INVITATION TO BIOLOGY

HELENA CURTIS

N. SUE BARNES

Invitation to Biology

THIRD EDITION

WORTH PUBLISHERS, INC.

INVITATION TO BIOLOGY, THIRD EDITION

PRINTED IN THE UNITED STATES OF AMERICA

LIBRARY OF CONGRESS CATALOG CARD NO. 80-54231

ISBN: 0-87901-131-9

FIRST PRINTING, FEBRUARY 1981

EDITOR: SALLY ANDERSON

ILLUSTRATOR: SHIRLEY BATY

PICTURE EDITOR: ANNE FELDMAN

PRODUCTION: KENNETH EKKENS

DESIGN: MALCOLM GREAR DESIGNERS

TYPOGRAPHER: NEW ENGLAND TYPOGRAPHIC SERVICE, INC.

PRINTING AND BINDING: R. R. DONNELLEY & SONS

COVER: SCANNING ELECTRON MICROGRAPH OF

HUMMINGBIRD FEATHER, ×510, JOHN MAIS

WORTH PUBLISHERS, INC.

444 PARK AVENUE SOUTH

NEW YORK, NEW YORK 10016

This book is dedicated to Caroline Rogers

Preface to the Teacher

The Third Edition of *Invitation to Biology* is in part a shorter version of *Biology,* Third Edition, in part a revision of *Invitation to Biology,* Second Edition, and in part a wholly new book. As you have doubtless noticed from the cover, with this edition *Invitation to Biology* also has a new coauthor. An experienced science editor and biology teacher, Sue Barnes made significant contributions to the Third Edition of *Biology* and has played a major role in this revision of *Invitation.*

Those familiar with the previous books can see from the Table of Contents that this edition of *Invitation* follows the same general plan. However, as we discovered in working on previous editions of this shorter book, it is not possible just to cut and paste, inserting a graceful transition or two. The entire book must be reconsidered, sentence by sentence, rewritten to a large extent, and once again submitted to the icy gaze of reviewers (a helpful but often merciless crew).

Writing a short book is harder, in many ways, than writing a longer one. One has to face, even more relentlessly, the question of what to put in and, much more difficult, what to leave out. What is modern biology? What do students *need* to know? What should students remember 10 years from now? What will still be true? The difficulties are compounded by the great variety of courses in which *Invitation* is used. Before embarking on this Third Edition, we wrote to many teachers, asking, among other questions, whether they would like the discussion of respiration and photosynthesis simplified, the treatment of classical and/or molecular genetics condensed, and a different balance between human physiology and comparative physiology. To each of these questions, the answers were about half and half. Rather than try to find a compromise that was likely to satisfy no one, we have sought in this edition to provide greater flexibility—to organize the material in such a way that by picking and choosing portions of the text to be assigned, a teacher can treat a particular topic in more or less detail, without encountering the repetition that leads to boredom or, on the other hand, discovering that a basic idea needed for understanding a later portion of the course is covered in optional material earlier in the text.

The Introduction to the Third Edition begins with a discussion of the signs of life, accompanied by a selection of photographs that we hope will whet the student's appetite for knowledge of the marvelous organization, functioning, and interactions of living organisms. We then venture to state the principles of biology. We believe that the principles of biology are what students have to understand and what they will remember if we—teachers and authors—have served them well. Throughout, we have tried to organize the text in such a way that the details and examples reinforce and provide greater understanding of the basic principles.

The overall organization of the book remains basically the same. The levels-of-biological-organization approach is again followed, beginning with the atoms and molecules that make up living systems. Part I deals with cells, Part II with individual organisms, and Part III with groups of organisms. The first two parts each have three sections, and the third has two sections. It is in the sequence of chapters within sections and within the chapters themselves that a significant amount of restructuring has occurred, resulting, we believe, in a book that is more coherent and satisfying for students and teachers alike.

In Section 1, we did not even consider eliminating the material that students woefully refer to as "the chemistry." One of the things the student *has* to know is that living systems are made up of atoms and molecules, organized in particular ways, according to certain general laws. (It is for this reason also that we have continued to present structural formulas throughout the text. The students certainly should not be expected to remember these formulas, but they should be reminded of what forces are at work.) An important undertaking of this revision has been to modernize the basic chemistry in the early chapters and to cover *only* those concepts that are used and built upon in subsequent chapters of the text. The goal has been to make the chemistry more coherent, more relevant to biology, and easier for students to understand.

In Section 2, Energetics, we have organized the material to provide two clear options. For those who wish a quite thorough discussion of respiration and photosynthesis, with biochemical details, all of Chapters 6, 7, and 8 should be covered. For those who wish a less rigorous treatment, with biochemical details held to a minimum, the following sequence should be used: all of Chapter 6; in Chapter 7, pages 99–101, 107, and the summary; and in Chapter 8, from page 110 through the first paragraph on page 115, page 116, and the last paragraph on page 120 through the summary.

The third section, Genetics, has remained much the same except for some internal reorganization. All of the material on human genetics, previously scattered throughout the section, has been gathered into Chapter 13, following immediately after the four chapters covering classical genetics. For those who wish to place major emphasis on classical genetics, these five chapters form a coherent unit. Similarly, Chapters 14–17 form a coherent unit on molecular genetics. Chapters 14 and 15 contain basic material on the structure of DNA and the process of protein synthesis, while Chapters 16 and 17 present current and (to us) exciting work. These latter two chapters are optional.

Part II, Biology of Organisms, was extensively revised in the previous edition and we have made relatively minor changes for this edition. Section 4, previously one long chapter on the diversity of organisms, has been broken into two shorter, more manageable chapters. Section 5, covering the biology of plants, remains intact. We have continued to resist all pressures to "integrate" plant and animal physiology. We have never seen this done in such a way that the beginning student could have any notion whatsoever of a plant as a complete, functioning, and interesting organism.

In Section 6, the focus continues to be on human physiology, although illuminating comparisons are made with other animals. New material has been added on the skeleton, the skin, and the immune system, the integration and control chapter has been broken into separate chapters on the nervous system and the endocrine system, and the chapter on reproduction and development has been moved later in the section. We believe that these changes give a better balance and a more logical flow to the section as a whole.

Part III, Biology of Populations, covers what G. E. Hutchinson so aptly termed "the ecological theater and the evolutionary play." Modern evolutionary theory and ecology are so inextricably intertwined that any separation of the two is arbitrary. We believe, however, that the student's understanding of modern ecology is deepened and enriched if it is preceded by a knowledge of the mechanisms of evolution. For that reason, we have continued to present evolution before ecology.

As in the Second Edition, the evolution section begins with an introductory chapter that places evolutionary theory in an historical and cultural perspective. Between editions, we have continued to be reminded of the fact that there are literate, intelligent people, young and old, who are not convinced that evolution actually took place. With such persons in mind, we have added more of the evidence—such as the testimony of the fossil record—that Darwinian evolution is a fact.

The ecology section has been reorganized so that it moves from the dynamics of growth in a single population, through the interactions of populations with one another, to the structure of communities and ecosystems, and finally to the overall organization and distribution of life on earth. The book closes with a chapter on social behavior (perhaps the most fascinating of all interactions among organisms) and a chapter on human evolution and ecology. As in the past, we have not devoted extensive space to environmental problems, believing that these problems are essentially social, political, and ethical. We have endeavored, however, to provide students with a solid biological foundation that will enable them to weigh intelligently the questions that we, as a society, must face in the years ahead.

The end-of-chapter summaries have been expanded and strengthened to provide greater assistance to students in mastering the essential conceptual material of each chapter. Similarly, many new questions have been added, some to provide a more thorough review of the chapter and others to extend the student's vision beyond the material covered in the text. We would also like to call your attention to the excellent Study Guide prepared by Vivian Null to accompany the text.

At the end of each section is a short list of books for further reading. We have not included lists of *Scientific American* reprints, simply on the assumption that these excellent supplementary materials are widely known, widely available, and constantly increasing in number. We would like to urge that students be encouraged to explore on their own such journals as *Scientific American, Science,* and *American Scientist.* In this way, they will have the pleasure of discovering something for themselves and of seeing the process of the acquisition of scientific knowledge as it takes place.

As with previous texts, we have been deeply dependent on the advice of reviewers and consultants. Among those who have made major contributions to the Third Edition of *Invitation* are: J. Wesley Bahorik, Kutztown State College; Bertrand Berlin, Middlesex Community College; John Blamire, Brooklyn College; Peter Clason, Oakland Community College; Ray Evert, University of Wisconsin; Abraham Flexer; Edward Florance,

Lewis and Clark College; Douglas Fratianne, Ohio State University; Michael Gaines, University of Kansas; Michael Ghiselin, University of Utah; Adair Gould, University of Delaware; Richard Haas, California State University, Fresno; Holt Harner, Broward Community College; Jean Harrison, University of California, Los Angeles; George Hennings, Kean College of New Jersey; Charles E. Holt, Massachusetts Institute of Technology; Keith I. King, Oregon State University; John Kirsch, Harvard University; Lynne Kunze, Virginia Western Community College; Dorothy Luciano; Robert M. May, Princeton University; B. J. D. Meeuse, University of Washington; Douglas Morrison, Rutgers University, Newark; Steven N. Murray, California State University, Fullerton; Bette Nicotri, University of Washington; Karin Rhines; Anthony Russell, University of Calgary; Tom Scott, University of North Carolina.

As in the past, it has been a pleasure to work with all the talented people of Worth Publishers, whose assistance and support were unfailing. They produced this book with their usual care and high standards.

Finally, we must thank all of the many teachers who have written to us about either *Biology* or *Invitation* with criticisms and suggestions. They helped us greatly in preparing the present text and we earnestly hope they will continue.

East Hampton, New York — HELENA CURTIS

New York, New York — N. SUE BARNES

January, 1981

Contents in Brief

Contents

Dolphins

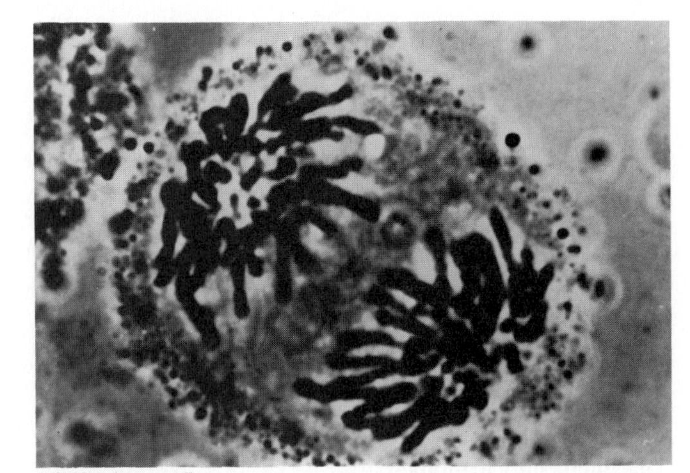
A dividing human cell

Zebras

Sedum

Axolotl

Sperm fertilizing egg

Water bug carrying eggs

Young baboons

INVITATION TO BIOLOGY

Introduction

I-1 From a biological point of view, a penguin is just an egg's way of making another egg. There are 18 different species of penguins, some of which live as far north as the equator, and all of which are, according to George Gaylord Simpson, "beautiful, interesting, inspiring, and funny."

Biology is defined as the "science of life." Its scope extends from the atoms and molecules that make up living matter through the organization of those particles in the body of an organism to the interactions of whole organisms and groups of organisms with one another and with the environment in which they live. That is also the scope of this text, which progresses from atoms and molecules through individual organisms and their component parts to a view of the biosphere, the entire world of living things.

When we speak of biology as the "science of life," what do we mean by "life"? Actually, there is no simple definition. Life does not exist in the abstract; there is no "life," only living things. Moreover, there is not any single, simple way to draw a sharp line between the living and the nonliving. There are, however, certain properties that, taken together, distinguish animate (that is, living) objects from inanimate ones.

The Signs of Life

The first characteristic of living things is that they are highly organized (Figure I–2). In living things, atoms—the particles of which all matter, both living and nonliving, is composed—are combined into a vast number of

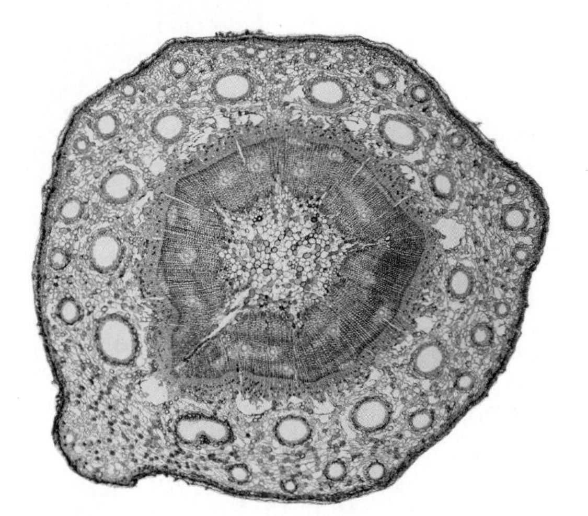

I-2 *Living things are highly organized, as in this cross section of a one-year-old pine stem. It reflects the complicated organization of many different kinds of atoms into molecules and of molecules into complex structures. Such complexity of form is never found in inanimate objects.*

very large molecules called macromolecules. Each type of macromolecule has a distinctive structure and a specific function in the life of the organism of which it is a part. Some macromolecules are linked with other macromolecules to form the structures of which the organism's body is composed. Others do not have a structural role but instead participate in the dynamic processes essential for the continuing life of the organism; for example, large molecules known as enzymes regulate virtually all of the processes occurring within living matter. The complex organization of both structures and processes is one of the most important properties by which an object can be identified as living.

The second characteristic is closely related to the first: Living systems maintain a chemical composition quite different from that of their surroundings (Figure I–3). The atoms present in living matter are the same as those in the surrounding environment, but they are present in different proportions and arranged in different ways. Although living systems constantly exchange materials with the environment, they maintain a stable and characteristic chemical composition. This important property is called homeostasis, which means simply "staying the same."

I–3 *Living organisms are homeostatic, which means "staying the same." Even this tiny, apparently fragile animal, a copepod, has a constant chemical composition that differs from its changing environment.*

A third characteristic of living things is the capacity to take in energy from the environment and to transform it and use it (Figure I–4). For example, in the process of photosynthesis, green plants take in light energy from the sun and use it to form complex molecules from water and the carbon dioxide in the air. The energy stored in these molecules is used by the plant to power its life processes and to build structures that add to its body. Animals can capture the stored energy by eating the plants; they, in turn, change the energy into other forms, such as heat, motion, and even electricity.

I–4 *Living things take energy from the environment and change it from one form to another. They are highly specialized at energy conversion. This young female Alaskan brown bear has just converted chemical energy stored in her body to energy of motion used in catching a salmon. After she has eaten and digested the salmon, the chemical energy stored in its body will be available for her use.*

I–5 *Living organisms respond to stimuli. For example, web-building spiders, such as this garden spider, are sensitive to the slightest vibrations of their webs. They can distinguish between vibrations caused by the wind and those caused by an intruder such as the grasshopper at the left. When the grasshopper became entangled in the web, the spider responded promptly, injecting it with venom and wrapping it in silk.*

Fourth, living systems can respond to stimuli (Figure I–5). Bacteria move toward or away from certain chemicals; green plants bend toward the light; mealworms congregate where it is damp; cats pounce on small moving objects. Although different organisms respond to widely varying stimuli, the capacity to respond is a fundamental and almost universal characteristic of life.

Fifth, and most remarkably, living things have the capacity to reproduce themselves with astonishing fidelity. Generation after generation, organisms produce more organisms like themselves. In each generation, however, there are slight variations between parents and offspring and among offspring, a fact of considerable importance, as we shall see later.

Some living things, primarily very small organisms such as bacteria, amoebas, and some types of algae, reproduce by simply dividing in two. The offspring are usually identical to the parent. Most organisms, however, grow and develop. For example, before hatching, the fertilized egg of a frog develops into the complex, but still immature, form that we recognize as a tadpole; after hatching, the tadpole continues to grow and undergoes further development, becoming a mature frog. Throughout the plant and animal kingdoms, similar striking patterns of growth and development occur (Figure I–6).

I–6 *Living things reproduce themselves, generation after generation, with astonishing fidelity. Although these two lion cubs will grow and develop into adults that closely resemble their mother (shown here) or father, they will not be identical to either. As we shall see, the slight variations between parents and offspring provide the raw material for evolution.*

A seventh characteristic of living things is that they are exquisitely suited to their environment. Moles, for instance, are furry animals that live underground in tunnels shoveled out by their large forepaws. Their eyes are small and almost sightless. Their noses, with which they sense the worms and other small animals that make up their diet, are fleshy and enlarged. This most important characteristic of living things is known as adaptation (Figure I–7, page 4).

(a) (b)

I-7 *Some unusual adaptations.*
(a) *This red-eyed tree frog of Central America, like many other animals that dwell in the canopy of the tropical rain forest, never descends to the forest floor. The frogs' eggs are laid on leaves that overhang water. After they hatch, the tadpoles drop into the water where they complete their development. The bulges at the ends of the toes are suction pads.*
(b) *Anteaters, which are found in both the Old World and the New World, are not closely related to one another genetically. However, they all have long noses, very long, wormlike tongues, and stout digging claws. This is the giant anteater of tropical America and, like most anteaters, it eats termites.*

These characteristics of living things are intimately interrelated, and each depends, to a large extent, on the presence of the others. At any given moment in its life, an organism is organized, maintains a constant internal environment, transforms energy, and is adapted to its external environment; the organism may or may not be responding to stimuli, reproducing, growing, and developing, but it possesses the capacity to do so.

The Nature of Science

As we noted earlier, biology is defined as the "science of life." Just as "life" does not exist in the abstract and can only be described in terms of the characteristics exhibited by living things, so "science" does not exist in the abstract. It is a particular human activity that can only be described in terms of the characteristics that distinguish it from other human activities.

Science, biological and other, is a way of seeking principles of order in the world of which we are a part. Art is another way, as are religion and philosophy. Science is distinguished from the other ways in which we seek understanding by the fact that it is limited to propositions about the natural world that can be verified objectively. Science has two separate but interacting components: (1) the collection of objective evidence—the data—based on observation or experiment or a combination of the two and (2) the structuring and interpretation of data by making meaningful connections among them. The great discoveries in science are not merely the addition of new data but the perception of new relationships among the available data—in other words, the development of new ideas.

The ideas of science are categorized in ascending order of validity as hypotheses, theories, and principles or laws. Lower on the scale than the hypothesis is the hunch, or educated guess, which is how most hypotheses begin. A hunch becomes a hypothesis when it is stated in such a way that it

Some Comments on Science and Scientists

Art and Science

The act of discovery in science engages the imagination (first of the man who makes it, and then of the man who appreciates it) as truly as does the act of creation in the arts.

J. Bronowski, *Identity of Man,* rev. ed., Natural History Press, Garden City, N.Y., 1971.

Science and Theory

The power of theories is that they combine many generalizations and other theories into networks of interlocking ideas that point to the future.

J. Bronowski, *The Common Sense of Science,* Random House, Inc., New York, 1959.

On principle, it is quite wrong to try founding a theory on observable magnitudes alone. It is the theory which decides what we can observe.

A. Einstein, from J. Bernstein, "The Secrets of the Old Ones, II," *The New Yorker,* March 17, 1973.

The Scientific Method

Indeed, scientists are in the position of a primitive tribe which has undertaken to duplicate the Empire State Building, room for room, without ever seeing the original building or even a photograph. Their own working plans, of necessity, are only a crude approximation of the real thing, conceived on the basis of miscellaneous reports volunteered by interested travelers and often in apparent conflict on points of detail. In order to start the building at all, some information must be ignored as erroneous or impossible, and the first constructions are little more than large grass shacks. Increasing sophistication, combined with methodical accumulation of data, make it necessary to tear down the earlier replicas (each time after violent arguments), replacing them successively with more up-to-date versions. We may easily doubt that the version current after only 300 years of effort is a very adequate restoration of the Empire State Building; yet, in the absence of clear knowledge to the contrary, the tribe must regard it as such (and ignore odd travelers' tales that cannot be made to fit).

E. J. DuPraw, *Cell and Molecular Biology,* Academic Press, Inc., New York, 1968.

Scientists at Work

Scientists at work have the look of creatures following genetic instructions; they seem to be under the influence of a deeply placed human instinct. They are, despite their efforts at dignity, rather like young animals engaged in savage play. When they are near to an answer their hair stands on end, they sweat, they are awash in their own adrenalin. To grab the answer, and grab it first, is for them a more powerful drive than feeding or breeding or protecting themselves against the elements.

It sometimes looks like a solitary activity, but it is as much the opposite of solitary as human behavior can be. There is nothing so social, so communal, so interdependent. An active field of science is like an immense intellectual anthill; the individual almost vanishes into the mass of minds tumbling over each other, carrying information from place to place, passing it around at the speed of light.

There are special kinds of information that seem to be chemotactic. As soon as a trace is released, receptors at the back of the neck are caused to tremble, there is a massive convergence of motile minds flying upwind on a gradient of surprise, crowding around the source. It is an infiltration of intellects, an inflammation.

There is nothing to touch the spectacle. In the midst of what seems a collective derangement of minds in total disorder, with bits of information being scattered about, torn to shreds, disintegrated, reconstituted, engulfed, in a kind of activity that seems as random and agitated as that of bees in a disturbed part of the hive, there suddenly emerges, with the purity of a slow phrase of music, a single new piece of truth about nature. . . .

There is something like aggression in the activity, but it differs from other forms of aggressive behavior in having no sort of destruction as the objective. While it is going on, it looks and feels like aggression: get at it, uncover it, bring it out, grab it, halloo! It is like a primitive running hunt, but there is nothing at the end of it to be injured. More probably, the end is a sigh. But then, if the air is right and the science is going well, the sigh is immediately interrupted, there is a yawping new question, and the wild, tumbling activity begins once more, out of control all over again.

Lewis Thomas, *The Lives of a Cell: Notes of a Biology Watcher,* Viking Press, Inc., New York, 1974.

is potentially testable, even if the test cannot be done immediately. When a hypothesis has survived a number of independent tests, it becomes, strictly speaking, a theory. A theory that has withstood repeated testing over a period of time becomes elevated to the status of a law or principle, although not always identified as such. The "theory" of evolution is an example. It is not "just a theory," in the common usage of the term, nor has it been for almost a hundred years. As far as scientists are concerned, it is a principle, just as is the cell "theory." However, our knowledge of many of the details of cellular structure and function or of the evolutionary process is still in the stage of theory, or even hypothesis.

This emphasis on the ideas of science is not to say that the facts are not important. It is the facts—the realities underlying the data accumulated through observation and experiment—that endure. It is for this reason that scientists stress objectivity; in all science, observations and experiments must always be reported in such a way that they can be repeated and verified, and they must always be repeated and verified before they are incorporated into the body of knowledge. The facts are the unyielding building blocks, the structural elements, stubborn and concrete, against which hypotheses are tested, with which theories are erected and revised, and against which dreams are sometimes shattered.

Unifying Principles of Modern Biology

Four principles of modern biology are so well established that biologists seldom discuss them among themselves. You could read widely in the literature of modern biology without seeing any of them explicitly mentioned. Yet it is impossible to understand either the ideas or the data of contemporary biology without being aware of the foundations. These four principles will be discussed in greater detail in the course of this text and will recur as major themes, but you should have all four in mind from the outset.

All Organisms Have Evolved

The key idea in the theory of evolution is that all species descended from other species; in other words, all living things share common ancestors in the distant past. Charles Darwin was not the first to propose this idea, but the credit is rightly his, for two reasons. First, his "long argument"—as *The Origin of Species* has been characterized—left little doubt that evolu-

(a)

(b)

I-8 *Recent biochemical tests have demonstrated conclusively the close evolutionary link between* (a) *the woolly mammoth, a creature that roamed North America, Asia, and Europe thousands of years ago, and* (b) *the modern elephant. Several years ago, a baby woolly mammoth that died about 40,000 years ago was found frozen in Siberia. Its tissues were so perfectly preserved that the exact structure of certain key molecules could be determined and compared with the structure of the same molecules in living elephants.*

tion had actually occurred and so marked a turning point in the history of biology. The second reason, which is closely related to the first, is that Darwin correctly perceived the mechanism by which evolution occurs.

Evolution is a two-stage process. The first stage is the occurrence, absolutely at random, of genetic variations among individuals—that is, variations that can be inherited. These variations have adaptive values; that is, they may be more or less useful to the organism as measured by its survival and reproduction. The second stage is natural selection, which is a process of interaction between an organism and its environment. As a result of this interaction, some organisms, because of their inherited characteristics, leave more offspring than organisms with other inherited characteristics. Given enormous amounts of time, evolution leads to the accumulation of changes that differentiate one group of organisms from another. The result is the great diversity of living things that now inhabit this planet.

An important corollary to the theory that all organisms have evolved is that life, as it now exists on this planet, arose only once. Impressive support is given to these concepts by the fact that the mechanism by which hereditary information is transmitted from one generation to another and the way in which the information is encoded are universal—essentially the same from the simplest bacterium to the most complex organism.

Evolution is viewed as the greatest unifying principle in biology.

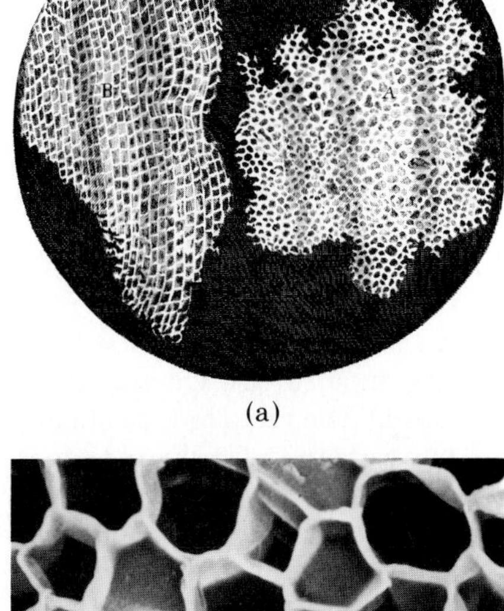

(a)

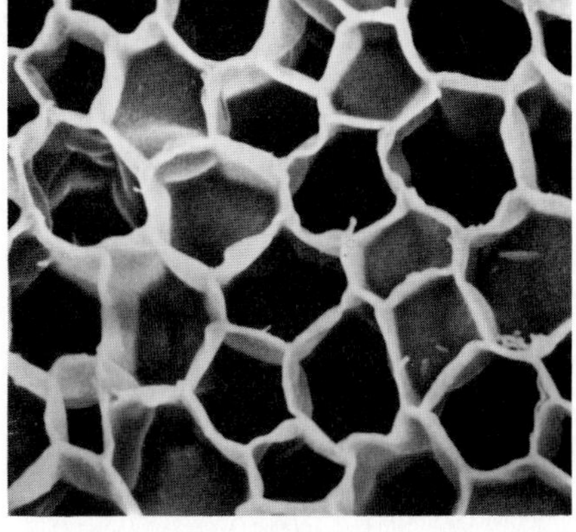

(b)

I–9 (a) *Robert Hooke's drawings of two slices of a piece of cork, reproduced from a book he published in 1665, and* (b) *an electron micrograph of a slice of cork. Hooke was the first to use the word "cells" to describe the tiny compartments that together make up an organism. The cells in these pieces of cork have died—all that remain are the outer walls. As we shall see in Section 1, the living cell is filled with a variety of substances, organized into distinct structures and carrying out a multitude of essential processes.*

All Organisms Are Made Up of Cells

A second important principle of biology is the cell theory. It states that the cell is the basic unit of living material and that all living organisms are composed of one or more of these fundamentally similar units.

The word "cell" was first used in a biological sense some 300 years ago. In the seventeenth century, Robert Hooke, using a microscope of his own construction, noticed that cork and other plant tissues are made up of small cavities separated by walls. He called these cavities "cells," meaning "little rooms." The word did not take on its present meaning, however, for more than 150 years.

In 1838, Matthias Schleiden, a German botanist, came to the conclusion that all plant tissues are organized in the form of cells. In the following year, zoologist Theodor Schwann extended Schleiden's observation to animal tissues and proposed a cellular basis for all life. The cell theory is of tremendous and central importance to biology because it emphasizes the basic sameness of all living systems. It therefore brings an underlying unity to widely varied studies involving many different kinds of organisms.

The cell theory took on an even broader significance when the great pathologist Rudolf Virchow generalized that cells can arise only from preexisting cells: "Where a cell exists, there must have been a preexisting cell, just as the animal arises only from an animal and the plant only from a plant. . . . Throughout the whole series of living forms, whether entire animal or plant organisms or their component parts, there rules an eternal law of continuous development."

In the perspective of evolution, Virchow's concept takes on an even larger significance. There is an unbroken continuity between modern cells—and the organisms that they compose—and the primitive cells that first appeared on earth more than 3 billion years ago.

All Organisms Obey the Laws of Physics and Chemistry

Until fairly recently, many prominent biologists believed that living systems are qualitatively different from nonliving ones, containing within them a "vital spirit" that enables them to perform activities that cannot be carried on outside the living organism. This concept is known as vitalism and its proponents as vitalists. (Do not dismiss this "foolish" idea too rapidly; you may discover that you too are a vitalist.)

In the seventeenth century, the vitalists were opposed by a group known as the mechanists. The French philosopher René Descartes (1596–1650) was a leading proponent of this group. The mechanists set about proving that the body worked essentially like a machine; the arms and legs move like levers, the heart like a pump, the lungs like a bellows, and the stomach like a mortar and pestle. By the nineteenth century, such simple mechanical models of living organisms had been abandoned, and the argument now centered on whether or not the chemistry of living organisms was governed by the same principles as the chemistry performed in the laboratory. The vitalists claimed that the chemical operations performed by living tissues could not be carried out experimentally in the laboratory, categorizing reactions as either "chemical" or "vital." The reductionists, as their opponents were now called (since they believed that the complex operations of living systems could be reduced to simpler and more readily understandable ones), achieved a partial victory when the German chemist Friedrich Wöhler (1800–1882) converted an "inorganic" substance (ammonium cyanate) into a familiar organic substance (urea). On the other hand, the claims of the vitalists were supported by the fact that, as chemical knowledge improved, many new compounds were found in living tissues that were never seen in the nonliving, or inorganic, world.

In the late 1800s, the leading vitalist was Louis Pasteur, who claimed that the changes that took place when fruit juice was transformed to wine were "vital" and could be carried out only by living cells—the cells of yeast. In spite of many advances in chemistry, this phase of the controversy lasted until almost the turn of the century.

I-10 *Over short distances, the cheetah is the fastest land animal in the world. It can attain speeds of 110 kilometers per hour but can maintain such speeds for little more than 30 seconds (covering a distance of about half a kilometer). The cheetah's motion—like all motion—is governed by the laws of physics, and the chemical reactions within its body that provide the necessary energy are governed by the laws of chemistry.*

However, in 1898, the German chemists Eduard and Hans Büchner showed that a substance extracted from the yeast cells could produce fermentation outside the living cell. (This substance was given the name enzyme, from *zyme,* the Greek word meaning "yeast" or "ferment.") A "vital" reaction was proved to be a chemical one, and the subject was eventually laid to rest. Today it is generally accepted that living systems "obey the rules" of chemistry and physics, and modern biologists no longer believe in a "vital principle."

Perhaps the greatest test for this principle of modern biology came about 30 years ago. As you know, one of the outstanding characteristics of living things is their capacity to reproduce, to generate faithful copies of themselves. By about 1950, this capacity had been shown to reside in a single type of chemical molecule, deoxyribonucleic acid (DNA). The race to discover the structure of this molecule began, and the question in everyone's mind was whether or not the structure of this one "simple" molecule could possibly explain the mysteries of heredity. As it turned out, and as we shall discuss in the course of this text, it could.

The next test for this concept seems to lie in the mechanisms of the brain. Even those of us who now willingly admit that the nature of heredity—so long a mystery—resides in a single molecule, an evolutionary accumulation of atoms organized in a particular way, are perhaps less willing to admit that our dreams and our ideals and our deepest emotions are based on "nothing but" the chemistry and organization of our brain cells.

Life Runs Downhill

Among the laws of physics that are pertinent to biology are the laws of thermodynamics. They state simply that (1) energy can be changed from one form to another but it cannot be gained or lost, that is, the total energy of the universe remains constant; and (2) all natural events proceed in such a way that concentrations of energy tend to dissipate or become random. A heated object, which is one example of concentrated energy, loses its heat to its surroundings.

A living system, which is a concentration of energy of another kind, can maintain itself only by a constant intake of energy. As we saw earlier, living organisms are experts at energy conversion; the energy they take in—whether in the form of sunlight or chemical energy stored in food—is transformed and used by each individual cell to do the work of the cell. In the course of this work, the energy may be further transformed to energy of motion, to heat energy, or even back to light energy again. It is ultimately dissipated, and the organisms must take in more energy.

I-11 *Because natural events proceed in such a way that concentrations of energy are ultimately dissipated, living systems can maintain themselves only by a constant intake of energy. These harvest mice are eating ripe wheat, in which energy from the sun has been stored.*

This flow of energy is the essence of life. Evolution may be viewed as a competition among organisms for the most efficient use of energy resources. Cells can be best understood as a complex of systems for transforming energy. At the other end of the biological scale, the structure of ecosystems and of the biosphere itself is determined by the relationships in terms of energy exchanges of the groups of organisms within it.

Most organisms budget their energy day to day, or even hour to hour, with no reserves beyond those stored in their own fat, liver, or other tissues. Some few, such as the honey bee and the squirrel, set aside some minimal provisions. Only we humans have been able to live on deficit financing; as you will see in Chapter 39, in advanced technological societies, it now costs 10 calories in energy for each calorie of food that reaches the table. This fact is not without relevance to world politics and economics.

Science and Human Values

Before we close this Introduction, we should mention one more characteristic of biology and other sciences, one that distinguishes science from art, religion, or philosophy. The raw materials of science are our observations of the phenomena of the natural universe. Science is limited to what is observable and measurable and, in this sense, is rightly categorized as materialistic. Hunches are abandoned, hypotheses superseded, theories shattered, but the observations endure, and, moreover, they are used over and over again, sometimes in wholly new ways. It is for this reason that scientists stress objectivity. (In the arts, by contrast, the emphasis is on subjectivity—experience as filtered through the individual consciousness.)

Because of this emphasis on objectivity, value judgments cannot be made in science in the way that such judgments are made in philosophy, religion, and the arts, and indeed in our daily lives. Whether or not something is good or beautiful or right in a moral sense, for example, cannot be determined by scientific methods. Such judgments, even though they may be supported by a broad consensus, are not subject to proof by science.

At one time, sciences, like the arts, were pursued for their own sake, for pleasure and excitement and satisfaction of the insatiable curiosity with which we are both cursed and blessed. In the twentieth century, however, the sciences have spawned a host of giant technological achievements—the hydrogen bomb, the Salk vaccine, DDT, indestructible plastics, nuclear energy plants, perhaps even ways to manipulate our genetic heritage—but have not given us any clues about how to use them wisely. It is thus little wonder that in the present generation there are many who are angry at science, as one would be angry at an omnipotent authority who apparently has the power to grant one's wishes but who refuses to do so.

The reason that science cannot and does not solve the problems we want it to is inherent in its nature. Most of the problems we now confront can be solved only by value judgments. For example, science gave us nuclear power and can give us predictions as to the extent of the biological damage that might result from accidents that allowed varying levels of radioactivity to escape into the environment. Yet it cannot help us, as citizens, in weighing the risk of damage from conceivable accidents against our energy needs. It can give us data to weigh in our judgments, but it cannot make those judgments for us.

Science has produced the knowledge that makes it possible to replace a diseased heart with a healthy heart. There are, however, many more patients who need heart transplant operations than there are healthy hearts available for transplanting. Scientific methods cannot help us decide who should receive heart transplants and who should be abandoned. Similarly, scientists can predict the possible extent of damage to the plants and animals of a particular area from the use of pesticides; they can also predict the reduction in food crops or the increase in malaria that would occur were pesticides prohibited. But scientists, in their capacity as scientists, cannot make the choice as to whether we should or should not use pesticides.

It is one of the ironies of this so-called "age of science and materialism" that probably never before have ordinary individual men and women, including scientists, been confronted with so many moral and ethical dilemmas. In this text, we shall discuss some of the dilemmas that have grown out of the achievements of modern science and technology. Our greater concern, however, is to provide you with the biological knowledge necessary to understand the relevant data as you make your own value judgments on the problems that confront us now and that will do so in the future.

The Study of Biology

Most texts, and this one is no exception, tend to stress what is known at the present time, rather than what is not known or how we came to know what we do. This tendency, although understandable, somewhat distorts the nature of biology, and indeed of science in general. A modern science is not a static accumulation of facts organized in a particular way, but a somewhat amorphous body of knowlege that not only constantly grows, developing new bulges and unpredictable appendages, but also may suddenly change its entire shape (as biology did in the nineteenth century with the acceptance of the theory of evolution). Science is dynamic, not static; consequently it cannot be contained within textbooks or libraries or information retrieval centers, but rather it is a process taking place in the minds of living scientists. In our enthusiasm for telling you what biologists have discovered thus far, do not let us convince you that all is known. Many questions are still unanswered. More important, many good questions have not yet been asked. Perhaps you may be the one to ask them.

You may have been persuaded to study biology because of the environmental problems now confronting us or because of a desire to know more about the mechanisms of your own body or an interest in genetic engineering or a career in medicine—in short, because it is "relevant." The study of biology is, indeed, pertinent to many aspects of our day-to-day existence, but do not make this your main reason for studying biology. Above all other considerations, study biology because it is "irrelevant"—that is, study it for its own sake, because, like art and music and literature, it is an adventure for the mind and nourishment for the spirit.

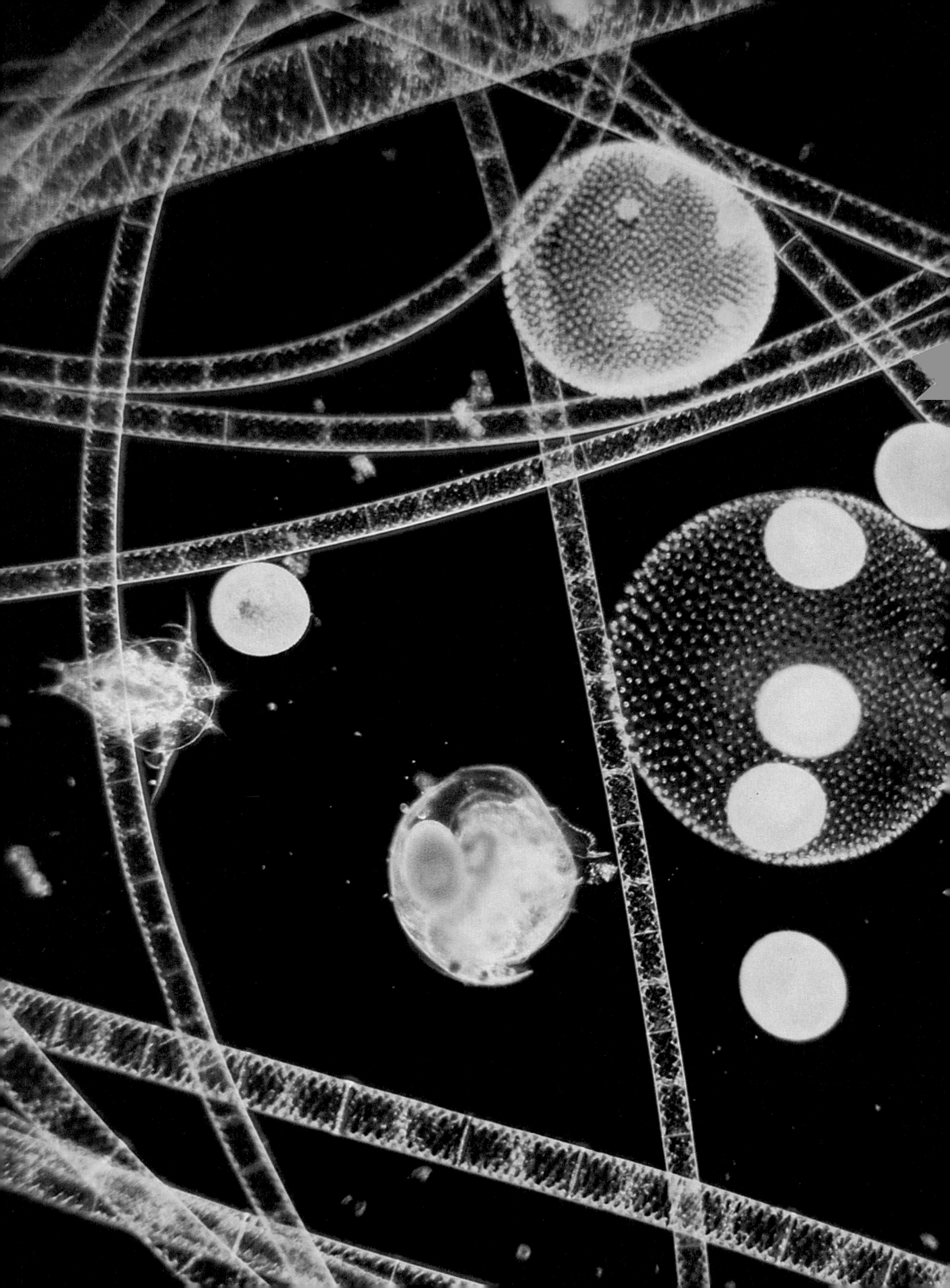

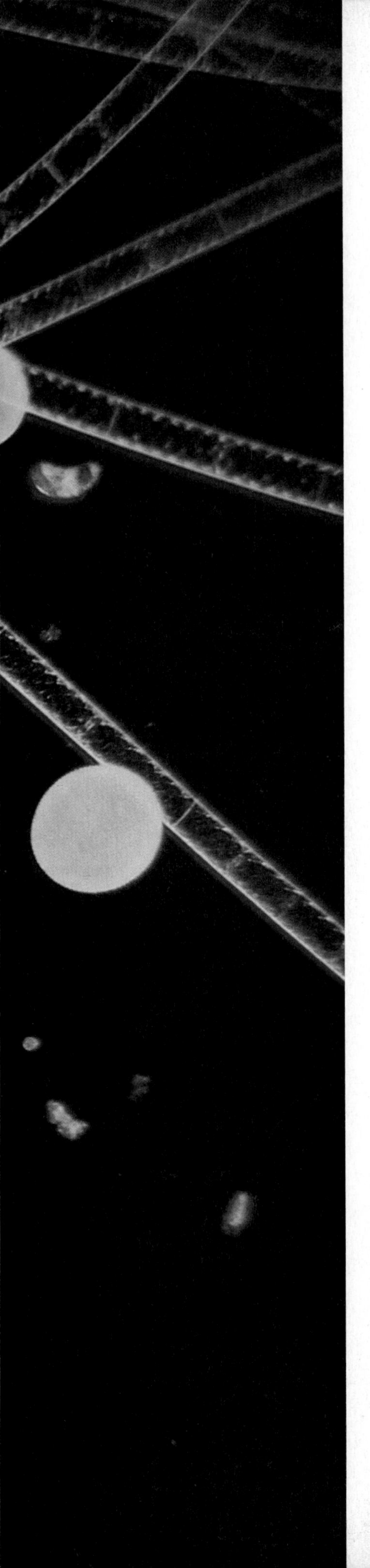

PART I

Biology of Cells

SECTION 1

The Unity of Life

All organisms are composed of cells. In Spirogyra, *a common freshwater alga, the cells are strung together in long filaments.* Volvox, *another freshwater alga, is a spherical colony containing hundreds (or even thousands) of individual cells. The large* Volvox *at the right has just burst, releasing daughter colonies into the water. Below and to the left of this* Volvox *is* Chydorus, *a small crustacean. The green mass in its body is an egg cell. At the far left is another crustacean, a copepod.*

1

Atoms and Molecules

The universe began, astronomers tell us, with an explosion that filled all space, with every particle of matter hurled away from every other particle. The temperature at the time of the explosion—some 20 billion years ago—was about 100,000,000,000 degrees Celsius (10^{11} °C). At this temperature, not even atoms could hold together; all matter was in the form of subatomic, elementary particles. Moving at enormous velocities, even these particles had fleeting lives. Colliding with great force, they annihilated one another, creating new particles and releasing more energy.

As the universe cooled, two types of stable particles, previously present only in relatively small amounts, began to assemble. (By this time, 100,000 years after the "big bang" is believed to have taken place, the temperature had dropped to a mere 2500°C, about the temperature of a white-hot wire in an incandescent light bulb.) These particles—protons and neutrons—are very heavy as subatomic particles go. Held together by forces that are still incompletely understood, they formed the central cores, or nuclei, of atoms.

These nuclei, with their positively charged protons, attracted small, light, negatively charged particles—electrons—which moved rapidly around them. Thus, atoms came into being. (Perhaps atoms existed before the explosion, but how will one ever know?)

According to current theory, it is from these atoms—blown apart, formed, and re-formed over 20 billion years—that all the stars and planets of the universe are formed, including our particular star and planet. And it is from the atoms present on this planet that living systems assembled themselves and evolved. Each atom in our own bodies had its origin in this explosion of some 20 billion years ago. We are, each of us, flesh and blood, but we are also stardust.

This text begins where life began, with the atom. At first, the universe aside, it might appear that lifeless atoms have little to do with biology. Bear with us, however. A closer look reveals that the activities we associate with being alive depend on combinations and exchanges between atoms, and the force that binds the electron to the atomic nucleus stores the energy that powers living systems.

1-1 *The Serpens Nebula, shown here, is composed of clouds of gases and dust. In the small, dark regions of the photograph, new clusters of stars are forming from dense concentrations of gas. The stars and planets of the universe have their origins in nebulae such as this.*

Atoms

All matter, including the most complex living organisms, is made up of combinations of *elements*, which are substances that cannot be broken down by ordinary chemical means. The smallest particle of an element is an *atom*. There are 92 naturally occurring elements, each differing from the others in the structure of its atoms (Table 1–1).

The atoms of each different kind of element have a characteristic number of positively charged particles, called *protons*, in their nuclei. For example, an atom of hydrogen, the lightest of the elements, has 1 proton in its nucleus; an atom of the heaviest element, uranium, has 92 protons in its nucleus. The number of protons in the nucleus of a particular atom is called the *atomic number*.

Outside the nucleus of an atom are negatively charged particles, known as *electrons*, which are attracted by the positive charge of the protons. The number of electrons in an atom equals the number of protons in its nucleus. The electrons determine the chemical properties of atoms, and chemical reactions involve changes in the numbers and energy of these electrons.

Atoms also contain *neutrons*, which are uncharged particles of about the same weight as protons. These, too, are found in the nucleus of the atom, where they seem to have a stabilizing effect. The *atomic weight* of an element is essentially equal to the number of protons plus neutrons in the nuclei of its atoms. (Electrons are so light by comparison that their weight is usually disregarded. When you weigh yourself, only about 30 grams—approximately 1 ounce—of your total weight is made up of electrons.)

The concept of the atom as the indivisible unit of the elements is almost 200 years old; however, our ideas about its structure have undergone many changes. These ideas, or hypotheses, are usually presented in the form of models, as are many scientific hypotheses.

The earliest model, emphasizing the indivisibility of the atom, resembled a billiard ball. When it was realized that electrons could be removed from the atom, the billiard-ball model gave way to the plum-pudding model, in which the atom was represented as a solid, positively charged mass with negatively charged particles, the electrons, embedded in it. Subsequently, however, physicists found that an atom is, in fact, mostly

Table 1–1 Atomic structure of some familiar elements

		Nucleus		
Element	Symbol	Number of Protons	Number of Neutrons	Number of Electrons
Hydrogen	H	1	0	1
Helium	He	2	2	2
Carbon	C	6	6	6
Nitrogen	N	7	7	7
Oxygen	O	8	8	8
Sodium	Na	11	12	11
Phosphorus	P	15	15	15
Sulfur	S	16	16	16
Chlorine	Cl	17	18	17
Calcium	Ca	20	20	20

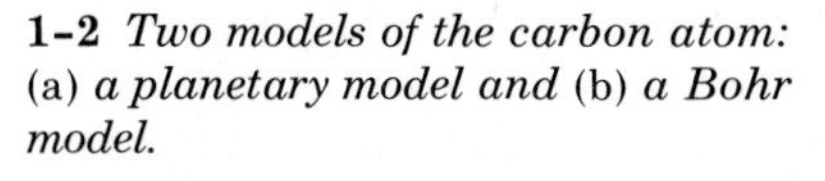
1-2 *Two models of the carbon atom:* (a) *a planetary model and* (b) *a Bohr model.*

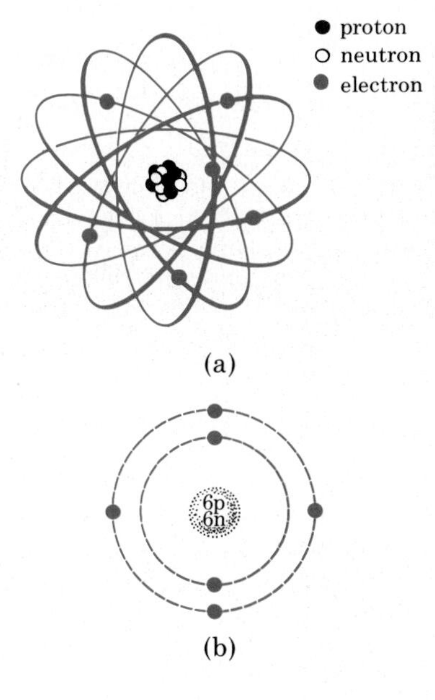

empty space. The distance from electron to nucleus, experiments indicated, is about 1,000 times the diameter of the nucleus; the electrons are so exceedingly small that the space is almost entirely empty. Thus the more familiar planetary model of the atom came into being, in which the electrons were depicted as moving in orbits around the nucleus (Figure 1-2a). Later, the Bohr model (named after physicist Niels Bohr) became the most popular one (Figure 1-2b). It emphasized the fact that different electrons of an atom have different amounts of energy and are at different distances from the nucleus. As we shall see, the Bohr model is not an accurate "picture" of an atom and has been superseded by another model (Figure 1-5, page 19). However, it can help us to understand certain properties of atoms that are of great importance in the chemistry of living systems.

Electrons and Energy

The distance of an electron from the nucleus is determined by the amount of *potential energy* (often called "energy of position") the electron possesses. The greater the amount of energy possessed by the electron, the farther it will be from the nucleus. Thus, an electron with a relatively small amount of energy is found close to the nucleus and is said to be at a low *energy level*; an electron with more energy is farther from the nucleus, at a higher energy level.

An analogy may be useful. A boulder on flat ground may be said to have no energy. If you change its position by pushing it up a hill, you give it potential energy. As long as it sits on the peak of the hill, the rock neither gains nor loses energy. If it rolls down the hill, however, it loses its potential energy (Figure 1-3a). The electron is like the boulder in that an input of energy can move it to a higher energy level—farther away from the nucleus. As long as it remains at the higher energy level, it possesses the added energy. And, just as the rock is likely to roll downhill, the electron also tends to go to its lowest possible energy level.

It takes energy to move a negatively charged electron farther away from a positively charged nucleus, just as it takes energy to push a rock up a hill. However, unlike the rock on the hill, the electron cannot be pushed partway up. With an input of energy, an electron can move from a lower energy level to any one of several higher energy levels, but it cannot move to an energy state somewhere in between. For an electron to move from one energy level to a higher one, it must absorb a discrete amount of energy, equal to the difference between the two particular energy levels. When the electron returns to its original energy level, that same amount of energy is released (Figure 1-3b).

1-3 (a) *The energy used to push a boulder to the top of a hill (less the heat energy produced by friction between the boulder and hill) becomes potential energy, stored in the boulder as it rests at the top of the hill. This potential energy is converted to energy of motion as the boulder rolls downhill.*
(b) *When an atom, such as the hydrogen atom diagrammed here, receives an input of energy, an electron may be boosted to a higher energy level. The electron thus gains potential energy, which is released when the electron returns to its previous energy level.*

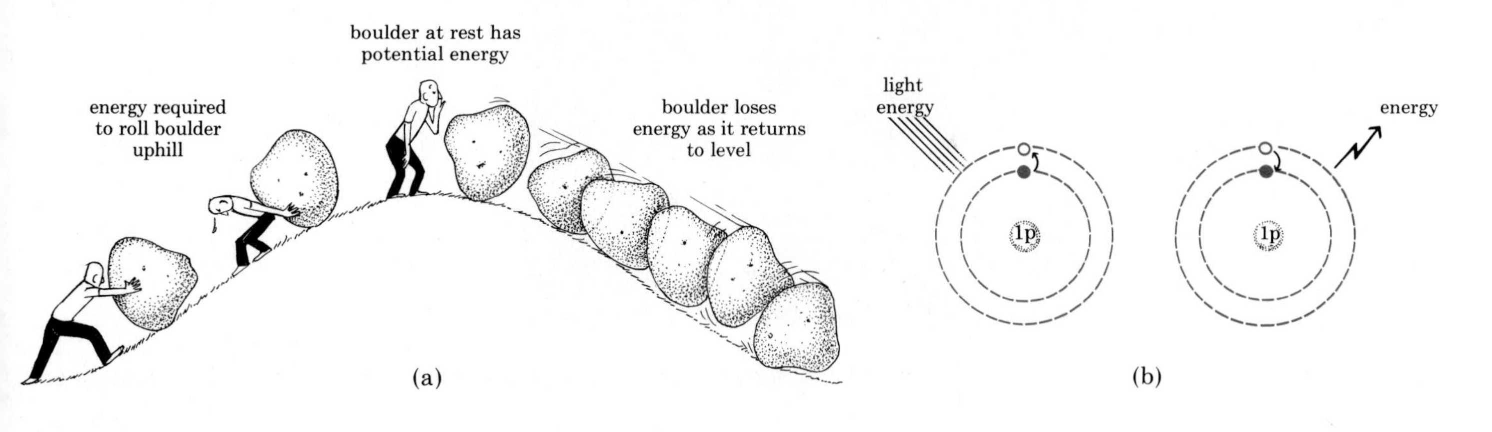

Isotopes

All atoms of a particular element have the same number of protons in their nuclei. Sometimes, however, different atoms of the same element contain different numbers of neutrons. These atoms differ from one another in their atomic weights but not in their atomic numbers. Such atoms are known as isotopes of the element. For example, three isotopes of hydrogen exist (Table 1-2). The common form of hydrogen, with its one proton, has an atomic weight of 1 and is symbolized as ^{1}H, or simply H. A second isotope of hydrogen contains one proton and one neutron and so has an atomic weight of 2; this isotope is symbolized as ^{2}H. A third, extremely rare isotope, ^{3}H, has one proton and two neutrons and so has an atomic weight of 3. The chemical behavior of the two heavier isotopes is essentially the same as that of ordinary hydrogen—all three isotopes have only one electron each, and the electrons determine chemical properties.

Most elements have several isotopic forms. The differences in weight, although very small, are sufficiently great that they can be detected with modern laboratory apparatus. Moreover, many, but not all, of the less common isotopes are radioactive. This means that the nucleus is unstable and emits energy as it changes to a more stable form. The energy given off by a radioactive isotope can be detected with a Geiger counter or on photographic film.

Isotopes have a number of important uses in biological research and in medicine. They can be used, for example, to determine the age of fossils (remnants of organisms long since dead) and of the rocks in which fossils are found. Each type of radioactive isotope emits energy and changes into another kind of isotope at a characteristic and fixed rate. As a result, the relative proportions of different isotopes in a rock sample give a good indication of how long ago that rock was formed.

Another use of radioactive isotopes is as "tracers." For example, an isotope of the element thallium, which is unreactive in the human body, can be used to detect blocked blood vessels in persons with symptoms of heart disease. The isotope is injected into the blood and its movement in the region of the heart is detected by a Geiger counter while the patient exercises on a treadmill. The Geiger counter is connected to a computer that produces a "picture" of the distribution of the isotope in the heart muscle. If a blood vessel is blocked by fatty deposits, the isotope cannot penetrate the region of heart muscle supplied by that blood vessel. This procedure, which has no known side effects, provides an extremely reliable indication of the presence or absence of a common type of heart disease (see page 381). Isotopes have a number of other diagnostic uses in medicine as well as an important role in the treatment of many forms of cancer.

Since isotopes of the same element all have the same chemical properties, a radioactive isotope will behave in an organism just as its more common nonradioactive isotope does. As a result, biologists have been able to use isotopes of a number of elements—especially carbon, nitrogen, and oxygen—to trace the course of many essential processes in living organisms.

Table 1-2 Isotopes of hydrogen

Isotope	Atomic Number	Atomic Weight	Number of Protons	Number of Neutrons	Number of Electrons
^{1}H	1	1	1	0	1
^{2}H	1	2	1	1	1
^{3}H	1	3	1	2	1

The dating of fossils using radioactive isotopes has established that the oldest known direct ancestors of human beings lived about 3½ million years ago. This fossil footprint, one of a trail of five found in Tanzania in east Africa, may have been left by one of those ancestors.

1-4 The leaves of these corn plants contain a pigment known as chlorophyll, which gives them their green color. When light strikes a molecule of chlorophyll, electrons in the molecule are raised to higher energy levels. As each electron returns to its previous energy level, the energy released is captured in the bonds of carbon-containing compounds.

In the green cells of plants and algae, the radiant energy of sunlight raises electrons to a higher energy level. In a series of reactions, which will be described in Chapter 8, these electrons are passed "downhill" from one energy level to another until they return to their original energy level. During these transitions, the radiant energy of sunlight is transformed into the chemical energy on which all life on earth depends.

The Arrangement of Electrons

At a given energy level, an electron moves around the nucleus at almost the speed of light. Since the location of an electron changes so rapidly, we usually speak about the pattern of the electron's motion rather than about its position. The volume of space in which the electron will be found 90 percent of the time is its *orbital.*

In any atom, the electrons at the lowest energy level—the first energy level—occupy a single spherical orbital, which can contain a maximum of two electrons (Figure 1-5a). Thus, for instance, hydrogen's single electron moves about the nucleus—90 percent of the time—within this single spherical orbital. Similarly, the two electrons of helium (atomic number 2) move within the single spherical orbital of the first energy level.

Atoms of higher atomic number than helium have more than two electrons. Since the first energy level is filled, these additional electrons must occupy higher energy levels, farther from the nucleus. At the second energy level, there are four orbitals, each of which can hold a maximum of two electrons (Figure 1-5b). Thus, the second energy level can contain a total of eight electrons—and so can the third.

The way an atom reacts chemically is determined by the number and arrangement of its electrons. An atom is most stable when all of its electrons are at their lowest possible energy levels. Therefore the electrons of an atom fill the energy levels in order—the first is filled before the second, the second before the third, and so on. Moreover, an atom in which the outermost energy level is completely filled with electrons is more stable

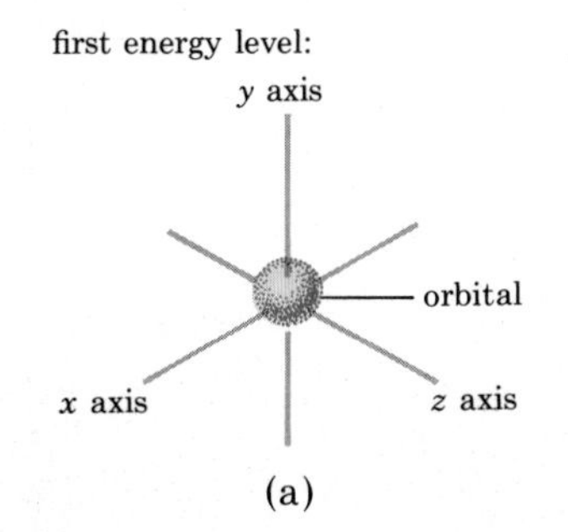

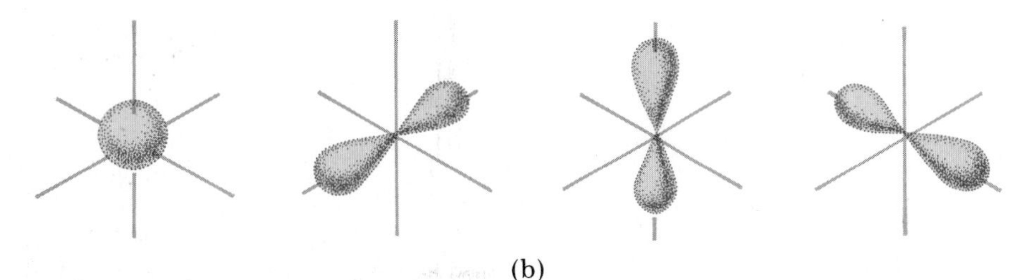

1-5 The most accurate representation of our current knowledge of atomic structure is provided by orbital models. (a) *The two electrons at the first energy level of an atom occupy a single spherical orbital. The nucleus is at the intersection of the axes (indicated by the gray lines).*

(b) *At the second energy level, there are four orbitals, each containing two electrons. One of these orbitals is spherical and the other three are dumbbell-shaped. The axes of the dumbbell-shaped orbitals are perpendicular to one another.*

For clarity, the orbitals are shown individually in this diagram. In reality, however, the spherical orbital of the second energy level surrounds the orbital of the first energy level, and portions of the dumbbell-shaped orbitals pass through the two spherical orbitals. The orbitals influence one another and determine the overall shape of the atom.

than one in which the outer energy level is only partially filled. For example, helium (atomic number 2) has two electrons at the first energy level, which means that its outer energy level (in this case, also its lowest energy level) is completely filled. Helium therefore tends to be unreactive. Similarly, neon (atomic number 10) has two electrons at the first energy level and eight at the second energy level; both energy levels are completely filled, and neon is unreactive.

In the atoms of most elements, however, the outer energy level is only partially filled (Table 1-3). These atoms tend to interact with other atoms in such a way that after the reaction both atoms have completely filled outer energy levels. Some atoms lose electrons; others gain electrons; and, in many of the most important chemical reactions that occur in living systems, atoms share their electrons with each other.

Bonds and Molecules

When atoms interact with one another, resulting in filled outer energy levels, new, larger particles are formed. These particles, consisting of two or more atoms, are known as *molecules*, and the forces that hold them together are known as *bonds*. There are two principal types of bonds: ionic and covalent.

Ionic Bonds

For many atoms, the simplest way to attain a completely filled outer energy level is either to gain or to lose one or two electrons. For example, chlorine (atomic number 17) needs one electron to complete its outer energy level (see Table 1-3). By contrast, sodium (atomic number 11) has a single electron in its outer energy level. This electron is strongly attracted by the chlorine atom and jumps from the sodium to the chlorine. As a result of this transfer, both atoms have completed outer energy levels and all the electrons are at the lowest possible energy levels. In the process, however, the original atoms have become electrically charged. Such charged atoms are known as *ions*. The chlorine atom, having accepted an electron from

Table 1-3 Electron arrangements in some familiar elements

Element	Atomic Number	Number of Electrons in Each Energy Level*		
		First	Second	Third
Hydrogen (H)	1	1	–	–
Helium (He)	2	2	–	–
Carbon (C)	6	2	4	–
Nitrogen (N)	7	2	5	–
Oxygen (O)	8	2	6	–
Neon (Ne)	10	2	8	–
Sodium (Na)	11	2	8	1
Phosphorus (P)	15	2	8	5
Sulfur (S)	16	2	8	6
Chlorine (Cl)	17	2	8	7

*The first energy level can hold a maximum of 2 electrons; the second and third energy levels can each hold a maximum of 8 electrons.

1-6 (a) *Oppositely charged ions, such as the sodium and chloride ions depicted here as spheres, attract one another. Table salt is crystalline NaCl, a latticework of alternating* Na^+ *and* Cl^- *ions held together by their opposite charges. Such bonds between oppositely charged ions are known as ionic bonds.*

(b) *The regularity of the latticework is reflected in the structure of salt crystals, magnified here about 14 times.*

sodium, now has one more electron than proton and is a negatively charged chloride ion: Cl^-. Conversely, the sodium ion has one less electron than proton and is positively charged: Na^+.

Because of their charges, positive and negative ions attract one another. Thus the sodium ion (Na^+) with its single positive charge is attracted to the chloride ion (Cl^-) with its single negative charge. The resulting substance, sodium chloride (NaCl), is ordinary table salt (Figure 1-6). Similarly, the calcium ion (Ca^{2+}), which has lost two electrons and thus has a double positive charge, can attract and hold two Cl^- ions. Calcium chloride is identified in chemical shorthand as $CaCl_2$, with the subscript 2 indicating that two chloride ions are present for each ion of calcium.

Bonds that involve the mutual attraction of ions of opposite charge are known as *ionic bonds*. Such bonds can be quite strong, but, as we shall see in the next chapter, many ionic substances break apart easily in water, producing free ions. Small ions such as Na^+ and Cl^- make up less than 1 percent of the weight of most living matter, but they play crucial roles. Potassium ion (K^+) is the principal positively charged ion in most organisms, and many essential biological processes occur only in its presence. Both Na^+ and K^+ are involved in the production and propagation of the nerve impulse. Calcium ion (Ca^{2+}) is required for the contraction of muscles, and magnesium ion (Mg^{2+}) forms a part of the chlorophyll molecule, the molecule in green plants and algae that traps radiant energy from the sun.

1-7 *This electron micrograph shows a nerve fiber resting in (and, at the right, pulled away from) a groove in a muscle fiber. The fibers are magnified about 1,900 times.*

The nerve fiber transmits impulses to the muscle that cause it to contract. Sodium and potassium ions are involved in producing and propagating impulses along the nerve fiber, while calcium ions are required for the contraction of the muscle fiber.

Covalent Bonds

Another way for an atom to complete its outer energy level is by sharing electrons with another atom. Bonds formed by shared pairs of electrons are known as *covalent bonds*. In a covalent bond, the shared pair of electrons forms a new orbital (called a molecular orbital) that envelops the nuclei of both atoms (Figure 1–8). In such a bond, each electron spends part of its time around one nucleus and part of its time around the other. Thus the electron sharing both completes the outer energy level and neutralizes the nuclear charge.

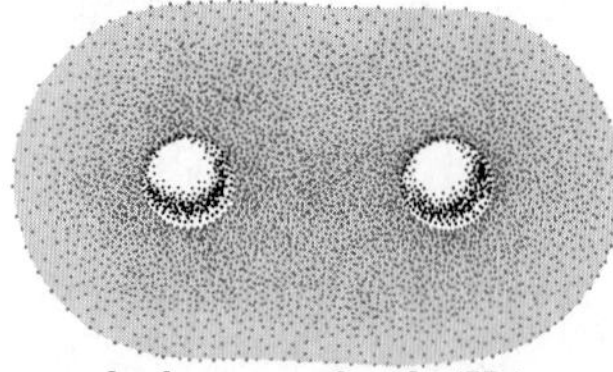

1–8 *In a molecule of hydrogen, each atom shares its single electron with the other atom. As a result, both atoms effectively have a filled first energy level, containing two electrons—a highly stable arrangement. This type of bond, in which electrons are shared, is known as a covalent bond.*

Atoms that need to gain electrons to achieve a filled, and therefore stable, outer energy level have a strong tendency to form covalent bonds. Thus, for example, a hydrogen atom forms a single covalent bond with another hydrogen atom. It can also form a covalent bond with any other atom that needs to gain an electron to complete its outer energy level.

Of extraordinary importance in living systems is the capacity of carbon atoms to form covalent bonds. A carbon atom has four electrons in its outer energy level (see Table 1–3). It can share each of those electrons with another atom, forming covalent bonds to as many as four other atoms (Figure 1–9). The covalent bonds formed by a carbon atom may be with different atoms (most frequently hydrogen, oxygen, and nitrogen) or with other carbon atoms. As we shall see in Chapter 3, this tendency of carbon atoms to form covalent bonds with other carbon atoms gives rise to the large molecules that form the structures of living organisms and that participate in essential life processes.

Polar Covalent Bonds

The electrons in covalent bonds are not always shared equally between the atoms involved. The nucleus of one kind of atom may have a greater attractive force for electrons than does the nucleus of another kind of atom. As a consequence, the shared electrons tend to spend more time around the nucleus with the greater attraction. The atom around which the electrons spend more time has a slightly negative charge, whereas the other atom has a slightly positive charge, since its nuclear charge is not entirely neutralized.

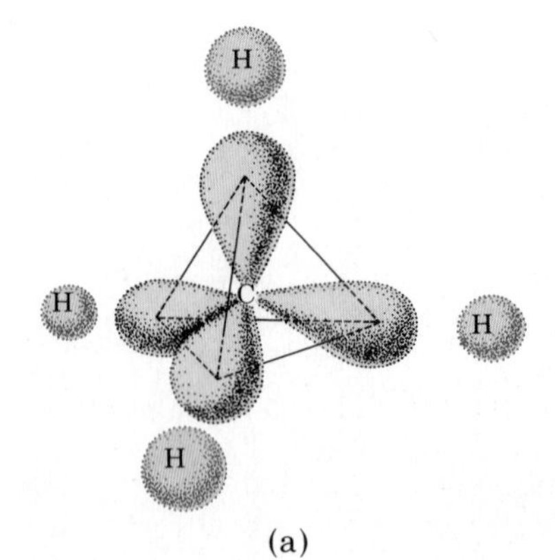

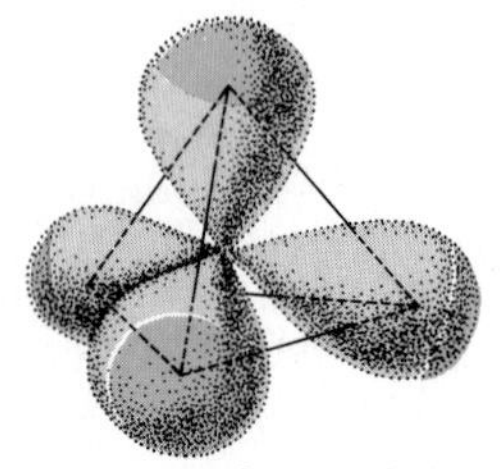

1–9 *A carbon atom, with four electrons in its outer energy level, can form covalent bonds with as many as four other atoms. Each electron occupies an orbital oriented toward one of the corners of a tetrahedron.* (a) *When a carbon atom reacts with four hydrogen atoms, each of its electrons forms a covalent bond with the single electron of one hydrogen atom, producing a methane molecule* (b). *Each pair of electrons moves in a new, molecular orbital. The molecule has the shape of a tetrahedron.*

Covalent bonds in which electrons are shared unequally are known as *polar covalent bonds*, and the molecules containing these bonds are said to be *polar molecules*. Such molecules often contain oxygen atoms, to which electrons are strongly attracted. The polar properties of many oxygen-containing molecules have very important consequences for living things. For example, many of the special properties of water (H_2O), upon which life depends, derive largely from its polar nature, as we shall see in the next chapter.

The Biologically Important Elements

Of the 92 naturally occurring elements, only six make up some 99 percent of all living tissue (Table 1–4). These six elements are carbon, hydrogen, nitrogen, oxygen, phosphorus, and sulfur, conveniently remembered as CHNOPS. These are not the most abundant of the elements of the earth's surface. Why, as life assembled and evolved from stardust, were these of such importance? One clue is that the atoms of all of these elements need to gain electrons to complete their outer energy levels (see Table 1–3). Thus they generally form covalent bonds. Because these atoms are small, the shared electrons in the bonds are held closely to the nuclei, producing very stable molecules. Moreover, with the exception of hydrogen, atoms of these elements can all form bonds with two or more atoms, giving rise to the complex molecules necessary for life.

Table 1–4 Atomic composition of three representative organisms

Element	Human	Alfalfa	Bacterium
Carbon	19.37%	11.34%	12.14%
Hydrogen	9.31	8.72	9.94
Nitrogen	5.14	0.83	3.04
Oxygen	62.81	77.90	73.68
Phosphorus	0.63	0.71	0.60
Sulfur	0.64	0.10	0.32
CHNOPS total:	97.90%	99.60%	99.72%

SUMMARY

Matter is composed of atoms, the smallest units of chemical elements. Atoms are made up of smaller particles. The nucleus of an atom contains positively charged protons and (except for hydrogen, ^1H) neutrons, which have no charge. The atomic number of an atom is equal to the number of protons in its nucleus. The atomic weight of an atom is the sum of the number of protons and neutrons in its nucleus. The chemical properties of an atom are determined by its electrons—small, negatively charged particles found outside the nucleus. The number of electrons in an atom equals the number of protons and thus the atomic number.

The electrons of an atom have differing amounts of energy. Electrons closer to the nucleus have less energy than those farther from the nucleus and thus are at a lower energy level. An electron tends to move to the lowest possible energy level, but with an input of energy, it can be boosted to a higher energy level. When the electron returns to a lower energy level, energy is released.

The chemical behavior of an atom is determined by the number and arrangement of its electrons. An atom is most stable when all of its electrons are at their lowest possible energy levels and those energy levels are completely filled with electrons. The first energy level can hold two electrons, and the second and third energy levels can each hold eight electrons. Chemical reactions between atoms result from the tendency of atoms to reach the most stable electron arrangement possible.

Particles consisting of two or more atoms are known as molecules, which are held together by chemical bonds. Two common types of bonds are ionic and covalent. Ionic bonds are formed by the mutual attraction of particles of opposite electric charge; such particles, formed when an electron jumps from one atom to another, are known as ions. In covalent bonds, pairs of electrons are shared between atoms; in some covalent bonds, known as polar covalent bonds, pairs of electrons are shared unequally, giving the molecule regions of positive and negative charge.

Six elements (CHNOPS) make up 99 percent of all living matter. The atoms of all of these elements are small and form tight, stable covalent bonds. With the exception of hydrogen, they can all form covalent bonds with two or more atoms, giving rise to the complex molecules that characterize living systems.

QUESTIONS

1. Describe the three types of particles of which atoms are composed. What is the atomic number of an atom? The atomic weight?

2. Although no model of the atom gives us an exact "picture," different models can help us to understand important characteristics of atoms. What characteristic of the atom was stressed by the planetary model? What important characteristic of electrons is emphasized by the Bohr model? What additional information about electrons is provided by the orbital model?

3. The street lights in many cities contain bulbs filled with sodium vapor. When electricity is passed through the bulb, a brilliant yellow light is given off. What is happening to the sodium atoms to cause this?

4. What is the difference between an energy level and an orbital? How many electrons can the first energy level of an atom hold? The second energy level? The third energy level?

5. Magnesium has an atomic number of 12. How many electrons are in its first energy level? Its second energy level? Its third energy level? How would you expect magnesium and chlorine to interact? Write the formula for magnesium chloride.

6. Explain the differences between ionic, covalent, and polar covalent bonds. What tendency of atoms causes them to interact with each other, forming bonds?

7. What six elements make up the bulk of living tissue? What characteristics do the atoms of these six elements share?

2

Water

In this chapter and the next, we are going to examine the molecules of which living things are composed. By far the most abundant of these molecules is water, which makes up 50 to 95 percent of the weight of any functioning living system.

Life on this planet began in water, and today, wherever liquid water is found, life is also present. There are one-celled organisms that eke out their entire existence in no more water than can cling to a grain of sand. Some kinds of algae are found only on the melting undersurfaces of polar ice floes. Certain bacteria can tolerate the near-boiling water of hot springs. In the desert, plants race through an entire life cycle—seed to flower to seed—following a single rainfall. In the rain forest, the water cupped in the leaves of a tropical plant forms a microcosm in which a myriad of small organisms are born, spawn, and die.

Water is the most common liquid on earth. Three-fourths of the surface of the earth is covered by water. In fact, if the earth's land surface were absolutely smooth, all of it would be $2\frac{1}{2}$ kilometers* under water. But

* About $1\frac{1}{2}$ miles. A metric table with English equivalents is in Appendix A.

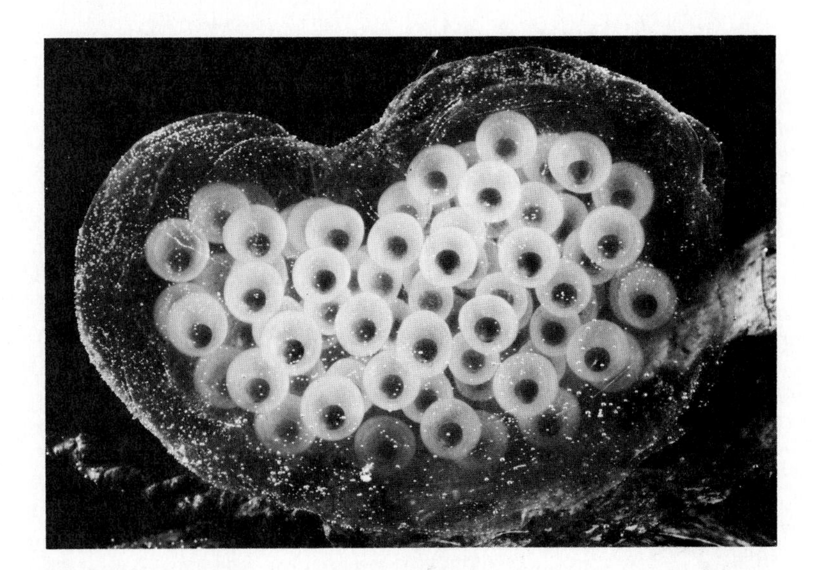

2-1 *The first living systems came into being, according to present hypotheses, in the warm primitive seas, and for many organisms, ourselves included, each new individual begins life bathed and cradled in water. These are frog eggs, encased in a jellylike sac.*

do not mistake "common" for "ordinary"; water is not in the least an ordinary liquid. Compared with other liquids it is, in fact, quite extraordinary. If it were not, it is highly unlikely that life on earth could ever have evolved.

Water and the Hydrogen Bond

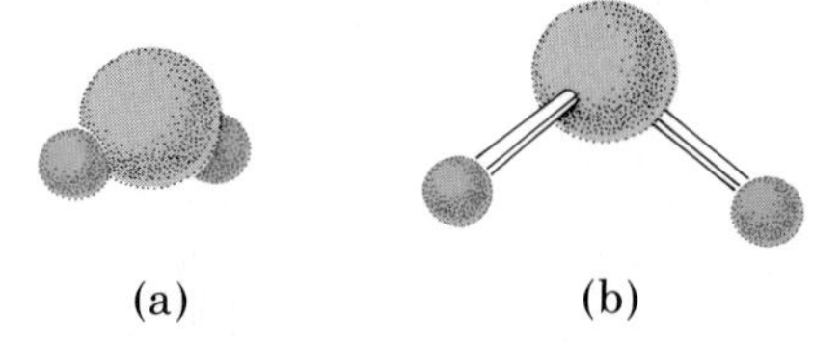

2-2 *The structure of the water molecule (H_2O) can be depicted in several different ways. In the space-filling model* (a), *the oxygen atom is represented by the gray sphere and the hydrogen atoms by the colored spheres. Because of its simplicity, this model is often used as a convenient symbol of the water molecule.*

The ball-and-stick model (b) *emphasizes that the atoms are joined by covalent bonds; it also gives some indication of the geometry of the molecule. A more accurate description of the molecule's shape is provided by the orbital model in Figure 2-3a.*

In order to understand why water is so extraordinary and how, as a consequence, it can play its unique and crucial role in relation to living systems, we must look at its molecular structure. Each water molecule is made up of two atoms of hydrogen and one atom of oxygen (Figure 2-2). Each of the hydrogen atoms is held to the oxygen atom by a covalent bond; that is, the single electron of each hydrogen atom is shared with the oxygen atom, which also contributes an electron to each bond.

The water molecule as a whole is neutral in charge, having an equal number of electrons and protons. However, the molecule is polar (page 23). Because of the very strong attraction of the oxygen nucleus for electrons, the shared electrons of the covalent bonds spend more time around the oxygen nucleus than they do around the hydrogen nuclei. As a consequence, the region near each hydrogen nucleus is a weakly positive zone. Moreover, the oxygen atom has four additional electrons in its outer energy level. These electrons are paired in two orbitals that are not involved in covalent bonding to hydrogen. Each of these orbitals is a weakly negative zone. Thus, the water molecule, in terms of its polarity, is four-cornered, with two positively charged "corners" and two negatively charged ones (Figure 2-3a).

When one of these charged regions comes close to an oppositely charged region of another water molecule, the force of the attraction forms a bond between them, which is known as a *hydrogen bond.* In water, a hydrogen bond forms between a negative "corner" of one water molecule and a positive "corner" of another. Every water molecule can establish hydrogen bonds with four other water molecules (Figure 2-3b).

Any single hydrogen bond is relatively weak and has an exceedingly short lifetime; on an average, each hydrogen bond lasts approximately

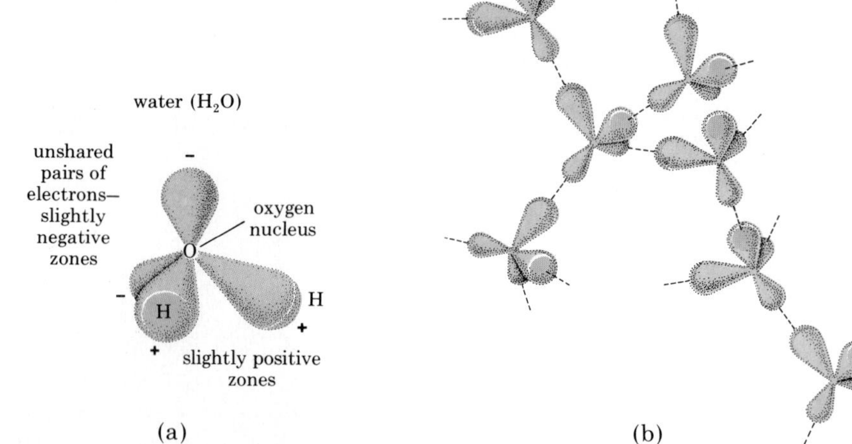

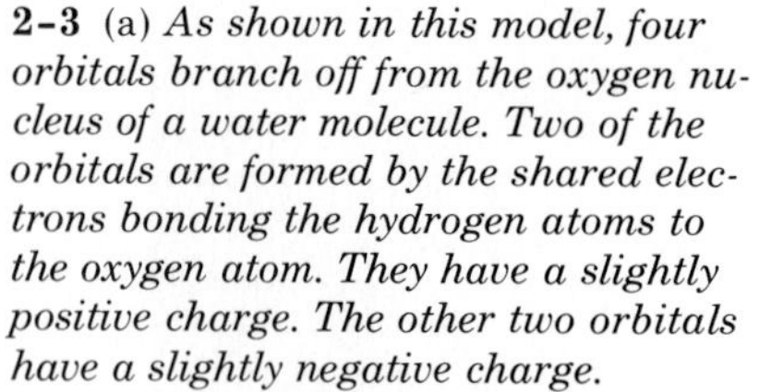

2-3 (a) *As shown in this model, four orbitals branch off from the oxygen nucleus of a water molecule. Two of the orbitals are formed by the shared electrons bonding the hydrogen atoms to the oxygen atom. They have a slightly positive charge. The other two orbitals have a slightly negative charge.*

(b) *As a result of these positive and negative zones, each water molecule can form hydrogen bonds (dashed lines) with four other water molecules. Under ordinary conditions of pressure and temperature, the hydrogen bonds are continually breaking and re-forming in a shifting pattern. Thus water is a liquid.*

2-4 *Droplets of dew on grass. The spherical structure of the droplets is a result of the attraction of water molecules for one another.*

1/100,000,000,000th of a second. But, as one is broken, another is made. All together, the hydrogen bonds have considerable strength, causing the water molecules to cling together in a liquid under ordinary conditions of temperature and pressure.

Now let us look at some of the consequences of these attractions among water molecules, especially as they affect living systems.

Surface Tension

Look at water dripping from a faucet. Each drop clings to the rim and dangles for a moment by a thread of water; then just as the tug of gravity breaks it loose, its outer surface is drawn taut, to form a sphere as the drop falls free. Gently place a needle or a razor blade flat on the surface of the water in a glass. Although the metal is denser than water, it floats. Look at a pond in spring or summer; you will see water striders and other insects walking on its surface almost as if it were solid. These phenomena are all the result of *surface tension*. Surface tension is the result of the cohesion, or clinging together, of the water molecules. (*Cohesion* is, by definition, the holding together of like substances. *Adhesion* is the holding together of unlike substances.)

2-5 *A water strider's legs dimple the surface of a pond. The water strider, which lives on this surface, has specialized hairs on its first and third pairs of legs that enable it to rest upon the surface film, depressing it but not penetrating. The second pair of legs, which penetrate the film, serve as propelling oars.*

The only liquid with a surface tension greater than that of water is mercury. Atoms of mercury are so greatly attracted to one another that they tend not to adhere to anything else. Water, however, because of its negative and positive charges, adheres strongly to any other charged molecules and to charged surfaces. The "wetting" capacity of water—that is, its ability to coat a surface—results from its polar structure, as does its cohesiveness.

Capillary Action and Imbibition

If you hold two dry glass slides together and dip one corner in water, the combination of cohesion and adhesion will cause water to spread upward between the two slides. This is *capillary action*. Capillary action similarly causes water to rise in very fine glass tubes, to creep up a piece of blotting paper, or to move slowly through the small spaces between soil particles and so become available to the roots of plants.

2-6 *The germination of seeds begins with changes in the seed coat that permit a massive uptake of water. The embryo and surrounding structures then swell, bursting the seed coat. In these acorns, photographed on the forest floor, the embryonic roots have emerged through the tough outer layers of the fruits.*

Imbibition ("drinking up") is the capillary movement of water molecules into substances such as wood or gelatin, which swell as a result. The pressures developed by imbibition can be astonishingly great. It is said that stone for the ancient Egyptian pyramids was quarried by driving wooden pegs into holes drilled in the rock face and then soaking the pegs with water. The swelling of the wood created a force great enough to break the stone slab free. Seeds imbibe water as they begin to germinate, swelling and bursting their seed coats (Figure 2-6).

Water and Temperature

Specific Heat of Water

The amount of heat a given amount of a substance requires for a given increase in temperature is its *specific heat* (also called heat capacity). One calorie* is defined as the amount of heat that will raise the temperature of 1 gram (1 cubic centimeter) of water 1°C. The specific heat of water is about twice the specific heat of oil or alcohol; that is, approximately 0.5 calorie will raise the temperature of 1 gram of oil or alcohol 1°C. It is four times the specific heat of air and 10 times that of iron. Only liquid ammonia has a higher specific heat. In other words, it requires a high input of energy to raise the temperature of water.

Heat is a form of energy—the *kinetic energy*, or energy of movement, of molecules. Molecules are always in motion; they vibrate, rotate, and shift position in relation to other molecules. Temperature, which is measured in degrees, reflects the *average* kinetic energy of the molecules. Heat, which is measured in calories, reflects the *total* energy in a collection of molecules; it includes both the molecular movement and the mass and number of moving molecules present. For example, a lake may have a lower temperature than does a bird flying over it, but the lake contains more heat because it has many more molecules in motion.

The high specific heat of water is a consequence of hydrogen bonding. The hydrogen bonds in water tend to restrict the movement of the molecules. In order for the kinetic energy of water molecules to increase sufficiently for the temperature to rise 1°C, it is necessary first to rupture a number of the hydrogen bonds holding the molecules together. When you heat a pot of water, much of the heat energy added to the water is used in breaking the hydrogen bonds between the water molecules. Only a relatively small amount of heat energy is therefore available to increase molecular movement.

What does the high specific heat of water mean in biological terms? It means that for a given rate of heat input, the temperature of water will rise more slowly than the temperature of almost any other material. Conversely, the temperature will drop more slowly as heat is removed. Because so much heat input or heat loss is required to raise or lower the temperature of water, organisms that live in the oceans or large bodies of fresh water live in an environment where the temperature is relatively constant. Also, the high water content of terrestrial plants and animals helps them to maintain a relatively constant internal temperature. This constancy of temperature is critical because biologically important chemical reactions take place only within a narrow temperature range.

* Nutritional calories are actually kilocalories (kcal); 1,000 calories equal 1 kilocalorie.

2-7 *Ammonia is very similar to water in its molecular structure, and biologists have speculated about whether it might substitute for water in life processes. The ammonia molecule (NH_3) is made up of hydrogen atoms covalently bonded to nitrogen, which, like the oxygen in the water molecule, retains a slight negative charge. But because there are three hydrogens with slight positive charges to one nitrogen, ammonia does not have the cohesive power of water and evaporates much more quickly. Perhaps this is why no form of life based on ammonia has been found, although NH_3 was very common in the primitive atmosphere.*

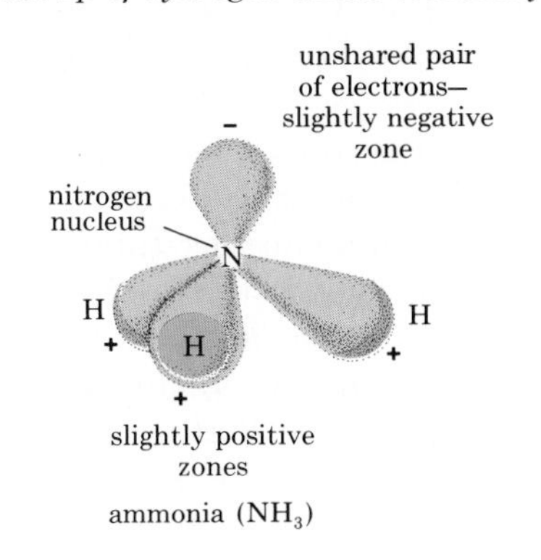

"Ammonia! Ammonia!"

[*Drawing by R. Grossman,* © *1962 The New Yorker Magazine, Inc.*]

Heat of Vaporization

Vaporization—or evaporation, as it is more commonly called—is the change from a liquid to a gas. Water has a high *heat of vaporization*. It takes more than 500 calories to change a gram of liquid water into vapor, 60 times as much as for ether and almost twice as much as for ammonia.

Hydrogen bonding is also responsible for water's high heat of vaporization. Vaporization comes about because some of the most rapidly moving molecules of a liquid break loose from the surface and enter the air. The hotter the liquid, the more rapid the movement of its molecules and, hence, the more rapid the rate of evaporation. But, whatever the temperature, so long as a liquid is exposed to air that is less than 100 percent saturated with the vapor of that liquid, evaporation will take place, right down to the last drop.

In order for a water molecule to break loose from its fellow molecules—that is, to vaporize—the hydrogen bonds have to be broken. This requires heat energy. At water's boiling point (100°C at a pressure of 1 atmosphere), 540 calories are needed to change 1 gram of liquid water into vapor. As a consequence, when water evaporates, as from the surface of your skin or a leaf, the escaping molecules carry a great deal of heat away with them. Thus evaporation has a cooling effect. Evaporation from the surface of a land-dwelling plant or animal is one of the principal ways in which these organisms "unload" excess heat and so stabilize their temperatures.

2-8 *Dogs unload heat by panting, which involves short, shallow breaths. When it is hot, a dog breathes at a rate of about 300 to 400 respirations a minute, compared with a rate of from 10 to 40 a minute in cool surroundings. As a result of panting, air passes rapidly over the large, moist tongue, evaporating saliva. Most of the air is inhaled through the nose and exhaled through the open mouth.*

Freezing

Water exhibits another peculiarity as it undergoes the transition from a liquid to a solid (ice). In most liquids, the density—that is, the weight of the material in a given volume—increases as the temperature drops. This greater density occurs because the individual molecules are moving more slowly and so the spaces between them decrease. The density of water also increases as the temperature drops, until it nears 4°C. Then the water molecules come so close together that every one of them can become hydrogen-bonded to four others. In the course of this bonding, the molecules move apart from each other, creating an open latticework (Figure 2-9a) that is the most stable structure for an ice crystal. Thus water as a solid takes up more volume than water as a liquid. Ice is less dense than liquid water and therefore floats in it.

This increase in volume has occasional disastrous effects on water pipes but, on the whole, turns out to be enormously beneficial for life forms. If water continued to contract as it froze, ice would be heavier than liquid water. As a result, lakes and ponds and other bodies of water would freeze from the bottom up. Once ice began to accumulate on the bottom, it would tend not to melt, season after season. Spring and summer might stop the freezing process, but laboratory experiments have shown that if ice is held to the bottom of even a relatively shallow tank, water can be boiled on the top without melting the ice. Thus if water did not expand when it froze, it would continue to freeze from the bottom up, year after year, and never melt again. Eventually, the body of water would freeze solid and life in it would be destroyed. By contrast, the layer of floating ice that actually forms tends to protect the organisms in the water. The ice layer effectively seals off the liquid water beneath it, keeping its temperature at or above the freezing point of water (0°C).

2-9 (a) *The crystalline structure of ice is an open latticework in which each water molecule is hydrogen-bonded to four other water molecules. In this structure, which is responsible for the beautiful patterns seen in snowflakes and frost, the water molecules are actually farther apart than they are in liquid water.*

(b) *When water freezes in the cracks and crevices of rock, the force created by its expansion splits the rock. Over long periods of time, this process breaks up masses of rock and contributes to the formation of soil.*

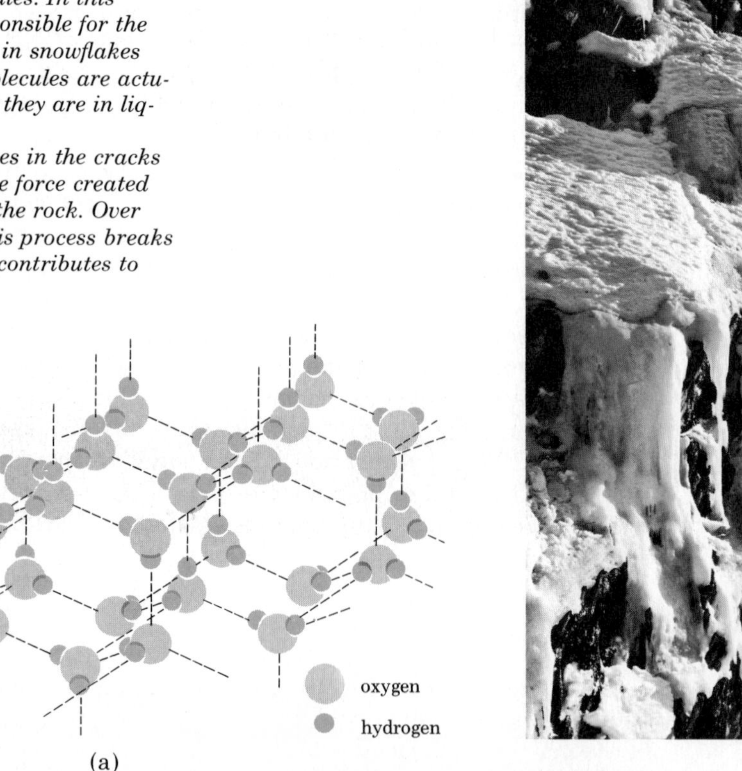

(a)

(b)

When ice melts, it draws heat from its surroundings. For example, ice cubes in a glass of water gradually melt, cooling the water in the process. The heat energy absorbed by the ice breaks the hydrogen bonds of the latticework. Conversely, as water freezes, it releases heat into its surroundings. In this way, ice and snow also serve as temperature stabilizers, particularly during the transition periods of fall and spring. Moderation of sudden changes in temperature gives organisms time to make seasonal adjustments essential to survival.

Water as a Solvent

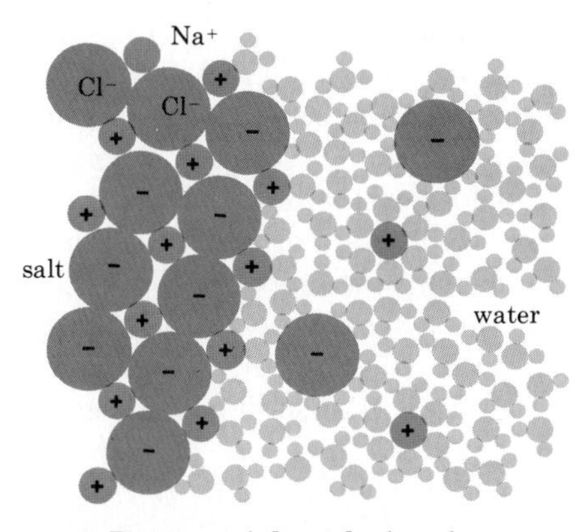

2-10 *Because of the polarity of water molecules, water can serve as a solvent for ionic substances and polar molecules. This diagram shows sodium chloride (NaCl) dissolving in water as the water molecules cluster around the individual sodium and chloride ions, separating them from each other.*

Many substances within living systems are found in solution. (A *solution* is a uniform mixture of the molecules of two or more substances. The substance present in the greatest amount—usually a liquid—is called the solvent, and the substances present in lesser amounts are called solutes.) The polarity of water molecules is responsible for water's capacity as a solvent. The polar water molecules tend to separate ionic substances, such as sodium chloride (NaCl), into their constituent ions. Then, as shown in Figure 2-10, the water molecules cluster around and segregate the charged ions. Many of the molecules important in living systems—such as sugars—also have areas of positive and negative charge. These molecules therefore attract water molecules and so dissolve in water.

Polar molecules that readily dissolve in water are often called *hydrophilic* ("water-loving"). Such molecules slip into solution easily because their partially charged regions attract water molecules and so compete with the attraction between the water molecules themselves.

Molecules, such as fats, that lack polar regions tend to be very insoluble in water. The hydrogen bonding between the water molecules acts as a force to exclude the nonpolar molecules. As a result of this exclusion, nonpolar molecules tend to cluster together in water, just as droplets of fats tend to coalesce, for example, on the surface of chicken soup. Such molecules are said to be *hydrophobic* ("water-fearing"), and the clusterings are known as hydrophobic interactions.

We will encounter these properties of hydrophobic and hydrophilic molecules again in later chapters. These weak forces—hydrogen bonds and hydrophobic forces—play very important roles in determining the shapes and properties of biologically important molecules.

Ionization: Acids and Bases

In liquid water, there is a slight tendency for a hydrogen atom to jump from the oxygen atom to which it is covalently bonded to the oxygen atom to which it is hydrogen-bonded (Figure 2-11). In this reaction, two ions are produced: the hydronium ion (H_3O^+) and the hydroxide ion (OH^-).

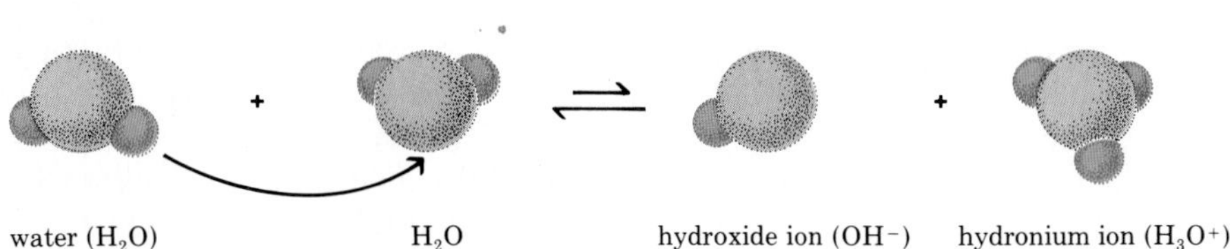

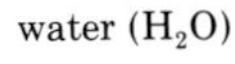

2-11 *When water ionizes, a hydrogen atom shifts from the oxygen atom to which it is covalently bonded to the oxygen atom to which it is hydrogen-bonded. In this diagram, the large spheres represent oxygen and the small spheres represent hydrogen.*

In any given volume of pure water, a small but constant number of water molecules will be ionized in this way. The number is constant because the tendency of water to ionize is exactly offset by the tendency of the ions to reunite; thus even as some molecules are ionizing, an equal number of others are forming, a state known as dynamic *equilibrium*.

Although the hydrogen ion does not exist in a separate form, by convention the ionization of water is expressed by the equation:

$$HOH \rightleftharpoons H^+ + OH^-$$

The arrows indicate that the reaction can go in either direction. The fact that the arrow pointing toward HOH is longer indicates that, at equilibrium, most of the H_2O is not ionized. As a consequence, in any sample of pure water, only a small fraction exists in ionized form.

In pure water, the number of H^+ ions exactly equals the number of OH^- ions. This is necessarily the case since neither ion can be formed without the other when only H_2O molecules are present. However, when an ionic substance or a substance with polar molecules is dissolved in water it may change the relative numbers of H^+ and OH^- ions. For example, when hydrochloric acid (HCl) dissolves in water, it is almost completely ionized into H^+ and Cl^- ions; as a result, an HCl solution contains more H^+ ions than OH^- ions. Conversely, when sodium hydroxide (NaOH) dissolves in water, it forms Na^+ and OH^- ions; thus, in a solution of sodium hydroxide in water, there are more OH^- ions than H^+ ions.

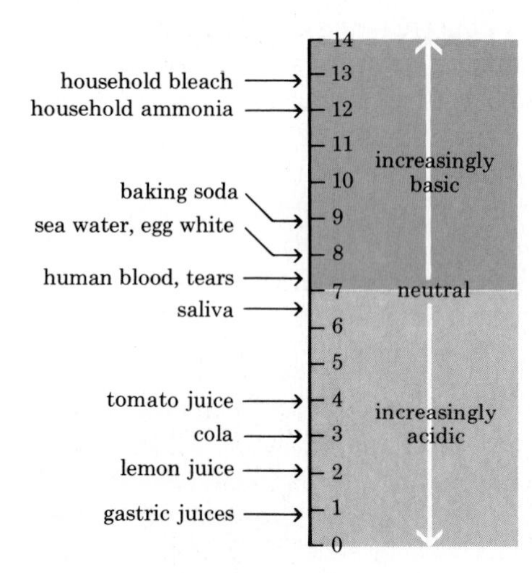

2-12 *pH values of various common solutions. A difference of one pH unit reflects a tenfold difference in the H^+ ion concentration. A cola drink, for instance, is 10 times as acidic as tomato juice, and gastric juices are about 100 times more acidic than cola drinks.*

A solution acquires the properties we recognize as acidic when the number of H^+ ions exceeds the number of OH^- ions; conversely, a solution is basic (alkaline) when the number of OH^- ions exceeds the number of H^+ ions. Thus, an *acid* is a substance that causes an increase in the relative number of H^+ ions in a solution, and a *base* is a substance that causes an increase in the relative number of OH^- ions. Strong acids and bases are substances, like HCl and NaOH, that ionize almost completely in water. Weak acids and bases are those that ionize only slightly.

Chemists define degrees of acidity by means of the pH scale. At pH 7.0, the concentrations of H^+ and OH^- ions are exactly the same, as they are in pure water. This is a neutral state. Any pH below 7.0 is acidic, and any pH above 7.0 is basic. Figure 2-12 shows the pH values of some familiar solutions. A difference of one pH unit represents a tenfold difference in the concentration of H^+ ion. Most of the chemical reactions of living systems take place within a narrow range of pH that hovers around neutrality.

The Water Cycle

Most of the water on earth—almost 98 percent—is in liquid form, in the oceans, lakes, and streams. Of the remaining 2 percent, some is frozen in polar ice and glaciers, some is in the soil, some is in the atmosphere in the form of vapor, and some is in the bodies of living organisms.

Water is made available to land organisms by processes powered by the sun. Solar energy evaporates water from the oceans, leaving the salt behind. Water is also evaporated, but in much smaller amounts, from moist soil surfaces, from the leaves of plants, and from the bodies of other organisms. These molecules—now water vapor—are carried up into the atmosphere by air currents. Eventually they fall to the earth's surface again as snow or rain. Most of the water falls on the oceans, since these cover most of

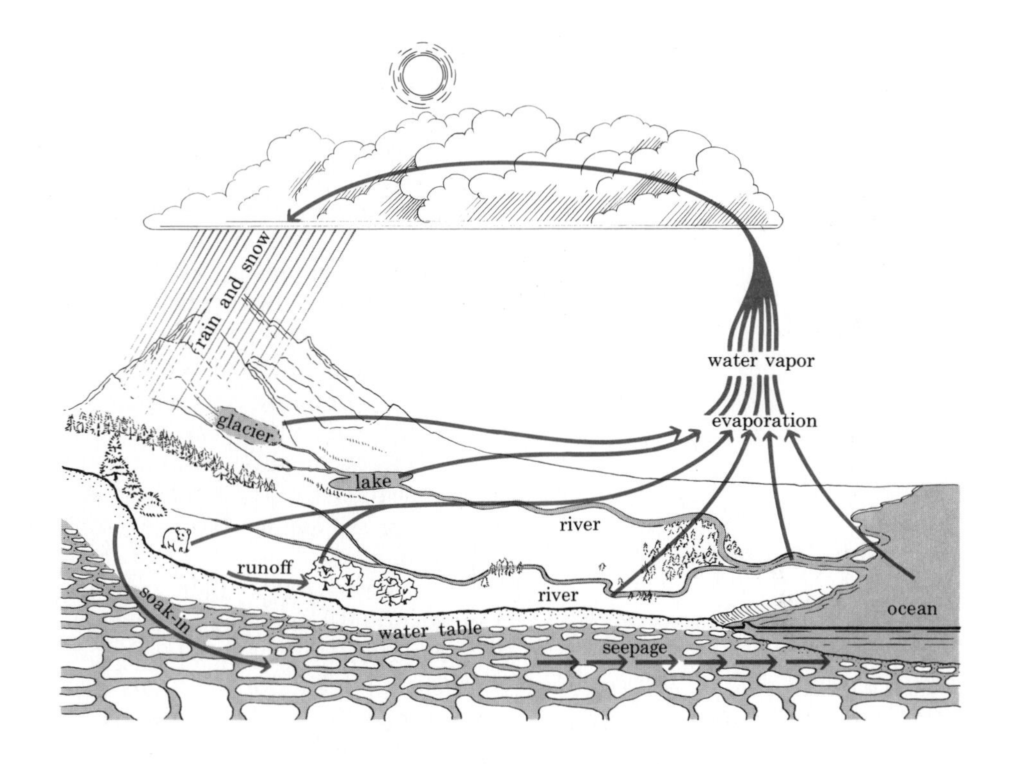

2-13 *The water cycle.*

the earth's surface. The water that falls on land is pulled back to the oceans by the force of gravity. Some of it, reaching low ground, forms ponds or lakes and streams or rivers, which pour water back into the oceans.

Some of the water that falls on the land percolates down through the soil until it reaches a zone of saturation. In the zone of saturation, all pores and cracks in the rock are filled with water (groundwater). The upper surface of the zone of saturation is known as the water table. Below the zone of saturation is solid rock, through which the water cannot penetrate. The deep groundwater, moving extremely slowly, eventually also reaches the ocean, thereby completing the water cycle.

As we have seen in this chapter, water, essential for life, is a most extraordinary substance. The earth's supply of water is the permanent possession of our planet, held to its surface by the force of gravity. Through the movements of the water cycle, it is perpetually available to living organisms.

SUMMARY

Water, the most common liquid on the earth's surface and the major component, by weight, of all living things, has a number of remarkable properties. These properties are a consequence of its molecular structure and are responsible for water's "fitness" for its roles in living systems.

Water is made up of two hydrogen atoms and one oxygen atom held together by covalent bonds. The water molecule is polar, with two weakly negative zones and two weakly positive zones. As a consequence, weak bonds form between water molecules. Such bonds, which link a somewhat positively charged hydrogen atom that is part of one molecule to a somewhat negatively charged oxygen atom that is part of another molecule, are known as hydrogen bonds. Each water molecule can form hydrogen bonds with four other water molecules. Although individual bonds are weak and constantly shifting, the total strength of the bonds holding the molecules together is very great.

Because of the hydrogen bonds holding the water molecules together (cohesion), water has a high surface tension and a high specific heat (the amount of heat that a given amount of the substance requires for a given increase in temperature). It also has a high heat of vaporization (the heat required to change a liquid to a gas). Just before it freezes, water expands; thus ice has a lower density and a larger volume than liquid water. As a result, ice floats on water.

The polarity of the water molecule is responsible for water's adhesiveness and hence its tendency for capillary movement. Similarly, water's polarity makes it a good solvent for ions and polar molecules. Molecules that dissolve readily in water are known as hydrophilic. Water molecules, as a consequence of their polarity, actively exclude uncharged molecules from solution. Molecules that are excluded from water solution are known as hydrophobic.

Water has a slight tendency to ionize, that is, to separate into H^+ ions and OH^- ions. In pure water, the number of H^+ ions and OH^- ions is equal. A solution that contains more H^+ ions than OH^- ions is acidic; one that contains more OH^- ions than H^+ ions is basic. The pH scale reflects the proportion of H^+ ions to OH^- ions. An acidic solution has a pH less than 7.0; a basic solution has a pH more than 7.0. Almost all of the chemical reactions of living systems take place within a narrow range of pH around neutrality.

Through the water cycle, the water above, on, and below the earth's surface is recirculated. As a result, it is continuously available to living organisms.

QUESTIONS

1. (a) Sketch the water molecule and label the areas of positive and negative charge. (b) What are the major consequences of the polarity of the water molecule? (c) How are these effects important to living systems?

2. The trick with the razor blade works better if the blade is a little greasy. Why?

3. Surfaces such as glass or raincoat cloth can be made "nonwettable" by application of silicone oils or other substances that cause water to bead up instead of spread flat. What do you suppose is happening, in molecular terms, when a surface becomes nonwettable?

4. What is vaporization? Describe the changes that take place in water as it vaporizes. What is heat of vaporization? Why does water have an unusually high heat of vaporization?

5. Generally, coastal areas have more moderate temperatures (not as cold in the winter, nor as hot in the summer) than inland areas at the same latitude. What explanation can you give for this phenomenon?

6. Describe the properties of water that are exhibited and suggested by the photograph at the left.

3

Organic Molecules

In this chapter, we present some of the types of *organic molecules*—that is, molecules containing carbon—that are found in living things. As you will see, the molecular drama is a grand spectacular with, literally, a cast of thousands. A single bacterial cell contains some 5,000 different kinds of molecules, and an animal or plant cell has about twice that many. These thousands of molecules, however, are composed of a relatively few elements (CHNOPS). Similarly, relatively few kinds of molecules play the major roles in living systems. Consider this, if you will, an introduction to the principal characters; the plot begins to unfold in Chapter 4.

The Central Role of Carbon

As we noted previously, water makes up from 50 to 95 percent of a living system, and ions, such as sodium (Na^+) and potassium (K^+), account for no more than 1 percent. Almost all the rest, chemically speaking, is composed of organic molecules.

Four different kinds of organic molecules are found in large quantities in organisms. These four are carbohydrates (which are composed of sugars), lipids (including fats and waxes), proteins (which are composed of amino acids), and nucleotides (complex molecules that can combine to form very large molecules known as nucleic acids). In this chapter, we shall consider the first three of these. Nucleotides will be taken up in later chapters when we discuss their key roles in energy exchanges and in genetic systems.

All of these molecules—carbohydrates, lipids, proteins, and nucleotides—contain carbon, hydrogen, and oxygen. In addition, proteins contain nitrogen and sulfur, and nucleotides, as well as some lipids, contain nitrogen and phosphorus.

The Carbon Backbone

As you will recall from Chapter 1, a carbon atom can form four covalent bonds with as many as four different atoms. Methane (CH_4), which is natural gas, is an example (Figure 1–9, page 22). Even more important, in terms of carbon's biological role, carbon atoms can form bonds with each other. Ethane contains two carbons; propane, three; butane, four; and so on, forming long carbon chains (Figure 3-1). In general, an organic molecule derives its overall shape from the arrangement of the carbon atoms that form the backbone, or skeleton, of the molecule. The shape of the molecule, in turn, determines many of its properties and its function within living systems.

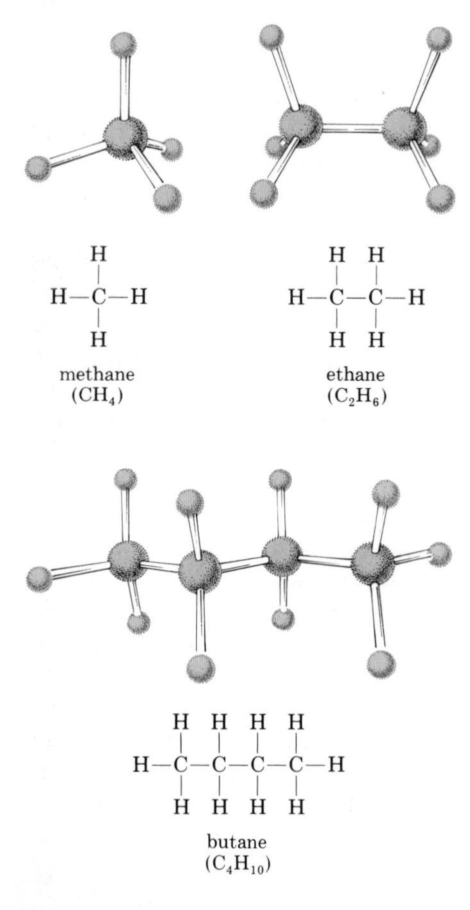

3-1 *Ball-and-stick models and structural formulas of methane, ethane, and butane. In the models, the gray spheres represent carbon atoms and the smaller colored spheres represent hydrogen atoms. The sticks in the models—and the lines in the structural formulas—represent covalent bonds, each of which consists of a pair of shared electrons.*

In the molecules shown in Figure 3-1, every carbon bond that is not occupied by another carbon atom is taken up by a hydrogen atom. Such compounds, consisting of only carbon and hydrogen, are known as *hydrocarbons*. Structurally, they are the simplest kind of organic compounds. Although hydrocarbons are derived from the remains of organisms that died millions of years ago (see page 253), they are relatively unimportant in living organisms. They are, however, of great economic importance. The bonds between carbon atoms and those between carbon and hydrogen atoms contain a great deal of stored energy. This energy is released when hydrocarbons are burned. The liquid fuels upon which we depend—gasoline, diesel fuel, and heating oil—are all hydrocarbons.

Carbohydrates

Carbohydrates are organic compounds composed of carbon, hydrogen, and oxygen. The proportion is roughly one carbon atom to two hydrogen atoms to one oxygen atom, as indicated by the shorthand formula CH_2O. Carbohydrates serve two extremely important functions in living systems. First, they are the principal energy-storage molecules in most organisms. Second, they are essential structural components in many organisms, particularly plants.

Carbohydrates are formed from small molecules known as *sugars*. There are three kinds of carbohydrates, classified according to the number of sugar molecules they contain. *Monosaccharides* ("single sugars"), such as ribose, glucose, and fructose, contain only one sugar molecule. *Disaccharides* ("two sugars") consist of two sugar molecules linked covalently. Familiar examples are sucrose (table sugar), maltose (malt sugar), and lactose (milk sugar). *Polysaccharides*, such as cellulose and starch, contain many sugar molecules linked together. A compound that is composed of many similar or identical subunits, as is a polysaccharide, is known as a *polymer* (from "many" and "part"). Each subunit is called a *monomer*.

Monosaccharides

Monosaccharides are the building blocks from which living cells construct polysaccharides and other essential molecules. Moreover, they are the principal energy source for most organisms.

In the process of *photosynthesis*, which we shall discuss in Chapter 8, the green cells of plants and algae capture radiant energy from sunlight and use it to combine carbon dioxide and water into sugar molecules. The captured energy is stored in the chemical bonds of the sugar molecules. This process, which involves many chemical reactions, can be summarized as follows:

$$\text{carbon dioxide} + \text{water} + \text{energy} \longrightarrow \text{sugar} + \text{oxygen}$$

In another series of reactions, living cells can break down (oxidize) the sugar molecules, releasing carbon dioxide, water, and energy:

$$\text{sugar} + \text{oxygen} \longrightarrow \text{carbon dioxide} + \text{water} + \text{energy}$$

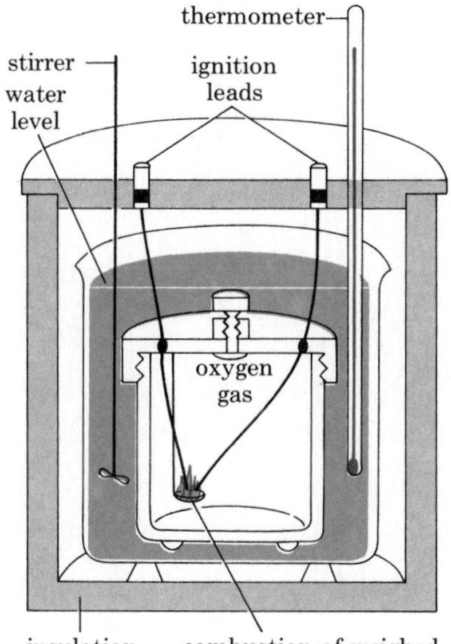

3-2 A calorimeter is used to measure the amount of energy stored in an organic compound. A known quantity of glucose or some other material is ignited electrically. As it burns, the rise in the temperature of the surrounding water is measured. Using the specific heat of water and the known weight of water in the calorimeter, one can then calculate the number of calories released by the burning of the sample.

Other organic substances, such as wood, gasoline, and alcohol, also release their stored energy when they are oxidized, as occurs when they are burned for fuel.

The amount of stored energy an organic compound contains is measured in terms of the amount of heat energy (expressed in calories) released when the compound is oxidized (Figure 3-2). In such a reaction, existing chemical bonds are broken and new bonds form. Further, the new bonds—those in water (H_2O) and carbon dioxide (CO_2) molecules—have less energy than the bonds that held the sugar or wood or gasoline molecules together; that is, the electrons of the newly formed molecules are at lower energy levels. As a result, there is some energy "left over" when the new bonds form. In the laboratory, a fireplace, or an automobile engine, this energy is released as heat. In a living organism, much of it is used to form chemical bonds among other atoms and molecules that are also present. Consequently, there is comparatively little release of heat, which, in any large amount, would be harmful to the organism.

The monosaccharide glucose is the principal energy source for most living cells. When a mole* of glucose is broken down to carbon dioxide and water, 686 kilocalories of energy are released. The yield is the same whether the reaction takes place in a living cell or in the laboratory. Inside the cell, almost 40 percent of this energy is repackaged in other chemical compounds, as we shall see in Chapter 7. In the laboratory, all of the energy is given off as heat.

Disaccharides

Disaccharides consist of two monosaccharide units covalently bonded together. Sucrose, our common table sugar, is a disaccharide made up of a glucose molecule and a fructose molecule. It is the form in which sugar is transported from the photosynthetic cells of plants (mostly in the leaf), where it is produced, to other parts of the plant body. Another common disaccharide is lactose, a sugar that occurs in milk. Lactose is made up of glucose combined with another monosaccharide, galactose. Sugar is transported through the blood of many insects in the form of another disaccharide, known as trehalose, which consists of two glucose units linked together.

Disaccharides are formed by the removal of a molecule of water from two monosaccharide molecules, a type of chemical reaction known as *condensation* (Figure 3-3). The addition of a molecule of water to a disaccharide splits the molecule back into two monosaccharide molecules. This splitting is known as *hydrolysis*, from *hydro*, meaning "water," and *lysis*, "breaking apart."

Hydrolysis is an energy-releasing reaction. The hydrolysis of sucrose yields 5.5 kilocalories per mole. Conversely, the formation of sucrose from glucose and fructose requires an energy input of 5.5 kilocalories per mole of sucrose.

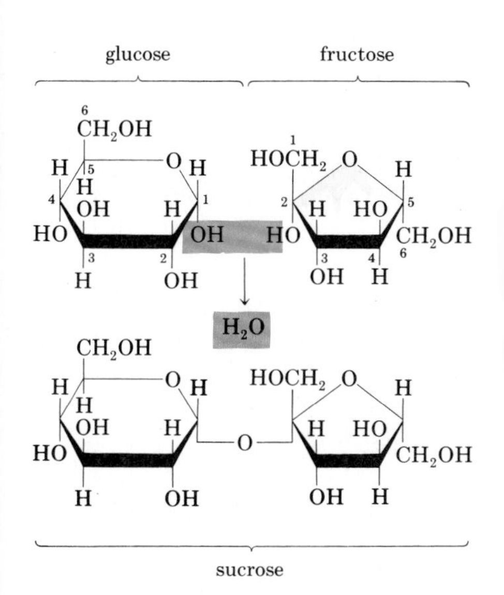

3-3 Sucrose is a disaccharide made up of two monosaccharides, glucose and fructose. The formation of sucrose involves the removal of a molecule of water (condensation). Splitting sucrose requires, conversely, the addition of a water molecule (hydrolysis).

* The mole is the principal unit of measure for quantities of substances involved in chemical reactions. The number of particles (whether ions, atoms, or molecules) in 1 mole of any substance is always exactly the same: 6.023×10^{23}. For example, 1 mole of hydrogen ions contains 6.023×10^{23} ions, 1 mole of carbon atoms contains 6.023×10^{23} carbon atoms, and 1 mole of glucose contains 6.023×10^{23} glucose molecules.

Representations of Molecules

As we saw in Chapters 1 and 2, chemists have developed various models to represent the structures of atoms and molecules. Each of these models is a way of organizing a particular set of scientific data and of focusing attention on particular characteristics of atoms and molecules.

Because the properties of a molecule depend on its three-dimensional characteristics, physical models are often the most useful. For example, ball-and-stick models of the kind shown in Figure 3–1 emphasize the geometry of a molecule and, in particular, the bonds between atoms. But these models fail to suggest the overall shape of the molecule created by the movement of electrons within their orbitals.

Space-filling models of the monosaccharides glucose and fructose. The darkest spheres, almost completely hidden in the center of each molecule, represent the carbon atoms. The large dark spheres at the surface of each molecule represent oxygen atoms, while the white spheres represent hydrogen atoms.

A closer approximation of molecular shape is provided by space-filling models. Each atom is represented by the edge of its outermost orbitals. Space-filling models are misleading, however, in that molecules do not fill space in the same way that we think of a table or a rock as filling space. The atoms that make up molecules consist mostly of empty space. If the outer orbitals of the electrons in an oxygen atom were the size of the perimeter of the Astrodome in Houston, the nucleus would be a ping-pong ball in the center of the stadium. What "fills" the space in molecules are areas of charge, produced by the movements of the electrons around the nuclei. One molecule "sees" another molecule in terms of those areas of charge. As a consequence, for instance, an enzyme that breaks down a glucose molecule will not have any effect on a fructose molecule because of the differences in the shape of those areas of charge. All the intricate biochemistry that goes on in the cell is based on this ability of molecules to "recognize" one another.

Ball-and-stick and space-filling models are often used in the laboratory, but they are less useful on paper because it is necessary to see them from all angles to see all of the atoms and their bonds. The most accurate two-dimensional representations of molecular structure are orbital models, such as those shown in Figure 2–3 (page 26). For molecules containing more than a few atoms, however, orbital models become extremely complicated. Thus, when representing complex molecules, such as those found in living systems, chemists usually use molecular formulas or structural formulas. A molecular formula indicates the number of atoms of each kind within the molecule, while a structural formula shows how the atoms are bonded to one another.

Glucose, for example, has 6 carbon atoms, 12 hydrogen atoms, and 6 oxygen atoms. Its molecular formula is $C_6H_{12}O_6$. However, fructose, another monosaccharide, also contains 6 carbons, 12 hydrogens, and 6 oxygens, and has a similar structure—a chain of carbon atoms to which hydrogen and oxygen atoms are attached. The differences between glucose and fructose are determined by which carbon atoms the other atoms are attached to. The molecules can therefore be distinguished by drawing structural formulas:

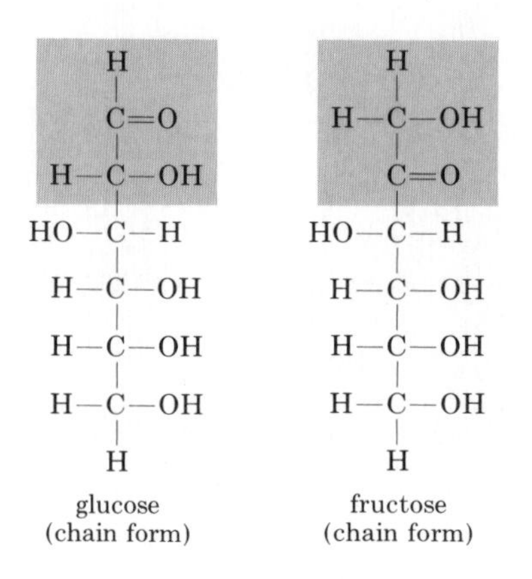

In these formulas, the symbol — represents a covalent bond, and the symbol = represents a *double* covalent bond (a bond formed by the sharing of four rather than two electrons between two atoms).

When glucose and fructose are in solution, however, they tend to form rings and so are more accurately represented by these structural formulas:

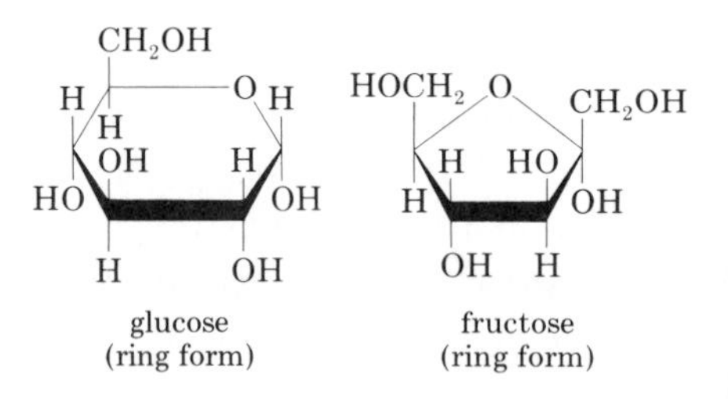

The lower edges of the rings are made thicker to hint at a three-dimensional structure. By convention, the carbon atoms at the intersections of the links in an organic ring structure are "understood" and not labeled.

Although structural formulas give us less information than physical models, they are a convenient tool as we examine the molecules involved in the structures and processes of living systems.

Polysaccharides

Polysaccharides are made up of monosaccharides linked together in long chains. They constitute storage forms for sugars. The principal storage polysaccharide in plants is starch (Figure 3–4), and in animals and fungi it is glycogen (Figure 3–5). Both starch and glycogen are built up of many glucose units. The differences between them are in the length of the polysaccharide chains (starch is made up of longer chains) and in the ways in which the glucose molecules are linked. Polysaccharides must be hydrolyzed to monosaccharides or disaccharides before they can be used as energy sources or transported through living systems.

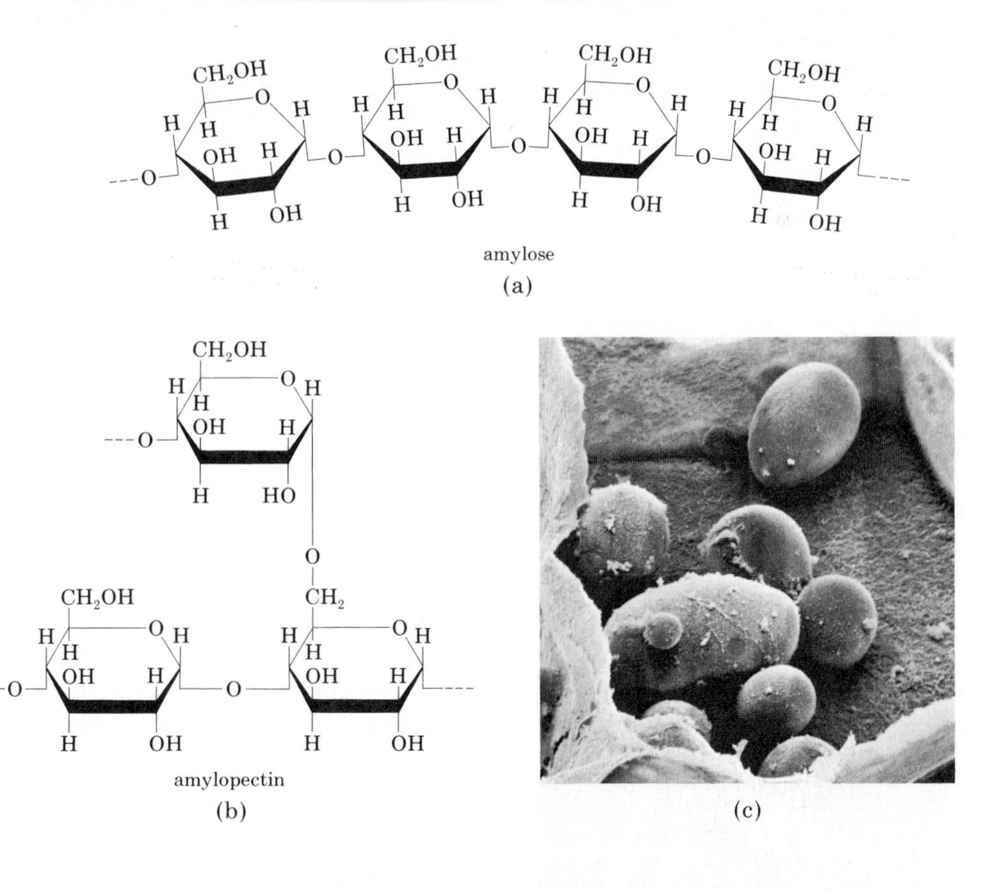

3–4 *In plants, sugars are stored in the form of starch. Starch is composed of two different types of polysaccharides, amylose* (a) *and amylopectin* (b). *A single molecule of amylose may contain 1,000 or more glucose units in a long unbranched chain, which winds to form a coil. A molecule of amylopectin is made up of about 48 to 60 glucose units arranged in shorter, branched chains.*

(c) *Starch molecules, perhaps because of their coiled nature, tend to cluster into granules. In this electron micrograph of a single storage cell of a potato, the spherical and egg-shaped objects are starch granules. They are magnified about 1,000 times.*

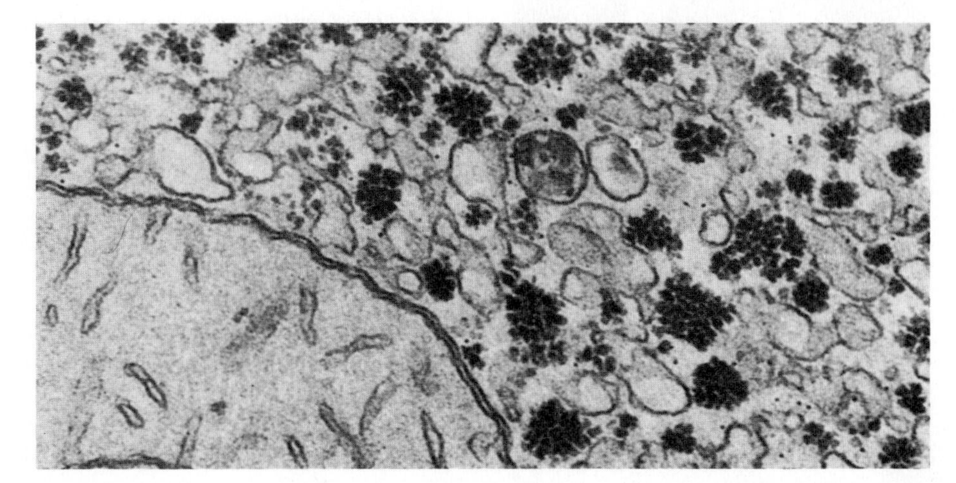

3–5 *Glycogen, which is the common storage form for sugar in humans and other vertebrates, resembles amylopectin in its general structure except that each molecule contains only 16 to 24 glucose units. The dark granules in this cross section of a liver cell are glycogen. When glucose is needed, it is provided by the conversion of glycogen.*

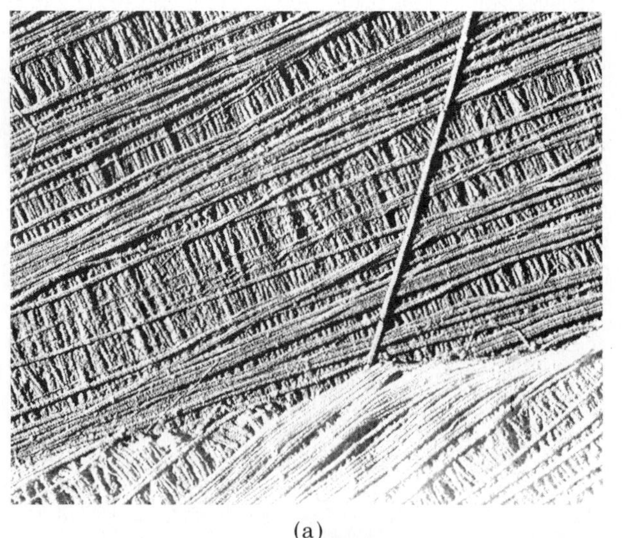

(a)

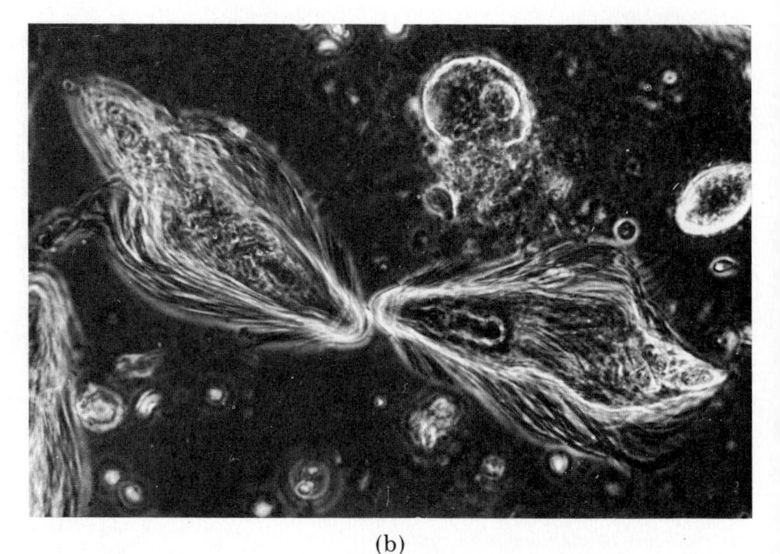

(b)

3-6 (a) *The plant cell wall is composed largely of cellulose. Each of the microfibrils you can see here (magnified about 30,000 times) is a bundle of hundreds of cellulose strands, and each strand is a chain of glucose monomers. The microfibrils, as strong as an equivalent amount of steel, are embedded in other polysaccharides.*

(b) *Because the orientation of the glucose units in cellulose is different from those in starch or glycogen, very few organisms can digest cellulose and so utilize its sugar monomers for energy. One such organism is the protist* Trichonympha, *of which two are shown here at a magnification of about 200 times.* Trichonympha *lives in the gut of termites and is entirely responsible for their well-known proficiency at digesting wood.*

3-7 *Green darner molting. The shells, or exoskeletons, of insects contain chitin, a nitrogen-containing polysaccharide. Some types of insects, after molting, recycle their sugars and nitrogen by thriftily eating their discarded exoskeletons.*

Polysaccharides also play structural roles. In plants, the principal structural polysaccharide is cellulose (Figure 3-6). Although cellulose, like starch, is a polymer of glucose, the bonds linking the glucose units in cellulose are slightly different from those in starch or glycogen. This small difference has a profound effect on the three-dimensional structure of the molecules and thus on their properties. Rather than forming granules, as do starch and glycogen molecules, cellulose molecules form long, rigid bundles. As a result, the biological role of cellulose is very different. In fact, cellulose can be hydrolyzed by only a few microorganisms. Cows and other ruminants, termites, and cockroaches can use cellulose for energy only because of microorganisms that inhabit their digestive tracts.

Chitin, which is a major component of the exoskeletons of insects and other arthropods and also of the cell walls of many fungi, is another tough, resistant, common polysaccharide (Figure 3-7).

Sugar as a Starting Material

In addition to serving as the building blocks of polysaccharides, glucose and other monosaccharides also serve as starting materials for the construction of more complex molecules. For example, although some of the sugar in our diet is converted to glycogen for storage in the liver (about four hours' worth), any excess is converted to fat for storage. The change from sugar to fat involves a partial breakdown and then a rebuilding of the molecule. These processes take place within the cell. The building blocks of proteins and of nucleotides are also synthesized within the cell, frequently using glucose or another monosaccharide as a starting material.

Why Is Sugar Sweet?

Actually it is not. Sugars have a number of objectively definable properties, such as molecular structure, melting point, and calorie content. Sweetness is not such a property. Just as beauty is in the eye of the beholder, sweetness resides not in the molecule itself but in the perception systems that have evolved to detect it.

Houseflies, for instance, have sugar detectors in their feet. If a housefly lights on a droplet of weak sugar solution, its proboscis (a tubular mouthpart through which it feeds) will automatically be extended. This useful response evolved over the millennia as a food-detecting mechanism. We similarly have sensory receptors, although our receptors, in keeping with our eating habits, are in our tongues. Because of sugar's value as an energy source, many other organisms have sugar-detecting mechanisms. A housefly reacts positively to about the same types of sugars that we do, although its detection mechanism, as gauged by the proboscis extension test, is about 10 million times more sensitive.

The capacity to detect sugars, and the fact that the sensation is pleasurable, probably evolved in insects and other animals because it promoted their ingestion of these energy-rich food molecules and, by extension, their survival. This taste for sugar has been exploited by plants, particularly the flowering plants. They have evolved nectaries, dripping with sugary syrup, by which they lure pollinators to their flowers, and fruits that turn sweet just as the seeds become mature and are ready to be transported to a germination site. Animals that consume these sugary products obtain not only energy-rich sugar molecules but also other essential nutrients, such as plant proteins and lipids, vitamins, and minerals.

Today, our taste for sugar is being further exploited by manufacturers of prepared foods. Many dry cereals are more than 50 percent sugar, and sugar is an added hidden ingredient in countless other products. We consume increasing amounts of sugar in place of other, equally essential nutrients. Sugar has thus become what behavioral scientists call a "supernormal stimulus." Baby cuckoos provide an example of a supernormal stimulus. Born in the nests of other birds, they gape so widely at the sight of food that the adult birds, all their parental instincts focused on this one oversized craw, let their own offspring starve. As another example, herring gulls abandon their own eggs for an artificial egg 1.5 times its size. And following a cue designed to lead us to a nutritious, vitamin-rich food supply, we let our children and ourselves grow fat, toothless, and malnourished, and die before our time.

The surface on which the feet of this housefly are resting has been covered with a weak solution of sugar water. Because its feet have "tasted" the sugar, its proboscis is extended.

Lipids

Lipids are fatty or oily substances. They have two principal distinguishing characteristics. First, they are nonpolar and so are generally insoluble in water. Second, they contain a larger proportion of carbon-hydrogen bonds than do any other organic compounds except hydrocarbons. As a consequence, lipids store a larger amount of energy than do most other organic compounds. Fats, for example, yield more than twice as many calories as an equivalent amount of carbohydrate or protein. On the average, fats yield about 9.3 kilocalories (kcal) per gram* as compared to 3.79 kcal per gram of carbohydrate or 3.12 kcal per gram of protein. The fact that lipids are insoluble and energy-rich determines their roles in living systems.

Fats

Lipids are stored in the cell chiefly in the form of fat, which cells synthesize from sugars. A fat molecule consists of three fatty acid molecules joined to one glycerol molecule (Figure 3–8). As with the disaccharides and polysaccharides, each bond is formed by the removal of a molecule of water (condensation).

The physical nature of the fat is determined by the length of the carbon chains in the fatty acids and by whether the acids are saturated or unsaturated. In saturated fatty acids, such as stearic acid, every carbon atom in the chain (except the last one) holds two hydrogen atoms, which completes the bonding possibilities of the carbon atom. Unsaturated fatty acids, such as oleic acid, contain carbon atoms joined by double bonds. Such carbon atoms are able to form additional bonds with other atoms (hence the term "unsaturated").

Unsaturated fats, which tend to be oily liquids, are more common in plants than in animals; examples are olive oil, peanut oil, and corn oil. Animal fats, such as butter and lard, contain saturated fatty acids and usually have higher melting temperatures.

* 1,000 grams = 1 kilogram = 2.2 pounds, so oxidation of a pound of fat would yield about 4,700 kilocalories, more than the 24-hour requirement for moderately active adults.

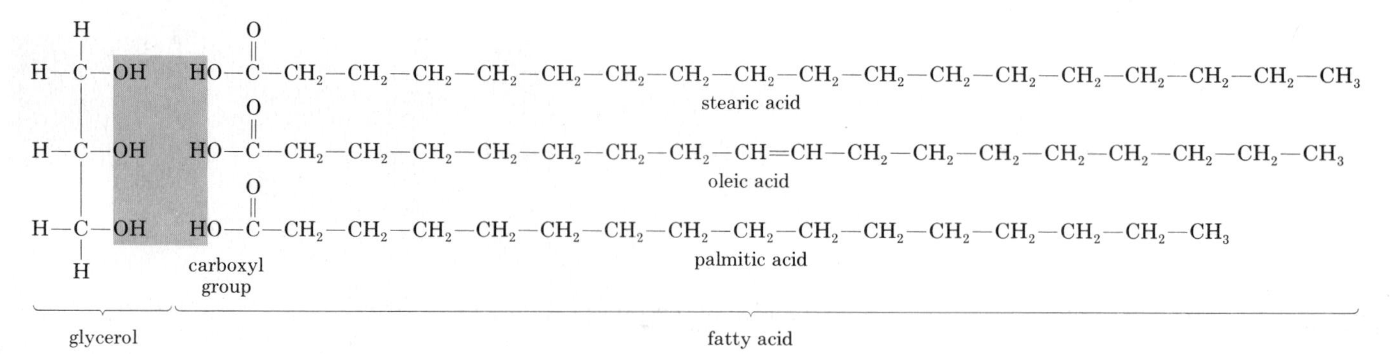

3–8 *A fat molecule consists of three fatty acids joined to a glycerol molecule. The long hydrocarbon chains of which the fatty acids are composed terminate in carboxyl (—COOH) groups, which become covalently bonded to the glycerol molecule. Each bond is formed by the removal of a molecule of water. The physical properties of the fat—such as its melting point—are determined by the lengths of the chains and by whether its component fatty acids are saturated or unsaturated. Three different fatty acids are shown here. Stearic acid and palmitic acid are saturated, and oleic acid is unsaturated, as you can see by the double bond in its structure.*

polar head — nonpolar tails

R—O—P(O⁻)(=O)—O—³CH₂

H—²C—O—C(=O)—CH₂CH₂CH₂CH₂CH₂CH₂CH₂CH=CHCH₂CH₂CH₂CH₂CH₂CH₂CH₂CH₃

H—¹C—O—C(=O)—CH₂CH₂CH₂CH₂CH₂CH₂CH₂CH₂CH₂CH₂CH₂CH₂CH₂CH₂CH₂CH₂CH₃

H

glycerol

3-9 *A phospholipid molecule consists of two fatty acids linked to a glycerol molecule, as in a fat, and a phosphate group (indicated by color) linked to the glycerol's third carbon. It also usually contains an additional chemical group, indicated by the letter R. The fatty acid "tails" are nonpolar and therefore insoluble in water (hydrophobic); the polar "head" containing the phosphate and R groups is soluble (hydrophilic).*

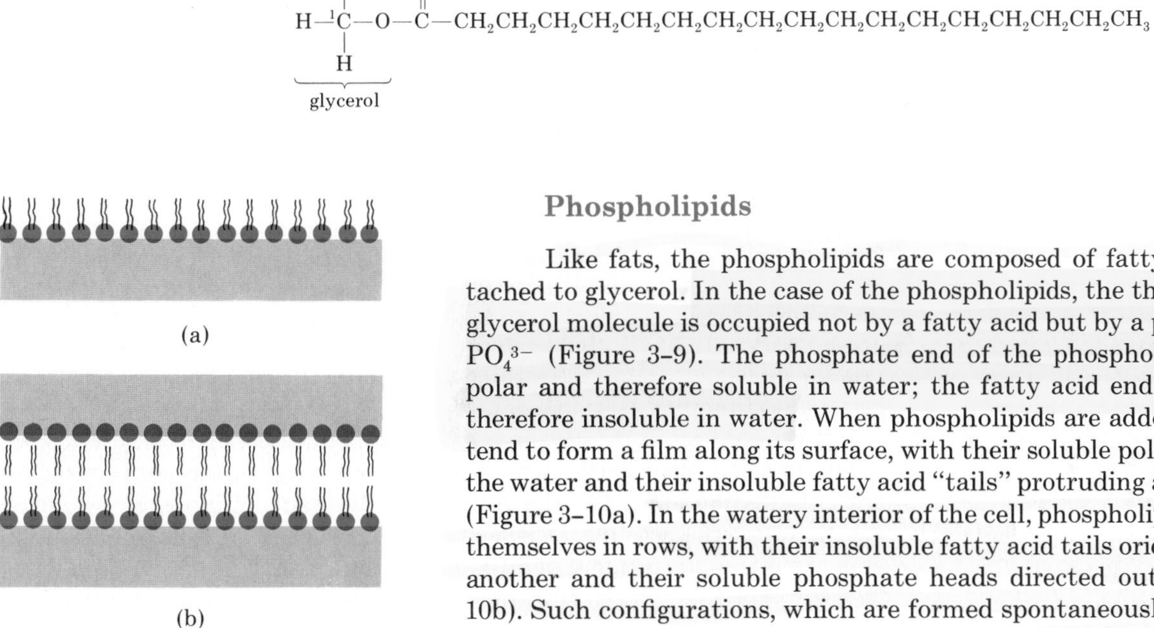

3-10 (a) *Because phospholipids have water-soluble heads and water-insoluble tails, they tend to form a thin film on a water surface with their tails extending above the water.* (b) *Surrounded by water, they spontaneously arrange themselves in two layers with their heads extending outward and their hydrophobic (water-fearing) tails inward. This arrangement is believed to be important in the structure of the cell membrane.*

Phospholipids

Like fats, the phospholipids are composed of fatty acid chains attached to glycerol. In the case of the phospholipids, the third carbon of the glycerol molecule is occupied not by a fatty acid but by a phosphate group, PO_4^{3-} (Figure 3-9). The phosphate end of the phospholipid molecule is polar and therefore soluble in water; the fatty acid end is nonpolar and therefore insoluble in water. When phospholipids are added to water, they tend to form a film along its surface, with their soluble polar "heads" under the water and their insoluble fatty acid "tails" protruding above the surface (Figure 3–10a). In the watery interior of the cell, phospholipids tend to align themselves in rows, with their insoluble fatty acid tails oriented toward one another and their soluble phosphate heads directed outward (Figure 3–10b). Such configurations, which are formed spontaneously by these molecules, are important in cellular structures, particularly membranes.

Waxes

Waxes are also a form of lipid. They are found as protective coatings on skin, fur, feathers, on the leaves and fruits of land plants (Figure 3–11), and on the exoskeletons of many insects.

3-11 *Waxes are also lipids. This electron micrograph shows waxy deposits on the upper surface of a eucalyptus leaf. The deposits are magnified 10,800 times. All groups of land plants synthesize waxes, which protect exposed plant surfaces from water loss.*

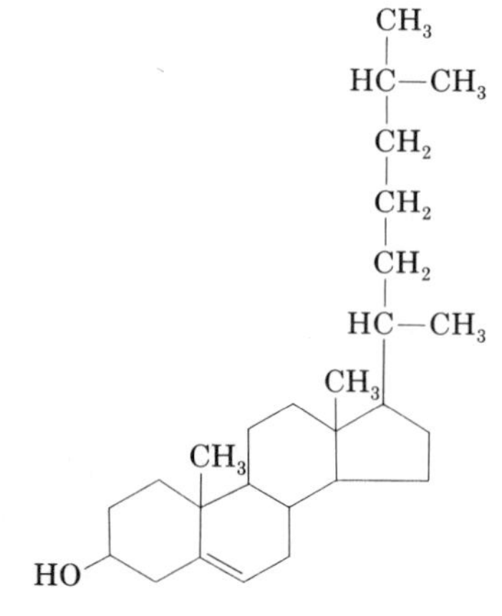

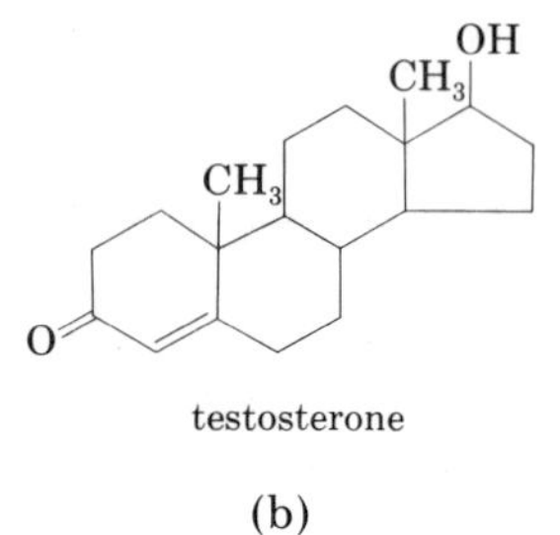

3-12 *Two examples of steroids.* (a) *The cholesterol molecule consists of four carbon rings and a hydrocarbon chain.* (b) *Testosterone, a male sex hormone synthesized from cholesterol by cells in the testes, also has the characteristic four-ring structure but lacks the hydrocarbon tail.*

Cholesterol and Other Steroids

The steroids do not resemble the other lipids structurally, but they are grouped with them because of their properties—particularly their insolubility in water. Structurally, they are characterized by four carbon rings (Figure 3-12).

The steroids are an important group of compounds. Cholesterol, along with the phospholipids, is found in cell membranes of animal cells. Sex hormones, such as estrogen and testosterone, and the hormones of the adrenal cortex, such as cortisol, are steroids; in the body they are synthesized from cholesterol. Cholesterol is synthesized in the liver from saturated fatty acids. It is also taken in in the diet, principally in meat, cheese, and egg yolks. High concentrations of cholesterol in the blood are associated with arteriosclerosis ("hardening of the arteries"), in which cholesterol is found in fatty deposits on the interior lining of diseased blood vessels (see page 381).

Proteins

Proteins, substances that play many essential and complicated roles in living systems, are made up of one or more chains of nitrogen-containing molecules known as *amino acids*. The sequence in which the amino acids are arranged in these chains determines the biological character of the protein molecule; even one small variation in this sequence may alter or destroy its function.

Proteins are composed of varying combinations of 20 different kinds of amino acids. Because protein molecules are large, often containing several hundred amino acids, the number of different amino acid sequences, and therefore the possible variety of protein molecules, is enormous—about as enormous as the number of different sentences that can be written with a 26-letter alphabet. Organisms, however, have only a very small fraction of the possible proteins. The single-celled bacterium *Escherichia coli,** for example, contains 600 to 800 different kinds of proteins at any one time; a cell in a plant or animal probably has several times that number. Each kind of protein has a special function and each, by its unique chemical nature, is specifically fitted for that function.

Amino Acids

Amino acids are made up of carbon, hydrogen, and oxygen, as are sugars. They also contain nitrogen. Every amino acid contains an amino group ($-NH_2$) and a carboxyl group ($-COOH$) bonded to the same carbon atom:

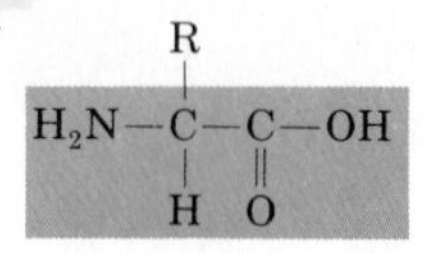

* Biologists use a binomial ("two-name") system for designating organisms. Every different kind of organism has a unique two-part name. The first part of the name refers to the genus (plural, genera) to which the organism belongs, and the second part refers to the species, a subdivision of the genus category. In this name, for example, *Escherichia* denotes the genus, while *coli* designates a particular kind, or species, of *Escherichia,* distinguished from all others by certain characteristics.

3-13 *Eight of the 20 different kinds of amino acids found in proteins. As you can see, their basic structures are the same, but they differ in their R groups. The other 12 amino acids—not shown here—are alanine (ala), glutamine (gln), histidine (his), leucine (leu), asparagine (asn), methionine (met), isoleucine (ile), aspartic acid (asp), lysine (lys), threonine (thr), tyrosine (tyr), and proline (pro).*

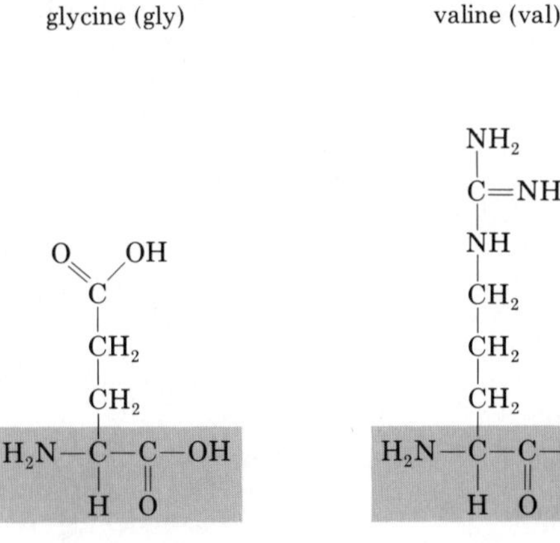

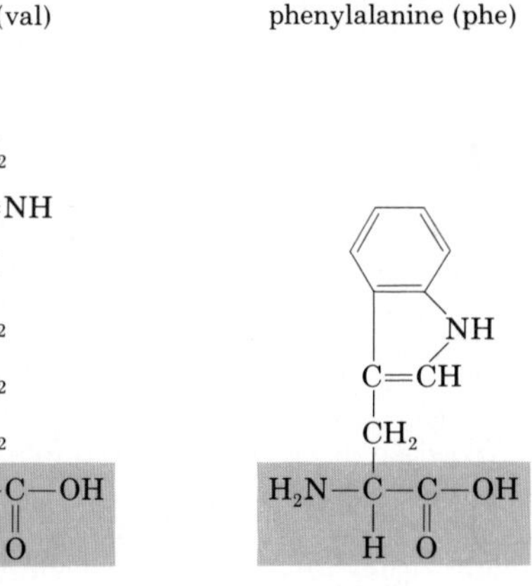

phenylalanine (phe)

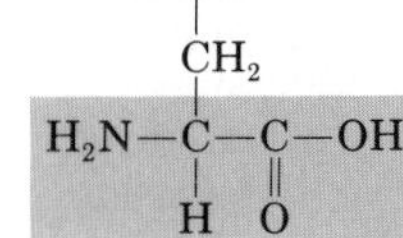

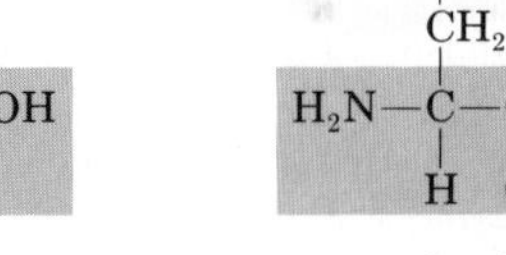

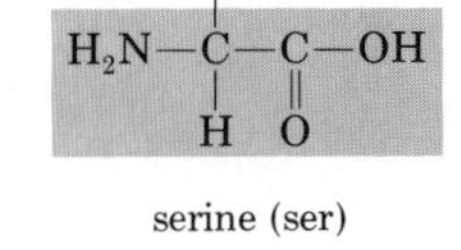

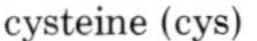

tryptophan (trp)

This is the basic structure of the molecule, and it is the same in all amino acids. The "R" stands for the rest of the molecule, which has a different chemical structure in each different kind of amino acid (Figure 3-13). These differences are very important because they determine the different biological properties of the individual amino acids and, as a consequence, of the various kinds of proteins.

The amino "head" of one amino acid can be linked to the carboxyl "tail" of another by the removal of a molecule of water, another example of condensation (Figure 3-14a). The linkage that is formed is known as a peptide bond, and the molecule that is formed by the linking of many amino acids is called a *polypeptide* (Figure 3-14b).

In order to assemble amino acids into proteins, a cell must have not only a large enough quantity of amino acids but also enough of every kind. This fact is of great importance in human nutrition (see essay on next page).

3-14 (a) *A peptide linkage is a covalent bond formed by condensation.* (b) *Polypeptides are polymers of amino acids linked together by peptide bonds, with the amino group of one acid joining the carboxyl group of its neighbor. The polypeptide chain shown here contains six different amino acids, but some chains may contain as many as 300 linked amino acid monomers.*

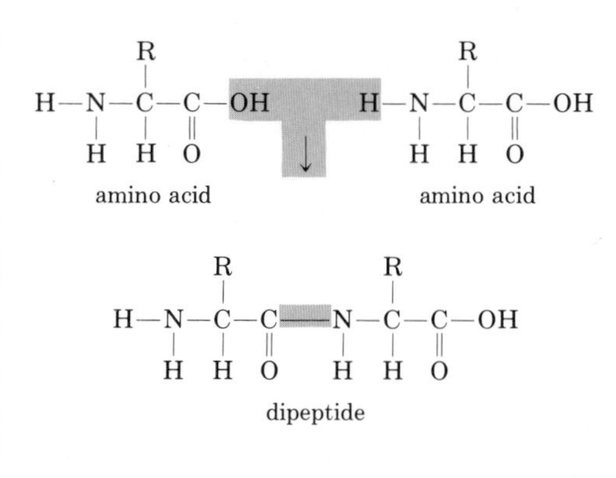

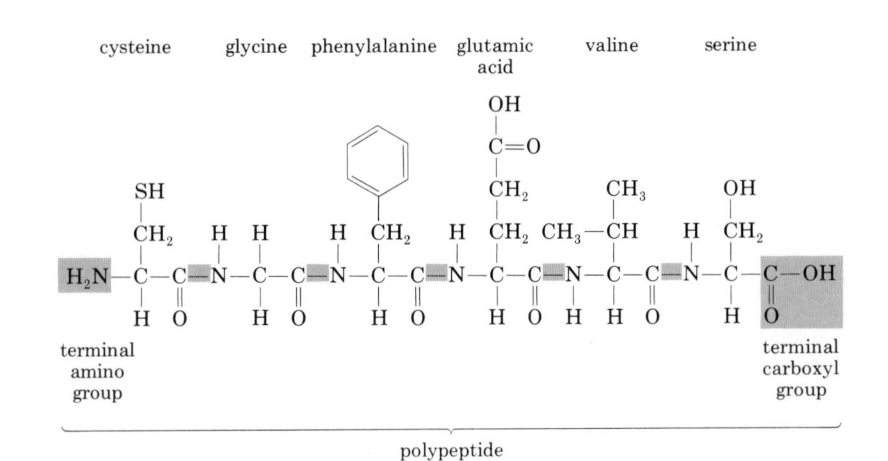

Amino Acids and Nitrogen

Like fats, amino acids are formed within living cells using sugars as starting materials. But while fats are made up only of carbon, hydrogen, and oxygen atoms, all available in the sugar and water of the cell, amino acids also contain nitrogen. Most of the earth's supply of nitrogen exists in the form of gas in the atmosphere. Only a few organisms, all microscopic, are able to incorporate nitrogen from the air into compounds—nitrites and nitrates—that can be used by living systems. Hence the proportion of the earth's nitrogen supply available to the living world is very small.

Plants incorporate the nitrogen in nitrites and nitrates into carbon-hydrogen compounds to form amino acids. Animals are able to synthesize some of their amino acids, using ammonia as a nitrogen source. The amino acids they cannot synthesize, the so-called essential amino acids, must be obtained either directly or indirectly from plants. For adult human beings, the essential amino acids are lysine, tryptophan, threonine, methionine, phenylalanine, leucine, valine, and isoleucine.

People who eat meat usually get enough protein and the correct balance of amino acids. People who are vegetarians, for either philosophical, esthetic, or economic reasons, have to be careful that they get enough protein and, in particular, the essential amino acids.

Until recently, agricultural scientists concerned with the world's hungry people concentrated on developing plants with a high caloric yield. Increasing recognition of the role of plants in supplying amino acids to the animal world has led to emphasis on the development of high-protein strains of food plants and of plants with essential amino acids, such as "high-lysine" corn.

Another approach to the right balance of amino acids is to combine certain foods. Beans, for instance, are likely to be deficient in tryptophan and in the sulfur-containing amino acids, but they are a good-to-excellent source of isoleucine and lysine. Rice is deficient in isoleucine and lysine but provides an adequate amount of the other essential amino acids. Thus rice and beans in combination make just about as perfect a protein menu as eggs or steak, as some nonscientists seem to have known for quite a long time.

The Levels of Protein Organization

In a living system, a protein is assembled in a long polypeptide chain, one amino acid at a time. In this process (to be described in Chapter 15), the head of one amino acid is linked to the tail of another, like a line of boxcars, always in a particular sequence. This sequence, which is dictated by the hereditary information in the cell for that particular protein, determines the structural features of the molecule as a whole and thus its biological function. The linear sequence of amino acids is known as the *primary structure* of the protein (Figure 3–15).

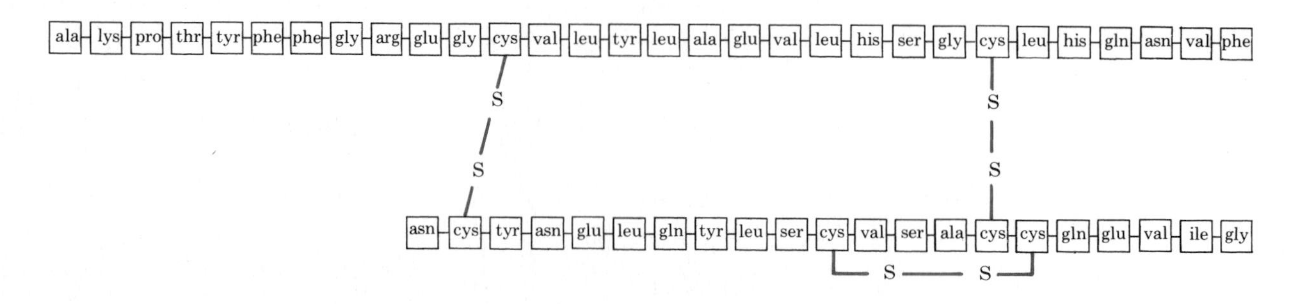

3–15 *The primary structure of a relatively small protein, the hormone insulin. This was one of the first proteins for which the primary structure was determined. Molecules of the sulfur-containing amino acid cysteine (cys) can form covalent bonds with each other. Such bonds are known as disulfide bridges. Insulin, as you can see, consists of two polypeptide chains held together by two disulfide bridges.*

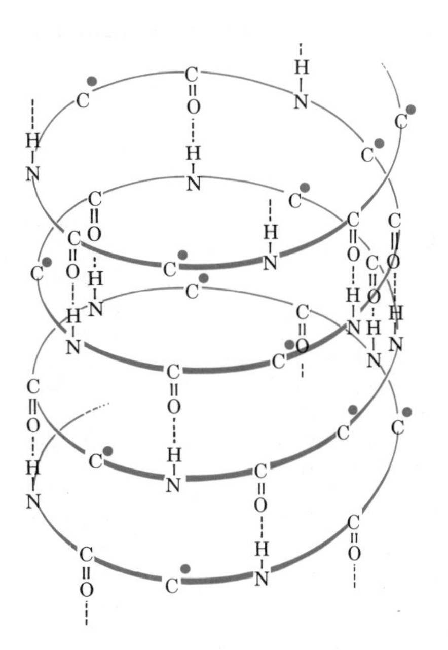

3-16 *A common secondary structure of a protein. The helix is held in shape by hydrogen bonds, indicated by the dashed lines. In this particular helix (the alpha helix), hydrogen bonds form between the double-bonded oxygen atom in one amino acid and the amino group in another amino acid that occurs four amino acids farther along the chain. The R groups, not shown in this diagram, are attached to the carbons indicated by the colored dots. The groups extend out from the helix.*

As the chain is assembled, it tends to fold in a simple pattern, known as its *secondary structure*. One common secondary structure is a coil or, as it is more commonly called by biochemists, a *helix*. The helix is held in shape by hydrogen bonds (Figure 3-16). The protein molecules of hair and of myosin (a muscle protein) tend to form helices and, as a consequence, are stretchable (elastic) because the hydrogen bonds break and re-form easily. (Straightening curly hair or curling straight hair involves breaking hydrogen bonds and making new ones.)

Other proteins, such as silk, are made of extended polypeptide chains lined up in parallel and linked to one another by hydrogen bonds. Polypeptides assembled in this way are smooth and supple but nonelastic.

Collagen, which is a principal component of cartilage, bones, and tendons, has another type of secondary structure. Collagen molecules are made by specialized cells called fibroblasts; as the long polypeptide chains are synthesized, three of them wrap around each other to form a cablelike fiber (Figure 3-17). Wool, fingernails, and horns are also structural proteins

3-17 *The structural protein collagen makes up about one-third of the protein in the human body and is probably the most abundant protein in the animal kingdom. Collagen is composed of long polypeptide chains wrapped around each other to form fibrils, such as those shown here, which are magnified 23,500 times. These fibers are a major constituent of skin, tendon, ligament, cartilage, and bone.*

(a) (b)

3-18 *The structural protein keratin is found in all vertebrates. It is the chief component of scales, wool, nails, and feathers.* (a) *These are cross sections of human hairs, magnified 173 times.* (b) *A feather is made up of a shaft to which thousands of barbs—each with many tiny barbules—are attached. The barbules on the lower half of each feather have tiny hooks on them that catch on the barbules of the adjacent feather, forming a solid, though flexible, structure for flight. When a bird preens itself, it is "zipping" its feathers back together. This feather, from a hummingbird, is magnified 170 times.*

whose texture and other characteristics vary because of differences in the arrangements of the polypeptide chains (Figure 3-18).

In other proteins, the long helix folds back on itself to make a complex tertiary structure. Such proteins are called globular proteins. Their tertiary structures form spontaneously as a result of attractions and repulsions among amino acids with different charges on their R groups. Enzymes—proteins that regulate chemical reactions in living systems—are globular proteins. As we shall see in Chapter 6, the three-dimensional structure of an enzyme is the critical factor in determining its biological function. Antibodies, important components of the immune system, are also globular proteins.

Often two or more globular proteins will fit together in a quaternary structure. Hemoglobin, for example, the oxygen-carrying molecule of the blood, is made up of four intertwined protein chains, each about 150 amino acids long. (A diagram of hemoglobin is shown on page 401.)

In every protein, the primary structure determines the secondary, tertiary, and quaternary structures.

3-19 (a) *The primary structure of a protein is the linear sequence of its amino acids.* (b) *Because of interactions among these amino acids, the molecule spontaneously forms a secondary structure, such as the alpha helix shown here, and* (c) *a tertiary structure, such as a globule.* (d) *Many globular proteins, including hemoglobin and some enzymes, are made up of more than one amino acid chain. This structure is known as a quaternary structure.*

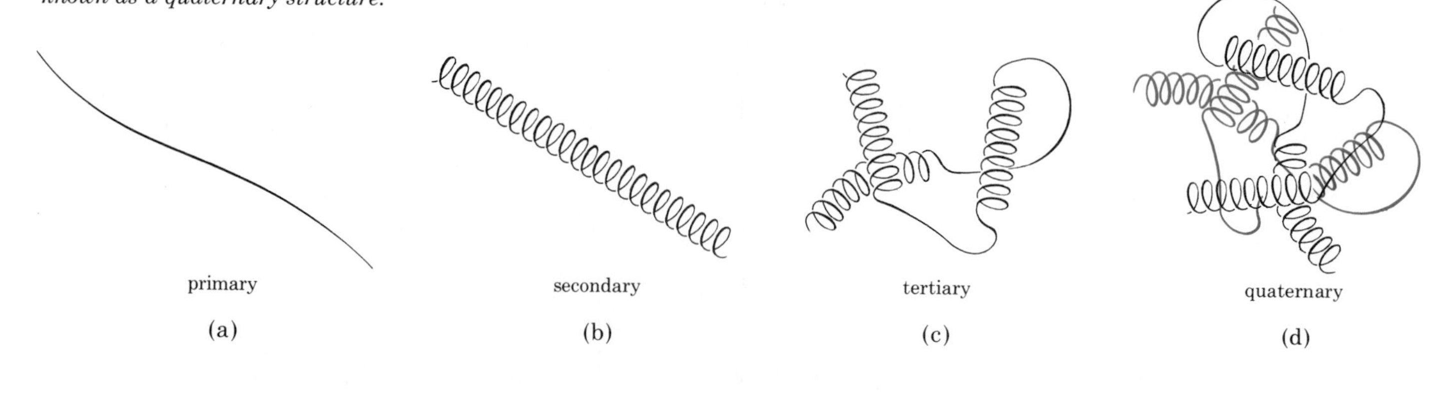

Organization and Living Systems

In the last three chapters, we have progressed from subatomic particles through atoms and molecules to large proteins, among the most complex of all molecules. Although the neutrons, protons, and electrons of which atoms are composed are all the same, the elements differ greatly from one another. Mercury is a heavy metallic liquid, chlorine is a green gas, sulfur is a yellow powder, pure carbon can take the form of a hard solid—a diamond—and so on. The differences lie not in the nature of the subatomic particles, but rather in their number and arrangement—that is, in their organization.

Just as subatomic particles combine to form atoms, atoms, as we have seen, combine to form molecules. At each level of organization, new properties appear. Water, for instance, is not the sum of the properties of hydrogen and oxygen; it is something more and also something different. In proteins, amino acids become organized into polypeptides, and polypeptide chains are arranged in a new level of organization, the tertiary or quaternary structure of the complete protein molecule. Only at this level of organization do the complex properties of the protein emerge, and only then can the molecule assume its function.

Living systems, as we noted earlier, obey the laws of physics and chemistry. Are we then "nothing but" a collection of atoms and molecules? There are recognizable differences between "life" and "nonlife." What is the basis of these differences? According to biologists, the differences are to be found not in the atoms and molecules themselves, but in their organization. The characteristics of living systems—like those of atoms or of molecules—do not emerge gradually as the degree of organization increases. They appear quite suddenly and specifically, in the form of the living cell—something that is more than and different from its constituent atoms and molecules.

No one knows exactly when or how this new level of organization—the living cell—first came into being. However, increasing knowledge of the conditions prevailing on our planet early in its history has provided strong evidence for the hypothesis that living cells spontaneously self-assembled from molecules present in the primitive seas. In the next chapter, we shall first examine this evidence and then move on to consider the intricate organization that characterizes the cells of which we and all other modern living systems are composed.

SUMMARY

The chemistry of living organisms is, in essence, the chemistry of organic compounds—that is, compounds containing carbon. Carbon is uniquely suited to this central role by its capacity to form covalent bonds with four other atoms, including other carbon atoms. Organic molecules derive their three-dimensional shapes primarily from their carbon skeletons.

Among the major types of organic molecules important in living systems are carbohydrates, lipids, and proteins. (Nucleotides, to be discussed later, constitute a fourth important type.)

Carbohydrates serve as a primary source of energy for living systems and also play important structural roles. The simplest carbohydrates are the monosaccharides ("single sugars"), such as glucose and fructose. Monosaccharides can be combined to form disaccharides ("two sugars"), such as sucrose, and polysaccharides (chains of many monomers). The polysaccharides starch and glycogen are storage forms for sugar, whereas cellulose, another polysaccharide, is an important structural material in plants. Disaccharides and polysaccharides are formed by condensation reactions in which monosaccharide units are covalently bonded with the removal of a molecule of water. They can be broken apart again by hydrolysis, with the addition of a water molecule.

Lipids are another source of stored energy and structural material for living systems. Compounds in this group, which includes fats and waxes, are generally insoluble in water. Phospholipids are major components of cellular membranes.

Proteins are very large molecules composed of long chains of amino acids; these are known as polypeptide chains. There are 20 different amino acids in proteins, and from these, an enormous number of different protein molecules are built. Many proteins form fibers and have important structural roles. Other proteins are globular and have regulatory and protective functions. The principal levels of protein organization are (1) primary structure, the linear amino acid sequence; (2) secondary structure, often a coiling of the polypeptide chain; (3) tertiary structure, the folding of the coiled chain into various shapes; and (4) quaternary structure, the folding of two or more polypeptide molecules around each other.

The properties of a molecule depend upon the organization of the atoms within the molecule. Similarly, the properties of a living system depend upon the organization of molecules within the cell.

QUESTIONS

1. Distinguish among the following: hydrocarbon/carbohydrate; glucose/fructose/sucrose; polysaccharide/polymer/polypeptide; glycogen/starch/cellulose.

2. Many of the synthetic reactions in living systems take place by condensation. What is a condensation reaction? What types of molecules undergo condensation reactions to form disaccharides and polysaccharides? To form fats? To form proteins?

3. Disaccharides and polysaccharides, as well as lipids and proteins, can be broken down by hydrolysis. What is hydrolysis? What two types of products are released when a polysaccharide such as starch is hydrolyzed? How are these products important for the living cell?

4. Plants usually store energy reserves as polysaccharides, whereas, in most animals, lipids are the principal form of energy storage. Why is it advantageous for animals to have their energy reserves stored as lipids rather than as polysaccharides? (Think about the differences in "life style" between plants and animals.) What kinds of storage materials would you expect to find in seeds?

5. Sketch the arrangement of phospholipids when surrounded by water.

6. In pioneer days, soap was made by boiling animal fat with lye (sodium hydroxide). The boiling hydrolyzed the bonds linking the fatty acids to the glycerol molecule, and the sodium hydroxide reacted with the fatty acid to produce soap. A typical soap is sodium stearate. In water, it ionizes to produce sodium ions (Na^+) and stearate ions:

$$CH_3CH_2CH_2CH_2CH_2CH_2CH_2CH_2CH_2CH_2CH_2CH_2CH_2CH_2CH_2CH_2C\begin{matrix} {}^{\diagup\!\!\diagup O} \\ {}_{\diagdown O^-} \end{matrix}$$

Explain how soap functions to trap and remove particles of dirt and grease.

4

Cells: Their Origin and Organization

4-1 *The tremendous amounts of energy released by the thermonuclear reactions at the heart of the sun give rise to an envelope of extremely hot gases surrounding its surface. These glowing gases are normally invisible, but during a solar eclipse they become strikingly apparent. The layer of gases may extend as far as 40,000 miles from the surface of the sun—a distance about five times greater than the diameter of the earth.*

The universe was perhaps 15 billion years old when the star that is our sun came into being. According to current hypotheses, it formed, like other stars, from an accumulation of dust and hydrogen and helium gases whirling in space among the older stars.

The immense cloud that was to become the sun condensed gradually as the hydrogen and helium atoms were pulled toward one another by the force of gravity, falling into the center of the cloud and gathering speed as they fell. As the cluster grew denser, the atoms moved more rapidly. More atoms collided with each other, and the gas in the cloud became hotter and hotter. As the temperature rose, the collisions became increasingly violent until the hydrogen atoms collided with such force that their nuclei fused, forming additional helium atoms and releasing nuclear energy. This thermonuclear reaction is still going on at the heart of the sun and is the source of the energy radiated from its glowing surface.

The planets, according to current theory, formed from gas and dust moving around the newly formed star. At first, particles would have collected at random, but as each mass grew larger, other particles began to be attracted by the gravity of the largest masses. The whirling dust and forming spheres continued to revolve around the sun until finally each planet had swept its own path clean, picking up loose matter like a giant snowball. The orbit nearest the sun was swept by Mercury, the next by Venus, the third by earth, the fourth by Mars, and so on out to Neptune and Pluto, the most distant of the planets. The planets, including earth, are calculated to have come into being about 4.6 billion years ago.

The Earth and the Biosphere

During the time earth and the other planets were being formed, the release of energy from radioactive materials kept their interiors very hot. When earth was still so hot that it was mostly liquid, the heavier materials collected in a dense core whose diameter is about half that of the planet. As soon as the supply of stellar dust, stones, and larger rocks was exhausted, the planet ceased to grow. As earth's surface cooled, an outer crust, a skin as thin by comparison as the skin of an apple, was formed. The oldest known rocks in this layer are about 3.98 billion years old.*

Only 50 kilometers below its surface, the earth is still hot—a small fraction of it is even still molten. We see evidence of this in the occasional volcanic eruption that forces lava (molten rock) through weak points in the earth's skin, or in the geyser, which spews up boiling water that has trickled down to the earth's interior.

The *biosphere* is the part of the planet within which life exists. It forms a thin film on the outermost layer, extending only about 8 or 10 kilometers up into the atmosphere and about as far down into the depths of the sea.

Why on Earth?

In our solar system, earth among all the planets is most favored for the production of life. A major factor is that earth is neither too close nor too distant from the sun. The chemical reactions on which life—at least as we know it—depends virtually cease at very low temperatures. At high temperatures, complex chemical compounds are too unstable for life to form or survive.

Earth's size and mass are also important factors. Planets much smaller than earth do not have enough gravitational pull to hold a protective atmosphere, and any planet much larger than earth is likely to have so dense an atmosphere that light from the sun cannot reach its surface.

Many biologists contend that given certain conditions—such as an energy source, water, a temperature range in which the water can exist in liquid form, and a long enough time—the evolution of some forms of life is inevitable. The discovery on another planet of a living organism, no matter how primitive, whose origin was independent of earth, would strongly support this hypothesis. There are approximately 10^{20} (the number 1 followed by 20 zeros) stars in the universe like our own sun that can provide energy

* The process by which these dates are estimated is described in Chapter 33.

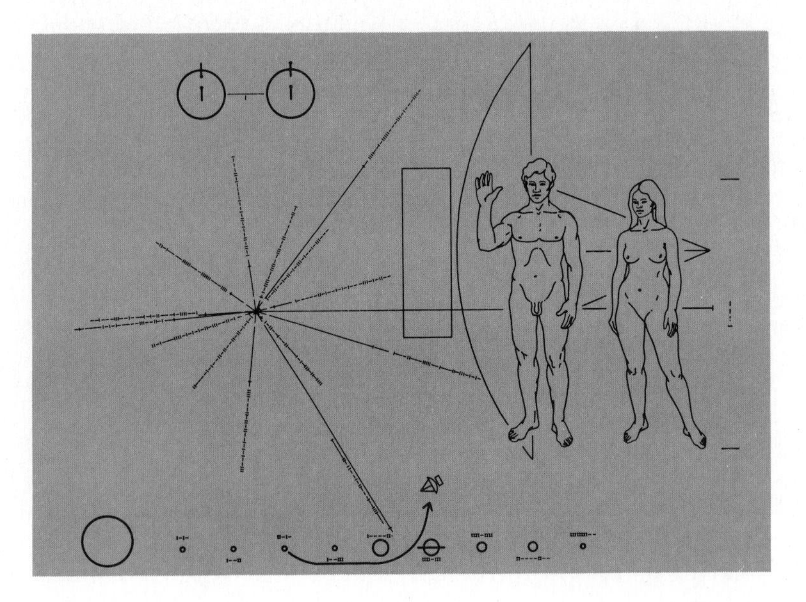

4-2 *Many astronomers and biologists believe that intelligent life exists in outer space. This small metal plaque, designed by Carl and Linda Sagan, was sent off in Pioneer 10, the first spacecraft to leave our solar system. It contains a message to the inhabitants of some far-distant planet: two members of the species* Homo sapiens, *with a hand raised in peaceful greeting, and their home address. Pioneer 10, launched in 1973, will reach the planetary system of another star in about 10 billion years.*

for living things. At least 10 percent of these, according to astronomers, are likely to be surrounded by planetary systems such as our own. If only 1 percent were to have planets with environments roughly similar to those of earth, that would offer some 10^{18} possibilities for the existence of life in other solar systems.

The Beginning of Life

Sometime between the time earth formed and the date of the earliest fossils discovered so far—an interval of about a billion years—life began. The chief raw materials for life were to be found in the atmosphere of the young earth. The principal component of the sun and hence of the solar system is hydrogen, and earth, when it formed, was probably surrounded by a cloud of hydrogen gas. Under conditions of an abundance of hydrogen, the available nitrogen, carbon, and oxygen atoms also present would tend to combine with hydrogen to form ammonia (NH_3), methane (CH_4) and other gases made up of carbon and hydrogen, and water (H_2O). These were the raw materials for living systems. As we have seen, combinations of these four elements available in the primitive atmosphere—hydrogen, oxygen, carbon, and nitrogen—make up more than 95 percent of the tissues of all living things.

In addition to these raw materials, energy abounded on the young earth. There was heat energy, both boiling (moist) heat and baking (dry) heat. Water vapor spewed out of the primitive seas, cooled in the upper

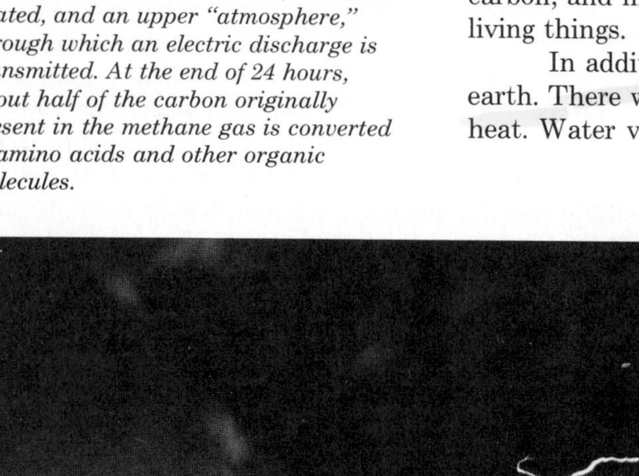

(a)

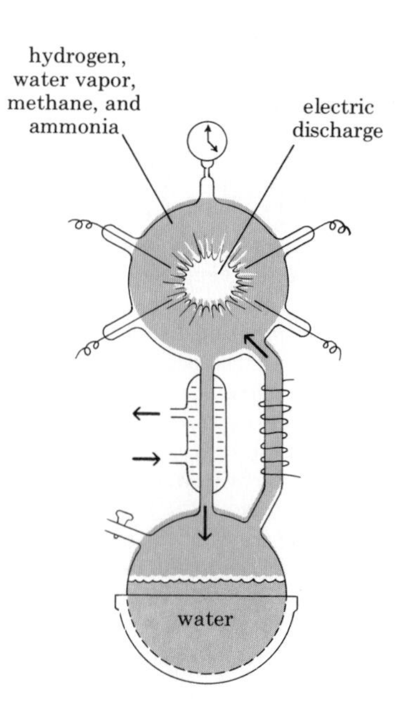

(b)

4-3 (a) *Bolts of lightning in the steam boiling up from a volcanic crater. Such sources of energy would have been present on the primitive earth and might have contributed to the formation of organic molecules. This photograph, taken in 1963, shows the birth of the island of Surtsey off the coast of Iceland.*

(b) *Conditions believed to exist on the primitive earth are simulated in the apparatus diagrammed here. Methane and ammonia are continuously circulated between a lower "ocean," which is heated, and an upper "atmosphere," through which an electric discharge is transmitted. At the end of 24 hours, about half of the carbon originally present in the methane gas is converted to amino acids and other organic molecules.*

atmosphere, collected into clouds, fell back on the crust of the earth, and steamed up again. Violent rainstorms were accompanied by lightning, which provided electrical energy. The sun bombarded the earth's surface with high-energy particles and ultraviolet light, another form of energy. Radioactive elements within the earth released their energy into the atmosphere.

These conditions can be simulated in the laboratory (Figure 4–3b), and scientists have now shown that under such conditions, organic molecules are produced from the simple gases ammonia, methane, and water vapor. The heat and electrical energy break apart the simple gas molecules, which then re-form into organic molecules. With various modifications of the experimental conditions, almost all of the common amino acids and other biologically important molecules have been produced.

Biochemists believe that, given the conditions existing on the young earth, chemical reactions producing such organic compounds were inevitable. As they reconstruct the process, these compounds were washed out of the atmosphere by driving rains. Gradually the compounds began to collect in ponds and shallow seas, becoming concentrated in an organic "soup." As we have seen, some organic molecules have a tendency to group together, forming boundaries between themselves and the surrounding solution, similar to the droplets formed by oil in water. Such aggregations may have been the forerunners of the first primitive cells.

The exact moment at which living systems, capable of reproducing themselves, first came into being is not known. The earliest fossils found so far, which are about 3.4 billion years old, resemble bacteria (Figure 4–4). The fossilized cells are sufficiently complex that it is clear that some little aggregation of chemicals had moved through the twilight zone separating the living from the nonliving millions of years before.

4–4 *This bacterium-like microfossil was found in South Africa in a deposit of a flintlike rock called black chert. It is about 3.4 billion years old, one of the oldest fossils now known.*

The short, straight line at the bottom of this micrograph and those that follow provides a reference for size; a micrometer, abbreviated μm, is 1/10,000 centimeter. The same system is used to indicate distances on a road map.

Heterotrophs and Autotrophs

The energy that produced the first organic molecules came from the heat, ultraviolet rays, and electrical disturbances that abounded in the primitive atmosphere. These organic molecules, in turn, became the source of energy for the earliest forms of life. The primitive cells or cell-like structures were able to use organic compounds, which were abundant, to meet their energy needs, just as our cells use sugars, fatty acids, and amino acids to meet their energy needs.

These early cells were *heterotrophs*, a category of organisms that today includes all animals and fungi, as well as many single-celled organisms. *Hetero* comes from the Greek word meaning "other," and *troph* comes from *trophos*, "one that feeds." A heterotrophic organism is one that is dependent upon others—that is, upon an outside source of organic molecules—for both its energy and its small building-block molecules.

As the primitive heterotrophs increased in number, they began to use up the complex molecules on which their existence depended and which had taken millions of years to accumulate. As the supply of these molecules decreased, competition began. Cells that could make efficient use of the limited energy sources now available were more likely to survive and leave offspring than cells that could not. In the course of time, cells evolved that were able to make their own energy-rich organic molecules out of simple inorganic materials. Such organisms are called *autotrophs*, "self-feeders."

Without the evolution of autotrophs, which became a new source of organic molecules for heterotrophs, life on earth would soon have come to an end.

The most successful autotrophs (that is, those that left the most offspring) were those that evolved a system for making direct use of the sun's energy—the process of photosynthesis. With the advent of photosynthesis, the flow of energy in the biosphere came to assume its modern form: radiant energy from the sun channeled through photosynthetic autotrophs to all other forms of life.

Prokaryotes and Eukaryotes

Living systems, as we noted on page 7, consist of small units, or compartments, called *cells*. There is an unbroken continuity between modern cells—and the organisms they compose—and the primitive cells that first appeared on earth more than 3 billion years ago.

One essential feature of all cells is an outer membrane, the *cell membrane* (sometimes called the plasma membrane), which separates the cell from its external environment. Another is the genetic material—the hereditary information—that directs a cell's activities and enables it to reproduce, passing on its characteristics to its offspring. The other contents of a cell (that is, everything within the cell membrane except the genetic material) are known as the *cytoplasm*. The cytoplasm contains a large variety of molecules as well as formed bodies called *organelles*. These specialized structures carry out particular functions within the cell, for example, assembling protein molecules.

There are, however, two fundamentally distinct kinds of cells, *prokaryotes* and *eukaryotes*. Prokaryotes and eukaryotes differ most notably in the organization of their genetic material. In prokaryotic cells, the genetic material is in the form of a large, single molecule of a chemical known as DNA (deoxyribonucleic acid); in eukaryotic cells, the DNA is associated with proteins in complex structures known as *chromosomes*. Moreover, in eukaryotes, the chromosomes are surrounded by a double membrane, the *nuclear envelope*, that separates them from the cytoplasm in a distinct *nucleus* (hence their name, *eu*, meaning "true," and *karyon*, meaning "nucleus" or "kernel"). By contrast, in prokaryotes ("before a nucleus"), the DNA is not contained within a membrane-bound nucleus.

Modern prokaryotes include the bacteria (Figure 4–5) and the blue-green algae (Figure 4–6), which many experts hold to be species of bacteria. The cell membrane is surrounded by an outer *cell wall* that is manufactured by the cell itself. The cytoplasm of prokaryotes contains very small organelles called *ribosomes*, on which protein molecules are assembled.

According to the fossil record, the earliest living organisms were comparatively simple cells, resembling present-day prokaryotes. Prokaryotes were the only forms of life on this planet for more than 1½ billion years until eukaryotes evolved. (The evolutionary relationship between prokaryotes and eukaryotes—quite an interesting subject—will be explored in Chapter 18.) Eukaryotic cells are usually larger than prokaryotic cells, and their organelles are more numerous and more complex, often enclosed within membranes. Some eukaryotic cells, including plant cells and fungi, have a cell wall; others, including the cells of our own bodies and those of other animals, do not. All multicellular organisms are made up of eukaryotic cells.

4-5 *Cells of* Escherichia coli, *a prokaryote that is a common inhabitant of the human digestive tract. The hereditary material (DNA) is in the lighter-appearing area in the center of each cell. The small, dense bodies in the cytoplasm are ribosomes. The two cells in the center have just finished dividing and have not yet separated completely.*

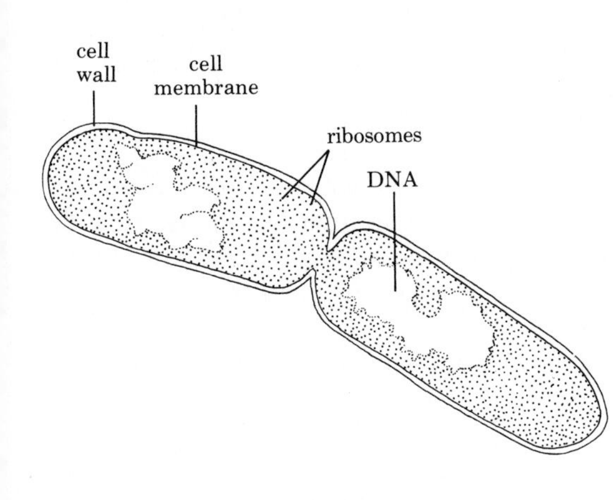

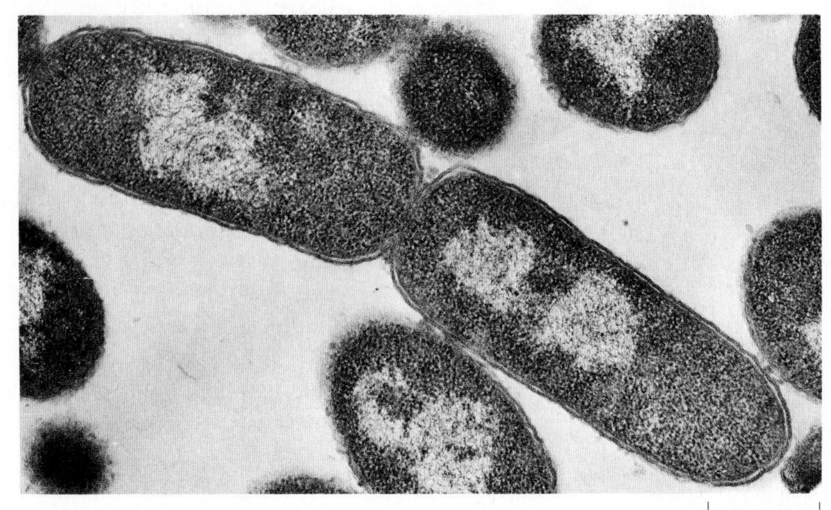

4-6 *Electron micrograph and diagram of a photosynthetic prokaryotic cell, a blue-green alga* (Anabaena azollae). *In addition to the hereditary material, this cell contains a series of membranes in which chlorophyll and other photosynthetic pigments are embedded. This cell can synthesize its own energy-rich organic compounds in chemical reactions powered by the radiant energy of the sun.*

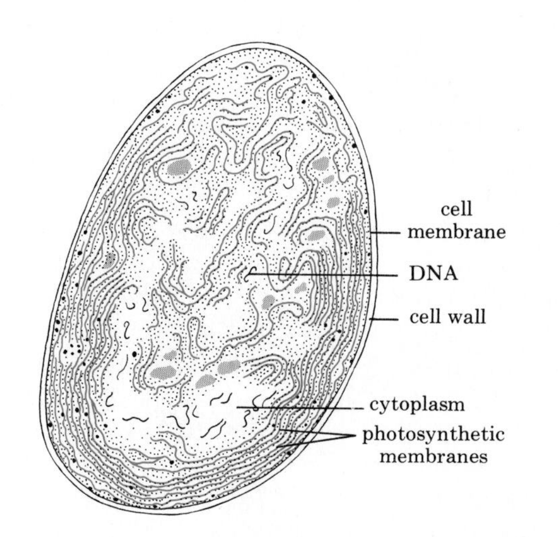

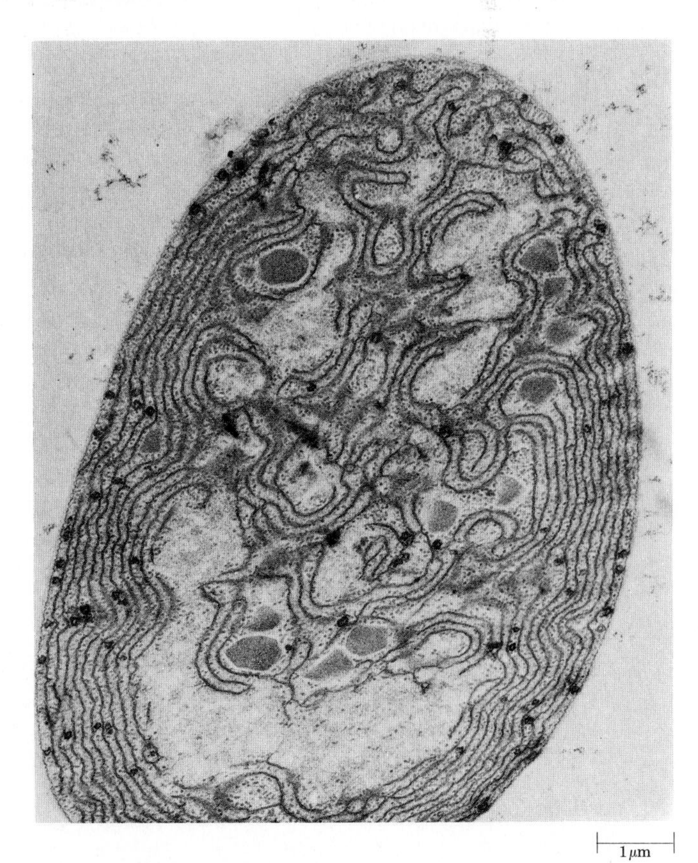

Figure 4–7 gives an example of a single-celled photosynthetic eukaryote, *Chlamydomonas*. It is a common inhabitant of freshwater ponds. These organisms are small, bright green (because of their chlorophyll) and move very quickly with a characteristic darting motion. Being photosynthetic, they are usually found near the water's surface. *Chlamydomonas* is believed to resemble the sort of cell from which the plants evolved.

The Origins of Multicellularity

The first multicellular organisms, as far as can be told by the fossil record, made their appearance a mere 750 million years ago (Figure 4–8). The various major groups of multicellular organisms—such as the plants, the animals, and the fungi—presumably evolved from different types of single-celled eukaryotes.

The cells of modern multicellular organisms closely resemble those of single-celled organisms (see Figure 4–9). They are bound by a cell membrane identical in appearance to the cell membrane of single-celled organisms. Their organelles are constructed according to the same design. The cells of multicellular organisms differ from single-celled organisms in that each type of cell is specialized to carry out a relatively limited function in the life of the organism. However, each remains a remarkably self-sustaining unit. The human body, made up of trillions of individual cells, is composed of at least 100 different types of cells, each specialized for its particular function but all working as a cooperative whole.

4–7 *Electron micrograph of* Chlamydomonas, *a photosynthetic eukaryotic cell, which contains a membrane-bound ("true") nucleus and numerous organelles. The most prominent organelle is the single, irregularly shaped chloroplast that fills most of the cell. It is surrounded by a double membrane and is the site of photosynthesis. Other membrane-bound organelles, the mitochondria, provide energy for cellular functions, including the flicking movements of the two flagella (one of which is visible in the micrograph). These movements propel the cell through the water. The cytoplasm is surrounded by a cell membrane, outside of which is a cell wall.*

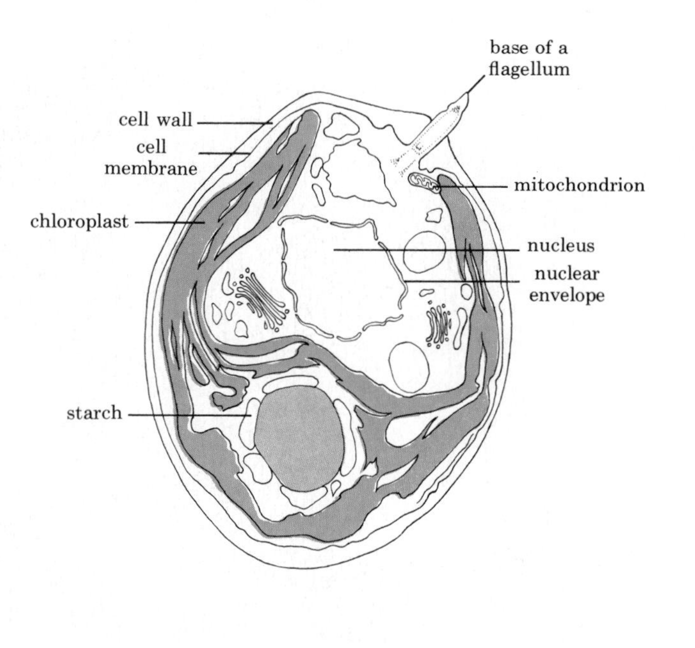

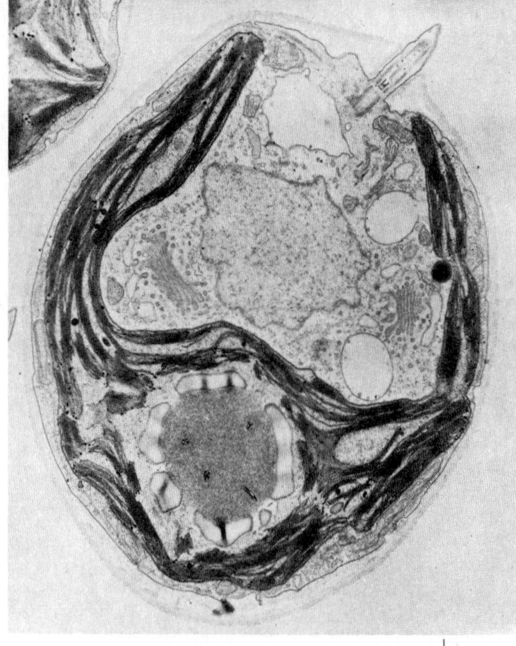

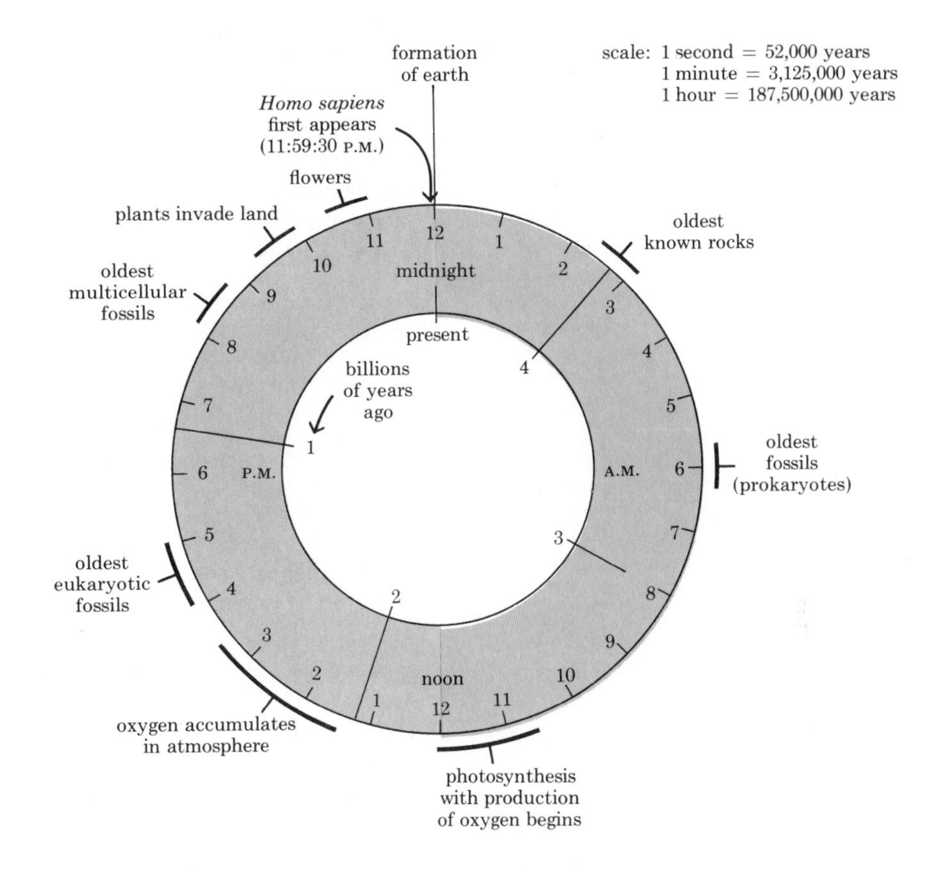

4-8 *The clockface of biological time. Life first appears relatively early in the earth's history, before 7:00* A.M. *on a 24-hour time scale. The first multicellular organisms do not appear until the twilight of that 24-hour day, and* Homo sapiens *is a late arrival—at about 30 seconds to midnight.*

4-9 *Electron micrograph of cells from the leaf of a corn plant. The nucleus can be seen on the right side of the central cell. The granular material within the nucleus is chromatin; it contains the hereditary material. Note the many mitochondria and chloroplasts, all enclosed by membranes. The vacuole and cell wall are characteristic of plant cells but are generally not found in animal cells. As you can see, this cell closely resembles* Chlamydomonas, *shown in Figure 4-7.*

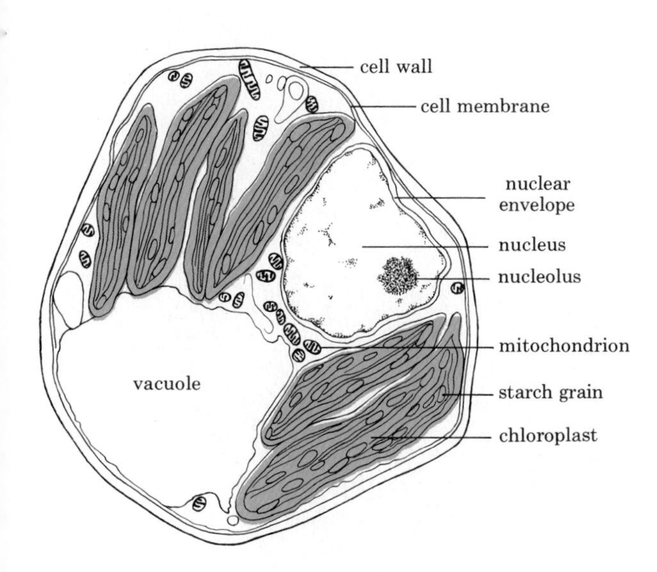

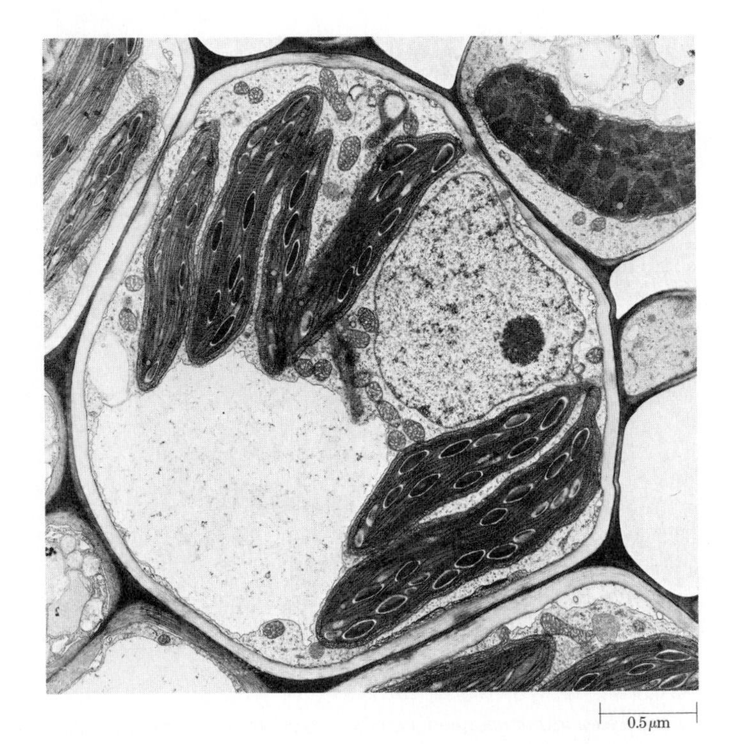

Viewing the Microscopic World

Unaided, the human eye has a resolving power of about 1/10 millimeter, or 100 micrometers. This means that if you look at two lines that are less than 100 micrometers apart, they merge into a single line. Similarly, two dots less than 100 micrometers apart look like a single blurry dot. To separate structures closer than this, optical instruments such as microscopes are used. The best light microscope has a resolving power of 0.2 micrometer, or 200 nanometers, and so improves on the naked eye about 500 times. It is theoretically impossible to build a light microscope that will do better than this.

Notice that resolving power and magnification are two different things; if you take a picture through the best light microscope of two lines that are less than 0.2 micrometer, or 200 nanometers, apart, you can enlarge that photograph indefinitely, but the two lines will continue to blur together. By using more powerful lenses, you can increase magnification, but this will not improve resolution. The limiting factor is the wavelength of light.

With the electron microscope, resolving power has been increased about 400 times over that provided by the light microscope. This is achieved by using "illumination" of a much shorter wavelength, consisting of electron beams instead of light rays. Areas in the specimen that permit the transmission of more electrons—"electron-transparent" regions—show up bright, and areas that absorb or deflect electrons—"electron-dense" regions—are dark. Electron microscopy at present, under the very best conditions, affords a resolving power of about 0.5 nanometer, roughly 200,000 times greater than that of the human eye. (A hydrogen atom is about 0.1 nanometer in diameter.)

Electrons, which have a very small mass, can pass through specimens only when the specimens are exceedingly thin. As a consequence, living matter has to be fixed, dehydrated, embedded in hard materials, and sliced by special cutting instruments before it can be examined by transmission electron microscopy. Moreover, most specimens are electron-transparent and have to be stained with heavy metals that bind differentially to different subcellular components. Thus only dead cells can be studied, and, moreover, it is sometimes difficult to determine whether or not the specimens have been changed by the preparation methods.

Recently, electron microscopy has been expanded in scope by the development of the scanning electron microscope. In scanning electron microscopy, the electrons whose imprints are recorded come from the surface of the specimen. The electron beam is focused into a fine probe, which is rapidly passed back and forth over the specimen. Complete scanning from top to bottom usually takes a few seconds. As a result of the electron bombardment from the probe, the specimen emits low-energy secondary electrons. Variations in the surface of the specimen alter the number of sec-

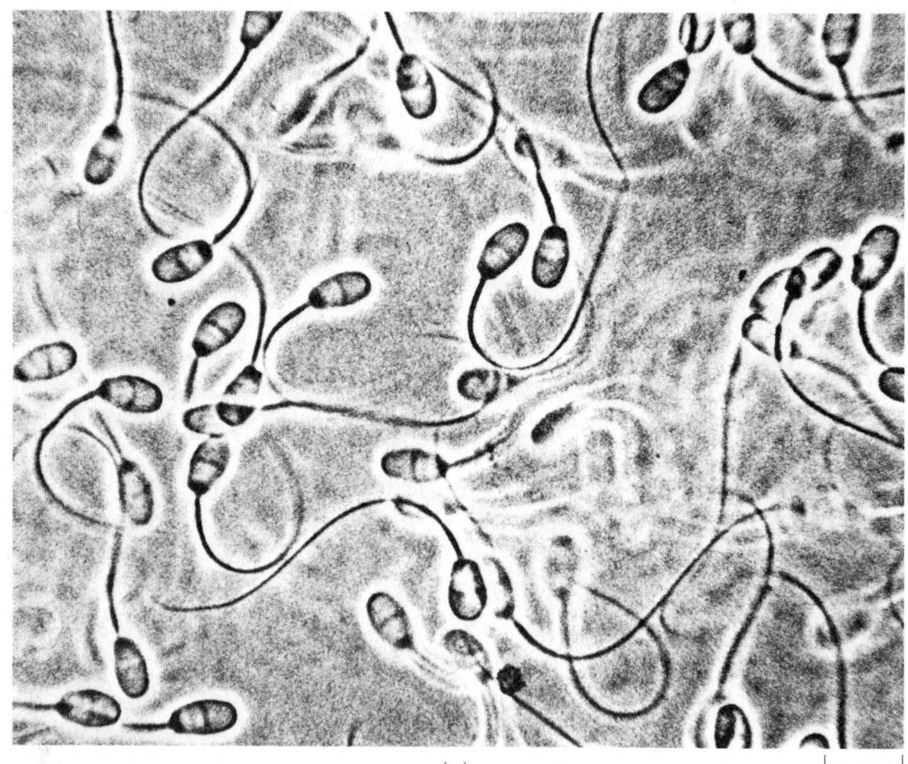

(a)

Rabbit sperm cells in (a) *a light micrograph;* (b) *an electron micrograph;* (c) *a scanning electron micrograph.*

ondary electrons emitted. Holes and fissures appear dark, and knobs and ridges are light. Electrons scattered from the surface plus secondary electrons are amplified and transmitted to a screen, which is scanned in synchrony with the electron probe.

Table 4-1 Measurements used in microscopy

1 centimeter (cm) = 1/100 meter = 0.4 inch*
1 millimeter (mm) = 1/1,000 meter = 1/10 cm
1 micrometer (μm)† = 1/1,000,000 meter = 1/10,000 cm
1 nanometer (nm) = 1/1,000,000,000 meter = 1/10,000,000 cm
or
1 meter = 10^2 cm = 10^3 mm = 10^6 μm = 10^9 nm

* A metric-to-English conversion table is found in Appendix A.

† Micrometers were formerly known as microns (μ).

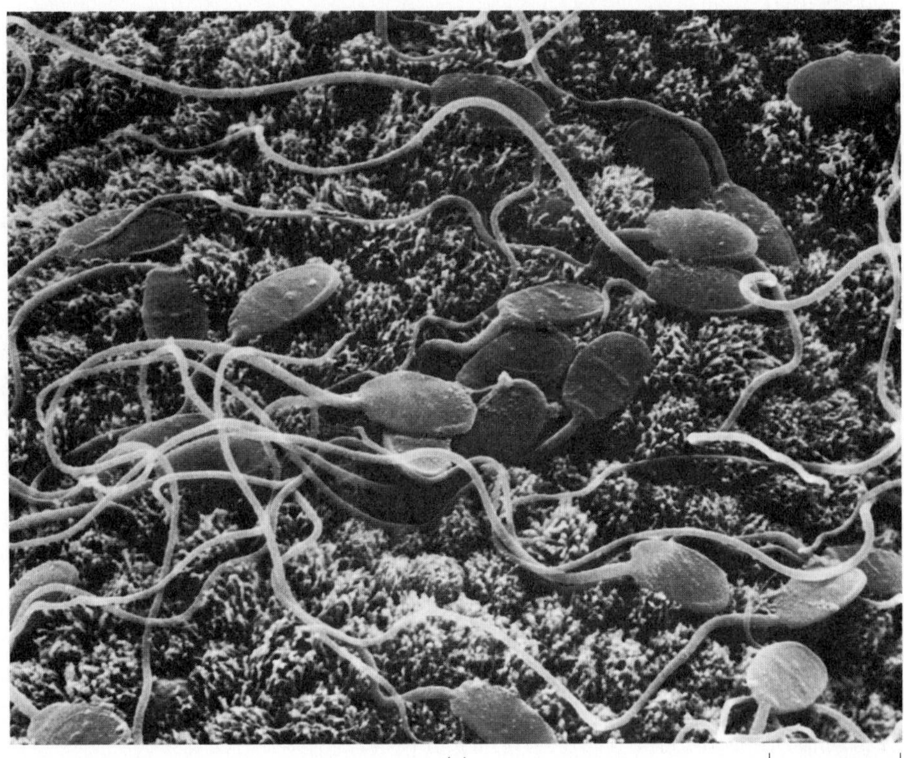

(c)

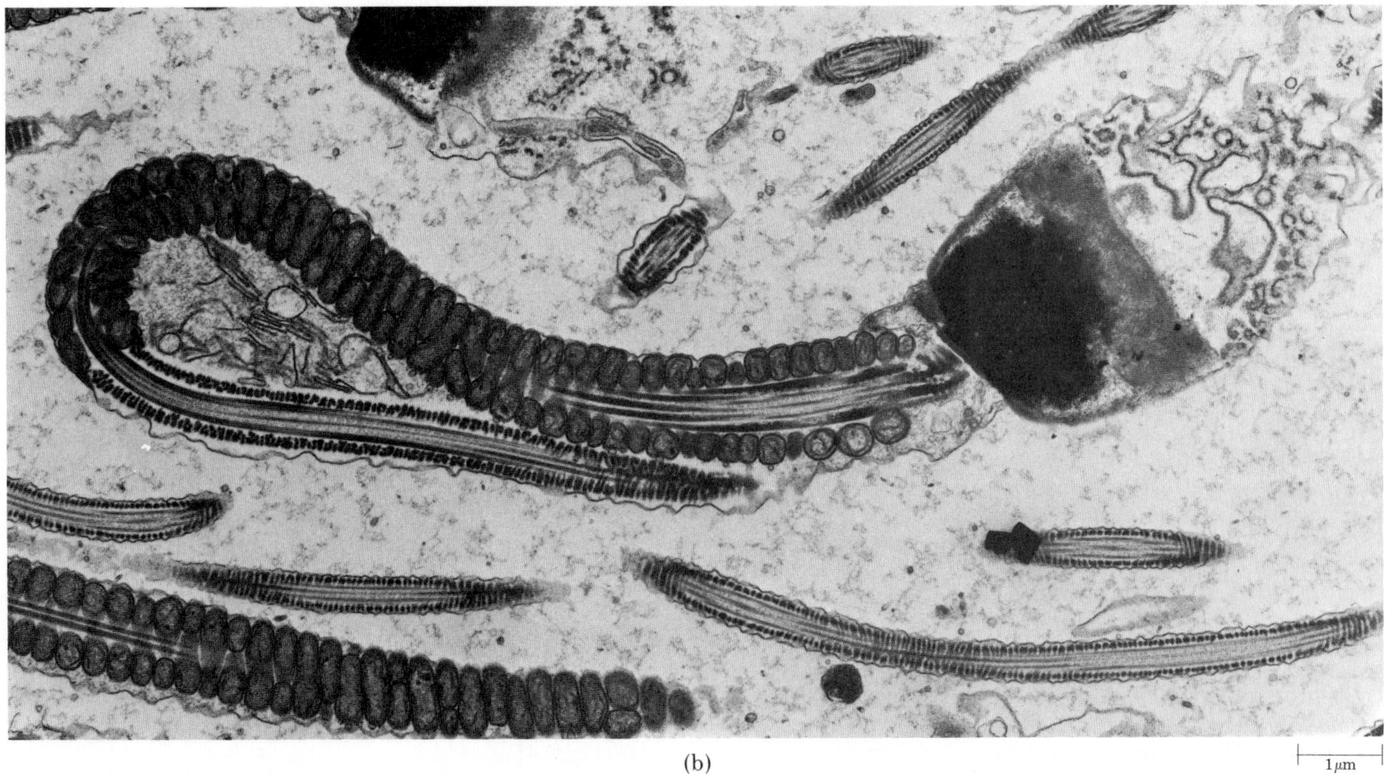

(b)

The Organization of the Cell

In the remainder of this chapter, we shall glimpse the structure and some of the functions of that amazing entity—the living cell. We use the word "glimpse" because, although much is still unknown about the life of the cell, what *is* known about cell structure and physiology would fill several large volumes. Our narrative is greatly abbreviated and therefore incomplete.

Although we can look at only one structure or process at a time, remember that most activities of a cell go on simultaneously and influence one another. *Chlamydomonas,* for instance, is swimming, photosynthesizing, absorbing nutrients from the water, building its cell wall, making proteins, converting sugar to starch (or vice versa), and oxidizing food molecules for energy, all at the same time. It is also likely to be orienting itself in the sunlight, it is probably preparing to divide, it is possibly "looking" for a mate, and it is undoubtedly carrying out at least a dozen or more other important activities.

Cell Size

Most of the cells that make up a plant or animal body are between 10 and 30 micrometers in diameter (see page 61). A principal restriction on cell size seems to be the relationship between volume and surface area. As you can see in Figure 4–10, as volume increases, surface area decreases rapidly in proportion to volume. Materials—such as oxygen, carbon dioxide, ions, food molecules, and waste products—entering and leaving the cell must move through its membrane-bound surface. The more active the metabolism* of a cell is, the more rapidly these materials must be exchanged with the environment if the cell is to continue to function. In small cells, the proportion of surface area to volume is greater than in large cells; thus, materials can move faster into, out of, and through small cells.

A second limitation on cell size appears to involve the capacity of the nucleus, the cell's control center, to regulate the cellular activities of a large, metabolically active cell. The exceptions seem to "prove" the rule. In

* Metabolism is simply the total of all of the chemical activities of a living system.

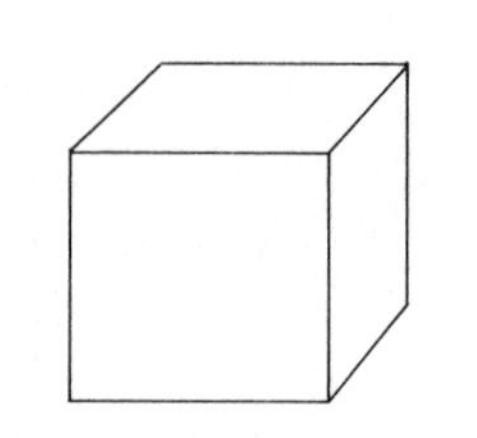
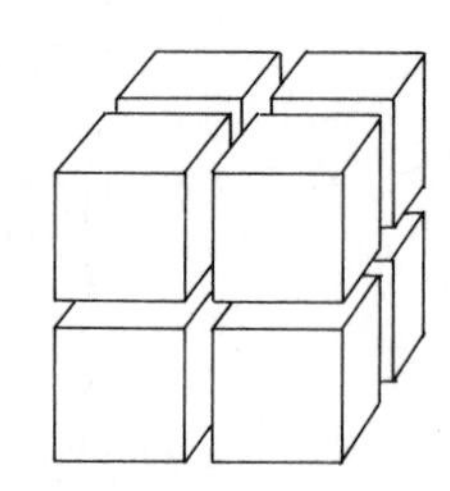
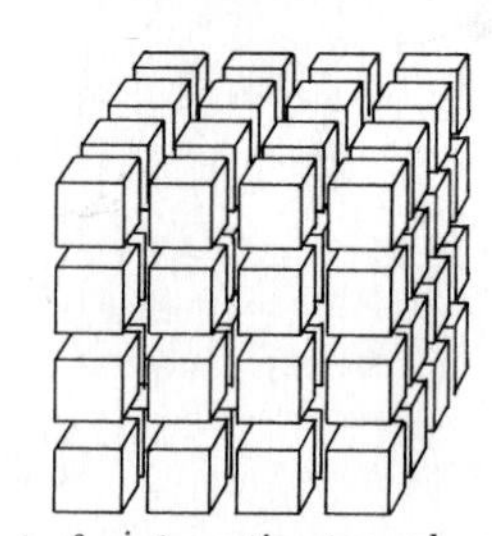

	one 4-centimeter cube	eight 2-centimeter cubes	sixty-four 1-centimeter cubes
surface area (square centimeters)	96	192	384
volume (cubic centimeters)	64	64	64
surface/volume	$1\frac{1}{2}$:1	3:1	6:1

4–10 *The single 4-centimeter cube, the eight 2-centimeter cubes, and the sixty-four 1-centimeter cubes all have the same total volume. As the volume is divided up into smaller units, however, the amount of surface area increases (for example, the single 4-centimeter cube has only one-fourth the total surface area of the sixty-four 1-centimeter cubes). Similarly, smaller cells have more surface area (more membrane surface through which materials can move into or out of the cell) and less volume (shorter distances through which materials must move within the cell).*

certain large, complex one-celled organisms—the ciliates, of which *Paramecium* is an example—each cell has two or more nuclei, the additional ones apparently copies of the original.

It is not surprising, therefore, that the most metabolically active cells are usually small. The relationship between cell size and metabolic activity is nicely illustrated by egg cells. Many egg cells are very large. A frog's egg, for instance, is 1,500 micrometers in diameter. Some egg cells are several centimeters across—for example, the egg cell, or yolk, of a chicken's egg. Most of this mass consists of stored nutrients for the developing embryo. When the egg cell is fertilized and begins to be active metabolically, it first divides many times before there is any actual increase in volume. Thus the cellular units are cut down to an efficient metabolic size.

Cell Boundaries

A cell can exist as a separate entity because of the cell membrane, which regulates the passage of materials into and out of the cell. The cell membrane is only about 9 nanometers thick and cannot be resolved in the light microscope. Now, with the electron microscope, it can be visualized as a continuous, thin double line (Figure 4–11).

The basic structure of all cell membranes is formed by two layers of phospholipid molecules, arranged with their hydrophobic tails pointing inward (Figure 3–10, page 43). It is essentially the same in all living cells, whether prokaryotic or eukaryotic. However, differences in the proteins and carbohydrates associated with the phospholipid molecules give the membranes of different types of cells unique properties. In Chapter 5, we shall examine the molecular structure of cell membranes and the ways in which they perform their essential tasks.

The Cell Wall

A principal distinction between plant and animal cells is that the former are surrounded by a cell wall. The wall is outside the membrane and is constructed by the cell. As a plant cell divides, a thin layer of gluey

4–11 *Electron micrograph showing a cross section of the cell membrane of a red blood cell. The cell membrane is indicated by the arrows. The basic structure of the membrane is believed to consist of two electron-dense (dark) layers of phospholipid molecules arranged with their hydrophobic tails pointing inward, forming the electron-transparent (light) inner "filling."*

The darker material at the left of the micrograph is hemoglobin, which fills the red blood cell.

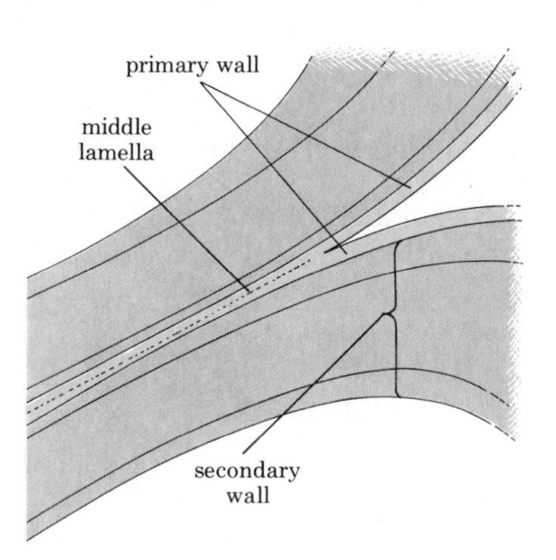

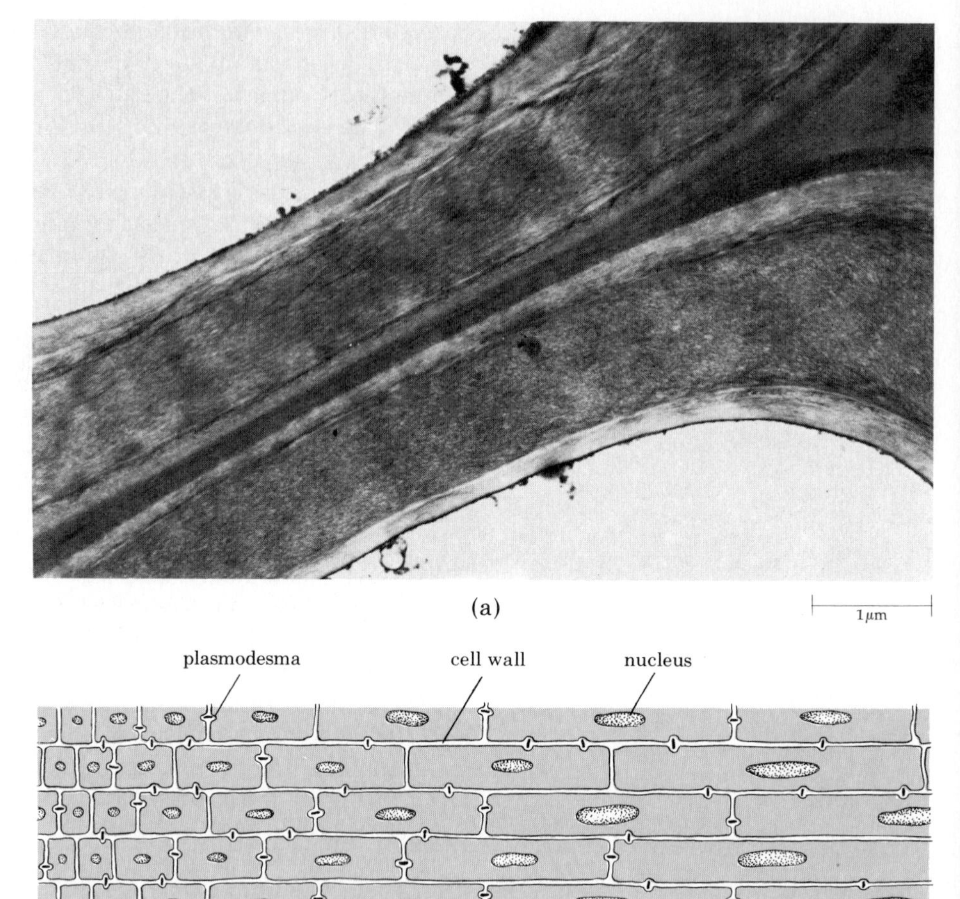

4–12 (a) *Electron micrograph of two adjacent cell walls of vessel elements, the cells that form the tubes through which water is conducted in plants. You can see the middle lamella, the primary walls, and the layered secondary walls, deposited inside the primary wall. The cells, which are from the wood of a ground hemlock, have died.*

(b) *Growth of plant cells is limited by the rate at which the cell walls expand. The walls control both the rate of growth and its direction; they do not expand in all directions but elongate in a single dimension. Cells at the left are newly formed; cells farther to the right are older and have started to elongate. Plasmodesmata (singular, plasmodesma) are channels connecting adjacent cells.*

material forms between the two new cells; this becomes the *middle lamella* (Figure 4–12a). Composed of pectins (the compounds that make jellies gel) and other polysaccharides, it holds adjacent cells together. Next, under the middle lamella, the plant cell constructs its primary cell wall. This wall is composed, to a large extent, of cellulose molecules wound together like wires in a cable and laid down in a matrix of gluey polymers. As you can see in Figure 3–6a on page 40, successive layers of microfibrils are oriented at right angles to one another in the completed cell wall. (Those of you familiar with building materials will note that the cellulose cell wall thus combines the structural features of both fiber glass and plywood.)

In plants, growth takes place largely by cell elongation. Studies have shown that the cell adds new materials to its walls throughout this elongation process. The cell, however, does not simply expand in all directions; its final shape is determined by the structure of its cell wall (Figure 4–12b).

As the cell matures, a secondary wall may be constructed. This wall is not capable of expansion, as is the primary wall. It often contains other molecules that have stiffening properties. In such cells, the living material of the cell often dies, leaving only the outer wall, a monument to the cell's architectural abilities.

Cellulose cell walls are also found in many algae; fungi and prokaryotes also have cell walls, but they are not made of cellulose.

4-13 *The cell nucleus fills the upper half of this electron micrograph. What you see here is the outer surface of the nuclear envelope. Clearly visible on this surface are pores, through which the interior of the nucleus can be seen. It is believed that the nucleus and the cytoplasm are able to "communicate" through the pores.*

The Nucleus

In eukaryotic cells, the nucleus is a large, often spherical body, usually the most prominent structure within the cell. It is surrounded by two membranes, which together make up the nuclear envelope (Figure 4-13). These two membranes are fused together at frequent intervals to create small pores that appear to form channels between the nucleus and the cytoplasm.

The chromosomes are found within the nucleus. When the cell is not dividing, they are visible only as a tangle of fine threads, called *chromatin*. The most conspicuous body within the nucleus is the *nucleolus*, the site at which the ribosomes are assembled.

The Functions of the Nucleus

The nucleus carries the hereditary information for the cell, the instructions that determine whether a particular cell will be an amoeba, part of a leaf, or part of a human liver. Each time a cell divides, this information is passed on to the two new cells. The nucleus exerts its influence by directing the ongoing activities of the cell, ensuring that the various complex molecules the cell requires are synthesized in the numbers needed. The way in which the nucleus performs these functions will be described in Section 3.

Some Organelles

Not long ago, the cell was visualized as a bag of fluid containing enzymes and other dissolved molecules along with the nucleus and a few organelles. With the development of electron microscopy, however, an increasing number of structures have been identified within the cytoplasm, which is now known to be highly organized and crowded with organelles. A representative array of them is seen in Figure 4-14 on the next page.

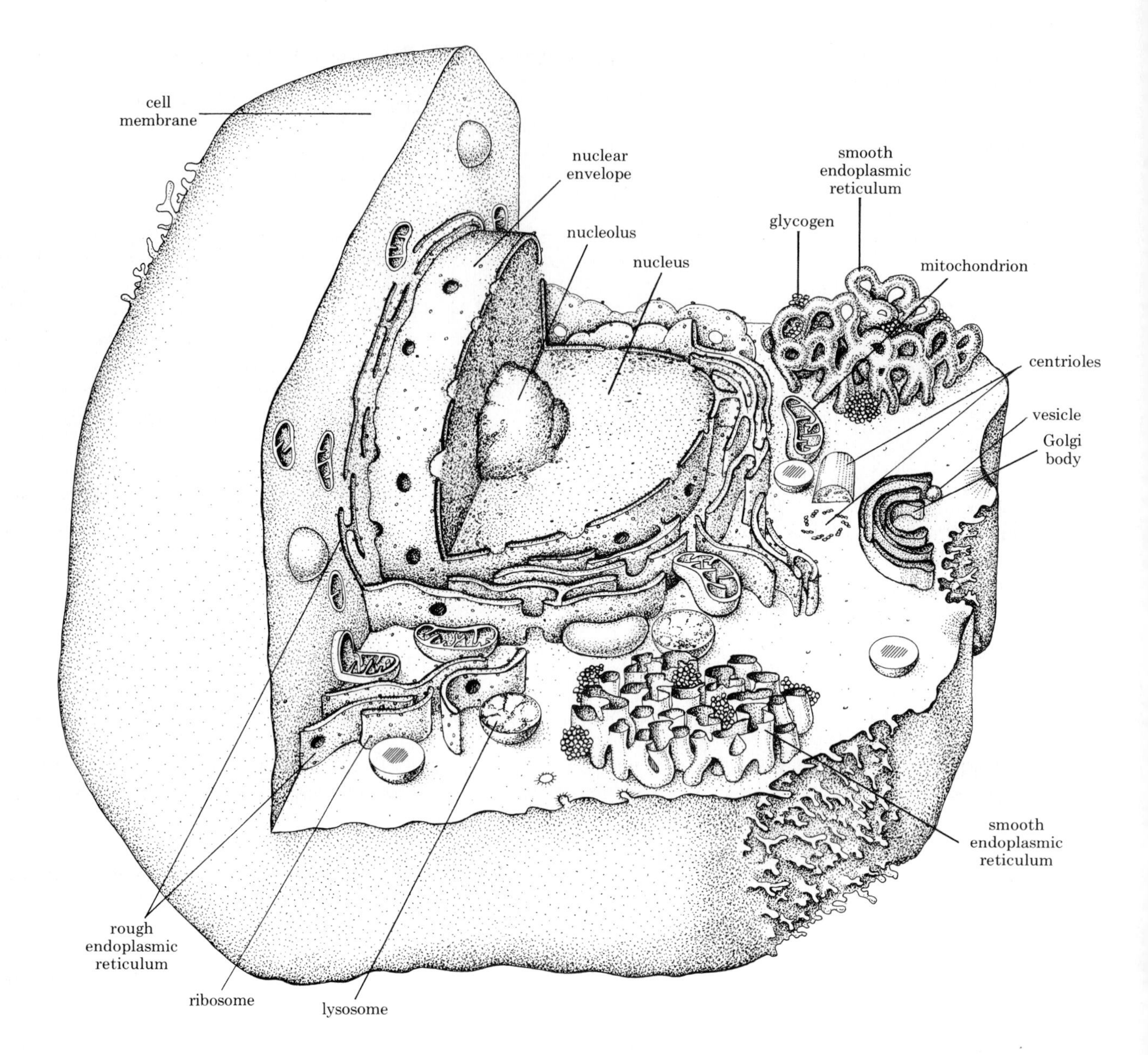

4-14 *An animal cell, as interpreted from electron micrographs. Like all cells, this one is bounded by an outer cell membrane, which acts as a selectively permeable barrier to the surrounding environment. All materials that enter or leave the cell, including food, wastes, and chemical messages must pass through this barrier.*

Within the membrane is found the cytoplasm, which contains the enzymes and other solutes of the cell. The cytoplasm is traversed and subdivided by an elaborate system of membranes, the endoplasmic reticulum. In some areas, the endoplasmic reticulum is covered with ribosomes, the special structures on which amino acids are assembled into proteins. Ribosomes are also found free in the cytoplasm.

Golgi bodies are packaging centers for molecules synthesized within the cell. The mitochondria are the sites of the chemical reactions that provide energy for cellular activities.

The largest body in the cell is the nucleus. It is surrounded by a double membrane, the nuclear envelope, which is continuous with the endoplasmic reticulum. Within the nuclear envelope is a nucleolus, the site where the ribosomes are formed, and the chromatin, which is the material of the chromosomes in an extended form.

Ribosomes and Endoplasmic Reticulum

Ribosomes, the most numerous of the cell's many organelles, are the sites at which amino acids are assembled into proteins. The more protein a cell is making, the more ribosomes it has.

The way in which the ribosomes are distributed in the cell seems to be related to the way the proteins are utilized. In cells, such as embryonic cells, that are making proteins for their own use, ribosomes tend to be distributed in the cytoplasm. In cells that are making digestive enzymes or other proteins for export, the ribosomes are found attached to a complex system of internal membranes called the *endoplasmic reticulum*. Endoplasmic reticulum with ribosomes attached to it is known as rough endoplasmic reticulum (Figure 4–15). It is continuous with the outer layer of the nuclear envelope. There is evidence that rough endoplasmic reticulum is involved both in producing proteins and in preparing them for shipment out of the cell.

Cells also contain smooth endoplasmic reticulum, that is, endoplasmic reticulum with no ribosomes on it. It is found largely in the form of tubules and plays a role in transporting substances from the interior of the cell to the surface and in synthesizing lipids.

4–15 *Rough endoplasmic reticulum, which fills most of this micrograph, is a system of membranes that separates the cell into channels and compartments and provides surfaces on which chemical activities take place. The dense objects on the membrane surfaces are ribosomes. This cell is from a pancreas, an organ extremely active in the synthesis of digestive enzymes, which are "exported" to the upper intestine, where most digestion takes place. In the lower right-hand corner of the micrograph are one mitochondrion and, above it, a portion of another.*

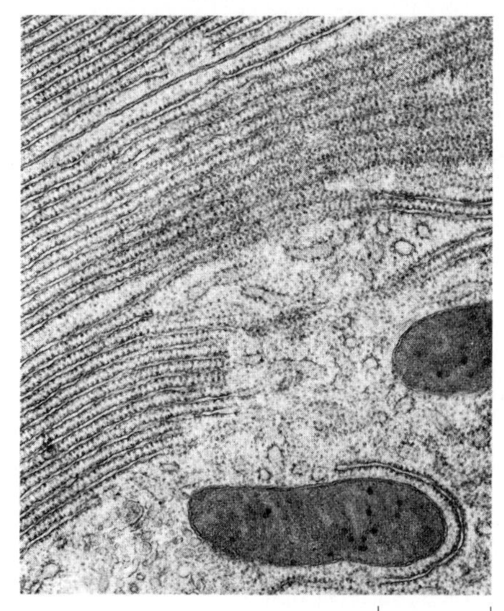

Golgi Bodies

A *Golgi body* consists of a group of flattened sacs composed of membranes stacked loosely on one another and surrounded by tubules and vesicles (very small membrane-enclosed sacs). Golgi bodies serve as packaging and distribution centers, especially for substances formed on the endoplasmic reticulum. Also, they are the sites of assembly of some complex molecules—for instance, combinations of sugars and proteins (glycoproteins)—that are found on the surfaces of cell membranes. In plant cells, they bring together the various components of plant cell walls. Golgi bodies are found in almost all eukaryotic cells. Animal cells usually contain 10 to 20 Golgi bodies, and plant cells may have several hundred.

4–16 *Golgi bodies are composed of special arrangements of membranes. Materials are packaged in membrane-enclosed vesicles at the Golgi bodies and are distributed within the cell or shipped to the cell surface. Note the vesicles pinching off from the edges of the flattened sacs.*

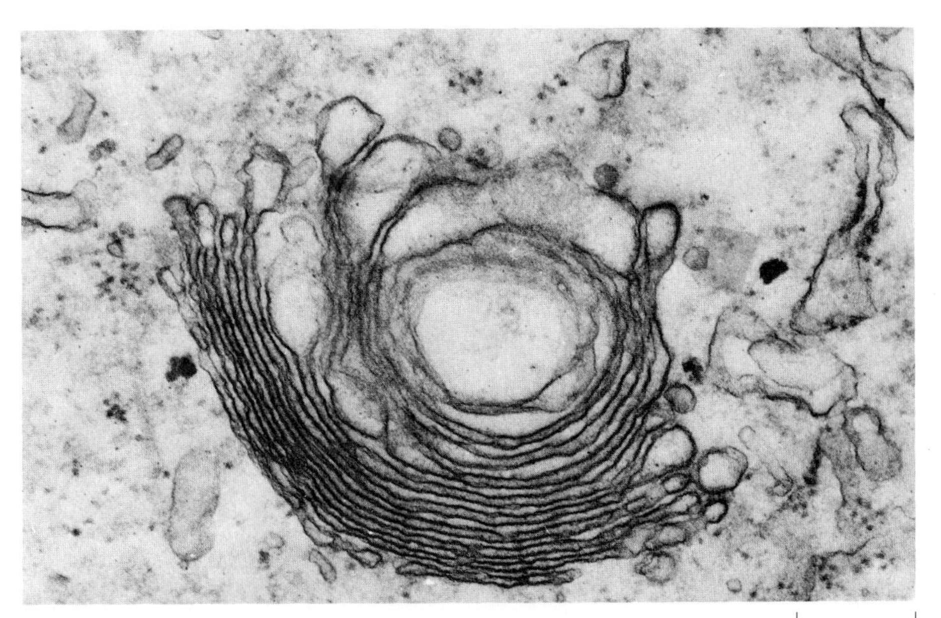

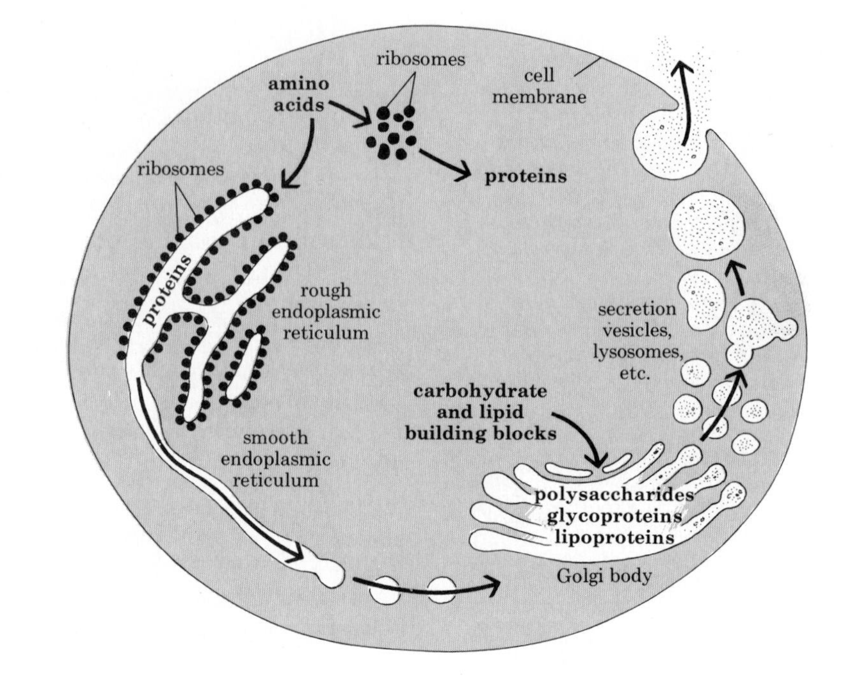

4-17 *Diagram illustrating the interaction of cytoplasmic organelles in the synthesis of proteins and in the packaging of macromolecules for export. Glycoproteins are combinations of sugars and proteins, and lipoproteins are combinations of fats and proteins. They are common components of membranes.*

Figure 4-17 summarizes the ways in which the rough and smooth endoplasmic reticulum and the Golgi body and its vesicles may interact to produce macromolecules (very large molecules) for export.

Lysosomes

One type of vesicle formed in the Golgi body is a *lysosome.* Lysosomes are essentially membranous bags that enclose destructive enzymes, thereby separating the enzymes from the rest of the cell. If the lysosomes break open, the cell itself is destroyed since the enzymes they carry (for example, digestive enzymes) are capable of breaking down all the major compounds found in a living cell.

An example of the function of lysosomes is given by white blood cells, which engulf bacteria in the human body. As the bacteria are taken up by the cell, they are wrapped in a membrane-enclosed sac, a vacuole. (Vacuoles are similar to vesicles but are larger.) When this occurs, the lysosomes within the cell fuse with the vacuoles containing the bacteria, releasing their destructive enzymes into the vacuole. These enzymes then digest the contents of the vacuoles. Why the enzymes do not destroy the membranes of the lysosomes that carry them is a pertinent question yet to be answered.

Chloroplasts and Mitochondria

The activities of a cell require energy. As we have seen, some cells (autotrophs) manufacture their own energy-rich organic compounds from inorganic molecules. Other cells (heterotrophs) must obtain organic molecules from outside sources.

Photosynthetic autotrophs capture radiant energy from the sun and transform it to chemical energy stored in organic molecules. This process, photosynthesis, requires special pigments, of which chlorophyll is the most common. Photosynthesis takes place, however, only when the chlorophyll molecules are embedded in a membrane. In all photosynthetic eukaryotes, the chlorophyll-bearing membranes are organized within a membrane-bound organelle, the *chloroplast.* (In photosynthetic prokaryotes, such as the blue-green alga in Figure 4–6, the chlorophyll is also contained in membranes, but these membranes are not separated from the rest of the cytoplasm by an outer membrane.)

Virtually all eukaryotic cells (including photosynthetic ones) have *mitochondria* (singular, mitochondrion), which are also membrane-bound organelles. In the process of cellular *respiration,* which occurs in the mitochondria, energy-rich molecules are broken down. The process uses oxygen and releases the energy needed for cellular activities. (When we breathe, we are working for our mitochondria, supplying them with the required oxygen.)

Chloroplasts and mitochondria are the essential power generators of eukaryotic cells. Without the energy these organelles make available, most other cellular functions could not be carried out. We shall examine the structure of these organelles when we consider the processes of photosynthesis and respiration in Section 2.

How Cells Move

All cells exhibit some form of movement. Even plant cells, encased in a rigid cell wall, exhibit active movement of the cytoplasm within the cell as well as chromosomal movements and changes in shape during cell division. Embryonic cells migrate in the course of development. Amoebas pursue and engulf their prey. Even little *Chlamydomonas* cells dart toward a light source.

Two different mechanisms of cellular movement have been identified. The first consists of assemblies of fibrous proteins, usually referred to as muscle protein because they were first identified and studied in muscle tissue. However, it has now been found that these proteins are present in a great variety of cells and appear to be associated with movement within the cells. The second mechanism involves long, thin structures—cilia and flagella—extending from the surface of many types of eukaryotic cells.

4–18 (a) *An isolated cell in motion. This cell, from a mouse embryo, was probably moving to the left when it was prepared for examination with the scanning electron microscope.* (b) *A similar cell, also moving to the left, was examined with a special microscopic technique that revealed the muscle proteins that make its movement possible. The protein visible here is actin, which, as we shall see in Chapter 24, is one of the proteins involved in the movement of skeletal muscle.*

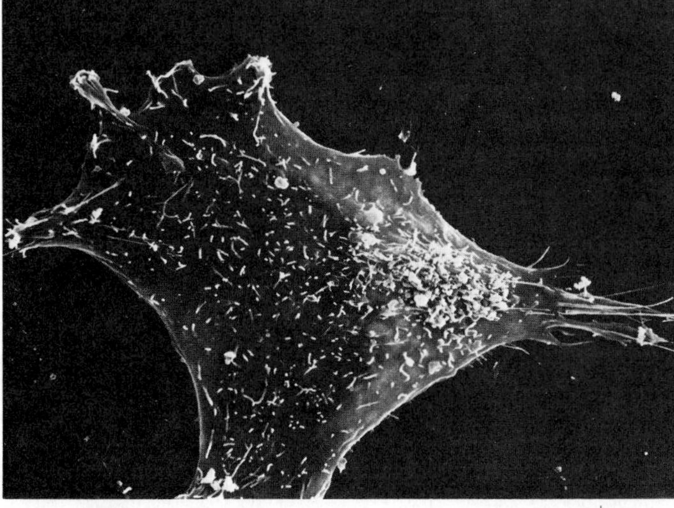

(a)

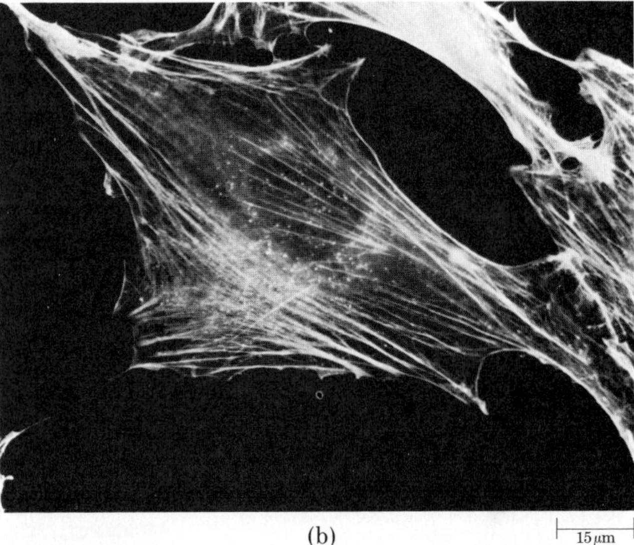

(b)

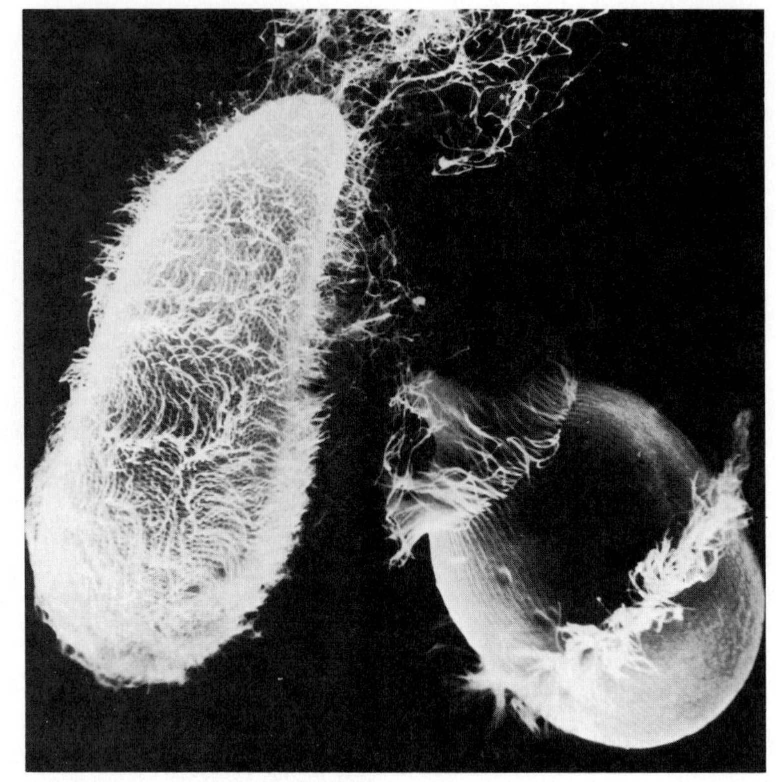

4-19 *Two ciliates, one-celled organisms distinguished by their many cilia. On the left,* Paramecium; *on the right,* Didinium. Didinium *is stalking* Paramecium. Paramecium, *in defense, has discharged a barrage of barbs (visible as a cloud at the top of the micrograph).* Didinium *is about to eject a bundle of slender, poisonous strands (not visible), which will paralyze* Paramecium *in a matter of seconds. In* Paramecium, *the cilia are distributed fairly evenly over the cell surface. In* Didinium, *they form two wreaths that circle the organism's barrel-shaped body.*

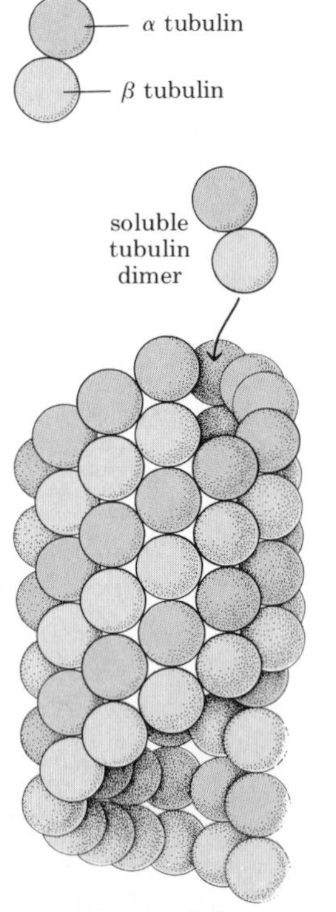

4-20 *Microtubules are hollow tubes, so small that they cannot be visualized by a light microscope. They are composed of subunits, each of which is a globular protein. The subunits are of two types, alpha tubulin and beta tubulin, which first come together to form a soluble dimer ("two parts"). The dimers then self-assemble into hollow tubules. Among their many functions, microtubules make up the internal structure of eukaryotic cilia and flagella.*

Cilia and Flagella

Cilia (from the Latin word for "eyelash") and *flagella* (singular, flagellum) are two names for essentially the same structure in eukaryotic cells. (The names were given before the basic similarity was realized.) When they are shorter and occur in larger numbers, the structures are more likely to be called cilia; when they are longer and fewer, they are usually called flagella. Thus we say that a *Paramecium* is covered with cilia, and *Chlamydomonas* has two flagella.

Many one-celled eukaryotes and also some very small multicellular ones, such as flatworms, are propelled by cilia. Similarly, the motile power of the human sperm cell comes from its single powerful flagellum, or "tail."

Many of the cells that form the tissues of our bodies are also ciliated. These cilia do not move the cells, but rather serve to sweep substances across the cell surface. For example, cilia on the surface of cells of the respiratory tract beat upward, propelling bits of soot, dust, pollen, tobacco tar—whatever foreign substances we have inhaled either accidentally or on purpose—to the backs of our throats, where they can be removed by swallowing.

Only a few large groups of eukaryotic organisms—most notably the flowering plants—have no cilia or flagella in any cells. Some bacteria move by means of flagella, but these prokaryotic flagella are so different in construction from those of eukaryotes that it would be useful if they had a different name.

All eukaryotic cilia and flagella have a similar structure. The basic unit of this structure is the microtubule (Figure 4-20). In each cilium or

4–21 (a) *Diagram of a cilium with its underlying basal body. All eukaryotic cilia and flagella, whether they are found on one-celled organisms or on the surfaces of cells within our own bodies, have this same internal structure, which consists of an outer ring of nine pairs of microtubules surrounding two additional microtubules in the center. The basal bodies from which they arise have nine outer triplets, with no microtubules in the center. The structural basis of the wheel-like formation in the basal body is not known. The "hub" of the wheel is not a microtubule, although it has about the same diameter.*
(b) *Cross section of flagella. These are from* Trichonympha, *the one-celled organism shown in Figure 3–6b on page 40.*

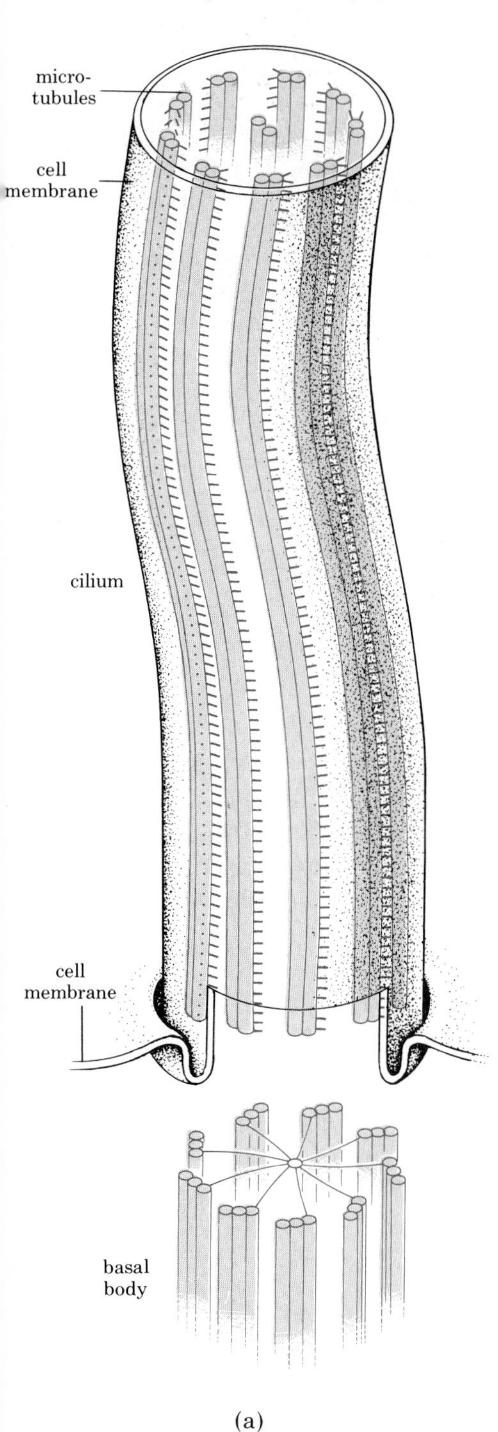

(a)

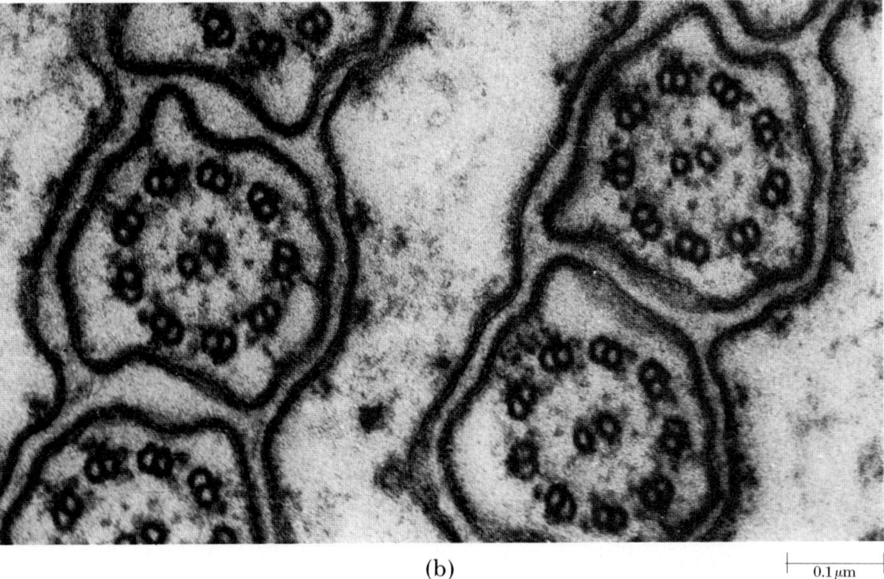

(b)

flagellum, nine pairs of fused microtubules form a ring that surrounds two additional, solitary microtubules in the center. The movement of cilia and flagella comes from within the structures themselves; if cilia are removed from cells and placed in a medium containing energy-rich chemicals, they twitch. The movement, according to one hypothesis, is caused by one outer pair moving tractor-fashion over its nearest neighbor. The "arms" that you can see on one of each pair of outer tubules (Figure 4–21a) have been shown to be enzymes involved in energy-releasing chemical reactions.

Cilia and flagella arise from *basal bodies*, which are also made up of microtubules. Their number and arrangement are somewhat different, however, as you can see in Figure 4–21a. The basal bodies are believed to keep the cilia or flagella supplied with fuel molecules and perhaps other substances as well. Eukaryotic cells with cilia and flagella also contain structures identical to basal bodies that are known as *centrioles*. The distribution of centrioles within the cell is different from that of basal bodies, and, until recently, it appeared that their function was also different. Thus they were given a different name, long before electron microscopy revealed their identical structure. Centrioles appear to have an important role in the movements of the chromosomes during cell division (Chapter 9).

The discovery of the complex internal structure of cilia and flagella, repeated over and over again throughout the living world, was one of the spectacular revelations of electron microscopy. For biologists, it is another glimpse down the long corridor of evolution, providing overwhelming evidence, once again, of the basic unity of earth's living things.

SUMMARY

The sun and its planets were formed about 4.6 billion years ago—the sun probably from the condensation of a hydrogen gas cloud, and the planets as accumulations of interstellar debris. Of the nine planets in this solar system, only earth is known to support life, but there are likely to be other planets in the universe with some form of life.

The primitive atmosphere of earth held the raw materials of living matter—hydrogen, oxygen, carbon, and nitrogen—combined in water vapor and gases. The energy required to break apart the simple gases in the atmosphere and re-form them into more complex molecules was present in heat, lightning, radioactive elements, and high-energy radiation from the sun. Laboratory experiments have shown that the types of molecules characteristic of living organisms—that is, organic molecules—can be formed under the conditions that probably prevailed on the young earth. These molecules gradually accumulated to form cells, the first living things.

The earliest cells were probably heterotrophs, that is, organisms that depend on outside sources for their energy-rich organic molecules. Organisms that can make their own organic molecules from inorganic substances are known as autotrophs. The most familiar of the modern autotrophs are the green plants, which use the sun's energy to power their synthetic reactions.

The earliest fossil cells resemble modern bacteria and blue-green algae. Such cells are known as prokaryotes. They are distinguished from the larger, more complex cells of plants and animals—known as eukaryotic cells—chiefly by the fact that eukaryotes have a membrane-bound nucleus and many specialized structures (organelles) not found in prokaryotes (see Table 4-2).

Table 4-2 Comparison of prokaryotic, animal, and plant cells

	Prokaryote	Animal	Plant
Cell membrane	Present	Present	Present
Cell wall	Present (noncellulose polysaccharide plus protein)	Absent	Present (cellulose)
Nucleus	No nuclear envelope	Surrounded by nuclear envelope	Surrounded by nuclear envelope
Chromosomes	Single, containing only DNA	Multiple, containing DNA and protein	Multiple, containing DNA and protein
Endoplasmic reticulum	Absent	Usually present	Usually present
Mitochondria	Absent	Present	Present
Chloroplasts	Absent	Absent	Present in photosynthetic cells
Ribosomes	Present (smaller)	Present	Present
Golgi bodies	Absent	Present	Present
Lysosomes	Absent	Often present	Similar structures (lysosomal compartments) present
Vacuoles	Absent	Small or absent	Usually large single vacuole in mature cell
9 + 2 cilia or flagella	Absent	Often present	Absent (in higher plants)
Centrioles	Absent	Present	Absent (in higher plants)

Multicellular organisms evolved comparatively recently—only about 750 million years ago. They are composed of eukaryotic cells specialized to perform particular functions.

Cells are the basic units of biological structure and function. Most cells are between 10 and 30 micrometers in diameter. One limit on the size of cells is the surface-to-volume ratio. Smaller units have a greater surface area in proportion to their volume than large ones.

All cells are bounded by an outer membrane, the cell membrane. The cells of plants, most algae, fungi, and prokaryotes have, in addition, a cell wall, constructed by the cell itself.

The nucleus of eukaryotic cells is surrounded by a double membrane, the nuclear envelope. The nucleus contains the hereditary information of the cell and, interacting with the cytoplasm, helps to regulate the cell's ongoing activities.

Cells have a variety of organelles, which carry on a vast number of biochemical activities. Proteins are assembled on ribosomes, small cytoplasmic particles present in both prokaryotic and eukaryotic cells. The cytoplasm of eukaryotic cells is subdivided by a network of membranes known as the endoplasmic reticulum, which serves as a work surface for many of the cell's biochemical activities and as a system of channels for moving materials through the cell. Cells that produce protein for export have extensive systems of endoplasmic reticulum with ribosomes attached, known as rough endoplasmic reticulum.

Golgi bodies, also composed of membranes, are packaging centers for materials being moved into and out of the cell. Lysosomes are sacs of destructive enzymes. They may fuse with other vesicles or vacuoles and digest their contents.

Chloroplasts and mitochondria are membrane-bound organelles found in eukaryotic cells. They are involved in energy capture and energy release, respectively. Photosynthesis occurs within the chloroplasts, and cellular respiration within the mitochondria.

Muscle proteins are associated with internal cellular movement, while cilia and flagella are associated with the external movement of cells or the movement of materials along cell surfaces. These hairlike structures are found on the surfaces of many types of eukaryotic cells. They have a highly characteristic 9 + 2 structure, with nine pairs of microtubules forming a ring surrounding two central microtubules.

QUESTIONS

1. Distinguish between the following: heterotroph/autotroph; prokaryote/eukaryote; cell membrane/cell wall; nucleus/nucleolus; chloroplasts/mitochondria.

2. As we noted, it is believed that many planets in the universe might contain some form of life. Suppose you were seeking such a planet. What characteristics would you look for?

3. (a) Sketch an animal cell. Include the principal organelles and label them. (b) Prepare a similar labeled sketch of a plant cell. (c) What are the major differences between the animal cell and the plant cell?

4. (a) Sketch a cross section of a cilium. (b) Sketch a cross section of the basal body of a cilium. (c) What are the differences between the two structures?

5. (a) Return to Chapter 2 and add approximate scale markers to Figures 2–4, 2–5, and 2–6. (b) Use a ruler and the scale marker at the bottom of each micrograph on pages 63 and 71 to determine: (1) the thickness (roughly) of a cell membrane, (2) the diameter of a flagellum, and (3) the diameter of a microtubule within a flagellum. (This is how the sizes of cellular components are determined by microscopists.) (c) Would a flagellum be resolvable in a light microscope (that is, is its diameter more than 0.2 micrometer)?

6. Explain the functions of each of the following structures: endoplasmic reticulum, ribosomes, Golgi bodies, lysosomes.

7. Why is the secondary cell wall of a plant *inside* the primary cell wall? Where is the cell membrane in relation to the two cell walls?

8. On the basis of what you know of the functions of each of the structures in Table 4–2 (page 72), what components would you expect to be most prominent in each of the following cell types: muscle cells, sperm cells, green leaf cells, red blood cells, white blood cells?

9. Two brothers were under medical treatment for infertility. Microscopic examination of their semen showed that the sperm were immotile and that the little "arms" were missing from the microtubular arrays. The brothers also had chronic bronchitis and other respiratory difficulties. Can you explain why?

5

How Things Get into and out of Cells

Cells are able to regulate the passage of materials across cell membranes. This is an important capacity. One way we identify living systems is that, although surrounded on all sides by nonliving matter, living matter contains different kinds and amounts of chemical substances than does nonliving matter. Without this difference, living matter would be unable to maintain the organization on which its existence depends.

The cell membrane is not simply an impenetrable barrier, however. Living matter constantly exchanges substances with the nonliving world around it. Control of these exchanges is essential in order to protect the cell's integrity and to maintain the conditions at which metabolic activities can take place. Cell membranes thus have a complex double function of keeping certain things out and letting others in. Similarly, internal membranes, such as those surrounding mitochondria, chloroplasts, and the nucleus, regulate the passage of materials among compartments within the cell and so regulate the internal environment.

Control of these exchanges across membranes depends on physical and chemical properties of the membranes and of the molecules that move through them. Let us first look at the membrane itself.

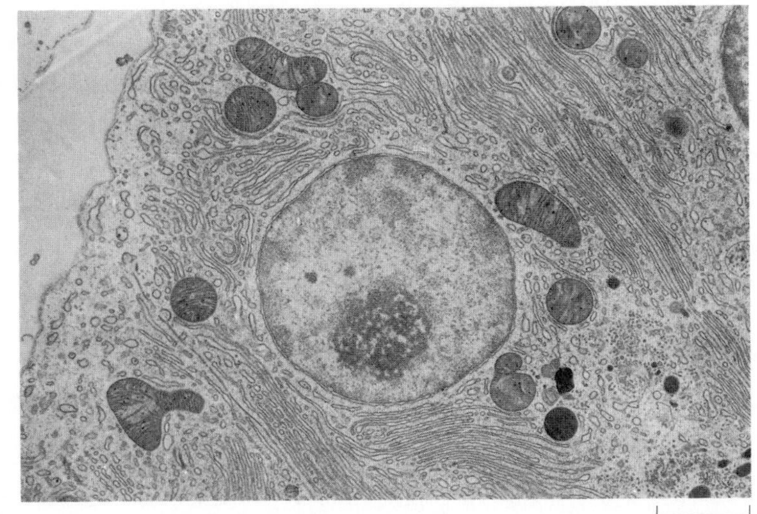

5-1 *Electron micrograph of a portion of a cell from the pancreas, an organ of the digestive system. The cell has a large, central nucleus with scattered chromatin, many mitochondria, large quantities of rough endoplasmic reticulum, and many small vesicles. Not only is the cell itself surrounded by a membrane and the nucleus by a double membrane system (the nuclear envelope), but its organelles are also surrounded by membranes. The membranes of the endoplasmic reticulum further divide the cell into membrane-bound channels. These membranes regulate the movements of substances into and out of cells and restrict their passage from one part of the cell to another.*

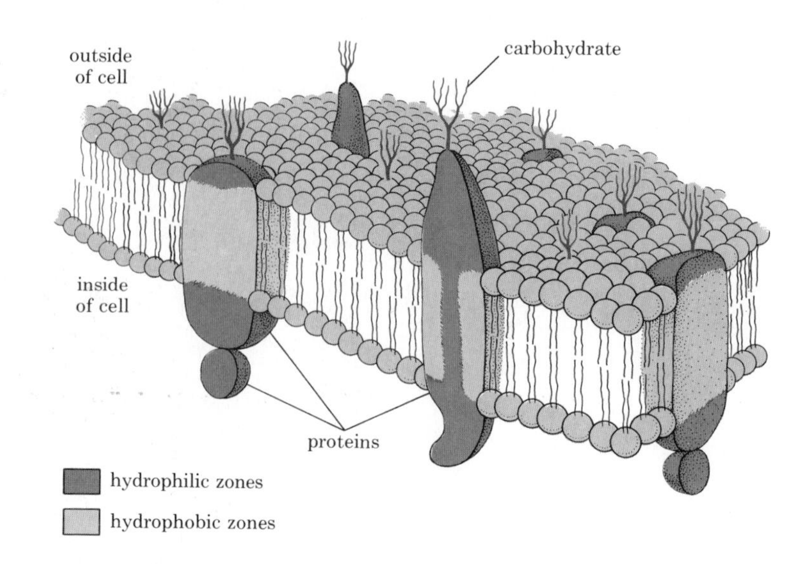

5-2 *Model of a cell membrane. The membrane is composed of phospholipid molecules and large protein molecules. The phospholipid molecules are arranged with their hydrophobic tails forming an inner layer and their hydrophilic phosphate heads on either side. The surface of the portion of a protein molecule embedded in the lipid bilayer is hydrophobic. The surface of a protein molecule exposed outside the lipid bilayer is hydrophilic. It is believed that pores with hydrophilic surfaces may pass through some of the protein molecules. The short carbohydrate chains attached to the outside of the membrane may be involved in the adhesion of cells to each other and with the "recognition" of molecules at the membrane surface.*

The Structure of the Cell Membrane

Figure 5-2 shows a current model of the structure of the cell membrane. As we saw in the last chapter, the membrane is composed of a phospholipid bilayer, that is, a double layer of phospholipid molecules arranged with their hydrophobic fatty acid tails pointed inward. Numerous large globular proteins are embedded in the lipid bilayer. All the membranes of a cell, including those surrounding the various organelles, have this same basic structure. There are, however, local differences in the types of lipids and, particularly, in the number and types of proteins.

The two surfaces of the cell membrane differ considerably in chemical composition. On the cytoplasmic side of the membrane, additional protein molecules are attached to some of the proteins protruding from the lipid bilayer. On the outside of the membrane, short carbohydrate chains are attached to the protruding proteins and to some of the lipid molecules. These carbohydrates are believed to play a role in the adhesion of cells to one another and in the "recognition" of molecules that are to be transported into or excluded from the cell.

According to this model, the proteins vary from membrane to membrane, depending on cell function, and also from place to place on the same membrane. The whole structure is generally quite fluid, and the proteins can be thought of as floating in a lipid sea. Some of the proteins are enzymes, which regulate particular chemical reactions, while others are carriers involved in the transport of molecules across the cell membrane.

Principles of Water Movement

Of the many kinds of molecules moving into and out of cells, by far the most important is water. Further, the other ions and molecules moving in and out are dissolved in water. Let us therefore look again at water, focusing our attention this time on how water moves.

Water molecules move from one place to another because of differences in potential energy, usually referred to as the *water potential*. Water moves from a region where water potential is greater to a region where water potential is lower, regardless of the reason for the water potential. A

5-3 *Water at the top of a falls, like a boulder on a hilltop, has potential energy. The movement of water molecules as a group, as from the top of the falls to the bottom, is referred to as bulk flow.*

simple example is water running downhill in response to gravity. Water at the top of a hill has more potential energy (that is, a greater water potential) than water at the bottom of a hill. As the water runs downhill, its potential energy is converted to mechanical energy if, for example, a watermill is placed in the path of the moving water.

Pressure is another source of water potential. If we fill a rubber bulb with water and then squeeze, water will squirt out of the nozzle. Like water at the top of a hill, this water has been given a high water potential and will move to a lower one. Can we make the water that is running downhill run uphill by means of pressure? Obviously we can. But only so long as the water potential produced by the pressure exceeds the water potential produced by gravity.

In solutions, water potential is affected by the concentration of dissolved particles (solutes). As the concentration of solute particles—that is, the number of solute particles per unit volume of solution—increases, the concentration of water molecules decreases, and vice versa. The water potential is directly related to the concentration of water molecules—the higher the concentration of water molecules, the greater the water potential. Conversely, the higher the concentration of solute particles, the lower the water potential. Individual water molecules move from regions of high water potential to regions of low water potential, a fact of great importance for living systems.

The concept of water potential is a useful one because it enables us to predict the way that water will move under various combinations of circumstances. Measurements of water potential are usually made in terms of the pressure required to stop the movement of water—that is, the hydrostatic (water-stopping) pressure—under the particular circumstances. The unit usually used to measure this pressure is the atmosphere. One atmosphere is the average pressure of the air at sea level.

Three mechanisms are involved in water movement: bulk flow, diffusion, and osmosis. In living systems, bulk flow moves substances from one part of a multicellular organism to another part, whereas diffusion and osmosis move substances across the cell membrane.

Bulk Flow

Bulk flow is the overall movement of a fluid. The molecules move all together and in the same direction. For example, water runs downhill by bulk flow in response to the differences in water potential at the top and the bottom of a hill. Blood moves through your body by bulk flow as a result of the water potential (blood pressure) created by the pumping of your heart. Sap—a concentrated solution of sucrose in water—moves by bulk flow from the leaves of a plant to other parts of the plant body.

Diffusion

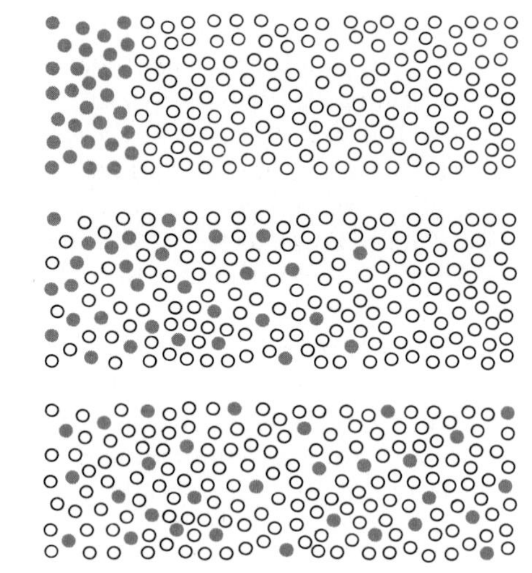

5-4 *Diagram of the diffusion process. Diffusion is the result of the random movement of individual molecules, which produces a net movement from a more concentrated to a less concentrated area. Notice that as one type of molecule (indicated by color) diffuses to the right, the other diffuses in the opposite direction. The result will be an even distribution of both types of molecules. Can you see why the net movement of molecules will slow down as equilibrium is reached?*

Diffusion is a familiar phenomenon. If you sprinkle a few drops of perfume in one corner of a room, the scent will eventually permeate the entire room even if the air is still. If you put a few drops of dye in one end of a glass tank full of water, the dye molecules will slowly become distributed evenly throughout the tank. The process may take a day or more, depending on the size of the tank, the temperature, and the relative size of the molecules.

Why do the dye molecules move apart? If you could observe the individual dye molecules in the tank (Figure 5-4), you would see that each one of them moves individually and at random. Looking at any single molecule—at either its rate of motion or its direction of motion—gives you no clue at all about where the molecule is located with respect to the others. So how do the molecules get from one side of the tank to the other? Imagine a thin section through the tank, running from top to bottom. Dye molecules will move in and out of this section, some moving in one direction, some moving in the other. But you will see more dye molecules moving from the side of greater dye concentration. Why? Simply because there are more dye molecules at that end of the tank. Since there are more dye molecules on the left, more dye molecules, moving at random, will move to the right, even though there is an equal probability that any one molecule of dye will move from right to left. Consequently, the net movement of dye molecules will be from left to right. Similarly, if you could see the movement of the individual water molecules in the tank, you would see that their net movement is from right to left.

What happens when all the molecules are distributed evenly throughout the tank? The even distribution does not affect the behavior of the molecules as individuals; they still move at random. And, since the movements are random, just as many molecules go to the left as to the right. But because there are now as many molecules of dye and of water on one side of the tank as on the other, there is no net movement of either. There is, however, just as much random motion as before, provided the temperature has not changed.

Substances that are moving from a region of higher concentration of their own molecules to a region of lower concentration are said to be moving *down a gradient.* Diffusion occurs only down a gradient. The steeper the downhill gradient—that is, the larger the difference in concentration—the more rapid the diffusion. (A substance moving in the opposite direction, toward a higher concentration of its own molecules, moves *against a gradient,* which is analogous to being pushed uphill.) In our imaginary tank, there are two gradients; the dye molecules are moving down one of them, and the water molecules are moving down the other. When the molecules have reached a state of equal distribution, that is, when there are no more gradients, they are said to be in dynamic equilibrium.

An aqueous solution in which diffusion is occurring is another example of water potential. The region of the tank in which there is pure water has a higher water potential than the region containing water plus dye or some other solute. Thus the water molecules are moving from a region of higher water potential to a region of lower water potential.

The essential characteristics of diffusion are (1) that each molecule moves independently of the others and (2) that these movements are random. The net result of diffusion is that the diffusing substance becomes evenly distributed.

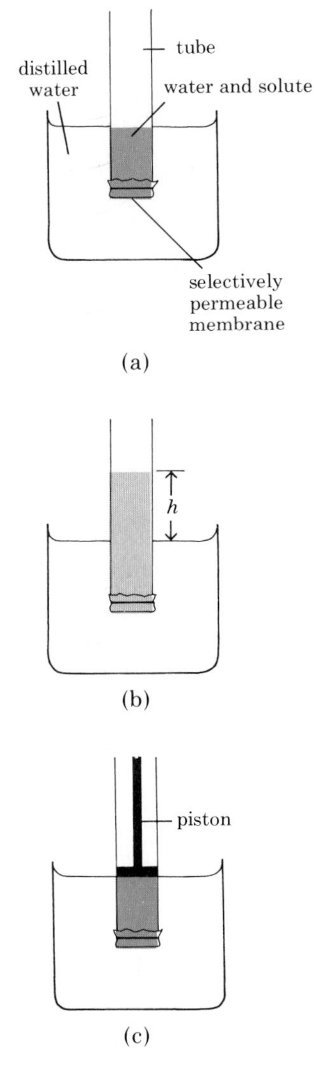

5-5 *Osmosis and osmotic pressure.* (a) *The tube contains a solution and the beaker contains distilled water.* (b) *The selectively permeable membrane permits the passage of water but not of solute. The movement of water into the solution causes the solution to rise in the tube until the osmotic pressure resulting from the tendency of water to move into a region of lower water concentration is counterbalanced by the height,* h, *and density of the column of solution.* (c) *The force that must be applied to the piston to oppose the rise in the tube of the solution is a measurement of the osmotic pressure. It is proportional to the height and density of the column of solution.*

Cells and Diffusion

Water, oxygen, carbon dioxide, a few other simple molecules, and some ions diffuse freely across cell membranes. For example, carbon dioxide gas is constantly produced as the cell oxidizes fuel molecules for energy. As a result, there is more carbon dioxide inside the cell than out. Thus a concentration gradient is maintained between the inside of the cell and the outside, and carbon dioxide diffuses out of the cell down this gradient. Conversely, oxygen gas is used up by the cell in the course of its internal activities, so oxygen present in air or water or blood tends to move into the cell by diffusion, again down a gradient.

Diffusion is also a principal way in which substances move within cells. The dependence upon diffusion is one of the factors limiting cell size. As you can see by studying Figure 5-4, the process becomes increasingly slower and less efficient as the distance "covered" by the diffusing molecules increases.

Osmosis

While permitting the passage of water, cell membranes block the passage of many of the materials dissolved in it. Such a membrane is said to be *selectively permeable*, and the movement of water through it is known as *osmosis*. Osmosis carries a net flow of water from a solution that has higher water potential to a solution that has lower water potential. In the absence of other factors that influence water potential (such as pressure), the movement of water in osmosis will be from a region of lower solute concentration (and therefore of higher water concentration) into a region of higher solute concentration (lower water concentration). The presence of solute decreases the water potential and so creates a gradient of water potential along which water moves (Figure 5-5).

The movement of water is not affected by *what* is dissolved in the water, only by *how much* is dissolved—that is, the concentration of particles of solute (molecules or ions) in the water. The word *isotonic* was coined to describe two or more solutions that have equal numbers of dissolved particles and therefore the same water potential. There is no net movement of water across a membrane separating two solutions that are isotonic to one another, unless, of course, pressure is exerted on one side. In comparing solutions of different concentration, the solution that has less solute (and therefore a higher water potential) is known as *hypotonic*, and the one that has more solute (a lower water potential) is known as *hypertonic*. (Note that *iso* means "the same"; *hyper* means "more"—in this case, more particles of solute; and *hypo* means "less"—in this case, fewer particles of solute.) In osmosis, water molecules move through a selectively permeable membrane into a hypertonic solution until the water potential is equal on both sides of the membrane.

Osmosis and Living Organisms

The movement of water across the cell membrane from a hypotonic to a hypertonic solution causes some crucial problems for living systems. These problems vary according to whether the cell or organism is hypotonic, isotonic, or hypertonic in relation to its environment. One-celled organisms that live in salt water, for example, are usually isotonic with the medium they inhabit, which is one way of solving the problem. Similarly, the cells of higher animals are isotonic with the blood and lymph that constitute the watery medium in which they live.

Many types of cells live in a hypotonic environment. In all single-celled organisms that live in fresh water, such as *Paramecium,* the interior of the cell is hypertonic to the surrounding water; consequently water tends to move into the cell by osmosis. If too much water were to move into the cell, it could dilute the cell contents to the point of interfering with function and could even eventually rupture the cell membrane. This is prevented by a specialized organelle known as a contractile vacuole, which collects water from various parts of the cell and pumps it out with rhythmic contractions (Figure 5–6). As you might expect, this bulk transport process requires energy.

5–6 *A* Paramecium *is hypertonic in relation to its environment, and hence water tends to move into the cell by osmosis. Excess water is expelled through its contractile vacuoles.* (a) *Photomicrograph of a* Paramecium *showing the position of its two rosette-like contractile vacuoles.* (b) *Collecting tubules converge toward the vacuole, filling it.* (c) *Then it contracts, emptying outside the cell membrane by way of a small central pore.*

Turgor Plant cells are usually hypertonic to their surrounding environment, and so water tends to move into them. This movement of water into the cell creates pressure within the cell against the cell wall. The pressure causes the cell wall to expand and the cell to enlarge. The elongation that occurs as a plant cell matures (Figure 4–12b) is a direct result of water uptake.

If the water potential on both sides of the plant cell membrane becomes equal, the net movement of water ceases. (We say "net movement" here because water molecules continue to move back and forth across the membrane. However, these movements are in equilibrium; that is, as many water molecules are going in as are coming out.)

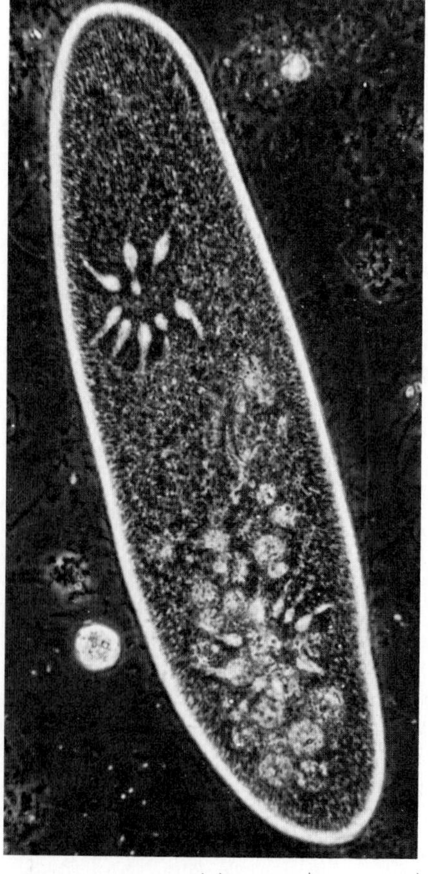

(a) 25 μm

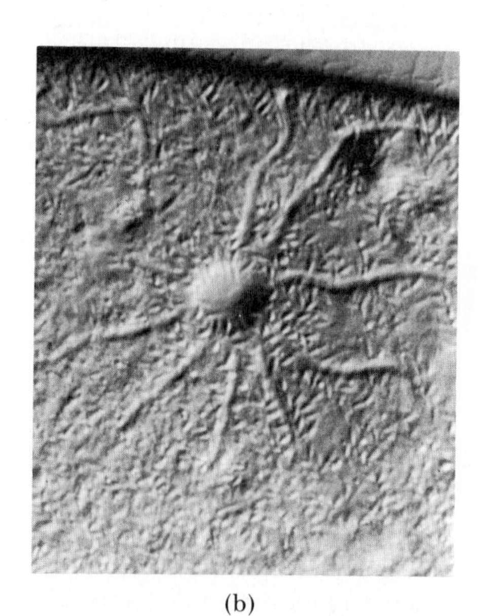

(b)

(c) 10 μm

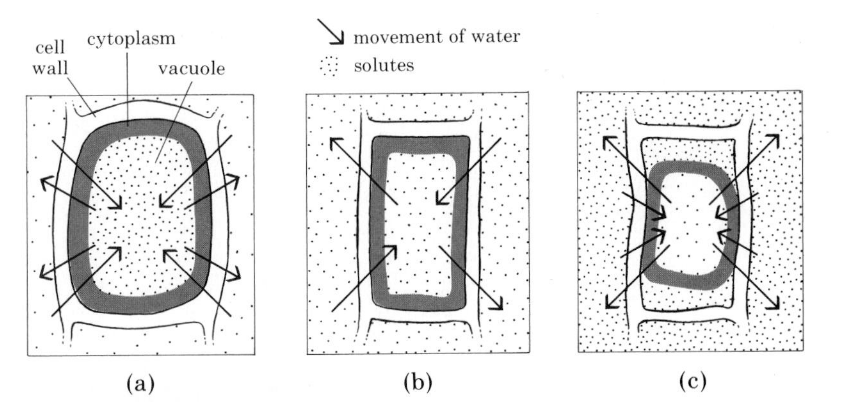

5-7 (a) *A turgid plant cell. The central vacuole is hypertonic in relation to the fluid surrounding it and so gains water. The expansion of the cell is held in check by the cell wall.* (b) *A plant cell begins to wilt if it is placed in an isotonic solution, so that water pressure no longer builds up within the vacuole.* (c) *A plant cell in a hypertonic solution loses water to the surrounding fluid and so collapses, with its membrane pulling away from the cell wall. Such a cell is said to be plasmolyzed.*

As the plant cell matures, the cell wall stops growing. Moreover, mature plant cells typically have large central vacuoles. These vacuoles often contain solutions of salts and other materials. (In citrus fruits, for example, they contain the acids that give the fruits their characteristic sour taste.) Because of these concentrated solutions, water continues to "try" to move into the cells. In the mature cell, however, the cell wall does not expand further, and equilibrium of salt concentration is not reached. As a consequence, the cell wall remains under constant pressure (Figure 5–7). This water pressure exerted on the cell wall from inside is known as *turgor.* Turgor, which keeps the cell walls stiff and the plant body crisp, results from an equilibrium of water potential. When turgor pressure is reduced, as a consequence of water loss, the plant wilts.

Crossing the Membrane

In order to get into or out of a cell (or a membrane-bound organelle, such as a mitochondrion), substances must pass through a membrane. As we have seen, cell membranes are permeable to such substances as water, oxygen, and carbon dioxide, which move in and out by diffusion or osmosis.

Other molecules used or produced by the cell are unable to diffuse through the membrane, because of either their size or their polarity. Most ions and polar molecules, for example, cannot pass through the hydrophobic interior of the lipid bilayer. Such substances are transported through the membrane by the embedded protein molecules. The tertiary structure (page 48) of a particular membrane protein determines what molecules it can transport. Many of these protein "carriers" can even move substances against a concentration gradient, a process known as *active transport* (Figure 5–8). For instance, carrier proteins move glucose into liver cells, where it

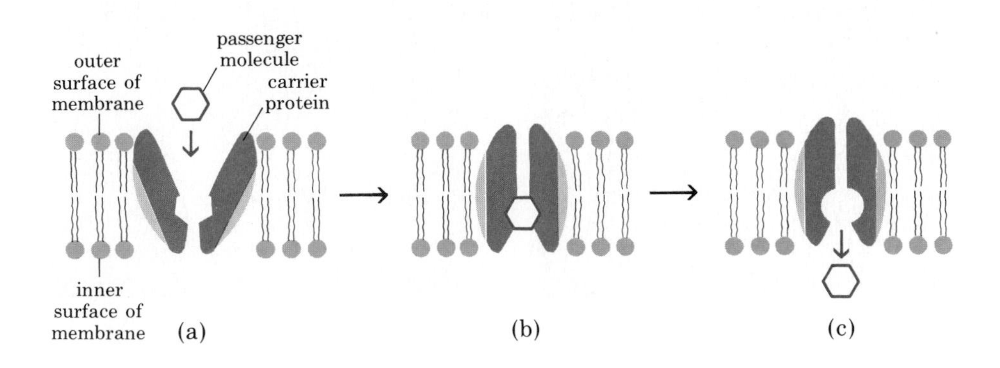

5-8 *A model of active transport.* (a) *A molecule to be transported into the cell fits precisely into the carrier protein* (b). *A chemical reaction then triggers a change in the shape of the protein* (c), *causing the molecule to be released to the interior of the cell.*

Another chemical reaction is required to return the carrier protein to its original shape, ready to transport another molecule into the cell.

Active transport processes require an expenditure of energy by the cell. They play an important role in maintaining its unique internal environment.

5-9 (a) *Phagocytosis. A food particle is enveloped in a portion of the cell membrane, which pinches off to become a separate vacuole. Lysosomes spill their contents of digestive enzymes into the vacuole.* (b) *The indigestible remains are eliminated by exocytosis as the vacuole membrane fuses with the cell membrane.*

phagocytosis

lysosomes

(a)

exocytosis

(b)

is stored as glycogen, even though the concentration of glucose is higher inside the liver cells than in the bloodstream.

Another type of transport into cells is *phagocytosis* ("cell eating"). This process is diagrammed in Figure 5-9a. It begins with the attachment of a solid particle to the cell membrane and is followed by the enveloping of the particle within a segment of the membrane. Thus the particle enters the cell within a sac, or vacuole. Often lysosomes fuse with these vacuoles, emptying their enzymes into them and so digesting or destroying the vacuole contents. Amoebas engulf their prey by phagocytosis, as shown in Figure 18-9 on page 239. As we noted in the last chapter, white blood cells engulf bacteria and other invaders in this way.

Pinocytosis ("cell drinking") is a process similar to phagocytosis. The only difference is that molecules in solution, rather than solid particles, are taken into the cells.

5-10 *Phagocytosis of a* Paramecium *by a* Didinium. *(See Figure 4-19 for the preamble to their encounter.)*

(a) *Ingestion of the* Paramecium *has begun. Because the* Paramecium *is larger than the* Didinium, *folding helps.*

(b) *The* Paramecium *is half "swallowed"; the part that is within the* Didinium *is surrounded by a membrane composed of the cell membrane of the* Didinium. *The process of compression has begun, as you can see in the tip of the* Paramecium *protruding from the oral rim of the* Didinium. *This compression is largely a matter of squeezing out water.*

(c) *Once the* Paramecium *is completely inside, the cell membrane of the* Didinium *will fuse over it, forming a food vacuole. A* Didinium *can eat a dozen* Paramecium, *each larger than itself, in a single day. Moreover, the* Paramecium *must also provide the means for its own demolition: The* Didinium *apparently lacks a crucial digestive enzyme that the* Paramecium *supplies.*

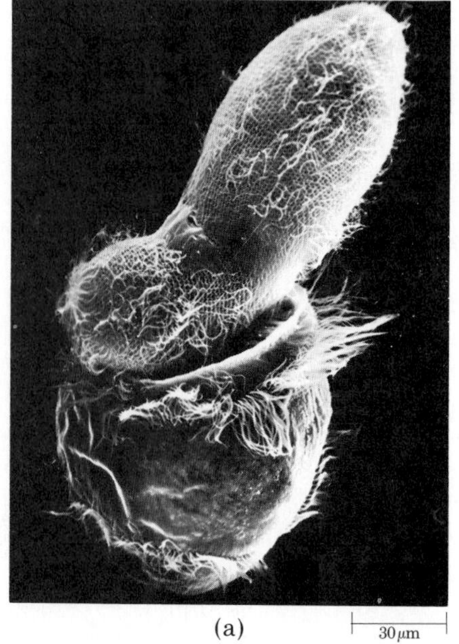

(a)

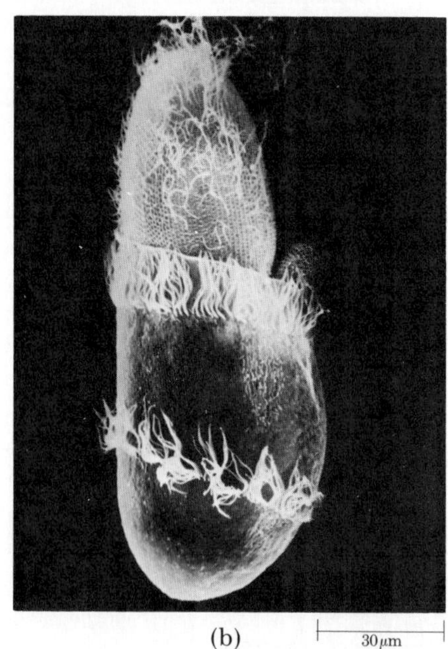

(b)

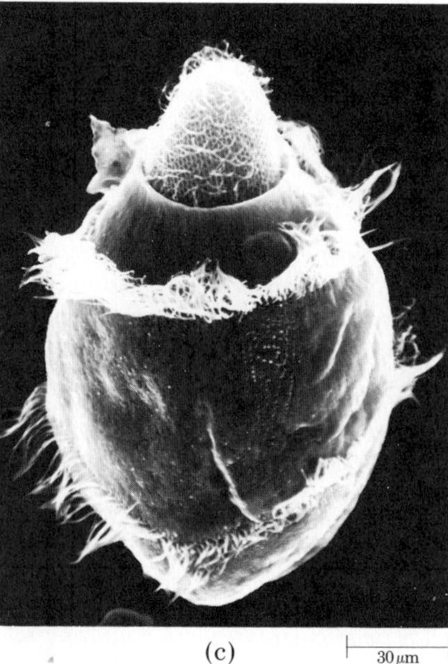

(c)

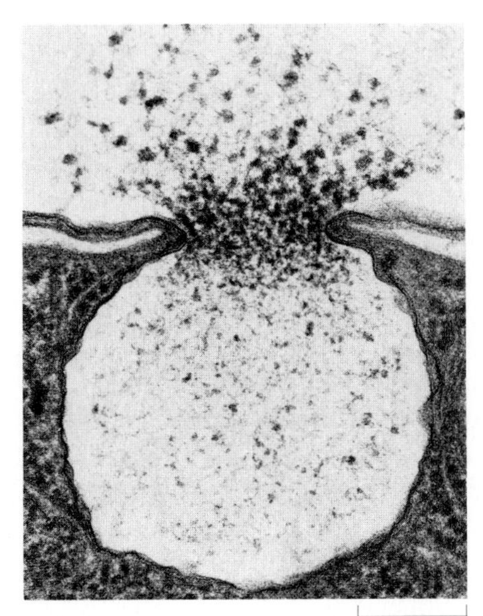

5-11 Exocytosis. A secretory vesicle, formed within the cell, discharges its contents. Notice how the membrane enclosing the vesicle has fused with the cell membrane.

Phagocytosis and pinocytosis can also work in reverse. A particle to be exported from the cell may be wrapped in a vacuole and transported to the cell surface. Here the membrane of the vacuole fuses with the outer membrane of the cell, expelling the vacuole's contents to the outside (Figure 5-9b). This process is called *exocytosis.*

The diffusion of molecules into the cell and the movement of water by osmosis are spontaneous processes, in which the molecules move from a state of high potential energy to one of lower potential energy. During these processes, the cell is essentially passive. Other forms of transport across membranes, however, require an energy expenditure on the part of the cell. As we noted earlier, it takes energy for a *Paramecium* to expel excess water through its contractile vacuoles. Similarly, the cell must expend energy to move substances against a concentration gradient by active transport or to form the membranes involved in phagocytosis and exocytosis. A steady supply of energy is required for these processes regulating the internal environment, as well as for all the other activities of the cell. In the next section, we shall see how cells provide themselves with this essential energy.

SUMMARY

The cell membrane regulates the passage of materials into and out of the cell, a function that makes it possible for the cell to maintain its structural and functional integrity. This regulation depends on interaction between the membrane and the materials that pass through it.

According to the current model, cell membranes are lipid bilayers in which globular proteins are embedded. Some of these proteins act as carriers, ferrying molecules through the membrane.

One of the principal substances passing into and out of cells is water. Water potential determines the direction in which the water moves; that is, water moves from where the water potential is higher to where it is lower. Water movement takes place by bulk flow, diffusion, and osmosis.

Bulk flow is the overall movement of water molecules as a group, as when water flows in response to gravity or pressure. The circulation of blood through the human body is an example of bulk flow.

Diffusion involves the random movement of molecules and results in net movement down a concentration gradient. Carbon dioxide and oxygen are two important molecules that move into and out of cells by diffusion across the membrane.

Osmosis is the movement of water through a membrane that permits the passage of water but inhibits the movement of most solutes; such a membrane is called a selectively permeable membrane. In the absence of other forces, the movement of water in osmosis is from a region of lower solute concentration (a hypotonic medium), and therefore of higher water potential, to one of higher solute concentration (a hypertonic medium), and so of lower water potential. Turgor in plant cells is a direct consequence of osmosis.

Molecules that are unable to cross the cell membrane by diffusion or osmosis are transported by carrier proteins in the membrane. When such movement is against a concentration gradient, it is known as active transport. Controlled movement into a cell may also occur by phagocytosis (solid particles) or pinocytosis (dissolved particles). Substances may be transported out of cells by exocytosis. All of these processes require a steady supply of energy.

QUESTIONS

1. Distinguish among the following: bulk flow/diffusion/osmosis; hypotonic/hypertonic/isotonic; phagocytosis/pinocytosis/exocytosis.

2. What is a concentration gradient? How does a concentration gradient affect diffusion? How does a concentration gradient affect osmosis?

3. When diffusion of dye molecules in a tank of water is complete, random movement of molecules continues (as long as the temperature remains the same). However, net movement stops. How do you reconcile these two facts?

4. Why is diffusion more rapid in gases than in liquids? Why is it more rapid at higher temperatures than at lower temperatures?

5. Three funnels have been placed in a beaker containing a solution (see the figure below). What is the concentration of the solution? Explain your answer.

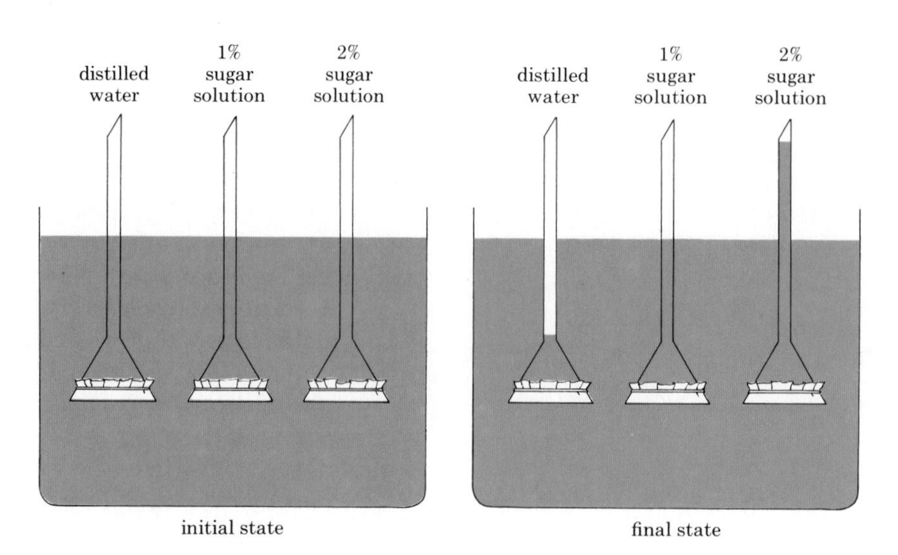

6. Imagine a pouch with a selectively permeable membrane containing a saltwater solution. It is immersed in a dish of fresh water. Which way will the water move? If you add salt to the water in the dish, how will this affect the water movement? What living systems exist under analogous conditions? How do you think they maintain water balance?

7. When you forget to water your house plants, they wilt and the leaves (and sometimes the stems) become very limp. What has happened to the plants to cause this change in appearance and texture? Within a few hours after you remember to water your plants, they resume their normal, healthy appearance. What has occurred within the plants to cause this restoration? Sometimes, if you wait too long to water your plants, they never revive. What do you suppose has happened?

8. In what three ways does active transport differ from diffusion or osmosis?

9. When you cut or scrape your skin, white blood cells rapidly converge on the site of the injury. What are they doing when they reach the site? Why is this important to your well-being?

SUGGESTIONS FOR FURTHER READING

BAKER, J. J., and G. E. ALLEN: *Matter, Energy and Life: An Introduction for Biology Students,* 4th ed., Addison-Wesley Publishing Co., Inc., Reading, Mass., 1981.*

A book for students who have had no previous chemistry or physics, which deals with such topics as the structure of matter, the formation of molecules, the course and mechanism of chemical reactions, as well as with the chemistry of living systems.

BRONOWSKI, J.: *The Ascent of Man,* Little, Brown and Company, Boston, 1973.*

An informal and illuminating history of the sciences, originally prepared as a television series. The emphasis is on science's relation to human culture. Well designed and illustrated.

DE ROBERTIS, E. D. P., and E. M. F. DE ROBERTIS, *Cell and Molecular Biology,* 8th ed., Saunders College/Holt, Rinehart and Winston, Philadelphia, 1980.

An up-to-date, well-illustrated text integrating the fundamental roles of nucleic acids and proteins with cell structure and function.

DYSON, ROBERT D.: *Cell Biology: A Molecular Approach,* 2d ed., Allyn & Bacon, Inc., Boston, 1978.

Comprehensive, clear, up-to-date account of cell structure and how it relates to metabolism and transport functions.

LEDBETTER, M. C., and KEITH R. PORTER: *Introduction to the Fine Structures of Plant Cells,* Springer-Verlag, New York, 1970.

An excellent atlas of electron micrographs of plant cells, with detailed explanations.

LOEWY, A. G., and P. SIEKEVITZ: *Cell Structure and Function,* 2d ed., Holt, Rinehart and Winston, Inc., New York, 1970.

An outstanding elementary text on cell structure and function.

PORTER, KEITH R., and MARY A. BONNEVILLE: *An Introduction to the Fine Structure of Cells and Tissues,* 4th ed., Lea & Febiger, Philadelphia, 1973.

An atlas of electron micrographs of animal cells; detailed commentaries accompany each. These are magnificent micrographs, and the commentaries describe not only what the pictures show but also the experimental foundations of our knowledge of cell ultrastructures.

THOMAS, LEWIS: *The Lives of a Cell: Notes of a Biology Watcher,* Viking Press, Inc., New York, 1974.*

THOMAS, LEWIS: *The Medusa and the Snail: More Notes of a Biology Watcher,* Viking Press, Inc., New York, 1979.*

Thomas, a physician and medical researcher, reveals the extent to which science can tune our intellectual antennae, broaden our perceptions, and extend our appreciation of ourselves and of the world around us. Anyone who wants to refute the contention that science destroys human values need look no further than these short, sensitive essays.

WEINBERG, STEVEN: *The First Three Minutes: A Modern View of the Origin of the Universe,* Basic Books, Inc., New York, 1977.*

A wonderful story, written for the intelligent nonscientist (characterized by the author as a smart old attorney who expects to hear some convincing arguments before he makes up his mind).

* Available in paperback.

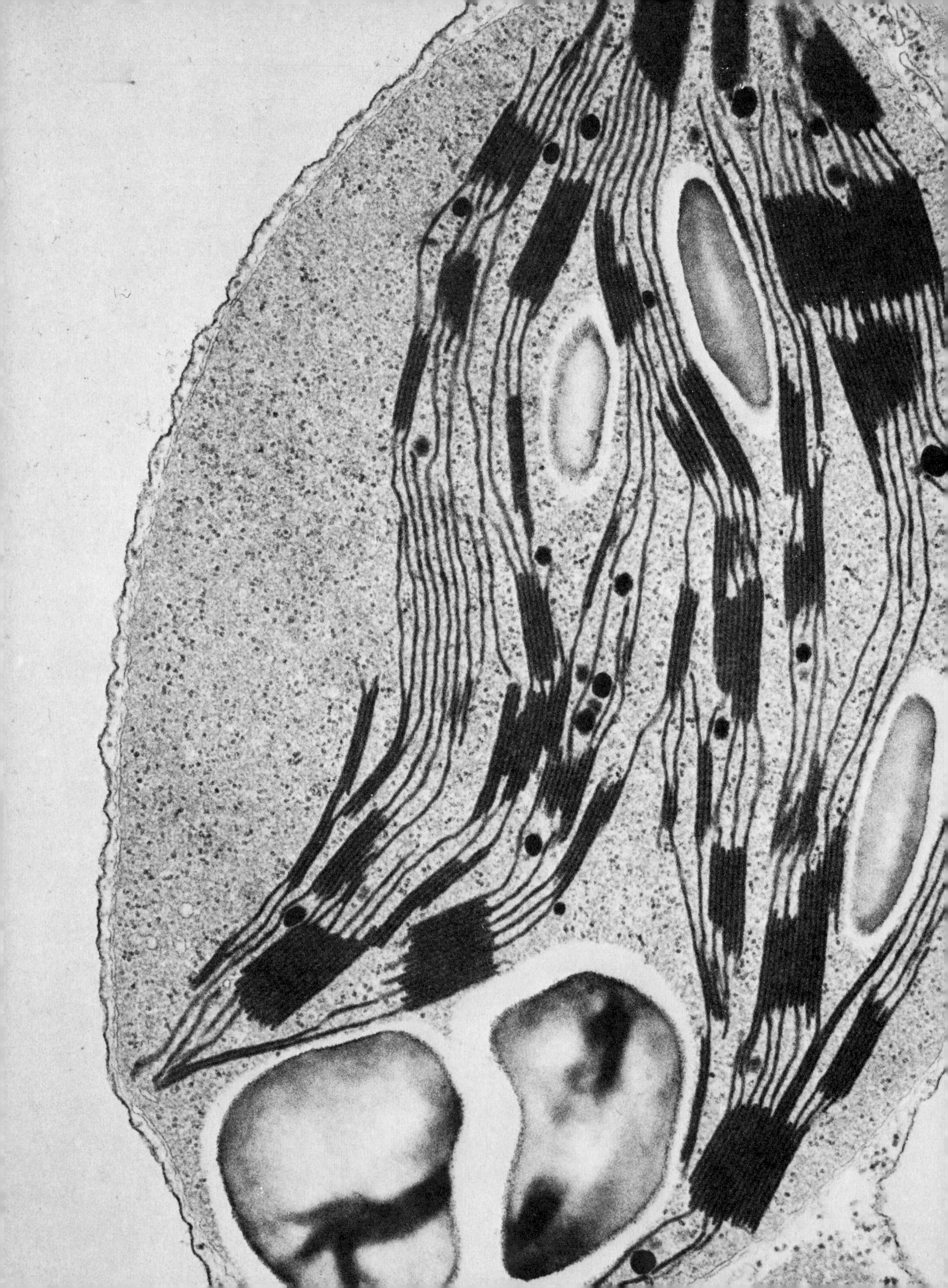

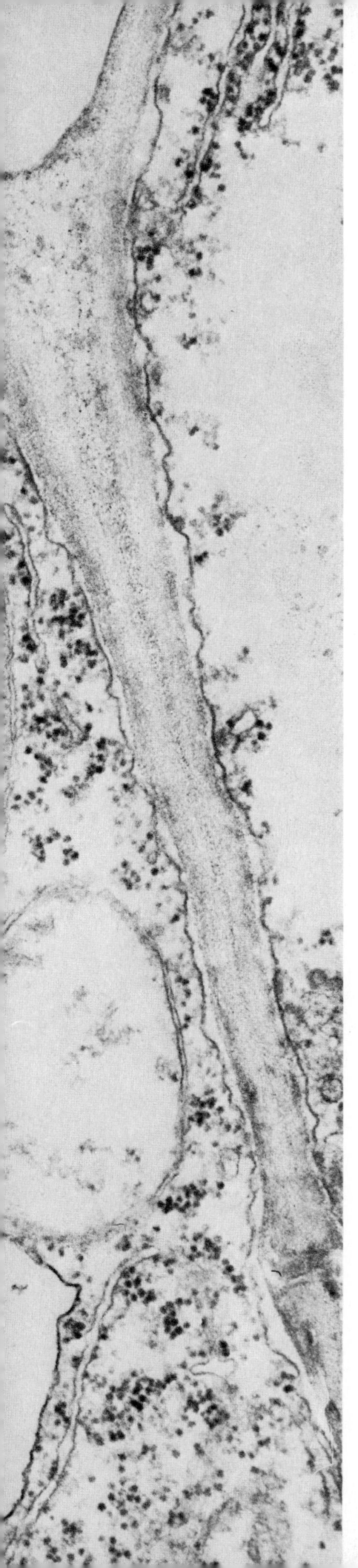

SECTION 2

Energetics

Chloroplasts—organelles that are found in eukaryotic algae and in the photosynthetic cells of green plants—capture light energy and convert it to chemical energy stored in organic molecules. Interconnected, stacked membranes within the chloroplast are, in effect, miniature solar cells. This chloroplast, from a blade of timothy grass, is magnified 52,700 times. The large, globular objects are starch grains.

6

The Flow of Energy

Life here on earth depends on the flow of energy from the thermonuclear reactions taking place at the heart of the sun. The amount of energy delivered to the earth by the sun is about 10^{24} (the number 1 followed by 24 zeros) calories per year. It is a difficult quantity to imagine. For example, the amount of solar energy striking the earth every day is about 1½ billion times greater than the amount of electricity generated in the United States each year.

About one-third of this solar energy is immediately reflected back into space as light (as it is from the moon). Much of the remaining two-thirds is absorbed by the earth and converted to heat. Some of this absorbed heat energy serves to evaporate the waters of the oceans, producing the clouds that, in turn, produce rain and snow. Solar energy, in combination with other factors, is also responsible for the movements of air and of water that help set patterns of climate over the surface of the earth.

A small fraction—less than 1 percent—of the solar energy reaching the earth becomes, through a series of operations performed by the cells of plants and other photosynthetic organisms, the energy that drives all the processes of life. Living systems change energy from one form to another, transforming the radiant energy from the sun into the chemical and mechanical energy used by everything that is alive (Figure 6–1).

6–1 *The flow of energy in a biological system. Radiant energy from the sun is transformed to chemical energy by the grasses of the African plains. The zebras, eating the grasses, use this stored chemical energy for growth and convert some of it to kinetic energy, which may help to keep them from participating in another energy transfer involving the waiting lions. Of the radiant energy falling on the grass, less than 10 percent is converted to chemical energy. Less than 10 percent of the chemical energy stored in the grass is converted to chemical energy stored in the zebra, and less than 10 percent of the zebra's stored energy is converted to chemical energy stored in the body of its predator, the lion.*

In this chapter, we shall look first at the general principles governing all energy transformations and then at the characteristic ways in which cells regulate the energy transformations that take place within living systems. In the chapters that follow, the principal and complementary processes of energy flow through the biosphere will be examined—glycolysis and respiration in Chapter 7 and photosynthesis in Chapter 8.

The Laws of Thermodynamics

Energy is such a common term today that it is surprising to learn that the word was coined less than 200 years ago, at the time of the development of the steam engine. It was only then that scientists and engineers began to understand that heat, motion, light, electricity, and the forces holding atoms together in molecules are all different forms of the same capacity to cause change, or, as it is most often expressed, to do work. This new understanding led to the study of thermodynamics—the science of energy transformations—and to the formulation of its laws.

The *first law of thermodynamics* states, quite simply: *Energy can be changed from one form to another, but it cannot be created or destroyed.* The total energy of any system plus its surroundings thus remains constant, despite any changes in form.

Electricity is a form of energy, as is light. Electrical energy can be changed to light energy (for example, by letting it flow through the tungsten wire in a light bulb). Conversely, light energy can be changed to electrical energy, a transformation that is the essential first step of photosynthesis, as we shall see in Chapter 8.

Energy can be stored in various forms and then changed into other forms. In automobile engines, for example, the energy stored in the chemical bonds of gasoline is converted to heat, which is then partially converted to mechanical movements (kinetic energy). Some of the energy is converted back to heat by the friction of these movements, and some of it leaves the engine in the exhaust products. Similarly, when organisms oxidize carbohydrates, they convert the energy stored in chemical bonds to other forms. On a summer evening, for example, a firefly converts chemical energy to kinetic energy, to flashes of light, and to electrical impulses that travel along the nerves of its body. Birds and mammals convert chemical energy into the heat necessary to maintain their body temperature, as well as into kinetic energy, electrical energy, and other forms of chemical energy. The first law of thermodynamics states that in these energy conversions, and in all others, the total energy of the system (in these examples, the individual organism) and its surroundings after the conversion is equal to the total energy before the conversion.

In all energy conversions, however, some useful energy is converted to heat and dissipated. Heat is simply the random motion of atoms and molecules. In a gasoline engine, about 75 percent of the energy originally present in the fuel is transferred to the surroundings in the form of heat—that is, it is converted to increased motion of atoms and molecules in the air. Similarly, the heat produced by "warm-blooded" animals, such as ourselves, is dissipated into the surrounding air or water.

Energy dissipated as heat has not been destroyed—it is still present in the random motion of atoms and molecules—but it has been "lost" for all practical purposes. It is no longer available to do useful work. This brings us to the *second law of thermodynamics*, which says that *in all natural processes, energy in a form to do work is eventually dissipated into the surroundings.* Another way of stating this is that the disorder, or randomness, of every natural system tends to increase.

Living Systems and the Second Law

The second law of thermodynamics, unlike the first, shows the direction that natural processes take. As a consequence, it is sometimes known as "time's arrow." The universe, according to the present model, is a closed system. The matter and energy present in it 20 billion years ago, at the time of the "big bang," are all the matter and energy it will ever have. Moreover, in every energy transformation, the final state has less useful energy and more disorder than the initial one. In this view, of course, the universe is running down. The stars will flicker out, one by one; life—any form of life on any planet—will come to an end. Finally, even the motion of

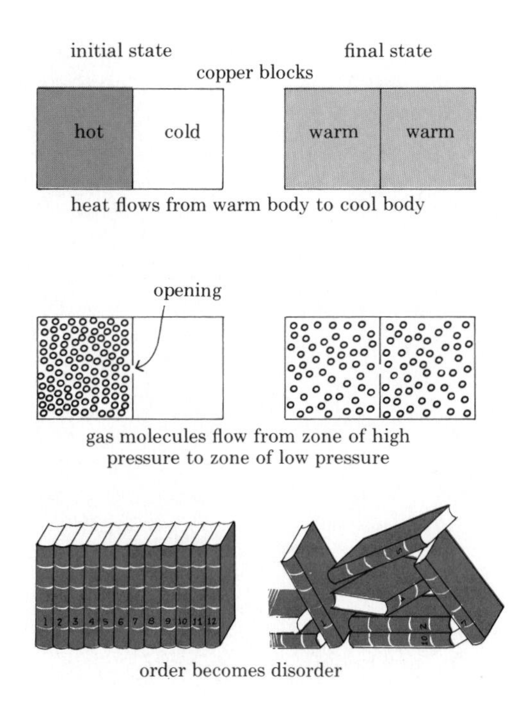

6-2 *Some illustrations of the second law of thermodynamics. In each case, a concentration of energy—in the hot copper block, in the gas molecules under pressure, and in the neatly organized books—is dissipated. In nature, processes tend toward randomness, or disorder. Only an input of energy can reverse this tendency and reconstruct the initial state from the final state. Ultimately, however, disorder will prevail, since the total amount of energy in the universe is finite.*

individual molecules will cease. However, even the most pessimistic among us do not believe this will occur for another 20 billion years or so.

In the meantime, life can exist *because* the universe is running down. Photosynthetic organisms are specialists at capturing light energy released by the sun as it slowly burns itself out. They use this energy to organize small, simple molecules (water and carbon dioxide) into larger, more complex molecules (sugars). In the process, the captured light energy is stored in the chemical bonds of sugars and other organic molecules. Living cells—including photosynthetic cells—can convert this stored energy into motion, electricity, light, and, by shifting the energy from one type of chemical bond to another, more convenient forms of chemical energy. At each transformation, of course, energy is lost to the surroundings as heat. But before the energy captured from the sun is completely dissipated, organisms use it to create and maintain the complex organization of structures and activities that we know as life.

Oxidation-Reduction

You will recall from Chapter 1 that electrons possess differing amounts of potential energy depending on their distance from the atomic nucleus and the attraction of the nucleus for electrons. An input of energy will boost an electron to a higher energy level, but without added energy an electron will move to the lowest energy level available to it.

Chemical reactions are essentially energy transformations in which energy stored in chemical bonds is transferred to other, newly formed chemical bonds. In such transfers, electrons shift from one energy level to another. In many reactions, electrons pass from one atom or molecule to another. These reactions, which are of great importance in living systems,

(a)

(b)

(c)

(d)

6-3 (a) *Life on earth can exist because photosynthetic cells, such as those in leaves, capture the radiant energy of the sun and convert it to chemical energy stored in the bonds of organic compounds. Other cells, in both plants and animals, use the chemical energy packaged by photosynthetic cells in various ways. Hummingbirds* (b), *for example, convert large amounts of chemical energy to the kinetic energy of flight. Their energy requirements are so high that, by a special adaptation, their body temperatures drop while they are at rest, which conserves stored chemical energy.* (c) *The skunk cabbage, a common plant in bogs and marshy areas of the northeastern United States, converts stored chemical energy to heat. The plant produces enough heat to melt surrounding snow or ice, while maintaining a nearly constant internal temperature of about 22°C (72°F). Gram for gram, the petals of the skunk cabbage are using energy at almost the same rate as the hummingbird.* (d) *Certain organisms, such as these bioluminescent mushrooms, transform some of their chemical energy into light energy, thereby glowing in the dark.*

are known as oxidation-reduction (or redox) reactions. The loss of an electron is known as *oxidation*, and the atom or molecule that loses the electron is said to be oxidized. The reason electron loss is called oxidation is that oxygen, which attracts electrons very strongly, is most often the electron acceptor.

Reduction is, conversely, the gain of an electron. Oxidation and reduction take place simultaneously because an electron that is lost by the oxidized atom is accepted by another atom, which is reduced in the process.

Redox reactions may involve only a solitary electron, as when sodium loses an electron and becomes oxidized to Na^+, and chlorine gains an electron and is reduced to Cl^-. Often, however, the electron travels with a proton, that is, as a hydrogen atom. In such cases, oxidation involves the removal of hydrogen atoms, and reduction the gain of hydrogen atoms. For example, when glucose is oxidized, hydrogen atoms are lost by the glucose molecule and gained by oxygen:

$$\underset{\text{glucose}}{C_6H_{12}O_6} + \underset{\text{oxygen}}{6O_2} \longrightarrow \underset{\text{carbon dioxide}}{6CO_2} + \underset{\text{water}}{6H_2O} + \text{energy}$$

The electrons are moving to a lower energy level, and energy is released.

Conversely, in the process of photosynthesis hydrogen atoms are transferred from water to carbon dioxide, thereby reducing the carbon dioxide to form glucose:

$$6CO_2 + 6H_2O + \text{energy} \longrightarrow C_6H_{12}O_6 + 6O_2$$

In this case, the electrons are moving to a higher energy level, and an energy input is required to make the reaction occur.

In living systems, the energy-capturing reactions (photosynthesis) and energy-releasing reactions (glycolysis and respiration) are oxidation-reduction reactions. As we noted earlier, the complete oxidation of a mole of glucose releases 686 kilocalories of energy (conversely, the formation of a mole of glucose from carbon dioxide and water requires 686 kilocalories). If this energy were to be released all at once, most of it would be dissipated as

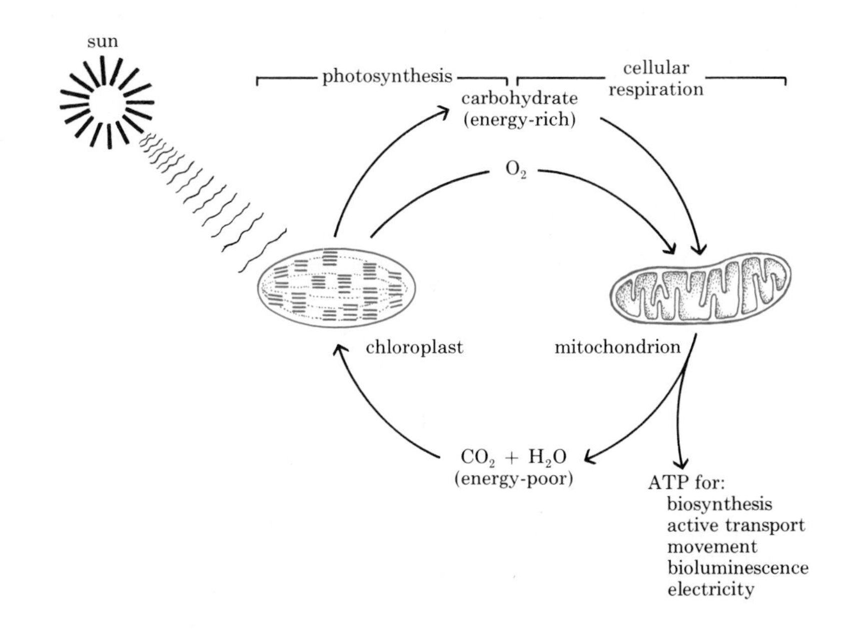

6-4 *The flow of biological energy. Chloroplasts, present in all photosynthetic eukaryotic cells, capture the radiant energy of sunlight and use it to convert water and carbon dioxide into carbohydrates, such as glucose, starch, and other foodstuff molecules. Oxygen is released into the air as a product of the photosynthetic reactions.*

Mitochondria, present in all eukaryotic cells, carry out the final steps in the breakdown of these carbohydrates and capture their stored energy in ATP molecules. This process, cellular respiration, consumes oxygen and produces carbon dioxide and water, completing the cycle.

heat. Not only would it be of no use to the cell, but the resulting high temperature would also be lethal. However, mechanisms have evolved in living systems that regulate these chemical reactions in such a way that energy is stored in particular chemical bonds from which it can be released in small amounts as the cell needs it. These regulating mechanisms, which require only a few kinds of molecules, enable cells to use energy efficiently, without disrupting the delicate balances that characterize a living system. To understand how these mechanisms work, we must look more closely at the proteins known as enzymes and at one small energy-storage molecule, adenosine triphosphate, abbreviated ATP.

Enzymes

Most chemical reactions require an input of energy to get started. This is true even for energy-releasing reactions such as the oxidation of glucose or the burning of wood. Existing chemical bonds must be broken before new bonds can be formed. This initial energy input is known as the *energy of activation.*

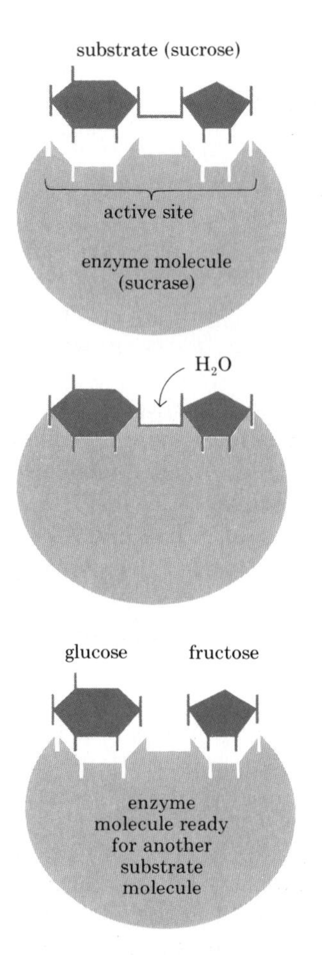

6-5 *A model of enzyme action. Sucrose, a disaccharide, is hydrolyzed to yield a molecule of glucose and a molecule of fructose. The enzyme involved in this reaction, sucrase, is specific for this process; as you can see, the active site of the enzyme fits the opposing surface of the sucrose molecule. The fit is so exact that a molecule composed, for example, of two subunits of glucose would not be affected by this enzyme.*

In the laboratory, the energy of activation is usually supplied as heat. But in a cell, hundreds of different reactions are going on at the same time, and heat would affect all of these reactions indiscriminately. Moreover, heat would break hydrogen bonds and would have other generally destructive effects on the cell. Cells get around this problem by the use of *enzymes,* globular proteins that are specialized to serve as catalysts. A catalyst is a substance that lowers the activation energy for a reaction by forming a temporary association with the molecules that are reacting. This temporary association weakens existing chemical bonds and makes it easier for new ones to form. As a result, very little added energy is needed to start the reaction, and it goes more rapidly than it would in the absence of the catalyst. The catalyst itself is not permanently altered in the process, and so it can be used over and over again.

Because of enzymes, cells are able to carry out chemical reactions at great speed and at comparatively low temperatures. A single enzyme molecule may catalyze the reaction of tens of thousands of identical molecules in a second. Thus enzymes are typically effective in very small amounts.

Over a thousand different enzymes are now known, each of them capable of catalyzing a specific chemical reaction. However, different types of cells are able to manufacture different enzymes—no cell contains all the known enzymes. The particular enzymes that a cell can manufacture are a major factor in determining the biological activities and functions of that cell. A cell can carry out a given chemical reaction at a reasonable rate only if it has a specific enzyme that can catalyze that reaction. The molecule (or molecules) on which an enzyme acts is known as its *substrate.* For example, in the reaction diagrammed in Figure 6-5, sucrose is the substrate, and sucrase is the enzyme (note that enzyme names typically end in "-ase").

Enzyme Structure and Function

The catalytic properties of an enzyme are the result of the tertiary structure of the protein. On the surface of the globular enzyme molecule, there is a region, known as the *active site,* into which the substrate molecule fits. Only a few amino acids of the enzyme are involved in any particular active site. Some of these may be adjacent to one another in the primary

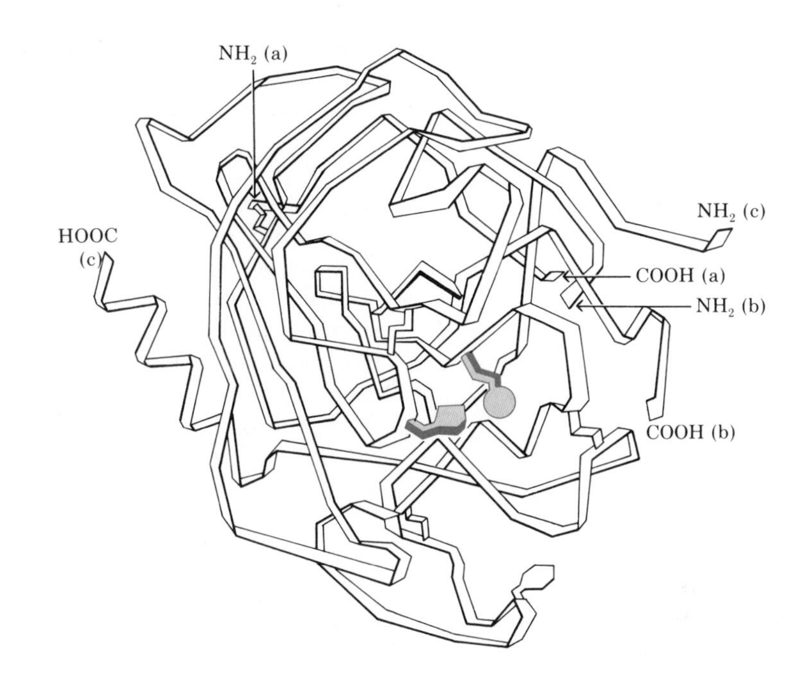

6–6 *Model of an enzyme. This enzyme (the digestive enzyme chymotrypsin) is composed of three polypeptide chains. The amino (NH_2) and carboxyl (COOH) ends of each chain are labeled. The three-dimensional shape of the molecule is a result of a combination of disulfide bonds and of interactions among the chains and between the chains and the surrounding water molecules.*

As a result of this bending and twisting of the polypeptide chains, particular amino acids come together in a highly specific configuration to form the active site of the enzyme. Two amino acids known to be part of the active site are shown in color.

structure, but often the amino acids of the active site are brought close to one another by the intricate folding of the amino acid chain that produces the tertiary structure. In an enzyme with a quaternary structure, the amino acids of the active site may even be on different polypeptide chains, as shown in Figure 6–6.

Within the last several years, studies of enzyme structure have suggested that the binding between enzyme and substrate alters the shape of the enzyme molecule. It is believed that this may put some strain on the reacting molecules and thus facilitate the reaction.

Enzymes characteristically work in series, like workers on an assembly line, each performing one small step in, for example, the oxidation of glucose. In many cases, they also work in conjunction with other molecules known as *coenzymes*. Coenzymes are nonprotein organic molecules that often function as electron carriers. In some oxidation-reduction reactions, electrons are passed to a coenzyme that serves as an electron acceptor. There are several different kinds of coenzymes in any given cell, each able to hold the electrons at a slightly different energy level. A coenzyme can accept electrons at one particular step in a reaction series, and then release them at another step in the same series or in a different series. This combination—enzymes working in series and electron-carrier molecules holding electrons at different energy levels—enables the cell to capture energy efficiently as electrons move from high energy levels to lower ones. Many vitamins—the compounds that humans and other animals cannot synthesize and so must obtain in their diets—are coenzymes or parts of coenzymes.

The Cell's Energy Currency: ATP

Glucose and other carbohydrates are storage forms of energy and also forms in which energy is transferred from cell to cell and organism to organism. In a sense, they are like money in the bank. *ATP* (adenosine triphosphate), however, is like the change in your pocket—it is the cell's immediately spendable energy currency. Almost every energy-requiring transaction in the cell involves ATP. For this reason, ATP is often referred to as the "universal currency" of the cell.

To understand the role of ATP, we must return briefly to the concept of the chemical bond. Because a chemical bond is a stable configuration of electrons, it takes energy to break an old bond and form a new one. This energy is the energy of activation. Because of enzymes, which greatly reduce the energy of activation required, the reactions essential to life are able to proceed at an adequate rate. Additional energy is required, however, in synthetic reactions, such as the formation of a disaccharide from two monosaccharide molecules. In such a reaction, the electrons forming the chemical bonds of the product are at a higher energy level than they were in the bonds of the starting materials. Just as energy must be supplied to push a boulder uphill, energy must be supplied to raise the electrons to the higher energy level. In the living cell, this energy is usually supplied by ATP.

How does ATP perform its function? To answer this question, we must look at its structure. The ATP molecule is composed of three subunits. The first of these is adenine, which is categorized as a nitrogenous base—that is, it has the properties of a base and contains nitrogen:

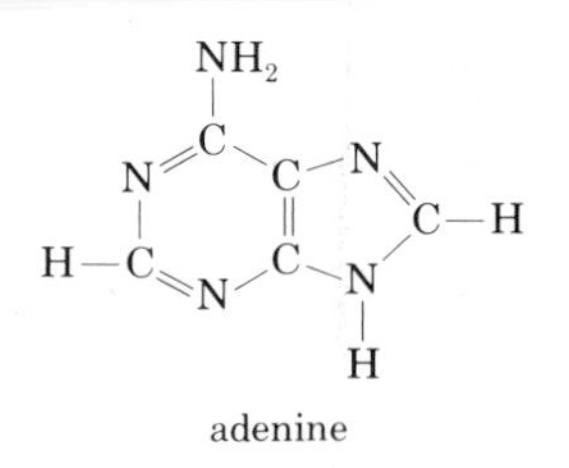

adenine

Adenine is a compound we shall be mentioning frequently in Section 3 because, in addition to its role as part of the ATP molecule, it is one of the principal components of the genetic material. (The use of adenine for these two quite different purposes is an example of the economy with which the cell operates.)

The second subunit of ATP is a five-carbon sugar, ribose, similar in structure to the sugars described in Chapter 3:

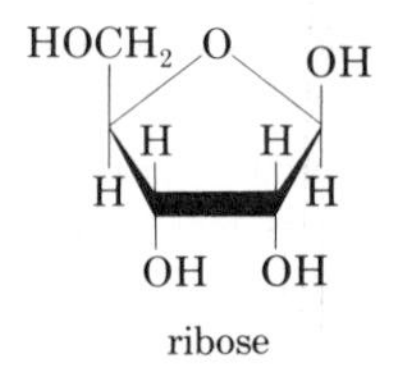

ribose

The third subunit is phosphate, which is an atom of phosphorus combined with four oxygen atoms. A compound consisting of a nitrogenous base plus a sugar plus a phosphate group is known as a *nucleotide*. Nucleotides are one of the important multipurpose chemical combinations found in the cell.

Adenine plus ribose plus one phosphate group make up the nucleotide known as adenosine monophosphate (AMP). In ATP, however, as the name implies, there are *three* phosphates, as shown in Figure 6–7. The symbol $\sim$ indicates a so-called "high-energy" bond. This is a somewhat confusing term. It does not mean a strong bond, such as the covalent bond between carbon and hydrogen. Instead, it means a bond that is easily broken and that releases, on a cellular scale, a relatively large amount of energy.

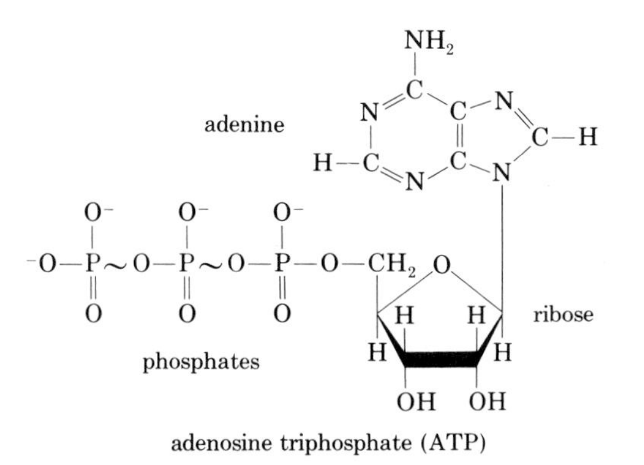

6-7 *Adenosine triphosphate (ATP) is the cell's chief energy currency. The bonds between the three phosphate groups in the molecule are important in ATP function.*

One of the principal characteristics of phosphate groups—and the reason for their important role in cellular reactions—is their capacity to form high-energy bonds. In ATP, two such bonds link the three phosphates together. Usually, in the course of cellular transactions, only one bond is broken. With the loss of one phosphate group, the ATP molecule becomes ADP, adenosine diphosphate. About 7 kilocalories of energy per mole are released. The ADP molecule can then be "recharged" with an input of 7 kilocalories per mole—regaining a third phosphate group and again becoming ATP.

An Example of ATP in Action

Let us look at a simple example of an energy exchange involving ATP. The formation of sucrose, a disaccharide synthesized in plants from the monosaccharides glucose and fructose, requires an input of energy to raise electrons to the higher energy level of the new chemical bond. This energy can be supplied by coupling the synthesis of sucrose to the removal of a phosphate group from the ATP molecule.

First, the terminal phosphate group of ATP is transferred to the glucose molecule:

$$\text{ATP} + \text{glucose} \longrightarrow \text{glucose phosphate} + \text{ADP}$$

In this reaction, some of the 7 kilocalories available from the conversion of ATP to ADP are conserved by the transfer of the phosphate group to the glucose molecule, which thus becomes "energized."

Next, glucose phosphate reacts with fructose to form sucrose:

$$\text{glucose phosphate} + \text{fructose} \longrightarrow \text{sucrose} + \text{phosphate}$$

In this second step, the phosphate group is released from the glucose. Most of the energy made available by its release (energy originally derived from ATP) is used to form the bond between glucose and fructose, producing sucrose. The free phosphate is then available, with an input of energy, to recharge an ADP molecule to ATP.

As we noted in Chapter 3, the formation of sucrose requires 5.5 kilocalories per mole. The conversion of ATP to ADP releases 7 kilocalories per mole. Thus most of the energy in the phosphate bond has been effectively utilized to carry out the work of the cell.

Where does the ATP come from? As we shall see in the next chapter, energy released in the cell's oxidation reactions, such as the breakdown of glucose, is used to recharge the ADP molecule. Thus the ATP/ADP system serves as a universal energy-exchange system, shuttling between energy-releasing reactions and energy-requiring ones.

SUMMARY

Living systems convert energy from one form to another as they carry out essential functions of maintenance, growth, and reproduction.

The laws of thermodynamics govern transformations of energy. The first law states that energy can be converted from one form to another but cannot be created or destroyed. The second law of thermodynamics states that in all natural processes, energy in a form to do work is eventually dissipated into the surroundings. Stated differently, the disorder, or randomness, of every natural system tends to increase. To maintain the organization on which life depends, living systems must have a constant supply of energy. The sun is the original source of this energy.

The energy transformations in living cells involve the movement of electrons from one energy level to another and, often, from one atom or molecule to another. Reactions in which electrons move from one atom or molecule to another are known as oxidation-reduction reactions. An atom or molecule that loses electrons is oxidized; one that gains electrons is reduced.

Enzymes are the catalysts of biological reactions, enormously increasing the rate at which reactions occur but remaining themselves unchanged in the process. Enzymes are large protein molecules folded in such a way that particular groups of amino acids form an active site. The reacting molecules, known as the substrate, fit precisely into this active site.

Enzymes generally work in series, with each enzyme performing one small step in a reaction sequence. Many enzymes work with other molecules, known as coenzymes, that function as electron carriers. Different coenzymes hold electrons at slightly different energy levels.

ATP supplies the energy for most of the activities of the cell. The ATP molecule consists of a nitrogenous base, adenine; a five-carbon sugar, ribose; and three phosphate groups. The three phosphate groups are linked by two high-energy bonds—bonds that release a relatively large amount of energy when they are broken. ATP participates as an energy carrier in most series of reactions that take place in living systems.

QUESTIONS

1. Distinguish among the following terms: first law of thermodynamics/second law of thermodynamics; oxidation/reduction; active site/substrate; ATP/ADP/AMP.

2. Why, in Figure 6–1, are there more plants than zebras and more zebras than lions?

3. Explain why it is that living systems, despite appearances, are not in violation of the second law of thermodynamics.

4. What is there about the orderliness of a living organism that most significantly distinguishes it from the orderliness of a machine, such as an IBM computer or the Bell Telephone System?

5. What is the basis for the specificity of enzyme action? What is the advantage to the cell of such specificity? What might be its disadvantages?

6. When an animal is deprived of a particular vitamin in its diet, it is likely to become ill and may die. What is a likely explanation?

7. Some human societies use the barter system for exchange of goods and services. However, all complex societies have some form of monetary exchange. What are the advantages of a monetary exchange? Relate your answer to the ATP/ADP system.

7

How Cells Make ATP: Glycolysis and Respiration

ATP is the principal energy carrier in living systems. It participates in a great variety of cellular events, from chemical biosyntheses, to the flick of a cilium, the twitch of a muscle, the active transport of a molecule across a cell membrane, or the propagation of an electrical impulse along a nerve. In the following pages, we shall show in some detail how a cell breaks down carbohydrates and captures and stores the released energy in the terminal phosphate bonds of ATP. The oxidation of glucose (or other carbohydrates) is complicated in detail but simple in its overall design.

The process begins with the glucose molecule, the form in which carbohydrates usually reach the cell. This molecule is split, and the hydrogen atoms (that is, protons and electrons) are removed from the carbon atoms and combined with oxygen. The summary equation for the oxidation of glucose is

glucose + oxygen ⟶ carbon dioxide + water + energy

or, $C_6H_{12}O_6 + 6O_2 \longrightarrow 6CO_2 + 6H_2O + 686$ kilocalories

About 40 percent of the energy released by this process is used to convert ADP to ATP. As you will recall, about 75 percent of the energy in gasoline is "lost" as heat in a gasoline engine, and only 25 percent is converted to useful forms of energy. The living cell is significantly more efficient.

7-1 *The chemical energy of ATP—provided by the oxidation of glucose and other organic compounds—moves these dolphins through the water. It also powers the many chemical activities taking place in their tissues and provides the heat that keeps their body temperatures well above that of the water around them.*

The Oxidation of Glucose: An Overview

The oxidation of glucose takes place in two major stages. The first is known as *glycolysis*. The second is respiration, which, in turn, consists of two stages: the *Krebs cycle* and the *electron transport chain*. Glycolysis occurs in the cytoplasm of the cell, and the two stages of respiration take place within the mitochondrion (Figure 7–2).

As we mentioned previously, mitochondria are surrounded by two selectively permeable membranes. The outer one is smooth, and the inner one folds inward. The folds are called cristae. The more active a cell, the more numerous are both its mitochondria and the cristae within them. Within the inner compartment of the mitochondrion, surrounding the cristae, is a dense solution containing enzymes, water, phosphates, and other molecules. This dense solution is the site of some of the reactions of the Krebs cycle. Other Krebs cycle reactions and the reactions of the electron transport chain occur in the internal membranes, where the necessary enzymes and electron-carrier molecules are present. The mitochondrion is thus a self-contained chemical plant. Its outer membrane lets most small molecules in or out freely, but the inner one permits the passage of only certain molecules.

In glycolysis and the Krebs cycle, hydrogen atoms are removed from the carbon skeleton of glucose and are passed to coenzymes that function as electron carriers. The first of these is nicotinamide adenine dinucleotide, abbreviated NAD^+. (As you can tell from its name, the molecule consists of two—"di"—nucleotides, each containing a nitrogenous base, a sugar, and a phosphate.) NAD^+ can accept a proton and two electrons, becoming reduced to NADH. The second molecule is flavin adenine dinucleotide, abbreviated FAD. FAD can accept two hydrogen atoms (that is, two protons and two electrons), being reduced to $FADH_2$. In glycolysis and the Krebs cycle, NAD^+ and FAD accept electrons and protons and are reduced.

In the final stage of respiration, NADH and $FADH_2$ give up their electrons to the electron transport chain. The electrons are passed "downhill" along a series of electron carriers known as cytochromes, all of which have similar structures (Figure 7–3). Although similar, the structures of the

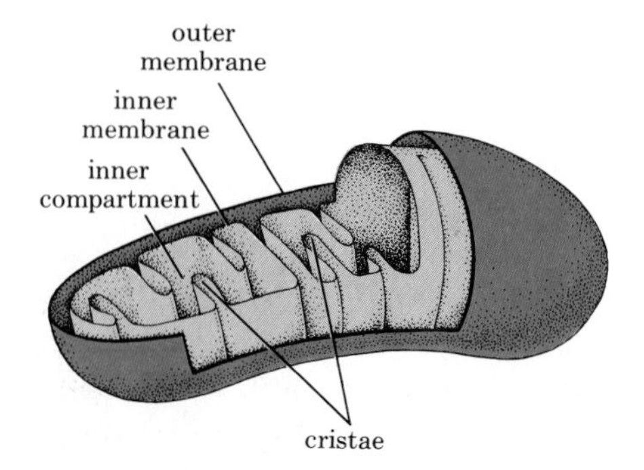

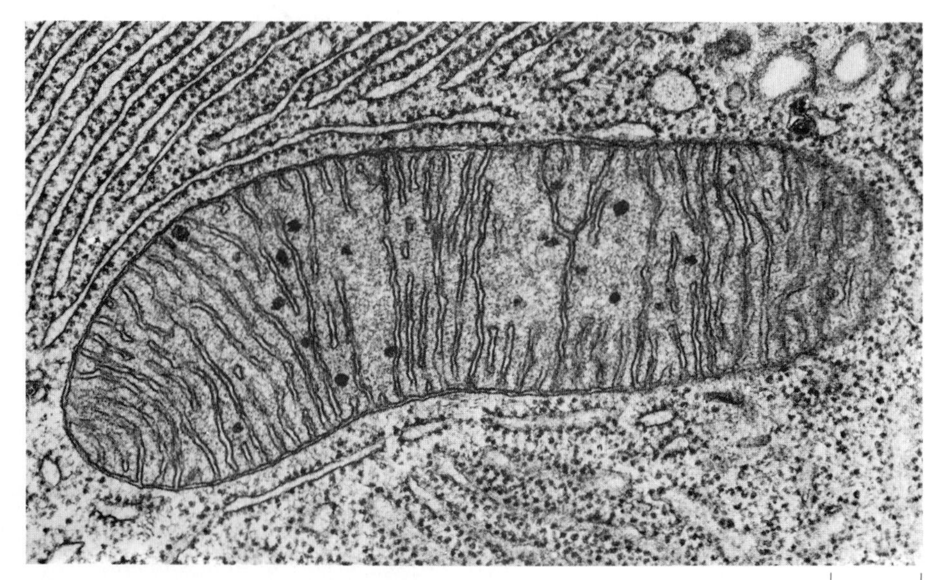

7–2 *Mitochondria are surrounded by two membranes. The inner membrane folds inward to make a series of shelves, or cristae. The enzymes and electron carriers involved in the final stage of respiration are built into these internal membranes.*

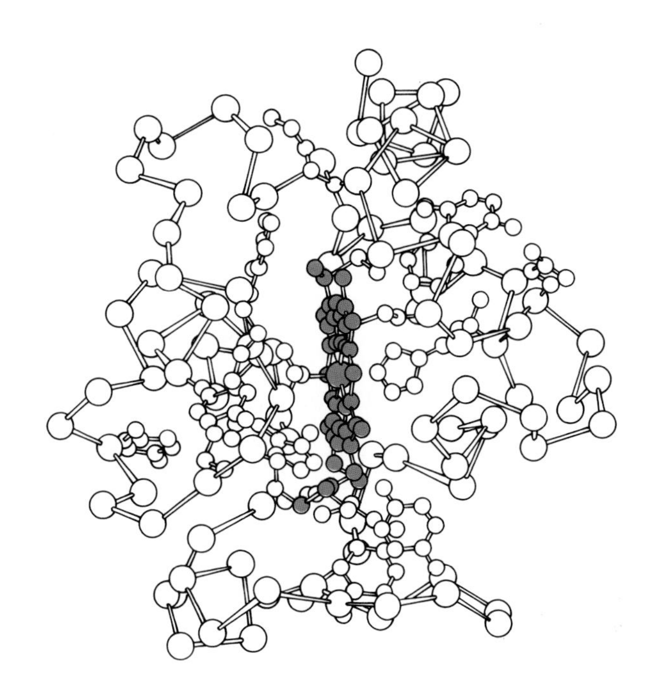

7–3 Cytochromes are molecules in which a heme group is held in an intricate protein structure. In the cytochrome c *structure shown here, the heme group is in color. An atom of iron is at the center of the heme group. Heme groups also occur in hemoglobin, the oxygen-carrying molecule in red blood cells.*

Cytochromes are involved in electron transfer. It is the iron within the molecule that actually combines with the electrons.

individual cytochromes differ enough to enable them to hold electrons at different energy levels. As the electrons drop, step by step, to lower energy levels, the energy released is used to form ATP from ADP and phosphate. When they reach their lowest energy level, the electrons and their accompanying protons combine with oxygen to form water.

In the presence of oxygen (an aerobic environment), the complete oxidation of a molecule of glucose produces 36 molecules of ATP. These convenient packages of energy are then available to power the many activities of the cell. In the absence of oxygen (an anaerobic environment), however, respiration cannot occur. Such a situation exists, for example, in muscle cells depleted of oxygen by strenuous exercise and in air-tight kegs containing yeast cells and fruit juices. Glycolysis, however, can occur, in conjunction with a process known as *anaerobic fermentation*. The energy yield is only two molecules of ATP for each molecule of glucose, but this is enough to supply the immediate needs of muscle cells until the oxygen supply can be replenished. And, it is enough to keep yeast cells alive and well while their enzymes carry out the fermentation reactions that produce wine, enjoyed by *Homo sapiens* (and perhaps other animals as well) since the dawn of history.

Now that you are familiar with the general outlines of glycolysis and respiration, let us look at the details of these processes and of anaerobic fermentation.

7–4 In glycolysis, the six-carbon glucose molecule is split into two three-carbon molecules of a compound known as pyruvic acid.

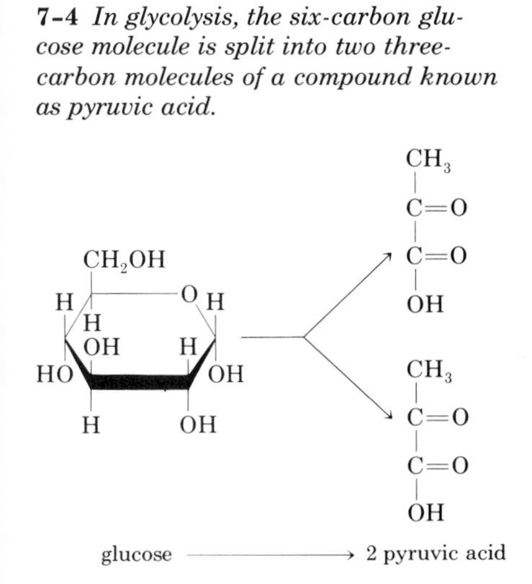

Glycolysis

In glycolysis, the six-carbon glucose molecule is split into two three-carbon molecules of a compound known as pyruvic acid (Figure 7–4). This is carried out in a series of nine reactions, each of which is catalyzed by a specific enzyme. About 143 kilocalories are released per mole of glucose during glycolysis; the remaining 543 kilocalories are stored in the bonds of the pyruvic acid molecules.

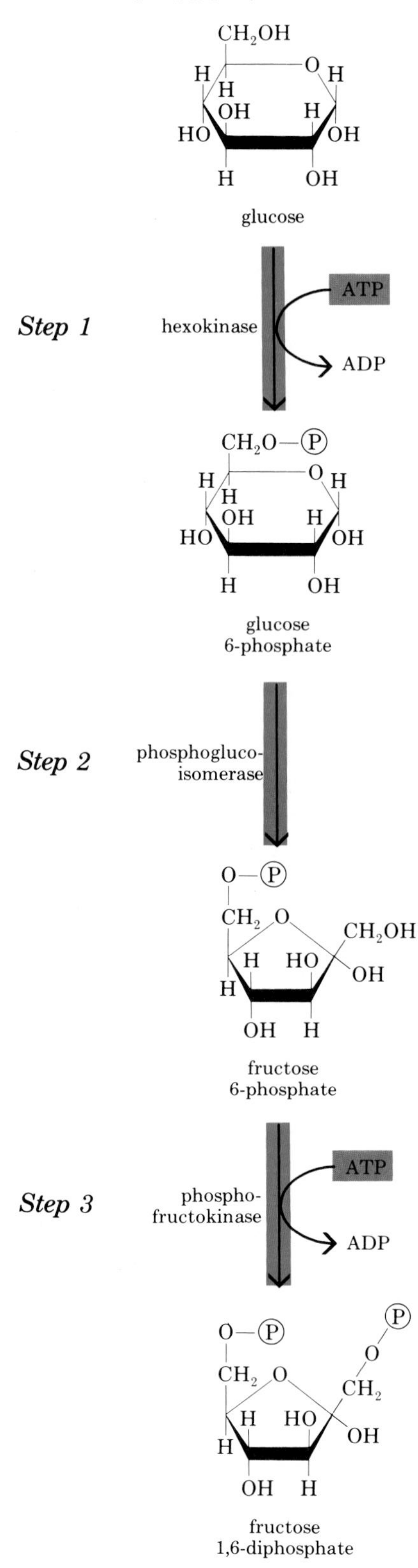

7-5 *The steps of glycolysis.*

Glycolysis exemplifies the way the biochemical processes of a living cell proceed in small sequential steps, each catalyzed by a specific enzyme. Do not try to memorize these steps, but follow them closely. Notice how the carbon skeleton of the molecule is dismembered and its atoms rearranged step by step. Note especially the formation of ATP from ADP and of NADH from NAD^+. ATP and NADH represent the cell's net energy harvest from glycolysis.

Step 1. The first steps in glycolysis require an input of energy. This energy is supplied by coupling these steps to the ATP/ADP system. The terminal phosphate group is transferred from an ATP molecule to the glucose molecule, to make glucose 6-phosphate. (This is also the first step, you may recall, in the biosynthesis of sucrose, page 97.) The reaction of ATP with glucose to produce glucose 6-phosphate and ADP is an energy-yielding reaction. Some of the energy released is conserved in the chemical bond linking the phosphate to the glucose molecule. This reaction is catalyzed by a specific enzyme (hexokinase), and each of the reactions that follows is similarly regulated by a specific enzyme.

Step 2. The molecule is reorganized, again with the help of a particular enzyme. The six-sided ring characteristic of glucose becomes the five-sided fructose ring. As you know, glucose and fructose both have the same number of atoms—$C_6H_{12}O_6$—and differ only in the arrangements of these atoms.

Step 3. In this step, which is similar to Step 1, fructose 6-phosphate gains a second phosphate by the investment of another ATP. The added phosphate is bonded to the first carbon, producing fructose 1,6-diphosphate, that is, fructose with phosphates in the 1 and 6 positions. Note that in the course of the reactions thus far, two molecules of ATP have been converted to ADP and no energy has been recovered.

Step 4. The six-carbon sugar molecule is split into two three-carbon molecules, dihydroxyacetone phosphate and glyceraldehyde phosphate. The two molecules are interconvertible by the enzyme isomerase. However, because the glyceraldehyde phosphate is used up in subsequent reactions, all of the dihydroxyacetone phosphate is eventually converted to glyceraldehyde phosphate. Thus, all subsequent steps must be counted twice to account for the fate of one glucose molecule. With the completion of Step 4, the preparatory reactions are complete.

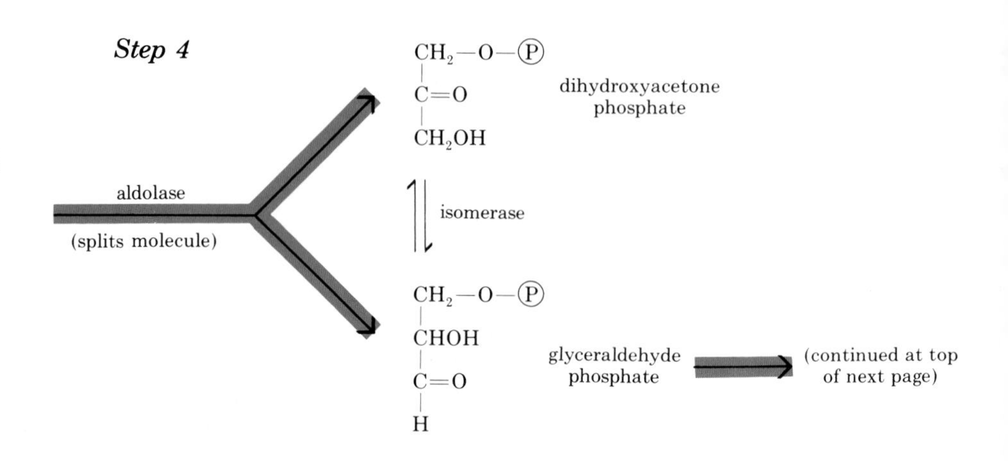

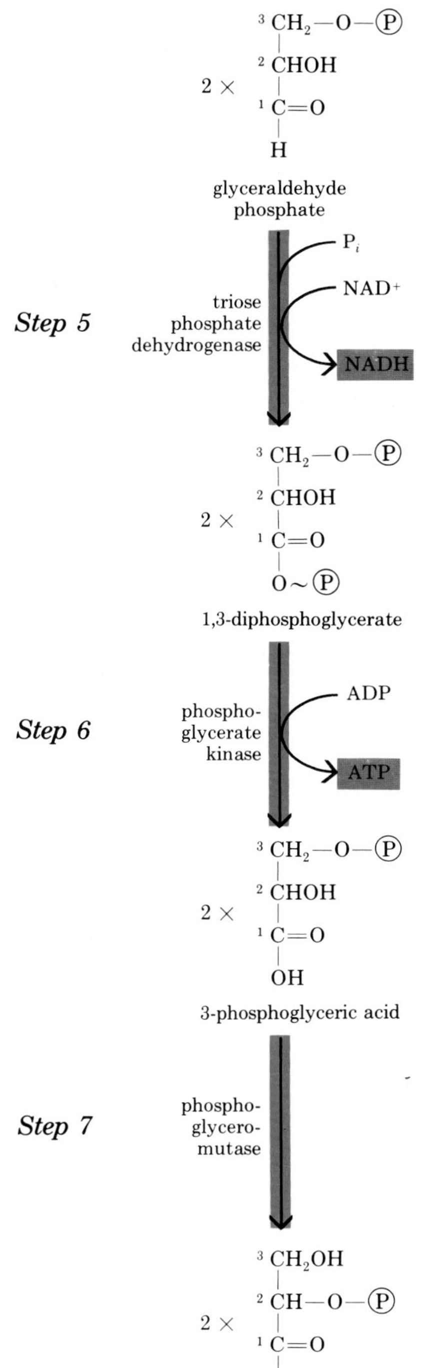

Step 5. Glyceraldehyde phosphate molecules are oxidized—that is, hydrogen atoms with their electrons are removed—and NAD^+ is reduced to NADH. This is the first reaction from which the cell harvests energy. Some of the energy from this oxidation reaction is also used to attach a phosphate group to what is now the 1 position of the glyceraldehyde phosphate molecule. (The designation P_i indicates inorganic phosphate, which is available in the cytoplasm.) Note that a "high-energy" bond is formed.

Step 6. The high-energy phosphate is released from the diphosphoglycerate molecule and used to recharge a molecule of ADP (a total of two molecules of ATP per molecule of glucose).

Step 7. The remaining phosphate group is transferred from the 3 position to the 2 position.

Step 8. In this step, a molecule of water is removed from the three-carbon compound. This internal rearrangement of the molecule concentrates energy in the vicinity of the phosphate group, forming a "high-energy" bond.

Step 9. The high-energy phosphate is transferred to a molecule of ADP, forming another molecule of ATP (again, a total of two molecules of ATP per molecule of glucose).

Summary of Glycolysis

The complete sequence begins with one molecule of glucose. Energy is invested at Steps 1 and 3 by the transfer of a phosphate group from an ATP molecule—one at each step—to the sugar molecule. The six-carbon molecule splits at Step 4, and from this point onward, the sequence yields energy. At Step 5, a molecule of NAD^+ takes energy from the system and is reduced to NADH. At Steps 6 and 9, molecules of ADP take energy from the system, form additional phosphate bonds, and become ATP.

To sum up: The energy from the phosphate bonds of two ATP molecules is needed to initiate the glycolytic sequence. Subsequently, two NADH molecules are produced from two NAD^+ and four ATP molecules from four ADP:

$$\text{glucose} + 2\text{ATP} + 4\text{ADP} + 2P_i + 2\text{NAD}^+ \longrightarrow 2\text{ pyruvic acid} + 2\text{ADP} + 4\text{ATP} + 2\text{NADH} + 2H^+ + 2H_2O$$

Thus one glucose molecule has been converted to two molecules of pyruvic acid. The net harvest—the energy recovered—is two molecules of ATP and two molecules of NADH per molecule of glucose. This series of reactions is carried out by virtually all living cells—from prokaryotes to the eukaryotic cells of our own bodies.

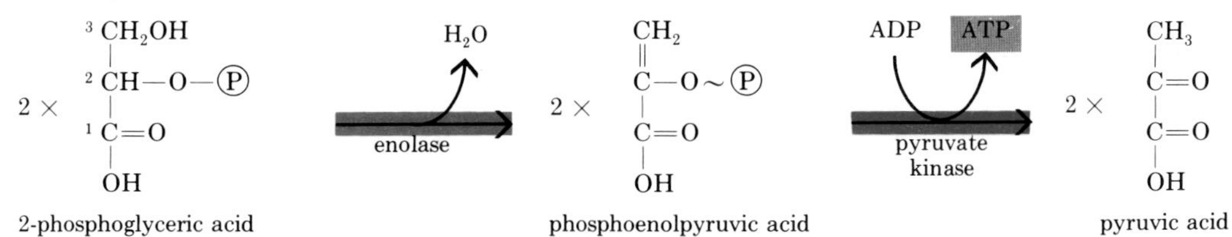

Respiration

Pyruvic acid passes from the cytoplasm, where it is produced by glycolysis, and crosses the outer and inner membranes of the mitochondria. Before entering the Krebs cycle, which takes place in the inner compartment of the mitochondrion, the three-carbon pyruvic acid molecule is oxidized (Figure 7-6). The third carbon is removed in the form of carbon dioxide, and a two-carbon acetyl group (CH_3CO) remains. In the course of this reaction, a molecule of NADH is formed from NAD^+. Each acetyl group is momentarily accepted by a compound known as coenzyme A. Like other coenzymes, it is a large molecule, a portion of which is a nucleotide. The combination of an acetyl group and coenzyme A is known simply as acetyl CoA. Fats and amino acids can also be converted to acetyl CoA and enter the respiratory sequence at this point.

7-6 (a) *The three-carbon pyruvic acid molecule is oxidized to the two-carbon acetyl group, which is combined with coenzyme A to form acetyl CoA. The oxidation of the pyruvic acid molecule is coupled to the reduction of NAD^+. Acetyl CoA enters the Krebs cycle.* (b) *Electron micrograph showing the enzymes involved in the oxidation of pyruvic acid to acetyl CoA. Each of the complexes visible here represents multiple copies of three different enzymes.*

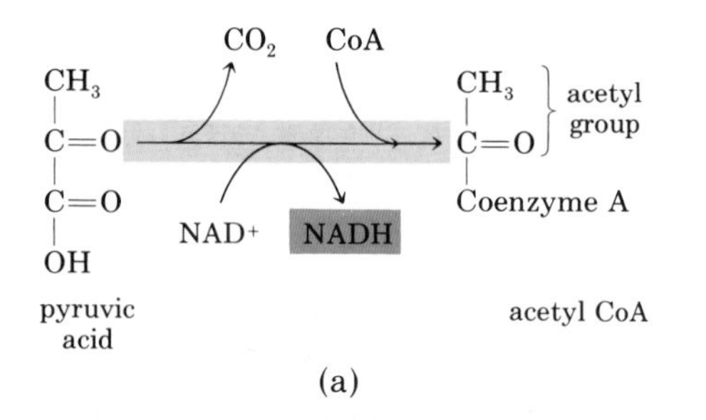

(a)

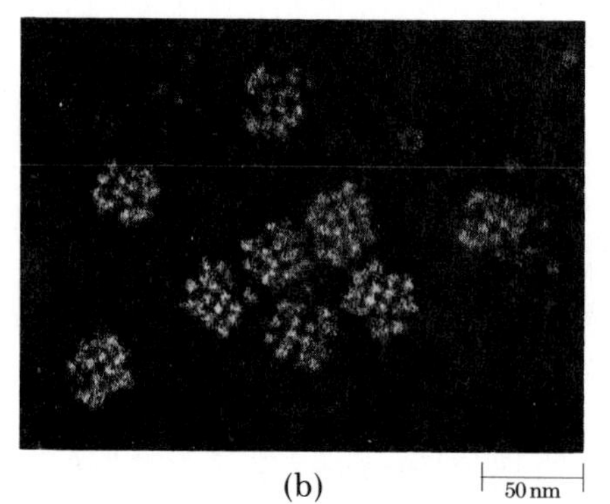

(b)

The Krebs Cycle

In the Krebs cycle (Figure 7-7), the two-carbon acetyl group is combined with a four-carbon compound (oxaloacetic acid) to produce a six-carbon compound (citric acid). In the course of the cycle, two of the six carbons are oxidized to carbon dioxide, and oxaloacetic acid is regenerated. Each turn around the cycle uses up one acetyl group and regenerates a molecule of oxaloacetic acid, which is then ready to begin the sequence once again.

In the course of these steps, some of the energy released by the oxidation of the carbon atoms is used to convert ADP to ATP (one molecule per cycle), some is used to produce NADH from NAD^+ (three molecules per cycle), and some is used to produce $FADH_2$ from FAD (one molecule per cycle).

$$\text{oxaloacetic acid} + \text{acetyl CoA} + \text{ADP} + P_i + 3NAD^+ + \text{FAD} \longrightarrow \text{oxaloacetic acid} + 2CO_2 + \text{CoA} + \text{ATP} + 3\text{NADH} + FADH_2 + 3H^+ + H_2O$$

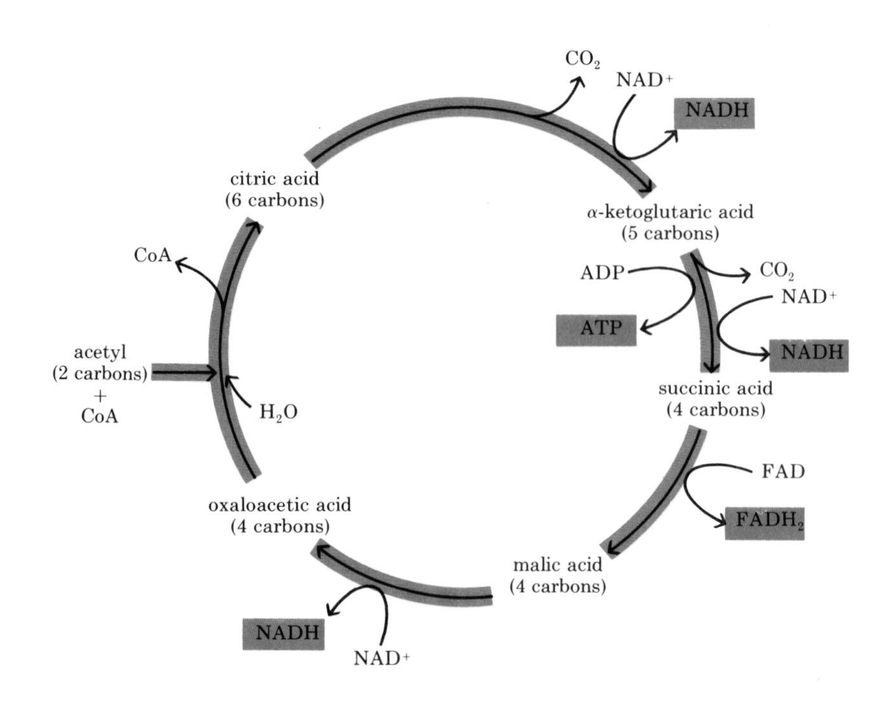

7-7 *Summary of the Krebs cycle. In the course of the cycle, the carbons donated by the acetyl group are oxidized to carbon dioxide and the hydrogen atoms are passed to electron carriers. As in glycolysis, a specific enzyme is involved at each step. One molecule of ATP, three molecules of NADH, and one molecule of $FADH_2$ represent the energy yield of the cycle.*

The Electron Transport Chain

The glucose molecule is now completely oxidized. Some of its energy has been used to produce ATP from ADP. Most of it, however, still remains in electrons removed from the carbon atoms as they were oxidized and passed to the electron carriers NAD^+ and FAD. These electrons are at a high energy level. As we noted earlier, in the course of the electron transport chain, these electrons are passed "downhill" and the energy released is used to form additional ATP molecules from ADP (Figure 7–8). At the end of the chain, the electrons are accepted by oxygen and combine with protons (hydrogen ions) to produce water.

Each time one pair of electrons passes from NADH to oxygen, three molecules of ATP are formed from ADP and phosphate. Each time a pair of electrons passes from $FADH_2$, two molecules of ATP are formed.

Overall Energy Harvest

We are now in a position to see how much of the energy originally present in the glucose molecule has been recovered in the form of ATP.

Glycolysis, in the presence of oxygen, yields two molecules of ATP directly and two molecules of NADH. The total gain, however, is not 8 ATP as you might calculate, but 6 ATP because there is a "cost" of 2 ATP to transport the electrons held by the two molecules of NADH across the mitochondrial membranes.

The conversion of pyruvic acid to acetyl CoA yields two molecules of NADH (inside the mitochondrion) for each molecule of glucose and so produces six molecules of ATP.

The Krebs cycle yields, for each molecule of glucose, two molecules of ATP, six of NADH, and two of $FADH_2$, or a total of 24 ATP.

As a balance sheet, Table 7-1, shows, the complete yield from a single molecule of glucose is 36 molecules of ATP. Note that all but two of the 36 molecules of ATP have come from reactions taking place in the mitochondrion, and all but four involve the oxidation of NADH or $FADH_2$ along the electron transport chain.

7-8 *Summary of glycolysis and respiration. Glucose is first broken down to pyruvic acid, with a yield of two ATP molecules and the reduction (dashed arrows) of two NAD^+ molecules to NADH. Pyruvic acid is oxidized to acetyl CoA, and one molecule of NAD^+ is reduced. (Note that this and subsequent reactions occur twice for each glucose molecule; this electron passage is indicated by solid arrows.) In the Krebs cycle, the acetyl group is oxidized and the electron acceptors, NAD^+ and FAD, are reduced. NADH and $FADH_2$ then transfer their electrons to the series of cytochromes and other electron carriers (FMN, flavin mononucleotide, and CoQ, coenzyme Q) that make up the electron transport chain. As the electrons are passed "downhill," the energy released is used to make ATP from ADP.*

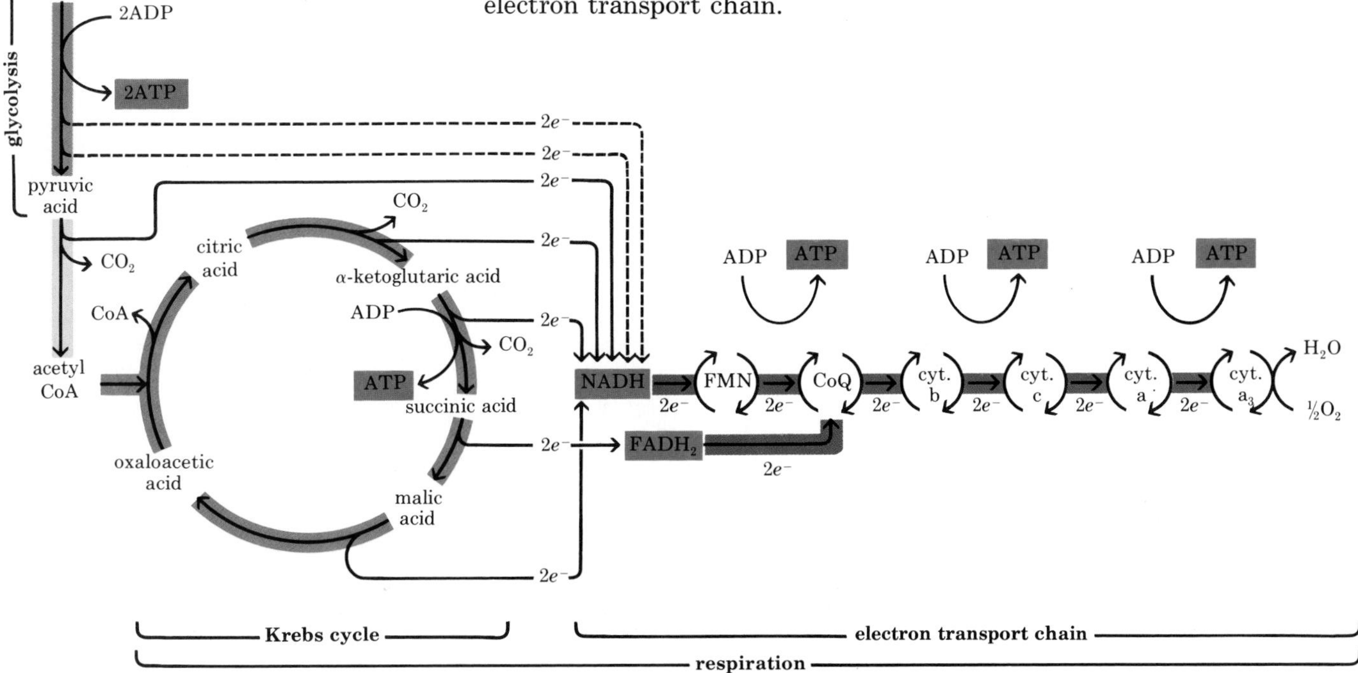

Table 7-1 Summary of energy yield from one molecule of glucose

Glycolysis:	2 ATP 2 NADH ⟶ 4 ATP*		⟶ 6 ATP
Respiration:			
Pyruvic acid ⟶ acetyl CoA:	1 NADH ⟶ 3 ATP	(× 2)	⟶ 6 ATP
Krebs cycle:	1 ATP 3 NADH ⟶ 9 ATP 1 $FADH_2$ ⟶ 2 ATP	(× 2)	⟶ 24 ATP

*The cell must use 2 ATP to transport the 2 NADH formed in glycolysis across the mitochondrial membranes. Thus, the net yield from these 2 NADH is only 4 ATP.

In the course of glycolysis and respiration, 686 kilocalories are released per mole of glucose. Almost 40 percent of this, about 252 kilocalories (7 kilocalories per mole of ATP × 36 moles of ATP), has been captured in the "high-energy" bonds of ATP.

The ATP molecules, once formed, are exported across the membranes of the mitochondrion by a shuttle system that simultaneously brings in one molecule of ADP for each ATP exported.

Anaerobic Fermentation

Pyruvic acid, formed in glycolysis, can follow one of several pathways. The pathway of respiration, which we have just examined, is the principal pathway of energy metabolism for most cells in the presence of oxygen.

In the absence of oxygen, pyruvic acid can be converted either to ethanol (ethyl alcohol) or to lactic acid, depending on the type of cell. For example, yeast cells, present as a "bloom" on the skin of grapes, can grow either with or without oxygen. When the glucose-filled juices of grapes are extracted and stored in air-tight kegs, the yeast cells turn the fruit juice to wine by converting glucose into ethanol (Figure 7-9). Yeast, like all living things, have a limited tolerance for alcohol, and when a concentration of about 12 percent is reached, the yeast cells cease to function.

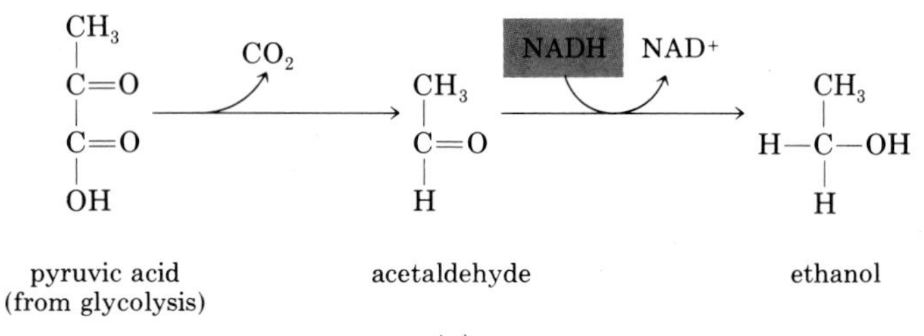

(a)

(b)

7-9 (a) *The steps by which pyruvic acid, formed by glycolysis, is converted anaerobically to ethanol (ethyl alcohol). In the first step, carbon dioxide is released. In the second, NADH is oxidized, and acetaldehyde is reduced. Most of the energy of the glucose remains in the alcohol, which is the end product of the sequence. However, by regenerating NAD^+, these steps allow glycolysis to continue, with its small but sometimes vitally necessary yield of ATP.*
(b) *Consequences of anaerobic glycolysis. Yeast cells visible on the grapes as dustlike "bloom," mix with the juice when the grapes are crushed. Storing the mixture under anaerobic conditions causes the yeast to break down the glucose in the grape juice to alcohol.*

Ethanol and the Liver

The human body can dispose fairly readily of most toxic products of its own manufacture, such as carbon dioxide and nitrogenous wastes. In contrast, most ingested toxic substances, such as ethanol (beverage alcohol) must first be broken down by the liver, which possesses special enzymes not present in other tissues.

It has been known for many years that heavy drinkers are at great risk for severe, and often fatal, liver disease. Recent studies conducted by Charles S. Lieber and his colleagues at the Bronx Veterans Administration Hospital and the Mount Sinai School of Medicine in New York City have demonstrated that the origin of the problem lies in the simple chemical steps involved in the breakdown of ethanol. Enzymes in the liver first oxidize ethanol to acetaldehyde by transferring two hydrogen atoms to NAD^+:

$$\underset{\text{ethanol}}{CH_3CH_2OH} + 2NAD^+ \longrightarrow \underset{\text{acetaldehyde}}{CH_3CHO} + 2NADH$$

The acetaldehyde is oxidized to acetic acid, which is, in turn, oxidized to carbon dioxide and water and eliminated from the body.

The chief culprits in the development of liver disease are the hydrogen atoms transferred from ethanol to NAD^+. These "extra" hydrogen atoms—carried in NADH molecules—follow two principal pathways within the cell. Most go directly to the electron transport chain, where they are oxidized, producing water and ATP. Because of the high levels of NADH present in the cell from the oxidation of ethanol, the production of NADH by glycolysis and the Krebs cycle is reduced. As a result, sugars, amino acids, and fatty acids are not broken down but are instead converted to fats. The fats accumulate in the liver. The mitochondria also become swollen and bloated, presumably as a result of the distortion of their normal function—the electron transport chain is doing very heavy duty, while the Krebs cycle is effectively shut down.

Other hydrogen atoms are used in the synthesis of glycerophosphates and fatty acids from the carbohydrate skeletons that are not being processed in glycolysis and the Krebs cycle. More fats accumulate. It does not take long. In human volunteers fed a good high-protein, low-fat diet, six drinks (about 10 ounces) a day of 86 proof alcohol produced an eight-fold increase in fat deposits in the liver in only 18 days. Fortunately, these early effects are completely reversible.

The liver cells work hard to get rid of the excess fats. The fats are not soluble in water (or in blood plasma). Before being released into the bloodstream, they are coated with a thin layer of protein. This coating and secretion process is carried out on the membranes of the endoplasmic reticulum. The liver cells of heavy drinkers show enormous proliferation of the endoplasmic reticulum.

As we noted, the fat deposits are initially reversible, but after a few years—depending on how much alcohol is consumed—liver cells, engorged with fat, begin to die, triggering the inflammatory process known as alcoholic hepatitis. Liver function becomes impaired. Cirrhosis is the next step; it is the formation of scar tissue, which interferes with the function of the individual cells and also with the supply of blood to the liver. This leads to the death of more cells. The liver can no longer carry out its normal activities—such as breaking down nitrogenous wastes—which is why cirrhosis is a cause of death. In fact, cirrhosis of the liver is the seventh leading cause of death in the nation, and the third leading cause of death between the ages of 25 and 65 in New York City and some other urban areas.

Not so long ago, it was commonly believed that a good diet was all that was required to protect even a heavy drinker from the deleterious effects of alcohol. In fact, if one were just to add a few vitamins to the alcohol itself, some sophisticates maintained, most of the long-term physical damage of alcohol would disappear. This new evidence refutes these comforting notions, and it comes at a time when alcohol is enjoying a resurgence of popularity among persons of high school and college age. (In populations of postgraduate age, as in other human societies the world over, it never lost its status as the drug abuse of choice.)

(a) *Liver tissue from a rat fed a balanced liquid diet for 24 days. This is the normal appearance of rat liver tissue.* (b) *In this liver tissue from another rat fed a liquid diet in which ethanol provided 36 percent of the total calories, many globular fat droplets have accumulated. This rat was also maintained on its special diet for 24 days.*

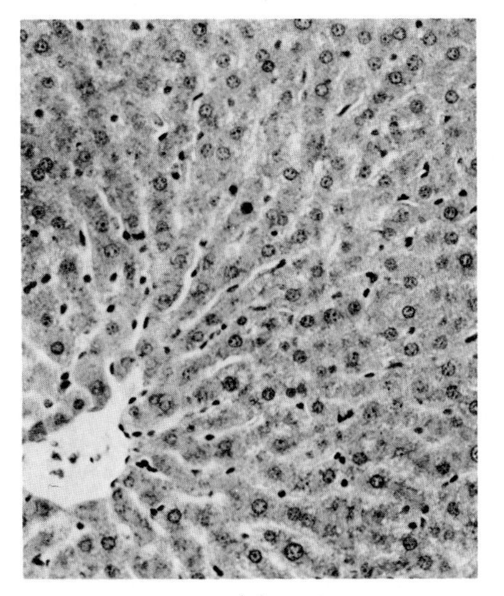

(a)

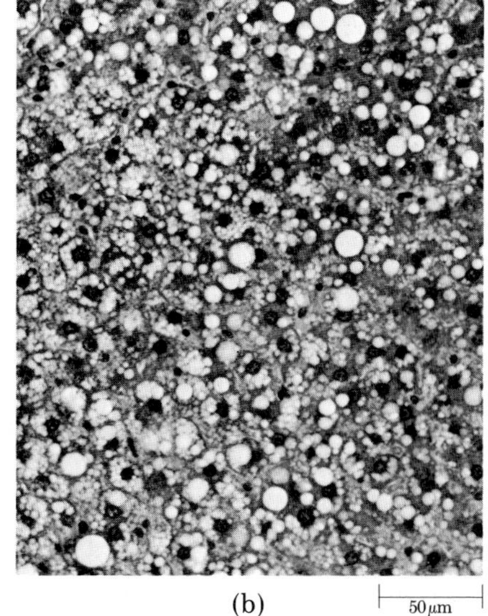

(b) 50μm

7-10 The enzymatic reaction that produces lactic acid from pyruvic acid anaerobically in muscle cells. In the course of this reaction, NADH is oxidized and pyruvic acid is reduced. The NAD+ molecules produced in this reaction and the one shown in Figure 7-9 are recycled in the glycolytic sequence. Without this recycling, glycolysis cannot proceed. Lactic acid accumulation results in muscle soreness and fatigue.

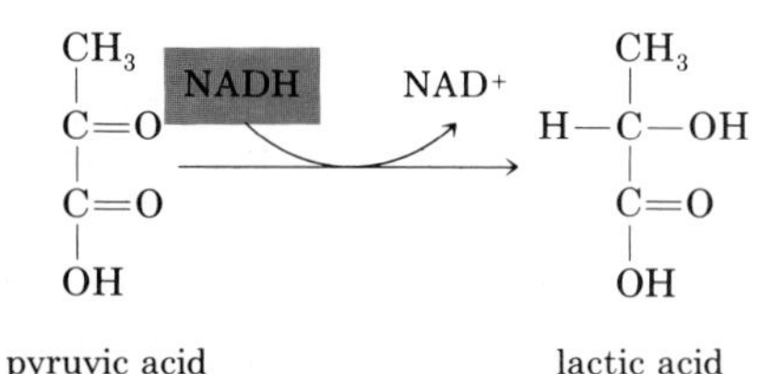

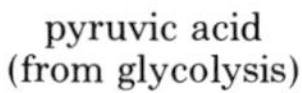

Lactic acid is formed from pyruvic acid by a variety of microorganisms and also by some animal cells when oxygen is scarce or absent (Figure 7-10). As we noted earlier, it is produced in muscle cells during strenuous exercise, as by an athlete during a sprint. We breathe hard when we run fast, thereby increasing the supply of oxygen, but even this increase may not be enough to meet the immediate needs of the muscle cells. These cells, however, can continue to work by accumulating what is known as an oxygen debt. Glycolysis continues, using glucose released into the bloodstream from glycogen stored in the liver, but the resulting pyruvic acid does not enter the Krebs cycle. Instead, it is converted to lactic acid, which, as it accumulates, produces the sensations of muscle fatigue. The lactic acid is ultimately carried by the blood to the liver where it is resynthesized to pyruvic acid and back to glucose or glycogen.

Why is pyruvic acid converted to lactic acid, only to be converted back again? The function of this reaction is simple: It regenerates the NAD^+ without which glycolysis cannot go forward (see Step 5, page 103). Even though this step seems to be wasteful in terms of energy consumption, it may be vitally important in the economy of the organism, spelling the difference between life and death when an animal "out of breath" needs one last burst of ATP to escape from a predator or catch a prey.

The fact that glycolysis does not require oxygen suggests that the glycolytic sequence evolved early, before free oxygen was present in the atmosphere. Presumably primitive one-celled organisms used glycolysis (or something very much like it) to extract energy from the organic compounds they absorbed from their watery surroundings. Carbon dioxide was formed as a by-product, and its increasing concentration in the primitive atmosphere made possible the evolution of photosynthesis—that all-important process, which we shall examine in the next chapter.

SUMMARY

The oxidation of glucose is a chief source of energy in most cells. As the glucose is broken down in a series of small enzymatic steps, the energy in the molecule is packaged in the form of "high-energy" bonds in molecules of ATP.

The first phase in the breakdown of glucose is glycolysis, in which the six-carbon glucose molecule is split into two three-carbon molecules of pyruvic acid, and two molecules of ATP and two of NADH are formed. This reaction takes place in the cytoplasm of the cell.

The second phase in the breakdown of glucose and other fuel molecules is respiration. It requires oxygen and, in eukaryotic cells, takes place in the mitochondria. It occurs in two stages: the Krebs cycle and the electron transport chain.

In the course of respiration, the three-carbon pyruvic acid molecules are broken down to two-carbon acetyl groups, which then enter the Krebs cycle. In the Krebs cycle, the two-carbon acetyl group is broken apart in a series of reactions to carbon dioxide. In the course of the oxidation of each acetyl group, four electron acceptors (three NAD^+ and one FAD) are reduced, and another molecule of ATP is formed.

The final stage of breakdown of the fuel molecule is the electron transport chain, which involves a series of electron carriers and enzymes embedded in the inner membranes of the mitochondrion. Along this series of electron carriers, the high-energy electrons accepted by NADH and $FADH_2$ during the Krebs cycle pass downhill to oxygen. Each time a pair of electrons passes down the electron transport chain, ATP molecules are formed from ADP and phosphate. In the course of the breakdown of the glucose molecule, 36 molecules of ATP are formed, most of them in the mitochondrion.

In the absence of oxygen, the pyruvic acid produced by glycolysis is converted to either ethanol or lactic acid by the process of anaerobic fermentation. NAD^+ is regenerated, allowing glycolysis to continue, producing a small but vital supply of ATP for the organism.

QUESTIONS

1. Distinguish among the following: oxidation of glucose/glycolysis/respiration/fermentation; NAD^+/NADH; Krebs cycle/electron transport chain; aerobic pathways/anaerobic pathways.

2. Sketch the structure of a mitochondrion. Describe where the various stages in the breakdown of glucose take place in relation to mitochondrial structure. What molecules cross the mitochondrial membranes during these processes?

3. (a) As we have seen, a cell obtains 36 molecules of ATP from each molecule of glucose that is completely oxidized. Account for the production of each molecule of ATP. (b) In the course of glycolysis, the Krebs cycle, and the electron transport chain, 40 molecules of ATP are actually formed. Why is the net yield for the cell only 36 molecules of ATP?

4. If oxygen-breathing organisms are so much more successful in converting energy than anaerobes, why are there any anaerobes left on this planet? Why didn't they all become extinct long ago?

5. Why is the alcohol content of naturally fermented wine never greater than 12 percent?

6. Cyanide can combine with—and deactivate—cytochrome *a* and cytochrome a_3. In our bodies, however, cyanide tends to react first with hemoglobin and to make it impossible for oxygen to bind to the hemoglobin. Either way, cyanide poisoning has the same effect: It inhibits the synthesis of ATP. Explain how this is so.

8

Photosynthesis, Light, and Life

8-1 *Cells of a green alga* (Cosmarium botrytis) *that lives in fresh water. Each cell is an individual, self-sufficient photosynthetic organism. The green color is due to chlorophyll, in which the radiant energy of sunlight is converted to chemical energy.*

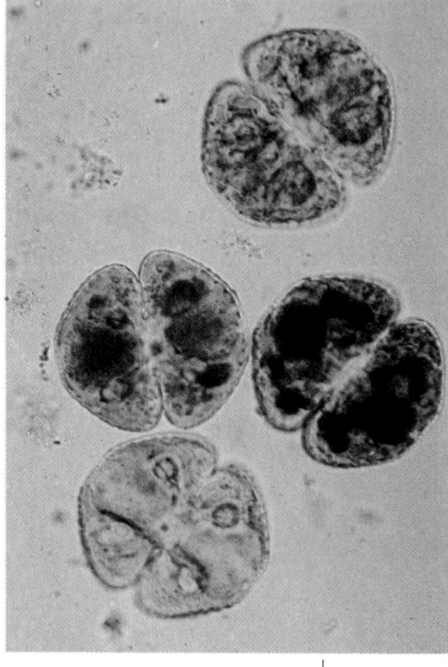

40 μm

The first photosynthetic organism probably appeared almost 3 billion years ago. Before that, the physical characteristics of earth and its atmosphere were the most powerful forces in determining which organisms flourished and which died out. With the evolution of photosynthesis, however, organisms began to change the face of our planet and, as a consequence, to exert strong influences on each other. Organisms have continued to change the environment, at an ever-increasing rate, up to the present day.

As we saw earlier, the atmosphere in which the first cells evolved probably consisted largely of ammonia (NH_3), water (H_2O), and methane (CH_4) and other carbon gases. Gradually, methane and ammonia were used up in the formation of the organic molecules on which the first cells depended. These earliest organisms lived in an environment without free oxygen, and, in fact, oxygen, with its powerful electron-attracting capacities, would have been poisonous to them. Their energy probably came from anaerobic glycolysis, using the organic molecules in the seas as fuel. This would have resulted not only in a decrease in the concentration of organic molecules but also in the accumulation of carbon dioxide in the atmosphere.

Then, it is hypothesized, there slowly evolved photosynthetic organisms that used carbon dioxide as their carbon source and released oxygen, as do most modern photosynthetic forms. As these photosynthetic organisms multiplied, they provided a new supply of organic molecules and free oxygen began to accumulate. In response to these changing conditions, cell species arose for which oxygen was not a poison but a requirement for existence.

As we saw in the last chapter, oxygen-consuming organisms have an advantage over those that do not use oxygen. A higher yield of energy can be extracted per molecule from the oxidation of carbon-containing compounds than from anaerobic processes, in which fuel molecules are not broken down completely. Energy released in cells by reactions using oxygen made possible the development of increasingly active, increasingly complex organisms. Without oxygen, the complex forms of life that now exist on earth could not have evolved.

Life on earth continues to be dependent on photosynthesis both for its oxygen and for its carbon-containing fuel molecules. Photosynthetic organisms capture light energy and use it to form carbohydrates and free

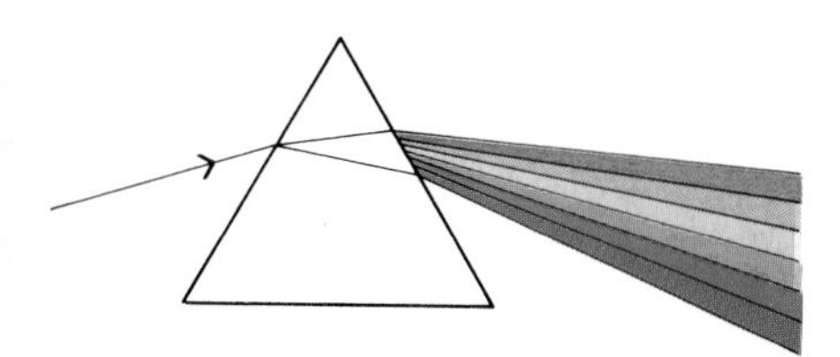

8-2 *White light is actually a mixture of different colors, ranging from violet at one end of the spectrum to red at the other. It is separated into its component colors when it passes through a prism—"the celebrated phaenomena of colors," as Newton referred to it.*

oxygen from carbon dioxide and water, in a complex series of reactions. The overall equation for photosynthesis is

carbon dioxide + water + light energy ⟶ glucose + oxygen

or, $6CO_2 + 6H_2O + 686 \text{ kilocalories} \longrightarrow C_6H_{12}O_6 + 6O_2$

To understand how organisms are able to capture light energy and convert it into stored chemical energy, we must first look at the characteristics of light.

The Nature of Light

Over 300 years ago, the English physicist Sir Isaac Newton (1642–1727) separated visible light into a spectrum of colors by passing it through a prism (Figure 8-2). Then by passing the light through a second prism, he recombined the colors, producing white light once again. By this experiment, Newton showed that white light is actually made up of a number of different colors, ranging from violet at one end of the spectrum to red at the other. Their separation is possible because light of different colors is bent at different angles in passing through the prism.

In the nineteenth century, through the genius of James Clerk Maxwell (1831-1879), it became known that what we experience as light is in truth a very small part of a vast continuous spectrum of radiation, the electromagnetic spectrum (Figure 8-3). All the radiations included in this spectrum act as if they travel in waves. The wavelengths—that is, the distances from one wave peak to the next—range from those of gamma rays, which are measured in nanometers, to those of low-frequency radio waves, which are measured in kilometers. Within the spectrum of visible light, red light has the longest wavelength, violet the shortest. Radiation of each particular wavelength has a characteristic amount of energy associated with it. The amount of energy is inversely proportional to the wavelength—the longer the wavelength, the lower the energy. Violet light, for example, has almost twice the energy of red light, the longest visible wavelength.

8-3 *Visible light is only a small portion of the vast electromagnetic spectrum. For the human eye, the visible spectrum ranges from violet light, which is made up of comparatively short rays, to red light, the longest visible rays.*

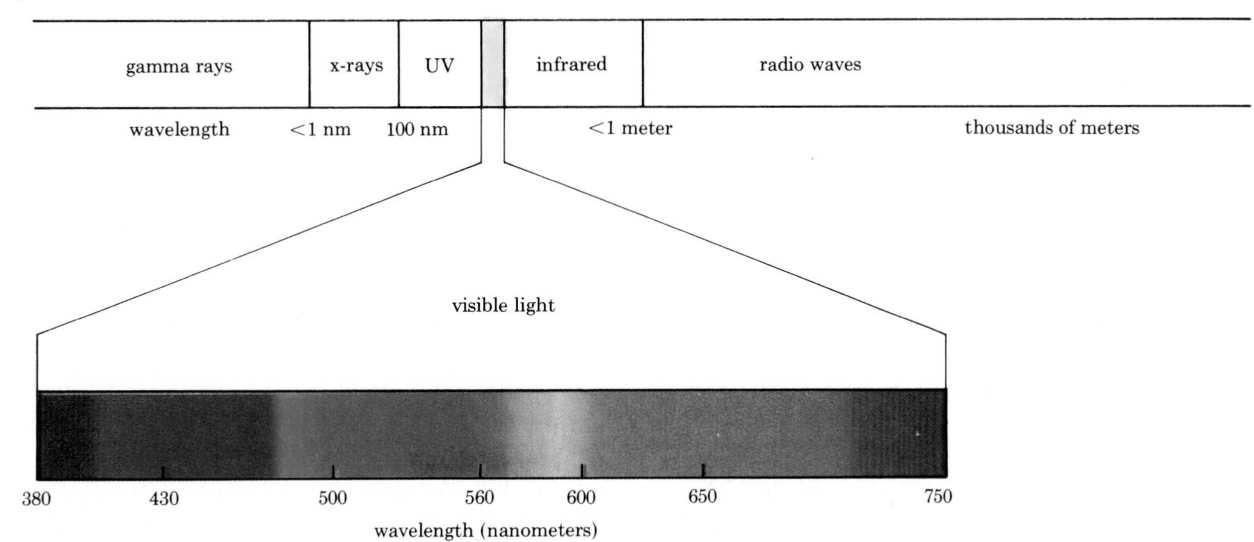

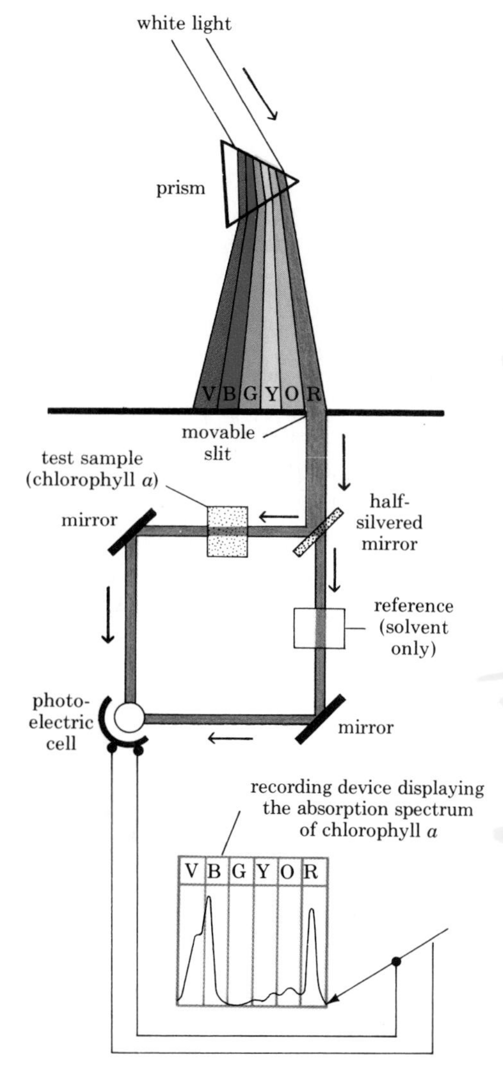

8-4 *The absorption spectrum of a pigment is measured with a spectrophotometer. This device directs a beam of light of each wavelength at the object to be analyzed and records what percentage of light of each wavelength is absorbed by the pigment sample as compared to a reference sample. Because the mirror is lightly (half) silvered, half of the light is reflected and half is transmitted. The photoelectric cell is connected to an electronic device that automatically records the percentage absorption at each wavelength.*

The Fitness of Light

From the physicist's point of view, the difference between radiations we can see and radiations we cannot see—so dramatic to the human eye—is only a few nanometers of wavelength, or, expressed differently, a small amount of energy. Why does this particular group of radiations, rather than some other, make the leaves grow and the flowers burst forth, cause the mating of fireflies and palolo worms, and, when reflecting off the surface of the moon, excite the imagination of poets and lovers? Why is it that this tiny portion of the electromagnetic spectrum is responsible for vision, for the rhythmic day-night regulation of many biological activities, for the bending of plants toward the light, and also for photosynthesis, on which all life depends? Is it an amazing coincidence that all these biological activities are dependent on these same wavelengths?

George Wald of Harvard, an expert on the subject of light and life, says no. He thinks that if life exists elsewhere in the universe, it is probably dependent on the same fragment of the vast spectrum. Wald bases this conjecture on two points. First, living things, as we have seen, are composed of large, complicated molecules held in special configurations and relationships to one another by hydrogen bonds and other weak bonds. Radiation of even slightly higher energies than the energy of violet light breaks these bonds and so disrupts the structure and function of the molecules. Radiations with wavelengths less than 200 nanometers—that is, with still higher energies—drive electrons out of atoms. On the other hand, light of wavelengths longer than those of the visible band—that is, with less energy than red light—is absorbed by water, which makes up the great bulk of all living things on earth. When this light reaches molecules, its lower energies cause them to increase their motion (increasing heat) but do not trigger changes in their electron configurations. Only those radiations within the range of visible light have the property of exciting molecules—that is, of moving electrons into higher energy levels—and so of producing chemical and, ultimately, biological changes.

The second reason for the visible band of the electromagnetic spectrum being "chosen" by living things is that it, above all, is what is available. Most of the radiation reaching the earth from the sun is within this range. Higher-energy wavelengths are screened out by the oxygen and ozone high in the modern atmosphere. Much infrared radiation is screened out by water vapor and carbon dioxide before it reaches the earth's surface.

This is an example of what has been termed "the fitness of the environment"; the suitability of the environment for life and that of life for the physical world are exquisitely interrelated. If they were not, life could not, of course, exist.

Chlorophyll and Other Pigments

In order for light energy to be used by living systems, it must first be absorbed. A pigment is any substance that absorbs light. Some pigments absorb all wavelengths of light and so appear black. Some absorb only certain wavelengths, transmitting or reflecting the wavelengths they do not absorb. Chlorophyll, the pigment that makes leaves green, absorbs light in the violet and blue wavelengths and also in the red; because it reflects green light, it appears green. Different pigments absorb light energy at different wavelengths. The absorption pattern of a pigment is known as the *absorption spectrum* of that substance (Figure 8-4).

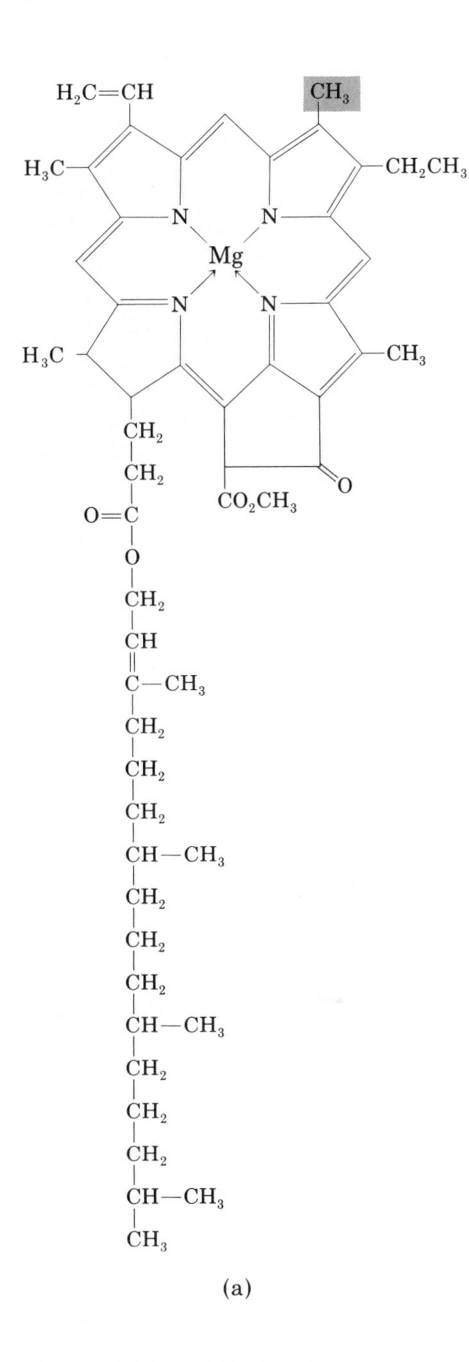

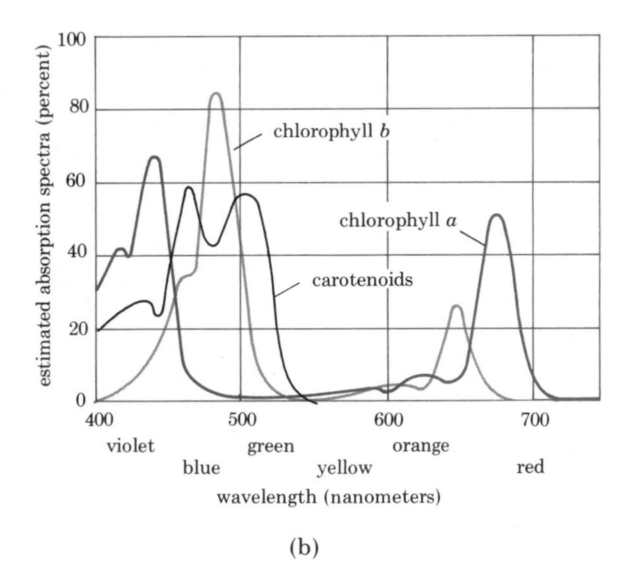

8-5 (a) *Chlorophyll* a *is a large molecule with a central atom of magnesium held in a ring structure that contains nitrogen. Attached to the ring is a long, hydrophobic carbon-hydrogen chain that may help to anchor the molecule in the internal membranes of the chloroplast. Chlorophyll* b *differs from chlorophyll* a *in having a CHO group in place of the CH_3 group indicated by color. Alternating single and double bonds, such as those in the chlorophylls, are common in pigments.*

(b) *The absorption spectra of chlorophyll* a, *chlorophyll* b, *and the carotenoids found in chloroplasts of plants. (Prepared by Govindjee.)*

Different groups of plants and algae use various pigments in photosynthesis. There are several different kinds of chlorophyll that vary slightly in their molecular structure. In plants, chlorophyll *a* is the pigment directly involved in the transformation of light energy to chemical energy. Most photosynthetic cells also contain a second type of chlorophyll—in plants, it is chlorophyll *b*—and a representative of another group of pigments called the carotenoids. The carotenoids are red, orange, or yellow pigments. In the green leaf, their color is masked by the chlorophylls, which are more abundant. In some tissues, however, such as those of a ripe tomato or the root of a carrot plant, the carotenoid colors predominate, as they do also when leaf cells stop synthesizing chlorophyll in the fall.

Chlorophyll *b* and the carotenoids are able to absorb light at wavelengths different from those absorbed by chlorophyll *a*. They apparently can pass the energy on to chlorophyll *a*, thus extending the range of light available for photosynthesis (Figure 8-5).

When pigments absorb light, electrons are boosted to a higher energy level. Three possible consequences are: (1) The energy may be dissipated as heat; (2) it may be re-emitted immediately as light energy of a longer wavelength; (3) the energy may trigger a chemical reaction, as happens in photosynthesis. Whether or not a particular pigment can participate in a chemical reaction depends not only on its structure but also on its relationship with neighboring molecules. Chlorophyll can convert light energy to chemical energy only when it is associated with certain proteins and embedded in a specialized membrane.

Chloroplasts

The structural unit of photosynthesis is the *thylakoid*, which usually takes the form of a flattened sac, or vesicle. In eukaryotes, the thylakoids form a part of the internal membrane structure of specialized organelles, the chloroplasts (Figure 8-6, on the next page). The alga *Chlamydomonas*, for instance, has a single very large chloroplast; the cell of a leaf characteristically has 40 to 50 chloroplasts, and there are often 500,000 chloroplasts per square millimeter of leaf surface.

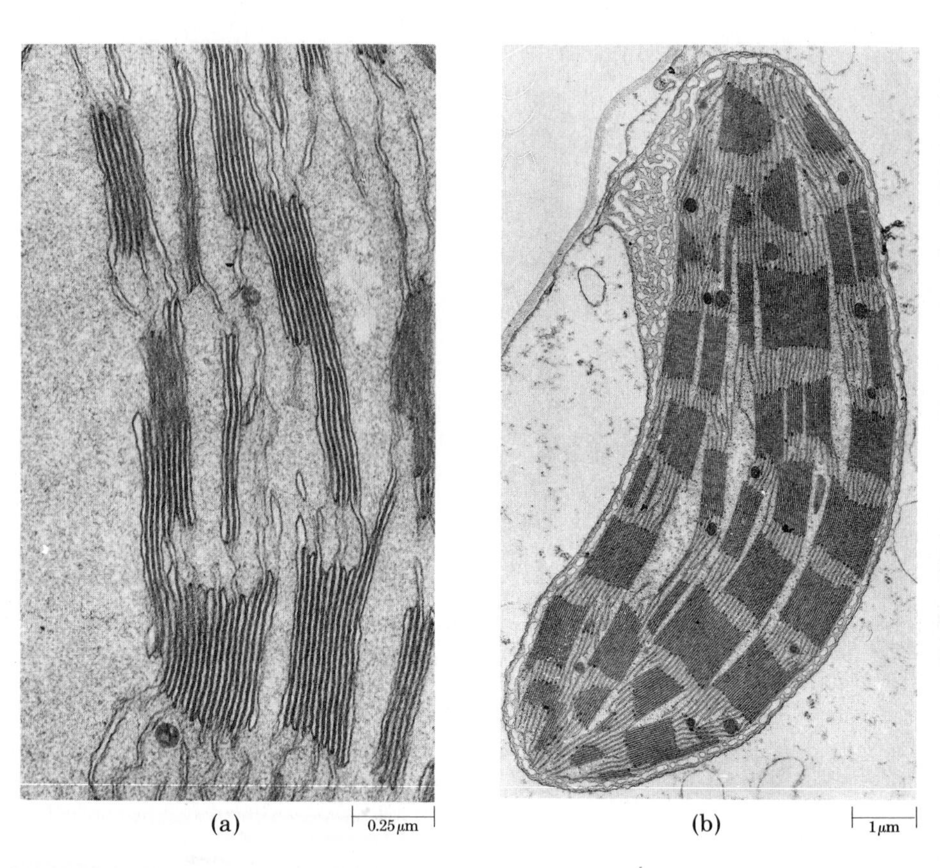

8–6 *Photosynthesis takes place in the thylakoid, a flattened sac, whose membranes contain chlorophyll and other pigments. In plants and algae, thylakoids are part of an elaborate membrane system enclosed in a special organelle, the chloroplast.* (a) *Stacks of thylakoids (grana) from a plant cell. Notice that some of the thylakoid membranes have extensions that interconnect the grana through the stroma that separates them.* (b) *A chloroplast, showing the elaborate system of internal membranes comprising interconnected stacks of thylakoids.*

Chloroplasts, like mitochondria, are surrounded by two membranes. The interior of the chloroplast is filled with a dense solution, the *stroma*, which is different in composition from the cytoplasm. With the light microscope, it is possible to see little spots of green within the chloroplasts of leaves. The early microscopists called these green specks *grana* ("grains"), and this term is still in use. Under the electron microscope, it can be seen that the grana are stacks of thylakoids.

All the thylakoids in a chloroplast are oriented parallel to each other. Thus, by swinging toward the light, the chloroplast can simultaneously aim all of its millions of pigment molecules for optimum reception, as if they were miniature electromagnetic antennae (which, of course, they are).

Photosynthesis: Light and Dark Reactions

The reactions of photosynthesis, it is now known, take place in two stages. In the first stage, known as the *light reactions*, light strikes chlorophyll *a* molecules that are packed in a special way in the thylakoid membranes. Electrons from the chlorophyll *a* molecules are boosted to higher energy levels, and, in a series of reactions, their added energy is used to form ATP from ADP and to reduce an electron-carrier molecule known as $NADP^+$. $NADP^+$ closely resembles NAD^+, and it too is reduced by the addition of two electrons and a proton, forming NADPH. Water molecules are also broken apart in the light reactions, supplying electrons that replace those boosted from the chlorophyll *a* molecules.

In the second stage of photosynthesis, the energy stored in ATP and NADPH is used to reduce carbon from carbon dioxide to a simple sugar. Thus the chemical energy of the electron-carrier molecules is transferred to molecules suitable for transport and storage in the algal cell or plant body. At the same time, a carbon skeleton is formed from which other organic molecules can be built. The reactions of this second stage, which occurs in the stroma of the chloroplast, are known as the *dark reactions*. They do not require light, but they do not require darkness either. Their only requirements are carbon dioxide and the products of the light reactions—ATP and NADPH. The binding of carbon dioxide into organic compounds in the dark reactions is known as the *fixation of carbon*.

Model of the Light Reactions

In the thylakoids, chlorophyll and other molecules are, according to the present model, packed into units called photosystems. Each unit contains from 250 to 400 molecules of pigment, which serve as light-trapping antennae. Once light energy is absorbed by one of the antenna pigments, it is bounced around (like a hot potato) among the other pigment molecules of the photosystem until it reaches a special form of chlorophyll *a*, which is the reaction center.

Present evidence indicates that there are two different photosystems. In Photosystem I, the reactive chlorophyll *a* molecule is known as P_{700} because one of the peaks of its absorption spectrum is at 700 nanometers. The reactive chlorophyll *a* molecule of Photosystem II is P_{680}. It is believed that these chlorophyll molecules have unusual properties because of their association with special proteins in the membrane.

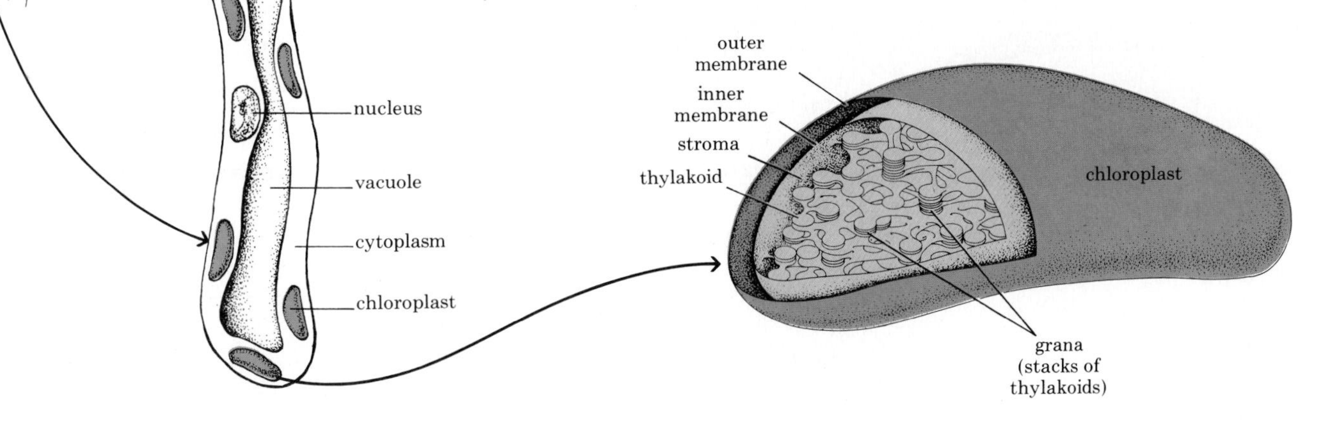

8-7 *Journey into a chloroplast. The plant shown is a geranium, which you may recognize by the characteristic shape of its leaves. The inner tissues of the leaf are completely enclosed by transparent epidermal cells that are coated with a waxy layer, the cuticle. Oxygen, carbon dioxide, and other gases enter the leaf largely through special openings, the stomata (singular, stoma). Gases and water vapor fill the spaces between cells in the spongy layer, leaving and entering cells by diffusion. Water, taken up by the roots, enters the leaf by way of the vascular bundle, and sugars, the products of photosynthesis, leave the leaf by this route, traveling to nonphotosynthetic parts of the plant. Much of the photosynthesis takes place in the palisade cells, elongated cells directly beneath the upper epidermis. They have a large central vacuole and numerous chloroplasts that move within the cell, orienting themselves with respect to the light. Light is captured in the membranes of the disk-shaped thylakoids within the chloroplast.*

The Carbon Cycle

By photosynthesis, living systems incorporate carbon dioxide from the atmosphere into organic, carbon-containing compounds. In respiration, these compounds are broken down again into carbon dioxide and water. These processes, viewed on a worldwide scale, result in the carbon cycle. The principal photosynthesizers in the cycle are plants and the phytoplankton, the marine algae. They synthesize carbohydrates from carbon dioxide and water and release oxygen into the atmosphere. About 75 billion metric tons of carbon per year are bound into carbon compounds by photosynthesis.

Some of the carbohydrates are used by the photosynthesizers themselves: Plants release carbon dioxide from their roots and leaves, and marine algae release it into the water where it maintains an equilibrium with the carbon dioxide of the air. Some 500 billion metric tons of carbon are "stored" in the seas, and some 700 billion metric tons in the atmosphere. Some of the carbohydrates are used by animals that feed on the live plants, on algae, and on one another, releasing carbon dioxide. The dead bodies of the plants and other organisms plus discarded leaves and shells, feces, and other waste materials settle into the soil or sink to the ocean floors where they are consumed by decomposers—small invertebrates, bacteria, and fungi. Carbon dioxide is also released by these processes into the reservoir of the air and oceans. Another, even larger store of carbon lies below the surface of the earth in the form of coal and oil, deposited there some 300 million years ago.

The natural processes of photosynthesis and respiration balance one another out. For many millions of years, the carbon dioxide of the atmosphere, as far as we can tell, has remained constant. By volume it is a very small proportion of the atmosphere, only about 0.03 percent. It is important, however, because carbon dioxide, unlike most other components of the atmosphere, absorbs heat from the sun's rays. Since 1850, carbon dioxide concentrations in the atmosphere have been increasing, owing in part to our use of fossil fuels, to our plowing of the soil, and to our destruction of forest land, particularly in the tropics. Some environmentalists predict that this increase in the carbon dioxide "blanket" will increase the temperature here on earth, with a consequential increase in the great deserts of the world. Others, looking on the sunnier side, foresee an increase in photosynthetic activity because of the increased carbon dioxide that will be available to plants and algae. Most, however, feel alarmed by the fact that although we do not know the consequences of what we are doing, we keep right on doing it.

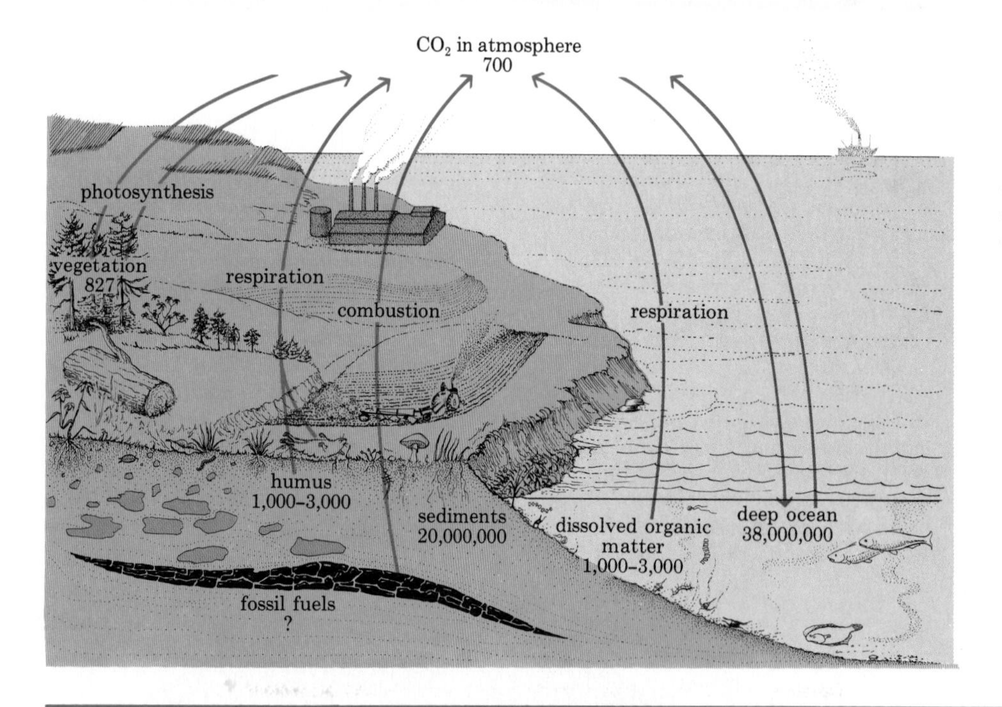

The carbon cycle. The arrows indicate the movement of carbon atoms. The numbers are all estimates of the amount of carbon stored, expressed in billions of metric tons. The amount of carbon released by respiration and combustion has, it is believed, begun to exceed the amount fixed by photosynthesis.

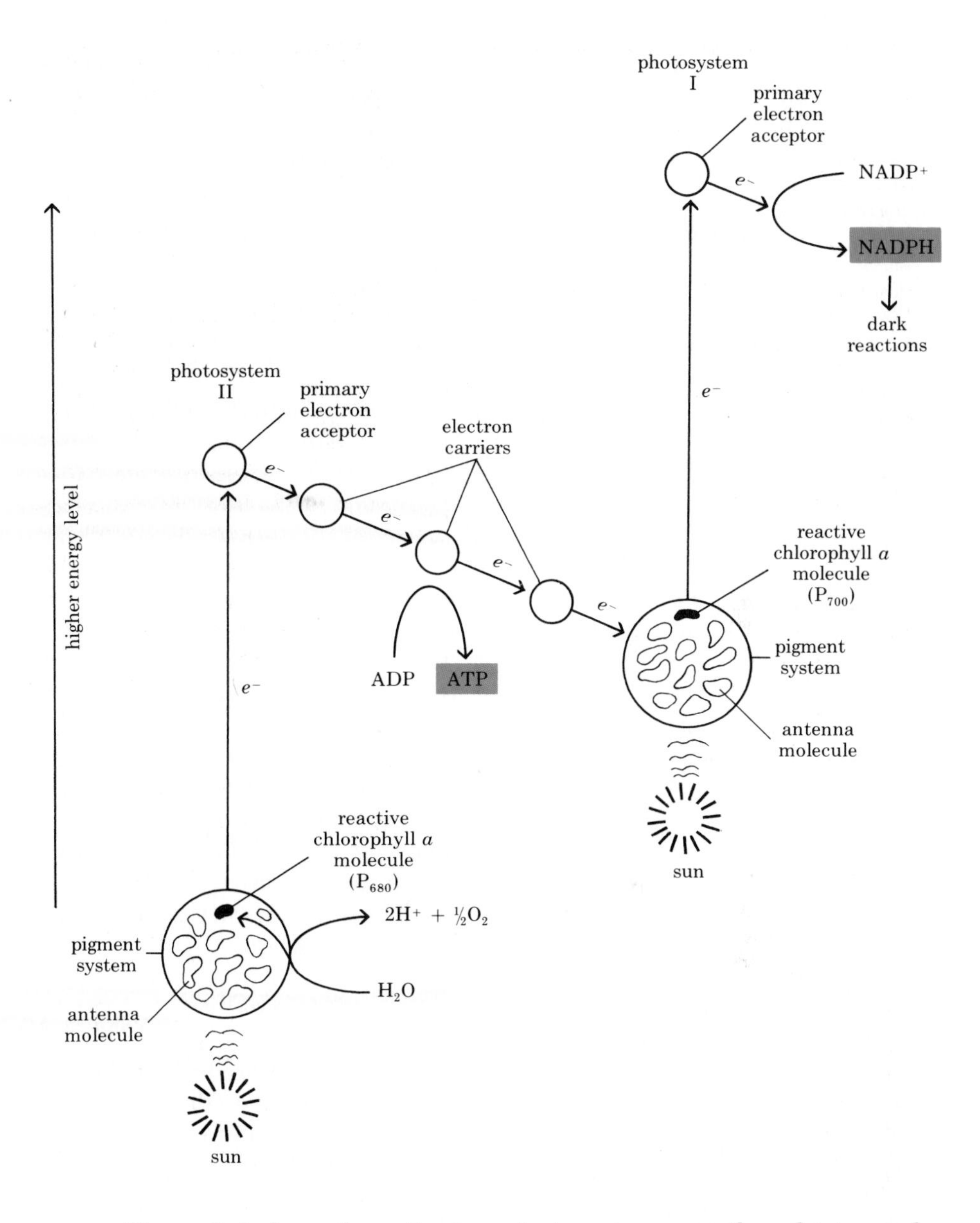

8-8 *Light energy trapped in the reactive chlorophyll* a *molecule of Photosystem II boosts electrons to a higher energy level. These electrons are replaced by electrons pulled away from water molecules, releasing protons (H^+) and oxygen gas. The electrons are passed from the electron acceptor along an electron transport chain to a lower energy level, the reaction center of Photosystem I. As they pass along this electron transport chain, some of their energy is packaged in the form of ATP. Light energy absorbed by Photosystem I boosts the electrons to another primary electron acceptor. From this acceptor, they are passed via other electron carriers to $NADP^+$ to form NADPH. The electrons removed from Photosystem I are replaced by those from Photosystem II. ATP and NADPH represent the net gain from the light reactions.*

Figure 8-8 shows how the two photosystems are thought to work together in photosynthesis. Light energy enters Photosystem II, where it is trapped by the reactive chlorophyll *a* molecule P_{680}. Electrons from the P_{680} molecule are boosted to a higher energy level, from which they are transferred to an electron-acceptor molecule. The electrons then pass downhill along an electron transport chain to Photosystem I. As the electrons pass along this chain, ATP is formed from ADP, in a process similar to that of the electron transport chain of mitochondria.

Three other events are taking place simultaneously:

1. The P_{680} chlorophyll molecule, having lost its electron, is avidly seeking a replacement. It finds it in the water molecule, which is dissociated into protons and oxygen gas.

2. Additional light energy is trapped in the reactive chlorophyll molecule (P_{700}) of Photosystem I. The molecule is oxidized, and an electron is boosted to a primary electron acceptor from which it goes downhill to $NADP^+$.

3. The electron removed from the P_{700} molecule is replaced by the electron from Photosystem II.

Thus in the light there is a continuous flow of electrons from water to Photosystem II to Photosystem I to $NADP^+$. In the words of Nobel laureate Albert Szent-Györgyi: "What drives life is . . . a little electric current, kept up by the sunshine."

The Dark Reactions

The Calvin Cycle: The Three-Carbon Pathway

The reduction of carbon takes place in a cycle named after its discoverer, Melvin Calvin. The Calvin cycle is analogous to the Krebs cycle (page 104) in that, in each turn of the cycle, the starting compound is regenerated. The starting (and ending) compound is a five-carbon sugar with two phosphates attached, ribulose diphosphate (RuDP).

The cycle begins when carbon dioxide is bound to RuDP, which then splits to form two molecules of phosphoglycerate, or PGA (Figure 8–9). (Each PGA molecule contains three carbon atoms, hence the name, the three-carbon pathway.) This reaction is catalyzed by a specific enzyme, RuDP carboxylase. Each step of the cycle is similarly regulated by a specific enzyme.

The complete cycle is diagrammed in Figure 8–10. At each full turn of the cycle, a molecule of carbon dioxide enters the cycle, is reduced, and a molecule of RuDP is regenerated. Three turns of the cycle introduce three molecules of carbon dioxide, the equivalent of one three-carbon sugar. Six revolutions of the cycle, with the introduction of six molecules of carbon dioxide, are necessary to produce a six-carbon sugar, such as glucose. The overall equation is

$$6RuDP + 6CO_2 + 18ATP + 12NADPH + 12H^+ + 12H_2O \longrightarrow 6RuDP + glucose + 18P_i + 18ADP + 12NADP^+$$

The immediate product of the cycle itself is glyceraldehyde phosphate, a three-carbon sugar. This same three-carbon sugar molecule is formed when the fructose diphosphate molecule is split at the fourth step in glycolysis (see page 102).

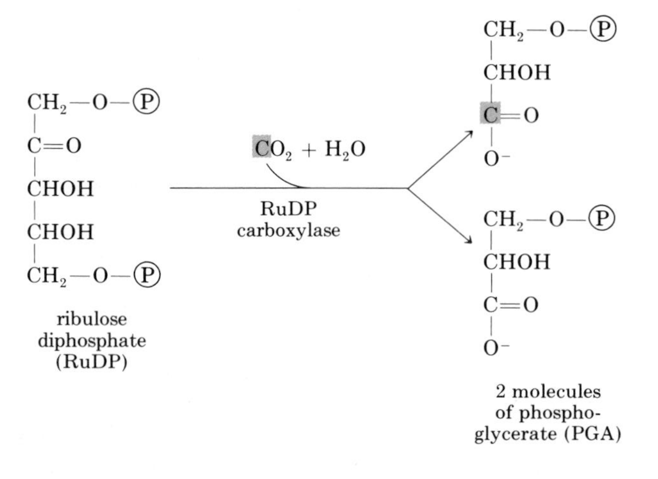

8–9 *Calvin and his collaborators briefly exposed photosynthesizing algae to radioactive carbon dioxide ($^{14}CO_2$). They found that the radioactive carbon is first incorporated into ribulose diphosphate (RuDP), which then immediately splits to form two molecules of phosphoglycerate (PGA). The radioactive carbon atom, indicated in color, appears in one of the two molecules of PGA. This is the first step in the Calvin cycle.*

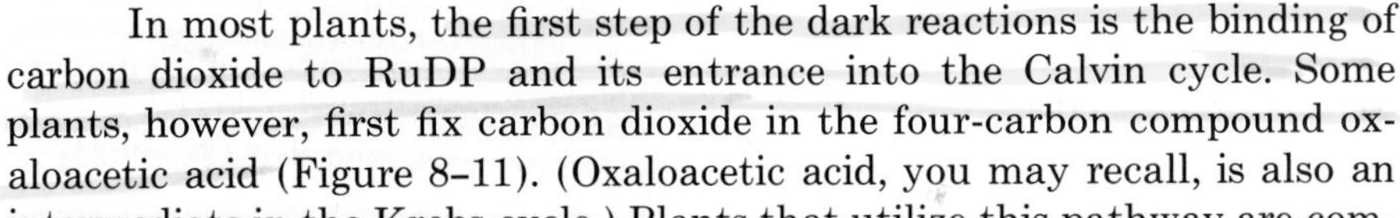

8-10 *Summary of the Calvin cycle. At each full "turn" of the cycle, one molecule of carbon dioxide enters the cycle. Three turns are summarized here—the number required to make one molecule of glyceraldehyde phosphate. Three molecules of ribulose diphosphate (RuDP), a five-carbon compound, are combined with three molecules of carbon dioxide, yielding six molecules of phosphoglycerate, a three-carbon compound. These are converted to six molecules of glyceraldehyde phosphate. Five of these three-carbon molecules are combined and rearranged to form three five-carbon molecules of RuDP. The "extra" molecule of glyceraldehyde phosphate represents the net gain from the Calvin cycle. The energy that "drives" the Calvin cycle is in the form of ATP and NADPH, produced by the light reactions.*

The Four-Carbon Pathway

In most plants, the first step of the dark reactions is the binding of carbon dioxide to RuDP and its entrance into the Calvin cycle. Some plants, however, first fix carbon dioxide in the four-carbon compound oxaloacetic acid (Figure 8-11). (Oxaloacetic acid, you may recall, is also an intermediate in the Krebs cycle.) Plants that utilize this pathway are commonly called C_4, or four-carbon plants, as distinct from the C_3 plants, in which carbon is fixed first in the three-carbon compound phosphoglycerate (PGA).

The carbon dioxide fixed in oxaloacetic acid by C_4 plants ultimately enters the Calvin cycle. After the oxaloacetic acid molecules are formed, they are transferred to adjacent cells where they are oxidized and carbon dioxide is released. This carbon dioxide then combines with RuDP and enters the Calvin cycle.

Why has evolution favored such a seemingly cumbersome method of providing carbon dioxide to the Calvin cycle? This question can be answered only by considering the function of the leaf as a whole. Carbon dioxide is not continuously available to the photosynthesizing cells. It enters the leaf by way of the stomata, specialized pores that open and close depending on, among other things, water stress.

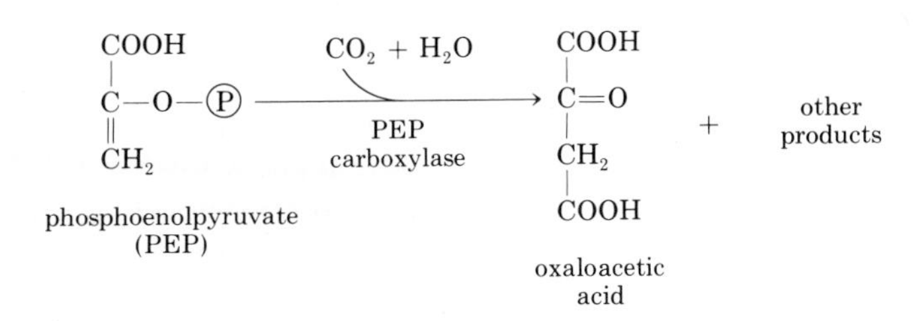

8-11 *In C_4 plants, carbon dioxide is combined with a compound known as phosphoenolpyruvate (PEP) to yield the four-carbon oxaloacetic acid. The reaction is catalyzed by the enzyme PEP carboxylase. The oxaloacetic acid is then broken down again to CO_2, which enters the Calvin cycle. Sugarcane, corn, and sorghum are among the best-known C_4 plants.*

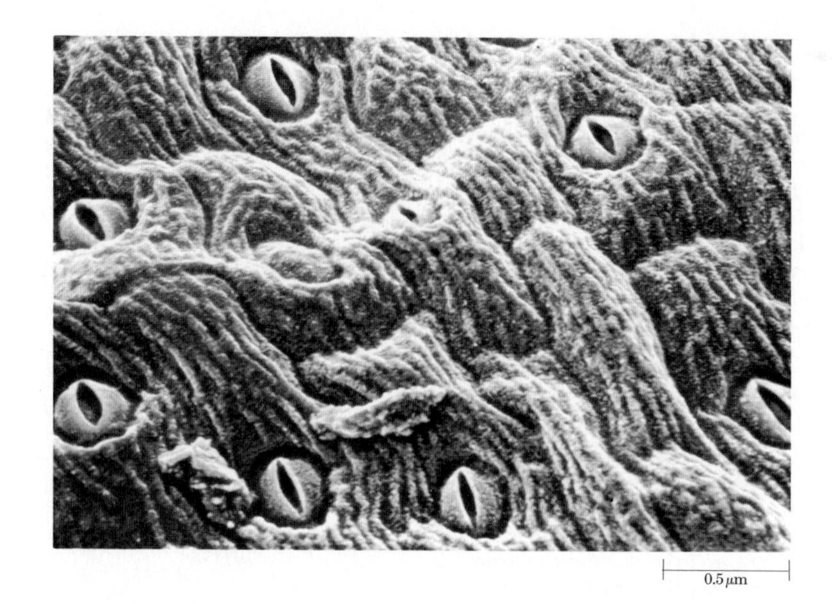

8-12 *Scanning electron micrograph of open stomata on the lower surface of a leaf. The carbon dioxide used in photosynthesis reaches the photosynthetic cells through these openings.*

PEP carboxylase, the enzyme that catalyzes the formation of oxaloacetic acid, has a high affinity for carbon dioxide. Even at low concentrations of carbon dioxide, the enzyme works rapidly to bind carbon dioxide to PEP. As a result, the carbon dioxide concentration within the leaf remains low, which maximizes the gradient of carbon dioxide between the cells and the outside air. Thus, when the stomata are open, carbon dioxide readily diffuses down the concentration gradient into the leaf. If the stomata must be closed much of the time—as they must be to conserve water in a hot, dry climate—the plant with C_4 metabolism will take up more carbon dioxide with each gasp (so to speak) than the plant that has only C_3 metabolism. Therefore, the C_4 plant is at a distinct advantage in drought-ridden areas.

Perhaps the most familiar example of the competitive capacity of C_4 plants is seen in lawns in the summertime. In most parts of the United States, lawns consist mainly of C_3 grasses such as Kentucky bluegrass. As the summer days become hotter and drier, these dark green, fine-leaved grasses are often overwhelmed by rapidly growing crabgrass, which disfigures the lawn as its yellowish-green, broader-leaved plants slowly take over. Crabgrass, you will not be surprised to hear, is a C_4 plant.

The list of plants known to utilize the four-carbon pathway has grown to over 100 genera, at least a dozen of which include both C_3 and C_4 species. This pathway has undoubtedly arisen independently many times in the course of evolution and is another example of the exquisite adaptation of living systems to their environment.

The Products of Photosynthesis

Glyceraldehyde phosphate, the three-carbon sugar produced by the Calvin cycle, may seem an insignificant reward, both for all the enzymatic activity on the part of the cell and for our own intellectual stress. However, this molecule and those derived from it provide (1) the energy source for all living systems, and (2) the basic carbon skeleton for all organic molecules. Carbon has been fixed—that is, it has been brought from the inorganic world into the organic one.

Molecules of glyceraldehyde phosphate may flow into a variety of different metabolic pathways, depending on the activities and requirements of the cell. Often they are built up to glucose or fructose, following a sequence that is in many of its steps the reverse of the glycolysis sequence described in the previous chapter. Plant cells use these six-carbon sugars to make starch and cellulose for their own purposes and sucrose for export. Animal cells store them as glycogen. All cells use sugars, including glyceraldehyde phosphate and glucose, as the starting point for the manufacture of other carbohydrates, fats and other lipids, and, with the addition of nitrogen, amino acids and nitrogenous bases. Finally, as we saw in the preceding chapter, the carbon fixed in photosynthesis is the source of ATP energy for heterotrophic cells.

SUMMARY

In photosynthesis, light energy is converted to chemical energy, and carbon is fixed into organic compounds.

Light energy is captured by the living world by means of pigments. The pigments involved in photosynthesis in eukaryotes include the chlorophylls and the carotenoids. Light absorbed by the pigments boosts their electrons to higher energy levels. Because of the way the pigments are packed into membranes, they are able to transfer this energy to reactive molecules, probably chlorophyll *a* packed in a particular way.

Photosynthesis takes place within cellular organelles known as chloroplasts. These organelles are surrounded by two membranes. Contained within the membranes of the chloroplast are a solution of organic compounds and ions known as the stroma and a complex internal membrane system consisting of fused pairs of membranes that form sacs called thylakoids. The pigments and other molecules responsible for capturing light are located in and on these membranes.

Photosynthesis takes place in two stages: (1) the light reactions, in which light energy is captured by chlorophyll and converted to the chemical energy of ATP and NADPH; and (2) the dark reactions, in which carbon atoms are reduced and carbohydrates formed.

In the currently accepted model of the light reactions in photosynthesis, light energy strikes antennae pigments of Photosystem II. Electrons are boosted uphill from the reactive chlorophyll *a* molecule P_{680} to an electron acceptor. As the electrons are removed, they are replaced by electrons from water molecules, with the simultaneous production of free oxygen. The electrons then pass downhill to Photosystem I along an electron transport chain, in the course of which ATP is generated. Light energy absorbed in antennae pigments of Photosystem I and passed to chlorophyll P_{700} results in the boosting of electrons to another electron acceptor. The electrons removed from P_{700} are replaced by the electrons from Photosystem II. The electrons are ultimately accepted by the electron carrier $NADP^+$. The energy yield from the light reactions is contained in molecules of NADPH and ATP.

In the dark reactions, which take place in the stroma, NADPH and ATP produced in the light reactions are used to reduce carbon dioxide to organic carbon. This is accomplished by means of the Calvin cycle. In the Calvin cycle, a molecule of carbon dioxide is combined with the starting material, a five-carbon sugar called ribulose diphosphate. At each turn of the cycle, one carbon atom enters the cycle. Three turns of the cycle produce a three-carbon molecule, glyceraldehyde phosphate. Two molecules of glyceraldehyde phosphate (six turns of the cycle) can combine to form a glucose molecule. At each turn of the cycle, RuDP is regenerated. The glyceraldehyde phosphate can also be used as a starting material for other organic compounds needed by the cell.

In C_4 plants, carbon dioxide is initially accepted by a compound known as PEP to yield the four-carbon oxaloacetic acid. The oxaloacetic acid is then oxidized, and carbon dioxide is transferred to the RuDP of the Calvin cycle. Under conditions of drought, C_4 plants are more efficient than C_3 plants.

QUESTIONS

1. Distinguish among the following: stroma/grana/thylakoid; light reactions/dark reactions; C_3 photosynthesis/C_4 photosynthesis.

2. Sketch a chloroplast and label its structure. Compare your sketch with Figure 8–7.

3. Why is it plausible to argue, as the Nobel laureate George Wald does, that wherever in the universe we find living organisms, we will find them (or at least some of them) to be colored?

4. Predict what colors of light might be most effective at stimulating plant growth. (Such is the principle of the special light bulbs used for plants.)

5. Describe in general terms the events of photosynthesis.

6. What is the source of the oxygen released in photosynthesis?

7. The experiment shown in Figure 4–3b on page 54 was attempted in Stockton, California, in 1973, but instead of making amino acids, the electric sparks caused the apparatus to explode. (No one was hurt, fortunately.) What is present in today's atmosphere that was not present in the primitive atmosphere and that would account for the explosion?

8. Over 100 genera of plants have acquired C_4 photosynthesis in the course of evolution. How is this adaptation advantageous to these plants? Many other plants, however, have not evolved C_4 photosynthesis. Why is it advantageous to such plants *not* to have C_4 photosynthesis?

SUGGESTIONS FOR FURTHER READING

CONANT, JAMES BRYANT (ed.): *Harvard Case Histories in Experimental Science,* vol. 2, Harvard University Press, Cambridge, Mass., 1964.

Case #5, Plants and the Atmosphere, *edited by Leonard K. Nash, describes the early work on photosynthesis, presented often in the words of the investigators themselves. The narrative illuminates the historical context in which the discoveries were made.*

LEHNINGER, ALBERT L.: *Bioenergetics: The Molecular Basis of Biological Energy Transformations,* 2d ed., The Benjamin/Cummings Publishing Company, Menlo Park, Calif., 1971.

A thorough account; Lehninger is one of the foremost experts on cellular energetics.

LEHNINGER, ALBERT L.: *First Course in Biochemistry,* Worth Publishers, Inc., New York, 1981.

This introductory text is outstanding both for its clarity and for its consistent focus on the living cell. There are numerous medical and practical applications throughout.

RABINOWITCH, EUGENE, and GOVINDJEE: *Photosynthesis,* John Wiley & Sons, Inc., New York, 1969.*

A lucid introduction, suitable for undergraduate students, to the processes of photosynthesis and to related physical and chemical concepts, such as entropy and free energy.

STRYER, LUBERT: *Biochemistry,* 2d ed., W. H. Freeman and Company, San Francisco, 1981.

A good introduction, handsomely illustrated, to cellular energetics.

*Available in paperback.

SECTION 3

Genetics

Hereditary characteristics are passed from generation to generation. Seventeen of the people in this photograph, which was taken in 1894, are direct descendants of Queen Victoria (seated center). Although she herself was not afflicted, Queen Victoria was a carrier of hemophilia, a group of diseases in which the blood does not clot properly. One of her sons and three of her grandsons had hemophilia, and two of her daughters and four of her granddaughters were carriers. Two of these granddaughters were Princess Irene of Prussia, standing to the left of Victoria and wearing a feather boa, and, to the right of Victoria, Alexandra (also wearing a boa), the future Tsarina of Russia. Nicholas II, to become the last Tsar of Russia, is standing beside Alexandra.

Alexis, the only son of Nicholas and Alexandra, inherited hemophilia from his mother. His parents' great concern for his health was one of the factors in the turbulent events surrounding the Russian revolution.

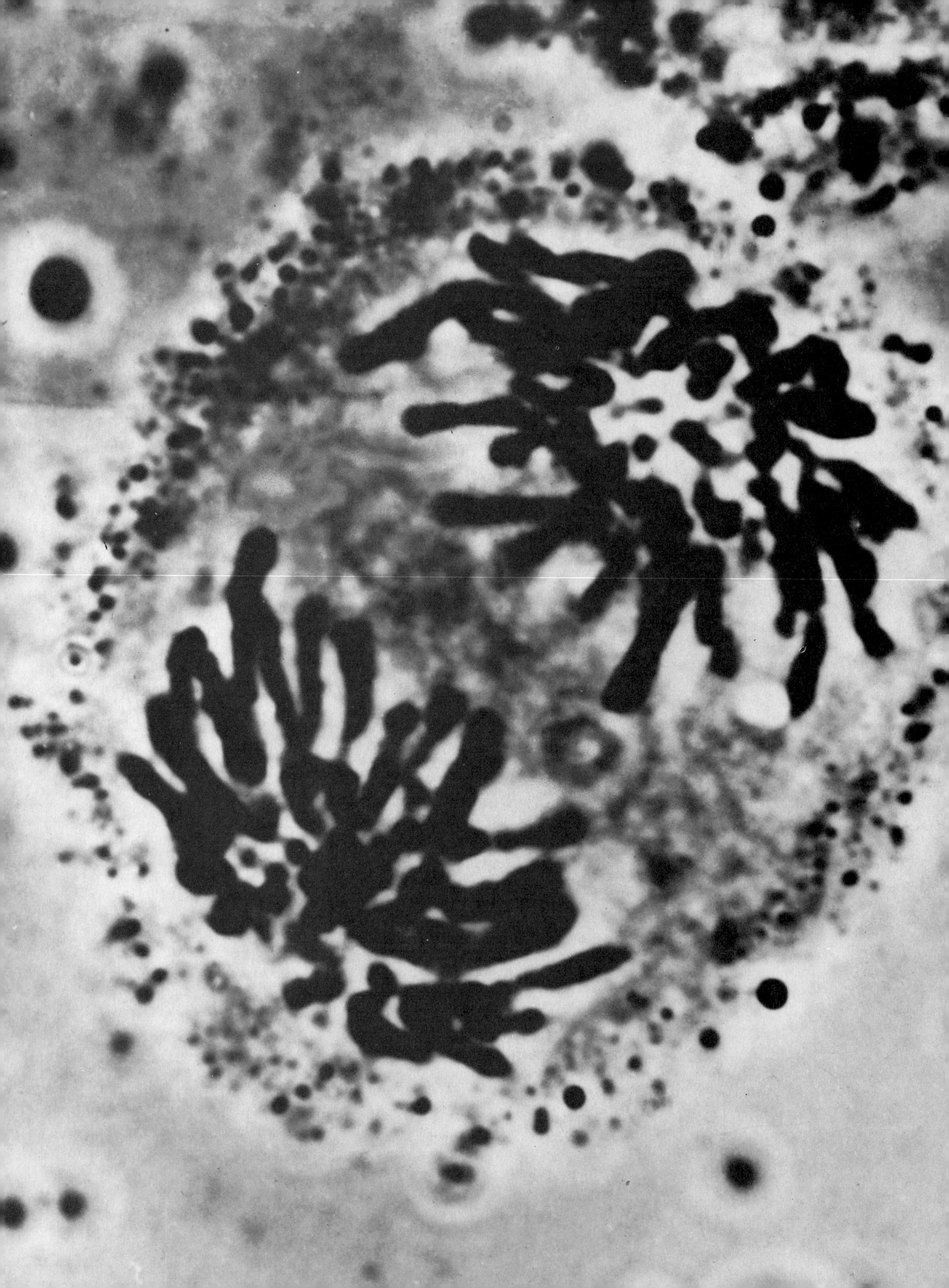

9

The Nucleus and Mitosis

Genetics is the branch of biology that deals with heredity, the process by which traits are passed from parents to offspring. As we saw in the Introduction, the capacity to reproduce is one of the special characteristics of living matter. For more than 3 billion years, living things have been reproducing, generation after generation, with each parent passing on to its offspring all the biological instructions necessary for the offspring to develop into the same kind of organism as the parent. A primary question of genetics is how this hereditary information is transmitted by the parent to its young. Or to put it more simply, why do cats have kittens, dogs have puppies, and hens have chicks? Why do you resemble your father or mother? What is the basis of this hereditary information? How is it passed from one generation to another? How is it translated into the specific characteristics of a particular organism?

These questions are among the most fundamental in biology. Our present understanding of the answers is based upon the work of some of the most brilliant scientists our civilization has ever known. In this chapter and the ones that follow, we are going to trace their discoveries.

The Cell's Nucleus

Every organism—whether a *Paramecium,* an oak, or a human—starts as a single cell. Thus the hereditary information must be carried somewhere, somehow in that single cell. More than a century ago, biologists, working with greatly improved microscopes, began to suspect that the key to the mystery of heredity was to be found within the cell's nucleus. As it turned out, they were quite right.

The nucleus of the eukaryotic cell is a prominent, roughly spherical body. Unlike the cytoplasm, which contains many distinct structures, the nucleus contains little that can be seen with any kind of microscope (Figure 9–2). The only things usually visible are its outer membranes (the nuclear envelope), a tangle of threadlike material (the chromatin), and one or two spherical bodies (the nucleoli). Yet despite its simple appearance, the nucleus plays a crucial role in the life of the organism. It is the repository of the hereditary information and is also a major factor in the control of cellular processes.

9–1 *A human cell divides. The long, dark bodies are chromosomes, the carriers of the hereditary information. The chromosomes have replicated and moved apart. Each set of chromosomes is an exact copy of the other. Thus the two new cells, the daughter cells, will contain the same hereditary material.*

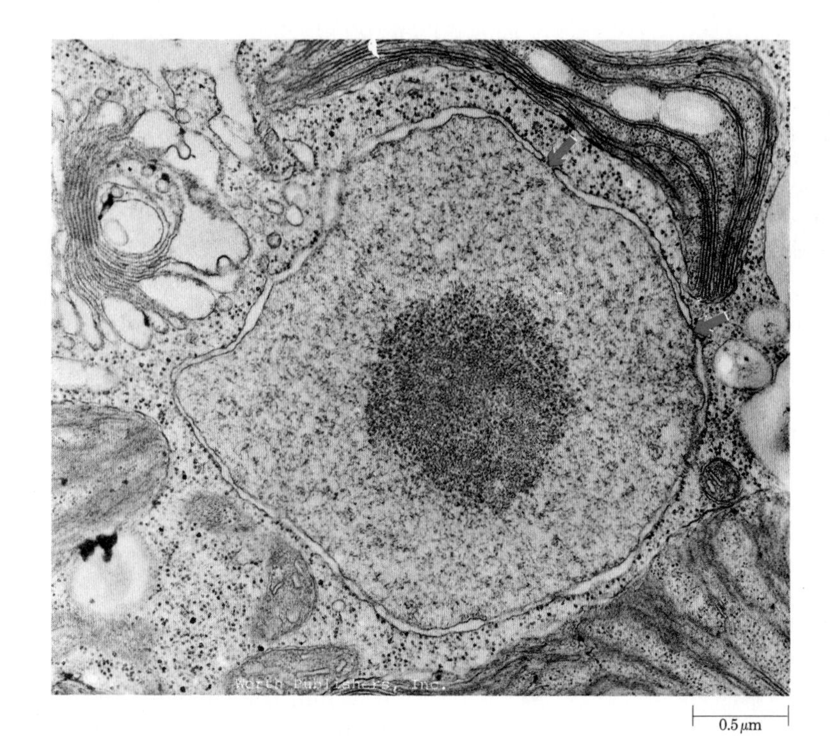

9–2 *Electron micrograph of the nucleus of the alga* Chlamydomonas. *The dark body in the center of the nucleus is the nucleolus. The major components of ribosomes are produced in the nucleolus; you can see partially formed ribosomes around its periphery. Notice also the nuclear envelope with its many pores, two of which are indicated by arrows. Above the nucleus is a portion of a chloroplast containing starch grains. A Golgi body is to the left of the nucleus, and mitochondria are visible below.*

Some Experimental Evidence

One of the most important early observations of the role of the nucleus in the life of the cell was made about a hundred years ago. A German embryologist, Oscar Hertwig, was observing the eggs and sperm of a sea urchin (Figure 9–3). Sea urchins produce eggs and sperm in great numbers. The eggs are relatively large and so are easy to observe. They are fertilized in the open water, rather than internally, as is the case with land-dwelling vertebrates such as ourselves. Watching the eggs being fertilized under his microscope, Hertwig observed that only a single sperm cell was required. Further, when the sperm cell penetrated the egg, its nucleus was released and fused with the nucleus of the egg. This observation, confirmed by other scientists and in other kinds of organisms, was important in establishing the fact that the nucleus is the carrier of heredity: The only link between father and offspring is the nucleus of the sperm.

20 μm

9–3 *This egg from a sea urchin is surrounded by sperm cells. Despite the great differences in size of egg and sperm, both contribute equally to the hereditary characteristics of the individual. Since the nucleus is approximately the same size in both cells, the early microscopists postulated that this part of the cell must be the carrier of the hereditary information. Sea urchin eggs and sperm cells have been used in many studies because sea urchins are relatively easy to obtain and fertilization, which is external, can be easily observed in the laboratory.*

Since Hertwig's time, a number of experiments have explored the role of the nucleus in the cell. In one simple experiment, the nucleus was removed from an amoeba by microsurgery. The amoeba stopped dividing and, in a few days, it died. If, however, a nucleus from another amoeba was implanted within 24 hours after the original one was removed, the cell survived and divided normally.

In the early 1930s, Joachim Hämmerling studied the comparative roles of the nucleus and the cytoplasm by taking advantage of some unusual properties of the marine alga *Acetabularia.* The body of *Acetabularia* consists of a single huge cell 2 to 5 centimeters in height. Individuals have a cap, a stalk, and a "foot," all of which are differentiated portions of the single cell. If the cap is removed, the cell will rapidly regenerate a new one. Different species of *Acetabularia* have different kinds of caps. *Acetabularia mediterranea,* for example, has a compact umbrella-shaped cap, and *Acetabularia crenulata* has a cap of petal-like structures.

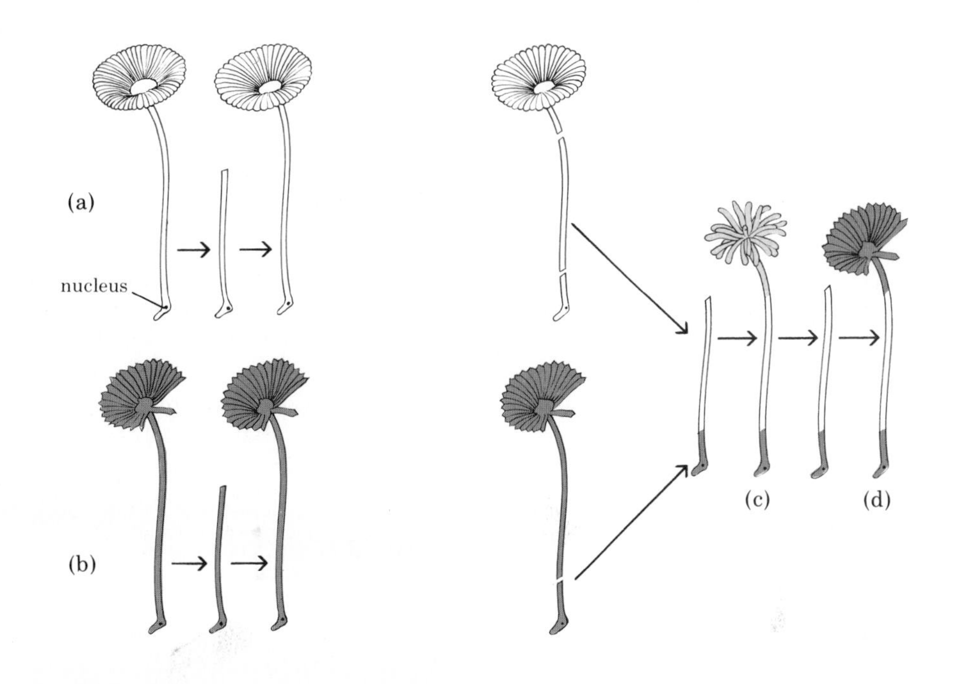

9–4 (a) *One species of* Acetabularia *has an umbrella-shaped cap, and* (b) *another has a ragged, petal-like cap. If the caps are removed, new caps form, similar in appearance to the amputated one. However, if the "foot" (containing the nucleus) is removed at the same time as the cap and a new nucleus from the other species is transplanted, the cap* (c) *that forms will have a structure with characteristics of both species. If this cap is removed, the next cap* (d) *that grows will be characteristic of the cell that donated the nucleus, not of the cell that donated the cytoplasm.*

Hämmerling took the "foot," which contains the nucleus, from a cell of *A. crenulata** and grafted it onto a cell of *A. mediterranea,* from which he had first removed the "foot" and the cap. The cap that then formed had a shape intermediate between those of the two species. When this cap was removed, the next cap that formed was completely characteristic of *A. crenulata* (Figure 9–4).

Hämmerling interpreted these results as meaning that certain cap-determining substances are produced under the direction of the nucleus. These substances accumulate in the cytoplasm, which is why the first cap that formed after nuclear transplantation was of an intermediate type. By the time the second cap formed, however, the cap-determining substances present in the cytoplasm before the transplant had been exhausted, and the form of the cap was completely under the control of the new nucleus.

Conclusion: The Functions of the Nucleus

We can see from these experiments that the nucleus performs two crucial functions for the cell. First, it carries the hereditary information for the cell, the instructions that determine whether a particular organism will develop as a *Paramecium,* an oak, or a human—and not just any *Paramecium,* oak, or human, but one that resembles the parent or parents of that particular, unique organism. Second, as Hämmerling's work indicated, the nucleus exerts a continuing influence over the ongoing activities of the cell, ensuring that the complex molecules that cells require are synthesized in the number and of the kind needed.

* By convention, with the second mention of a scientific binomial, it is permissible to abbreviate the first (genus) name. This is fortunate, particularly when dealing with such names as *Acetabularia.* It is also permissible to use the genus name alone when one is referring to all members of a particular genus. The second (species) name cannot stand alone, however.

Cell Division

Living organisms increase both in number and in size by means of cell division. One-celled organisms grow by assimilating materials from the environment and synthesizing these materials into new structural and functional molecules. When such a cell reaches a certain size, it divides. The two daughter cells, each of which has received about half of the mass of the parent cell, then begin growing again. In many-celled plants and animals, cell division is the means by which the organism grows, starting from one single cell, and also by which injured or worn-out tissues are repaired and replaced.

In all of these instances, the new cells produced are structurally and functionally similar both to the parent cell and to one another. They are similar, in part, because each new cell receives about half of the parent cell's cytoplasm and organelles. More important, in terms of structure and function, each new cell inherits an *exact* replica of the hereditary information of the parent cell. How cells accomplish this impressive feat is the subject of the rest of this chapter.

At the time of cell division, the nucleus changes its appearance. The threadlike material, the chromatin, condenses into a group of structures, the chromosomes, which appear rodlike under the light microscope. Because of the behavior of these bodies at the time of cell division, it was long suspected that they were the carriers of the hereditary information.

Cell division in eukaryotes consists of two overlapping stages, *mitosis* and *cytokinesis*. In mitosis, the nuclear envelope breaks down; the chromosomes, which have previously been replicated, are apportioned equally; and two new identical nuclei are formed. In cytokinesis, the cytoplasm of the parent cell is divided and packaged into two parts, each containing one of the nuclei.

Prior to mitosis, the chromosomal material is replicated, so that at the beginning of mitosis each chromosome consists of two replicas called *chromatids*. The two identical chromatids remain joined together at a constricted area, the *centromere* (Figure 9–5a). During mitosis, the chromatids are separated, and each of the two daughter cells receives one complete set of the chromosomes that were present in the parent cell. Cytokinesis is not as precise as mitosis. The various cytoplasmic organelles are apportioned more or less equally between the daughter cells.

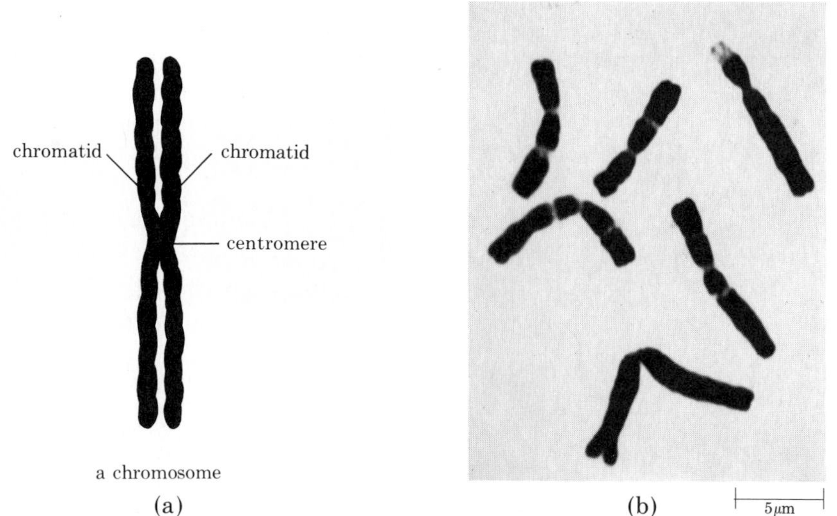

9–5 (a) *A chromosome at the beginning of mitosis. The chromosomal material has replicated, and each chromosome now consists of two identical parts, called chromatids. The centromere, the constricted area at the center, is the site of attachment of the two chromatids.* (b) *If you look carefully, you can see that each of these chromosomes actually consists of two longitudinal halves. Some chromosomes have several constrictions. The centromere, the area of attachment, is usually the constriction nearest the center.*

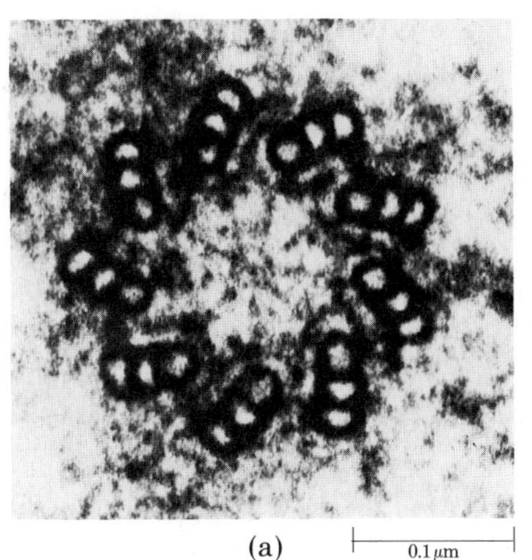

(a)

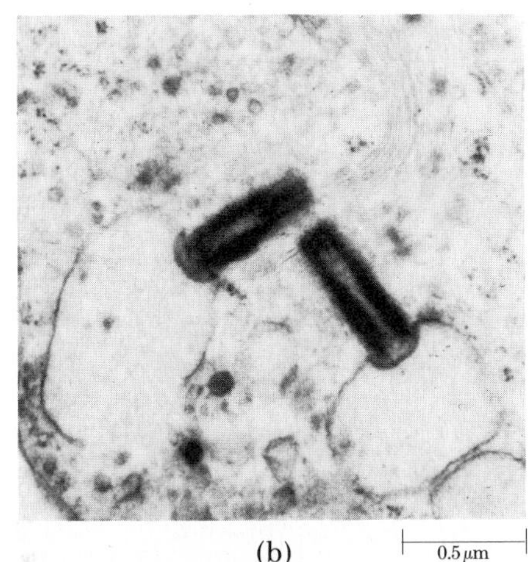

(b)

9–6 (a) *Cross section of a centriole. As we saw in Chapter 4, centrioles and basal bodies are structurally identical. It is hypothesized that basal bodies organize microtubules into cilia and flagella and that, similarly, centrioles organize the microtubules of spindle fibers. However, plant cells that do not have centrioles are also able to form spindles.*

(b) *A centriole pair. Centrioles are usually—but not always—formed from preexisting centrioles, with the newly formed centriole appearing at a right angle to the previously existing one.*

The Phases of Mitosis

The process of mitosis is conventionally divided into four phases: prophase, metaphase, anaphase, and telophase. Of these, prophase is usually by far the longest. If a mitotic division takes 10 minutes (which is about the minimum time required), prophase will last about 6 minutes. The schematic drawings that follow show mitosis as it occurs in an animal cell.

Between divisions, the cell is said to be in interphase, which is not considered one of the stages of mitosis. Although little can be seen in the nucleus during interphase, an essential step in the preparation for mitosis occurs during this period: The chromosomes replicate.

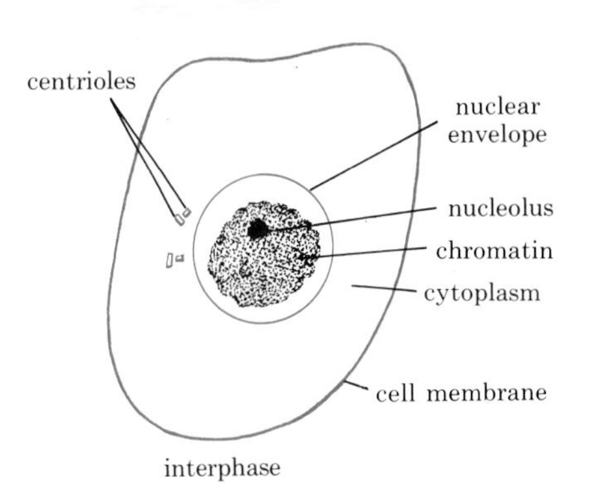

interphase

In early *prophase*, the chromatin condenses, and the individual chromosomes begin to become visible. Each chromosome consists of two duplicate chromatids pressed closely together longitudinally and connected at the centromere (Figure 9–5b). In most cells (higher plants are the principal exception), two pairs of centrioles can be seen at one side of the nucleus, outside the nuclear envelope. Each pair consists of one mature centriole and a smaller, newly formed centriole lying at right angles to the first.

During prophase, the centriole pairs move apart. Between the centriole pairs, forming as they separate (or perhaps separating them as they form), are *spindle fibers* made up of microtubules (see page 70) and other proteins. Additional fibers, known collectively as the aster, radiate outward from the centrioles. The nucleolus usually disperses, disappearing from view. The nuclear envelope breaks down, marking the end of prophase.

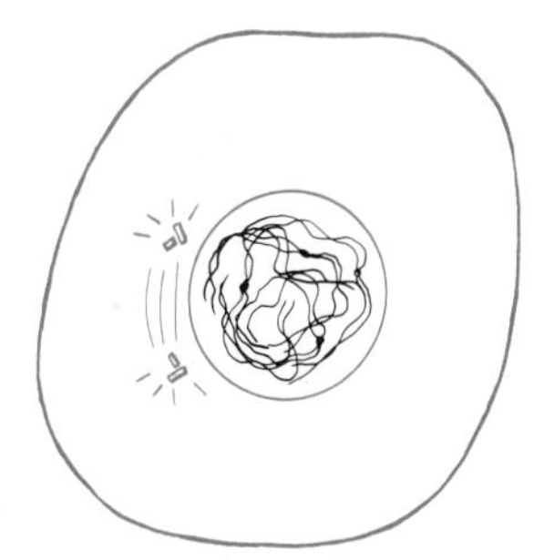
early prophase

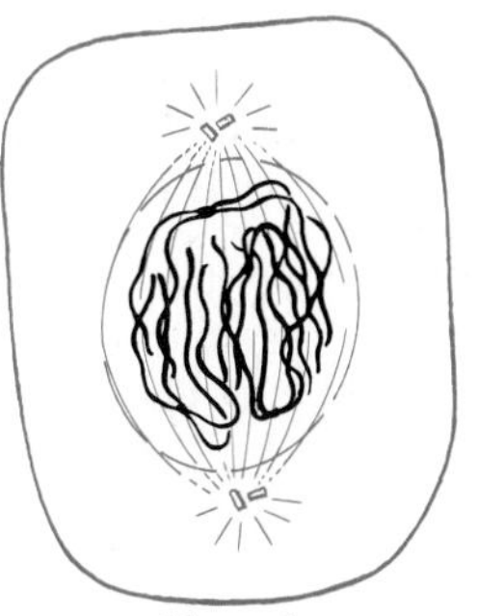
mid-prophase

By the end of prophase, the centriole pairs are at opposite ends of the cell, and the members of each pair are of equal size. The spindle is fully formed. It is a three-dimensional football-shaped structure consisting of three groups of microtubules: (1) astral rays; (2) continuous fibers reaching from pole to pole; and (3) shorter fibers that are attached to disk-shaped structures, called *kinetochores*, within the centromere of each chromatid pair (Figure 9–7).

During early *metaphase* the chromatid pairs move back and forth within the spindle, apparently maneuvered by the spindle fibers. They appear to be tugged first toward one pole and then the other. Finally they become arranged precisely at the midplane (equator) of the cell. This marks the end of metaphase.

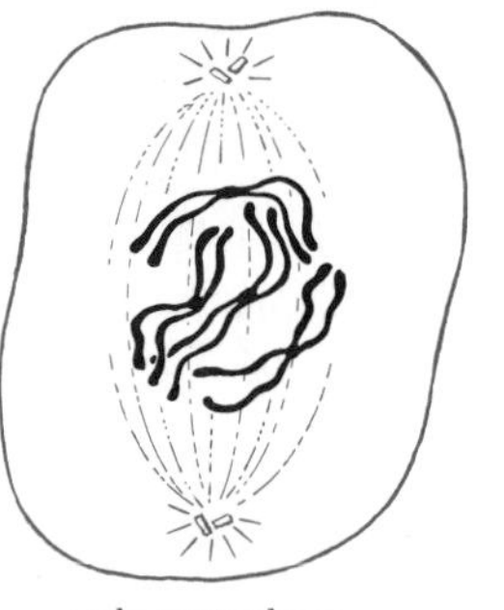

early metaphase

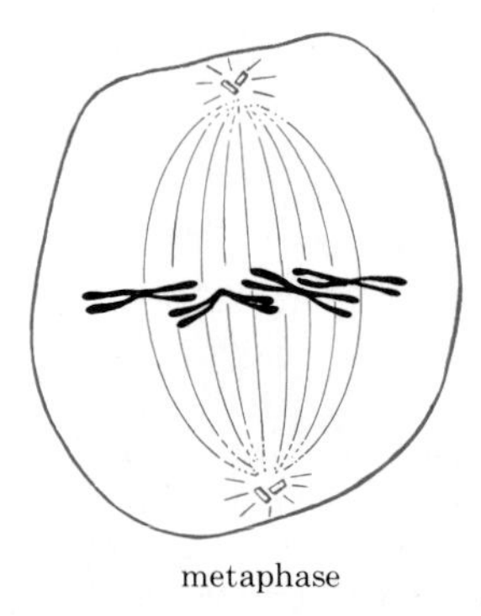

metaphase

At the beginning of *anaphase*, the chromatid pairs separate into individual chromosomes. The two duplicate chromosomes move apart, apparently drawn toward the poles by the spindle fibers. The centromeres move first, while the arms of the chromosomes seem to drag behind. As anaphase continues, the two identical sets of newly separated chromosomes move toward the opposite poles of the spindle.

By the beginning of *telophase*, the chromosomes have reached the opposite poles and cytokinesis has begun. The spindle disperses. During late telophase, nuclear envelopes form around the two sets of chromosomes, which once more become diffuse. Often, a new centriole forms adjacent to each of the previous ones, so that each daughter cell has a new centriole pair. In each nucleus, the nucleolus re-forms.

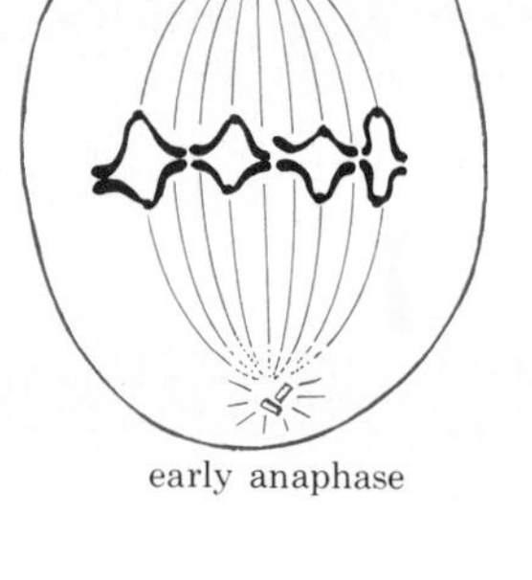

early anaphase

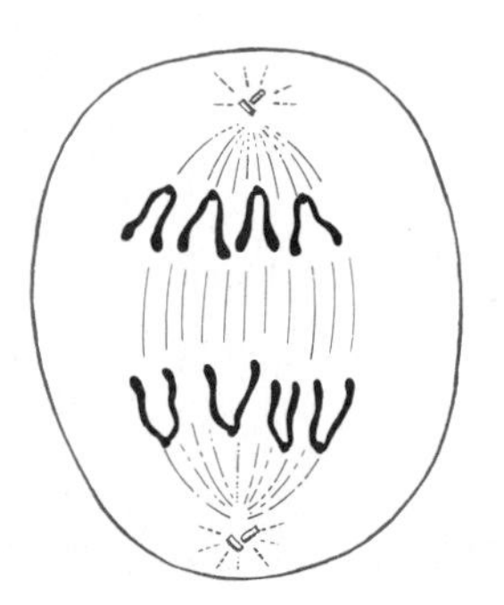

late anaphase

early telophase

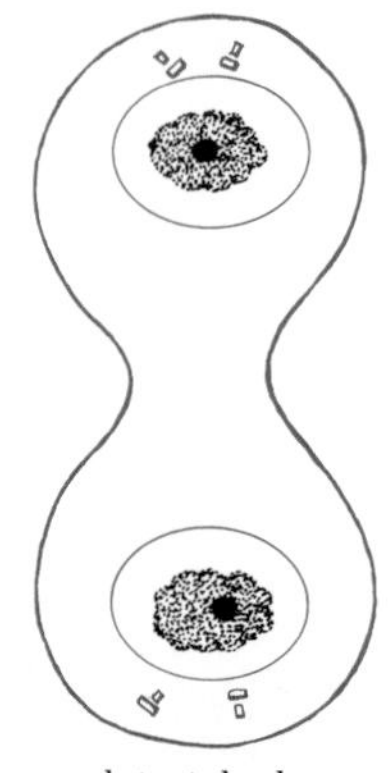

late telophase

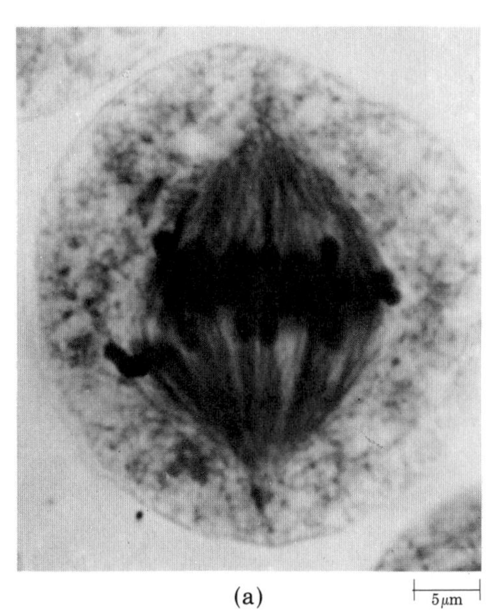

(a)

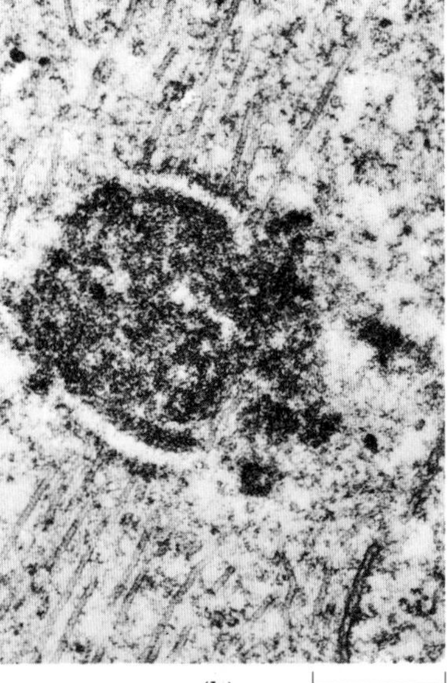

(b)

9-7 (a) *This micrograph of the spindle in a wheat cell illustrates the spindle's three-dimensional quality. The gray streaks are the spindle fibers. The chromosomes are the large dark bodies near the equator of the spindle.*

(b) *In this electron micrograph, spindle fibers can be seen extending to the kinetochores of a green alga. The dark material is the metaphase chromosome, most of which is out of the plane of the thin section prepared for this micrograph. The kinetochores are the two arc-shaped areas at either side of the chromosome.*

(a)

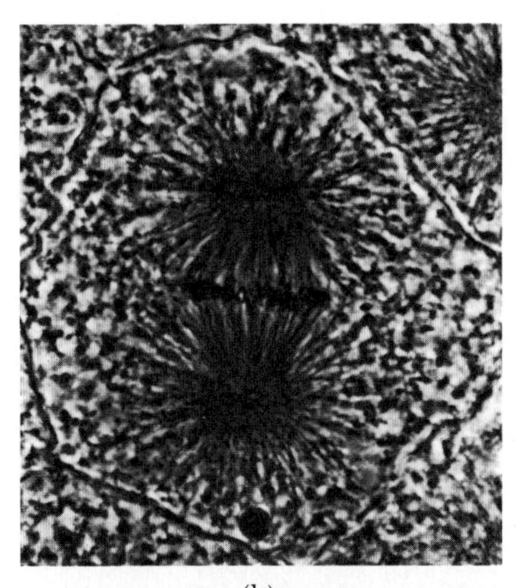

(b)

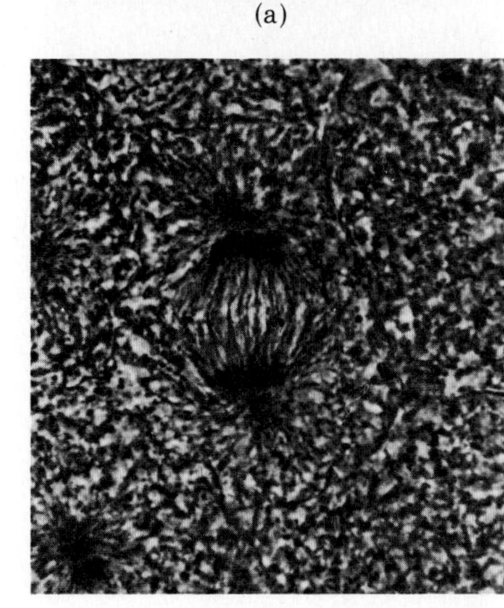

(c)

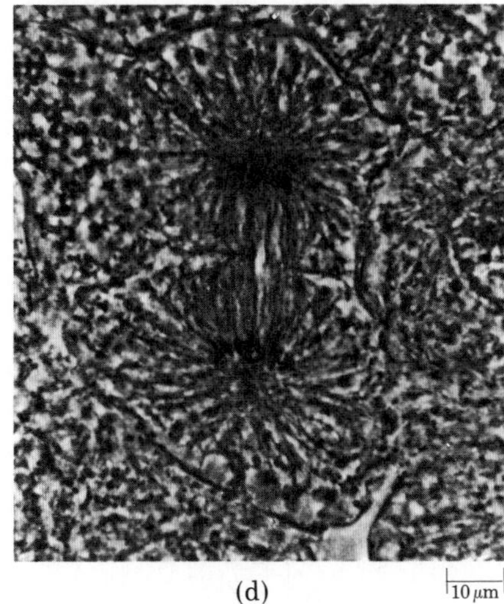

(d)

9-8 *Mitosis in embryonic cells of a whitefish.* (a) *Prophase. The chromosomes have become visible, the nuclear envelope has broken down, and the spindle apparatus has formed.* (b) *Metaphase. The chromatid pairs, perhaps guided by the spindle fibers, are lined up at the equator of the cell. Some of them appear to have begun to separate.* (c) *Anaphase. The two sets of chromosomes are moving apart.* (d) *Telophase. The chromosomes are completely separated, the spindle apparatus is disappearing, and a new cell membrane is forming that will complete the separation of the two daughter cells.*

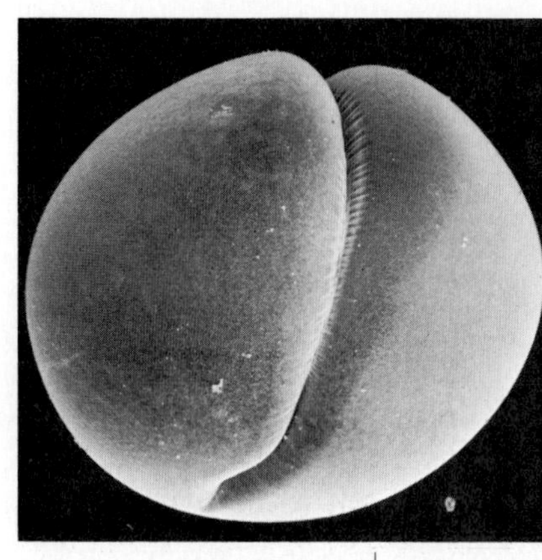

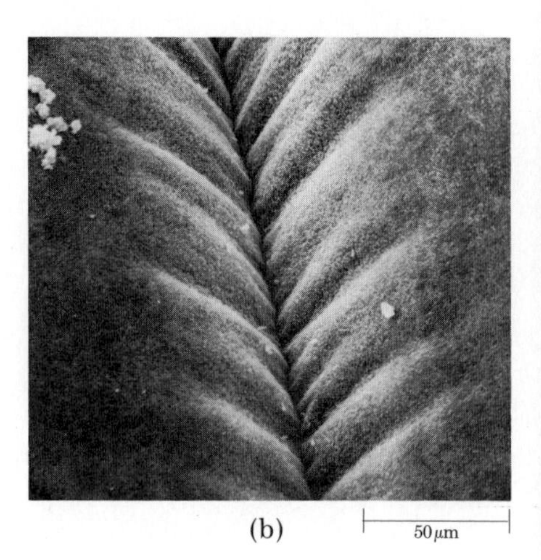

9-9 (a) *Cytokinesis in an animal cell, the egg of a frog.* (b) *Note the constriction furrows.*

Cytokinesis

Cytokinesis usually begins during telophase of mitosis. It differs in some respects in plant and animal cells. In animal cells, the cell membrane constricts along the surface of the cell in the area where the equator of the spindle used to be. At first a furrow appears on the surface, and this gradually deepens into a groove (Figure 9-9). Eventually the connection between the daughter cells dwindles to a slender thread, which soon parts.

In plant cells, the cytoplasm is divided by the formation of a *cell plate* (Figure 9-10), which is formed from a series of vesicles produced from Golgi bodies. The vesicles eventually fuse to form a flat, membrane-bound space, the cell plate. As more vesicles fuse, the edges of the growing plate fuse with the membrane of the cell. In this way, a space is established between the two daughter cells, completing their separation. Cellulose is then laid down against the membranes, each new cell forming its own cell wall. The space itself eventually is impregnated with pectins and becomes the middle lamella (see page 64).

When cell division is complete, two daughter cells are produced, indistinguishable from each other and from the parent cell.

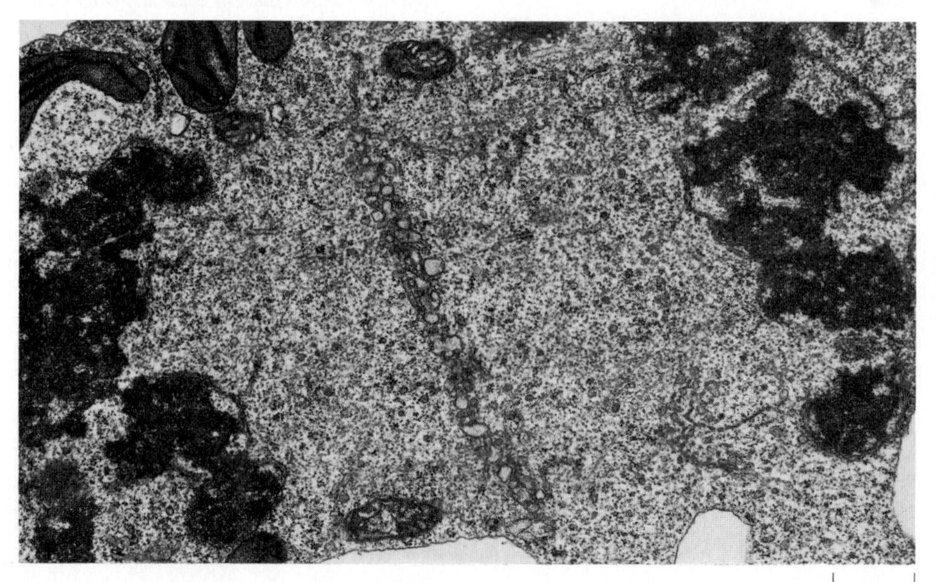

9-10 *In plants, the separation of the two cells is effected by the formation of a structure known as the cell plate. Vesicles appear across the equatorial plane of the cell and gradually fuse, forming a flat, membrane-bound space, the cell plate, which extends outward until it reaches the wall of the dividing cell. The large, dark forms on either side of the micrograph are the chromosomes.*

The Cell Cycle

Dividing cells pass through a regular sequence of cell growth and division, known as the *cell cycle* (Figure 9–11). Completion of the cycle requires varying periods of time from a few hours to several days, depending on both the type of cell and external factors, such as temperature or available nutrients.

The S (synthesis) phase of the cell cycle is the period during which the chromosomes are replicated. G (gap) phases precede and follow the S phase. The G_1 phase occurs after mitosis and precedes the S phase; the G_2 phase follows the S phase and occurs before mitosis. These are periods of intensive biochemical activity, during which the cell doubles in size and its enzymes, ribosomes, mitochondria, and other molecules and organelles approximately double in number. The G and S phases together are referred to as interphase. Mitosis occupies the rest of the cell cycle.

Some cells pass through successive cell cycles throughout the life of the organism. Single-celled organisms and certain cells in growth centers of both plants and animals fall into this category. An example is the cells in the human bone marrow that give rise to red blood cells. The average red blood cell lives only about 120 days, and there are about 25 trillion (2.5×10^{13}) of them in an adult. To maintain this number, about 2.5 million new red blood cells must be produced by cell division each second. At the other extreme, some highly specialized cells, such as nerve cells, lose their capacity to replicate once they are mature. A third group of cells retains the capacity to divide but does so only under special circumstances. Cells in the human liver, for example, do not ordinarily divide, but if a portion of the liver is removed surgically, the remaining cells (even if as few as a third of the total remain) continue to replicate until the liver reaches its former size. Then they stop.

When cells stop growing, they stop in the G_1 phase. It is hypothesized that substances are synthesized during the G_1 phase that either inhibit or stimulate the S stage and the rest of the cycle, thus determining whether or not cell division will occur. Knowledge of the control mechanisms involved would not only be interesting biologically but also might be of great importance in the control of cancer. Cancer cells differ from normal cells largely in that the cancer cells keep on dividing at the expense of the host tissues.

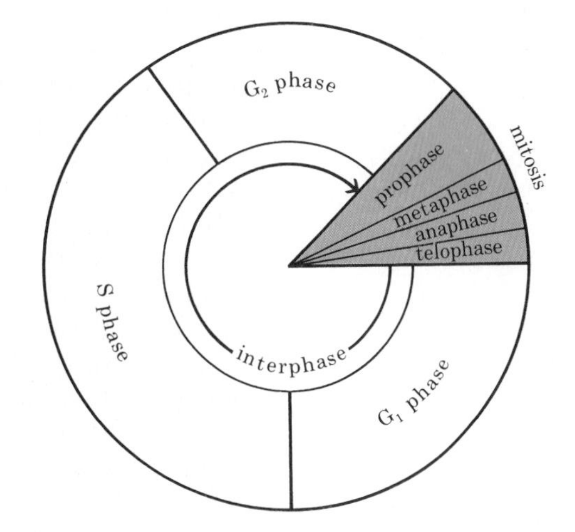

9–11 *The cell cycle. Dividing cells go through four principal phases, including the stages of mitosis (shown in color) and the S (synthesis) phase, during which the chromosomal material is replicated. Separating mitosis and the S phase are two G phases. The first of these (G_1) is a period of general growth and replication of cytoplasmic organelles. During the second (G_2), structures directly associated with mitosis, such as the spindle fibers, are synthesized. After the G_2 phase comes mitosis, which is, in turn, divided into four stages. In cells of different species, the different phases occupy different proportions of the total cycle.*

9-12 *In this multicellular animal, a* Hydra, *a new organism is developing on the body of the parent. It will eventually break away and live independently. The chromosomes in its cells are identical to those in the cells of the parent* Hydra. Hydra *is a common inhabitant of freshwater lakes and ponds.*

Mitosis and Reproduction

In one-celled eukaryotic organisms, mitosis is the key event in reproduction—it is the means by which exact replicas of the chromosomes are transmitted from parent to offspring. Mitosis plays the same essential role in the reproduction of some large organisms. Sea anemones and sponges, for example, may break apart to form new sea anemones and sponges. Plants may produce roots or runners from which new individuals arise. And in some organisms, such as *Hydra* (Figure 9-12), a new individual may develop on the body of the parent. Reproduction in which exact replicas of the chromosomes are transmitted from parent to offspring by mitosis is known as *asexual reproduction.* It is also sometimes called vegetative reproduction, since it is particularly common among plants.

Although all multicellular organisms produce new cells—and thus grow—by mitosis and some produce entirely new individuals by mitosis, most multicellular organisms arise from a unique single cell—the fertilized egg. This cell is formed by the fusion of two cells, the sperm and the egg. How are these cells formed? And what process ensures that the new organism receives a full complement of chromosomes, with exactly one half from the father and one half from the mother? To answer these questions, in the next chapter we shall look at another kind of nuclear division, meiosis.

SUMMARY

The nucleus is the most prominent structure within most eukaryotic cells. It is surrounded by a double membrane, the nuclear envelope. Within it are the nucleolus and the chromosomes.

Cell division is an orderly process that includes division of the nucleus (mitosis), during which the chromosomes are apportioned between the two daughter cells, and division of the cytoplasm (cytokinesis). When the cell is in interphase (not dividing), the chromosomes are visible only as thin strands of threadlike material (chromatin). During this period, if mitosis is to take place, the chromosomal material is replicated. As mitosis begins, the chromatin condenses into pairs of identical chromatids held together at the centromere. At the end of mitosis, the chromatids separate, and each chromatid, now an independent chromosome, moves to one of the daughter cells. In this way, the chromosomes are equally divided between the two new cells.

Cytokinesis in animal cells results from constrictions in the cell membrane between the two nuclei. In plant cells, the cytoplasm is divided by the coalescing of vesicles to form the cell plate. In both cases, the result is the production of two new, separate cells. As a result of mitosis, each has received an exact copy of the chromosomes of the parent cell, and, as a result of cytokinesis, approximately half of the cytoplasm and organelles.

Dividing cells pass through a regular sequence of cell growth and cell division known as the cell cycle. It consists of two G (gap) phases, an S (synthesis) phase, and mitosis.

In multicellular organisms, mitosis is the key event in the production of new cells for growth and replacement. In one-celled organisms and some many-celled organisms, it is also the means by which exact replicas of the chromosomes are transmitted from parent to offspring in reproduction.

QUESTIONS

1. Distinguish among the following: mitosis/cytokinesis; cell division/cell cycle; centriole/centromere/kinetochore; chromatid/chromosome.

2. The drawings below show stages in mitosis in a plant cell. Identify each stage and describe what is happening.

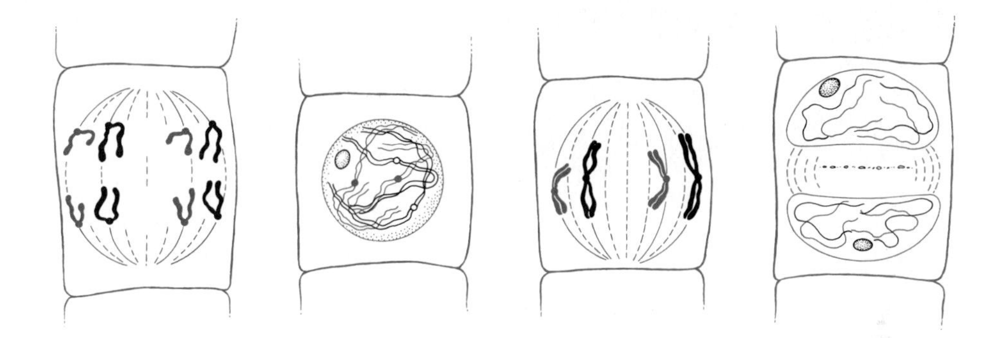

3. What is a chromosome? How is it related to chromatin?

4. Why do we often refer to chromatids as "sister chromatids"? When are sister chromatids formed? How? When do they first become visible under the microscope?

5. In what ways does cell division in plant cells differ from that in animal cells?

6. What is the function of cell division in the life of an organism? Suppose you, as an organism, were made up of a large single cell rather than trillions of small ones. How would you differ from your present self?

10

Meiosis and Sexual Reproduction

Most multicellular organisms, including, for example, pea plants, sea urchins, and human beings, reproduce sexually. Sexual reproduction generally requires two parents, each of which contributes genetic information to the new offspring. It always involves two events: *fertilization* and *meiosis.* Fertilization is the means by which the different genetic contributions of the two parents are brought together to form the new genetic identity (genome) of the offspring. Meiosis is a special kind of nuclear division that evolved from mitosis and uses much of the same machinery. However, as you will see, meiosis differs from mitosis in some important respects.

Haploid and Diploid

To understand meiosis, we must look again at the chromosomes. Every organism has a chromosome number characteristic of its particular species. A mosquito has 6 chromosomes per cell; a cabbage, 18; corn, 20; a cat, 38; a human being, 46; a plum, 48; a dog, 78; and a goldfish, 94.

However, the *gametes* (from the Greek word *gamos,* "to marry")—that is, the male and female sex cells, the sperm and the eggs—have exactly half the number of chromosomes that is characteristic of the body (somatic) cells of an organism. The number of chromosomes in the gametes is referred to as the *haploid* ("single") number and the number in the somatic cells as the *diploid* ("double") number.

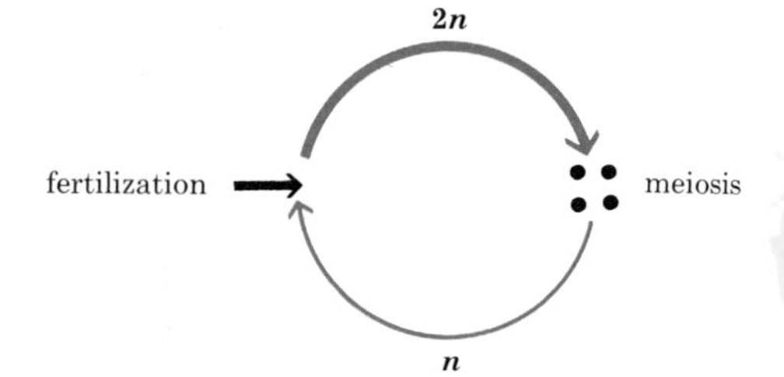

10–1 *Sexual reproduction is characterized by two events: the coming together of the sex cells, or gametes (fertilization), and meiosis. Following meiosis, the number of the chromosomes is single, or haploid* (n). *Following fertilization, the number is double, or diploid* (2n).

For brevity, the haploid number is designated n and the diploid number as $2n$. In humans, for example, $n = 23$ and $2n = 46$. When a sperm fertilizes an egg, the two haploid nuclei fuse, $n + n = 2n$, and the diploid number is restored (Figure 10–1). The diploid cell produced by the fusion of two gametes is known as a *zygote.*

In every diploid cell, each chromosome has a partner. These pairs of chromosomes are known as homologous pairs, or *homologues.* The two resemble each other in size and shape. Normally, one homologue is derived from one parent, and its partner is derived from the other parent.

In the special kind of nuclear division called meiosis, the diploid number of chromosomes is reduced to the haploid number. If this reduction did not occur, the chromosome number would double each generation. Moreover, as we shall see, meiosis is in itself a source of new chromosome combinations.

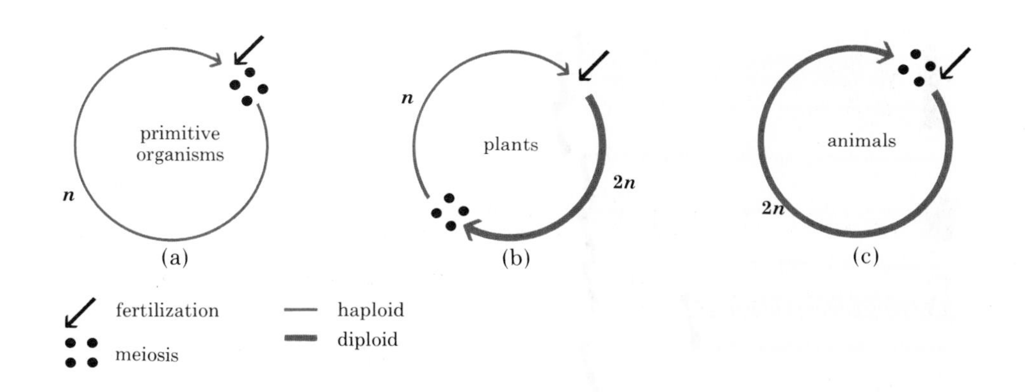

10-2 *Fertilization and meiosis occur at different points in the life cycle of different organisms.* (a) *In some one-celled organisms, meiosis occurs immediately after fertilization, and most of the life cycle is spent in the haploid state (signified by the thin line).* (b) *In plants, fertilization and meiosis are separated, and the organism characteristically has a diploid phase and a haploid phase.* (c) *In animals, meiosis is immediately followed by fertilization. As a consequence, during most of the life cycle the organism is diploid (signified by the thick line).*

Meiosis and the Life Cycle

Meiosis occurs at different times during the life cycle of different organisms (Figure 10-2). In the alga *Chlamydomonas* it occurs immediately after fusion of the mating cells (Figure 10-3). The cells are ordinarily haploid, and meiosis restores the haploid number.

In plants, such as ferns, a haploid phase typically alternates with a diploid phase (Figure 10-4). The common and conspicuous form of a fern is the *sporophyte*, the diploid organism. By meiosis, fern sporophytes produce spores, usually on the undersides of their fronds. These spores have only the haploid number of chromosomes. They germinate to form much smaller plants (*gametophytes*), typically only a few cell layers thick. In these plants all the cells are haploid. The small, haploid plants produce gametes, which fuse and then develop into a new, diploid sporophyte. This process, in which a haploid stage is followed by a diploid stage and again by a haploid stage, is

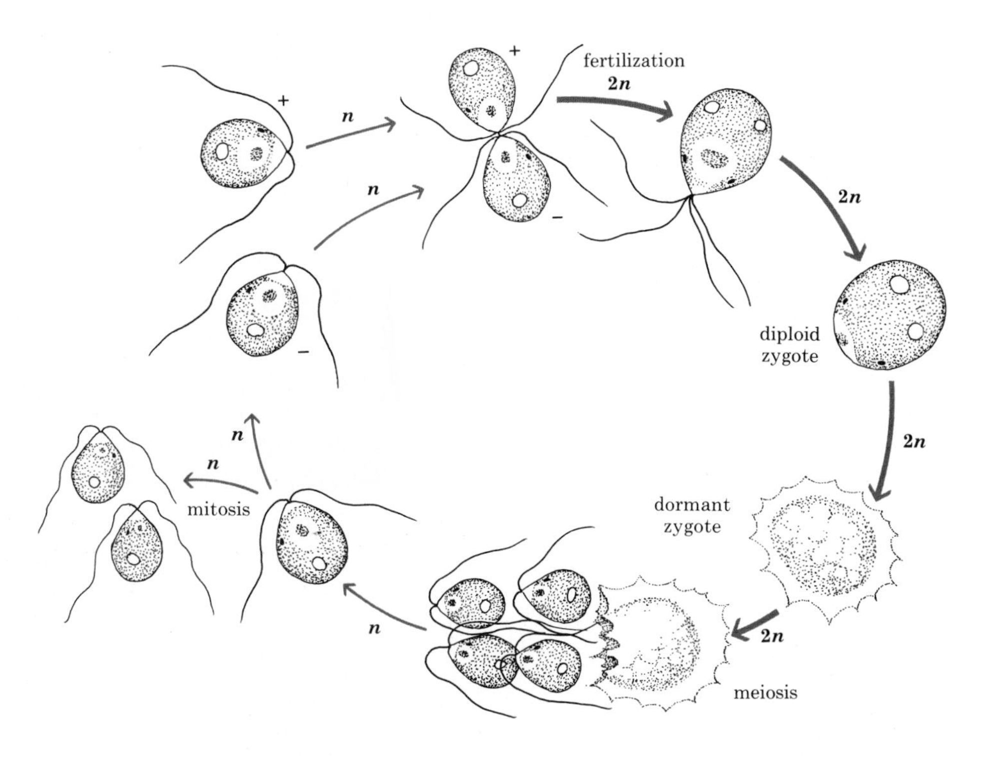

10-3 *The life cycle of* Chlamydomonas *is of the type shown in Figure 10-2a. The organism is haploid for most of its life cycle (thin arrows). Fertilization, the fusing of cells of different mating strains (indicated here by + and −), temporarily produces the diploid state (thick arrows). The diploid zygote divides meiotically, forming four new haploid cells, which will probably divide repeatedly by mitosis before entering another sexual cycle.*

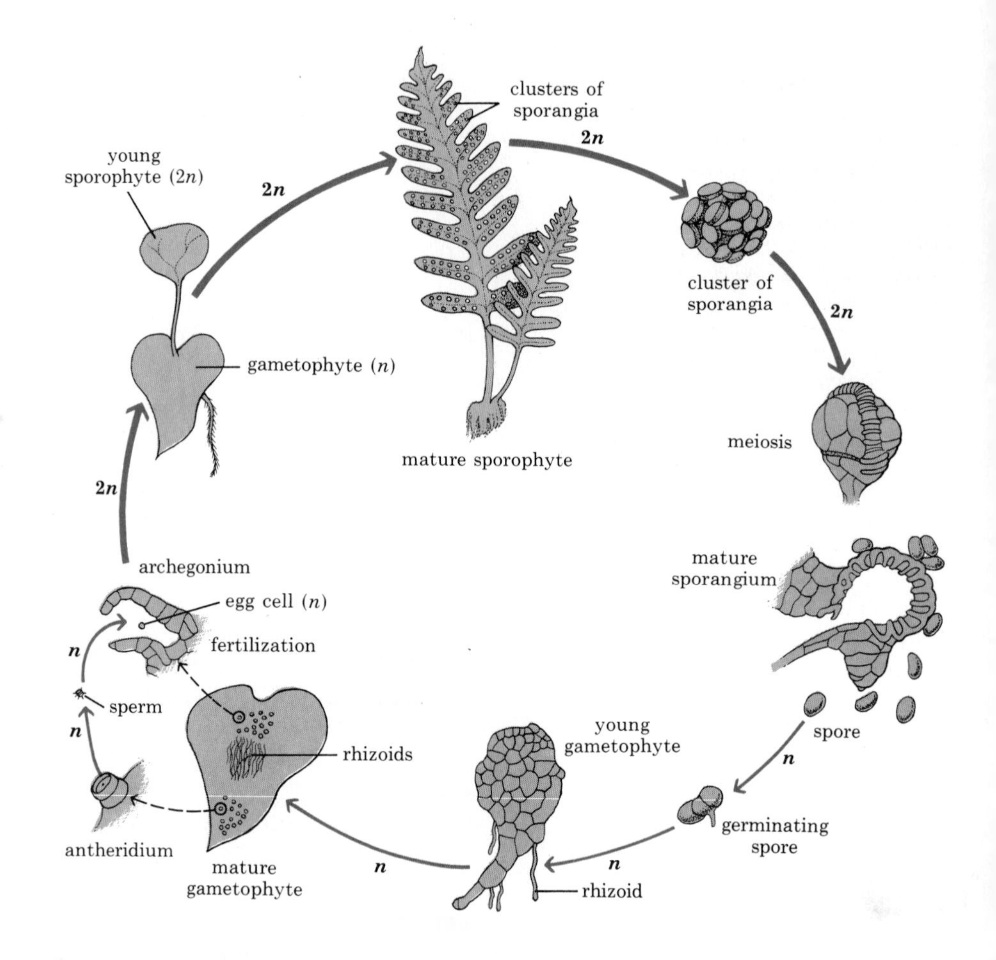

10–4 *The life cycle of a fern is of the type shown in Figure 10–2b. Following meiosis, spores, which are haploid, are produced in the sporangia and then are shed (extreme right). The spores develop into haploid gametophytes. In many species, the gametophytes are only one layer of cells thick and are somewhat heart-shaped, as shown here (bottom). From the lower surface of the gametophyte, filaments, the rhizoids, extend downward into the soil.*

On the lower surface of the gametophyte are borne the flask-shaped archegonia, which enclose the egg cells, and the antheridia, which enclose the sperm. When the sperm are mature and there is an adequate supply of water, the antheridia burst, and the sperm cells, which have numerous flagella, swim to the archegonia and fertilize the eggs. From the fertilized (2n) *egg, the zygote, the* 2n *sporophyte grows out of the archegonium within the gametophyte. After the young sporophyte becomes rooted in the soil, the gametophyte disintegrates. When it becomes mature, the sporophyte develops sporangia, in which meiosis occurs, and the cycle begins again.*

known as *alternation of generations.* As we shall see in Chapter 19 (page 251), alternation of generations occurs in all plants, although not always in the same form.

Human beings have the typical animal life cycle, in which meiosis immediately precedes fertilization (Figure 10–5). Virtually all of the life cycle is spent in the diploid state.

The Phases of Meiosis

The events that take place during meiosis somewhat resemble those that take place during mitosis. But there are some important differences:

1. A single cell undergoes two successive nuclear divisions—designated meiosis I and meiosis II—producing a total of four cells.
2. Each of the four cells contains half the number of chromosomes present in the original cell.
3. The haploid cells produced by meiosis contain new combinations of chromosomes.

In the following discussion we shall describe meiosis in a plant cell in which the diploid number is 8 ($n = 4$). Four of the eight chromosomes were inherited from one parent and four from the other parent. For each chromosome from one parent there is a homologue from the other parent—a chromosome of similar size and shape. During interphase the chromosomes are replicated, so that by the beginning of meiosis I each chromosome consists of two identical chromatids held together at the centromere.

10-5 *The life cycle of* Homo sapiens. *Gametes, which are haploid, are produced by meiosis. At fertilization, they fuse, restoring the diploid number in the fertilized egg. The fertilized egg develops into a mature man or woman, who again produces haploid gametes. As is the case with most other animals, the life cycle is almost entirely diploid, the only exception being the gametes. This is the type of life cycle diagrammed in Figure 10-2c.*

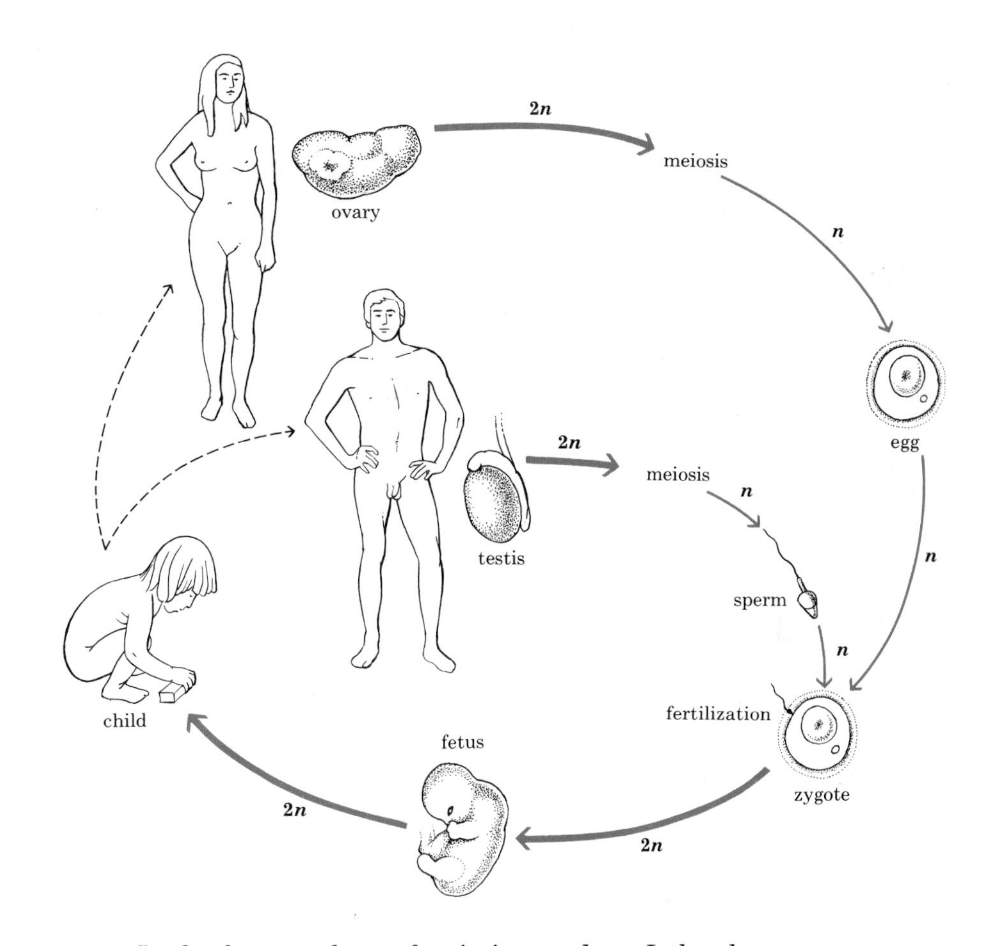

In the first prophase of meiosis, *prophase I*, the chromosomes come into view and the homologous chromosomes come together in pairs. Since each chromosome consists of two identical chromatids, the pairing of the homologous chromosomes actually involves four chromatids (Figure 10-6a). Thus a homologous pair is also known as a tetrad (from the Greek *tetra*, meaning "four"). *Crossing over*, an interchange of segments of one chromosome with corresponding segments from its homologue, can take place at this time (Figure 10-6b). As we shall see in Chapter 12, such exchanges of chromosomal material are important sources of variation in the hereditary material.

10-6 (a) *During prophase I of meiosis, chromosomes become arranged in homologous pairs. Each homologous pair consists of four chromatids and is therefore also known as a tetrad.*
(b) *Crossing over. During meiosis I, homologues, paired up as tetrads, connect at crossover points, where exchanges of segments of the chromosomes—crossing over—take place.*

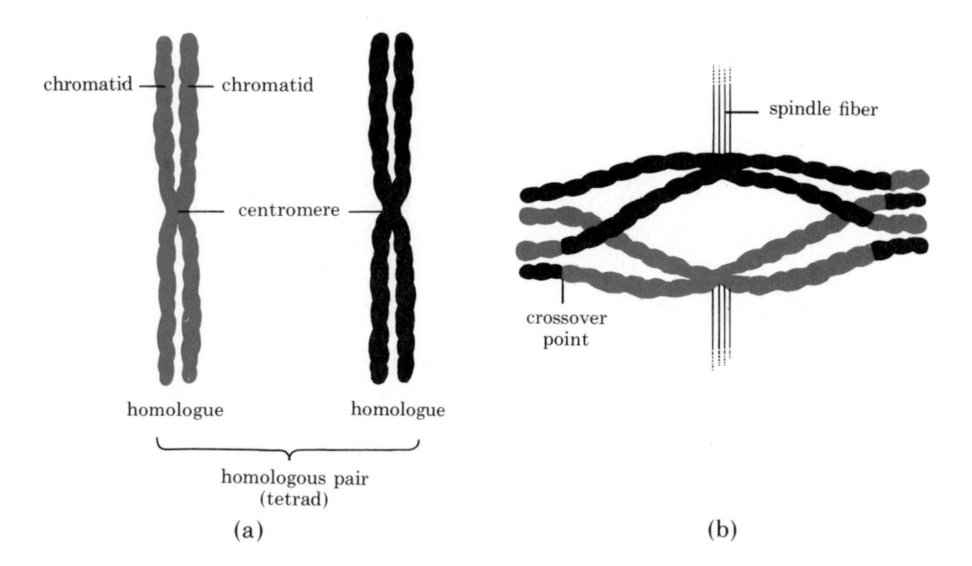

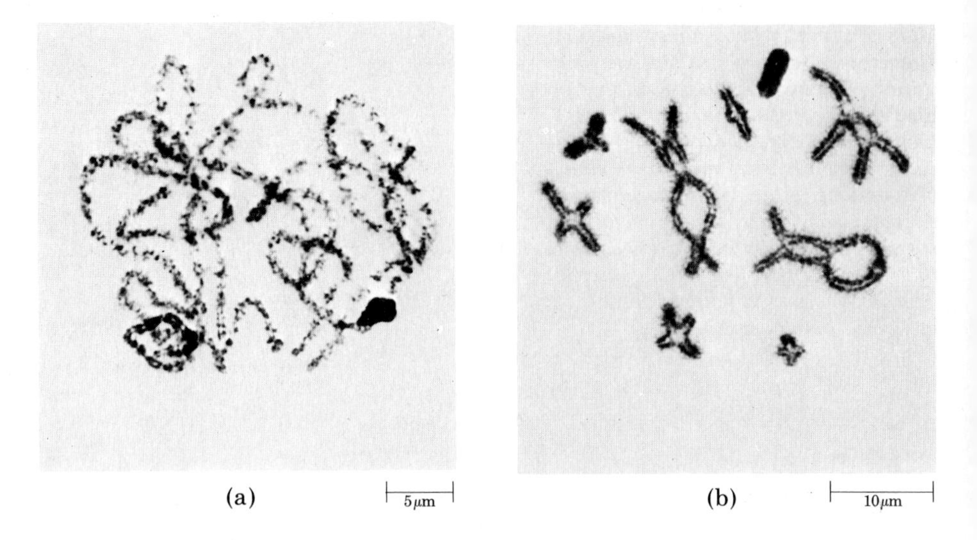

10-7 (a) *Early prophase I in the formation of a sperm cell in a grasshopper. The homologous chromosomes are now paired; the individual chromatids are not visible, however, so each chromosome appears as a single structure, and the tetrads appear double-stranded (rather than four-stranded).*

(b) *Late prophase I. All four chromatids can be seen in most of the tetrads.*

If you remember the arrangement of the homologous chromosomes at this stage of meiosis, you will be able to remember all the subsequent events with little difficulty.

Toward the end of prophase I, the spindle apparatus is formed, the nucleolus disperses, and the nuclear envelope breaks down.

In *metaphase I*, the four homologous pairs line up along the equatorial plane of the cell. Each pair consists of four chromatids.

At *anaphase I*, the homologues, each consisting of two chromatids, separate, as if pulled apart by the spindle fibers attached to their kinetochores. However, the two sister chromatids do not separate as they did in mitosis.

By the end of the first meiotic division, *telophase I*, the homologues have separated. As the cell enters interphase II, the chromosomes disappear from view and nuclear envelopes may re-form.

Each nucleus now contains only half the number of chromosomes as the original cell.* Further, these chromosomes may be different from any

* In counting, it is often difficult to know whether to count a chromosome that has replicated but has not divided as 1 or 2. It is customary to count such a chromosome as 1. The trick is to count centromeres.

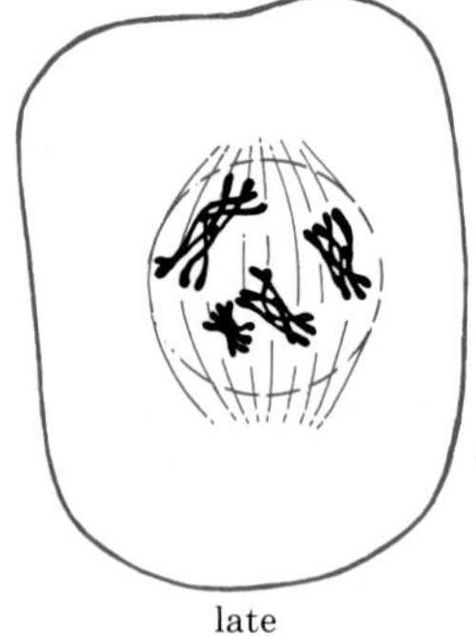

late prophase I

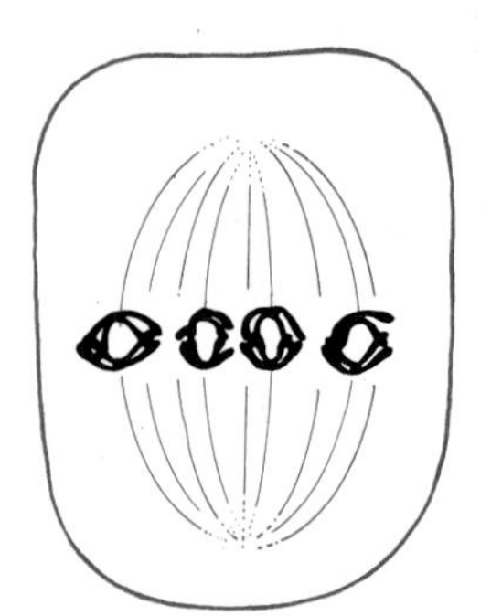

metaphase I

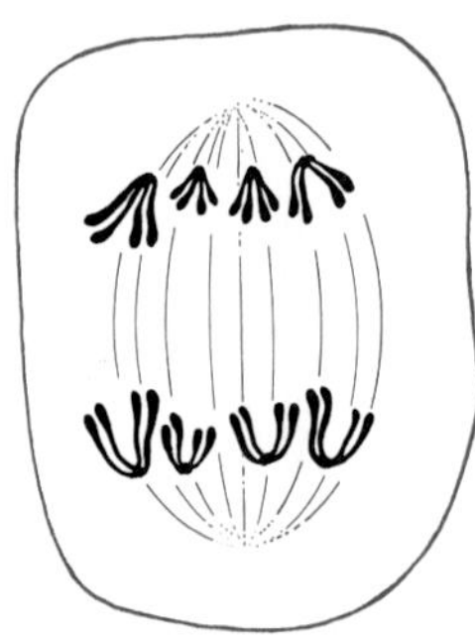

anaphase I

telophase I

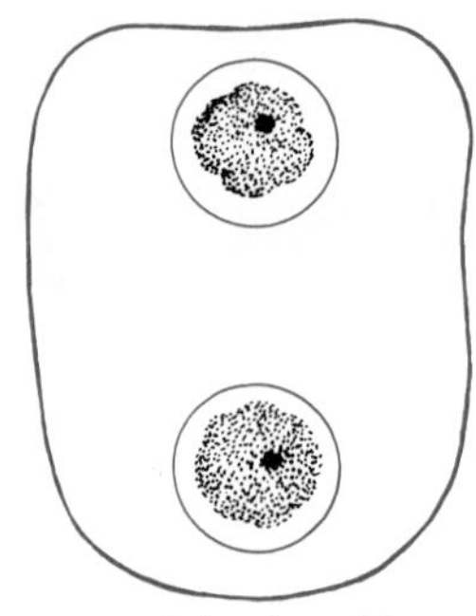

interphase II

10-8 *Anaphase II in wake-robin* (Trillium erectum), *a plant that flowers in the early spring. The chromosomes are almost completely separated. Each diploid nucleus has divided to form four sets of chromosomes.*

Following telophase II and cytokinesis, each of the haploid cells will divide mitotically and differentiate to form a pollen grain. The haploid pollen grain is a microscopic gametophyte and contains sperm cells.

25 μm

one that was present in the original cell because of exchanges taking place during crossing over.

Meiosis II resembles mitosis except that it is not preceded by replication of the chromosomal material. During *prophase II*, the chromosomes condense again. There are four in each nucleus (the haploid number), and they are still in the form of two chromatids held together at the centromere. The nuclear envelopes, if present, dissolve, and new spindle fibers begin to appear.

During *metaphase II*, the four chromatid pairs in each nucleus line up on the equatorial plane. At *anaphase II*, as in mitosis, the sister chromatids separate, and each resulting chromosome moves toward one of the poles.

During *telophase II*, a nuclear envelope forms around each set of chromosomes. There are now four nuclei in all, each containing the haploid number of chromosomes. Cell division (cytokinesis) proceeds as it does following mitosis. The cells then begin to differentiate into spores (in plants) or gametes (in animals).

So, beginning with one cell containing eight chromosomes (four homologous pairs), we end with four cells, each with four chromosomes (no homologous pairs).

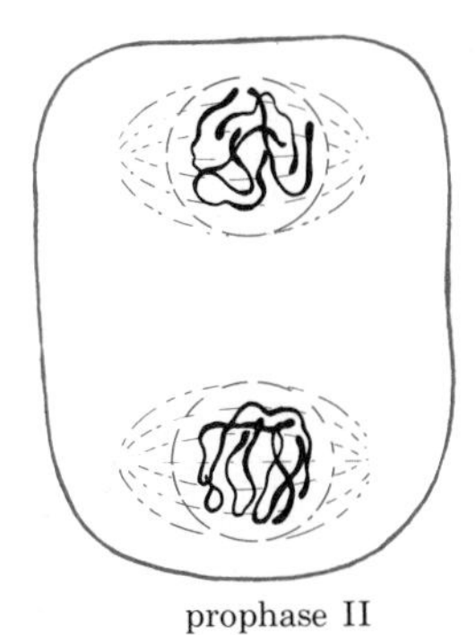

prophase II

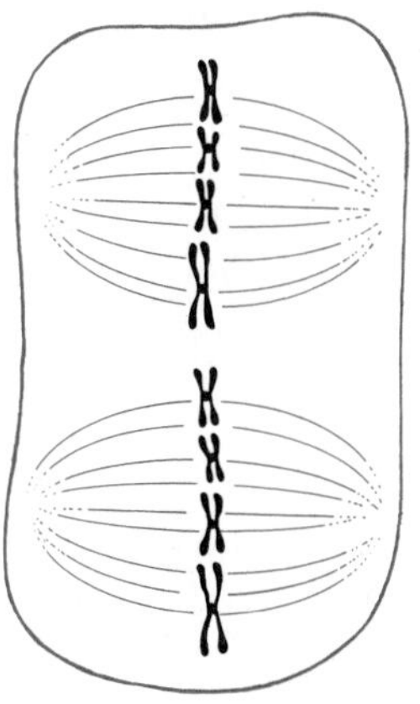

metaphase II

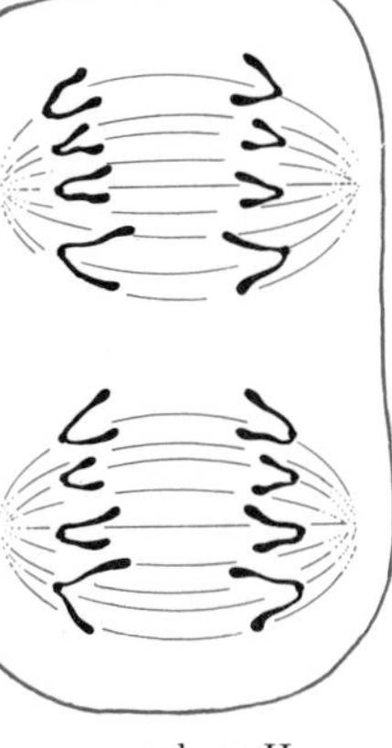

anaphase II

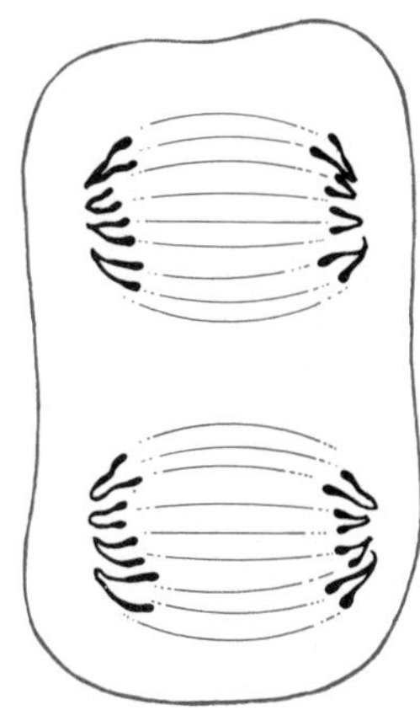

telophase II

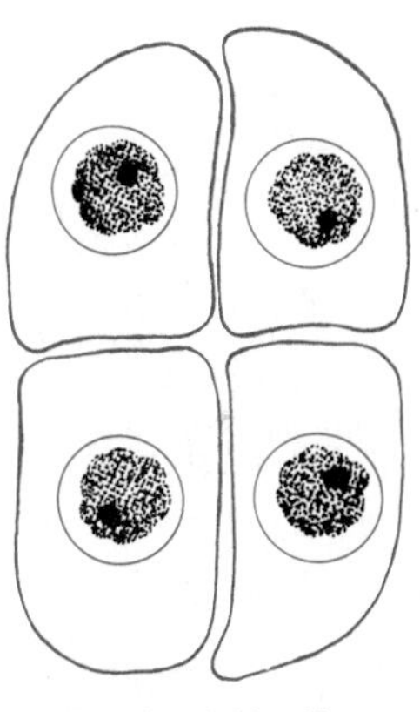

four haploid cells

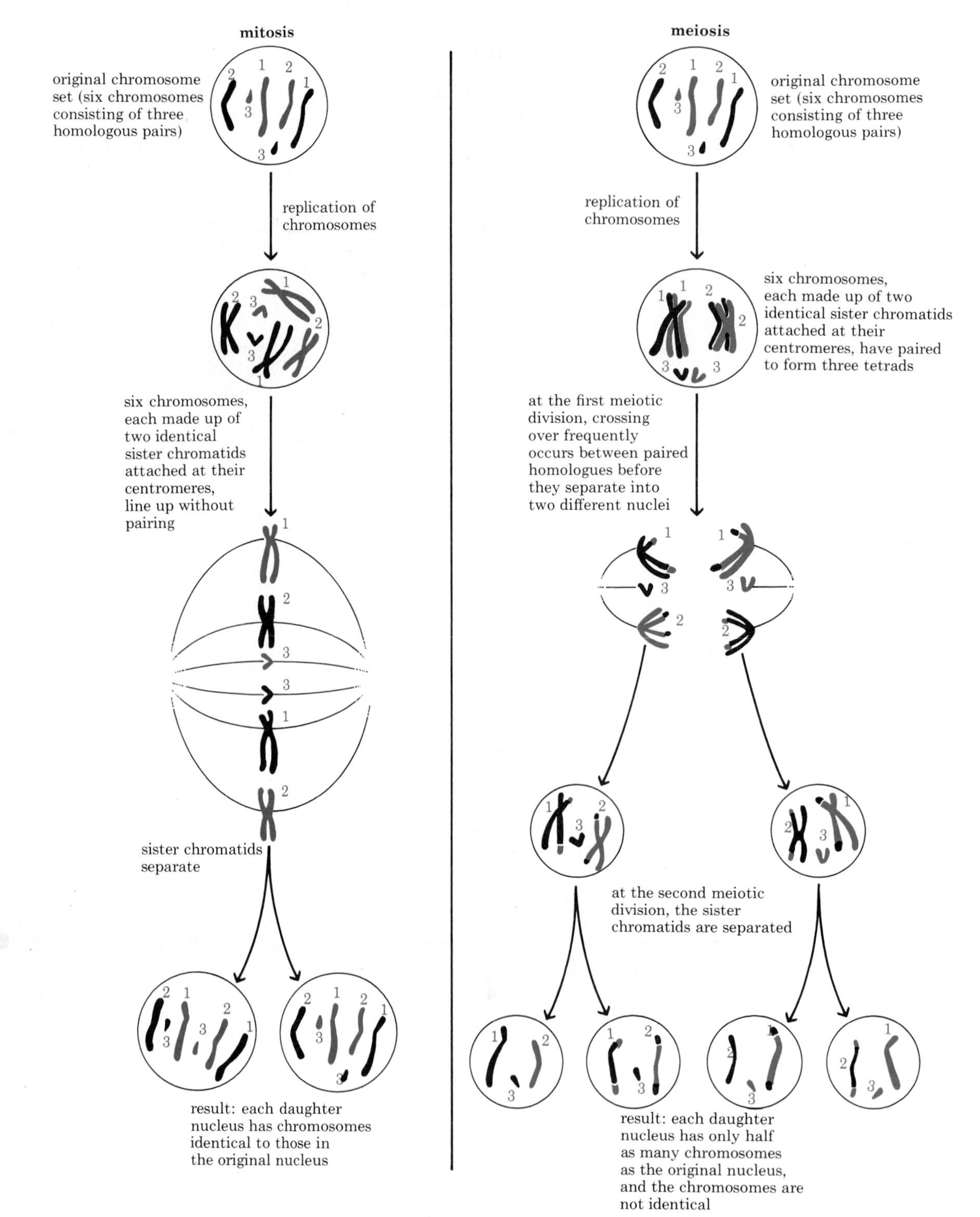

10-9 *A comparison of mitosis and meiosis. In these examples, the organism has six chromosomes* (2n = 6).

Meiosis in the Human Species

In our own species, meiosis takes place in the reproductive organs, the testes of the male and the ovaries of the female. In the male, a cell known as a primary spermatocyte undergoes the two divisions of meiosis to produce four spermatids, which then differentiate into sperm (see Figure 10–10).

In the female, the meiotic divisions are equal in terms of the chromosomes, but unequal in the way the cytoplasm is apportioned (Figure 10–11). One egg cell (ovum) is produced along with two or three polar bodies, which contain the other nuclei formed by meiosis. These then disintegrate. As a result of this unequal division, the ovum is well supplied with essential materials for the developing embryo.

10–10 *The series of changes resulting in the formation of sperm cells begins with the growth of spermatogonia into large cells known as primary spermatocytes. At the first meiotic division, each primary spermatocyte divides into two haploid secondary spermatocytes. The second meiotic division results in the formation of four haploid spermatids. The spermatids differentiate into functional sperm. For simplicity, only six* (n = 3) *chromosomes are shown.*

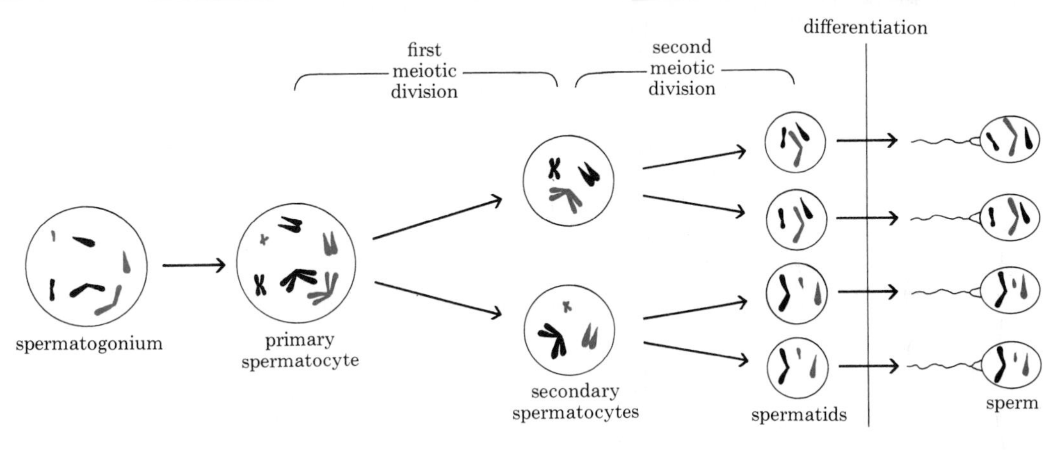

10–11 *Formation of the ovum. A primary oocyte undergoes a meiotic division to produce a secondary oocyte and a polar body. This first meiotic division begins in the human female during the third month of fetal development and ends at ovulation, which may take place 50 years later. The second meiotic division, which produces the egg cell and a second polar body, does not take place until after fertilization. The first polar body may also divide.*

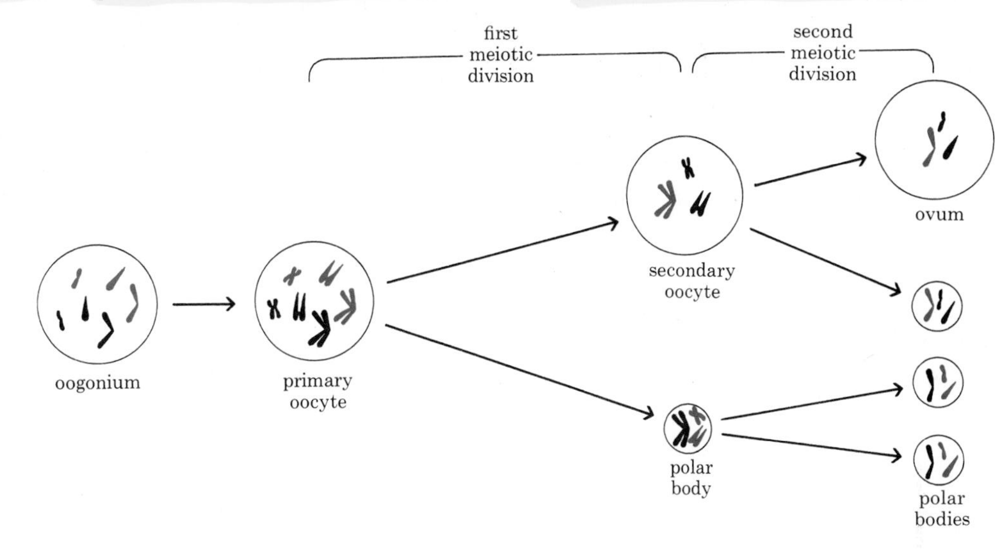

The Function of Sexual Reproduction

Sexual reproduction demands a tremendous expenditure of energy on the part of the organisms involved. Male animals often produce thousands or even millions of sperm cells for each one that reaches an egg. Plants cover themselves with flowers. Birds invest in bright, improbable plumage. Members of the human species write poems, sing songs, and renounce kingdoms, all in the cause of sexual reproduction. Yet, as we saw in the last chapter, organisms can reproduce very adequately by asexual means (mitosis). What, then, is the function of sexual reproduction?

The answer that biologists offer to this question is that sexual reproduction provides a source of variations in a population. Because of the reshuffling and crossing over of chromosomes that takes place at meiosis, as well as the coming together of gametes at fertilization, sexually produced individuals are different from their parents. The resulting variations provide the rich store of material on which natural selection can operate. The fact that all higher animals—the most rapidly and recently evolved—reproduce sexually appears to confirm the evolutionary advantages of this method of reproduction.

SUMMARY

Sexual reproduction involves a special kind of nuclear division called meiosis. Meiosis is the process by which the chromosomes are reassorted and cells are produced that have the haploid chromosome number (n). The other principal component of sexual reproduction is fertilization, the coming together of haploid cells, which restores the diploid number ($2n$). There are characteristic differences among major groups of organisms as to where in the life cycle these events take place.

At the start of meiosis, the chromosomes arrange themselves in pairs. The members of the pairs are known as homologues. One homologue of each pair is of maternal origin, and one is of paternal origin. Each homologue consists of two identical chromatids. Early in meiosis, crossing over occurs between homologues, resulting in exchanges of chromosomal material.

In the first stage of meiosis, the homologues are separated. Two nuclei are produced, each with a haploid number of chromosomes, which, in turn, consist of two chromatids each. The cell enters interphase, but the chromosomal material is not replicated. In the second stage of meiosis, the sister chromatids separate as in mitosis. When the two nuclei divide, four haploid cells are formed.

Each of the haploid cells produced by meiosis contains a unique assortment of chromosomes due to crossing over and random assortment of homologues. Thus meiosis is a source of the variation upon which evolution depends.

QUESTIONS

1. Distinguish between the following terms: haploid/diploid; sporophyte/gametophyte; gamete/zygote; meiosis I/meiosis II.

2. Draw a diagram of a cell with six chromosomes ($n = 3$) at meiotic prophase I. Label each pair of chromosomes differently (for example, label one pair A^1 and A^2, and another B^1 and B^2, etc.).

3. Diagram the eight possible gametes resulting from meiosis in a plant cell with six chromosomes ($n = 3$). Label each chromosome differently, as in Question 2. Neglect crossing over.

4. Identify the stages of meiosis in crested wheat grass shown in the micrographs below.

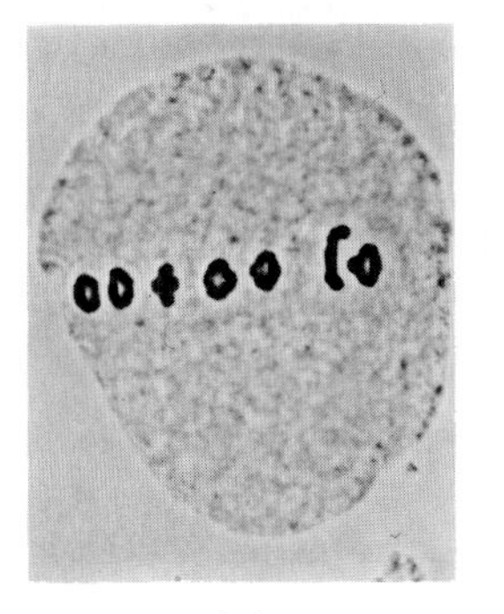

(a)

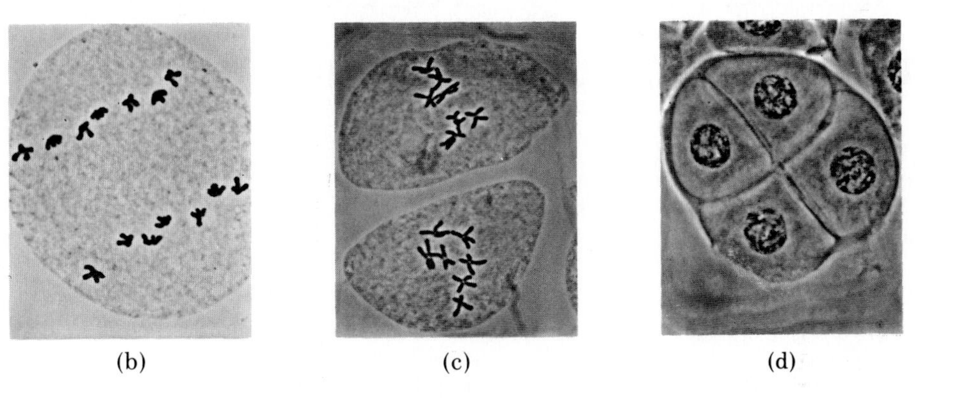

(b) (c) (d)

5. Compare and contrast the processes and the consequences of meiosis and mitosis.

6. In our bodies and in those of other higher animals, both mitosis and meiosis occur. Describe the functions of these two processes in our bodies. Why is it advantageous to have both processes?

7. When does sexual reproduction *not* require two parents? (If you can't answer this question, wait until you have read the next chapter and try again.)

11

From an Abbey Garden: The Beginning of Genetics

In Chapter 9, we saw that the nucleus carries the information that is passed from parent to offspring and that controls cellular structure and function. Moreover, each time a cell divides, the chromosomes are equally apportioned between the daughter cells by the complex process of mitosis. Similarly, when gametes are produced by meiosis, each gamete receives a haploid set of chromosomes.

Long before developments in microscopy made these discoveries possible, biologists were attempting to answer some puzzling questions about heredity. For example, why does a child resemble its mother in certain features and its father in other features? Why do some features seem to "skip a generation," with the result that a child resembles a grandparent more closely than either parent? Such questions were of considerable practical importance to breeders of plants and animals, who were attempting to develop varieties with particular desirable characteristics.

At about the same time that Hertwig was observing the fusion of sea urchin egg and sperm, Gregor Mendel was conducting the first experiments that provided useful answers to these basic questions about heredity. His work, carried out in a quiet monastery garden in what is now Brno, Czechoslovakia (then known as Brünn), and ignored until after his death, marks the beginning of modern genetics.

The Concept of the Gene

By Mendel's time, breeding experiments with domestic plants and animals had shown that both parents contribute to the characteristics of their offspring. Further, it was known that these contributions are carried in the gametes, that is, the sperm and the eggs.

Mendel's great contribution was to demonstrate that inherited characteristics are carried as discrete units, which are parceled out in different ways—reassorted—in each generation. These discrete units later came to be known as *genes.*

For his experiments in heredity, Mendel chose the common garden pea. It was a good choice. The plants were commercially available and easy to cultivate. Different varieties had clearly different characteristics that "bred true," reappearing in crop after crop. Further, the sexual organs of

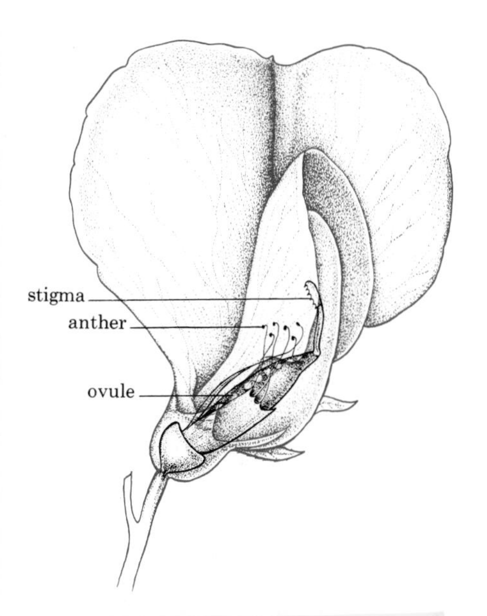

11-1 *In a flower, pollen develops in the anther and the egg cells in the ovule. Pollination occurs when pollen grains, trapped on the stigma, germinate and grow down to the ovule, where they release sperm cells. Egg and sperm unite, and the fertilized egg develops within the ovule. In the garden pea, the ovule and embryo form the peas (the seeds).*

Pollination in most species involves the pollen from one plant (often carried by an insect) being caught on the stigma of another plant. This is called cross-pollination.

In the pea flower, however, the stigma and anthers are completely enclosed by petals, and the flower, unlike most, does not open until after fertilization has taken place. Thus the plant normally self-pollinates. In his crossbreeding experiments, Mendel pried open the bud before the pollen matured and removed the anthers with tweezers. Then he artificially pollinated the flower by dusting the stigma with pollen collected from other plants.

the pea flower are entirely enclosed by petals, even when they are mature (Figure 11-1). Consequently, the flower normally self-pollinates, that is, sperm cells from the flower's own pollen fertilize its egg cells. Although the plants could be crossbred experimentally, accidental crossbreeding could not occur to confuse the experimental results. As Mendel said in his original paper, "The value and utility of any experiment are determined by the fitness of the material to the purpose for which it is used."

Mendel's choice of the pea plant for his experiments was not original. However, he was successful in formulating the basic principles of heredity—where others had failed—because of his approach to the problem. First, he tested a very specific model in a series of logical experiments. He planned his experiments carefully and imaginatively, choosing for study clear-cut hereditary differences and avoiding characteristics that could be "more or less" apparent in the offspring. Second, he studied the offspring of not only the first generation but also the second generation. Third, and most important, he counted the offspring and then analyzed the results mathematically. Even though his mathematics was simple, the idea that a biological problem could be studied quantitatively was startlingly new. Finally, he organized his data in such a way that his results could be evaluated simply and objectively. The experiments themselves were described so clearly that other scientists could repeat and check them, as eventually they did.

The Principle of Segregation

Mendel began with 32 different types of pea plants, which he studied for several years before he began his quantitative experiments. As a result of these observations, he selected for study seven traits that appeared as conspicuously different characteristics in different varieties of plants. One variety of plant, for example, always produced yellow peas, or seeds, while another always produced green ones. In one variety, the seeds, when dried, had a wrinkled appearance; in another variety, they were smooth. The complete list of alternate traits is given in Table 11-1.

Mendel performed experimental crosses, removing the anthers from flowers and dusting the stigmas with pollen from a flower of another variety. He found that in every case in the first generation (now known in biological shorthand as the F_1, for "first filial generation"), one of the alternate traits disappeared completely without a sign. All the seeds produced as a result of a cross between yellow-seeded plants and green-seeded plants were as yellow-seeded as the yellow-seeded parent. All of the flowers produced by plants resulting from a cross between a purple-flowered plant and a white-flowered plant were purple. Traits that appeared in the F_1 generation, such as yellow seeds and purple flowers, Mendel called *dominant.*

The interesting question was: What happened to the second trait—the greenness of the seed or the whiteness of the flower—which had been passed on so faithfully for generations by the parent stock? Mendel let the pea plant itself carry out the next stage of the experiment by permitting the F_1 plants to self-pollinate. The traits that had disappeared in the first generation reappeared in the second, or F_2, generation. In Table 11-1 are the results of Mendel's actual counts. These traits, which were present in the parent generation and reappeared in the F_2 generation, must also have been present somehow in the F_1 generation, although not apparent there. Mendel called these traits *recessive.*

Table 11-1 Results of Mendel's experiments with pea plants

	Original Crosses			(F_2) Second Generation		
Trait	Dominant	×	Recessive	Dominant	Recessive	Total
Seed form	Round	×	Wrinkled	5,474	1,850	7,324
Seed color	Yellow	×	Green	6,022	2,001	8,023
Flower position	Axial	×	Terminal	651	207	858
Flower color	Purple	×	White	705	224	929
Pod form	Inflated	×	Constricted	882	299	1,181
Pod color	Green	×	Yellow	428	152	580
Stem length	Tall	×	Dwarf	787	277	1,064

11-2 *Garden pea plant with seeds (peas) in the approximate ratio of three yellow seeds for each green seed, as observed by Mendel.*

If you analyze the results in Table 11-1 as Mendel did, you will notice that the dominant and recessive traits appear in the second, or F_2, generation in ratios of about 3:1. How do the recessives disappear so completely and then appear again, and always in such constant proportions? It was in answering this question that Mendel made his greatest contribution. He saw that the appearance and disappearance of traits and their constant proportions could be explained if hereditary characteristics are determined by discrete factors (genes). These factors, Mendel saw, must have occurred in the F_1 plants in pairs, one factor inherited from each parent. These pairs of factors separated again when the mature F_1 plants produced sex cells, resulting in two kinds of gametes, with one gene of the pair in each.

The hypothesis that every individual carries pairs of genes for each trait and that the members of a pair segregate during the formation of gametes is known as Mendel's first law, or the *principle of segregation.*

Consequences of Segregation

According to Mendel's hypothesis, the two genes in a pair might be the same, in which case the self-pollinating plant would breed true. Or the two factors might be different; such different, or alternative, forms came to be known as *alleles.* Yellow-seededness and green-seededness, for instance, are determined by alleles, different forms of the gene for seed color. When the alleles of a gene pair are the same, the organism is said to be *homozygous* for that particular trait; when the alleles of a gene pair are different, the organism is *heterozygous* for that trait.

When gametes are formed, genes are passed on to them, but each gamete contains only one of two possible alleles for any given trait. When the two gametes combine in the fertilized egg, the genes occur in pairs again. If the alleles are the same, both will be expressed. If the alleles are different, one may be dominant over the other; in this case, the organism will appear as if it had only the dominant allele. This outward appearance of a trait is known as the *phenotype.* However, in the genetic makeup, or *genotype,* each allele still exists independently and as a discrete unit, even though it may not be expressed in the phenotype. The recessive allele will separate from its dominant partner when gametes are again formed. Only if two recessive alleles come together—one from the female gamete and one from the male gamete—will the phenotype then show the recessive trait.

When pea plants homozygous for purple flowers are crossed with pea plants having white flowers, only pea plants with purple flowers are produced, although each plant in the F_1 generation will carry a gene for purple and a gene for white. Figure 11–3 shows what happens in the F_2 generation if the F_1 generation self-pollinates. Notice that the result would be the same if an F_1 individual were cross-fertilized with another F_1 individual, which is how these experiments are performed with animals and with plants that are not self-pollinating.

In order to test his hypothesis, which is diagrammed in Figure 11–4, Mendel performed two additional experiments. He crossed white-flowering

11–3 *A pea plant homozygous for purple flowers is represented as* WW *in genetic shorthand. The gene for purple flowers is designated* W *because of a convention by which geneticists, in indicating a pair of alleles, use the first letter of the less common form (white). The capital indicates the dominant, the lowercase the recessive. A* WW *plant can produce gametes with only a purple-flower* (W) *gene. The female symbol* ♀ *indicates that this flower contributed the egg cells, or female gametes.*

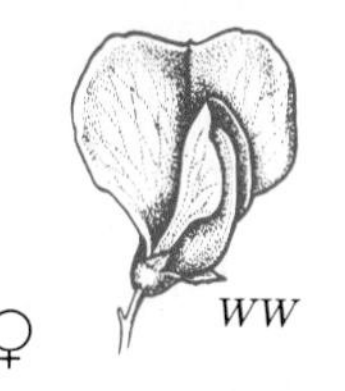

A white pea plant (ww) *can produce gametes with only a white-flower* (w) *gene. The male symbol* ♂ *indicates that this flower contributed the sperm cells, or male gametes.*

When a w *sperm cell fertilizes a* W *egg cell, the result is a* Ww *pea plant, which, since the* W *gene is dominant, will produce purple flowers. However, this* Ww *plant can produce gametes with either a* W *or a* w *allele.*

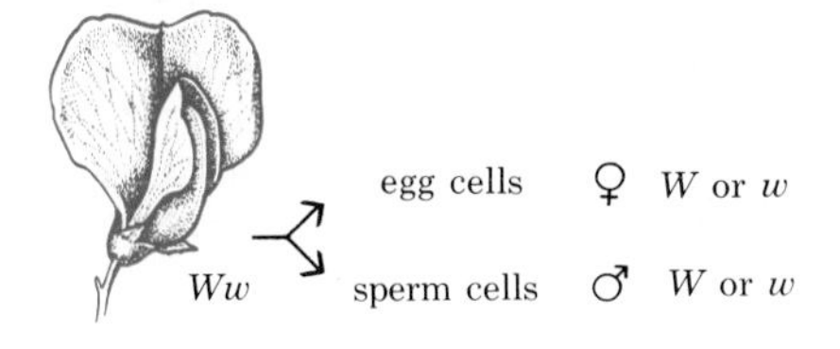

And so, if the plant self-pollinates, four possible crosses can occur:

♀ W × ♂ W → *purple flowers*
♀ W × ♂ w → *purple flowers*
♀ w × ♂ W → *purple flowers*
♀ w × ♂ w → *white flowers*

These results are summarized in Figure 11–4.

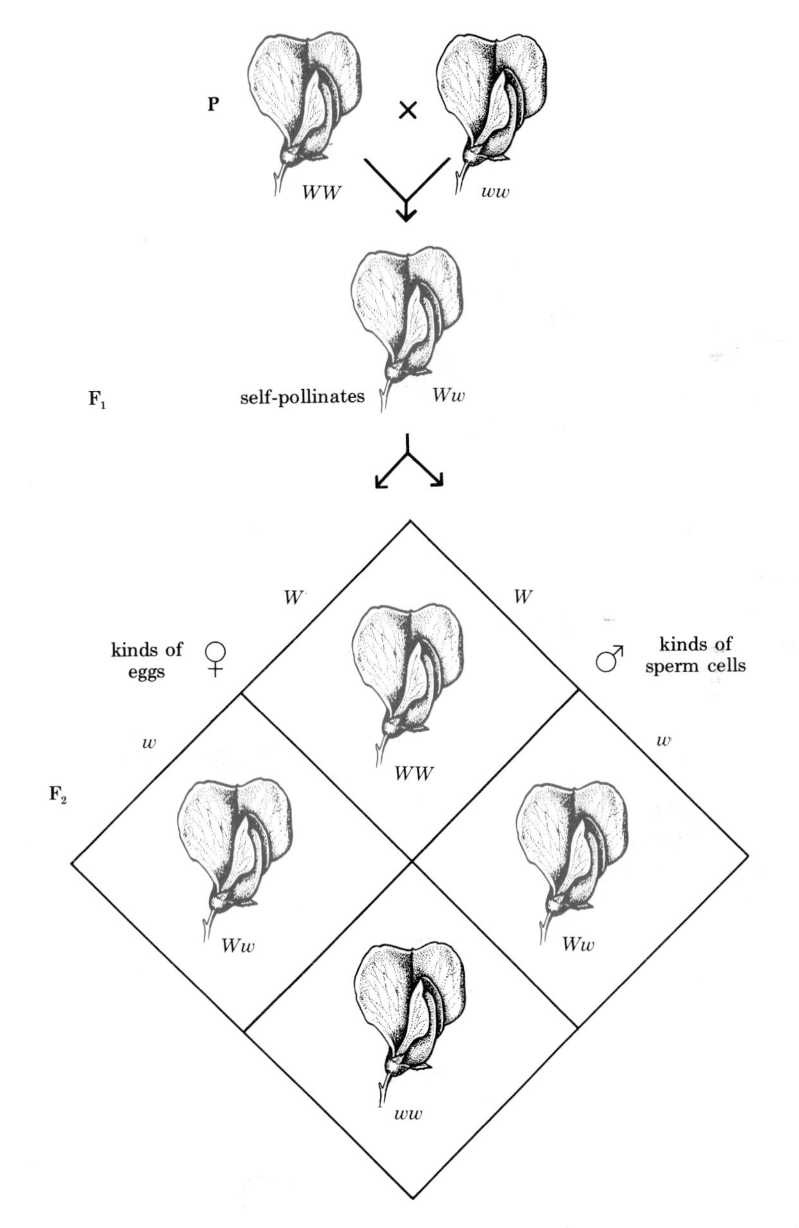

11–4 *A cross between a parent* (P) *pea plant with two dominant genes for purple flowers* (WW) *and one with two recessive genes for white flowers* (ww). *The phenotype of the offspring in the* F_1 *generation is purple, but note that the genotype is* Ww. *The* F_1 *heterozygote self-pollinates, producing four kinds of gametes,* ♀ W, ♂ W, ♀ w, *and* ♂ w, *in equal proportions. The* W *and* w *sperm cells and eggs combine randomly to form, on the average,* ¼ WW *(purple),* ½ Ww *(purple), and* ¼ ww *(white) offspring. It is this underlying 1:2:1 genotypic ratio that accounts for the phenotypic ratio of 3 dominants (purple) to 1 recessive (white). The distribution of traits in the* F_2 *generation is shown by a Punnett square, named after the English geneticist who first used this sort of checkerboard diagram for the analysis of genetically determined traits.*

Mendel and the Laws of Chance

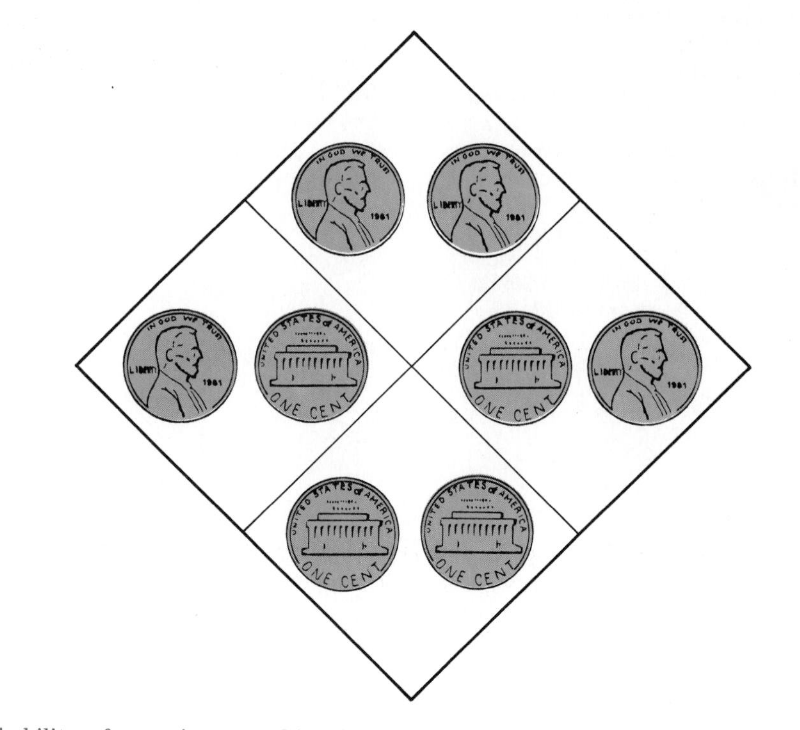

In applying mathematics to the study of genetics, Mendel was stating that the laws of chance apply to biology as they do to the physical sciences. Toss a coin. The chance that it will turn up heads is fifty-fifty, or $\frac{1}{2}$. The chance that it will turn up tails is also fifty-fifty, or $\frac{1}{2}$. Now toss two coins. The chance that one will turn up heads is again $\frac{1}{2}$. The chance that the second will turn up heads is also $\frac{1}{2}$. The chance that both will turn up heads is $\frac{1}{2} \times \frac{1}{2}$, or $\frac{1}{4}$. The probability of two independent events occurring together is simply the probability of one occurring alone multiplied by the probability of the other occurring alone. The chance of both turning up tails is similarly $\frac{1}{2} \times \frac{1}{2}$. The chance of the first turning up tails and the second turning up heads is $\frac{1}{2} \times \frac{1}{2}$, and the chance of the second turning up tails and the first heads is $\frac{1}{2} \times \frac{1}{2}$.

We can diagram this in a Punnett square (see figure), which indicates that the combination in each square has an equal chance of occurring. It was undoubtedly the observation that one-fourth of the offspring in the F_2 generation showed the recessive phenotype that indicated to Mendel that he was dealing with a simple case of the laws of probability.

If there were three coins involved, the probability of any given combination would be simply the product of all three fifty-fifty possibilities: $\frac{1}{2} \times \frac{1}{2} \times \frac{1}{2}$, or $\frac{1}{8}$. Similarly, with four coins, the probability of any given combination is $\frac{1}{2} \times \frac{1}{2} \times \frac{1}{2} \times \frac{1}{2}$, or $\frac{1}{16}$. The Punnett square on page 154 expresses the probability (or chance) of each of any one of four possible combinations.

Notice that in planning his experiments, Mendel made several assumptions: (1) of the male gametes produced, one half contain one paternal allele and one half contain the other allele; (2) of the female gametes produced, one half contain one maternal allele and one half contain the other allele; (3) the male and female gametes combine at random. Thus, the laws of probability could be employed.

If you toss two coins 4 times, it is unlikely that you will get the precise results diagrammed above. However, if you toss two coins 100 times, you will come close to the results predicted in the Punnett square, and if you toss two coins 1,000 times, you will be very close indeed. As Mendel knew, the relationship between dominants and recessives might well not have held true if he had been dealing with a small sample. The larger the sample, the more closely it will conform to results predicted by the laws of chance.

plants with white-flowering plants and confirmed that they bred true—that only white-flowering plants were produced. Next he crossed one of his F_1 individuals, the result of a cross between purple- and white-flowering plants, with a true-breeding white-flowering plant. To the outside observer, it would appear as if Mendel were simply repeating his first experiment, crossing plants having purple flowers with plants having white flowers. But he knew that if his hypothesis were correct, his results would be different from those of his first experiment. In fact, he predicted the results of such a cross before he made it. Can you? Stop a moment and think about it.

The easiest way to analyze the possible result of such a cross is to diagram it, as in Figure 11–5. This experiment, which reveals the genotype of the parent with the dominant trait, is known as a *testcross*. A testcross is

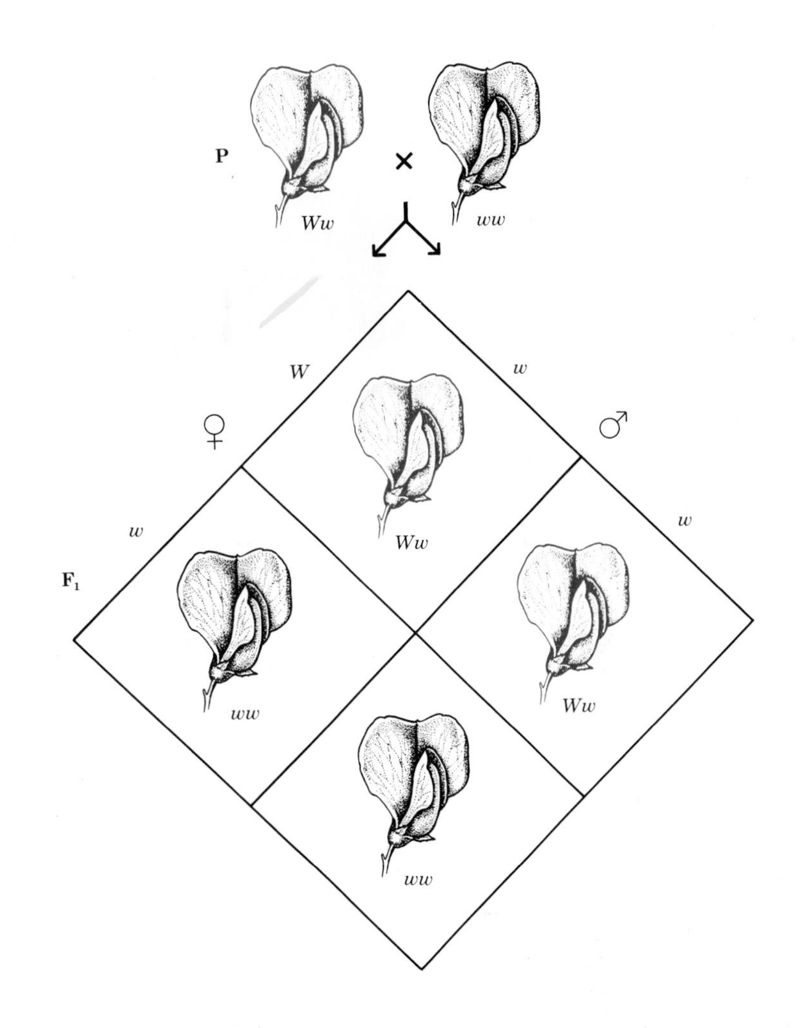

11–5 *A testcross. In order for a pea flower to be white, the plant must be homozygous for the recessive gene* (ww). *But a purple pea flower can come from a plant with either a* Ww *or a* WW *genotype. How could you tell such plants apart? Mendel solved this problem by breeding such plants with homozygous recessives. This sort of experiment is known as a testcross. As shown here, a phenotypic ratio in the* F_1 *generation of 2 purple to 2 white indicates a heterozygous purple-flowering parent. What would have been the result if the plant being tested had been homozygous for purple flowers?*

an experimental cross between an individual with the dominant phenotype for a given trait with another individual that is homozygous recessive for the trait. The resulting ratio of offspring will indicate whether the individual with the dominant phenotype was homozygous or heterozygous for the trait being studied. In the testcross shown in Figure 11–5, the cross revealed that the genotype of the plant being tested was *Ww* rather than *WW*.

The Principle of Independent Assortment

In a second series of experiments, Mendel studied crosses between pea plants that differed in two characteristics. For example, one parent plant had peas that were round and yellow, and the other had peas that were wrinkled and green. The round and yellow traits, you will recall (see Table 11–1), are dominant, and the wrinkled and green are recessive. As you would expect, all the seeds produced by a cross between the parental types were round and yellow. When these F_1 seeds were planted and the flowers allowed to self-pollinate, 556 seeds were produced. Of these, 315 showed the two dominant characteristics, round and yellow, but only 32 combined the recessive traits, green and wrinkled. All the rest of the seeds were unlike either parent; 101 were wrinkled and yellow, and 108 were round and green. Totally new combinations of characteristics had appeared.

This experiment did not contradict Mendel's previous results. Round and wrinkled still appeared in the same 3:1 proportion (423 round to 133 wrinkled), and so did yellow and green (416 yellow to 140 green). But the round and the yellow traits and the wrinkled and the green ones, which had originally combined in one plant, behaved as if they were entirely independent of one another. From this, Mendel formulated his second law, the *principle of independent assortment.* This principle states that members of each pair of genes are distributed independently when the gametes are formed.

Figure 11–6 diagrams Mendel's interpretation of these results. It shows why, in a cross involving two gene pairs, each pair with one dominant and one recessive allele, the ratio of distribution of the phenotypes will be, on the average, 9:3:3:1. The 9 represents the proportion of F_2 offspring that will show the two dominant traits, 1 the proportion that will show the two recessive traits, and 3 and 3 the proportions that will show the two alternative combinations of dominants and recessives. The ratios apply when one of the original parents is homozygous for both recessive traits and the other homozygous for both dominant ones, as in the experiment just described *(RRYY × rryy),* as well as when each original parent is homozygous for one recessive and one dominant trait *(rrYY × RRyy).*

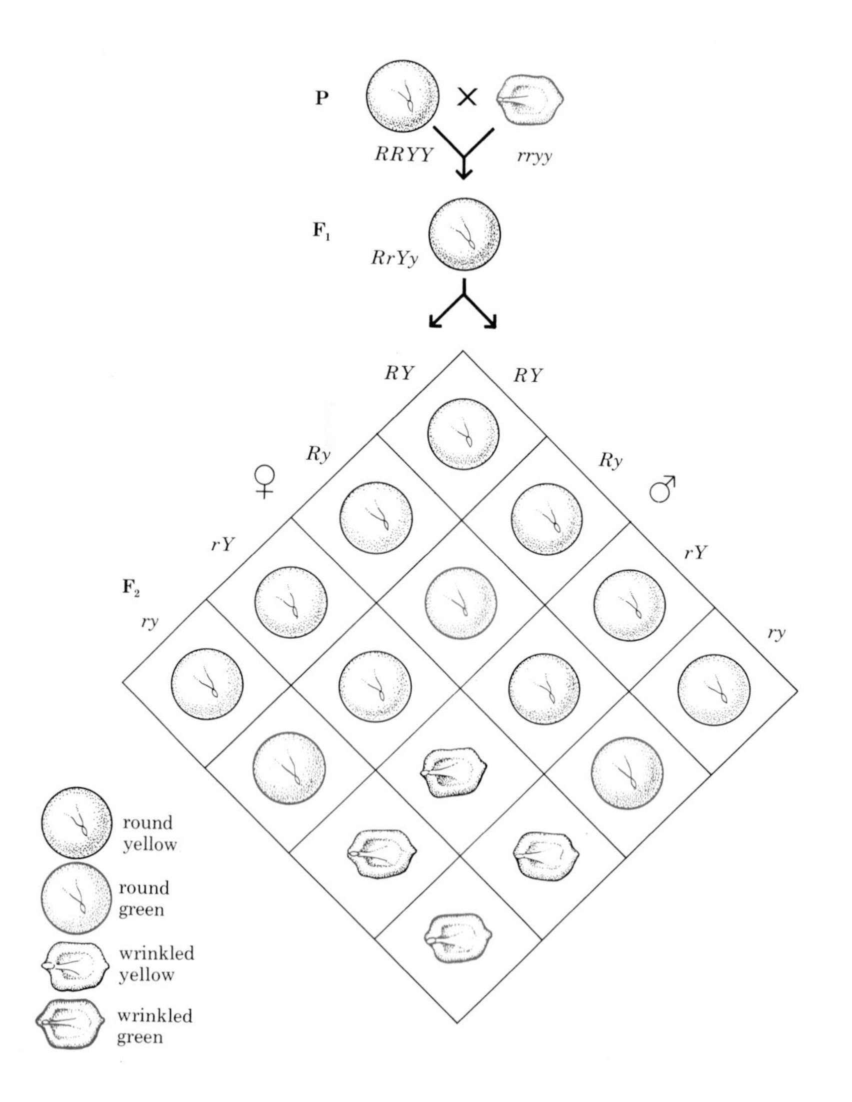

11–6 *One of the experiments from which Mendel derived his principle of independent assortment. A plant homozygous for round* (RR) *and yellow* (YY) *peas is crossed with a plant having wrinkled* (rr) *and green* (yy) *peas. The* F_1 *generation are all round and yellow, but notice how the traits will, on the average, appear in the* F_2 *generation if the* F_1 *is allowed to self-pollinate. Of the 16 kinds of offspring, 9 show the two dominant traits* (RY), *3 show one combination of dominant and recessive* (Ry), *3 show the alternate combination* (rY), *and 1 shows the two recessives* (ry). *This 9:3:3:1 distribution is always the expected result from a cross involving two pairs of independent dominant-recessive alleles. (The letters* R *and* Y *are used because round and yellow are the less common forms in nature.)*

11-7 *Gregor Mendel, holding a plant, is third from the right in this photograph of members of the Augustinian monastery in Brünn in the 1860s. In his experiments carried out in the monastery garden, Mendel showed that hereditary determinants are carried as separate units from generation to generation.*

The Influence of Mendel

Mendel's experiments were first reported in 1865 before a small group of people at a meeting of the Brünn Natural History Society. None of them, apparently, understood what Mendel was talking about. His paper was, however, published the following year in the *Proceedings* of the Society, a journal that was circulated to libraries all over Europe. In spite of this, his work was ignored for 35 years, during most of which he devoted himself to the duties of an abbot, and he received no scientific recognition until after his death. (He was, to use DuPraw's phrase, page 5, an odd traveler whose tale could not be made to fit.)

It was not until 1900 that biologists were finally prepared to accept Mendel's findings. Within a single year, his paper was independently rediscovered by three scientists working in three different European countries. Each of them had done similar experiments and was searching the scientific literature to seek confirmation of his results. And each found, in Mendel's brilliant analysis, that much of his own work had been anticipated. The rediscovery of Mendel's work, in conjunction with studies already under way at the turn of the century, produced an extraordinary period in the history of genetics, as we shall see in the next chapter.

SUMMARY

How is the hereditary information transmitted from cell to cell and generation to generation? Gregor Mendel provided the first answers to this question. According to Mendel's principle of segregation, hereditary characteristics are determined by discrete factors (now called genes) that appear in pairs, one of each pair inherited from each parent.

The genetic makeup of an organism is known as its genotype. Its appearance, or outward characteristics, constitute the organism's phenotype. Both genes in a pair may be alike (the homozygous condition), or they may be different (heterozygous). Different genes of a gene pair are called alleles (alternative forms). In a heterozygous pair, only one allele may be evident in the phenotype. An allele that is expressed in the phenotype to the exclusion of the other is a dominant allele; one that is concealed in the phenotype is a recessive allele. When two organisms that are heterozygous for different alleles of the same gene are crossed, the ratio of dominant to recessive in the phenotype of the offspring is 3:1.

Mendel's other great principle, the principle of independent assortment, applies to the behavior of two or more gene pairs. This law states that the members of each pair of genes segregate independently of one another. In crosses involving two independent gene pairs, the expected phenotypic ratio in the F_2 generation is 9:3:3:1.

QUESTIONS

1. Distinguish between the following terms: gene/allele; dominant/recessive; homozygous/heterozygous; genotype/phenotype; the F_1/the F_2.

2. State Mendel's two principles in your own words.

3. In the experiments summarized in Table 11–1, which of the alternate traits appeared in the F_1 generation?

4. Why is a homozygous recessive always used in a testcross?

5. (a) In a pea plant that breeds true for tall, what possible gametes can be produced? (Use the symbol D for tall, d for dwarf.) (b) In a pea plant that breeds true for dwarf, what possible gametes can be produced? (c) What will be the genotype of the F_1 generation produced by a cross between these two types? (d) What will be the phenotype of the F_1 generation produced by a cross between these two types? (e) What will be the probable distribution of traits in the F_2 generation? Illustrate with a Punnett square.

6. Suppose you have just flipped a coin five times and it has turned up heads every time. What are the chances that the next time you flip it, it will be tails?

7. The ability to taste a bitter chemical, phenylthiocarbamide (PTC), is due to a dominant gene. In terms of tasting ability what are the possible phenotypes of a man both of whose parents are tasters? What are his possible genotypes?

8. If the man in Question 7 marries a woman who is a nontaster, what proportion of their children could be tasters? Suppose one of the children is a nontaster. What would you know about the father's genotype? Explain your results by drawing Punnett squares.

9. A taster and a nontaster have four children, all of whom can taste PTC. What is the probable genotype of the parent who is a taster? Is there another possibility?

10. A pea plant that breeds true for round, green seeds *(RRyy)* is crossed with a plant that breeds true for wrinkled, yellow seeds *(rrYY)*. Each parent is homozygous for one dominant trait and for one recessive trait. What is the genotype of the F_1 generation? What is the phenotype?

The F_1 seeds are planted and their flowers are allowed to self-pollinate. Draw a Punnett square to determine the ratios of the phenotypes of the F_2 generation. How do the results compare with those of the experiment shown in Figure 11–6?

11. Mendel did not know of the existence of chromosomes. Had he known, what change might he have made in his second principle?

12

Mendel Rediscovered: The Golden Age of Genetics

Between the publication of Mendel's experiments in 1865 and their rediscovery in 1900, several important developments occurred in biology. Darwin's evolutionary theory had become accepted, but it was still a mystery as to how small variations persist in populations. Hertwig's experiments on sea urchin reproduction had demonstrated that the nuclei of sperm and egg cells are the hereditary links between parents and offspring. The steps in mitosis and meiosis had been pieced together. But the relationships among these new discoveries were not known. Many biologists were working on problems of evolution and heredity, and the rediscovery of Mendel's work catalyzed other new ideas and experiments. The results of this work led, in turn, to modifications of Mendel's two principles.

In this chapter, we shall describe some of the major ideas that resulted from what has been rightly called the "Golden Age of Genetics." In the next chapter, we shall examine how these ideas explain some puzzling aspects of human heredity.

12-1 *Hugo de Vries is shown standing next to* Amorphophallus titanum, *which has the largest flower cluster of any of the flowering plants. De Vries, a Dutch botanist, was one of the rediscoverers of Mendel's work. He was also the first to recognize the nature of mutations and their role in evolutionary processes.*

Mutations and Evolution

Mendel's work filled an important gap in Darwin's theory by explaining why small variations persist in populations. However, Mendelian principles presented new problems to the early evolutionists, because they appeared to offer no possibility for major changes in the genetic makeup of organisms. Segregation of traits explained how variations were maintained from generation to generation. Independent assortment explained how individuals could have characteristics in combinations not present in either parent and so be better adapted, in evolutionary terms, than either parent. But if all hereditary variations were to be explained by the reshuffling process proposed by Mendel, there would be little or no opportunity for major changes in organisms.

A solution to the problem was proposed by one of Mendel's rediscoverers, a Dutch botanist named Hugo de Vries. De Vries was studying the evening primrose. Heredity in the primrose, he found, was generally orderly and predictable, as in the garden pea. Occasionally, however, a characteristic appeared that was not present in either parent or indeed anywhere in the lineage of that particular plant. De Vries hypothesized that this characteristic came about as the result of an abrupt change in a gene and that the trait produced by the changed gene was then passed along like any other hereditary trait. De Vries spoke of this hereditary change as a *mutation* and of the organism that carried it as a *mutant.* Different alleles of the same gene, de Vries proposed, arise as a result of mutations. Mutations are the source of new genetic characteristics, and these are the sources of the variations on which evolution depends.

In 1927, H. J. Muller found that treatment of gametes by x-rays greatly increases the rate at which mutations occur. It was soon discovered that other radiations, such as ultraviolet light, and also some chemicals can act as *mutagens,* agents that produce mutations. Studies have shown that whether mutations are produced by external factors, such as radiations and chemicals, or by unknown factors within the cell, they are usually detrimental to the organism. In fact, many of them are lethal. This is not surprising. If one were to change a word at random in a Shakespearean sonnet or a wire at random in a television set, an improvement would be unlikely, and the results might well be disastrous. It is reasonable to assume that the genetic makeup of an individual is as precise in its engineering and as delicate in its balance as a sonnet or a television set. However, as a result of mutations, there is a wide range of variability in a natural population. In a complex or shifting environment, a particular variation might give an individual or its offspring a slight edge.

When mutations occur in eggs or sperm, their consequences are concealed until the next generation. It is, in part, because of the effects on the gametes that many people are concerned about toxic chemicals in our environment and the increase in radiation exposure caused by increased use of nuclear energy or by unwarranted or careless use of x-rays. Another reason for concern is that mutations may also occur in body cells. As we shall see (page 220), there is evidence that such mutations are a principal cause of cancer.

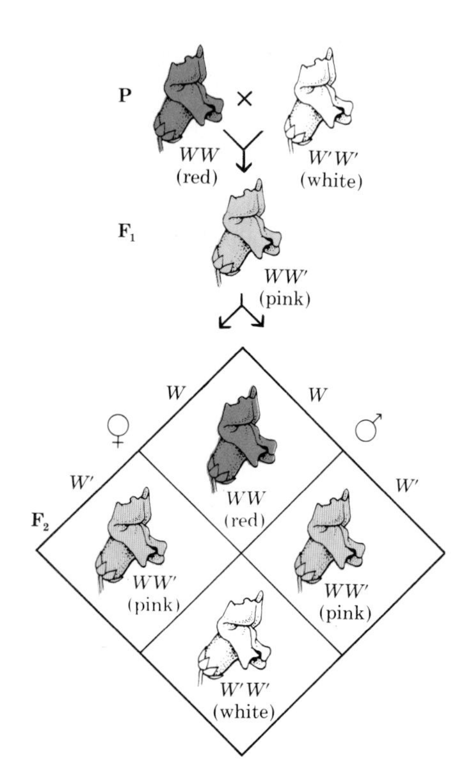

12–2 *A cross between a red* (WW) *snapdragon and a white* (W′W′) *snapdragon. This looks very much like the cross between a purple- and a white-flowering pea plant shown in Figure 11–4, but there is a significant difference because neither allele is dominant. The flower of the heterozygote is a blend of the two colors.*

Modifying Mendel

Incomplete Dominance

During the decade that followed the rediscovery of Mendel's work, many studies were carried out that, while confirming his work in principle, showed that the action of genes is more complex than it had first appeared to be. An important example is *incomplete dominance.*

Dominant and recessive traits are not always so clear-cut as in the pea plant. Some traits appear to blend. For instance, a cross between a red-flowering snapdragon and a white-flowering snapdragon produces heterozygotes that are pink (Figure 12–2). But when members of this generation self-pollinate, the traits begin to sort themselves out once again, showing that the alleles themselves, as Mendel had affirmed, remain discrete and unaltered. As we shall see, what occurs in the snapdragon and in similar crosses is actually a result of the combined effect of gene products.

Broadening the Concept of the Gene

Mendel's experiments seemed to suggest that each gene affects a single characteristic in a one-to-one relationship. It was soon discovered, however, that a single gene sometimes affects many traits in an organism *(pleiotropy),* and that, conversely, a single trait is often affected by many genes *(polygenic inheritance).*

Pleiotropy

A striking example of pleiotropy is provided by a gene in rats that produces a protein involved in the formation of cartilage. A mutation of this one gene causes a whole complex of congenital deformities, including thickened ribs, a narrowing of the tracheal passage (through which air moves to and from the lungs), a loss of elasticity in the lungs, blocked nostrils, a blunt snout, and a thickening of the heart muscle. Needless to say, the mutation leads to a greatly increased mortality rate. Since cartilage is one of the most common structural substances of the body, the widespread effects of such a gene are not difficult to understand. In fact, it is very likely that Mendel's allele for wrinkled peas, for example, also affected other structural characteristics of the pea plant. The frizzle trait in fowl (Figure 12–3) is another example of pleiotropy.

Polygenic Inheritance

A trait affected by a number of genes does not show a clear-cut difference between groups of individuals—such as the differences tabulated by Mendel—but rather shows a gradation of small differences, which is

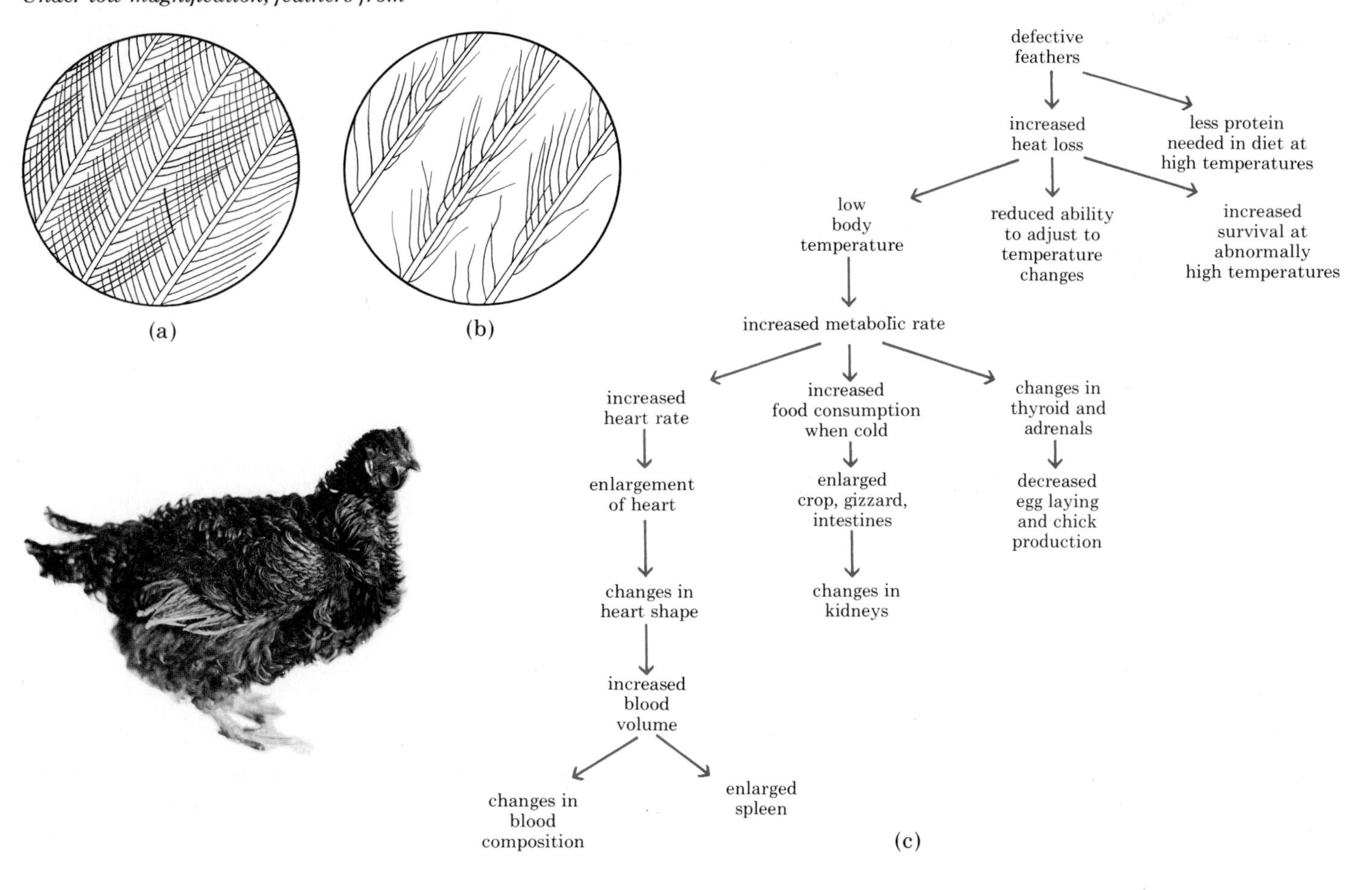

12–3 *The frizzle trait in fowl is an example of pleiotropy, the capacity of a gene to produce a variety of phenotypic effects. "Frizzle" is manifested primarily by differences in the feathers.* (a) *Under low magnification, feathers from normal homozygous birds show a closely interwebbed structure.* (b) *Feathers from birds homozygous for the frizzle trait are weak and stringy and provide poor insulation. Some of the consequences of the manifestation of this single gene are shown in* (c). *Notice that the frizzle trait is a liability at low temperatures but may increase survival at high temperatures.*

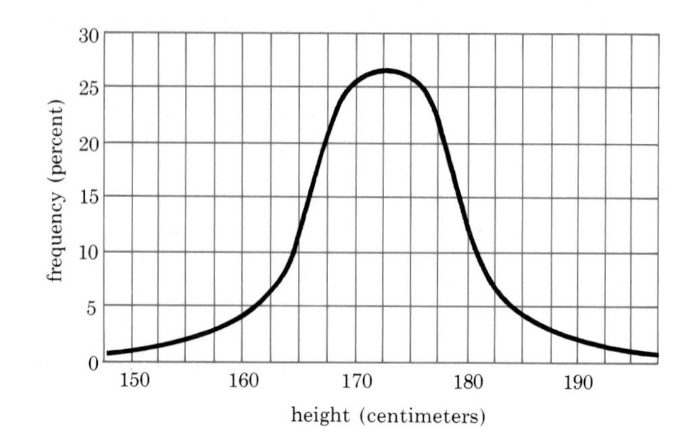

12-4 *Height distribution of males in the United States. Height is an example of polygenic inheritance; that is, it is affected by a number of genes. Such genetic traits are characterized by small gradations of difference. A graph of the distribution of such traits takes the form of a bell-shaped curve, as shown, with the mean, or average, usually falling in the center of the curve. The larger the number of genes involved, the smoother the curve.*

known as *continuous variation.* If you make a chart of genetic differences among individuals that are affected by a number of genes, you get a curve such as that shown in Figure 12-4.

Fifty years ago, the average height of males in the United States was less than it is now but the shape of the curve was the same; in other words, the great majority fell within the middle range and the extremes in height were represented by only a few individuals. Some of these height variations are produced by environmental factors, such as diet, but even if all the men in a population were maintained from birth on the same type of diet, there would still be a continuous variation in height in the population. This is due to genetically determined differences in hormone production, bone formation, and numerous other factors.

Table 12-1 illustrates a simple example of polygenic inheritance, color in wheat kernels, which is controlled by two pairs of genes. Human skin color is under a similar kind of genetic control, as are many other human characteristics.

Table 12-1 The genetic control of color in wheat kernels (polygenic inheritance*)

Parents:	$R_1R_1R_2R_2 \times r_1r_1r_2r_2$ (Dark Red) (White)		
F_1:	$R_1r_1R_2r_2$ (Medium Red)		
F_2:	Genotype	Phenotype	
1	$R_1R_1R_2R_2$	Dark red	15 red to 1 white
2 } 4	$R_1R_1R_2r_2$	Medium-dark red	
2	$R_1r_1R_2R_2$	Medium-dark red	
4 } 6	$R_1r_1R_2r_2$	Medium red	
1	$R_1R_1r_2r_2$	Medium red	
1	$r_1r_1R_2R_2$	Medium red	
2 } 4	$R_1r_1r_2r_2$	Light red	
2	$r_1r_1R_2r_2$	Light red	
1	$r_1r_1r_2r_2$	White	

* Two genes are involved, each of which has two alleles: R_1 and r_1 for gene 1 and R_2 and r_2 for gene 2.

Genotype and Phenotype

There is a crucial difference between genotype and phenotype. From the moment of its conception, every organism is acted upon by the environment, and the expression of any gene is always the result of the interaction of gene and environment. To take a simple, familiar example, a seedling may have the genetic capacity to be green, to flower, and to fruit, but it will never turn green if it is kept in the dark, and it may not flower and fruit unless certain precise environmental requirements are met. Himalayan rabbits are all white if they are raised at high temperatures (above 35°C). However, rabbits of the same genotype, when raised at room temperature, have black ears, forepaws, noses, and tails.

In humans, also, similar genotypes are expressed quite differently in different environments. A simple example is found in height, which is influenced by a group of genetic factors (see facing page). This influence is indicated by the fact that tall parents generally have tall children and short parents tend to have short children. On the other hand, adult height is also clearly influenced by environmental factors in childhood, such as diet, sunshine, and incidence of disease. The population of the United States grew taller with each generation for four generations, presumably as a result of general improvements in the standard of living and in the average diet (this trend has just come to an end; grown children in this country are now no taller, on the average, than their parents).

As René Dubos reminds us,* children are growing faster; a boy now reaches full height at about 19 years, but 50 years ago, maximum stature was not usually attained until age 29. Of more social consequence is the fact that puberty is also being reached earlier. In Norway, for example, the mean age of the onset of menstruation fell from 17 in 1850 to 13 in 1960. Historical evidence indicates, however, that also in Imperial Rome and Western Europe in Shakespeare's time, teen-agers reached puberty at an early age (Juliet, remember, was not yet 14). Apparently, the slowing of the growth and maturation rate of the general population was a consequence of the industrial revolution and increasing urbanization, which reduced the standards of nutrition and health. Thus we have a situation in which the genetic potential has remained apparently unaltered over the centuries, but the phenotypic expression has undergone fluctuations. A side effect of this change in phenotypic expression is that teen-agers are now reaching physical maturity early in a society in which childhood and dependency have become greatly prolonged.

* René Dubos, *So Human an Animal,* Charles Scribner's Sons, New York, 1968.

The water buttercup, Ranunculus peltatus, *grows with half the plant body submerged in water. Although the leaves are genetically identical, the broad, floating leaves differ markedly in both form and physiology from the finely divided leaves that develop under water. Such plants illustrate dramatically that the final form of an organism is the result of an interaction between heredity and environment.*

Genes and Chromosomes

In 1902, two years after the rediscovery of Mendel's work, Walter Sutton, then a graduate student at Columbia University, was observing meiosis in the production of sperm cells in grasshoppers. He was struck by the parallels between what he was seeing and the first principle of Mendel: Chromosomes come in pairs (Figure 12–5), and so do Mendelian factors (genes). Chromosome pairs (homologues) separate when the gametes are formed; so do genes. And genes and chromosomes come together again in pairs in the offspring.

What about Mendel's second principle in relation to the movement of the chromosomes at meiosis? This principle, as you will recall, states that members of different pairs of genes assort independently. We can see that this can be true if—and this is an important point—the genes are on different pairs of chromosomes, as shown in Figure 12–6.

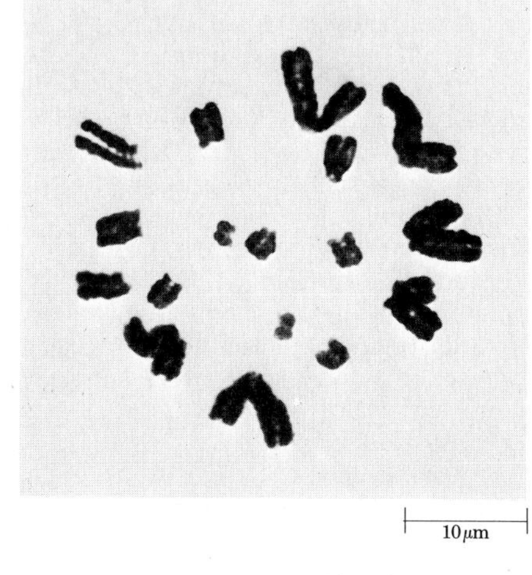

12–5 *Chromosomes from a diploid cell of a grasshopper, during metaphase of a mitotic division. Each chromosome consists of two closely aligned chromatids. Note that even though these chromosomes are not paired, it is possible to pick out some of the homologues. The observation that chromosomes come in homologous pairs was one of Sutton's clues to the meaning of meiosis.*

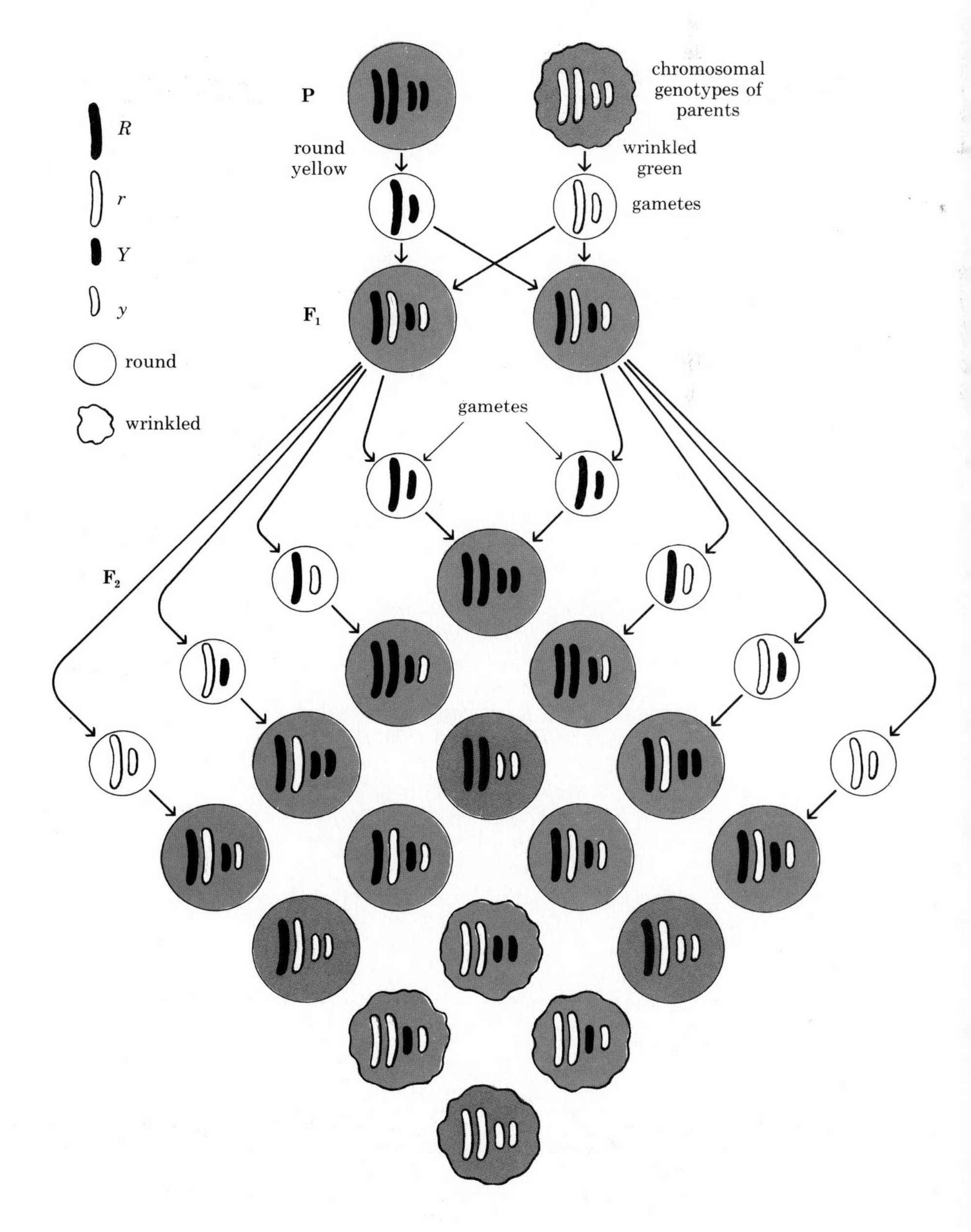

12–6 *The chromosome distributions in Mendel's cross of round yellow and wrinkled green peas, according to Sutton's hypothesis. Although the pea has 14 chromosomes (*n = 7*), only 4 are shown here, the two carrying the genes for round or wrinkled and the two carrying the genes for yellow or green. (This is analogous to what Mendel did when he selected the two traits to study.) As you can see, one parent is homozygous for the recessives, one for the dominants. Therefore, the only gametes they can produce are* RY *and* ry. *(Remember,* R *now stands not just for the gene but also for the chromosome carrying the gene, as do the other letters.) The* F_1 *generation, therefore, must be* Rr *and* Yy. *When a mother cell of this generation undergoes meiosis,* R *is separated from* r *and* Y *from* y *when the respective homologues separate at anaphase I. Four different types of haploid egg cells are possible, as the diagram reminds us, and also four different types of sperm nuclei. These can combine in 4 × 4, or 16, different ways, as illustrated in the Punnett square. (Yellow is indicated by the color, green by gray.)*

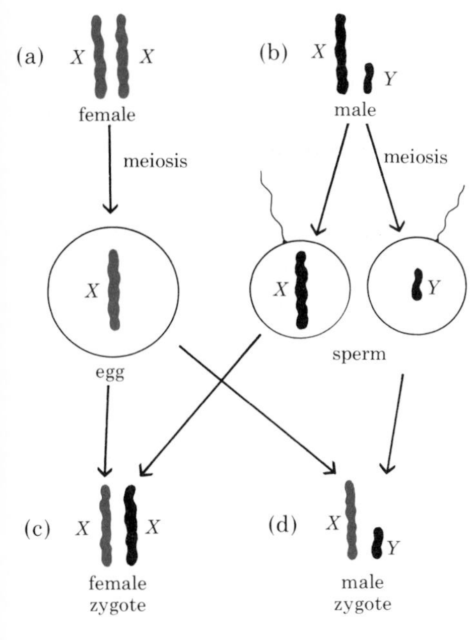

12-7 *How the sperm cell determines the sex of human offspring.* (a) *At meiosis, every egg cell receives an* X *chromosome from the mother.* (b) *A sperm cell may receive either an* X *chromosome or a* Y *chromosome.* (c) *If a sperm cell carrying an* X *chromosome fertilizes the egg, the offspring will be female* (XX). (d) *If a sperm cell carrying a* Y *chromosome fertilizes the egg, the offspring will be male* (XY).

On the basis of these parallels, Sutton proposed that the factors described by Mendel are carried on the chromosomes. Although Sutton's proposal was widely accepted as a working hypothesis, proof of the physical location of the gene depended on a series of studies that we shall describe in the remainder of this chapter.

Sex Determination

Strong evidence in support of Sutton's hypothesis was provided by the observation that male and female animals often show chromosomal differences. One pair of chromosomes differs between the sexes, and these are known as the *sex chromosomes.* All the other pairs of chromosomes, which are the same whether the animal is male or female, are known as *autosomes.* In the females of many species, the two sex chromosomes are structurally the same. These are, by convention, termed *X* chromosomes, and so the female is designated *XX*. The sex chromosomes of the male consist of one *X* chromosome, which is the same as the female *X* chromosome, and one *Y* chromosome. Thus the males of these species are designated *XY*. However, there are some animal groups, such as birds, in which the females are *XY* and the males *XX*.

In *XY* males, when the gametes are formed by meiosis, half of the sperm carry an *X* chromosome and half carry a *Y* chromosome. All the gametes formed by a female carry the *X*. The sex of the offspring is determined by whether the female gamete is fertilized by a male gamete carrying an *X* chromosome or a male gamete carrying a *Y* chromosome (Figure 12-7). Since equal numbers of *X* and *Y* sperm are produced, there is theoretically an even chance of male or female offspring. In humans, however, the ratio of male to female births is about 106 to 100. The reason for this is not known, but it has been suggested that the male-determining sperm may have an advantage in getting to the egg.

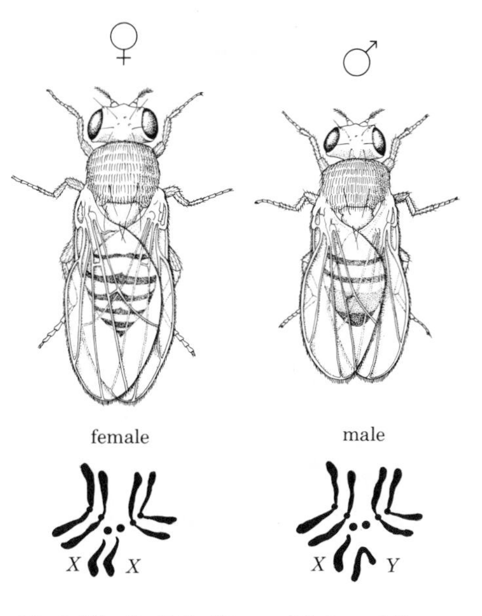

12-8 *The fruit fly* (Drosophila) *and its chromosomes. Fruit flies have only four pairs of chromosomes, a fact that simplified Morgan's experiments. Six of the chromosomes (three pairs) are autosomes (including the two small spherical chromosomes in the center) and two are sex chromosomes.*

The Golden Age of *Drosophila*

Early in the 1900s, T. H. Morgan began a study of genetics at Columbia University, founding what was to be the most important laboratory in the field for several decades. By a remarkable combination of insight and good fortune, he selected the fruit fly *Drosophila* as his experimental material. *Drosophila* means "lover of dew," although actually this little fly is not attracted by dew but feeds on the fermenting yeast that it finds in rotting fruit. The fruit fly was a likely choice for a geneticist since it is easy to breed and maintain. These tiny flies, each only 3 millimeters long, produce a new generation every two weeks. Each female lays hundreds of eggs at a time, and an entire population can be kept in a half-pint bottle, as they were in Morgan's laboratory. Also, *Drosophila* has only four pairs of chromosomes (Figure 12-8), a feature that turned out to be particularly useful, although Morgan could not have foreseen that.

Biologists have often used for their experiments "insignificant" plants and animals—such as pea plants, sea urchins, and fruit flies. Underlying this approach is the assumption that basic biological principles are universal, applying equally to all living things. As it turned out, the little fruit fly proved to be a "fit material" for a wide variety of genetic investigations. In the decades that followed, *Drosophila* was to become famous as the biologist's principal tool in studying animal genetics.

Calico Cats, Barr Bodies, and the Lyon Hypothesis

A dark spot of chromatin—called a Barr body—can be seen at the outer edge of the nucleus of female mammalian cells in interphase. According to the Lyon hypothesis—named after Mary Lyon, who proposed it—this dark spot is an inactivated *X* chromosome. Early in embryonic life, according to Lyon, this inactivation occurs in one or the other *X* chromosome in each cell of the female mammal (except for those cells from which egg cells will form). Thus all the somatic cells of female mammals are not identical but are one of two types, depending on which of the *X* chromosomes is active.

Calico cats (also sometimes known as tortoise-shell cats) have coats that are both black and yellow (actually an orange-yellow). They are also almost always female. In cats, the alleles for black or yellow coat color are carried on the *X* chromosome, so calico cats neatly fit the Lyon hypothesis. There are, however, occasional male calicos. These males have an extra *X* chromosome and are almost always sterile.

Sex-Linked Characteristics

The investigators in Morgan's laboratory were, at first, looking for genetic differences among individual flies that they could study in breeding experiments similar to those Mendel had carried out in the pea plant. One of the first differences to turn up was a mutant male fly with white eyes. (Red is the usual color of a fruit fly's eyes.) The white-eyed male was mated with a red-eyed female, and all the offspring had red eyes. Apparently the white-eyed male was a homozygous recessive; when it was crossed with a red-eyed female, presumably homozygous dominant, all of the F_1 offspring were red-eyed heterozygotes.

Morgan then crossbred the F_1 offspring, just as Mendel had done in his pea experiments. However, instead of the expected 3:1 ratio of dominant to recessive phenotypes (that is, of red-eyed to white-eyed individuals, regardless of sex), this is what resulted:

Red-eyed females	2,459
White-eyed females	0
Red-eyed males	1,011
White-eyed males	782

The startling thing about these results was that all of the white-eyed flies of the F_2 generation were males. Why were there no white-eyed females? To explore the situation further, Morgan crossed the original white-eyed male with one of the F_1 (red-eyed) females. The following results were obtained from this testcross:

Red-eyed females	129
White-eyed females	88
Red-eyed males	132
White-eyed males	86

Morgan and his co-workers examined these figures and on the basis of them proposed the following hypothesis: The gene for eye color is carried only on the *X* chromosome. (In fact, as it was later shown, the *Y* chromosome carries very little genetic information.) The allele for white eyes must indeed be recessive, since all of the F_1 flies had red eyes. Thus a heterozygous female would have red eyes—which is why there were no white-eyed females in the F_2 generation. However, a male that received an *X* chromosome carrying the allele for white eyes would always be white-eyed since no other allele would be present.

Further experimental crosses proved Morgan's hypothesis to be right (Figure 12–9). They also showed that white-eyed fruit flies are more likely to die before they hatch than are red-eyed fruit flies, which explains their lower-than-expected numbers in the F_2 generation and the testcross.

These experiments introduced the concept of sex-linked traits. As we shall see, such traits are important in human genetics. The experiments also established what Sutton had hypothesized: Genes *are* on chromosomes.

Linkage

Mendel showed that certain pairs of alleles, such as those for round and wrinkled peas, assort independently of other pairs, such as those for yellow and green peas. However, as we noted previously, two pairs of alleles assort independently only if they are on different pairs of homologous

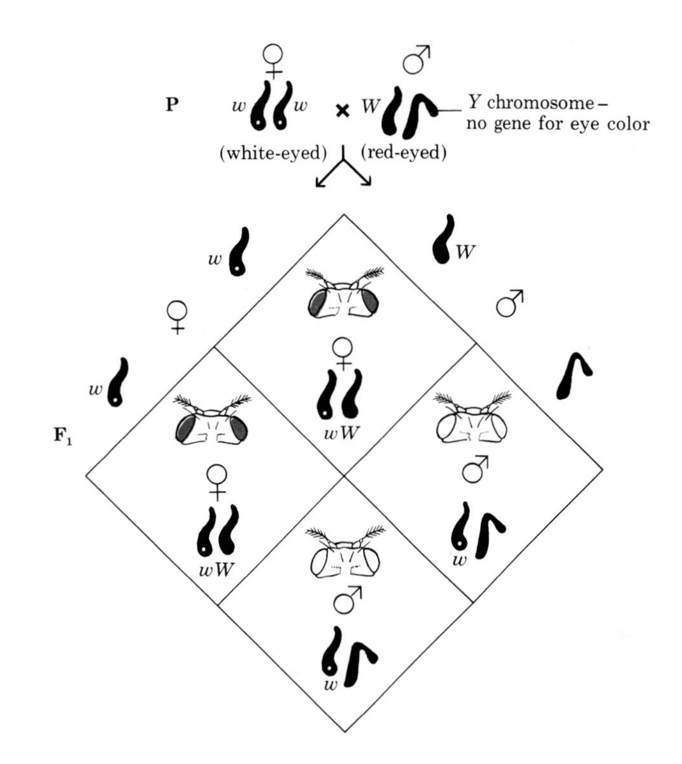

12-9 *Offspring of a cross between a white-eyed female fruit fly and a red-eyed male fruit fly, illustrating what happens when a recessive gene, indicated by the white dot, is carried on an X chromosome. The* F_1 *females, with one X chromosome from the mother and one from the father, are heterozygous* (Ww) *and so will be red-eyed. But the* F_1 *males, with their single X chromosome received from the mother carrying the recessive* (w) *trait, will be white-eyed because the* Y *chromosome carries no gene for eye color. Thus the recessive allele on the X chromosome inherited from the mother will be expressed.*

chromosomes. If two pairs of alleles are on the same pair of homologous chromosomes, then segregation of one pair of alleles will not be independent of the segregation of the other pair. In other words, if the alleles of two different genes are on the same chromosome, they should both be transmitted to the same gamete at meiosis. Genes that tend to stay together because they are on the same pair of homologous chromosomes are said to be in the same *linkage group.*

As increasing numbers of *Drosophila* mutants were isolated and analyzed at Columbia University, it became clear that the mutations fell into four linkage groups, in accord with the four pairs of chromosomes visible in the cells. Indeed, in all organisms that have been studied in sufficient genetic detail, the number of linkage groups and the number of pairs of chromosomes have been the same, providing further support for Sutton's hypothesis.

Crossing Over

Large-scale studies of linkage groups soon revealed some unexpected difficulties. For instance, most fruit flies have gray bodies and long wings, both dominant traits. When individuals homozygous for these traits were bred with mutant fruit flies having black bodies and short wings (both recessive traits), all the F_1 offspring had gray bodies and long wings, as would be expected. Then the F_1 generation was inbred. Two outcomes seemed possible:

1. The genes for body color and wing length would be assorted independently, giving rise to Mendel's 9:3:3:1 ratios in the phenotypes and indicating that the genes for these traits were on different chromosomes.

2. The genes for the two traits would be linked. In this case, 75 percent of the flies would be gray with long wings and 25 percent, homozygous for both recessives, would be black with short wings.

In the case of these particular traits, the results most closely resembled the second possibility, but they did not conform exactly. In a few of the offspring, the genes for these traits seemed to assort independently; that is, some few flies appeared that were gray with short wings, and some that were black with long wings. How could this be? Somehow genes that were usually linked together had become separated.

To find out what was happening, Morgan tried a testcross, breeding a member of the F_1 generation with a homozygous recessive. If black and gray, long and short assorted independently—that is, if they were on different chromosomes—25 percent of the offspring of this cross should be black with long wings, 25 percent gray with long wings, 25 percent black with short wings, and 25 percent gray with short wings. On the other hand, if the genes for color and wing size were on the same chromosome and so moved together, half of the offspring of the testcross should be gray with long wings and half should be black with short wings. But actually, as it turned out, over and over, in counts of hundreds of fruit flies resulting from such crosses, 42 percent were gray with long wings, 42 percent were black with short wings, 8 percent were gray with short wings, and another 8 percent were black with long wings.

Morgan was convinced by this time that genes are located on chromosomes. It now seemed clear that the genes controlling the two traits were located on a single pair of homologous chromosomes since the traits did not show up in the 25:25:25:25 percentage ratios of independently assorted gene pairs. The only way in which the observed figures could be explained was if the members of one pair of alleles could be exchanged between the two homologous chromosomes without disturbing the other pair of alleles.

As we noted in Chapter 10, it now has been established that exchange of portions of homologous chromosomes—crossing over—takes place at the beginning of meiosis. Presumably, the actual exchange takes place when the homologues are closely grouped early in the first meiotic prophase. As the homologues begin to separate, the points at which crossing over has occurred become visible.

12-10 *Homologous chromosomes of a grasshopper, as seen in prophase I. All four chromatids are visible. Crossing over—the exchange of genetic material—has probably occurred at the points at which these chromatids intersect. The arrows indicate the position of the centromeres.*

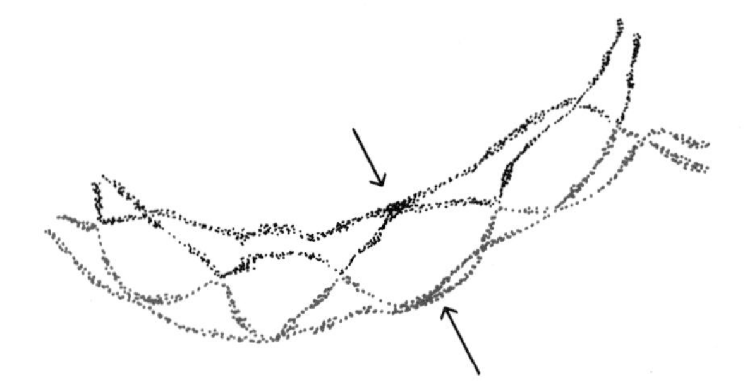

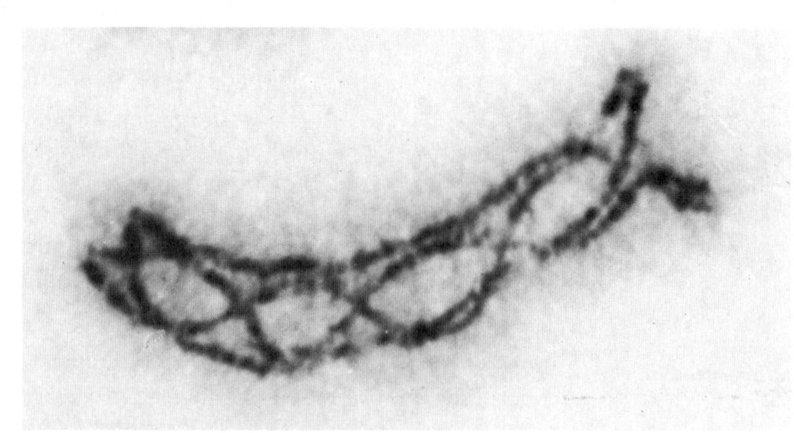

2 μm

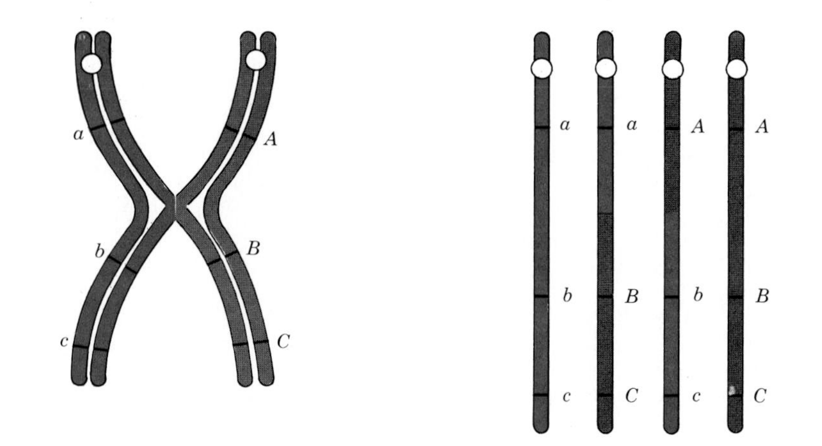

12–11 *Crossing over takes place when breaks occur in the chromatids of homologous chromosomes during prophase I, when the chromosomes are paired. The broken end of each chromatid joins with the chromatid of a homologous chromosome. In this way, alleles are exchanged between chromosomes. The white circles symbolize centromeres.*

Mapping the Chromosome

With the discovery of crossovers, it began to seem clear not only that the genes are carried on the chromosomes, but also that they must be positioned at particular spots, or *loci* (singular, locus), on the chromosomes. Furthermore, the alleles of any given gene must occupy corresponding loci on homologous chromosomes. Otherwise, exchange of sections of chromosomes would result in genetic chaos rather than in an exact exchange of alleles.

As other traits were studied, it became clear that the percentage of separations, or crossovers, between any two genes, such as those for gray body and long wing, was different from the percentage of crossovers between two other genes, such as those for gray body and long leg. In addition, as Morgan's experiments had shown, these percentages were fixed and predictable. It occurred to A. H. Sturtevant, one of the many brilliant young geneticists attracted to Morgan's laboratory, that the percentage of crossovers probably had something to do with the distances between genes, or in other words, with their spacing along the chromosome. This simple concept opened the way to the "mapping" of chromosomes.

Sturtevant postulated (1) that genes are arranged in a linear series on chromosomes; (2) that genes that are close together will be separated by crossing over less frequently than genes that are farther apart; and (3) that it should therefore be possible, by determining the frequencies of crossovers, to plot the sequence of the genes along the chromosome and the relative distances between them. In Figure 12–11, for example, you can see that in a crossover, the chances of a strand's breaking and recombining with its homologous strand somewhere between *B* and *C* should be less likely than this happening somewhere between *A* and *C*.

In 1913, Sturtevant began constructing chromosome maps using data from crossover studies in fruit flies. As a standard unit of measure, he arbitrarily took the distance that would give (on the average) one crossover per 100 gametes. The genes with 10 percent crossover would be 10 units apart; those with 8 percent crossover would be 8 units apart. By this method, he and other geneticists constructed maps locating a wide variety of genes and their mutants in *Drosophila* (Figure 12–12). The distances on such maps cannot be absolute because the distance unit is an arbitrary one.

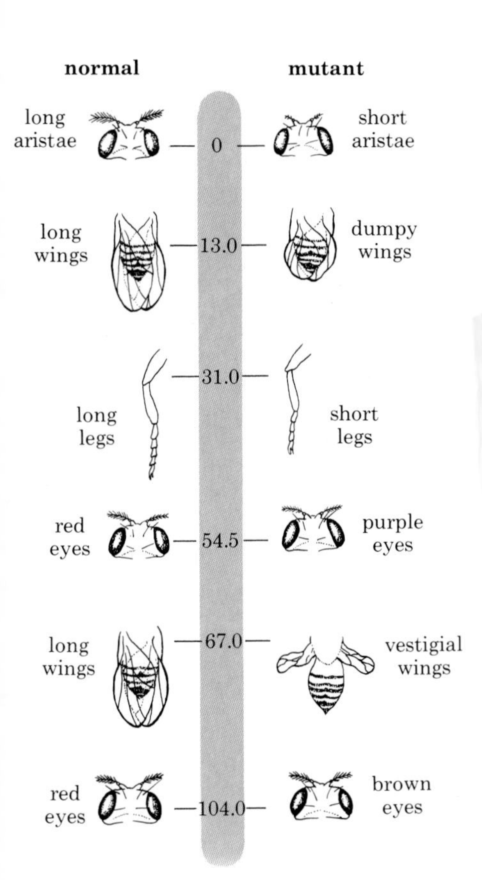

12–12 *A portion of the genetic map of* Drosophila melanogaster, *showing some of the genes on chromosome 2 and their relative positions, as calculated by the frequency of crossovers. As you can see, more than one gene may affect a single characteristic, such as eye color.*

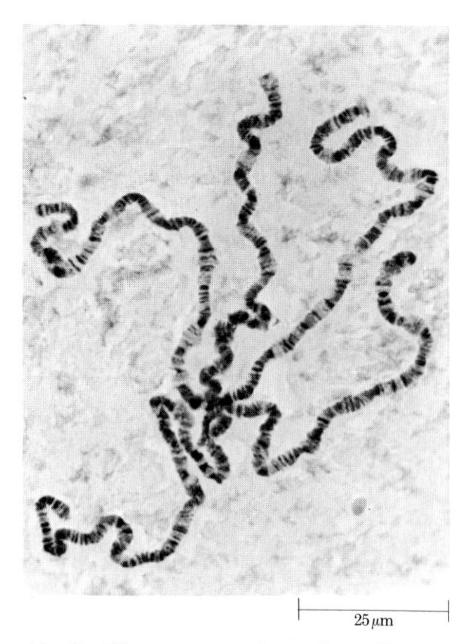

12–13 *Chromosomes from the salivary gland of a* Drosophila *larva. These chromosomes are 100 times larger than the chromosomes in ordinary body cells, and their details are therefore much easier to see. (This micrograph was taken with a light microscope, not an electron microscope.) Because of the distinctive banding patterns, it is possible in some cases to assign genes to specific bands on particular chromosomes.*

The relative distances may also be in error in some cases because breaking and rejoining of chromosome segments may be more likely to occur at some sites on the chromosome than others.

These studies confirmed that genes are arranged linearly on chromosomes. However, the question of *how* homologous chromosomes break in exactly corresponding sites and so exchange equal amounts of genetic material at crossing over is still under investigation.

Giant Chromosomes

In *Drosophila,* as in many other insects, certain cells do not divide during the larval (immature) stage of the insect. In such cells, however, the chromosomes continue to replicate, over and over again, but since daughter chromosomes do not separate after replication, they simply become larger and larger. In 1933, such giant chromosomes were reported in the salivary glands of the larvae of *Drosophila.* As you can see in Figure 12–13, these giant chromosomes are marked by very distinctive dark and light bands. The dark bands are areas in which the chromatin is more tightly packed. Changes in these banding patterns were found to correlate with observed genetic changes in the flies. In addition, the results of crossovers and breaks could actually be *seen.* Thus further confirmation of Sturtevant's brilliant hypothesis came 20 years later from an unexpected source.

SUMMARY

The rediscovery of Mendel's work in 1900 was the catalyst for many new discoveries in genetics, leading to the modification of some of Mendel's conclusions. In this chapter we examined some of the major concepts that resulted from the "Golden Age of Genetics."

Mutations are abrupt changes in the genotype. They are the ultimate source of the genetic variations that Mendel studied and that provide the raw material for evolution.

Incomplete dominance is a situation in which the effects of both alleles are apparent in the heterozygote.

Some genes affect two or more superficially unrelated characteristics; this property of a gene is known as pleiotropy.

Many characteristics are under the control of a number of separate genes. This phenomenon is known as polygenic inheritance. Such traits typically show continuous variation, as represented by a bell-shaped curve.

Sutton was among the first to notice the analogy between the behavior of the chromosomes at meiosis and the segregation and assortment of genes described by Mendel. On the basis of this observation, Sutton proposed that genes are carried on chromosomes. This hypothesis led to the recognition of an important exception to Mendelian principles: Independent assortment is modified if the particular genes involved are on the same chromosome.

The fruit fly *Drosophila* has been used in a wide variety of genetic studies. It has four pairs of chromosomes; three pairs are structurally the same in both sexes, but the fourth pair, the sex chromosomes, is different. In fruit flies, as in many other species (including humans), the two sex chromosomes are *XX* in females and *XY* in males.

At the time of meiosis, the sex chromosomes are segregated. Each egg cell receives an *X* chromosome, but half the sperm cells receive *X* chromosomes and half receive *Y* chromosomes. Thus it is the sperm cell that determines the sex of the embryo in species with *XY* males.

In the early 1900s, experiments with mutations in the fruit fly showed that certain characteristics are sex-linked. Recessive genes carried on the *X* chromosomes appear in the phenotype far more often in males than in females (the *Y* chromosome carries much less genetic information than the *X* chromosome); a female heterozygous for a sex-linked characteristic will show the dominant trait, whereas a single recessive gene in the male, if carried on the *X* chromosome, will result in a recessive phenotype since no other allele is present.

Some genes assort independently in breeding experiments, and others tend to remain together. Genes that tend to travel together are said to be in the same linkage group; a linkage group consists of a pair of homologous chromosomes.

Genes are arranged in a fixed linear array along the length of the chromosomes, and the alleles of a given gene are at corresponding sites (loci) on homologous chromosomes. Alleles are sometimes exchanged between the chromatids of homologous chromosomes at meiosis (crossing over). Measurements of the frequency of exchange permit the construction of maps showing the relative positions of gene loci along the chromosome.

QUESTIONS

1. The so-called "blue" (really gray) Andalusian variety of chicken is produced by a cross between the black and white varieties. Only a single pair of alleles is involved. What color chickens (and in what proportions) would you expect if you crossed two blues? If you crossed a blue and a black? Explain.

2. In one strain of mice, skin color is determined by five different pairs of alleles. The colors range from almost white to dark brown. Would it be possible for any given pair of mice to produce offspring darker or lighter than either parent? Explain.

3. Height and weight in animals follow a distribution similar to that shown in Figure 12–4. By inbreeding large animals, breeders are usually able to produce some increase in size among their stock. But after a few generations, increase in size characteristically stops. Why?

4. Draw a diagram similar to Figure 12–7 indicating sex determination in a robin.

5. A couple has three girls. What are the chances that the next child will be a boy?

6. Suppose you would like to have a family consisting of two girls and a boy. What are your chances, assuming you have no children now? If you already have one boy, what are your chances of completing your family as planned?

7. A diploid organism has 42 chromosomes per cell. How many linkage groups does it have?

8. The first two pairs of genes studied simultaneously in crosses were *Rr* and *Yy* in pea plants. As it turned out, they were on different chromosomes. How do you think the development of the principles of genetics would have been affected had they been linked?

9. The genes for coat color in cats are carried on the X chromosome. Black *(b)* is the recessive and yellow *(B)* is the dominant. What coat colors would you expect in the offspring of a cross between a black female and a yellow male? What coat colors would you expect in the sons of a calico female, regardless of the coat color of the father?

10. In a series of breeding experiments, a linkage group composed of genes A, B, C, D, and E was found to show approximately the following crossover frequencies:

Gene	A	B	C	D	E
A	–	8	12	4	1
B	8	–	4	12	9
C	12	4	–	16	13
D	4	12	16	–	3
E	1	9	13	3	–

Crossovers per 100 Gametes

Using Sturtevant's standard unit of measure, "map" the chromosomes.

11. You and a geneticist are looking at a mahogany-colored Ayrshire cow with a newly born red calf. You wonder if it is male or female, and the geneticist says it is obvious from the color which sex the calf is. He explains that in Ayrshires the genotype AA is mahogany and aa is red, but the genotype Aa is mahogany in males and red in females. What is he trying to tell you—that is, what sex is the calf? What are the possible phenotypes of the calf's father?

13

Human Genetics

The principles of genetics are, of course, the same for humans as they are for members of any other species. There are some important differences, however, in the methodology of genetic studies. Breeding experiments, so readily performed with fruit flies and garden peas, are not possible with humans. Most knowledge of patterns of human heredity is based on family pedigrees, and the pedigrees are usually worked out only when a medical problem is apparent. The human propensities for speech and record-keeping, however, often make it possible to obtain information that stretches back several generations. As a result, a large quantity of data has been accumulated on certain traits and their transmission.

Mendelian Inheritance

Transmission by Dominant Genes

Many human characteristics are governed by simple Mendelian inheritance of dominant genes. Among these are cleft chin, polydactylism (extra fingers), and brown teeth. Certain diseases, such as Huntington's disease, which afflicted the folk singer Woody Guthrie, are also caused by dominant genes. As you can rapidly calculate, each child of a parent heterozygous for one of these traits has a fifty-fifty chance of inheriting the gene. In the case of Huntington's disease, the usual age of onset of the disease is between 35 and 45. By that time, individuals with the disease have usually had children, who must then spend their lives waiting to see whether or not the disease will occur.

13-1 *Woody Guthrie, the well-known folk singer and songwriter, inherited the gene for Huntington's disease from his mother. The disease, a progressive deterioration of the nervous system, usually causes death 5 to 15 years after onset.*

Transmission by Recessive Genes

A number of other diseases are inherited as simple Mendelian recessives. One of the most familiar examples is phenylketonuria, or PKU. It is caused by a lack of function of the enzyme that normally breaks down the amino acid phenylalanine. When this enzyme is missing or deficient, phenylalanine and its abnormal breakdown products accumulate in the bloodstream. These breakdown products are harmful to the developing cells of the brain and can result in mental retardation. PKU is caused by a

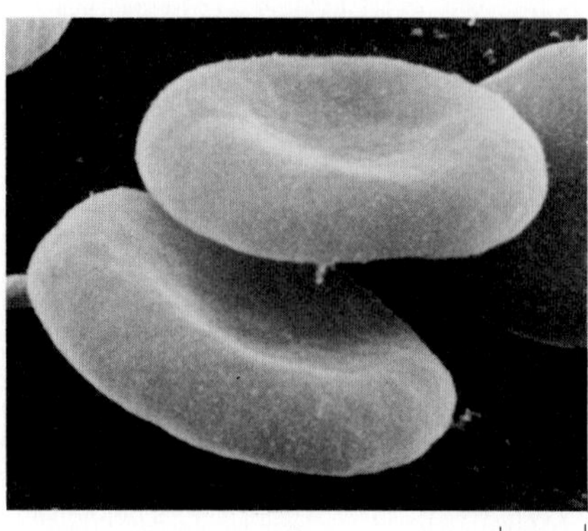

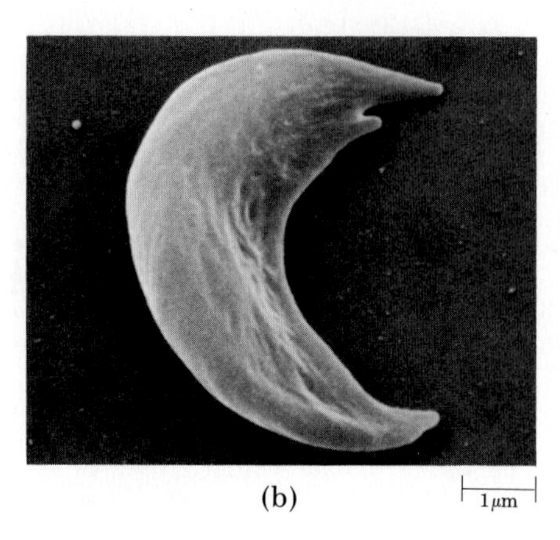

13–2 *Scanning electron micrographs of* (a) *red blood cells containing normal hemoglobin, and* (b) *red blood cell containing the abnormal hemoglobin associated with sickle cell anemia. The abnormal hemoglobin molecules stick together, distorting the shape of the cells. As a result, the cells cannot pass readily through the capillaries.*

single recessive gene in the homozygous state. The disease can, however, be detected shortly after birth and successfully treated (see essay on page 178).

Another disease caused by a homozygous recessive gene is sickle cell anemia. This gene causes the production of structurally abnormal hemoglobin. In the heterozygote, the sickle cell gene is usually detectable only by laboratory tests. In persons homozygous for the sickling gene, however, a large proportion of the red blood cells "sickle"—that is, form a sickle shape—and clog the small capillaries. This causes often painful blood clots and deprives vital organs of their full supply of blood, resulting in intermittent illness and often a shortened life span. This hereditary disease occurs with a high frequency among blacks. In the United States about 9 percent of blacks are heterozygous for the sickle cell gene, and about 0.2 percent are homozygous and therefore have the symptoms of sickle cell anemia.

Multiple-Allele Inheritance

Although a single individual can have only two alleles for any single gene, many alleles can exist within a population. Probably the most familiar characteristic in human beings that is determined by a group of alleles of a single gene is the ABO blood series.

A, B, and O are all alleles of the same gene, and all affect the surface of the red blood cell. Individuals with at least one A allele have a polysaccharide called A on the surface of their blood cells. Individuals with at least one B allele have another polysaccharide, called B. Individuals with one A allele and one B allele have both polysaccharides. Individuals with two O alleles have neither polysaccharide. If you receive a transfusion of blood cells containing foreign polysaccharide—that is, one not present in your own body—you will have an immune reaction against it. This immune reaction causes the foreign blood cells to clump together, or agglutinate (Figure 13–3), and can be severe enough to cause death.

The genetic basis of the ABO blood types is shown in Table 13–1. Alleles A and B are codominants; that is, the effects of one are not masked by the effects of the other. If you have AB blood, it means that one of your parents is type A or AB and the other is type B or AB. If you have A blood, it means that you inherited an A allele from one parent and either an A allele or an O allele from the other. If you have O-type blood, you must have had parents who each carried an allele for O, although they may have been, phenotypically, either A or B.

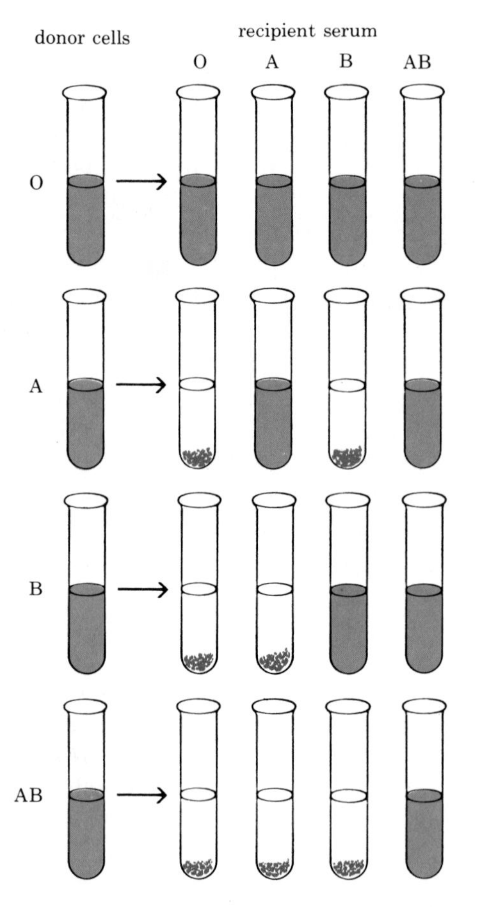

13–3 *Severe and sometimes fatal reactions can occur following transfusions of blood of a different type from the recipient's. These reactions are the result of agglutination of the blood cells caused by specific proteins, called antibodies, present in the recipient's serum. Blood-group reactions can be demonstrated equally well in test tubes, as shown here. For example, when type A donor cells are mixed with type O recipient serum, the donor cells agglutinate and settle to the bottom of the test tube. Persons with type O blood used to be called universal donors and those with type AB blood, universal recipients. Now other factors are checked as well.*

Table 13-1 Genetic basis of the ABO blood types

Phenotype (Polysaccharides Made)	Genotype (Alleles Present)	Reaction with Antibodies		Antibodies in Blood Plasma
		Antibody A	Antibody B	
O	O/O	No	No	Antibody A, antibody B
A	A/A, O/A	Yes	No	Antibody B
B	B/B, O/B	No	Yes	Antibody A
AB	A/B	Yes	Yes	None

Discovery of the ABO blood groups, which has made blood transfusions safe and practical, ranks among the great medical advances of the twentieth century. A number of additional blood factors controlled by other genes have also been found, and they too are checked in modern blood banks to determine blood compatibility.

Sex-Linked Characteristics

Color Blindness

As in *Drosophila*, the *Y* chromosome of a human male carries much less genetic information than the *X* chromosome. Genes for color vision, for example, are carried on the *X* chromosome in humans but not on the *Y* chromosome. Color blindness is produced by a recessive allele of the normal gene. The allele for complete color vision is dominant; a woman with one *X* chromosome with the normal allele and one *X* chromosome with the allele for color blindness will have normal color vision. If she transmits the *X* chromosome with the recessive allele to a daughter, the daughter will also have normal color vision if she receives an *X* chromosome with the normal allele from her father (that is, if he is not color-blind). If, however, the *X* chromosome with the recessive allele is transmitted from mother to son, he will be color-blind since, lacking a second *X* chromosome, he has only the recessive allele (Figure 13-4). Nonsexual characteristics, such as color blindness, that are controlled by alleles on the *X* chromosome are said to be *sex-linked.*

13-4 *Color blindness in humans is determined by a recessive allele on the* X *chromosome. In the pedigree shown here, the mother has inherited one normal and one defective allele. The normal allele will be dominant, and she will have normal color vision. However, half her eggs (on the average) will carry the defective allele and half will carry the normal allele—and it is a matter of chance which kind is fertilized. Since her husband's* Y *chromosome, the one that determines a son rather than a daughter, carries no gene for color discrimination, the single allele the wife contributes (even though it is a recessive allele) will determine whether or not the son is color-blind. Therefore, half her sons (on the average) will be color-blind. Assuming that her children marry individuals with* X *chromosomes with the normal alleles, the expected distribution of the trait among her grandchildren will be as shown on the chart.*

normal female

normal male

carrier female

color-blind male

X normal *X* chromosome

X chromosome with gene for color blindness

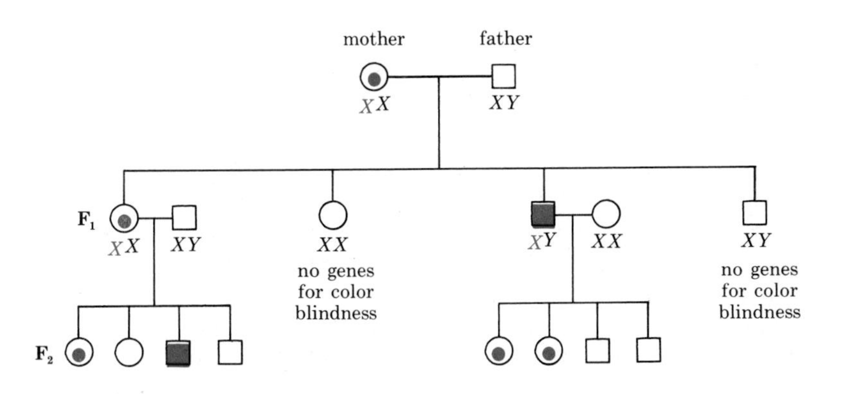

Hemophilia

A classic example of a recessive allele transmitted on the X chromosome is the type of hemophilia that has afflicted some royal families of Europe since the nineteenth century. Hemophilia is a group of diseases in which the blood does not clot normally. In some kinds of hemophilia, even minor injuries carry the risk of the patient's bleeding to death.

Queen Victoria was probably the original carrier in her family (Figure 13–5). Because none of her relatives other than her descendants were affected, we conclude that the mutation occurred on an X chromosome in one of her parents or in the cell line from which her own eggs were formed. Prince Albert, Victoria's husband, could not have been responsible because male-to-male inheritance of the disease is impossible. (Why?) One of her sons, Leopold, Duke of Albany, died of hemophilia at the age of 31. At least two of Victoria's daughters were carriers (see photograph on pages 124–125), since a number of *their* descendants were hemophiliacs. And so, through various intermarriages, the disease spread from throne to throne across Europe. In Tsarevitch Alexis, son of the last Tsar of Russia, the gene for hemophilia, inherited from Victoria, had considerable political consequences.

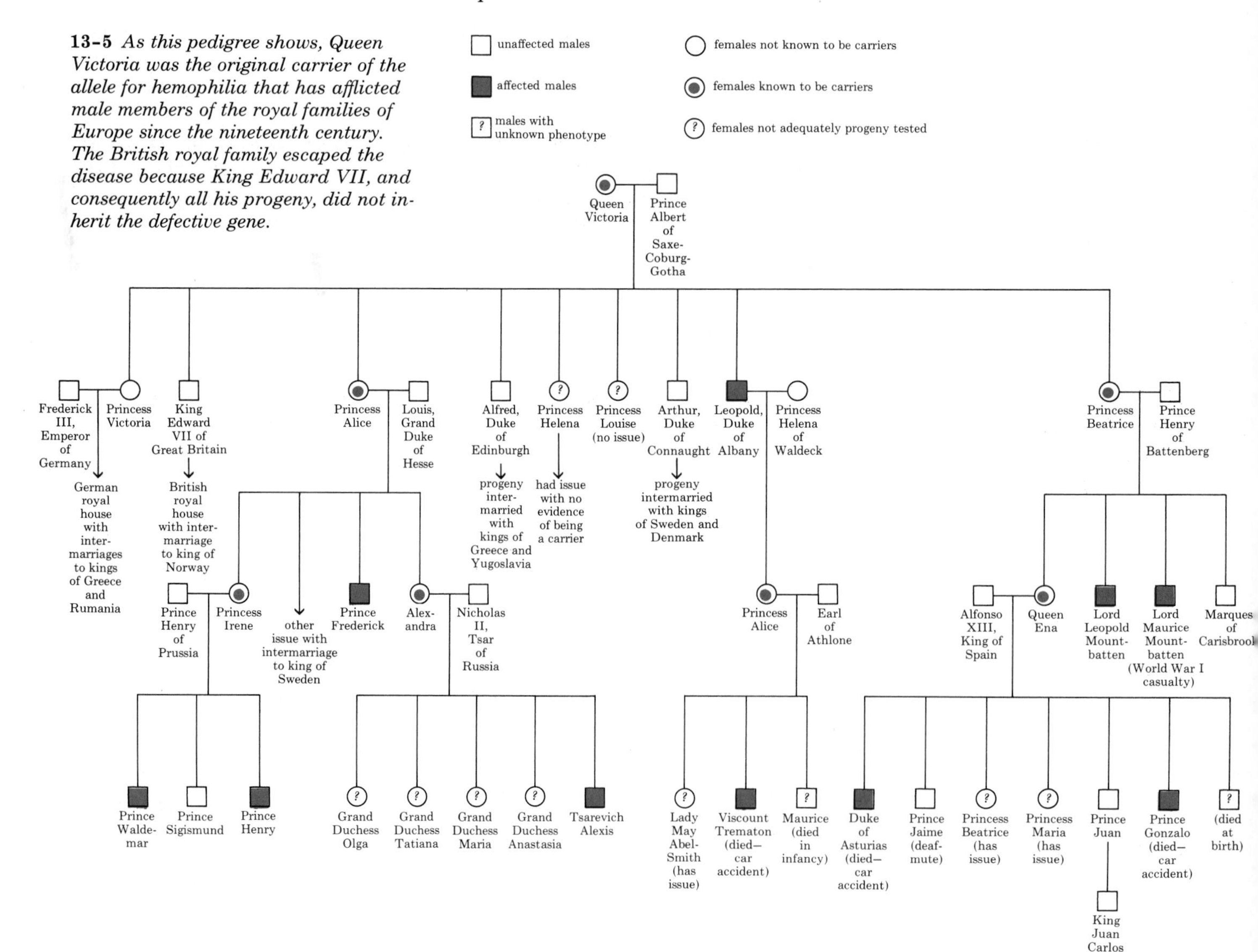

13–5 *As this pedigree shows, Queen Victoria was the original carrier of the allele for hemophilia that has afflicted male members of the royal families of Europe since the nineteenth century. The British royal family escaped the disease because King Edward VII, and consequently all his progeny, did not inherit the defective gene.*

13-6 *Gregory Efimovitch Rasputin (1871–1916). Nicholas II and Alexandra of Russia fell under the sinister influence of Rasputin because of the hemophilia of their son Alexis, on whom Rasputin exerted apparently mystical healing powers. The gene for hemophilia had been inherited from Queen Victoria. Alexis did not die of his disease but was executed with other members of the royal family in 1918.*

Chromosome Abnormalities

From time to time, because of a "mistake" during meiosis, chromatids may not separate properly. This phenomenon is known as *nondisjunction.* The results of nondisjunction are sex cells with one chromosome too many and other sex cells with one chromosome too few. A cell with one too few (unless it is a *Y*) cannot produce a viable embryo, but a cell with one too many sometimes can. The result is an individual with an extra chromosome in every cell of his or her body.

The presence of additional chromosomes often produces widespread abnormalities. Many infants with such abnormalities are stillborn; among those who survive, many are mentally deficient. These individuals frequently have abnormalities of the heart and other organs as well.

Down's Syndrome

One of the most familiar chromosomal abnormalities is the form of mental deficiency known as Down's syndrome, after the physician who first described it. (The disease is also sometimes called mongolism.) Down's syndrome usually involves more than one defect and so is referred to as a syndrome, a group of disorders that occur together. The syndrome includes, in most cases, not only mental deficiency but also a short, stocky body type with a thick neck and, often, abnormalities of other organs, especially the heart.

Down's syndrome arises when a child inherits three, rather than two, copies of chromosome 21. In about 95 percent of the cases, the cause of the genetic abnormality is nondisjunction, which results in an extra copy of chromosome 21 in the cells of the child (Figure 13–7b).

13-7 (a) *The normal diploid chromosome number of a human being is 46, 22 pairs of autosomes and the two sex chromosomes. In this karyotype (see essay on next page), the autosomes are grouped by size* (A, B, C, *etc.*), *and then the probable homologues are paired. A normal woman has two* X *chromosomes and a normal man, shown here, an* X *and a* Y.

(b) *The karyotype of a male with Down's syndrome caused by nondisjunction. Note that there are three chromosomes 21.*

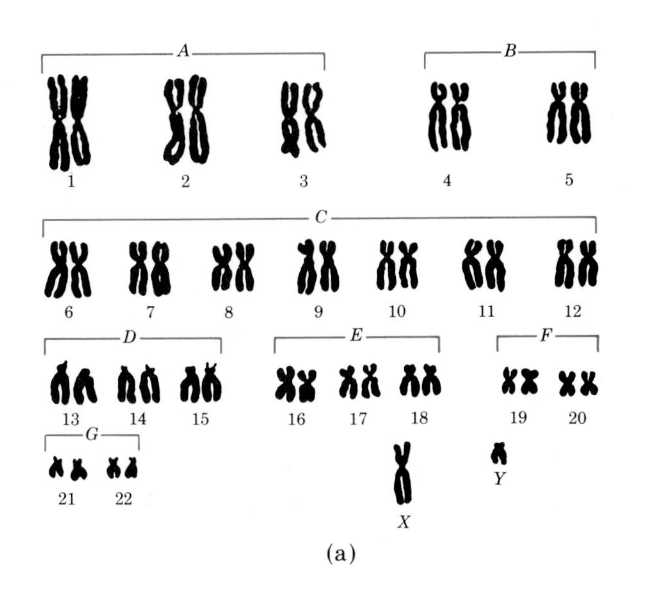

A
B
1 2 3 4 5
C
6 7 8 9 10 11 12
D
E
F
13 14 15 16 17 18 19 20
G
21 22
X
Y
(b)

Preparation of a Karyotype

Chromosome typing for the identification of hereditary defects is being carried out at an increasing number of genetic counseling centers throughout the United States. The result of the procedure is known as a karyotype. The chromosomes shown in a karyotype are metaphase chromosomes, each consisting of two sister chromatids held together at their centromeres. White blood cells in the process of dividing have been interrupted at metaphase by the addition of colchicine, which prevents the subsequent steps of mitosis from taking place. After treating and staining, the chromosomes are photographed, enlarged, cut out, and arranged according to size. Certain abnormalities, such as an extra chromosome or piece of a chromosome, can be detected.

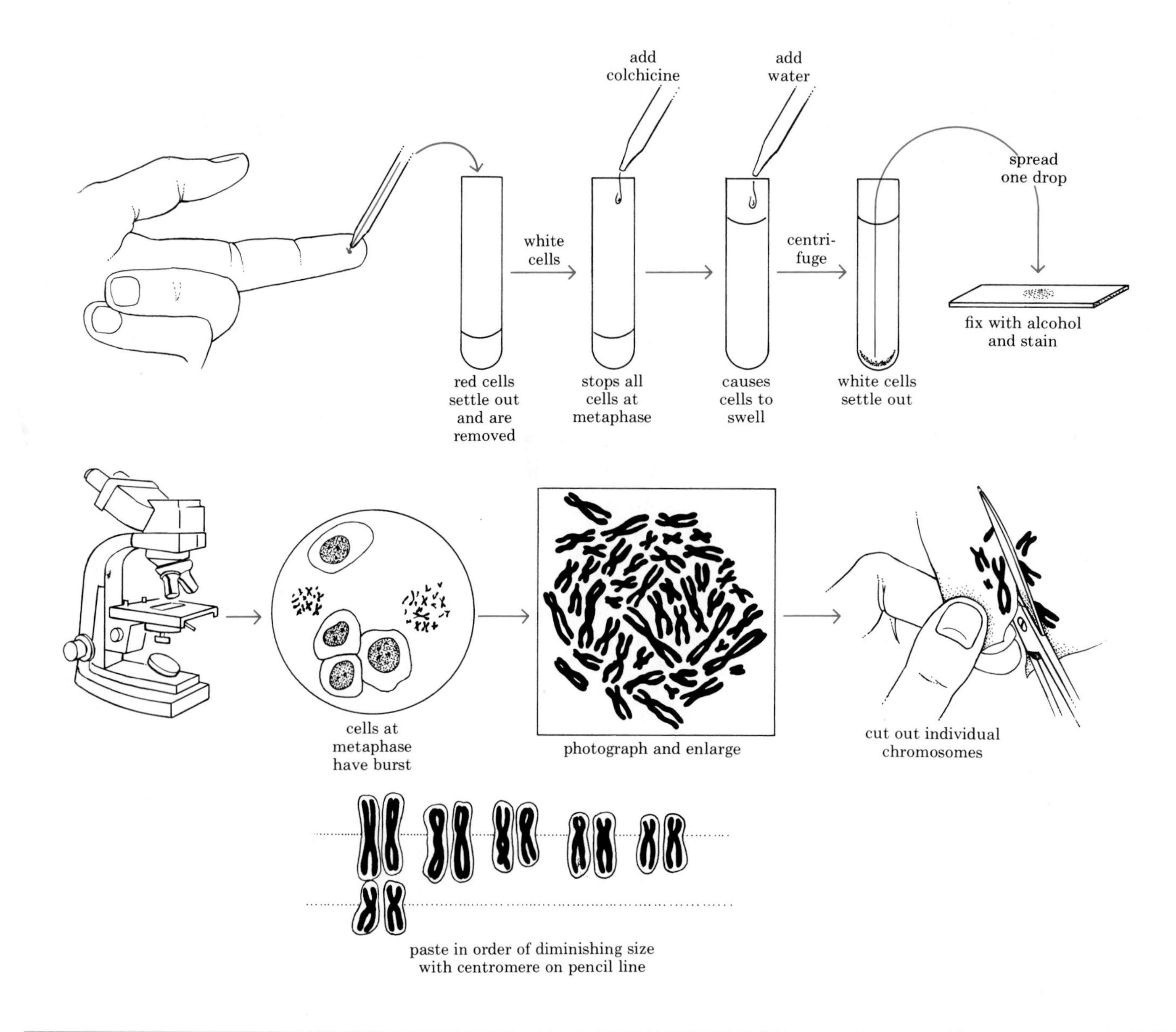

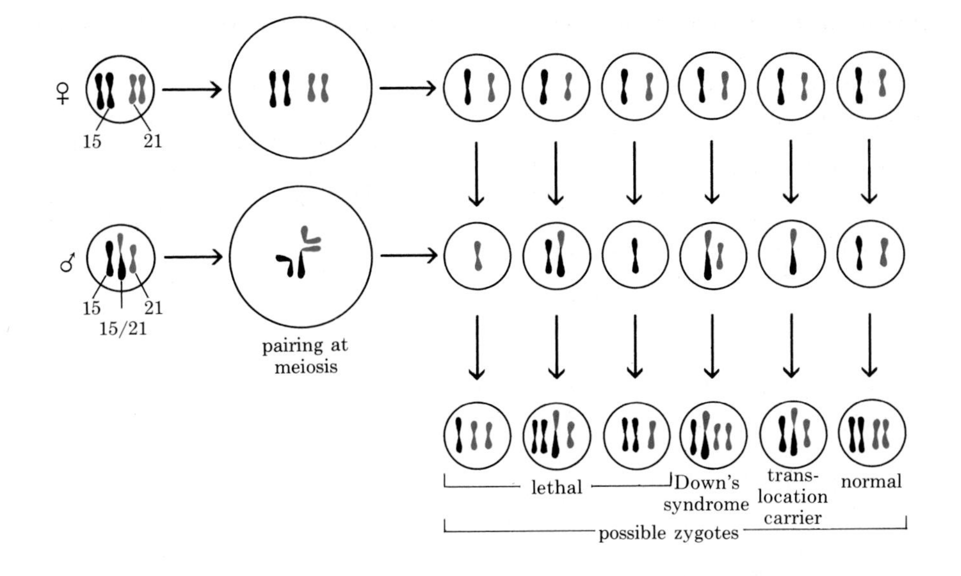

13-8 *Transmission of a translocation defect. The mother, top row, has normal pairs of chromosomes 21 and 15, and each of her egg cells will contain a normal 21 and a normal 15. The father (more frequently, although not always, the translocation carrier) has one normal 15, one normal 21, and a translocation 15/21. He himself appears normal, but his chromosomes cannot pair normally at meiosis. There are six possibilities for the offspring of these parents: The infant will (1) die before birth (three of the six possibilities), (2) have Down's syndrome, (3) be a translocation carrier like the father, or (4) be normal. Tests for the chromosomal abnormality can be made in prospective parents and in the fetus before birth.*

Down's syndrome may also be the result of an abnormality, known as *translocation,* in the chromosomes of one of the parents. Translocation occurs when a portion of a chromosome is broken off and becomes attached to another chromosome. The person with Down's syndrome caused by translocation usually has a third chromosome 21 (or, at least, most of it) attached to a larger chromosome, such as chromosome 15.

When cases of Down's syndrome due to translocation are studied, it is usually found that one parent, although phenotypically normal, has only 45 separate chromosomes. One chromosome is composed of most of chromosomes 15 and 21 joined together. The possible genetic makeups of the offspring of this parent are diagrammed in Figure 13-8. Three out of the six possible combinations are lethal. One of the remaining three will produce Down's syndrome, one will be normal, and one will be a carrier.

Thus parents who have had a child with Down's syndrome are advised to have their karyotypes prepared. If either parent shows an abnormality, they are warned that they may transmit the condition. If the karyotypes of both parents are normal, they do not run a greater-than-average risk of having another congenitally defective child.

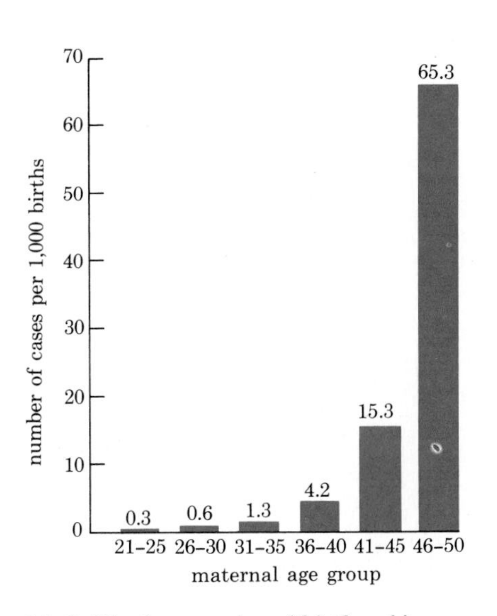

13-9 *The frequencies of births of infants with Down's syndrome in relation to the ages of the mothers. The number of cases shown for each age group represents the occurrence of Down's syndrome in every 1,000 births by mothers in that group. As you can see, the risk of having a child with Down's syndrome increases rapidly after the mother's age exceeds 40. A similar increased risk occurs after the father's age exceeds 55.*

It has been known for many years that Down's syndrome and a number of other defects involving nondisjunction are more likely to occur among infants born to older women (Figure 13-9). The reasons for this are not known, but the formation of egg cells is well under way in the human female before she is born, so the increasing incidence of abnormalities may be correlated in some way with the aging of the mother's reproductive cells. Recent studies, however, indicate that in about 25 percent of Down's syndrome cases, the extra chromosome comes from the father rather than the mother. Moreover, the frequency of Down's syndrome increases with increasing paternal age as well as with increasing maternal age. Paradoxically, very young parents are also at a high risk of having children with Down's syndrome.

Abnormalities in the Sex Chromosomes

Nondisjunction may also produce individuals with extra sex chromosomes. An *XY* combination in the twenty-third pair, as you know, produces maleness, but so do *XXY*, *XXXY*, and even *XXXXY*. These latter

The Social Cost of Genetic Disease

As a conservative estimate, some 120,000 infants in the United States are born each year with genetic diseases. The financial cost to society is tremendous. The emotional cost is, of course, incalculable. At present, our capacity to reduce these costs is very limited.

PKU is an example of a genetic disease that can be treated. If it is detected early and the child is kept on a carefully controlled diet during the first six years of life, he or she will develop into a normal adult. The only method of detection presently available is to routinely test the urine or blood of all newborn infants for abnormally high levels of phenylalanine. Although this is a very costly procedure, it is now being done in most hospitals nationwide. Is it worth the cost? For this particular disease, it is indeed financially worth the expense, quite apart from humanitarian considerations.

A "cost-benefit" analysis shows that the costs of *not* diagnosing and treating PKU children are greater than those of providing them with treatment. The test costs $1.25 per child. That comes to about $4,000,000 to test the 3,000,000 children born each year in the United States. The incidence of PKU is 1 in 15,000 births, so the testing should uncover about 200 afflicted children each year who can then receive the special diet. The cost of administering the diet is about $1,000 a year per child. The cost of treatment is thus $1,200,000 (200 children at $1,000 a year for six years). The total annual cost of this program nationwide thus comes to $5,200,000. This is a considerable sum to spend on 200 children, but what would be the cost to society if the program were not carried out? Each of those 200 children would have to be maintained in a mental institution for an average of approximately 30 years at a cost of about $5,000 per child per year, or a grand total of $30,000,000 per year. This simple type of cost-benefit analysis shows that carrying out the PKU program nationwide, while costing $5,200,000, actually saves almost $25,000,000 a year. Thus, for PKU at least, the effort is clearly worth the cost, and the same is probably true for other genetic disorders as well.

With other genetic diseases, it is possible to routinely screen prospective parents to detect "carriers." Such screening is now being carried out almost entirely on a voluntary basis, and the decisions made on the basis of the results are private ones. Few would argue against the expansion of such programs and of encouraging members of any "at risk" groups to take advantage of them. Sickle cell anemia, some types of Down's syndrome, and a few other diseases can be detected in the carriers, but, unfortunately, many hereditary diseases cannot.

For some of these diseases, amniocentesis (see figure) can be used to detect afflicted fetuses. It is now becoming possible not only to detect conditions involving gross chromosomal abnormalities, such as Down's syndrome, but also diseases based on enzyme deficiencies. However, the only "treatment" for diseases diagnosed by amniocentesis is abortion. Financial assistance is available in some states for women seeking abortion in order to avoid the birth of a handicapped child. In financial terms, such programs are undoubtedly cost effective, given the cost to society of caring, for example, for a person with Down's syndrome for a lifetime. Even those who would personally choose abortion under such circumstances would find it abhorrent, however, to insist—or even to suggest—that all women elect this alternative.

Finally, and most tragic, are those diseases for which virtually nothing can be done at this time. In the case of Huntington's disease, for example, early detection is not possible by any known means and so it follows that it is impossible to predict who will or who will *not* transmit the disease. The costs of long-term care for Huntington's disease patients are so high that health agency personnel sometimes advise the healthy husband or wife to get a divorce so that the patient can become a ward of the state and so become eligible for medical benefits. There is no "cost benefit" here, to society, the patient, or the family. What are our responsibilities? Even were biologists able to trace the origins of the disease to specific chemical defects in the gene, biology cannot answer these social and moral questions.

Amniocentesis. A sharp-pointed needle is inserted into the amniotic cavity, and fluid containing cells from the fetus is withdrawn into a syringe. The cells can then be analyzed for genetic defects. The procedure cannot be done until the fourth month of pregnancy.

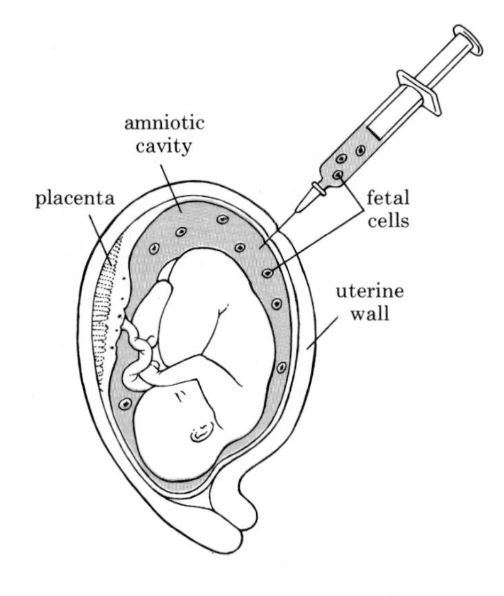

males are usually sexually underdeveloped and sterile, however. *XXX* combinations sometimes produce normal females, but many of the *XXX* women and all *XO* women (women with only one *X* chromosome) are sterile. Many, but not all, individuals with sex chromosome abnormalities are mentally retarded.

Nature and Nurture

A little more than half a century ago, a leading American psychologist, John Watson, threw out a challenge: "Give me a dozen healthy infants, well-formed, and my own specified world to bring them up in, and I'll guarantee to take any one at random and train him to become any type of specialist I might select—doctor, lawyer, merchant, chief, and yes, even beggarman and thief, regardless of his talents, penchants, tendencies, abilities, vocations, and the race of his ancestors."*

The pendulum now seems to have swung entirely in the other direction. No longer, in the popular view, is an infant completely plastic, capable of being molded into any shape by his or her environment. Rather, a member of the human species is seen as an untidy collection of behavioral souvenirs, acquired during our ascendancy from the apes (or perhaps even on the long route from the amoeba).

This change in attitude has little to do with studies actually carried out in humans, but is rather a product of a new and extremely interesting discipline known as ethology. Ethology is the study of animal behavior under natural conditions. Ethologists are particularly concerned with species differences in behavior and their evolutionary origins (as contrasted, for example, with studying a solitary pigeon shut up in a box).

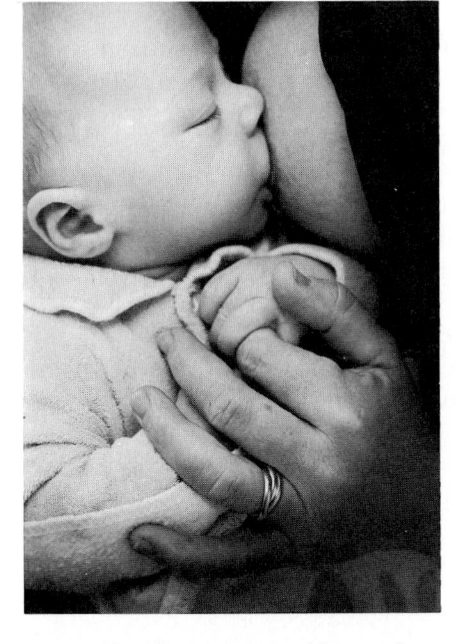

13-10 *Very few patterns of human behavior are clearly based solely on instinct. One such, of obvious survival value, is the suckling response, present in infants at birth.*

In animals, some patterns of behavior are clearly genetically determined. A spider, for instance, although raised in solitary confinement, can build an extremely complex web on its first try. Among vertebrates, many species of fish and birds have elaborate courtship and territory-defending rituals, the basic patterns of which are part of their genetic equipment—that is, instinctive rather than learned.

Mammals also show inherited behavior. For example, the first time a flying squirrel raised in a bare laboratory cage isolated from other squirrels is given nuts or nutlike objects, it will "bury" them in the bare floor. It makes scratching movements as if to dig out the earth, pushes the nuts down into the "hole" with its nose, covers them over with imaginary earth, and stamps on them. Another example is domestic animals, which are bred for personality traits and disposition as well as for physical characteristics.

Are humans naturally aggressive? Do we inherit genes for territoriality? Are some races more intelligent than others? At this time, very, very little is known about the answers to these questions. Schizophrenia has recently been shown to have a strong genetic basis, as indicated by studies of siblings separated at birth (by being adopted into different families). These studies reveal that if one sibling developed schizophrenia, the other was far more likely to be schizophrenic than an unrelated person of the same age. Some forms of mental deficiency have been associated with chromosomal abnormalities, as previously noted, and with genetic "errors," as in cases of PKU. However, studies purporting to show genetically based

*J. B. Watson, *Behaviorism,* Peoples Institute, New York, 1925, p. 82.

differences in IQ between races or ethnic groups have been under attack, largely because of environmental inequities or the cultural bias of the test itself. The single factor that correlates most consistently with low IQ is low birth weight—a subject explored further in Chapter 31.

Beyond such scraps of evidence as we have cited, there are very few data. The nature of one's nature is, of course, an interesting subject and generates arguments and discussions in which anyone can participate, since, even if special knowledge is required, it is not available. Consequently, at present, arguments about these issues must be regarded as social, ethical, and political, rather than as scientific. However, because they have reverberating consequences, they may not be regarded as trivial.

SUMMARY

The principles of genetics are the same for humans as they are for other species. For example, some traits, such as cleft chin and Huntington's disease, are transmitted by simple dominant genes.

Some other human diseases are due to simple Mendelian recessives. In such diseases, the genetic trait appears in the phenotype of the offspring only if he or she is homozygous for the recessive allele—that is, if each parent carries a recessive allele. PKU and sickle cell anemia are due to Mendelian recessives in the homozygous state.

Some traits, such as the ABO blood groups, are the result of multiple alleles of a single gene.

Humans have two sex chromosomes (*X* and *Y*) and 22 pairs of autosomes. For study, the chromosomes are arranged in a karyotype.

In humans (and all other mammals), the *Y* chromosome determines maleness. The *Y* chromosome carries far fewer genes than the *X* chromosome. As a consequence, if a male receives an *X* chromosome carrying recessive alleles, those alleles (which in the female would be dominated by the normal dominant alleles) will usually be expressed. Characteristics resulting from the expression of such genes are said to be sex-linked. Color blindness and hemophilia are examples of sex-linked characteristics that are carried by females but are seen chiefly among males.

Nondisjunction is the failure of chromatids to separate properly at the time of meiosis. As a consequence of nondisjunction, children may be born with an extra chromosome. The presence of an extra chromosome can produce abnormalities such as Down's syndrome. Extra sex chromosomes can also result from nondisjunction; this is often, but not always, associated with sterility and mental retardation.

There are very few data available on the existence of hereditary influences on human behavior or intelligence.

QUESTIONS

1. PKU, phenylketonuria, is a disease caused by the presence of two recessive alleles. If two healthy parents have a child with PKU, what are their genotypes with respect to PKU? What are their chances of having another child with the same disease?

2. If two healthy parents have a child with sickle cell anemia, what are their genotypes with respect to this allele? Having had one such child, what are their chances of having another child with the same disease?

3. What proportion of the children of the parents in Question 2 will be carriers of the sickle cell gene (that is, heterozygous)? What proportion of their children will not carry the gene? (Draw a Punnett square to diagram this problem.)

4. If you are of blood type O and neither of your parents is, what are their possible genotypes? What is the probability that one of your siblings is also of type O? (A Punnett square may again be helpful.)

5. Why is male-to-male inheritance of color blindness impossible? Under what conditions would color blindness be found in a woman? If she married a man who was not color-blind, would her sons be color-blind? Her daughters?

6. A woman whose maternal grandfather was hemophiliac has parents who seem to be normal. She too seems normal, as does her husband. What are the chances that her first son will be normal? (Hint: Determine the genotype of the woman's mother and then the possible genotypes of the woman herself.)

7. Describe the two types of chromosomal abnormality that can cause Down's syndrome. Is it possible to identify prospective parents who are at a higher-than-average risk? If so, how?

14

The Chemical Basis of Heredity: The Double Helix

The work discussed in the previous chapters of this section is often referred to as "classical genetics." In some ways, this is an unfortunate term because "classical" carries a sense of something that happened a long time ago and that is interesting today only for historical reasons. Classical genetics provides the framework for all modern genetics, and classical studies continue today. As we shall see in Section 7, they are the basis for population genetics, one of the most dynamic of the modern branches of biology.

Classical genetics showed that genes are carried on chromosomes, that they come in alternate forms (alleles), that alleles occupy corresponding positions (loci) on homologous chromosomes, and that chromosomes with their alleles are reassorted at meiosis. Classical geneticists discovered that segments of chromatids are exchanged during meiosis, and they made use of these exchanges to construct chromosome maps. They introduced the concept of mutations and of interactions of genes with other genes, as in polygenic inheritance, to produce the phenotype. All this work—and more—was accomplished in less than half a century.

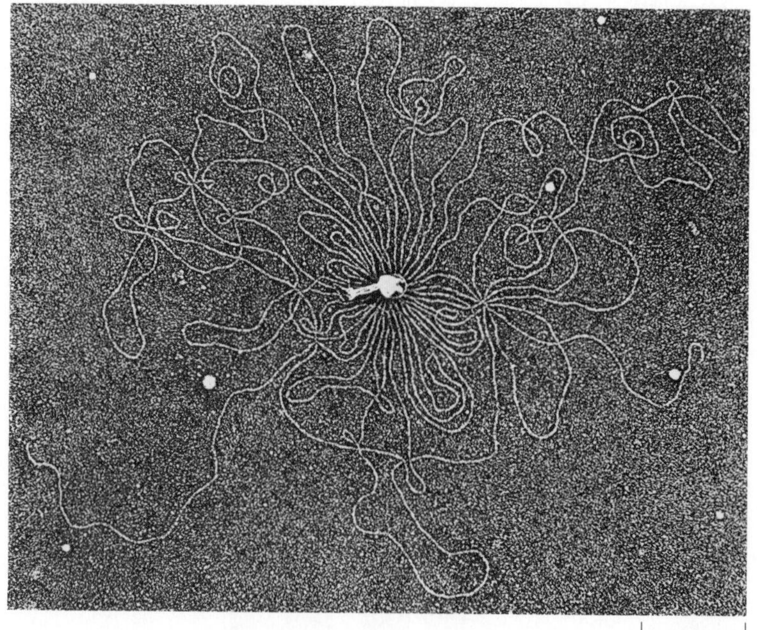

14-1 *Modern genetics is concerned with the chemical basis of heredity. The genetic information has been shown to be contained in a large, complex molecule known as deoxyribonucleic acid (DNA). This electron micrograph shows the genetic material of a bacteriophage, a virus that attacks bacterial cells. It is in the form of one long, continuous molecule of DNA. (The molecule shown here, however, has been broken apart at one point—note the free ends at the right and left.) In the center of the micrograph is the outer coat of the virus, made of protein, from which the DNA has been released.*

A turning point in genetics came when scientists began to focus on the question of how it was possible for these little lumps of matter—the chromosomes—to be the bearers of what they had come to realize must be an enormous amount of complex information. The chromosomes, like all other parts of the living cell, are composed of atoms arranged into molecules. Some scientists, a number of them quite eminent in the field of genetics, thought it would be impossible to understand the complexities of genetics in terms of the structure of "lifeless" chemicals. (Although they did not use the term to describe themselves, they were, in fact, the vitalists—page 8—of the twentieth century.) Others thought that if the chemical structure of the chromosomes were understood, we could then come to understand how chromosomes function as the bearers of the genetic information. The work of the latter group is now termed "molecular genetics."

Genes and Proteins

In the decade following the rediscovery of Mendel, it had been suggested that genes act by influencing the production of enzymes—the highly specific catalysts of biochemical reactions. The problem was to devise a test for this hypothesis. Traits such as wrinkledness in peas—or indeed the visible features of any organism—are usually the end product of a vast number of chemical reactions, most of which are still unknown.

In 1941, George Beadle and Edward Tatum found a way to test the hypothesis. Their experimental organism, the red bread mold *Neurospora,* was another "fit material." Its life cycle is brief, it can be grown in large quantities in the laboratory, and, most important, it is haploid during most of its life cycle. Thus recessive mutations are not masked by normal, dominant alleles. Another important feature of *Neurospora* is that it is able to make for itself all of the amino acids and nearly all of the vitamins. As a result, it can grow on a "minimal medium" containing only sugars, one vitamin, and a few minerals.

The making of an amino acid requires a series of chemical reactions, each of which is controlled by a particular enzyme. If, as a result of a mutation, *Neurospora* were to lose any one of the enzymes involved, for example, in making the amino acid proline, it could no longer grow on the minimal medium. The mutant could, however, grow on a medium supplemented with proline. Beadle and Tatum x-rayed *Neurospora* spores to increase the mutation rate, allowed them to germinate and grow on a medium enriched with all of the amino acids, and then collected the next generation of spores for their experiments. This ensured that any changes observed were indeed genetic ones. On the basis of their experiments, summarized in Figure 14–2, Beadle and Tatum formulated the one-gene–one-enzyme hypothesis. It states that genes control a cell's activities by controlling enzymes, with each gene responsible for one particular enzyme.

Although all enzymes are proteins, not all proteins are enzymes. Some, for instance, are hormones, like insulin, and others are structural proteins, like collagen. These proteins, too, are under genetic control. Thus, the theory was later generalized to "one gene, one protein." In other words, enzymes and other protein molecules are the immediate products of genes.

The Structure of Hemoglobin

Linus Pauling was one of the first to see some of the implications of these new ideas. Perhaps, Pauling reasoned, human diseases involving hemoglobin, such as sickle cell anemia, could be traced to a slight variation

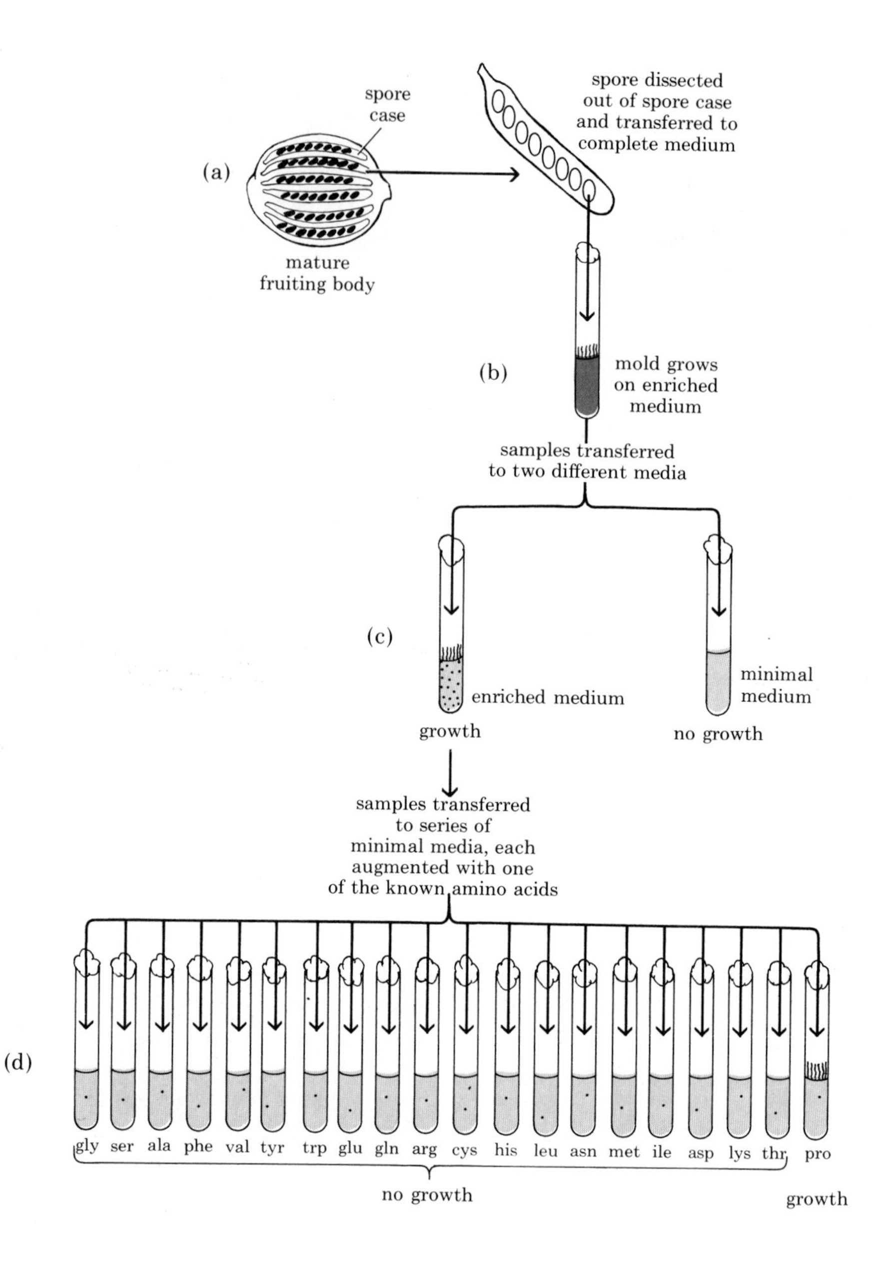

14-2 *How Beadle and Tatum tested the mutants of* Neurospora. *By these experiments, they were able to show that a change in a single gene results in a change in a single enzyme.* (a) *Spore cases are removed from fruiting bodies of* Neurospora *and the spores dissected out.* (b) *Each spore is transferred to an enriched medium, containing all* Neurospora *normally needs for growth plus supplementary amino acids.* (c) *A fragment of the mold is tested for growth in the minimal medium. If no growth is observed on the minimal medium, it may mean that a mutation has occurred that renders this mutant incapable of making a particular amino acid, and so tests are continued.* (d) *Subcultures of mold that grow on the enriched medium but not on the minimal one are tested for their ability to grow in minimal media supplemented with only one of the amino acids. As in the example shown here, a mold that has lost its capacity to synthesize the amino acid proline is unable to survive in a medium that lacks that amino acid. Further tests are then made to discover, in each case, which enzymatic step has been impaired.*

from normal in the protein structure of the hemoglobin molecule. To test this hypothesis, he took samples of hemoglobin from people with sickle cell anemia, from others heterozygous for the gene, and from others free of it. To try to detect differences in these proteins, he used a process known as electrophoresis, in which organic molecules are dissolved in a solution and exposed to a weak electric field. Very small differences, even in very large molecules, may be reflected in the electric charges of the molecules, and the molecules will move differently in the electric field. Pauling found that a person who has sickle cell anemia makes a different sort of hemoglobin than a person who does not have the disease. A person who is heterozygous (carrying one copy of the recessive gene for sickling and one copy for normal hemoglobin) makes both kinds of hemoglobin molecules.

A few years later, it was found that the actual difference between the normal and the sickle cell hemoglobin lies in two of the molecule's 600 amino acids. The hemoglobin molecule is composed of four polypeptide chains—two identical alpha chains and two identical beta chains. Each

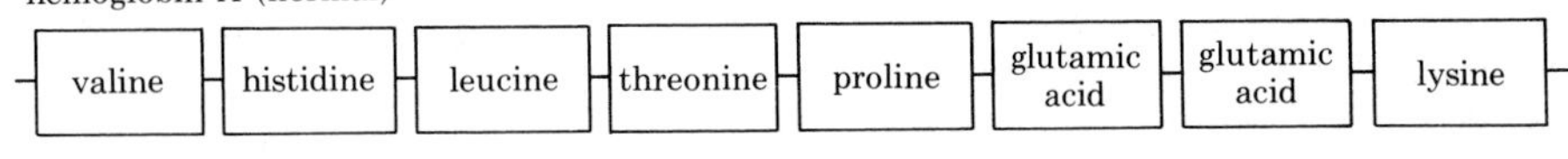

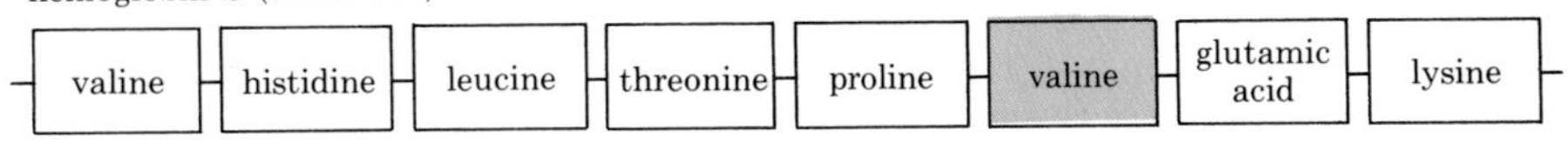

14-3 *An example of the remarkable precision of the "language" of proteins. Portions of the beta chains of the hemoglobin A (normal) molecule and the hemoglobin S (sickle cell) molecule are shown. The entire structural difference between the normal molecule and the sickle cell molecule (literally, a life-and-death difference) consists of this single substitution of one valine for one glutamic acid.*

chain consists of about 150 amino acids. In a precise location in each beta chain, one glutamic acid is replaced by one valine (Figure 14-3). This was one of the earliest demonstrations that a protein has a unique series of amino acids.

Speculative thinkers in the field of biology were quick to see that the amino acids, the number of which is so provocatively close to the number of letters in our own alphabet, could be arranged in a variety of different ways. It appeared that these different arrangements might account for both the great diversity of proteins and their great specificity. The proteins were seen as making up a sort of language—"the language of life"—that spelled out the directions for all the many activities of the cell.

The Gene: Protein or DNA?

During the 1930s and 1940s, scientists became more and more concerned with the question, "Exactly what is a gene?" The attempt to identify the chemical nature of the gene gave rise to two opposing schools of thought, and the controversy lasted for a number of years—until, as we shall see, it was conclusively resolved in the 1950s.

The chromosomes of eukaryotes are composed of protein and a chemical known as deoxyribonucleic acid, abbreviated as DNA. Many prominent investigators, particularly those who had been studying proteins, believed that the genes themselves were proteins. They thought that the chromosomes contained master models of all the proteins that would be required by the cell and that enzymes and other proteins active in cellular life were copied from these master models. This was a logical hypothesis, but as it turned out, it was wrong.

The Transforming Factor

To trace the beginning of the other hypothesis—the one that proved right—it is necessary to go back to 1928 and pick up an important thread in modern biological history. In that year, an experiment was performed that seemed at the time very remote from either biochemistry or genetics. Frederick Griffith, a public health bacteriologist, was studying the possibility of developing vaccines against pneumococci, the bacteria that cause one kind of pneumonia.

Pneumococci, as Griffith knew, come in either virulent (disease-causing) forms with polysaccharide capsules or nonvirulent (harmless) forms without capsules (Figure 14-4). The existence of the capsule and its composition are both genetically determined—that is, they are inherited

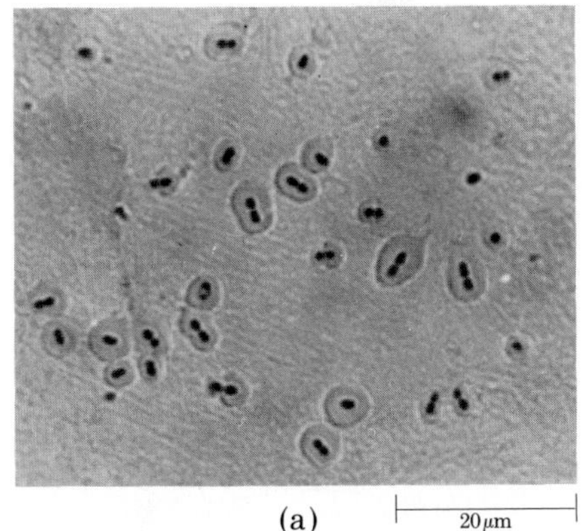
(a) 20 μm

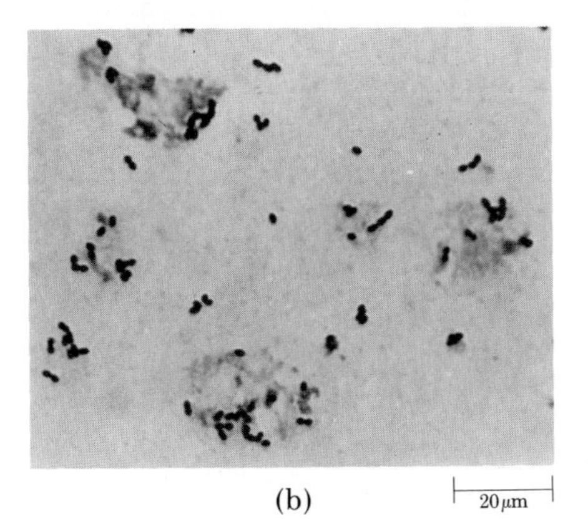
(b) 20 μm

14-4 (a) *Encapsulated and* (b) *nonencapsulated forms of pneumococci. The encapsulated form, which is resistant to phagocytosis by white blood cells, produces pneumonia; the nonencapsulated form is harmless.*

properties of the bacteria. Griffith thought that injections of heat-killed virulent bacteria or of live, closely related but harmless strains (ones without capsules) might immunize mice to live virulent pneumococci. As a result of his experiments, which are outlined in Figure 14–5, Griffith found that something can be passed from dead to live bacteria that changes their hereditary characteristics. This phenomenon was known as *transformation*, and the "something" that caused it was called the *transforming factor*.

In 1943, O. T. Avery and his co-workers at Rockefeller University demonstrated that the transforming factor is DNA. Subsequent experiments showed that a variety of genetic traits can be passed from members of one strain of bacterial cells to those of another, similar strain by means of isolated DNA.

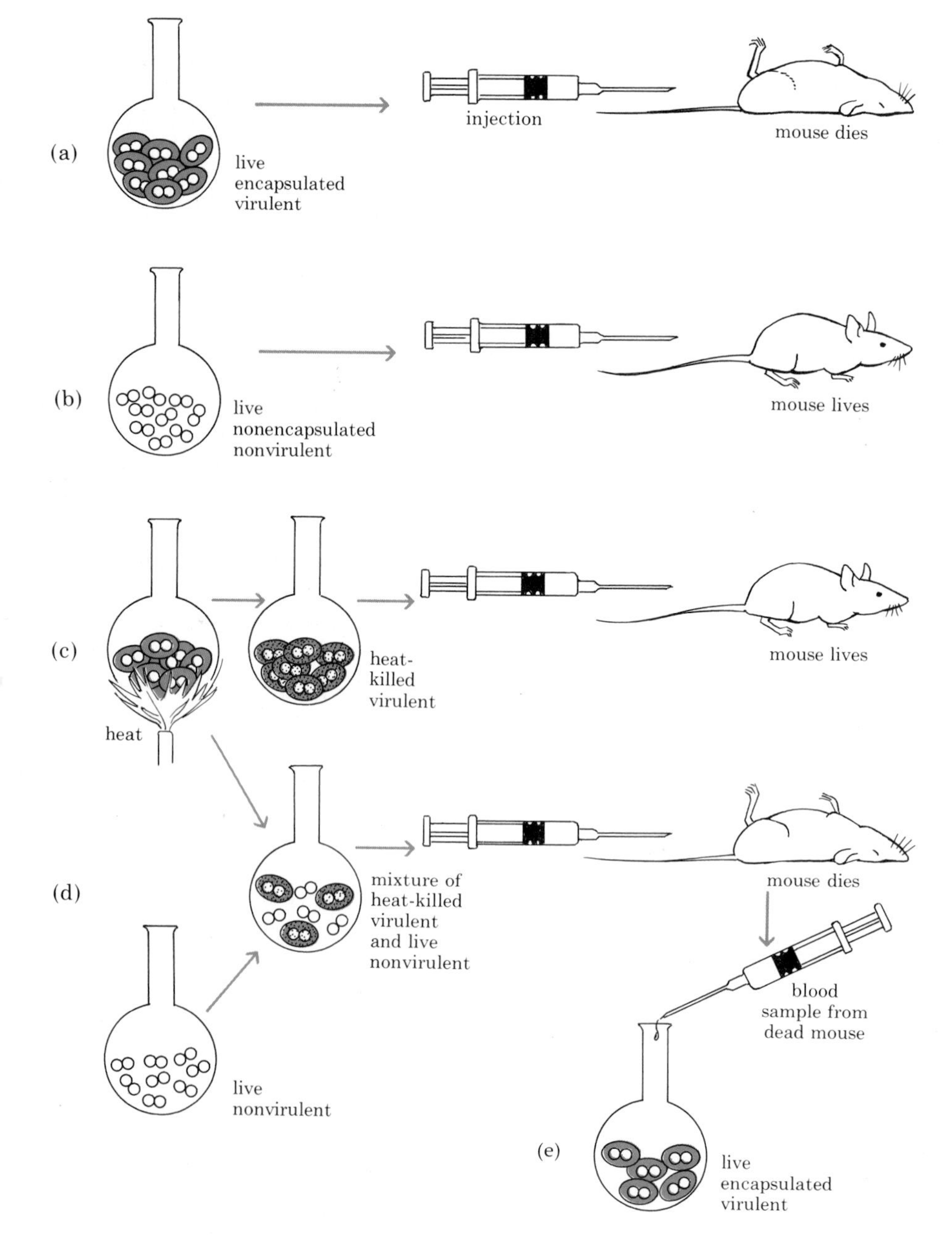

14–5 *Discovery of the transforming factor, a substance that can transmit genetic characteristics from one cell to another, resulted from studies of pneumococci, pneumonia-causing bacteria. One strain of these bacteria has capsules (protective outer sheaths); another does not. The capacity to make capsules and cause disease is an inherited characteristic, passed from one bacterial generation to another as the cells divide.* (a) *Injection into mice of encapsulated pneumococci killed the mice.* (b) *The nonencapsulated strain was harmless.* (c) *If the encapsulated strain was heat-killed before injection, it too was harmless.* (d) *If, however, heat-killed encapsulated bacteria were mixed with live nonencapsulated bacteria and the mixture was injected into mice, the mice died.* (e) *Blood samples from the dead mice revealed live encapsulated pneumococci. Something had been transferred from the dead bacteria to the live ones that endowed them with the capacity to make capsules and cause pneumonia. This "something" was later isolated and found to be DNA.*

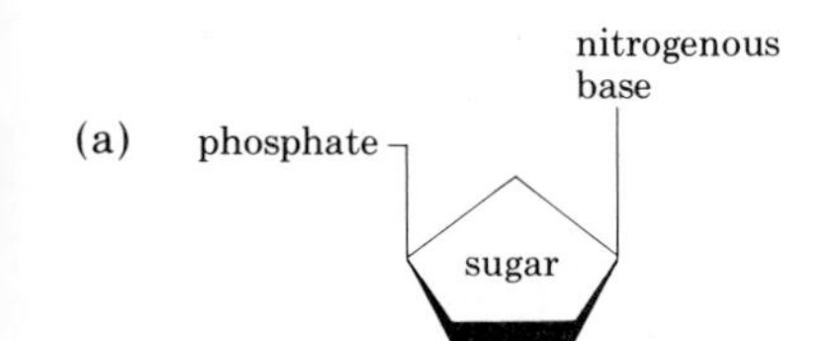

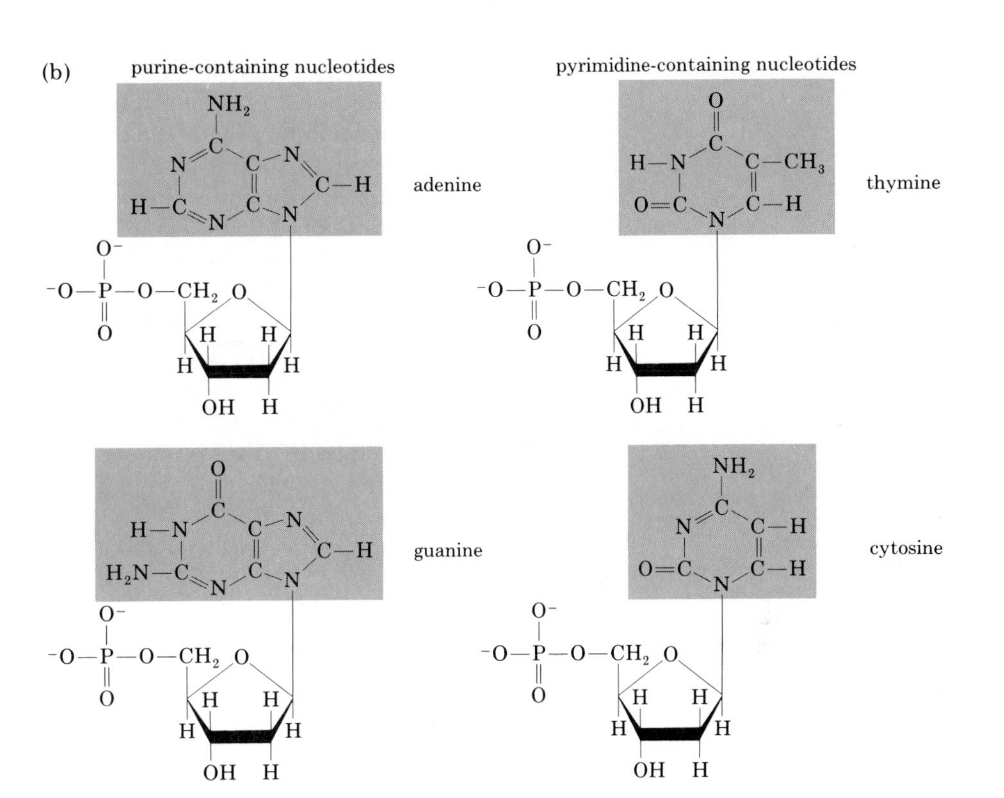

14-6 (a) *A nucleotide is made up of three different parts: a nitrogenous base, a sugar, and a phosphate.* (b) *Each nucleotide in DNA contains one of the four possible nitrogenous bases, a deoxyribose sugar, and a phosphate.*

DNA had first been isolated by a German chemist named Friedrich Miescher in 1869—in the same remarkable decade in which Darwin published *The Origin of Species* and Mendel presented his results to the Brünn Natural History Society. The substance Miescher isolated was white, sugary, slightly acidic, and contained phosphorus. Since he found it only in the nuclei of cells, he called it nucleic acid. This name was later amended to deoxyribonucleic acid, to distinguish it from a similar chemical also found in cells, ribonucleic acid (RNA).

By the time of Avery's discovery, it was known that DNA is made up of nucleotides (Figure 14-6). Each nucleotide consists of a nitrogenous base, a deoxyribose sugar, and a phosphate. The nitrogenous bases are of two kinds: purines, which have two rings, and pyrimidines, which have one ring. There are two kinds of purines found in DNA, adenine (A) and guanine (G), and two kinds of pyrimidines, cytosine (C) and thymine (T). So DNA is made up of four types of nucleotides, differing only in their nitrogen-containing purine or pyrimidine.

Avery's results offered evidence for DNA as the genetic material, but his discovery was slow to gain full recognition. This was partly because bacteria, which are, of course, prokaryotes, were considered "lower" and "different" and partly because the DNA molecule—made up of only four components—seemed too simple for the enormously complex task of carrying the hereditary information.

The Phage Experiments

Another series of crucial experiments was made with an even "lower" organism, a virus that attacks bacteria. These viruses, known as *bacteriophages* ("bacteria eaters"), were originally chosen for study because they reproduce at a phenomenal rate. Moreover, the bacteria they attack are *Escherichia coli,* the familiar bacteria found in the healthy human intestine. Bacteriophages invade bacterial cells, multiply within them, and then escape—usually bursting (lysing) the cell. The entire infection cycle can take place in 25 minutes, during which time several hundred virus progeny can be produced from a single virus.

Another interesting thing about bacteriophages is that they consist only of DNA and protein (see Figure 14–1), the two leading contenders for the role as carrier of the genetic information. The question of which type of molecule carries the viral genes—the hereditary information by which new viral particles are made—was answered in 1952 by Alfred D. Hershey, working with Martha Chase. Their experiments are summarized in Figure 14–7. Remember that protein contains sulfur and no phosphorus and DNA contains phosphorus and no sulfur. You will then see why these experiments demonstrated that only the DNA of the bacteriophages was involved in the replication process and that the protein could not be the hereditary material.

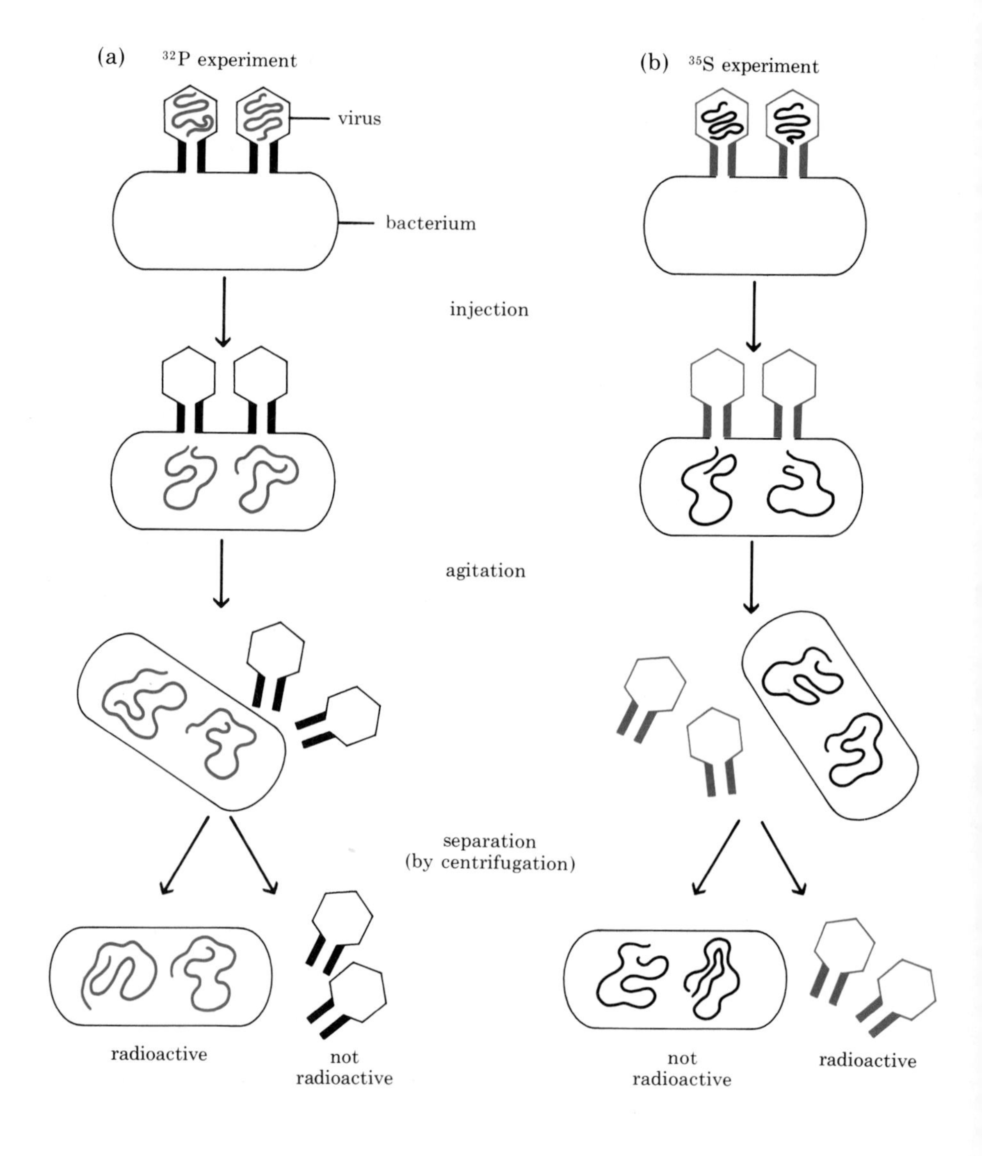

14–7 *A summary of the Hershey-Chase experiments. Two separate samples of viruses were prepared, one in which the DNA was labeled with the radioactive isotope ^{32}P and the other in which the protein was labeled with the radioactive isotope ^{35}S. One culture of bacteria* (a) *was infected with ^{32}P-labeled phage and another* (b) *with ^{35}S-labeled phage. After the infectious cycle had begun, the cells were agitated in a blender to separate them from any viral material remaining outside the cells. The two samples—one containing extracellular material and the other intracellular material—were then tested for radioactivity. Hershey and Chase found that the ^{35}S had stayed outside the bacterial cells with the empty viral coats and the ^{32}P had entered the cells, infected them, and caused the production of new virus progeny. It was therefore concluded that the genetic material of the virus is DNA rather than protein.*

14-8 *Electron micrograph of bacteriophages attacking a cell of* E. coli. *The viruses are attached to the bacterial cell by their tails. Some of the viruses have apparently injected their DNA into the cell, as indicated by their empty heads. A complete cycle of virus infection takes only about 25 minutes. At the end of that period, several hundred new virus particles are released from the cell.*

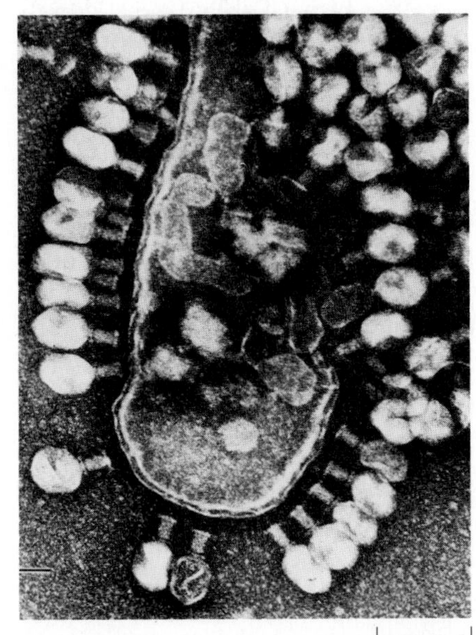

Electron micrographs later confirmed that this type of bacteriophage attaches to the bacterial cell wall by its tail and injects its DNA into the cell, leaving the empty protein coat (the "ghost") on the outside (Figure 14-8). In short, the protein is just a container for the DNA, which carries all the hereditary information of the virus and directs the synthesis both of new DNA and of new protein for the viral progeny.

Further Evidence for DNA

The role of DNA in transformation and in viral infection formed very convincing evidence that DNA is the genetic material. Two other lines of work also helped to lend weight to the argument. Alfred Mirsky, in a long series of careful studies, showed that, in general, the somatic cells of any given species contain equal amounts of DNA. The principal exceptions are the gametes, which regularly contain just half as much DNA as other cells.

A second important series of contributions was made by Erwin Chargaff. Chargaff analyzed the purine and pyrimidine content of the DNA of many different kinds of living things and found that the nitrogenous bases do not occur in equal proportions. The proportions of the four nitrogenous bases are the same in all cells of a given species, but they vary from one species to another. In other words, the bases, according to Chargaff's analysis, do not occur in a regular sequence–for instance, ATCG, ATCG–because if they did, the proportions would all be the same. Rather, they must be arranged irregularly–for instance, ATCG, TTAC. Therefore, these variations could very well provide a "language" in which the instructions for controlling cell growth could be written. Some of Chargaff's results are reproduced in Table 14-1. Can you, by examining these figures, notice anything interesting about the proportions of purines and pyrimidines?

The Watson-Crick Model

In the early 1950s, a young American scientist, James Watson, went to Cambridge, England, on a research fellowship to study problems of molecular structure. There, at the Cavendish Laboratory, he met physicist Francis Crick. Both were interested in DNA, and they soon began to work together to solve the problem of its molecular structure. They did not do experiments in the usual sense but rather undertook to examine all the data about DNA and unify them into a meaningful whole.

Table 14-1 Composition of DNA in several species

	Purines		Pyrimidines	
Source	Adenine	Guanine	Cytosine	Thymine
Human being	30.4%	19.6%	19.9%	30.1%
Ox	29.0	21.2	21.2	28.7
Salmon sperm	29.7	20.8	20.4	29.1
Wheat germ	28.1	21.8	22.7	27.4
E. coli	24.7	26.0	25.7	23.6
Sea urchin	32.8	17.7	17.3	32.1

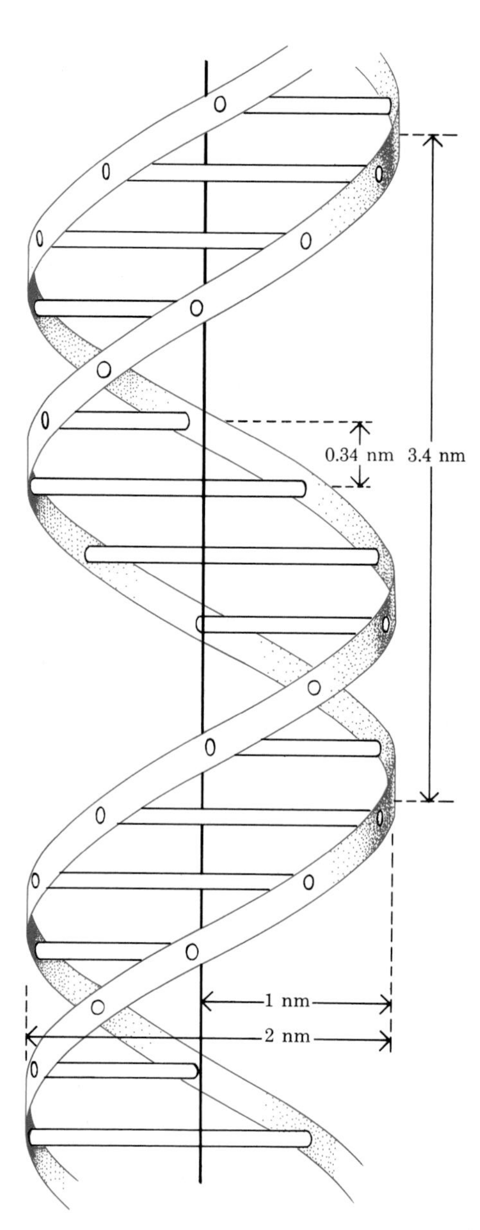

14–9 *The double-stranded helical structure of DNA, as first presented in 1953 by Watson and Crick. The framework of the helix is composed of the sugar-phosphate units of the nucleotides. The rungs are formed by the four nitrogenous bases adenine and guanine (the purines) and thymine and cytosine (the pyrimidines). Each rung consists of two bases. Knowledge of the distances between the atoms, determined from x-ray diffraction pictures, was crucial in establishing the structure of the DNA molecule.*

The Known Data

By the time Watson and Crick began their studies, quite a lot of information on the subject had already accumulated. It was known that DNA contains nucleotides, each consisting of either one purine or one pyrimidine plus a deoxyribose sugar and a phosphate group. X-ray studies showed that the molecule takes the form of a helix (coil). In 1950, Pauling had shown that proteins sometimes take this form (see page 47) and that the helical structure is maintained by hydrogen bonding between successive turns in the helix. Pauling had suggested that the structure of DNA might be similar. And, finally, there were Chargaff's data (Table 14–1), which indicated that the ratio of DNA nucleotides containing thymine to those containing adenine is approximately 1:1 and that the ratio of nucleotides containing guanine to those containing cytosine is also approximately 1:1.

Building the Model

From these data, Watson and Crick attempted to construct a model of DNA that would fit the known facts and explain the biological role of DNA. In order to carry the vast amount of genetic information, the molecules should be heterogeneous and varied. Also, there must be some way for them to replicate readily and with great precision so that faithful copies could be passed from cell to cell and from parent to offspring, generation after generation.

Let us see what the Watson-Crick model looked like. If you were to take a ladder and twist it into a helix, keeping the rungs perpendicular, you would have a crude model of the molecule (Figure 14–9). The two longitudinal railings are made up of alternating sugar and phosphate molecules. The rungs of the ladder are formed by the nitrogenous bases—adenine (A), thymine (T), guanine (G), and cytosine (C)—one base bonded to each sugar-phosphate and two bases forming each rung. The paired bases meet across the helix and are joined together by hydrogen bonds, the relatively weak, common, and very important chemical bonds that Pauling had demonstrated in his studies of protein structure.

The distance between the two sides, or railings, according to x-ray measurements, is 2 nanometers. Two purines in combination would take up more than 2 nanometers, and two pyrimidines would not reach all the way across. But if a purine paired in each case with a pyrimidine, there would be a perfect fit. The paired bases—the "rungs" of the ladder—would therefore always be purine-pyrimidine combinations.

As Watson and Crick worked their way through the data, they assembled actual tin-and-wire models of the molecule (see essay, page 193), testing where each piece would fit into the three-dimensional puzzle. First, they noticed, the nucleotides along any one strand of the double helix could be assembled in any order: ATGCGTACATTGCCA, and so on (Figure 14–10a). Since a DNA molecule may be thousands of nucleotides long, there is a possibility for great variety.

DNA Replication

The most exciting discovery came, however, when Watson and Crick set out to construct the matching strand. They encountered an interesting and important restriction. Not only could purines not pair with purines and pyrimidines not pair with pyrimidines, but also, because of the configura-

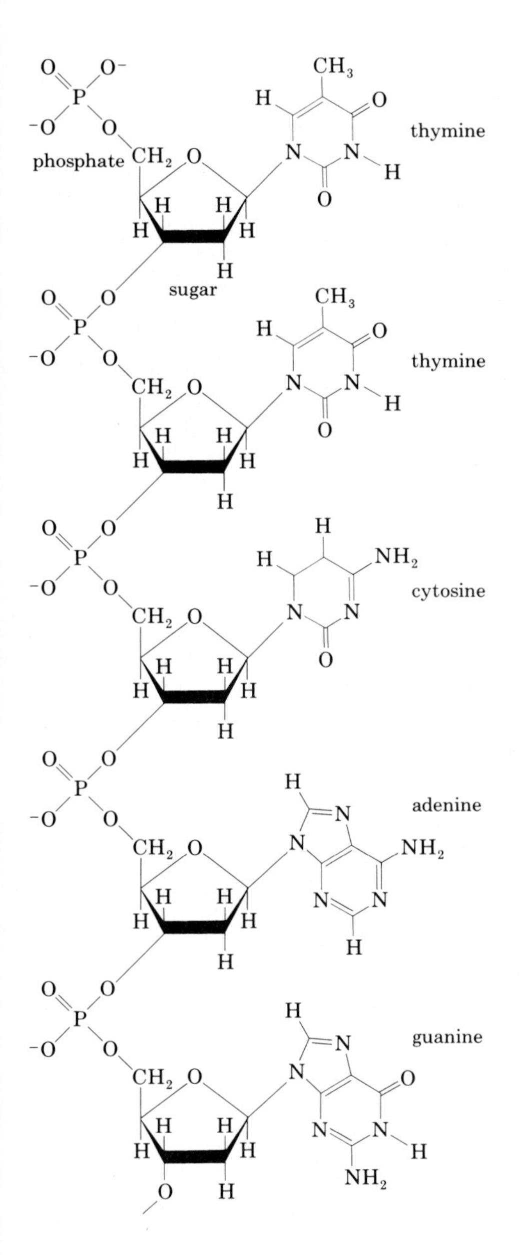

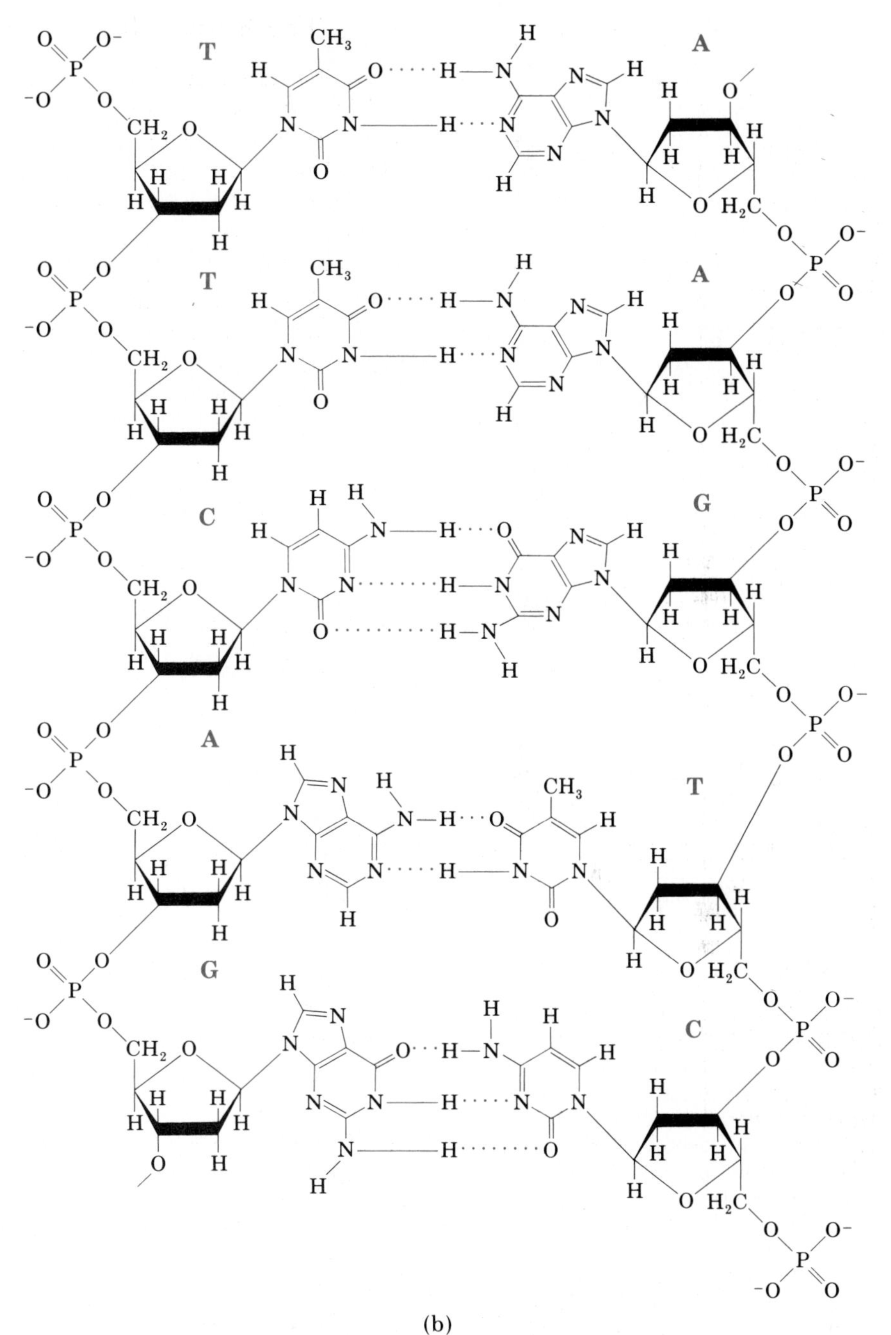

14-10 (a) *The structure of a portion of one strand of a DNA molecule. Each nucleotide consists of a sugar, a phosphate group, and a purine or pyrimidine base. Note the repetitive sugar-phosphate-sugar-phosphate sequence that forms the backbone of the molecule. The sequence of bases varies from one DNA molecule to another. In the figure, the order of nucleotides (from the top) is TTCAG.*

(b) *The double-stranded structure of a portion of the DNA molecule. The strands are held together by hydrogen bonds between the bases. Because of bonding requirements, adenine can pair only with thymine and guanine only with cytosine. Thus the order of bases along one strand determines the order of bases along the other.*

tions of the molecules, adenine could pair only with thymine and guanine only with cytosine (Figure 14–10b). Look at Table 14–1 again and see how well these chemical requirements confirm Chargaff's data.

This fact about the structure of the DNA molecule immediately suggested the method by which it reproduces itself. At the time of chromosome replication, the molecule opens up, bit by bit, the bases separating at the hydrogen bonds. The two strands separate, and new strands form along each old one, using nucleotides available in the cell. If a T is present on the old strand, only an A can fit into place on the new strand; a G will pair only with a C; and so on. In this way, each strand forms a copy of its original partner strand, and two exact replicas of the molecule are produced (Figure 14–11). The age-old question of how hereditary information is duplicated and passed on, generation after generation, had in principle been answered.

DNA as the Carrier of Genetic Information

The Watson-Crick model showed clearly how the DNA molecule is able to carry the genetic information. The information is coded in the sequence of the bases, and *any* sequence of bases is possible. Since the number of paired bases ranges from about 5,000 for the simplest known virus up to an estimated 5 billion in the 46 human chromosomes, the possible variations are astronomical. The DNA from a single human cell—which if extended in a single thread would be almost 2 meters long—can contain information equivalent to some 600,000 printed pages of 500 words each, or a library of about a thousand books. Obviously, the DNA structure can well account for the endless diversity among living things.

Once the structure of DNA was revealed, there was little doubt as to its genetic role.

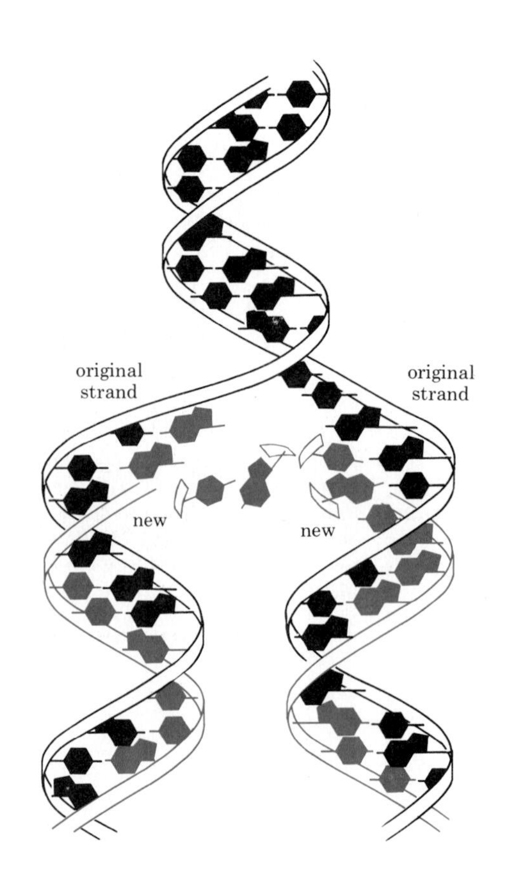

14-11 *The DNA molecule shown here is in the process of replication, separating down the middle as the paired bases separate at the hydrogen bonds. (For clarity, the bases are shown out of plane.) Each of the original strands then serves as a template along which a new, complementary strand forms from nucleotides available in the cell.*

Who Might Have Discovered It?

Then there is the question, what would have happened if Watson and I had not put forward the DNA structure? This is "iffy" history which I am told is not in good repute with historians, though if a historian cannot give plausible answers to such questions I do not see what historical analysis is about. If Watson had been killed by a tennis ball I am reasonably sure I would not have solved the structure alone, but who would? Olby has recently addressed himself to this question. Watson and I always thought that Linus Pauling would be bound to have another shot at the structure once he had seen the King's College x-ray data, but he has recently stated that even though he immediately liked our structure it took him a little time to decide finally that his own was wrong. Without our model he might never have done so. Rosalind Franklin was only two steps away from the solution. She needed to realise that the two chains must run in opposite directions and that the bases, in their correct tautomeric forms, were paired together. She was, however, on the point of leaving King's College and DNA, to work instead on TMV [tobacco mosaic virus] with Bernal. Maurice Wilkins had announced to us, just before he knew of our structure, that he was going to work full time on the problem. Our persistent propaganda for model building had also had its effect (we had previously lent them our jigs to build models but they had not used them) and he proposed to give it a try. I doubt myself whether the discovery of the structure could have been delayed for more than two or three years.

There is a more general argument, however, recently proposed by Gunther Stent and supported by such a sophisticated thinker as Medawar. This is that if Watson and I had not discovered the structure, instead of being revealed with a flourish it would have trickled out and that its impact would have been far less. For this sort of reason Stent had argued that a scientific discovery is more akin to a work of art than is generally admitted. Style, he argues, is as important as content.

I am not completely convinced by this argument, at least in this case. Rather than believe that Watson and Crick made the DNA structure, I would rather stress that the structure made Watson and Crick. After all, I was almost totally unknown at the time and Watson was regarded, in most circles, as too bright to be really sound. But what I think is overlooked in such arguments is the intrinsic beauty of the DNA double helix. It is the molecule which has style, quite as much as scientists. The genetic code was not revealed all in one go but it did not lack for impact once it had been pieced together. I doubt if it made all that difference that it was Columbus who discovered America. What mattered much more was that people and money were available to exploit the discovery when it was made. It is this aspect of the history of the DNA structure which I think demands attention, rather than the personal elements in the act of discovery, however interesting they may be as an object lesson (good or bad) to other workers.

Francis Crick: "The Double Helix: A Personal View," *Nature,* vol. 248, pages 766–769, 1974.

(a)

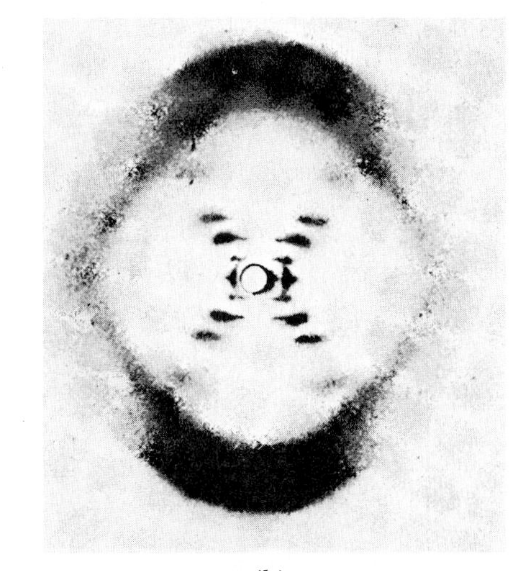

(b)

(a) *Watson (left) and Crick in 1953 examining one of their models of DNA.*

(b) *X-ray diffraction photograph of DNA taken by Rosalind Franklin in the laboratories of Maurice Wilkins, who shared the Nobel Prize with Watson and Crick. The reflections crossing in the middle indicate that the molecule is a helix. The heavy dark regions at the top and bottom are due to the closely stacked bases perpendicular to the axis of the helix.*

SUMMARY

Classical genetics is concerned with the mechanics of inheritance—how the units of hereditary material are passed from one generation to the next and how changes in the hereditary material are expressed in individual organisms. In the 1930s, a new question arose and geneticists began to explore the nature of the gene.

In 1941, working with the bread mold *Neurospora,* Beadle and Tatum established the "one-gene–one-enzyme" principle. This principle was later expanded to "one gene, one protein." Proteins are large and complex molecules, with very specific structures. Even small differences in structure can produce large differences in function, as in the case of hemoglobin.

DNA (deoxyribonucleic acid) is the genetic material of the cell. Investigations showing that both the transforming factor in bacteria and the carrier of the genetic information in bacteriophages are DNA provided some of the first evidence for this hypothesis.

Further support for the genetic role of DNA came from two more findings: (1) Almost all tissue cells of any given species contain equal amounts of DNA, and (2) the proportions of nitrogenous bases are the same in the DNA of all cells of a given species, but they vary in different species.

In 1953, Watson and Crick proposed a structure of DNA. The DNA molecule, according to their model, is a double-stranded helix, shaped somewhat like a twisted ladder. The two sides of the ladder are composed of repeating groups of a phosphate and a five-carbon sugar. The "rungs" are made up of paired bases, one purine base pairing with one pyrimidine base. There are four bases—adenine (A), guanine (G), thymine (T), and cytosine (C). A can pair only with T, and G only with C. The four bases are the four "letters" used to spell out the genetic message. The paired bases are joined by hydrogen bonds.

The DNA molecule is self-replicating. The two strands come apart down the middle, breaking at the hydrogen bonds, and each strand forms a new complementary strand from nucleotides available in the cell.

On the basis of this structure, as revealed by Watson and Crick, the role of DNA as the carrier and transmitter of the genetic information became widely accepted.

QUESTIONS

1. The bread mold with which Beadle and Tatum worked is haploid. Would haploid organisms follow Mendel's principles? Why would it be easier and faster to conduct genetic experiments with haploid organisms rather than diploid organisms?

2. What advantages does *Neurospora* have over *Drosophila* as an organism for genetic experiments?

3. In pea plants, a cross between a purple-flowered plant and a white-flowered plant produces a purple-flowered plant. In snapdragons, a cross between a red-flowered plant and a white-flowered plant produces a pink-flowered plant. Explain how the Beadle-Tatum hypothesis might account for these differences.

4. What are the steps by which Griffith demonstrated the existence of the transforming factor? Can you think of any implications of Griffith's discovery for modern medicine?

5. One of the chief arguments for the erroneous theory that proteins constitute the genetic material is that proteins are heterogeneous. Explain why the genetic material must have this property. What feature of the Watson-Crick DNA model is important in this respect?

6. What characteristics of bacteriophages made them such a useful experimental tool in resolving the question of whether the genetic material was DNA or protein?

7. Suppose you are talking to someone who has never heard of DNA. How would you support an argument that DNA is the genetic material? List at least five of the strong points in such an argument.

15

Protein Synthesis and the Genetic Code

Like most important scientific discoveries, the Watson-Crick model raised more questions than it answered. Given that genes are made of DNA and that the products of genes are specific proteins, what is the link between them? How does DNA influence protein synthesis?

One early hypothesis was that the DNA somehow formed a template for protein production. Templates are patterns, or guides, and the word "template" is usually associated with the metalwork patterns used in industry. By extension, the word came to be applied to a biological molecule that, by its shape, directs or molds the structure of another molecule. During replication of the DNA molecule prior to cell division, for example, each original strand of DNA serves as a template for the formation of its new partner strand.

There was, however, simply no way that a strand of protein could be matched up against a strand of DNA in a templatelike fashion. The relationship between DNA and protein had to be a more indirect one. If the proteins, with their 20 amino acids, were the "language of life," the DNA molecule, with its four nitrogenous bases, could be envisioned as a sort of code for the language. So the term *genetic code* came into being.

The Triplet Code

As it turned out, the idea of a "code of life" was useful not only as a dramatic metaphor but also as a working analogy. Scientists, seeking to understand how the DNA so artfully stored in the nucleus could order the quite dissimilar structures of protein molecules, approached the problem with methods used by cryptographers in deciphering codes.

As we saw in Chapter 3, there are 20 biologically important amino acids. The primary structure of each particular kind of protein molecule consists of a specific linear arrangement of these 20 different amino acids. Similarly, there are four different nucleotides, arranged in a specific linear sequence in a DNA molecule. If each nucleotide "coded" one amino acid, only four amino acids could be provided for. If two nucleotides specified one amino acid, there could be a maximum number, using all possible arrangements, of 4^2, or 16—still not quite enough. Therefore, following the code analogy, at least three nucleotides in sequence must specify each amino acid. This would provide for 4^3, or 64, possible combinations, or *codons*.

The three-nucleotide, or triplet, codon was widely and immediately adopted as a working hypothesis, although its existence was not actually demonstrated until a decade after the Watson-Crick discovery. Proof depended on answering yet another question: How is the information encoded in the DNA molecule translated into protein?

15-1 Chemically, RNA is very similar to DNA, but there are two differences in its nucleotides. One difference is in the sugar component; instead of deoxyribose, RNA contains ribose, which has an additional oxygen atom. The other difference is that instead of thymine, RNA contains the closely related pyrimidine uracil (U). A third, and very important, difference between the two nucleic acids is that most RNA does not possess a regular helical structure and is usually single-stranded.

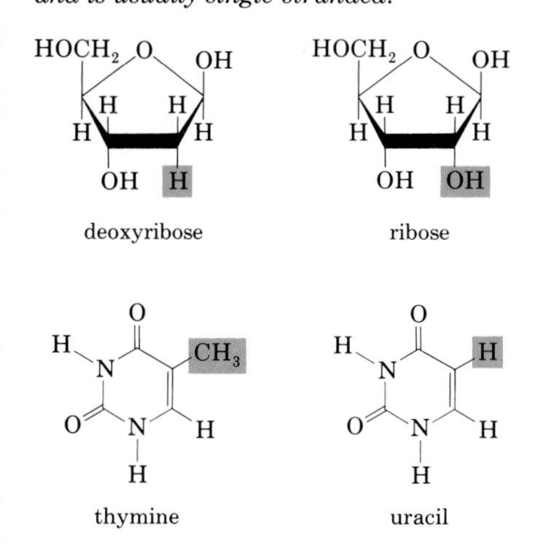

The RNAs

The search for the answer to this question led to the examination of another kind of molecule, ribonucleic acid (RNA), a close chemical relative of DNA (Figure 15-1). There were clues that RNA might play a role in the translation of genetic information into a sequence of amino acids. For instance, cells that are making large amounts of protein are rich in RNA. The RNA is found mostly in the cytoplasm, and it is there that most protein synthesis takes place. Also, cells making large amounts of protein have numerous ribosomes, and ribosomes are rich in RNA.

The role of RNA molecules in translating the code was studied by breaking apart *Escherichia coli* cells, separating their contents into various fractions, and seeing which fractions and (finally) which components of which fractions were essential for protein biosynthesis in a test tube. It was found that the machinery is complex and involves, among other things, three forms of RNA.

Messenger RNA (mRNA) is a long molecule consisting of a single strand of 1,000 to 10,000 nucleotides. This molecule forms along a strand of DNA, using it as a template (Figure 15-2), in a way similar, in principle, to that in which a new DNA strand forms. The reaction is catalyzed by a special enzyme, RNA polymerase.

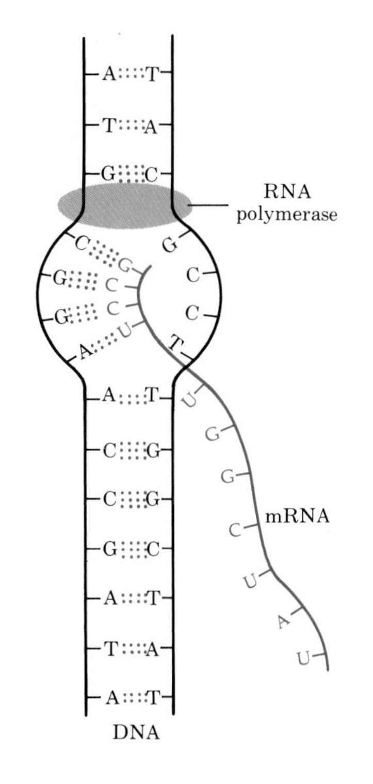

15-2 The beginning of protein biosynthesis is the formation of messenger RNA (mRNA). At the chromosome, a strand of RNA is copied from a sequence of DNA. At the point of attachment of the enzyme RNA polymerase, the DNA opens up. Nucleotide building blocks are assembled into RNA using one strand of the DNA as a template. As the RNA polymerase moves along the DNA molecule, the hydrogen bonds between the two strands of the DNA re-form and the newly formed RNA strand is released. The strand of mRNA is a "negative print" of the sequence of nucleotides in the DNA.

Transfer RNA and ribosomal RNA are also formed by this same copying process from DNA.

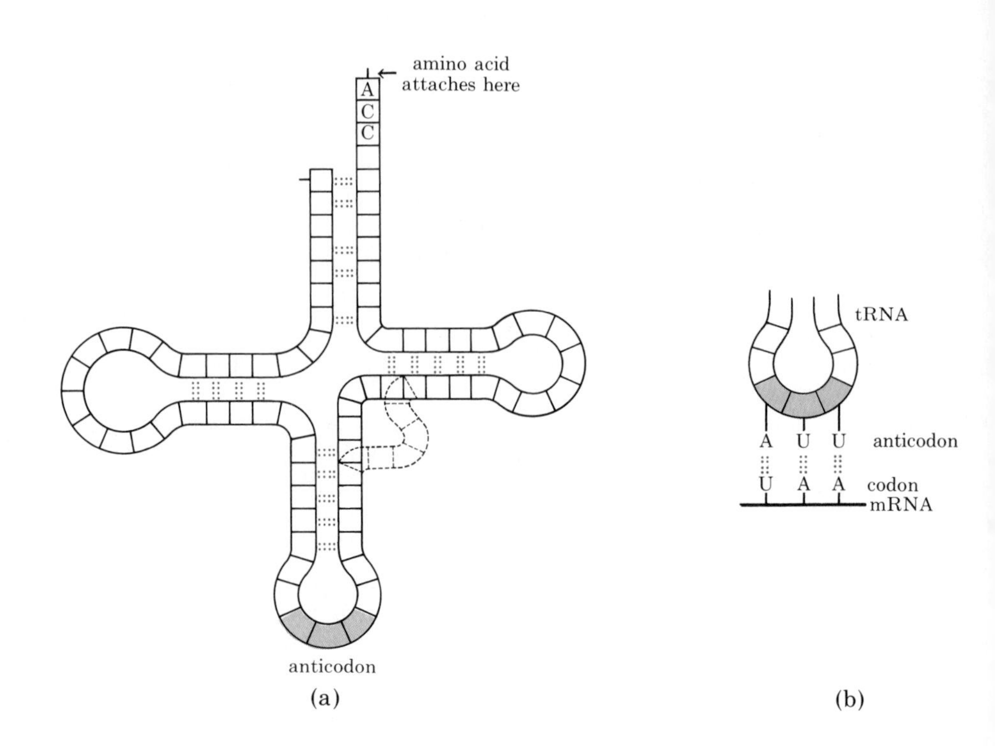

15-3 (a) *General structure of tRNA molecules. Such molecules consist of about 80 nucleotides linked together in a single chain. The chain always terminates in a CCA sequence. An amino acid can link to its specific tRNA molecule at this CCA end. The other nucleotides vary according to the particular tRNA.*

All tRNA molecules appear to have the configuration shown here; in some, however, there is an additional bulge, indicated by the dashed lines. Some of the bases are hydrogen-bonded to one another, as indicated by the colored dots. The unpaired bases at the bottom of the diagram (indicated in color) are known as the anticodon. They serve to "plug in" the tRNA molecule to an mRNA codon, as shown in (b).

Transfer RNA (tRNA) molecules are relatively small and have a characteristic cloverleaf shape (Figure 15-3). There are more than 20 kinds of tRNA molecules—at least one for each kind of amino acid found in proteins.

Ribosomes are about half RNA and half protein. The form of RNA they contain is called *ribosomal RNA* (rRNA). As shown in Figure 15-4, each ribosome is composed of two subunits, each with its characteristic RNAs and proteins.

As you might expect, investigators found that protein synthesis requires—in addition to the three types of RNA—a supply of amino acids, a number of enzymes, and ATP as an energy source. Now, with this rather large cast of characters in mind, let us look at what happens.

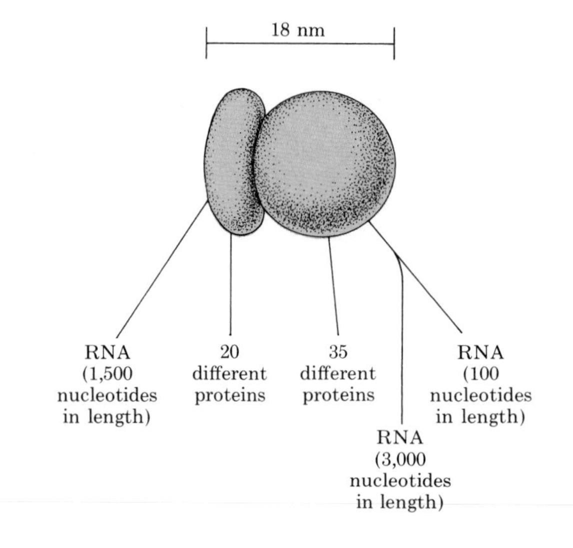

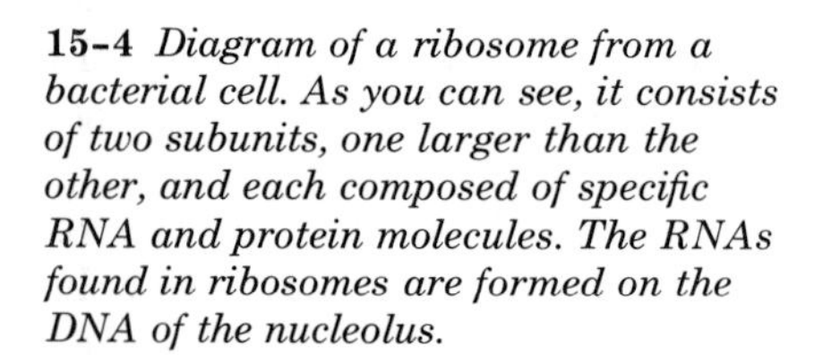
15-4 *Diagram of a ribosome from a bacterial cell. As you can see, it consists of two subunits, one larger than the other, and each composed of specific RNA and protein molecules. The RNAs found in ribosomes are formed on the DNA of the nucleolus.*

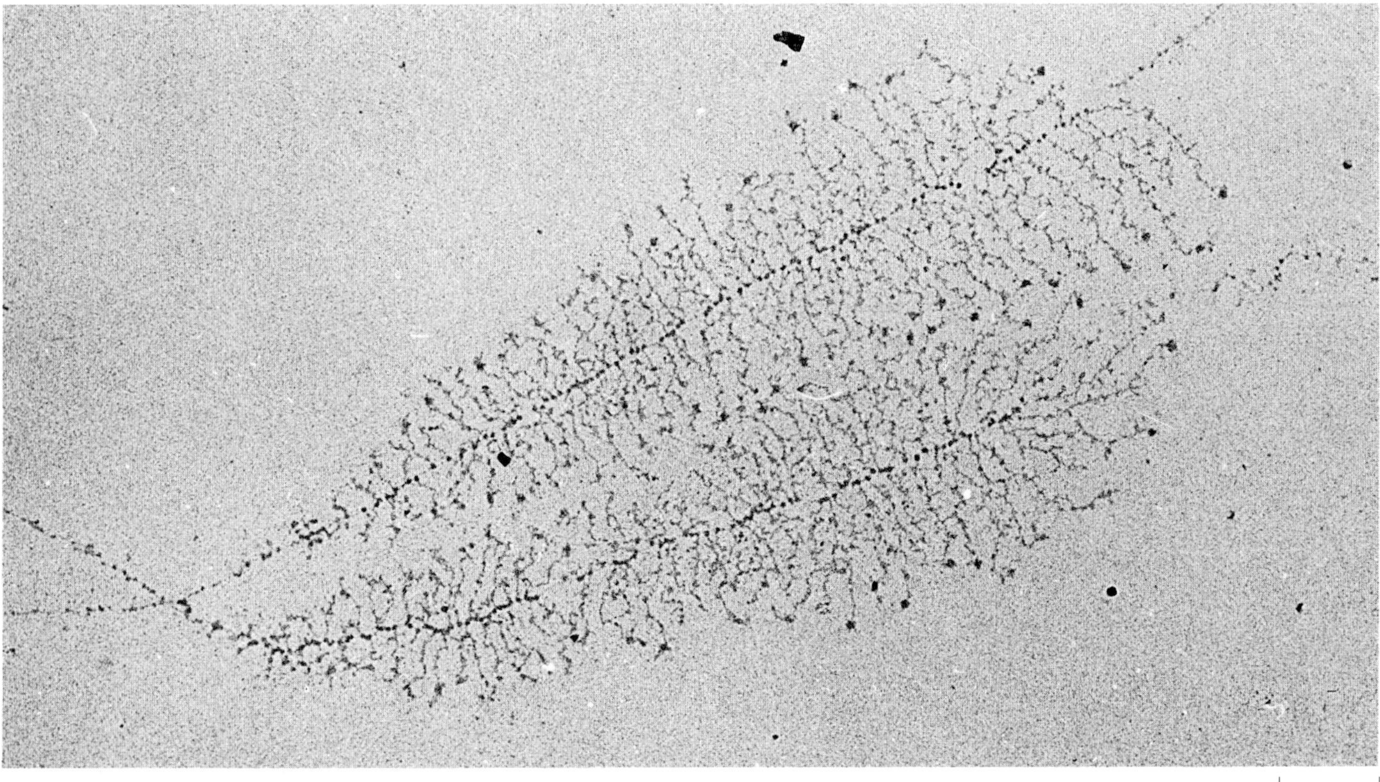

15-5 *Transcription. This electron micrograph shows two different segments of DNA from the nucleolus of an amphibian egg cell. The fine fibrils perpendicular to the DNA strands are RNA molecules that are being transcribed from the DNA. Transcription begins at the left of each of these DNA strands and ends at the upper right; the longer RNA molecules at the right are completely transcribed.*

The Making of a Protein

The manufacture of a protein is one of the most intricate processes carried out by living organisms. It begins when a strand of mRNA, with the help of the enzyme RNA polymerase, forms along a segment of one strand of the DNA helix. The mRNA forms along the DNA strand according to the same base-pairing rules as those that govern the formation of a DNA strand. The only difference is that in mRNA uracil substitutes for thymine. Because of the copying mechanism, the completed mRNA strand carries an inverse copy of the DNA message. This stage of protein biosynthesis—the formation of a coded strand of RNA—is known as transcription (see Figure 15-5).

In eukaryotic cells, the mRNA undergoes a certain amount of "editing" (which we shall discuss in the next chapter) and then moves into the cytoplasm. Here are the amino acids, enzymes, ATP molecules, ribosomes, and molecules of tRNA. Each type of tRNA attaches by one end to its particular amino acid. This reaction requires ATP and a special enzyme for each amino acid–tRNA combination. Energy from the ATP molecule is transferred to the amino acid–tRNA combination and is used in the subsequent steps of protein synthesis.

Once in the cytoplasm, the mRNA molecule attaches to a ribosome. It seems to fit between the two subunits of the ribosome, which is then able to move along the mRNA molecule, somewhat like a railroad car moving along a track. As the ribosome moves along, codons of the mRNA are exposed at the ribosome surface. As each codon is exposed, a molecule of

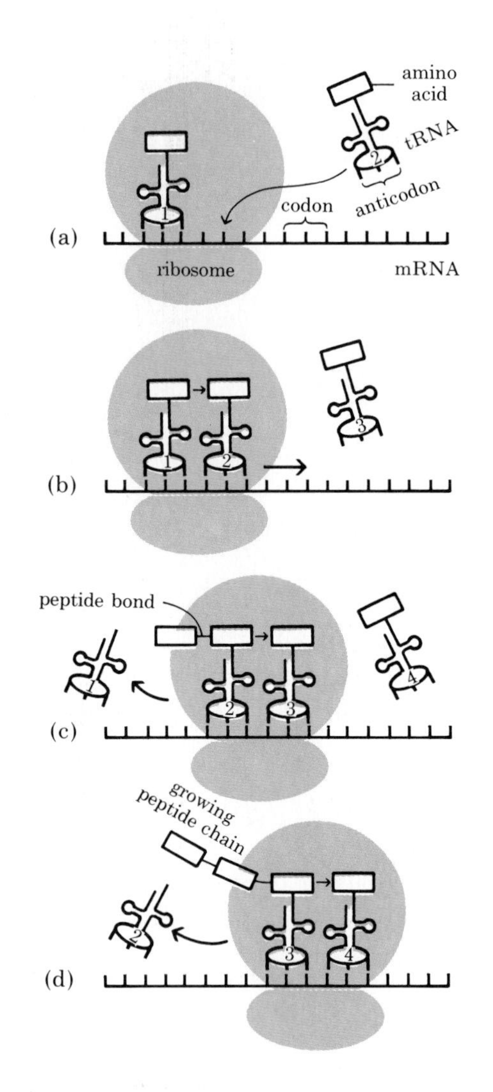

15-6 *The process of chain elongation in protein synthesis.* (a) *After the ribosome and the first tRNA have complexed with the mRNA,* (b) *the second tRNA with its amino acid associates with the complex.* (c) *A peptide bond is formed between the two amino acids and the first tRNA is released. The ribosome then moves along the messenger so that the next codon is exposed within the ribosome. The appropriate tRNA with its attached amino acid complexes with the exposed mRNA codon.* (d) *A new peptide bond is formed and the second tRNA is released. The ribosome moves once again as in* (c) *and the process continues.*

tRNA, with its particular amino acid in tow, zeroes into position. Only the "correct" tRNA can attach to the mRNA; that is, its special nucleotide triplet—the *anticodon*—must be able to pair with the exposed codon on the mRNA molecule (see Figure 15-3b).

After the first tRNA molecule has bonded to the mRNA, the ribosome moves along, exposing the next codon (Figure 15-6). The next tRNA molecule with its amino acid then moves into place. The energy that was stored in the bond between the first tRNA molecule and its amino acid is now used to forge a peptide bond between the two amino acids. The first tRNA molecule is then released from the mRNA molecule, leaving the dipeptide (two amino acids) in place, attached to the second tRNA molecule. The ribosome moves again, and the process is repeated over and over until the complete protein molecule is formed. This stage in the biosynthesis of proteins—the assembly of amino acids in a specific sequence—is known as *translation*. The information *transcribed* from the nucleotide sequence of DNA to the nucleotide sequence of mRNA has been *translated* into the amino acid sequence of the protein (Figure 15-7).

When one ribosome moves along an mRNA molecule, one copy is formed of the specific protein molecule for which the mRNA codes. In order to produce the quantities of a particular protein needed by a cell, many rounds of protein synthesis may be required. Numerous molecules of a

15-7 *How a protein is made. At least 20 different kinds of tRNA molecules are formed on the DNA in the nucleus of the cell. These molecules are so structured that each can be attached (by a special enzyme) at one end to a specific amino acid. Each contains an anticodon that fits an mRNA codon for that particular amino acid.*

The process begins when an mRNA strand is formed on the DNA template in the nucleus and travels to the cytoplasm. At the point of attachment to the ribosome, the matching tRNA molecule, with its amino acid, plugs in momentarily to the codon in the mRNA. As the ribosome moves along the mRNA strand, another tRNA linked to its particular amino acid fits into place. The first tRNA molecule is then released, leaving behind its amino acid, which is bonded to the second amino acid by a peptide bond. As the process continues, the amino acids are brought into line one by one, following the exact order laid down by the DNA code.

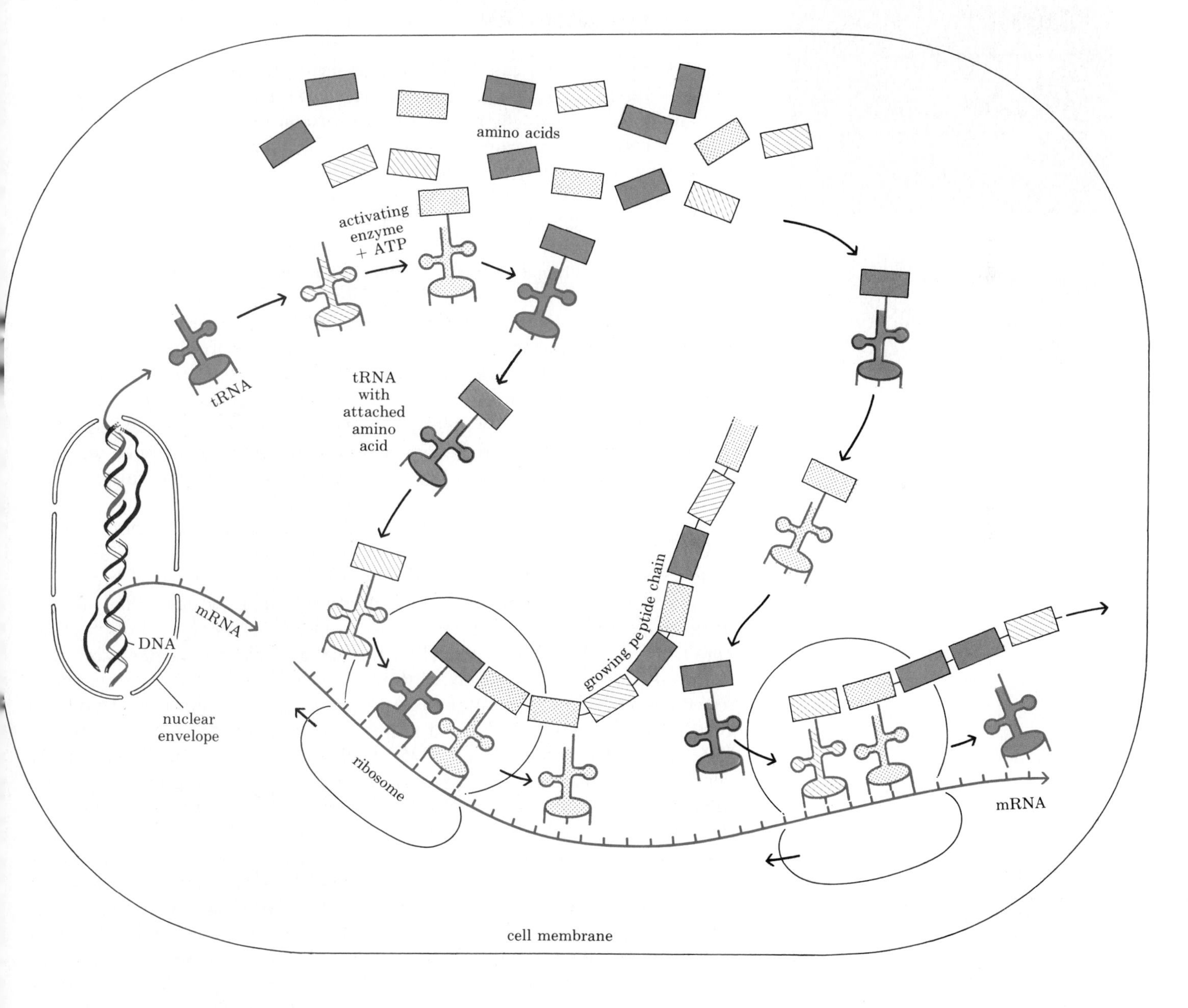

15-8 *Groups of ribosomes "reading" the same mRNA strand are called polyribosomes. These are from yeast cells.*

particular mRNA can be transcribed from a given sequence of DNA. Multiple copies of the various tRNA and rRNA molecules are also formed. The mRNA molecules, the various tRNAs, and the ribosomes are then used over and over. Further, a number of ribosomes can be moving along the mRNA molecule at the same time, making several identical copies of the particular protein molecule (Figure 15-8).

Breaking the Code

Long before all the details of transcription and translation became known, Marshall Nirenberg of the National Institutes of Health set out to test the messenger RNA hypothesis. He prepared extracts from *Escherichia coli* cells that contained ribosomes, ATP, tRNA, and other necessary factors. To these extracts he added radioactively labeled amino acids and crude samples of RNA from a variety of sources. All of the RNAs stimulated protein synthesis; the amounts of protein produced were small but definite. In other words, the cellular material would start producing protein molecules even when the RNA "orders" it received were from a complete "stranger."

Nirenberg and his associate, Heinrich Matthaei, next tried an artificial RNA. Perhaps if the cell extracts could read a foreign message and translate it into protein, they could read a totally synthetic message, one dictated by the scientists themselves. Severo Ochoa of New York University had developed a process for linking nucleotides into a long strand of RNA. By this process, he had produced an RNA molecule that contained only one nitrogenous base, uracil, repeated over and over again. It was called "poly-U." What kind of protein would poly-U dictate?

Nirenberg and Matthaei prepared 20 different test tubes, each of which contained extracts of *E. coli* with ribosomes, tRNA, ATP, other necessary factors, and all the amino acids. In each test tube, one of the amino acids, and only one, carried a radioactive label. Synthetic poly-U was added to each test tube. In 19 of the test tubes, no radioactive polypeptides were formed, but in the twentieth one, to which radioactive phenylalanine had been added, the investigators were able to detect newly formed, radioactive polypeptide chains. When the polypeptides were analyzed, they were found to consist only of phenylalanines, one after another. Nirenberg and Matthaei had dictated the message "uracil . . . uracil . . . uracil . . . ," and a clear answer had come back, "phenylalanine . . . phenylalanine . . . phenylalanine." The experiment not only defined the first code word (UUU = phe) but made available methods for defining the others.

It was not long before the codes were worked out for all the amino acids, using synthetic mRNA. Of the 64 possible triplet combinations, 61 of them specify particular amino acids. With 61 combinations coding for 20 amino acids, you can see that there must be more than one codon for many of the amino acids. As shown in Figure 15-9, codons specifying the same amino acid often differ only in the third nucleotide, leading to the speculation that the first two may be sufficient to hold the tRNA to the mRNA in most instances.

Biological Implications

Some of the biological implications of these findings are strikingly clear. For example, let us take another look at sickle cell anemia in the light of Figure 15-9. Normal hemoglobin contains glutamic acid; sickle cell he-

first letter	U (second letter)	C (second letter)	A (second letter)	G (second letter)	third letter
U	UUU phe	UCU ser	UAU tyr	UGU cys	U
U	UUC phe	UCC ser	UAC tyr	UGC cys	C
U	UUA leu	UCA ser	UAA stop	UGA stop	A
U	UUG leu	UCG ser	UAG stop	UGG trp	G
C	CUU leu	CCU pro	CAU his	CGU arg	U
C	CUC leu	CCC pro	CAC his	CGC arg	C
C	CUA leu	CCA pro	CAA gln	CGA arg	A
C	CUG leu	CCG pro	CAG gln	CGG arg	G
A	AUU ile	ACU thr	AAU asn	AGU ser	U
A	AUC ile	ACC thr	AAC asn	AGC ser	C
A	AUA ile	ACA thr	AAA lys	AGA arg	A
A	AUG met	ACG thr	AAG lys	AGG arg	G
G	GUU val	GCU ala	GAU asp	GGU gly	U
G	GUC val	GCC ala	GAC asp	GGC gly	C
G	GUA val	GCA ala	GAA glu	GGA gly	A
G	GUG val	GCG ala	GAG glu	GGG gly	G

15-9 *The genetic code, consisting of 64 triplet combinations (codons) and their corresponding amino acids (see page 45). Since 61 triplets code 20 amino acids, there are "synonyms," as many as six for leucine, for example. Most of the synonyms, as you can see, differ only in the third nucleotide. Of the 64 codons, only 61 specify particular amino acids. The other three codons are stop signals, which cause the chain to terminate. The code is shown here as it would appear in the mRNA molecule.*

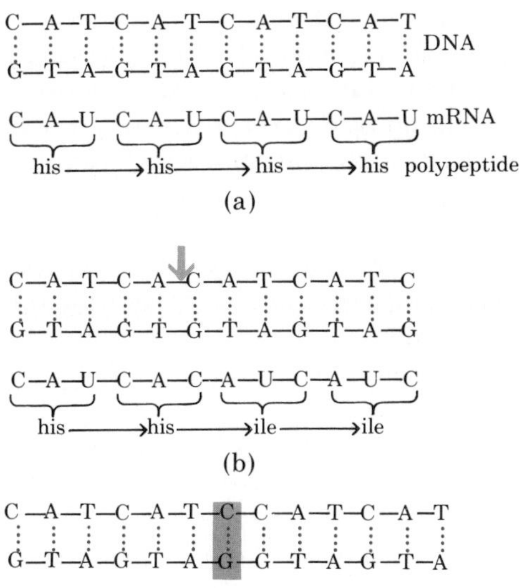

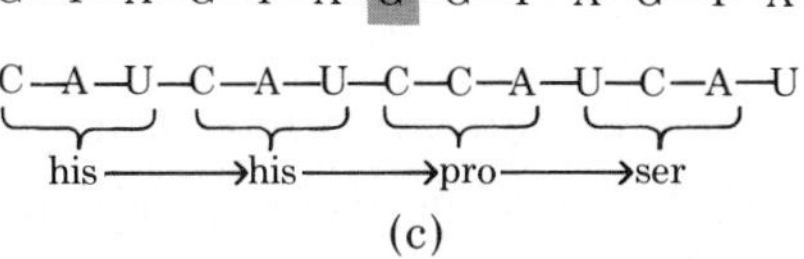

15-10 *Deletion or addition of nucleotides within a gene leads to changes in the protein produced. The original DNA molecule, the mRNA transcribed from it, and the resulting polypeptide are shown in* (a). *In* (b) *we see the effect of the deletion of a nucleotide pair (T-A), as indicated by the arrow. The reading frame for the gene is altered, and a different sequence of amino acids occurs in the polypeptide. A similar change results from the addition of a nucleotide pair (color), as shown in* (c).

moglobin contains valine. In mRNA, GAA or GAG specifies glutamic acid (glu), and GUU, GUC, GUA, or GUG specifies valine (val). So the difference between the two is the replacement of one adenine by one uracil in a molecule that, since it dictates a protein that contains more than 150 amino acids, must contain more than 450 nitrogenous bases. In other words, the tremendous functional difference–literally a matter of life and death–can be traced to a single "misprint" in over 450 nucleotides.

De Vries, some 80 years ago, defined mutation in terms of traits appearing in the phenotype. In the light of current knowledge, the definition is somewhat different: A mutation is a change in the sequence or number of nucleotides in a nucleic acid that can be transmitted from parent to offspring.

Many mutations involve only a small number of nucleotides. For example, substitution of one nucleotide for another can lead to changes in the protein produced by a gene, as in the case of sickle cell anemia. Other changes can result from deletion or addition of nucleotides within a gene (Figure 15-10). When this occurs, the reading frame of the gene shifts–that is, the way in which the nucleotides are grouped into triplets changes–resulting in the production of an entirely new protein. Similar shifts in reading frame can also result from mutations involving relatively large segments of a chromosome. If, for example, a chromosome breaks in two places, the broken piece may be removed, turned around, and replaced "backwards." Such a change reverses the sequence of nucleotides, drastically changing the coded "message."

Although the discovery of the mechanism of protein synthesis and the working out of the genetic code answered many questions and shed new light on mutations, they raised many new questions. The genetic material, the DNA, is undifferentiated in form, being an enormously long sequence of nucleotides. Yet the information it contains is compartmentalized in the units we call genes. How is this accomplished? And, how does the cell regulate its genes so they produce the right amounts of the right proteins at the right times? To answer these questions, we shall turn our attention in the next chapter to the structure of the chromosome.

15-11 *A representation of the information flow from DNA to protein. Replication of the DNA occurs only once in the life of a cell, during the S phase of the cell cycle (page 135) prior to mitosis or meiosis. Transcription and translation, however, occur repeatedly throughout the cell's lifetime.*

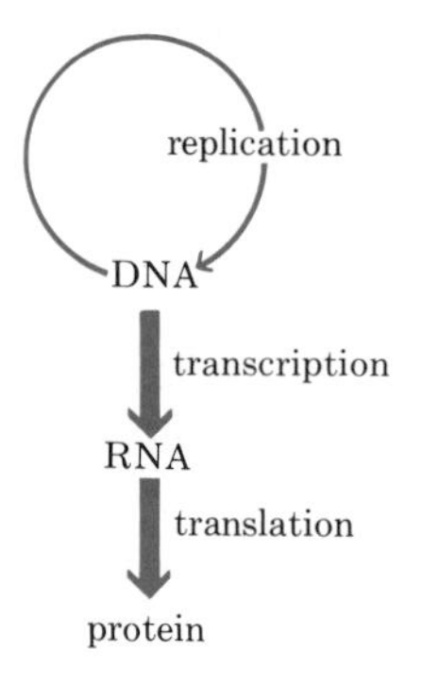

SUMMARY

Genetic information is coded in the molecules of DNA, and these, in turn, determine the sequence of amino acids in molecules of protein. A gene is a segment of a DNA molecule that specifies the complete sequence of one protein.

The way in which the gene directs the production of a protein, according to current theory, is as follows. Each series of three nucleotides along a DNA strand is the DNA code for a particular amino acid. The information is transcribed from the DNA to a long, single strand of RNA (ribonucleic acid). This type of RNA molecule is known as messenger RNA, or mRNA. The mRNA forms along one of the strands of DNA, following the principles of base pairing first suggested by Watson and Crick. The mRNA therefore is complementary to the DNA strand.

The mRNA strand leaves the cell's nucleus and becomes attached to a ribosome. At the point where the strand of mRNA is in contact with the ribosome, small molecules of another type of RNA, known as transfer RNA (tRNA), which serve as adapters between the mRNA and the amino acids, are bound temporarily to the mRNA strand. This bonding takes place by complementary base pairing between the mRNA codon and the tRNA anticodon. Each tRNA molecule carries the specific amino acid called for by the mRNA codon to which the tRNA attaches. Thus, following the sequence originally dictated by the DNA, the amino acid units are brought into line one by one and are formed into a polypeptide chain.

The genetic code has now been "broken"; that is, it is now known which amino acid is called for by a given mRNA codon. Of the 64 possible triplet combinations of the four-letter nucleotide code, 61 combinations have been identified with one of the 20 amino acids that make up protein molecules. The other three triplets serve as "punctuation marks," terminating protein synthesis.

Mutations are now defined as changes in the sequence or number of nucleotides in the nucleic acid of a cell or organism. They may take the form of substitutions of one nucleotide for another, deletions or additions of nucleotides, or reversal of the nucleotide order.

QUESTIONS

1. Explain the term "genetic code." In what respects is it a useful analogy?

2. Describe the structure and function of each of the three types of RNA involved in protein biosynthesis.

3. Define and distinguish between transcription and translation.

4. In a hypothetical segment of a DNA molecule, the sequence of bases is ... AAGTTTGGTTACTTG .... What would be the sequence of bases in an mRNA strand transcribed from this DNA segment? What would be the sequence of amino acids coded by the mRNA? Does it matter where the transcription from DNA to mRNA begins? Show why.

5. Fill in the missing letters.

T G T	_ _ _	_ _ _	DNA
_ _ A	C _ _	_ _ _	DNA
U _ _	_ C A	_ _ _	mRNA codons
_ _ _	_ _ _	G C A	tRNA anticodons

Specify the amino acid that each triplet codes for.

16

The Structure of the Chromosome and the Regulation of Gene Expression

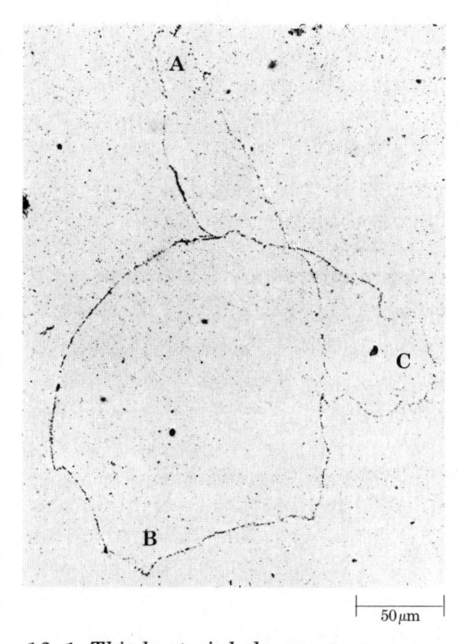

16-1 *This bacterial chromosome was caught in the act of replication by a technique known as autoradiography. An* E. coli *cell was grown for a short time in a medium containing thymine labeled with a radioactive isotope (^3H). The cell was broken open and the DNA, which was allowed to spread out, was placed on a photographic plate for two months. During this period, decay of the radioactive isotope left spots on the photographic emulsion.*

Replication proceeds from the starting point in opposite directions around the circular DNA molecule. A and B are believed to represent already replicated portions and C the unreplicated portion. This autoradiograph, taken by John Cairns, established that bacterial DNA is circular (a single molecule with no end).

The genetic code—three nucleotides coding for a specific amino acid—is universal. It is the same for *Escherichia coli, Homo sapiens,* and all other organisms—awesome evidence that all living things are descended from a common ancestor. However, recent work in molecular genetics has shown that, despite the basic similarities of bacterial chromosomes and eukaryotic chromosomes, there are important differences between them.

The Bacterial Chromosome

The bacterial chromosome is a single, circular loop of DNA (Figure 16-1) with little protein attached to it. Not all of a bacterium's genes are necessarily in its chromosome. In addition to the DNA in the chromosome, many bacteria also contain much smaller DNA molecules, called *plasmids,* which replicate along with the cell. Plasmids sometimes become part of the larger DNA molecule and, as we shall describe in the next chapter, are an important tool in much current research.

Control of Gene Expression in Bacteria

The chromosome of *E. coli* is calculated, on the basis of its weight, to consist of about 4 million nucleotide pairs, enough for more than 4,000 different genes. Thus far, about 1,000 genes have been definitely identified, each coding for a specific protein.

Not all of a bacterial cell's genes are active at the same time, however. The cell needs different proteins at different times, depending on the situation. For example, when the disaccharide lactose is available to *E. coli* cells, the cells need the enzyme beta-galactosidase in order to split lactose into glucose and galactose, releasing energy for the cell's use (Figure 16-2). When lactose is not available, the cells do not need beta-galactosidase. Experiments have shown that when *E. coli* cells are growing on lactose, approximately 3,000 molecules of beta-galactosidase are present in every normal cell (about 3 percent of all the protein in the cell). If lactose is not present, there is only about one molecule of this enzyme per cell. Substances, such as lactose, that increase the amount of an enzyme produced by a cell are known as *inducers,* and the enzymes they influence are known as *inducible enzymes.*

16–2 The splitting of lactose (milk sugar) to galactose and glucose requires the enzyme beta-galactosidase. Normal E. coli *cells synthesize beta-galactosidase only when lactose is present.*

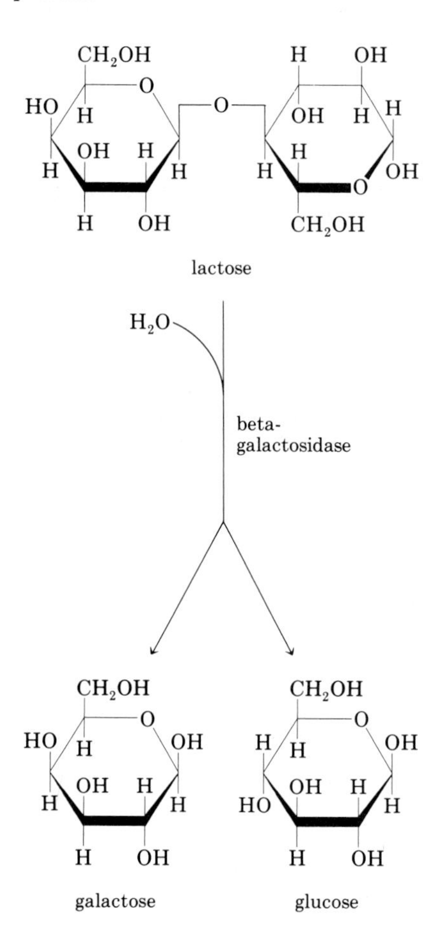

Other substances act to repress enzyme production. For example, *E. coli* can make all of its own amino acids. If, however, a particular amino acid—histidine, for example—is present in a sufficient amount, the cell will then stop making all of the enzymes associated with the biosynthesis of histidine. Such enzymes are known as *repressible enzymes*, and the substances that inhibit their production are known as *corepressors*.

The capacity to turn enzyme production on and off is genetically determined, as shown by mutants of *E. coli* that make beta-galactosidase even in the absence of lactose. These mutants are at a disadvantage compared to normal cells, since by making an enzyme in the absence of its substrate they are using their energies and resources uneconomically.

The Operon

To explain how inducers and corepressors function in the regulation of enzyme biosynthesis in bacterial cells, the French scientists François Jacob and Jacques Monod developed the hypothesis of the *operon*. An operon is a group of structural genes (genes that code for enzymes or other proteins) that are aligned along a single segment of DNA and that are regulated as a unit. The lactose operon, for example, includes the gene that codes for beta-galactosidase and two other genes, which also code for enzymes involved in the breakdown of lactose. These genes are adjacent to one another and are transcribed one after the other, along the DNA, forming a single mRNA molecule. The mRNA molecules produced are active for only a very short time, after which they are broken down. Thus, if a bacterial cell needs a continuous supply of enzymes, it must make a continuous supply of the specific mRNA.

Preceding the operon, as shown in Figure 16–4a, are two overlapping regions: (1) the *promoter*, a short segment of DNA to which RNA polymerase attaches at the beginning of transcription, and (2) the *operator*, which serves as the on-off switch for the operon. If the operator is "on," the RNA polymerase can move along the DNA and transcribe a molecule of mRNA. If the operator is "off," the RNA polymerase cannot move along the DNA and no mRNA will be made.

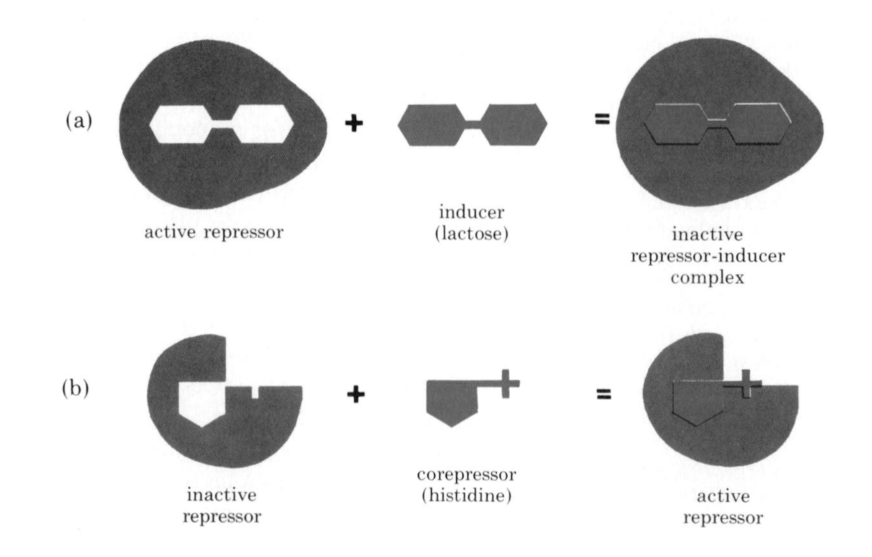

16–3 According to current theory, the synthesis of proteins is regulated by interactions involving either an inducer or a corepressor and another molecule known as a repressor.

(a) In the case of inducible enzymes, such as beta-galactosidase, the repressor molecule actively represses production of the enzyme until the repressor combines with the inducer (in this case, lactose). The repressor is then inactivated, and production of the inducible enzyme can begin.

(b) In the case of a repressible enzyme, the repressor becomes active only when it combines with the corepressor. Thus, in the absence of the corepressor, the repressible enzyme is actively produced.

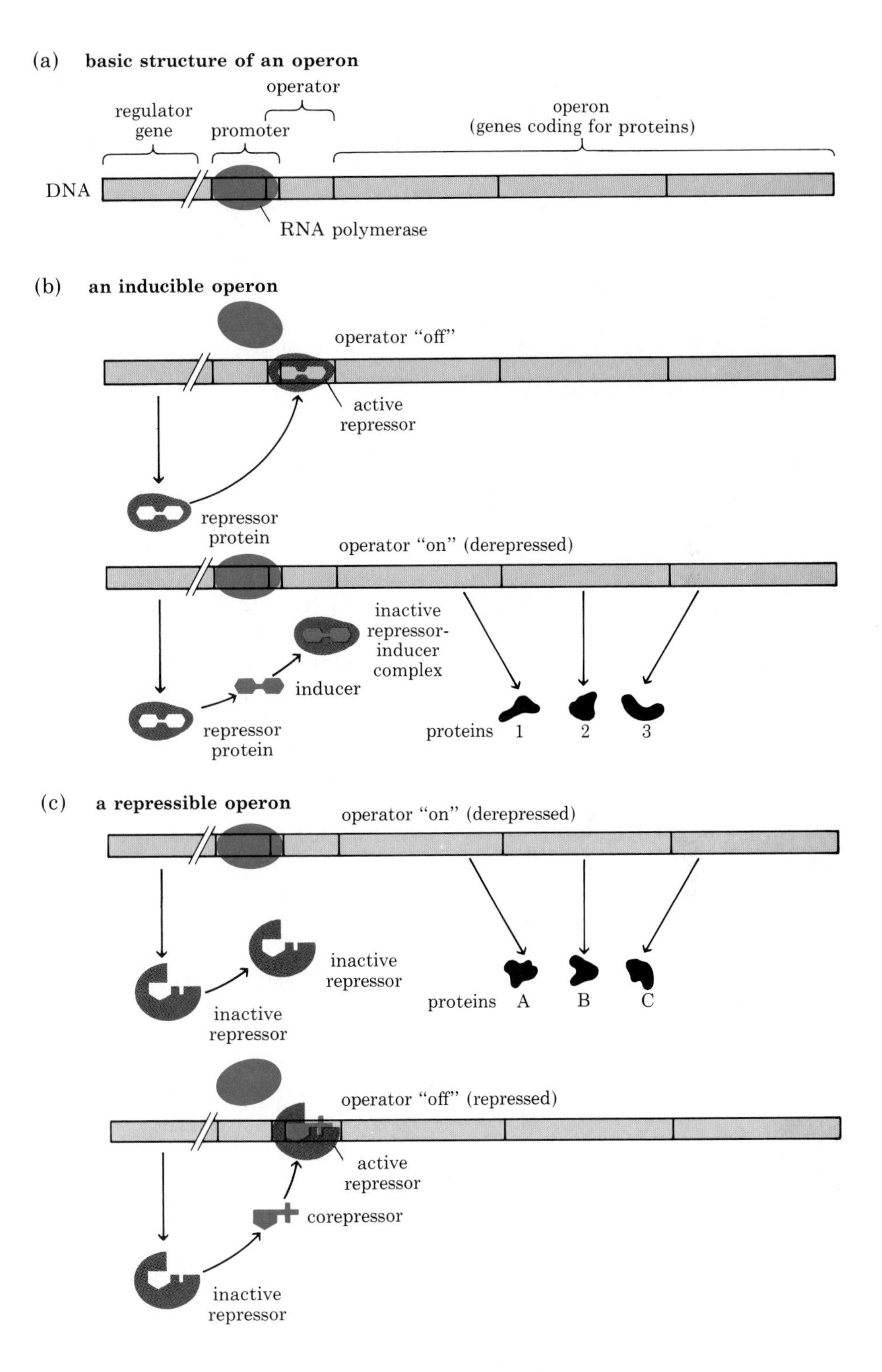

16-4 (a) *An operon is a group of structural genes that code for proteins, often enzymes that work sequentially in a particular reaction pathway. Adjacent to the operon on the bacterial chromosome is a segment of DNA that contains the promoter and the operator. The operator slightly overlaps the promoter. The promoter is the site at which RNA polymerase attaches; this is the enzyme responsible for the transcription of mRNA. Another gene involved in operon function is the regulator. The regulator gene, which is not necessarily adjacent to the other genes (as indicated by the diagonal lines), produces a regulatory protein known as a repressor. Some operons are inducible, others repressible.*

(b) *In operons activated by inducers, the regulator gene codes for a protein that binds to DNA at the operator and so prevents the RNA polymerase from initiating transcription. The inducer counteracts the effect of the repressor by binding with it and maintaining it in an inactive form. Thus, when the inducer is present, the repressor can no longer attach to the operator, and synthesis of mRNA proceeds.*

(c) *In operons regulated by corepressors, the repressor can bind to the operator only when it is combined with a corepressor. Thus transcription proceeds until a corepressor is produced by the enzymes for which the operon codes.*

Whether the operator is on or off is controlled by yet another gene, the *regulator*. The regulator codes for a specific protein known as the *repressor*, which may bind to the operator. If the repressor is bound to the operator, the operator is "off," and no mRNA can be transcribed from the operon. With no repressor bound to it, the operator is "on," mRNA is transcribed, and the enzymes coded for by the operon are synthesized.

The repressor is controlled by another "signal" compound. In the case of inducible enzymes, this compound is the inducer. The inducer binds

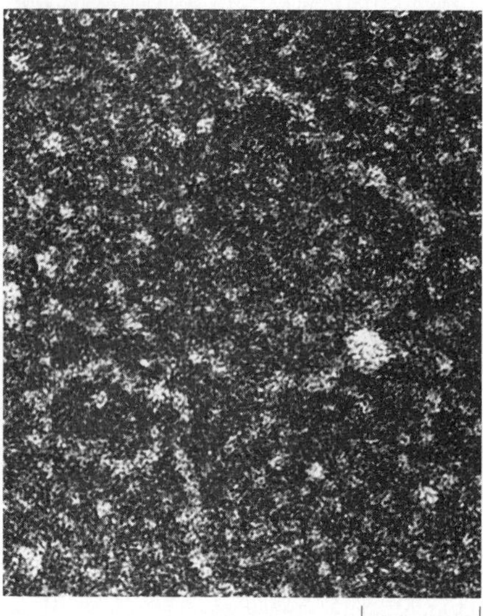

16–5 *A repressor protein (the white spherical form at the middle right in the micrograph) attached to the operator of the lactose operon.*

with the repressor molecule and changes its shape so that it can no longer attach itself to the operator (Figure 16–4b). In the absence of the active repressor, mRNA molecules are formed along the structural genes, and from these molecules proteins are produced. When the supply of the inducer is exhausted (as a result of the action of the enzymes produced), the repressor once again assumes control, and mRNA production and protein formation cease. For the lactose operon, the inducer is lactose.

In the case of repressible enzymes, the "signal" compound is a corepressor. The repressor is active only when bound with the corepressor (Figure 16–4c).

In further confirmation of the operon hypothesis, mapping studies show that genes for enzymes with related activities occur in clusters. The operon system is now widely accepted as a principal genetic control mechanism in bacteria. It is a simple and efficient means of coordinating the production of enzymes required by the cell in the same amounts at the same time.

The Eukaryotic Chromosome

The eukaryotic chromosome is far more complex in structure. Unlike the bacterial chromosome, it is more than half protein. These proteins are believed to play a major role in maintaining the structure of the chromosomes and in the events that take place during mitosis and meiosis. They may also be involved in regulating gene expression in eukaryotic cells.

The regulation of gene expression is a much different problem for a multicellular eukaryote than for *E. coli.* A multicellular organism usually starts life as a fertilized egg, the zygote. The zygote undergoes multiple mitotic divisions, producing many cells. At some stage these cells begin to differentiate, becoming muscle cells, nerve cells, blood cells, intestinal cells, and so forth. Each cell type, as it differentiates, begins to produce characteristically different proteins that distinguish it from other types of cells. This is nicely illustrated by mammalian red blood cells. In the early stages of fetal life, red blood cells produce one type of fetal hemoglobin; at later stages, a second type of fetal hemoglobin is produced; then, sometime after the birth of the organism, the red blood cells begin to produce the alpha and beta chains characteristic of adult hemoglobin. Thus the genes are expressed in a carefully controlled sequence, one after the other. The DNA segments that code for all of these hemoglobin molecules are expressed only in the red blood cells.

It is generally believed, however, that all of the genetic information originally present in the zygote is also present in every diploid cell of the organism. In other words, the DNA segments that code for hemoglobin (both the fetal types and the adult type) are present in skin cells and heart cells and liver cells and nerve cells and, indeed, in every one of the more than 100 different types of cells in the body. Similarly, the DNA sequence that codes for the hormone insulin is present not only in the special cells of the pancreas that manufacture insulin but also in all the other cells. Since each type of cell produces only its characteristic proteins—and not the proteins characteristic of other cell types—it becomes apparent that in any cell in any multicellular organism, most of the DNA has been selectively "silenced."

Cloning

Since each cell of a multicellular organism seems to carry copies of all of the organism's genes, it is theoretically possible to generate a whole organism from a single somatic cell. This form of genetic manipulation, which is known as cloning, has been the subject of much public attention.

A clone is an organism or group of organisms that have been produced from a single parent by a series of mitotic divisions and that are therefore genetically identical. Clones are found in nature only among one-celled organisms and in a few invertebrates and plants that reproduce asexually. When you root an African violet leaf, the new plant is genetically identical to the parent plant and is, strictly speaking, a clone. It has been discovered that it is also possible to produce a whole carrot from an isolated mature carrot cell.

The first clones of a vertebrate animal were produced as the result of experiments performed by J. B. Gurdon of Oxford University in England. Gurdon removed nuclei from the intestinal cells of tadpoles—that is, frog larvae—and placed them into frogs' egg cells in which the nuclei had been destroyed. Some of the egg cells then developed into normal frogs (see illustration).

Artificial cloning is proving to be a valuable technique for the production of uniform, disease-free varieties of plants. Its potential usefulness in the breeding of animals is less clear. Animals in which cloning might be worthwhile, economically speaking, are those in which the fertilized egg, unlike the eggs of frogs, develops inside the body of the female. The technical problems of removing and reimplanting such an egg are soluble—as the birth of a "test-tube baby" in Britain has demonstrated. It is not clear, however, that the delicate procedures involved could be carried out on a large enough scale to make them economically feasible.

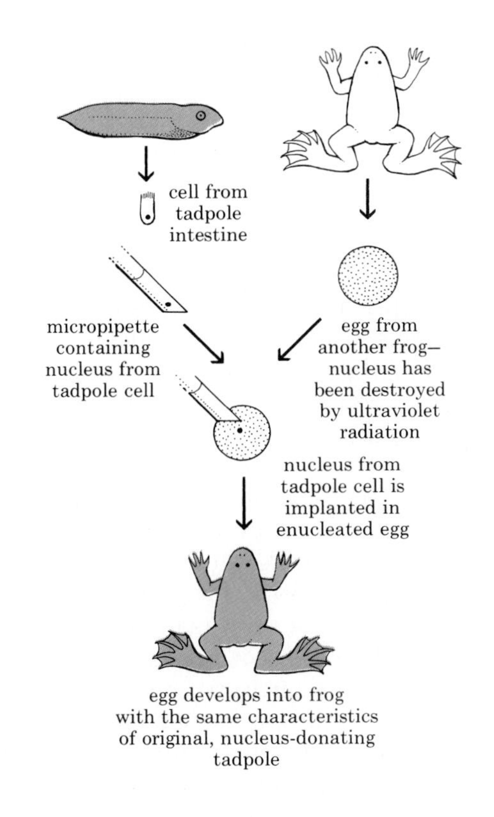

In experiments by J. B. Gurdon, the nucleus was removed from an intestinal cell of a tadpole and implanted into an egg cell in which the nucleus had been destroyed. In some cases, the egg developed normally, indicating that the tadpole intestinal cell nucleus contained all the information required for all the cells of the organism.

The application of cloning to humans, although more publicized, is less clear. One possibility that has been mentioned is the production of organs for transplant. There are two main obstacles to the widespread use of transplanted organs to replace diseased ones. One is the lack of suitable replacement organs, which need to be healthy, relatively young, and undamaged. The other, more serious obstacle is that an animal body reacts more or less strongly against any cells that are not genetically identical to its own cells, rejecting the transplant. For this reason, tissues can be exchanged readily only between identical twins. Currently transplants are done whenever possible between genetically similar persons, often members of the same family. Also, drugs are given to the patient to reduce his or her reactions to the foreign tissue, but since these drugs reduce the reactions to *all* foreign substances, they greatly increase the dangers of infection. One possible way to overcome these two obstacles might be the production of new organs from the patient's own cells growing in tissue culture; such organs, clearly, would be acceptable to his or her body.

Before such a feat could even be attempted, far more would have to be known about the control of human development, differentiation, and gene expression. Thus, at the moment, the cultivation of human replacement tissues is a very remote possibility. It has also been suggested—although, so far, not in public by scientists in the field—that in this way a family could "reincarnate" a loved one simply by removing a cell from his or her body; a society could reproduce its leaders in politics, science, and the arts; and a government could produce large numbers of people of a proven useful type. Beyond the overwhelming technical difficulties that would be involved, the complex interaction of heredity, environment, and personal experience that produces each unique human being makes the likelihood of cloning an identical human being virtually impossible. There have been great financial rewards from such ideas, but only to writers of science fiction.

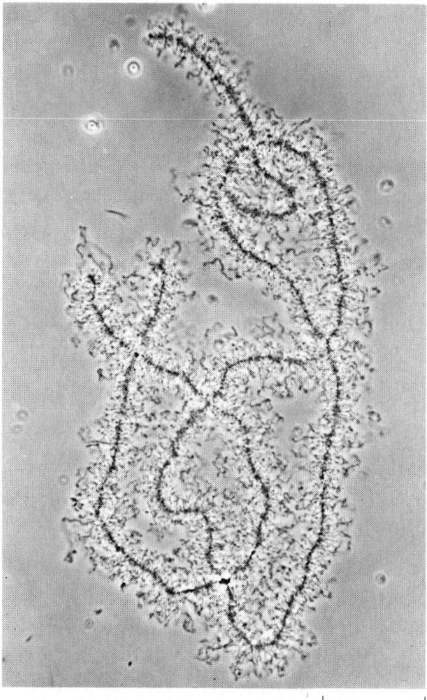

16-6 *Two homologous lampbrush chromosomes. The individual chromatids are closely paired and so cannot be distinguished.*

Chromosomal Proteins and Gene Regulation

Gene expression in eukaryotic cells seems to be related to the way the DNA and protein are packed together in the chromosome. The primary evidence is that RNA is transcribed only when the nucleus is in interphase, never when the chromatin is tightly coiled, as it is during mitosis.

A second type of evidence comes from studies of the giant chromosomes of insects (see page 168). At different stages of larval growth in insects, it is possible to observe diffuse thickenings, or "puffs," in various regions of these chromosomes. The puffs are open loops of DNA, and studies with radioactive isotopes indicate that they are the sites of rapid RNA synthesis (transcription). When ecdysone, a hormone that produces molting in insects, is injected, the puffs occur in a definite sequence that can be related to the developmental stage of the animal.

Somewhat similar observations have been made with another type of material, lampbrush chromosomes, so called because they resemble the brushes used to clean kerosene lamps. They are found in the nuclei of egg cells of fish, birds, reptiles, and amphibians during the first prophase of meiosis. These cells are very actively engaged in the RNA and protein syntheses required for the rapid growth of the mature egg. As you can see in Figure 16-6, each chromatid has a number of lateral loops, which branch out from the main axis. These loops appear to be reeled in and out from the body of the chromosome. Each of these loops consists of a thin fiber. If this thin fiber is exposed to an enzyme that digests away the associated protein, all that remains is a thin filament about the diameter of a single DNA molecule. These loops, like the chromosome puffs in giant chromosomes, are also the sites of active RNA synthesis.

Thus, according to present information, mRNA is not transcribed from tightly coiled DNA, and, conversely, mRNA transcription is correlated with the unfolding of DNA.

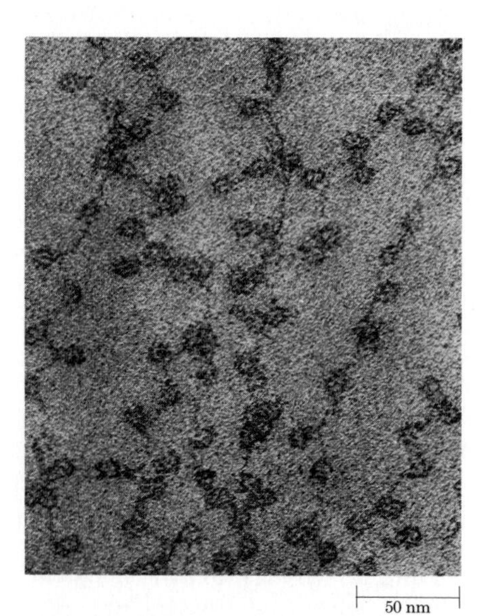

16-7 *Chromatin from a red blood cell of a chicken. Each bead is composed of about 140 base pairs of DNA and an assembly of histone molecules. The string between the beads is DNA.*

The Proteins of the Chromosome

There are two types of proteins in the chromosome. One group is composed of large molecules that are quite varied in structure and are believed to play a regulatory role. They make up 50 to 70 percent of the total protein of chromatin, but they are difficult to analyze. As yet, little is known about them.

The second type of protein is a group of relatively small molecules known as histones. They are always present in chromatin, are very similar from species to species, and are synthesized in large numbers during the S phase of the cell cycle (see page 135), when DNA is also being synthesized. There are at least five distinct types of histones. They are believed to be important and universal structural elements of chromatin.

Much recent attention has been focused on the way in which DNA and histones might fit together. The structure that has emerged from biochemical studies and electron micrographs resembles beads on a string (Figure 16-7). The beads are tight clusters of DNA and histones, and the string is short stretches of DNA. When a fragment of DNA is tied up in a bead, it is about one-sixth the length it would be if it were fully extended. Presumably the beads and strings are further folded in some way.

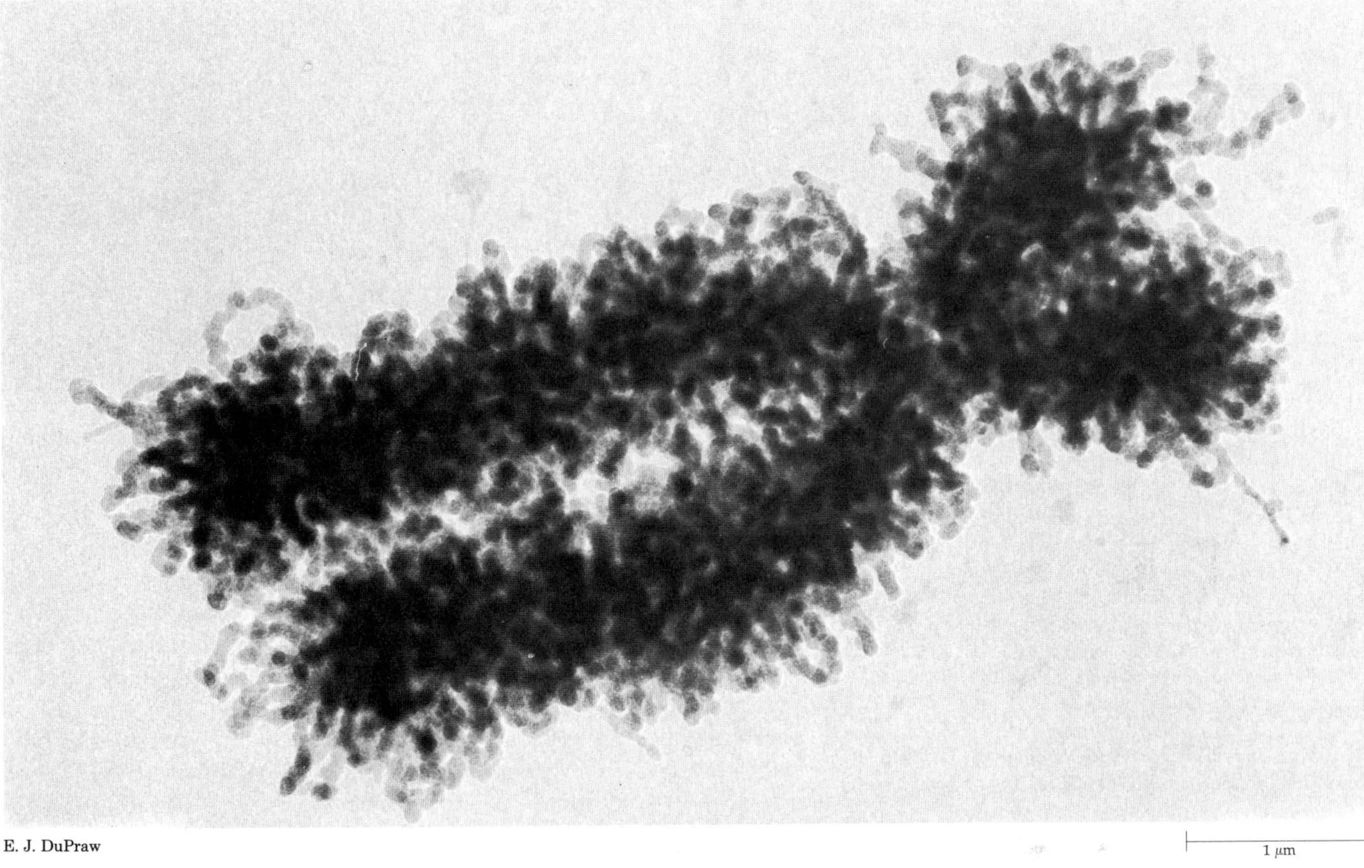

E. J. DuPraw

16-8 *Human chromosome 12, shown at metaphase, before the chromatids have separated. The chromatin is highly condensed. Each chromatid, according to present evidence, contains a single DNA molecule.*

When Watson and Crick deduced the structure of DNA, the meaning of the structure—in terms of the replication of DNA—was immediately obvious to them. The structure of chromatin—the way the DNA and protein are packaged together—remains, functionally speaking, quite mysterious. And, recent discoveries about the DNA itself have contributed to the mystery.

The DNA of the Eukaryotic Chromosome

DNA is an "exquisitely thin filament," in the words of E. J. DuPraw, who calculates that a length sufficient to reach from the earth to the sun would weigh only half a gram. The DNA of each chromosome is believed to be in the form of a single molecule. In a human chromosome, each of these molecules is believed to be from 3 to 4 centimeters long. Thus each human cell contains between 1 and 2 meters of DNA, and the entire human body contains some 25 billion kilometers of DNA double helix.

The technical problems of investigating a molecule as large as the DNA of a eukaryotic cell are enormous. Because of the base-pairing properties of DNA, it is possible, however, to compare structures of different DNA molecules and different sequences on the same molecule without actually identifying them. If DNA is heated gently, the hydrogen bonds holding the two strands together are broken and the strands separate.

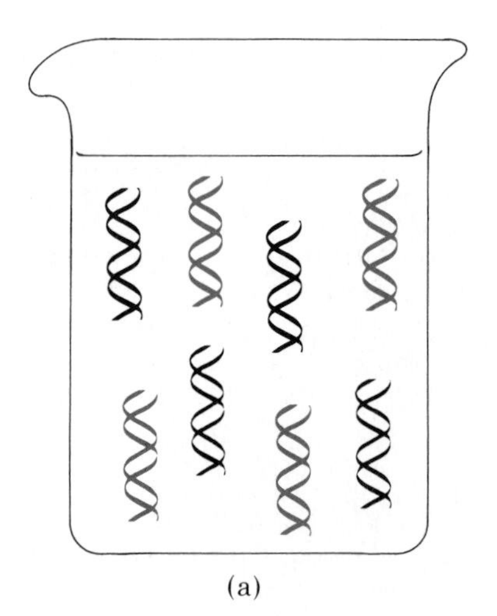

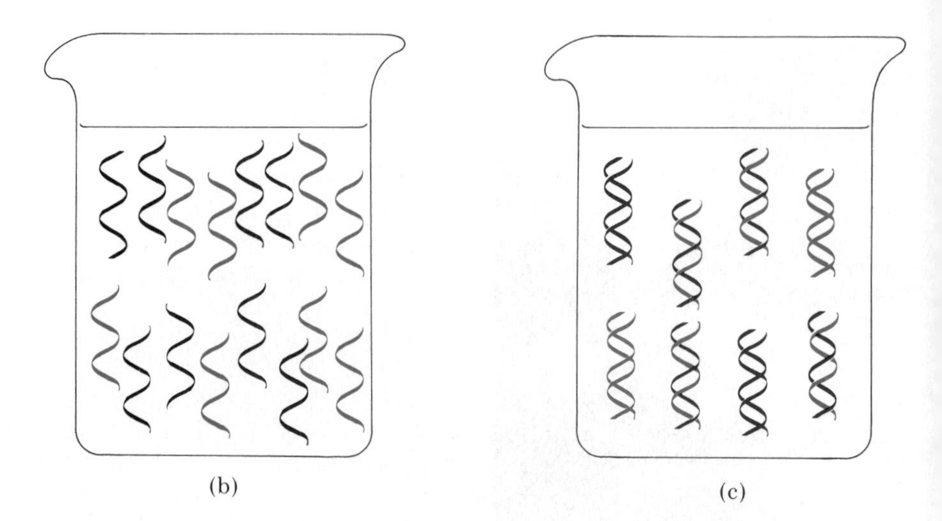

16–9 (a) *Two types of DNA are mixed in a beaker. One (indicated in color) is labeled with radioactive phosphorus.* (b) *When heated, the double-stranded DNA separates into two single strands.* (c) *When cooled, the DNA strands recombine in double helices. The double-stranded molecules are then separated from any single-stranded molecules present, and the amount of radioactivity in the newly formed duplexes is analyzed. The extent to which radioactive and nonradioactive strands are able to form duplexes in a given length of time is an indicator of the similarities of the DNAs.*

When the solution is cooled slightly, the bonds re-form (Figure 16–9). The extent to which segments from two samples of DNA reassociate and the speed with which they do so provides an estimate of the similarity between the DNAs. Similarly, the extent and rate of combination of an mRNA molecule with a DNA molecule make it possible to determine whether or not the mRNA could have been transcribed from that particular DNA.

Studies using this method and others have revealed that the DNA of the eukaryotic cell is composed of three distinct kinds of sequences. One type of DNA is made up of very short (about 100 base pairs), simple, and highly repetitive sequences. (The sequence in a hermit crab, for instance, is ATCCATCCATCC.) Such sequences may be repeated as many as 10^7 times per cell and may make up as much as 10 percent of all the DNA. Both structural and regulatory roles have been hypothesized for these repetitive sequences.

A second type of DNA consists of gene-sized (that is, about 1,000 base pairs) sequences repeated over and over again, from a few times to hundreds of thousands of times.

The third type of DNA consists of gene-sized segments that are unique sequences. About 70 percent of the total DNA of a eukaryotic cell is made up of this unique-sequence DNA. This much DNA could code for about half a million genes. With the number of genes in a mammalian cell estimated at about 50,000, it would appear that the unique-sequence DNA contains about 10 times more DNA than is required.

Intervening Sequences

A related, particularly startling discovery was made in 1978. As more DNAs and more RNAs were analyzed, direct comparisons could be made between chromosomal DNA and the corresponding mRNA. It was discovered that right in the middle of a DNA segment coding for a protein, a long sequence of bases would sometimes occur that did not appear in the mRNA molecule in the cytoplasm. These *intervening sequences*, as they are called, may be several hundred nucleotides in length, making up as much as a third of the original mRNA molecule transcribed in the nucleus. Before the mRNA moves from the nucleus into the cytoplasm, the intervening se-

16–10 (a) *New evidence indicates that the DNA of a gene (color) often includes "silent" intervening sequences (white). These sequences are transcribed into mRNA in the nucleus* (b), *but are then snipped out of the molecule by special enzymes* (c). *Other enzymes splice together the remaining, code-carrying portions of the mRNA molecule, which then enters the cytoplasm* (d) *and directs the synthesis of the protein for which it codes* (e). *The function of the intervening sequences is not known.*

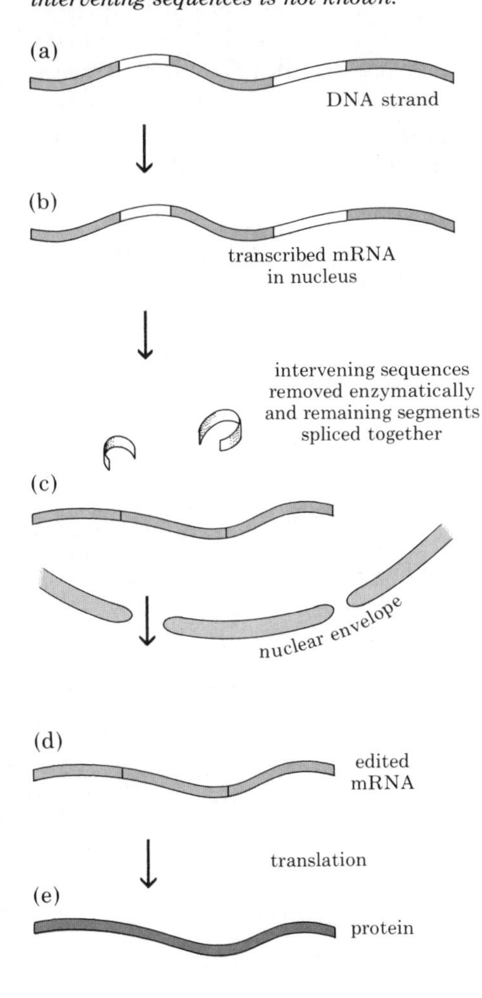

quences are removed enzymatically and the code-carrying portions of the mRNA are rejoined (Figure 16–10).

Genes coding for such intervening sequences have now been found in *Drosophila,* yeast, chick, rabbit, mouse, and a virus. They are apparently not present in prokaryotes. Given the extraordinary efficiency of the transcription-translation process, the fact that intervening sequences are so widespread leads one to believe that they have some meaning. Do they have a role in regulating gene expression? Are they a factor in crossovers? Do they make possible the introduction of new variations in the genetic material, beyond those resulting from the reshuffling of complete genes that is known to occur in meiosis? And, are they thus an important—but previously unsuspected—part of the mechanism of evolution?

The discovery of intervening sequences has given rise to questions that, when answered, should further extend our knowledge of the regulation of gene expression.

SUMMARY

There are major structural differences between prokaryotic and eukaryotic chromosomes, primarily related to the differences in gene regulation in the two types of organisms.

The bacterial chromosome is a single, circular loop of DNA. In addition, a bacterial cell may contain a number of much smaller, also circular, molecules of DNA called plasmids.

A principal means of genetic regulation in bacteria is the operon system. This system is made up of a promoter, the site at which the RNA polymerase molecule binds; an operator, which is the site of regulation; and the operon, one or more structural genes that code for proteins. The operator is under the influence of another gene, known as the regulator. The regulator codes for a protein, known as a repressor, that attaches to the DNA molecule at the operator and blocks mRNA transcription. The repressor does not act alone but rather in conjunction with another compound. This compound may be an inducer, in which case it inactivates the repressor, or it may be a corepressor, in which case it enables the repressor to function.

The eukaryotic chromosome, which is far more complex in structure, consists of DNA packed together with two types of protein molecules. One is a group of large and varied proteins that are believed to have specific regulatory functions. The other is a group of relatively simple proteins known as histones. Recent evidence indicates that the histone-DNA complexes resemble beads on a string.

Each chromosome, according to present evidence, contains a single, very long DNA double helix. Three types of DNA sequences have been discovered. In one type, the sequences are short and highly repetitive. In the second type, the sequences are gene-sized and highly repetitive. In the third type, the sequences are also gene-sized but are heterogeneous and nonrepetitive. Only this third type, which comprises about 70 percent of the total DNA, is believed to contain the genetic message.

In a number of organisms, silent regions, known as intervening sequences, have been found within the genes coding for protein. These sequences are transcribed into mRNA but are edited out before the mRNA is translated into protein. Their function is not known.

QUESTIONS

1. Distinguish among the following: regulator/promoter/operator/operon; inducer/repressor/corepressor; inducible enzyme/repressible enzyme.

2. In what ways are the chromosomes of prokaryotes and eukaryotes similar? In what ways are they different?

3. Compounds can be used by cells in two different ways. One type is broken down (usually as an energy source). Another is used as a building block for a larger molecule. Which type of compound would you expect to function as a corepressor? As an inducer? Do the examples given in the text conform to these expectations?

4. Gene expression theoretically can be regulated at the level of transcription, translation, or activation of the protein. In the latter case, the polypeptide produced by protein synthesis is an inactive form. It undergoes some structural modification (controlled by enzymes) before it can perform its function in the cell. What would be the advantages of each type of regulation, in terms of the cell? Under what circumstances might one type be more useful than another? Which is the more economical?

17

Recombinant DNA

The discoveries about the gene made by molecular biologists during the past 30 years clearly rank among the great scientific achievements of all time. These discoveries have given us not only a greatly increased understanding of the workings of heredity but also the tools with which to extend that understanding. Moreover, within the past few years it has become possible to deliberately manipulate the composition of DNA in living cells, creating new, artificial combinations of hereditary information. This work, in which portions of DNA molecules from different sources are recombined (hence the name "recombinant DNA"), is providing new knowledge and new practical possibilities, particularly in medicine and agriculture, at a stunning pace. To understand this work, however, we must look first at the ways in which DNA molecules recombine without human intervention.

Natural Recombinations

Every time a eukaryotic organism reproduces sexually, DNA is recombined. Crossing over and the random assortment of homologous chromosomes during meiosis ensure that the exact composition of the DNA in a particular gamete—whether sperm or egg—will be different from that in the egg and sperm that originally gave rise to the parent organism. The DNA composition of the gamete will also be different from that of other gametes produced by meiosis in the same parent organism. Thus, a totally new combination of DNA—of hereditary information—occurs in the zygote when egg and sperm unite in fertilization. As we have seen, this natural recombination maintains the variations on which evolution depends.

Natural recombinations of DNA occur in prokaryotes through three separate processes. In transformation, as we saw in Chapter 14, bacterial cells release DNA (transforming factor) into the environment, where it can be taken up by other cells and recombined with the DNA originally present in those cells. In the second process, bacterial conjugation, DNA is transferred directly from one cell to another. Transduction, the third process, involves the transfer of DNA from one cell to another by means of viruses. These naturally occurring processes are essential tools in current recombinant DNA work.

Bacterial Conjugation

Bacteria mate by a process called conjugation. When this event occurs, a donor cell passes some or all of its circular chromosome—that is, its DNA—to another cell (Figure 17–1). Like many other phenomena, conjugation has been studied most extensively in *Escherichia coli.*

17-1 *Electron micrograph of conjugating* E. coli *cells. A slender connecting bridge has formed between the elongated donor cell at the top of the micrograph and the more rotund recipient. DNA enters the recipient cell through this bridge.*

In *E. coli,* cells are either donors or recipients of DNA. Whether or not a cell can function as a donor depends on its possession of a particular plasmid (see page 205) known as a fertility, or F, factor. A donor cell is thus known as F^+ and a recipient as F^-. The F factor plasmid can be passed from one cell to another. When this happens, the recipient cell, in turn, becomes a donor cell (Figure 17–2a).

Like some other plasmids, the F factor can exist independently in the cell or can be part of the chromosome. Transfer of a portion of the bacterial chromosome from one cell to another occurs only if the F factor has become integrated into the chromosome. The integrated F factor, like other genes, occupies a particular position on the chromosome. During conjugation, the chromosome opens and begins to replicate at the point of attachment of the F factor. As it replicates, the free end of the chromosome moves into the F^- cell (Figure 17–2b). Usually only part of the chromosome is transferred. This fragment can then be integrated into the chromosome of the F^- cell.

There are numerous plasmids besides the F factor, and they may contain as few as two genes or as many as several hundred. Plasmids readily undergo recombination with other plasmids. This natural capacity of plasmids is the source of potentially serious public health problems (see essay).

Transduction

Transduction is the transfer of genes from one cell to another by means of viruses. It has been observed with bacterial viruses, which, as we noted in Chapter 14, are composed of nucleic acid wrapped in a protein coat. Transduction occurs when some of the nucleic acid of the host cell becomes wrapped in the protein coat of a virus and is carried to a new host cell.

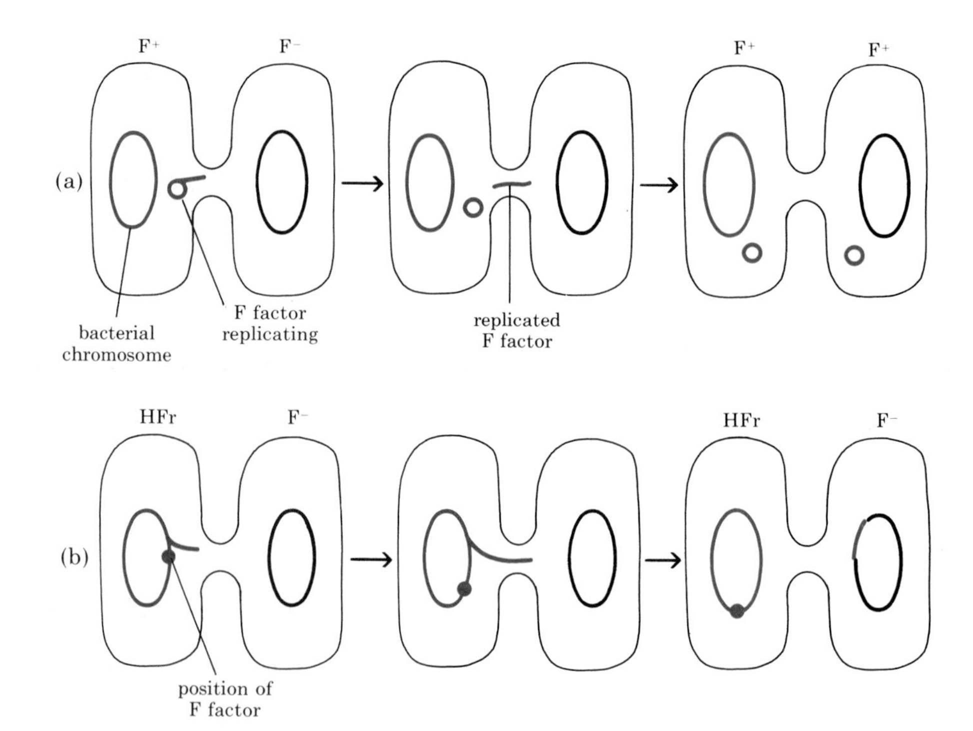

17-2 (a) *The F factor is a plasmid that confers upon its host cell the capacity to be a donor cell and transfer an F factor to a recipient (F^-) cell.*
(b) *When the F factor is part of the chromosome, the cell is referred to as HFr (for high frequency of recombination). From such donor cells, the chromosome itself, or a portion of it, enters the recipient cell. That portion may undergo recombination with the chromosome of the recipient cell, exchanging genes with it. The recipient cell usually remains an F^- cell because the F factor is not usually transferred.*

Infectious Drug Resistance

For many years, transfers of genetic information among bacterial cells have been regarded mainly as laboratory phenomena, chiefly of interest to research scientists engaged in studies such as gene mapping. However, it is now clear that such transfers also occur frequently in nature among certain groups of bacteria.

In a large bacterial population, such as that of the human intestines, there may be a few cells that, as a result of mutation, are resistant to a particular drug—streptomycin, for example. If the bacterial population is exposed to streptomycin—as when a patient is being treated with the drug—susceptible bacterial cells are destroyed and the resistant ones remain. These multiply and produce a population of cells made up entirely of resistant individuals. This is one reason why responsible physicians advocate restraint in the use of antibiotics. (A second reason is that with each treatment with antibiotics, the chance of the patient's developing an allergic reaction to that drug increases.)

It is now known that bacteria can become drug-resistant by the transfer of plasmids from resistant to nonresistant cells. Like F factors, these plasmids are readily passed from cell to cell. Under experimental conditions, 100 percent of a population of nonresistant cells can become resistant within an hour after being mixed with suitable resistant bacteria. Such transfers were first discovered in a strain of *Shigella,* a bacterium that causes dysentery. Further studies showed that not only can *Shigella* transfer drug resistance to other *Shigella* organisms but also the innocuous *E. coli* can transfer resistance factors to this and other, unrelated groups of bacteria. Infectious resistance is now found among more types of bacteria, including those that cause typhoid fever, gastroenteritis, and undulant fever as well as dysentery.

These findings cast considerable doubt on the advisability of the current widespread use of antibiotics. For example, antibiotics are ineffective against viruses, yet many physicians (sometimes at their patients' insistence) routinely prescribe them for the common cold, a virus-caused disease. Antibiotics are routinely added to animal feed because they promote growth. In members of one farm family whose chickens were being fed antibiotics, 36 percent of the bacteria present in their intestines were resistant to three or more antibiotic compounds, compared with 6 percent for a neighboring control family.

The problem is, of course, that if these practices continue, an increasingly larger percentage of disease-causing bacteria will be resistant to any treatment now available for them. Thus, by our own lack of wisdom, we are threatening to negate some of the greatest medical advances of our century.

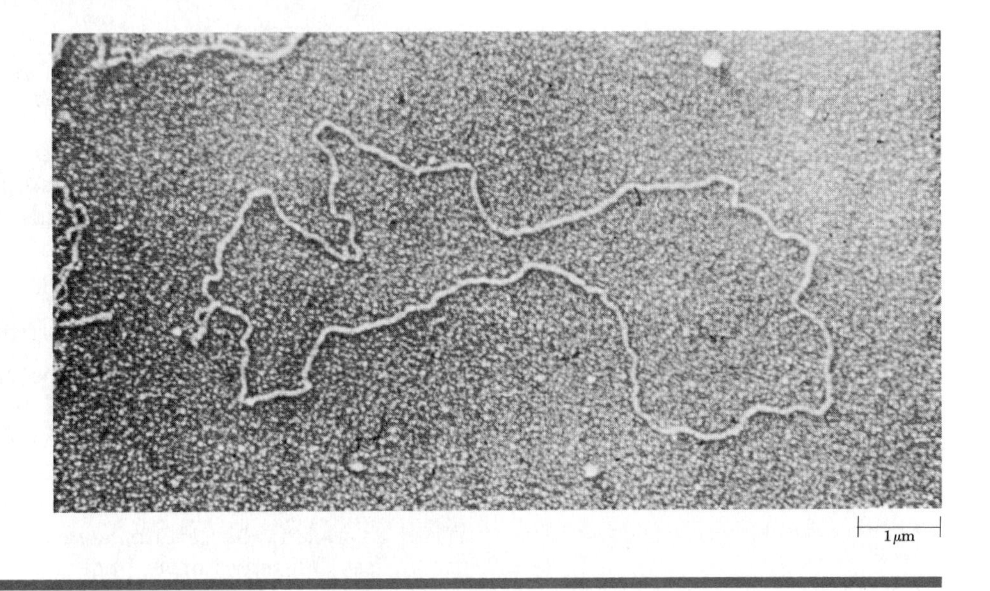

This plasmid, known as R6, carries genes that confer resistance to six different drugs, including the antibiotics tetracycline, neomycin, and streptomycin.

Two forms of transduction are known: general and restricted. General transduction occurs when fragments of the DNA of a bacterial cell, broken down in the course of infection, are caught up in a bacteriophage head and are carried to a new host cell. These particles carry little or no viral DNA and so they cannot set up an active infection in the recipient cell. Instead, the portion of the bacterial chromosome carried into the new host may become a part of the host chromosome.

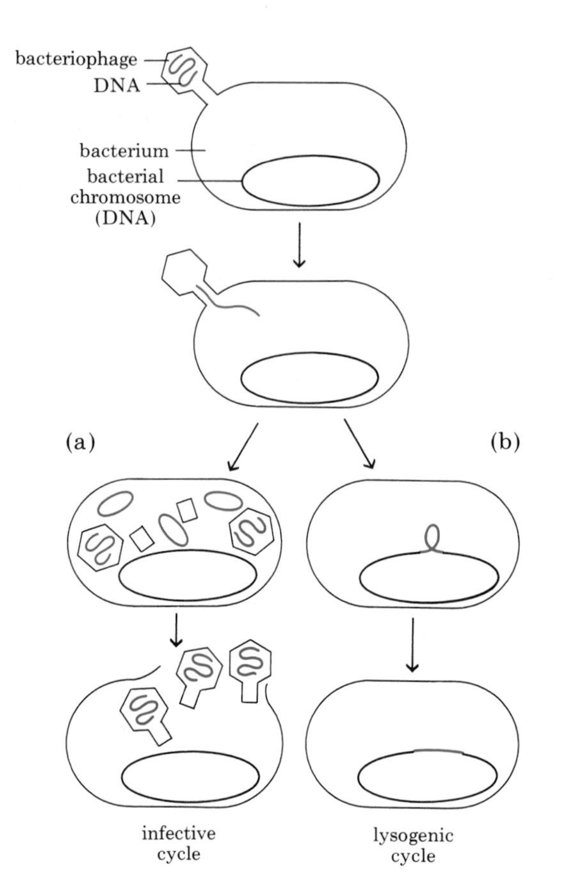

17-3 *When certain types of bacterial viruses infect bacterial cells, one of two events may occur.* (a) *The virus DNA may enter the cell and set up an infection, such as we described in Chapter 14, or* (b) *the DNA of the virus may simply lie latent in the cell. In the latter case it may become part of the bacterial chromosome, replicating with it. Bacteria harboring such viruses are known as lysogenic. From time to time such a virus, known as a prophage, becomes activated and sets up a new infective cycle.*

Restricted transduction occurs only with certain types of bacterial viruses. These bacteriophages do not necessarily set up an infective cycle when they enter a host cell but may instead exist as a molecule of naked DNA in the host cell—in other words, as a plasmid. Like the F factor and some other plasmids, they may become part of the host cell chromosome (Figure 17-3). They are then called prophages and are duplicated faithfully each time the bacterial cell divides. Bacteria that harbor prophages are known as *lysogenic bacteria*.

From time to time the relationship between a lysogenic bacterium and its resident prophage may shift. The viral DNA breaks loose from the bacterial chromosome and sets up an infective cycle. Exposure to x-rays, ultraviolet light, or certain chemicals (all generally the same agents that cause mutations) increases the rate of activation of viruses in host bacteria. When a lysogenic bacterium releases a burst of newly synthesized viruses, the viruses can infect other nearby bacterial cells.

When the viruses leave a host bacterial cell chromosome, they sometimes take a fragment of the chromosome with them (Figure 17-4). This fragment is replicated with them through any infective cycle. If the DNA of one of these viruses becomes part of the chromosome of a new host, the genes from the previous host may be inserted into the new host's chromosome and become a part of its genetic equipment. For instance, the bacteriophage lambda can carry the bacterial genes that produce the enzymes concerned with the breakdown of galactose—in other words, the galactose operon. When a lambda prophage carrying the galactose genes becomes a prophage in a bacterial cell that cannot synthesize one or more of these enzymes, the infected cell gains the capacity to utilize galactose for growth.

It has long been suspected that certain viruses can enter into relationships with eukaryotic cells similar to those relationships between the lysogenic bacteria and their viruses. Herpes simplex, the virus that causes fever blisters, is an example. If you suffer from fever blisters or know someone who does, you know that the lesions tend to break out repeatedly in the same person. Further, they characteristically occur in times of stress, such as when that person has a fever or eats certain food or gets a sunburn (ultraviolet exposure). It is believed that certain cells constantly harbor the herpes simplex virus, that it multiplies when the cells multiply, and that it causes cell damage—the fever blister—only when it is activated.

17-4 (a) *Like the F factor, the DNA of a virus may become part of a bacterial chromosome.*
(b) *Spontaneously, or as a result of x-rays or chemical treatment, the viral DNA may break loose from the chromosome, setting up a new infective cycle.*
(c) *Sometimes the virus takes a fragment of adjacent host chromosome with it. This fragment may be large enough to include several host genes.*
(d) *When and if the viral DNA combines with the chromosome of a new host cell, the cell acquires the new genetic information. This process is known as transduction.*

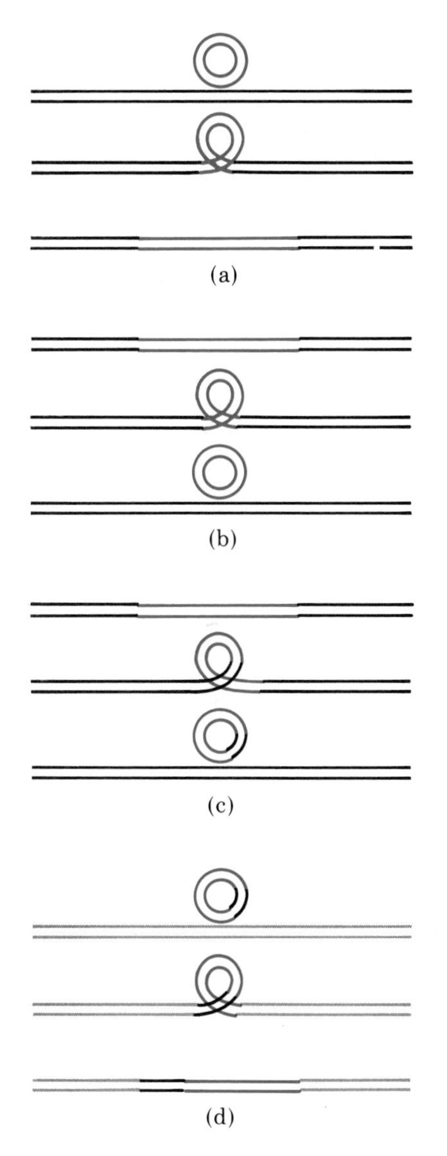

Gene Splicing

Transformation, bacterial conjugation, and transduction are not new discoveries. There are, however, some new developments that make it possible to manipulate these phenomena to some extent. They involve principally the discovery of enzymes that operate on DNA. One group of enzymes, known as DNA ligases, mend breaks in DNA molecules. A second group of enzymes, known as restriction enzymes, cleave DNA molecules at specific sites. These enzymes apparently have the function of recognizing and destroying DNA that is foreign to the cell. In the laboratory, they can be used to cut DNA molecules into relatively small, uniform fragments.

Some restriction enzymes cut through both strands of the DNA molecule at the same site. Others cleave the molecule by cutting through the two strands a few bases apart. One enzyme, for instance, known as Eco RI (isolated from a plasmid-containing *E. coli*), cleaves DNA only at the sequence GAATTC. This can have surprising consequences. The double-stranded molecule would be

...GAATTC...
...CTTAAG...

Eco RI cleaves each strand between the guanine and the adenine:

↓
...GAATTC...
...CTTAAG...
↑

What is left, as you will readily see, is

...G and AATTC...
...CTTAA and G...

The protruding ends—TTAA on one strand and AATT on the other—can be joined again with each other by the appropriate DNA ligase and also—and this is most important—with any other segment of DNA that has been cleaved by the same restriction enzyme. Thus it is now possible to splice together segments of DNA from, apparently, a limitless variety of sources.

These segments of DNA can be carried by plasmids. For instance, Stanley Cohen of Stanford University Medical School and a group of co-workers isolated a small plasmid of *E. coli,* designated pSC101, that makes bacteria resistant to the antibiotic tetracycline. pSC101 has only one

Genes, Viruses, and Cancer

Cancer is a group of diseases in which particular cells in the body cease to respond to whatever controls growth under normal circumstances. These cells multiply autonomously—crowding out, invading, and destroying other tissues. There have been many theories about the causes of cancer. One of the older hypotheses, now gaining renewed support, is that cancer is the result of a somatic mutation. Such a mutation occurring in a single somatic cell would be passed on by mitosis to all of its progeny.

More recently, accumulating information about viruses has led many scientists to believe that viruses are the cause of at least some forms of human cancer. One of the reasons for this belief is the fact that many different kinds of cancers in other animals are caused by viruses, including cancers in frogs and chickens and in mice, rats, and other mammals. It has been impossible, however, to establish conclusively that any human cancers are caused by viruses.

With the discovery of viruses—such as bacterial prophages—that can alter the genetic material of living cells, the somatic theory of mutation and the virus theory appeared to merge. It has recently been discovered that at least one cancer-causing virus, the polyoma virus, does become part of the chromosomes of cells growing in tissue culture. When the viral DNA is inserted into the host chromosome, the cell becomes cancerous. By all criteria, its offspring are also cancerous. They have the capacity to cause cancer in the animal or strain of animal from which the cells were originally taken.

Discovery of the cause of cancer, although a great intellectual achievement, would not necessarily mean control of the disease. We have, for instance, the example of sickle cell anemia, in which the cause is known to the last nucleotide, yet for which, at this writing, there is no known cure. Nor are there any cures for viral diseases, although some of them can be prevented very effectively before infection occurs. On the other hand, the cause of at least one very common type of cancer, cancer of the lung, is well established, but cigarette smoking has not stopped. Nevertheless, we continue to hope that such research will hasten the day when a control will be found for what is probably the most dreaded of human diseases.

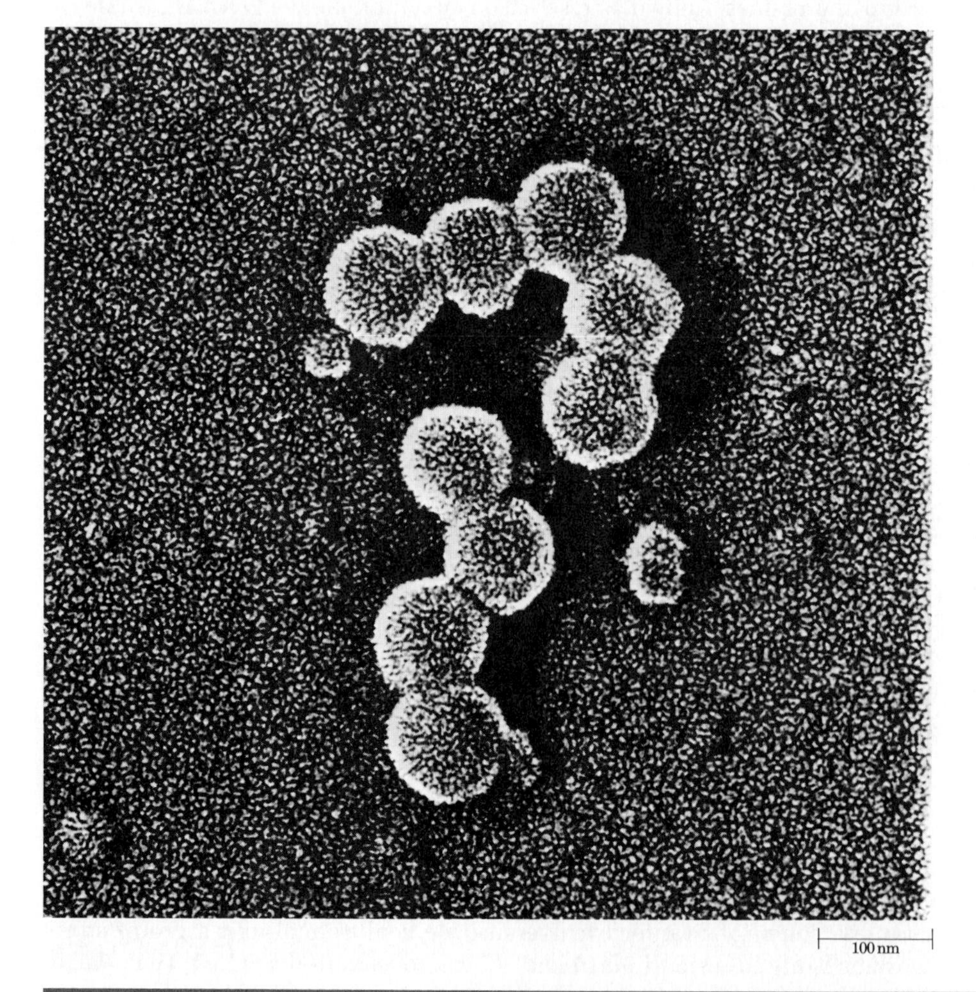

SV40 (simian virus 40), originally isolated from monkeys, has been shown to produce cancers in baby hamsters and other laboratory animals. It is small, even for a virus; each particle is only about 45 nanometers in diameter. It is composed of a molecule of DNA surrounded by an icosahedral (twenty-sided) protein coat.

GAATTC sequence in the entire molecule and, as a consequence, is cleaved at only one site by Eco RI. These investigators have shown that the insertion of an additional segment of a DNA molecule into the plasmid (as shown in Figure 17-5) does not affect the uptake of the plasmid by *E. coli,* its capacity to make the recipient cells tetracycline-resistant, or its ability to replicate.

At present, biochemical geneticists have a variety of restriction enzymes, each of which cleaves DNA molecules at a different site, and of DNA ligases with which to splice the segments together again. They also have two different types of hosts for newly spliced genes—bacteriophages and bacterial plasmids—and a means to introduce the genes into new hosts.

17-5 *An enzyme (Eco RI) has been found that opens DNA molecules at specific sites, leaving "sticky" ends exposed. These ends, consisting of TTAA and AATT sequences, can rejoin or can join with any other fragment of DNA that has been cleaved by the same enzyme. Thus it is possible to splice together DNA from different sources; for example, a foreign gene can be inserted into a bacterial plasmid, as shown here.*

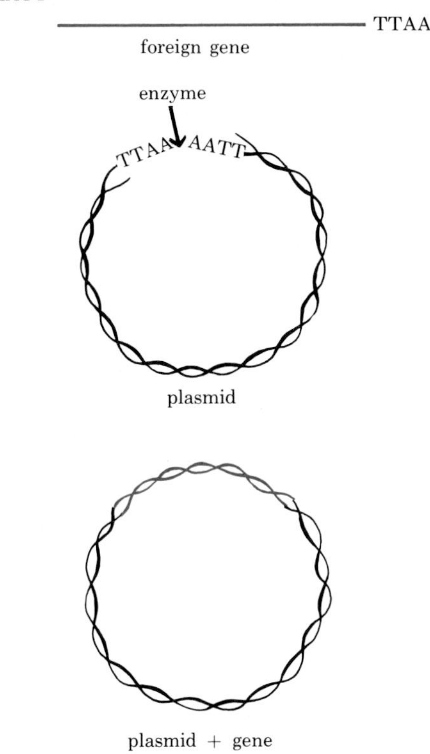

The Possibilities

What can be done with this new technology? When it first became possible to recombine DNA molecules from different sources, scientists suggested a number of applications. For example, viruses might be tailor-made to carry particular genes to particular cells. Such a procedure could provide a cure for hereditary diseases such as sickle cell anemia or PKU.

Other scientists pointed out that if segments of human DNA that code for proteins important in medical treatment—such as insulin and other hormones—were inserted into bacterial plasmids, one could have a low-cost, virtually limitless source of these proteins.

Another particularly beneficial practical application might be the production of agriculturally important plants with combinations of new characteristics—such as the capacity both to carry out C_4 photosynthesis (page 119) and to fix nitrogen (page 559). Some investigators are working with four strains of the bacterium *Pseudomonas,* each of which has a capacity to break down a component of fuel oil. The attempt is being made to combine them genetically into one super oil-eater that, in powdered form, can simply be sprinkled on oil spills.

Perhaps most important and intriguing of all is the possibility of being able to isolate a eukaryotic gene and study it outside of the enormously complex eukaryotic cell and in the well-understood and relatively simple *E. coli* host. A new field of research is now open to molecular biology.

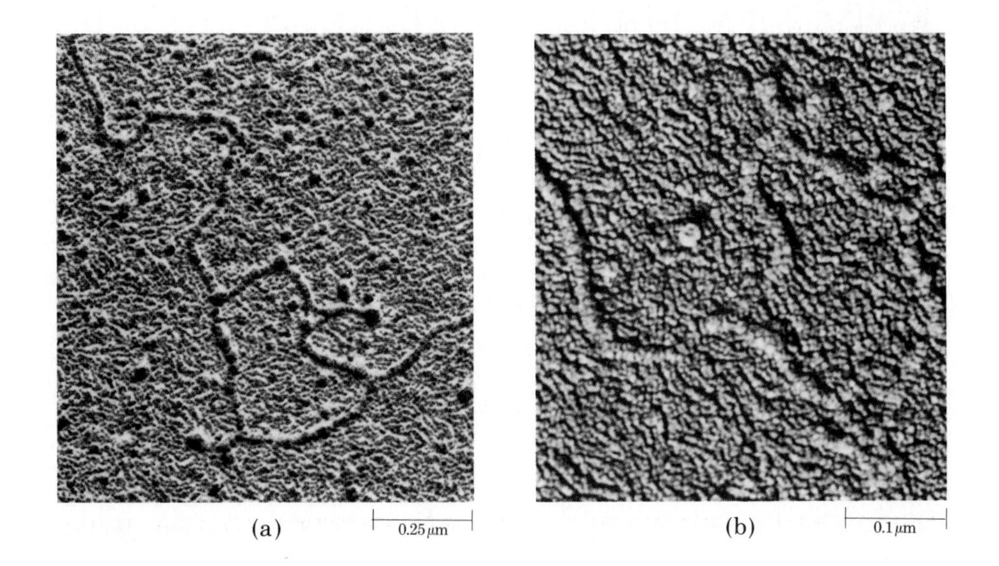

17-6 *Recombinant DNA techniques have been used by scientists at the University of Michigan to isolate a human gene. The gene that codes for ribosomal RNA is shown here at two different magnifications and under different conditions. In each micrograph, the human gene is the region of the double-stranded DNA molecule where the strands have split apart to form a loop. The DNA on either side of this loop is bacteriophage DNA.*

The Accomplishments

In the past few years, the possibilities of recombinant DNA work have begun to become realities. The first success was achieved by Keiichi Itakura and his associates at the City of Hope Medical Center. In late 1976, they constructed a plasmid containing the lactose operator gene, which, you will recall, is the on-off switch for the lactose operon (page 206). The next step was to select a gene that could be spliced to the operator gene and then turned on by the operator in the presence of lactose. They selected the gene for the mammalian hormone somatostatin because (1) it was a small protein (only 14 amino acids), (2) it could be detected in small amounts, and (3) it was a potentially useful compound.

The investigators knew the sequence of amino acids in somatostatin and thus could determine what the sequence of nucleotides in DNA should be. They then synthesized the gene, bonding together the nucleotides one by one. They spliced the somatostatin gene into the lactose operon and supplied this plasmid—naked DNA—to *E. coli* cells. Only a few bacteria were transformed, but these multiplied, the plasmid along with them, until several colonies of bacteria were obtained, all containing the somatostatin gene as part of the lactose operon. When the operator was turned on by the addition of lactose, proteins were synthesized. Chemical treatment of these proteins released pure somatostatin. The somatostatin, tested in laboratory animals, was found to have the biological activity of the natural hormone.

Since the initial success with somatostatin, a number of other biologically useful proteins have been prepared using recombinant DNA techniques. Perhaps the most important is human insulin. At present, millions of diabetics depend on daily injections of insulin. The molecule is too complex to synthesize commercially. It is presently extracted from the pan-

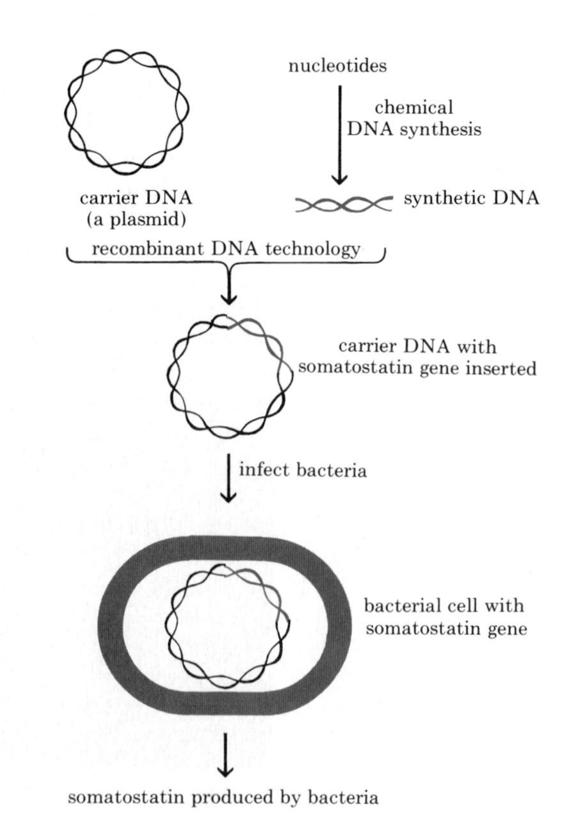

17-7 *How cells were bacterial made to produce the hormone somatostatin.*

creatic glands of pigs and cattle slaughtered for food, but the demand, due to the increasing number of diabetics, is threatening to exceed the supply. Moreover, many patients are allergic to these nonhuman proteins. The insulin that has been produced by *E. coli,* in a process similar to that used for somatostatin, is identical to human insulin. Initially, the amounts of insulin prepared by recombinant DNA techniques were extremely small. By mid-1980, however, *E. coli* "factories" began to produce sufficiently large quantities of human insulin that plans could be made for clinical testing.

Recombinant DNA techniques are also being used to produce human growth hormone and a substance known as interferon. Interferon, which is produced by the immune system, has an important role in fighting viral infections and may be of some value against some forms of cancer. Until recently, however, interferon could be obtained only from white blood cells; the quantities available were so minute that very little testing could be done. Bacterial production of human interferon should yield enough of the substance for researchers to thoroughly explore its role in the body's natural defenses against viruses and its potential uses in treatment.

Recently, a group of investigators in Edinburgh, Scotland, succeeded in transferring genes from the virus causing one form of hepatitis to *E. coli.* The viral DNA replicates with the bacterial genes and can be passed to the progeny. In the past, it has been impossible to grow the hepatitis virus successfully in the laboratory, and thus studying it has been extremely difficult. This new achievement will make it possible to produce enough viral DNA for analysis. And, there is hope that eventually it will be possible to produce enough viral protein that a hepatitis vaccine can be developed.

In the span of one scientific generation, molecular geneticists not only have identified the genetic material in chemical terms and demonstrated how it is replicated, transcribed, and translated, but they have also produced genes of their own making and been able to make them function. This work, with its many potential applications, is still in its infancy. Many new understandings, and probably surprises, lie ahead.

17-8 *Gene splicing.* (a) *A plasmid has been cut open with a restriction enzyme, leaving two "sticky" ends. A small segment of foreign DNA (lower right) also has "sticky" ends that can join with the ends of the plasmid by base pairing.* (b) *With the aid of DNA ligases, the foreign DNA has been spliced into the plasmid.* (c) *The plasmid now containing the foreign DNA is about to enter a bacterial cell.* (Huntington Potter and David Dressler, LIFE Magazine, 1980, Time, Inc.)

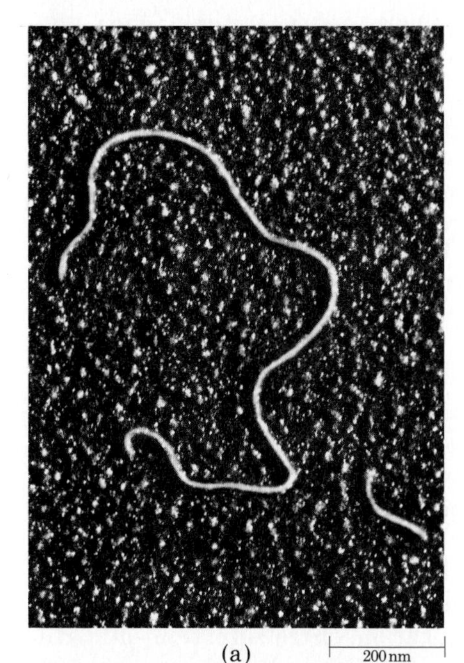

(a) 200 nm

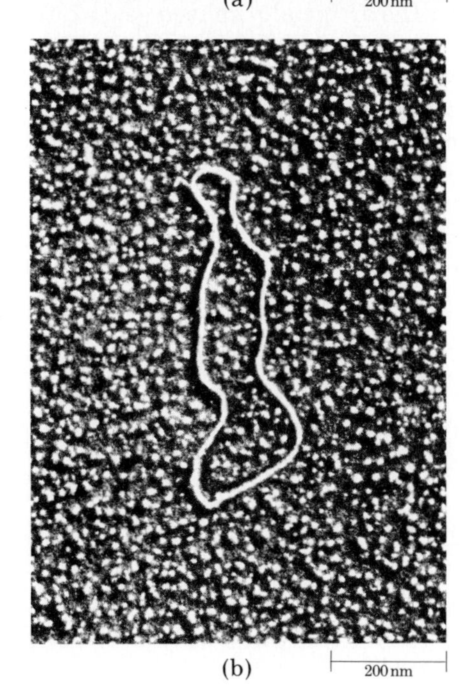

(b) 200 nm

(c) 200 nm

SUMMARY

The DNA of eukaryotic organisms is recombined naturally in each generation as a result of meiosis and crossing over, followed by fertilization. In prokaryotes, recombination can occur as a result of transformation (discussed in Chapter 14), bacterial conjugation, or transduction.

Bacterial conjugation is a form of mating in which DNA passes from one cell to another. Donor cells are characterized by the presence of the F (fertility) factor. The F factor is a plasmid and, like many other plasmids, it can exist independently or as part of the host chromosome. When the F factor is integrated into the chromosome, the chromosome can later break at that point and replicate, and all or part of the daughter chromosome may be passed to a recipient cell.

In transduction, DNA is transferred from one host cell to another by way of viral infection. In general transduction, fragments of the DNA of a bacterial cell are carried to a new host cell by a bacteriophage that has, however, lost its capacity to set up a new infection. These DNA fragments may be integrated into the chromosome of the new host cell.

Restricted transduction occurs when the DNA of a bacteriophage does not set up an infection in the host cell, but rather remains latent in the cell as a plasmid. This viral DNA may become part of the chromosome of the cell and is then called a prophage. Bacteria that contain prophages are known as lysogenic bacteria. Certain agents (generally those that cause mutations) can activate the viral DNA, causing it to break loose from the bacterial chromosome and start a new infective cycle. If some of the bacterial chromosome is carried along with the viral DNA, it will be replicated with it and incorporated into the new virus particles. Later it may be transferred to the chromosome of a new host cell.

Two groups of enzymes have been discovered that operate on DNA molecules. Restriction enzymes cut DNA molecules apart at specific sites, whereas DNA ligases mend breaks in the molecules. These enzymes can be used in the laboratory to splice specific genes into plasmids or into the DNA of bacteriophages. Transformation, conjugation, or transduction—all natural processes—can then be used to transfer the artificially spliced genes into new host cells.

Recombinant DNA studies are proceeding at a very rapid pace. They have opened up entirely new fields of research and have many potential applications in medicine and agriculture. Genes for human insulin, growth hormone, and interferon have been successfully incorporated into bacterial cells, which have, in turn, been induced to synthesize the protein coded for by the human gene.

QUESTIONS

1. Distinguish among the following: transformation/bacterial conjugation/transduction; donor cell/recipient cell; general transduction/restricted transduction; prophage/lysogenic bacterium; restriction enzyme/DNA ligase.

2. In what way does bacterial mating resemble fertilization? How does it differ from it?

3. As bacterial mating indicates, it is possible to separate the production of new genetic material from reproduction. Why do you think these two processes are combined in eukaryotic cells?

4. Describe two possible outcomes of infection of a bacterial cell by a bacterial virus.

5. *Escherichia coli* cells used in recombinant DNA studies are "disabled"; that is, they lack the capacity to synthesize their cell walls, for example, or to make a nitrogenous base, such as thymine. Consequently, they are able to survive only in an enriched laboratory medium. Why is such a precaution necessary?

SUGGESTIONS FOR FURTHER READING

BODMER, WALTER F., and L. L. CAVALLI-SFORZA: *Genetics, Evolution, and Man,* W. H. Freeman and Company, San Francisco, 1976.

An excellent undergraduate text with emphasis on human genetics.

GOODENOUGH, URSULA: *Genetics,* 2d ed., Holt, Rinehart and Winston, Inc., New York, 1978.

An up-to-date, general introductory text with emphasis on molecular genetics.

JACOB, FRANÇOIS: *The Logic of Life: A History of Heredity,* Pantheon Books, New York, 1973.

Jacob's principal theme concerns the changes in the way people have looked at the nature of living beings. These changes, which are part of our total intellectual history, determine both the pace and direction of scientific investigation. The opening chapters are particularly brilliant.

JUDSON, HORACE F.: *The Eighth Day of Creation: Makers of the Revolution in Biology,* Simon and Schuster, New York, 1979.*

A comprehensive study of the human and scientific aspects of molecular biology from the 1930s through the mid-1970s. As events unfold, the story is told from each participant's point of view. We are treated to an enlightening and personal glimpse at the development of scientific thought.

OLBY, ROBERT: *The Path to the Double Helix,* University of Washington Press, Seattle, Wash., 1975.

An account, written by a professional historian of science, of twentieth-century genetics. Olby is interested not only in the scientific concepts and experiments but also in the various personalities involved and their effects on one another and on the course of scientific discovery.

PETERS, JAMES A. (ed.): *Classic Papers in Genetics,* Prentice-Hall, Inc., Englewood Cliffs, N. J., 1959.*

Includes papers by most of the scientists responsible for the important developments in genetics: Mendel, Sutton, Morgan, Beadle and Tatum, Watson and Crick, and so on. You should find this book very interesting; the authors are surprisingly readable, and the papers give a feeling of immediacy that no modern account can achieve.

STRICKBERGER, MONROE W.: *Genetics,* 3d ed., The Macmillan Company, New York, 1975.

An up-to-date and cohesive account of the science, this book is much broader in coverage than most texts at this level.

WATSON, J. D.: *The Double Helix,* Atheneum Publishers, New York, 1968.*

"Making out" in molecular biology. A brash and lively book about how to become a Nobel laureate.

WATSON, J. D.: *Molecular Biology of the Gene,* 3d ed., The Benjamin/Cummings Publishing Company, Menlo Park, Calif., 1976.*

For the student who wants to go more deeply into the questions of molecular biology, this is a detailed and authoritative account.

* Available in paperback.

PART II

Biology of Organisms

SECTION 4

The World of Living Things

Representatives of two major classes of organisms. This African grasshopper is eating euphorbia, African plants that—although unrelated—are similar in many ways to the cacti of the American deserts.

18

The Diversity of Life: Microorganisms

In Part II of this book we are going to be concerned with organisms, the living things that populate the world around us. The chief focus will be on the plants, particularly the flowering plants, and on the more complex animals, particularly humans. These are the subjects of Sections 5 and 6, respectively. In this chapter and the next, we are going to introduce these organisms by examining the tremendous diversity of life on our planet today. We shall also trace the outlines of evolutionary history.

The Classification of Organisms

Most people have a limited awareness of the natural world and are concerned chiefly with those organisms that influence their own lives. For example, gauchos, the cowboys of Argentina, who are famous for their horsemanship, have some 200 names for different colors of horses but divide all the plants known to them into four groups: *pasto,* or fodder; *paja,* bedding; *cardo,* wood; and *yuyos,* everything else.

Most of us are like the gauchos. Once beyond the range of common plants and animals, and perhaps a few uncommon ones that are of special interest to us, we usually run out of names. Biologists, however, face the problem of systematically investigating, identifying, and exchanging information about well over a million different kinds of organisms. In order to do this, they must have a system for naming all these organisms and for grouping them together in orderly and logical ways. The problems of developing such a system are immensely complicated. Some 700,000 different kinds of insects, for instance, have been named and classified to date.

Every individual organism is a member of a particular *species*, which, in turn, belongs to a larger group, a *genus* (plural, genera). In the eighteenth century, the binomial system of nomenclature became generally accepted for representing these facts, and this system is still in use today. As we saw earlier (page 44), the binomial consists of two parts, the name of the genus and a descriptive word that indicates the particular species included within that genus. The wolf, for example, is classified in the genus *Canis* (the Latin word for "dog"), and its scientific name is *Canis lupus.* The use of this unique combination of words indicates that we are referring to the wolf rather than such near relations as the coyote *(Canis latrans)* and the domestic dog *(Canis familiaris).*

18-1 *This unusual micrograph, taken by scientists at the Bronx-Lebanon Hospital Center in New York, shows a bacterial cell exploding. This bacterium is a member of the species* Staphylococcus aureus, *the cause of many human infections. The contents of the cell were under higher pressure than the surrounding environment, and when the cell was treated with an antibiotic that damaged the cell wall, it exploded. The cell is magnified 228,000 times in this micrograph.*

(a)

(b)

(c)

18-2 *These cats are all members of the genus* Felis: (a) *a cougar,* Felis concolor, (b) *an ocelot,* Felis pardalis, *and* (c) *a domestic cat,* Felis catus.

According to the modern system of classification, related genera are grouped into *families*, related families into *orders*, related orders into *classes*, related classes into *phyla* (singular, phylum), and related phyla into *kingdoms*. The classification of a particular organism is determined by a number of criteria, such as structural features, details of biochemistry, and patterns of reproduction and development. Table 18–1 (page 232) shows the classification of two different types of organisms. Notice how much information is used in classifying an organism.

Until very recently, most biologists recognized only two kingdoms of organisms: Plantae and Animalia. The fungi, algae, and bacteria were grouped with the plants, and the protozoans—one-celled organisms that eat and move—were classified with the animals. As knowledge of cell structure and biochemistry has accumulated, the number of groups recognized as constituting different kingdoms has increased. The most recent proposals recommend five kingdoms: Prokaryotae, Protista, Fungi, Plantae, and Animalia. We shall follow this system in the text and in Appendix C, where a more complete classification of organisms is found.

The five-kingdom system is not the only one in modern use. For example, some systems group the eukaryotic algae and the fungi with the plants, and the protozoans with the animals, ending up with three kingdoms: prokaryotes, plants, and animals. Ideally, the different categories should reflect degrees of evolutionary relationship. In truth, no system is really satisfactory in this regard.

In the remainder of this chapter, we shall look at the first three kingdoms: the prokaryotes, the protists, and the fungi. Although they are dissimilar in many important ways, they are conventionally studied together in the science known as microbiology. Viruses will also be described briefly. While they do not fit comfortably in any of the kingdoms, they too are traditionally a part of microbiology.

The Prokaryotes

The only living members of the kingdom Prokaryotae are the bacteria and the blue-green algae. As we noted in Chapter 4, prokaryotes are by far the simplest of all cells. Their genetic material is in the form of a long, circular molecule of DNA. The DNA is not combined with protein, the genetic material is not isolated from the rest of the cell by a special mem-

18-3 *Representatives of the five kingdoms.* (a) *Prokaryotes. Bacteria,* Neisseria gonorrhoeae *(dark spheres), the causative agent of gonorrhea, being ingested by white blood cells.*

(b) *Protists.* Vorticella, *like most protists, is a single cell. It attaches itself to a substrate by a long stalk. A contractile fiber, visible in this micrograph, runs through the stalk.*

(c) *Fungi. Inky cap mushrooms. Fungi are characterized by a multicellular underground network and also spores and sporangia, which, in mushrooms, are borne in these familiar structures.*

(d) *Plants. Marsh marigolds. The flowers of angiosperms, attractive to pollinators, are among the principal reasons for their evolutionary success.*

(e) *Animals. A luna moth. A nervous system with a variety of sense organs is a chief characteristic of the animal kingdom.*

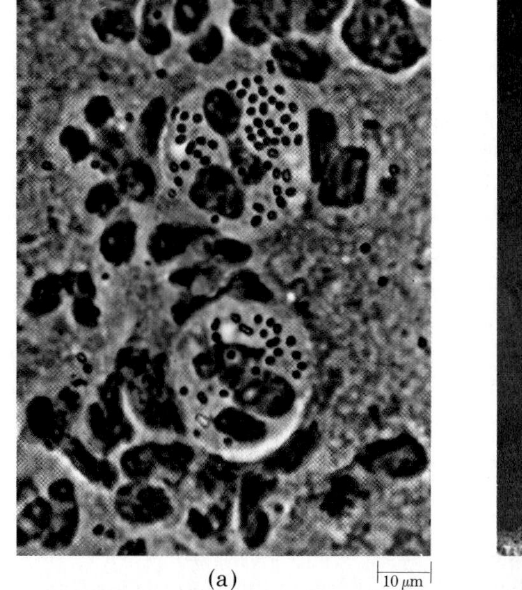

(a) 10 μm

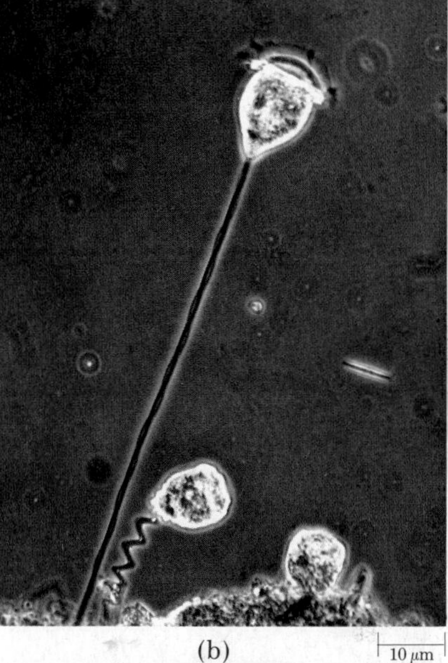

(b) 10 μm

(c)

(e)

(d)

Table 18-1 Biological classifications

Red Maple *(Acer rubrum)*

Category	Name	Characteristics
Kingdom	Plantae	Multicellular organisms that usually have rigid cell walls and usually possess chlorophyll
Subkingdom	Embryophyta	Plants forming embryos
Phylum	Tracheophyta	Vascular plants (plants with conducting tissues)
Subphylum	Pterophytina	Generally large, conspicuous leaves; complex vascular pattern
Class	Angiospermae	Flowering plants, seed enclosed in ovary
Subclass	Dicotyledoneae	Embryo with two seed leaves (cotyledons)
Order	Sapindales	Soapberry order, usually woody plants
Family	Aceraceae	Maple family; characterized by watery, sugary sap; chiefly trees of temperate regions
Genus	*Acer*	Maples and box elder
Species	*Acer rubrum*	Red maple

Human *(Homo sapiens)*

Category	Name	Characteristics
Kingdom	Animalia	Multicellular organisms requiring complex organic substances for food
Phylum	Chordata	Animals with notochord, dorsal hollow nerve cord, gills in pharynx at some stage of life cycle
Subphylum	Vertebrata	Spinal cord enclosed in a vertebral column, body basically segmented, skull enclosing brain
Superclass	Tetrapoda	Land vertebrates, four-limbed
Class	Mammalia	Young nourished by milk glands, skin with hair or fur, body cavity divided by diaphragm, red blood cells without nuclei, high body temperature
Order	Primates	Tree dwellers or their descendants, usually with fingers and flat nails, sense of smell reduced
Family	Hominidae	Flat face; eyes forward; color vision; upright, bipedal locomotion
Genus	*Homo*	Large brain, speech, long childhood
Species	*Homo sapiens*	Prominent chin, high forehead, sparse body hair

brane, and cell division is not accompanied by mitosis. The cytoplasm contains ribosomes but no complex membrane-bound organelles. The cell is surrounded by a cell membrane and a cell wall. This wall, which is a combination of a unique polysaccharide and protein, is different in composition from the cell walls of any other type of organism. In fact, the unusual cell-wall composition shared by the bacteria and the blue-green algae was one of the clues to their close evolutionary ties.

Bacteria

Bacteria are not only the oldest but also the most abundant group of organisms in the world. They can live in places and under conditions that support no other forms of life. They have been found in the icy wastes of Antarctica, in the near-boiling waters of natural hot springs, and in the dark depths of the oceans. When conditions are unfavorable, many kinds can take the form of hard, resistant spores, which may lie dormant for years until conditions become more favorable for growth.

Bacteria obtain energy in a great variety of ways. Some are autotrophs; that is, they are able to synthesize energy-rich organic compounds from simple inorganic substances. Among the autotrophs are photosynthetic bacteria. Like green plants, they capture light energy with the aid of specialized pigments, which are, however, different from the pigments found in algae or plants. Unlike algae and plants, photosynthetic bacteria

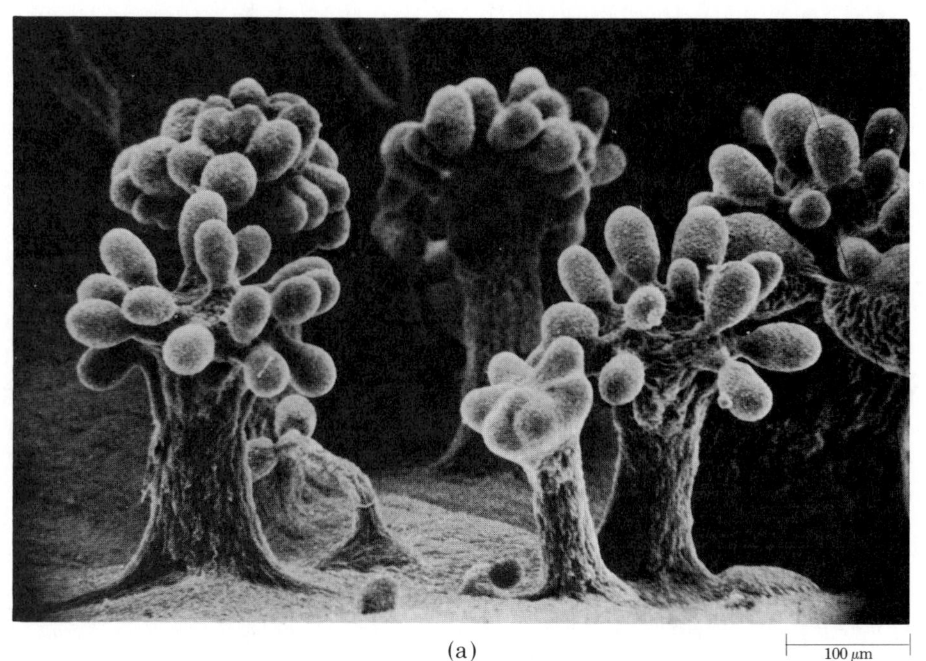

(a) 100 μm

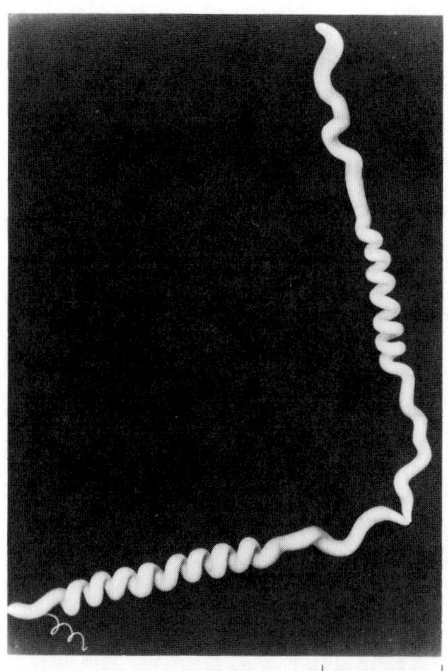

(b) 2 μm

18-4 *Examples of three classes of bacteria.* (a) *The myxobacteria are distinguished by their slime tracks. Also, some types form fruiting bodies, such as those in this scanning electron micrograph of* Chondromyces crocatus.

(b) *The spirochetes are spiral bacteria. The members of some species are 500 micrometers long, which is an enormous size for prokaryotes. They move by means of a rapid spinning or whirling of the cell around its long axis.* Treponema pallidum, *two of which are shown here, is the causative agent of syphilis.*

(c) *Rickettsiae are the smallest known cells. Typhus is caused by* Rickettsia typhi, *shown here. It is spread from rats to humans by fleas. It can then be transmitted by body lice, and under crowded conditions, large numbers of people can be infected in a very short time. More human lives have been taken by rickettsial diseases than by any other infection except malaria. At the siege of Granada in 1489, 17,000 Spanish soldiers were killed by typhus, but only 3,000 in combat. In the Thirty Years War, the Napoleonic campaigns, and the Serbian Campaign during World War I, typhus was also the decisive factor.*

(c) 0.5 μm

split hydrogen sulfide (H_2S) and other compounds rather than water and do not release oxygen gas. Other autotrophic bacteria are chemoautotrophs; they obtain their energy from the oxidation of inorganic molecules such as certain compounds of nitrogen, sulfur, and iron.

Most bacteria are heterotrophs, and it is this group that is of the greatest significance to us. Some of the heterotrophs obtain their energy from the tissues or body fluids of other living organisms. These are the disease-causing (pathogenic) bacteria (see essay, page 238). By far the largest numbers of heterotrophic bacteria, however, live on dead organic matter. These bacteria are of the utmost importance to us and to all other living things because they are the principal decomposers of the biosphere. Through the action of the bacteria (and the fungi, which are also decomposers), materials incorporated into the bodies of once-living organisms are released and made available to successive generations of living things. We shall discuss the subject of recycling in greater detail in Chapter 39.

18-5 (a) *A gelatinous colony of the blue-green alga* Nostoc. (b) Oscillatoria, *a filamentous blue-green alga. These algae, now widely regarded as types of photosynthetic bacteria, have chlorophyll* a *and produce oxygen during photosynthesis.*

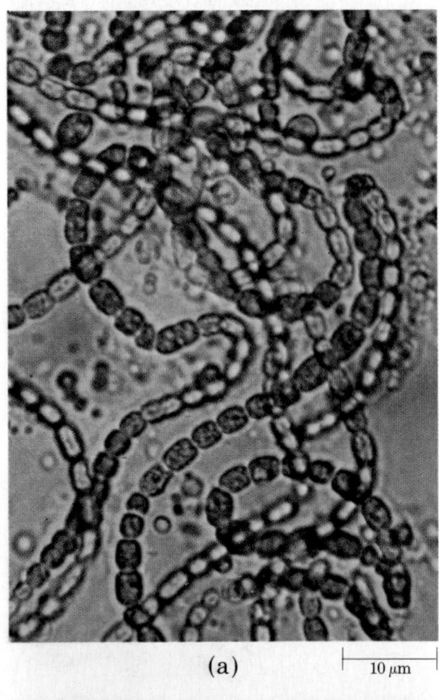

(a) 10 μm

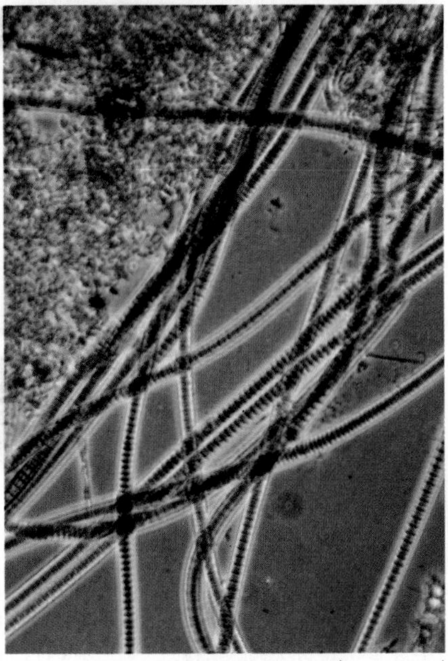

(b) 100 μm

Blue-Green Algae

Blue-green algae resemble bacteria in their small size, in their simple, prokaryotic organization, and in the unusual composition of their cell walls. Most experts believe that they should be classified as another type of photosynthetic bacteria.

The blue-green algae resemble eukaryotic algae and also plants in their photosynthesis. In all these groups, the principal photosynthetic pigment is chlorophyll *a,* and photosynthesis involves the splitting of the water molecule (page 114). In blue-green algae, however, chlorophyll and the other photosynthetic pigments are not enclosed in chloroplasts, as they are in eukaryotic cells; rather, they are embedded in membranes distributed around the periphery of the cell (see Figure 4-6, page 57).

Blue-green algae grow for the most part in fresh water. They are sometimes found as isolated cells, but more often they form clusters, threads, or chains (Figure 18-5). Many species of blue-green algae are able to incorporate atmospheric nitrogen into organic compounds. In Southeast Asia, blue-green algae growing on the surface of rice paddies are important contributors of nitrogen to the rice plants.

Viruses: A Case Apart

Viruses do not fit easily into any of the kingdoms of living organisms. Because of their simplicity, however, they are usually studied with the bacteria and blue-green algae. Many biologists believe that viruses should not be regarded as living organisms at all but rather as parts of cells that have set up a partially independent existence. S. E. Luria, who was awarded the Nobel Prize for his work with bacterial viruses, has called them "bits of heredity looking for a chromosome." (As we noted in Chapter 17, some viruses do indeed "find" a chromosome.)

All viruses consist of nucleic acid—either DNA or RNA—and protein. The protein forms an outer coat, which protects the nucleic acid and determines what sort of cell the virus can infect. Usually, particular viruses attack only particular cells. For example, influenza virus invades the lining of the human respiratory tract, tobacco mosaic virus infects the leaves of the tobacco plant, and bacteriophages attack bacterial cells.

Viruses can multiply only within a living cell. In some virus infections, the protein coat is left outside the cell while the nucleic acid enters (see Figure 14-8, page 189); in others, the intact virus enters the cell, but once inside, the protein is destroyed by enzymes, freeing the viral nucleic acid from its coat. In the case of the DNA viruses, the DNA of the virus replicates and also codes for messenger RNA. The mRNA, in turn, produces enzymes and coat protein needed by the virus. The virus uses the equipment of the host cell, including ribosomes, transfer RNA molecules, amino acids, and nucleotides, for its synthetic activities. In the case of the RNA viruses, the nucleic acid both replicates and serves directly as messenger RNA, producing the enzymes and coat protein needed by the virus. The end product of either type of infection is hundreds or thousands of new viral particles. These are produced and assembled within the infected cell, which is often broken apart and destroyed as the particles are released. Characteristically, it is this lysing of the host cell that causes the symptoms associated with a virus infection.

18-6 (a) *Adenovirus, one of the many viruses that cause colds in humans. This virus is an icosahedron. Each of its 20 sides is an equilateral triangle composed of identical protein subunits. Many viruses, and also Buckminster Fuller's geodesic domes, are constructed on this principle. There are 252 subunits in all. Within the icosahedron is a core of double-stranded DNA.* (b) *A model of the adenovirus, made up of 252 tennis balls.*

(c, d) *Electron micrograph and diagram of influenza virus. The virus is composed of a core of RNA surrounded by a lipoprotein envelope through which protrude stubby protein spikes. The influenza virus, for reasons that are not understood, mutates frequently. The changes in its nucleic acid alter its proteins and, hence, previously formed antibodies no longer "recognize" it. New strains of influenza virus are likely to arise more rapidly than new vaccines can be produced to combat them.*

(e, f) *Electron micrograph and diagram of a bacteriophage, showing the many different structural components of the protein coat. The DNA of the virus codes for all these structural proteins.*

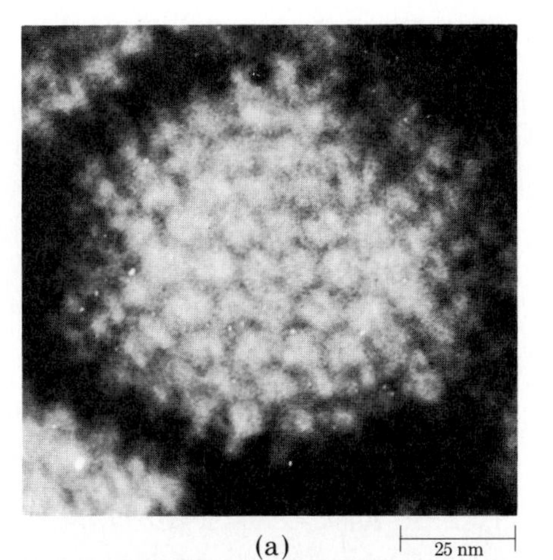

(a)

(b)

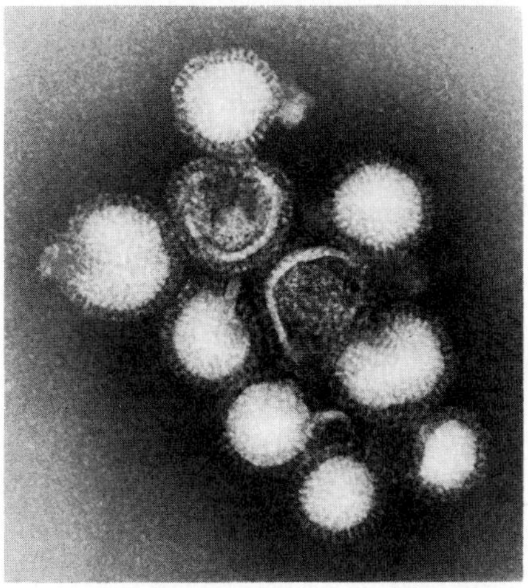

(c)

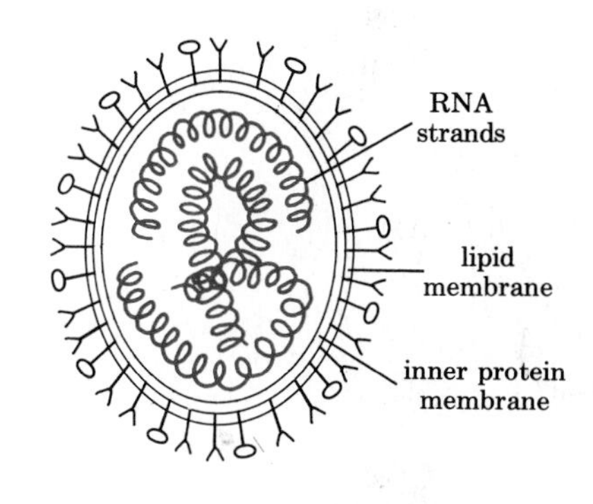

(d)

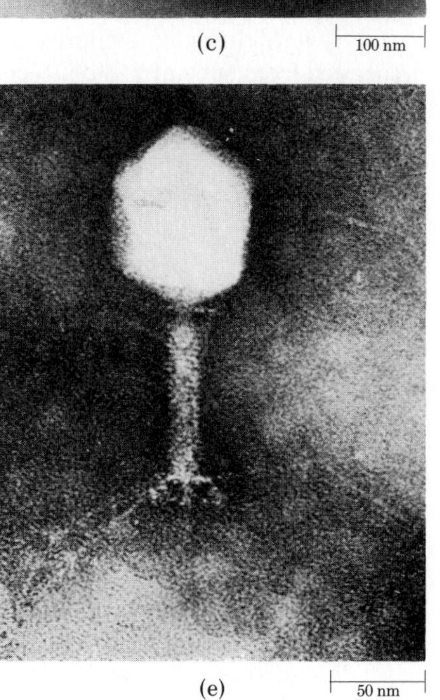

(e)

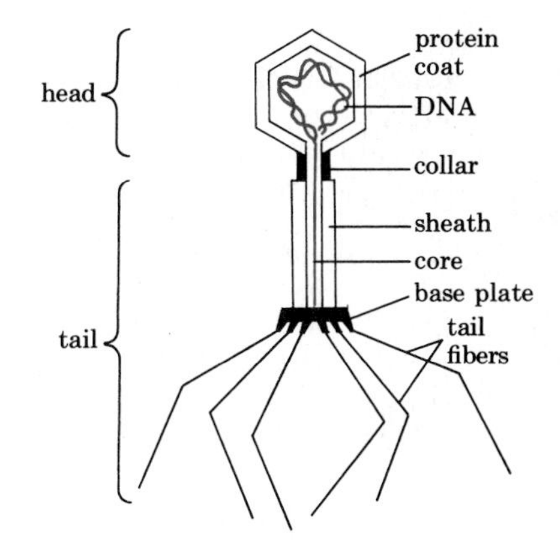

(f)

The Protists

The kingdom Protista includes all eukaryotic one-celled heterotrophs and also all eukaryotic algae, some of which are multicellular. Among protists, the DNA of the nucleus is combined with protein, organized into chromosomes, and enclosed within the nuclear envelope. Protists have mitochondria, and photosynthetic protists have chloroplasts as well. Many of these organisms have cilia or flagella with the characteristic 9 + 2 structure (see page 71).

Many protists have a sexual cycle, such as that of *Chlamydomonas* (page 139). In those that do not, biologists tend to believe that some form of sexual reproduction once existed and was lost.

There are eight major phyla of protists (Table 18–2), six of which are algae. Biologists generally agree (1) that the protists represent quite a number of different evolutionary lines, and (2) that all other eukaryotic organisms–fungi, plants, and animals–originated from primitive protists.

Origin of the Protists

The step from the prokaryotes to the first eukaryotes (the protists) was one of the big evolutionary transitions, second only to the origin of life. The question of how it came about is a matter of current and lively discussion. One interesting hypothesis is that larger, more complex cells evolved when certain prokaryotes took up residence inside other prokaryotic cells.

About 2½ billion years ago, oxygen began slowly to accumulate in the atmosphere as a result of the photosynthetic activity of blue-green algae. Those bacteria that were able to use oxygen in ATP production gained a strong advantage, and so such forms began to prosper and increase. Some of these evolved into modern forms of bacteria. Others, according to the hypothesis, moved into larger cells and evolved into mitochondria.

Several lines of evidence support the idea that mitochondria are descended from specialized bacteria. Mitochondria contain their own DNA, and this DNA is present in a single, continuous molecule, like the DNA of

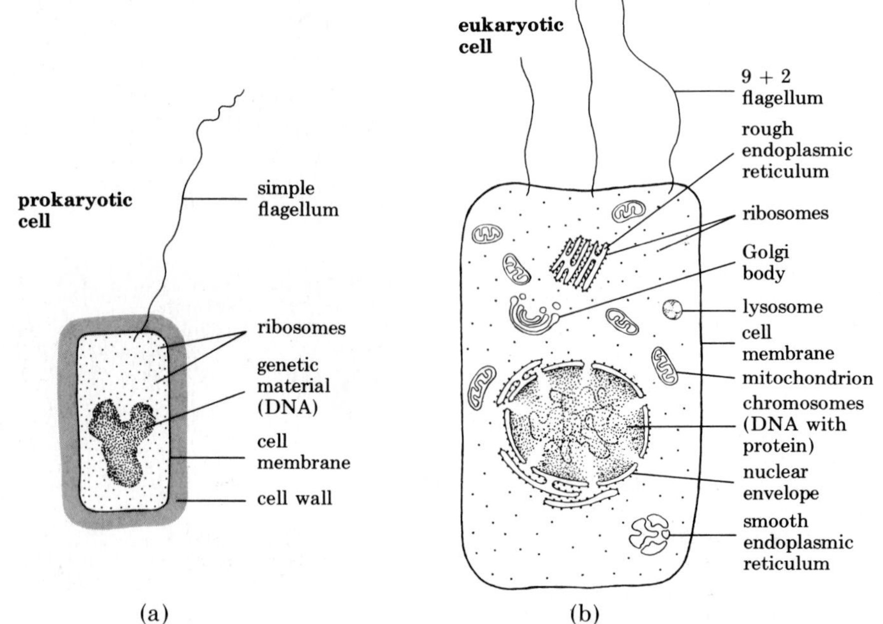

18–7 *Comparison between* (a) *a prokaryotic cell and* (b) *a eukaryotic cell. The eukaryote contains a membrane-bound nucleus; complex membrane-bound organelles, such as mitochondria and chloroplasts; endoplasmic reticulum; and Golgi bodies. Its chromosomes are a complex of DNA and special proteins. Flagella, when present, have a characteristic 9 + 2 structure. All these features are lacking in prokaryotes.*

Table 18-2 The kingdom Protista

Phylum Protozoa	Unicellular heterotrophs, including flagellates (mastigophores), amoeba-like organisms (sarcodines), ciliates, and sporozoans
Phylum Euglenophyta	*Euglena* and related algae, all unicellular, mostly freshwater
Phylum Chrysophyta	Diatoms and related algae, all unicellular, mostly marine
Phylum Pyrrophyta	Dinoflagellates and related algae, all unicellular, marine and freshwater
Phylum Chlorophyta	Green algae, including unicellular, colonial, and multicellular forms; mostly freshwater
Phylum Phaeophyta	Brown algae, including the kelps; all multicellular, almost all marine
Phylum Rhodophyta	Red algae, all multicellular, mostly marine
Phylum Gymnomycota	Slime molds

bacteria. Many of the enzymes contained in the cell membranes of bacteria are found in mitochondrial membranes. Mitochondria contain ribosomes that resemble those of bacteria both in their small size and in some details of their chemical composition. Further, mitochondria appear to be produced only by other mitochondria, which divide within their host cell. (However, to make the situation more complex, there is not nearly enough DNA in mitochondria to code for all the mitochondrial proteins. Host-cell DNA is also required for mitochondrial structures.)

We know little about the original cells in which these bacteria first set up housekeeping—or, indeed, if they actually existed. But if such cells did exist, they probably had no means of using oxygen for cellular respiration. Thus they were dependent entirely on anaerobic fermentation as an energy-releasing process, which, as we have seen (page 101), is relatively inefficient. Cells with oxygen-utilizing respiratory assistants would have been more efficient than those lacking them and so would have outreproduced them.

In an analogous fashion, photosynthetic prokaryotes ingested by larger, nonphotosynthetic cells are believed to be the forerunners of chloroplasts. By this symbiosis ("living together"), the smaller cells gained nutrients and protection, and the larger cells were given a new energy source.

This hypothesis accounts for the presence in eukaryotic cells of complex organelles not found in the far simpler prokaryotes. It gains plausibility from the fact that many modern organisms contain bacteria and algae within their cells (Figure 18-8), indicating that such associations are not difficult to establish.

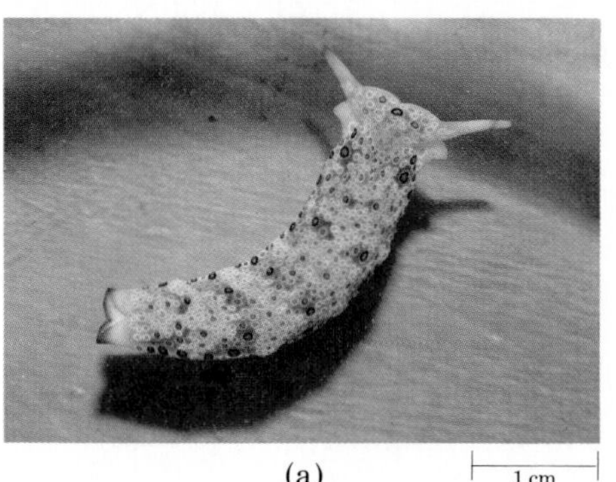

(a)

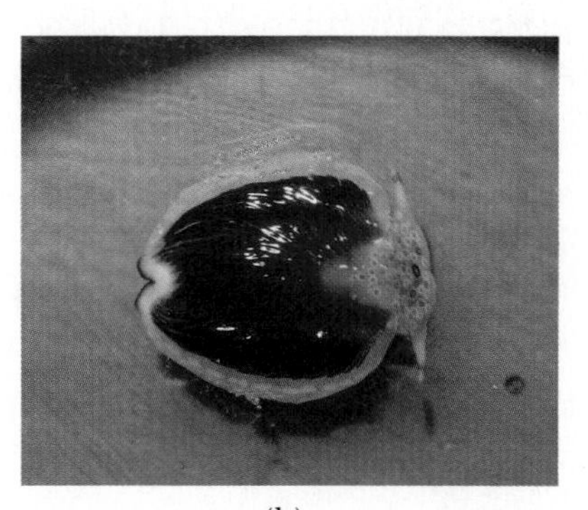

(b)

18-8 Placobranchus, *a marine mollusk. The tissues of this animal contain chloroplasts, which it obtains by eating certain green algae.*

(a) *Ordinarily, structures called parapodia are folded over the animal's back, hiding the chloroplast containing tissues.* (b) *When the parapodia are spread apart, however, the deep green tissues become visible. The chloroplasts carry on photosynthesis so efficiently that within a 24-hour cycle of light and darkness some individuals produce more oxygen than they consume.*

Infectious Disease: Its Causes, Prevention, and Control

Infectious disease can be caused by many types of microorganisms: bacteria, viruses, protists, and fungi. Most of the time, potentially pathogenic microbes live within their host organism without apparent ill effect. Disease is likely to be the result of a sudden change in the microbe, in the host, or in their relationship. For instance, many people harbor small numbers of *Mycobacterium tuberculosis* without any symptoms of disease; however, factors such as malnutrition, fatigue, or other diseases may weaken host defenses so that the signs of tuberculosis appear. Similarly, as we saw earlier, herpes simplex virus may remain latent for months or years at a time, producing fever blisters only in response to some change in the condition of the host.

The pathogenic effects of microbes are produced in a variety of ways. The viruses, as we have seen, enter particular types of cells and often destroy them. Bacteria may also produce cell destruction. Frequently, however, the effects we recognize as disease are caused not by the direct action of the pathogens but by toxins, or poisons, produced by them. For instance, diphtheria is caused by *Corynebacterium diphtheriae*. The organisms are inhaled and establish infection in the upper respiratory tract. They produce a powerful toxin that is transported through the bloodstream to body cells, where it inhibits protein synthesis.

Some diseases are the result of the body's reaction to the pathogen. In pneumonia caused by *Streptococcus pneumoniae,* the infection causes a tremendous outpouring of fluid and cells into the lungs, thus interfering with breathing. The symptoms caused by fungus infections of the skin similarly result from inflammatory responses.

About 100 years ago, microbes were recognized as disease agents. This opened the way to control measures, among the most important of which was the introduction of sterile procedures in hospitals. Even more important was the institution of public health measures—for example, the eradication of fleas, lice, mosquitoes, and other agents that carry disease; disposal of sewage and other wastes; protection of public water supplies; pasteurization of milk; and quarantine.

Many infectious diseases can be prevented by immunization. Vaccines have been developed against a large number of diseases, including mumps, measles, German measles, and polio (all virus-caused) and diphtheria, whooping cough, and tetanus (all bacteria-caused). A vaccine for a particular pathogen may be made from a related microorganism, a killed strain of the pathogen itself, or a strain that has been weakened by growing in another host organism. The vaccine stimulates the immune system to produce antibodies against the pathogen. (Antibodies and immunity will be discussed in Chapter 27.)

In 1935, sulfanilamide, the first of the "wonder drugs" for the control of bacterial infection, was discovered in Germany. In 1940, the effects of penicillin were reported from England. Penicillin was the first known antibiotic—by definition, a chemical that is produced by a living organism and is capable of inhibiting the growth of microorganisms. Many antibiotics are produced by bacteria, and some are formed by fungi. Many, including penicillin, can now be synthesized in the laboratory.

Antibiotics and other chemotherapeutic agents are effective because they interfere with some essential process of the pathogen without affecting the cells of the host. Penicillin, for example, blocks a key reaction in the synthesis of the cell wall in many types of bacteria.

For the past 40 years, we have enjoyed relative freedom from infectious disease. However, our success in combating infectious disease has led to a complacency that could result in serious outbreaks of disease in the future. For example, many parents are now neglecting to have their children immunized against diseases for which safe and effective vaccines are available. And, as we noted earlier (page 217), increasing numbers of bacterial strains are becoming resistant to antibiotics.

Until the 1950s, poliomyelitis was one of the most feared of all childhood diseases. It was brought under control first by a killed-virus vaccine (Salk vaccine), which was subsequently replaced by a live-virus vaccine (Sabin vaccine), the one now in wide general use. The possible aftereffects of polio are depicted in this Egyptian hieroglyph, dating from the nineteenth dynasty (1320–1200 B.C.*).*

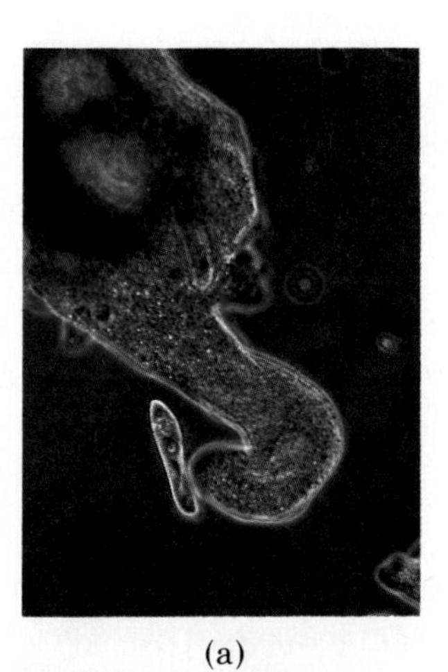
(a)

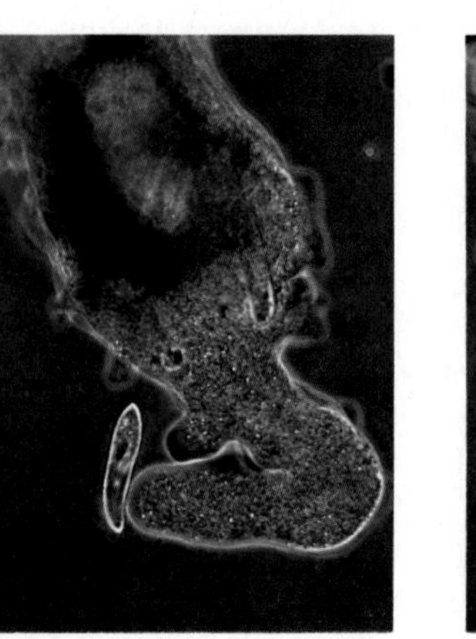
(b)

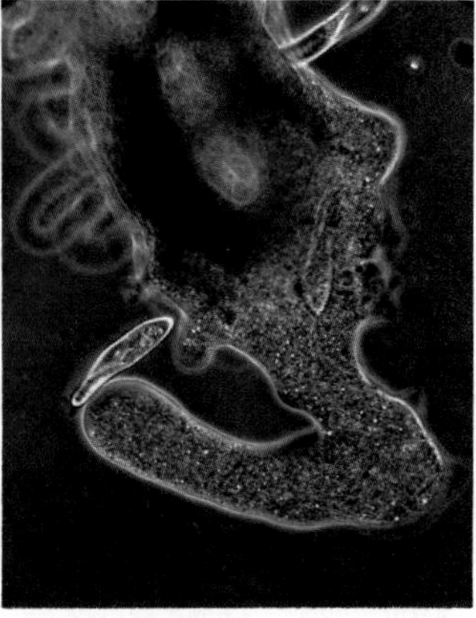
(c)

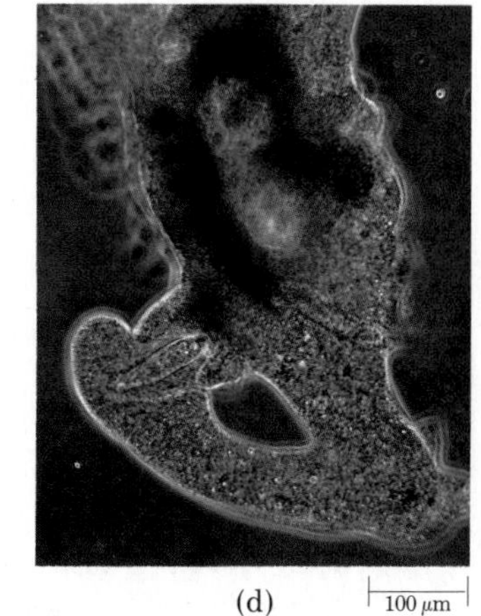
(d)

18–9 *The giant amoeba* Chaos chaos *captures a* Paramecium. *Although seemingly disorganized,* Chaos *is able to sense its prey, move toward it, and send out a pseudopod of the right size and shape to envelop it.*

Types of Protists

Protozoa

The Protozoa—"first animals"—are a collection of varied and interesting organisms, all of which are unicellular heterotrophs. There are four main classes in the phylum Protozoa: Mastigophora, Sarcodina, Ciliophora, and Sporozoa.

The Mastigophora are characterized by their flagella; most have one or two, but some have many. Most are free-living, but some live in or on other organisms from which they obtain nutrients. *Trichonympha* (Figure 3–6b, page 40) and *Trypanosoma gambiense,* a flagellate that causes African sleeping sickness, are examples of such parasites.

The Sarcodina include the amoebas and amoeba-like organisms. Generally, they move and feed by the formation of pseudopodia, which are extensions of the cytoplasm (Figure 18–9).

The Ciliophora, or ciliates, are the most specialized and complicated of the Protozoa and probably are the most complex of all cells. They are characterized by cilia, all of which have the 9 + 2 structure. *Paramecium* and *Didinium,* shown in Figure 4–19 (page 70), are examples of ciliates.

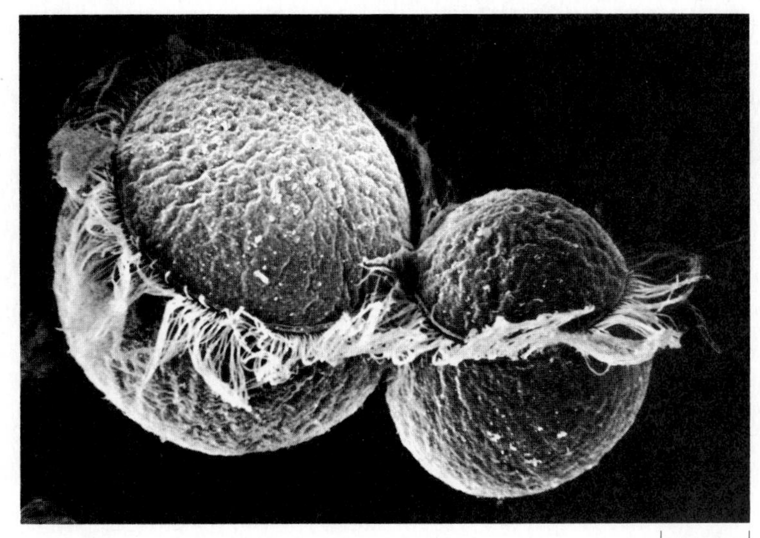

18–10 *These two ciliates, members of the genus* Opisthonecta, *are undergoing conjugation, a complex form of sexual reproduction in which the two cells exchange genetic information.*

The Sporozoa are all parasites. The most infamous are the members of the genus *Plasmodium* that cause malaria, which is, on a worldwide basis, still a principal cause of death among humans.

Algae

The six phyla of algae listed in Table 18–2 resemble one another in that they are all made up principally of photosynthetic organisms. (Several phyla, however, include a few forms that are clearly related although not photosynthetic.) These algae all have chlorophyll *a* as their principal photosynthetic pigment. They are unicellular or have multicellular forms that are relatively simple as compared with the plants.

The six phyla can be distinguished from one another in a variety of ways. These include their accessory photosynthetic pigments, the presence or absence of a cell wall, and the composition of the cell wall, if present. Accessory pigments that mask the green of the chlorophylls are responsible for the different colors of the members of the various phyla.

Plankton Many different types of one-celled algae live in the seas and in bodies of fresh water, generally in the upper levels where light is abundant. Together with small invertebrate animals and the immature forms of larger animals, they form the *plankton* (from *planktos,* the Greek word for "wandering"). The photosynthetic members of the plankton community—sometimes called *phytoplankton*—carry out most of the photosynthesis that takes place in the oceans. They are the food source for most of the oceans' inhabitants (Figure 18–11).

Seaweeds The larger, multicellular algae make up the seaweeds. These include the Rhodophyta (red algae), Phaeophyta (brown algae), and some members of the Chlorophyta (green algae). These forms are found near the shores, where the waters are rich in minerals such as nitrogen and phosphorus washed down from the land. The shortage of such minerals in the open ocean is the limiting factor in plans to farm the seas.

18–11 *Some prominent members of the plankton community.* (a) Noctiluca scintillans, *the "shining light of the night," is bioluminescent, as its name implies. Yellow-brown diatoms that have been ingested can be seen inside the cell.* (b) *A diatom, showing the characteristic intricately marked shell, which contains silicon.* (c) *A dinoflagellate* (Ceratium tripos). *These organisms often have thick cellulose walls and bizarre shapes. You can see one flagellum in motion.*

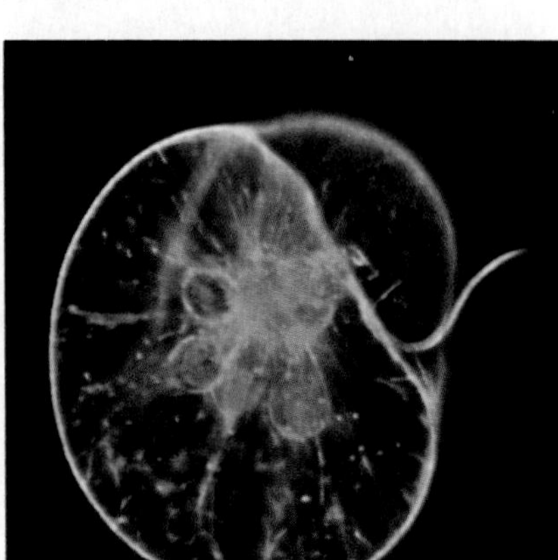

(a) 100 μm

(b) 50 μm

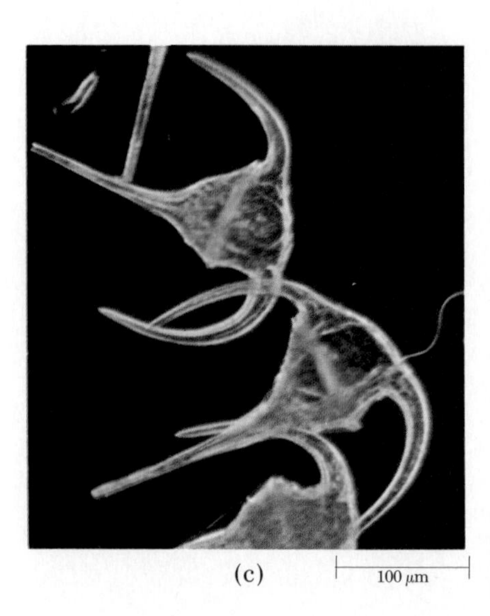

(c) 100 μm

(a)

(b)

18-12 (a) *A brown alga,* Laminaria, *showing the holdfasts.* (b) *Unlike the brown algae, most red algae are made up of filaments. The branched filaments of this red alga are hooked, enabling it to cling to other seaweeds.*

The seaweeds that evolved along the shores were under different evolutionary pressures than the planktonic algae. Chief factors in this environment are the waves and tides. One important adaptation found among certain coastal-dwelling seaweeds is the *holdfast*, a special anchoring structure that adheres to the rocks (Figure 18–12a). The upper portions of the seaweed characteristically are thin and spread out; they produce enough sugars to nourish the other parts of the organism, which may be 30 meters or more below the surface of the water and overshadowed by the upper portions of the alga. Some of the large red algae and brown algae have specialized conducting tissues that transport the products of photosynthesis downward.

Not only did the marine algae take on specialized shapes, but they also developed specialized accessory pigments. Water filters out the light, removing first the longer wavelengths, the reds and oranges; at the deeper levels, only a faint blue light penetrates. The red algae capture the energy of these blue rays. The brown algae absorb blue-green light, and the green algae, whose principal pigments are the chlorophylls, make fullest use of the longer-waved red light available near the surface. As a consequence of the evolution of specialized pigments, the various seaweeds use virtually the entire spectrum of light as well as the living space along the rocky shore.

Chlorophyta: Ancestors of the Plants The Chlorophyta (green algae), of which *Chlamydomonas* (page 58) is an example, are of particular interest to biologists because they are believed to be the group from which the plants arose. Evidence for this relationship is based on a comparison between modern green algae and modern plants. Like the plants, the green algae contain chlorophylls *a* and *b* and carotenoids as their photosynthetic pigments, they accumulate their food as starch, and their cell walls are composed of cellulose.

Many species of green algae are colonial or multicellular. In colonial organisms, the cells making up the colony may be interconnected, but there is no division of labor among them. In a multicellular organism, different

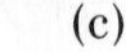

18-13 *It is possible to identify several evolutionary lines among the Chlorophyta. Each presumably started with a single-celled organism resembling* Chlamydomonas, *and each represents a different trend toward multicellularity.*
(a) Valonia, *which is about the size of a hen's egg, has many nuclei but no partitions separating them.*
(b) *In* Spirogyra, *a freshwater alga, the cells all elongate and then are divided by cross walls so that they are strung together in long, fine filaments. The chloroplasts form spirals that look like strips of green tape within each cell.*
(c) *In* Ulva, *or sea lettuce, the cells divide both longitudinally and laterally, with a single division in the third plane. This produces a broad body two cells thick.*
(d) *In* Volvox, *hundreds or thousands of* Chlamydomonas-*like cells are united by tiny strands of protoplasm into a hollow sphere. New daughter colonies can be seen forming inside the mother colonies.*

cells are specialized for different functions. Green algae have achieved multicellularity in several different ways (Figure 18–13).

Multicellular green algae typically have a life cycle that includes alternation of generations. As we saw in Chapter 10 (page 139), in this type of life cycle a diploid form that reproduces asexually (by spores) alternates with a haploid form that reproduces sexually (by gametes). The diploid, spore-producing form is known as the sporophyte, and the structures in which the spores are produced are known as sporangia. The haploid, gamete-producing form is known as the gametophyte, and the structures in which gametes are produced are known as gametangia. In some algae, the gametes are identical in appearance; such gametes are known as isogametes. In other algae, one gamete is larger than the other; the larger gamete is the egg and the smaller the sperm. This is where sexual differentiation, with its mixed blessings and complications, all began.

The life cycle of *Ulva,* a green alga that produces isogametes, is shown in Figure 18–14. This reproductive pattern is another link between the green algae and plants, in all of which some form of alternation of generations is found.

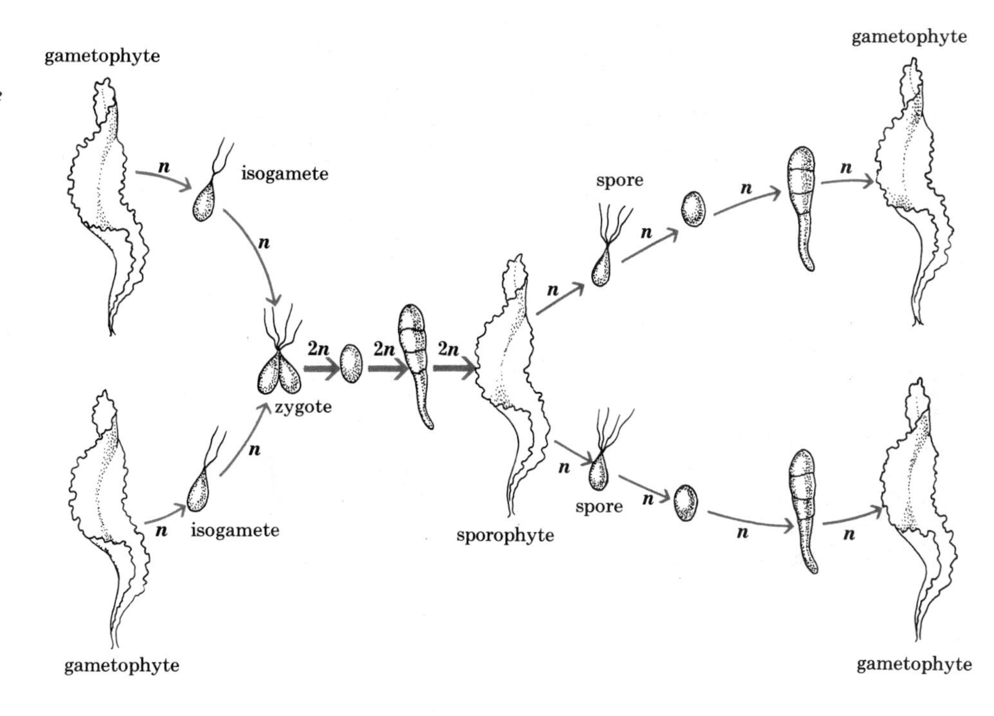

18-14 *In the sea lettuce,* Ulva, *we can see the reproductive pattern known as alternation of generations, in which one generation produces spores, the other gametes. The haploid* (n) *gametophyte produces haploid isogametes, and the gametes fuse to form a diploid* (2n) *zygote. A sporophyte, in which all the cells are diploid, develops from the zygote. The sporophyte produces haploid spores by meiosis. The haploid spores develop into haploid gametophytes, and the cycle begins again.*

In Ulva, *the two generations look alike. In some other species of green algae, the sporophyte and the gametophyte do not resemble one another.*

Slime Molds

The slime molds (phylum Gymnomycota) are a group of curious organisms that resemble amoebas. However, during some or all of their life cycle, the slime molds are multicellular or, at least, multinucleate—that is, they have many nuclei but are not partitioned by cell membranes.

Slime molds reproduce by means of spores. A spore, by definition, is a cell that is capable of developing into an adult organism without fusion with another cell. Thus it contrasts with a gamete, which must unite with another gamete to form a zygote, which then develops into an adult individual. The spores of slime molds, like the spores of many other organisms, are produced by meiosis in sporangia (Figure 18-15). Sporangia often elevate the spores, which facilitates their dispersal by air currents.

18-15 (a) *The plasmodium—a streaming mass of protoplasm—of a slime mold. Such a plasmodium, with its multiple nuclei, can pass through a piece of silk or filter paper and come out the other side apparently unchanged.* (b) *Sporangia of a plasmodial slime mold on a rotting log.*

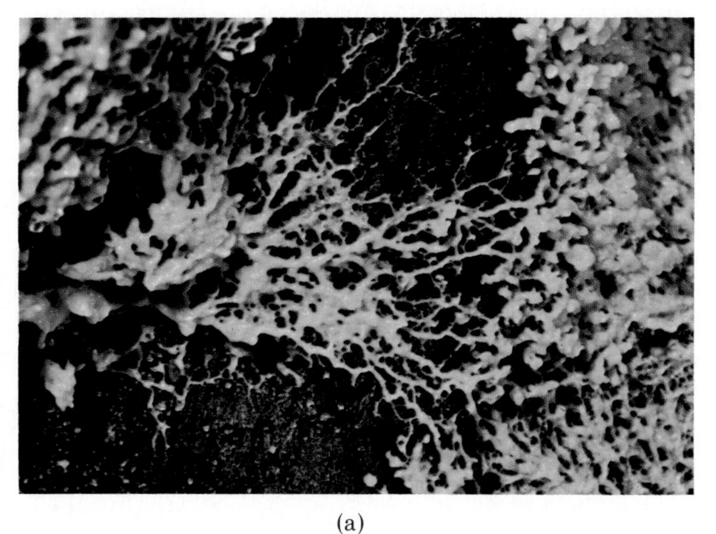

(a)

(b)

Ergotism, the Witches of Salem, and LSD

Many fungi are parasitic on plants. The disease of plants called ergot is caused by *Claviceps purpurea,* a parasite of rye and other grasses. Although ergot seldom causes serious damage to a crop of rye, it is dangerous because a small amount mixed with rye grains is enough to cause severe illness among domestic animals or among the people who eat bread made with the flour.

Ergotism, the toxic condition caused by eating grain infected with ergot, is often accompanied by gangrene, nervous spasms, psychotic delusions, and convulsions. It occurred frequently during the Middle Ages, when it was known as St. Anthony's fire. In one epidemic in the year 994, more than 40,000 people died. In 1722, ergotism struck down the cavalry of Tsar Peter the Great on the eve of battle for the conquest of Turkey and thus changed the course of history.

Recent evidence suggests that the adolescent girls who denounced the "witches" of Salem, Massachusetts, were victims of ergotism. As recently as 1951, there was an outbreak in a small French village in which 30 people became temporarily insane, imagining that they were pursued by demons and snakes; five of the villagers died.

Ergot, which causes muscles to contract and blood vessels to constrict, has various medical uses. It is also the initial source for the psychedelic drug lysergic acid diethylamide (LSD).

The Fungi

The fungi are a group of organisms so unlike any others that, although they were long classified with the plants, it now seems appropriate to assign them to a separate kingdom. Except for some one-celled forms, such as the yeasts, the fungi are basically composed of masses of filaments. A fungal filament is called a *hypha,* and all the hyphae of a single organism are collectively called a *mycelium.* The walls of the hyphae usually contain chitin, a polysaccharide that is never found in plants. (It is, however, the principal component of the exoskeleton—the hard outer covering—of insects and other arthropods.) Cytoplasm flows within the mycelium. The complex reproductive structures of fungi, such as mushrooms, are composed of tightly packed hyphae.

All fungi are heterotrophs. They obtain food by absorbing dissolved organic molecules. Typically a fungus secretes digestive enzymes onto a food source and then absorbs the smaller molecules released. Consequently, fungi live in soil, in water, or in some other medium containing organic substances. Growth is their only form of mobility, except for spores, which may travel through air or water.

Yeasts are of great economic importance in the production of wine, beer, and bread. Other fungi are important in the preparation of cheeses, such as Roquefort and Camembert, and antibiotics, including penicillin. Fungi cause many plant diseases, including potato blight, which was responsible for the great potato famines in Ireland, and downy mildew of grapes, which threatened the entire French wine industry during the latter part of the nineteenth century. Fungi also cause a number of diseases in humans, such as ringworm and thrush, and are important as destroyers of food and clothing (especially cotton and leather).

Together with the bacteria, fungi are the decomposers of the world. As we shall see in Chapter 39, their activities are as vital to the continued survival of other forms of life as are those of the food producers.

Reproduction in the Fungi

Fungi reproduce both asexually and sexually. Asexual reproduction takes place either by the fragmentation of the mycelium (with each fragment becoming a new individual) or by the production of spores. In some of the fungi, spores are produced in sporangia. The bright colors and powdery textures associated with many types of molds are the colors and textures of spores and sporangia, which are often elevated above the mycelium on sporangiophores.

Sexual reproduction is often initiated by the coming together of hyphae of different mating strains. Either before or after they come in contact, the touching tips of the hyphae may develop into specialized reproductive structures called gametangia. The gametangia fuse, the nuclei unite, forming a zygote, and meiosis takes place. Reproduction in one type of fungus, the black bread mold, is shown in Figure 18–17.

The Lichens

A lichen is a specific combination of a fungus and an alga. The organisms resulting from these associations are completely different from either the alga or the fungus growing alone, as are the physiological conditions under which the lichen can survive.

18-16 *Three fungi.* (a) *A common morel. These (and truffles) are among the most prized of the edible fungi.* (b) *A shelf fungus, which grows on decaying wood.* (c) Amanita muscaria. *Members of this genus include the most beautiful and also the most poisonous of the mushrooms. The ring near the top of the stalk is one of the identifying characteristics of this group. One mushroom has been picked to show the gills, on which sexual spores are formed.*

(a)

(b)

(c)

18-17 *Asexual and sexual reproduction in the common black bread mold* Rhizopus. *The mold consists of branched hyphae, including rhizoids, which anchor the mycelium; stolons, which run above the surface of the bread; and sporangiophores.*

(a) *At maturity, the fragile wall of the sporangium disintegrates, releasing the asexual spores, which are carried away by air currents. Under suitable conditions of warmth and moisture, the spores germinate, giving rise to new masses of hyphae.*

(b) *Sexual reproduction occurs when two hyphae from different mating strains come together, forming gametangia, which fuse to form a thick-walled, resistant zygote. After a period of dormancy, the zygote undergoes meiosis and germinates, producing a new sporangium.*

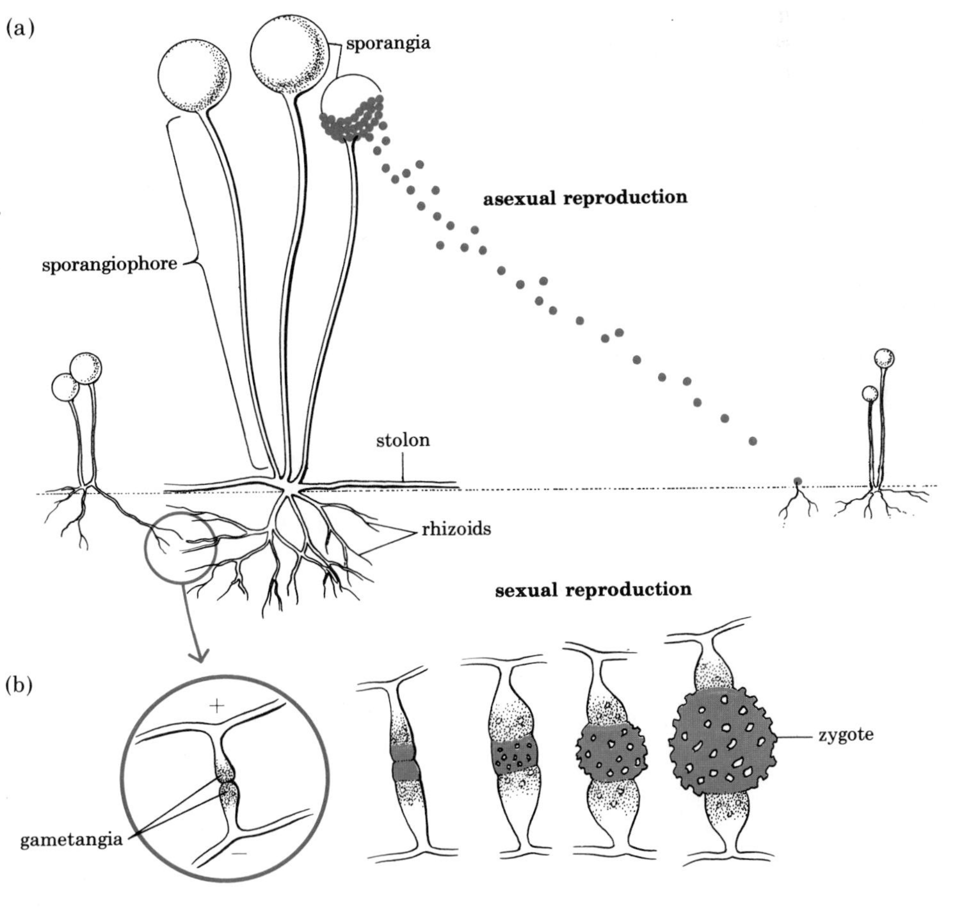

(a)

18-18 (a) *Lichen growing on a dead cedar in a North Carolina salt marsh.* (b) *Crustose lichens on a rock.*

The lichens are widespread in nature, and they are often the first colonists of bare rocky areas. Lichens do not need an organic food source, as do their component fungi, and unlike free-living algae, they can remain alive even when dried out. Because they absorb substances from rainwater, they are particularly susceptible and sensitive to airborne toxic compounds. Thus, the presence or absence of lichens is a sensitive index of air pollution.

Lichens reproduce most commonly by the breaking off of fragments containing both fungal hyphae and algae. New lichens are formed by the capture of an alga by a fungus. Sometimes the alga is destroyed by the fungus, in which case the fungus also dies. If the alga survives, a lichen is produced.

SUMMARY

Organisms are named according to a binomial system, which designates the genus and species of the organism. Genera, in turn, are grouped into families, families into orders, orders into classes, classes into phyla, and phyla into kingdoms. Five kingdoms are recognized in this text: Prokaryotae, Protista, Fungi, Plantae, and Animalia.

The kingdom Prokaryotae comprises the bacteria and the blue-green algae. Prokaryotes are small, relatively simple cells without a membrane-bound nucleus and organelles. Their genetic material is in the form of a circular DNA molecule. They have cell walls of unique composition. The bacteria are a diverse group of organisms, including autotrophs (both chemoautotrophs and photosynthetic autotrophs) and heterotrophs. Among the heterotrophs are the decomposers, which are of great ecological importance in the recycling of organic matter, and the pathogenic (disease-causing) bacteria. All blue-green algae are photosynthetic. Some species of bacteria and blue-green algae can incorporate atmospheric nitrogen into organic compounds.

Viruses do not fit in any of the five kingdoms of organisms, and, in fact, some biologists do not consider them to be living organisms. A virus consists essentially of nucleic acid (either DNA or RNA) enclosed in an outer coat of protein. They can multiply only within a living cell, usually only one particular type of cell.

The other four kingdoms of organisms are all composed of eukaryotes. In eukaryotic cells, DNA, the genetic material, is associated with proteins and organized into chromosomes. These are contained within a nucleus bounded by a double membrane, the nuclear envelope. Eukaryotic cells also have membrane-bound organelles, including mitochondria and, in the case of photosynthetic cells, chloroplasts. They have cilia and flagella with the characteristic 9 + 2 structure. It is hypothesized that eukaryotes evolved as certain prokaryotes began living inside other prokaryotic cells.

The kingdom Protista is a highly diverse group, comprising unicellular eukaryotic heterotrophs, the Protozoa; the slime molds; and six phyla of algae. Unicellular algae are the photosynthetic components of the plankton, the food source for many marine organisms. The common marine seaweeds are multicellular algae, including brown algae, red algae, and some green algae.

The green algae are believed to be the group from which the plants arose. Both green algae and plants have (1) chlorophylls *a* and *b* and carotenoids as the photosynthetic pigments; (2) cellulose cell walls; and (3) a life cycle in which a haploid form, the gametophyte, alternates with a diploid form, the sporophyte (alternation of generations).

The Fungi are heterotrophs and, with a few exceptions, are simple multicellular organisms. They absorb organic materials from the substrate on or in which they live. Most have cell walls that contain chitin. The body of a multicellular fungus is the mycelium, which is made up of a mass of filaments, called hyphae. Many fungi reproduce by spores that are often produced within sporangia. Fungi are ecologically important as decomposers. They are also economically useful in the production of foodstuffs and antibiotics. They are the causes of a number of serious plant diseases as well as some diseases in animals.

Lichens are associations of fungi and algae that are adapted to conditions—such as lack of water and of organic matter—under which neither form could survive alone.

QUESTIONS

1. Distinguish between the following terms: blue-green algae/green algae; Protista/Protozoa; pathogen/toxin; flagellate/ciliate; colonial organism/multicellular organism; slime mold/bread mold; hypha/mycelium; sporangia/gametangia.

2. Describe the principal differences between prokaryotes and eukaryotes.

3. What are the distinguishing features of the kingdoms Prokaryotae, Protista, and Fungi?

4. Many microorganisms produce antibiotics. What do you think their function might be for the organisms that produce them?

5. Some biologists consider viruses to be living organisms. By what criteria might viruses be considered alive?

6. Explain alternation of generations, using as your example the sea lettuce, *Ulva*.

19

The Diversity of Life: Plants and Animals

In the previous chapter, we considered the three kingdoms of microorganisms: the prokaryotes, the protists, and the fungi. In this chapter, we shall briefly survey the two kingdoms most familiar to us: the plants and the animals.

Table 19–1 gives the major events in the evolution of plants and animals, as well as information about the climate and geology of the different time periods. You may find it helpful to refer to this table as you proceed through the chapter. Notice that the table begins with the present and goes back in time; thus the earliest events are at the bottom of the table.

19–1 *The water fern,* Marsilea, *is a plant that lives in the water. However, like whales and porpoises, it bears the unmistakable traces of an ancestral sojourn on the land.*

The Plants

Plants are multicellular photosynthetic organisms adapted for life on land. As we saw in the last chapter, the common ancestor of all the modern plants seems to have been a relatively complex multicellular green alga. It invaded the land some 500 million years ago. It may have been "prepared" for its invasion of land by its differentiation into a holdfast and an upper photosynthetic area, such as are seen among the modern coastal seaweeds. This "preparation" is known as *preadaptation.* Preadaptation is the evolution in one environment of a structure or function that turns out to be advantageous when later generations shift to a new environment.

Once the ancestral plant had made the transition to land, some new evolutionary developments took place. (These were postadaptations, which evolved *after* the new environment had been invaded.) These adaptations must have occurred early in the evolutionary history of plants since most modern plants, even though very diverse, share them. They include a fatty coating, the *cuticle,* that covers the aboveground parts of the plant body and retards water evaporation, and *stomata,* specialized openings in leaves and green stems through which carbon dioxide enters and oxygen exits. Another adaptation is the surrounding of the gametangia and sporangia by protective layers of cells that keep the reproductive cells inside from drying out. Finally, as we noted previously, all plants have a reproductive cycle that involves alternation of generations.

Table 19-1 Major physical and biological events in geologic time

Millions of Years Ago	Period	Epoch	Life Forms	Climates and Major Physical Events
Cenozoic Era				
	Quaternary	Recent Pleistocene	Planetary spread of *Homo sapiens;* extinction of many large mammals. Deserts on large scale.	Fluctuating cold to mild. Four glacial advances and retreats (Ice Age); uplift of Sierra Nevada.
1½–7	Tertiary	Pliocene	Large carnivores. First known appearance of hominids (humanlike primates).	Cooler. Continued uplift and mountain building, with widespread extinction of many species.
7–26		Miocene	Whales, apes, grazing mammals. Spread of grasslands as forests contract.	Moderate uplift of Rockies.
26–38		Oligocene	Large, browsing mammals. Apes appear.	Rise of Alps and Himalayas. Lands generally low. Volcanoes in Rockies.
38–53		Eocene	Primitive horses, tiny camels, modern and giant types of birds.	Mild to very tropical. Many lakes in western North America.
53–65		Paleocene	First known primitive primates and carnivores.	Mild to cool. Wide, shallow continental seas largely disappear.
Mesozoic Era				
65–136	Cretaceous		Extinction of dinosaurs. Marsupials, insectivores, and angiosperms become abundant.	Lands low and extensive. Last widespread oceans. Elevation of Rockies at end of period.
136–195	Jurassic		Dinosaurs' zenith. Flying reptiles, small mammals. Birds appear. Gymnosperms and ferns.	Mild. Continents low. Large areas in Europe covered by seas. Mountains rise from Alaska to Mexico.
195–225	Triassic		First dinosaurs. Primitive mammals appear. Forests of gymnosperms and ferns.	Continents mountainous. Large areas arid. Eruptions in eastern North America. Appalachians uplifted and broken into basins.
Paleozoic Era				
225–280	Permian		Reptiles evolve. Origin of conifers and possible origin of angiosperms; earlier forest types wane.	Extensive glaciation in southern hemisphere. Appalachians formed by end of Paleozoic; most of seas drain from continent.
280–345	Carboniferous Pennsylvanian Mississippian		Age of amphibians. First reptiles. Variety of insects. Sharks abundant. Great swamps, forests of ferns, gymnosperms, and horsetails.	Warm. Lands low, covered by shallow seas or great coal swamps. Mountain building in eastern U.S., Texas, Colorado. Moist, equable climate, conditions like those in temperate or subtropical zones, little seasonal variation, root patterns indicate water plentiful.
345–395	Devonian		Age of fish. Amphibians appear. Shellfish abundant. Lunged fish. Extinction of primitive vascular plants. Origin of modern subclasses of vascular plants.	Europe mountainous with arid basins. Mountains and volcanoes in eastern U.S. and Canada. Rest of North America low and flat. Sea covers most of land.
395–440	Silurian		Earliest vascular plants. Rise of fish and reef-building corals. Shell-forming sea animals abundant. Modern groups of algae and fungi.	Mild. Continents generally flat. Again flooded. Mountain building in Europe.
440–500	Ordovician		First primitive fish. Invertebrates dominant. Invasion of land by plants?	Mild. Shallow seas, continents low; sea covers U.S. Limestone deposits; microscopic plant life thriving.
500–600	Cambrian		Age of marine invertebrates. Shelled animals.	Mild. Extensive seas. Seas spill over continents.
Precambrian Era				
Over 600			Earliest known fossils. Soft-bodied marine invertebrates.	Dry and cold to warm and moist. Planet cools. Formation of earth's crust. Extensive mountain building. Shallow seas. Accumulation of free oxygen.

The Bryophytes

The two major groups of plants, the bryophytes and the tracheophytes, separated early in evolutionary history, probably more than 400 million years ago. The *bryophytes*, which include the mosses, liverworts, and hornworts, are plants that did not develop elaborate conducting systems. They are comparatively simple in their structure and are relatively small, usually less than 15 centimeters in height. Bryophytes absorb moisture through leaflike structures as well as through rootlike structures known as rhizoids. (The terms "root" and "leaf" are reserved for structures with vascular, or conducting, tissues.) As you would imagine, bryophytes are most abundant in moist areas.

In the bryophytes, unlike any other plants, the gametophyte (haploid form) is larger than the diploid sporophyte (Figure 19–2). The sperm are released when sufficient moisture is present, enabling them to swim by means of their flagella to the egg-containing archegonium, which attracts them chemically. Fusion of egg and sperm takes place within the archegonium. The resulting zygote develops into a sporophyte, which remains attached to the gametophyte and is often nutritionally dependent on it. Typically, the sporophyte consists of a single, large capsule elevated on a stalk, from which the spores are discharged.

Asexual reproduction, often by fragmentation, is also common.

19–2 *A bryophyte, a haircap moss with spore capsules. The lower green structures are the gametophytes, which are haploid* (n); *the stalks and capsules are the diploid* (2n) *sporophytes.*

(a)

(b)

(c)

19–3 (a) *In the club mosses,* Lycopodium, *the sporangia are borne on specialized leaves, sporophylls, which are aggregated into a cone at the apex of the branches, as shown in this club moss, a running ground pine. The airborne, waxy spores give rise to small, independent, subterranean gametophytes. The sperm, which are biflagellated, swim to the archegonium, and the young sporophyte, or embryo, develops there.*

(b) *The horsetails, of which there is only one living genus* (Equisetum), *are easily recognized by their jointed, finely ribbed stems, which contain silicon. At each node, there is a circle of small, scalelike leaves. Spore-bearing leaves are clustered into a cone at the apex of the stem. The gametophytes are independent, and the sperm are coiled, with numerous flagella.*

(c) *A cinnamon fern. Among ferns, the leaf, or frond, is commonly divided into leaflets, or pinnae. Spores are borne in sporangia, either on the underside of leaves or on separate stalks, as shown here.*

The Tracheophytes

The *tracheophytes,* or vascular plants, which dominate the modern landscape, are characterized by their efficient systems for the transport of water and sugars. Modern tracheophytes include the club mosses, horsetails, ferns, gymnosperms (the most familiar of which are the conifers), and angiosperms (flowering plants).

Trends among the Vascular Plants

Better Conducting Systems Among the vascular plants, there have been three marked evolutionary trends. The first is the development of increasingly efficient conducting systems. One conducting system, the *xylem,* transports water and ions from the roots to the leaves, often more than 100 meters. The other conducting system, the *phloem,* carries sucrose and other products of photosynthesis from the leaves to the nonphotosynthetic cells of the plant. The structure and function of the vascular system in modern angiosperms will be described in Chapter 21.

Reduction of the Gametophyte The second pronounced trend is the reduction in size of the gametophyte generation. Among the ferns, as we saw in Chapter 10, the gametophyte is separate from the sporophyte but is much smaller. It is nutritionally independent of the sporophyte. In a few species of fern, the gametophytes exist as separate sexes, with some gametophytes producing male gametes and others producing female gametes. Most fern gametophytes, however, are hermaphroditic, producing both male and female gametes (see Figure 10–4, page 140).

19-4 (a) *Male cones of jack pine* (Pinus banksiana) *shedding pollen. The pollen grains are immature male gametophytes, which complete their maturation when they reach the ovules, embedded in the female cones. There they release sperm cells that fuse with the egg cells.*

(b) *Female cone. The female gametophytes develop in ovules on the base of a scale of the cone, and the eggs are fertilized there. Each scale contains two ovules. When the seeds are mature, they drop from the cone.*

Among the gymnosperms, there are two types of gametophytes, one male and one female. Both are entirely heterotrophic, being dependent on the parent sporophyte for their nutrition and development. The female gametophyte, in fact, never leaves the protection of the sporophyte, and the egg cell is fertilized there by sperm from the male gametophyte (Figure 19-4).

In the angiosperms, the most recently evolved of the vascular plants, the gametophytes are microscopic. The female gametophyte, or embryo sac, consists of seven cells with a total of eight haploid nuclei, and the male gametophyte consists of three cells, each of which contains a haploid nucleus. And yet the ancestral pattern persists, and it is not possible to understand reproduction among this most important group of modern plants without knowledge of its legacy from the past.

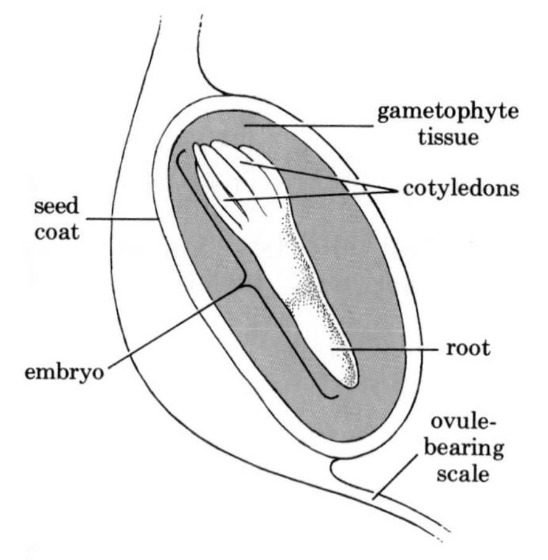

19-5 *Pine seed. The outer layers of the ovule have hardened into a seed coat, enclosing the female gametophyte and the embryo, which now consists of an embryonic root and a number of embryonic leaves, the cotyledons.*

When the seed germinates, the root will emerge from the seed coat and penetrate the soil. When the root absorbs water, the tightly packed cotyledons will elongate and swell with the moisture, rising above ground on the lengthening stem and forcing off the seed coat. During this period, the cotyledons absorb nutrients that are stored in the gametophyte tissue and are essential for the growth of the embryo into a seedling.

The Seed The third major development among the vascular plants, and perhaps the most important to survival on land, is the seed. The seed is a complex structure in which the young sporophyte, or embryo, is contained within a protective outer covering, the seed coat (Figure 19-5). The seed coat, which is derived from tissues of the parent sporophyte, protects the embryo from drying out while it remains dormant, sometimes for years, until conditions are favorable for its germination.

Seeds became important toward the close of the Paleozoic era, which ended some 225 million years ago. Earlier in this era, in the Carboniferous period, the temperature was warm and the lands were low, covered by shallow seas or great swamps. Water was plentiful, and there was little seasonal variation in temperature. This was the period in which our fossil fuel deposits were formed from the masses of vegetation in the swamps. At the close of the Carboniferous period, there were worldwide changes of climate, with widespread glaciers and droughts. The seed was in existence at this time; according to the fossil record, some of the fernlike plants and even some of the club mosses had seedlike structures. But it was not until the close of the Permian period, when the land became colder and drier, that plants that had seeds gained a major evolutionary advantage and came to be the dominant plants of the land.

Fuels: Past, Present, and Future

In the process of photosynthesis, plants and algae use the energy of the sun to convert carbon dioxide and water into carbohydrates. Most of these carbohydrates are then oxidized by the plants and algae themselves, or by heterotrophs that eat the plants and algae, by decomposers such as fungi and bacteria, or, less frequently, by burning (as in a forest fire). A small percentage of the carbohydrates, those stored in the bodies of dead organisms, are buried in sediment or mud under conditions in which decay is only partial. This partially decayed material is known as peat. The peat may eventually become covered with sedimentary rock, placing it under pressure. Depending on time, temperature, and other factors, peat may become compressed and converted into coal, petroleum, or natural gas—the so-called fossil fuels.

During certain periods in the earth's history, the rate of fossil fuel formation was greater than at other times. One such period was the Carboniferous, some 300 million years ago. The lands were low, covered by shallow seas or swamps, and, in what are now the temperate regions, conditions were favorable for growth year round. The principal large plants were ferns, tree-sized club mosses and horsetails, and primitive gymnosperms. Their fossils are found near the large coal seams of this period.

People have burned peat and coal, as well as wood, for domestic use for several thousand years. The great fossil fuel deposits, however, have been mined and consumed on a large scale only since about 1900. The growth of industries, of cities, of automobile use, and of the human population closely parallels the great increase in fuel consumption. Fossil fuels are still being formed, but the rate of formation is so slow that, for all practical purposes, supplies are not renewable and are finite.

We use fossil fuels both directly and indirectly for a variety of purposes. When such fuels are burned, they release heat, which can be used to warm our buildings and cook our food; to power automobile, diesel, and jet engines; or to produce steam from water. The steam, in turn, can be used to power machinery or to turn turbines that generate electricity. As the supply of fossil fuels diminishes, alternative sources of energy are being sought for these functions. Many have been suggested, including the direct use of solar energy for heat and to power solar cells to generate electricity; harnessing wind and water (actually an indirect use of solar energy) and the surge of tides to generate electricity; tapping the heat energy beneath the surface of the earth; and preparing "synthetic fuels" from low-grade coal and oil-bearing shale. For some of these possibilities the technology is available now, but for others it is primitive. For a number of years, nuclear power, which can be used to generate electricity, seemed the most plausible new energy source. However, many people, both scientists and nonscientists, are deeply concerned about the hazards of nuclear power, and its future role remains to be seen.

An exciting new prospect is the use of renewable fuels—the carbohydrates produced by photosynthesis—to meet at least some of our energy needs. There are several possibilities, all of which involve growing crops specifically for fuel. In some parts of the world, such as rural India, the most pressing need is fuel for domestic use. Attempts are being made to replant forest regions that have been stripped over the centuries. The goal is a very basic one: to provide local populations with a continually renewing source of firewood.

Another possibility—important for both the developed and the developing world—is the biological production of alcohol. As you know, alcohol is produced by the anaerobic fermentation of sugars by yeast. In Brazil, sugarcane plantations like the one shown in the photo are being converted to "energy plantations." Parts of the sugarcane plant with a low sugar content are burned, providing energy for the sugar mill and to extract alcohol from the fermentation mixture; enough heat is left over to generate electricity as well. The resulting alcohol can be used alone as fuel in properly converted automobile engines, mixed with gasoline to provide gasohol, or used as a starting material in the production of synthetic chemicals. Experiments are underway elsewhere with the use of other crops, including cassava, sorghum, pineapple, and corn, as sources of starch for alcohol production.

Another potential biological fuel source is plants that produce hydrocarbons. There are a number of plants in different parts of the world that produce a liquid quite similar to crude oil. Experimental work has begun to determine the conditions under which these plants can be cultivated to produce a maximum energy yield.

Our technological civilization has been built with solar energy captured by photosynthetic organisms hundreds of millions of years ago. Its future may well depend on solar energy captured by photosynthetic organisms now and in the years to come.

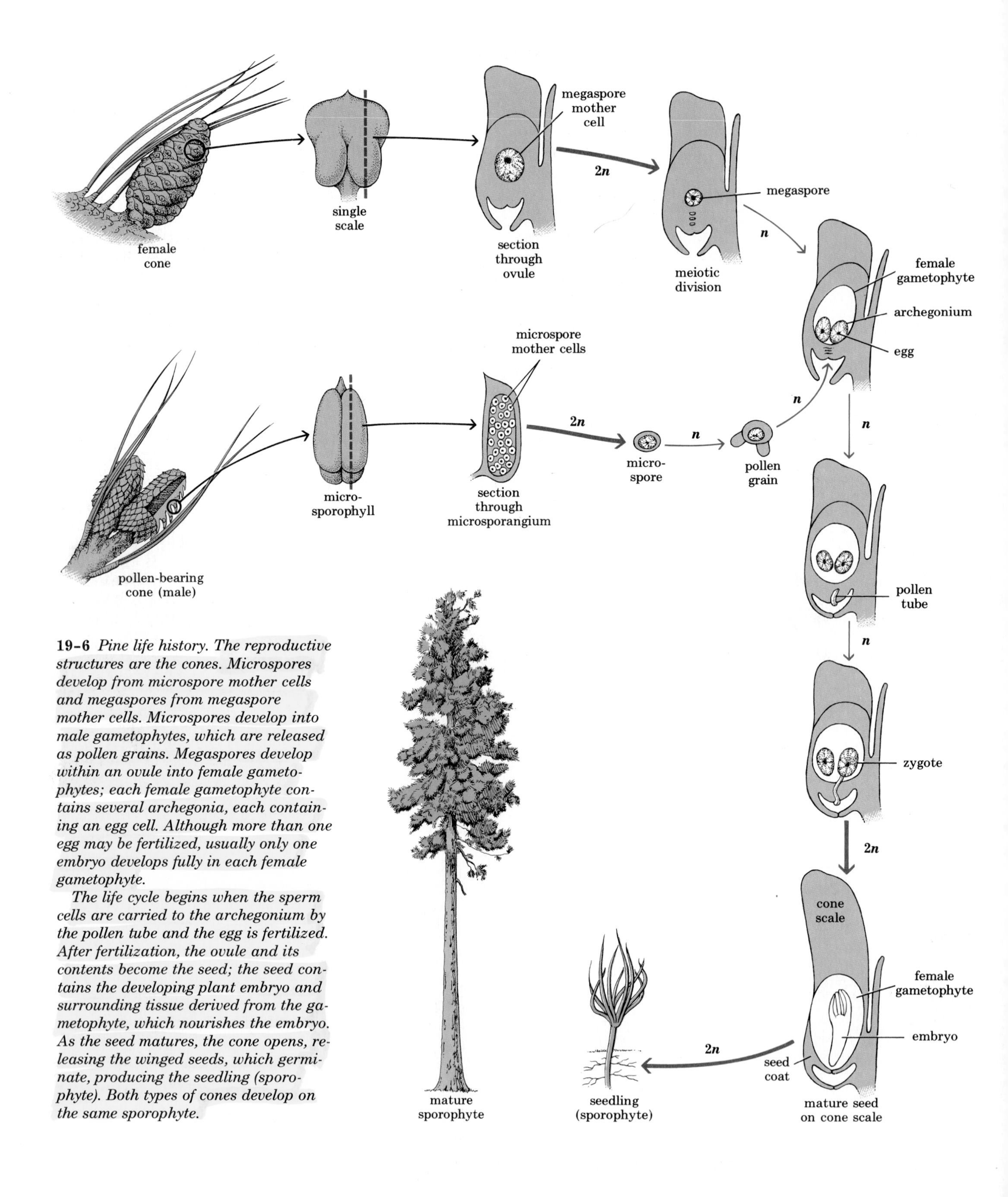

19–6 *Pine life history. The reproductive structures are the cones. Microspores develop from microspore mother cells and megaspores from megaspore mother cells. Microspores develop into male gametophytes, which are released as pollen grains. Megaspores develop within an ovule into female gametophytes; each female gametophyte contains several archegonia, each containing an egg cell. Although more than one egg may be fertilized, usually only one embryo develops fully in each female gametophyte.*

The life cycle begins when the sperm cells are carried to the archegonium by the pollen tube and the egg is fertilized. After fertilization, the ovule and its contents become the seed; the seed contains the developing plant embryo and surrounding tissue derived from the gametophyte, which nourishes the embryo. As the seed matures, the cone opens, releasing the winged seeds, which germinate, producing the seedling (sporophyte). Both types of cones develop on the same sporophyte.

The Seed Plants

The two major modern groups of plants are both seed plants: the *gymnosperms*, "naked seed" plants, and the *angiosperms* (from the Greek word *angio*, meaning "vessel"—literally, a seed borne in a vessel). The largest and most familiar group of gymnosperms are the conifers, "cone-bearers," in which the female gametophyte develops on the scale of a cone and is fertilized there. The life cycle of a familiar conifer, a pine, is shown in Figure 19–6.

Among the angiosperms, the female gametophyte, which contains a single egg cell, develops within the flower, where the egg is fertilized. The mature seed is not exposed, as it is in the gymnosperms, but is enclosed in a fruit, which develops from a part of the flower called the carpel. Sometimes other parts of the flower, in addition to the carpel, contribute to the fruit (Figure 19–7). Thus the angiosperm seed has an additional covering that encloses it, protects it, and often aids in its dispersal (Figure 19–8).

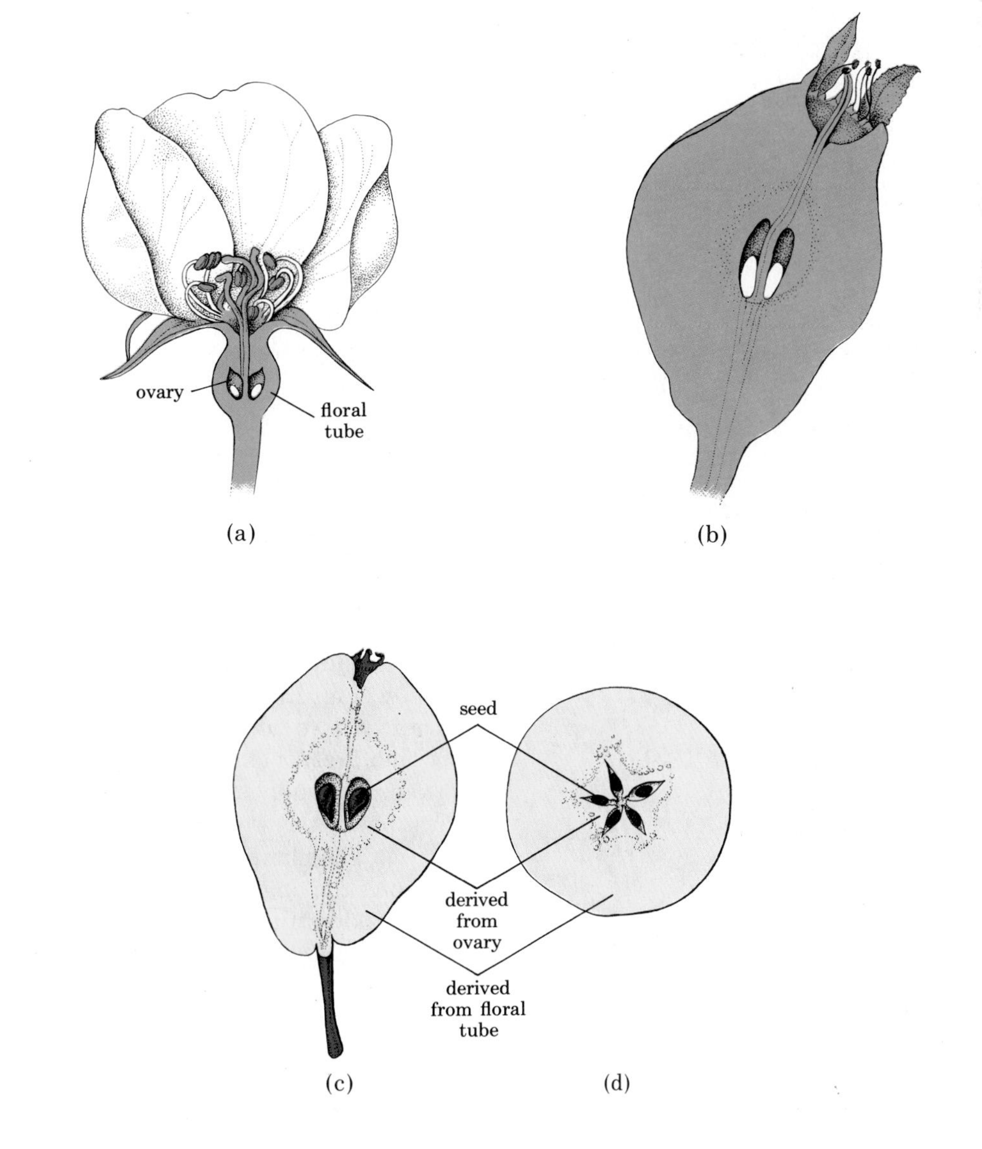

19–7 *Development and structure of the pear.* (a) *Flower of the pear. The ovary is the basal portion of the carpel.* (b) *Older flower, after petals have fallen.* (c) *Longitudinal section and* (d) *cross section of the mature fruit. The core of the pear is the ripened ovary wall. The fleshy, edible part develops from the floral tube.*

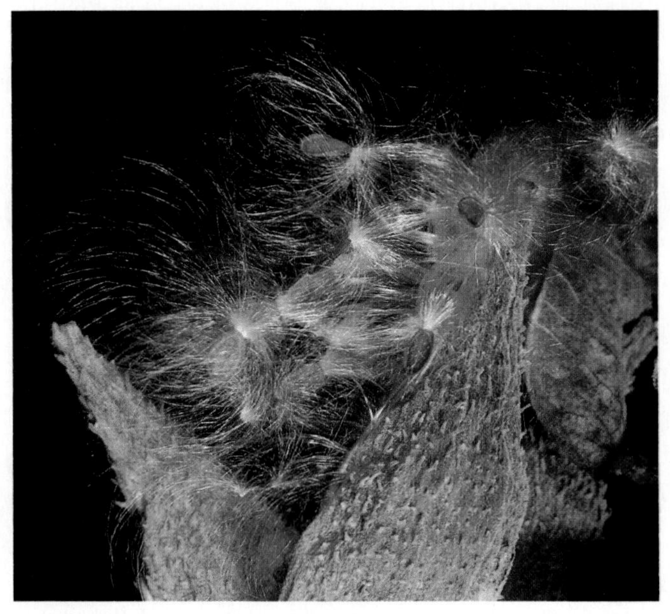

(a)

(b)

19-8 *Angiosperms are characterized by fruits. A fruit is a mature ovary and, often, accessory parts of the flower, enclosing the seed or seeds. The fruit aids in dispersal of the seed. Some fruits are borne on the wind, some are carried from one place to another by animals, some float on water, and some are even forcibly ejected by the parent plant.*

(a) *In milkweed, the fruit bursts open when it is ripe, releasing seeds with tufts of silky hair that aid in their dispersal.*

(b) *The tough seeds of these blackberries will pass unharmed through the digestive tract of the dormouse. If they are deposited in a suitable environment, they will germinate, forming new blackberry plants.*

The Role of Plants

The only forms of life on land that do not depend on plants for their existence are a few kinds of autotrophic prokaryotes. For all other land organisms, the chloroplast of the plant cell is the "needle's eye" through which the sun's energy is channeled into the biosphere. Even those animals that eat only other animals—the carnivores—could not exist if their prey, or their prey's prey, had not been nourished by plants.

Moreover, plants are the channels by which many of the simple inorganic substances vital to life enter the biosphere. Carbon is taken from the carbon dioxide of the atmosphere and incorporated into organic compounds during photosynthesis. Elements such as nitrogen and sulfur are taken from the soil in the form of simple inorganic compounds and incorporated into proteins, vitamins, and other essential organic compounds within green plant cells. Animals cannot make these organic compounds from inorganic materials and so are entirely dependent on plants for these compounds as well as for their energy supply.

The plants on which we depend are almost exclusively angiosperms. Among the angiosperms, the cereal grasses, such as wheat, rice, and corn, are of particular importance. Their dry fruits (grains) are relatively easy to store and are high in calories and protein. The ability to stockpile food in the form of grain (and also in the form of domesticated animals that ate the grasses and their grain) was an essential component in the agricultural revolution of some 11,000 years ago. Without the angiosperms, it is unlikely that modern civilization could have come into being.

The Animals

Animals are many-celled heterotrophs. They depend directly or indirectly for their nourishment on plants or algae. Typically they digest their food in an internal cavity and store food as glycogen or fat. Their cells do not have walls. Generally animals move by means of contractile cells (muscle cells) containing characteristic proteins. Reproduction is usually sexual. The more recently evolved animals—the arthropods and the vertebrates—are the most complex of all organisms. They have many kinds of specialized tissues, including elaborate sensory and neuromotor mechanisms not found in any of the other kingdoms.

For most of us, animal means mammal, and mammals are, in fact, the chief focus of attention in Section 6. However, the mammals, or even the vertebrates as a whole, represent only a small fraction of the animal kingdom. More than 90 percent of the species of animals are *invertebrates*—that is, animals without backbones—and most of these are insects.

Origins of the Animals

Animals, like plants, presumably had their origins among the protists. In the case of animals, however, we have fewer clues about which particular group of protists most closely resemble the ancestral ones. The fossil record indicates that by the Cambrian period, which ended some 500 million years ago, most, if not all, of the invertebrate phyla with modern representatives were already in existence. Although numerous efforts have been made to trace the evolutionary relationships of these earlier forms, the Precambrian fossil record is inadequate to verify any hypotheses about the chronological order of their appearance. Most of the evidence for charts such as Figure 19–26, on page 269, comes from studies of modern forms.

There are about 30 phyla of invertebrates, each distinguished from all the others by a particular type of body plan. Of these, we are going to discuss only a few, concentrating on the most "successful," as judged on the basis of the number of surviving species and individuals.

Sponges: Phylum Porifera

Sponges, the most primitive of the modern animals, represent a level of organization somewhere between a colony of cells and a true multicellular organism. As with colonial forms, if the individual cells of the sponge are separated, such as by pressing the tissue through a fine cloth, and then mixed together again, the cells will reassemble, resuming their previous organization.

In contrast to true colonial forms, however, the cells of the sponge are to a certain extent differentiated and specialized. The different cell types include feeding cells (the flagellated collar cells); epithelial cells, some of which contain contractile fibers; and amoebocytes, which serve several functions, including carrying food from the collar cells to the epithelial cells and other nonfeeding types. The amoebocytes also produce stiffening structures, either inorganic spicules or organic fibers, which form the skeleton, the only part of the sponge that remains when it is dried out and cleaned.

19–9 *A purple sponge.*

Because all the digestive processes of sponges are carried out within single cells, even a giant sponge—and some stand taller than a man—can consume nothing larger than microscopic particles. These are obtained by

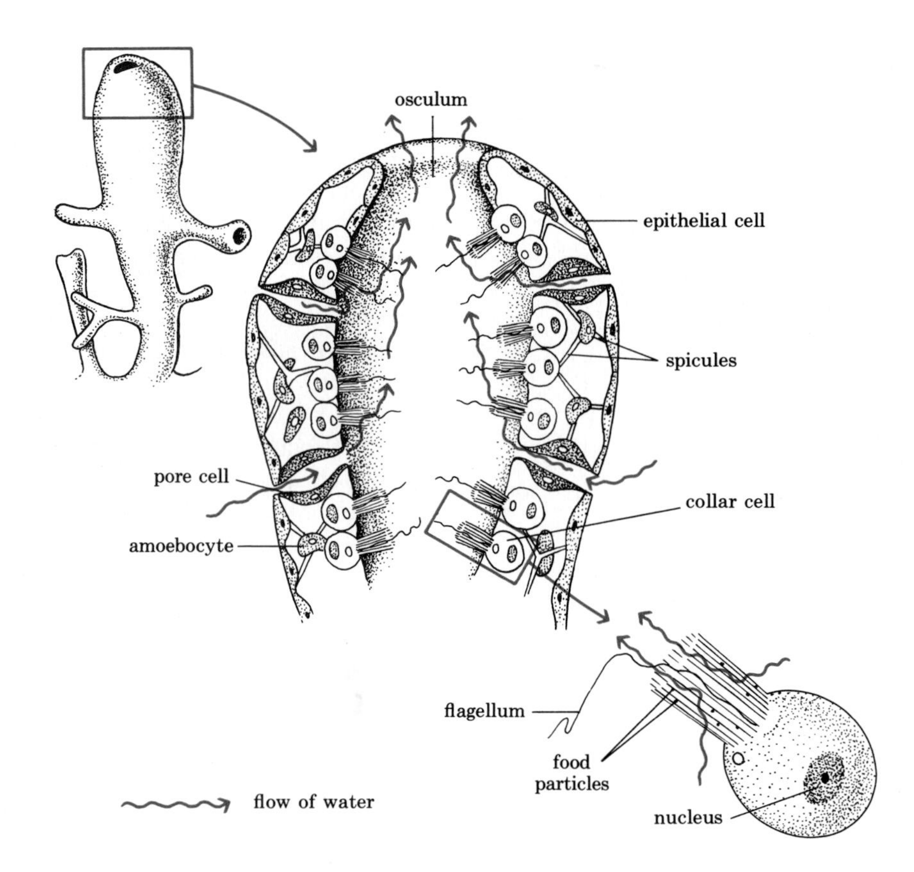

19-10 *The body of a simple sponge is dotted with tiny pores, from which the phylum derives its name (Porifera, or "pore bearer"). Water containing food particles is drawn into the internal cavity of the sponge through these pores and is forced out the osculum. The water is moved by the beating of the flagella protruding from the collar cells and by the sucking effect of flow of the local currents across the osculum. Each collar cell has about 20 retractile filaments and a single flagellum, the lashing of which directs a current of water through the filaments. Minute particles are filtered out and cling to one or more filaments and are then drawn into the cell. A sponge 10 centimeters high filters more than 20 liters of water a day.*

filtering the water that is driven through the body of the sponge by the flagella of the collar cells. Sponges are well suited to a slow life on the ocean floor. Sunlight is not necessary to their existence, and they have been found at depths that support few other living creatures. However, they are generally regarded as an evolutionary dead end; that is, no other groups were derived from them.

19-11 *This jellyfish, the medusa form of the coelenterate* Gonionemus murbachii, *is swimming actively, its bell contracted by muscles around its margin. The specialized muscle cells contain assemblies of contractile protein fibers.*

Coelenterates: Phylum Coelenterata

The coelenterates are a large group of aquatic animals, including jellyfish, *Hydra,* corals, and sea anemones. They are characterized by a hollow body made up of two layers of tissue, the ectoderm and the endoderm. Between these two layers is a jellylike filling, most conspicuous in the jellyfish. The interior cavity of the body is known as the *coelenteron,* from which the coelenterates get their name. There are two basic body plans: polyp and medusa (Figure 19-12a).

The life cycle of coelenterates characteristically includes both polyp and medusa. There is also an immature larval form, known as the planula, which is a small, free-swimming ciliated organism. In the typical life cycle, the planula settles down to give rise to a polyp, which reproduces asexually and may form extensive colonies. From the polyp, young medusas (jellyfish) may bud off. These are sexually reproducing forms that give rise to planulas again (Figure 19-13). All adult forms are radially symmetrical; that is, their body parts are arranged around an axis, like spokes around the hub of a wheel.

19-12 (a) *Among coelenterates, there are two basic body plans: the vase-shaped polyp (left) and the bowl-shaped medusa (right). The coelenteron, characteristic of the phylum, is a digestive cavity with a single opening. The coelenterate body has two tissue layers, ectoderm and endoderm, with gelatinous mesoglea between them.*

(b) *Cnidoblasts, specialized cells located in the tentacles and body wall, are a distinguishing feature of coelenterates. The interior of the cnidoblast is filled by a nematocyst, which consists of a capsule containing a coiled tube, as shown on the left. A trigger on the cnidoblast, responding to chemical or mechanical stimuli, causes the tube to shoot out, as shown on the right. The capsule is forced open and the tube turns inside out, exploding to the outside. The cnidoblast cannot be "reloaded"; it is absorbed and a new cell grows to take its place.*

19-13 *The life cycle of the coelenterate* Aurelia. *Sperm and egg cells are released from adult medusas into the surrounding water. Fertilization takes place, and the resulting zygote develops first into a hollow sphere of cells, the blastula. It then elongates and becomes a ciliated larva called a planula. The planula eventually settles to the bottom, attaches by one end to some object, and develops a mouth and tentacles at the other end, thus transforming into the polyp stage. The body of the polyp grows and, as it grows, begins to form medusas, stacked upside down like saucers. These bud off, one by one, and grow into full-sized jellyfish.*

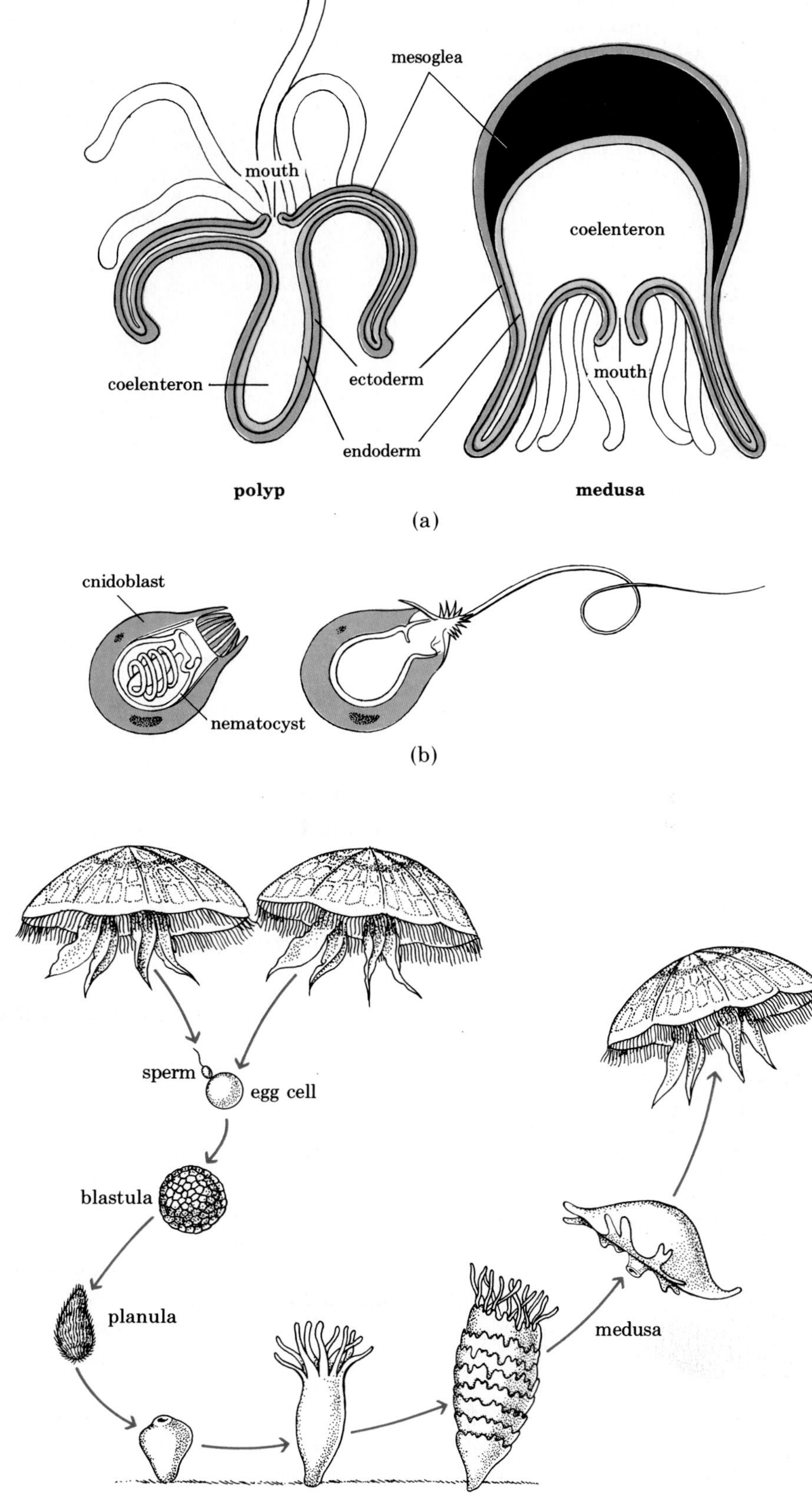

19-14 *The sea anemone is a coelenterate in which the medusa phase has been lost. It feeds by tentacles, moving food into its coelenteron, which is divided longitudinally by partitions. These are pink anemones, photographed near Point Loma, California.*

In both polyps and medusas, food is captured by means of tentacles and pushed into the coelenteron, which opens wide and engulfs it. Within the coelenteron the food is digested to some extent by enzymes secreted by cells lining the cavity. Digestion is then completed within food vacuoles in the cells. The indigestible remains are ejected by the same opening through which the food enters. The coelenterates are able to eat almost any prey they can stuff into their remarkably expandable body cavities.

Coelenterates have primitive nervous systems. The medusa form—exemplified by the jellyfish—has two rings of nerve cells that circle the margin of the bell. The polyp form—as in *Hydra* (page 136)—has a continuous network of nerve cells just below the outer surface. This network links the body into a functional whole and makes possible coordinated movements. Highly specialized stinging cells, cnidoblasts, are found only among the coelenterates. They are used for defense and to capture prey (Figure 19-12b).

The most ecologically important members of this group are the coral builders. They are responsible for the formation of great land masses in the sea where ordinarily no land would exist. The 2,000-kilometer-long Great Barrier Reef off the northeast shores of Australia is an example of such a coral-created land mass. A coral reef is composed primarily of the accumulated limestone skeletons of coral coelenterates, covered by a thin crust occupied by the living colonial animals.

Many biologists believe that a primitive coelenterate is to be found on the evolutionary pathway leading from the protists to the higher forms. A principal clue in this evolutionary puzzle, they point out, is the ciliated larva. It is easy to imagine a gradual transition between such a form and the simplest of the flatworms, the next big step in the order of complexity. As shown in Figure 19-26, on page 269, a form corresponding to the planula larva may have been the starting point for wormlike animals.

Flatworms: Phylum Platyhelminthes

The flatworms are the simplest animals, in terms of body plan, to show bilateral symmetry (Figure 19–15). In bilaterally symmetrical animals, the body plan is organized along a longitudinal axis, with the right half an approximate mirror image of the left half. A bilaterally symmetrical animal can move more efficiently than a radially symmetrical one (which is better adapted to a sedentary existence). It also has a top and a bottom, or in more precise terms (applicable even when it is turned upside down or, as in the case of humans, standing upright), a dorsal and a ventral surface. Like most bilateral organisms, the flatworm also has a distinct "headness" and "tailness." Apparently, when one end goes first, it is advantageous to collect the sensory cells into that end. With the aggregation of sensory cells came a gathering of nerve cells; this gathering is a forerunner of the brain.

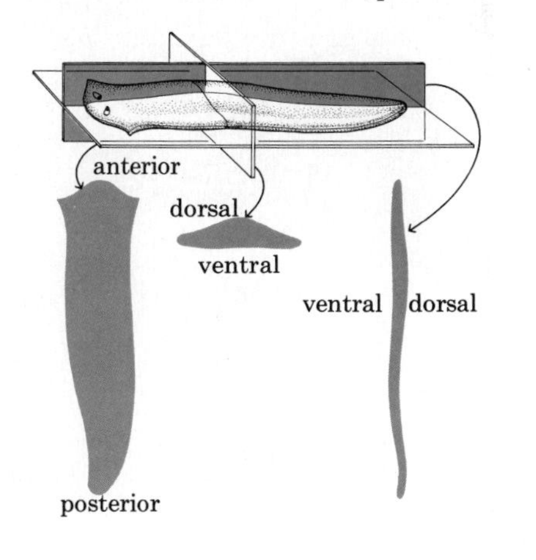

19–15 *In a bilaterally symmetrical organism, such as the planarian shown here, the right and left halves of the body are mirror images of one another. The upper and lower (or back and front) surfaces are known as dorsal and ventral. The end that goes first is termed anterior and the rear, posterior.*

Flatworms have the three distinct tissue layers—ectoderm, mesoderm, and endoderm—characteristic of all animals above the coelenterate level of organization. Moreover, not only are their tissues specialized for various functions, but also two or more types of tissue cells may combine to form organs. Thus while coelenterates are largely limited to the tissue level of organization, flatworms can be said to have gained the organ level of complexity.

The free-living flatworms, of which the freshwater planarians are familiar examples, suck bits of dead animals into their highly branched digestive cavities, where they are digested by the cells lining the cavity. The indigestible residue is ejected through the mouth. Two light-sensitive spots at the anterior end resemble a pair of crossed eyes, and two projections on either side of the head are areas sensitive to chemical stimuli such as emanate from meat or other food. Within the mesoderm is a fairly complex musculature, which enables the animal, when disturbed, to move rapidly with a sort of loping motion. Planarians have an extensive excretory system and complex reproductive organs.

The phylum also includes the parasitic flukes and tapeworms, which often have complex life cycles involving several hosts in succession.

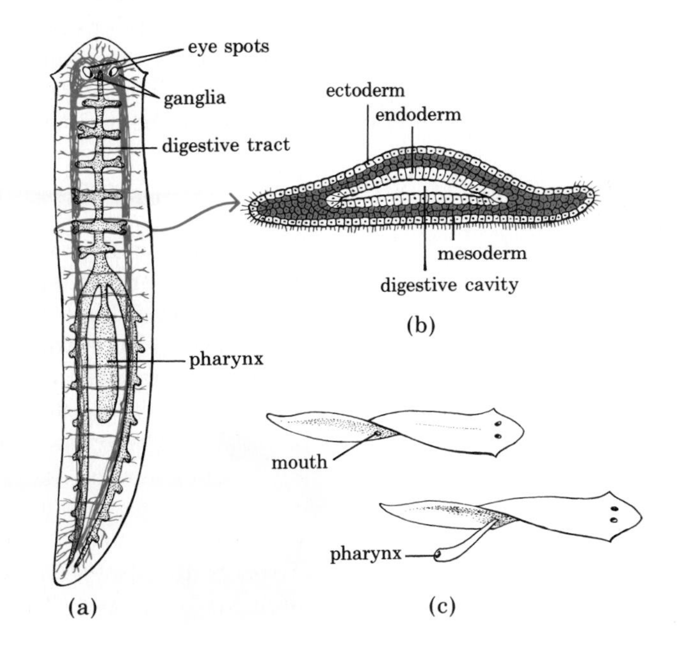

19–16 *Example of a flatworm: the freshwater planarian.* (a) *The nervous system is indicated in color. Note that some of the fibers are aggregated into two cords, one on each side of the body. Clusters of nerve cells in the head, known as ganglia, are the beginnings of a brain. The eye spots are light-sensitive areas.*

(b) *Planaria, like other flatworms, but unlike coelenterates, have three layers of body tissues.*

(c) *A carnivore, the planarian feeds by means of its extensible pharynx.*

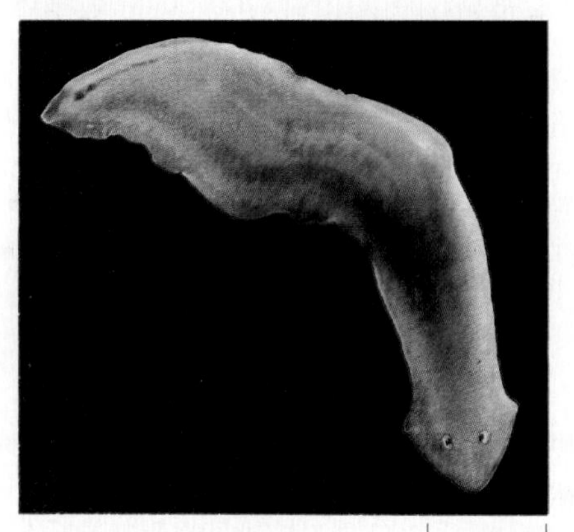

(a)

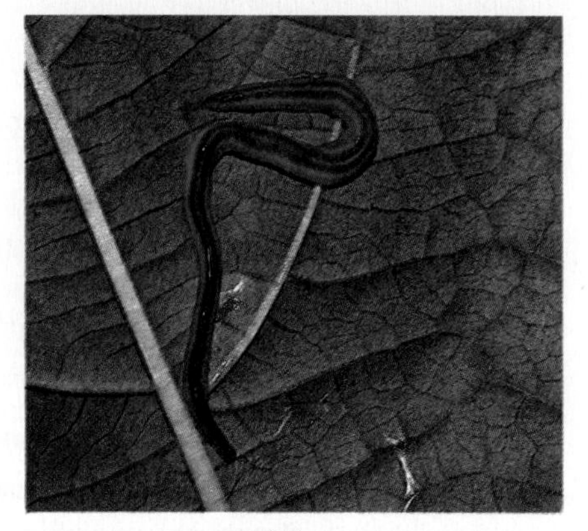

(b)

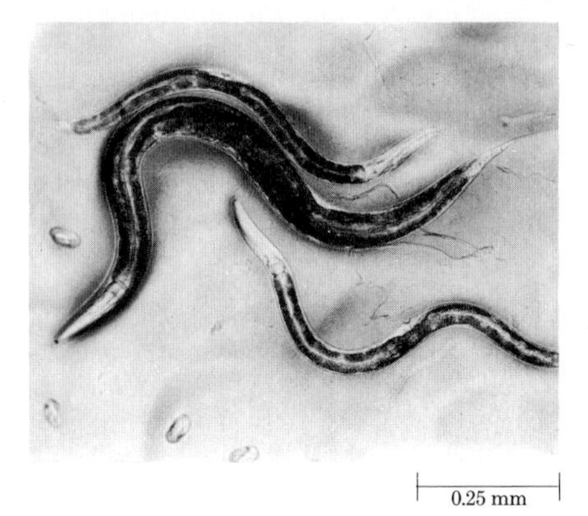

(c)

19–17 (a) *A planarian. The two eye spots are actually light-sensitive patches. A planarian can see only about as well as you can with your eyes closed.*

(b) *A ribbon worm. These worms range in length from less than 2 centimeters to 30 meters and are virtually all the colors of the spectrum. All members of the phylum, however, have thin bodies (seldom more than ½ centimeter thick), a mesoderm, a mouth-to-anus digestive tract, and a long, muscular tube that can be thrust out to grasp prey.*

(c) *An adult and two immature nematodes. Strands of shed cuticle can be seen extending from the right-hand side of the adult nematode. Four eggs are visible at the lower left of the micrograph.*

Ribbon Worms: Phylum Rhynchocoela

The ribbon worms, although a small phylum, are of special interest to biologists attempting to reconstruct the evolution of the invertebrates. They appear to be closely related to the flatworms, but with an important difference: They have a one-way digestive tract beginning with a mouth and ending with an anus. This is a far more efficient arrangement than the one-opening digestive system of the coelenterates and flatworms. In the one-way tract, food moves assembly-line fashion, with the consequent possibility that various segments of the tract can be specialized for different stages of digestion.

This phylum is called Rhynchocoela ("beak" plus "hollow") because these worms are characterized by a long, retractile, slime-covered tube (proboscis). The proboscis, sometimes armed with a barb, seizes prey and draws it to the mouth, where it is engulfed.

Roundworms: Phylum Nematoda

There are probably several hundred thousand species of roundworms, but only a small percentage of these have been described and named. Most are free-living microscopic forms. It has been estimated that a spadeful of good garden soil usually contains about a million nematodes. Some are parasites; all species of plants and animals are parasitized by at least one species of nematode. Humans are hosts to about 50 species.

Nematodes have a three-layered body plan and a tubular gut with a mouth and an anus. Between the endoderm and the mesoderm, they have a body cavity known as a pseudocoelom. Six other minor phyla, mostly small, wormlike animals, have body plans based on the pseudocoelom. Nematodes are unsegmented and are covered by a thick, continuous cuticle, which is molted periodically as they grow.

Parasitic Worms

Throughout the world, parasitic worms are a major cause of death and disability. Estimates of the numbers of people affected are staggering: 200 million people infected by the blood fluke *Schistosoma,* a flatworm; 650 million people infected by *Ascaris* and 450 million people by *Ancylostoma* (hookworm), both intestinal nematodes; 250 million people infected by *Filaria* and 50 million by *Onchocerca,* both nematodes with minute larvae known as microfilariae. The diseases caused by these organisms and others are epidemic in parts of China, the Middle East, Africa, Latin America, and the Caribbean. Pockets of these diseases, as well as diseases caused by other parasitic worms, also exist in the United States.

All of these parasitic worms have a life cycle with two or more hosts, one of which is the human being. *Schistosoma,* the cause of schistosomiasis, for example, has a life cycle involving aquatic snails and humans. The worms emerge from the snails and enter a human host, for example, through the bare feet of a person working in a rice paddy. They work their way to the blood vessels of the intestine and liver, where they produce vast numbers of eggs. The eggs have sharp spines that damage the tissues in which they are laid and through which they pass as they leave the body in the feces. If the egg-contaminated human feces reach water in which the snails live, the eggs enter the snails for the next stage of their life cycle. There they hatch, develop, and emerge to infect another human host. The damage inflicted by *Schistosoma* on the human host can be fatal, but not before the parasite has ensured its return passage to a snail and, ultimately, to its next human host.

Onchocerca volvulus, the cause of a disease that occurs in fertile regions of west Africa and in the coffee-growing highlands of Mexico and Guatemala, does not kill but rather maims its victims. The parasite is transmitted by the blackfly *Simulium.* When the fly bites, it transmits infective larvae known as filariform larvae. These larvae lodge in the tissues under the skin, mature, and mate. The eggs hatch within the female parasite and are released as microfilariae (microscopic larvae). The female can produce more than 2,000 microfilariae per day, and she can live in the host human for 15 to 20 years. When a blackfly bites an infected human, it ingests microfilariae, which then pass through a series of developmental stages within the blackfly, resulting in the filariform larvae. When the fly bites its next victim, the cycle begins again. The human victim soon develops a chronic inflammation of the skin, which becomes so changed in texture that it is known as "pachyderm skin." The most serious consequence, however, is that the microfilariae invade the eye, producing blindness.

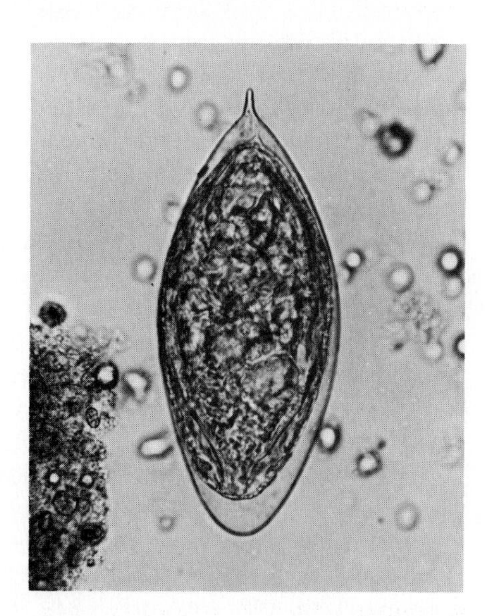

A Schistosoma *egg. The eggs hatch in aquatic snails, from which the immature worms emerge. They grow to maturity in the blood vessels of a human host. The adult worms live in permanent male-female pairs, with the very small female occupying a special groove in the body of the much larger male. Each pair of worms produces several thousand fertilized eggs a day.*

Combating the diseases caused by parasitic worms is not an easy task. In principle, these diseases could be eliminated by blocking transmission of the worms from host to host (through, for instance, improved sanitation), by eradicating the nonhuman hosts, or by killing the parasites themselves. In practice, the investments required for education, sanitation projects, eradication of alternate hosts, and treatment of human victims are often beyond the reach of the countries where the diseases are most prevalent. Even if money were no problem, the technical difficulties would remain formidable. For example, the worms often have a number of hosts (*Schistosoma* infects not only humans but also livestock and various wild animals), a host such as the blackfly or the aquatic snail is widely distributed, and the few drugs that are effective against the worms are quite toxic to humans.

You may recall (page 238) that chemotherapeutic approaches to bacterial diseases are effective because bacteria have various enzymes not found in eukaryotic cells. Compounds that block unique and essential bacterial enzymes can kill the bacteria without harm to the host cells. New research is revealing that parasitic worms also have unique enzymes and metabolic pathways. Most of these are directly related to the parasitic life style, such as migration within the body of the host, the mechanisms by which the parasites attach themselves to the host, and the massive production of eggs. And, it turns out that some parasitic worms have their own unique enzymes for such an essential process as glycolysis. There is a new hope that, with increased understanding of the basic biology of the parasitic worms, it may be possible to develop specific drugs that will act only on particular aspects of worm metabolism and that will not have severe side effects on the human victims.

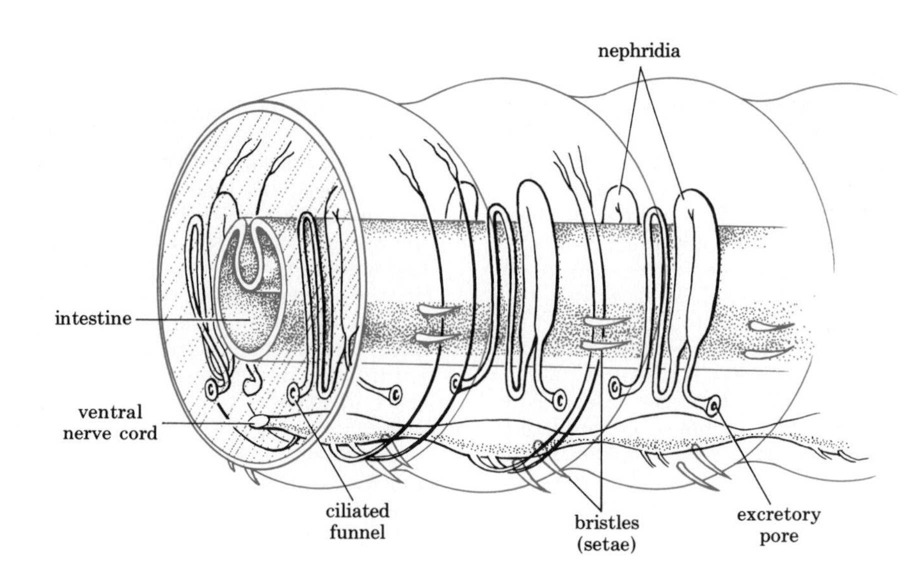

19-18 *Three segments of the earthworm, an annelid. On each segment are four pairs of bristles, which are extended and retracted by special muscles. These are used by the worm to anchor one part of its body while it moves another part forward. Two excretory tubes, or nephridia, are in each segment (except the first three and the last). Each nephridium really occupies two segments since it opens externally by a pore in one segment and internally by a ciliated funnel in the segment immediately in front of it. The intestine, nephridia, and other internal organs are suspended in a large fluid-filled cavity, the coelom, which also serves as a hydrostatic skeleton.*

Segmented Worms: Phylum Annelida

This phylum contains almost 9,000 species of marine, freshwater, and soil worms, including the familiar earthworm. The term annelid means "ringed" and refers to the most distinctive feature of this group: the division of the body into segments, visible as rings on the outside and with partitions on the inside (Figure 19–18). This segmented pattern is found in a modified form in arthropods, too, such as dragonflies, millipedes, and lobsters, which are thought to have evolved from the same ancestors that gave rise to modern annelids. Although the earthworm is probably the most familiar of the annelids, it is atypical because of its much-reduced appendages (Figure 19–19).

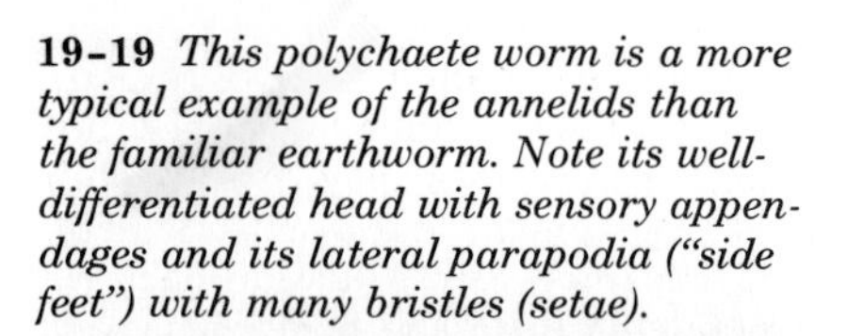

19-19 *This polychaete worm is a more typical example of the annelids than the familiar earthworm. Note its well-differentiated head with sensory appendages and its lateral parapodia ("side feet") with many bristles (setae).*

The annelids have a three-layered body plan, a tubular gut, and a well-developed circulatory system that transports oxygen (diffused through the skin) and food molecules (from the gut) to all parts of the body (Figure 19–20). The excretory system is made up of specialized paired tubules, nephridia, which occur in each segment of the body except the first three and the last. Annelids have a nervous system and a number of special sensory cells, including touch cells, taste receptors, light-sensitive cells, and cells concerned with the detection of moisture.

In the flatworms and ribbon worms, the mesoderm is packed solid with muscle and other tissues, but in the annelids there is a fluid-filled cavity, the *coelom* (pronounced see-loam), in this middle layer (Figure 19–21d). (Note that the term coelom, although it sounds like "coelenteron" and comes from the same Greek root, meaning "cavity," refers quite specifically to a cavity *within* the mesoderm, whereas the coelenteron is a digestive cavity lined by endoderm.) Within the coelom, the gut is suspended by double layers of mesoderm known as mesenteries. The fluid in the coelom constitutes a hydrostatic skeleton for the annelid, stiffening the body in somewhat the same way water pressure stiffens and distends a fire hose.

Although the opening of a cavity within the mesoderm may seem less dramatic than other evolutionary innovations, it is extremely important. Within such a space, organ systems can bend, twist, and fold back on

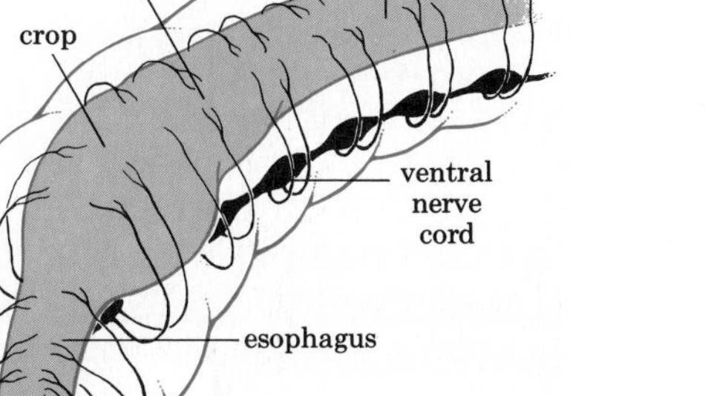

19-20 (a) *The digestive tract of an earthworm. The mouth leads into a muscular pharynx, which sucks in decaying vegetation and other material. These are stored in the crop and ground up in the gizzard with the help of soil particles. The rest of the tract is a long intestine (gut) in which food is digested and absorbed.*

(b) *The circulatory system of the earthworm is made up of longitudinal vessels running the entire length of the animal, one dorsal and several ventral. Smaller vessels (the parietal vessels) in each segment collect the blood from the tissues and feed it into the muscular dorsal vessel, through which it is pumped forward. In the anterior segments are five pairs of hearts—muscular pumping areas in the blood vessels—whose irregular contractions force the blood downward through the parietal vessels to the ventral vessels, from which it returns to the posterior segments. The arrows indicate the direction of blood flow.*

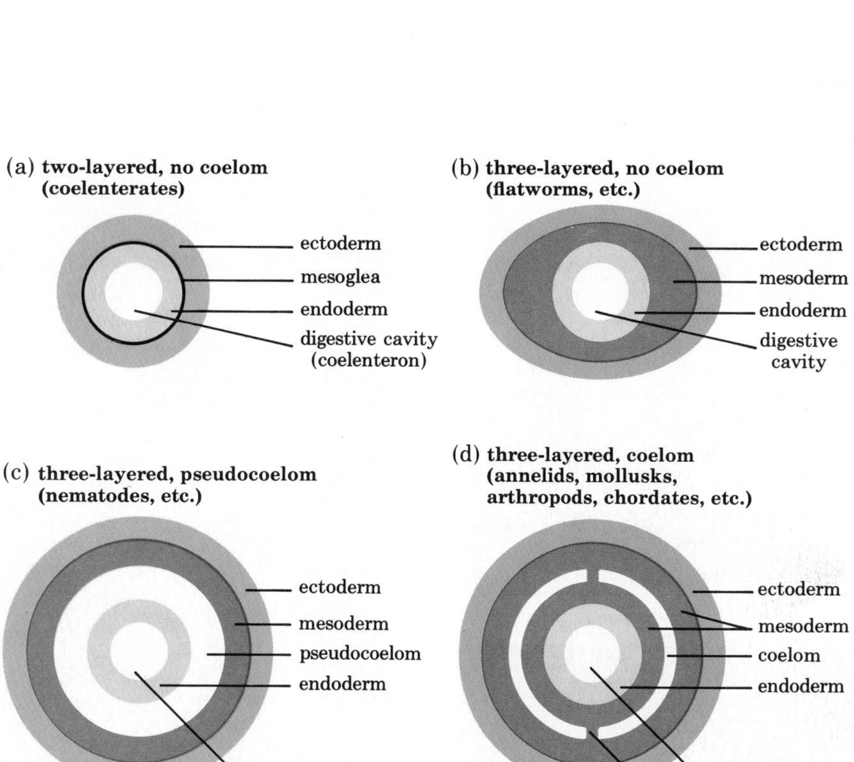

19-21 *Basic body plans of the animal phyla, as shown in cross section.* (a) *A body that consists of only two tissue layers is characteristic of coelenterates.*

(b) *Flatworms and ribbon worms have three-layered bodies, with the layers closely packed on one another.*

(c) *Nematodes have three-layered bodies with a pseudocoelom between the endoderm and mesoderm.*

(d) *Annelids and most other animals, including vertebrates, have bodies that are three-layered with a cavity, the coelom, within the middle layer (mesoderm). The mesodermal mesenteries suspend the gut within the body wall.*

themselves, increasing their functional surface areas and filling, emptying, and sliding past one another, surrounded by lubricating coelomic fluid. Consider the human lung, constantly expanding and contracting in the chest cavity, or the 6 or 7 meters of coiled human intestine; neither of these could have evolved until the coelom made room for them.

Mollusks: Phylum Mollusca

The mollusks constitute one of the largest phyla of animals, both in number of species and in number of individuals. They are characterized by soft bodies within a hard, calcium-containing shell. In some forms, however, the shell has been lost in the course of evolution, as in slugs and octopuses, or greatly reduced in size, as in squids. There are three major classes of mollusks: (1) the gastropods, such as the snails, whose shells are generally in one piece; (2) the bivalves, including the clams, oysters, and mussels, which have two shells joined by a hinge ligament; and (3) the cephalopods, the most active and intelligent of the mollusks, including the cuttlefish, squids, and octopuses.

Although the mollusks are diverse in size and shape, they all have the same fundamental body plan. There are three distinct body zones: a head-foot, which contains both the sensory and the motor organs; a visceral mass, which contains the organs of digestion, excretion, and reproduction; and a mantle, which hangs over and enfolds the visceral mass and which secretes the shell. The mantle cavity, a space between the mantle and the visceral mass, houses the gills; the digestive, excretory, and reproductive systems discharge into it.

(a)

(b)

(c)

(d)

19-22 *Representative mollusks.* (a) *A bivalve. The blue eyes of this scallop are visible among its tentacles.*
(b) *A land-dwelling gastropod. The shell, which is secreted by the mantle and grows as the soft body grows, covers and protects the visceral mass. The head contains sensory organs, including two eyes at the tips of the longer tentacles.*
(c) *A squid. Jet-propelled, the squid is moving to the left. If you look closely, you can see its siphon just below and to the right of its eye.*
(d) Octopus macropus. *Note its well-developed eyes and the suckers on the undersurface of the arms.*

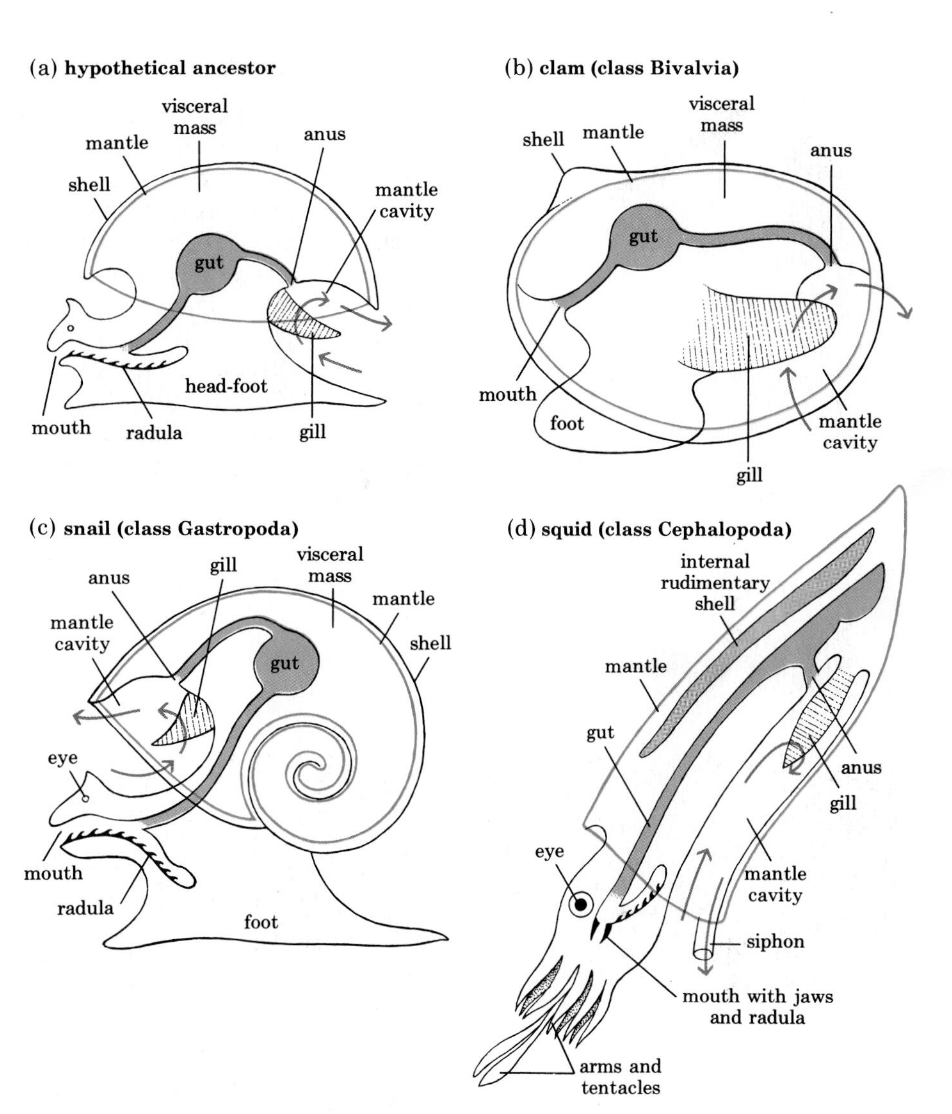

19-23 *Mollusks are characterized by soft bodies composed of a head-foot, a visceral mass, and a mantle, which can secrete a shell. They exchange gases with the surrounding water through gills, except for the land snails, in which the mantle cavity has been modified for air breathing. A hypothetical primitive mollusk is shown in* (a).

The three major modern classes are the bivalves, such as the clam (b), *which are generally sedentary and feed by filtering water currents, created by beating cilia, through large gills; the gastropods, exemplified by the snail* (c), *in which the visceral mass has become coiled and rotated through 180° so that mouth, anus, and gills all face forward and the head can be withdrawn into the mantle cavity; and the cephalopods, such as the squid* (d), *in which the head is modified into a circle of arms and part of the head-foot forms a tubelike siphon through which water can be forcibly expelled, providing for locomotion by jet propulsion. The arrows indicate the direction of water movement.*

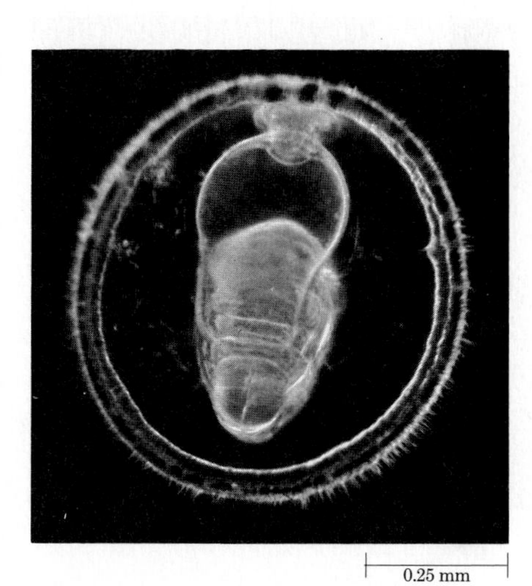

19-24 *The trochophore larva. Although their adult forms are very different, certain annelids, mollusks, and arthropods have larvae of this type. This particular larva will develop into a polychaete worm.*

A characteristic organ, found only in this phylum, is the radula. This tooth-bearing, tonguelike organ is variously used by different kinds of mollusks to scrape off algae, to drill holes in the shells of barnacles or of other mollusks, or to aid in the ingestion of prey animals. It is present in all mollusks except the bivalves.

The molluscan circulatory system, which includes a chambered heart, is the most efficient circulatory system of the invertebrates. The nervous systems of the gastropods and bivalves are simple, but those of the cephalopods are more complex. The octopus, in particular, has a highly developed nervous system and a brain and pair of eyes that rival those of vertebrates in complexity and are strikingly similar in both structure and function.

Although the annelids and the mollusks are quite different in their basic body plans, there are some evolutionary links between them. One of these is the trochophore larva (Figure 19-24).

(a)

(b)

(c)

(d)

(e)

(f)

19-25 *Some arthropods.* (a) *A millipede. Arthropods are segmented, like the annelids from which they presumably arose. However, unlike annelids, adult arthropods have rigid jointed exoskeletons and appendages.* (b) *This crayfish, a familiar crustacean, is eating a worm. Note the compound eyes and the many appendages.* (c) *A wolf spider. These common spiders have eight eyes, four small ones in a row beneath four large ones. Note that the segments of the body have become fused into two regions—cephalothorax ("head-thorax") and abdomen.*

Examples of insects include (d) *a soldier beetle;* (e) *a giant mesquite bug; and* (f) *everybody's favorite beetle, a ladybug.*

Arthropods: Phylum Arthropoda

The arthropods, "joint-footed" animals, are divided into three major classes: crustaceans (crabs and lobsters), arachnids (spiders, ticks, mites, and scorpions), and insects. Smaller classes include such organisms as centipedes, millipedes, and horseshoe crabs.

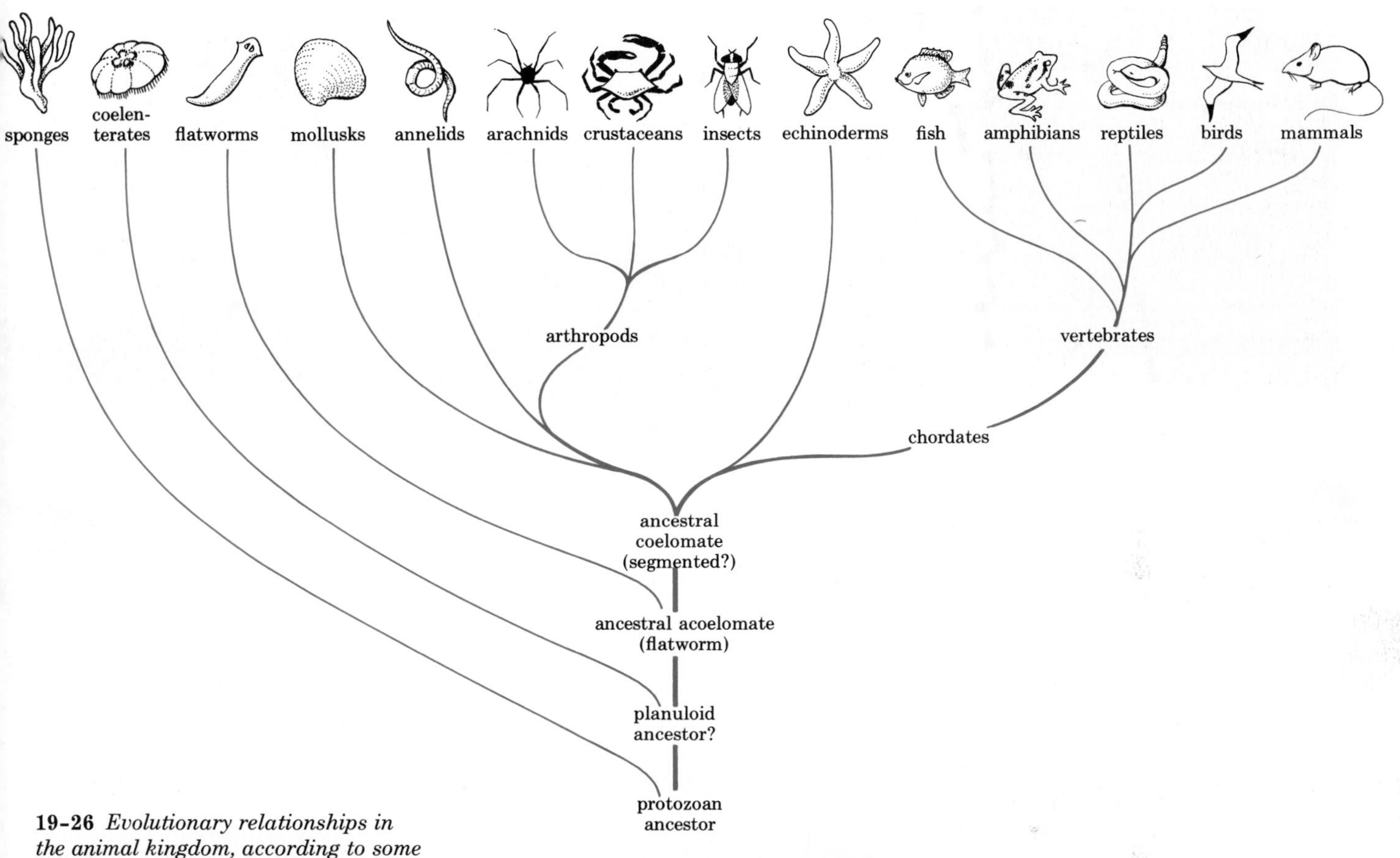

19-26 *Evolutionary relationships in the animal kingdom, according to some hypotheses. Relationships are based on structural similarities, such as the segmentation in annelids and arthropods, and also on resemblances among larval forms and in developmental patterns. The first fossils available are from the Cambrian period and, by this time, the major groups of invertebrates had already diverged.*

Arthropoda is by far the largest phylum of animals, with almost 1 million species classified to date. There are more species of insects than of all other animals combined; of beetles alone, some 275,000 have been classified. Some experts estimate that, including the unclassified forms, there are more than 10 million living insect species.

The enormous success of the insects appears to be related to the high degree of specialization of the individual species. Insects are usually so selective about where they live and how they eat that many different species can live noncompetitively in a very small area—such as on a single plant or in or on one small animal.

Arthropod Characteristics

All arthropods are segmented, a characteristic that suggests a common ancestry with the annelids (Figure 19-26). In the evolution of the arthropods, the various segments of the body have become specialized in different ways. The most anterior segments form the head, the next constitute the thorax, and the final segments are the abdomen (Figure 19-27, page 270). In the more complex arthropods, the segments of the head and thorax have become fused, but the basic segmented pattern is clearly evident in the adult forms of more primitive arthropods—such as centipedes and millipedes—and the immature stages of many (witness the caterpillar).

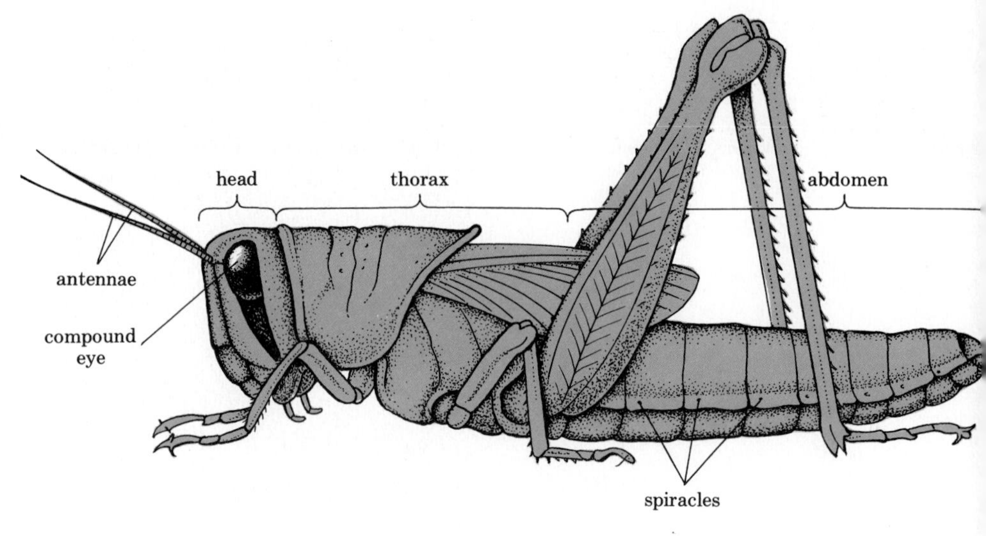

19–27 *In the grasshopper, an insect, the head consists of six fused segments that have appendages specialized for tasting and biting. Each of the three segments of the thorax carries a pair of legs (three pairs in all), and two of them carry wings (in the grasshopper, the forewings are hardened as protective covers). The spiracles in the abdomen open into a network of chitin-lined tubules through which air circulates to various tissues of the body. This sort of tubular breathing system is found only among insects and some other land-dwelling arthropods.*

In some of the arthropods, such as the centipedes, which have a pair of legs on almost every segment of the clearly wormlike body, the jointed appendages are uniform in size and structure. More typically, arthropods have highly differentiated appendages, specialized for walking, swimming, feeling, feeding, chewing, biting, and other such functions, depending on the species (Figure 19–28).

Arthropods have a hard outer covering, or *exoskeleton*, containing chitin. It deters predators and, in the land forms, protects the animals from drying out. The exoskeleton is many-jointed and has muscles attached to it. As different parts of the animal move, different parts of the exoskeleton also move. The exoskeleton, however, covers the animal completely and does not grow. As a consequence, arthropods must molt, a process by which the old exoskeleton is discarded and a new and larger one is formed (see Figure 3–7, page 40). At molting time, the arthropod secretes an enzyme that dissolves the inner layer of the exoskeleton, and a new skeleton, not yet hardened, is formed beneath the old one. The animal wriggles out from the old skeleton, which splits open. After emerging, the arthropod expands rapidly by taking in air or water, stretching the new exoskeleton before it hardens.

The anterior section of the arthropod gut is a chitin-lined extension of the exoskeleton. It serves primarily to grind food into smaller particles. Large digestive glands secrete enzymes into the midsection, where digestion takes place.

The circulatory system of arthropods is an open one, in which the tubular heart pumps blood through vessels into blood sinuses, or cavities. Blood returns from the sinuses to the heart through valved openings.

The insects and some arachnids have an unusual respiratory system consisting of chitin-lined tracheae (air ducts) that pipe air directly into various parts of the body. Air flow is regulated by the opening and closing of special pores (spiracles) in the exoskeleton. Terrestrial arthropods that do not have tracheae have book lungs, so called because they are constructed in a series of leaflike plates. Both tracheae and book lungs are found only in this phylum.

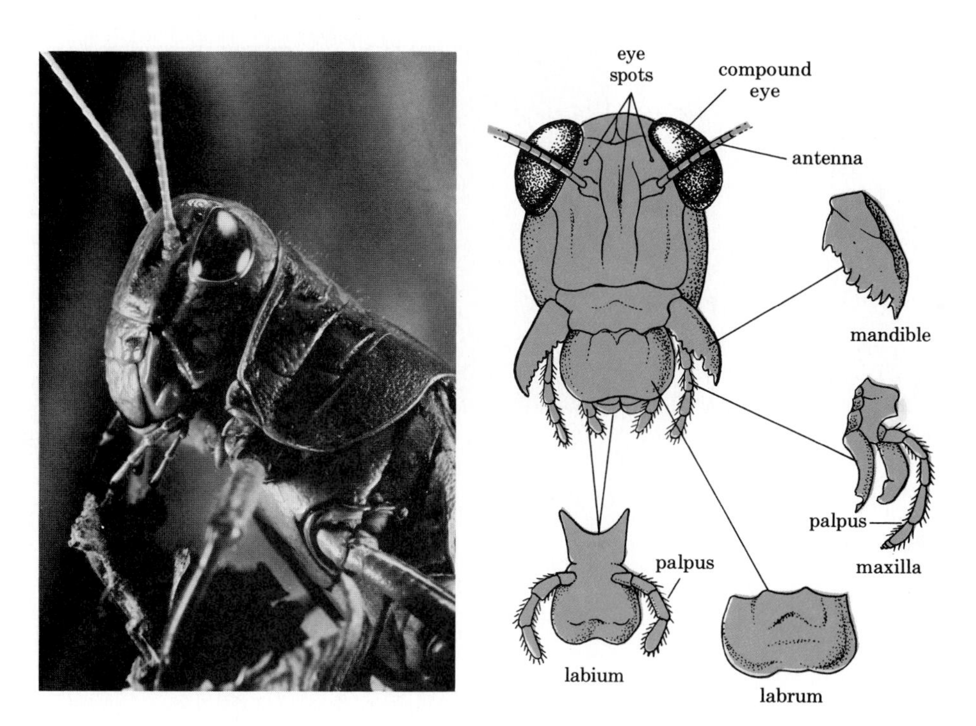

19-28 *Mouthparts of a grasshopper. The mandibles are crushing jaws. The labium and the labrum are the lower and upper lip. The maxillae move food into the mouth, and the palpi assist in tasting.*

Excretion in terrestrial forms is by means of tubes, called Malpighian tubules, attached to and emptying into the hindgut. Malpighian tubules are also an exclusively arthropod characteristic.

The nervous system of the arthropods is complex, making possible the intricate and finely tuned movements involved in activities such as flight, mating in midair, and building webs and hives. Arthropods have a number of extremely sensitive sensory organs, such as the compound eye (Figure 19-29) found among crustaceans and insects. They also have a complex hormone-producing system, which plays a major role in molting.

Some arthropods, especially insects such as ants, honey bees, and termites, live in highly organized societies. The social organization of honey bees is described in Chapter 41.

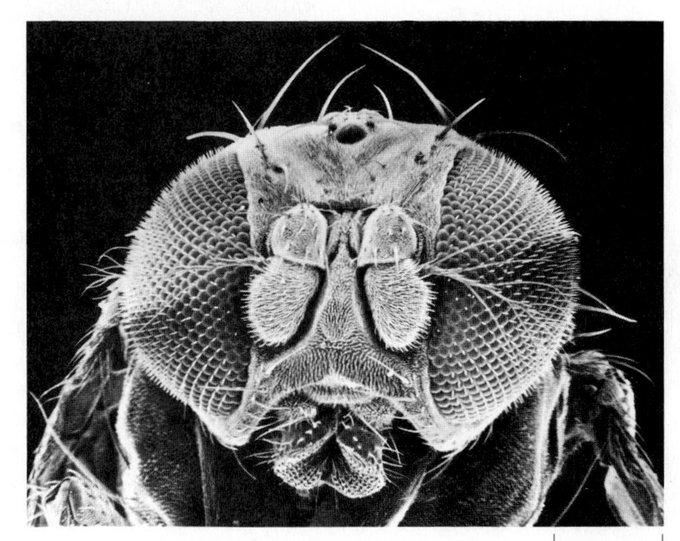

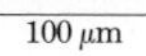

19-29 *The head of* Drosophila, *as shown by the scanning electron microscope. Note the large compound eyes on either side of the head. Although insect eyes cannot change focus, they can define objects only a millimeter from the lens, a useful adaptation for an insect.*

(a)

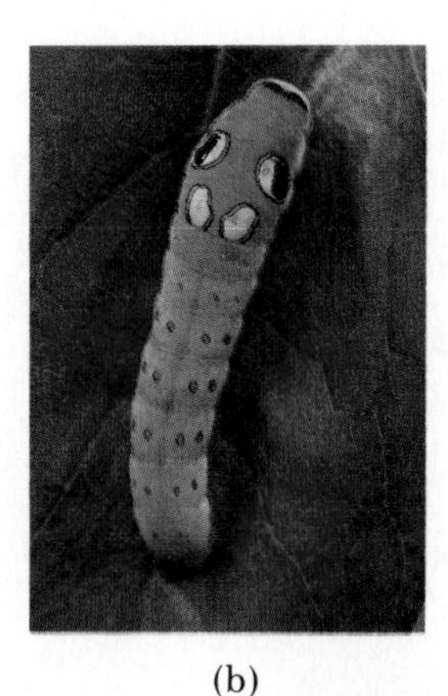
(b)

(c)

(d)

19-30 *Developmental stages of a spicebush swallowtail butterfly:* (a) *the egg;* (b) *caterpillar, the larval form;* (c) *pupa, the resting form in which metamorphosis takes place;* (d) *a butterfly, the adult, reproductive form.*

Insect Development

Most insects go through definite developmental stages. In some species the young, although sexually immature, looks like a small copy of the adult; it grows larger by a series of molts until it reaches full size. In others, such as the grasshopper, the newly hatched young is wingless and somewhat different in proportions from the adult, but it is otherwise similar.

Almost 90 percent of all insects, however, undergo a complete metamorphosis, so that the adult is entirely different from the immature form. The immature feeding forms are all properly referred to as *larvae*, although they are also commonly known as caterpillars, grubs, or maggots, depending on the species. Following the larval period, the insect undergoing complete metamorphosis enters an outwardly quiescent pupal stage, in which extensive remodeling of the organism occurs. The adult insect emerges from the pupa. Thus the insect that undergoes complete metamorphosis exists in four different forms in the course of its life history: the egg, the larva, the pupa, and the adult (Figure 19-30). The larvae and adults are so different that they usually do not compete for food or other resources, another example of the extreme specialization found in the insect world.

19-31 *Sand dollar. Note the five-part body plan. Like other echinoderms, the sand dollar has a water vascular system and tube feet. The mouth is on the lower surface.*

Echinoderms: Phylum Echinodermata

The echinoderms ("prickly skins") include the sea urchins, sand dollars, sea cucumbers, and sea lilies. Adult echinoderms are radially symmetrical, like the coelenterates. The most familiar of the echinoderms are the starfish, which have the spiny skin from which the phylum derives its name. Many species of starfish have five arms, but some have more. Other members of the phylum also have bodies arranged in fives. The water vascular system (Figure 19-32) supports the clinging and pulling activities of the tube feet, with which many members of the phylum cling, move, and attack their prey.

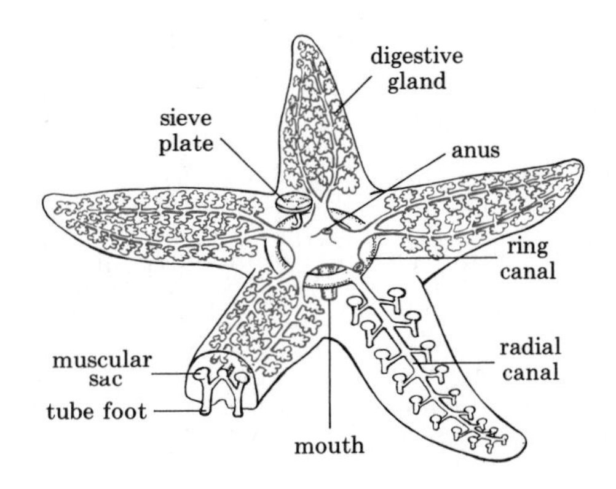

19-32 *The water vascular system of the starfish supports its means of locomotion. Water enters through minute openings in the sieve plate and is drawn down a tube to the ring canal. Five radial canals, one for each arm, connect the ring canal with many pairs of tube feet, which are hollow, thin-walled cylinders ending in suckers. Each tube foot connects with a rounded muscular sac. When the sac contracts, the water in it, prevented by a valve from flowing back into the radial canal, is forced under pressure into the tube foot. This stiffens the tube, making it rigid enough to walk on and extends the foot until it attaches to the substrate by its sucker. The muscles of the foot then contract, forcing the water back into the sac and creating the suction that holds the foot to the surface. If the tube feet are planted on a hard surface, such as a rock or a clamshell, the collection of tubes will exert enough force to pull the starfish forward or to open the clam.*

Although the adults are radially symmetrical, echinoderm larvae are bilaterally organized. The fossil record shows that radial symmetry is a late, secondary development in the evolution of the group. In the development of these larvae, the first opening of the digestive tract of the embryo becomes the anus, while the mouth breaks through secondarily at the other end of the larval gut. Animals with such a pattern of early development are known as deuterostomes ("second the mouth"), in contrast to the protostomes such as mollusks, annelids, and arthropods, in which the mouth develops first. The primitive chordates from which the vertebrates arose also are deuterostomes. For this reason, among others, the vertebrates are believed to have a closer evolutionary relationship with this group than with the other invertebrate phyla (see Figure 19–26, page 269).

Chordates: Phylum Chordata

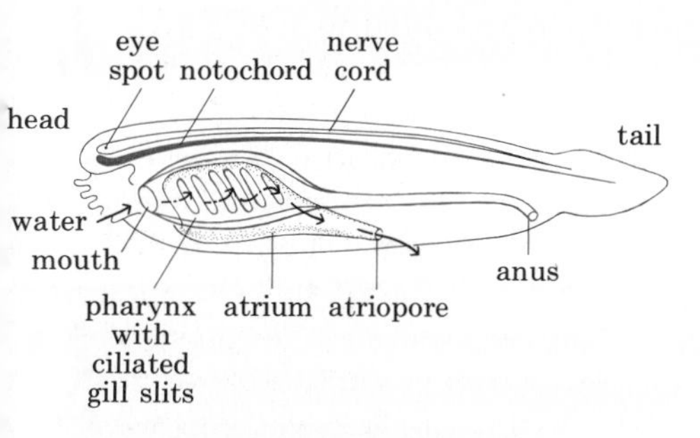

19-33 Branchiostoma, *the lancelet, exemplifies three distinctive chordate characteristics: (1) a notochord, the dorsal rod that extends the length of the body; (2) a dorsal, tubular nerve cord; and (3) pharyngeal gill slits.*

Mammals and other vertebrates belong to a large phylum of animals comprising the chordates. *Branchiostoma,* the lancelet, is a good example of a chordate (Figure 19–33). The lancelet is a small, blade-shaped, semi-transparent animal found in shallow marine waters all over the warmer parts of the world. Although it can swim very efficiently, it spends most of its time buried in the sandy bottom, with only its mouth protruding above the surface.

This animal has three features that identify it as a member of the chordate phylum. The first is the *notochord,* a rod that extends the length of the body and serves as a firm but flexible axis. The notochord is a structural support. Because of it, *Branchiostoma* can swim with strong undulatory motions that move it through the water with a speed unattainable by the flatworms or aquatic annelids.

The second chordate characteristic is the hollow nerve cord, a tube that runs beneath the dorsal surface of the animal, above the notochord. (The principal nerve cords in the invertebrates, by contrast, are solid and are almost always near the ventral surface.)

The third characteristic is a pharynx with gill slits. The pharyngeal apparatus becomes highly developed in fish, in which it serves a respiratory function, and traces of the gill slits remain even in the human embryo. In *Branchiostoma,* the pharynx serves primarily for collecting food. The cilia on the sides of the pharyngeal gill slits pull in a steady current of water, which passes through the slits into a chamber known as the atrium and then exits through the atriopore. Food particles are collected in the sieve-like pharynx.

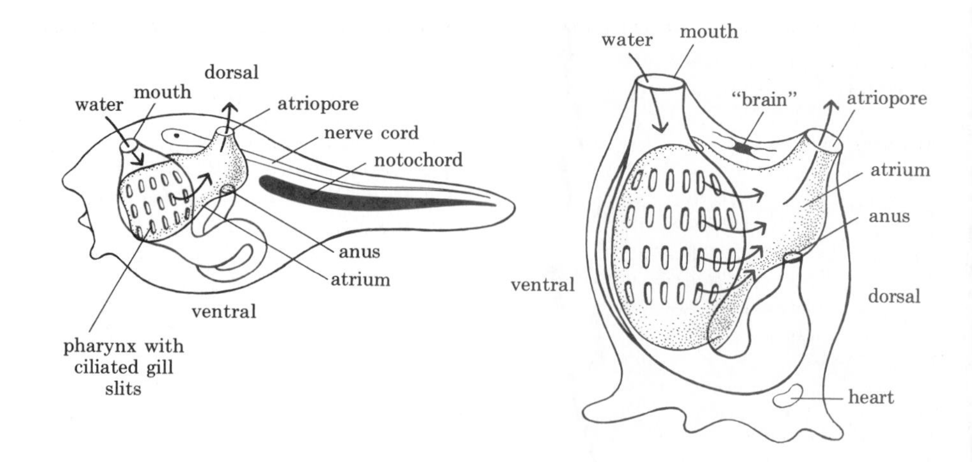

19-34 *Two stages in the life of a tunicate. At the left is the larva; the adult form is shown on the right. After a brief free-swimming existence, the larva settles to the bottom and attaches at the anterior end. Metamorphosis then begins. The larval tail, with the notochord and dorsal nerve cord, disappears, and the animal's entire body is turned 180°. The mouth is carried backward to open at the end opposite that of attachment, and all the other internal organs are also rotated back.*

The chordates are believed to have arisen from a group of organisms that resembled the modern tunicates. Tunicates, in their larval form, have chordate characteristics (Figure 19-34).

The Principal Chordates: The Vertebrates

The vertebrates constitute the largest and most familiar group of chordates. All vertebrates have a vertebral column, or backbone, as their structural axis. This is a flexible bony support that develops around the notochord, supplanting it entirely in most species. The backbone is made up of bony elements, the vertebrae, which encircle the nerve cord along the length of the spine. The brain is similarly enclosed and is protected by bony skull plates. Between the vertebrae are cartilaginous disks, which give the vertebral column its flexibility. Associated with the vertebrae are a series of muscle segments by which sections of the vertebral column can be moved separately. This segmented pattern persists in the embryonic forms of higher vertebrates but is largely lost in the course of development.

One of the great advantages of a bony endoskeleton, as compared with the exoskeletons of invertebrates, is that it is composed of living tissue that can grow with the animal. In the developing vertebrate embryo, the skeleton is largely cartilaginous, with the bone gradually replacing cartilage in the course of maturation (Figure 19-35). The growing portions of the bones characteristically remain cartilaginous until the animal reaches its full size. In addition to its powers of growth, bone can also repair itself.

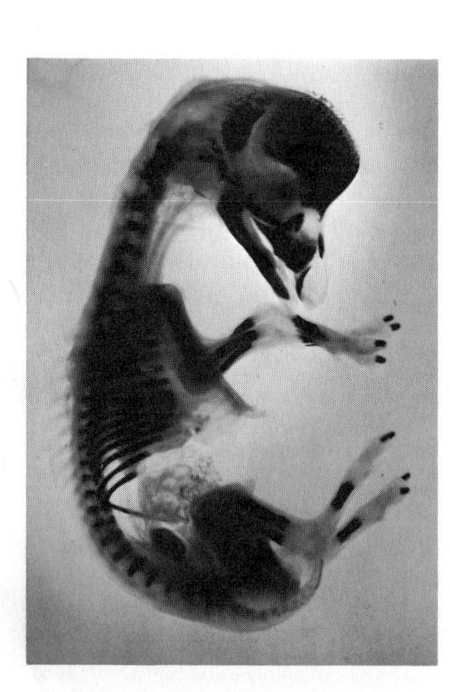

19-35 *An elk fetus. The bones have been stained to show them more clearly, so that you can see the extent to which the skeleton is still cartilaginous. Notice the legs, for example. Only the dark areas are bone; these will gradually grow and replace the cartilage as the animal matures.*

There are seven classes of vertebrates: the fish (comprising three classes), the amphibians, the reptiles, the birds, and the mammals.

Fish The first fish were jawless and had a strong notochord running the length of their bodies. Today these jawless fish (class Agnatha), once a large and diverse group, are represented only by the hagfish and the lampreys. They have a notochord throughout their lives, like *Branchiostoma,* and a cartilaginous skeleton.

The sharks (including the dogfish) and skates, the second major class of fish, also have a completely cartilaginous skeleton. Their skin is covered with small, pointed teeth (denticles), which resemble vertebrate teeth structurally and give the skin the texture of coarse sandpaper.

The third major class is the osteichthyans, the fish with bony skeletons. This group includes the trout, bass, salmon, perch, and many others—most of the familiar freshwater and saltwater fish.

Animal Locomotion

By and large, animals move and plants sit still. The reason is that animals are heterotrophic and must search out and capture their food, whereas plants are autotrophic and can live wherever there is sufficient sunlight, water, and air.

That animals must be mobile has many consequences for their structure and is one of the main reasons that most animals are more complex than members of the kingdom Plantae. One of the best ways of appreciating the evolutionary increase in the complexity of animal phyla is in terms of the adaptations that permit animals to get about and find their food (or to escape animals that covet *them*). The simplest kind of motion directed by an organism involves the cilia and flagella of some very small animals, for example, the planula larva of the coelenterates. Such a method is not practical for larger organisms, however, and bundles of specialized muscle fibers are the chief organs of locomotion in most animals. Still, animals as large as flatworms rely in part on cilia to creep along.

In addition to pushing against the ground, the water, or the ocean floor, muscles must also have something to work against in the animal's body—otherwise the body would just bend in the direction of contraction and a floppy, uncoordinated motion would result. (It is rather like using a tool such as a screwdriver. No matter how much the head were pushed against the slot in the screw, this would accomplish nothing if the screwdriver itself did not also resist twisting.) There are two basic systems for giving animals rigidity: the hydrostatic skeleton and stiffened appendages. The first, which is typical of wormlike organisms, is simply a fluid-filled tube or space that resists bending. This resistance causes the body to return to its original shape after the muscles have contracted. The primary significance of the pseudocoelom and the coelom is that they create just such a bend-resistant tube within an animal's body. The pseudocoelom, which is found in nematodes, makes possible a great advance over the simple, rather flaccid movements of flatworms and most coelenterates. Since nematodes have only one sort of muscle fiber, running the length of the body, their motion is a thrashing one, resulting from contractions first on one side and then on the other.

The ultimate in the hydrostatic skeleton is that found in the annelids. These worms have segmented muscles that run both lengthwise and around the body and that can be contracted independently to lengthen or shorten the body. Moreover, in each segment the coelom is divided into left and right compartments, which are separated from those ahead and behind. This arrangement allows exquisite control over the movements of small parts of the body. In earthworms, lengthening the body places the bristles of the extended part of the body ahead of the rest of the animal; the bristles anchor the worm while contractions that shorten the posterior segments pull the rest of the body forward. Earthworm locomotion is a special case among the annelids, however. Many of the marine annelids, or polychaetes, augment hydrostatic movements with swimming motions of the lateral appendages of their bodies, the parapodia.

Movements can be more precise when force is brought to bear on a small area, as it is by the much more elaborate legs of arthropods and chordates. But appendages pose a real problem: They must be stiffened by some substance if they are to support weight and resist bending. The parapodia of polychaetes are laced with blood vessels that give some hydrostatic stiffening. In the arthropods, it is the chitinous exoskeleton that gives firmness and something for internal muscles to attach to and pull against. In chordates, the skeleton is brought inside the animals, but the principle is the same. It is interesting to note, however, that the notochord, the basis of the vertebrate endoskeleton, is really a tube of fluid-filled cells—a hydrostatic skeleton buried deep within the primitive chordate's body.

Other variations in means of locomotion are mostly modifications of the hydrostatic principle. The methods bivalves have for getting about, the jet propulsion of cephalopods and some coelenterates, and the water vascular system of echinoderms all involve a fluid-filled chamber that provides resistance to muscle contractions.

Of course, no statement in biology is ever more than a generalization, and a great variety of animals do not move at all, but adopt the opposite strategy of allowing the food to come to them. In the last century, Louis Agassiz characterized a barnacle as "nothing more than a little shrimp-like animal, standing on its head in a limestone house and kicking food into its mouth." It is an accurate description, and many mollusks, sponges, and some colenterates and annelids simply filter passing food items out of the water. On the other hand, some plants certainly move, suggesting that there may be reasons for plants and animals to get about other than to search for food. Dispersal is one reason and involves the need to colonize new territories and to seek out mates.

According to present evidence, the bony fish went through most of their evolution in fresh water and spread to the seas at a much later period. The fresh water in which they lived was shallow and often stagnant (devoid of oxygen). It is believed that under these conditions some primitive osteichthyans obtained oxygen by swallowing air. Gradually, in some species, simple lungs evolved that served as accessory organs to the gills. (Lungs and gills and their comparative functions will be described in more detail in Chapter 28.)

19-36 *Rays, like skates and sharks, are cartilaginous fish (class Chondrichthyes) that have existed in their present form for about 350 million years. Their flattened body is an adaptation to bottom living.*

Lunged fish were the most common fish in the later Devonian period, a time of recurring drought. Some of these lunged fish, ancestors of modern lungfish, would bury themselves in the mud during periods when the water dried up completely. In others, skeletal supports evolved that served to prop up the thorax of the fish. These fish could gulp air even when their bodies were not supported by water. Eventually, it is believed, these lobe-finned fish gained the capacity to waddle to a new body of water. Thus the transition to land began as an attempt to remain in the water.

19-37 *Many frogs are clearly fishlike in their larval (tadpole) stages. As adults, they require water to reproduce and their moist skins are an important accessory respiratory organ. Frogs, like all adult amphibians, are carnivores. They catch insects with a flick of their long tongues, which are attached at the front of their mouths and which have a sticky, flypaper-like surface.*

Amphibians Amphibians descended from air-breathing lunged fish. Modern amphibians include frogs and toads (which are tailless as adults) and salamanders (which have tails throughout their lives). They can readily be distinguished from the reptiles by their thin, usually scaleless skins, which serve as respiratory organs. Frogs also have lungs, into which they gulp air, but some salamanders respire entirely through their skins and the mucous membranes of their throats. Because water evaporates rapidly through their skins, amphibians can die of desiccation (drying out) in a dry environment.

Most frogs in cold climates have two life stages, one in water and the other on land (hence their name, from *amphi* and *bios,* meaning "two lives"). The eggs are laid in water and are fertilized externally. They hatch into gilled larvae (tadpoles). The tadpoles later develop into adults that lose their gills and develop lungs. The adults live out of the water, at least in the summer. However, there are many variations on this theme. Some of the American salamanders fertilize their eggs on land; the males deposit sperm packets that are picked up by the females. Many modern amphibians also skip the free-living larval stage. The eggs, which may be laid on land, in a hollow log or cupped leaf, or even carried by the parent, hatch into miniature versions of the adult. Thus, in their own way, the modern amphibians are as far removed from their ancestral forms as we are in our way.

19-38 *Many reptiles lay eggs in which the embryos can develop on land. These newly hatched black snakes are sunning themselves, a behavior characteristic of ectotherms—animals that take in heat from the environment. Black snakes, like all other snakes, are carnivorous, devouring their prey whole.*

Reptiles As you will recall, the vascular plants were freed from the water by the evolution of the seed. Analogously, the vertebrates became truly terrestrial with the evolution in the reptiles of the amniote egg, an egg that retains its own water supply and so can survive on land. The reptilian egg, which is much like the familiar hen's egg in basic design, contains a large yolk, the primary food supply for the developing embryo, and abundant albumen (egg white), which supplies additional nutrients and water. A membrane, the amnion, surrounds the developing embryo with a liquid-filled space that substitutes for the ancestral pond. The gilled state is passed in a shelled egg. (In mammals also, although their eggs typically develop internally, the embryos are enclosed in water within the enveloping amnion and pass through a gilled stage before birth.)

Reptiles are characteristically four-legged, although the legs are absent in snakes and some lizards. In keeping with their terrestrial existence, reptiles have a dry skin, usually covered with protective scales. Modern reptiles, of which there are 5,400 species, include lizards, snakes, turtles, and crocodiles.

Until recently the dinosaurs, which are clearly of reptilian descent, were assumed to have been *ectothermic*—that is, to have regulated their body temperatures within broad limits by taking in heat from the environment or giving it off to the environment. Several lines of evidence indicate, however, that at least some groups of dinosaurs were *endothermic*—that is, their body temperatures were regulated internally.

Birds Birds are essentially reptiles specialized for flight (Figure 19-39). Their bodies contain air sacs, and their bones are hollow. The frigate bird, a large seagoing bird with a wingspread of more than 2 meters, has a skeleton that weighs only 110 grams (about 4 ounces). The most massive bone in the bird skeleton is the keel, or breastbone, to which are attached the huge muscles that operate the wings. Flying birds have jettisoned all extra weight; the female's reproductive system has been trimmed down to a single ovary, and even this becomes large enough to be functional only in the mating season.

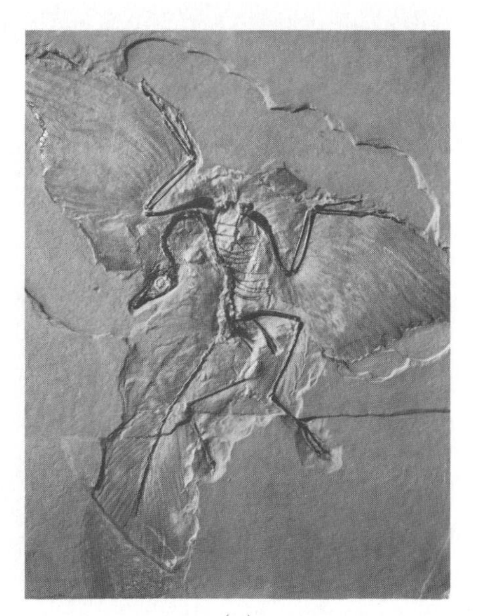

(a)

(b)

19-39 (a) *The oldest known fossil bird,* Archaeopteryx, *dates from the middle Jurassic period, about 150 million years ago. It still had many reptilian characteristics. The teeth and the long, jointed tail are not found in modern birds. The clearly evident feathers may have been related as much to endothermy as to flight.*

(b) *A modern bird, a common tern, landing at its nest. If you look closely, you can see the speckled, well-camouflaged eggs in the nest.*

Birds have feathers, which is their outstanding, unique physical characteristic. They maintain a high and constant body temperature, which distinguishes them from most of the modern reptiles (although a few, such as leatherback turtles, show some degree of endothermy). In modern birds, feathers make possible flight and also serve as insulation. (Only animals that are endothermic require insulation; insulation would be a disadvantage for animals that warm their bodies by exposure to the environment.) Birds also have scales, a reminder of their reptilian ancestry. Many birds are born at an immature stage, and virtually all birds require a long period of parental care.

Mammals Mammals also descended from the reptiles. Characteristics distinguishing mammals from other vertebrates are that mammals (1) have hair, (2) provide milk for their young from specialized glands (mammae), and (3) like birds, but unlike other vertebrates, maintain a high body temperature by generating heat metabolically.

In nearly all mammalian species, the young are born alive, as they are in some fish and reptiles, which retain the eggs in their bodies until they hatch. Some very primitive mammals, however, the *monotremes*, such as the duckbilled platypus, lay eggs with shells but nurse their young after hatching. The *marsupials*, which include the opossum and the kangaroo, also bear their young alive. They differ from the major group of mammals, however, in that the infants are born at a tiny and immature stage and are often kept in a special protective pouch in which they suckle and continue their development (Figure 19–40). Most of the familiar mammals are *placentals*, so called because they have an efficient nutritive connection, the placenta, between the uterus and the embryo. As a result, the young develop to a much more advanced stage before birth. Thus the young are afforded protection during their most vulnerable period. The earliest placentals were small, shy, and probably nocturnal, thus avoiding the carnivorous dinosaurs. They undoubtedly lived mostly on insects, grubs, worms, and eggs.

19–40 *Marsupial infants are born at an immature stage and continue their development attached to a nipple in a special protective pouch of the mother.* (a) *This tiny kangaroo accidentally became dislodged from its mother's pouch. As you can see, it is still attached to the nipple. After the picture was taken, the baby was restored to the pouch, with no apparent ill effects from its premature introduction to the outside world.* (b) *Opossum infants spend about two weeks in the womb and about three months in their mother's pouch. A newborn opossum is much smaller than a honey bee.*

(a)

(b)

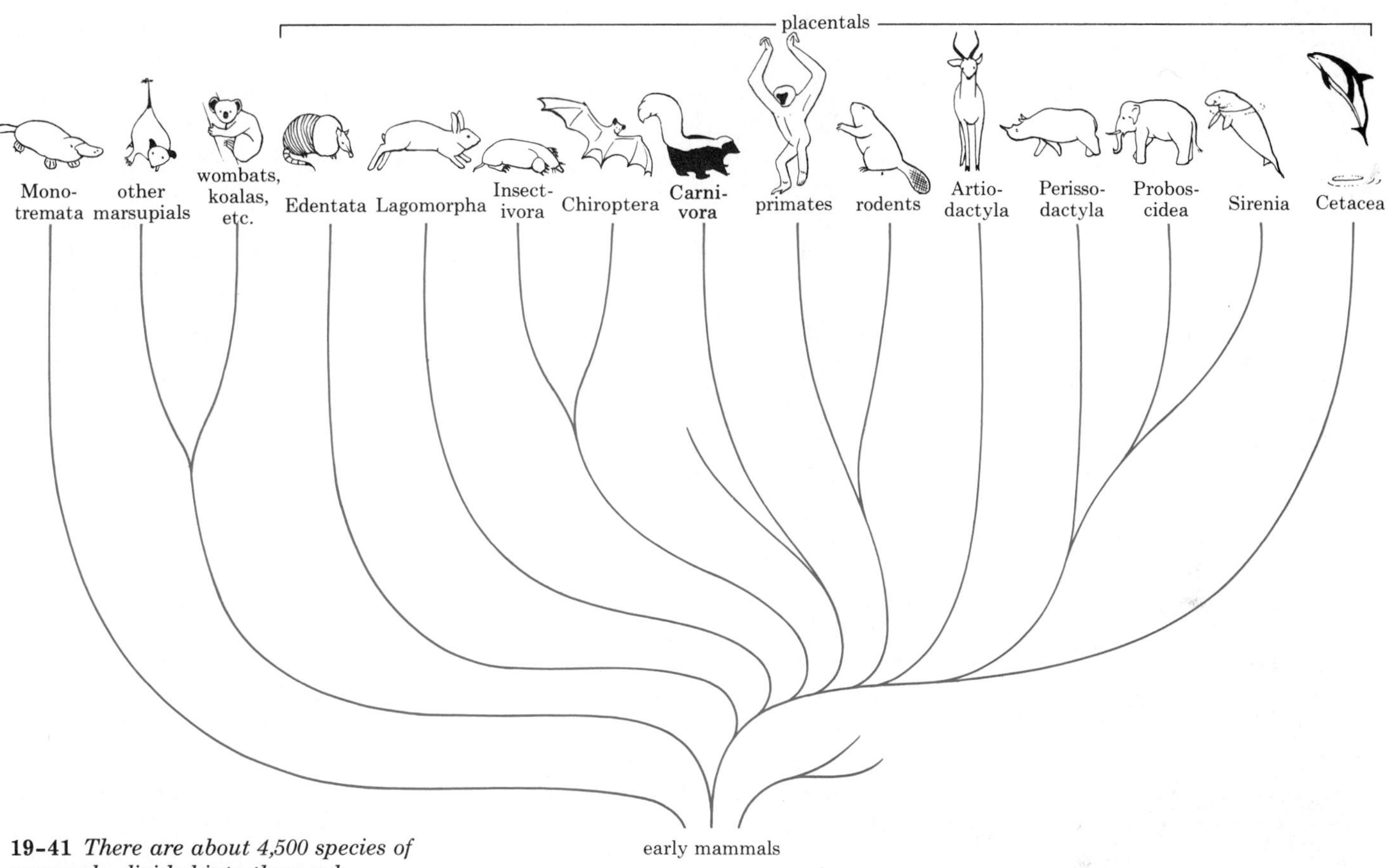

19–41 *There are about 4,500 species of mammals, divided into three subclasses: the monotremes, the marsupials, and the placentals. There are twelve major orders of placentals.*

The major evolutionary lines of mammals are summarized in Figure 19–41. The primates, the order to which we belong, are placental mammals characterized by four kinds of teeth (canines, incisors, premolars, and molars), opposable first digits (thumbs and usually big toes), two pectoral mammae, an expanded brain (particularly the cerebral cortex), and a tendency toward single births. We are distinguished from the other primates by our upright posture, long legs and short arms, high forehead and small jaw, and sparse body hair.

Among all the mammals, humans are among the least specialized. We are omnivores. Unlike the carnivores, which are meat eaters, and the several orders of herbivores, we eat a wide variety of fruits, vegetables, and other animals. Our hands closely resemble those of a primitive reptile, in contrast to the highly specialized forelimbs developed by, for example, whales, bats, and horses. We cannot see as well as an eagle. Our sense of smell is much less keen than a dog's, and our sense of taste far less sensitive than a housefly's. Many animals can run faster, swim more powerfully, and climb trees with more agility (though few can do all three). Humans have, however, one area of extreme specialization: the brain. Because of our brain, we are unique among all the other animals in our capacity to reason, to speak, to plan, to learn, and so, to some extent, to control our own future and that of the other organisms with which we share this planet.

(a) (b) (c) (d) (e) (f)

19–42 *An assortment of mammals.* (a) *Lagomorphs, such as the jack rabbit shown here, have two pairs of upper incisors whereas rodents, such as beavers* (b), *have only one. In both lagomorphs and rodents, the teeth grow continually.*

Carnivores, such as this walrus bull (c) *and leopard* (d), *are adapted to hunt and kill for food. The leopard's prey is a Thomson's gazelle, which is an artiodactyl (a two-toed ungulate). Zebras* (e), *which are perissodactyls (odd-toed ungulates), are herbivores adapted for grazing.*

(f) *Elephants, the Proboscidea, are the largest land mammals living today; some reach a weight of 7.5 metric tons.*

SUMMARY

Plants are multicellular photosynthetic organisms adapted for life on land. Their adaptations include a fatty outer coating, the cuticle; stomata; and protective layers of cells surrounding the reproductive cells. All plants also have a life cycle with alternation of generations. The ancestor of plants is believed to have been a green alga. From this common ancestor, two major groups evolved: the bryophytes and the tracheophytes.

The bryophytes are relatively small plants lacking elaborate vascular systems. They include the mosses, liverworts, and hornworts. In the bryophytes, the gametophyte (haploid) generation is the larger, dominant form.

The tracheophytes are the vascular plants. The largest groups of these are the gymnosperms and the angiosperms. The gymnosperms are "naked seed" plants in which the ovules and seeds are exposed. The angiosperms are the flowering plants in which the seed (the mature ovule and its contents) is enclosed in an outer protective layer, the fruit, formed from parts of the flower. Fruits, with their enclosed seeds, are often dispersed by animals.

Animals are multicellular heterotrophs that depend directly or indirectly on plants or algae as their source of food energy. Almost all digest their food in an internal cavity. Most are motile. Reproduction is usually sexual. More than 90 percent of the animal species are invertebrates, animals without backbones.

Major phyla of invertebrates include: (1) Porifera (sponges), which have a unique body plan, simple cellular organization, and food-trapping collar cells; (2) Coelenterata, characterized by radial symmetry, a two-layered body plan, a coelenteron, and cnidoblasts; (3) Platyhelminthes (flatworms), with a three-layered body and bilateral symmetry; (4) Rhynchocoela (ribbon worms), with a three-layered body plan, a one-way digestive tract, and a retractile proboscis; (5) Nematoda (roundworms), a very large group characterized by a three-layered body, a one-way digestive tract, and a pseudocoelom; (6) Annelida (segmented worms), in which the three-layered body is divided externally and internally into segments and has a coelom and a circulatory system; (7) Mollusca, characterized by a muscular head-foot and a mantle that secretes a shell; (8) Arthropoda, segmented animals with a hard, jointed exoskeleton, the largest phylum in the animal kingdom, including crustaceans, arachnids, and insects; and (9) Echinodermata, with radial symmetry, a spiny exoskeleton, and a water vascular system.

The vertebrates are the principal members of the phylum Chordata. They are distinguished from the invertebrates by a vertebral column and an internal skeleton. The major classes of vertebrates include fish, amphibians, reptiles, birds, and mammals.

QUESTIONS

1. Distinguish among the following terms: cuticle/stomata; xylem/phloem; vertebrate/invertebrate; coelenteron/coelom/pseudocoelom; radial symmetry/bilateral symmetry; exoskeleton/endoskeleton; larva/pupa; ectotherm/endotherm; marsupial/placental.

2. Bryophytes, among the plants, and amphibians, among the animals, often live in habitats intermediate between fresh water and dry land, rather than between salt water and land. Propose an argument (referring back to Chapter 5) to explain why invasions of the land were more likely by organisms previously adapted for life in fresh water.

3. In order for a characteristic to be selected for, in evolutionary terms, it must be advantageous at the time selection is taking place. How, then, can you explain preadaptation?

4. A coelom provides hydrostatic support for an animal. Why is a coelenteron less likely to fill this same function?

5. In which phylum do you find each of the following and what is its function: collar cell, cnidoblast, mantle, radula, siphon, spiracle, tube foot?

6. Label the drawing below.

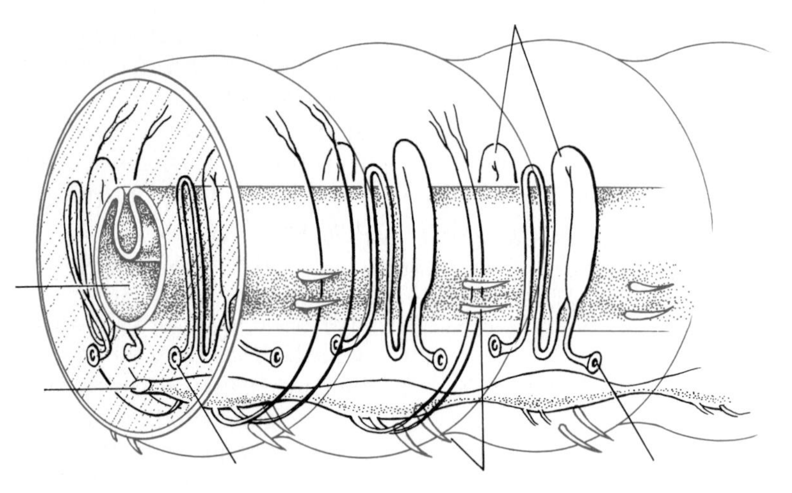

7. Many mollusks have lost or may be in the evolutionary process of losing their shells. What are the advantages to an organism of having a shell? Of losing one?

8. Smallness may also be an advantage to an organism. What are some advantages of smallness?

9. Consider the graceful swimming of a fish or squid. What is the role of the semirigid beam running the length of the animal, whether the "pen" of a squid, the notochord of a *Branchiostoma,* or the vertebral column of a fish?

10. Describe the identifying characteristics of the phylum Chordata. What is the functional significance of each?

SUGGESTIONS FOR FURTHER READING

ALEXOPOULOS, C. J.: *Introductory Mycology,* 3d ed., John Wiley & Sons, Inc., New York, 1979.

A general, brief introduction to the fungi.

BARNES, ROBERT D.: *Invertebrate Zoology,* 4th ed., Saunders College/Holt, Rinehart and Winston, Philadelphia, 1980.

One of the best general introductions to protozoans and invertebrates.

BONNER, J. T.: *The Cellular Slime Molds,* 2d ed., Princeton University Press, Princeton, N.J., 1967.

A record of experimental work with a small but fascinating group of organisms.

BUCHSBAUM, RALPH, and LORUS J. MILNE: *The Lower Animals: Living Invertebrates of the World,* Doubleday & Company, Inc., Garden City, N.Y., 1962.

A collection of handsome photos of the invertebrates accompanied by a text prepared by two noted zoologists but directed toward the general reader.

CURTIS, HELENA: *The Marvelous Animals,* Natural History Press, Garden City, N.Y., 1968.

An informal introduction to one-celled organisms.

DAWSON, E. Y.: *Marine Botany: An Introduction,* Holt, Rinehart and Winston, Inc., New York, 1966.

A short, lively text that covers seaweeds, marine bacteria, fungi, phytoplankton, and sea grasses.

DESMOND, ADRIAN J.: *The Hot-Blooded Dinosaurs: A Revolution in Paleontology,* The Dial Press, Inc., New York, 1976.*

Desmond, a historian of science, describes the development of evolutionary theories, particularly as they were influenced by the discovery of dinosaur fossils, and presents a lively review of the evidence that is leading an increasing number of paleontologists to the conclusion that dinosaurs were endothermic. (If you are not interested in the history of paleontology, you may want to begin in the middle.) The text is well written and the illustrations are wonderful.

EVANS, HOWARD E.: *Life on a Little-Known Planet,* E. P. Dutton & Co., Inc., New York, 1978.*

Professor Evans is the author of many popular articles and books on insects. This book profits from his wide knowledge, clarity, and humor.

HICKMAN, CLEVELAND P.: *Biology of the Invertebrates,* 2d ed., The C. V. Mosby Company, St. Louis, 1973.

Probably the best general text on the invertebrates.

KLOTS, ALEXANDER B., and ELSIE B. KLOTS: *Living Insects of the World,* Doubleday & Company, Inc., Garden City, N.Y., 1975.

A spectacular gallery of insect photos. The text is informal but informative, written by experts for the general public.

LARGE, E. C.: *The Advance of the Fungi,* Dover Publications, Inc., New York, 1962.*

A fascinating popular account of the closely interwoven histories of fungi and humans, first published in 1940.

MARGULIS, LYNN: *Origin of Eukaryotic Cells,* Yale University Press, New Haven, Conn., 1970.

A fascinating discourse on the origin of eukaryotic cells by serial symbiotic events, beautifully illustrated and well reasoned.

RAVEN, PETER H., RAY F. EVERT, and HELENA CURTIS: *Biology of Plants,* 3d ed., Worth Publishers, Inc., New York, 1981.

This general botany text contains an excellent presentation of the evolution of plants and related organisms.

ROMER, ALFRED: *The Procession of Life,* World Publishing Co., Cleveland, 1968.*

A history of evolution, written by an expert but as readable as a novel.

RUSSELL-HUNTER, W. D.: *A Biology of Lower Invertebrates,* The Macmillan Company, New York, 1968.*

RUSSELL-HUNTER, W. D.: *A Biology of Higher Invertebrates,* The Macmillan Company, New York, 1969.*

Short, authoritative accounts, with emphasis on function.

SLEIGH, M.: *The Biology of Protozoa,* University Park Press, Baltimore, Md., 1973.*

A general biology of the Protozoa, including chapters on structure, metabolism, reproduction, and ecology, with many excellent illustrations.

SMITH, A. H.: *The Mushroom Hunter's Field Guide,* The University of Michigan Press, Ann Arbor, Mich., 1980.

A clear, concise, well-illustrated guide to edible mushrooms, enlivened with good advice and pertinent anecdotes.

STANIER, R. Y., et al.: *The Microbial World,* 4th ed., Prentice-Hall, Inc., Englewood Cliffs, N.J., 1976.

An introduction to the biology of microorganisms, with special emphasis on the properties of bacteria.

* Available in paperback.

SECTION 5

Plants and the Land

About 11,000 years ago, people discovered that the energy-rich grains of certain grasses not only could be eaten but also could be stored and then planted the next season to produce new supplies of food. Ever since, the human population has depended heavily on agriculture. This field in Colorado is filled with barley plants on which the grains are ripening. Barley grains are used to make malt beverages, in breakfast foods, and for livestock feed.

20

Introducing the Plant: The Leaf

For most of earth's history, the land was bare. A billion years ago, seaweeds clung to the shores at low tide and perhaps some gray-green lichens patched a few inland rocks. But, had anyone been there to observe it, the earth's surface would generally have appeared as barren and forbidding as the bleak Martian landscape. According to the fossil record, plants first began to invade the land a mere half billion years ago. Not until then did the earth's surface truly come to life. As a film of green spread from the edges of the waters, other forms of life—heterotrophs—were able to follow. The shapes of these new forms and the ways in which they lived were determined by the plant life that preceded them. Plants supplied not only their food—their chemical energy—but also their nesting, hiding, stalking, and breeding places.

And so it is today. In all terrestrial communities except those created by human activities, the character of the plants still determines the character of the animals and other forms of life that inhabit a particular area. Even we members of the human species, who have seemingly freed ourselves from the life of the land and even, on occasion, from the surface of the earth, are still dependent on the photosynthetic events that take place in the green leaves of plants.

Plants, as defined in this text (see page 248), are multicellular photosynthetic organisms adapted for life on land. In this section, we shall focus on the largest and most abundant class of plants—the angiosperms. This class is characterized by specialized structures—flowers—in which sexual reproduction takes place, in which the seed is formed, and from which the fruit develops. There are about 235,000 species of angiosperms. The group is divided into two large subclasses, the *dicots* (170,000 species) and the *monocots* (65,000 species). The differences between the two groups are summarized in Figure 20–2 on the next page.

20–1 *This young bean seedling has just pushed through the soil, beginning its process of development into a mature bean plant. The first foliage leaves are emerging from between the cotyledons, structures that we know more familiarly as the two halves of the bean seed. The cotyledons contain the food reserves on which the young seedling is dependent; gradually the food will be used up, and the cotyledons will wither and fall from the plant. By that time, the seedling will be able to supply its own food through photosynthesis.*

The Plant Body

One can best understand the structure of the plant body by comparing the photosynthetic cells of an angiosperm with single-celled algae. Look again at the green alga *Chlamydomonas* shown in Figure 4–7 (page 58) and the corn leaf cell shown in Figure 4–9 (page 59). In both cells, the principal

20-2 *The angiosperms are divided into two broad groups: the dicots and the monocots. The names refer to the fact that the embryo in the dicots has two cotyledons ("seed leaves") and in the monocots has one. In the dicots, the vascular tissues are arranged around a central core in the stem; in monocots, they are scattered. The veins of dicot leaves are usually fanlike or featherlike; those of monocot leaves are usually parallel. Dicots characteristically have taproots, and monocot roots are often fibrous. These characteristics are discussed in more detail in the text.*

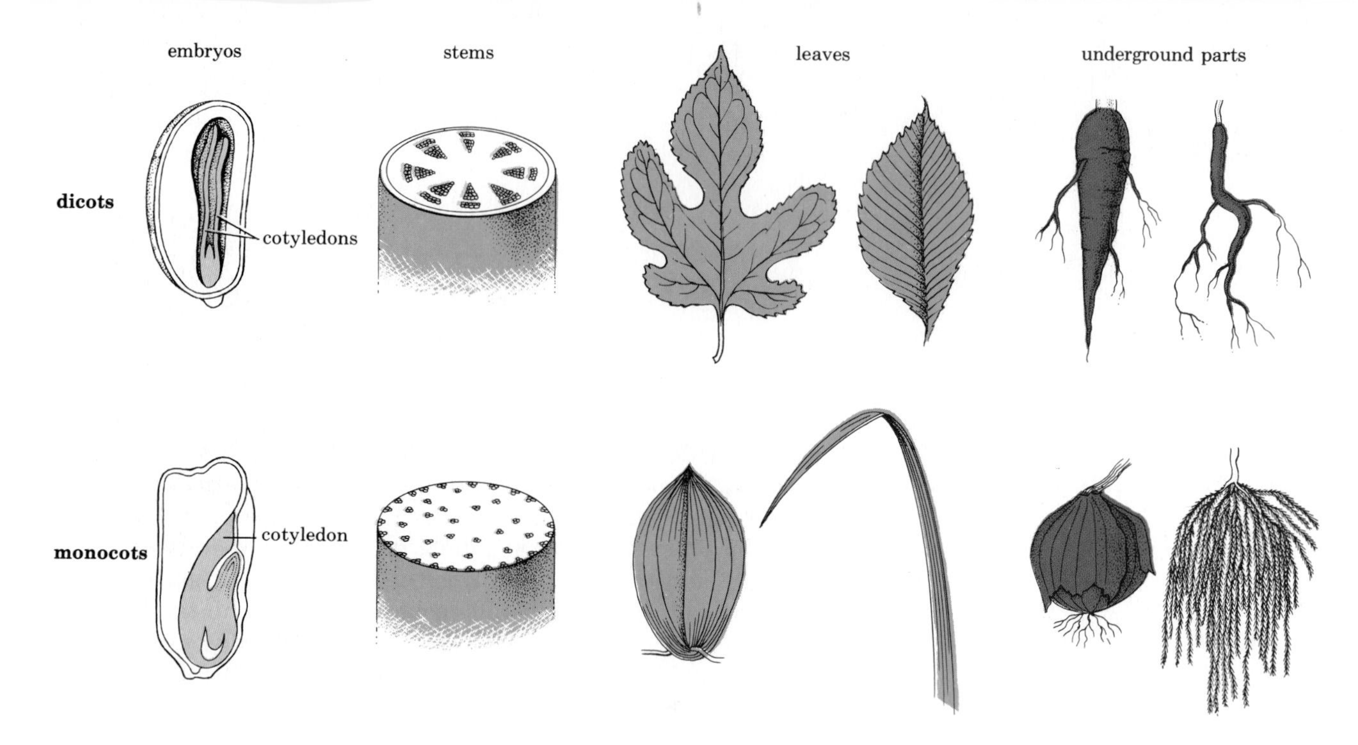

organelles are the light-gathering chloroplasts. These chloroplasts, like those of other green algae and of all plants, contain chlorophylls *a* and *b* and carotenoids. Both cells are surrounded by a cell membrane outside of which is a cell wall composed mostly of cellulose. In addition to sunlight, both types of cell require water, carbon dioxide, and a few minerals. The algal cells can obtain these from the water in which they live. The photosynthetic cells of plants, however, live in a very different environment and require a complex life-support system. The plant body is, in effect, that life-support system.

Figure 20-3 diagrams the external structure of a familiar angiosperm, a geranium. This plant, like other vascular plants, is characterized by a root system that anchors the plant in the ground and collects water and minerals from the soil; a stem or trunk that raises the photosynthetic parts of the plant toward the sun; and structures highly specialized for light capture and photosynthesis, the leaves. These are all interconnected by a complicated and efficient system for the transport of food, water, and minerals. Such a system is the distinguishing feature of all vascular plants, a group that also includes the gymnosperms and the ferns (see page 251).

These structures—leaves, stems, and roots—are all adaptations to a photosynthetic life on land. In the remainder of this chapter, we shall examine the leaf; in the next chapter, we shall consider roots, stems, and the conducting systems that connect the different parts of the plant body.

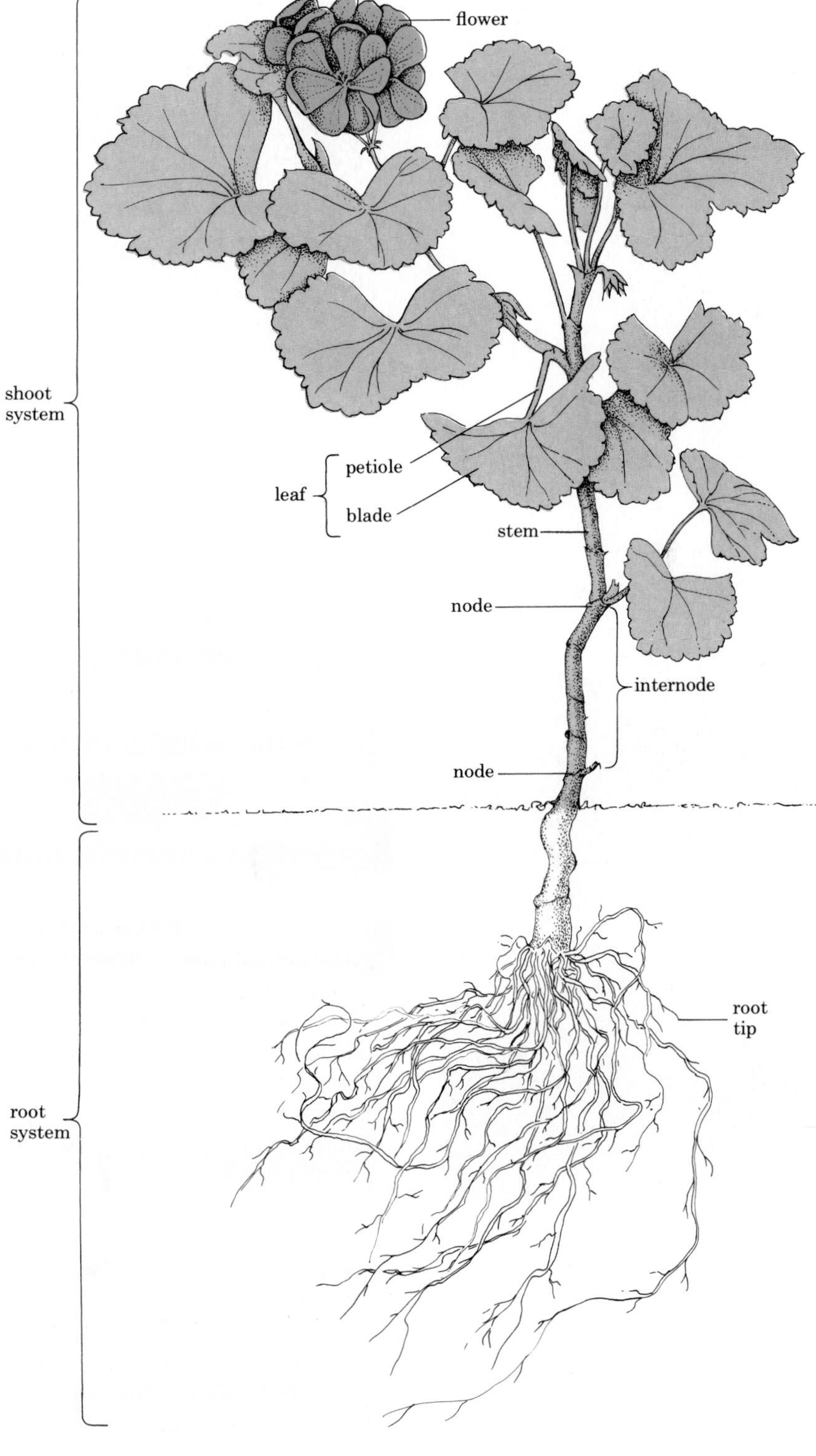

20-3 *The body plan of a flowering plant. The aboveground shoot system consists of the stem, the leaves, whose primary function is photosynthesis, and the flowers, the reproductive structures. Leaves appear at regions on the stem known as nodes. The portions of the stem between nodes are called internodes. The belowground structures, the roots, supply water and minerals to the stem, leaves, and flowers.*

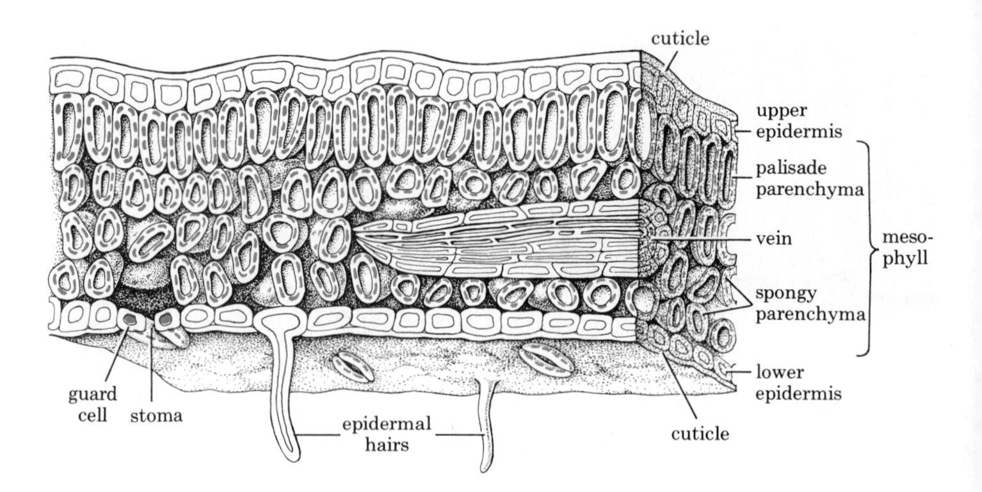

20–4 *Diagram of the interior of a leaf. Chloroplasts are indicated in color.*

The Leaf and Photosynthesis

The blade of the leaf is characteristically a thin, broad light-collecting surface. Its primary function is the exposure of the photosynthetic cells to sunlight. In the leaf, the photosynthetic cells are sandwiched between two layers of small, stiff-walled epidermal cells that are transparent and so permit the passage of light to the photosynthetic cells. The outer surfaces of the epidermal cells are coated with a fatty layer, the cuticle, which retards the escape of water from the interior of the leaf (Figure 20–4).

The photosynthetic cells are of a general type known as *parenchyma cells.* These are the cells from which all other cells of the plant body have evolved. In the leaf, parenchyma cells make up the tissue called *mesophyll*, or "middle leaf." There are two types of parenchyma cells in the mesophyll: *palisade parenchyma*, column-shaped cells near the upper surface of the leaf in which most of the photosynthesis takes place, and *spongy parenchyma*, irregularly shaped cells that underlie the palisade cells.

20–5 *Veins in the leaf of a maple. Continuous with the vascular tissue of the stem and root, they branch and divide into finer and finer bundles, reaching within a short distance of every photosynthetic cell. As you can tell from the pattern of the veins, the maple is a dicot.*

The supply of water to the photosynthetic cells is the most crucial problem that plants had to solve, evolutionarily speaking, in their transition to land. The success of the various groups of modern plants is in direct proportion to the efficiency of their water transport systems. Water is brought into the leaf through the vascular tissues; the vascular bundles of the leaf are called *veins* (Figure 20–5). The vascular tissues of the leaf pass through the petiole (leaf stalk) and are continuous with the vascular tissues of the stem and root. Within the leaf, water moves from the cells of the xylem—the water-conducting tissue—into the open spaces surrounding the spongy parenchyma. As we noted previously, its escape from the leaf is retarded by the fatty cuticle of the epidermal cells.

Sugars, produced by photosynthesis, are removed from the leaf by the veins and transported to other parts of the plant body in the phloem—the food-conducting tissue.

Stomata

The leaf, as we have described it so far, appears to be a waterproof, airtight package in which photosynthetic cells are held aloft and advantageously displayed to sunlight. However—and this is a serious difficulty—photosynthesis also requires carbon dioxide, which, on land, is available

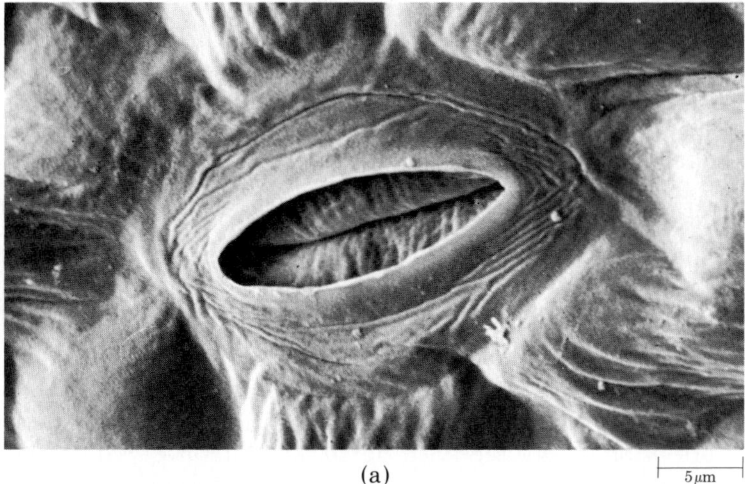

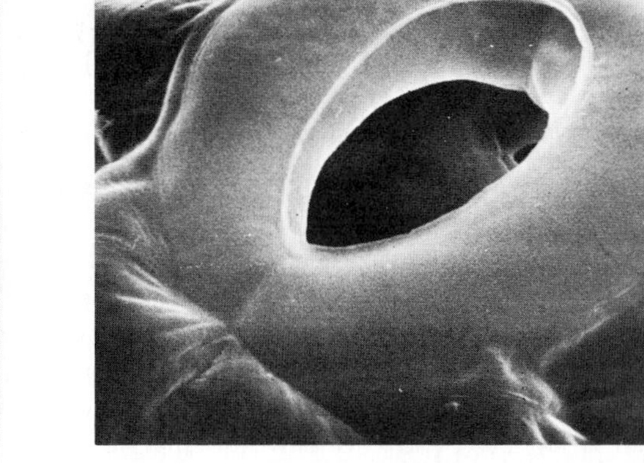

20-6 (a) *Portion of a parsley leaf showing a closed stoma.* (b) *View of an open stoma of cucumber. The stomata lead into air spaces within the leaf that surround the thin-walled mesophyll cells. The air in these spaces, which make up 15 to 40 percent of the total volume of the leaf, is saturated with water vapor that has evaporated from the photosynthetic cells.*

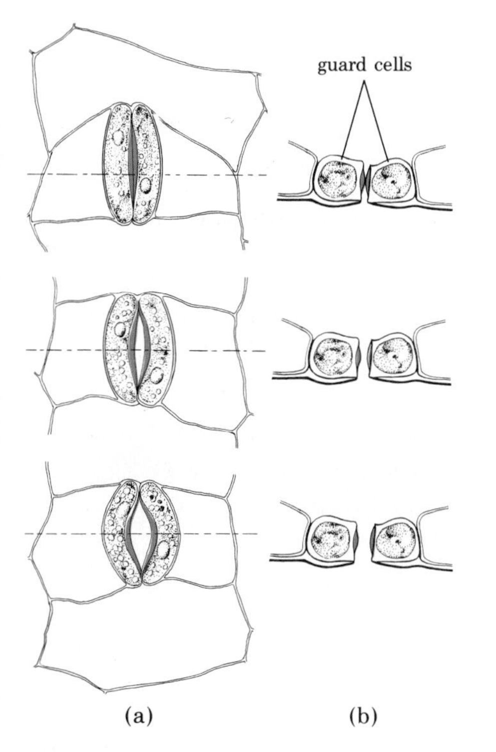

20-7 *Mechanism of stomatal movement. A surface view is shown in* (a) *and a cross section in* (b). *Each stoma has two guard cells that open the stoma when they are turgid and close it when they lose turgor. In many species, the guard cells have thickened walls adjacent to the stomatal opening. As turgor pressure increases, the thinner parts of the cell wall are stretched more than the thicker parts, causing the cells to bow out and the stoma to open.*

only in air. As plants made the transition to land, a crucial problem was getting carbon dioxide from the air to the photosynthetic cells of the mesophyll without simultaneously losing large quantities of water from the leaf.

The process of evolution has not produced a perfect solution to this problem, but it has produced an effective compromise in the form of stomata (from *stoma,* the Greek word for "mouth"). A stoma consists of two specialized cells in the leaf epidermis, called guard cells, and a small opening, or pore, that they surround. The guard cells open and close the pore (Figure 20-6), thus regulating the loss of water and the exchange of gases. About 90 percent of the water lost by the plant body escapes through the stomata; the rest is lost through the cuticle.

Stomata are commonly most abundant on the undersurface of leaves. They may be very numerous. For example, on the lower surface of tobacco leaves there are about 19,000 stomata per square centimeter. On the upper surface there are about 5,000 stomata per square centimeter.

Figure 20-7 shows the mechanism of stomatal movement. The movement is controlled by the turgor pressure (see page 80) of the guard cells. When the water available to the leaf drops below a certain point, the turgor pressure of the guard cells drops and the stomata close, thus conserving the remaining water.

Factors other than water loss—including light, carbon dioxide concentration, temperature, and ion concentrations—also affect the stomata. For example, in many species the stomata close regularly in the evening and open in the morning, even though there may be no change in the water available to the plant. In most species, an increase in carbon dioxide concentration in the mesophyll causes the stomata to close. Temperatures higher than 35°C (95°F) can cause stomatal closing in some species. In hot climates, the stomata of many plants close regularly at midday. The way in which these factors affect the turgor pressure of the guard cells—and thus regulate the opening and closing of the stomata—is still poorly understood.

In a number of succulents (fleshy plants with water-filled tissues), the stomata are closed throughout the day, opening only at night. Such plants include cacti, pineapples, and members of the stonecrop family. At night, when the stomata are open, the plants take in carbon dioxide and convert it to four-carbon acids. During the day, the carbon dioxide is released from the acids and used immediately in photosynthesis. This pathway of photosynthesis is known as CAM (Crassulacean acid metabolism, from Crassulaceae, the plant family to which the stonecrops belong). It resembles, in principle, the C_4 photosynthesis described on page 119.

20-8 *Modified leaves.* (a) *Spines on a prickly-pear cactus.* (b) *Succulent leaves, adapted for water storage* (Sedum). (c) *An onion bulb, which is composed of leaves modified for food storage.*

Leaf Shapes and Adaptations

Leaves come in various shapes and sizes, ranging from broad fronds to tiny scales. Some of these differences can be correlated with the environments in which the plants live. Large leaves with broad surfaces are often found in plants that grow under the canopy in a tropical rain forest, where water is plentiful but where there is intense competition for light. Small, leathery leaves are often associated with harsh, dry climates, where the light-capturing surface has to be sacrificed for water conservation.

In some plants, such as desert cacti, leaves are modified as spines, which are hard, dry, and nonphotosynthetic. (The terms "spine" and "thorn" are often used interchangeably; however, thorns are technically modified branches.) In other plants, for example, the garden pea, leaves are modified as tendrils. In many plants, leaves are succulent; that is, they are modified for water storage. Other leaves are specialized for food storage. A bulb, such as the onion, is a large bud consisting of a short stem with many modified leaves attached to it. The "head" of a cabbage also consists of a stem bearing numerous thick, overlapping leaves. In some plants, the petioles become thick and fleshy: Celery and rhubarb are two of the most familiar examples.

(a)

(b)

(c)

Plants and Air Pollution

One of the primary functions of the leaf is the exchange of gases with the surrounding air—taking in the carbon dioxide used in photosynthesis and releasing the oxygen produced. As a consequence, the leaf, like the human lung, is exceedingly sensitive to air pollution.

Air pollution has many forms. Some pollutants are in the form of particles. The particles may be organic, such as those present in the smoke produced by burning fossil fuels, leaves, and garbage, or inorganic, such as the lead compounds released in the combustion of leaded gasoline. As a major component of smog, these particles reduce the amount of sunlight reaching the earth's surface and thus the amount available to the photosynthesizing leaf. Such particles have other, more direct effects on plants. They may clog the stomata and prevent them from functioning. Moreover, some metallic particles act as plant poisons, presumably by inhibiting essential enzymes.

Fluorides, which are released into the air as waste products from the manufacture of phosphates, steel, and aluminum, act as cumulative poisons. Fluorides enter the plant through the stomata and cause leaf tissue to collapse. It is believed that they inhibit enzymes involved in cellulose synthesis. Thousands of acres of Florida citrus groves have been damaged by fluorides discharged from phosphate fertilizer plants.

When ores containing sulfur are processed, sulfur dioxide is produced. Sulfur oxides are also released by the burning of fossil fuels containing sulfur, such as high-sulfur coal. Acids derived from them are a major component of acid rain (see page 572). In several parts of the United States, virtual deserts have been created by the emission of sulfur dioxide combined with metals. For example, a large copper smelter was opened at the turn of the century in a mountainous area in Tennessee. Within a few years, all vegetation had been killed in the formerly luxuriant forest surrounding the smelter. Although air pollution controls were instituted in 1905, the area remains barren to this day.

Photochemical smog, most familiar in southern California, is produced by the action of sunlight on automobile exhaust gases. One of the principal ingredients of photochemical smog is nitrogen dioxide (NO_2), which is a by-product of gasoline combustion. In the presence of light, NO_2 is split into NO and atomic oxygen. The latter reacts very rapidly with the normal molecular oxygen (O_2) of the air to produce ozone (O_3). Ozone damages the thin-walled palisade cells of the plant leaf, apparently by altering the permeability of the membranes of both the cells and their chloroplasts. Another component of photochemical smog is a nitrogen-containing organic compound called PAN. Photochemical smog concentrations as low as 0.25 part per million reduce photosynthesis by 66 percent.

Although progress has been made in recent years in reducing air pollution, the problem is far from solved. There is also concern on the part of many scientists that large-scale development of synthetic fuels and increased use of coal to meet our energy needs will reverse the progress that has been made. Our concern is usually with the effects of air pollution on ourselves. It would be wise if we broadened that concern to include the plants upon which we and all terrestrial animal life are dependent.

(a) *These blackberry leaves have been damaged by sulfur dioxide. Areas of injured leaf tissue are surrounded by healthy tissue.* (b) *The effects of ozone on this tobacco leaf are visible as flecks of dead tissue on the upper surface of the leaf. When ozone injury is more severe, large lesions are visible on both surfaces of the leaf.*

(a)

(b)

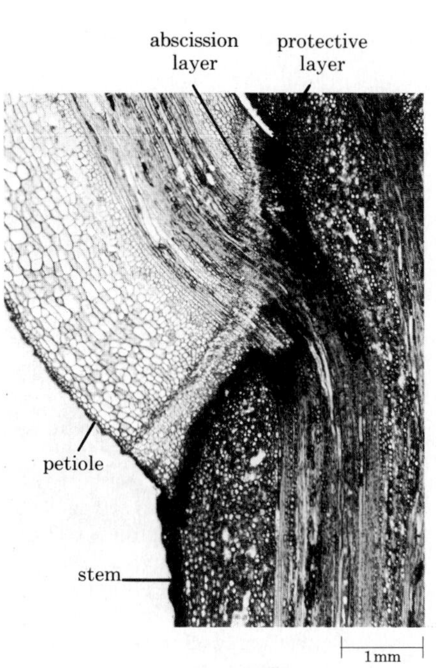

20-9 *Abscission layer in a maple leaf, as seen in a longitudinal section through the base of the petiole.*

Adaptations to Seasonal Water Shortages

The first angiosperms evolved in the tropics and were probably *perennials*: plants in which vegetative structures survive year-round and from year to year. As we have seen, plants lose water through the leaves. Many groups of early angiosperms adapted to seasonal water shortages by dropping their leaves during periods of drought; such plants are said to be *deciduous*. Because of this adaptation, angiosperms were able to spread north and south from the tropics to areas where seasonal water shortages result from the cold.

The fall of the leaf comes about when special enzymes dissolve the middle lamellae (see page 64) in a layer of weak, thin-walled cells in the petiole. These cells compose the *abscission layer*, and the process is known as *abscission* (Figure 20-9). As abscission proceeds, the cells separate. Before the leaf falls, a protective layer develops and ultimately becomes the leaf scar.

As the leaf dies, new chlorophyll can no longer be formed. The green color gradually disappears, unmasking the yellows and oranges of the carotenoids and producing the blaze of colors characteristic of the deciduous forest in the fall. The red and purple colors seen in some fall leaves are due to another group of pigments, the anthocyanins. These pigments, which are in solution in the vacuoles of the leaf cells, typically form when the temperature drops.

The dropping of leaves in deciduous trees comes about not as a direct reaction to cold or drought. Rather, in response to an environmental cue—the relative length of day and night—the plant produces chemicals that cause leaf abscission. (The way this and similar responses come about is the subject of Chapter 23.)

In other groups of angiosperms, changes in the life cycle evolved in response to seasonal water shortages. In *annual plants*, the leaves and all other vegetative parts die at the end of a single growing season. Only the seeds survive, and they do not germinate until the period of water shortage is past. In *biennial plants*, the cycle from seed to seed spans two growing seasons. The first season of growth ends with the formation of a root, a stem, and a rosette of leaves close to the soil. In the second growing season, stem elongation, flowering, and seed formation take place, followed by the death of the vegetative parts of the plant. Many common food plants are biennials, including cabbage and related plants (see page 485), the sugar beet, and parsley.

SUMMARY

Plants are multicellular photosynthetic organisms adapted to life on land. The largest and most recently evolved class of plants is the angiosperms, the flowering plants. Angiosperms are divided into two subclasses: monocots and dicots.

The plant body has specialized photosynthetic areas (leaves), conducting and supporting structures (stems), and structures that anchor the plant in the soil and absorb water and minerals from it (roots).

The photosynthetic cells of leaves are parenchyma cells. The "middle leaf," or mesophyll, is composed of palisade parenchyma and spongy parenchyma. It is sandwiched between two layers of epidermal cells that are covered by a fatty, water-resistant cuticle. Veins, the vascular bundles of the leaf, conduct water and minerals to the leaf cells and remove sugars from the leaf.

Stomata, minute pores in the epidermis, regulate the exchange of gases in the leaf. Most of the water lost by the plant body escapes through the stomata. The opening and closing of the guard cells of the stomata is controlled by the turgor pressure of the cells, which, in turn, may be regulated by a number of factors.

In some species, leaves are modified as spines or tendrils or for water or food storage. Some perennial plants (plants in which vegetative parts persist year-round and from year to year) have adapted to periods of seasonal drought by becoming deciduous. Other plants have adapted by completing the growing cycle—seed to seed—in a single season (annuals) or in two seasons (biennials).

QUESTIONS

1. Distinguish among the following terms: epidermis/cuticle; palisade parenchyma/spongy parenchyma; annual/biennial/perennial; deciduous/abscission.

2. Describe the distinguishing characteristics of monocots and dicots.

3. Sketch the interior of a leaf, labeling the principal cells and tissues. Describe the function of each of the labeled parts.

4. Superimpose on this sketch arrows showing the pathway of water, of oxygen, and of carbon dioxide into or out of the leaf.

5. How is it physically possible for an increase in the turgor of the guard cells to open the stomata? Consider what would happen if you partially inflated a cylindrical balloon, applied a strip of adhesive tape along its length, and then inflated it further. What does the experiment suggest about the role of wall thickenings in the guard cells?

6. What are the major adaptations that enable plants to survive the periodic drought (winter) of temperate regions?

7. With the exception of spines, all leaves are specialized for photosynthesis. Some leaves are specialized for other functions as well. List three other functions for which leaves may be specialized, and give an example of each specialization.

21

Roots, Stems, and Transport Systems

In the previous chapter, we saw that the plant body consists of three parts—leaves, stems, and roots—which are connected by systems that transport water, minerals, and food. For the plant to live, water and minerals must move from the roots through the stem to the leaves, and sugars must move from the photosynthetic cells of the leaves to the nonphotosynthetic parts of the plant. In this chapter, we shall first examine the characteristics of roots and stems. Then we shall consider the processes by which substances are transported throughout the plant body.

Roots

Roots are specialized structures that anchor the plant and take up water and essential minerals. Characteristically, the root system makes up more than half of the plant body. The lateral spread of tree roots is usually greater than the spread of the crown of the tree. For example, in a study made on a four-month-old rye plant, the total surface area of the root system was calculated to be 639 square meters, 130 times the surface area of the leaves and stem.

Types of Roots

The first root of a plant, which originates in the embryo, is called the primary root. In dicots, this root develops into a *taproot*, which, in turn, gives rise to lateral or branch roots (Figure 21–1a). In monocots, the primary root is usually short-lived and the final root system develops from the base of the stem; roots of this kind are called *adventitious roots*. ("Adventitious" describes any structure growing from an unusual place.) These adventitious roots and their branches develop into a fibrous root system (Figure 21–1b).

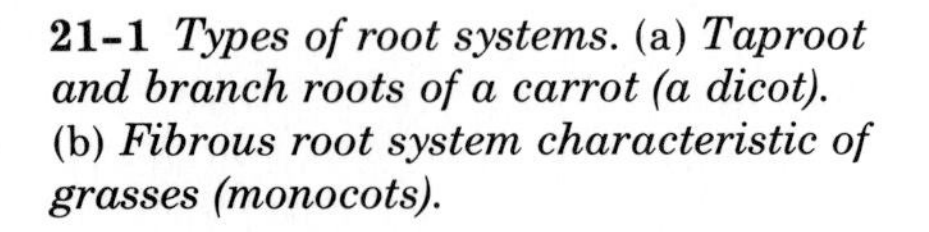

21–1 *Types of root systems.* (a) *Taproot and branch roots of a carrot (a dicot).* (b) *Fibrous root system characteristic of grasses (monocots).*

Special Adaptations of Roots

Aerial roots are adventitious roots produced from aboveground structures. In some plants, such as corn, aerial roots serve as prop roots (Figure 21–2a). Trees that grow in swamps, such as the red mangrove and

21-2 (a) *Prop roots of corn. These are adventitious roots, arising from the stem.*
(b) *Aerial roots of bald cypress, a conifer. Such roots serve to anchor the plant in a marshy habitat and probably provide aeration as well. The bald cypress drops its needles at the end of each growing season—hence, its name. The only other deciduous conifer in North America is the larch.*

the bald cypress (Figure 21-2b), often have aerial roots. In swampy areas the soil is low in oxygen, and, in some species, the aerial roots are believed not only to anchor the plant but also to supply the root cells with the oxygen needed for respiration.

Most roots are storage organs, and in some species the roots are highly specialized for this function. Beets, carrots, and sweet potatoes are examples of such roots.

The Structure of the Root

The internal structure of the root is comparatively simple. There are three concentric layers: the *epidermis*, the *cortex*, and the *vascular cylinder* (Figure 21-3).

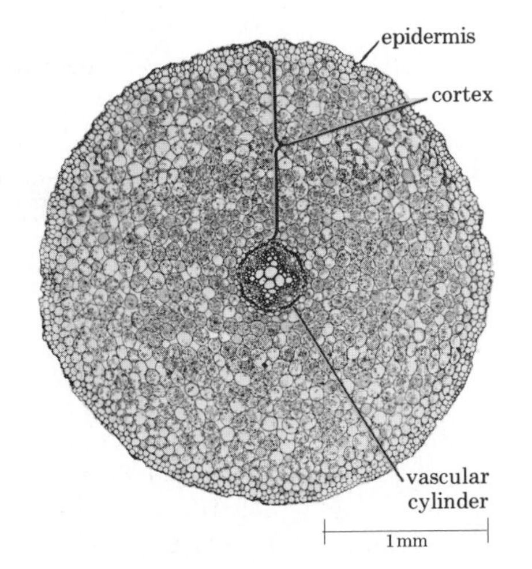

21-3 *Root of a buttercup in cross section. The vascular cylinder is shown in more detail in Figure 21-7.*

21-4 *Root hairs of a radish seedling. Most of the uptake of water and minerals takes place through the root hairs, which begin to form just behind the growing tip of the root.*

The Epidermis

The epidermis, which encloses the entire surface of the root, absorbs water and minerals from the soil. As you would expect, its cuticle is very thin compared with that found on the surface of the leaf epidermis.

The epidermal cells of the root are characterized by fine, tubular outgrowths, known as root hairs (Figure 21-4). Root hairs are slender extensions of the epidermal cells; in fact, the nucleus of the epidermal cell is often found within the root hair. Much of the water and minerals that enter the root is taken up through these delicate outgrowths of the epidermis. In the study of the rye plant previously mentioned, the roots were estimated to have some 14 billion root hairs. Placed end to end, they would have extended more than 10,000 kilometers.

The Cortex

As you can see in Figure 21-3, the cortex occupies by far the greatest volume of the young root. The cells of the cortex are parenchyma cells, like those of the mesophyll of the leaf; however, as you would expect, the parenchyma cells of the root lack chloroplasts. They store starch and other organic substances. The tissue of the cortex contains many air spaces. Oxygen-containing air from the soil enters these spaces through the epidermal cells and is used by the cortical cells in respiration.

Unlike the rest of the cortex, the cells of the innermost layer, the *endodermis*, are compact and have no spaces between them. Each endodermal cell is encircled by a *Casparian strip*, a fatty band within the cell wall (Figure 21-5). The strip is continuous and is not permeable to water. Therefore, water and dissolved substances, which move freely around the other cortical cells and through their cell walls, must pass through the cell membranes of endodermal cells. As you will recall (page 79), water, oxygen, and carbon dioxide pass easily through cell membranes, but many ions and

21-5 *Diagrammatic cross section of a root, showing the two pathways of uptake of water and minerals. Along pathway A, water moves by osmosis and solutes by active transport through the cell membranes of a series of living cells. Along pathway B, water flows through the cell walls and along their surfaces, and the solutes flow with the water or by diffusion. Notice the location of the Casparian strip and how it blocks off pathway B all around the vascular cylinder of the root. In order to pass the Casparian strip, the solutes must be transported through the cell membranes of the endodermal cells, as in pathway A.*

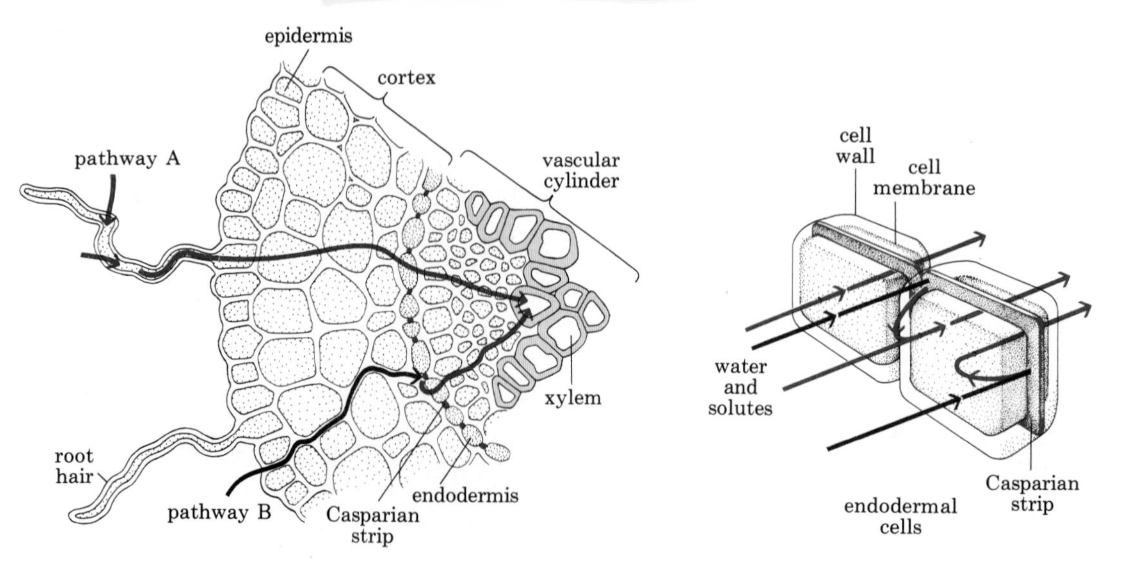

21-6 *Cross section of the root of a corn plant, showing the vascular cylinder enclosing the pith. Part of a branch root can be seen at the lower right.*

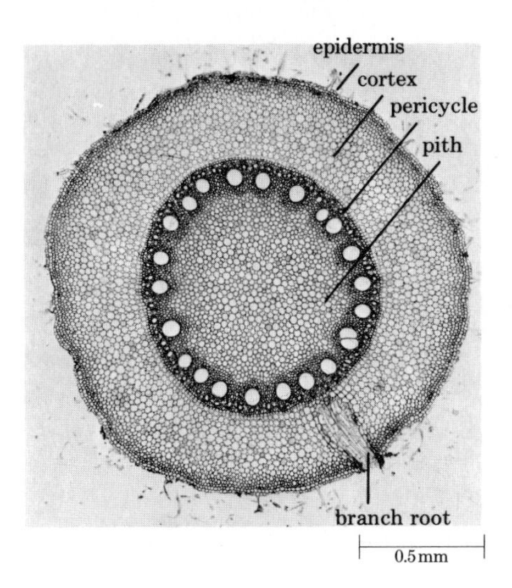

other substances do not. Thus, the membranes of the endodermal cells regulate the passage of such substances into the vascular tissues of the root and so determine what enters the plant body.

The Vascular Cylinder

The vascular cylinder of the root consists of the vascular tissues (xylem and phloem) completely surrounded by one or more layers of cells, the *pericycle*. Branch roots arise from the pericycle. In most species, the vascular tissues are grouped in a solid cylinder, as shown in Figure 21-3. In some, however, they form a hollow cylinder around a *pith*, a central core of tissue (Figure 21-6). Figure 21-7 shows the details of the vascular cylinder of a buttercup root. The vascular tissues of the root are continuous with those of the stem.

Stems

The Structure of the Stem

The stem holds the leaves up to the light and provides for the transportation of substances to and from the leaves. The outer surface of a young green stem, like that of the leaf and the root, is made of epidermal cells. Like the leaf, the green stem is covered with a fatty cuticle, contains stomata, and is photosynthetic.

Ground Tissue

The bulk of the tissue of a young stem is known as the *ground tissue*. Like the mesophyll of the leaf and the cortex of the root, ground tissue is composed mostly of parenchyma cells. The turgor of these cells provides the chief support for young green stems.

Stems also have supporting tissues formed from other types of cells. These cells generally have secondary walls (see page 64) containing lignin, a complex macromolecule that impregnates the cellulose and toughens and hardens it. Two types of cells are of particular importance: fibers and

21-7 *Details of the vascular cylinder shown in Figure 21-3. The endodermis is considered part of the cortex. The outermost layer of the vascular cylinder is the pericycle, from which branch roots arise. The conducting tissues, the xylem and the phloem, are enclosed by the pericycle.*

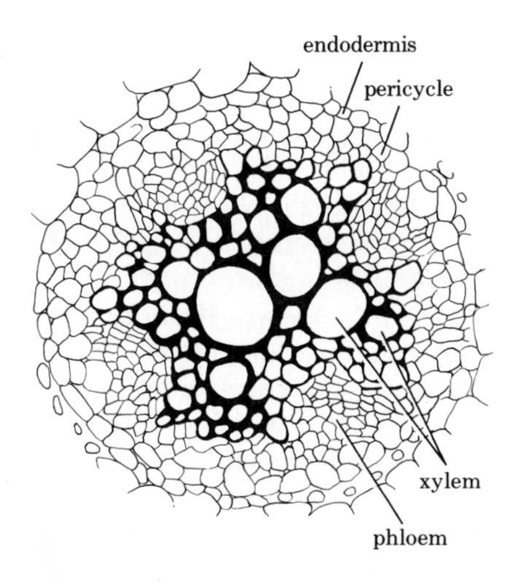

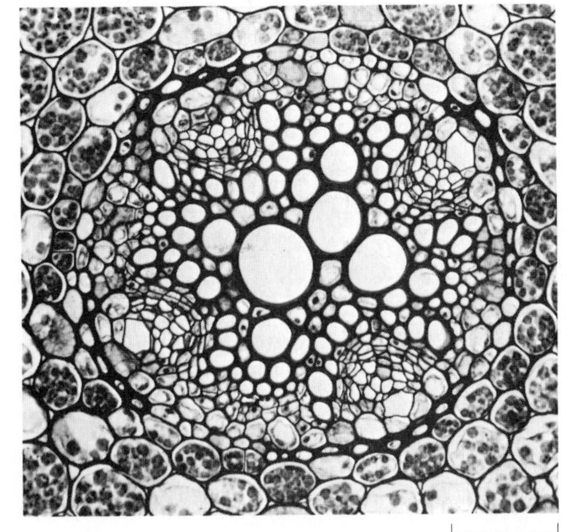

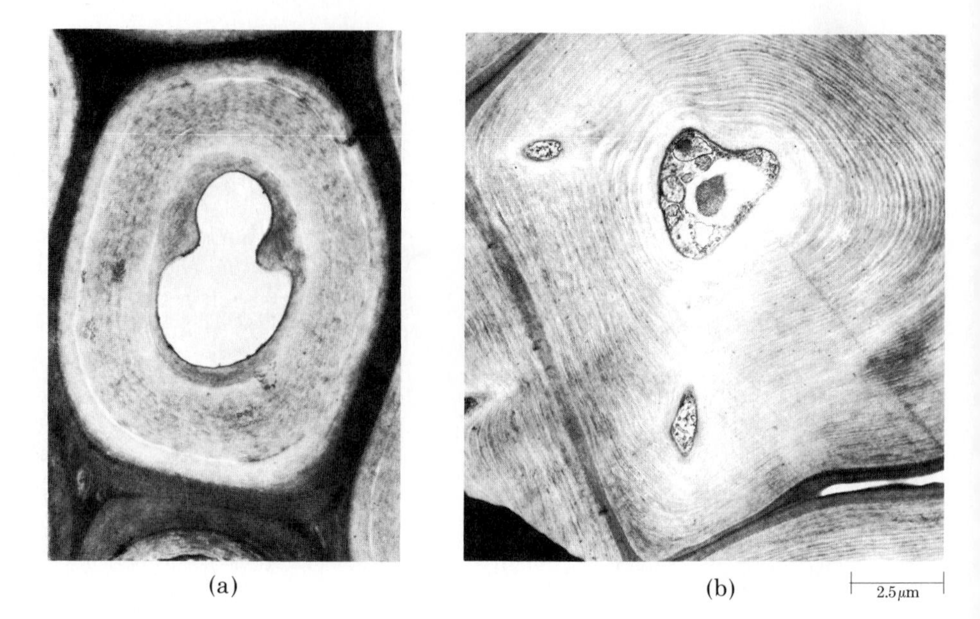

21-8 *Two types of cells found in the ground tissue of stems.* (a) *Fiber. Its specialization is thickened, often lignified, cell walls that give it strength and rigidity. Many fibers, but not all, are dead at maturity, like the cell shown here.* (b) *Sclereid. This type of cell, which has very thick lignified walls, is often found in seeds and fruits, as well as in stems. This sclereid is from a pear; sclereids give the fruit its characteristic gritty texture. This particular cell is alive, but many sclereids are dead at maturity.*

sclereids. Fibers, which are elongated, somewhat elastic cells (Figure 21–8a), typically occur in strands or bundles arranged in patterns characteristic of the plant. They are often associated with the vascular tissues. Plant fibers such as flax, hemp, jute, sisal, and raffia have long been used in human artifacts, including baskets, rope, and cloth. Sclereids, or stone cells (Figure 21–8b), which are variable in form, are also common in stems. Layers of sclereids are found in seeds, nuts, and fruit stones as well, where they form the hard outer coverings.

Conducting Tissues

The conducting tissues of the stem—the xylem and the phloem—are embedded in the ground tissue. Xylem conducts water and minerals from the roots to other parts of the plant body. Phloem conducts the products of photosynthesis, chiefly in the form of sucrose, from the leaves to the nonphotosynthetic cells of the plant. It is customary to think of xylem as transporting water up and phloem as transporting sugars down, but if you think of the various shapes of plants you can see that water must also often be transported laterally, as along a tendril, or even down, as to the branches of a weeping willow. Conversely, sugars must often go upward, as into a flower or fruit.

The conducting cells of the xylem are *tracheids* and *vessel elements*. Both of these cell types have thick secondary walls containing lignin. Also, both are dead at maturity. Tracheids are long, thin cells that overlap one another on their tapered ends (Figure 21–9a). These overlapping surfaces contain thin areas, pits, where no secondary wall has been deposited. Water passes from one tracheid to the next through the pits. Vessel elements differ from tracheids in that their end walls contain perforations or are broken down entirely (Figure 21–9b and c). Thus, the vessel elements form a continuous *vessel*, which is a more effective conduit than a series of tracheids. Most gymnosperms have only tracheids, but angiosperms may have both tracheids and vessels.

The conducting cells of the phloem are called *sieve-tube elements* (Figure 21–10). A *sieve tube* is a vertical column of sieve-tube elements

21-9 Tracheids and vessel elements are the conducting cells of the xylem in angiosperms. (a) *Tracheids are a more primitive and less efficient type of conducting cell. Water moving from one tracheid to another passes through pits. Pits are not perforations but areas in which there is no secondary cell wall. Water moving from one tracheid to another passes through two primary cell walls and the middle lamella.*

Vessel elements differ from tracheids in that the primary walls of vessel elements are perforated at the end where they are joined with other vessel elements. (b) *There may be numerous small perforations in adjoining walls of vessel elements, or* (c) *the adjoining walls may break down completely as the cells mature, forming a single opening. Vessel elements are also characteristically shorter and wider than tracheids and their adjoining walls are less oblique.*

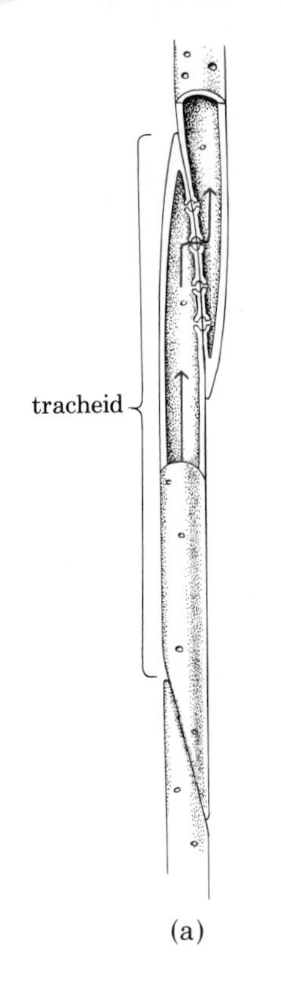

(a)

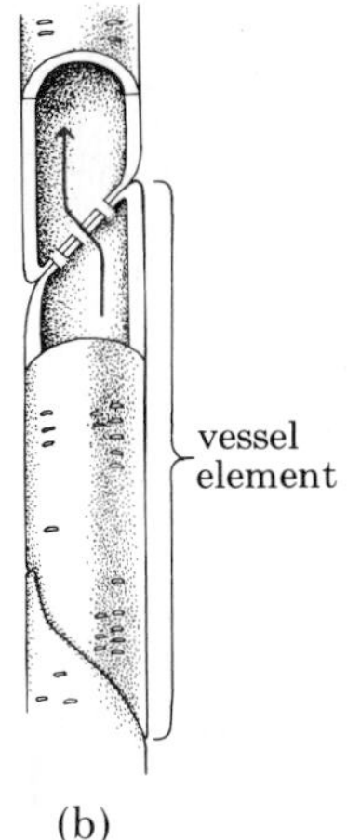

(b)

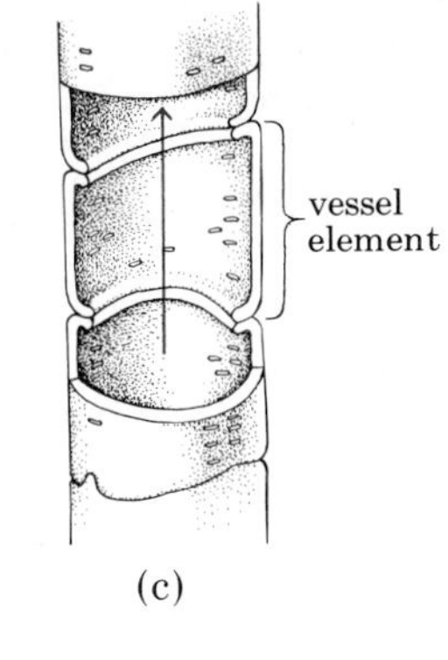

(c)

joined by their end walls. These end walls, called sieve plates, have openings leading from one sieve-tube element to the next.

The sieve-tube elements, which are alive at maturity, are filled with a watery fluid, called sieve-tube sap. They also contain a protein substance that forms a film along the longitudinal surface of the cell. The nucleus of a sieve-tube element disintegrates as the cell matures, as do many of the organelles. Sieve-tube elements are associated with *companion cells*, which are thought to provide nuclear control and energy for the sieve tubes. A sieve-tube element can function only if its cell membrane is intact, and it is likely that the companion cell helps to maintain the membrane.

Both phloem and xylem tissues contain parenchyma cells, which store food and water, and supporting fibers.

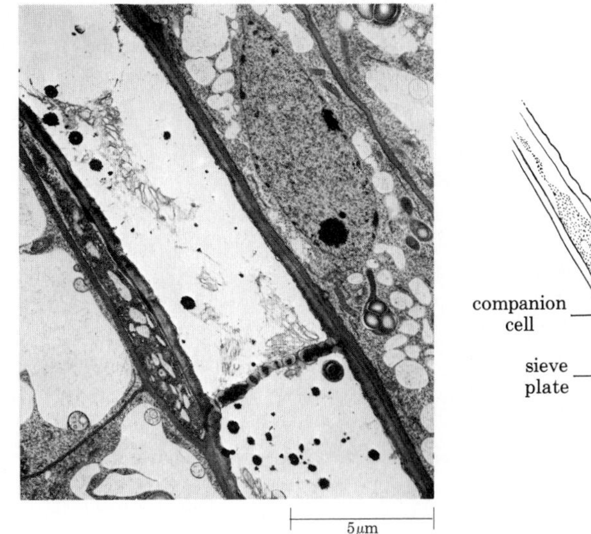

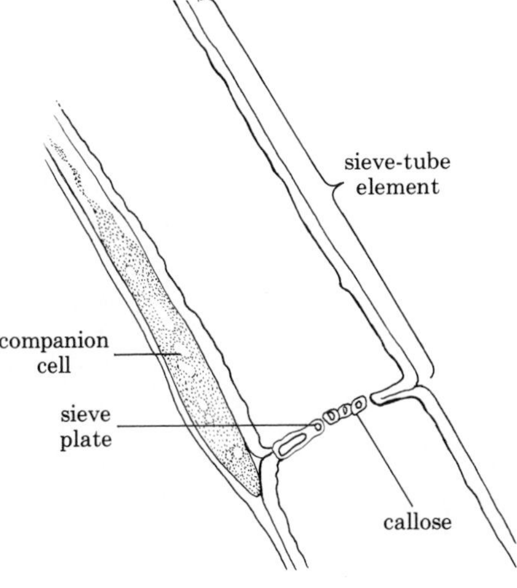

21-10 In angiosperms, the conducting elements of the phloem are sieve tubes, made up of individual cells, the sieve-tube elements. These cells, which lack nuclei at maturity, are usually found in close association with companion cells, which do have nuclei. Sieve-tube elements are joined by a sieve plate. The light-colored (electron-transparent) material near the pore is callose, a polysaccharide characteristically associated with sieve plates, especially in old or injured cells. Callose plugs the pores when a sieve-tube element is injured, thus preventing leakage. Immediately to the left of the upper sieve-tube element, identifiable by its dense cytoplasm, is a portion of a companion cell. A parenchyma cell is to the right.

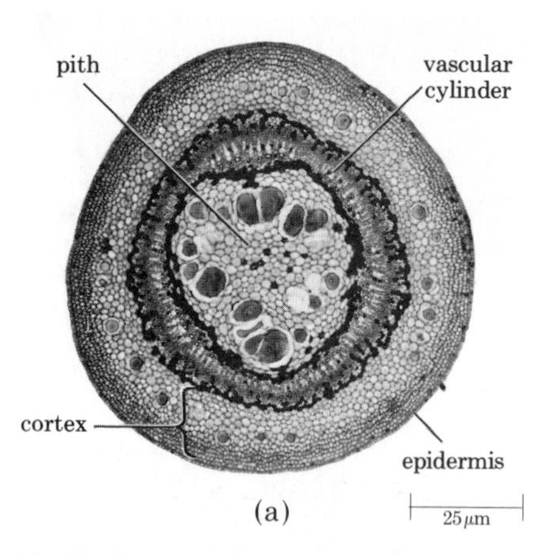

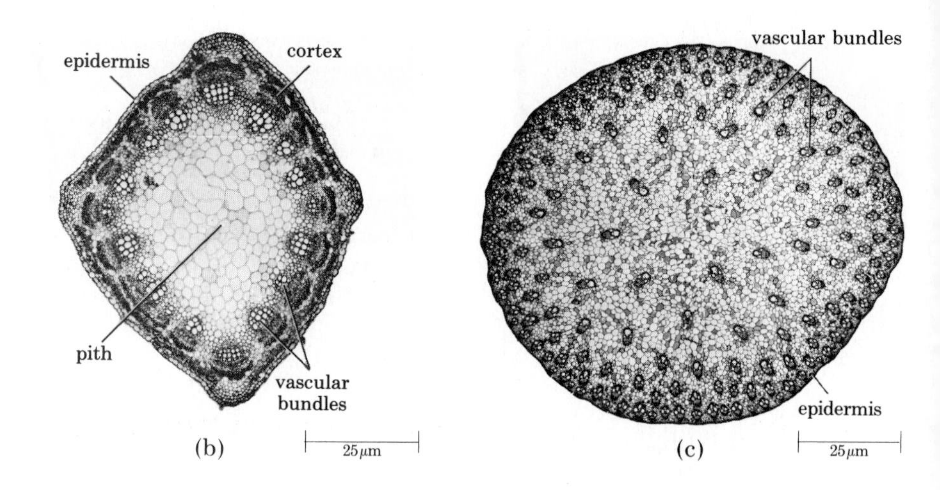

21-11 *Cross sections of two dicot stems and one monocot stem.* (a) *In the linden, a dicot, the vascular tissue forms a continuous cylinder.* (b) *In alfalfa, also a dicot, the cylinder is made up of separate vascular bundles.* (c) *In corn, a monocot, numerous vascular bundles are scattered throughout the ground tissue.*

Stem Patterns

In green stems, the xylem and the phloem are usually arranged in longitudinal parallel strands, the vascular bundles. In dicots, the vascular bundles form a ring, the vascular cylinder, around a central area of ground tissue called the pith (Figures 21-11a and b). The cylinder of ground tissue outside the vascular bundles is called the cortex. Within each bundle, the xylem is characteristically on the inside, adjacent to the pith, and the phloem is on the outside, adjacent to the cortex. In monocots, the vascular bundles are scattered throughout the ground tissue (Figure 21-11c).

Special Adaptations of Stems

The stems of some climbing plants coil themselves around the structures on which they are growing. Others produce modified branches in the form of tendrils. (As we saw previously, tendrils may also be modified leaves.) The tendrils of grape, English ivy, and Virginia creeper are all modified stems.

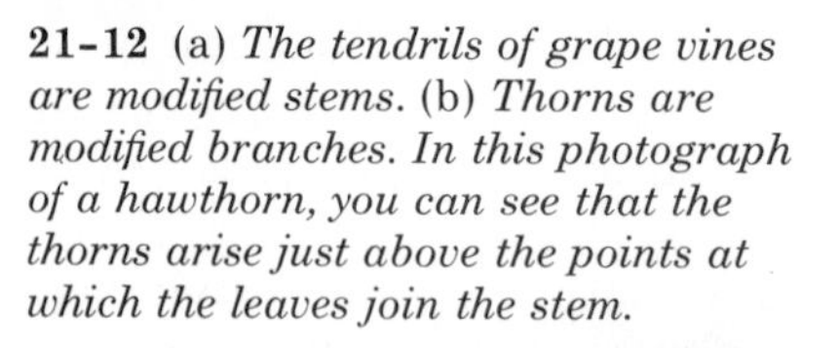

21-12 (a) *The tendrils of grape vines are modified stems.* (b) *Thorns are modified branches. In this photograph of a hawthorn, you can see that the thorns arise just above the points at which the leaves join the stem.*

(a)

(b)

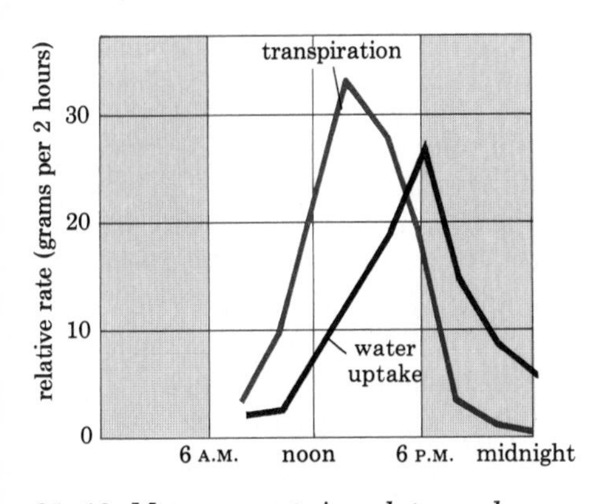

21-13 *Measurements in ash trees show that a rise in water uptake follows a rise in transpiration. These data suggest that the loss of water generates forces for its uptake.*

Runners, such as those found in most varieties of strawberry, are long, slender stems that grow along the surface of the soil. Rhizomes are underground stems that often, as in the grasses, produce upright stems bearing leaves and flowers.

Stems may also be adapted for food storage. White potatoes are enlarged rhizomes known as tubers. In some species of plants that grow in arid environments, stems are adapted for water storage. The water-storing tissues consist of large parenchyma cells that lack chloroplasts; the stem of a cactus may be 98 percent water by weight.

The Movement of Water: Transpiration

In order to carry out photosynthesis, plants must exchange gases with the surrounding air. In the course of this exchange, as we noted in Chapter 20, they lose water through the stomata and, to a lesser extent, through the cuticle. The loss of water vapor from the plant body is known as *transpiration.*

The quantity of water passing through a plant is enormous—far greater than that used by an animal of comparable weight. An animal requires less water because most of its water remains in its body and recirculates; in vertebrates, the water is in the form of blood plasma. By contrast, in plants, more than 90 percent of the water that enters the roots is given off into the air as water vapor. A single corn plant needs 160 to 200 liters of water from seed to harvest, and a hectare (2.47 acres) of corn requires almost 5 million liters of water a season.

Water enters the body of most plants almost entirely through the roots. During periods of rapid transpiration, water may be removed from around the roots so quickly that the soil in the vicinity of the roots becomes depleted. Water will then move slowly by diffusion and capillary action through the soil toward the depleted region near the roots. (For a review of the principles of water movement, see pages 27 to 28 and 76 to 79. An understanding of these processes is necessary for understanding the rest of this chapter.)

21-14 *Guttation droplets on the edge of a rose leaf. Guttation, the loss of liquid water, is a result of root pressure. The water escapes through specialized pores located near the ends of the principal veins of the leaf. Guttation, which is restricted to relatively small plants, usually occurs at night when the air is moist.*

Root cells, like other living parts of the plant, contain a higher concentration of solutes (both organic and inorganic) than does soil water. As a result, water from the soil enters the roots by osmosis. Osmosis is sufficient to move water a short distance up the stem, a phenomenon known as root pressure. Guttation (Figure 21-14) is a visible consequence of root pressure. But how can water reach 20 meters high to the top of an oak tree, travel three stories up the stem of a vine, or move 125 meters up in a tall redwood?

One important clue is the observation that during times when the most rapid transpiration is taking place—which is, of course, when the flow of water up the stem must be the greatest—xylem pressures are characteristically negative (less than atmospheric pressure). The existence of negative pressure can be demonstrated readily. If you peel a piece of bark from a transpiring tree and make a cut in the xylem, no sap runs out. In fact, if you place a drop of water on the cut, the drop will be drawn in.

What is the pulling force? It is not simple suction, as the negative pressure might indicate. Suction simply removes air from a system so that the water (or other liquid) is pushed up by atmospheric pressure. But atmospheric pressure is only enough to raise water (against no resistance) about 11 meters at sea level, and many trees are much taller than 11 meters.

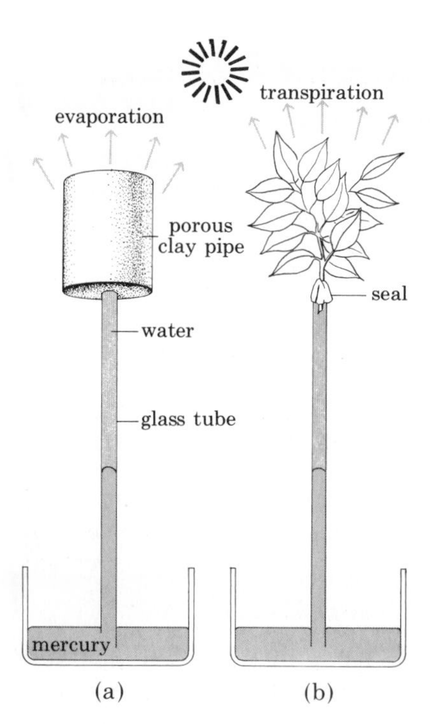

21-15 (a) *A simple model that illustrates the cohesion-tension theory. A piece of porous clay pipe, closed at both ends, is filled with water and attached to the end of a long, narrow glass tube also filled with water. The water-filled tube is placed with its lower end below the surface of a volume of mercury contained in a beaker. As water molecules evaporate from the pores in the pipe, they are replaced by water "pulled up" through the narrow glass tube in a continuous column. As the water evaporates, mercury rises in the tube to replace it.* (b) *Transpiration from plant leaves results in sufficient water loss to create a similar negative pressure.*

The Cohesion-Tension Theory

According to the now generally accepted theory, the explanation is to be found not only in the properties of the plant but also in the remarkable properties of water, to which the plant has become exquisitely adapted. As we pointed out in Chapter 2, in every water molecule, two hydrogen atoms are covalently bonded to a single oxygen atom. Each hydrogen atom is also held to the oxygen atom of a neighboring water molecule by a hydrogen bond. The cohesion resulting from this secondary attraction can produce a tensile, or holding, strength of as much as 140 kilograms per square centimeter (2,000 pounds per square inch) in a thin column of water.

In the leaf, water evaporates, molecule by molecule, from the cells into the intercellular spaces within the leaves. The water potential of the leaf cell falls, and the water from the vessels or tracheids moves, molecule by molecule, into the leaf cell. But each molecule in the xylem vessel is linked to other molecules in the vessel. They, in turn, are linked to others, forming one long, narrow, continuous thread of water reaching right down to a root tip. As the molecule of water moves through the stem and into the leaf, it tugs the next molecule along behind it.

Because the diameter of the vessels is very small and because the water molecules adhere to the walls, even as they are cohering to one another, gas bubbles, which could rupture the column, do not usually form. The pulling action, molecule by molecule, causes the negative pressure observed in the xylem. The technical term for a negative pressure is *tension*, and this theory of water movement is called the *cohesion-tension theory*.

The principle of cohesion-tension is illustrated in Figure 21-15. As indicated in the diagram, the power for this process comes not from the plant, which plays only a passive role in transpiration, but directly from the energy of the sun.

The Uptake of Minerals

Minerals—naturally occurring inorganic substances—are taken up in solution by the roots and travel through plants with the transpiration stream. As you saw in Figure 21-5, the cells of the endodermis play a major role in determining which substances enter the xylem.

The need of plants for some elements can be determined by an analysis of the molecules of which plant cells are composed. Other requirements are identified by studying the capacity of plants to grow in distilled water to which small amounts of various minerals are added. This sounds easier than it is. Sometimes, it has been found, a substance—boron, for example—is needed in such small amounts that it is almost impossible to set up experimental conditions that exclude it. Thus, it is difficult to prove that its absence is lethal.

The greatest requirement of plants is for nitrogen. Although nitrogen is the most abundant element in air, most plants cannot use gaseous nitrogen. They are dependent upon nitrogen-containing ions—ammonium (NH_4^+) and nitrate (NO_3^-)—from the soil. Plant cells reduce the nitrate ions to ammonium ions, and the ammonium ions are then combined with carbon-containing compounds to form amino acids, nucleotides, chlorophyll, and other nitrogen-containing compounds.

Table 21-1 lists the mineral elements required by plants, the form in which they are usually absorbed, and some of the uses plants make of them. Six elements (macronutrients) that plants require in relatively large

Table 21-1 A summary of mineral elements required by plants

Element	Form in Which Absorbed	Approximate Concentration in Whole Plant (as % of Dry Weight)	Some Functions
Macronutrients			
Nitrogen	NO_3^- (or NH_4^+)	1–3%	Component of amino acids, proteins, nucleotides, nucleic acids, chlorophyll, and coenzymes.
Potassium	K^+	0.3–6%	Involved in osmosis and ionic balance and in opening and closing of stomata. Activator of many enzymes.
Calcium	Ca^{2+}	0.1–3.5%	Component of cell walls. Enzyme cofactor. Involved in cell membrane permeability.
Phosphorus	$H_2PO_4^-$ or HPO_4^{2-}	0.05–1.0%	Formation of "high-energy" phosphate compounds (ATP and ADP). Component of nucleic acids and of several essential coenzymes.
Magnesium	Mg^{2+}	0.05–0.7%	Part of the chlorophyll molecule. Activator of many enzymes.
Sulfur	SO_4^{2-}	0.05–1.5%	Component of some amino acids, proteins, and coenzyme A.
Micronutrients			
Iron	Fe^{2+}, Fe^{3+}	10–1,500 parts per million (ppm)	Required for chloroplast development. Component of cytochromes.
Chlorine	Cl^-	100–10,000 ppm	Involved in osmosis and ionic balance. Probably essential in photosynthesis in the reactions in which oxygen is produced.
Copper	Cu^{2+}	2–75 ppm	Activator of some enzymes.
Manganese	Mn^{2+}	5–1,500 ppm	Activator of some enzymes. Required for oxygen release in photosynthesis.
Zinc	Zn^{2+}	3–150 ppm	Activator of many enzymes.
Molybdenum	MoO_4^{2-}	0.1–5.0 ppm	Required for nitrogen metabolism.
Boron	BO^{3-} or $B_4O_7^{2-}$ (borate or tetraborate)	2–75 ppm	Functions unknown. Possibly involved in carbohydrate transport.
Elements Essential to Some Plants or Organisms			
Cobalt	Co^{2+}	Trace	Required by nitrogen-fixing microorganisms.
Sodium	Na^+	Trace	Involved in osmosis and ionic balance, probably for many plants not essential. Required by some desert and salt-marsh species and may be required by all plants that utilize C_4 photosynthesis.

amounts and seven (micronutrients) that are needed in smaller quantities have been identified.

You might anticipate that organisms make use of what is most readily available; indeed, they seem to have done this when life originated from elements in the gases of the primitive atmosphere. But the table reveals some findings that you might not expect. Sodium, for instance, which is one of the most abundant of the elements, is apparently not required at all by most plants. The fact that plants evolved with no functions requiring sodium is even more striking when you consider that sodium is vital to animals. In the seas, where both plant and animal life seem to have originated, sodium is the most abundant mineral element and is far more readily available than potassium, which it closely resembles in its essential properties. Similarly, although silicon and aluminum are almost always present in large amounts in soils, few plants require silicon and apparently none requires aluminum. On the other hand, most plants need molybdenum, which is relatively rare.

Carnivorous Plants

Many species of plants are meat eaters; their diets include insects, other invertebrates, and even some vertebrates, such as small birds and frogs. Unlike carnivorous animals, they do not utilize their prey for energy but rather as a source of mineral elements, particularly nitrogen and phosphorus. As Darwin noted in his book on this subject, published in 1875, these plants are usually found in swamps, bogs, and peat marshes where acids leach the soil of nutrients.

(a) Yellow pitcher plants attract insects into their flowerlike tubular leaves by means of nectar. Following the nectar over the rim, the insect finds itself on a carpet of fine, transparent hairs. When the nectar trail ends, and the insect turns to exit, it encounters the points of these fine hairs, pointing downward and blocking its way. If it moves inward, it encounters a slick waxed surface from which it slides or drops into a foul-smelling broth of rainwater, digestive enzymes, and decomposing bacteria at the pitcher's base.

(b) The sundew is a tiny plant, often only 2 to 5 centimeters across, with club-shaped tentacles on the upper surface of its leaf. These tentacles secrete a clear sticky liquid that attracts insects. When an insect is caught by one tentacle, the others bend inward toward it until the insect drowns, its air passages filled with the mucilage-like fluid. The tentacles also secrete digestive enzymes.

(c, d) The Venus flytrap has leaves that close around any insect moving on the surface of the leaf. The closing of the leaf is triggered by touching one or two of the three trigger hairs in the middle of each leaf lobe. Venus flytraps are found in nature only on the coastal plain of North and South Carolina, usually on the edges of wet depressions and pools. It has long been believed that the Venus flytrap lures its victims by exuding nectar, but recent studies indicate that ants and other insects—despite its name, its usual diet in the wild consists of crawling invertebrates—visit the leaves by chance.

(e) The common bladderwort is a free-floating aquatic plant. The traps are tiny, flattened, pear-shaped bladders. Each bladder has a mouth guarded by a hanging door. The tripping mechanism consists of four stiff bristles near the lower free edge of the door. When a small animal brushes against these bristles, the hairs distort the lower edge of the door, causing it to spring open. Water then rushes into the bladder, carrying the animal with it, and the door closes behind. The animals decay by bacterial action, and the minerals and organic compounds released are taken up by the cellular walls of the trap. The undigested exoskeletons remain in the bladders.

(a) (b) (c) (d) (e)

"Fungus-Roots": Mycorrhizae

In recent years, it has been found that a particular kind of association between fungi and the roots of plants plays a crucial role in mineral uptake. If seedlings of many forest trees are grown in nutrient solutions and then transplanted to prairie and other grassland soils, they fail to grow. Eventually they die from malnutrition, despite the fact that soil analysis shows that there are abundant nutrients in the soil. If a small amount (0.1 percent by volume) of forest soil containing fungi is added to the soil around the roots of the seedlings, however, they will grow promptly and normally. The restoration of normal growth is caused by the functioning of mycorrhizae—associations between roots and fungi. Mycorrhizae are now thought to occur in more than 90 percent of all families of plants.

In mycorrhizal associations, the fungus may form a sheath of hyphae around the root. Only the cortex is actually invaded by the fungus. Roots with mycorrhizae usually lack root hairs, the role of water and mineral uptake from the soil evidently being at least partially assumed by the fungus.

The exact relationship between roots and fungi is not known. Apparently the roots secrete sugars, amino acids, and possibly some other organic substances that are used by the fungi. Although the evidence is just now accumulating, it seems that the chief contribution of the fungi is in water uptake and in converting minerals in the soil and in decaying material into an available form.

A study of the fossils of early vascular plants has revealed that mycorrhizae were as frequent in them as they are in modern vascular plants. This has led to the interesting suggestion that the evolution of mycorrhizal associations may have been the critical step allowing plants to make the transition to the bare and relatively sterile soils of the then-unoccupied land.

21-16 *Effects of mycorrhizae on tree nutrition. Nine-month-old seedlings of white pine were grown for two months in a sterile nutrient solution and then transplanted to prairie soil. The seedlings on the left were transplanted directly. The seedlings on the right were placed in forest soil for two weeks before being transplanted to the prairie.*

The Movement of Sugars: Translocation

The photosynthetic cells of the plant, which are most abundant in the leaves, provide the food energy for all of the other cells of the plant. The process by which the products of photosynthesis are transported to other tissues is known as *translocation.*

Translocation has proved a very difficult process to study because the sieve-tube elements of the phloem are quite delicate. When they are disturbed, callose and the protein lining the interior of the cells plug the pores in the sieve plates (see Figure 21–10, page 301), preventing the movement of substances through them.

Two investigative techniques have proved useful in circumventing this difficulty. One technique is to expose plants to radioactive carbon dioxide ($^{14}CO_2$). The sugars made from this carbon dioxide are thus labeled with a radioactive tag, and so their passage through the plant can be traced. Another technique involves the use of aphids for taking samples of sieve-tube sap (Figure 21–17).

Data gathered by these techniques indicate that sieve-tube sap contains (by weight) 10 to 25 percent solutes, more than 90 percent of which are sugars, mostly sucrose. Low concentrations of amino acids and other nitrogen-containing substances are also present. The rate of movement of these substances along the sieve tube is remarkably fast: In one set of experiments, for example, it was estimated that the sap was moving at a rate of about 100 centimeters per hour, far faster than could be accounted for by diffusion.

The Pressure-Flow Hypothesis

The most popular current explanation for translocation is the *pressure-flow hypothesis.* According to this hypothesis, the solutes move in solutions that, in turn, move because of differences in water potential caused by concentration gradients of sugar.

(a)

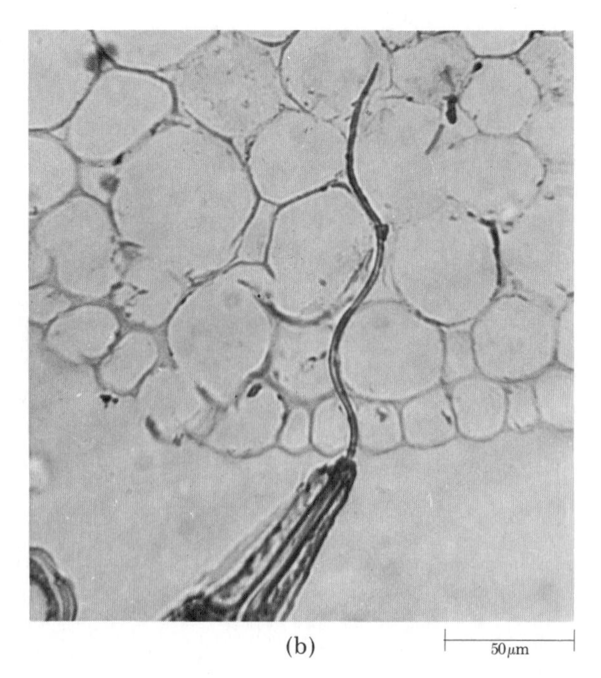

(b)

21–17 *Assistance by aphids.* (a) *Aphids are very small insects that feed on plants, sucking out their juices; you probably have seen them on rose bushes.* (b) *The aphid, as the micrograph reveals, drives its sharp mouthparts, or stylets, like a hypodermic needle between the epidermal cells and then taps the contents of a single sieve-tube element. If the aphid is anesthetized, it is possible to sever the stylets and leave them undisturbed in the cell. The fluid may continue to exude through the stylets from the sieve tube for several days, and pure samples of the fluid flowing through the sieve tube can be collected for analysis without damaging the sieve tube or interfering with its function.*

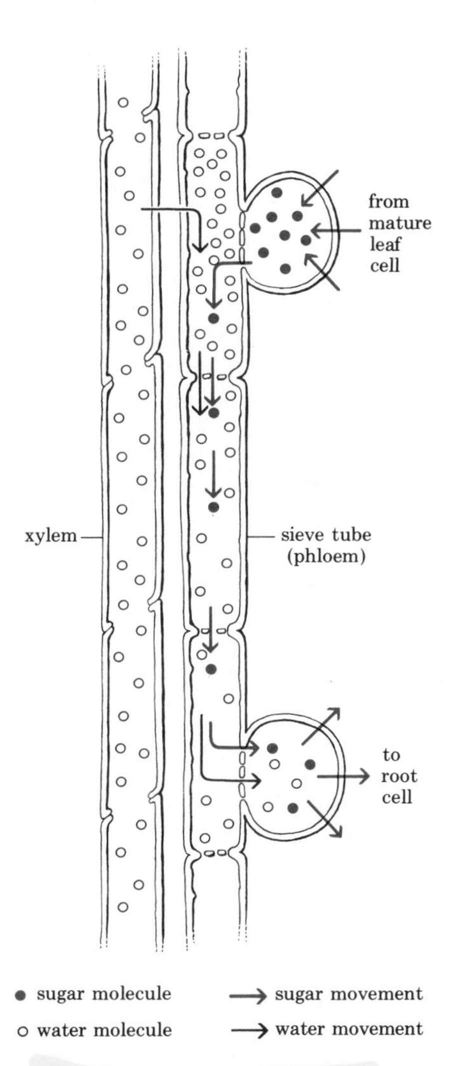

21-18 A model of the pressure-flow mechanism. Sugar molecules are pumped into a sieve tube by active transport. As a consequence of the increased concentration of sugar, water enters the sieve tube by osmosis. Sugar molecules, required by the root cells, are removed from the phloem by active transport, and the sugar concentration of the sieve tube falls. As a consequence of the lowered sugar concentration, water moves out of the sieve tube by osmosis. Because of the active secretion of sugar molecules into the sieve tube at one end and the active absorption of sugar molecules from the sieve tube at the other end, a flow of sugar solution takes place along the tube.

It is believed that sugar manufactured in the photosynthetic cells of the leaf is pumped by active transport into the sieve tubes. The energy required for this active transport is expended not by the sieve tubes but by other phloem cells (such as companion cells) bordering the sieve tubes. The added sugar decreases the water potential in the sieve tube and causes water to move into the sieve tube from the xylem by osmosis. At a nutrient-utilizing tissue—for example, a root tip—the sugar is removed from the sieve tube, again as the result of active transport by neighboring cells. The water molecules follow the sugar molecules out, again by osmosis. Thus the water flows in at one end of the sieve tube and out at the other. Between these two points, the water and its solutes, including sugar, move passively by bulk flow (Figure 21-18).

Translocation carries sugars and other nutrients not only to the cells of the root but also to the other nonphotosynthetic parts of the plant, including the flowers. As we shall see in the next chapter, sugars are often concentrated in the flower—a fact of considerable importance in the evolution of angiosperms.

SUMMARY

Roots are specialized for anchoring the plant in the soil, for absorbing water and minerals, and for transporting them to other parts of the plant body. Structurally, the root consists of three concentric layers: the epidermis, the cortex, and the vascular cylinder. The epidermis is the protective and absorbing tissue. Much of the absorption takes place through fine extensions of the epidermal cells known as root hairs. The cortex consists largely of parenchyma cells adapted to the storage of water and sugars. The innermost layer of the cortex is the endodermis, a single layer of specialized cells whose walls contain a waterproof zone, the Casparian strip. The Casparian strip helps the plant regulate its uptake of minerals. The vascular cylinder consists of xylem and phloem, surrounded by a layer of pericycle, from which branch roots arise.

Stems hold the photosynthetic structures—the leaves—up to the light, conduct water to the photosynthetic cells, and transport sugars from them. The outer surface of a green stem is made up of epidermal cells. The bulk of the stem is ground tissue, which may be divided into an outer cylinder (the cortex) and an inner core (the pith). The ground tissue is largely composed of parenchyma cells but also may contain fibers and sclereids.

The conducting tissues consist of xylem and phloem. In angiosperms, the conducting tissues of the xylem are made up of a series of tracheids and vessel elements. Tracheids and vessel elements characteristically have thick secondary walls and are dead at maturity. The conducting cells of the phloem are the sieve-tube elements, living cells with their end walls perforated, which form continuous sieve tubes. Associated closely with each sieve-tube element are one or more companion cells. Xylem and phloem also contain parenchyma cells and fibers.

Transpiration is the loss of water vapor by plants. As a consequence of transpiration, plants require large amounts of water. Water enters the plant from the soil through the roots and travels through the plant body in the conducting cells of the xylem. According to the cohesion-tension theory, water moves through the tracheids and vessels under negative pressure. Because the molecules of water cling together (cohesion), a continuous column of water molecules is pulled from the root, molecule by molecule, by the evaporation of water above.

Minerals enter the plant body in solution in the water from the soil. Nitrogen, an essential element, is taken up in the form of nitrate and ammonium ions. Mycorrhizae, "fungus-roots," are associations between plant roots and fungi that facilitate the uptake of minerals by the roots.

The movement of organic compounds from the photosynthetic parts of the plant is known as translocation. It takes place in the phloem. According to the pressure-flow hypothesis, sugars are pumped into the sieve tubes in the leaf by active transport and are pumped out of them in various parts of the plant body where they are needed for energy. Water moves into and out of the sieve tubes by osmosis, following the sugar molecules. These processes create a difference in water potential along the sieve tube, and thus water and the sugars dissolved in it move by bulk flow along the sieve tube.

QUESTIONS

1. Distinguish among the following: xylem/phloem; tracheid/vessel element/sieve-tube element; transpiration/translocation; cohesion-tension theory/pressure-flow hypothesis.

2. Sketch cross sections of (a) a root, (b) a dicot stem, and (c) a monocot stem. Label the principal cells or tissues in each sketch, and describe the function of each of the labeled parts.

3. What two cell layers are present in roots but absent in stems? What functions do the cells of these layers perform in the roots, and why are they necessary for the well-being of the plant?

4. What properties of water discussed in Section 1 are important to the movement of water and solutes through plants?

5. We have seen that the tissue of the root cortex contains many air spaces. Also, we noted that some plants growing in swampy areas have aerial roots. You may have discovered that overwatering can kill house plants. What do these data indicate about roots?

6. Transpiration has often been described as a "necessary evil" to the plant. Why is it necessary? Why is it evil?

7. Gardeners advise removing many leaves of a plant after transplanting. Why does this help the plant to survive?

8. Consider a tree transpiring most rapidly at midday and an investigator with a sensitive instrument for measuring changes in the diameter of the trunk. If water is pulled up from the top (cohesion-tension theory), what change in diameter should be observed from night to day? (The change was, in fact, one of the early bits of evidence for this theory.)

9. Most plants cannot live in areas in which there is a high salt concentration, such as salt marshes. Explain.

10. (a) Experts in flower arranging advise recutting the stems of flowers while holding them under water. Explain why. (b) Some florists advise adding ordinary table sugar to the water in which cut flowers are placed. When this is done, some types of flowers will remain fresh for several weeks. What is the explanation?

11. Identify the tissues and cells through which a molecule of water travels from the time it enters the root until it is used in photosynthesis.

22

Plant Reproduction, Development, and Growth

Unlike the higher animals, whose reproduction is almost exclusively sexual, many plants reproduce both sexually and asexually. (Asexual reproduction is often referred to as vegetative reproduction.) Organisms produced by asexual reproduction are genetically identical to their single parent, whereas organisms produced by sexual reproduction, which involves meiosis and fertilization, are genetically different from both parents.

Asexual Reproduction

There are many forms of asexual reproduction among plants. One of the most familiar occurs by means of horizontal stems growing either above ground (runners) or below ground (rhizomes). Strawberries are a common example of plants that propagate by runners, as are spider plants, commonly grown as hanging plants. Plants that reproduce by rhizomes include potatoes, many flowering garden perennials, such as lilies-of-the-valley and irises, and the sod-forming grasses of lawns and pastures. Both runners and rhizomes develop adventitious roots.

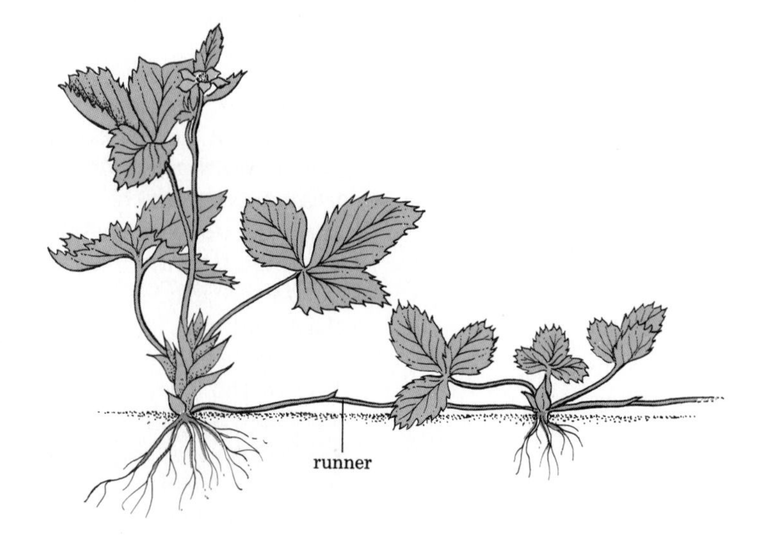

22-1 *Wild strawberry plants reproduce asexually by means of aboveground stems, or runners. Roots and leaves develop at every second node along the modified stems. These plants also form flowers and can reproduce sexually.*

The Seedless Orange

In 1869, the U.S. Department of Agriculture received a letter from Mrs. F. E. C. Schneider, the wife of the first Presbyterian missionary to the state of Bahia in Brazil. She described a remarkable new fruit—a seedless orange—and offered to send some of the trees on which this amazing contradiction grew. The next year, 12 trees arrived in Washington, D.C., where they were carefully nurtured. In 1873, two of the trees were sent to Riverside, California, and it is from these two trees that the entire navel orange industry, in California and in the world, has developed.

At one time, the navel orange reproduced like any other citrus tree—that is, by sexual means—and its fruits contained seeds. Citrus seeds have been found in the ruins of southern Babylon and dated at 4000 B.C. Around 1200 A.D., Arabian traders introduced citrus trees into the Mediterranean area, where the fruits quickly became a valuable supplement to the diet. Although the reasons were not understood, it was soon discovered that eating citrus fruits could prevent scurvy among sailors at sea for long periods of time. (British sailors ate limes—hence the nickname "limeys.") Columbus had orange seeds with him when he landed in Haiti in 1493. Oranges first came to the Brazilian coast in 1549 and took root in Bahia. It was there that the first seedless orange arose through a freak mutation.

The commercial value of an orange with no seeds at all is obvious. But, how can new orange trees be produced if the plant is sterile and its fruits contain no seeds? The answer lies in asexual reproduction. To propagate this mutation, the bark of a young, normal orange tree is carefully slit open. A stem cutting from the mutant variety is placed into the opening, and the wound is sealed with tape. If the stem cutting and the rooted tree have not been damaged too much by the surgery, the grafted shoot will "take" and begin growing. (Plants do not have an immune system that can reject the transplant.) Over a period of time, all of the nonmutant branches are removed from the normal orange tree and replaced with grafts from the mutant variety. A "new" tree has been produced that grows only seedless oranges. Asexual reproduction has totally replaced sexual reproduction, but only with a little help from the horticulturist and Mrs. Schneider.

On the left, an orange with seeds; on the right, a navel orange, which is seedless.

Many members of the lily family, which includes onions and tulips as well as lilies, reproduce asexually by bulbs. Some species of plants with arching stems, such as raspberries, develop new roots from stem tips that touch the soil. New plants may form if the stem is subsequently broken and separated from the parent plant. The leaves of some plants, such as African violets, develop adventitious roots when they are detached from the parent plant and may give rise to new individuals in that way. Some species of *Kalanchoë* produce plantlets along the edges of leaves; the plantlets later drop to the ground and develop into separate plants. If the taproot of a dandelion is injured or broken near the soil line, a callus (a mass of unspecialized cells) forms, plugging the wound. Eventually several new plants grow from this callus tissue. Thus both lawn mowers and grazing animals have highly beneficial effects on the dandelion population.

22-2 Flower of a round-leaved hepatica. Primitive flowers are believed to have been—like the hepatica—radially symmetrical, with numerous separate floral parts. The insect is a pollen-eating beetle.

The capacity of many species of plants to reproduce asexually has been exploited in the development of cultivated varieties of plants for food or ornamental use. Such plants are, of course, genetically identical to the parent stock, and so vegetative reproduction is a way of preserving uniformity. Many plants are reproduced by stem cuttings, which simply involves sticking young stems in the ground and protecting them from drying out until adventitious roots appear. Rooting can often be facilitated by hormone treatment (see page 331). Another artificial form of plant propagation is grafting, in which a stem cutting is attached to the main stem of a rooted woody plant. Most fruit trees and roses are propagated in this way.

Many economically important plants are sterile and can only be propagated vegetatively; these include pineapples, bananas, seedless grapes, navel oranges, and numerous ornamental plants.

Sexual Reproduction: The Flower

The flower is the structure of sexual reproduction in the angiosperms. It is often, by happy coincidence, extraordinarily beautiful to the human eye. Unlike the reproductive organs of animals, which are permanent structures that develop in the embryo, flowers are transitory, developing seasonally. A complete flower consists of four sets of floral appendages, which grow in spirals or whorls. Each floral part, evolutionarily speaking, is a modified leaf.

The outermost parts of the flower are the *sepals*, which are commonly green and obviously leaflike in structure. The sepals enclose and protect the flower bud. Next are the *petals*, collectively called the *corolla*; these are also usually leaf-shaped but are often brightly colored. They advertise the presence of the flower among the green leaves, attracting insects or other animals that visit flowers for their nectar (a sugary liquid) or for other edible substances. These animals, as they forage for food, may carry pollen from flower to flower.

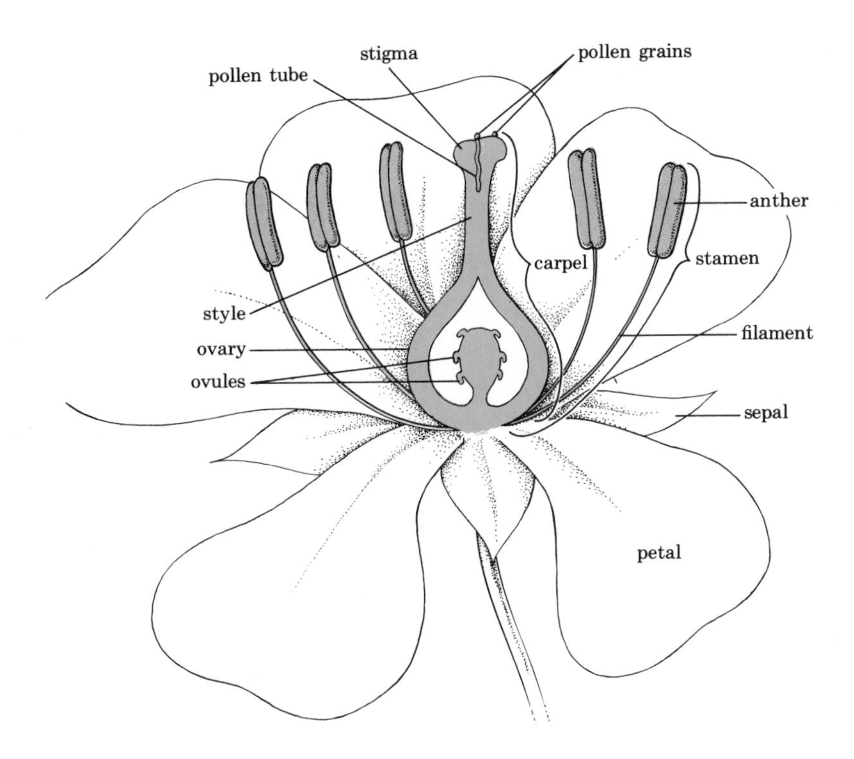

22-3 *Structure of a flower. The sex organs are shown in color. Some flowers have only the male reproductive structures (stamens), some only the female (carpels). A flower that possesses both stamens and carpel, such as this flower, is known as a complete flower.*

Pollen grains have been deposited on the sticky surface of the stigma, and the pollen tube of one is growing down the style to an ovule.

22-4 *Corn is monoecious; that is, it has separate male and female flowers borne on the same plant. The tassels, at the top of the stem, are the male (pollen-producing) flowers. Each thread of "silk," seen emerging from the ear of corn, is the combined stigma and style of a female flower.*

Within the corolla are the *stamens*. Each stamen consists of a single elongated stalk, the *filament*, and at the end of the filament, the *anther*. The pollen grains, formed within the anther, are immature male gametophytes. (For a review of the concept of alternation of generations, see page 139). When ripe, the pollen grains are released, often in large numbers, through narrow slits or pores in the anther.

The centermost appendages of the flower are the *carpels*, which contain the female gametophytes. Typically a carpel consists of a *stigma*, which is a sticky surface specialized to receive the pollen; a slender stalk, the *style*, down which the pollen tube grows; and a base, the *ovary*. Within the ovary are the *ovules*. Each ovule encloses a female gametophyte with a single egg cell. When the egg cell is fertilized, the ovule develops into a seed and the ovary develops into the surrounding fruit.

In some species, flowers are either male or female. Male and female flowers may be present on the same plant, as in corn, squash, oaks, and birches, or on different plants, such as the tree of heaven (*Ailanthus),* the date palm, and the American mistletoe. Species in which separate male and female flowers are borne on the same plant are said to be monoecious ("in one house"); species in which the male and female flowers are on separate plants are known as dioecious ("in two houses").

Evolution of the Flower

As we noted in Chapter 19, the angiosperms are believed to have evolved from the gymnosperms. Gymnosperms, like angiosperms, are vascular plants, although their vascular systems are less highly developed. However, unlike the angiosperms, they do not produce a flower and their seeds are not enclosed in fruit (hence the name "naked seed").

The early gymnosperms from which the angiosperms evolved were probably wind-pollinated, as are modern gymnosperms (see Figure 19-4, page 252). And, as in the modern gymnosperms, the ovule probably exuded droplets of sticky sap in which pollen grains were caught and drawn to the female gametophytes. Insects, probably beetles, feeding on plants must have come across the protein-rich pollen grains and the sticky, sugary droplets. As they began to depend on these new-found food supplies, they inadvertently carried pollen from plant to plant.

Beetle pollination must have been more efficient than wind pollination for some plant species because, clearly, selection began to favor those plants that had insect pollinators. The more attractive the plants were to the beetles, the more frequently they would be visited and the more seeds they would produce. Any chance variations that made the visits more reliable or that made pollination more efficient thus offered immediate advantages; more seeds would be formed, and more offspring would be likely to survive. Nectaries (nectar-secreting organs) evolved, which lured the pollinators. Plants developed white or brightly colored flowers that called attention to the nectar and other food supplies. The carpel, originally a leaf-shaped structure, became folded on itself. As a result, the ovule was enclosed and protected from hungry pollinators. By the beginning of the Cenozoic era, some 65 million years ago, the first bees, wasps, butterflies, and moths had appeared. These are long-tongued insects for which flowers are often the only source of nutrition. From this time on, flowers and insects have had a profound influence on one another's history, each shaping the other as they evolved together.

(a)

(b)

(c)

(d)

(e)

22-5 *Pollinators.* (a) *A honey bee foraging in a flower of* Salvia. *Notice that the anthers are depositing pollen grains on the bee's thorax.*

(b) *A longhorn beetle pollinating a lily. The pollen-covered head of the beetle is brushing against a stigma of the flower.*

(c) *Female rufous hummingbird probing for nectar in a columbine. Her head is collecting pollen, which she will carry to another flower. Flowers pollinated by birds are scentless, bright red or orange, and have copious nectar that makes the visit worthwhile.*

(d)*By thrusting its face into the tubular corolla of an organ-pipe cactus flower, a* Leptonycteris *bat is able to lap up nectar with its long, bristly tongue. Pollen grains clinging to its face and neck are transferred to the next flower visited by the bat. Bat-pollinated flowers have dingy colors and a musty scent (similar to that produced by bats to attract one another), and they open at night.*

(e) *A gossamer-winged butterfly sipping nectar in a daisy. Notice the long, sucking tongue of the butterfly.*

22-6 *Orchidaceae, with about 20,000 species, is the largest family of flowering plants. Its flowers are highly specialized.* (a) *The parts of an orchid flower. The lip is a modified petal that serves as a landing platform for insects.* (b) Cymbidium *orchids. This genus of Asian orchids contains about 30 species, most of which grow in tropical or subtropical regions.*

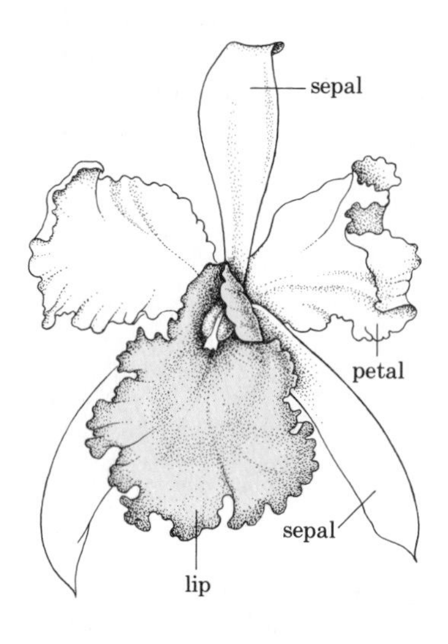

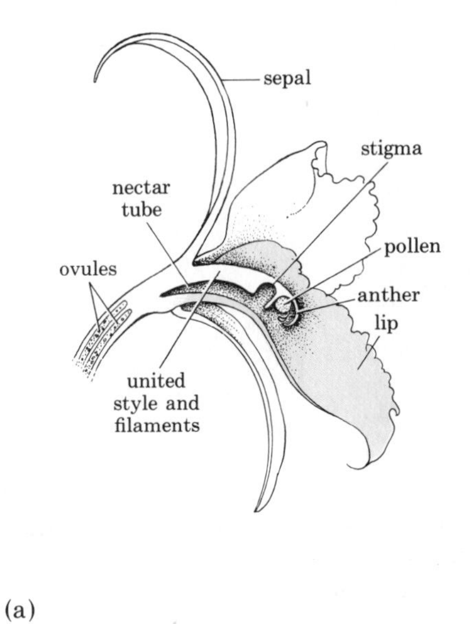

(a)

(b)

22-7 *Composites (family Asteraceae), with some 13,000 species, are the second largest family of flowering plants. Their flowers are also highly specialized.*

(a) *The organization of the head of a composite. The individual flowers are subordinated to the overall effect of the head, which acts as a large, single flower in attracting insects. The internal structure of the disk flower is indicated in color.*

(b) Helianthus annuus, *"annual flower of the sun," is composed of numerous separate florets, each comprising a carpel and fused anthers enclosed in a small corolla of fused petals. The ray flowers (with yellow petals) are often sterile.*

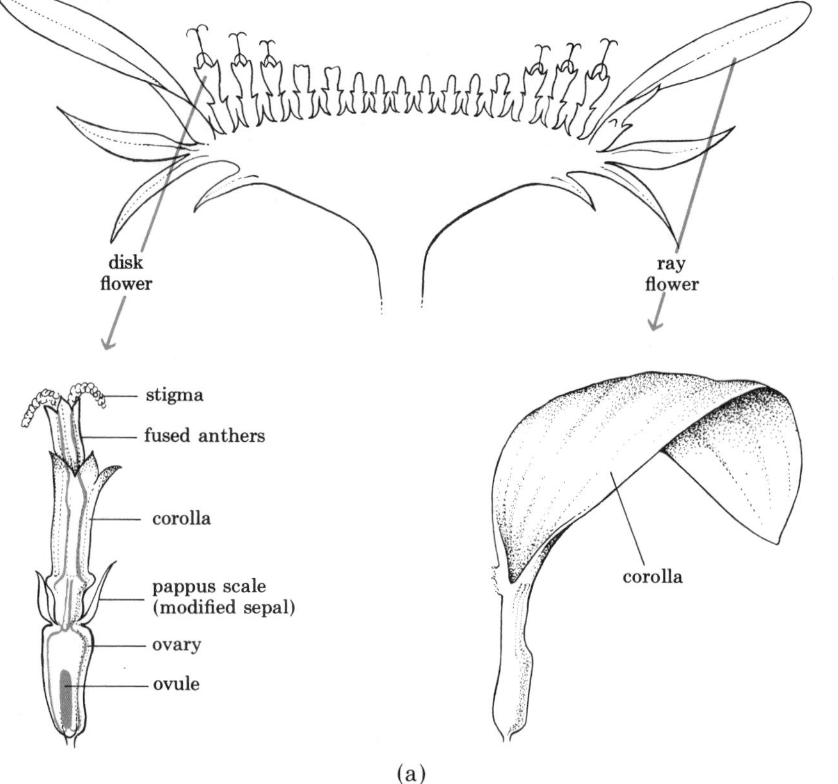

(a)

(b)

A flower that attracts only a few kinds of animal visitors and attracts them regularly has an advantage over flowers visited by more promiscuous pollinators: Its pollen is less likely to be wasted on a plant of another species. In turn, it is an advantage for the insect to have a "private" food supply that is relatively inaccessible to competing insects. Most of the distinctive features of modern flowers are special adaptations that encourage regular visits (constancy) by particular pollinators. The varied colors and odors are "brand names" for guiding pollinators. The diverse shapes, such as deep nectaries and complex landing platforms that are found, for example, in orchids, snapdragons, and irises, represent ways of excluding indiscriminate pollinators.

Pollination

For flowering plants, a new cycle of life begins when a pollen grain comes into contact with the stigma of a flower of the same species. Pollen is commonly produced in great quantities; the probability that any particular pollen grain will reach the stigma of an appropriate flower is very small.

By the time a pollen grain is released from its parent flower, it usually consists of three haploid cells—two sperm cells contained within a larger cell (known as a tube cell), which is, in turn, enclosed by the thickened outer wall of the pollen grain. The pollen grain contains its own nutrients and has so tough an outer coating that intact grains thousands of years old have been found in peat bogs.

As you will recall, in multicellular algae and more primitive plants, such as ferns, there is a distinct cycle of alternation of generations in which the sporophyte produces spores that produce gametophytes that produce gametes, with gametophyte and sporophyte having separate existences. In the course of plant evolution, the gametophyte stage has been steadily reduced in size. In the angiosperms, all that remains of the male gametophyte is the tough, tiny pollen grain. The sperm cells are the gametes.

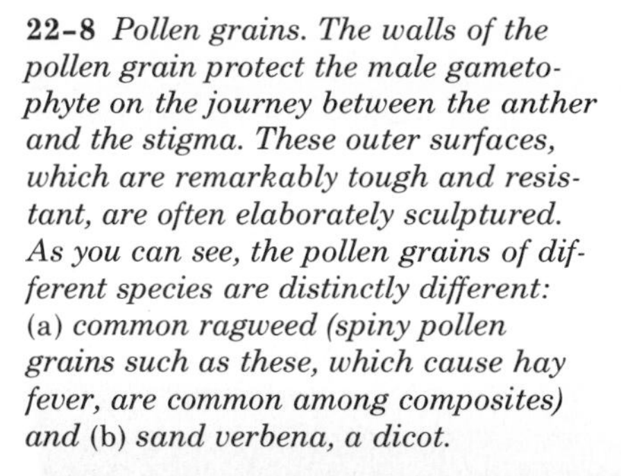

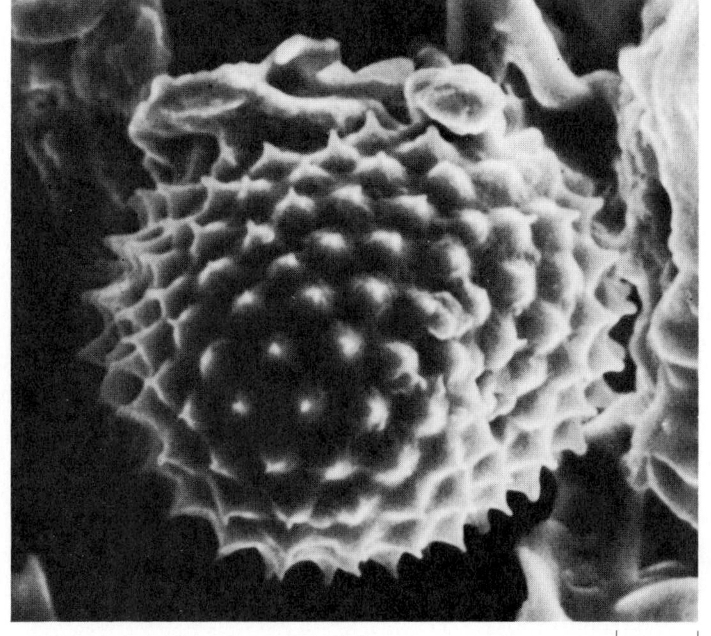

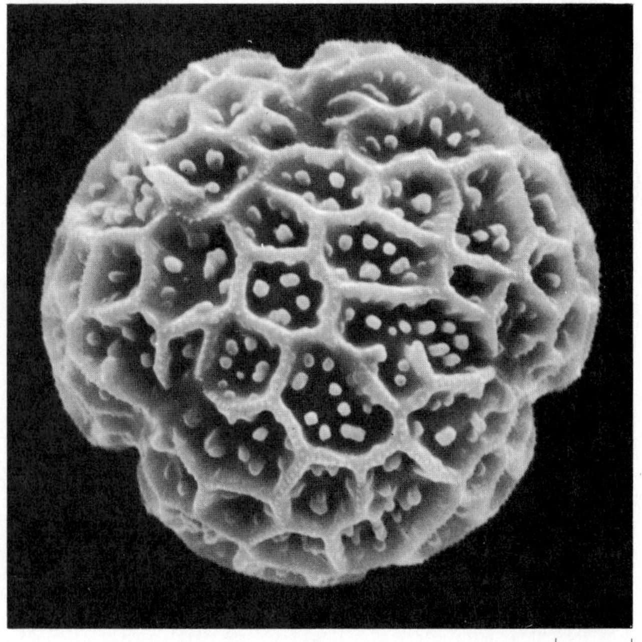

22-8 *Pollen grains. The walls of the pollen grain protect the male gametophyte on the journey between the anther and the stigma. These outer surfaces, which are remarkably tough and resistant, are often elaborately sculptured. As you can see, the pollen grains of different species are distinctly different:* (a) *common ragweed (spiny pollen grains such as these, which cause hay fever, are common among composites) and* (b) *sand verbena, a dicot.*

22–9 *Fertilization in angiosperms. The pollen tube of the male gametophyte, or pollen grain, grows down through the style and enters the ovule, in which the female gametophyte has developed to a seven-cell stage. One of the sperm nuclei unites with the egg nucleus, and the zygote is formed. The other sperm nucleus fuses with the two polar nuclei that are present in a single large cell (which in the drawing fills most of the ovule). From the resulting triploid* (3n) *cell, the endosperm will develop. The carpel shown here contains a single ovule.*

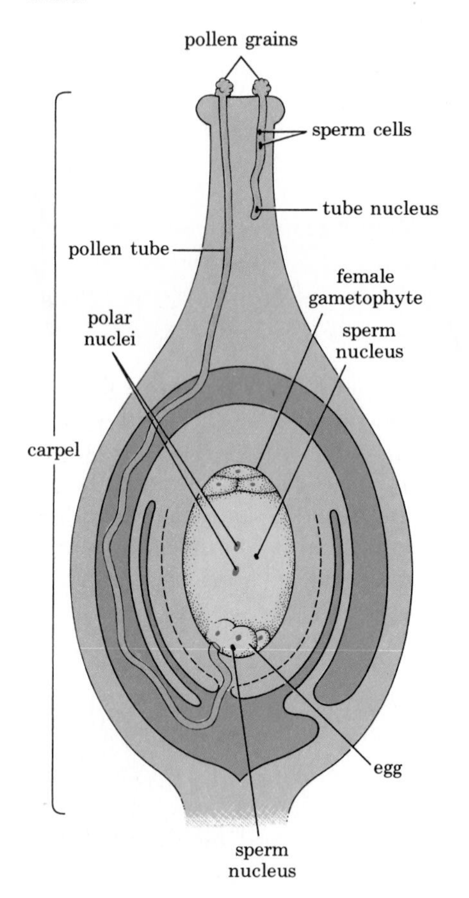

Fertilization

Once on the stigma, the pollen grain germinates, and, under the influence of the tube nucleus, a pollen tube grows down through the style into an ovule (Figure 22–9). The ovule contains the female gametophyte, which has also become reduced in size in the course of evolution. In many species it consists of seven cells, with a total of eight haploid nuclei. One of the smaller cells is the egg, containing a single haploid nucleus. The large central cell contains two haploid nuclei, called the polar nuclei because they move to the center from each end, or pole, of the gametophyte.

The nucleus of one of the sperm cells carried by the pollen tube unites with the egg. This fertilized cell, the zygote, develops into the embryo—the young sporophyte. The nucleus of the second sperm cell unites with the two nuclei of the central cell. From the resultant $3n$ cell, a specialized tissue called the *endosperm* develops. It surrounds and nourishes the embryo. These extraordinary phenomena of fertilization and triple fusion—together called "double fertilization"—take place, in all the natural world, only among the flowering plants.

The Embryo

The zygote (the fertilized egg cell) divides mitotically, and as the embryo grows, its cells begin to *differentiate*—become different from one another. The embryo begins to take on a characteristic shape, a process known as *morphogenesis*. At the same time, the $3n$ cell divides mitotically to produce the endosperm.

Growth Areas

In the earliest stages of embryonic growth, cell division takes place throughout the body of the young plant. As the embryo grows older, however, the production of new cells by mitosis becomes gradually restricted to certain parts of the plant body: the *apical meristems* of the root and the shoot. During the rest of the life of the plant, all the primary growth—which chiefly involves the elongation of the plant body—originates in these meristems.

The existence of such meristematic areas, which add to the plant body throughout the life of the plant, is one of the principal differences between plants and animals. Higher animals stop growing when they reach maturity, although the cells of certain "turnover" tissues, such as skin or the lining of the intestine, continue to divide. Plants, however, continue to grow during their entire life span. Growth in plants is the counterpart, to some extent, of mobility in animals. By growth, a plant modifies its relationship with the environment, for example, in turning toward the light and extending its roots.

The Seed and the Fruit

The seed consists of the embryo, which develops from the fertilized egg; the stored food, which consists of or derives from the endosperm; and the seed coat, which develops from the outermost layers of the ovule. The fruit develops from the wall of the ovary. In a peach, for instance, which

22-10 *Seeds.* (a) *In dicots such as the common bean, the endosperm is digested as the embryo (shown in color) grows and the food reserve is stored in the fleshy cotyledons.* (b) *In corn and other monocots, the single cotyledon, known as the scutellum in corn and other grains, absorbs food reserves from the endosperm. The coleoptile is a sheath that encloses the apical meristem of the shoot.*

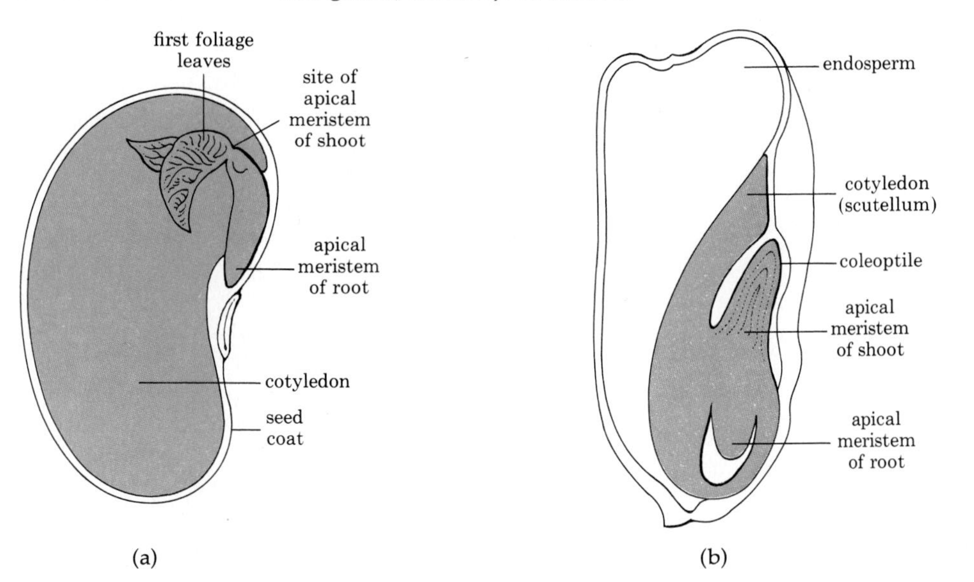

contains only one ovule in each ovary, the skin, the succulent edible portion of the fruit, and the stone are three distinct layers of the wall of the mature ovary (the base of the carpel). The almond-shaped structure within the stone is the seed. In a pea, the pod is the mature ovary wall and the peas are the seeds (the mature ovules and their contents). A raspberry is an aggregate of many fruits from a single flower, each fruit containing a single seed and each formed from a separate carpel.

As the ovary ripens into fruit and the seed forms, the petals, stamens, and other floral parts of the parent plant may fall away (see Figure 19-7, page 255).

Seed Dormancy

The seeds of most wild plants require a period of dormancy before they will germinate. This genetic requirement ensures that the seed will "wait" at least until the next favorable growth period. Some seeds can remain dormant and yet viable—with the embryo in a state of suspended animation—for hundreds of years.

The seed coat apparently plays a major role in maintaining dormancy. In some species, the seed coat seems to act primarily as a mechanical barrier, preventing the entry of water and gases, without which growth is not possible. In these cases, growth is initiated by the seed coat's being worn away in various ways—such as being washed by rainfall, abraded by sand or soil, burned away in a forest fire, or partially digested as it passes through the digestive tract of a bird or other animal. In other species, dormancy seems to be maintained chiefly by chemical inhibitors in the seed coat. These inhibitors undergo chemical changes in response to various environmental factors, such as light or prolonged cold or a sudden rise in temperature, which neutralize their effects, or they may be washed or eroded away. Eventually, the embryo resumes growth.

The Staff of Life

Grains are the small, one-seeded fruits of grasses. Because they are dry, they can be stored as food. The collecting and storing of grains from wild grasses is believed to have been an important impetus to the agricultural revolution of some 11,000 years ago (see Chapter 42). Today we are heavily dependent on cultivated wheat, rice, corn, rye, and other grains. In many countries, they constitute the chief food resource.

The fruit of wheat, sometimes known as the kernel, is made up of the embryo, the endosperm, and the surrounding seed coat. More than 80 percent of the bulk of the wheat kernel and 70 to 75 percent of its protein are in the endosperm. White flour is made from the endosperm. Wheat germ, the embryo, forms about 3 percent of the kernel. It is usually removed as wheat is processed because it contains oil, which makes the grain more likely to spoil. Bran is the seed coat plus the aleurone layer (outer part of endosperm); it constitutes about 14 percent of the kernel. The bran is also removed when wheat is milled to make white flour. Actually, the presence of the bran somewhat decreases the caloric value of the wheat kernel. We are unable to digest bran because it is mostly cellulose. Bran therefore tends to speed the passage of food through our intestinal tracts, resulting in decreased absorption. The wheat germ and the bran, which contain most of the vitamins, are sometimes used for human consumption but more often are fed to livestock.

Wheat is about 9 to 14 percent protein, most of which, as we noted, is contained in the endosperm. Its protein value is diminished, however, by its lack of certain essential amino acids, notably lysine.

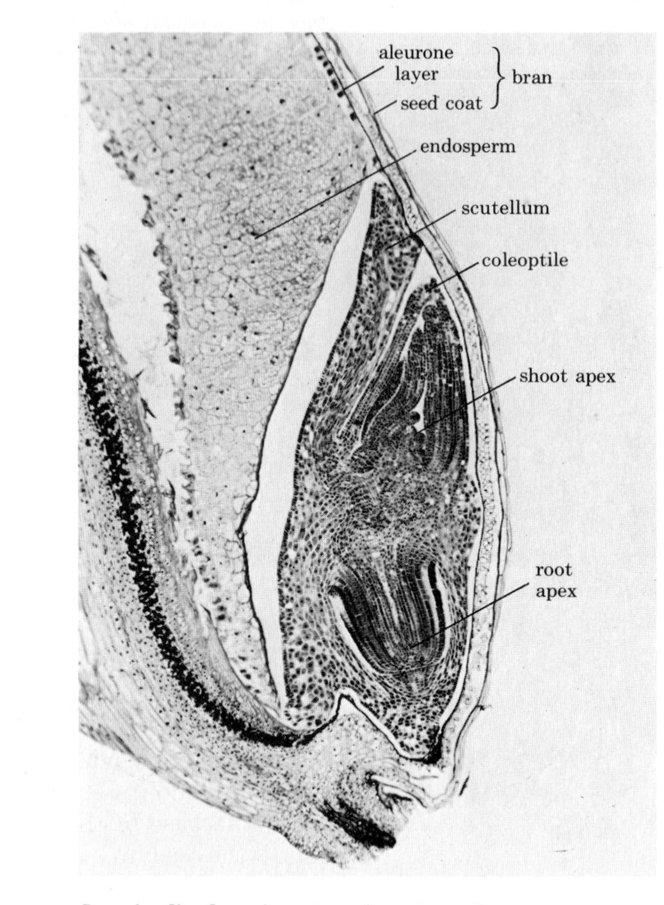

Longitudinal section of a wheat kernel.

Primary Growth

A dormant seed characteristically contains very little moisture (only about 5 to 10 percent of its total weight). Germination begins with a massive entry of water into the seed. The seed coat ruptures and the young sporophyte emerges.

Primary growth begins immediately. It involves the elongation of stems and roots, the formation of branches, and the differentiation of the conducting tissues and other specialized tissues of the young shoot and root. All primary growth originates in the apical meristems of the shoot and root.

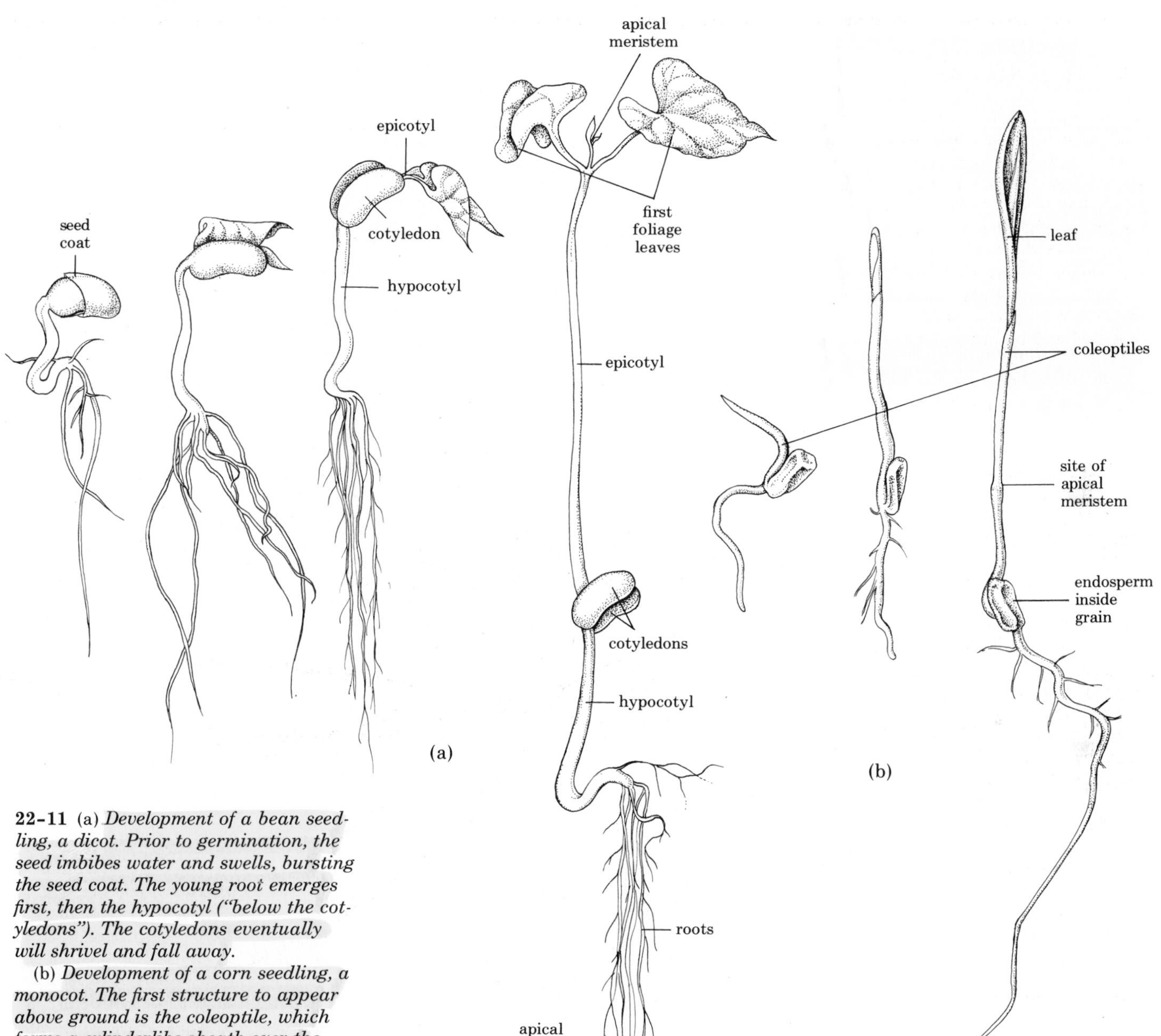

22-11 (a) *Development of a bean seedling, a dicot. Prior to germination, the seed imbibes water and swells, bursting the seed coat. The young root emerges first, then the hypocotyl ("below the cotyledons"). The cotyledons eventually will shrivel and fall away.*

(b) *Development of a corn seedling, a monocot. The first structure to appear above ground is the coleoptile, which forms a cylinderlike sheath over the growing shoot of the plant. Typically, the shriveled endosperm, with the scutellum buried in it, is still present in the young seedling.*

Primary Growth of the Root

The first part of the embryo to break through the seed coat, in nearly all seed plants, is the root. Figure 22-12 (page 322) diagrams the growing zone of the root of a dicot. At the very tip is the root cap, which protects the apical meristem as the root is pushed through the soil. The cells of the root cap wear away and are constantly replaced by new cells from the meristem.

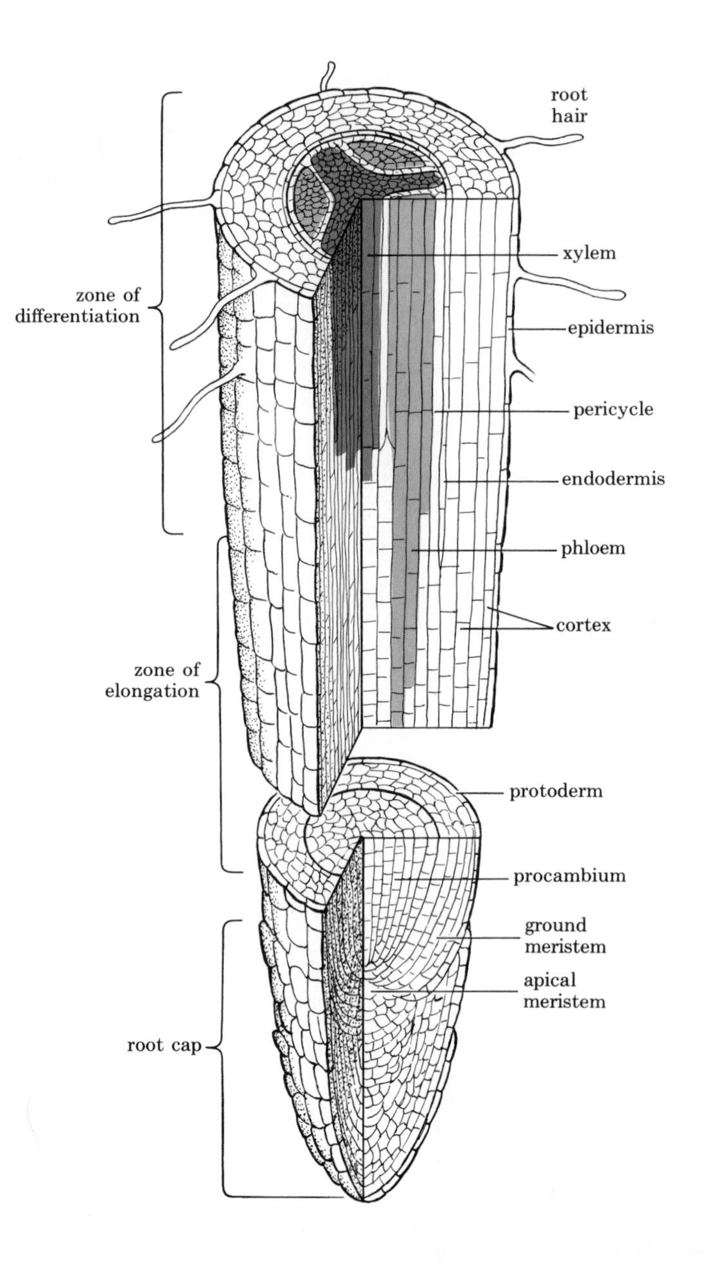

22-12 *The growth regions of a dicot root. New cells are produced by the division of cells within the apical meristem. The cells above the meristem undergo a characteristic series of changes as the distance increases between them and the root tip. First, there is a maximum rate of cell division, followed by cell elongation without further division. The latter accounts for most of the lengthening of the root. As the cells elongate, they differentiate into various specialized tissues of the root. The protoderm becomes the epidermis, the ground meristem becomes the cortex, and the procambium becomes the xylem and phloem. Some of the cells produced in the apical meristem form the protective root cap.*

Certain cells in the meristem retain the capacity to produce new cells and thus perpetuate the meristem. All the other cells in the root—which are the progeny of these relatively few meristematic cells—eventually differentiate, some becoming cells of the root cap and others forming the complex tissues of the root. Just above the point where cell division ceases, the cells gradually elongate, growing to 10 or more times their previous length, often within the span of a few hours.

As the cells elongate, they begin to differentiate. The conducting cells of the phloem mature first and then the conducting cells of the xylem. In the region of the root where the xylem first forms, the endodermis also takes shape. To the inside of the endodermis, the pericycle forms. In this same general region of the root, the epidermal cells differentiate and begin to extend root hairs into the crevices between the soil grains.

This same basic pattern of growth is seen in the first root of a seedling and in the growing root tips of a tall tree.

Primary Growth of the Shoot

The shoot includes all the aboveground parts of the plant. The organization of the developing shoot tip is somewhat similar to that seen in the root: first, a zone in which most of the cell division takes place; next, a zone of cell elongation; and finally, a zone of differentiation. However, because of the regular occurrence of nodes and their appendages—the leaves and buds—these zones are not as distinct in the shoot as they are in the root.

As in the root, the outermost layer of cells develops into the epidermis. In the shoot, these cells have a relatively conspicuous fatty cuticle. Other cells differentiate to form ground tissue and the primary vascular tissues—the primary xylem and the primary phloem. The pattern of development is more complicated, however, than in the root tip since the apical meristem of the shoot is the source of tissues that give rise to new leaves, branches, and flowers.

Figure 22–13 shows the shoot tip of a lilac. Here you can see the apical meristem, which is very small, and the beginnings—primordia—of leaves. As you can see, leaves are formed in an orderly sequence at the shoot tip. The vascular tissue begins to differentiate in the leaf primordia, eventually becoming part of the general vascular system that connects the plant from root to leaf tip. As the internodes elongate, the young leaves become separated so that the leaves clustered so tightly together around the apex in Figure 22–13 will eventually be spaced out along the stem of the plant.

As the growing tip of the shoot elongates, small masses of meristematic tissue (buds) are left just above the point at which the leaf joins the stem (the leaf axil). These buds may remain dormant or they may give rise to new leafy shoots or flowers.

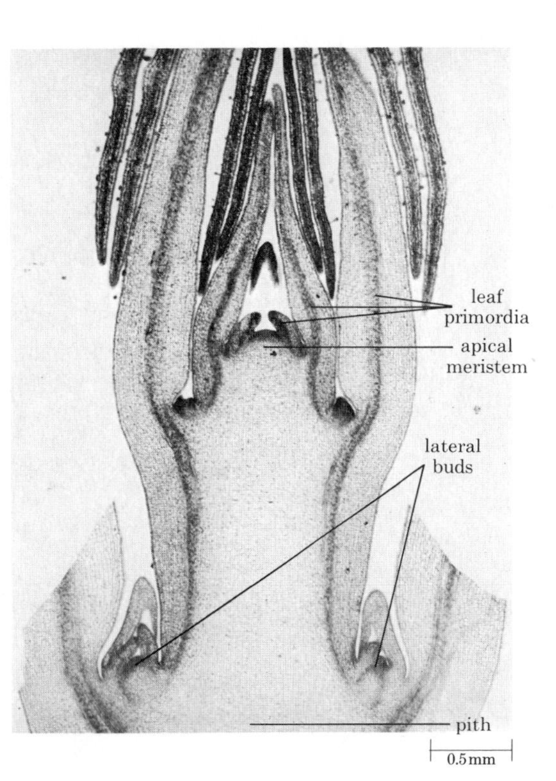

22–13 *Longitudinal section of the shoot tip of a lilac* (Syringa vulgaris). *At the tip of the stem is the apical meristem, the central zone of cell division. The leaf primordia, from which new leaves will form, originate along the sides of the shoot apex. In this picture, two leaf primordia can be seen arising from the apical meristem. Successively older leaf primordia had formed on both sides. The developing stem below the apical meristem and the lateral buds on either side are also regions of active cell division.*

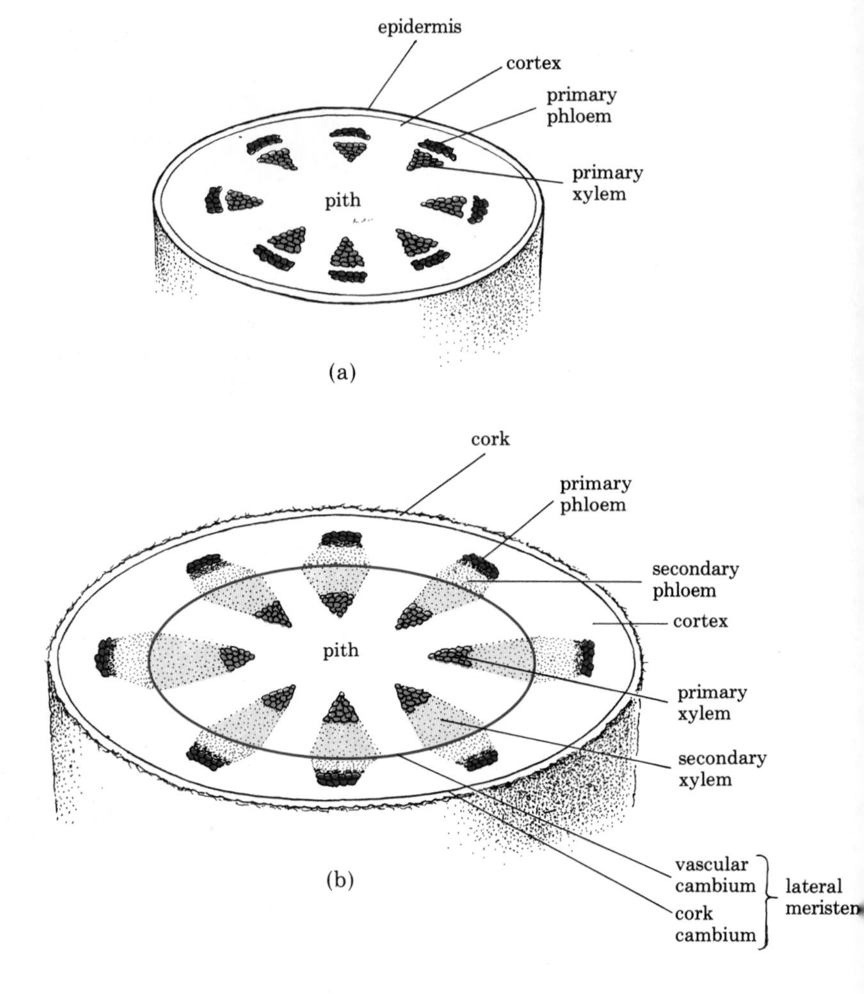

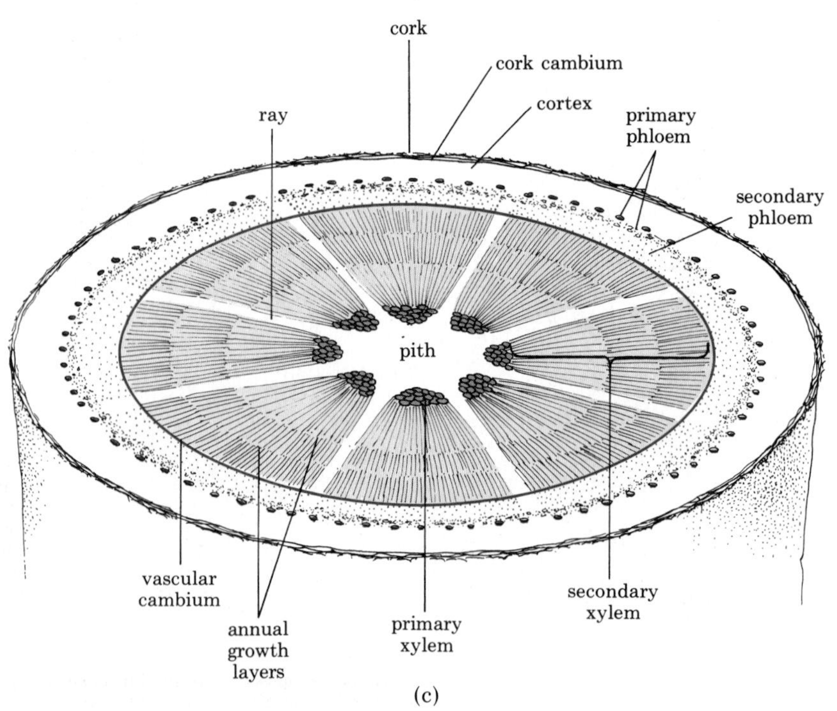

22-14 (a) *Stem of a dicot before the onset of secondary growth.*

(b) *Beginnings of secondary growth. Secondary xylem and secondary phloem are produced by the vascular cambium, a meristematic tissue formed late in primary growth. As the trunk increases in diameter, the epidermis is stretched and torn. This apparently triggers the formation of the cork cambium, from which cork is formed, replacing the epidermis.*

(c) *Cross section of a three-year-old stem, showing annual growth layers. Rays are strands of living cells that transport nutrients and water laterally (across the trunk). On the perimeter of the outermost growth layer of xylem is the vascular cambium, encircled by a band of secondary phloem. The primary phloem and also the cortex will eventually be sloughed off. In an older stem, the thin cylinder of active secondary phloem is immediately adjacent to the vascular cambium. The tissues outside the vascular cambium, including the phloem, constitute the bark.*

Secondary Growth

Secondary growth is the process by which woody plants increase the thickness of trunks, stems, branches, and roots after primary growth has ceased (Figure 22–14). The so-called "secondary tissues" are not derived from the apical meristems; they are the result of the production of new cells by *lateral meristems* that are known as the vascular cambium and the cork cambium.

The *vascular cambium* is a thin, cylindrical sheath of tissue between the xylem and the phloem. In plants with secondary growth, the cambium cells divide continually during the growing season, adding new xylem cells—that is, secondary xylem—toward the inside of the cambium, and secondary phloem toward the outside. Some daughter cells remain as a cylinder of undifferentiated cambium. As the tree grows older, the living cells of the xylem in the center of the trunk die, and their neighboring vessels cease to function. This nonfunctioning xylem is called heartwood, as distinct from sapwood, which consists of living cells and functional vessels.

As the girth of stems and roots increases by secondary growth, the epidermis becomes stretched and torn. In response to this tearing process, a new type of cambium, the *cork cambium,* forms. Cork, which is a dead tissue at maturity, is produced from the cork cambium.

22–15 *A tree trunk showing the relationships of the successive concentric layers. The heartwood is composed entirely of dead xylem cells. The cortex, which is outside the phloem in a green stem, is sloughed off in the process of secondary growth.*

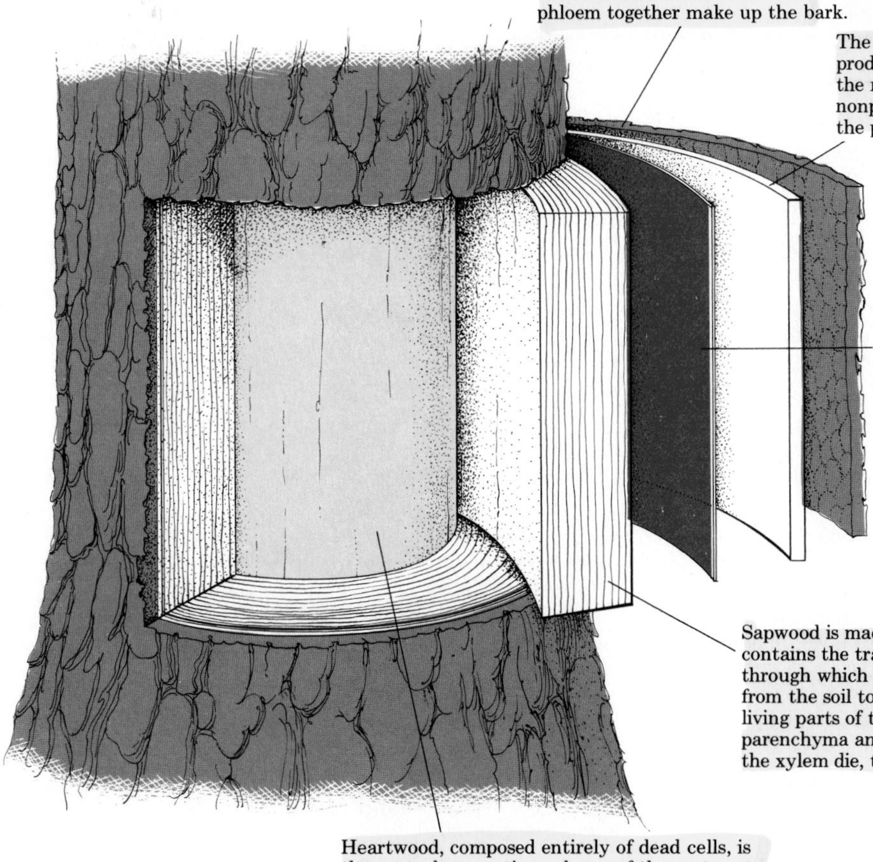

Cork, which is a dead tissue, protects the inner tissues from drying out, from mechanical injury, and from insects and other herbivores. Cork and phloem together make up the bark.

The phloem conducts the sugars produced by photosynthesis to the roots and other living, nonphotosynthetic parts of the plant.

The cambium layer produces secondary xylem and secondary phloem.

Sapwood is made up of xylem tissue, which contains the tracheids and vessels through which water and minerals move from the soil to the leaves and other living parts of the tree. As the parenchyma and supporting cells of the xylem die, they become heartwood.

Heartwood, composed entirely of dead cells, is the central supporting column of the mature tree.

By the continuous formation of secondary layers of xylem and (to a small extent) phloem, tree trunks increase their diameter as primary growth increases their height. Season after season the new xylem forms visible growth layers, or rings. These visible growth rings are the result of a difference in the density of the wood produced early in the growing season—when the cells are larger and have thinner walls—and late in the season—when the cells are smaller and have thicker walls. Each growing season leaves its trace, so that the age of a tree can be estimated by counting the number of growth rings in a section near its base. Since the rate of growth of a tree depends on climatic conditions, it is also possible to estimate from the width of the annual growth layers of ancient trees fluctuations in temperature and rainfall that occurred hundreds of years ago.

A Brief Review

As we have seen in the past three chapters, the different parts of the plant body are specialized for particular functions. These functions include photosynthesis, absorption of water and minerals, transport of solutions, support, reproduction, and continuing growth. The specialized tissues that perform these functions include numerous cell types, which are summarized in Table 22–1.

Table 22–1 Summary of main cell types in angiosperms

Cell Type	Location	Characteristics	Function
Apical meristem	Apices of shoots and roots	Many-sided, small, thin-walled cells; vacuoles usually small	Origin of primary meristematic tissues
Epidermis	Surface of entire primary plant body	Flattened, variable in shape, overlaid by cuticle; specialized guard cells	Protective covering
Parenchyma	Everywhere, usually dominant in pith, cortex, mesophyll	Many-sided, usually thin-walled; abundant air spaces between cells	Photosynthesis, storage, wound healing, among others
Sclereid	Anywhere; in pith and cortex of stems; in leaves and flesh of fruits; seed coats	Irregular; massive secondary wall; alive or dead at maturity	Produces hard texture, mechanical support
Fiber	Primary and secondary xylem and phloem; cortex	Very long, narrow cell, with secondary cell wall; alive or dead at maturity	Support
Tracheid	Primary or secondary xylem	Elongate, with pits in walls; dead at maturity	Conduction of water and solutes
Vessel element	Interconnected series (= vessels) in primary or secondary xylem	Elongate, with pits in walls and end walls perforated; dead at maturity	Conduction of water and solutes
Sieve-tube element	Primary or secondary phloem, usually with companion cells; form interconnected series (= sieve tubes)	Elongate, with specialized sieve plates; nucleus lacking at maturity	Conduction of organic solutes
Vascular cambium	Lateral, between secondary phloem and xylem tissues	Elongate, spindle-shaped and short, with rays	Produces secondary xylem and phloem
Cork	Surface of stems and roots with secondary growth	Flattened cells, compactly arranged; dead at maturity, the cells often air-filled	Restricts gas exchange and water loss

It is essential for the well-being of the plant that the different functions performed by the various cells and tissues be integrated. Moreover, the plant, like all other organisms, must respond appropriately—and in a timely fashion—to its environment if it is to survive. In the next chapter, we shall consider some ways in which the functions of plant tissues and their response to the environment are regulated.

SUMMARY

The flower is the reproductive structure of the angiosperms. The anthers of the flower produce pollen grains, the male gametophytes. Each anther is attached to a slender filament; the entire structure, anther plus filament, is known as the stamen. The carpel typically consists of the stigma, an area on which the pollen grains germinate, a slender style, and, at its base, the ovary. The ovary contains one or more ovules. Within each ovule, a female gametophyte containing an egg cell develops. The ovule later becomes the seed, and the ovary becomes the fruit. The various shapes and colors of flowers evolved under selection pressures for more efficient pollinating mechanisms.

In flowering plants, a new life cycle begins when a pollen grain germinates upon the stigma of a flower of the same species, sending a pollen tube down the style and into an ovule. Within the ovule is the female gametophyte, which usually consists of seven cells, with a total of eight haploid nuclei, including the egg nucleus. One of the three haploid nuclei of the pollen grain (a sperm nucleus) unites with the egg cell nucleus; the other sperm nucleus unites with the two polar nuclei of the female gametophyte. The embryo develops from the first union, and the nutritive endosperm, which is a triploid ($3n$) tissue, from the second. The phenomena of fertilization and triple fusion together are found only among the flowering plants and are known as "double fertilization."

The angiosperm seed, or mature ovule, consists of the embryo, the seed coat, and stored food, which may be present in the form of endosperm.

The embryo has two apical meristems—the apical meristem of the shoot and the apical meristem of the root—and either one or two cotyledons ("seed leaves"). From the endosperm, the cotyledons absorb nutrients that nourish the growing embryo. Early in embryonic life, cell division becomes confined to limited areas in the plant body: the apical meristems of the root and the shoot.

When the seed germinates, growth of the root and shoot proceeds from the apical meristems of the embryo. Certain cells within the meristems divide continuously. Some of these cells continue to divide. Others elongate and then differentiate, forming, according to their position, various specialized cells, including those of the xylem and the phloem. In the root, the apical meristem forms a root cap, which protects the root tip as it is pushed through the soil. Root hairs, which are the principal pathways of absorption by the roots, are outgrowths of the epidermal cells.

Leaf primordia arise from the apical meristem of the shoot. As the nodes are separated by elongation of the internodes, small masses of meristem (buds) are formed in the axils of the leaves. These buds may remain dormant or may give rise to branches or specialized shoots.

Secondary growth is the process by which the woody plants increase their girth. Such growth arises primarily from the vascular cambium, a sheath of meristematic tissue completely surrounding the xylem and completely surrounded by phloem. The cambium cells divide during the growing season, adding new xylem cells (secondary xylem) on their inner surfaces and new phloem cells (secondary phloem) on their outer surfaces. As the trunk increases in girth, the epidermis is eventually ruptured and destroyed and is replaced by cork.

QUESTIONS

1. Distinguish among the following terms: runner/rhizome; ovary/ovule/egg cell/seed; pollination/fertilization; epidermis/endodermis/endosperm; apical meristem/lateral meristem; primary growth/secondary growth.

2. Sketch a flower. Give the function of each floral part.

3. In many plants, pollen production in the anthers occurs prior to or after the full development of the carpel of the same flower. What are the consequences of this shift in time frames?

4. The dark spots on the foxglove flowers at the left are called honey guides. What do you suppose their function is?

5. The jack pine produces seeds with very tough seed coats. These seed coats rupture, and the seeds germinate, only after exposure to intense heat, as in a forest fire. What might be the advantages of such seeds for the jack pine?

6. Sketch a dicot embryo at the time of seed release. Identify each part in terms of the future development of the plant body.

7. Suppose you carve your initials 1 meter above the ground on the trunk of a mature tree that is growing vertically at an average of 15 centimeters a year. How high will your initials be at the end of 2 years? At the end of 20 years?

8. Sketch a tree trunk with secondary growth. Compare your sketch with Figure 22–14c.

9. Rabbits often eat all of the bark off young trees, at the height they can reach. When they do this, the tree dies. Why?

10. What plant part are you eating when you eat each of the following: carrots, eggplant, lima beans, corn on the cob, green beans, celery, tomatoes, spinach, brussels sprouts, broccoli, white potatoes?

23

Plant Hormones and Plant Responses

We saw in the previous chapter that as a plant grows it does far more than increase its mass and volume. It differentiates, forming a variety of cells, tissues, and organs, and undergoes morphogenesis, taking on the shape characteristic of the adult sporophyte. Moreover, many of its activities are finely tuned to its environment and to the changing patterns of the seasons. Many of the details of how these processes are regulated are not known, but it is clear that plant development and growth depend on the interplay of a number of internal and external factors. Chief among the internal factors are the plant hormones.

Hormones and the Regulation of Plant Growth

A *hormone*, by definition, is a substance that is produced in one tissue and transported to another, where it exerts one or more highly specific effects. Hormones help the plant integrate the growth, development, and metabolic activities of its various tissues. Typically they are active in very small quantities. In the shoot of a pineapple plant, for example, only 6 micrograms of auxin, a common growth hormone, are found per kilogram of plant material.

The term "hormone" comes from the Greek word meaning "to excite." It is now clear, however, that many hormones also have inhibitory influences. So, rather than thinking of hormones as stimulators, it is perhaps more useful to consider them as chemical messengers. But this term also needs qualification. As we shall see, the response to the particular "message" depends not only on its content but also upon how it is "read" by its recipient.

23-1 *An example of phototropism. Here a bean seedling turns toward the light.*

Auxin

The first plant hormone to be isolated was *auxin*. The effects of this hormone were observed by Charles Darwin and his son Francis and first reported in *The Power of Movement in Plants,* published in 1881. The Darwins were studying the bending toward light, or *phototropism*, of grass seedlings. They noted that the bending takes place below the tip, in the lower part of the coleoptile (the sheath that surrounds the shoot tip). Then they showed that if they covered the terminal portion of the coleoptile with a cylinder of metal foil or a hollow tube of glass blackened with India ink and exposed the plant to a light coming from the side, the characteristic bending of the seedling did not occur. If, however, the tip was enclosed in a transparent glass tube, bending occurred normally. Bending also occurred normally when the lightproof cylinder was placed below the tip (Figure 23–2). "We must therefore conclude," they stated, "that when seedlings are freely exposed to a lateral light some influence is transmitted from the upper to the lower part, causing the material to bend."

23–2 *The Darwins' experiment.* (a) *Light striking a growing coleoptile (such as the tip of this oat seedling) causes it to bend toward the light.* (b) *Placing an opaque cover over the tip of the seedling inhibits this bending response, but* (c) *an opaque collar placed below the tip does not. These experiments indicate that something produced in the tip of the seedling and transmitted down the stem causes the bending.*

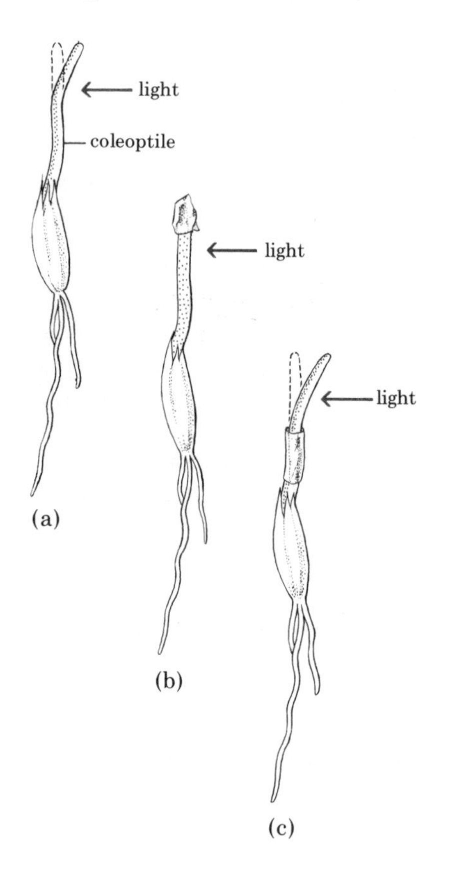

In 1926, the Dutch plant physiologist Frits W. Went succeeded in separating this "influence" from the plants that produced it. Went cut off the coleoptile tips from a number of oat seedlings. He placed the tips on a slice of agar (a gelatinlike substance), with their cut surfaces in contact with the agar, and left them there for about an hour. He then cut the agar into small blocks and placed a block off-center on each stump of the decapitated plants, which were kept in the dark during the entire experiment. Within one hour, he observed a distinct bending *away* from the side on which the agar block was placed (Figure 23–3).

Agar blocks that had not been in contact with a coleoptile tip produced either *no* bending or only a slight bending *toward* the side on which the block had been placed. Agar blocks that had been exposed to a section of coleoptile lower on the shoot produced no physiological effect.

Went interpreted these experiments as showing that the coleoptile tip exerted its effects by means of a chemical stimulus (in short, a hormone) rather than a physical stimulus, such as an electrical impulse. This chemical stimulus came to be known as auxin, a term coined by Went from the Greek word *auxein,* "to increase." Several different substances with activity similar to that of auxin have now been isolated from plant tissues, and others have been synthesized in the laboratory.

The phototropism observed by the Darwins results from the fact that under the influence of light, auxin migrates from the light side to the dark side of the tip. The cells on the dark side, having more auxin, elongate

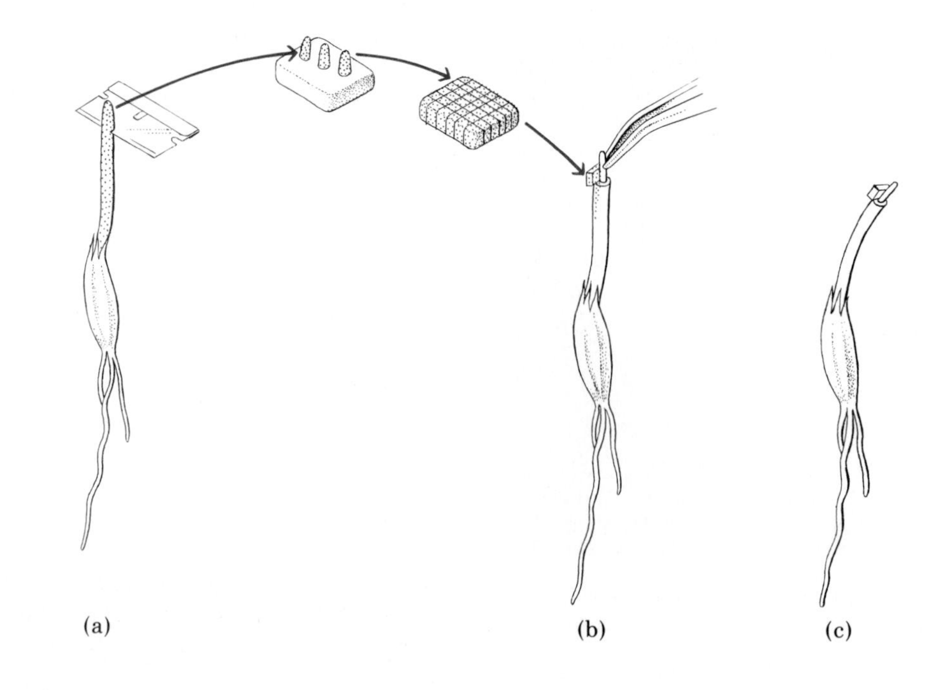

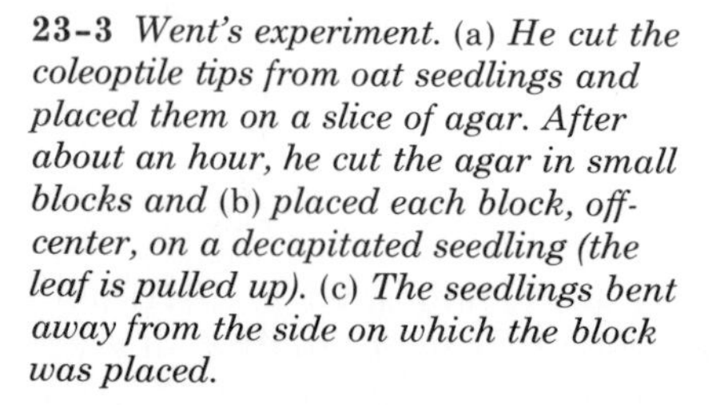

23–3 *Went's experiment.* (a) *He cut the coleoptile tips from oat seedlings and placed them on a slice of agar. After about an hour, he cut the agar in small blocks and* (b) *placed each block, off-center, on a decapitated seedling (the leaf is pulled up).* (c) *The seedlings bent away from the side on which the block was placed.*

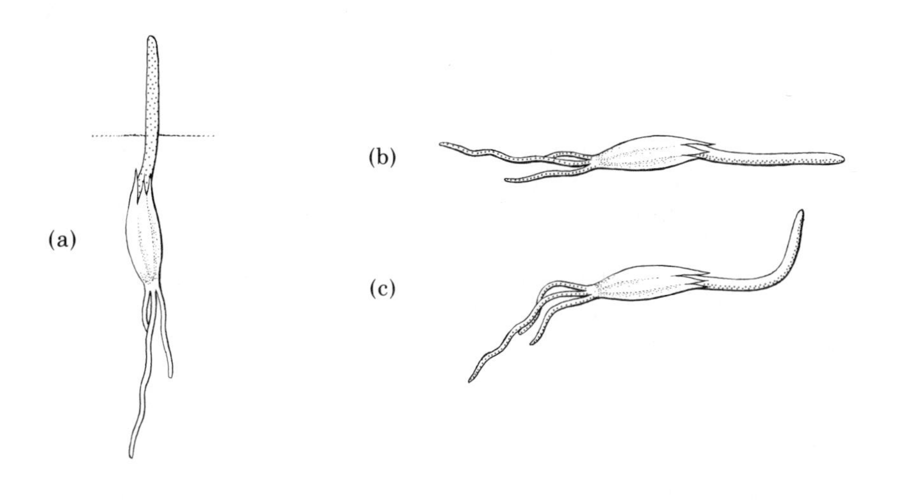

23-4 *Why shoots grow upward and roots downward.* (a) *When a seedling is vertical, the auxin produced in the tip is distributed evenly and the plant grows erect.* (b) *However, when a seedling is placed on its side, auxin is transported laterally and accumulates in cells on the lower side of the shoot;* (c) *these cells then grow more rapidly and so the shoot grows upward. It has been hypothesized that auxin similarly accumulates in the cells on the lower side of the root, where it has an inhibitory effect. As a result, the cells grow more slowly, so that the root tip turns down. However, many investigators believe that other inhibiting factors are involved.*

more rapidly than those on the light side, causing the plant to bend toward the light. This response has high survival value for young plants.

Other Responses to Auxin

Various plant tissues show other responses to auxin. Shoots elongate in response to auxin produced in the apical meristem. Apparently, under the influence of auxin the cell wall becomes more plastic. The cell is then more able to take up water and elongate.

In low concentrations, auxin stimulates the growth of branch roots and adventitious roots; rooting preparations used by gardeners contain auxin. In higher concentrations, auxin inhibits the growth of the main roots, which may explain why seedlings can orient themselves in the ground, a phenomenon known as geotropism (Figure 23-4).

In most dicot species, the growth of lateral buds is inhibited by auxin. If you pinch off the growing tip (meristem) of the stem of the house plant *Coleus,* for example, the lateral buds begin to grow vigorously, producing a plant with a bushier, more compact body. If you treat the "eyes" (actually lateral buds) of a potato with auxin, they will be inhibited from sprouting. As a result, the tubers can be stored longer. The formation of the abscission layer (page 294) has been correlated with diminished production of auxin in the leaf. Auxin is also involved in the growth of fruits (Figure 23-5).

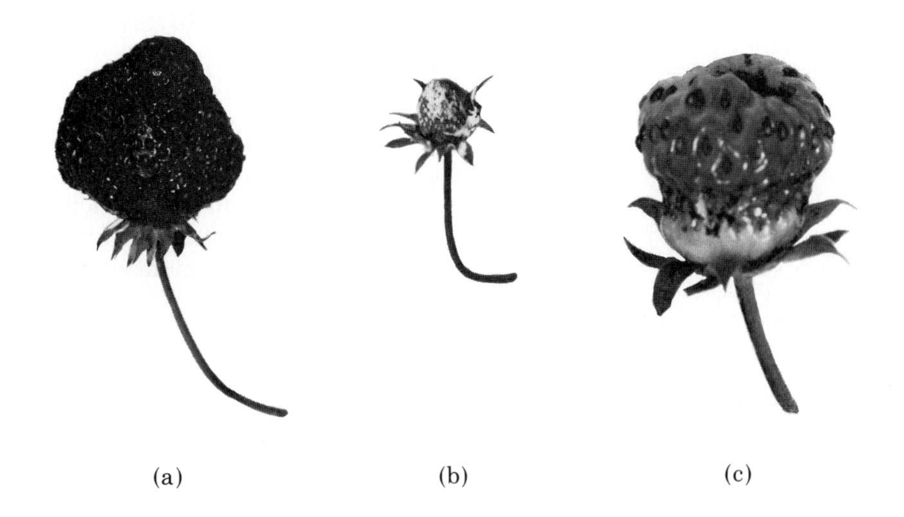

23-5 *Auxin, apparently produced by developing seeds, promotes the growth of fruit.* (a) *Normal strawberry,* (b) *strawberry from which all seeds have been removed, and* (c) *strawberry in which three horizontal rows of seeds were left. If a paste containing auxin is applied to* (b), *the strawberry grows normally.*

23-6 *Bolting and flowering in cabbage plants following gibberellin treatment. The plants on the left were not treated.*

The mechanism or mechanisms by which one chemical can produce so many varied and even opposite effects is not understood. In fact, exposing a plant tissue to a hormone has been compared to putting a coin in a vending machine. You may get your morning newspaper, a candy bar, or four minutes of country and western music. It depends not so much on the coin as on the machine in which you put it. Similarly, the effects of auxin and other plant hormones depend not only on the hormone but also on the target tissues and the surrounding chemical environment.

The Gibberellins

The *gibberellins* were first discovered by a Japanese scientist who was studying a disease of rice plants called "foolish seedling disease." The diseased plants grew rapidly but were spindly and tended to fall over. The cause of the symptoms, it was found, was a chemical produced by a fungus, *Gibberella fujikuroi,* which infected the seedlings. The substance, which was named gibberellin, and many closely related substances were subsequently isolated not only from the fungus but also from many species of plants.

The most remarkable results are seen when gibberellins are applied to certain plants that are genetic dwarfs. Under gibberellin treatment, these dwarfs become indistinguishable from normal tall plants. This suggests that the dwarf plants lack the gene for an enzyme essential in their own production of gibberellins.

Some plants —lettuce and cabbage are common examples—first grow as rosettes; the leaves develop but the internodes do not elongate until just before flowering. At that time the stem elongates rapidly, a phenomenon known as bolting. In the case of cabbage and other biennials, bolting and flowering do not normally occur until after a period of cold. However, bolting and flowering can be induced earlier by the application of gibberellin (Figure 23-6).

Gibberellins have also been shown to play a role in the growth of plant embryos and seedlings. In grass seeds, there is a specialized layer of cells, the aleurone layer, just inside the seed coat (see page 320). These cells are rich in protein. During the early stages of germination, the embryo produces gibberellin, which diffuses to the aleurone layer. In response to the gibberellin, the aleurone cells produce enzymes that hydrolyze the starch in the endosperm, converting it to sugars that the embryo and then the seedling can use (Figure 23-7). In this way the embryo itself calls forth the substances needed for its survival and growth at the time they are required.

The Cytokinins

The *cytokinins* are a group of hormones originally detected in coconut "milk," which is a liquid endosperm. Their principal action is to promote cell division, hence their name (from the term "cytokinesis"). They have now been found in numerous plants, largely in actively dividing tissues, including germinating seeds, fruits, and roots.

Responses to Cytokinin and Auxin Combinations

Studies of responses to combinations of auxin and cytokinins are helping physiologists understand how plant hormones work together to produce the total growth pattern of the plant. Apparently, the undifferentiated plant cell—such as the meristematic cell—has two courses open to it: Either it can enlarge, divide, enlarge, and divide again, or it can elongate without cell division. The cell that divides repeatedly remains essentially undifferentiated, whereas the elongating cell tends to differentiate and become specialized. In studies of tobacco stem tissue, the addition of auxin to the culture medium in which the tissue was growing produced rapid cell expansion, so that giant cells were formed. A cytokinin alone had little or no effect. Auxin plus cytokinin resulted in rapid cell division, so that large numbers of relatively small cells were formed.

By slight alterations in the relative concentrations of auxin and cytokinin, investigators have been able to affect the development of undifferentiated cells growing in tissue culture. When a high concentration of auxin is present, undifferentiated tissue gives rise to organized roots. With higher concentrations of cytokinins, buds appear (Figure 23-8).

However, lest you think this is simple, we shall describe another tissue culture study, in which tuber tissue of the Jerusalem artichoke was used. In this study, it was shown that a third factor, the calcium ion, can modify the action of the auxin-cytokinin combination. Auxin plus low concentrations of cytokinin was shown to favor cell enlargement, but as calcium ion was added to the culture, there was a steady shift in the growth pattern from cell enlargement to cell division. High concentrations of calcium ion apparently prevent the cell wall from expanding, and at such concentrations the cell switches course and divides. Thus, not only do hormones modify the effects of hormones, but these combined effects may be, in turn, modified by nonhormonal factors, such as calcium ions and, undoubtedly, many others.

23-7 *An experiment investigating the action of gibberellin in barley seeds. Forty-eight hours before the picture was taken, each of these seeds was cut in half and the embryo removed. The seed at the bottom was treated with plain water, the seed in the center was treated with a solution of 1 part per billion of gibberellin, and the seed at the top was treated with 100 parts per billion of gibberellin. As you can see, digestion of the starchy storage tissue has begun to take place in the seeds treated with gibberellin.*

Other Hormones

Soon after the discovery of the growth-promoting hormones, plant physiologists began to speculate that growth-inhibiting hormones would be found, since it is clearly advantageous to the plant not to grow at certain times and in certain seasons. Not long afterward, an inhibitory hormone was isolated from dormant buds. Subsequently, the same hormone was discovered in leaves, where it was found to promote abscission. The hormone is called *abscisic acid*. In addition to promoting abscission, it generally inhibits the growth-promoting effects of other hormones.

Ethylene is an unusual growth regulator in that it is a gas, a simple hydrocarbon, $H_2C{=}CH_2$. Its effects have been known for a long time. In the early 1900s, many fruit growers made a practice of improving the color and flavor of citrus fruits by "curing" them in a room with a kerosene stove. (Long before this, the Chinese used to ripen fruits in rooms where incense was being burned.) It was long believed that it was the heat that ripened the fruits. Ambitious fruit growers, who went to the expense of installing more modern heating equipment, found to their sorrow that this was not the case. As experiments showed, the incomplete combustion products of kerosene were actually responsible for ripening the fruits. The most active gas was identified as ethylene. As little as 1 part per million of ethylene in the air will speed the ripening process. Subsequently, it was found that ethylene is produced by plants, as well as by kerosene stoves, that it appears just before and also during fruit ripening, and that it is responsible for a number of changes in color, texture, and chemical composition that take place as fruits mature. Ethylene is also believed to interact with auxin in a number of other responses.

23-8 *Two buds forming on undifferentiated tissue (callus) from a geranium following treatment with both auxin and a cytokinin. Callus from some types of plants will continue to grow as undifferentiated tissue, or roots, or buds, depending on the relative proportions of auxin and cytokinins.*

Another hormone that has been postulated to exist, but never isolated, is a flowering hormone. Experiments that indicate the existence of such a hormone are outlined in Figure 23-9 on the next page.

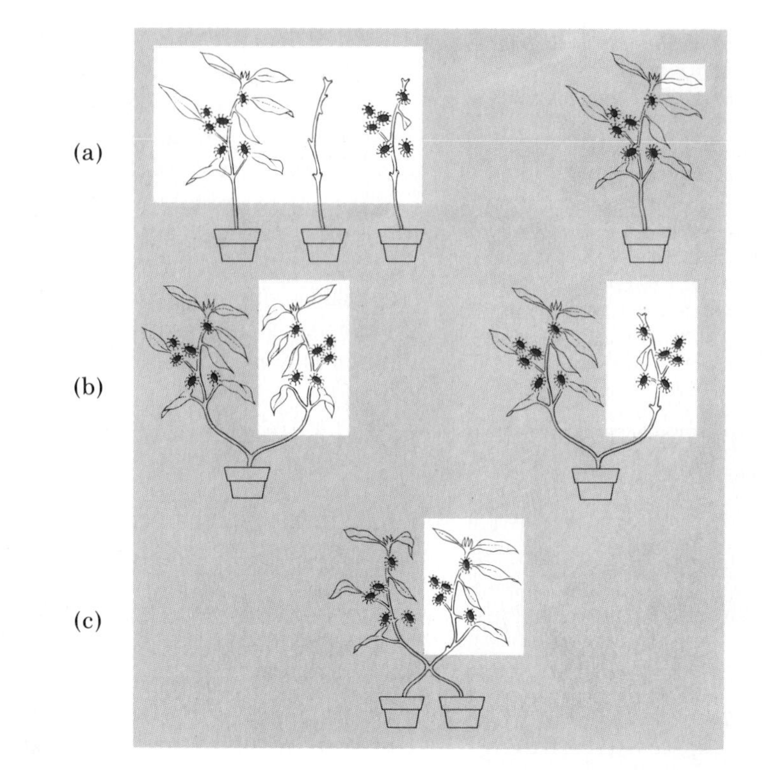

23-9 *Experiments that indicate the existence of a flowering hormone.* (a) *When certain plants, such as the cocklebur shown here, are exposed to an appropriate light cycle, those with leaves flower and those without leaves do not. When even one-eighth of a leaf remains on a plant, flowering occurs, and the illumination of a single leaf—not necessarily the whole plant—suffices. These experiments indicate that a chemical originating in the leaves causes the plant to flower.* (b) *This conclusion is supported by experiments on branched plants. Exposure of one branch to the light induces flowering on the other branch as well, even when only a portion of a leaf is present on the lighted branch.* (c) *When two plants are grafted together, exposure of one of the plants to the light cycle induces flowering in both the lighted plant and the grafted one.*

Plants and Photoperiodism

In many regions of the biosphere, the most important environmental changes affecting plants (and, indeed, land organisms in general) are those that result from the changing seasons. Plants are able to accommodate to these changes because of their capacity to sense and, more important, to anticipate the yearly calendar of events: the first frost, the spring rains, long dry periods, long growing spells, and even the time that nearby plants of the same species will be in flower. For many plants, all of these determinations are made in the same way: by measuring the relative periods of light and darkness. This phenomenon is known as *photoperiodism*.

Photoperiodism and Flowering

The effects of photoperiodism on flowering are particularly striking (Figure 23–10). Plants are of three general types: day-neutral, short-day, and long-day. Day-neutral plants flower without regard to day length. Short-day plants flower in early spring or fall; they must have a light period shorter than a critical length—for instance, the cocklebur flowers when exposed to 15½ hours or *less* of light. Other short-day plants are poinsettias, strawberries, primroses, ragweed, and some chrysanthemums.

Long-day plants, which flower chiefly in the summer, will flower only if the light periods are *longer* than a critical length. Spinach, potatoes, clover, henbane, and lettuce are examples of long-day plants.

The discovery of photoperiodism explained some puzzling data about the distribution of common plants. Why, for example, is there no ragweed in northern Maine? The answer, investigators found, is that ragweed starts producing flowers when the day is less than 14½ hours long. The long summer days do not shorten to 14½ hours in northern Maine until August, and then there is not enough time for ragweed seed to mature

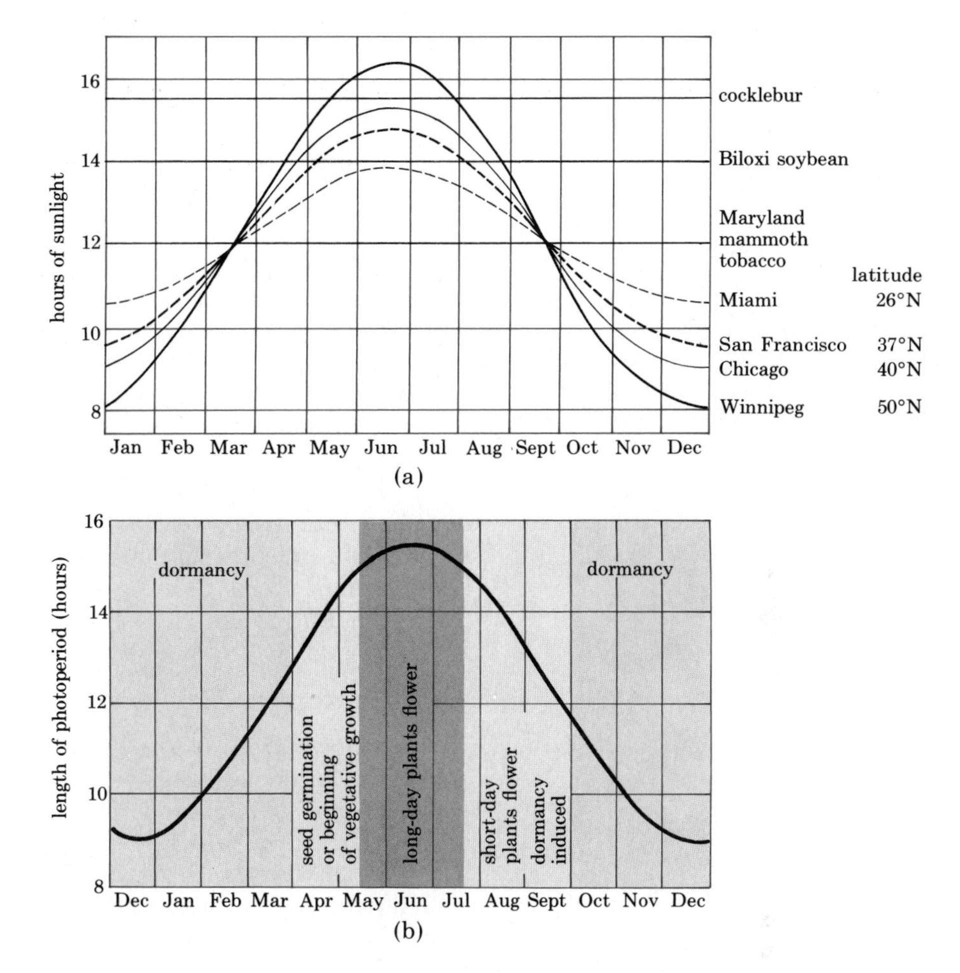

23-10 (a) *The relative length of day and night determines when plants flower. The four curves depict the annual change in day length in four North American cities at four different latitudes. The black lines indicate the effective photoperiod of three different short-day plants. The cocklebur, for instance, requires 15½ hours or less of light. In Chicago, it can flower as soon as it matures, but in Winnipeg the buds do not appear until early in August, so late that frost usually kills the plants before the seed is mature.*
(b) *Relationship between day length and the developmental cycle of plants in the temperate zone.*

23-11 *The cocklebur, a short-day plant, has been important in experimental studies of photoperiodism.*

before the frost. For somewhat similar reasons, spinach cannot produce seeds in the tropics. Spinach needs 14 hours of light a day for a period of at least two weeks in order to flower, and days are not this long in the tropics.

Note that ragweed and spinach will both bloom if exposed to 14 hours of daylight, yet one is designated as short-day and one as long-day. The important factor is not the absolute length of the photoperiod but rather whether it is longer or shorter than a particular critical interval for that variety. And in some varieties 5 or 10 minutes' difference in exposure can determine whether or not a plant will flower.

The first field studies on photoperiodism in plants were carried out early in this century by scientists at the U.S. Department of Agriculture. These scientists have since become known as the Beltsville group, for the small town in Maryland where they carried out their studies.

Measuring the Dark

Subsequently, other investigators, Karl C. Hamner and James Bonner, began a laboratory study of photoperiodism. They used the cocklebur as the experimental organism. As we mentioned previously, the cocklebur is a short-day plant, requiring 15½ hours or less of light per 24-hour cycle to flower. It is particularly useful for experimental purposes because a single exposure under laboratory conditions to a short-day cycle will induce flowering two weeks later, even if the plant is immediately returned to long-day conditions. Also, as we saw in Figure 23-9, the cocklebur can withstand a good deal of rough treatment.

In the course of these studies, in which they tested a variety of experimental conditions, the investigators made a crucial and totally unexpected discovery. If the period of darkness is interrupted by as little as a 1-minute exposure to the light of a 25-watt bulb, flowering does not occur. Interruption of the light period by darkness has no effect whatsoever on flowering. Subsequent experiments with other short-day plants showed that they, too, required periods not of uninterrupted light but of uninterrupted darkness (Figure 23–12).

On the basis of the findings of the Beltsville group, commercial growers of chrysanthemums had found that they could hold back blooming in the short-day plants by extending the daylight with artificial light. Now, on the basis of the new experiments by Hamner and Bonner, they were able to economize on their electric bill and still have flowers for late-season football games.

What about long-day plants? They also measure darkness. A long-day plant that will flower if it is kept in a laboratory in which there is light for 16 hours and dark for 8 hours will also flower on 8 hours of light and 16 hours of dark if the dark is interrupted by even a brief exposure to light (Figure 23–12).

23–12 Experiments on photoperiodism showed that plants measure the darkness rather than the light. Short-day plants flower only when the darkness exceeds some critical value. Thus, the cocklebur, for instance, will flower on 8 hours of light and 16 hours of darkness. If the 16-hour period of darkness is interrupted even very briefly, as shown on the right, the plant will not flower.

The long-day plant, on the other hand, which will not flower on 16 hours of darkness, will *flower if the darkness is interrupted. Long-day plants flower only when the darkness* is less *than some critical value.*

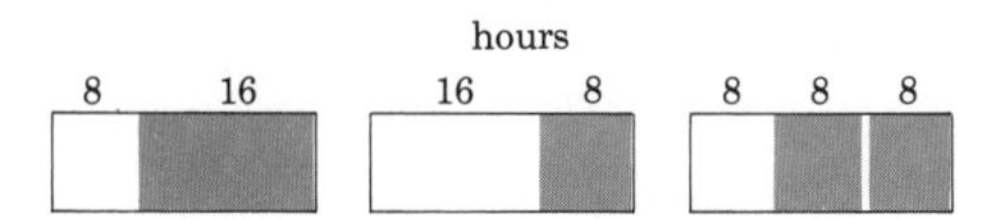

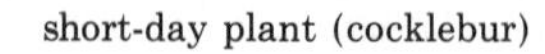

long-day plant (henbane)

Photoperiodism and Phytochrome

Following up on the clues from the Hamner and Bonner experiments, the Beltsville group was able to detect and eventually to isolate the pigment involved in photoperiodism. This pigment, which they called phytochrome, exists in two different forms. One form, known as P_r, absorbs red light with a wavelength of 660 nanometers; the other form, P_{fr}, absorbs far-red light with a wavelength of 730 nanometers. P_{fr}, which is the active form of the pigment, promotes flowering in long-day plants and inhibits flowering in short-day plants.

When P_r absorbs red light, it is converted to P_{fr}. This conversion takes place in daylight or in incandescent light; in both of these types of light, red wavelengths predominate over far-red. When P_{fr} absorbs far-red light, it is converted back to P_r. The P_{fr} to P_r conversion can also take place in the dark, which is how it usually occurs in nature (Figure 23–13).

Other Phytochrome Responses

Many types of small seeds, such as lettuce, germinate only when they are in loose soil, near the surface. This ensures that the seedling will reach the light before it runs out of stored food. Red light, a sign that sunlight is present, stimulates seed germination by converting phytochrome to the active form.

Phytochrome is also involved in the early development of seedlings. When a seedling develops in the dark, as it normally does underground, the stem elongates rapidly, pushing the shoot up through the soil layers. Any seedling grown in the dark will be elongated and spindly with small leaves (Figure 23–14). It will also be almost colorless, because the chloroplasts do not synthesize chlorophyll until they are exposed to light. Such a seedling is said to be etiolated. When the seedling tip reaches the light, normal growth begins. Phytochrome is involved in the switching from etiolated to normal growth. If a dark-grown bean seedling is exposed to only 1 minute of red (660 nanometers) light, it will respond with normal growth. If, however, the exposure to red light is followed by a 1-minute exposure to far-red light, etiolated growth continues.

23–13 P_r changes to P_{fr}, the active form, when exposed to red light, which is present in sunlight. P_{fr} reverts to P_r when exposed to far-red light. In darkness, P_{fr} slowly reverts to P_r.

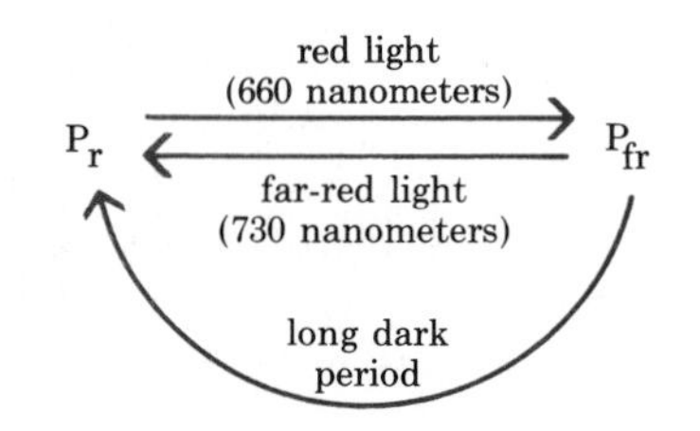

23-14 *Dark-grown seedlings, such as the ones on the right, are thin and pale with longer internodes and smaller leaves than the normal seedling on the left. This group of characteristics, known as etiolation, has survival value for the seedling because it increases its chances of reaching light before its stored energy supplies are used up.*

The way in which phytochrome acts is not known. One recent suggestion is that it alters the permeability of the cell membrane, permitting particular substances to enter the cell, or, perhaps, inhibiting their entry, and that these substances, which probably include hormones, regulate the cell's activities.

Circadian Rhythms

Some species of plants have flowers that open in the morning and close at dusk. Others spread their leaves in the sunlight and fold them toward the stem at night (Figure 23-15). As long ago as 1729, a French scientist noticed that these daily movements continue even when the plants are kept in constant dim light. More recent studies have shown that less evident activities, such as photosynthesis, auxin production, and the rate of cell division, also have daily rhythms. These regular day-night cycles are called *circadian rhythms*, from the Latin words *circa*, meaning "about," and *dies*, "day." They have been found throughout the plant and animal kingdoms.

Biological Clocks

It is now generally agreed that circadian rhythms are internal—that is, caused by factors within the plant—or animal—itself. The mechanism by which they are controlled is known as a *biological clock*.

Biological clocks are believed to play an essential role in many aspects of plant and animal physiology. They are clearly involved in photoperiodism. In order for plants to detect changes in day length—and remember that, in some cases, they are accurate to within 10 minutes—they must have some constant, a clock, with which to compare them. The chemical nature of the biological clock—or indeed whether there is just one kind of clock or many—is still not known.

Some plants secrete nectar or perfume at certain specific times of the day. As a result, insects—which have their own biological clocks—become programmed to visit these flowers at these times, thereby ensuring maximum rewards for both the insects and the flowers.

23-15 *Leaves of the wood sorrel, day* (a) *and night* (b). *One hypothesis concerning the function of such "sleep" movements is that they protect the leaves from absorbing moonlight on bright nights, thus protecting photoperiodic phenomena. Another theory, proposed by Darwin almost 100 years ago, is that the folding protects against heat loss from the leaves by night.*

(a)

(b)

23-16 *Tendrils of* Smilax *(greenbrier). Twisting is caused by varying growth rates on different sides of the tendril.*

Touch Responses in Plants

Many plants respond to touch. One of the most common examples is seen in tendrils. They wrap around objects with which they come in contact (Figure 23-16) and so enable the plant to cling and climb. The response can be rapid; a tendril may wrap around a support one or more times in less than an hour. Cells touching the support shrink slightly and those on the other side elongate. There is some evidence that auxin plays a role in this response.

A more spectacular response is seen in the sensitive plant, *Mimosa pudica,* in which the leaflets and sometimes entire leaves droop suddenly when touched. This response is a result of a sudden change in the turgor pressure of cells at the base of the petioles. There is some controversy about its survival value to the plant. *Mimosa pudica* often grows in dry, exposed areas where it may be subjected to drying winds. Strong winds may shake the leaves enough to make them fold up, so conserving water. Another suggestion is that the wilting response makes the plant unattractive to grazing animals or that it startles chewing insects.

Turgor changes triggered by touch are also involved in the capture of prey by the carnivorous Venus flytrap (see essay, page 306). The leaves of the Venus flytrap have two lobes, and each leaf half is equipped with three sensitive hairs. When an insect alights on one of these leaves, it brushes against the hairs, setting off an electrical impulse that triggers the closing of the leaf. The toothed edges mesh, the leaf halves gradually squeeze closed, and the insect is pressed against digestive glands located on the inner surface of the trap.

The trapping mechanism is so specialized that it can distinguish between living prey and inanimate objects, such as pebbles and small sticks, that may fall on the leaf by chance. The leaf will not close unless two of its hairs are touched in succession or one hair is touched twice.

Studies made by investigators at Washington University in St. Louis have shed new light on how the sundew traps insects. As you saw on page 306, the club-shaped leaves of the plant are covered with tiny tentacles. A sticky droplet surrounding the tip of each tentacle attracts insects. When

23-17 *Sensitive plant* (Mimosa pudica). (a) *Normal position of leaves and leaflets.* (b, c) *Successive responses to touch.*

(a)

(b)

(c)

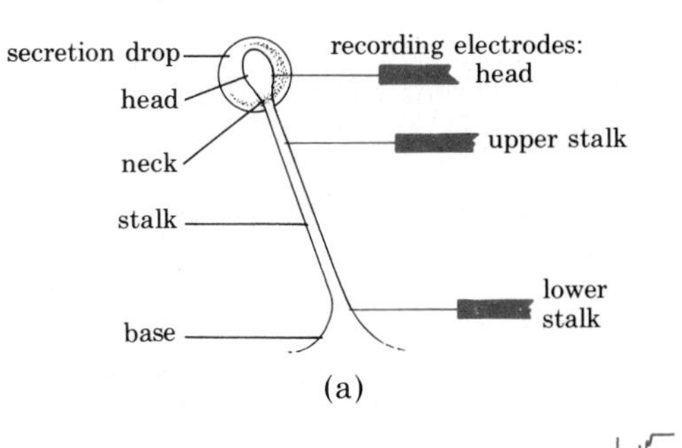

fly introduced

50 mV

20 sec

(b)

23-18 (a) *Diagram of a sundew tentacle, with electrodes in place.* (b) *A record of electrical impulses produced by positioning a fruit fly so that its feet stroked the head of a tentacle. The recording ended when the contact was broken as the tentacle bent.*

an insect is caught on the tip of a tentacle, the tentacle bends in rapidly, carrying the prey to the center of the leaf, where it is digested by enzymes. Microelectrodes placed in the tentacles (Figure 23-18) have revealed that this rapid response, like the triggering of the Venus flytrap, is accompanied by an electrical impulse that moves down the tentacle. These impulses are the same, in principle, as the nerve impulse in animals (see page 358).

The investigators predict that electric signals will be found to coordinate a variety of functions in plants.

SUMMARY

Plants respond to factors in both their internal and external environments. Such responses enable the plant to develop normally and to remain in touch with changing external conditions.

Hormones are important factors in plant response. A hormone is a chemical that is produced in particular tissues of an organism and carried to other tissues of the organism, where it exerts one or more specific influences. Characteristically, it is active in extremely small amounts.

Auxin is a hormone that is produced principally in rapidly dividing tissues, such as coleoptile tips and apical meristems. It causes lengthening of the shoot and the coleoptile, chiefly by promoting cell elongation. It often inhibits growth of lateral buds, thus restricting growth principally to the apex of the plant. The same concentration of auxin that promotes growth in the stem may inhibit growth in the main root system. Auxin promotes the initiation of branch roots and adventitious roots; however, it retards abscission in leaves and fruits. In developing fruits, auxin produced by seeds stimulates growth of the ovary wall. The capacity of auxin to produce such varied effects appears to result from different responses of the various target tissues.

The gibberellins were first isolated from a parasitic fungus that causes abnormal growth in rice seedlings. They were subsequently found to be natural growth hormones also present in plants. Gibberellins induce bolting and flowering in many plants. In some dwarf plants, application of gibberellins restores normal growth. Gibberellins are also involved in embryo and seedling growth. They stimulate the production of hydrolyzing enzymes that act on the stored starch of the endosperm, converting it to sugar, which nourishes the seedling.

The cytokinins, a third class of hormone, promote cell division. It is possible by altering the concentrations of auxin and cytokinins to alter patterns of growth in undifferentiated plant tissue in tissue culture.

Abscisic acid, a growth-inhibiting hormone, induces dormancy in vegetative buds and accelerates abscission. Ethylene is a gas that is produced by fruits during the ripening process. Ethylene promotes the ripening of fruits and is now considered a natural growth regulator. There is evidence for the existence of a flowering hormone (or hormones), but it has not yet been isolated.

Plants respond to a number of environmental stimuli. Photoperiodism is the response of organisms to changing periods of light and darkness in the 24-hour day. Such a response controls the onset of flowering in many plants. Some plants will flower only when the periods of light exceed a critical length. Such plants are known as long-day plants. Other plants, short-day plants, flower only when the periods of light are less than some critical period. Day-neutral plants flower regardless of photoperiod. Interruption of the dark phase of the photoperiod, even by a brief exposure to light, can serve to reverse the photoperiod effects, indicating that the dark period rather than the light period is the critical factor.

Phytochrome, a pigment commonly present in small amounts in the tissues of plants, is the receptor molecule for transitions between light and darkness. The pigment can exist in two forms, P_r and P_{fr}. P_r absorbs red light of a wavelength of 660 nanometers, whereas P_{fr} absorbs far-red light of a 730-nanometer wavelength. P_{fr} is the active form of the pigment; among its many known effects, it promotes flowering in long-day plants, inhibits flowering in short-day plants, promotes germination in lettuce seeds, and promotes normal growth in seedlings. Its mechanism of action is unknown.

Circadian rhythms are regular cycles of growth and activity occurring approximately on a 24-hour basis. Many of these rhythms have been shown to be independent of the organism's environment and to be controlled by some internal regulator—a biological clock. Its chemical nature is not known.

Some species of plants respond to touch. Examples include the winding of tendrils, the collapse of the leaves of the sensitive plant *(Mimosa),* the triggering of the carnivorous Venus flytrap, and the bending of the tentacles of the sundew.

QUESTIONS

1. Describe Went's experiments and the conclusions that can be drawn from them.

2. Describe the principal roles played by each of the following hormones: auxin, gibberellin, cytokinin, abscisic acid, and ethylene.

3. For many years it has been known that oranges hasten the ripening of bananas. Can you think of an explanation for this phenomenon?

4. Speculate on why—in the course of evolution—a gas was selected to mediate a process such as fruit ripening.

5. Distinguish between the following: phototropism/photoperiodism; circadian rhythm/biological clock.

6. Plants must, in some manner, synchronize their activities with the seasons. Of the clues that they might use, day (or night) length seems to have been selected. What might be the advantage of using photoperiod rather than, say, temperature as a season detector?

7. Photoperiodic systems are not nearly as sensitive to low levels of illumination as are many visual systems. Why might extreme sensitivity be a disadvantage in a photoperiodic system? What havoc might be wrought by the widespread use of bright lights for street illumination?

8. Suppose you were given a chrysanthemum plant, in bloom, one autumn, and you decided to keep it indoors as a house plant. What precautions would you need to take the following autumn to ensure that it would bloom again?

SUGGESTIONS FOR FURTHER READING

GALSTON, A. W., and P. DAVIES: *Life of the Green Plant,* 3d ed., Prentice-Hall, Inc., Englewood Cliffs, N.J., 1980.*

A convenient and concise summary of plant growth and development.

GREULACH, VICTOR: *Plant Structure and Function,* The Macmillan Company, New York, 1973.

A short, lucid introduction to botany, with emphasis on physiology.

LEDBETTER, M. C., and KEITH PORTER: *Introduction to the Fine Structure of Plant Cells,* Springer-Verlag, New York, 1970.

An excellent atlas of electron micrographs of plant cells, with a detailed explanation of each.

RAVEN, PETER H., RAY F. EVERT, and HELENA CURTIS: *Biology of Plants,* 3d ed., Worth Publishers, Inc., New York, 1981.

An up-to-date and handsomely illustrated general botany text.

RAY, PETER M.: *The Living Plant,* 2d ed., Holt, Rinehart and Winston, Inc., New York, 1972.*

An outstanding, short text.

SALISBURY, FRANK B., and CLEON ROSS: *Plant Physiology,* 2d ed., Wadsworth Publishing Co., Inc., Belmont, Calif., 1978.

A good, modern plant physiology text for more advanced students.

TORREY, JOHN G.: *Development in Flowering Plants,* The Macmillan Company, New York, 1967.*

How flowering plants develop, with an emphasis on the underlying physiological process.

* Available in paperback.

SECTION 6

Human Physiology

Genus, Homo; *species,* sapiens. *An immature form.*

24

The Human Animal: An Introduction

24-1 *In most species of animals, each new individual arises from the union of a sperm cell with an egg cell. This egg cell, which is being fertilized by a sperm cell, is from a hamster. Hamsters, like humans, are vertebrate animals belonging to the class Mammalia.*

The animal kingdom is made up of a vast array of organisms, ranging from microscopic aquatic forms to extremely complicated and highly organized multicellular creatures that are, by far, the most complex of all living things. In this section of the book, we shall examine *Homo sapiens* as an example of an animal organism. This one species, however, will not command our full attention. Because we are part of a continuum of nature, it is logical to study other animals in order to understand the human animal; it is equally logical to study the human animal (now one of the best understood) in order to gain a deeper understanding of animal life in general.

The human being is a vertebrate and, as such, has a bony, articulated (jointed) internal skeleton that supports the body and grows as it grows. The dorsal nerve cord (known as the spinal cord) is surrounded by bony segments, the vertebrae, and the brain is enclosed in a protective casing, the skull. The skeleton is moved by means of muscles.

As in other vertebrates, and most invertebrates as well, the human body contains a cavity, or coelom (page 264). In humans, the coelom is divided into two parts, the thoracic cavity and the abdominal cavity (Figure 24-2). These are separated by a dome-shaped muscle, the diaphragm. The thoracic cavity contains the heart, lungs, and esophagus (part of the digestive tract). The abdominal cavity contains the stomach, intestines, liver, and a number of other organs.

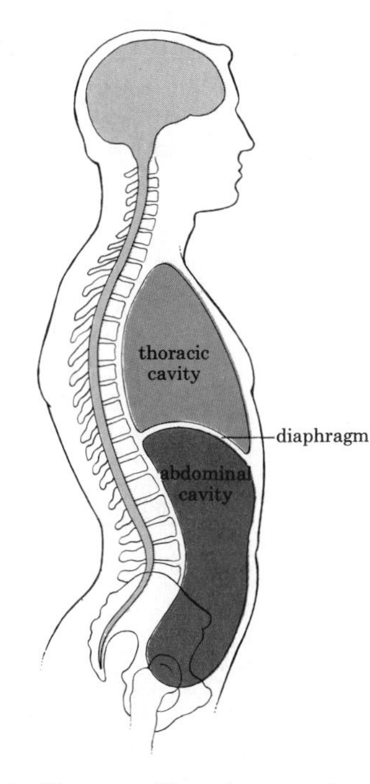

24-2 *Humans, like other vertebrates, are characterized by a dorsal nervous system enclosed in vertebrae and the skull. As in other mammals, a muscular diaphragm divides the coelom into the thoracic and abdominal cavities.*

Cells and Tissues

The human body, like that of all other complex animals, is made up of a variety of different cells. These cells are organized into *tissues*, which are groups of cells carrying out a unified function.

Although experts can distinguish more than 100 different types of cells in the human body, they customarily classify them in terms of only four tissue types: (1) epithelial, (2) connective, (3) muscle, and (4) nerve.

Epithelial Tissue

Epithelial tissue consists of a continuous sheet of cells that provides a protective covering for the whole body and contains sensory nerve endings. It also forms the lining membranes of internal organs, cavities, and passageways. Hence everything that goes into and out of the body must pass through epithelial cells.

Epithelium is classified according to the shape of the individual cells as squamous, cuboidal, or columnar (Figure 24-3). It may consist of only a single layer of cells (simple epithelium) or several layers (stratified epithelium). One surface of the epithelial sheet is always attached to an underlying layer, called the basement membrane. This membrane is composed of a fibrous polysaccharide material produced by the epithelial cells themselves.

The epithelium of the body cavities and passageways frequently contains modified epithelial cells that secrete mucus, which lubricates the surfaces. *Glands* are special types of epithelial tissue; gland cells produce specific substances, such as perspiration, saliva, hormones, or digestive enzymes. Glandular epithelium is composed of cuboidal or columnar epithelial cells.

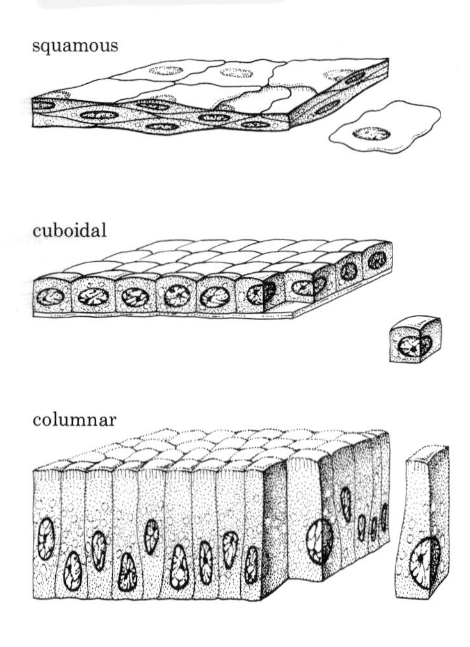

24-3 *The three types of epithelial cells that cover the inner and outer surfaces of the body. Squamous cells, which usually perform a protective function, make up the outer layers of the skin and the lining of the mouth and other mucous membranes. There are usually several layers of these flat cells piled on top of one another. Cuboidal and columnar cells, in addition to lining various passageways, perform much of the chemical work of the body.*

Connective Tissue

Connective tissue binds together and supports the other three kinds of tissue. Unlike epithelial tissue cells, the cells of connective tissue are widely separated from one another by large amounts of intercellular substances. The intercellular substances include: (1) connecting and supporting fibers, such as collagen (Figure 3-17, page 47), which is a major component of skin, tendons, ligaments, and bones; (2) elastic fibers, which are often found in the walls of hollow, distensible organs, such as the stomach and the uterus; and (3) reticular fibers, which form networks inside solid organs, such as the liver and the pancreas. All of these fibers are embedded in a ground substance that is more or less fluid and amorphous (without any shape or form).

Bone, like other connective tissues, consists of cells, fibers, and ground substance (Figure 24-5a). It is unlike the others in that the collagen fibers are impregnated with hard crystals. Bone tissue is amazingly strong and light; our bones make up only about 18 percent of our weight.

Blood and lymph are connective tissues in which the ground substance is plasma. The characteristic cells of blood and lymph are described in Chapter 27.

24-4 *The outer surface, or epidermis, of the skin is composed of stratified squamous epithelium. In the lower portion of this scanning electron micrograph, the layers of epithelium are clearly visible.*

On the skin surface, some of the older cells are being sloughed off. This is a continuous process in epithelial tissue that is subjected to wear and tear. Replacement cells are produced by mitosis in the underlying layers of the skin. The rodlike structures at the left are hairs emerging from the skin surface.

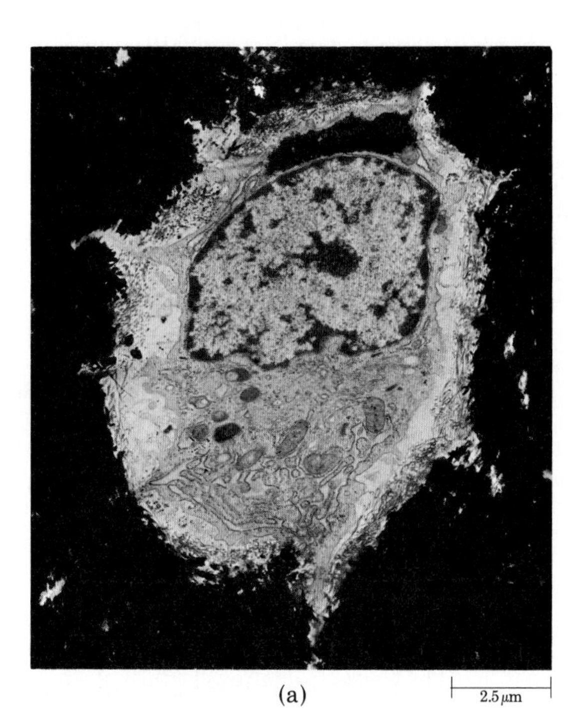

(a)

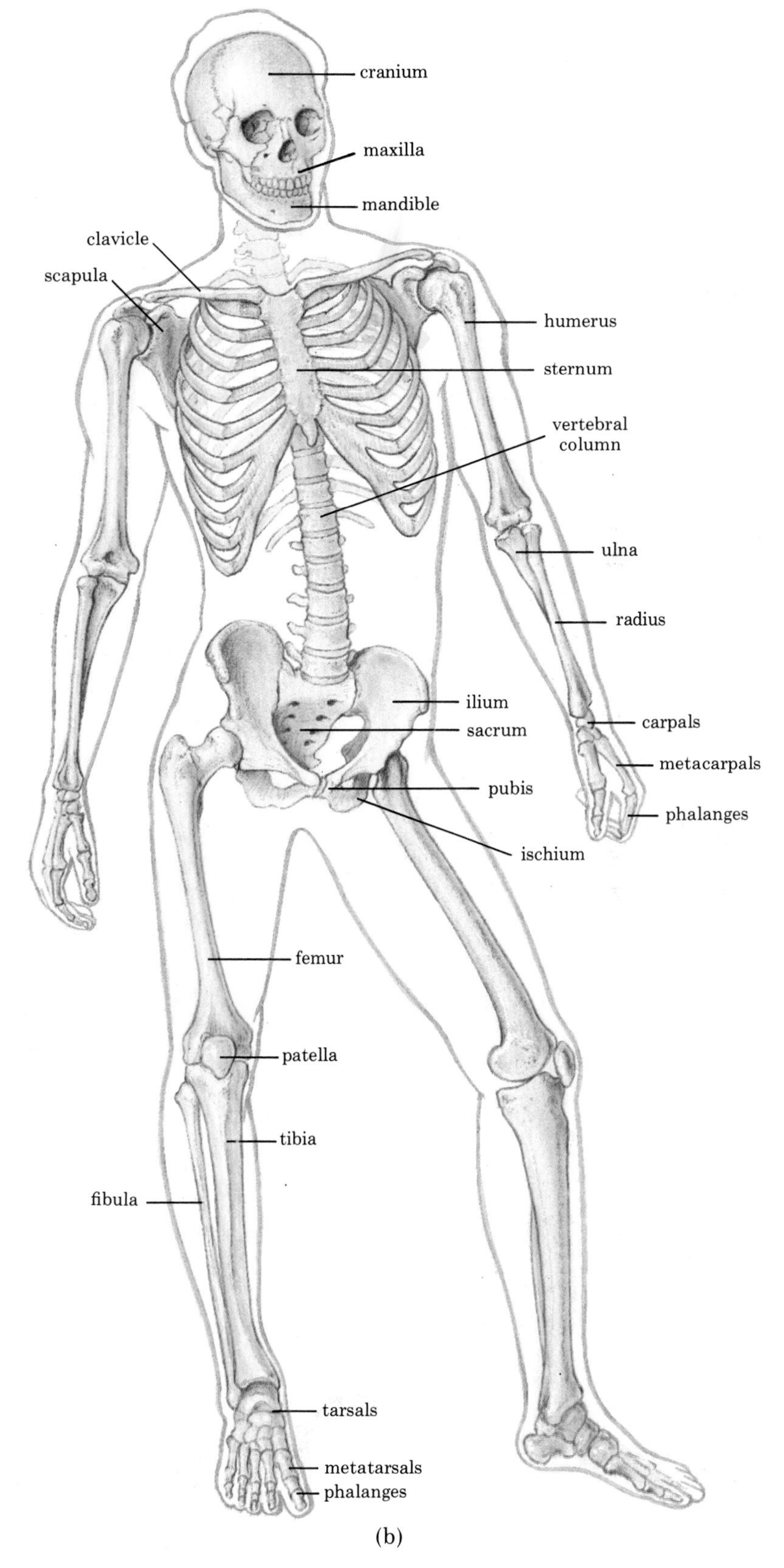

(b)

24-5 (a) *Electron micrograph of a bone cell (osteocyte). The large nucleus occupies the upper half of the cell. The light area just outside the cell membrane contains collagen fibrils, and the dark area surrounding the cell and the collagen fibrils is hardened bone. Young bone cells produce the extracellular ground substance, which gradually hardens as calcium compounds in the ground substance crystallize in and on the collagen fibrils. Cytoplasmic processes extend from the cells through channels in the hardened bone.*

(b) *The skeleton of a human adult contains 206 bones. Twenty-nine are in the skull, including 14 face bones and six small bones of the ears. There are 27 bones in each hand and 26 in each foot. Bones are living organs, made up not only of connective tissue but of the other tissue types as well. They are surrounded by a fibrous sheath, which contains blood vessels that nourish the bone tissues.*

The Injury-Prone Knee

A joint is a structure, usually movable, at which two bones are opposed. The knee is the hinge joint between two strong, inflexible bones, the femur and the tibia. It is covered by the "kneecap," or patella (removed in this drawing, but visible in Figure 24-5b).

As in all joints, the bones of the knee are bound together by ligaments. The collateral ligaments bind the joint externally; they are slack when the knee is flexed and taut when the knee is extended. The cruciate ligaments cross within the joint, stabilizing it. The menisci are C-shaped wedges of cartilage, resting on the upper surface of the tibia and cushioning the joint.

Knees are particularly susceptible to injury, as Joe Namath, Bobby Orr, Wilt Chamberlain, and others will sadly attest. The reason for the knee's vulnerability becomes clear when you contemplate the analogous problem of fastening together two match sticks with several rubber bands, to produce a hinge that is both strong and flexible and can not only swing back and forth but also twist, bend, and rotate. Common knee injuries are torn ligaments, especially the collateral ligaments (as a result of the knee's being hit from the side), and crushed menisci. When the shock-absorbing wedges of menisci are lost, bone chips begin to accumulate in the joint, making it less flexible and much more painful.

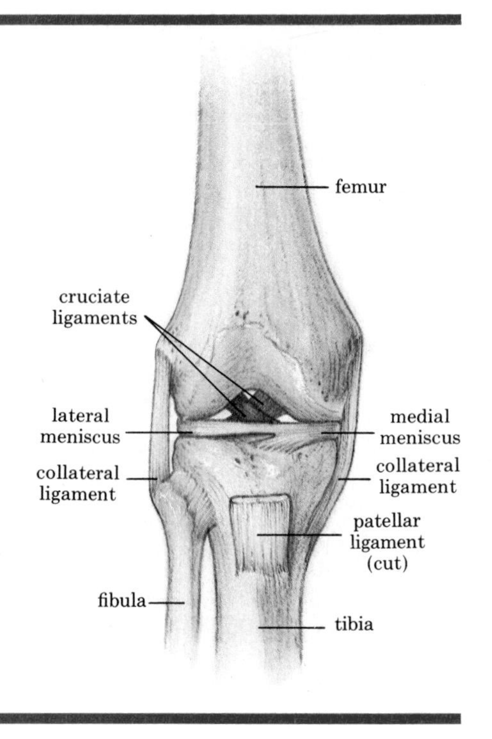

The right knee, viewed from the front, with the patella removed.

Muscle Tissue

About 40 percent of our body's weight is made up of muscle tissue. There are two general types: smooth muscle and striated muscle, so called because of its striped appearance (Figure 24-6). Smooth muscle surrounds the walls of the internal organs, such as the digestive organs and the uterus; it is sometimes called involuntary muscle since it is not under our conscious control. The muscles that move the skeleton are striated muscle and are sometimes called voluntary muscles since we can move them at will. A special type of striated muscle, cardiac muscle, makes up the wall of the heart and is involuntary.

24-6 *Photomicrograph of skeletal muscle fibers, showing striated pattern. The arrow indicates a small blood vessel, a capillary. Within it, you can see red blood cells, which carry oxygen to the muscle fibers, each of which is a single cell.*

In this discussion, we shall focus on skeletal muscle because it is the subject of current research, and much of what we know about the other forms of muscle is inferred from what has been learned about skeletal muscle. Also, studies of the machinery of skeletal muscle, involving as they do both electron microscopy and biochemistry, have brought scientists tantalizingly close to visualizing the precise role of individual molecules.

Muscle Action

Skeletal muscles, like all muscles, act by contracting. A skeletal muscle is typically attached to two or more bones, either directly or by means of the tough strands of connective tissue known as tendons. Some of these tendons, such as those that connect the finger bones and their muscles in the forearm, may be very long. When the muscle contracts, the bones move at a joint, which is held together by ligaments and contains a lubricating fluid. Most of the skeletal muscles of the body work in antagonistic pairs, one muscle flexing, or bending, the joint, and the other extending, or straightening, it (Figure 24-7). Sometimes when both muscles of an antagonistic pair contract simultaneously, the joint does not move but is stabilized. This is particularly important in maintaining upright posture.

A muscle, such as the biceps, consists of bundles of muscle fibers—often hundreds of thousands of fibers—held together by connective tissue. Each fiber is a single cell with many nuclei. These fibers are very large cells—50 to 100 micrometers in diameter and, often, several centimeters long. Each cell is surrounded by an outer cell membrane that has been given the special name of sarcolemma.

Embedded in the cytoplasm of each muscle cell (fiber) are some 1,000 to 2,000 smaller structural units, the *myofibrils*. They run parallel for the length of the cell. Each myofibril is, in turn, composed of units called *sarcomeres*. The repetition of these units gives the muscle its characteristic striated pattern.

Figure 24-8 shows a sarcomere as seen in a longitudinal section of muscle. As the diagram shows, each sarcomere is composed of two types of filaments running parallel to one another. The thicker filaments in the central portion of the sarcomere are composed of a protein known as myosin; the thinner filaments are made primarily of actin, which is also a protein. When viewed in cross section, each thick filament is surrounded by six thin filaments.

24-7 *Muscles attached to bone move the vertebrate skeleton. They often work in antagonistic pairs, with one relaxing as the other contracts. Muscles cannot lengthen spontaneously; they lengthen only when pulled by antagonistic muscles. For example, when you move your hand toward your shoulder, as shown here, the biceps contracts and the triceps relaxes. When you move your hand down again, the triceps contracts, while the biceps relaxes. The muscles that move the skeleton, such as those diagrammed here, are known as skeletal muscles. They are striated, as shown in Figure 24-6.*

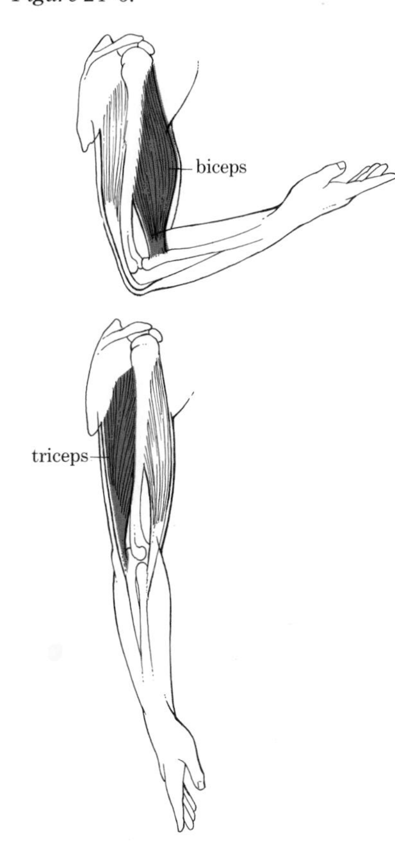

24-8 *Electron micrograph and diagram of a sarcomere, the contractile unit of muscle. Each sarcomere is composed of an array of thick and thin protein filaments arranged longitudinally. The way the thin and thick filaments overlap makes the muscle fiber appear striated (banded). The Z line is where thin filaments from adjoining sarcomeres interweave. The I band is a region that contains only thin filaments. The A band marks the extent of the thick filaments. The part of the A band where there are no thin filaments is called the H zone. The thick filaments are interconnected and held in place at the M line. Muscle contraction involves the sliding of the thin filaments between the thick ones.*

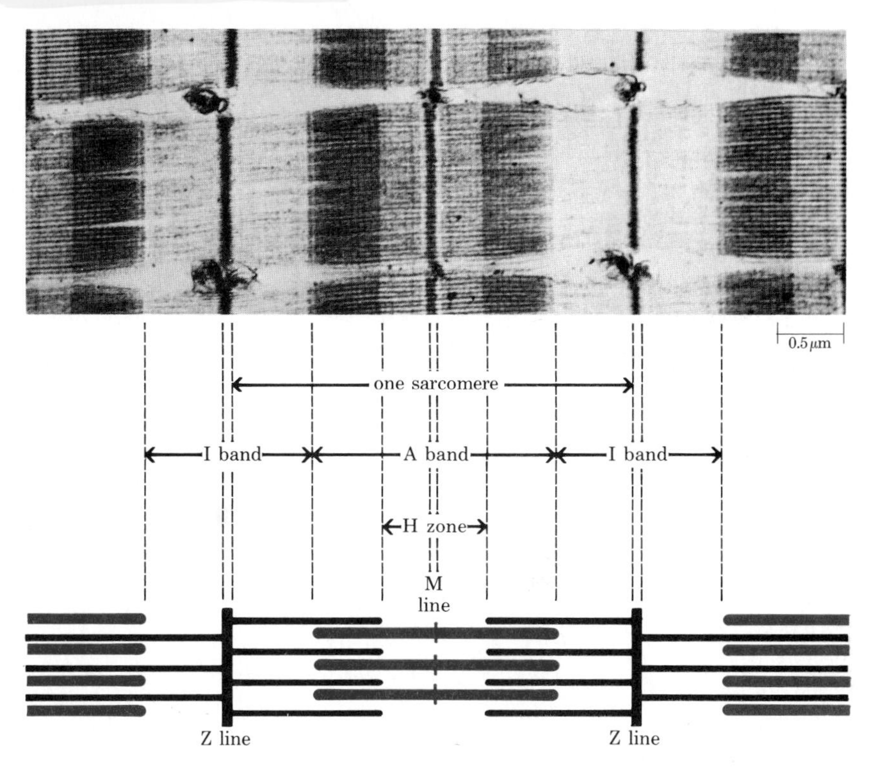

24-9 (a) *Skeletal muscle is composed of individual muscle cells, the muscle fibers. These are cylindrical cells, often many centimeters long, with numerous nuclei.* (b) *Each muscle fiber is made up of many cylindrical subunits, the myofibrils. These rods run from one end of the cell to the other and contain the contractile proteins.* (c) *The myofibril is divided into segments, sarcomeres, by thin, dark partitions, the Z lines. The Z lines of adjacent myofibrils are in line with each other and appear to run from myofibril to myofibril across the fiber, giving the muscle cell its striated appearance.* (d) *Each sarcomere is made up of thick and thin filaments. Chemical analysis shows that the thick filaments consist of bundles of a protein called myosin. Each individual myosin molecule is composed of two protein chains wound in a helix; the end of each chain is folded into a globular structure.* (e) *Each thin filament consists primarily of two actin strands coiled about one another in a helical chain. Each strand is composed of globular actin molecules.* (f) *According to current hypothesis, the globular myosin heads protruding from the thick filaments serve as levers, attaching to the actin molecules of the thin filament and pulling them toward the center of the H zone, shortening the sarcomere and contracting the myofibril. When the myofibrils contract, the fiber shortens, and when enough fibers shorten, the entire muscle shortens, producing skeletal movements.*

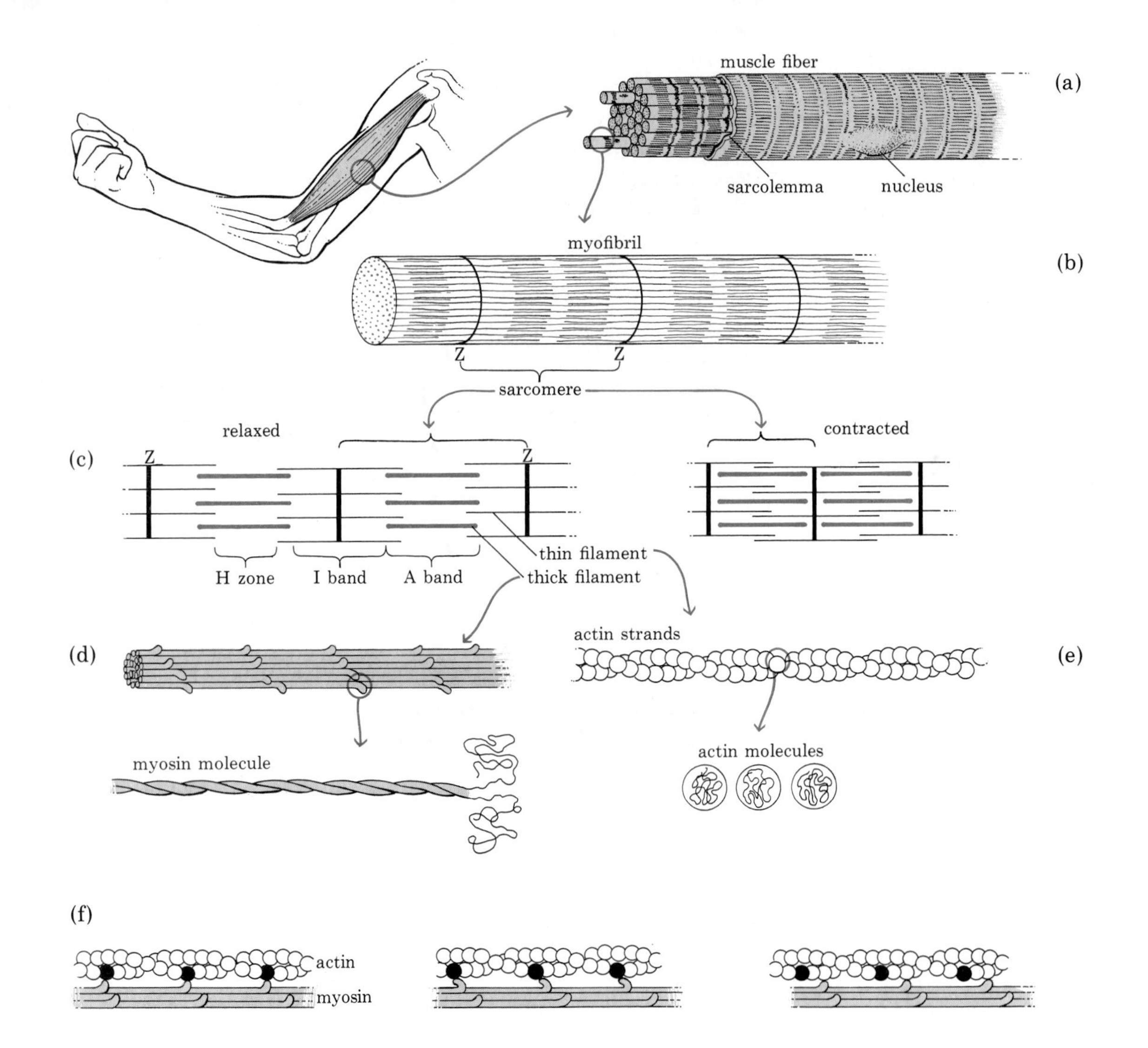

Muscles function by contracting. When muscle is stimulated, the thin (actin) filaments of the sarcomere slide past the thick (myosin) filaments. Since the thin filaments are anchored into the Z lines, this causes each sarcomere to shorten, and thus the myofibril as a whole contracts. According to the current hypothesis, cross bridges between the thick and thin filaments form, break, and re-form rapidly, as one filament "walks" along the other (Figure 24–9).

Actin and Myosin

The actin strands in muscle are composed of many smaller globular subunits assembled in a long chain. As shown in Figure 24–9e, each thin filament is composed primarily of two such actin chains wound around one another.

The myosin molecule has the longest protein chains known, each one consisting of some 1,800 amino acid units. Each chain has a globular structure at one end. The myosin molecule consists of two of these chains wound around each other, with the globular "heads" free. The thick filaments, in turn, are composed of bundles of myosin molecules. These globular heads have two crucial functions: They are the binding sites that link the actin and myosin molecules, and they also act as enzymes to split ATP to ADP, thus providing the energy for muscle contraction.

To sum up, when a muscle fiber is stimulated, the thin (actin) filaments slide between the thick (myosin) filaments, and ATP is hydrolyzed to ADP in the process. The sarcomeres shorten and the fibers contract. If enough fibers contract, a whole muscle contracts, and the organism or part of the organism moves. Thus the intricate machinery of the sarcomere is, in essence, a mechanism for the conversion of chemical energy into kinetic energy, the energy of motion.

24–10 *Electron micrograph of smooth muscle from the epididymis (part of the spermatic duct) of a mouse. Smooth muscle is made up of long, spindle-shaped cells containing contractile proteins. Portions of a number of smooth muscle cells can be seen in the picture. The nucleus of one cell, distinguishable by its granular texture, can be seen in the middle of the micrograph. Very thin fibrils run lengthwise through the cells and make up the bulk of the cytoplasm.*

Smooth Muscle

Smooth muscle, unlike striated muscle, consists of cells with single nuclei (Figure 24–10). These cells contain numerous fibrils that run lengthwise through the cell but are not arranged in any discernible pattern. Like striated muscle, smooth muscle cells contain large amounts of actin and myosin, which are presumed, but not proved, to be in the fibrils and to play a role in the contraction of smooth muscle. Smooth muscle also requires ATP for contraction.

Nerve Tissue

The fourth major tissue type is nerve tissue. The basic units of nerve tissue are the *neurons*, or nerve cells, which make up about 10 percent of nerve tissue cells. Neurons receive and transmit signals from either the external or the internal environment. They also process this information and are responsible for the complex functions of consciousness, memory, and thought. The other 90 percent of nerve tissue cells are *glial cells*, which not only support the neurons physically but also nourish, protect, and insulate them.

A typical neuron has three basic parts: the *cell body*, which contains the nucleus and most of the metabolic machinery of the cell; the *dendrites*, which receive stimuli and transmit impulses to the cell body (the cell body may also receive stimuli directly); and the *axon*, which transmits impulses

24-11 *Four examples of neurons. Dendrites, which are characteristically multiple, transmit signals toward the neuron cell body, whereas axons, of which there is only one per cell, transmit signals away from the cell body.*

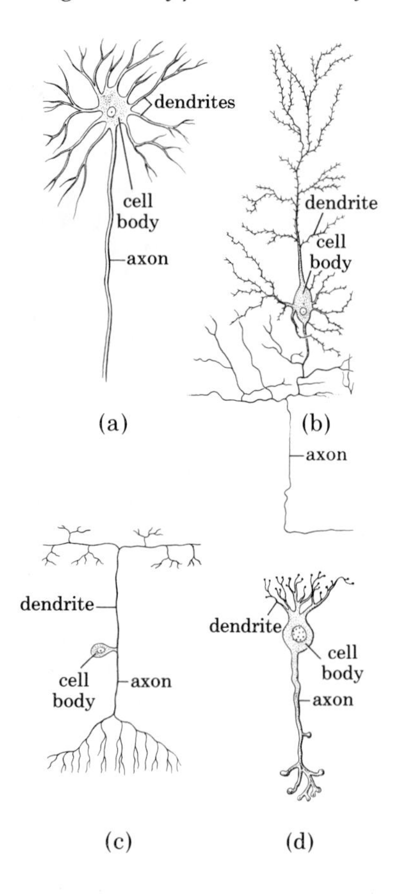

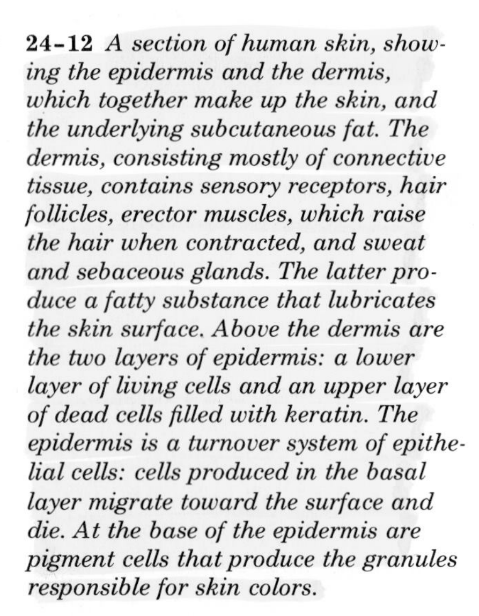

away from the cell body (Figure 24-11). A single neuron may have many dendrites, but it usually has only one axon, although the axon may be branched.

Axons and dendrites are commonly referred to as *nerve fibers*. Along these fibers, nerve impulses are transmitted from one part of the organism to another. A *nerve* consists of numerous nerve fibers and blood vessels bound together by connective tissue.

Neurons are the most spectacular of the body's cells. They may reach astonishing lengths. For example, the axon of a single motor neuron—a nerve cell that activates muscle—may extend from the spinal cord down the whole length of the leg to the toe. Or a sensory neuron—one that transmits sensations to the brain—based near the spinal cord may send a dendrite down to the toe and an axon up virtually the entire length of the spinal cord. In an adult human, such a cell might be close to 2 meters long (5 meters in a giraffe).

In the next chapter, we shall describe the structure of nerve cells in more detail and consider, in particular, the way in which they transmit the nerve impulse.

Organs and Organ Systems

Organs are composed of functionally related groups of tissue. The stomach, for example, is made up of layers of glandular epithelium (the stomach lining), connective tissue, nerves, and smooth muscle. The structure of the body's largest organ, the skin, is shown in Figure 24-12.

Organs, in turn, generally are part of *organ systems*. The stomach, for instance, is part of the digestive system. In the chapters that follow, we shall consider each of the major organ systems of the vertebrate body.

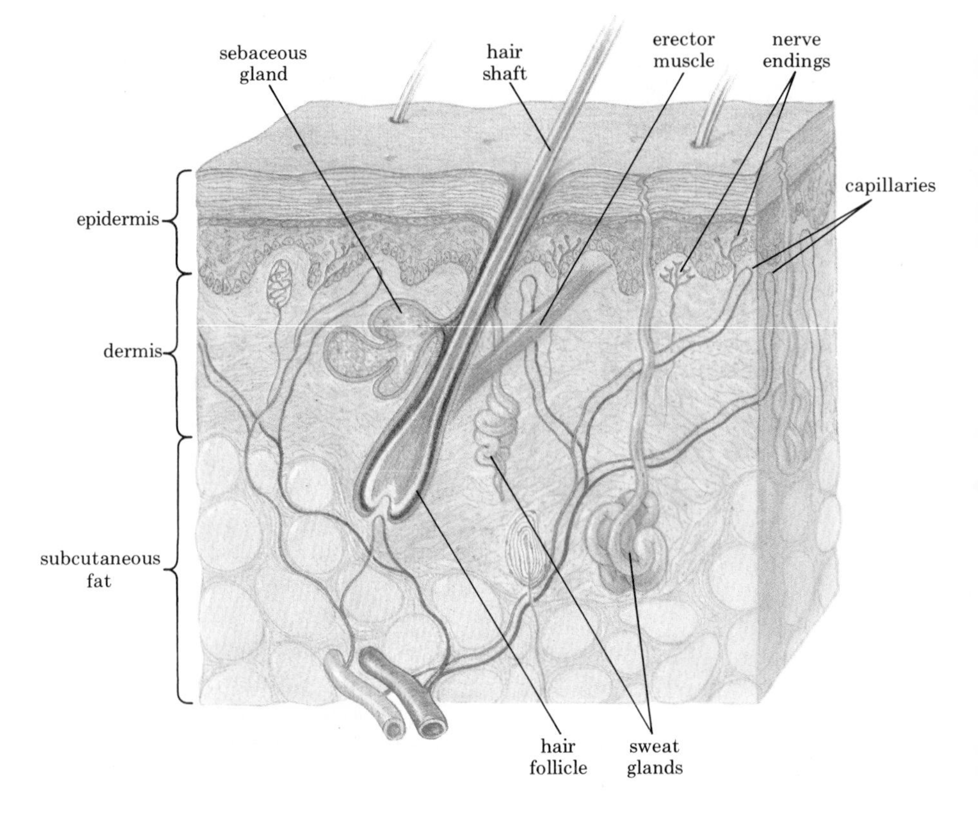

24-12 *A section of human skin, showing the epidermis and the dermis, which together make up the skin, and the underlying subcutaneous fat. The dermis, consisting mostly of connective tissue, contains sensory receptors, hair follicles, erector muscles, which raise the hair when contracted, and sweat and sebaceous glands. The latter produce a fatty substance that lubricates the skin surface. Above the dermis are the two layers of epidermis: a lower layer of living cells and an upper layer of dead cells filled with keratin. The epidermis is a turnover system of epithelial cells: cells produced in the basal layer migrate toward the surface and die. At the base of the epidermis are pigment cells that produce the granules responsible for skin colors.*

The Whole Organism

The adult human body is composed of trillions of cells of more than 100 kinds, grouped together in many different structural and functional units. Moreover, each cell is also working for itself, breaking down glucose, making ATP, building and maintaining membranes and organelles, producing enzymes and other protein molecules, dividing or not dividing. Yet in order for the organism to survive, all of these cells, tissues, and organs have to function as a whole.

For example, do you recall what it feels like to be terrified? The physical characteristics of fear result from a simultaneous discharge of many nerve fibers. Some of these cause the blood vessels in the skin and intestinal tract to contract; this contraction increases the return of the blood to the heart, raising the blood pressure and sending more blood to the muscles. The heart beats both faster and stronger, and the respiratory rate increases. The pupils dilate. The muscles attached to the hair follicles in the skin contract; this is probably a legacy from our furry forebears, which looked larger and more ferocious with their hair standing on end. The rhythmic movement of the intestine stops, and sphincters, the circular muscles at the end of the intestines and the opening of the bladder, relax; these reactions inhibit digestive operations, but the relaxing of the sphincters may also have the disconcerting consequence of causing involuntary defecation or urination. The adrenal glands pour out epinephrine (adrenaline), which, with other hormones, causes the release of large quantities of sugar from the liver into the bloodstream; this sugar provides an extra energy source for the muscles. As a consequence, the body as a whole is prepared for "fight or flight"—or, at least, action that would have been appropriate at some stage in our evolution.

However, response to the external environment is only a small part of the organism's need for integration of its many individual functional units. Equally complex and vital, although less obvious, activities are required for the constant regulation of the internal environment. This maintenance of a relatively constant internal environment is known as *homeostasis.* As we noted in the Introduction, homeostasis—"staying the same"—is one of the chief characteristics of living organisms. It is by regulating the internal environment that organisms have been able to free themselves to an increasing degree from the external environment, moving from salt water to fresh water, and from the water to the land and air.

Homeostasis is characteristic even of single cells and very simple organisms. In an organism as complex as *Homo sapiens,* it involves constant monitoring and regulation of a large number of different factors, such as temperature, pH, and the concentrations of oxygen, carbon dioxide, glucose and other nutrients, hormones, and a variety of ions. Despite changes in the external environment and the organism's response to it, body temperature and the composition of the body fluids remain relatively unchanged.

Virtually every cell, tissue, and organ in the body contributes in some way to this remarkable stability. The liver acts as a metabolic factory and warehouse, removing organic molecules or adding them as they are needed. Oxygen for ATP production may be used rapidly, as during muscle exertion, or slowly, as during sleep, but the supply in the bloodstream available to the individual cells remains constant because of variations in the rates at which the lungs take in oxygen and the heart pumps blood. The kidney processes and removes just the right amount of wastes, water, and salts—and so on, down the long list of organs, tissues, and cells. Therefore, although we shall examine the various organ systems of the body one by one, it is important to remember that all are acting in concert and that the function of each is affected by all the others. In the next two chapters we shall discuss the two major control agencies—the nervous system and the endocrine system—in more detail.

SUMMARY

As vertebrate animals, humans have a bony, jointed internal skeleton that includes a vertebral column enclosing the spinal cord and a skull enclosing the brain. The human body, like that of other mammals, contains a coelom that is divided by a muscular diaphragm into two compartments, the abdominal cavity and the thoracic cavity.

Our bodies are made up of cells of four principal tissue types: epithelial tissue, connective tissue, muscle, and nerve. Epithelium serves as a covering or lining for the body and its cavities. It also secretes mucus, perspiration, saliva, hormones, and digestive enzymes. Connective tissue strengthens and supports the other tissues of the body. It is characterized by large amounts of intercellular substances. Muscle is composed of assemblies of contractile fibers, which convert chemical energy to kinetic energy. Nerve tissue includes neurons and glial cells. Neurons carry signals from one part of the body to another.

Tissues are groups of cells that carry out a unified function. Various types of tissues are grouped in different ways to form organs, and organs are grouped to form organ systems.

The various cells, tissues, and organs of the body operate in an integrated manner so that the organism maintains a relatively constant internal environment (homeostasis) and is capable of responding as a whole to its external environment.

QUESTIONS

1. Distinguish among the following terms: diaphragm/coelom; smooth muscle/striated muscle; muscle cell/myofibril; actin/myosin; sarcomere/sarcolemma; nerve cell/neuron/glial cell; axon/dendrite.

2. What is the functional significance of each of the four tissue types? Give an example of each type.

3. What are the three types of cells found in epithelial tissue? What is the basis for classification of epithelial cells?

4. In addition to protection, what is the other major function of epithelial tissue?

5. What is the major structural difference between connective tissue and epithelial tissue? What functions of connective tissue account for this structural difference?

6. What is meant by antagonistic paired muscles?

7. Diagram and label a sarcomere that is in a relaxed state.

8. How does the structure you diagrammed in Question 7 enable muscle to perform its primary function?

9. Give some examples of homeostasis. Why is it important?

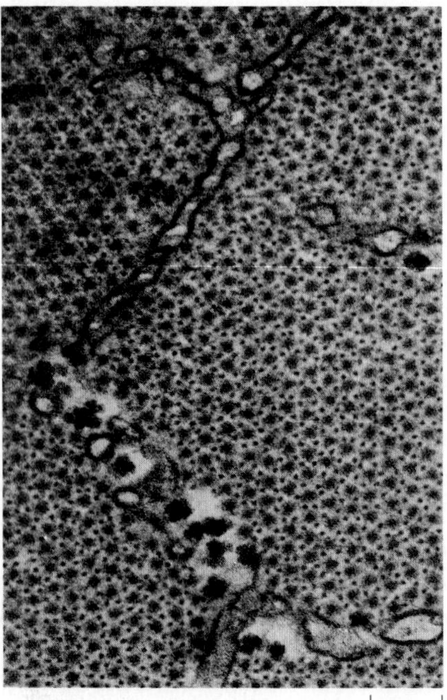

10. Can you interpret this electron micrograph? (You may be able to find the clue you need on page 349.)

25

Integration and Control: The Nervous System

The principal regulatory systems of the human body are the nervous system and the endocrine system (the hormone-secreting glands and their products). These two systems operate continuously to integrate and control the body's activities. In this chapter and the next, we are going to examine these vast communication networks in order to provide a background for understanding the regulation of such functions as circulation, respiration, digestion, temperature control, excretion, and reproduction, which are the subjects of the chapters that follow. Discussion of the human brain and of some of the current research on its functions is reserved for Chapter 32.

Nervous Systems

Functionally, a nervous system differs from an endocrine system chiefly in its capacity for rapid response—a nerve impulse can travel through an organism in a matter of milliseconds. Hormones, however, move at a somewhat slower rate (by way of the bloodstream) and, characteristically, elicit a slower response. Plants, which are not noted for their liveliness, rely largely on an elaborate interplay of hormones to coordinate their activities. But animals, while also relying on hormones, are generally characterized by nervous systems—networks of specialized nerve cells.

Invertebrate nervous systems range from the very simple to the complex (Figure 25–1, page 356). In the coelenterate *Hydra,* the nerve cells form a diffuse, continuous network. These neurons receive information from sensory receptor cells, which can be found among the epithelial cells on the inner (feeding) and outer surfaces of the animal. The neurons stimulate muscular cells that cause movements in the body wall.

The nervous system of the planarian, a flatworm, is more highly organized than that of *Hydra.* Some of the nerve net is condensed into two cords, and there are two clusters of neurons, known as *ganglia* (singular, ganglion), at the anterior end of the body.

In the earthworm, the two cords have come together in a single nerve cord, which runs along the ventral surface of the body. Along the nerve cord are ganglia, one for each segment. The nerve cord forks just below the pharynx and the two forks meet again, terminating in two large ganglia.

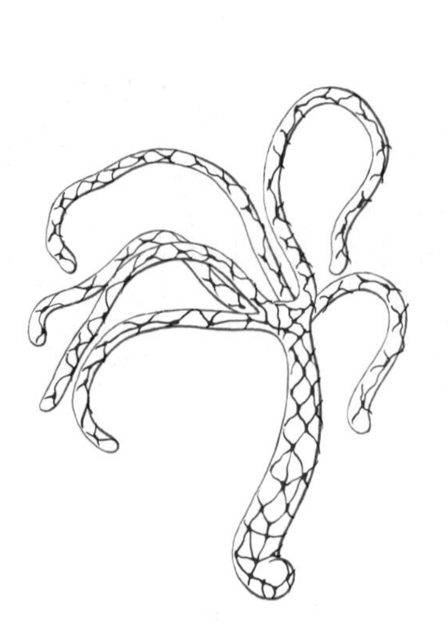

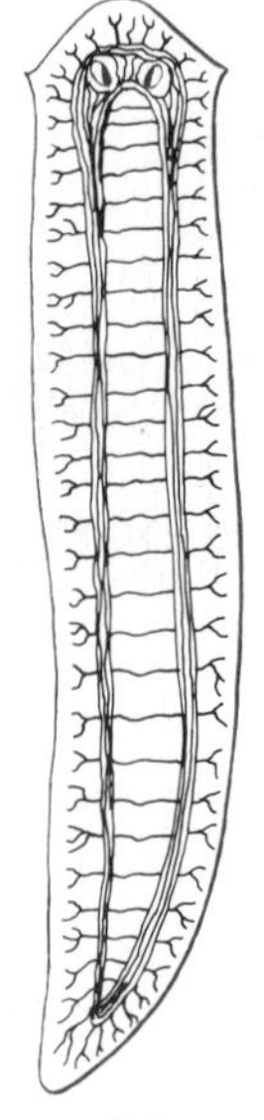

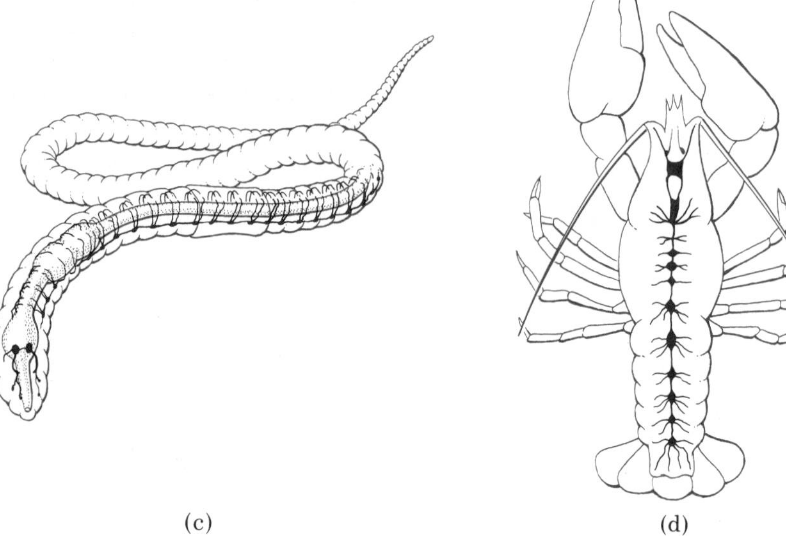

25-1 *Some invertebrate nervous systems.* (a) Hydra, *a coelenterate;* (b) *planarian, a flatworm;* (c) *earthworm, an annelid; and* (d) *crayfish, an arthropod.*

Arthropods, such as the crayfish, have sizable collections of neurons in the head region. These collections are large enough to be called a brain. As in the earthworm, the nervous system is characterized by a chain of additional ganglia interconnected by a bundle of nerve fibers that run along the ventral surface. As a consequence of this type of nervous system, many quite complicated arthropod activities—for example, the complex movements of the finely articulated appendages—can be coordinated by the nearest ganglion.

In vertebrates, the nervous system is greatly elaborated. Birds and mammals, in particular, have highly developed and finely tuned nervous systems that monitor internal and external changes and initiate responses to them. The functional unit of the nervous system, however, remains the same—the neuron.

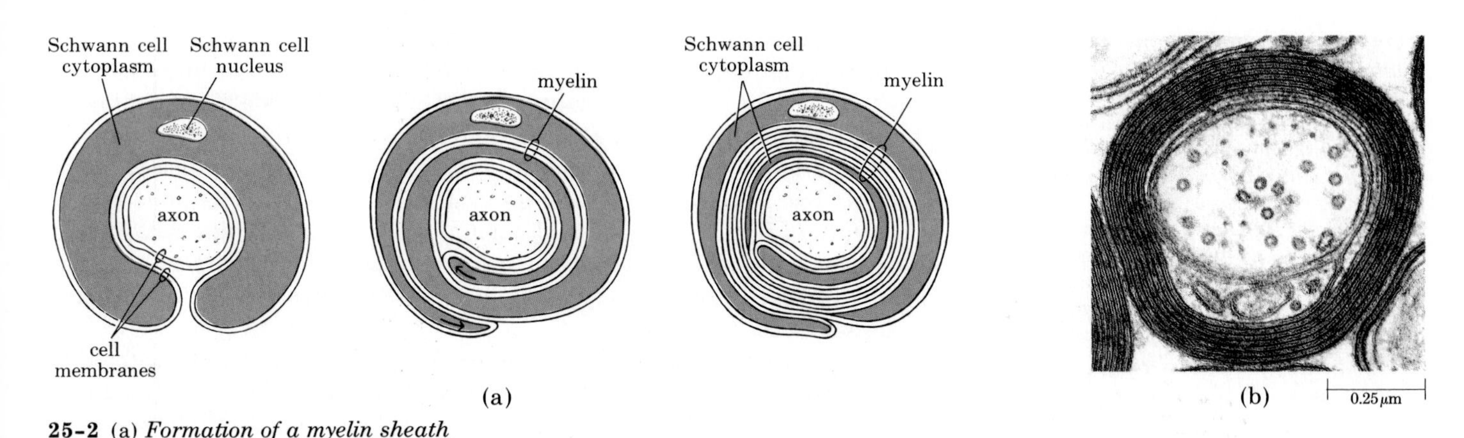

25-2 (a) *Formation of a myelin sheath by a Schwann cell. As the Schwann cell grows, it extends itself around and around the nerve fiber, in this case an axon, and gradually extrudes its cytoplasm from between the layers. The myelin sheath, which consists of layers of lipid-containing cell membrane, insulates the nerve fiber.* (b) *Electron micrograph of a cross section of a mature myelin sheath.*

Neurons

As we noted in the previous chapter, a neuron has three basic parts: the cell body, the dendrites, and the axon. Vertebrate nerve fibers (axons and dendrites) are often enveloped in a myelin sheath formed by Schwann cells (Figure 25-2), which are a type of glial cell. Myelinated fibers transmit messages more rapidly than fibers without a myelin sheath.

There are three types of neurons in the nervous system: sensory neurons, interneurons, and motor neurons. *Sensory neurons* are activated by sensory stimuli. These stimuli may be received by naked nerve endings, by nerve endings enclosed in specialized capsules (Figure 25-3), or by separate cells—sensory receptors—adjacent to sensory neurons. Often sensory

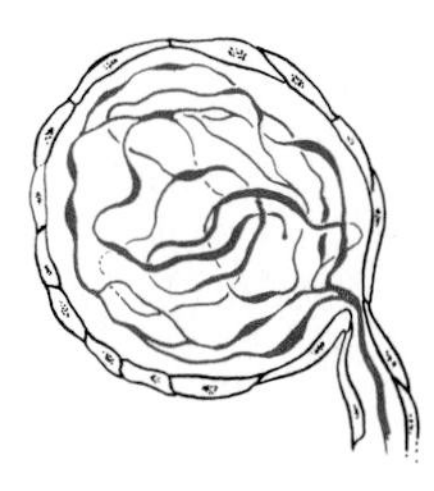

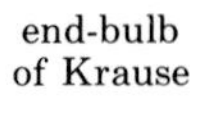

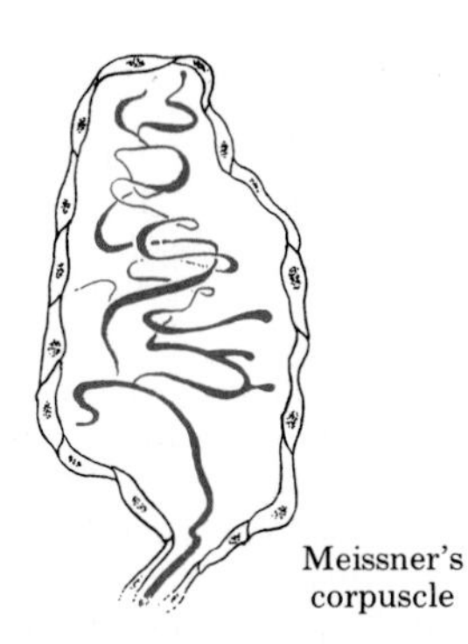

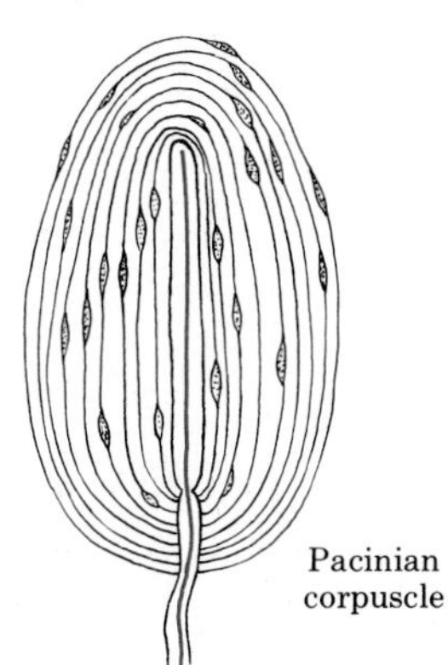

25-3 *Skin contains nerve endings—either naked or enclosed in specialized capsules—that are activated by different types of stimuli. Free nerve endings are pain receptors, end-bulbs of Krause are thought to be receptors for cold, Meissner's corpuscles may be receptors for touch, and the onion-shaped Pacinian corpuscles are for pressure and vibration. Nerve fibers are shown in color.*

In the Pacinian corpuscle, the best studied of the four, a single myelinated nerve fiber (dendrite) enters the corpuscle, which is composed of many concentric layers of connective tissue. Pressure on these outer layers stimulates the dendrite.

receptors are found within sensory organs, such as the eye or ear. The impulses generated in the nerve endings of the sensory neuron are carried along the dendrites to its cell body, which lies just outside the spinal cord. The impulse is then transmitted along the axon of the sensory neuron, which enters the spinal cord and, in turn, transmits the impulse to *interneurons.* The impulse may then be transmitted to other regions of the spinal cord or brain. In the case of a simple reflex, the impulse is transmitted to the dendrites or cell body of a *motor neuron* and is carried along its axon to an effector, which is typically a muscle (Figure 25–4). These events are summarized in Figure 25–5.

In the course of evolution, relatively few changes have occurred in the structure of the individual nerve cells, which are remarkably similar throughout the animal kingdom. Vast changes have occurred, however, in the number and arrangement of such cells and in the uses made of them. There are estimated to be 10^{11} (a hundred billion) neurons in the human body, the great majority of which are interneurons located within the brain. It has been further estimated that a typical interneuron in the brain may receive signals from as many as 1,000 other neurons. Let us look at the nature of these signals.

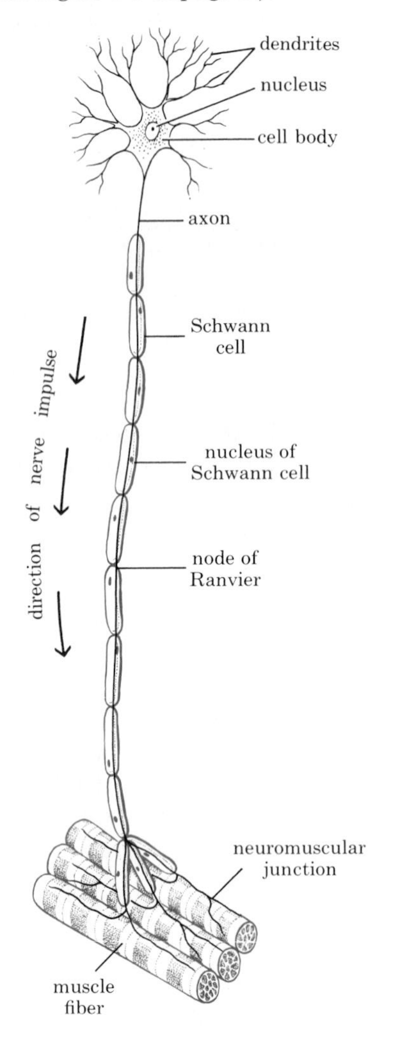

25–4 *Diagram of a motor neuron. The stimulus is received on the naked nerve surface, usually by the dendrites, which transmit the signal to the cell body and to the axon. The signal travels along the axon, which is insulated by a myelin sheath composed of Schwann cells. The nodes of Ranvier are gaps in the sheath that occur at the junctions of adjacent Schwann cells. This motor neuron branches to form neuromuscular junctions with muscle fibers (see also Figure 1–7 on page 21).*

The Nerve Impulse

For almost 200 years it has been known that nerve conduction is somehow associated with electrical phenomena. As you will recall, there are two types of electric charge, positive and negative. Like charges repel one another and unlike charges attract. Thus, negatively charged particles tend to move toward an area of positive charge, and vice versa.

The difference in the amount of electric charge between an area of positive charge and an area of negative charge is called the *electric potential.* An electric potential is a form of potential energy, like a boulder at the

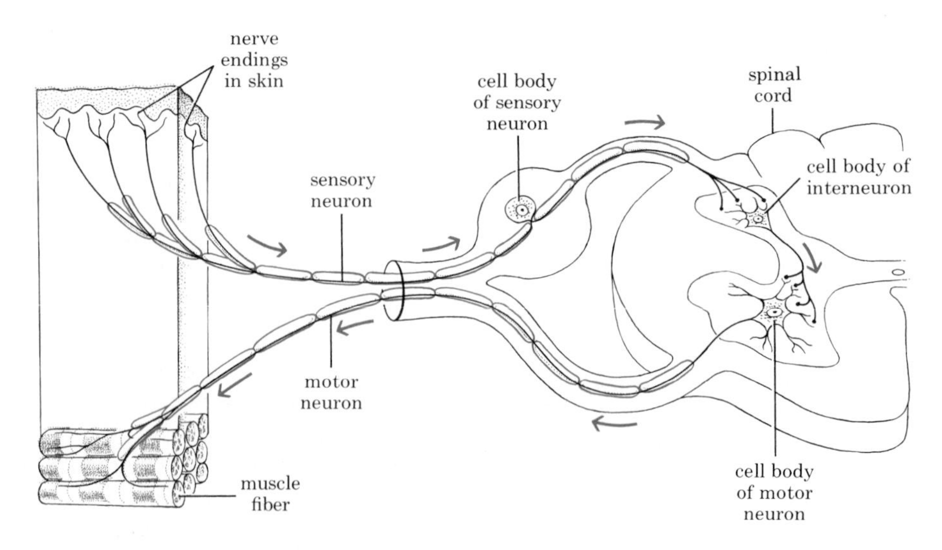

25–5 *Diagram showing the functions of the different types of neurons. In this example, nerve endings in the skin, when appropriately stimulated, transmit signals along the sensory neuron to an interneuron in the spinal cord. The interneuron transmits the impulse to a motor neuron. As a result of the stimulation of the motor neuron, muscle fibers contract. Additional interneurons, not shown here, may also be stimulated by the sensory neuron and carry the impulse to the brain.*

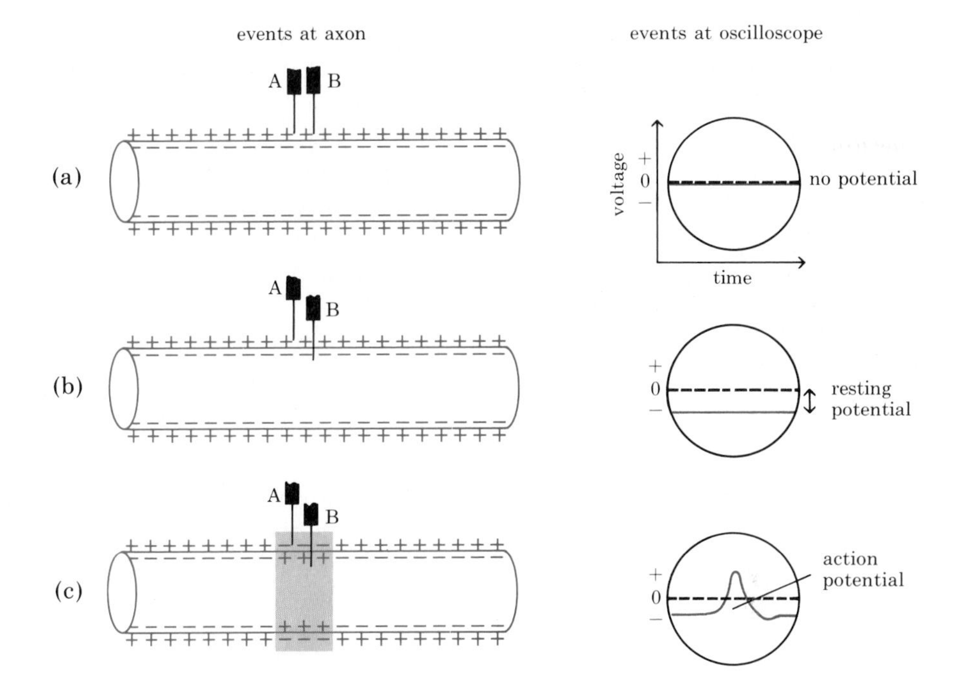

25-6 *The electric potential of the membrane of the axon is measured by microelectrodes connected to an oscilloscope.* (a) *When both electrodes are outside the membrane, no potential is recorded.* (b) *When one electrode penetrates the membrane, the oscilloscope shows that the interior is negative with respect to the exterior and that the difference between the two is about 70 millivolts. This is the resting potential.* (c) *When the axon is stimulated and a nerve impulse passes along it, the oscilloscope shows a brief reversal of polarity—that is, the interior becomes positive in relation to the exterior. This brief reversal in polarity is the action potential.*

top of a hill or water behind a dam. This potential energy is converted to electrical energy when charged particles are allowed to move through a solution or along a wire between the two areas of different charge. The force with which the charged particles move, which is analogous to the force of water running downhill, is measured in volts or, when it is very small, in millivolts.

Major advances in understanding the nerve impulse occurred when it became possible to measure electrical impulses in an individual neuron. Such measurements are made with microelectrodes tiny enough to penetrate a living cell without injuring it. The microelectrodes are connected to a very sensitive voltmeter, and the results are displayed on a device called an oscilloscope. The studies on how the nerve impulse is transmitted were carried out on the axon, and therefore we refer primarily to the axon in this discussion. Nerve impulses, however, travel along the long dendrites of sensory neurons in the same way.

When both electrodes are outside the axon, no voltage difference is recorded (Figure 25-6a). When one electrode penetrates the axon, a voltage difference of about 70 millivolts can be detected between the outside and the inside of the axon. The interior of the membrane is negative with respect to the exterior (Figure 25-6b). This is the *resting potential* of the membrane. When the axon is adequately stimulated, the oscilloscope records a very brief reversal of polarity—that is, the interior becomes positive in relation to the exterior (Figure 25-6c). This reversal in polarity is called the *action potential.* The action potential is the nerve impulse.

The nerve impulses from any one neuron are always the same. This is known as an all-or-nothing response; either the neuron responds to the stimulus with at least one full-sized action potential or it does not respond at all. The only variation is the frequency at which impulses are produced (Figure 25-7, page 360). The message transmitted by a neuron is, in effect, a Morse code with dashes but no dots.

25-7 (a) *Nerve impulses can be monitored by electronic recording instruments. The impulses from any one neuron are all the same; that is, each impulse has the same duration and voltage change as any other.*

(b) *Nerve impulses from a sensory neuron (touch receptor) in cat skin. The skin was touched and pressed in at varying depths, as indicated by the figures at the left. As you can see, the more deeply the skin was pressed in, the more rapidly nerve impulses were produced. The vertical lines represent individual action potentials on a compressed time scale.*

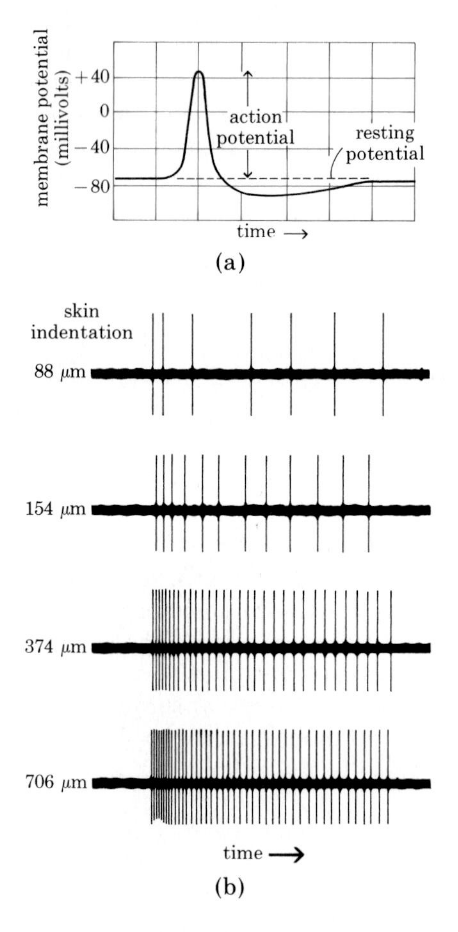

The Ionic Basis of the Action Potential

The electric potential of the axon is made possible by differences in the distribution of ions on either side of the membrane. The most important differences are in the concentrations of potassium ions (K^+) and sodium ions (Na^+). The concentration of K^+ ions is about 10 times higher inside the axon than in the fluid outside, and the concentration of Na^+ ions is about 10 times higher outside the axon than inside. These concentration differences are maintained by a protein carrier molecule in the membrane (see Figure 5–8, page 81). This carrier molecule, which is known as the sodium-potassium pump, transports Na^+ ions out of the axon and pumps in K^+ ions.

In its resting state, the axon membrane is impermeable not only to large, negatively charged ions but also to sodium ions. As a result, most of the Na^+ ions that are pumped out of the axon remain on the outside. The membrane is, however, relatively permeable to potassium ions, which tend to leak out, diffusing down the concentration gradient. According to the current hypothesis, it is this leaking out of the K^+ ions—while the negatively charged ions are trapped inside—that makes the inside of the membrane slightly negative, creating the resting potential (Figure 25–8).

When the membrane is stimulated, it suddenly becomes permeable to Na^+ ions. They rush down their concentration gradient, attracted also by the negatively charged ions inside. This influx of positively charged ions momentarily reverses the polarity of the membrane so that it becomes positive on the inside and negative on the outside, producing the action potential (Figure 25–9). This change in permeability lasts for only about half a millisecond; then the membrane regains its previous impermeability to Na^+. During this time, the permeability to K^+ increases, and there is an outward flow of K^+ ions due to the concentration gradient and also the positive charge inside the cell at the peak of the action potential. This outward flow of positive K^+ ions counteracts the previous inward flow of positive Na^+ ions, and the resting potential is quickly restored. (The actual number of ions involved is very small. Only a very few Na^+ ions need enter to reverse the polarity of the membrane, and only the same small number of K^+ ions need move out of the cell to restore the resting potential.) Subsequently, the sodium-potassium pump restores the Na^+ and K^+ concentrations to their original levels.

Propagation of the Impulse

An important feature of the nerve impulse is that, once initiated, it continues to move along the axon, renewing itself until it reaches the end of the axon. The action potential is self-propagating because at its peak, when the inside of the membrane at the active region is positive, positively charged ions move from this region to adjacent areas inside the axon, which are still negative. As a result, the adjacent area becomes depolarized—that is, the electric potential across the membrane decreases. This causes increased permeability to Na^+ ions, which then rush in across the momentarily depolarized membrane and create a new action potential. This new action potential, in turn, depolarizes the next adjacent area of the membrane (Figure 25–10). (The segment of the axon behind the action potential has a brief refractory period during which it cannot be reexcited and which keeps the action potential from going backwards.) As a consequence of this renewal process, an axon, which would be a very poor conductor of an ordinary electric current, is capable of transmitting a nerve impulse over a considerable distance with absolutely undiminished strength.

25-8 *Resting axon, with membrane polarized. There is a slight excess of positive charge outside the membrane caused by the outward diffusion of K^+ ions.*

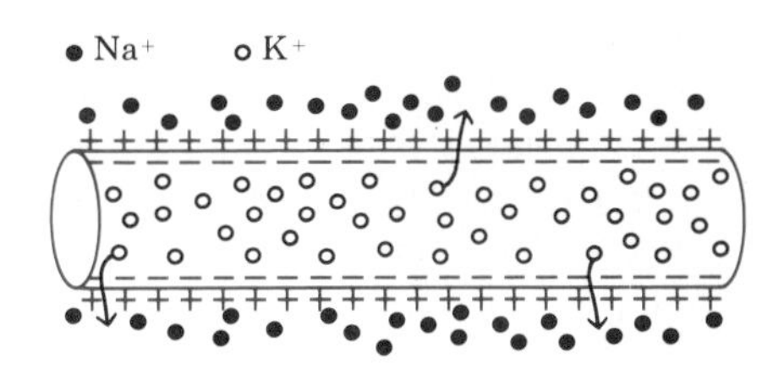

25-9 *Action potential. A portion of the membrane becomes momentarily permeable to Na^+ ions. Na^+ ions rush in, and the polarity of the membrane is reversed at that point.*

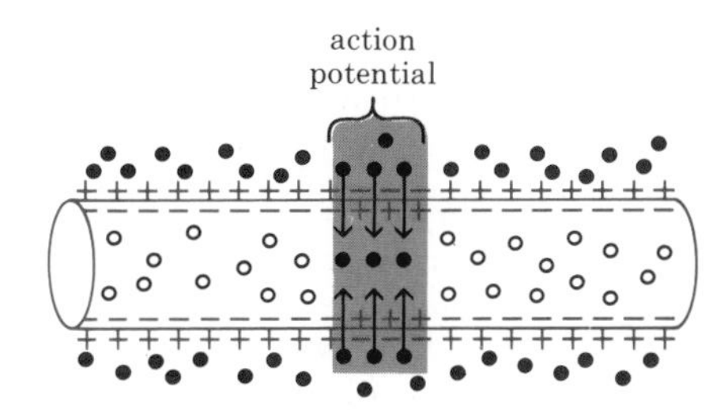

25-10 *Propagation of the action potential. In advance of the action impulse, a small segment of the membrane becomes slightly depolarized, owing to the movement of positively charged ions along the inside of the membrane. When the membrane becomes depolarized in this way, its permeability to Na^+ increases, and the Na^+ ions rush in, creating a new action potential and depolarizing another segment of the membrane.*

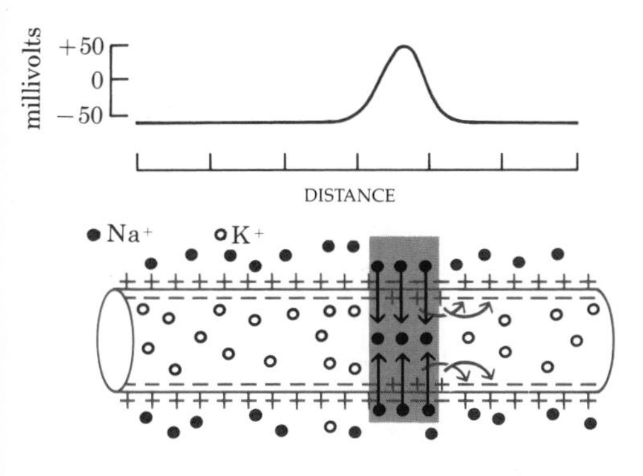

The Synapse

Signals travel from one neuron to another across a specialized junction known as a *synapse*. The neurons do not touch at the synapse; there is a space between them known as the synaptic cleft. Transmission across most synapses in mammals is by chemical means. In synaptic knobs at the end of the axon are numerous small vesicles, visible in electron micrographs (Figure 25-11), which contain a chemical transmitter substance. Arrival of the action potential at the axon terminal causes these vesicles to empty their contents into the synaptic cleft. The transmitter substance crosses the cleft and combines with receptor molecules on the membrane of the second cell, changing the permeability of the membrane and initiating an impulse in that cell. A few of these chemical transmitters have been identified. Two principal ones are acetylcholine and norepinephrine.

Unlike the nerve impulse along the fiber—which is an all-or-nothing proposition—signals transmitted by chemicals across a synaptic junction can modulate one another. That is, some may excite and some may inhibit. A single neuron may receive signals from hundreds, even thousands, of synapses, and, based on its processing of all the signals, it will or will not be excited sufficiently to initiate an impulse. Synapses are therefore relay and control points that are extremely important in the functioning of the nervous system.

After their release, the chemical transmitter molecules are rapidly destroyed by specific enzymes or are taken up again by the synaptic knob. Such removal of the transmitter puts a halt to its effect. This in itself is an essential feature in the control of the activities of the nervous system. The activity of many drugs, such as LSD and psilocybin, is thought to depend on their interference with chemical transmission across synapses (page 451).

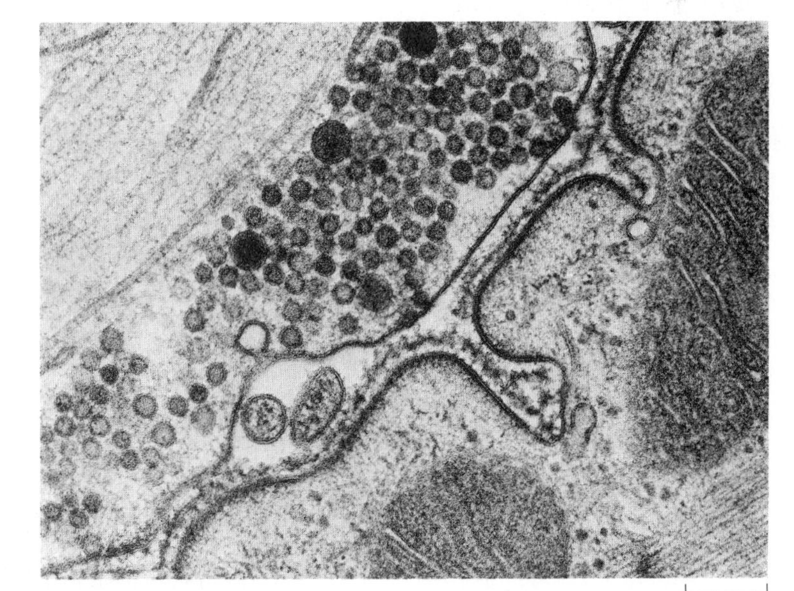

25-11 *In this electron micrograph, the area extending from the lower left-hand corner to the top center is occupied by an axon terminal of a motor neuron in a frog. Spherical vesicles containing transmitter molecules (in this case, acetylcholine) cluster near the axon membrane. To the right of the synaptic cleft is a muscle cell in which you can see two mitochondria and, in the lower right-hand corner, myofibrils. Release of transmitter molecules into the synaptic cleft stimulates the muscle cell and causes it to contract.*

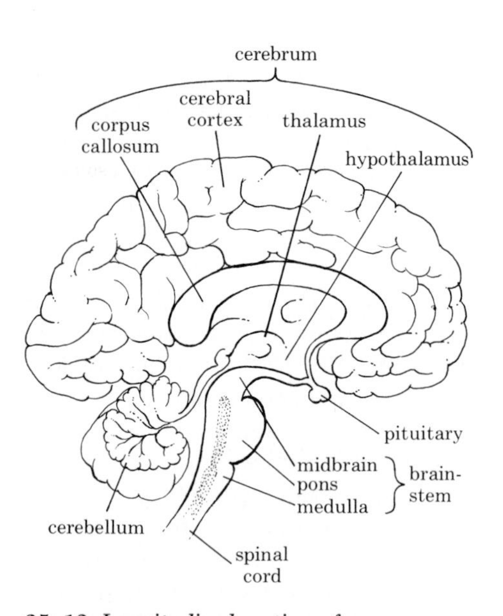

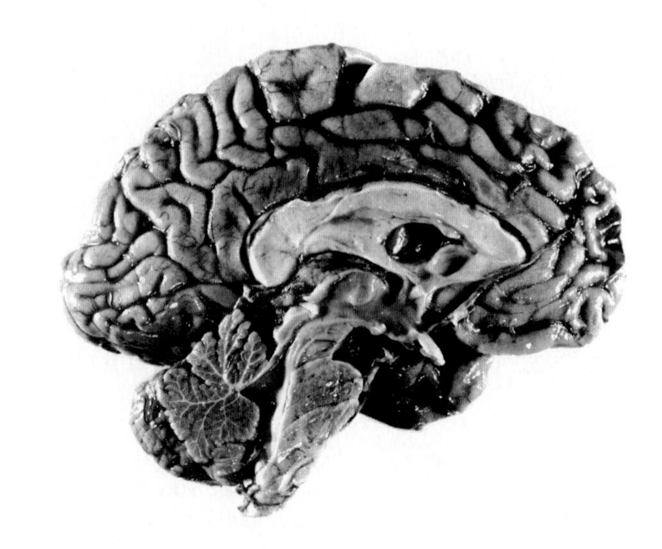

25-12 *Longitudinal section of a human brain. The gray matter, characteristic of the cerebral cortex, consists of nerve cell bodies, glial cells, and unmyelinated fibers. The white matter is made up predominantly of myelinated fibers.*

The Central Nervous System

The brain and the spinal cord form the *central nervous system.* In vertebrates, they are protected by the bones of the skull and the vertebrae.

Anatomically, the human brain consists of a number of different structures (Figure 25-12). The three principal divisions are the *brainstem,* the *cerebellum,* and the *cerebrum.* The brainstem is the old brain, evolutionarily speaking, and is surprisingly similar from fish to *Homo sapiens.* The cerebrum is divided into two hemispheres, which have, in humans, an outer layer known as the *cerebral cortex.* Both the brainstem and the cerebrum are made up of a number of anatomically distinct structures (see Table 25-1).

The brain and spinal cord are divided into gray matter and white matter. The *white matter* contains nerve fibers carrying information between various regions of the body and the central nervous system. Its whiteness is due to the fatty layers of the myelin sheaths. The *gray matter* contains interneurons, glial cells, and, in the brainstem and spinal cord, the cell bodies of motor neurons as well. Clusters of cell bodies within the brain and spinal cord are called nuclei; other clusters of cell bodies, found outside the brain and spinal cord, are called ganglia.

Figure 25-13 shows a segment of the human spinal cord, with pairs of spinal nerves entering and emerging from the cord through spaces between the vertebrae. Each of these pairs innervates the skeletal muscles and sensory receptors of a different and distinct area of the body. In mammals, there are 31 such pairs.

The Peripheral Nervous System

The neurons and fibers outside the central nervous system constitute the *peripheral nervous system.* (Note, however, that fibers of the peripheral nervous system extend into the central nervous system.) The neurons of the peripheral system include both motor neurons and sensory neurons. The fibers of motor and sensory neurons are bundled together into nerves, which are classified as either *cranial nerves,* nerves connecting directly to the brain (such as the optic nerve), or *spinal nerves,* those that

Table 25-1 Structural divisions of the human brain

Structure	Description	Function
Brainstem	Stalk of brain; enlarged, knobby extension of spinal cord; consists of medulla, pons (bridge), and midbrain. Contains all nerve fibers that pass between spinal cord and higher brain centers. Also contains nuclei involved with many autonomic reflexes.	Transmission of signals between spinal cord and higher brain centers; control of heartbeat, respiration, some other autonomic functions (see page 364)
Cerebellum	Bulbous, convoluted mass of nerve tissue that, in humans, lies in back of and under the cerebrum	Unconscious coordination of muscular activities; reaches largest size, proportionately, in birds
Cerebrum	Largest, most prominent part of human brain, divided by groove into right and left hemispheres	Receives and integrates incoming information; makes associations between new data and stored information in cortex; coordinates responses
Cerebral cortex	Thin (2 millimeter layer) of gray matter, wrinkled and folded, overlying surface of cerebrum; the cortex of each hemisphere consisting of four lobes: frontal, parietal, temporal, occipital (see page 447); reaches greatest development in higher primates, especially humans	Sensory, motor, and association functions (see page 448)
Cerebral white matter	Nerve tracts connecting areas of cortex to each other and to rest of brain	Transmission of information to and from cerebral cortex
Corpus callosum	Nerve fibers between hemispheres	Exchange of information between hemispheres
Thalamus	Two egg-shaped masses of gray matter below cerebral hemispheres	Main relay center between brainstem and cerebrum; nuclei sort and process sensory information before transmission to cortex
Hypothalamus	Small mass (about 4 grams) below thalamus, above pituitary	Controls pituitary; integration center for sex drive, anger, hunger, thirst, pleasure, temperature regulation; source of ADH and oxytocin (see page 370)

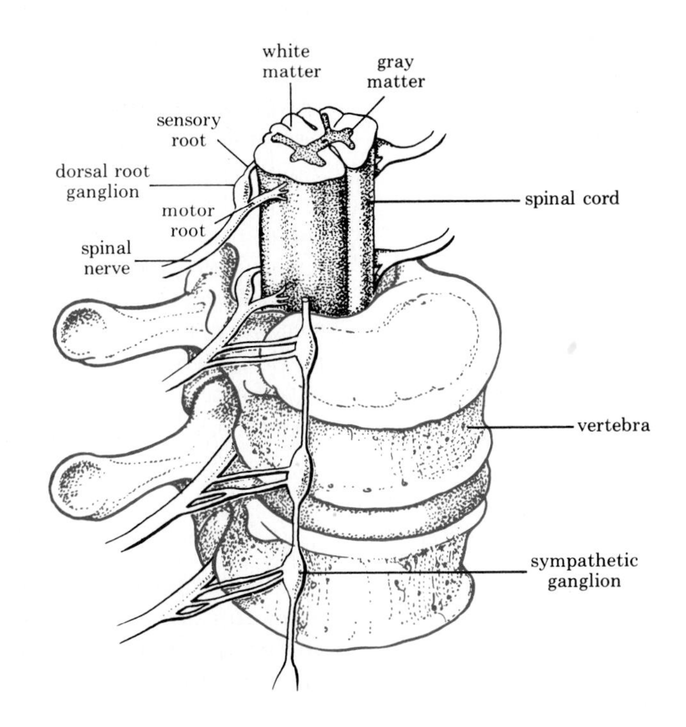

25-13 *A segment of the human spinal cord. Each spinal nerve divides into two fiber bundles, the sensory root and the motor root, at the vertebral column. The sensory bundle connects with the cord dorsally; the cell bodies of the sensory neurons are in the dorsal root ganglia. The motor bundle connects ventrally with the spinal cord. The sympathetic ganglia are part of the autonomic nervous system (page 364). The butterfly-shaped gray matter within the spinal cord is composed of interneurons, glial cells, and the cell bodies of motor neurons. The surrounding white matter consists of ascending and descending nerve fibers.*

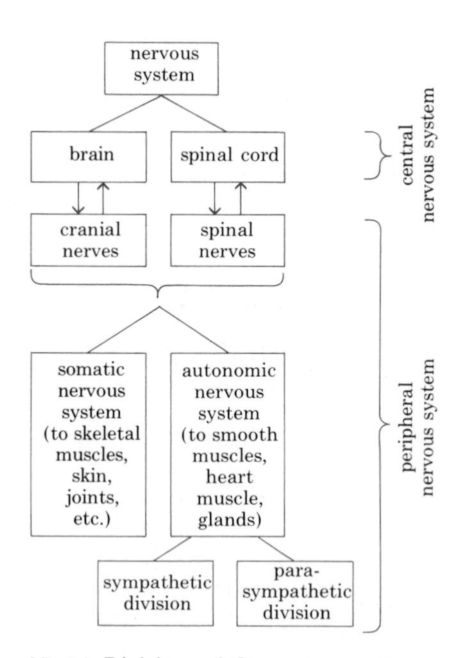

25-14 *Divisions of the nervous system.*

connect with the spinal cord. Each fiber within a nerve is capable of transmitting a separate message, like the wires in a telephone cable.

Near the central nervous system, the motor and sensory components of the spinal nerves separate from each other. The sensory fibers feed into the dorsal (back) side of the spinal cord. They may synapse immediately with interneurons or motor neurons as in a simple reflex, such as that shown in Figure 25-5 on page 358, or they may turn and ascend toward the brain. The cell bodies of the sensory neurons are in the dorsal root ganglia outside the spinal cord. Motor fibers emerge from the spinal cord on the ventral (front) side. The cell bodies of the motor neurons are in the spinal cord, where they receive impulses from interneurons or sensory neurons.

The Somatic and Autonomic Nervous Systems

As you can see in Figure 25-14, there are two divisions of the peripheral nervous system: the somatic and the autonomic. The *somatic nervous system* includes both motor and sensory neurons. The *autonomic nervous system* is primarily a motor system, consisting of the neurons that control heart muscle, glands, and smooth muscle (the type of muscle found in the walls of blood vessels and in the digestive, respiratory, and reproductive tracts). The autonomic nervous system is generally categorized as an "involuntary" system, in contrast to the "voluntary" somatic system, which controls the muscles that we can move at will, that is, the skeletal muscles.

You will readily recognize that the distinction here between "voluntary" and "involuntary" is not clear-cut. Skeletal muscles often move involuntarily, as in a reflex action, and it is reported that some individuals, such as practitioners of yoga, can control their rate of heartbeat and the contractions of some smooth muscle.

Anatomically, the motor neurons of the somatic system are distinct and separate from those of the autonomic system, although fibers of both types may be bundled together in the same nerve. The cell bodies of the motor neurons of the somatic system are located within the central nervous system, with long axons running without interruption all the way to the skeletal muscles. The motor fibers of the autonomic nervous system also originate in cell bodies inside the central nervous system, but they do not travel all the way to their target organs, or effectors. Instead, motor neurons of the autonomic system form a synapse with a second neuron in a ganglion outside the central nervous system. Such a ganglion is called a sympathetic ganglion (see Figure 25-13). This second neuron innervates the target muscle or gland.

The two-neuron pathway with synapses outside the central nervous system constitutes a characteristic difference between the autonomic and somatic systems. Another major difference is that, in vertebrates, the somatic system can only stimulate or not stimulate its effector, whereas the autonomic nervous system can stimulate or inhibit the activity of its target organ.

Divisions of the Autonomic Nervous System: Sympathetic and Parasympathetic

The autonomic nervous system itself has two divisions: the *sympathetic division* and the *parasympathetic division*. These two divisions are anatomically and functionally distinct. The major anatomical differences between them are:

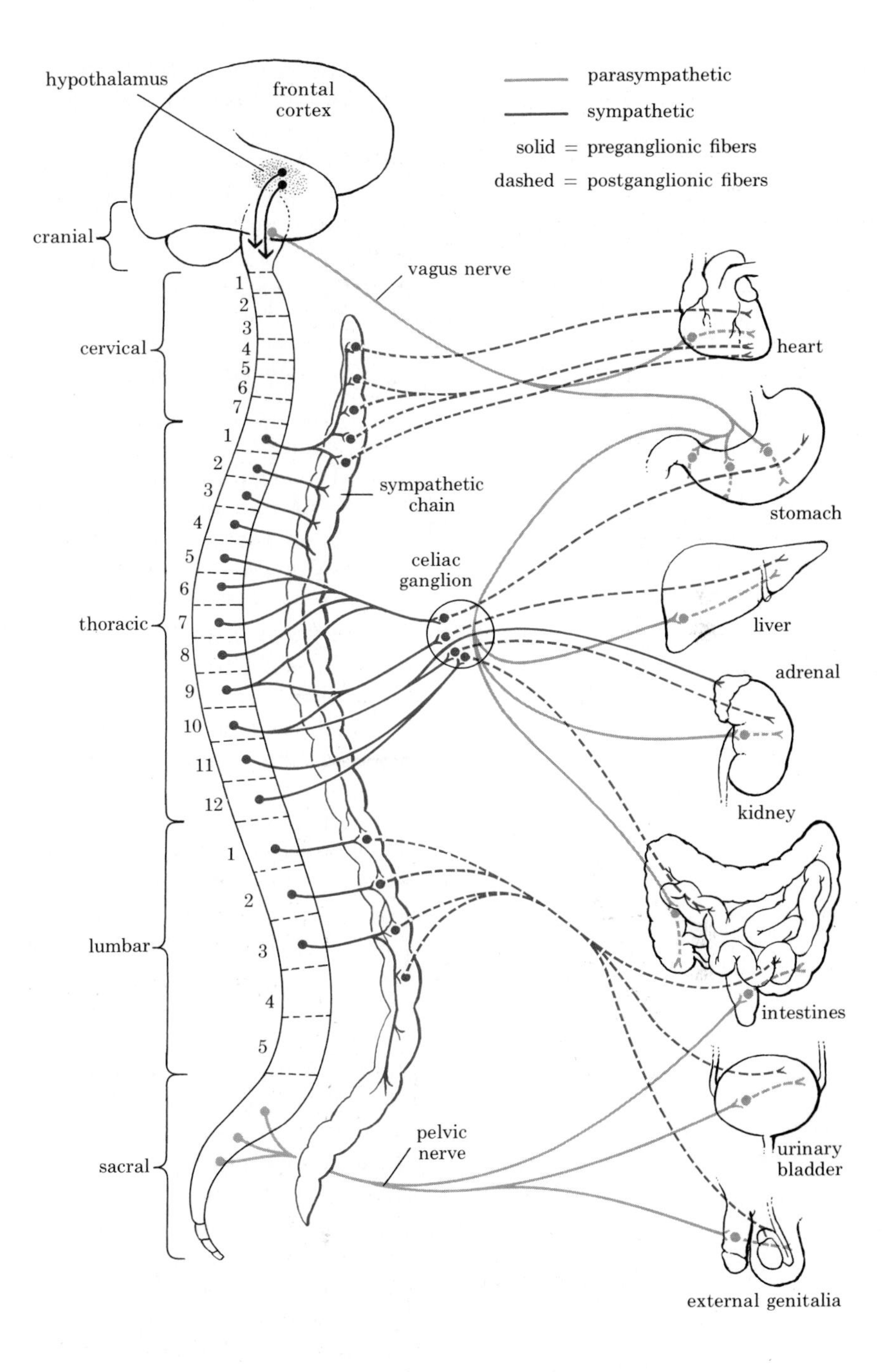

25-15 *The autonomic nervous system. It differs from the somatic system anatomically in that its motor neurons are entirely outside the central nervous system and synapse with nerve fibers (axons) emerging from the brain and spinal cord. The fibers emerging from the central nervous system are known as preganglionic fibers, and those terminating in the effector organs are known as postganglionic fibers.*

The autonomic nervous system consists of the sympathetic and the parasympathetic divisions. The preganglionic fibers of the parasympathetic division exit from the base of the brain and from the sacral region of the spinal cord and synapse with the postganglionic neurons at or near the target organs. The sympathetic division originates in the thoracic and lumbar regions; preganglionic fibers of the sympathetic division synapse with postganglionic neurons in the sympathetic chain or in other ganglia, such as the celiac ganglion, part of the solar plexus.

Most, but not all, internal organs are innervated by both divisions, which usually function in opposition to each other. In general, the sympathetic division produces the effect of exciting organs involved in "fight-or-flight" reactions, and the parasympathetic division stimulates more tranquil functions, such as digestion.

1. The parasympathetic division consists of nerve fibers from the brain and from the lower (sacral) region of the spinal cord. The sympathetic division originates in the middle portion—the thoracic and lumbar regions—of the spinal cord.

2. The ganglia in the parasympathetic division are near or in the target organ. In the sympathetic division, the ganglia are in a chain running parallel and close to the spinal cord.

Generally speaking, the effects of the two divisions are antagonistic. The symptoms of fear, described on page 353, are the results of increased discharge of the neurons of the sympathetic division, which mobilize the body for action. The parasympathetic division, on the other hand, is more concerned with the restorative activities of the body—that is, with rest and rumination. Parasympathetic stimulation slows down the heartbeat, increases the movements of the smooth muscles of the intestines, and stimulates secretions of the salivary glands and the digestive glands of the stomach. As you can see in Figure 25–15, most of the major visceral organs are innervated by fibers from both the sympathetic division and the parasympathetic division. The two divisions work in close cooperation for the ultimate homeostatic regulation of the body's functions.

SUMMARY

The nervous and endocrine systems provide the communication and integration necessary to coordinate the numerous functions that enable an animal to regulate its internal environment.

The unit of the nervous system is the neuron, or nerve cell, which consists of dendrites and a cell body, which receive signals, and an axon, which relays the signals to other cells. The signal travels along the axon or dendrite in the form of an action potential, which is a transient reversal in the electrical polarity of the membrane of the fiber. The message carried depends on the frequency with which action potentials are transmitted in a given fiber and on the particular fibers that are activated.

Nerve cells transmit signals to other cells across a junction called a synapse. The signal crosses the synaptic cleft in the form of a chemical transmitter that stimulates or inhibits the target cell.

The central nervous system consists of the brain and the spinal cord, which are encased, in vertebrates, in the skull and vertebral column. The nervous system outside the central nervous system constitutes the peripheral nervous system. Spinal nerves emerge from the vertebral column in pairs, each pair consisting of motor fibers and sensory fibers. The motor fibers of each pair innervate the muscles of a different area of the body, and the sensory fibers receive impulses from sensory receptors in the same area.

In vertebrates, the peripheral nervous system has two major divisions: (1) the somatic nervous system, which innervates the "voluntary" muscles, and (2) the autonomic nervous system, which controls the muscles and glands involved in the digestive, circulatory, urinary, and reproductive functions. In the autonomic system, the motor neurons arising from the spinal cord synapse outside the central nervous system with second neurons. These postganglionic neurons stimulate or inhibit the effectors. The autonomic system has two divisions: (1) the sympathetic division, which is largely responsible for mobilization for action; and (2) the parasympathetic division, which controls restorative activities, such as digestion and rest.

QUESTIONS

1. Distinguish among the following terms: sensory neuron/interneuron/motor neuron; resting potential/action potential; central nervous system/peripheral nervous system; gray matter/white matter.

2. Describe the way in which the nerve impulse propagates itself. Draw a diagram of this process.

3. What are three significant differences between the somatic and autonomic nervous systems?

4. What are three significant differences between the sympathetic and parasympathetic divisions of the autonomic nervous system?

26

Integration and Control: The Endocrine System

Hormones are secreted by glands—epithelial tissues specialized for secretion. Glands are classified as exocrine or endocrine. *Exocrine glands* secrete their products into ducts; examples are digestive glands and the sweat glands of the human skin. *Endocrine glands* secrete their products into the bloodstream (or, more precisely, into the extracellular fluids from which the substances diffuse into the bloodstream). They are sometimes referred to as "ductless" glands. "Endocrine" is generally used as a synonym for "hormone-secreting," and endocrinology is the study of hormones.

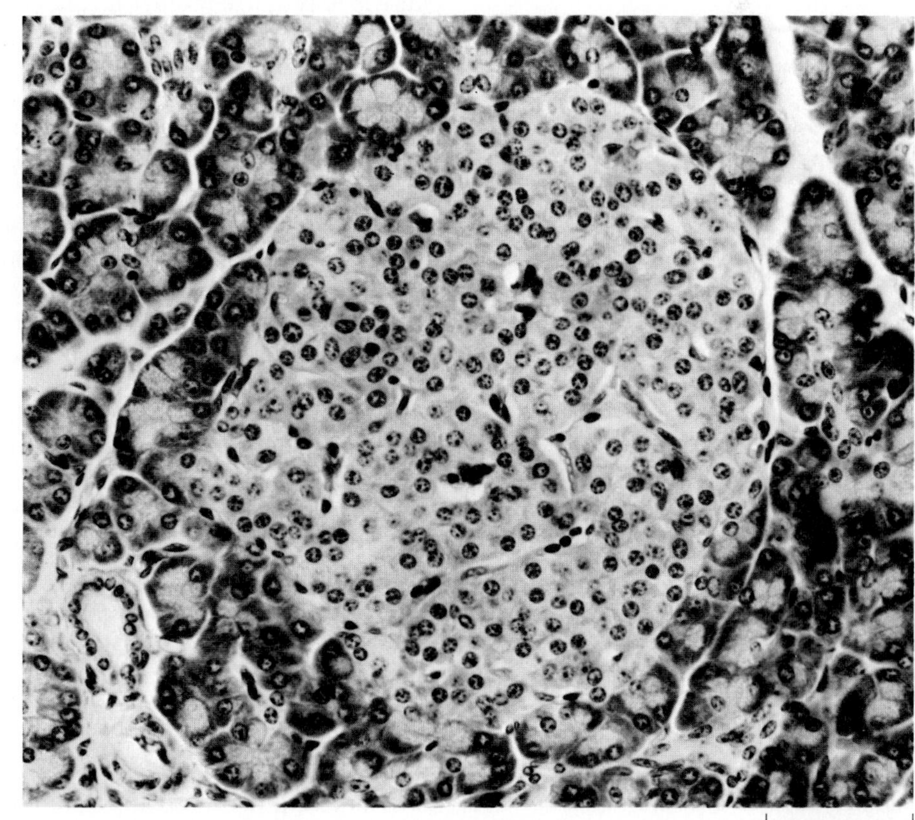

26–1 *A cross section of pancreatic tissue. The pancreas is both an exocrine and an endocrine gland. The group of small cells in the center of the micrograph are islet cells, endocrine cells that secrete the hormones insulin and glucagon into the bloodstream. The surrounding exocrine cells produce digestive enzymes, which are carried via the pancreatic duct to the small intestine.*

Table 26–1 summarizes the principal endocrine glands of vertebrates and the hormones they secrete. In the following pages we shall discuss most of these glands and hormones. We shall, however, defer discussion of the hormones and glands involved in reproduction until Chapter 31.

The Pituitary Gland

The pituitary gland is situated at the base of the brain in the geometric center of the skull. It is about the size of a kidney bean.

The anterior lobe of the pituitary is the source of at least six different hormones. One of these is growth hormone, sometimes called somatotropin. It promotes the growth of bone and muscle. As is the case with most of the hormones, growth hormone is best known by the effects caused by too much or too little. If there is a deficit in the production of growth hormone in childhood, a midget results, the so-called "pituitary dwarf." An excess of growth hormone during childhood results in a giant; most circus giants are the result of such an excess. Excessive growth hormone in the adult does not lead to giantism, since growth of the long bones has ceased. It leads instead to acromegaly, an increase in the size of the jaw and the hands and feet, adult tissues that are still sensitive to the effects of growth hormone. Growth hormone also affects glucose metabolism, inhibiting the uptake and oxidation of glucose by many types of cells. It also stimulates the breakdown of fatty acids, thus conserving glucose.

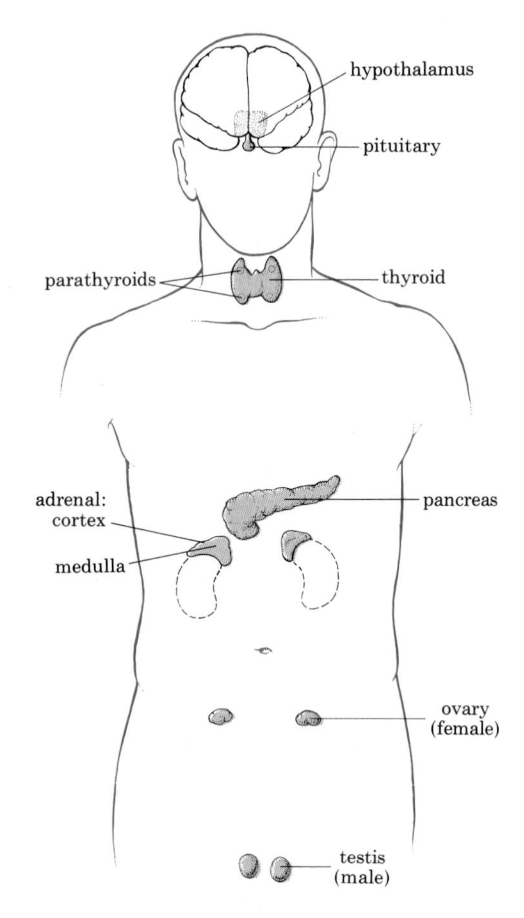

26–2 Some of the hormone-producing (endocrine) organs. The pituitary releases hormones that, in turn, regulate the hormone secretions of the thyroid, the adrenal cortex, and the sex glands. The pituitary is itself under the regulatory control of an area of the brain known as the hypothalamus, indicated in color. The hypothalamus thus is the major link between the nervous system and the endocrine system.

A second hormone produced by the anterior pituitary is prolactin, which stimulates secretion of milk in mammals. As long as an infant continues to nurse, the impulses produced by the suckling of the breast are transmitted by way of the central nervous system to the pituitary and cause it to produce prolactin. The prolactin, in turn, acts upon the breast to maintain the production of milk. Once suckling ceases, the synthesis and release of prolactin decrease and milk production stops. Thus supply is regulated by demand. In some birds—for example, ringdoves and the familiar parakeet—prolactin stimulates the production of a milklike fluid by the lining of the crop (a pouch in the bird's gullet). This crop "milk" is fed to the young birds for about two weeks after hatching.

Four of the hormones secreted by the anterior pituitary are *tropic hormones*—hormones that act upon other endocrine glands to regulate their secretions. One of these tropic hormones is TSH, the thyroid-stimulating hormone. TSH stimulates cells in the thyroid gland to increase their production and release of thyroxine, the thyroid hormone. The increased concentration of thyroxine, in turn, inhibits the release of TSH by the pituitary. Adrenocorticotropic hormone (ACTH) has a similar regulatory relationship with the production of cortisol, one of the hormones produced by the adrenal cortex (the outer layer of the adrenal gland). This type of regulatory system, in which the activity of a system is decreased as the products of its operation increase, is known as *negative feedback*. The most familiar example of a negative feedback system is a thermostat that turns off a furnace when the temperature rises. Many examples of negative feedback are known, both in biological systems and in those designed by engineers.

The other two tropic hormones secreted by the anterior pituitary are gonadotropins—hormones that act upon the *gonads*, or gamete-producing organs (the testes and ovaries). These hormones, follicle-stimulating hormone (FSH) and luteinizing hormone (LH), will be discussed in Chapter 31.

Table 26-1 Some of the principal endocrine glands of vertebrates and the hormones they produce

Gland	Hormone	Principal Action	Mechanism Controlling Secretion	Chemical Composition
Pituitary, anterior lobe	Growth hormone (somatotropin)	Stimulates bone and muscle growth, inhibits oxidation of glucose, promotes breakdown of fatty acids	Hypothalamic-inhibiting and -releasing hormones	Protein
	Prolactin	Stimulates milk production and secretion in "prepared" gland	Hypothalamic-inhibiting and -releasing hormones	Protein
	Thyroid-stimulating hormone (TSH)	Stimulates thyroid	Thyroxine in blood; hypothalamic-releasing and -inhibiting hormones	Glycoprotein
	Adrenocorticotropic hormone (ACTH)	Stimulates adrenal cortex	Cortisol in blood; hypothalamic-releasing and -inhibiting hormones	Polypeptide (39 amino acids)
	Follicle-stimulating hormone (FSH)*	Stimulates ovarian follicle, spermatogenesis	Estrogen in blood; hypothalamic-releasing and -inhibiting hormones	Glycoprotein
	Luteinizing hormone (LH)*	Stimulates corpus luteum and ovulation in female, interstitial cells in male	Progesterone or testosterone in blood; hypothalamic-releasing and -inhibiting hormones	Glycoprotein
Hypothalamus (via posterior pituitary)	Oxytocin	Stimulates uterine contractions, milk ejection	Nervous system	Peptide (9 amino acids)
	Antidiuretic hormone (ADH, vasopressin)	Controls water excretion	Osmotic concentration of blood; nervous system	Peptide (9 amino acids)
Thyroid	Thyroxine, other thyroxinelike hormones	Stimulate and maintain metabolic activities	TSH	Iodinated amino acids
	Calcitonin	Inhibits release of calcium from bone	Concentration of calcium in blood	Polypeptide (32 amino acids)
Parathyroid	Parathyroid hormone (parathormone)	Stimulates release of calcium from bone, promotes calcium uptake from gastrointestinal tract, inhibits calcium excretion	Concentration of calcium in blood	Protein
Adrenal cortex	Cortisol, other cortisol-like hormones	Affect carbohydrate, protein, and lipid metabolism	ACTH	Steroids
	Aldosterone	Affects salt and water balance	Renin from kidney, K^+ ions in blood	Steroid
Adrenal medulla	Epinephrine and norepinephrine	Increase blood sugar, dilate some blood vessels, increase rate of heartbeat	Nervous system	Catecholamines
Pancreas	Insulin	Lowers blood sugar, increases storage of glycogen	Concentration of glucose in blood	Protein
	Glucagon	Stimulates breakdown of glycogen to glucose in the liver	Concentration of glucose and amino acids in blood	Polypeptide (29 amino acids)
Ovary, follicle	Estrogens*	Develop and maintain sex characteristics in females, initiate buildup of endometrium	FSH	Steroids
Ovary, corpus luteum	Progesterone and estrogens*	Promote continued growth of endometrium	LH	Steroids
Testis	Testosterone*	Supports spermatogenesis, develops and maintains sex characteristics of males	LH	Steroid

* These hormones will be discussed in Chapter 31.

The Hypothalamus

The pituitary gland lies beneath an area of the brain known as the hypothalamus and is directly under its influence. The pituitary is also under the influence, by way of the hypothalamus, of other brain centers. Some nine hormones have now been isolated from the hypothalamus that act either to stimulate or inhibit the secretion of hormones by the anterior pituitary. These hypothalamic-releasing and -inhibiting hormones are sometimes called "brain hormones." They are small peptides—one of them is only three amino acids in length. Another, somatostatin, was the first mammalian hormone to be synthesized using recombinant DNA technology (see page 222). It inhibits the release of growth hormone from the pituitary.

Figure 26–3 summarizes the negative feedback system linking the hypothalamus and pituitary with the thyroid, adrenal cortex, and gonads. With the discovery of the close working relationship between the pituitary and the hypothalamus, it became clear that the nervous and endocrine systems are not separate methods of control but are, in fact, a single regulatory system integrating the organism's activities and promoting its homeostasis.

The hypothalamus is also the source of two hormones stored in and released from the posterior pituitary: oxytocin and antidiuretic hormone, or ADH. (The ADH of many vertebrates is called vasopressin because it increases blood pressure; it has no such effect in humans, however, except in very high concentrations.) Oxytocin accelerates childbirth by acting on the smooth muscle of the uterus and increasing contractions. It is also responsible for the "letting down" of milk that occurs when the infant begins to suckle. ADH, as we shall see in Chapter 30, regulates the excretion of water by the kidneys.

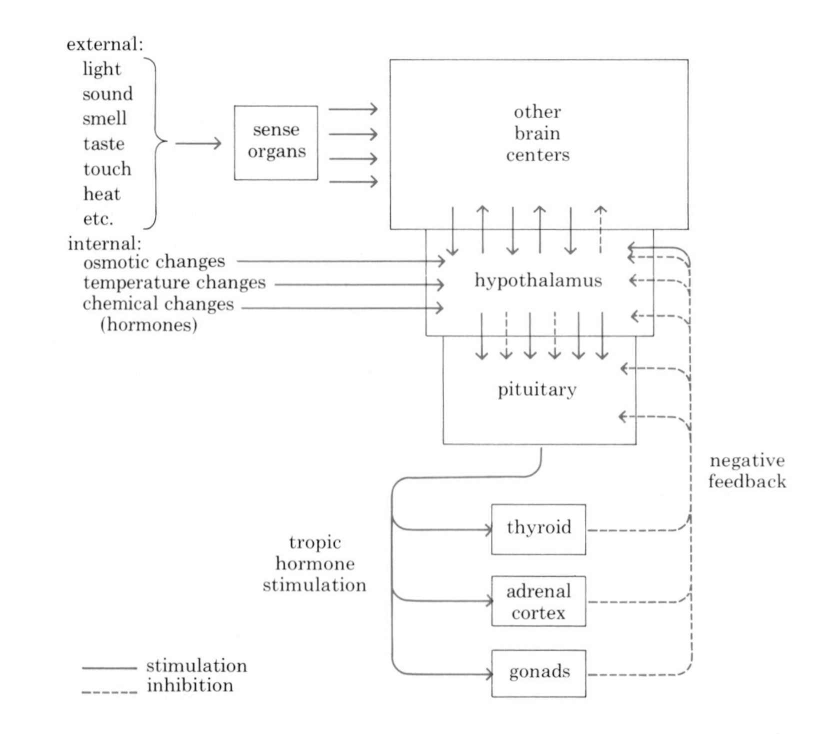

26–3 *The production of many major hormones is regulated by a complex negative feedback system involving the pituitary and the hypothalamus. The hypothalamus controls the pituitary's secretion of tropic hormones, and these, in turn, stimulate the secretion of hormones from the thyroid, adrenal cortex, and gonads (the testes or the ovaries). As the concentration of the hormones produced by these target glands rises in the blood, the hypothalamus decreases its production of stimulating hormones, the pituitary decreases its hormone production, and production of hormones by the target glands also slows. By way of the hypothalamus, which receives information from many other parts of the brain, hormone production is regulated also in response to changes in the external and internal environments.*

The Thyroid Gland

26–4 *Thyroxine, the principal hormone produced by the thyroid gland. Note the four iodine atoms in its structure. Because we need iodine for thyroxine, it is an essential component of the human diet. Where iodine is present in the soil, it is available in minute quantities in drinking water and in plants. In the United States, table salt is ordinarily iodized or must be specifically labeled as being uniodized.*

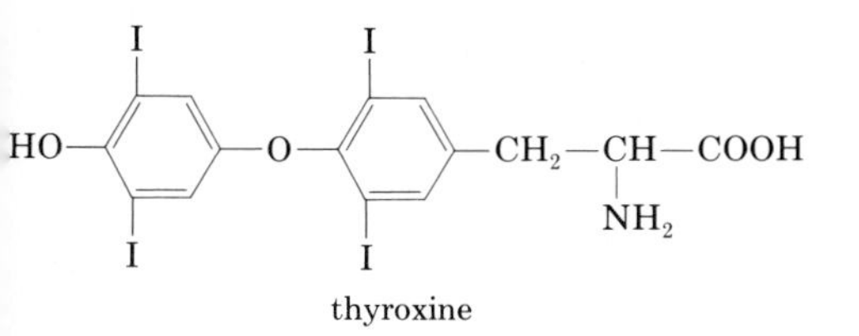

The thyroid, under the influence of the thyroid-stimulating hormone from the pituitary, produces thyroxine, which is an amino acid combined with four atoms of iodine (Figure 26–4). Thyroxine accelerates the rate of cellular respiration. Hyperthyroidism, the overproduction of thyroxine, produces nervousness and excitability, increased heart rate and blood pressure, and weight loss. Hypothyroidism (too little thyroxine) in infancy affects development, particularly of the brain cells, and if not treated in time, can lead to permanent mental deficiency and dwarfism. In adults, hypothyroidism is associated with dry skin, brittle hair, intolerance to cold, and lack of energy.

The thyroid also secretes calcitonin in response to rising calcium levels in the blood. This hormone inhibits the release of calcium from bone.

The Parathyroid Glands

26–5 *The pea-sized parathyroid glands, the smallest of the known endocrine glands, are located behind or within the thyroid gland. They produce parathyroid hormone (parathormone), which increases concentrations of blood calcium.*

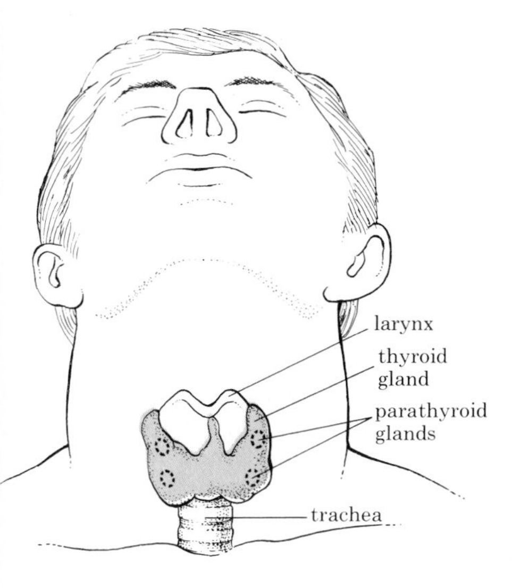

The pea-sized parathyroid glands, the smallest of the known endocrine glands, are located behind or within the thyroid gland. They produce parathyroid hormone (parathormone), which raises the concentration of calcium ion in the blood by increasing its absorption from the intestine and by reducing its excretion from the kidneys. Calcium ion is essential for muscle contraction, nerve conduction, and many other functions.

Parathyroid hormone also stimulates the release into the bloodstream of calcium from bone, which contains 99 percent of the body's total calcium. Thus, parathyroid hormone and calcitonin work as a fine-tuning mechanism, regulating blood calcium. The production of both hormones is regulated directly by the amount of calcium in the blood.

Adrenal Cortex

The adrenal glands are on top of the kidneys. The outer layer of the adrenal gland, the adrenal cortex, is the source of a number of steroids (see page 44) with hormonal activity. Cortisol and aldosterone (Figure 26–6) are among the most active.

Cortisol and cortisol-like hormones promote the formation of glucose from protein and fat. At the same time, they inhibit the uptake of glucose by most cells, with the notable exception of brain cells, thus favoring mental activities at the expense of other body functions. Their release increases during periods of stress such as facing new situations, engaging in athletic competitions, and taking final exams. Thus they work in concert with the sympathetic nervous system. In large amounts, they suppress inflammation and the production of antibodies.

26–6 *Cortisol and aldosterone are steroids, as you can tell from their characteristic four-ring structure. Both hormones are secreted by the adrenal cortex.*

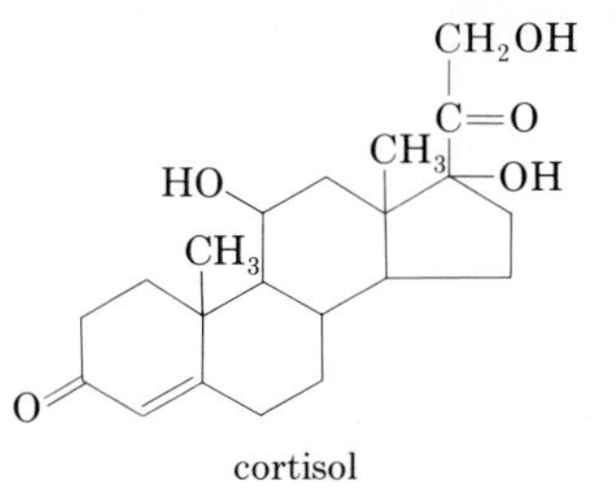

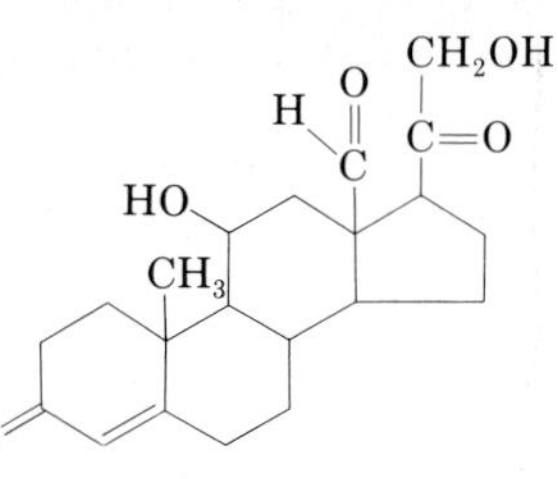

Pheromones

Pheromones (from *pher,* meaning "to carry," plus hormone) are chemical signals exchanged between organisms, usually members of the same species. They may have been the evolutionary forerunners of hormones, serving to coordinate the activities of one-celled organisms. Pheromones are usually produced in special glands and are discharged into the environment.

Among modern species, communication by pheromones seems to be most prevalent among the insects. Pheromones, for example, serve as sex attractants, drawing males to females. Among the best studied are the mating substances of moths. One female gypsy moth, by the emission of minute amounts of a pheromone commonly known as disparlure, can attract male moths that are several kilometers downwind. Since the male can detect as little as a few hundred molecules per milliliter of the attractant, disparlure is still potent even when it has become widely diffused.

Pheromones play a variety of other roles in insect communication. Ants lay down pheromones as trail markers, signposts to a food source. When a honey bee stings, it leaves not only the stinger in the victim's skin but also a chemical substance that recruits other honey bees to the attack. Similarly, worker ants of many species release pheromones as alarm substances when they are threatened by an invader; the pheromone spreads through the air to alarm and recruit other workers. If these ants, too, encounter the invader, they will release the pheromone, so the signal will either die out or build up, depending on the magnitude of the threat.

Pheromones have also been found among mammals. The most familiar are the scent-marking substances in the urine of male dogs and cats, which serve as a warning signal to other males. Male mice, it has been found, release substances in their urine that alter the reproductive cycles of the females. The smell of urine from a strange male, for example, can alter hormone balance and interrupt the pregnancy of a female, leaving her free to mate with the newcomer.

Sex attractant pheromones have also been found among the primates. It is suspected that sex attractants are exchanged among members of the human species, but no human substance has yet been isolated and definitely identified as a pheromone. However, the fact that the apocrine sweat glands, the sources of "body odor," begin to function at puberty strongly suggests that the chemicals they produce originally played such a role. Similarly, striking differences in the capacity of men and women to detect certain odors may be a clue to the existence of human pheromones.

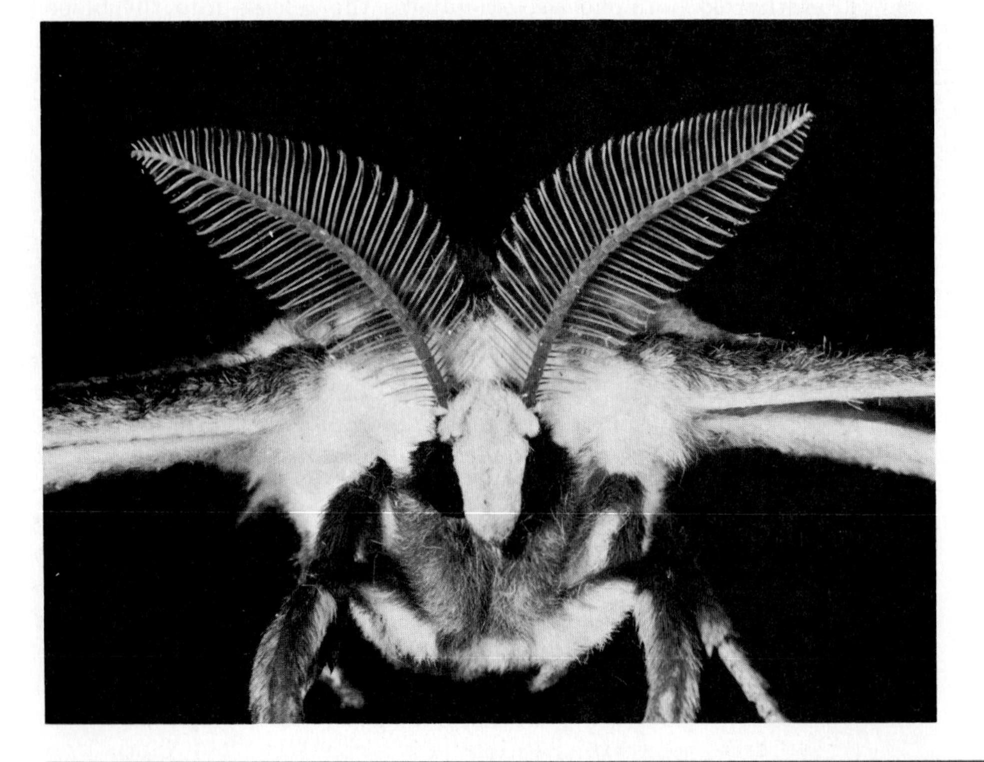

The lavishly plumed antennae of male moths, such as the luna moth shown here, can detect minute amounts of the pheromones emitted by females. The male moth characteristically flies upwind, and the pheromone, of course, disperses downwind. When a male moth detects the pheromone of a female of the same species, he will fly toward the source. If he loses the scent, he flies about at random until he either picks it up again or abandons the search. It is not until he is quite close to the female that he can fly "up the gradient" and use the intensity of the odor as a locating device.

Aldosterone and related hormones affect the concentration of ions, particularly sodium and potassium, in the blood.

The adrenal cortex is also the source of small amounts of male sex hormones in both men and women. An adrenal tumor may result in increased production of these hormones and, consequently, in women, the growth of facial hair and other masculine characteristics. Bearded ladies in circuses are often the victims of such tumors.

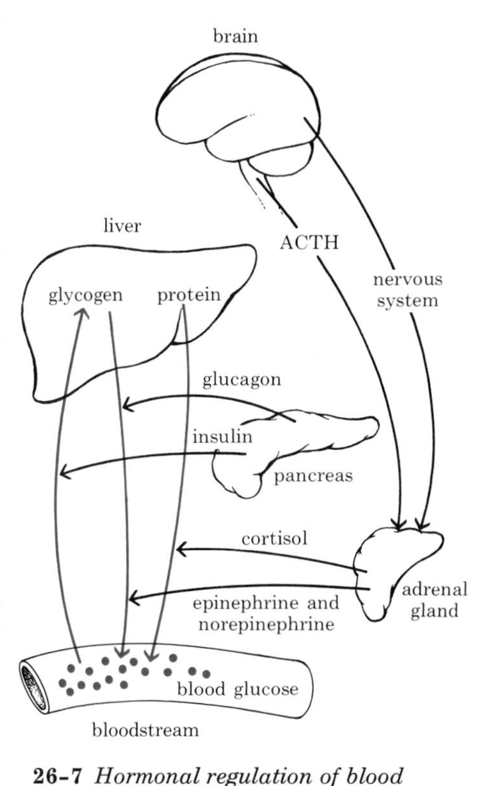

26-7 *Hormonal regulation of blood glucose. When blood sugar concentrations are low, the pancreas releases glucagon, which stimulates the breakdown of glycogen and the release of glucose from the liver. When blood sugar concentrations are high, the pancreas releases insulin, which removes glucose from the bloodstream by increasing its passage into cells and promoting its conversion into glycogen, the storage form. Under conditions of stress, ACTH, produced by the pituitary, stimulates the adrenal cortex to produce cortisol and related hormones, which increase the breakdown of protein and its conversion to glucose in the liver. At the same time, the adrenal medulla releases epinephrine and norepinephrine, which also raise blood sugar.*

Adrenal Medulla

The adrenal medulla, the central portion of the adrenal gland, actually does not conform to the definition of a gland in that it is not made up of glandular epithelium. It is composed, instead, of nervous tissue; it is, in essence, a large ganglion whose cells secrete epinephrine and norepinephrine into the bloodstream. These hormones raise blood pressure, stimulate respiration, dilate the respiratory passages, and increase the rate of heartbeat. They also increase the concentration of glucose in the bloodstream by promoting the activity of the enzyme that breaks down glycogen to glucose 6-phosphate. The adrenal medulla is stimulated by nerve fibers of the sympathetic division and acts as an enforcer of sympathetic activity. Epinephrine and norepinephrine are destroyed by enzymes in the blood within minutes after their release, which is another mechanism for tight regulatory control. (The steroid and protein hormones are also broken down, although more slowly and principally by the liver.)

The Pancreas

The islet cells of the pancreas (Figure 26–1, page 367) are the source of two hormones concerned with the metabolism of glucose: insulin and glucagon. Insulin is secreted in response to a rise in blood sugar or amino acid concentration (as after a meal). It lowers the blood sugar by stimulating cellular uptake of glucose and by stimulating the conversion of glucose to glycogen.

Glucagon, produced by different islet cells of the pancreas, increases blood sugar. It stimulates the breakdown of glycogen to glucose in the liver and the breakdown of fats and protein, which decreases glucose utilization.

Thus, as we have seen, at least six different hormones are involved in regulating blood sugar: growth hormone, cortisol, epinephrine and norepinephrine, insulin, and glucagon. This tight control over blood glucose is particularly important for brain cells. Unlike other cells in the body, which can derive energy from the breakdown of amino acids and fats, brain cells can utilize only glucose under most circumstances and so are immediately affected by low blood sugar.

Mechanism of Action of Hormones

Virtually all cells of the body are equally exposed to the hormones released into the bloodstream, yet not all respond. Recent research has indicated the mechanisms underlying the specificity of action of two major groups of hormones, the steroids and the protein hormones.

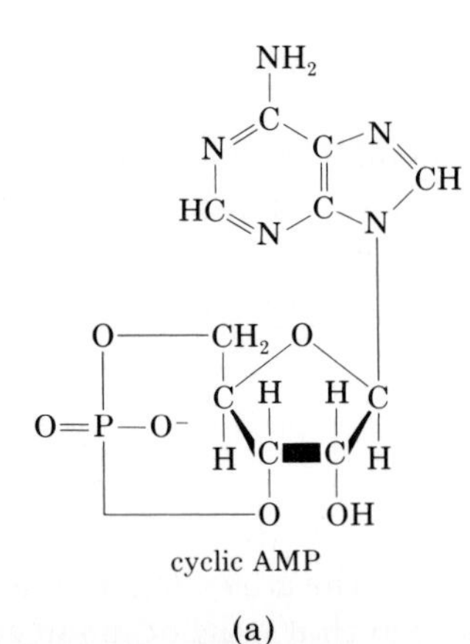

(a)

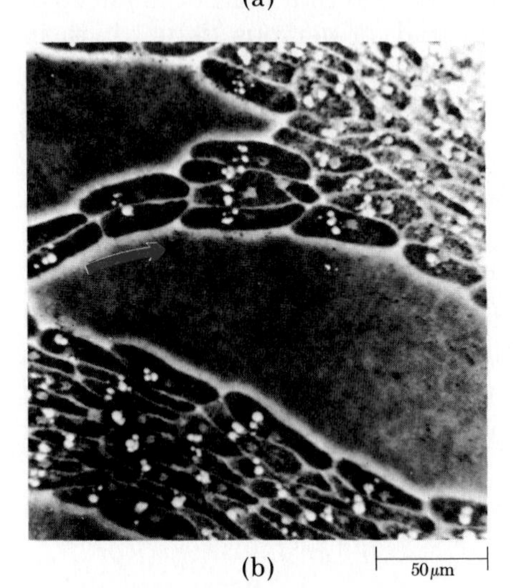

(b)

26-8 (a) *Cyclic AMP (adenosine monophosphate) acts as a "second messenger" within the cells of vertebrates. Following stimulation by various hormones—the "first messengers"—cyclic AMP is formed from ATP. "Cyclic" refers to the fact that the atoms of the phosphate group form a ring.* (b) *Cyclic AMP is also the chemical that attracts the amoebas of the cellular slime molds, causing them to aggregate into a sluglike body, which then behaves like a multicellular organism. The arrow shows the direction in which the cells are moving.*

Steroid hormones are relatively small, lipid-soluble molecules. Thus they pass easily through cell membranes and freely enter all the cells of the body. However, in their target cells and only in their target cells, the hormones encounter a specific receptor molecule that combines with them. The receptor, a protein, and the steroid together act directly on the DNA of the cell to promote the synthesis of messenger RNA and so of specific enzymes and other proteins. (Note that this is similar to the operation of the operon, described on page 206. However, it differs in that hormone and receptor apparently serve to stimulate RNA synthesis in these cells, rather than to remove a repressor molecule, as occurs in the operon.) These findings explain how the very slight differences in configuration among steroid molecules can be correlated with drastically different effects: The protein receptors, like enzymes, are highly specific in their combining properties.

By contrast, protein hormones, which are much larger molecules, do not enter cells but rather combine with receptor molecules on the membrane surface, setting in motion a "second messenger" that is responsible for the sequence of events inside the cell. These findings may have important medical implications. For example, it has long been known that juvenile diabetes—diabetes in young persons—is caused by a deficiency of the hormone insulin, which, as we noted, promotes the uptake of glucose by cells. It was assumed by analogy that the diabetes that arises in older persons and is associated with obesity had the same cause. It has now been found, however, that most adult diabetes results from a decrease in the number of insulin binding sites on receptor cells. Such patients are treated most effectively by diet.

Another group of studies, for which Earl W. Sutherland was awarded the Nobel Prize in 1971, has shown that a chemical known as cyclic AMP (Figure 26-8) is the second messenger in a number of target cells. Hormones that trigger the action of cyclic AMP include ACTH, TSH, LH, ADH, epinephrine, insulin, and glucagon. The differences in their effects are due to the presence of different enzyme systems within the cells that respond to cyclic AMP.

At almost the same time that cyclic AMP was identified by these research workers in mammalian physiology, biologists studying a peculiar group of organisms known as cellular slime molds isolated a chemical of great importance in this biological system. The cells of the cellular slime mold begin as individual amoebas and then come together to form a single organism. The chemical that calls them together was named acrasin, after Acrasia, the mythological siren who lured seamen in Homeric legend. Acrasin has now been identified as cyclic AMP, and this discovery is yet another example of the long thread of evolutionary history that links all organisms.

SUMMARY

The endocrine system consists of glands that secrete chemicals (hormones) that are taken up into the bloodstream. These chemicals exert specific effects on certain organs and tissues.

The principal endocrine glands of vertebrates include the pituitary, the hypothalamus, the thyroid, the parathyroids, the adrenal cortex and medulla, the pancreas (which is also an exocrine gland), and the gonads (ovaries or testes). Through the regulatory activities of the hypothalamus, the nervous and endocrine systems profoundly influence one another.

Hormone production is characteristically regulated by negative feedback systems. In the case of thyroxine and the steroid hormones, the concentration of a particular hormone in the blood affects the release from the pituitary of the tropic hormone that stimulates the gland producing that particular hormone. In the case of other hormones, the concentration in the bloodstream of other factors, such as ions, affects hormone release. For example, increased concentrations of calcium ion in the bloodstream act directly on the parathyroid glands to inhibit their secretion of parathyroid hormone. Such feedback systems are important mechanisms of homeostasis.

The effects of a hormone on its target tissues depend on the presence of specific receptors for that hormone either within the cell or on the surface of its cell membrane. Some hormones, such as steroid hormones, pass easily into the cell and combine with their receptors. Other hormones (for example, protein hormones) combine with receptor molecules on the membrane, activating a "second messenger," cyclic AMP, within the cell.

QUESTIONS

1. Distinguish between the following: endocrine/exocrine; pituitary/hypothalamus; thyroid gland/parathyroid glands; adrenal cortex/adrenal medulla; insulin/glucagon.

2. Draw a diagram of the negative feedback system regulating the production of thyroxine.

3. Describe how the concentration of calcium ion in the bloodstream is regulated by the thyroid and parathyroid glands.

4. Which hormones act to increase the level of blood glucose? To decrease it? How does each hormone exert its effects?

5. How does a steroid hormone exert its specific effects on a target cell? In what way is the action of a protein hormone different?

6. What types of functions would you expect to be controlled by hormones rather than by nerves? Does the reality, as described in this chapter, fulfill your expectations?

27

Circulation

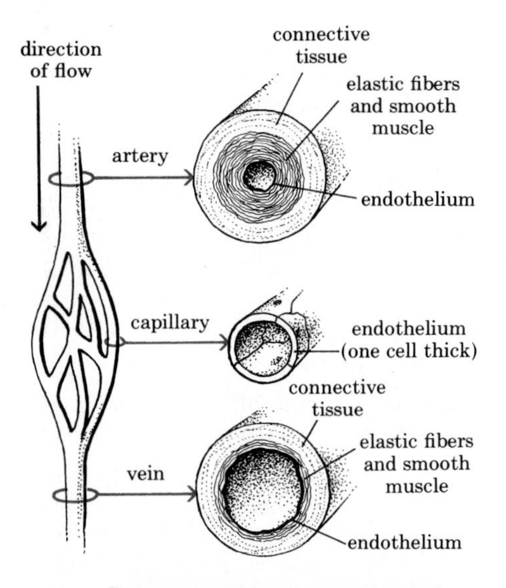

27–1 *Structure of blood vessels. Arteries have thick, tough, elastic walls that can withstand the pressure of the blood as it leaves the heart. Capillaries have walls only one cell thick. Exchange of gases, nutrients, and wastes between the blood and the cells of the body takes place through these thin capillary walls. Veins have larger lumens (passageways) and thinner, more readily extensible walls that minimize resistance to the flow of blood on its return to the heart.*

As we discussed earlier in this book, every cell in the body of a complex organism builds its own membranes and organelles, makes its own ATP, and assembles its own enzymes and other proteins. To engage in these activities, cells need oxygen and nutrients and must dispose of carbon dioxide and other wastes.

In single-celled organisms and very small multicellular ones, the needs of each cell are supplied directly by the medium in which the organism lives. In larger, more complex animals, these needs are supplied by a transport system. In humans and other vertebrates, gases (oxygen and carbon dioxide), nutrient molecules, hormones, and wastes are transported in the bloodstream. The blood circulates through a closed circuit of continuous vessels, propelled by contractions of a specialized muscle, the heart. This system is known as the *cardiovascular system* (from *cardio,* meaning "heart," and *vascular,* meaning "vessel").

In the cardiovascular system, the heart pumps blood into the large arteries, from which it travels to branching, smaller arteries (the *arterioles*) and then into very small vessels, the *capillaries.* From the capillaries, the blood passes into small veins, the *venules,* then into larger veins, and through them, back to the heart.

The Blood Vessels

In humans, the diameter of the opening of the largest artery, the *aorta,* is 2.5 centimeters, that of the smallest capillary only 6 micrometers, and that of the largest vein, the *vena cava,* 3 centimeters. Arteries, veins, and capillaries differ not only in their diameters but also in the structure of their walls (Figure 27–1). The walls of the capillaries consist of only one layer of endothelium, a special type of epithelial tissue (Figure 27–2). The walls of arteries and veins, which are lined with endothelium, also contain muscle and supporting tissues (Figure 27–3).

The Capillaries and Diffusion

The capillaries carry out the actual function of the cardiovascular system. Through their thin walls, nutrients, oxygen, carbon dioxide, and other molecules are exchanged between the blood and the fluids surround-

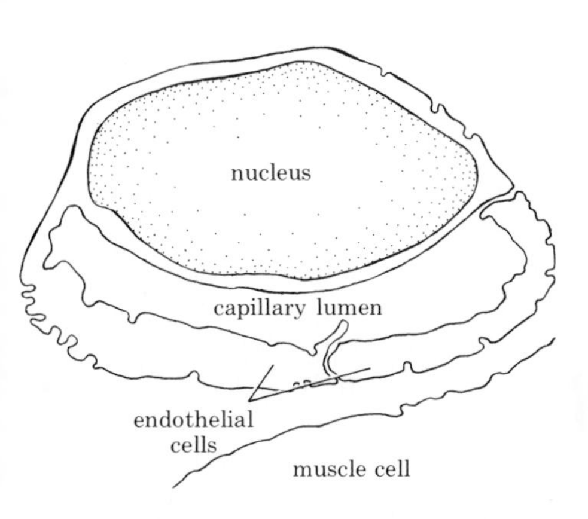

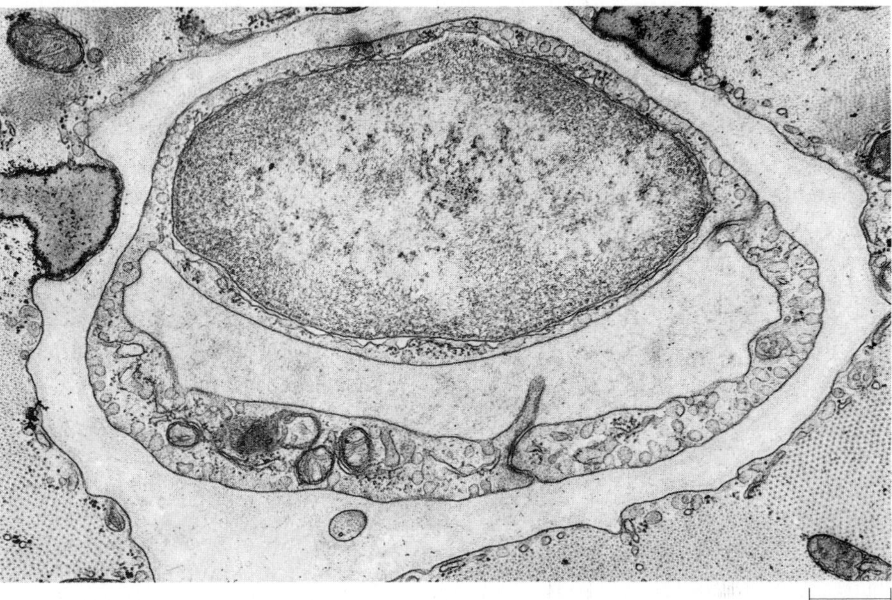

27-2 *Electron micrograph of a capillary from heart muscle. Portions of two endothelial cells can be seen, fitting together to form the capillary lumen, the passageway through which the blood flows. The space between the capillary and the surrounding muscle contains extracellular fluids.*

ing the body cells. To understand how this system works, we must return to the principles of water movement described in Chapter 5.

Most of the molecules that cross the capillary walls pass in or out by diffusion. Some additional molecules cross by bulk flow; the pressure of the blood within the capillaries tends to force fluid out through the capillary walls. At the same time, because of the protein molecules present in the blood, fluid tends to reenter the bloodstream by osmosis. Thus, there is normally a balance between inflow and outflow. When this balance is disturbed (for example, when the endothelium is damaged by a blow) and outflow is greater, excess fluid collects in the tissues, a condition known as edema.

Every cell in the human body is within rapid diffusion distance from a capillary—the total length of the capillaries in a human adult is more than 80,000 kilometers (50,000 miles). Even the cells in the walls of the large veins and arteries depend on the capillary system for their blood supply, as does the heart itself.

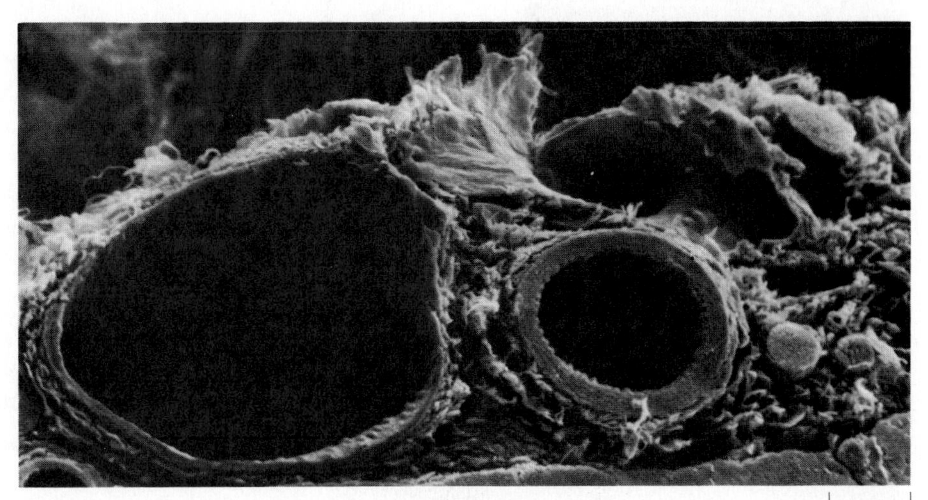

27-3 *In this scanning electron micrograph, a medium-sized vein is at the left and a smaller artery at the right. The artery can be identified by its thick, muscular wall. Connective tissue surrounds the two blood vessels and holds them in place.*

The Heart

Evolution of the Heart

In its simplest form—as in the earthworm (page 265)—the heart is a muscular contractile part of a blood vessel. In the course of vertebrate evolution the heart has undergone some structural adaptations, as shown in Figure 27–4.

Fish have a single heart divided into an *atrium*, which is the receiving area for the blood, and a *ventricle*, which is the pumping area from which the blood is expelled into the vessels. The ventricle of the fish heart pumps blood directly into the capillaries of the gills, where it picks up oxygen and releases carbon dioxide. From the gills, oxygenated blood is carried to the tissues. By this time, however, most of the propulsive force of the heartbeat has been dissipated by the resistance of the capillaries in the gills, so that the blood flow through the rest of the tissues (the systemic circulation) is relatively sluggish.

In amphibians, there are two atria; one receives oxygenated blood from the lungs, and the other receives partially deoxygenated blood from the systemic circulation. (As you will recall, the moist skin of amphibians is also an organ for gas exchange with the environment; oxygen diffuses into and carbon dioxide diffuses out of the capillaries in the skin.) Both atria empty into a single ventricle, which pumps the blood simultaneously through the lungs and the systemic circulation. By this arrangement, the blood enters the systemic circulation under high pressure.

In birds and mammals, the heart is functionally separated into two organs. The right heart receives blood from the tissues and pumps it into the lungs, where it becomes oxygenated. From the lungs, the oxygenated

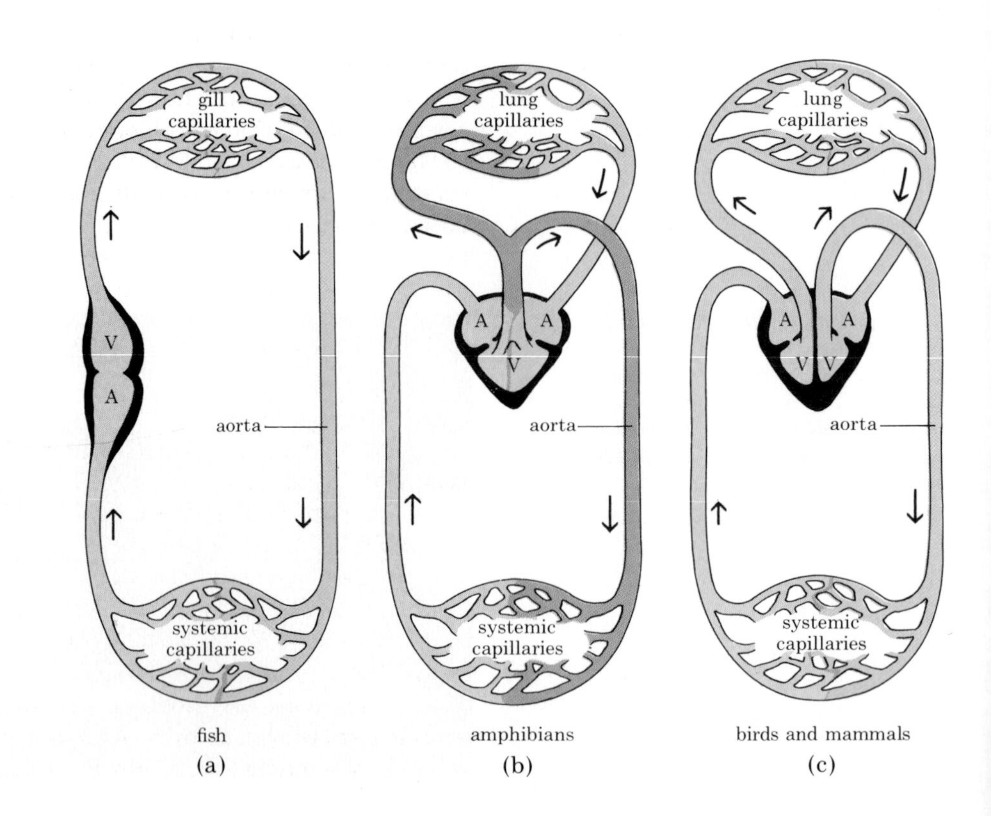

27–4 *Vertebrate circulatory systems. Oxygenated blood is indicated in color.* (a) *In the fish, the heart has only one atrium (A) and one ventricle (V). Blood oxygenated in the gill capillaries goes straight to the systemic capillaries without first returning to the heart.*

(b) *In amphibians, the single primitive atrium has been divided into two separate chambers. Oxygenated blood from the lungs enters one atrium, and oxygen-poor blood from the tissues enters the other. The blood is mixed somewhat in the single ventricle and then pumped again through the lungs and body tissues.*

(c) *In birds and mammals, the ventricle is divided into two separate chambers, so that there are, in effect, two hearts—one for pumping oxygen-poor blood through the lungs and one for pumping oxygen-rich blood through the body tissues.*

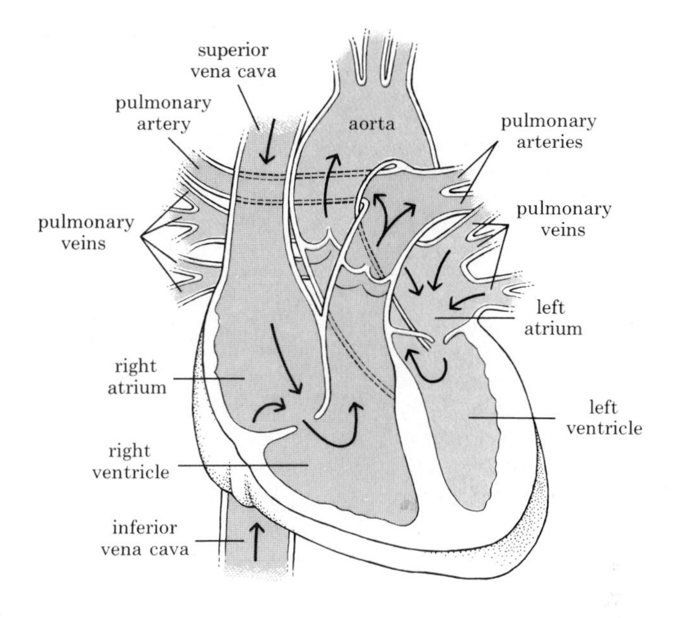

27-5 *The human heart. Blood returning from the systemic circulation through the superior and inferior venae cavae enters the right atrium and passes to the right ventricle, which propels it through the pulmonary arteries to the lungs, where it is oxygenated. Blood from the lungs enters the left atrium through the pulmonary veins, passes to the left ventricle, and then is pumped through the aorta to the body tissues.*

blood returns to the left heart, from which it is pumped into the body tissues. This efficient, high-pressure circulatory system is necessary to maintain the high metabolic rate of both birds and mammals, with their constant body temperature and their generally high level of activity.

The Heart and Regulation of Its Beat

Figure 27-5 shows a diagram of the human heart. Its walls are made up of a specialized muscle—the cardiac muscle. Blood returning from the body tissues enters the right atrium through two large veins, the superior and inferior venae cavae. Blood returning from the lungs enters the left atrium through the pulmonary veins. The atria, which are thin-walled compared with the ventricles, expand as they receive the blood. Both atria then contract simultaneously, assisting the flow of blood through open valves into the ventricles. Then the ventricles contract simultaneously, closing the valve between each atrium and ventricle. The right ventricle propels the blood into the lungs through the pulmonary arteries; the left ventricle propels oxygenated blood into the aorta, from which it travels to the other body tissues. In a healthy adult at rest, this rhythmic process takes place about 70 times a minute; under strenuous exercise, the rate more than doubles.

Most muscle contracts only when stimulated by a motor nerve, but the stimulation of cardiac muscle cells originates in the muscle itself. A vertebrate heart will continue to beat even after it is removed from the body if it is kept in a nutrient solution. In vertebrate embryos, the heart begins to beat very early in development, before the appearance of any nerve supply. In fact, isolated embryonic heart cells in a test tube will beat.

The beat of the cardiac muscle is initiated by a special area of the heart, the *sinoatrial node*, which is located in the right atrium (Figure 27-6). This node functions as the *pacemaker*. It is composed of cardiac muscle cells that depolarize spontaneously, initiating their own action potential and contraction. From the pacemaker the action potential spreads

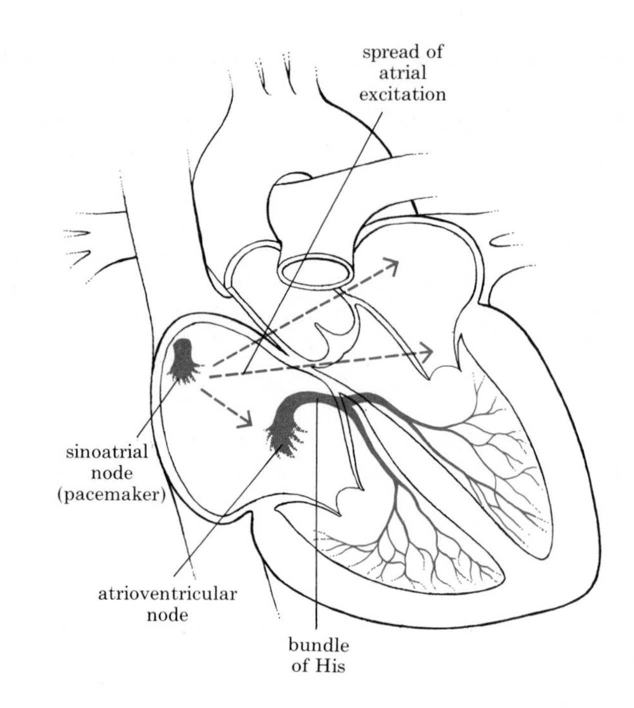

27-6 *The beat of the mammalian heart is controlled by a region of specialized muscle tissue in the right atrium, the sinoatrial node, which functions as the heart's pacemaker. Some of the nerves regulating the heart have their endings in this region. Excitation spreads from the pacemaker through the atrial muscle cells, causing both atria to contract almost simultaneously. When the wave of excitation reaches the atrioventricular node, its conducting fibers pass the stimulation to the bundle of His, which triggers simultaneous contraction of the ventricles. Because the fibers of the atrioventricular node conduct relatively slowly, the ventricles do not contract until after the atrial beat has been completed.*

throughout the right and left atria. As it passes along the surface of the individual cardiac muscle cells, it activates their contractile machinery, and they contract. The action potential travels very quickly, so many cells are activated almost simultaneously.

About 100 milliseconds after the pacemaker fires, impulses traveling through special conducting fibers stimulate a second area of nodal tissue, the *atrioventricular node*. The atrioventricular node is the only electrical bridge between the atria and the ventricles. It consists of slow-conducting fibers. Thus the atrioventricular node imposes a delay between the atrial and ventricular contractions, so that the atrial beat is completed before the beat of the ventricles begins. From the atrioventricular node, impulses are carried by special muscle fibers, the *bundle of His* (named after its discoverer), to the walls of the right and left ventricles, which then contract simultaneously.

If you listen to a heartbeat, you hear "lubb-dup, lubb-dup." The deeper, first sound ("lubb") is the closing of the valves between the atria and the ventricles; the second sound ("dup") is the closing of the valves leading from the ventricles to the arteries. If any one of the four valves is damaged, as from rheumatic fever, blood may leak back through the valve, producing the noise characterized as a "heart murmur" (a "ph-f-f-t" sound).

Although the autonomic nervous system does not initiate the vertebrate heartbeat, it does control its rate. Fibers from the parasympathetic division travel through the vagus nerve (a large nerve that runs through the neck) to the pacemaker. These nerve fibers secrete acetylcholine, which has a slowing effect on the pacemaker and thus decreases the rate of heartbeat. Conversely, fibers from the sympathetic division secrete norepinephrine, which stimulates the pacemaker, increasing the rate of heartbeat. Epinephrine and norepinephrine from the adrenal medulla affect the heart in the same way that the sympathetic nerves do.

Diseases of the Heart and Blood Vessels

Cardiovascular diseases cause more deaths in the United States than accidents and all other diseases combined. According to recent estimates, more than 40 million people in this country—18 percent of the total population—have some form of cardiovascular disease.

About 65 percent of the cardiovascular deaths are caused by heart attacks. A heart attack is the result of an insufficient supply of blood to an area of heart muscle; with their oxygen supply cut off, the muscle cells die. It can be caused by a blood clot—a thrombus—that forms in the blood vessels of the heart itself or, occasionally, by a clot that forms elsewhere in the body and travels to the heart and lodges in a vessel there. (A wandering clot such as this is known as an embolus.) A heart attack can also result from a blocking of a blood vessel due to atherosclerosis (see below). Recovery from a heart attack depends on how much of the heart tissue is damaged and whether or not other blood vessels in the heart can enlarge their capacity and supply these tissues, which then may recover to some extent. Angina pectoris is a related condition in which the heart muscle receives an insufficient blood supply, often a result of a narrowing of the vessels. The symptoms are a pain in the center of the chest and, often, in the left arm and shoulder.

A stroke is caused by interference with the blood supply to the brain. This may be the result of a thrombus (cerebral thrombosis), an embolus, or the bursting of a blood vessel in the brain. Its effects depend on the severity of the damage and where it occurs in the brain.

Atherosclerosis contributes to both heart attacks and strokes. In this disease, the linings of the arteries thicken and their surface becomes roughened by deposits of cholesterol, fibrin (clotting material), and cellular debris. The arteries, becoming inelastic, no longer expand and contract, and blood moves with increasing difficulty through the narrowed vessels. Hardening of the arteries—arteriosclerosis—may subsequently occur if calcium becomes deposited in the artery walls. Thrombi and emboli thus form more easily and are more likely to block the vessel. The causes are not clear-cut. Arteriosclerosis is associated with high levels of cholesterol (page 44) in the blood. Other important risk factors are cigarette smoking, high blood pressure, stress, diabetes, and possibly lack of exercise. Female hormones seem to protect against arteriosclerosis; the condition is much less common in premenopausal women than it is among men of the same age. This difference in susceptibility to arteriosclerosis is a major reason that, in the United States, women now live, on the average, almost 10 years longer than men.

High blood pressure—hypertension—contributes significantly to cardiovascular disease. It places additional strain on the arterial walls and increases the chances of emboli. Despite lack of knowledge about the causes, hypertension can be treated and blood pressure reduced to safer levels. In the United States, high blood pressure is twice as common among black males as among white males and almost twice as common among black females as among white females. Related to this disturbing statistic is the fact that the incidence of deaths among black men and women from cardiovascular diseases is even higher than in the white population.

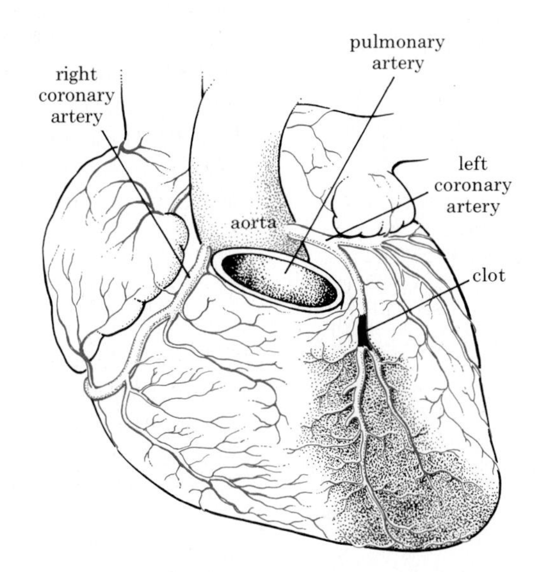

A heart attack. When a clot forms in a blood vessel, the cells in the area supplied by this vessel are deprived of oxygen and die. The severity of the heart attack depends, in part, on the extent of damage to the heart muscle.

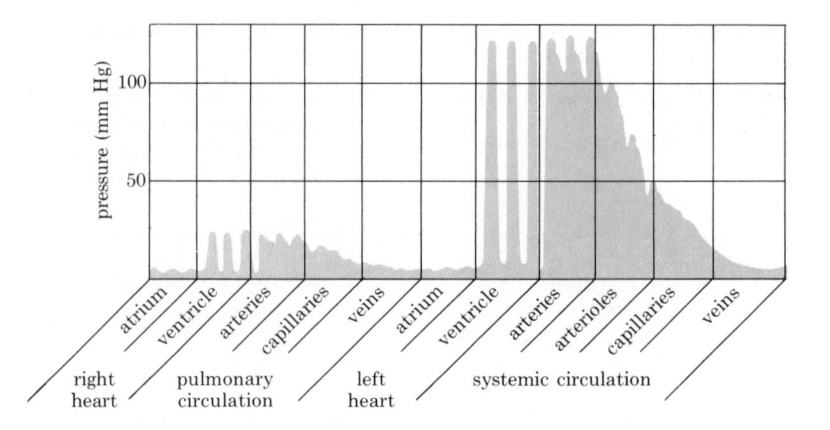

27–7 *In mammals, blood goes from the right heart to the lungs, from the lungs to the left heart, and, from the left heart, it enters the systemic circulation, moving from arteries to arterioles to capillaries to veins. Blood pressure varies in the different areas of the cardiovascular system. The fluctuations in blood pressure produced by three heartbeats are shown in each section of the diagram. Note the fall in pressure as blood traverses the arterioles of the systemic circulation.*

Blood Pressure

The contractions of the ventricles propel the blood into the arteries with considerable force. Blood pressure is a measure of the force per unit area with which blood pushes against the walls of the blood vessels. It is conventionally described in terms of how high it can push a column of mercury. For medical purposes, it is usually measured at the artery of the upper left arm. Normal blood pressure in a young adult is 120 millimeters of mercury (120 mm Hg) when the heart is contracting (the systolic blood pressure) and 80 mm Hg when the heart relaxes (diastolic pressure); this is stated as a blood pressure of 120/80.

The pressure is generated by the pumping action of the heart and changes with the rate at which it contracts. Blood flow is directly proportional to blood pressure; the greater the pressure, the greater the flow.

Blood pressure is not the same in the various parts of the cardiovascular system (Figure 27–7). As the blood gets farther along the system of vessels, the pressure drops, becoming lower in the veins and lowest in the right atrium. Blood in the veins, which are soft-walled, is helped back toward the heart by the contractions of the body muscles. For example, when you walk, your leg muscles bulge and squeeze the veins that lie between the contracting muscles. This raises the pressure within the veins and increases blood flow back to the heart. Valves in the veins prevent backflow (Figure 27–8).

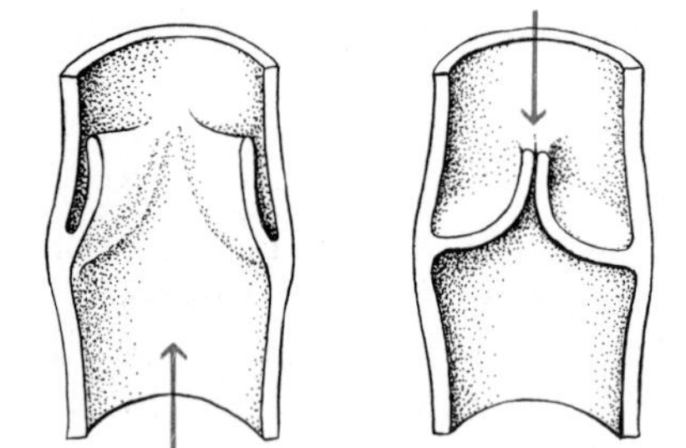

27–8 *Valves in the veins open to permit movement of blood toward the heart but close to prevent backflow.*

Regulating Blood Flow

The blood flow to different parts of the body changes as the activities of the organism change. For example, blood flow through muscle increases during exercise, flow to the stomach and intestines increases during digestion, and flow through the skin increases at high temperatures and decreases at low temperatures.

The regulation of blood flow depends on a very simple physical principle: Fluid flow through a tube is proportional to the fourth power of the radius of the tube (r^4). As you may recall, the radius of a tube is one-half of its diameter. The diameter of the arterioles, which directly supply the capillaries, can be altered by rings of smooth muscle in the vessel walls. As the smooth muscle contracts, the opening of the arteriole gets smaller, and blood flow through the arteriole (and the capillary bed it feeds) decreases

27-9 *Diagram of a capillary bed. The muscular wall of the arteriole controls the blood flow through the capillaries. The arteriolar muscles, which are innervated by the sympathetic division of the autonomic nervous system, make it possible to regulate the blood supply to different regions of the body depending on the physiological state of the organism at different times.*

(Figure 27-9). Conversely, when the smooth muscle relaxes, the arteriole opens wider, and blood flow into the capillaries increases. These smooth muscles are controlled by autonomic nerves (chiefly sympathetic nerves), the hormones epinephrine and norepinephrine, and the levels of other chemicals that are produced locally in the tissues themselves.

Cardiovascular Regulating Center

Activity of the nerves controlling the smooth muscle of the blood vessels is coordinated with the activity of nerves regulating heart rate and strength of heartbeat by the *cardiovascular regulating center*. This center is located in the medulla, a small region of the brain close to the junction of the brain and spinal cord (see Figure 25-12, page 362). It controls the sympathetic and parasympathetic nerves to the heart as well as the nerves to the smooth muscle in the arterioles. Thus, if blood flow through a particular body area is increasing due to dilation of blood vessels, the heart is simultaneously stimulated to develop greater pressure to support the greater flow.

The cardiovascular regulating center integrates the reflexes that control blood pressure. It receives information about existing blood pressure from specialized stretch receptors in large arteries in the neck (the carotid arteries), the venae cavae, the aorta, and the heart. The effector organs of the reflex are, as we have indicated, the heart and blood vessels.

The blood-pressure reflex is another example of negative-feedback control. When pressure falls, the activity of the heart is increased and the blood vessels are constricted, which raises the pressure again. Conversely, heart activity is decreased and the blood vessels dilated in response to high pressure.

The Blood

27-10 *In vertebrates, oxygen is transported in red blood cells, shown here in a scanning electron micrograph. Red blood cells, which are only about 7 or 8 micrometers in diameter, are flexible and so are able to twist and turn in their passage through the capillaries.*

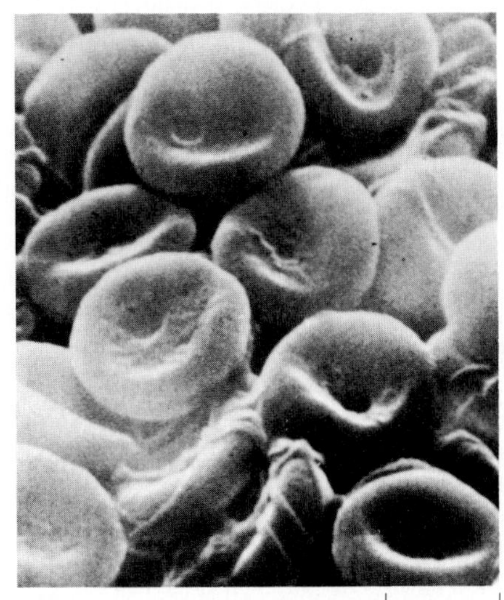

An individual weighing 75 kilograms (165 pounds) has about 5 liters of blood. About 60 percent of it is a straw-colored liquid called *plasma*. The plasma, which is mostly water, carries many different kinds of ions and molecules. They include fibrinogen, the protein from which blood clots are formed; nutrients, such as glucose, fats, and amino acids; various ions; antibodies, hormones, and enzymes; and waste materials, such as urea and uric acid. The other 40 percent of the blood is made up of red blood cells, white blood cells, and platelets.

Red Blood Cells

Red blood cells, or *erythrocytes* (Figure 27-10), transport oxygen to all the tissues of the body. They are among the most highly specialized of all cells. As a mammalian red blood cell matures, it extrudes its nucleus and mitochondria, and its other cellular structures dissolve. Almost the entire volume of a mature red blood cell is filled with hemoglobin, the molecule that combines with oxygen to transport it from the lungs to the tissues.

There are about 5 million red blood cells per cubic millimeter of blood—some 25 trillion (25×10^{12}) in the whole body. In human beings, an individual red blood cell has a life span of about 120 to 130 days. New ones are produced in the bone marrow of adults at the rate of about 2 million per second.

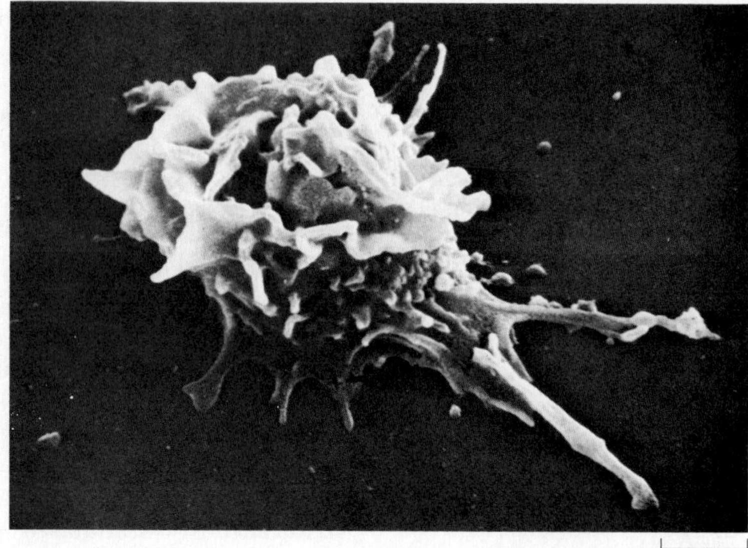

27-11 *A scanning electron micrograph of a phagocytic white blood cell from a rabbit. This type of white blood cell defends the body against foreign invaders by extending pseudopodia around the foreign material, incorporating it within a food vacuole, and destroying it with the help of enzymes from the lysosomes (page 68).*

Although there are several types of white blood cells, all are nearly colorless, are larger than red blood cells, contain no hemoglobin, and have a nucleus.

White Blood Cells

For every 1,000 red blood cells in the human bloodstream, there are one or two white blood cells, or *leukocytes*. The body contains several different types of white blood cells, which are classified according to their appearance under the microscope.

The chief function of the white blood cells is the defense of the body against invaders such as viruses, bacteria, and various foreign particles. Unlike red blood cells, white blood cells are not confined within the blood vessels but can migrate out into the tissues to the site of foreign invasion. They appear spherical in the bloodstream, but in the tissues they become flattened and amoeba-like. Like amoebas, they move by means of pseudopodia and many are phagocytic (Figure 27-11). Certain types of white blood cells are involved in immunity, as we shall see later in this chapter.

White blood cells are often destroyed in the course of fighting infection. Pus is composed largely of these dead cells. New white blood cells are formed constantly in the spleen, in bone marrow, and in certain other tissues to take the place of those cells that are destroyed.

Platelets

Platelets, cell fragments found in the bloodstream of mammals, are colorless round or biconcave disks less than half the size of red blood cells. They play an important role in the formation of clots and in plugging breaks in blood vessels.

The Lymphatic System

Not quite all of the fluid forced out of the capillaries by the pressure of the circulating blood (page 377) is returned to the capillaries by osmosis. In higher vertebrates, this fluid is collected by the *lymphatic system*, which routes it back to the bloodstream. The lymphatic system is like the venous system in that it consists of an interconnecting network of progressively larger vessels. An important difference, however, is that the lymph capillaries begin blindly in the tissues, rather than forming part of a continuous circuit. Fluid from the tissues seeps into the lymph capillaries, from which it travels to large ducts that empty into the superior vena cava (Figure 27-12). This fluid is known as *lymph.* Some nonmammalian vertebrates have lymph "hearts," which help to move the fluid. In mammals, lymph is moved by contractions of the body muscles with valves preventing backflow, as in the venous system. Also, recent studies have shown that lymph vessels contract rhythmically; these contractions may be the principal factor propelling the lymph.

Lymph nodes, which are masses of spongy tissue, are distributed throughout the lymphatic system. They have two functions: They remove cellular debris and foreign particles from the lymph before it enters the blood, and they are the sites of proliferation of *lymphocytes,* white blood cells important in immunity. The lymph nodes produce about 10 billion lymphocytes per day.

The spleen, thymus, tonsils, and adenoids are also composed of lymphoid tissue. The thymus, a spongy two-lobed organ that lies high in the chest, plays a critical role in the maturation of a particular type of lymphocyte known as the T-cell, the subject of much current research.

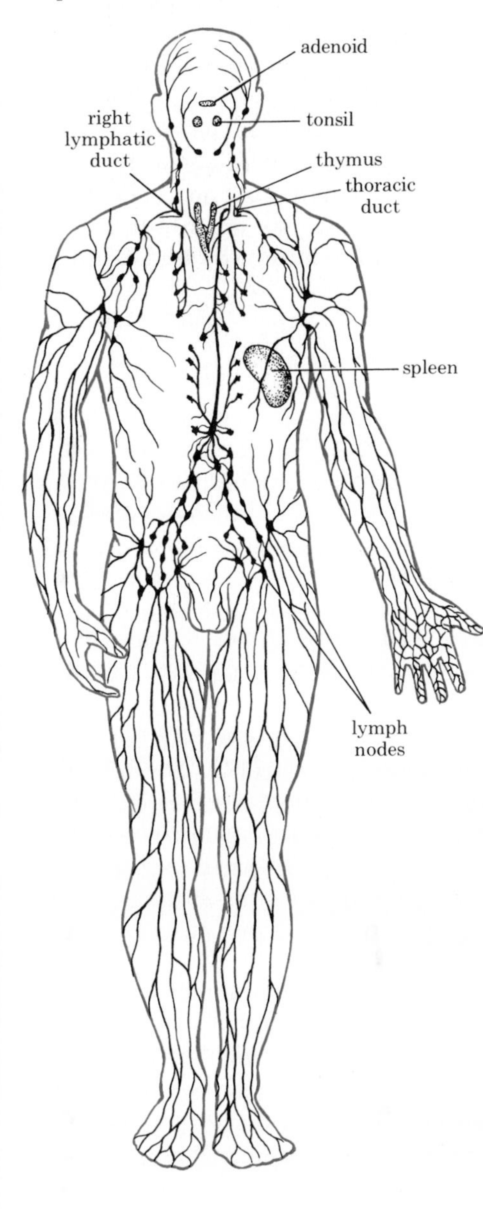

27-12 *Human lymphatic system. Lymphatic fluid reenters the bloodstream through the thoracic duct and right lymphatic duct, which empty into the superior vena cava.*

Defense Mechanisms

The body has three lines of defense against invading bacteria, viruses, and other foreign particles. The first line of defense is the outer wrapping of skin and mucous membranes. The skin is an effective barrier as long as it is intact. Mucous membranes are more fragile, but they are constantly flushed and cleansed with fluids, such as mucus and saliva, that contain antimicrobial substances.

Once a microorganism penetrates the skin or mucous membranes, it encounters a second line of defense, consisting primarily of phagocytic white blood cells. Suppose, for example, you cut your skin. The injured cells immediately release histamine and other chemicals that increase both blood flow into the area and the permeability of nearby capillaries. Circulating white blood cells move through the capillary walls, crowding into the site of the injury. These cells engulf the foreign invaders, literally eating themselves to death. Blood clots begin to form, walling off the injured area. Pus may accumulate. The local temperature of the area often rises, creating an environment unfavorable to the multiplication of microorganisms while accelerating the motion of white blood cells. This series of events is known as the *inflammatory response.*

If the inflammatory response is insufficient, a third defense mechanism—the *immune system*—comes into play.

The Immune Response

The immune system differs from the other defenses of the body in that it is highly specific, involving recognition of a particular invader and the tailoring of an attack against it.

The *immune response* consists of a primary response to the initial attack of an invader and a secondary response to subsequent attacks by the invader. In the primary response, the body recognizes an invading substance as foreign, or "not-self." Certain lymphocytes are then activated to multiply, and they eliminate the invader. The secondary response involves a "memory" of the lymphocytes for the invader, so that when that particular substance enters the body again, it is attacked immediately and usually

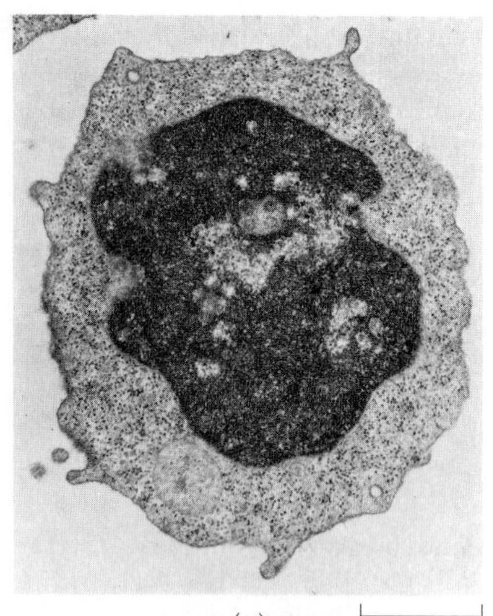

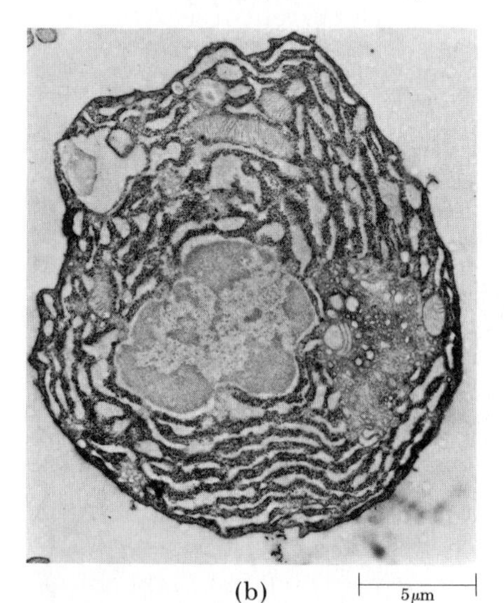

27–13 *As a B-cell* (a), *a type of lymphocyte, differentiates into a plasma cell* (b), *it becomes highly specialized for the manufacture and export of antibodies, which are complex globular proteins. It grows larger, its nucleus becomes relatively smaller and less dense, and there is a large increase in endoplasmic reticulum and ribosomes.*

destroyed before an infection can take hold. After you have been infected with a particular bacterium or virus, you are often protected against reinfection by that agent. Common examples of diseases that confer immunity, as this phenomenon is known, are measles, mumps, and chicken pox. Vaccines are agents that trigger the immune response of the body to a particular infectious agent without actually producing the disease; thus, the invader, when it does appear, is met with the secondary response and so the disease is prevented.

The study of the immune system is currently one of the most active areas of biological research. The intricate and precise workings of this defense system are still being unraveled, but the general outlines, as well as some of the details, are beginning to emerge. Biologists have found that there are two different kinds of immunity: humoral immunity and cellular immunity.

Humoral Immunity

Humoral immunity is mediated by lymphocytes known as B-cells. It involves the formation of *antibodies*, molecules that circulate in the "humors," or fluids, of the body. Like enzymes, antibodies are complex globular proteins. They are highly specific, and their specificity is based on their combining in a very precise way with another molecule. The molecule with which an antibody combines is known as an *antigen* (shorthand for "antibody-generating substance"). Virtually all proteins and some polysaccharides and lipids can act as antigens. A single cell, such as a bacterial cell, may carry a number of substances that act as antigens, each of which can elicit a specific antibody.

Antibodies act against invaders in one of three ways: (1) They may coat the foreign particles or cause them to clump together in such a way that they can be taken up by the phagocytic white blood cells; (2) they may combine with the invading particles in such a way that they interfere with some vital activity—for example, covering the protein coat of a virus at the site where the virus attaches to the cell membrane; or (3) they may themselves, in combination with another blood component known as complement, actually lyse and destroy the foreign cell.

Each type of B-cell has one—and only one—type of antibody on its cell surface, but there may be as many as 100,000 molecules of the antibody present. When the B-cell encounters an antigen with which its antibody combines, it begins to multiply. More B-cells with that particular antibody are produced, and some of them differentiate to form *plasma cells* (Figure 27–13). These cells are, in essence, antibody factories—a single plasma cell can produce about 2,000 molecules of its particular antibody per second. The antibodies are released into the bloodstream and circulate throughout the body to react with the invading material.

Division of a B-cell takes about 10 hours, and the process is repeated nine times in four or five days. At that point, antibody production usually catches up with the multiplication of the infectious organism and eliminates it. After the infection, the circulating antibodies produced by the plasma cells disappear, but the increased population of B-cells bearing that particular antibody on the cell surface persists indefinitely in the circulation. If the antigen with which the antibody combines enters the body again, these B-cells immediately begin their defensive measures against it. Such B-cells, known as "memory cells," are responsible for long-term immunity.

27-14 (a) *Diagram of an IgG antibody molecule. The two heavy chains are connected by two disulfide bridges, and each heavy chain has a flexible hinge region. The light and heavy chains both have constant (gray) and variable (colored) regions.* (b) *Because each antibody molecule has two antigen-binding sites, one antibody can bind to antigens on different cells or particles, thus causing them to stick together (agglutinate). Phagocytic white blood cells then consume these larger masses of foreign particles and antibodies.*

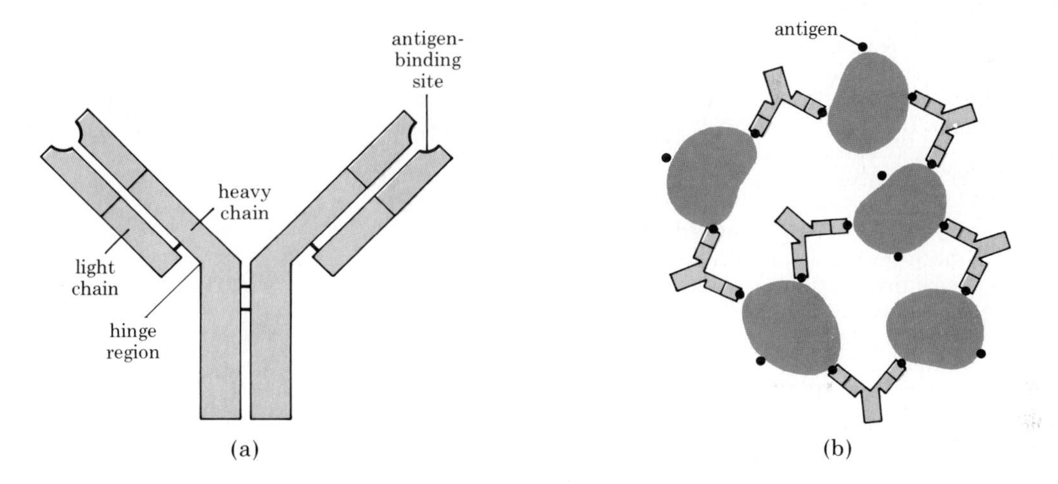

The Structure of Antibodies There are five major classes of antibodies, or immunoglobulins, as they are called by biochemists. One class, known as IgG, is the most abundant in the bloodstream and is the best understood.

Each IgG antibody consists of four subunits, two "light" chains and two "heavy" chains. Each of the chains has a region, known as the constant region, where the sequence of amino acids is apparently common to all IgG antibodies. Another region of each chain, called the variable region, has a specific sequence in each particular antibody. The combining sites, it is not surprising to find out, involve the variable regions. Each antibody has at least two combining sites, so it can attach to two antigens at the same time. The antigens may be on the same microorganism or on two different ones (Figure 27-14). Each heavy chain contains a flexible hinge region, so the antigen-binding sites need not be a fixed distance apart.

Cellular Immunity

Cellular immunity involves another kind of lymphocyte, known as a T-cell. Like B-cells, T-cells are highly specific in their recognition of particular antigens. T-cells, however, interact *directly* with the invading substance rather than indirectly through the production of antibodies. Several types of T-cells function in the regulation of the immune response. Others seem to be important in controlling both virus infections and cancer. Some of these T-cells are also involved in the rejection of tissues—such as skin, kidney, or heart—transplanted from one individual to another (except an identical twin). T-cells can be seen aggregating around a tissue transplant before and during its rejection. Surgeons and research workers concerned with extending the use of tissue transplants are seeking ways to suppress or paralyze cellular immunity in such a way that these foreign but potentially lifesaving tissues can survive.

Disorders Related to the Immune System

The immune response is a powerful bulwark against disease, but it sometimes goes awry. For example, the immune system can ordinarily distinguish between "self" and "not-self." Substances that are present during embryonic life, when the immune system is developing, will not be antigenic in later life. This recognition occasionally breaks down, however, and the immune system attacks the body. Certain disorders, among them one type of anemia and myasthenia gravis, have been identified as autoimmune diseases–that is, diseases in which an individual makes antibodies against his or her own cells. There is growing evidence that other disorders, such as some forms of rheumatoid arthritis, may have the same basis.

Hay fever and other allergies are the result of immune responses to pollen, dust, or other substances that are weak antigens to which most people do not react.

Another medical problem caused by the immune system is Rh disease of the newborn. During the last month before birth, the human baby usually acquires antibodies from its mother. Most of these antibodies are beneficial. An important exception, however, is found in the antibodies formed against a blood factor, the Rh factor. The Rh factor is a genetically determined antigen found on the surface of red blood cells. If a woman who lacks the Rh factor (that is, an Rh-negative woman) has children fathered by a man homozygous for the Rh factor, all the children will be Rh positive; if he is heterozygous, about half of the children will be Rh positive.

Before or during the birth of the mother's first Rh-positive child, fetal red blood cells–bearing the Rh antigen–are likely to enter her bloodstream. Her immune system then produces antibodies against the foreign antigen, and these antibodies persist in her blood. In subsequent pregnancies, they may be transferred to the fetus. If that fetus is Rh positive, the antibodies will react with its red blood cells, destroying them (see illustration). This reaction can be fatal either before or just after birth.

In 1968, two medical scientists at Columbia University developed a substance called RhoGAM, which contains antibodies against the Rh factor. If RhoGAM is injected into an Rh-negative woman at the birth of her first Rh-positive child, it destroys the fetal Rh-positive cells that have entered her bloodstream before they are able to trigger a reaction by her own immune system. As a result, her next Rh-positive child is protected against Rh disease. At the birth of each Rh-positive child, RhoGAM must be administered to protect the next child. The use of RhoGAM has reduced the death toll of Rh disease by more than 75 percent since its introduction.

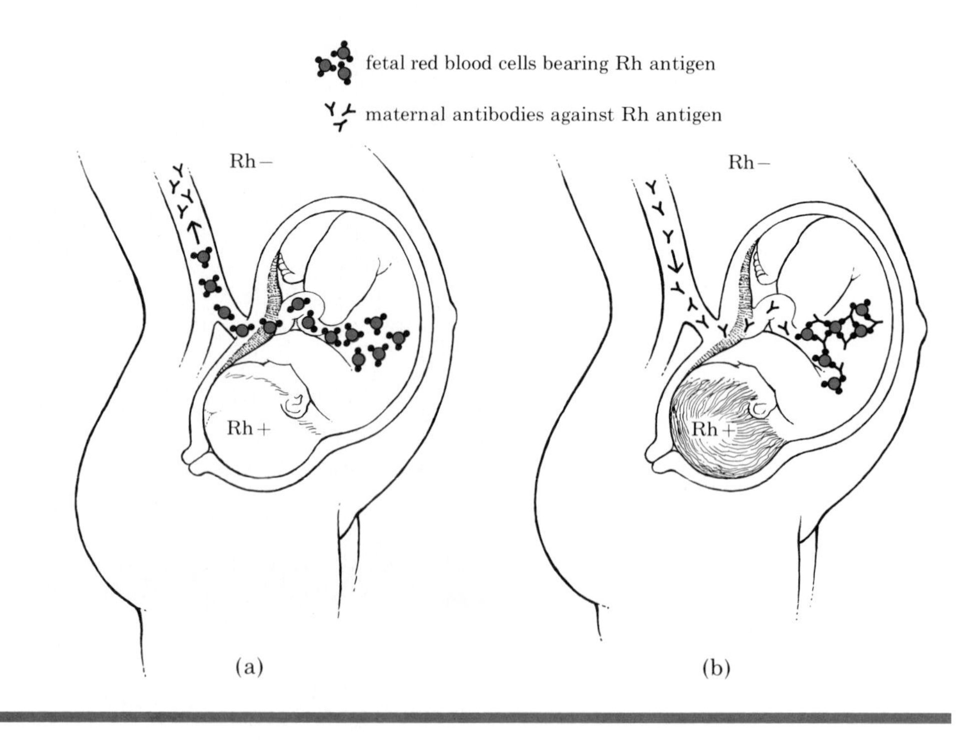

The events leading to Rh disease. (a) *Late in the first pregnancy–or during the actual birth of the child–fetal blood cells spill across the barrier that separates the maternal circulation from the fetal circulation. Antigens on the red blood cells of the Rh-positive fetus stimulate the production of antibodies by the mother's immune system. These antibodies remain in the mother's bloodstream indefinitely.* (b) *Late in the second pregnancy, the antibodies pass through the barrier from the maternal blood to the blood of the fetus. If the fetus is Rh positive, the antibodies react with the antigens on its red blood cells, destroying them.*

Immunity and Cancer

Cancer cells resemble the host's own cells in many ways. Yet, within the host, they act like foreign organisms, invading and "choking off" or competing with normal tissues. Moreover, they have antigens on their cell surfaces that can be distinguished from the antigens of the normal host cells. Does this mean that people can mount an immune response against their own cancers? A growing number of cancer researchers believe not only that cancer can induce an immune response but also that it usually does so. In fact, it usually does so successfully, overwhelming the cancer before it is ever detected by the patient or the physician. The cancers that are discovered represent occasional failures of the immune system. This conclusion suggests that bolstering the patient's immune system may provide a means for cancer prevention and control.

SUMMARY

Transport of oxygen, nutrients, hormones, and wastes is accomplished in vertebrates by closed circulatory systems in which the blood is pumped by the muscular contractions of the heart into a vast circuit of arteries, arterioles, capillaries, venules, and veins. This network ultimately services every cell in the body. The essential function of the circulatory system is performed by the capillaries, through which substances are exchanged with the fluids surrounding the individual cells.

Heart structure in vertebrates varies with the demands of their differing metabolic rates. The fish has a two-chambered heart; the atrium receives deoxygenated blood, and the ventricle propels blood through the oxygenating gill capillaries into the systemic circulation. In amphibians, oxygenated and partially deoxygenated blood are received in the two atria, are mingled somewhat in the ventricle, and are then pumped simultaneously through the lungs and the systemic circulation. Birds and mammals have a double circulatory system, made possible by a four-chambered heart that functions as two separate pumping organs. One side pumps deoxygenated blood to the lungs, and the other pumps oxygenated blood to the body tissues.

Synchronization of the heartbeat is controlled by the sinoatrial node (the pacemaker) located in the right atrium, and by the atrioventricular node, which delays the stimulation of ventricular contraction until the atrial contraction is completed. The rate of heartbeat is under neural and hormonal control. Acetylcholine slows the pacemaker; epinephrine and norepinephrine speed it up.

The blood is composed of plasma, red blood cells (erythrocytes), white blood cells (leukocytes), and platelets. The fluid part of the blood is plasma, which is chiefly water in which are dissolved or suspended nutrients, ions, antibodies, enzymes, waste substances, and other compounds. Red blood cells are the carriers of oxygen-bearing hemoglobin. White blood cells defend the body against invaders by phagocytosis or in the immune response. Platelets are involved in the formation of blood clots.

Fluid that seeps out of the capillaries can reenter the capillaries by osmosis or be returned to the blood by the lymphatic system. The lymph also picks up bacteria, cellular debris, and other foreign particles. These are filtered out of the lymph by the lymph nodes. The lymphatic system is also the site of maturation of lymphocytes, the white blood cells involved in the immune response. The spleen, thymus, tonsils, and adenoids are also components of the lymphatic system.

The immune response is a defensive reaction that involves recognizing a substance entering the circulatory system as foreign and developing a specific defense against it. Humoral immunity involves the production of antibodies by plasma cells, which are B-cells (a type of lymphocyte) that have differentiated after contact with an antigen. Antibodies are complex proteins that combine with specific antigens and inactivate them. Cellular immunity is mediated by T-cells, another type of lymphocyte, which interact directly with the foreign substance. There is evidence that a failure of the immune system may be involved in the development of cancer.

QUESTIONS

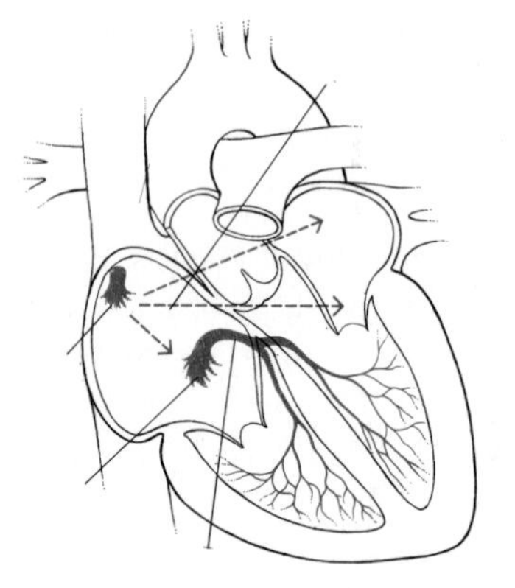

1. What is the function of the cardiovascular system? What are its principal components?

2. Distinguish between the following: blood/plasma; aorta/vena cava; atrium/ventricle; right heart/left heart; sinoatrial node/cardiovascular regulating center; humoral immunity/cellular immunity.

3. Label the diagram at the left.

4. Trace the course of a single red blood cell from the right ventricle to the right atrium in the mammal.

5. What is the advantage of the atrioventricular fibers being slow-conducting?

6. How does the radius of a blood vessel affect the blood flow through it?

7. When you are very frightened, you turn quite pale. What is occurring in the cardiovascular system to cause this change? Why is such an adaptation useful?

8. If a live bacterium enters the body, what are the ways in which it can be destroyed?

28

Respiration

Table 28-1 Composition of dry air

Gas	% of Volume
Oxygen	21
Nitrogen	77
Argon	1
Carbon dioxide	0.03
Other gases*	0.97

* Includes hydrogen, neon, krypton, helium, ozone, and xenon.

Respiration has two meanings in biology. At the cellular level, "respiration" refers to the oxygen-requiring chemical reactions that take place in the mitochondria and are the chief source of energy for the cell. At the level of a whole organism, "respiration" refers to the process of taking in oxygen from the environment and returning carbon dioxide to it. The latter process—which is, of course, essential for the former—is the subject of this chapter.

The means by which organisms obtain energy in the absence of oxygen are not very efficient, as we saw in Chapter 7. Before oxygen came to be present in the atmosphere, the only forms of life were one-celled organisms; the only present-day forms that can carry on life processes without oxygen are a few types of bacteria and yeasts. As we noted earlier (page 110), the transition to an atmosphere containing free oxygen was one of the giant steps in evolution. The modern atmosphere is about 21 percent oxygen (Table 28-1). From this reservoir in the air organisms obtain—either directly or indirectly—the oxygen that supports their life processes.

Air under Pressure

At sea level, the air around us exerts a pressure on our skin of 1 atmosphere (about 15 pounds per square inch). This pressure is enough to support a column of water about 10 meters high or a column of mercury 760 millimeters high (Figure 28-1, page 392). Atmospheric pressure is generally measured in terms of mercury simply because mercury is relatively heavy—so the column will not be inconveniently tall.

The pressure of a particular gas in the air is proportional to its concentration. Thus, 21 percent of the total air pressure, or 159 mm Hg, results from the pressure of oxygen in the air. It is because of this pressure that oxygen dissolves in water and is available to the many aquatic organisms that obtain their oxygen from the water. Similarly, it is as a result of this pressure that oxygen enters the bodies of air-breathing organisms.

We are so accustomed to the pressure of the air around us that we are unaware of its presence or of its effects on us. However, if you visit a place—such as Mexico City—that is at a comparatively high altitude and therefore has a lower atmospheric pressure, you will feel lightheaded at first and will tire easily. Regular inhabitants of such places breathe more deeply,

28-1 *Atmospheric pressure is usually measured by means of a mercury barometer.* (a) *To make a simple mercury barometer, fill a long glass tube, open at one end only, with mercury. Closing the tube with your finger,* (b) *invert it into a dish of mercury. Remove your finger. The mercury level will drop until the pressure of its weight inside the tube is equal to the atmospheric pressure outside. At sea level, the height of the column will be about 760 millimeters (29.91 inches).*

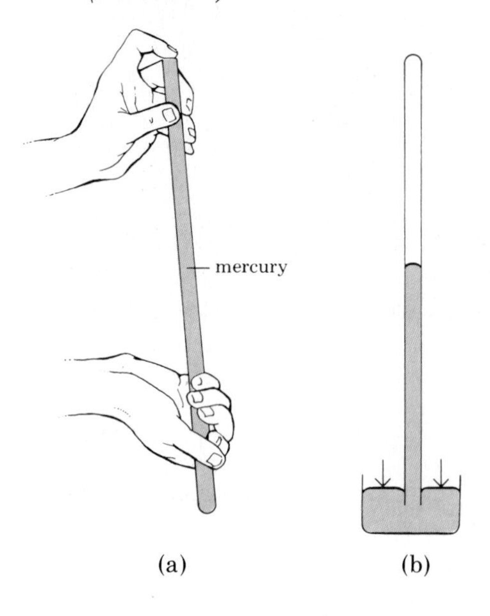

have enlarged hearts that circulate oxygen-carrying blood more rapidly, and have more red blood cells for oxygen transport.

The consequences of increased gas pressures are seen in deep-sea divers. Early in the history of deep-sea diving, it was found that when divers come up from the bottom too quickly, they get the "bends," which are always painful and sometimes fatal. The bends develop as a result of breathing compressed air. High pressures force more nitrogen from the air in the lungs into solution in the blood and tissues. If the body is rapidly decompressed, nitrogen bubbles out of the blood, like the carbon dioxide bubbles that appear in a bottle of soda when you first remove the top. The nitrogen bubbles lodge in the capillaries, stopping blood flow, or invade nerves or other tissues.

Evolution of Respiratory Systems

Oxygen enters and moves within cells by diffusion. Within the cell, it takes part in the oxidation of breakdown products of glucose and other carbon-containing compounds that serve as cellular energy sources. In this process, carbon dioxide is produced, which then diffuses out of the cell down the concentration gradient. This is true of any cell, whether an amoeba, a *Paramecium,* a liver cell, or a brain cell. However, substances can move effectively by diffusion only for very short distances (less than 1 millimeter). These limits pose no problem for very small animals, in which each cell is quite close to the surface, or for animals in which much of the body mass is not metabolically active—like the "jelly" of jellyfish. Many eggs and embryos also respire in this simple way, particularly in the early stages of development.

However, diffusion cannot possibly meet the needs of large organisms in which cells in the animal's interior may be many centimeters from the air or water serving as the oxygen source. As organisms increased in size in the course of evolution, there also evolved circulatory and respiratory systems (Figure 28-2) that transport large numbers of gas molecules by bulk flow. (Remember that diffusion is the result of the random movement of individual molecules; bulk flow is the overall movement of molecules in response to pressure or gravity.)

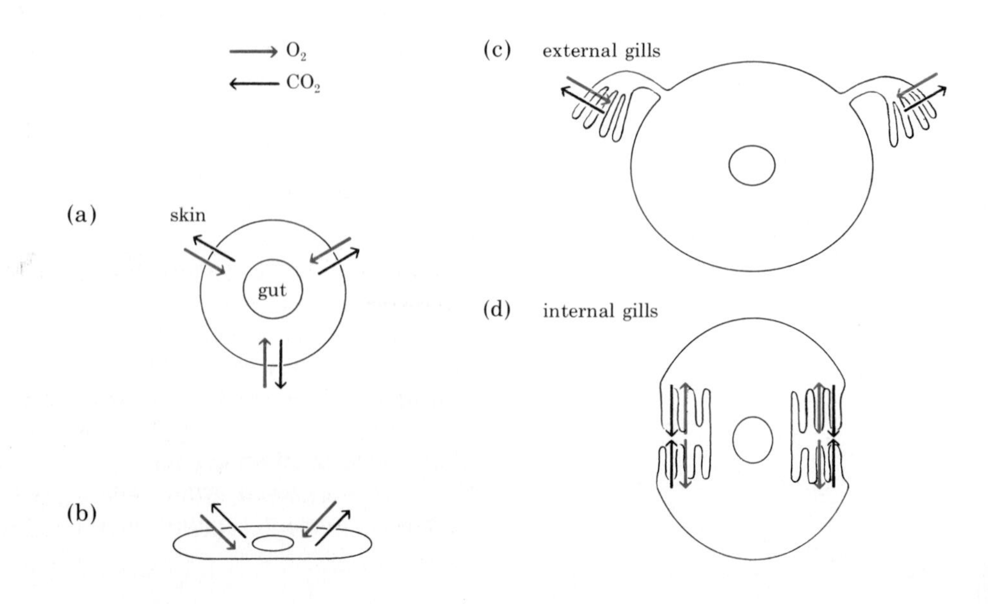

28-2 *Respiratory systems.* (a) *Gas exchange across the entire surface of the body is found in a wide range of small animals from protozoans to earthworms.* (b) *Gas exchange across the surface of a flattened body. Flattening increases the surface-to-volume ratio and also decreases the distance over which diffusion has to occur within the body. Flatworms are examples.*
(c) *External gills increase the surface area but are unprotected and therefore easily damaged. External gills are found in polychaete worms and some amphibians. Gas exchange usually takes place across the rest of the body surface as well as the gills.* (d) *With internal gills, a ventilation mechanism draws water over the highly vascularized gill surfaces, as in fish.*
(e) *Gas exchange at the terminal ends of fine tracheal tubes that branch through the body and penetrate all the tissues is characteristic of insects and some other arthropods.* (f) *Lungs are highly vascularized sacs into which air is drawn by a ventilation mechanism. Lungs are found in all air-breathing vertebrates.*

28–3 *Gills are essentially outpocketings of the epithelium that increase the surface area exposed to water. Often the gills are covered by an exoskeleton, as in crustaceans, or a flaplike gill cover, as in fish. In the axolotl, the amphibian shown here, the external nature of the gills is clearly evident. The source of their bright red color is blood flowing through dense capillary networks just one cell layer beneath the gill surface.*

An early stage in the evolution of gas-transport systems is exemplified by the earthworm (page 264). As is the case with most other kinds of worms, an earthworm has a network of capillaries just one cell layer beneath the surface of its body. Oxygen and carbon dioxide diffuse directly through the body surface into and out of the blood as it travels through these capillaries. The blood picks up oxygen by diffusion as it travels near the surface of the animal and releases oxygen by diffusion as it travels past the oxygen-poor cells in the interior of the earthworm's body. Conversely, blood picks up carbon dioxide from the cells and releases it by diffusion near the surface of the animal. Thus the gases move into and out of the earthworm by diffusion but are transported within the animal by bulk flow.

This system is particularly suitable for worms because their tube shape exposes a proportionately large surface area. Some worms can adjust their surface area in relation to oxygen supply. If you have an aquarium at home, you may be familiar with *Tubifex,* a type of worm frequently sold as fish food. When these worms are placed in water low in oxygen, such as a poorly aerated aquarium, they stretch out as much as 10 times their normal length and thus increase the surface area through which oxygen diffuses.

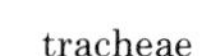
tracheae

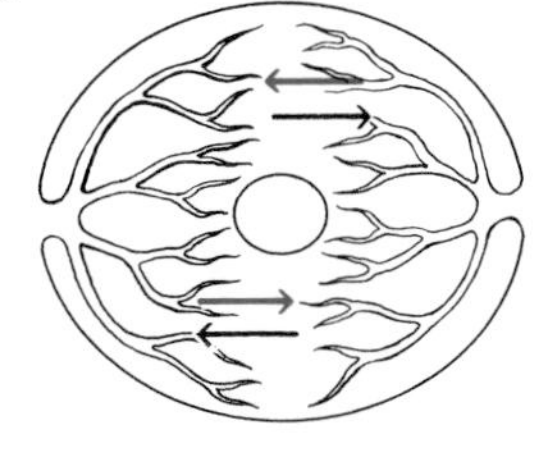

lungs

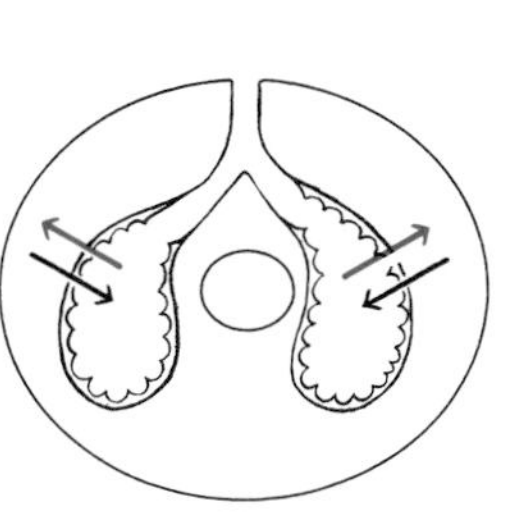

Evolution of the Gill

Gills and lungs are other ways of increasing the respiratory surface. Gills are outgrowths (Figure 28–3), whereas lungs are ingrowths, or cavities. The respiratory surface of the gill, like that of the earthworm, is a layer of cells, one cell thick, exposed to the environment on one side and to circulatory vessels on the other. The layers of gill tissue may be spread out flat, stacked, or convoluted in various ways. The gill of a clam, for instance, is shaped like a steam-heat radiator (which is also designed to provide a maximum surface-to-volume ratio).

The vertebrate gill is believed to have originated primarily as a feeding device. Primitive vertebrates breathed mostly through their skin. They filtered water into their mouths and out of what we now call their gill slits,

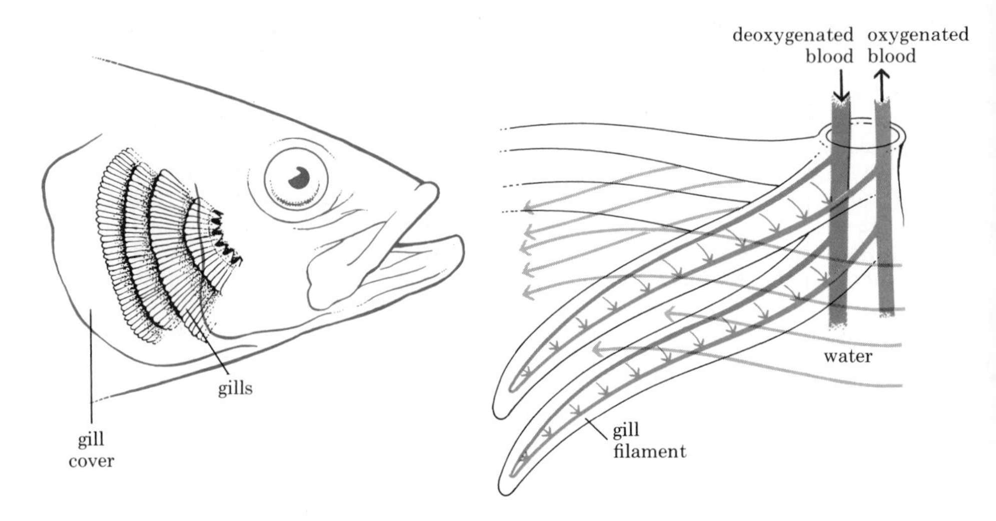

28-4 *Rates of diffusion are proportional not only to the surface areas exposed but also to differences in concentration of the diffusing molecules. The greater the difference in concentration of a molecule, the more rapid its diffusion. In the gill of the fish, the circulatory vessels are arranged so that the blood is pumped through them in a direction opposite to that of the oxygen-bearing water. This arrangement results in a far more complete transfer of oxygen to the blood than if the blood flowed in the same direction as the water.*

extracting bits of organic matter from the water as it went through. (*Branchiostoma*—page 273—which is believed to resemble closely the ancestral vertebrate, feeds in this way.)

In the course of time, numerous selection pressures, chiefly involved with predation, came into operation. As one consequence, there was a trend toward an increasingly thick skin, even one armored or covered with scales. Such a skin is not, of course, useful for respiratory purposes. At the same time, related forces were operating to produce animals that were larger and swifter and so more efficient at capturing prey and escaping predators. Such animals also had larger energy requirements—and consequently larger oxygen requirements. These problems were solved by the "capture" of the gill for a new purpose: respiratory exchange. The surface area and blood supply of the epithelium of the gill have slowly increased over the millennia. The modern gill is the result of this evolutionary process.

In most fish, the water (in which oxygen is dissolved) is pumped in at the mouth by movements of the bony gill cover and flows out across the gills (Figure 28-4). The fish can regulate the rate of water flow, and sometimes assist it, by opening and closing its mouth. Fast swimmers, such as mackerel, obtain enough oxygen to meet their needs by keeping their mouths open as they swim. As a result, water moves rapidly over the gills. Such fish have become so dependent on this method of respiration that if they are kept in an aquarium or any other space where their motion is limited, they will suffocate.

Evolution of the Lung

Lungs are internal cavities into which oxygen-containing air is taken. They have certain disadvantages as compared with gills; it is more efficient from the point of view of diffusion to have a continuous flow across the respiratory surface. Air, however, is a far better source of oxygen than is water; one-fifth of the modern atmosphere is free oxygen. Not only must more water be processed to obtain a given amount of oxygen, but water also weighs a great deal more. A fish spends up to 20 percent of its energy in the muscular work associated with respiration, whereas an air breather expends only 1 or 2 percent of its energy in respiration. Also, oxygen diffuses about 300,000 times more rapidly through air than through water, and so can be replenished much more quickly as it is used up by respiring organisms. All the higher vertebrates—the birds and the mammals—are air breathers, even those that live in the water.

Energy and Oxygen

Oxygen consumption is directly related to energy expenditure, and, in fact, energy requirements are usually calculated by measuring oxygen intake or the release of carbon dioxide. The energy expenditure at rest is known as basal metabolism. The basal metabolism of a warm-blooded (endothermic) animal is much higher than that of an ectotherm, with 10 to 30 percent of the energy budget being expended on maintaining a constant temperature.

Metabolic rates increase sharply with exercise. A person exercising consumes 15 to 20 times the amount of oxygen that he or she consumes sitting still. Oxygen consumption of a running animal increases linearly with its speed. If you are an aquatic animal, such as a fish or a porpoise, swimming is far less expensive, in terms of energy requirements, than is running. However, a human, whose body is basically unfit for moving through the water, uses five times as much energy—as measured by oxygen consumption—in swimming as in running.

In terms of energy consumption, flying is less expensive than walking and running. Many birds can fly more than 1,000 kilometers nonstop, whereas an animal of similar weight could not travel on the ground for that distance without stopping for food. (Migratory birds may also hitchhike on atmospheric currents.) However, flying, for a bird, is more expensive, in energy-consumption terms, than swimming is for a fish.

Oxygen is required for the energy-producing reactions that take place in the mitochondria. The oxygen consumption of a running animal—such as this male white-tailed deer—increases directly in proportion to its speed.

Lungs are not essential for air breathing. As we saw, earthworms are air breathers. The overwhelming advantage of lungs, however, is that the respiratory areas can be kept moist without a large loss of water by evaporation. Although lungs are largely a vertebrate "invention," they are found in some lower animals. Land-dwelling snails, for example, have independently evolved lungs that are remarkably similar to the lungs of some amphibians.

Some primitive fish had lungs as well as gills, although the lungs were not efficient enough to serve as more than accessory structures. These lungs were probably a special adaptation to life in fresh water, which, unlike ocean water, may stagnate (become depleted of oxygen) because of decay or algal bloom. A few species of lungfish still exist. By coming to the surface and gulping air into their lungs, they can live in water that does not have sufficient oxygen to support other fish life.

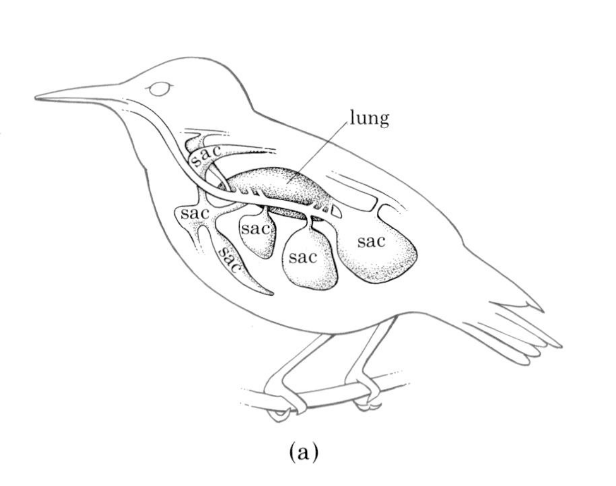

(a)

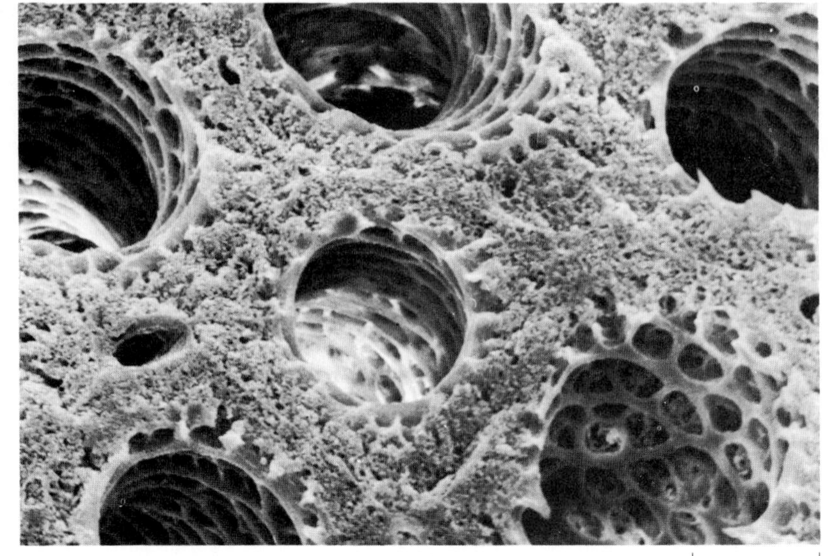

(b)

28-5 (a) *The lungs of birds are extraordinarily efficient. They are small and are expanded and compressed by movements of the body wall. Each lung has several air sacs attached to it, which empty and fill like balloons at each breath. No gas exchange takes place in the sacs. They appear rather to act as bellows, flushing the lung with fresh air at every breath. As a result, there is little residual "dead" air left in the lungs, as there is in mammals.*

(b) *Scanning electron micrograph of lung tissue from a domestic fowl. The tubes visible here are ventilated by air sacs, which serve to draw in fresh air. Gas exchange takes place in the broad meshwork of air capillaries and blood capillaries, which make up the sponge-like respiratory tissue seen here surrounding the tubes.*

Amphibians and reptiles have relatively simple lungs with small internal surface areas, although their lungs are far larger and more complex than those of the lungfish. The lungs of lungfish developed directly from the pharynx, the posterior portion of the mouth cavity, which leads to the digestive tract. In amphibians, reptiles, and other air-breathing vertebrates, we see the evolution of the windpipe, or *trachea*. It is guarded by a valve mechanism, the glottis, and nostrils, which make it possible for the animal to breathe with its mouth closed. Amphibians still rely largely on their skin for gas exchange, but reptiles breathe almost entirely through their lungs.

An important feature of all vertebrate lungs is that the exchange of air with the atmosphere takes place as a result of changes in lung volume. Such lungs are known as *ventilation lungs*. Frogs gulp air and force it into their lungs in a swallowing motion; then they open the glottis and let the air out again. In reptiles, birds, and mammals, air is drawn into the lungs as a consequence of changes in the size of the lung cavity, brought about by the activity of chest muscles.

Respiration in Large Animals: Some Principles

Let us stop for a moment to sum up this discussion. In large animals, both diffusion and bulk flow move oxygen molecules between the external environment and actively metabolizing tissues. This movement occurs in four stages:

1. Movement by bulk flow of the oxygen-containing external medium (air or water) to a thin membrane close to small blood vessels in lungs or gills.
2. Diffusion of the oxygen across this membrane into the blood.
3. Movement by bulk flow of the oxygen with the circulating blood to the tissues where it will be used.
4. Diffusion of the oxygen from the blood into the tissue cells.

Carbon dioxide, which is produced in the tissue cells, follows the reverse path as it is eliminated from the body.

The Human Respiratory System

The human respiratory system is shown in Figure 28–6a. Air enters the nasal passages, where it is warmed and cleaned, after which it passes through the trachea. The trachea is strengthened by rings of cartilage that prevent it from collapsing during inspiration (breathing in). The trachea leads into the *bronchi* (singular, bronchus), which subdivide into smaller and smaller passageways, the *bronchioles*. The bronchioles end in small air sacs known as *alveoli*, each only 1 or 2 millimeters in diameter (Figure 28–6b).

The alveoli have very thin walls that contain numerous capillaries. The barrier between an alveolus and its capillaries is only about 0.3 millimeter. Gases are exchanged between the air in the alveoli and the blood in the capillaries by diffusion. The other parts of the respiratory system serve mainly to transport air by bulk flow to and from the alveoli. A pair of human lungs has about 300 million alveoli, providing a respiratory surface of some 70 square meters—approximately 40 times the surface area of the entire human body.

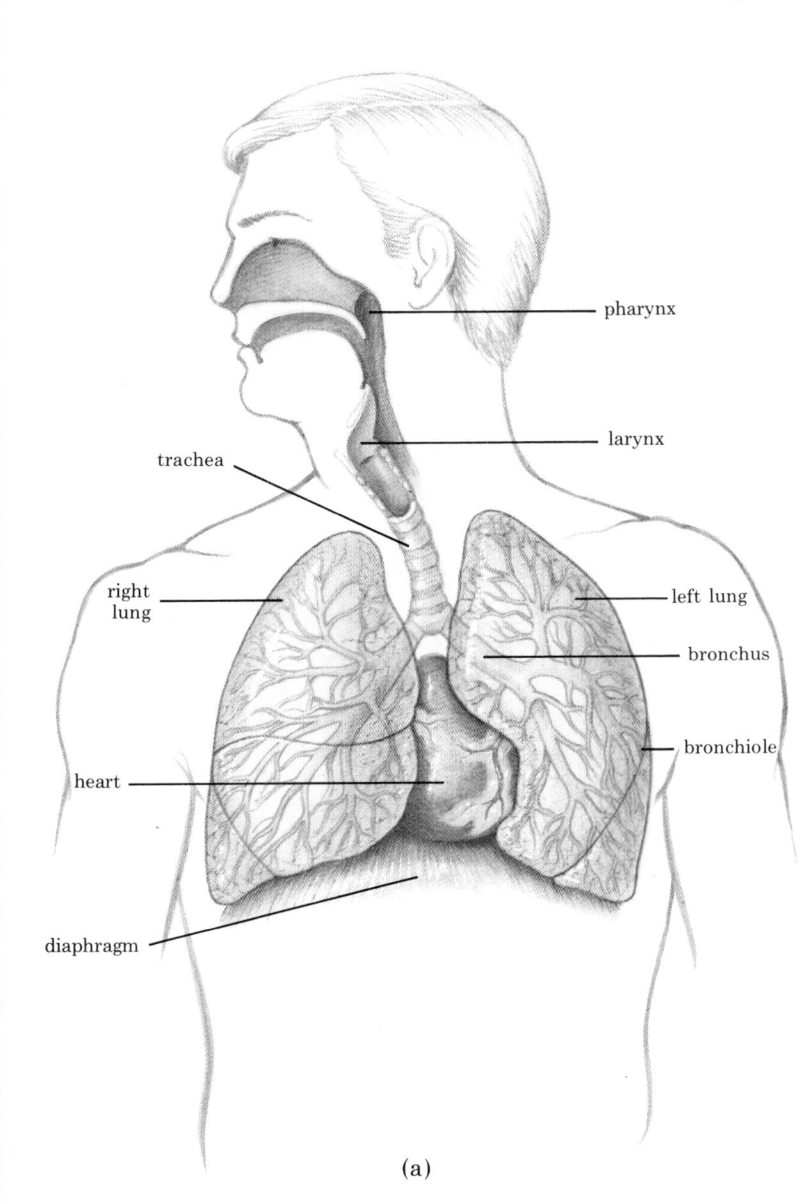

(a)

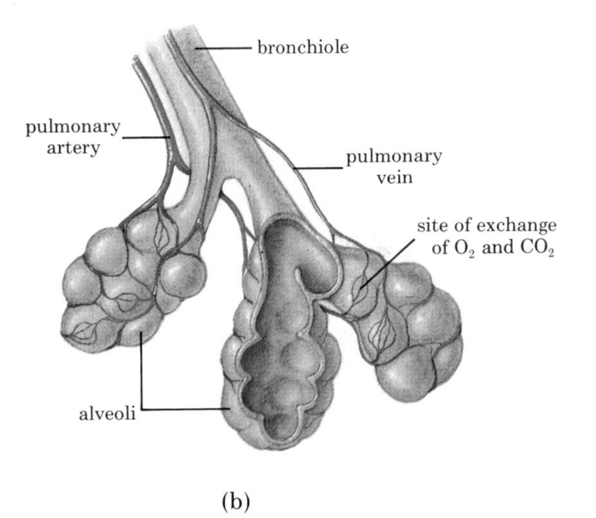

(b)

28–6 *The human respiratory system.* (a) *Air enters through the nose or mouth and passes into the pharynx and down the trachea, bronchi, and bronchioles to the alveoli* (b) *in the lungs. From each alveolus, of which there are some 300 million in a pair of lungs, oxygen and carbon dioxide diffuse into and out of the bloodstream, through the capillary walls.*

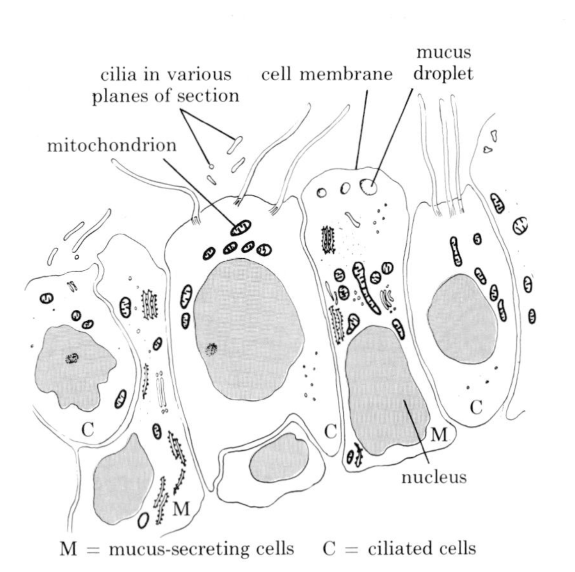

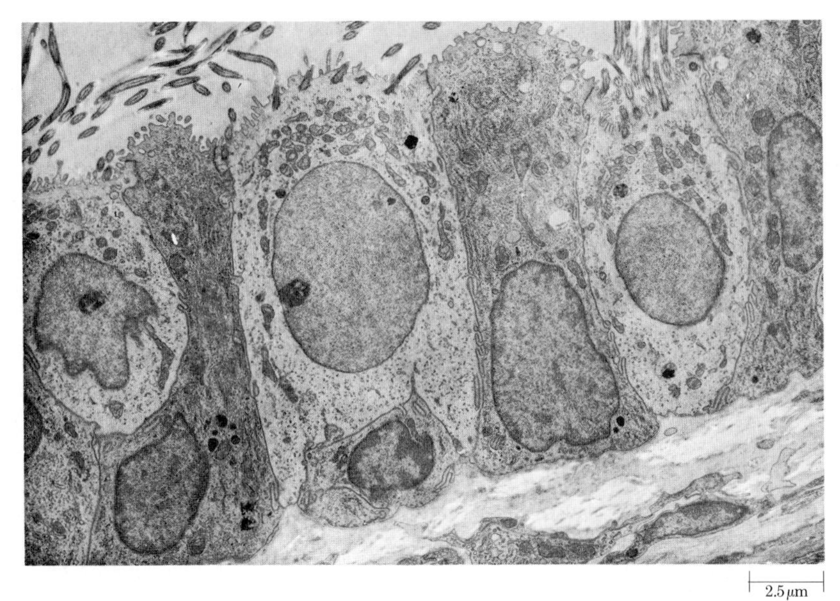

28-7 *Electron micrograph and diagram of tracheal epithelial cells. As you can see, the epithelium contains both ciliated cells (marked C in the diagram) and mucus-secreting cells (M), sometimes called goblet cells. The beating of the cilia distributes a protective coating of mucus and removes foreign particles from the trachea. In the micrograph, the white circles in the cytoplasm of the middle goblet cell are secretion droplets of mucus.*

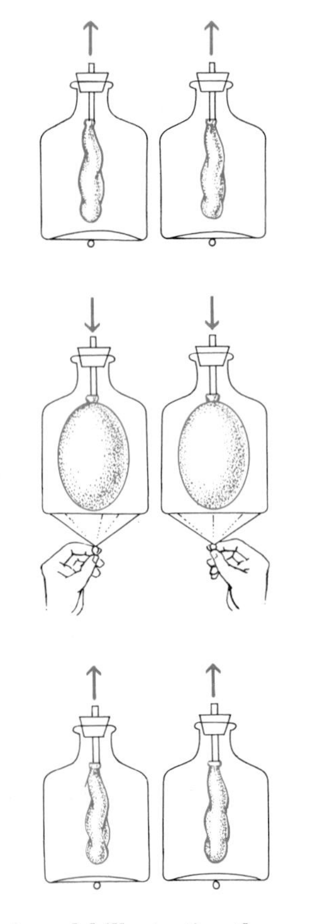

28-8 *A model illustrating the way air is taken into and expelled from the lungs.*

The trachea and bronchi are lined with epithelial cells, which include mucus-secreting cells and ciliated cells (Figure 28-7). The mucus coats the respiratory system and traps foreign particles that enter with the air. The cilia beat continuously, pushing mucus and the foreign particles embedded in it up toward the pharynx, from which it is generally swallowed. We are usually aware of this production of mucus only when it is increased above normal as a result of an irritation of the membranes.

Mechanics of Respiration

Air flows into or out of the lungs when the air pressure within the alveoli differs from the pressure of the external air (atmospheric pressure). When alveolar pressure is greater than atmospheric pressure, air flows out of the lungs, and expiration occurs. When alveolar pressure is less than atmospheric pressure, air flows into the lungs, and inspiration occurs. When the pressures are the same, as occurs momentarily between inspiration and expiration, there is no air flow.

The pressure in the lungs is varied as the result of changes in the volume of the thoracic cavity. These changes are brought about by the contraction and relaxation of the muscular diaphragm and of the intercostal ("between-the-ribs") muscles. We inhale by contracting the dome-shaped diaphragm, which flattens it and lengthens the thoracic cavity, and by contracting the intercostal muscles, pulling the rib cage up and out. These movements enlarge the thoracic cavity; the pressure within it falls, and air moves into the lungs. Air is forced out of the lungs as the muscles relax, reducing the volume of the thoracic cavity. Usually, only about 10 percent of the air in the lung cavity is exchanged at every breath, but as much as 80 percent can be exchanged by deliberate deep breathing.

Cancer of the Lung

Lung cancer is the most rapidly increasing form of cancer in the United States and the most common cause of death from cancer in men. About 117,000 men and women will develop lung cancer in the United States this year, and about 90 percent of them will die in less than three years. Most of these patients will be cigarette smokers.

The middle and lower lobes of a cancerous lung are shown in (a). The cancer is the solid grayish-white mass. Its rapid growth has replaced most of the normal lung tissue in the middle lobe. It has also probably begun to spread to other parts of the body. Notice the bronchial branch that leads into the cancer and is destroyed by it. The lung tissue remaining around the cancer is compressed, airless, and dark red. Away from the cancer, in the lower lobe, the lung is filled with air and is light pink.

The lung is normally protected by ciliated cells lining the trachea and bronchi. The cilia sweep out particles caught in the mucus secreted by the goblet cells. Cancer cells arising in the bronchus generally do not have cilia. The scanning electron micrographs show (b) the normal ciliated surface of the bronchus and (c) the surface of a bronchus with cancer.

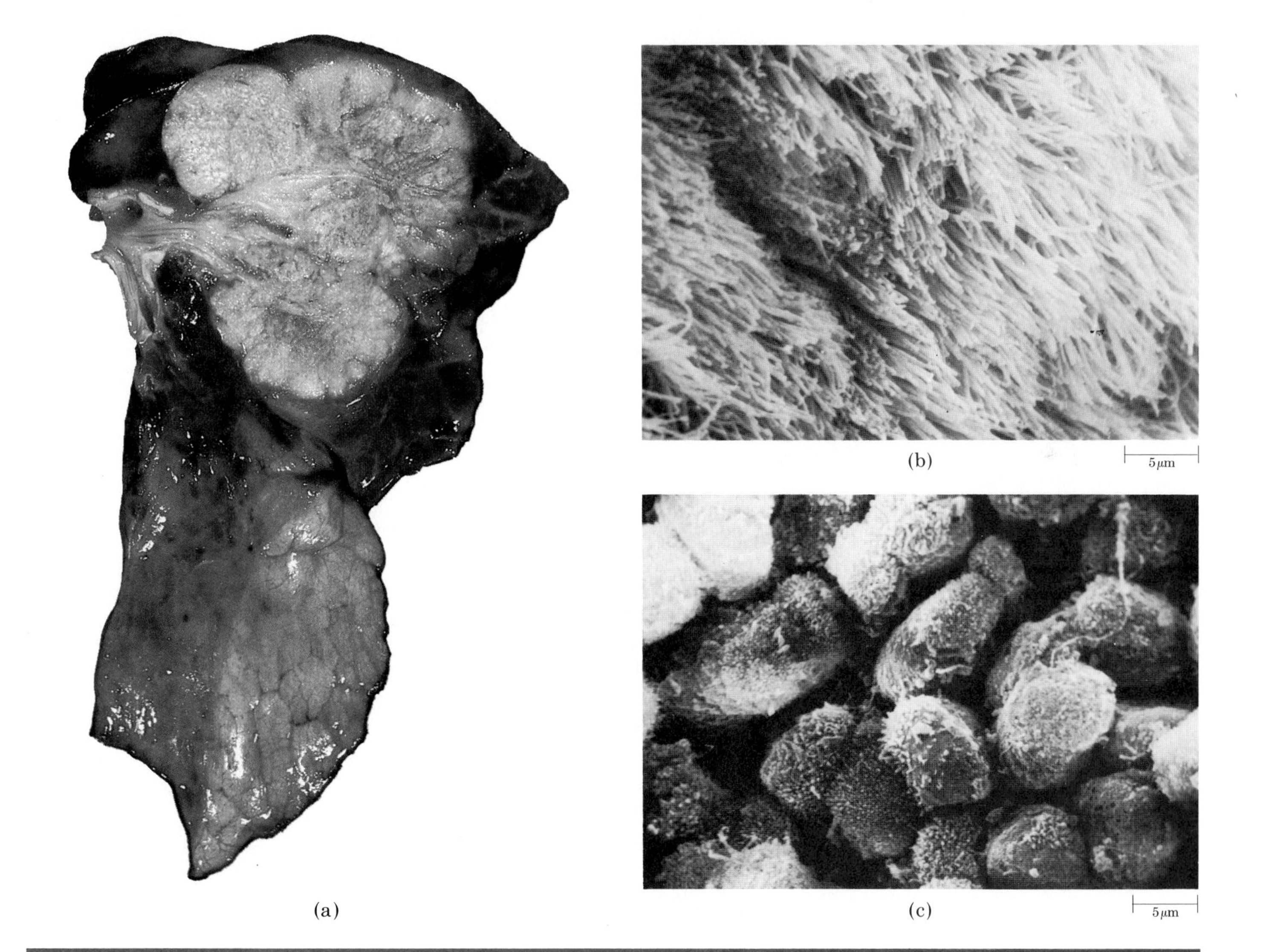

(a) (b) (c)

The Diving Reflex

Most mammals have about the same oxygen requirements as humans. Structurally, their respiratory organs are very similar. Yet marine mammals can dive to great depths and stay submerged for long periods of time. A sperm whale, for example, has been found at a depth of more than 1,000 meters and has been known to stay submerged for 75 minutes. A Weddell seal, a much smaller animal, can dive to 600 meters and stay submerged for 70 minutes, or can swim submerged for between 2,000 and 4,400 meters—1 to 2 miles. Are their lungs different from ours? Is their blood different? Do they have special oxygen reserves, like a submarine?

Studies in a variety of diving mammals have shown that none have lungs significantly larger, in proportion, than ours. In fact, seals (and perhaps other diving mammals) exhale before diving or early in the dive. Blood volume is increased, however. In humans, blood is about 7 percent of the body weight, whereas in diving marine mammals, it is 10 to 15 percent. The blood vessels are proportionately enlarged, and they appear to serve as a reservoir of oxygenated blood. Moreover, the proportion of red blood cells is higher. Still, physiologists calculate, these adaptations would not provide enough oxygen for long dives.

Studies of diving mammals have shown that a major survival factor is a group of automatic reactions known, collectively, as the diving reflex. During a dive, the heart rate slows and blood supply is reduced to tissues that are tolerant of oxygen deprivation, such as the digestive organs, skin, and muscles. Muscles obtain energy by anaerobic glycolysis, producing large amounts of lactic acid and building up an oxygen debt (see page 108). Most of the oxygen is shunted to the heart and brain, whose cells would begin to die after about four minutes without oxygen. (In a pregnant female, the fetus similarly is given high priority for the available oxygen.)

Recent studies by Martin Nemiroff of the University of Michigan Medical School indicate that the diving reflex is also present in human beings. In humans, its primary survival value is during birth, when the infant may be cut off from an oxygen supply during the later stages of labor.

The observations by Nemiroff came in the course of investigating 60 near-drownings. It has been generally assumed that a person who is submerged for four minutes or more will die or, at best, suffer irreversible brain damage. In the group of 60, 15 had been rescued after four minutes or more in the chilly waters of Michigan lakes. Of these 15, 11 survived without brain damage. One of these, a college student who crashed through the ice in an automobile accident, was under water for 38 minutes; he finished the semester with a 3.2 average. Another survivor resumed his career as a practicing physician. All the survivors had been submerged in cold (below 21°C) water, which slowed the metabolic demands of their cells. Some required assisted breathing for as long as 13 hours.

Moral: Even if a drowning victim is cold, blue, with no pulse, no heartbeat, and the eyes fixed in a glassy stare—as was the case with the student—if he or she has been in cold water, efforts at resuscitation should be started immediately and continued indefinitely. We are not seals, but owing to an ancient survival strategy, we have unexpected powers of underwater survival.

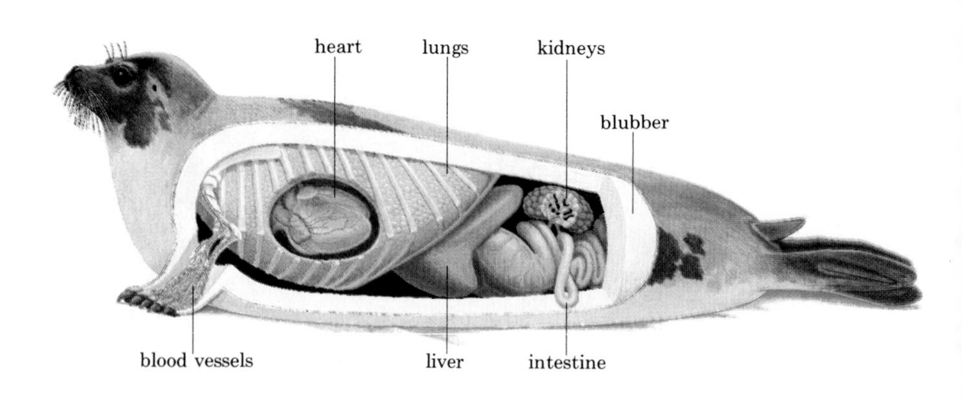

A seal exhales as it dives, reducing the likelihood of the "bends" (see page 392). It has $1\frac{1}{2}$ times as much blood as a land mammal of the same size and, when it dives, can reduce its heartbeat from 120 to 30 beats per minute. Other adaptations of the seal include an intestine more than twice as long as a human's, a kidney composed of lobules that can each perform all the functions of the whole kidney, special arrangements of blood vessels that allow regulation of body temperature, and a thick insulating layer of blubber.

Control of Respiration

The rate and depth of respiration are controlled by respiratory neurons in the medulla of the brain. These neurons receive and coordinate information about the carbon dioxide, hydrogen ion, and oxygen concentrations of the blood. Action potentials are relayed from these neurons to motor nerves that originate in the spinal cord and that control the muscles of the diaphragm and chest.

The system is so sensitive to even the smallest change in the chemical composition of the blood, particularly to the concentration of carbon dioxide, that only the slightest variations are permitted to occur. If the carbon dioxide concentration increases only slightly, breathing immediately becomes deeper and faster, permitting more carbon dioxide to leave the blood until the carbon dioxide level has returned to normal.

You can, of course, increase your breathing rate by voluntarily contracting and relaxing your diaphragm and chest muscles, but breathing is normally involuntary. It is impossible to commit suicide by deliberately holding your breath; as soon as you lose consciousness, the involuntary controls take over once more and breathing resumes. Deaths due to barbiturates and other drugs with a depressant activity are usually the result of a damping of the activity of the respiratory neurons. The effect can often be reversed if artificial respiration is administered quickly and for a long enough period of time.

Transport of Gases: The Role of Red Blood Cells

Oxygen is relatively insoluble in blood plasma—only about 0.3 milliliter (ml) of oxygen will dissolve in 100 ml of plasma at normal atmospheric pressure. The capacity of the blood to transport oxygen from the lungs to the body tissues is greatly increased by a special carrier molecule, hemoglobin. A hemoglobin molecule is made up of four subunits, each of which comprises a heme unit and a polypeptide chain (Figure 28–9). The heme unit consists of a complex nitrogen-containing ring structure with one atom of iron in the center. Each heme unit can combine with one molecule of

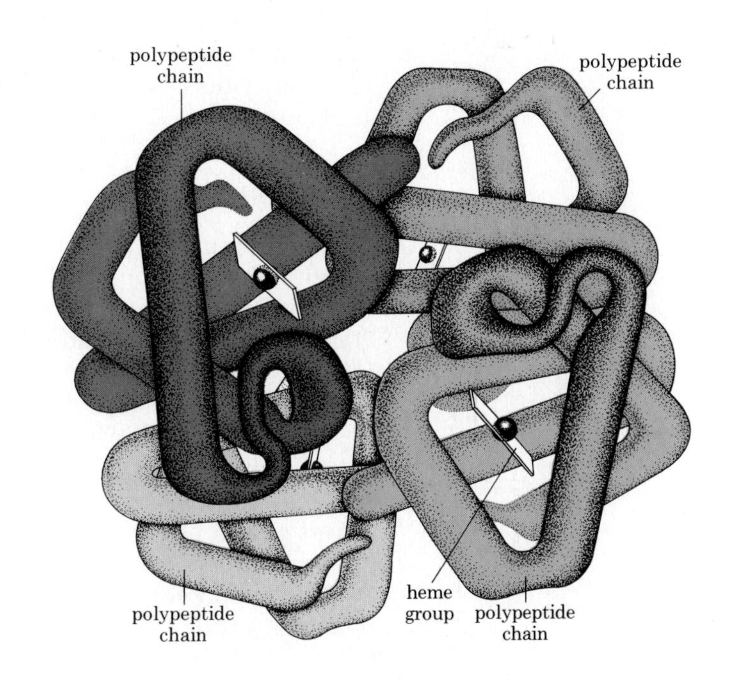

28–9 *The hemoglobin molecule consists of four heme groups with their polypeptide chains intertwined in a quaternary structure (page 48). Each molecule can hold up to four oxygen molecules. (Adapted with permission from R. E. Dickerson and I. Geis,* The Structure and Action of Proteins, *The Benjamin/Cummings Publishing Company, Menlo Park, Calif., 1969. Copyright 1969 by Dickerson and Geis.)*

oxygen; thus each hemoglobin molecule can carry four molecules of oxygen. Hemoglobin enables our bloodstreams to carry about 60 times as much oxygen as could be transported by an equal volume of plasma alone.

In many invertebrates, hemoglobin or other carrier molecules circulate freely in the bloodstream. In vertebrates, hemoglobin is transported in the red blood cells. As we saw in the last chapter, these cells are highly specialized for their transport function; a mature red blood cell carries some 265 million molecules of hemoglobin.

Hemoglobin is a red pigment that becomes a brighter red upon oxygenation. For this reason, blood flowing through arteries is a brighter red than blood flowing through veins. Whether oxygen combines with hemoglobin or is released from it depends on the oxygen concentration of the surrounding plasma. In the capillaries of the alveoli, where oxygen concentration is high, most of the hemoglobin is combined with oxygen. In the tissues, however, where the oxygen concentration is lower, oxygen is released from the hemoglobin molecules into the plasma and diffuses into the tissues. This system compensates automatically for the oxygen requirements of the tissues. Thus, because a highly active tissue uses more oxygen, the oxygen concentration in the tissue is lower and more oxygen is released from the hemoglobin.

Carbon dioxide is more soluble than oxygen, and some of it is simply dissolved in the blood. Most, however, is carried in combination with water as bicarbonate ion (HCO_3^-):

$$\underset{\text{carbon dioxide}}{CO_2} + \underset{\text{water}}{H_2O} \rightleftharpoons \underset{\text{bicarbonate ion}}{HCO_3^-} + \underset{\text{hydrogen ion}}{H^+}$$

As you can see by the arrows, this reaction can go in either direction. The direction it actually takes depends on the concentration of carbon dioxide. In the tissues, where carbon dioxide concentration is high, bicarbonate is formed. In the lungs, where carbon dioxide is low, bicarbonate dissociates to form carbon dioxide and water. Once it is released, the carbon dioxide diffuses from the plasma into the alveoli and flows out of the lung with the expired air.

The reaction of carbon dioxide and water is catalyzed by the enzyme carbonic anhydrase, which is present in red blood cells along with hemoglobin. Thus transport of carbon dioxide by the blood also depends on red blood cells.

SUMMARY

Organisms obtain most of their energy from oxidation of carbon-containing compounds. This process requires oxygen and releases carbon dioxide. Respiration is the means by which an organism obtains the oxygen required by its cells and rids itself of carbon dioxide.

Oxygen is available in both water and air. It moves into cells and body tissues by diffusion. However, efficient movement of oxygen by diffusion requires a relatively large surface area exposed to the source of oxygen and a short distance over which the oxygen has to diffuse. Selection pressures for increasingly efficient means of gas exchange led to the evolution in vertebrates of gills and lungs. Both gills and lungs present enormously increased surface areas for the exchange of gases. They also have a rich blood supply for transporting these gases to and from other parts of the animal's body.

Respiration in large animals involves both diffusion and bulk flow. Bulk flow brings air or water to the lungs or gills and circulates oxygen and carbon dioxide in the bloodstream. Gases are exchanged between the blood and the air in the lungs and between the blood and the tissues by diffusion.

In humans, air enters the lungs through the trachea, or windpipe, and goes from there into a network of increasingly smaller tubules, the bronchi and bronchioles, which terminate in small air sacs, the alveoli. Gas exchange actually takes place across the alveolar walls. Air moves into and out of the lungs as a result of changes in the pressure within the thoracic cavity, which, in turn, result from changes in the size of the thoracic cavity. The muscles that bring about respiration are under the control of nerves that originate in the spinal cord. These nerves, in turn, are under the control of respiratory neurons in the medulla of the brain that respond to signals caused by slight changes in the carbon dioxide, hydrogen ion, and oxygen concentrations of the blood.

In vertebrates, oxygen is carried by hemoglobin molecules, each of which can bind four molecules of oxygen. The hemoglobin is packed within red blood cells. Carbon dioxide is transported in the blood plasma largely in the form of bicarbonate ions.

QUESTIONS

1. Distinguish among the following: gills/lungs; bronchi/bronchioles/alveoli.

2. Sketch the human respiratory system. When you have finished, compare your drawing to Figure 28–6.

3. Trace the path of an oxygen atom from the air into a body cell.

4. What are the advantages and disadvantages of obtaining oxygen from air rather than from water? You may be able to think of several of each besides those mentioned in the text.

5. Carbon monoxide (CO), which is extremely poisonous, combines with hemoglobin more readily than does oxygen. The resulting compound, which is a brighter red than normal hemoglobin, can no longer combine with oxygen. From these facts, suggest how you might recognize and give assistance to a victim of carbon monoxide inhalation.

6. One of the results of long-term smoking is the loss of bronchial cilia. What effects would you expect this to have on normal lung function?

29

Digestion

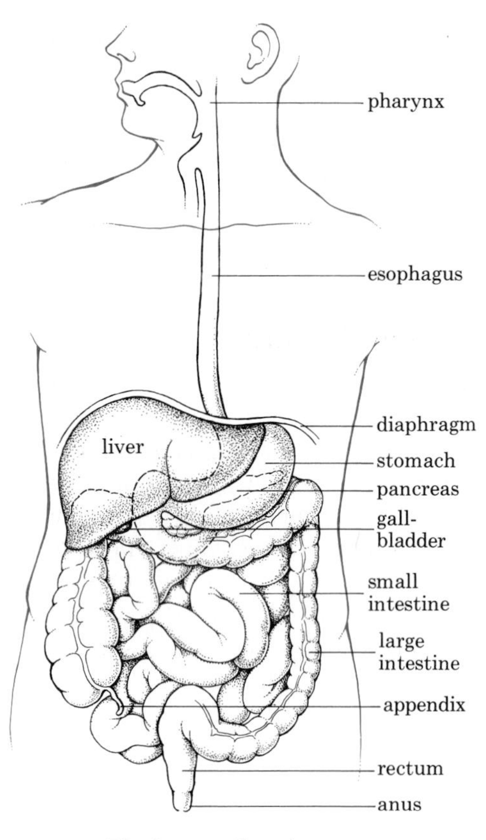

29-1 *The human digestive tract.*

Digestion is the process by which heterotrophs break down the tissues of other organisms to atoms and molecules that can be used by their own cells. Such atoms and molecules include monosaccharides (such as glucose), amino acids, fatty acids, and iron and other inorganic substances. Once within the cells, most of the molecules can be processed for entry into the Krebs cycle, where they are oxidized and their energy is eventually packaged in the high-energy bonds of ATP. Or, after suitable alterations, they can be synthesized into proteins and other molecules required for cellular structures and activities. Digestion thus provides both the building-block molecules needed by the individual cells and the cellular fuel for their various activities.

Digestive Tract in Vertebrates

In higher animals, as we saw in Chapter 19, digestion takes place in a long tube, running from mouth to anus, known as the digestive tract. It has different areas that are specialized for particular stages of the digestive process (Figure 29-1). Food is moved along this tube by successive waves of muscular contractions known as *peristalsis*.

The Mouth: Primary Processing

The mechanical breakdown of food begins in the mouth. Many vertebrates have teeth especially adapted for tearing or grinding. Some, such as birds and turtles, have horny beaks or bills. The tongue, also a vertebrate development, serves largely to move and manipulate food in mammals. However, some vertebrates, such as hagfish and lampreys, have tongues equipped with horny "teeth." The sticky tongues of frogs and toads, which are attached at the front rather than the back of the mouth, can flip out to catch insects. Mammalian tongues carry taste buds (Figure 29-2), which are stimulated by certain chemicals. In humans, the tongue has developed a secondary function of forming sounds for communication.

While food is being chewed in the mouth, it is moistened by saliva, a watery, slightly alkaline secretion produced by the salivary glands (Figure 29-3). The saliva, which also contains mucus, lubricates the food so that it

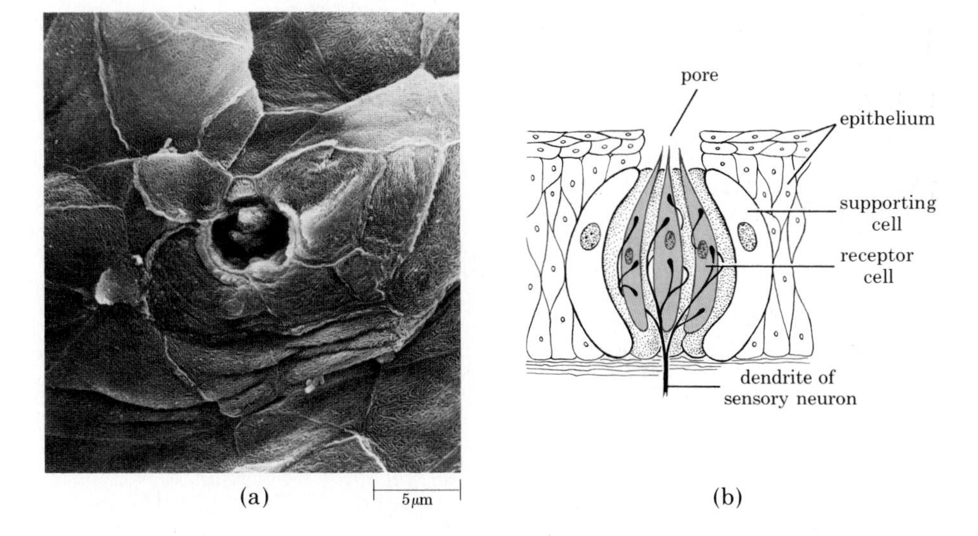

29–2 (a) *The outer pore of a taste bud. The receptor cells of the taste bud are just visible within the pore. This scanning electron micrograph shows the surface of the human tongue. Other taste buds are found on the roof of the mouth, the pharynx, and the larynx (voice box).* (b) *Longitudinal section of a taste receptor.*

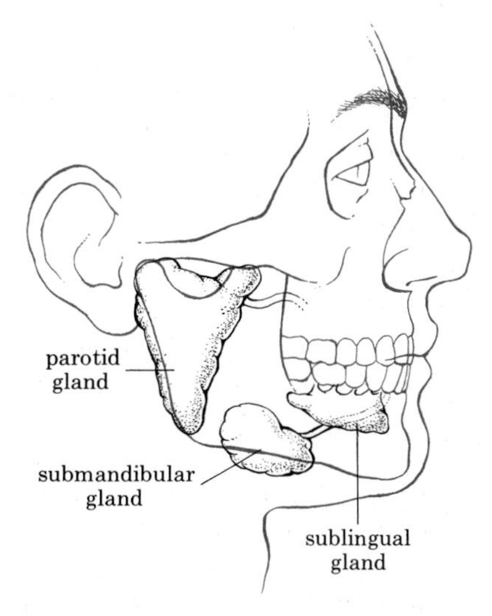

29–3 *The bulk of the saliva is produced by three pairs of salivary glands. Additional amounts are supplied by minute glands in the mucous membrane lining the mouth. The parotid glands are the sites of infection by the mumps virus.*

can be swallowed easily. In humans, saliva also contains a digestive enzyme, amylase, which begins the breakdown of starches. Carnivores, such as dogs, which characteristically tear and gulp their food, have no digestive enzymes in their saliva.

The secretion of saliva is controlled by the autonomic nervous system. It can be initiated by reflexes originating in taste buds and in the walls of the mouth, and also by the mere smell or anticipation of food. (Think hard, for a moment, about eating a lemon.)

The Pharynx and Esophagus: Swallowing

From the mouth, food is propelled backward toward the pharynx, a passageway at the back of the mouth that connects with the trachea and the esophagus. Swallowing is the passing of food to the esophagus, a muscular tube about 25 centimeters long, and through the esophagus to the stomach (Figure 29–4). Swallowing begins as a voluntary action and, once underway in humans, continues involuntarily. In humans, the upper part of the esophagus is striated muscle, but the lower part is smooth muscle. Dogs, cats, and other animals that gulp their food have striated muscle along the whole length of the esophagus.

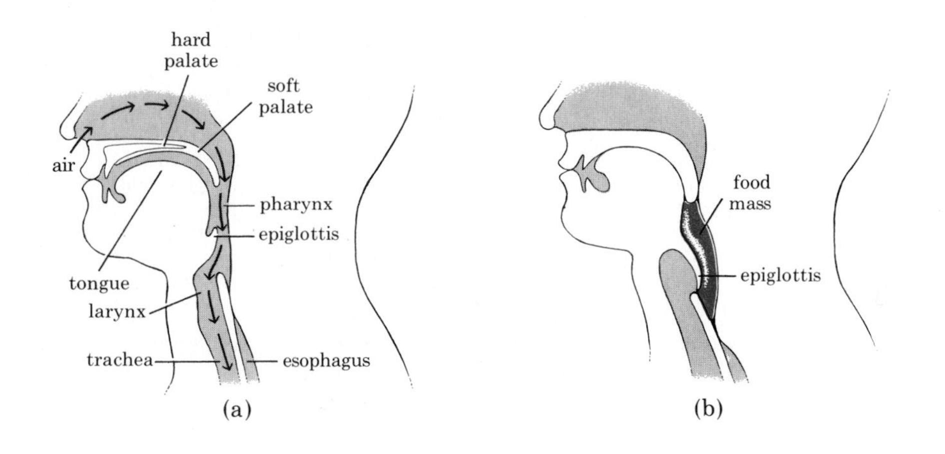

29–4 *Separation of the digestive and respiratory systems in mammals makes it possible to keep food out of the lungs. The pharynx, the common passageway of the two systems, is at the back of the mouth and connects with the trachea and the esophagus.* (a) *Face and neck, showing parts of the respiratory and digestive systems.* (b) *Swallowing. As the food mass descends, the epiglottis tips downward, and the food mass passes into the esophagus.*

The Heimlich Maneuver

Food strangulation claims the lives of almost 3,000 persons a year in this country alone, more than accidents involving firearms or airplanes. It occurs when a food mass enters the trachea rather than the esophagus (Figure 29–4). If the food becomes lodged, the victim cannot speak or breathe and, if the airway is completely blocked, will die in four or five minutes. (The fact that the victim cannot speak helps onlookers to distinguish it from a heart attack; although the symptoms are similar, heart attack sufferers can talk.)

The food can nearly always be dislodged by the Heimlich maneuver, a procedure so simple that it has been carried out successfully in at least two instances by eight-year-olds. There are three steps: (1) Stand behind the victim and wrap your arms around his or her waist. (2) Make a fist with one hand, grasp it with your other hand, and then place the fist against the victim's abdomen, slightly above the navel and below the rib cage. (3) Press your fist into the victim's abdomen with a quick upward thrust. The sudden elevation of the diaphragm compresses the lungs and forces air up the trachea, pushing the food out. Repeat several times if necessary.

If the victim is sitting, the procedure can be carried out in the same way. If the victim is lying on his back, face him, and kneel astride the hips. Put the heel of one hand on the abdomen above the navel and below the rib cage, put the other hand on top of the first hand, and press with a quick upward thrust. You can even do it on yourself: Place your hands at your waist, make a fist, and press quickly upward.

The victim should be seen by a physician immediately after the emergency treatment because it is possible to break a rib or cause other internal injuries, especially if the movements are performed incorrectly. But, considering the alternative, it is well worth the risk.

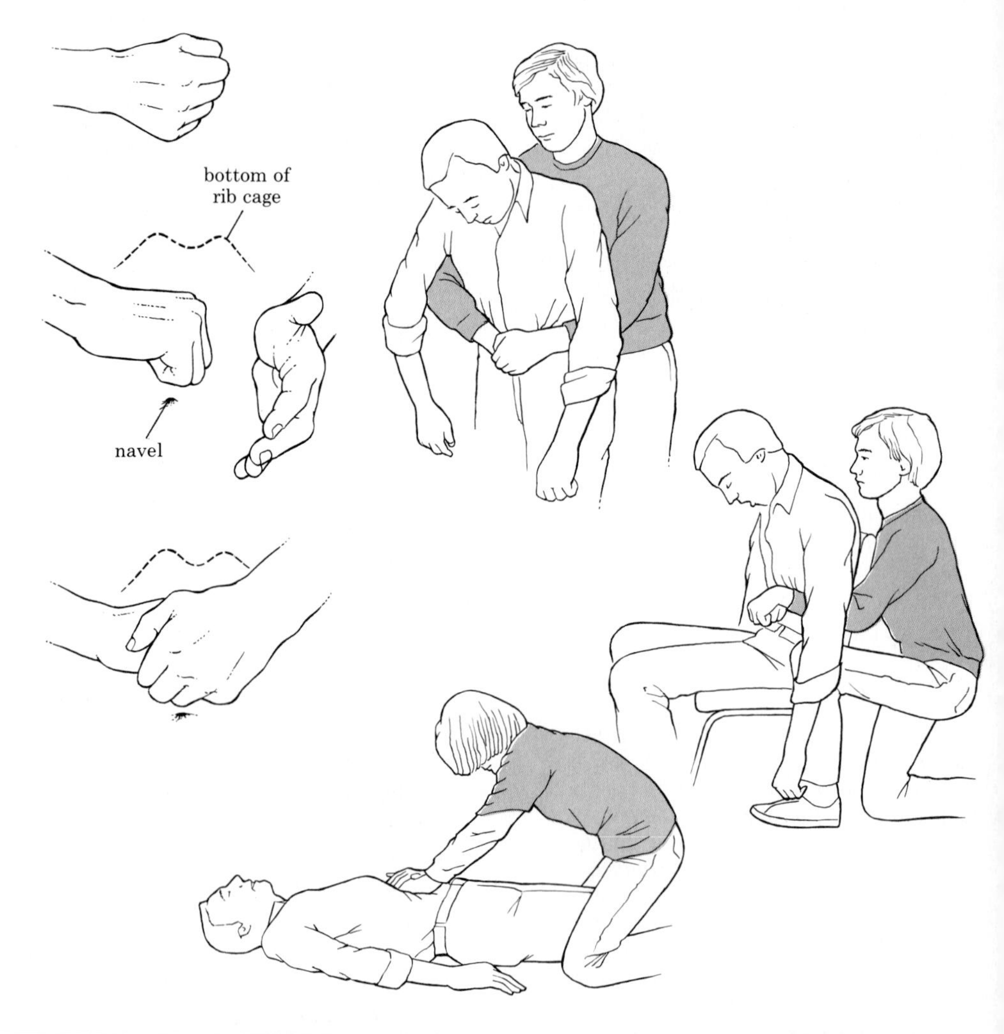

The maneuver can be carried out with the victim in a standing, sitting, or lying position. If the person is larger than you are, it is preferable to have him sitting or lying down. Note how the fist is made (you might want to practice this) and the position of the hands. These are both very important.

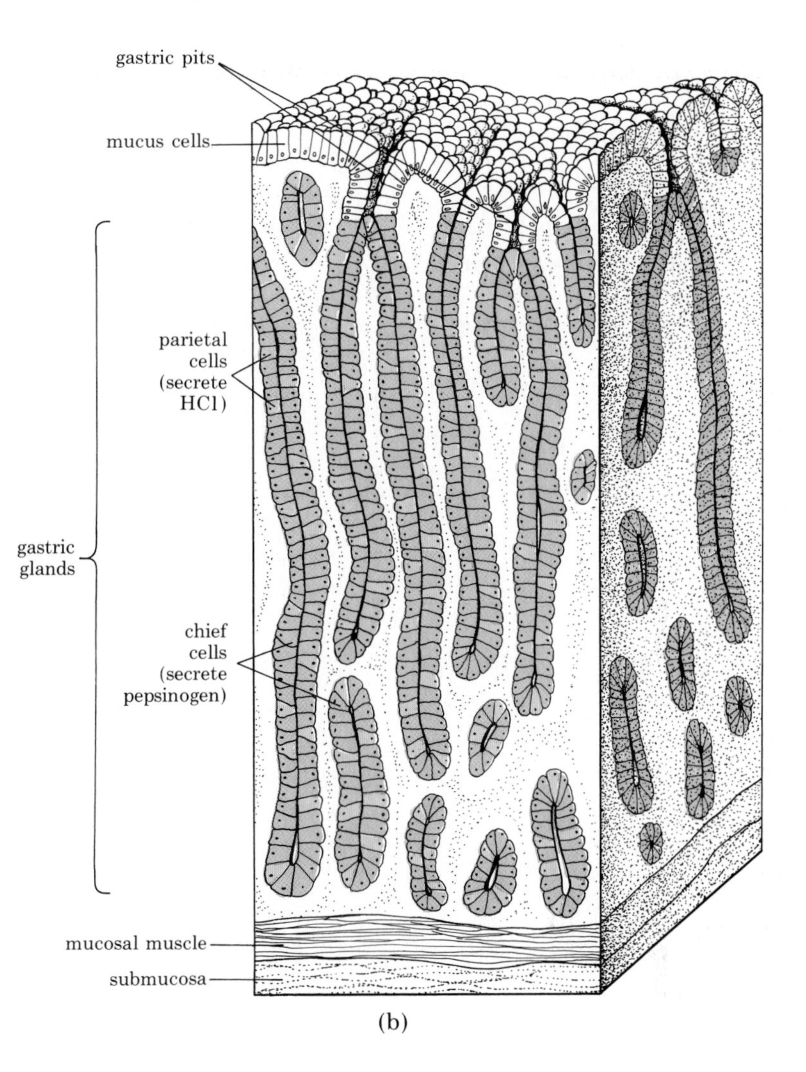

29–5 (a) *Surface of the stomach, as shown in a scanning electron micrograph. The numerous indentations are gastric pits.* (b) *A cross section of stomach mucosa. The parietal cells secrete hydrochloric acid and the chief cells produce pepsinogen. Mucus, secreted by other cells of the epithelium, coats the surface of the stomach and lines the gastric pits, protecting the stomach surface from digestion.*

The Stomach: Secondary Processing

The stomach is a collapsible, elastic bag, which, unless it is fully distended, lies in folds. Distended, it holds 2 to 4 liters of food. The walls of the stomach are lined by a layer of epithelial cells, which constitute the gastric mucosa (Figure 29–5). These gastric cells secrete mucus, pepsinogen, and hydrochloric acid (HCl), collectively called gastric juice.

As a consequence of the HCl secretion, the pH of gastric juice is normally between 1.5 and 2.5, far more acidic than any other body fluid. The burning sensation you feel if you vomit is caused by the acidity of gastric juice acting on unprotected membranes. The HCl, which kills most bacteria and other living cells in the ingested food, loosens the tough, fibrous components of tissues and erodes the cementing substances between cells. HCl also initiates the conversion of pepsinogen to its active form, pepsin. Pepsin, an enzyme that breaks proteins down into peptides, is active only at the low pH of the stomach.

The stomach is under both neural and hormonal control. Anticipation of food and the presence of food in the mouth stimulate churning movements of the stomach and the production of gastric juice. When protein-containing food reaches the stomach, its presence causes the release of a hormone, gastrin, from gastric cells into the bloodstream. This hormone acts on the cells of the stomach to increase their secretion of gastric juice.

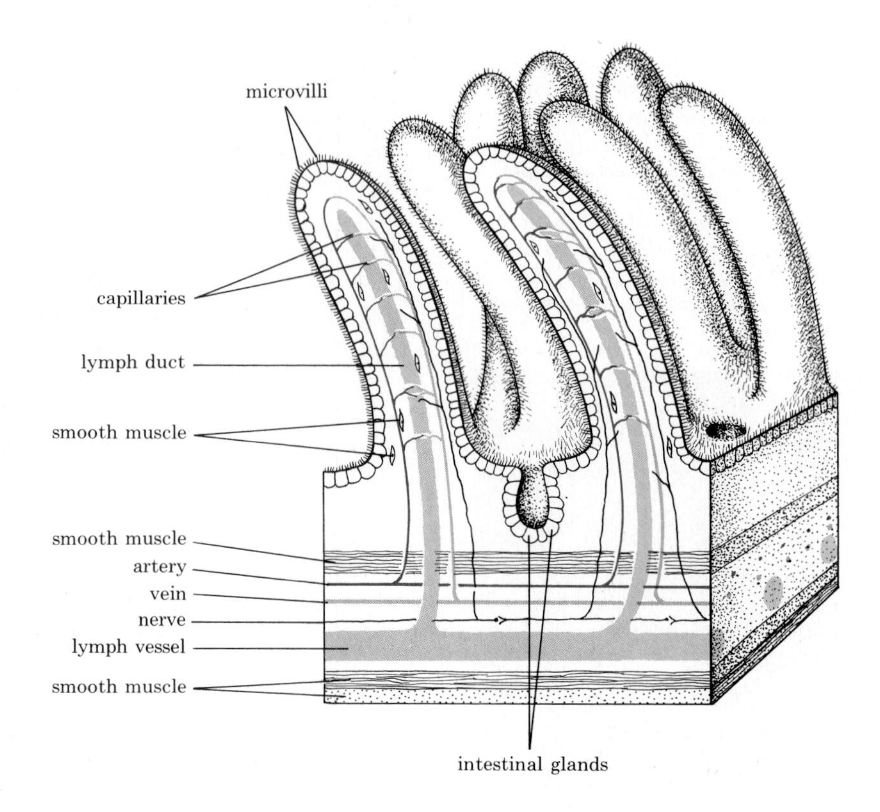

29-6 *Longitudinal section of villi of small intestine. Food molecules are absorbed through the walls of the villi and, with the exception of fat molecules, enter the bloodstream by means of the capillaries. Fats, hydrolyzed to fatty acids and glycerol and then resynthesized, are taken up into the lymphatic system. The villi can move independently of one another; their motion increases after a meal.*

In the stomach, food is converted into a semiliquid mass. It is gradually moved by peristalsis through the pyloric sphincter, a ring of muscle separating the stomach and small intestine. The stomach is usually empty four hours after a meal.

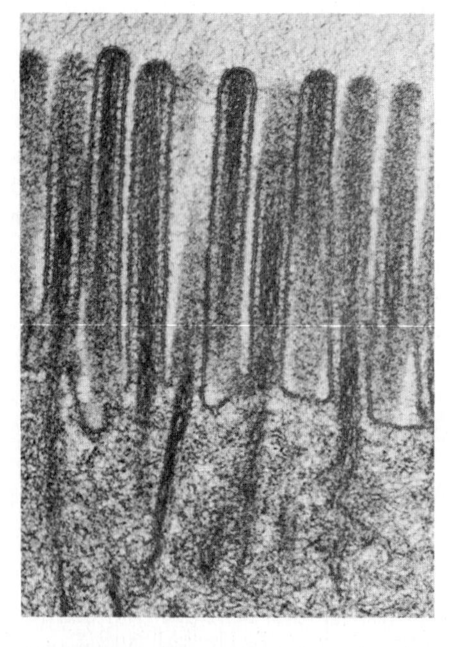

29-7 *Microvilli on the surface of an intestinal epithelial cell. These cellular extensions greatly increase the absorptive surface of the intestine.*

The Small Intestine: Digestion and Absorption

In the small intestine, the breakdown of food molecules begun in the mouth and stomach is completed. The food molecules are then absorbed from the digestive tract into the circulatory system of the body, from which they are delivered to the individual cells.

Anatomically, the lining of the small intestine is characterized by numerous fingerlike projections, *villi* (Figure 29-6), and tiny cytoplasmic projections, *microvilli*, on the surface of individual epithelial cells (Figure 29-7). Both of these structural features increase the surface area of the small intestine. The small intestine is 3 to 4 meters (10 to 13 feet) long in the adult; its total surface area is about 180 square meters, the area of a singles tennis court. The upper 20 centimeters of the small intestine, known as the *duodenum*, is the most active in the digestive process; the rest is principally concerned with absorption of food molecules.

A number of substances are involved in the digestive processes of the small intestine. Mucus is secreted by the epithelial cells of the intestinal lining. Digestive enzymes are produced both by the intestinal cells and by the pancreas. (The pancreatic enzymes enter the duodenum through the pancreatic duct.) In addition to enzymes, the small intestine receives an alkaline fluid from the pancreas, which neutralizes the stomach acid, and bile, which is produced in the liver and stored in the gallbladder. Bile contains a mixture of salts that, like laundry detergents, emulsify fats, breaking them apart into droplets. In this form they can be attacked by enzymes.

In the small intestine, amylases continue the breakdown of starch begun in the mouth, while lipases hydrolyze fats into glycerol and fatty acids. Three types of enzymes break down proteins. One group breaks apart the long protein chains. A second type of enzyme acts only on the end of a chain, splitting off dipeptides (pairs of amino acids), with some enzymes working on the amino end and some on the carboxyl end. A third group of enzymes comes into action to break the remaining dipeptides into single amino acids. The amino acids and dipeptides are absorbed through the epithelial cells of the intestine and enter the bloodstream.

The digestive activities of the small intestine are coordinated and regulated by hormones (Table 29-1). In the presence of acidic food from the stomach, the duodenum releases secretin, a hormone that stimulates the pancreas and liver to secrete alkaline fluids. Fats and amino acids in the food stimulate production of another hormone, cholecystokinin, which triggers the release of enzymes from the pancreas and the emptying of the gallbladder. At least eight other substances are suspected of being gastrointestinal hormones. A complex interplay of stimuli and checks and balances serves to activate and inactivate digestive enzymes and to adjust the chemical environment. In addition to hormonal influences, the intestinal tract is also regulated by the autonomic nervous system.

The Large Intestine: Resorption of Water and Elimination

The chief function of the large intestine is the resorption of water, sodium, and other minerals. In the course of digestion, large amounts of water—some 2 or 3 liters—enter the stomach and small intestine by osmosis from the body fluids or as secretions of the glands lining the digestive tract. When the resorption process is disrupted, as in diarrhea, severe dehydration can result. Infants dying from infectious diarrhea, still the chief cause of infant death in many countries, die principally as a result of water loss.

The large intestine harbors a considerable population of bacteria (including the familiar *Escherichia coli*), which break down additional food substances. Living on these food substances, largely materials we lack the

Table 29-1 Major gastrointestinal hormones

Hormone	Source	Stimulus for Production	Action
Gastrin	Stomach	Protein-containing food in stomach, also parasympathetic nerves to stomach	Stimulates secretion of gastric juice
Secretin	Duodenum	HCl in duodenum	Stimulates secretion of alkaline pancreatic fluids and bile
Cholecystokinin	Duodenum	Fats and amino acids in duodenum	Stimulates release of pancreatic enzymes and release of bile from gallbladder

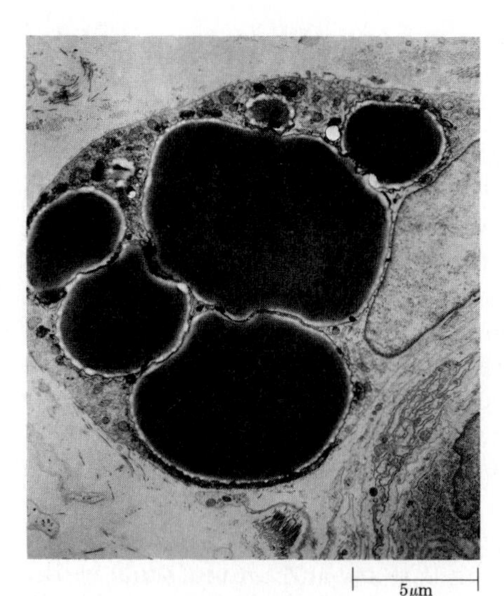

29-8 *A fat cell from the skin of a newborn rat. The nucleus is to the right. The large stored droplets of fat that nearly fill the cell have been stained for electron microscopy. When excess calories are taken in in the diet, fat accumulates in these specialized cells, and when caloric intake is less than sufficient, fat is mobilized, broken down to glycerol and fatty acids, and released into the bloodstream.*

enzymes to digest, they synthesize amino acids and vitamins, which are absorbed into the bloodstream. These bacteria are the chief source of vitamin K.

The bulk of the fecal matter consists of bacteria (mostly dead cells) and cellulose fibers, along with other indigestible substances. It is stored briefly in the rectum and eliminated through the anus as feces.

Regulation of Blood Glucose

As we noted at the beginning of this chapter, a major function of digestion is to provide an energy source for each of the individual cells of which the body is composed. Although vertebrates rarely eat 24 hours a day, their blood sugar—the cellular energy supply—remains extraordinarily constant. The liver plays a central role in this critical process. Glucose and other monosaccharides are absorbed into the blood from the intestinal tract and are passed directly to the liver by way of the hepatic portal vein. Some of the glucose is taken up by the liver cells, which convert it to glycogen (the liver stores enough glycogen to satisfy the body's needs for about four hours) and to fat. The fat is stored in fat cells (Figure 29-8), which also form fat from glucose. The liver similarly breaks down amino acids and converts them to glucose (excess amino acids are not stored). The nitrogen from the amino acids is excreted in the form of urea, and the glucose is stored as glycogen.

Whether the liver takes up or releases glucose and the amount it takes up or releases are determined primarily by the concentration of glucose in the blood. The concentration of glucose is, in turn, regulated by a number of hormones and is influenced by the autonomic nervous system (see Figure 26-7, page 373).

Some Nutritional Requirements

Because of the liver's activities in converting various types of food molecules into glucose and because most tissues can use fatty acids as an alternate fuel, the energy requirements of the body can be met by carbohydrates, proteins, or fats—the three principal types of food molecules. Energy requirements are ordinarily met by a combination of the three. Carbohydrates and proteins supply about the same number of calories per gram, and fats about twice as much as either of them.

In addition to calories, the cells of the body need the 20 different kinds of amino acids required for assembling proteins. Vertebrates are not able to synthesize all 20 amino acids. Humans can synthesize 12, either from a simple carbon skeleton or from another amino acid. The other eight, which must be obtained in the diet, are known as essential amino acids.* Plants are the ultimate source of these essential amino acids, but it is difficult (although by no means impossible) to obtain sufficient quantities of them by eating a completely vegetable diet, largely because plant proteins are relatively deficient in lysine and tryptophan (see essay, page 46).

* The essential amino acids are isoleucine, leucine, lysine, methionine, phenylalanine, threonine, tryptophan, and valine.

Two Troublemakers: The Appendix and the Gallbladder

Two digestive tract structures are notorious troublemakers. The most familiar is the appendix. This worm-shaped blind pouch, about 8 centimeters in length, is attached to the large intestine just below the junction of the small and large intestines (see Figure 29-1). It has no known function in humans.

The appendix is thought to be an evolutionary remnant from herbivorous ancestors. As you know, plant tissues contain large amounts of cellulose, a tough polysaccharide for which terrestrial animals have no digestive enzymes. We are completely unable to digest cellulose, and it passes out of our bodies in the feces. However, the digestive tracts of many herbivorous animals contain large populations of symbiotic bacteria that are able to ferment cellulose anaerobically, releasing nutrients for themselves and the host animal. This fermentation takes place in different areas of the digestive tract in different species. In the ruminants—a group that includes cattle, sheep, goats, camels, and giraffes—it occurs in the rumen, a large pouch formed from the esophagus. In the horse family and in rabbits and other lagomorphs, bacterial fermentation occurs in the cecum, a blind sac at the beginning of the large intestine. It is from this area of the human intestine that the appendix extends.

In the course of human evolution, the appendix—and whatever function it once performed—became superfluous. The structure remains, however, and bacteria and undigested food can enter it. If this material accumulates and hardens, the appendix may become irritated, inflamed, and then infected. If it ruptures, its contents are spilled into the abdominal cavity. Serious, and even fatal, infection can result. For this reason, the appendix is routinely removed in the course of abdominal surgery performed for other reasons.

The most frequent type of abdominal surgery among adults in this country is removal of the gallbladder. As we have seen, the gallbladder stores the bile produced by the liver and releases it into the small intestine in response to the hormone cholecystokinin. Bile is produced continually in the liver and travels through small ducts to the common bile duct, which leads to the small intestine (see illustration). The cystic duct connects the gallbladder to the common bile duct. Just before the common bile duct enters the small intestine, the main pancreatic duct merges with it. This duct carries enzymes and alkaline fluids from the pancreas. Between periods of digestion, a sphincter muscle just below the junction of these two ducts closes. As a result, bile from the liver is shunted into the gallbladder. When the products of fat and protein digestion enter the duodenum shortly after the next meal, the sphincter opens and bile and pancreatic juices begin flowing into the small intestine.

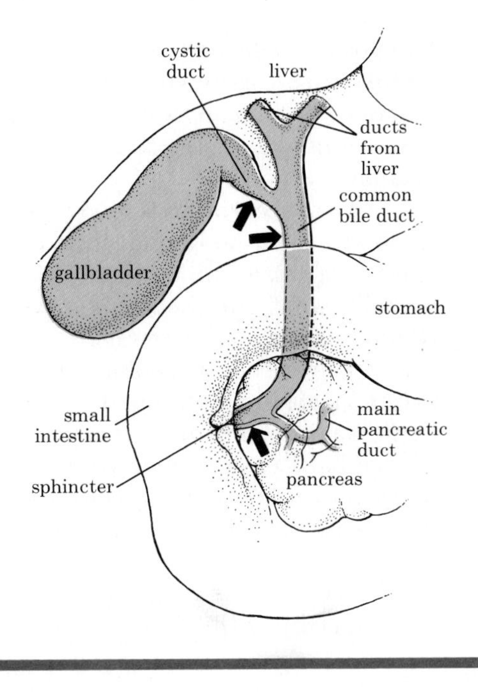

The ducts connecting the liver, the gallbladder, the pancreas, and the small intestine. The arrows indicate the different locations at which gallstones can lodge.

In the gallbladder, sodium ions are pumped from the bile and water follows by osmosis, making the bile more concentrated. It is here that the problem begins. Bile contains four main ingredients: bile salts, lecithin (a phospholipid), cholesterol, and various pigments produced by the breakdown of hemoglobin released from dead red blood cells. The relative concentrations of the bile ingredients are in a very delicate balance. If the concentration of cholesterol increases or that of bile salts or lecithin decreases, cholesterol can precipitate out, forming gallstones. If the stones lodge in the cystic duct, they block the release of bile from the gallbladder and can cause infection and acute pain. If they lodge in the common bile duct, they block the movement of bile from the liver into the intestine. As a result, fats cannot be digested. Moreover, bile accumulates in the bloodstream, resulting in jaundice. If gallstones lodge farther down the duct, after it has merged with the pancreatic duct, they cut off not only the supply of bile but also the supply of pancreatic fluids. As a consequence, practically no digestion or absorption of nutrients occurs in the small intestine.

Some 15 million Americans have gallstones, and about 500,000 require hospitalization for treatment each year. The most common form of treatment is surgical removal of the gallstones and gallbladder. New forms of treatment are now being tested and may be in general use within a few years. One of the most promising is the use of bile salts to dissolve the gallstones.

People who have had their gallbladders removed surgically have normal digestive functions. Their livers continue to produce bile, and the flow is increased during digestion. Although the gallbladder, unlike the appendix, has a function, it is apparently a nonessential function. Some bile-producing animals, including rats and horses, have no gallbladder at all.

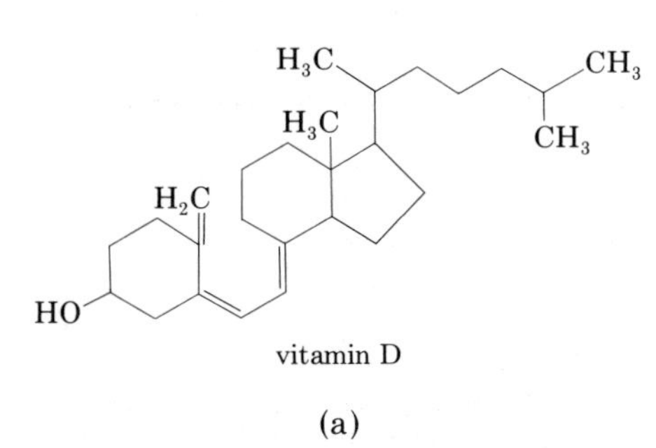

(a)

(b)

29-9 (a) *Vitamin D is a steroid that is produced in the skin by the action of ultraviolet rays from the sun on a cholesterol-like compound (the absorbed energy opens the second ring of the molecule).*

According to one current hypothesis, the ancestors of modern Homo sapiens *originated in the tropics and were all dark-skinned. In those populations that moved northward, however, the screening effects of the darker pigment, which inhibits the production of vitamin D, caused selection for lighter skin color, whereas no such selection pressures occurred in the sun-drenched areas nearer the equator. Nor did such selection occur among the Eskimos* (b), *who eat a diet rich in fish oils, a major source of vitamin D.*

Mammals also require but cannot synthesize certain polyunsaturated fats that provide fatty acids needed for fat synthesis. These can be obtained by eating plants or insects (or eating other animals that have eaten plants or insects).

Vitamins are an additional group of molecules required by living cells that cannot be synthesized by animal cells. They are characteristically required only in small amounts. Many of them function as coenzymes. Table 29–2 indicates some of the principal vitamins required in the human diet. There is no clear evidence that the ingestion of amounts of any particular vitamin in excess of the amounts available in well-balanced diets has any beneficial effects on a normally healthy individual. Some, including vitamins A, D, and K, are toxic in large doses.

The body also has a dietary requirement for a number of inorganic substances, or minerals. These include calcium and phosphorus for bone formation, iodine for thyroid hormone, iron for hemoglobin and cytochromes, and sodium, chloride, and other ions essential for ionic balance. Most of these are present in the ordinary diet or in drinking water. Like the vitamins, however, they must be given in supplementary form when the dietary intake is inadequate or when the individual is not able to assimilate them normally.

SUMMARY

Digestion is the process by which food is broken down into molecules that can be taken up by the bloodstream and so distributed to the individual cells of the body. It occurs in successive stages, regulated by an interplay of hormones and nervous stimuli. In mammals food is processed initially in the mouth, where the breakdown of starch begins. It moves through the esophagus to the stomach, where gastric juices destroy bacteria and begin to break down proteins.

Most of the digestion occurs in the duodenum, the upper portion of the small intestine. Digestive activity in the duodenum, which is performed by enzymes, is almost completely under hormonal regulation. The breakdown of starch by amylases continues, fats are hydrolyzed by lipases, and proteins are reduced to single amino acids. Hormones (secretin and cholecystokinin) are secreted by duodenal cells, and these, in turn, stimulate the functions of the pancreas and the liver. The pancreas releases an alkaline fluid containing digestive enzymes; the liver produces bile, which is also alkaline.

Most absorption of food molecules, minerals, and water occurs in the small intestine. Additional water is resorbed from the residue of the food mass as it passes through the large intestine. The large intestine contains bacteria, which are the source of certain vitamins. Undigested remains are eliminated through the anus.

The chief energy sources for cells in the mammalian body are glucose and fatty acids circulating in the blood. The organ principally responsible for maintaining a steady supply of glucose is the liver, which stores glucose (in the form of glycogen) when glucose levels in the blood are high and breaks down glycogen or amino acids, releasing glucose, when the levels drop. These activities of the liver are regulated by a number of hormones.

Requirements for good nutrition include molecules for fuel (which can be obtained from carbohydrates, fats, or proteins), essential amino acids, certain essential fatty acids, vitamins, and minerals.

Table 29-2 Vitamins

Designation: Letter and Name	Major Sources	Function	Deficiency Symptoms
A, carotene	Egg yolk, green or yellow vegetables, fruits, liver, butter	Formation of visual pigments, maintenance of normal epithelial structure	Night blindness; dry, flaky skin
B-complex vitamins:			
B_1, thiamine	Brain, liver, kidney, heart, whole grains	Formation of cofactor involved in Krebs cycle	Beri-beri, neuritis, heart failure, mental disturbance
B_2, riboflavin	Milk, eggs, liver, whole grains	Part of electron carrier FAD	Photophobia, fissuring of skin
B_3, niacin (nicotinic acid)	Whole grains, liver and other meats, yeast	Part of electron carriers NAD, NADP, and of CoA	Pellagra, skin lesions, digestive disturbances
B_5, pantothenic acid	Present in most foods	Forms part of CoA	Neuromotor and cardiovascular disorders, gastrointestinal distress
B_6, pyridoxine	Whole grains, liver, kidney, fish, yeast	Coenzyme for amino acid metabolism and fatty acid metabolism	Dermatitis, nervous disorder
B_{12}, cyanocobalamin	Liver, kidney, brain	Maturation of red blood cells, coenzyme in amino acid metabolism	Anemia, malformed red blood cells
Biotin	Egg yolk, synthesis by intestinal bacteria	Concerned with fatty acid synthesis, CO_2 fixation, and amino acid metabolism	Scaly dermatitis, muscle pains, weakness
Folic acid	Liver, leafy vegetables	Nucleic acid synthesis, formation of red blood cells	Failure of red cells to mature, anemia
C, ascorbic acid	Citrus fruits, tomatoes, green leafy vegetables	Vital to collagen and ground (intercellular) substance	Scurvy, failure to form connective tissue fibers
D_3, calciferol	Fish oils, liver, milk and other dairy products, action of sunlight on lipids in the skin	Increases Ca^{2+} absorption from gut, important in bone and tooth formation	Rickets (defective bone formation)
E, tocopherol	Green leafy vegetables	Maintains resistance of red cells to hemolysis, cofactor in electron transport chain	Increased red blood cell fragility
K, naphthoquinone	Synthesis by intestinal bacteria, leafy vegetables	Enables synthesis of clotting factors by liver	Failure of blood coagulation

QUESTIONS

1. Distinguish between the following: amylases/lipases; gastric juice/bile; gastrointestinal hormones/gastrointestinal enzymes; digestion/absorption; villi/microvilli; vitamins/essential amino acids.

2. Diagram the human digestive tract, including all the major organs. Check your drawing against Figure 29-1.

3. Trace the chemical processing of a hamburger on a bun as it passes through your digestive tract.

4. If you oxidize a pound of fat, whether butter or the fat in your own body cells, about 3,500 kilocalories are released. Suppose that you go on a weight-reducing diet, limiting yourself to 1,000 kilocalories a day. You spend most of your time sitting in a library (studying, of course) and so you expend only about 3,000 kilocalories a day. How many pounds will you lose each week?

5. Many animals are either herbivores or carnivores, and some are even specialists on a single organism as a food source. We, however, are omnivores. We also have an unusually large number of substances specifically required in our diets. How might these two facts be related?

30

Control of the Internal Environment

One of the identifying properties of living systems is that they are homeostatic: They are able to maintain a relatively constant internal environment despite continual exchanges with and changes in the external environment. This capacity exists even in the simplest living organisms.

In animals, many activities contribute to homeostasis. We have seen numerous examples in the previous chapters—the homeostatic regulation of blood sugar, for example, or the fine-tuning of hormone levels by feedback systems. In this chapter, we are going to examine three particularly noteworthy homeostatic functions: regulating temperature, regulating blood chemistry, and maintaining water balance.

Temperature Regulation

Some 200 years ago, Dr. Charles Blagden, then Secretary of the Royal Society of London, went into a room that had been heated to 126°C (260°F), taking with him a few friends, a small dog in a basket, and a raw steak. The entire group remained there for 45 minutes. Dr. Blagden and his friends emerged unaffected. So did the dog. (The basket had kept the dog's feet from being burned by the floor.) But the steak was cooked.

The control of body temperature is an important factor in regulation of the internal environment. Physiological processes involve a multitude of biochemical reactions, virtually all controlled by enzymes. The rate at which an enzyme functions is determined by a number of factors, one of the most important of which is temperature. In general, the rate of activity of an enzyme approximately doubles with every 10°C rise in temperature. Thus changes in body temperature can cause significant changes in the delicate chemical balances maintained within the living organism.

Temperature regulation, like water conservation, is a problem faced by animals as they become terrestrial. In general, large bodies of water, for the reasons discussed in Chapter 2, maintain a very stable temperature. The animals that live in them, with few exceptions, maintain a body temperature that is the same as the temperature of the water. The few exceptions include the aquatic mammals and certain large fish, such as tuna, that retain inside their bodies heat generated by muscular activities. Because the temperatures of large bodies of water are low (seldom exceeding 5°C), the pace of life in the water, physically and intellectually, is generally slow.

30-1 *In laboratory studies, the internal temperature of reptiles was shown to be almost the same as the temperature of the surrounding air. It was not until observations were made of these animals in their own environment that it was found that they have behavioral means for temperature regulation that give them a surprising degree of temperature independence. By absorbing solar heat, reptiles can raise their temperature well above that of the air around them. This Colorado collared lizard has become quite warm from the sun and has raised its body to allow cooling air currents to circulate across its belly.*

Terrestrial reptiles—snakes, lizards, and tortoises—are able to maintain remarkably stable body temperatures during their active hours by varying the amount of solar radiation they absorb. By careful selection of suitable sites, such as the slope of a hill facing the sun, and by orienting their bodies with a maximum surface exposed to sunlight, they can heat themselves rapidly. Such heating can occur as quickly as 1°C per minute, even on mornings when the air temperature is close to 0°C. By frequently changing position, these reptiles are able to keep their temperature within quite a narrow range as long as the sun is shining. When it is not, they seek the safety of their shelters; there they are not immobilized in an exposed position and are less vulnerable to attack. As we noted in Chapter 19, such animals, which obtain their heat from the external environment, are known as ectotherms.

"Warm-Blooded" Animals

The so-called "warm-blooded" animals, primarily mammals and birds, are endotherms (page 277). With some exceptions, endotherms are also *homeotherms*, animals that maintain a quite constant internal body temperature. The primary source of heat in endotherms is the oxidation of glucose and other energy-containing molecules within the body cells. Endotherms have a much higher metabolic activity than ectotherms, with a large proportion of their energy budget allocated to heat production. Also, small endotherms have a proportionally larger heat budget than large ones, because of surface-to-volume ratios (page 62). Endotherms are characterized by layers of insulating material—fur, feathers, fat—and by mechanisms for disposing of excess heat, such as panting in dogs and sweating in humans.

The human body maintains a remarkably constant temperature, deviating very little from 37°C. This constancy of temperature is maintained by an automatic system—a thermostat—in the hypothalamus, which precisely measures the body temperature, compares it with the thermostat "set point," and triggers the appropriate control mechanisms. Receptor cells in this center monitor the temperature of the blood flowing through the hypothalamus. Although the surface of the skin is covered with receptors for hot and for cold, these are not the principal agents in the regulation of internal temperature.

30-2 *The size of the extremities in a particular type of animal can often be correlated with the climate in which it lives.* (a) *The fennec fox of the North African desert has large ears that help it to dissipate body heat.* (b) *The red fox of the eastern United States has ears of intermediate size, and* (c) *the Arctic fox has relatively small ears. Similar correlations of characteristics such as size, weight, and color with environment can also sometimes be made among animals of a single species living over an extended geographic range.*

(a)

(b)

(c)

The elevation of temperature known as fever is due not to a malfunction of the thermostat but to a resetting. Thus, at the onset of fever, an individual typically feels cold and often has chills; although the body temperature is rising, it is still lower than the new thermostat setting. The substance primarily responsible for the resetting of the thermostat is a protein released by white blood cells in response to a foreign organism. The adaptive value of fever, if any—and we can only presume that there is one—is a subject of current research.

Regulation at High Temperatures

As the internal temperature rises, the blood vessels near the skin surface are dilated, and the supply of blood to the skin increases. If the air is cooler than the body surface, heat can be transferred directly to the air. Animals that live in hot climates characteristically have larger exposed surface areas than animals that live in the cold (Figure 30–2).

The rate at which heat is lost from the surface is increased by the evaporation of perspiration or saliva. In humans and most other large animals, when the external temperature rises above body temperature, perspiration begins. Evaporation requires heat energy (see page 29), which comes from the body surface, and the blood just below the skin surface cools as this heat energy is used. Dogs and some other animals pant. Horses and human beings sweat from all over their body surfaces. For animals that dissipate heat by water evaporation, temperature regulation at high temperatures necessarily involves water loss. This, in turn, stimulates thirst and, as we shall see, water conservation by the kidneys. Dr. Blagden and his friends were probably very thirsty.

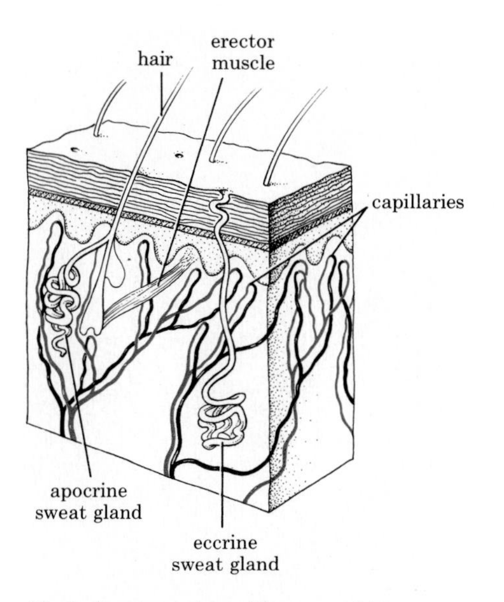

30–3 *Cross section of human skin showing structures involved in temperature regulation. In cold, the arterioles constrict, reducing the flow of blood through the capillaries, and the hairs, each of which has a small erector muscle under nervous control, stand upright. In animals with body hair (*Homo sapiens *is one of the few terrestrial mammals that is virtually hairless), air trapped between the hairs insulates the skin surface, conserving heat.*

In heat, the arterioles dilate and the eccrine glands secrete a salty liquid. Evaporation of this liquid cools the skin surface, dissipating heat (approximately 540 calories for every gram of H_2O). The apocrine glands are scent glands, producing a milky, odorous fluid. They are abundant in the armpit, navel, and anogenital areas. Beneath the skin is a layer of subcutaneous fat that serves as insulation, retaining the heat in the underlying body tissues.

Regulation at Low Temperatures

As temperatures go down, blood vessels near the skin surface are constricted, reducing heat loss through the skin. Metabolic processes increase. Part of this increase is due to increased muscle activity (shivering, shifting from foot to foot), and part is due to hormones that stimulate metabolism. Muscular activity raises heat production by increasing glucose oxidation and by friction.

Prime mobilizers of chemical energy are the autonomic nerves to fat tissue, which increase the metabolic breakdown of fat, and the hormone epinephrine, which is released from the adrenal medulla. In some animals, the thyroid gland increases its release of thyroxine in response to cold. Thyroxine increases the metabolic rate; it apparently exerts its effects directly on the mitochondria. Cortisol production by the adrenal cortex may also increase under the stress of exposure to cold. Both increased thyroxine and cortisol, you will recall (page 368), are mediated by the action of the hypothalamus on the pituitary, which, via tropic hormones, stimulates the target glands. So again we have an example of integration and control in which the nervous and endocrine systems function as a coordinated unit (Figure 30–4).

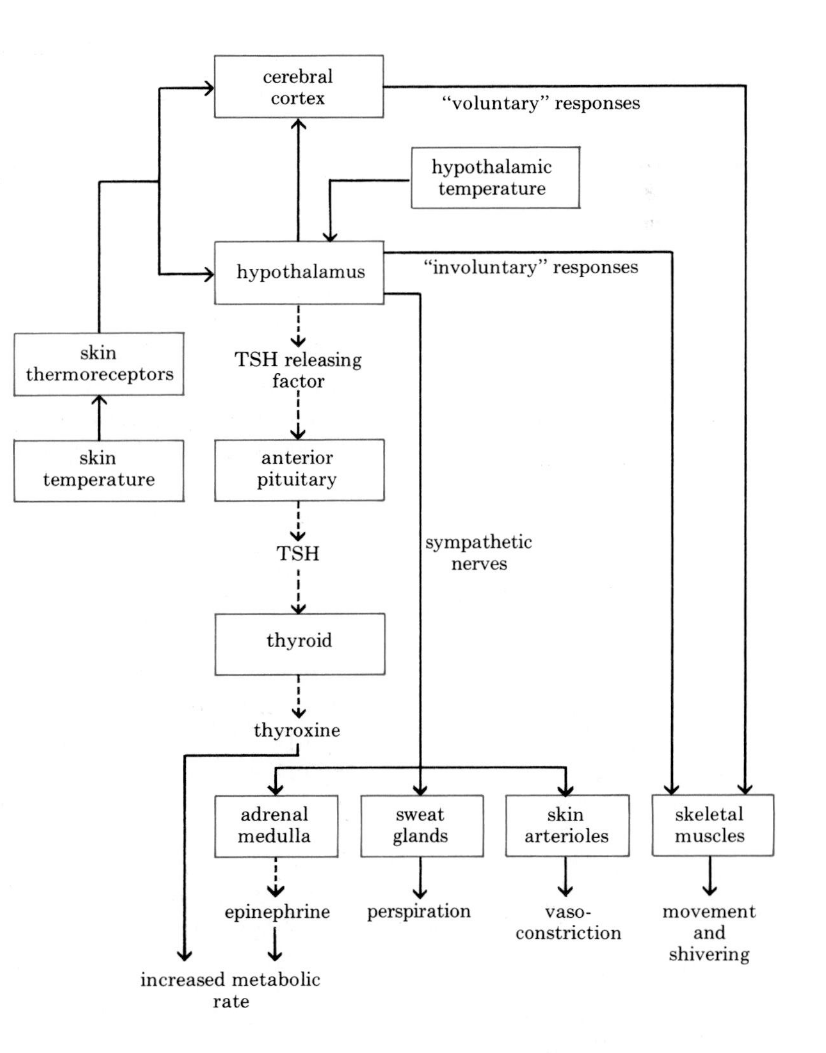

30–4 *Body temperature in mammals, regulated by a complex network of activities involving both the nervous and endocrine systems, is controlled largely by the hypothalamus. (The production of TSH is not thought to be a significant part of the response in humans.) Many behavioral responses are also involved.*

(a)

(b)

(c)

30–5 *Some adaptations to cold.* (a) *Prior to hibernating, mammals, such as the dormouse shown here, store up food reserves in body fat.* (b) *Note the heat-conserving reduction in surface-to-volume ratio in this hibernating chipmunk.* (c) *Bats have a number of special adaptations that enable them to conserve heat despite their small body size. Some hibernate. In others, the temperature drops when they are at rest (as it does in hummingbirds). Members of the species shown here also roll up their ears when they get cold.*

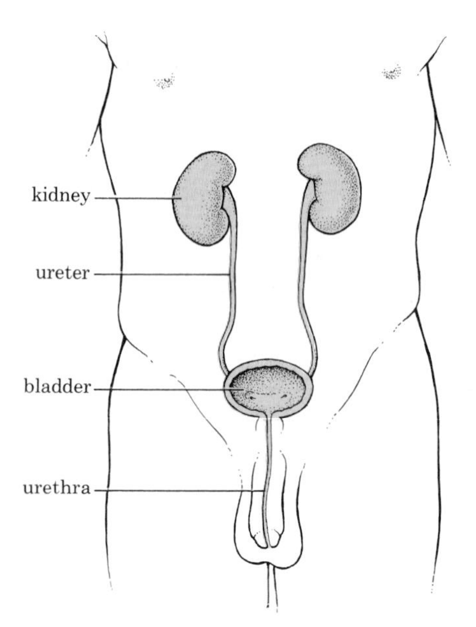

30–6 *The urinary system is one of the major organ systems of the body. Fluids are processed in the kidneys, and waste products and water—the urine—pass along a pair of tubes, the ureters, to the bladder, where they are stored. Urine leaves the body by way of the urethra.*

Regulation of Blood Chemistry

Animals are about 60 percent water. About two-thirds of this water is within the cells, and one-third, the extracellular fluid, surrounds, bathes, and nourishes the cells. Thus the extracellular fluid serves the same purpose for the body cells that the pre-Cambrian seas served for the earliest cells that arose in them. As they became multicellular in the course of evolution, animals began to produce their own extracellular fluid, similar in composition to the salty fluid of the sea. As they did so, they also evolved mechanisms for regulating its composition.

In many invertebrates and all vertebrates, the internal chemical environment is to a large extent regulated by a special organ system (Figure 30–6). Among terrestrial vertebrates, the most important components of this system are the kidneys. It is possible to relate major advances in vertebrate evolution—such as the transition to land—to increasing efficiency of kidney function. In fact, one can trace vertebrate evolution in terms of the evolution of the kidney, as Homer Smith did in his classic book *From Fish to Philosopher.*

Substances Regulated by the Kidneys

Regulation of the extracellular fluid by the kidneys involves the processing of three different categories of substances: waste products, ions, and water.

Human Circadian Rhythms

Human beings, like plants and other animals, are governed by circadian rhythms (page 337). It has long been known that we are more likely to be born between 3 and 4 A.M. and also to die in these same early morning hours. Body temperature fluctuates as much as 1°C (about 2°F) during the course of the day, usually reaching a high about 4 P.M. and a low about 4 A.M. Alcohol tolerance is greatest at 5 P.M. Many people have an automatic internal alarm system that wakes them at the same hour every morning—whether they want to or not. Secretion of some hormones, heart rate, blood pressure, and urinary excretion of potassium, sodium, and calcium all vary according to a circadian rhythm.

A study by the Federal Aviation Administration showed that pilots flying from one time zone to another—from New York to Europe, for instance—exhibit "jet lag," a general decrease in mental alertness and ability to concentrate and an increase in decision time and physiological reaction time. Measurements of circadian rhythms show that the body may be "out of sync" for as long as a week after such a flight. This brings into question present policies of diplomats speeding to foreign capitals at times of international crisis or of troops being air-transported into combat.

Elaborate studies in which individuals have been kept isolated in constant conditions for long periods of time—in underground bunkers, for example—have indicated that circadian rhythms are under the control of a biological clock or clocks. Such internal clocks presumably are involved in the homeostatic regulation of the internal environment.

The chief waste products that cells release into the bloodstream are carbon dioxide and the nitrogenous compounds, such as ammonia, produced by the breakdown of amino acids. Carbon dioxide is eliminated from the lungs, or it diffuses out into water through the skin or gills. In simple aquatic organisms, ammonia also diffuses out into the water. In larger, more complex organisms, nitrogenous wastes must be converted to some other form since ammonia is highly toxic, even in low concentrations.

Many birds, terrestrial reptiles, and insects excrete waste nitrogen in the form of crystalline *uric acid*. In birds, the uric acid is mixed with the undigested wastes in the cloaca (the common exit chamber for the digestive and urinary tracts), and the combination is dropped as a semisolid paste, familiar to frequenters of public parks and admirers of outdoor statuary. This nitrogen-laden substance forms a rich natural fertilizer. Guano, the excreta of seabirds, accumulates in such quantities on the small islands where the birds gather that at one time it was harvested commercially. Mammals excrete nitrogenous wastes mostly in the form of *urea* (Figure 30–7), which must be dissolved in water.

$$H_2N-\underset{\underset{O}{\|}}{C}-NH_2$$

urea

30–7 *Urea, the principal form in which nitrogen is excreted in most mammals. It is formed in the liver by the combination of two molecules of ammonia (NH_3) with one of carbon dioxide through a complex series of energy-requiring reactions. What would be the other product of this reaction?*

Chemical regulation also involves maintaining closely controlled concentrations of ions such as Na^+, K^+, H^+, Mg^{2+}, Ca^{2+}, Cl^-, and HCO_3^-. These ions are important for their specific roles in such processes as the maintenance of membrane permeability, transmission of action potentials, and muscle contraction.

The concentration of a particular substance in the body depends not only on the absolute amount of the substance but also on the amount of water in which it is dissolved. Thus, regulation of the water content of the blood is an important aspect of the regulation of blood chemistry. We shall return to this topic later in the chapter, after we have examined the structure and function of the kidney.

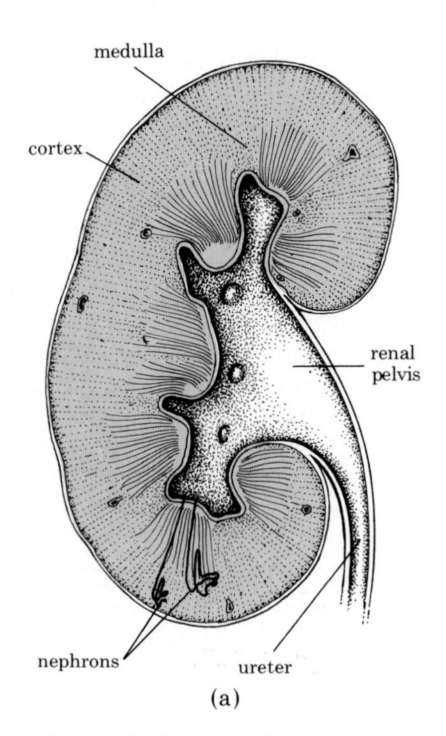

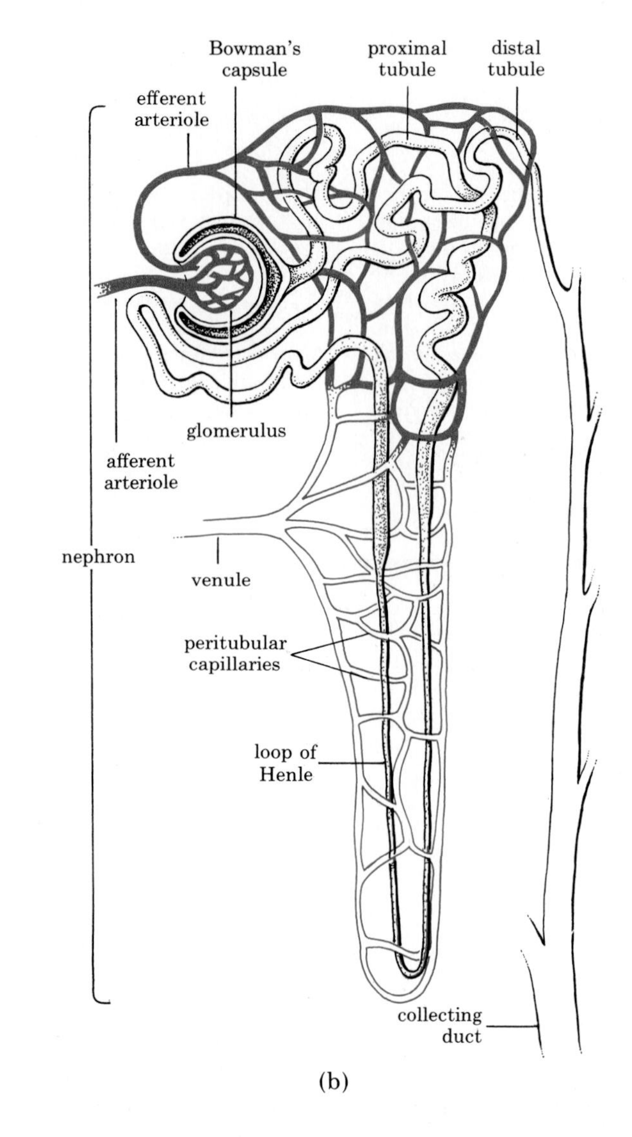

30-8 (a) *In longitudinal section, the human kidney is seen to be made up of an outer region, the cortex, which contains the fluid-filtering mechanisms, and an inner region, the medulla, through which collecting ducts carrying the urine merge and empty into the funnel-shaped renal pelvis, which, in turn, empties into the ureter.*
(b) *The nephron is the functional unit of the kidney. Blood enters the nephron through the afferent arteriole leading into the glomerulus. Fluid is forced out by the pressure of the blood through the thin capillary walls of the glomerulus into Bowman's capsule. The capsule connects with the long renal tubule, which has three regions: the proximal tubule; the loop of Henle, which extends into the medulla; and the distal tubule. As the fluid travels through the tubule, almost all the water, ions, and other useful substances are resorbed into the bloodstream through the peritubular capillaries. Other substances are secreted from the capillaries into the tubules. Waste materials and some water pass along the entire length of the tubule into the collecting duct and are excreted from the body as urine.*

Structure of the Kidney

In humans and other mammals, the functional unit of the kidney is the *nephron*. It consists of a cluster of capillaries known as the *glomerulus*, a bulb called *Bowman's capsule*, and a long, narrow tube, the *renal tubule* (Figure 30-8). Each of the two human kidneys contains about a million nephrons with a total length of some 80 kilometers (50 miles) in an adult.

Urine is formed in the nephrons and passed into the renal pelvis, which is, in essence, a funnel. From this funnel, the urine trickles continuously through the ureter to the bladder, which stores the urine until it is excreted through the urethra.

Function of the Kidney

Blood enters the kidney through the renal artery, which divides into progressively smaller branches, the arterioles, which each lead, in turn, to a glomerulus. Unlike other capillary beds, a glomerulus lies between two arterioles. The efferent arteriole—the one leading out—then divides again into capillaries, the peritubular ("around-the-tubes") capillaries, which then merge to form a venule that empties into the renal vein.

30-9 The section of kidney tissue shown in this scanning electron micrograph was prepared by the freeze-fracture technique. A glomerulus occupies the center of the micrograph, enclosed at the top and sides by Bowman's capsule. The space within the capsule is continuous with the proximal tubule, the beginning of which you can see at the upper right.

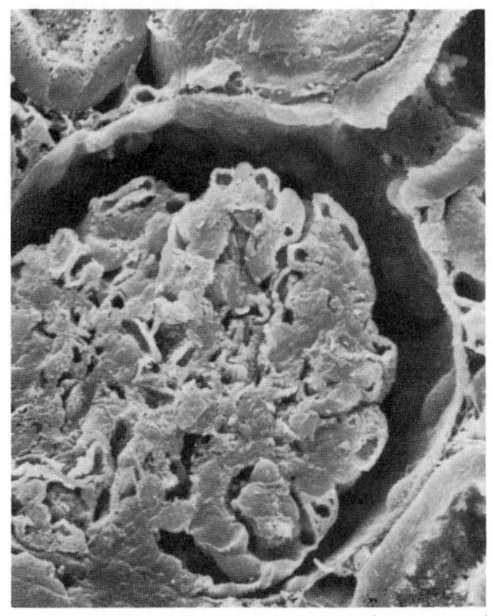

The blood within a glomerulus is at a pressure about twice that in other capillaries. As a consequence, about one-fifth of the blood plasma that enters the kidney is forced out of the glomerular capillaries into Bowman's capsule. This crucial first stage of kidney function is called *filtration,* and the fluid entering the capsule is called the filtrate. Except for the absence of large molecules, such as proteins, the filtrate has the same chemical composition as the plasma.

The filtrate then begins its long passage through the renal tubule, the walls of which are made up of a single layer of cells. These cells, which are specialized for active transport, accomplish the next two stages of kidney function. In the second stage, *secretion,* molecules left in the plasma are selectively removed from the peritubular capillaries and actively secreted into the filtrate. Penicillin, for example, is removed from the circulation in this way.

The third stage, *resorption,* occurs simultaneously. Most of the water and solutes that initially entered the tubule are transported back into the peritubular capillaries. For example, glucose, most amino acids, and most vitamins are returned to the bloodstream. Waste products such as urea, however, are not resorbed and remain in the fluid in the tubule.

Each day some 180 liters of fluid from the bloodstream are processed by the nephrons (remember, the total blood supply is only about 5 liters). Together, the processes of filtration, secretion, and resorption so carefully adjust the chemical composition of the fluid that blood leaving the kidneys can be considered to be completely remade.

In the fourth stage of kidney function, *excretion,* the fluid remaining in the renal tubule leaves the nephron. It is now the urine.

30-10 The four basic components of kidney function: (1) *filtration through the glomerular capillaries into Bowman's capsule;* (2) *secretion from the peritubular capillary into the tubule;* (3) *resorption from the tubule to the peritubular capillary;* (4) *excretion. The pressure in the glomerular capillaries is regulated by constriction of the efferent arteriole.*

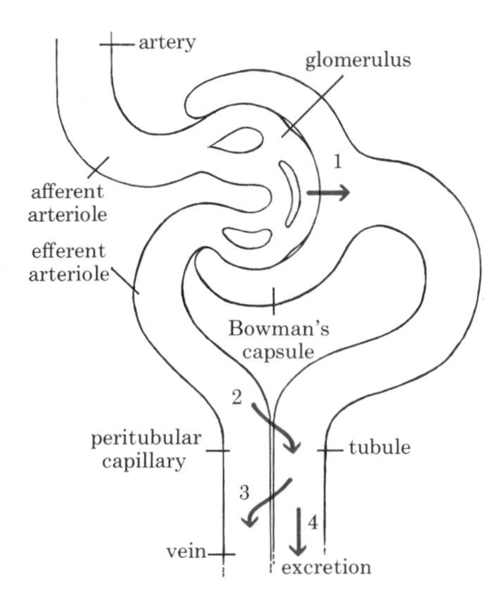

Water Balance

An Evolutionary Perspective

The earliest organisms probably had a salt and mineral composition much like that of the environment in which they lived. The early organisms and their surroundings were probably also isotonic; that is, each had the same total effective concentration of dissolved substances, so that there was no net movement of water into or out of the cell. When organisms moved to fresh water (a hypotonic—less concentrated—environment), they had to develop systems for "bailing themselves out," since fresh water tended to move into their bodies by osmosis. The contractile vacuole of *Paramecium* is an example of such a bailing device (see page 80).

If the vertebrates evolved in fresh water, as is generally believed, the first function of the kidneys was probably to pump water out and to keep salt and other desirable solutes, such as glucose, in. In freshwater fish, the kidney works primarily as a filter and resorber, and the urine is hypotonic—that is, it has a concentration of solutes lower than that of body fluids.

Saltwater fish have a different problem. Their body fluids are generally less concentrated than their environment, and so they tend to lose water by osmosis. Their need is to conserve water and thereby keep their body fluids from becoming too concentrated. This problem has been solved in different ways by different groups of fish. In hagfish, for example, body fluids are about as salty as the salt waters of the surrounding ocean and so are isotonic with them. Cartilaginous fish such as the shark are also isotonic

30-11 *Some marine animals, such as the turtle, have special glands in their heads that can secrete sodium chloride at a concentration about twice that of sea water. Since ancient times, turtle watchers have reported that these great armored reptiles come ashore, with tears in their eyes, to lay their eggs. It is only recently that biologists have learned that this is not caused by an excess of sentiment—as is the case with Lewis Carroll's mock turtle—but is, rather, a useful solution to the problem of excess salt. Marine birds have similar salt-secreting glands.*

to sea water, but they achieve their isotonicity in a different way. In the course of evolution, they developed an unusual tolerance for urea, so instead of constantly excreting it, as do most other fish, they retain a high concentration of it in the blood. This high concentration of urea makes their body fluids almost isotonic in relation to sea water; hence, they do not tend to lose water by osmosis.

The bony fish, the most abundant group of marine fish, have body fluids that are hypotonic in relation to the environment, having an osmotic concentration only about one-third that of sea water. Thus, like terrestrial animals, they have the potential problem of losing so much water to their environment that the solutes in the body fluids could become too concentrated and the cells could die. In bony marine fish, this problem has been solved by the evolution of special gland cells in the gills that excrete excess salt. Hence these fish can drink salt water and still remain hypotonic. (Freshwater fish, conversely, have salt-absorbing cells in their gills.)

Since terrestrial animals do not always have automatic access to either fresh or salt water, they must regulate water content in other ways, balancing off gains and losses.

Sources of Water Gain and Loss

Terrestrial animals gain water by drinking fluids, by eating water-containing foods, and as an end product of the oxidative processes that take place in the mitochondria, as we saw in Chapter 7. When 1 gram of glucose is oxidized, 0.6 gram of water is formed. When 1 gram of protein is oxidized, only about 0.3 gram of water is produced. Oxidation of 1 gram of fat, however, produces 1.1 grams of water because of the high hydrogen concentration in fat (the extra oxygen comes from the air).

On the average, a human takes in about 2.3 liters of water a day in food and drink and gains an additional 200 milliliters a day by oxidation of food molecules. Water is removed from the blood and excreted as urine, is lost from the lungs in the form of the moisture in exhaled air, is eliminated in the feces, and is lost by evaporation from the skin.

Excretion of Water

In humans, urine is usually the major route of water loss. In a normal adult, the rate of water excretion in urine averages 1.2 liters a day. Although the actual amount of urine produced may vary from a few hundred milliliters a day to several thousand, there will be a variation of less than 1 percent in the fluid content of the body. However, a minimum output of about 500 milliliters of water is necessary for health, since this much water is needed to remove potentially toxic waste products.

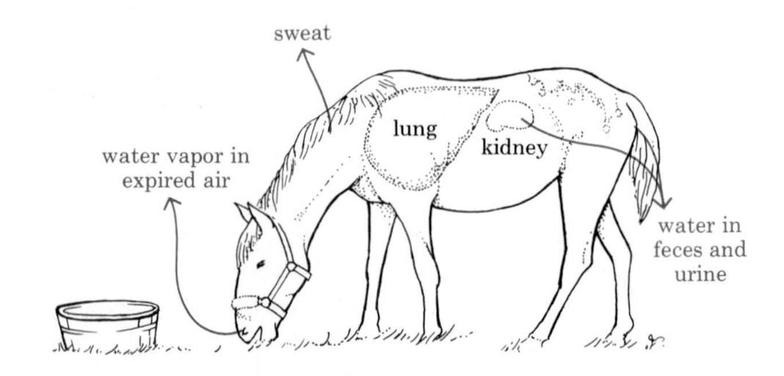

30-12 *A mammal is in water balance when the total amount of water lost in expired air, evaporation from skin, and in urine and feces equals the total amount of water gained by the intake of food and fluids and by the oxidation of food molecules.*

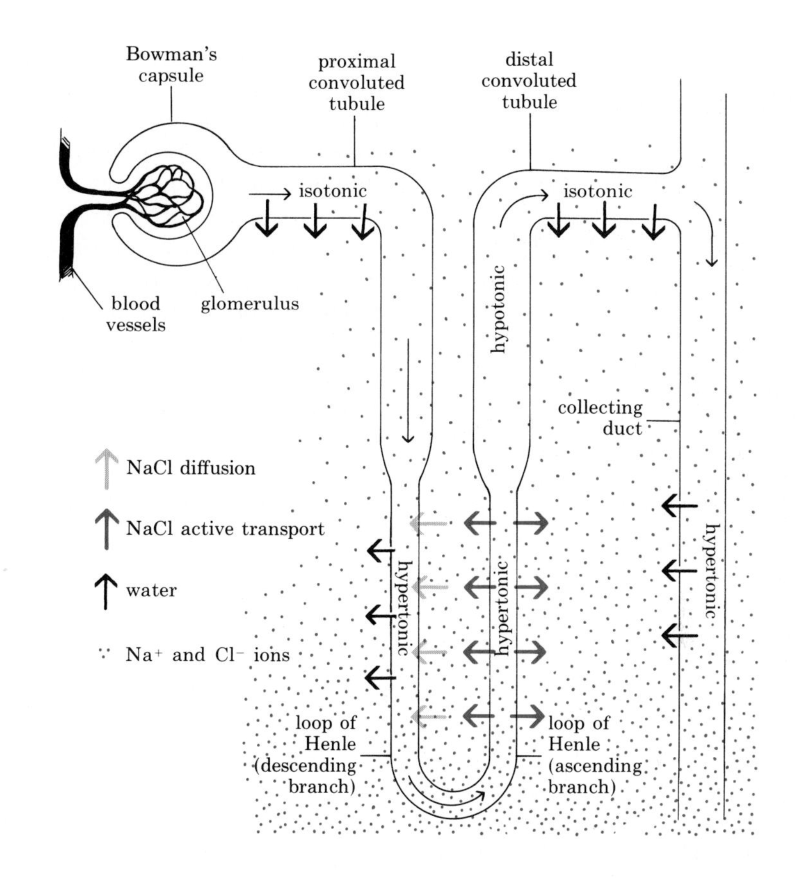

30–13 *The formation of hypertonic urine in the human nephron. As shown in this diagram, the fluid from Bowman's capsule first enters the proximal convoluted tubule, descends the loop of Henle, ascends it, and then passes through the distal convoluted tubule into the collecting duct.*

In the loop of Henle, water diffuses freely through the cells lining the walls of the descending branch while sodium chloride is actively pumped from the ascending branch back to the surrounding fluid. The ions pumped out of the ascending branch pass back into the descending branch by simple diffusion since the fluids surrounding the descending branch are more concentrated than those within it. Thus the salt recirculates to the ascending branch, where the sodium chloride is pumped out again. Water cannot follow the salt out of the ascending branch because its walls are impermeable to water.

This recirculation of salt has two consequences: (1) As the urine passes through the loop of Henle, much salt but little water is removed, and (2) the lower part of the loop of Henle and also the lower part of the collecting duct are bathed in a fluid containing many times the level of salt normally found in tissue or blood.

The fluid entering the proximal tubule is isotonic with the blood plasma; that is, it has the same solute concentration as does the blood plasma. As the fluid descends the loop of Henle, it becomes increasingly concentrated. In the ascending branch, it becomes less and less concentrated, and as it enters the distal tubule, it is hypotonic. By the end of the distal tubule, it is once more isotonic. The fluid then flows to the collecting duct, which traverses the zone of high salt concentration.

From this point onward, the urine concentration depends on the cells that form the walls of the collecting duct. If the duct walls are not permeable to water, the water entering the collecting duct from the distal tubule stays in the duct, and a less concentrated urine is excreted. If ADH (antidiuretic hormone) is present, however, the cells of the collecting duct are permeable to water, and water diffuses out into the surrounding salty fluid, as shown in the diagram. In this case a highly concentrated (hypertonic) urine is passed down the duct to the renal pelvis, the ureter, the bladder, and finally out the urethra.

Water Conservation: The Loop of Henle

The control of urinary water loss is a major mechanism by which body water is regulated. Animals with free access to fresh water typically excrete a copious urine that is hypotonic in relation to their blood. Most terrestrial animals cannot afford to be so profligate with water, however. In response to evolutionary pressures, birds and mammals have developed the ability to excrete hypertonic urines—that is, urines that are more concentrated than their own body fluids. This ability is associated with a hairpin-shaped section of the nephron known as the *loop of Henle*. By sampling fluids in and around the nephron and analyzing the ions present in each sample, physiologists have been able to determine how such a deceptively simple-looking structure makes this function possible. Their findings are summarized in Figure 30–13.

Aldosterone and ADH

In mammals, at least two hormones act on the nephron to affect the composition of the urine. One of these is aldosterone, which is produced by the adrenal cortex. Aldosterone stimulates resorption of sodium from the distal tubule and secretion of potassium into it. When the adrenal glands are removed, or when they function poorly (as in Addison's disease), sodium chloride and water are lost in the urine, and the tissues of the body become depleted of them. Generalized weakness results, and if a patient with Addison's disease is not given hormone replacement therapy, the fluid loss can eventually be fatal. Aldosterone production is controlled by a complex feedback circuit involving, as you would expect, sodium and potassium levels in the bloodstream.

The second hormone, ADH (antidiuretic hormone), is formed in the hypothalamus and released from the posterior pituitary. It affects the amount of water excreted in the urine. ADH acts on the membranes of the collecting ducts of the nephrons and increases their permeability to water, so more water moves, by diffusion, back into the blood from the nephron.

The amount of ADH released depends on the osmotic concentration of the blood and also on the blood pressure. Osmotic receptors that monitor the solute content of the blood are located in the hypothalamus. Pressure receptors that detect changes in blood volume are found in the walls of the heart, in the aorta, and in the carotid arteries. Stimuli received by these receptors are transmitted to the hypothalamus. Factors that increase the concentration of solutes in the blood or decrease blood pressure, or both, increase the production of ADH and conservation of water in the body. Such factors include dehydration and hemorrhage. Factors that decrease blood concentration, such as the ingestion of large amounts of water, or that increase blood pressure—epinephrine, for instance—signal the hypothalamus to decrease ADH production, and so more water is excreted. Alcohol suppresses ADH secretion and thus increases urinary flow, a phenomenon familiar to imbibers of beer and other alcoholic beverages. Pain and emotional stress trigger ADH secretion and thus decrease urinary flow.

Adaptations of Desert Animals

We have seen in the course of this chapter, first, that water evaporation is a major mechanism for dissipating heat and, second, that water balance is essential for the smooth functioning of the internal environment. How do animals in the desert resolve these two conflicting demands? There are a variety of answers to this question.

30-14 *The kangaroo rat, a common inhabitant of the American desert, may spend its entire life without drinking water. It lives on seeds and other dry plant materials, and, in the laboratory, it can be kept for months on a diet of only fatty seeds such as barley or rolled oats. Analysis shows the kangaroo rat is highly conservative in its water expenditures. It has no sweat glands, and, being nocturnal, it searches for food only when the external temperature is relatively cool. Its feces have a very low water content, and its urine is highly concentrated. Its major water loss is through respiration, and even this loss is reduced by the animal's long nose in which some cooling of the expired air takes place, with condensation of water from it.*

Some animals can derive all their water from food and do not require fluids. The kangaroo rat of the American desert, for example, can spend its entire life without drinking water if it eats the right type of food. It is not surprising that it selects a diet of fatty seeds. If it is fed high-protein seeds, such as soybeans—the oxidation of which produces a large amount of nitrogenous waste and a relatively small amount of water—the kangaroo rat will die of dehydration unless some other source of water is available.

Many desert animals regulate their temperature—and thus the demands on their water supply—by behavior, as do the reptiles mentioned earlier in the chapter. One of the most interesting examples is found in the architecture of the burrows built by prairie dogs, which follow some basic principles of aerodynamics to air-condition their tunnels.

As you may know, airplane wings are shaped so that air will move along the top surface faster than along the bottom surface. This creates lower pressure above the wings so that the plane is, in effect, pulled upward. A prairie dog constructs a burrow about 18 meters long, with an opening at either end. Then, following the principle illustrated by the airplane wing, it constructs a high mound around one end and a low mound around the other, often gathering dirt from some distance around and working and reworking the structure. Air moves faster over the higher mound, and the pressure is less than at the other, lower opening of the burrow. As a consequence, air is pulled through the burrow from the lower end. Duke University scientists, who have studied the prairie dog's engineering accomplishments, calculate that when air is moving over the mounded openings at the rate of 1.6 kilometers (1 mile) an hour, a prairie dog burrow gets a complete change of air every 10 minutes.

30-15 *By facing the sun, a camel exposes as small an area of body surface as possible to the sun's radiation. Its body is insulated by fat on top, which minimizes heat gain by radiation. The underpart of its body, which has much less insulation, radiates heat out to the ground cooled by the animal's shadow. Other adaptations of the camel to desert life include long eyelashes, which protect its eyes from the stinging sand, and flattened nostrils, which retard water loss.*

As a final example, let us consider the camel, for so long the companion of desert travelers. A camel has several advantages over a human in the desert. For one thing, the camel excretes a much more concentrated urine; in other words, it does not need to use so much water to dissolve its waste products. (In fact, we are very uneconomical with our water supply; even dogs and cats excrete a urine twice as concentrated as ours.)

Also, a camel can lose more water proportionally than a human and still continue to function. If a person loses 10 percent of the body weight in water, he or she becomes delirious, deaf, and insensitive to pain. If as much as 12 percent is lost, the individual is unable to swallow and so cannot recover without assistance. Laboratory rats and many other common animals can tolerate dehydration of up to 12 to 14 percent of body weight. Camels can tolerate the loss of more than 25 percent of their body weight. They can go without drinking for one week in the summer months, three weeks in the winter.

Probably most important, the camel can tolerate a fluctuation in internal temperature of as much as 6°C. This tolerance means that it can let its temperature rise during the daytime (which the human thermostat would never permit) and cool during the night. The camel begins the next day at below its normal temperature—storing up coolness, in effect. It is estimated that the camel saves as much as 5 liters of water a day as a result of being able to tolerate these internal temperature fluctuations.

Finally, a camel orients itself to the sun in such a way as to expose as small an area as possible to the sun's radiation. Its fatty hump on top insulates the animal against the sun's rays striking its back. The relatively uninsulated underpart of its body radiates heat out to the ground cooled by the animal's shadow. As you might guess, when a camel goes to sleep on cold desert nights, it curls up.

SUMMARY

Homeostasis—the capacity to maintain a constant internal environment—is a characteristic of living systems. The control of body temperature is an essential feature of homeostasis. Ectotherms are animals that adjust their body temperatures principally by regulating the amount of heat taken in from the external environment. For endotherms, such as *Homo sapiens,* the source of heat is the oxidation of glucose and other fuel molecules by the body cells and friction from muscular activity.

Heat is carried by the bloodstream from the core of the body to the surface, where it is dissipated. In hot weather, as body temperature rises, the blood vessels near the surface of the skin are dilated, increasing the flow of blood to the skin. As external temperatures rise above body temperatures, humans and most other large animals begin to sweat. Evaporation of sweat from body surfaces requires heat and cools the surface. In the cold, blood vessels supplying the skin surface are constricted. As body temperature drops, shivering begins; the increased cellular metabolism and muscular activity involved in shivering produce heat. Temperature is regulated by an automatic system, or thermostat, in the hypothalamus that measures the body temperature and sets in motion the response mechanisms.

The urinary system enables an animal to regulate its internal chemical environment. Regulation is accomplished by (1) the excretion of toxic waste products, especially the nitrogenous compounds produced in the breakdown of amino acids; (2) control of the ionic content of body fluids; and (3) maintenance of water balance.

The excretory unit of the kidneys is the nephron; each nephron consists of a long tubule attached to a closed bulb (Bowman's capsule), which encloses a twisted cluster of capillaries, the glomerulus. Fluids are forced from the glomerulus into Bowman's capsule (filtration). By a combination of filtration, secretion, and resorption, the circulating body fluids are processed during their passage along the tubules and excess water and waste products are converted into urine.

The problems of water balance are different for animals living in saltwater, freshwater, and terrestrial environments. Terrestrial animals generally need to conserve water. An important means of water conservation is the capacity for excreting a urine that is hypertonic in relation to the blood. The loop of Henle is the portion of the mammalian nephron that makes possible the production of a hypertonic urine.

Water and salt excretion are subject to hormonal regulation. Aldosterone, produced by the adrenal cortex, increases the resorption of sodium and secretion of potassium by the renal tubules. ADH, a hormone formed in the hypothalamus, acts on the collecting ducts of the nephrons, increasing their permeability to water and thus decreasing water loss as more water is returned to the circulating blood.

Desert animals have a number of solutions to the dual problem of temperature regulation and water conservation. These include dietary habits, the capacity to excrete a very concentrated urine, a comparatively wide tolerance for fluctuations in total water and temperature, and a number of behavioral adaptations.

QUESTIONS

1. Distinguish among the following: extracellular fluid/intracellular fluid; endotherm/ectotherm/homeotherm; Bowman's capsule/glomerulus; aldosterone/ADH.

2. Describe what happens to the human organism as the external temperature rises. As it falls.

3. Explain the following terms in relation to kidney function: filtration, secretion, resorption, and excretion.

4. Sketch a nephron, indicating the pathways of the blood and the glomerular filtrate.

5. Diagram the paths of glucose and urea through the kidney.

6. Why does a high-protein diet require an increased intake of water? (You should be able to think of two different reasons.) Why does a person lose some weight after shifting to a low-salt diet, even without reducing the caloric intake? Given the fact that amino acids in excess of the body's requirements are broken down by the liver, not stored, what is the advantage of a high-protein diet? What might be a disadvantage?

7. Viewed as a whole, the kidney selectively removes substances from the bloodstream. A closer look, however, reveals that it does so by filtering out all small molecules through the glomerulus and then resorbing those that are needed in the tubules. Thus, the system need not identify wastes as such; rather, it must identify useful substances. Why would such an arrangement have been beneficial to early mammals? Why is it especially beneficial to modern mammals, including ourselves?

8. Could a human being on a life raft survive by drinking sea water? By catching and eating bony saltwater fish? Explain your answers.

31

Reproduction and Development

In the preceding chapters of this section, we have been primarily concerned with the ways in which the individual human body maintains itself. In this chapter, we are going to discuss the process by which parents produce new individuals, maintaining the continuity of human life, and how these individuals develop to become reasonable facsimiles of their parents. In the following chapter, we shall consider the human brain—the structure that most distinguishes us from the other members of the animal kingdom.

Life began for each of us with the fusion of two remarkable cells, a sperm and an ovum. These cells are produced in the gonads (gamete-producing organs) of the male and female reproductive systems.

The Male Reproductive System

Formation of Sperm

Normal males from adolescence until old age produce an average of several hundred million sperm a day. These cells are produced in the *testes* (singular, testis). Each human testis contains about 125 meters of tightly coiled seminiferous ("seed-bearing") tubules (Figure 31–1). Cells in these

31–1 (a) *The testis is made of tightly packed coils of seminiferous tubules containing sperm cells in various stages of development. The entire developmental sequence takes eight to nine weeks.*
(b) *As shown in the idealized cross section of a seminiferous tubule, spermatogonia develop into cells known as primary spermatocytes. These divide (first meiotic division) into two equal-sized cells, the secondary spermatocytes. In the second meiotic division, four equal-sized spermatids are formed. These differentiate into functional sperm. The Sertoli cells support and nourish the developing sperm.*

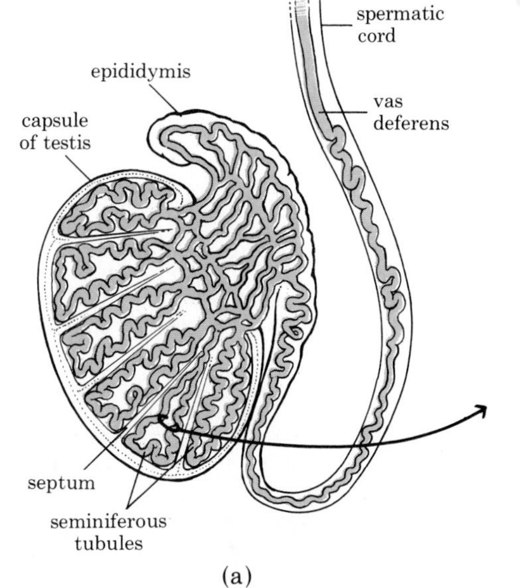

basement membrane
spermatogonium
primary spermatocyte
secondary spermatocyte
spermatids
Sertoli cell
sperm cells
(b)

31–2 *Diagram of a human sperm cell. The mature cell consists primarily of the nucleus, carrying the "payload" of tightly condensed DNA and associated protein, the very powerful tail (flagellum), and mitochondria, which provide the power for sperm movement. The acrosome is a specialized lysosome (see page 68).*

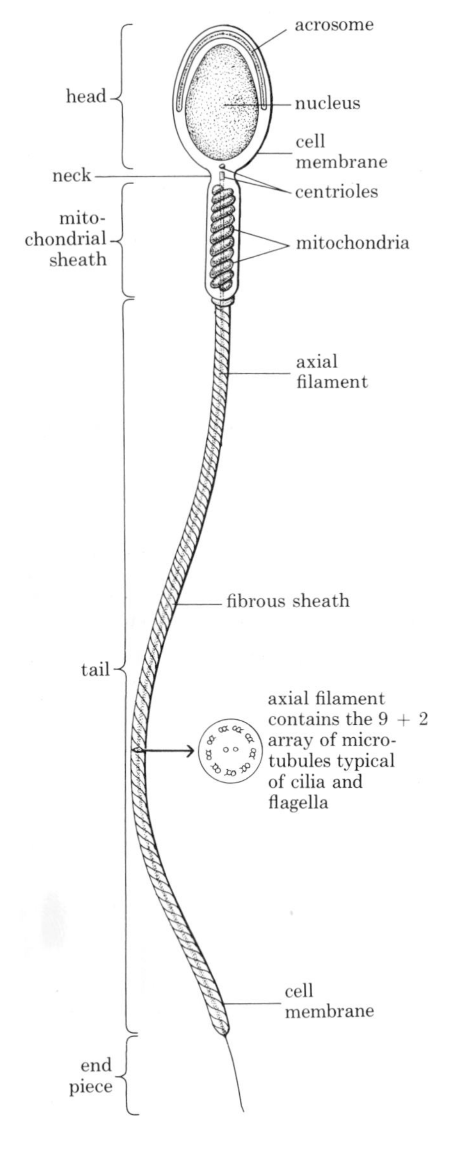

tubules, the spermatocytes, divide meiotically, each producing four haploid cells, the spermatids.* These cells then differentiate into the highly specialized sperm cells (Figure 31–2).

Pathway of the Sperm

The pathway of the sperm can be traced in Figure 31–3. From the testis, the sperm are carried to the *epididymis.* The sperm are immotile when they enter the epididymis and gain motility only after some 18 hours there. From the epididymis, the sperm pass to the *vas deferens* (plural, vasa deferentia), where most of them are stored. Vasectomy, an increasingly popular method of birth control, involves cutting and tying off the vasa deferentia just above the testes, as shown in Figure 31–4.

The vasa deferentia lead from the testes into the abdominal cavity. Within the abdominal cavity, they loop around the bladder, where they merge with ducts of the *seminal vesicles.* The vas deferens from each testis then enters the *prostate gland* and merges with the urethra, which extends the length of the *penis.* The penis serves both for the excretion of urine and the ejaculation of sperm.

* For a review of meiosis, see pages 140 to 144. The stages in the development of sperm cells are diagrammed in Figure 10–10 on page 145.

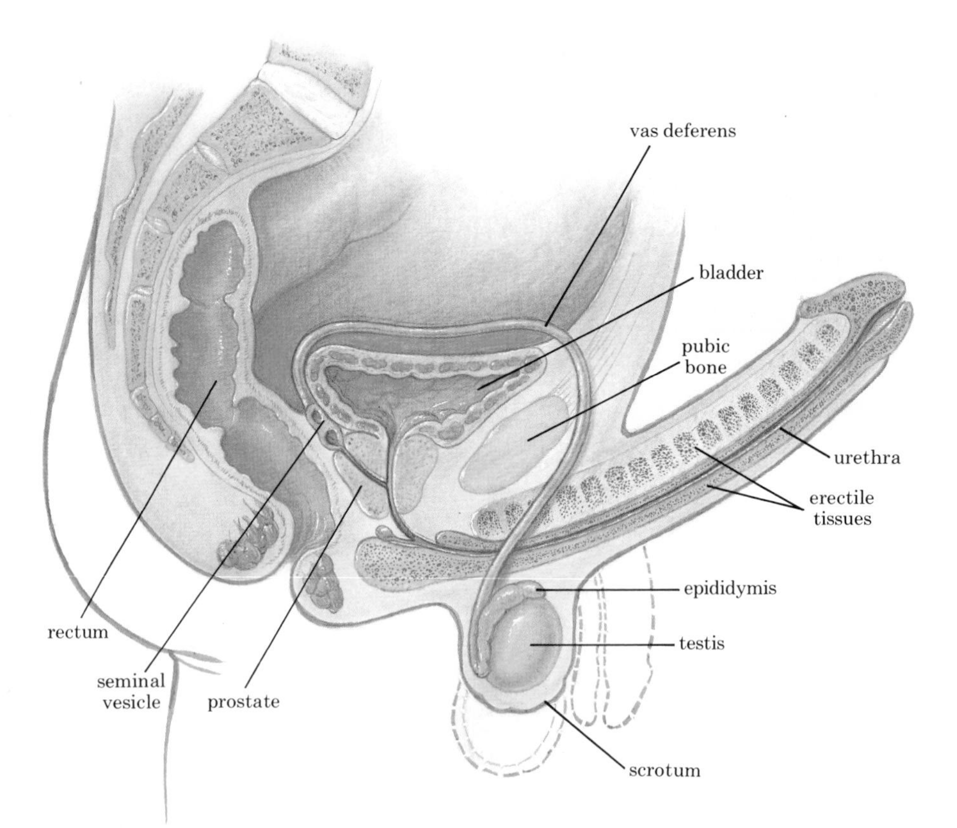

31–3 *Diagram of the human male reproductive tract, showing the penis and scrotum before (dashed lines) and during erection. Sperm cells formed in the seminiferous tubules of the testis enter the vas deferens, which connects with a duct from the seminal vesicle and then within the prostate gland joins the urethra. The sperm cells are mixed with fluids, mostly from the seminal vesicles and prostate gland. The resulting mixture, the semen, is released through the urethra of the penis. The urethra is also the passageway for urine, which is stored in the bladder.*

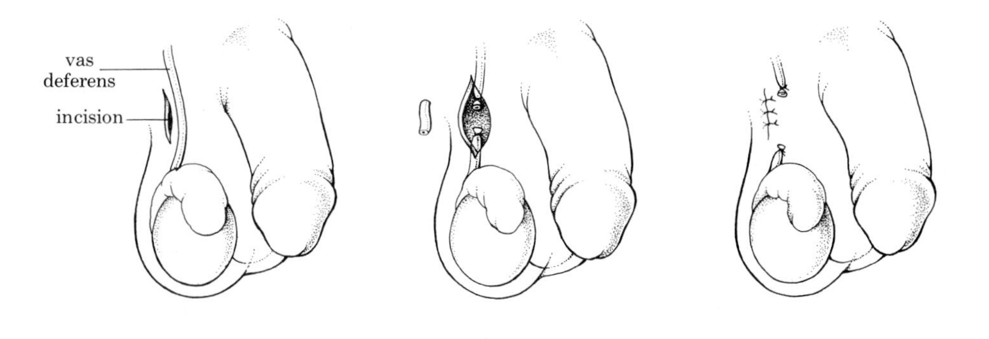

31-4 *During a vasectomy, the vas deferens on each side is severed, and the cut ends are folded back and tied off, preventing the release of sperm from the testis. The sperm cells are resorbed by the body, and the semen is normal except for the absence of sperm.*

This is a relatively safe and almost painless procedure that does not require a general anesthetic or hospitalization. Because the blood vessels between the testes and the rest of the body are left intact, the procedure does not affect hormone levels, sexual potency, or performance. Its chief drawback is that it is generally not reversible.

Erection of the penis is an adaptation that ensures deposit of sperm in the female reproductive tract. It occurs as a consequence of an increased flow of blood that fills the spongy, erectile tissues of the penis (Figure 31-5). The blood flow is controlled by nerve fibers to the blood vessels of the erectile tissues. Ejaculation of sperm from the penis is a reflex similar to the simple reflex described on page 358. Stimulation of the penis, such as may be produced by repeated thrusting in the vagina, results in the contraction of smooth muscle in the walls of the vas deferens, causing the sperm to empty into the urethra. Other muscles forcibly expel the sperm from the penis. These muscular spasms resulting in ejaculation account for many of the sensations associated with orgasm.

As the sperm move through the vas deferens, fluid is added from the seminal vesicles and prostate gland. The seminal fluid, which is alkaline and contains fructose, suspends and nourishes the sperm cells. It also helps neutralize the acidic fluids of the female reproductive tract. Sperm and fluid together form the semen. From 3 to 6 milliliters of semen are released per complete ejaculation.

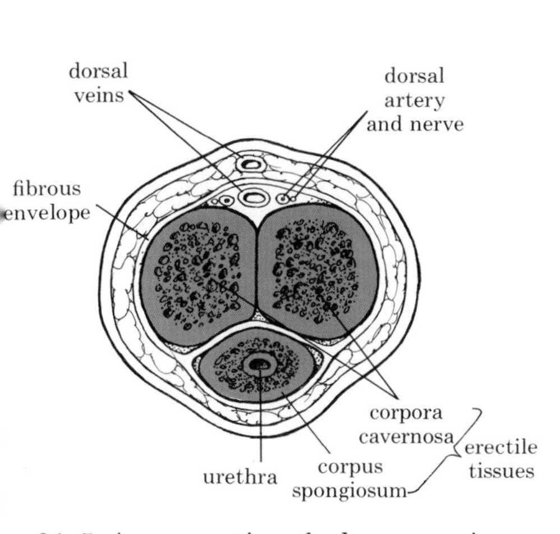

31-5 *A cross section of a human penis. The penis is formed of three cylindrical masses of spongy erectile tissue that contain a large number of small spaces, each about the size of a pinhead. Erection of the penis is caused by dilation of the blood vessels carrying blood to the spongy tissues.*

The Role of Hormones

In addition to producing the sperm cells, the testes are also the source of male hormones, known collectively as *androgens*. Androgens are produced by cells, known as interstitial cells, found in the connective tissue surrounding the seminiferous tubules. The principal androgen is testosterone.

Androgens produced in the male fetus are important in the development of the external sex organs. When the human male is about 10 years old, increased androgen production is associated with the enlargement of the penis and testes and also the prostate, the seminal vesicles, and other accessory organs.

Androgens also affect other parts of the body not directly involved in the production and deposition of sperm. In the human male, these effects include growth of the larynx and an accompanying deepening of the voice, muscle development, skeletal size, and distribution of body hair. Androgens stimulate the apocrine sweat glands, whose secretion attracts bacteria and so produces the body odors associated with sweat after puberty. They may cause the oil-secreting glands of the skin to become overactive, resulting in acne. These characteristics, associated with sexual development but not directly involved with reproduction, are known as secondary sex characteristics.

In other animals, the androgens are responsible for the lion's mane, the powerful musculature and fiery disposition of the stallion, the cock's

31–6 *Secondary sex characteristics among vertebrates. The male and female animals in each of these photos are actively courting.*

comb and spurs, and the bright plumage of many adult male birds (Figure 31–6). They are also responsible for a variety of behavior patterns such as the scent marking of dogs, the courting behavior of the sage grouse, and various forms of aggression toward other males found in a great many vertebrate species.

Regulation of Hormone Production

The production of testosterone is regulated by a gonadotropic hormone called luteinizing hormone (LH). LH is produced by the pituitary gland, under the influence of the hypothalamus. It acts on the interstitial cells of the testes to stimulate their output of testosterone. The increased concentration of testosterone, in turn, inhibits the release of LH by the pituitary (Figure 31–7). This is another example of regulation by negative feedback.

Another pituitary hormone, also under the control of the hypothalamus, is follicle-stimulating hormone (FSH). It acts on the seminiferous tubules, stimulating the development of sperm cells. Testosterone is also required for sperm development (Table 31–1).

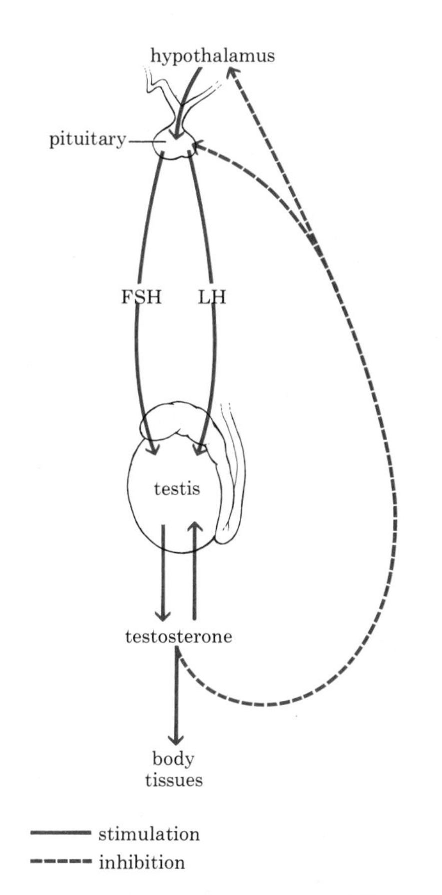

31-7 *Production of the male sex hormone testosterone is controlled by a negative feedback system. The hypothalamus, a brain center, regulates the pituitary gland's production of the gonadotropin LH. LH stimulates production of testosterone by the interstitial cells of the testes. As testosterone increases to a certain level of concentration in the blood, it inhibits the production of LH. Decreased levels of LH result in decreased production of testosterone. FSH, another gonadotropin secreted by the pituitary under the influence of the hypothalamus, stimulates sperm production.*

Table 31-1 Major mammalian gonadotropic and sex hormones in males

Hormone	Principal Source	Principal Effects	Control
FSH	Pituitary	Stimulates sperm production	Hypothalamus
LH	Pituitary	Stimulates testosterone production	Hypothalamus
Testosterone	Testes	Produces and maintains male sex characteristics, stimulates sperm production	LH

Environmental stimuli can also affect the production of testosterone. Studies on bulls, for example, have shown that after a bull sees a cow, his blood level of LH rises as much as seventeenfold; within about half an hour, the blood level of testosterone reaches its peak. Human testosterone production has also been shown to fluctuate in response to environmental stimuli. In many animals, hormone production is strictly seasonal. For example, in many species of deer testosterone is almost undetectable in adult males during the period from spring to midsummer, while new antlers are growing. As soon as testosterone levels begin to rise, the antlers die, losing their nerve and blood supply and their "velvet" cover. In the hard antler stages that follow, stags have pushing contests in which they lock antlers. Antler size is a determinant of hierarchy within a group. In early fall, testosterone levels begin to soar, the stag's voice deepens, and his neck and shoulder muscles thicken. A period of intense sexual activity, known as rutting, begins. Each stag sets out to corral a harem, which he defends until rutting season is over. Then he returns to the "stag line," his antlers drop off, and the testosterone in his blood becomes once more undetectable.

The Female Reproductive System

The female reproductive system is shown in Figure 31-8 (page 432). The gamete-producing organs are the *ovaries,* each a solid mass of cells about 3 centimeters long. The oocytes, from which the ova (eggs) develop, are in the outer layer of the ovary.

Other important structures include the uterus, the vagina, and the vulva. The *uterus* is a hollow, muscular, pear-shaped organ about 7.5 centimeters long and 5 centimeters wide. It is lined by the *endometrium,* which has two principal layers, one of which is shed at menstruation and another from which the shed layer is regenerated. The opening of the uterus is the *cervix,* through which the sperm pass on their way toward the egg cell and through which the fetus emerges at the time of birth.

The *vagina* is a muscular tube about 7.5 centimeters long that leads from the cervix of the uterus to the outside of the body. It is the receptive organ for the penis and is also the birth canal. Its opening is between the urethra, the tube leading from the bladder, and the anus.

The external genital organs of the female are collectively known as the *vulva.* The *clitoris,* most of which is embedded in the surrounding tissue, is about 2 centimeters long and corresponds to the penis in the male (in the early embryo, the structures are identical). The *labia* (singular, labium) are folds of skin. The labia majora are fleshy and, in the adult, covered with pubic hair. They enclose and protect the underlying, more delicate structures. The labia minora are thin and membranous.

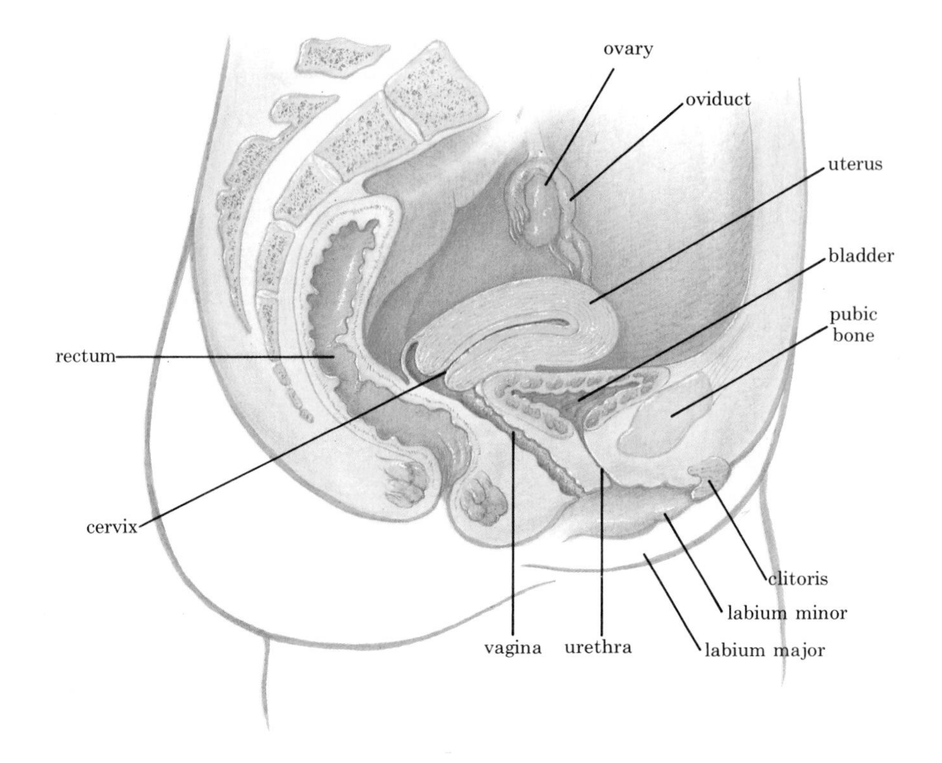

31-8 *The female reproductive organs. Notice that the uterus lies at right angles to the vagina. This is one of the consequences of the bipedalism and upright posture of* Homo sapiens *and one of the reasons that childbirth is more difficult for the human female than for other mammals.*

Formation of Ova

In human females, the primary oocytes begin to form during the third month of fetal development. By the time of birth, the two ovaries contain some 2 million primary oocytes, which have reached prophase of the first meiotic division. These primary oocytes remain in prophase until the female matures sexually. Then, under the influence of hormones, the first meiotic division of a primary oocyte resumes, resulting in a secondary oocyte and a polar body. The first meiotic division is completed at about the time of *ovulation* (the release of the oocyte from the ovary).

Of the original 2 million primary oocytes, about 300 to 400 reach maturity, usually one at a time about every 28 days after puberty. As many as 50 years may elapse between the beginning and the end of the first meiotic division in a particular oocyte. The second meiotic division, resulting in production of an ovum and a polar body from the secondary oocyte, begins at ovulation and is completed only after a sperm penetrates the oocyte.

Maturation of the oocyte involves both meiosis and a great increase in size. This size increase reflects the accumulation of stored food reserves and metabolic machinery, such as messenger RNA and enzymes, required for the early stages of development. As you can see in Figure 10–11 on page 145, oocytes do not divide into equal-sized cells, as spermatocytes do. Instead, one very large cell (100 micrometers in diameter in humans) is produced. The other nuclei are, in effect, discarded.

An oocyte and the cells surrounding it in the ovary are known as an *ovarian follicle* (Figure 31–9). The cells of the follicle nourish the oocyte and also secrete estrogens, hormones that initiate the buildup of the endometrium.

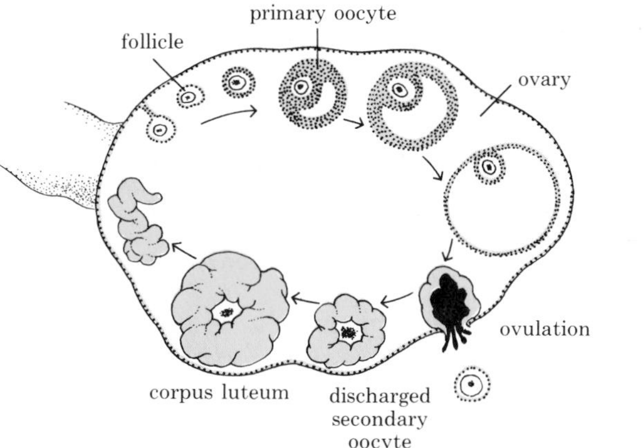

31-9 *Oocytes develop near the surface of the ovary within follicles. After an oocyte is discharged from a follicle (ovulation), the remaining cells of the ruptured follicle give rise to the corpus luteum, which secretes estrogens and progesterone. If the ovum is not fertilized, the corpus luteum is resorbed in two to three weeks. If the ovum is fertilized, the corpus luteum persists, sustaining the production of estrogens and progesterone, which maintain the uterus during pregnancy.*

Pathway of the Oocyte

During the final stages of its growth, the ovarian follicle moves to the surface and eventually bursts, releasing the oocyte. The oocyte is swept into the *oviduct* (sometimes called a Fallopian tube) by the beating of cilia on long, fingerlike projections at the opening of the tube (Figure 31-10). (This mechanism is so effective that women who have only one ovary and only one oviduct—and these on opposite sides of the body—have become pregnant.)

The oocyte then moves slowly down the oviduct, propelled by peristaltic waves produced by the smooth muscle of the walls. The journey from the oviduct to the uterus takes about three days. An unfertilized oocyte, however, lives only about 24 hours after it is ejected from the follicle. So, fertilization, if it is to occur, must occur in an oviduct.

Sterilization in women is usually carried out by severing the oviducts, thus preventing sperm from meeting the egg. Tubal ligation ("tying off the tubes"), as the procedure is called, usually does not produce hormone changes.

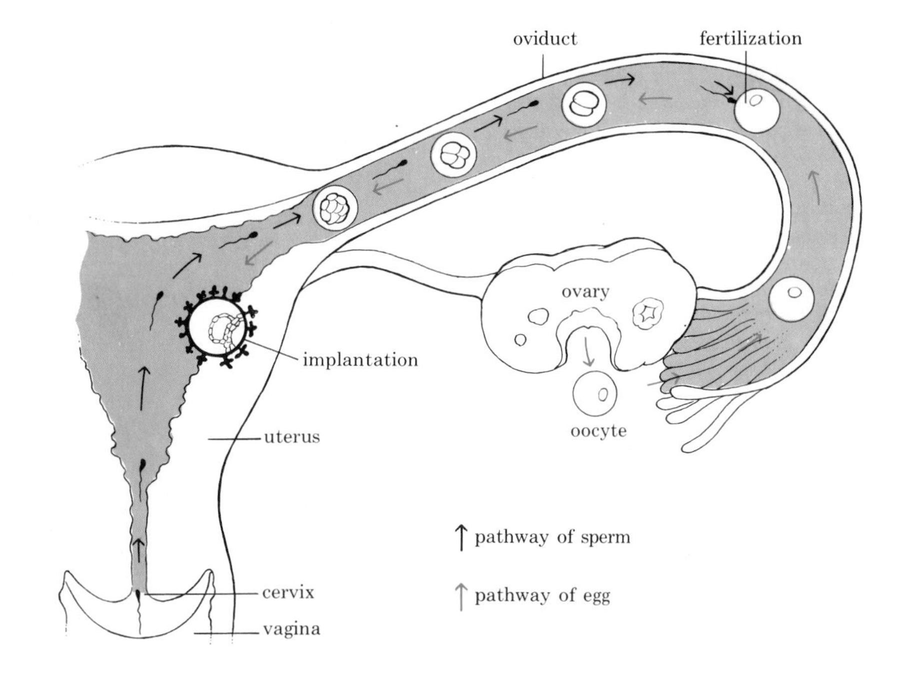

31-10 *Fertilization of the egg by sperm. About once a month in the nonpregnant female of reproductive age, an oocyte is ejected from the ovary and is swept into one of the oviducts. Fertilization, when it occurs, normally takes place within an oviduct, after which the embryo becomes implanted in the lining of the uterus. Muscular movements of the oviduct, plus the beating of the cilia that line it, propel the egg cell down the tube toward the uterus. If the oocyte is not fertilized, it dies, usually within 12 to 24 hours. A sperm cell has an average life of 48 hours within the female reproductive tract.*

Table 31-2 Major mammalian gonadotropic and sex hormones in females

Hormone	Principal Source	Principal Effects	Control
FSH	Pituitary	Stimulates growth of ovarian follicle, stimulates estrogen production	Hypothalamus
LH	Pituitary	Stimulates growth of ovarian follicle and release of oocyte, stimulates estrogen and progesterone production	Hypothalamus
Estrogens	Ovary, placenta	Produce and maintain female sex characteristics, thicken lining of uterus	FSH and LH
Progesterone	Ovary, placenta	Further prepares uterine lining for pregnancy, inhibits uterine movements, promotes development of milk ducts	LH

Hormonal Regulation in Females

The production of oocytes in all vertebrate females is cyclic. The cycle involves both the interplay of hormones—including estrogens, progesterone, and the two gonadotropic hormones, FSH and LH—and changes in the follicle cells and the lining of the uterus. The recurring pattern of varying hormone levels and tissue changes is known in humans as the *menstrual cycle* (Figure 31-11).

The onset of the menstrual cycle marks the beginning of puberty in the human female. The average age of puberty is 13½, but the normal range is very wide. The increased production of female sex hormones preceding puberty induces the development of secondary sex characteristics, such as pubic and axillary (underarm) hair and enlargement of hips and breasts.

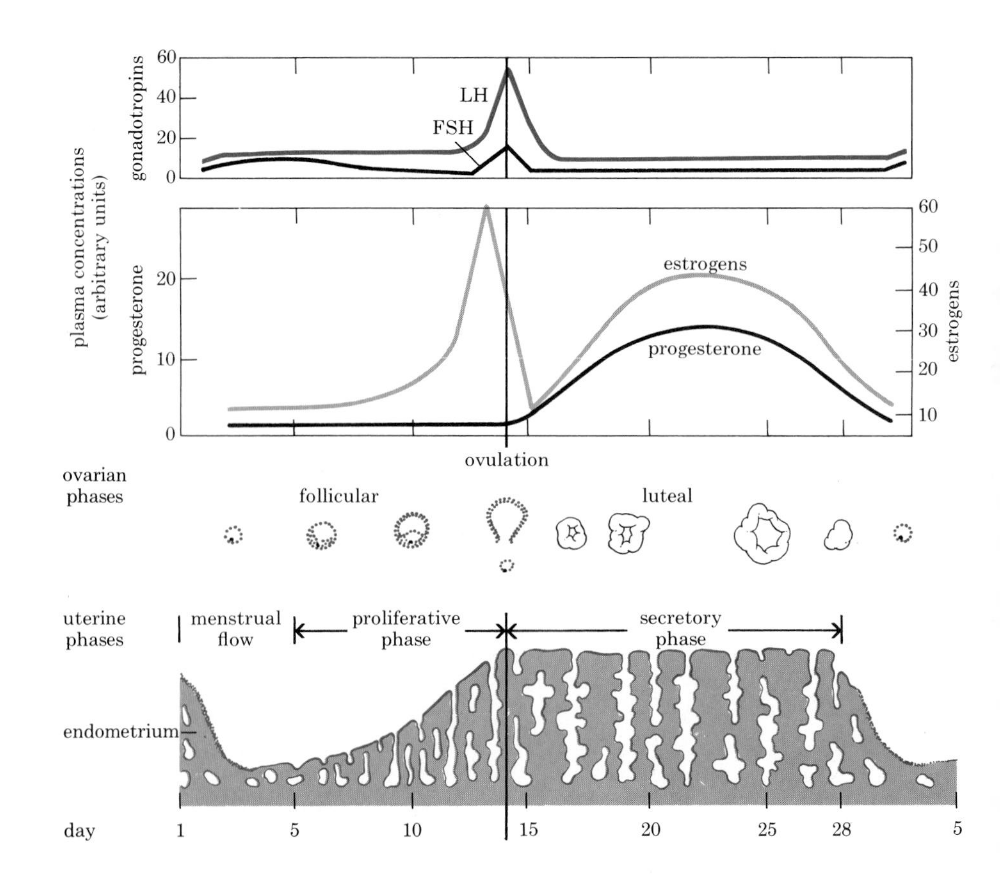

31-11 *Diagram of events taking place during the menstrual cycle. The cycle begins with the first day of menstrual flow, which is caused by the shedding of the endometrium, the lining of the uterine wall. The increase of FSH and LH during the first week promotes the growth of the ovarian follicle and its secretion of estrogens. Under the influence of estrogens, the endometrium regrows. A sharp increase of LH from the pituitary about midcycle stimulates the release of the oocyte (ovulation). (It is not known what role, if any, is played by the simultaneous increase in FSH.) Following ovulation, LH and FSH levels drop. The follicle is converted to the corpus luteum, which secretes estrogens and also progesterone. Progesterone further stimulates the endometrium, preparing it for implantation. If pregnancy does not occur, the corpus luteum degenerates, the production of progesterone and estrogens falls, the endometrium begins to slough off, FSH and LH concentrations increase once more, and the cycle begins anew.*

Orgasm in the Female

Under the influence of a variety of stimuli, the clitoris, the labia, and other tissues in the pelvic region of the female become engorged and distended with blood, as does the penis of the male. This process is somewhat slower in women than in men, largely because the valves trapping the blood in the female sexual structures tend to leak. The distension of the tissues is accompanied by the secretion into the vagina of a fluid that lubricates its walls.

Orgasm in the female, as in the male, is marked by rhythmic muscular contractions, followed by expulsion into the veins of the blood trapped in the engorged tissues. Homologous muscles produce orgasm in the two sexes, but in the female there is no ejaculation of fluid through the urethra or the vagina. Orgasm in the female is not necessary for conception.

Contraceptive Techniques

Using no contraceptive methods, 80 percent of child-bearing-age women having regular sexual intercourse become pregnant within a year. However, a variety of contraceptive techniques is now available for couples who wish to prevent or defer pregnancy. In Table 31-3 (page 436), they are rated in order of effectiveness, in terms of average number of pregnancies per year among the women of child-bearing age using them. In most cases, two figures are given for effectiveness. The first, lower, figure is an "ideal" figure, obtainable when the method is used consistently and correctly. The second figure is an average figure, reflecting actual experience.

Human Development

Fertilization

About 300 to 400 million sperm cells are released into the vagina at ejaculation. Of these, several hundred thousand make their way up the oviducts, moving against the beating of the cilia that line the tubes. Sperm cells can survive about 48 hours in the female reproductive tract.

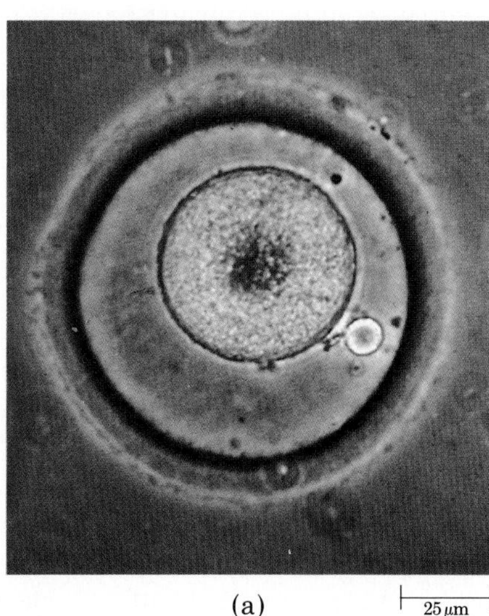

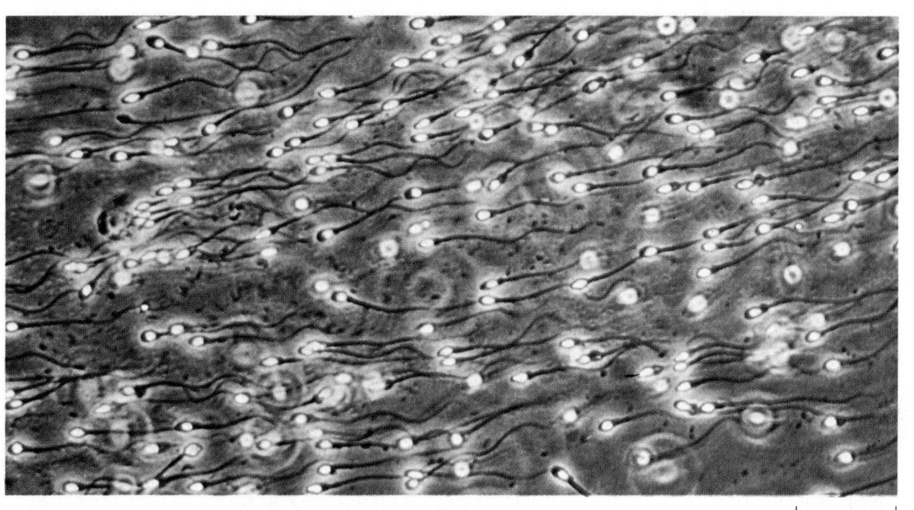

31-12 (a) *A human egg. The dark strands within the nucleus are chromosomes. A polar body is at the lower right.* (b) *Human sperm. Of the 300 to 400 million sperm cells released in an ejaculation, only one can fertilize each egg cell.* (a, *Roberts Rugh and Landrum B. Shettles, M.D.,* From Conception to Birth: The Drama of Life's Beginnings, *Harper & Row, Publishers, Inc., New York, 1971;* b, *Fritz Goro, Time-Life Picture Agency.)*

Table 31-3 Methods of birth control currently available

Method	Mode of Action	Effectiveness (Pregnancies per 100 Women per Year)	Action Needed at Time of Intercourse	Requires Instruction in Use	Possible Undesirable Effects
Vasectomy	Prevents release of sperm	0	None	No	Usually produces irreversible sterility
Tubal ligation	Prevents passage of egg cell to uterus	0	None	No	Usually produces irreversible sterility
"The pill" (estrogens and progesterone)	Prevents follicle maturation and ovulation or implantation	0–10	None	Yes, timing	Early—some water retention, breast tenderness, nausea; late—increased risk of cardiovascular disease
Intrauterine device (coil, loop, IUD)	Possibly prevents implantation	1–5	None	No	Menstrual discomfort, displacement or loss of device, uterine infection
"Minipill" (progesterone alone)	Probably prevents sperm from entering uterus	1–10	None	Yes, timing	?
"Morning-after pill" (50 × normal dose of estrogen)	Arrests pregnancy, probably by preventing implantation	?	None	Yes, timing	Breast swelling, nausea, water retention, cancer (?)
Condom (worn by male)	Prevents sperm from entering vagina	3–10	Yes, male must put on after erection	Not usually	Some loss of sensation in male
Diaphragm with spermicidal jelly	Prevents sperm from entering uterus, jelly kills sperm	3–17	Yes, insertion before intercourse	Yes, must be inserted correctly each time	None known
Vaginal foam, jelly alone	Spermicidal, mechanical barrier to sperm	3–22	Yes, requires application before intercourse	Yes, must use within 30 minutes of intercourse; leave in at least 6 hours after	Not usually, may cause irritation
Withdrawal	Removes penis from vagina before ejaculation	9–25	Yes, withdrawal	No	Frustration in some
Rhythm	Abstinence during probable time of ovulation	13–21	None	Yes, must know when to abstain	Requires abstinence during part of cycle
Douche	Washes out sperm that are still in the vagina	?–40	Yes, immediately after	No	None

Although only one sperm cell fertilizes the egg cell, a large number must reach it for fertilization to occur. The additional sperm cells apparently play an accessory role, perhaps by bringing about chemical changes necessary for fertilization. The membrane of one sperm cell fuses with the membrane of the secondary oocyte, and the sperm cell's contents are emptied into the oocyte's cytoplasm. After the second meiotic division of the oocyte is completed, the nuclei meet within the egg cell and merge. This event determines what particular individual—what genetic identity—will exist.

After fertilization, the egg continues its passage down the oviduct, where the first cell divisions take place. At about 36 hours after fertilization, the fertilized egg divides to form two cells. At 60 hours, the two cells divide to form four cells. At three days, the four cells divide to form eight. As the

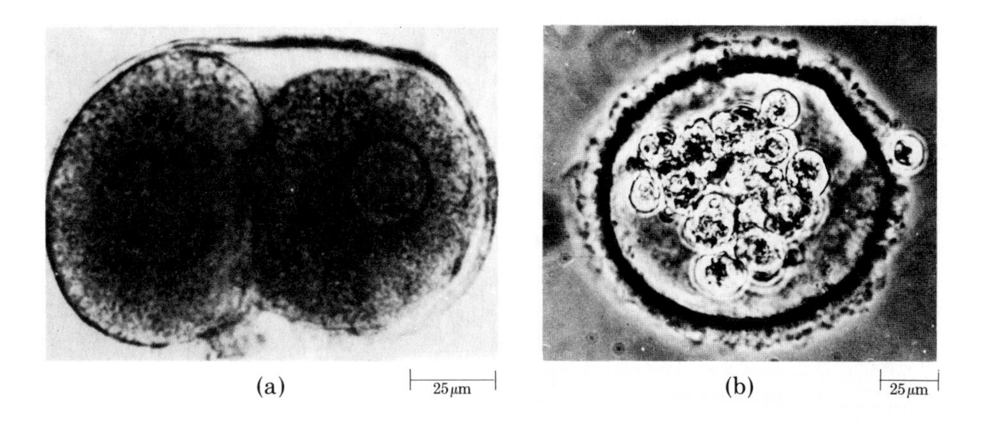

31–13 (a) *A human embryo at the two-cell stage. The cells are still surrounded by an outer membrane.* (b) *The cells continue to divide, but, since the volume of the embryo does not increase, they are still easily contained within the same membrane.*

cells divide further, an inner space develops, so that a hollow ball of cells is formed. At this stage, the ball of cells is called a *blastula*. One side of the ball of cells, which is thicker than the other side, will develop into the embryo itself; the remaining cells will develop into the membranes that enclose the embryo. Although the embryo soon consists of many cells, it is not significantly larger than the single egg cell from which it originated.

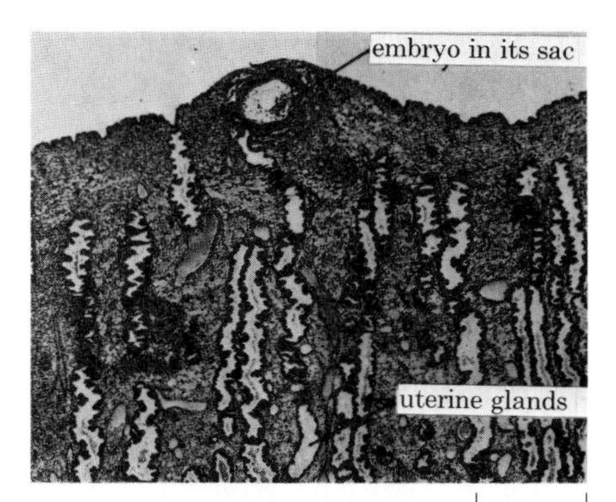

31–14 *Implantation. The tiny embryo invades the lining of the uterus within a week after fertilization. Subsequently, the placenta begins to form; this organ is the source of hormones that help to maintain pregnancy. Implantation usually occurs three to four days after the young embryo reaches the uterus.*

Implantation

About the sixth day after fertilization, the embryo makes contact with the tissues of the uterus. As a result of chemical interactions between the tiny mass of cells and the rich uterine lining, the outer epithelium of the lining breaks down and the embryo becomes embedded (implanted) in the nourishing tissue (Figure 31–14). As the embryo penetrates the tissues of the uterus, it becomes surrounded by ruptured blood vessels and the nutrient-filled blood escaping from them. By this stage, the embryo is developing extraembryonic membranes analogous to the membranes surrounding the chick embryo (see essay, page 438). The outermost membrane, the chorion, develops fingerlike projections that further invade the uterine tissues. Blood vessels develop in these tissues that connect with the blood vessels of the embryo as they develop. Ultimately, the placenta forms.

The placenta is a spongy tissue through which oxygen, food molecules, and wastes are exchanged between mother and embryo. It is formed from a maternal tissue, the endometrium, as well as from an extraembryonic membrane, the chorion, and has a rich blood supply from both. Within the placenta, the capillaries of the embryonic and maternal circulatory systems intertwine but are not directly connected. Molecules, including food and oxygen, diffuse from the maternal bloodstream through the placental tissue and into the blood vessels that carry them into the embryo. Similarly, carbon dioxide and other waste products from the embryo are picked up from the placenta by the maternal bloodstream and carried away for disposal through the mother's lungs and kidneys. As the embryo grows larger, it remains attached to the placenta by the long umbilical cord, which permits it to float in its sac of amniotic fluid.

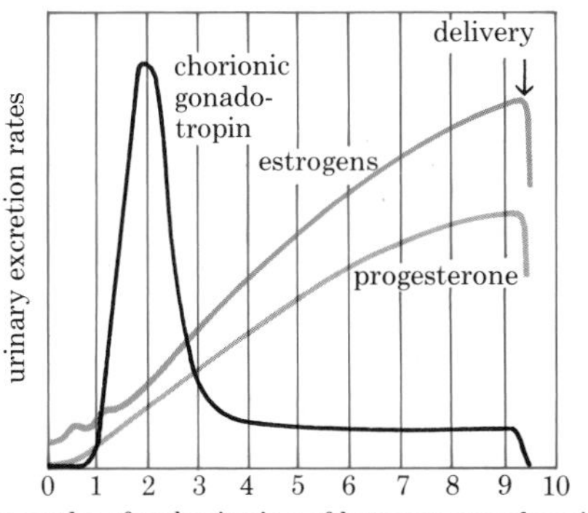

31–15 *Levels of estrogens, progesterone, and chorionic gonadotropin excreted in the urine during pregnancy. Urinary excretion rates indicate the hormone concentrations in the blood.*

The placenta is also a source of hormones. Early in pregnancy, it produces a hormone known as chorionic gonadotropin (Figure 31–15). Pregnancy tests involve the detection of this hormone in blood or urine. Chorionic gonadotropin stimulates the corpus luteum to continue its production of estrogens and progesterone. Later in pregnancy, the placenta itself produces estrogens and progesterone in large amounts.

The Amniote Egg

As we noted in Chapter 19, one of the most significant evolutionary steps among the animals was the development of the amniote egg, which contains its own water supply and so can be laid on land. In the amniote egg, the embryo is enveloped in a liquid-filled membrane, the amnion.

The amniote egg actually includes four membranes—the yolk sac, the allantois, the amnion, and the chorion—known collectively as the extraembryonic membranes. Each develops from the embryo itself. The figures show the development of these membranes in the chick egg, the most familiar example of the amniote egg.

In early development (a), body folds of the embryo begin to separate from the underlying yolk. One membrane, the yolk sac, grows around and almost completely encloses the yolk. A second, the allantois, arises as an outgrowth of the rear of the gut. The third and fourth are elevated over the embryo by a folding process during which the membrane is doubled. When the folds fuse (b), two separate membranes are formed. The inner one is the amnion, and the outer one is the chorion. The chorion eventually fuses with the allantois to form the chorioallantoic membrane (c), which, in the later stages of development, encloses embryo, yolk, and all the other structures.

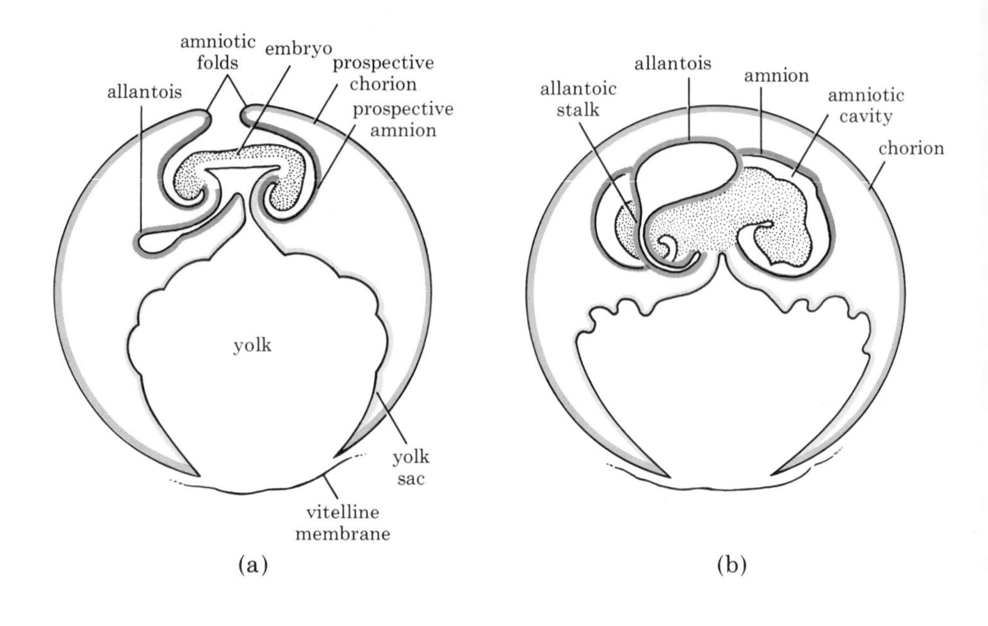

Although only a few mammals lay eggs (these are the monotremes, such as the duckbill platypus), the mammalian embryo develops within a system of membranes that closely resemble those found in the amniote egg. Even the yolk sac is still present, en-

The First Trimester

During the second week of its life, the embryo grows to 1½ millimeters in length, and its major body axis begins to develop. (In this and subsequent measurements, the embryo is measured from crown to rump.) As the body elongates, it can be seen to be divided into segments, known as somites.

During the third week, the embryo grows to 2⅓ millimeters long, and most of its major organ systems begin to form: the neural groove (Figure 31-16), which is the beginning of the central nervous system, the first organ system to develop; the heart and blood vessels; the primitive gut; and the muscle rudiments.

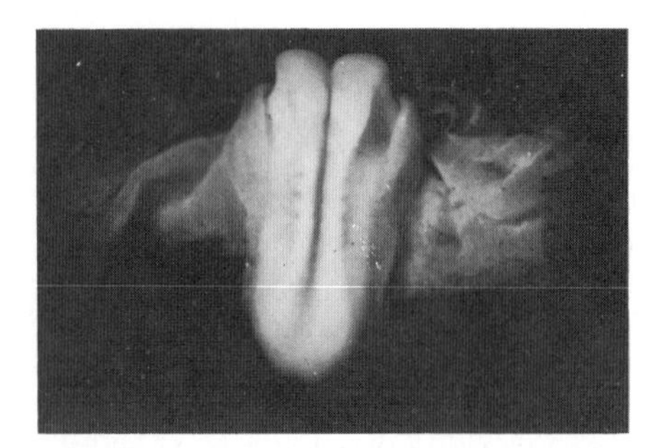

31-16 *As the embryo develops, two ridges form and enlarge. The groove between them is called the neural groove. From the neural groove, the dorsal nerve cord—one of the "trademarks" of the chordates—develops. These two folds soon close. This embryo is 18 days old.*

By 22 days, the very rudimentary heart, still only a tube, begins to flutter and then to pulsate. From this time on, the heart will not stop its 100,000 beats per day until death of the individual. Soon after, the eyes begin to form. Also, by this time, about 100 cells have been set aside (in the yolk sac) as germ cells, from which the ova or sperm cells of the individual will eventually develop.

By the end of the first month, the embryo is 5 millimeters in length and has increased its mass 7,000 times. The neural groove has closed, and the embryo is now C-shaped (Figure 31-17). At this stage, it can be clearly seen that the tissues lateral to the notochord (page 273) are arranged in paired somites. Each embryo has 40 pairs of somites, from which muscles,

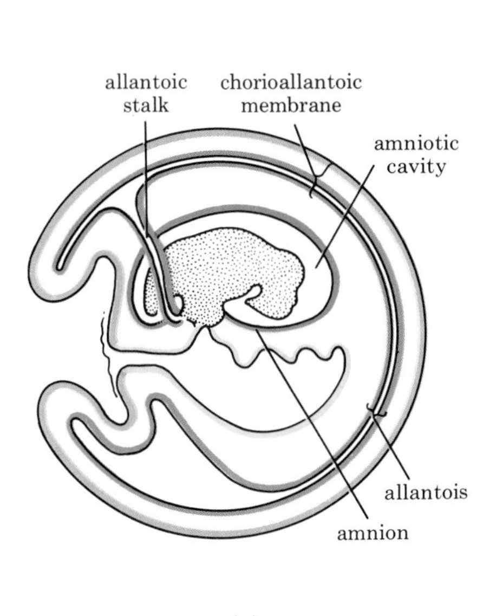

closing a space where the yolk "used to be." Among the placental mammals, the outermost extraembryonic membrane, the chorion, has become specialized to form part of the placenta, through which the embryo receives its nourishment.

bones, and connective tissues will develop. This segmentation of the muscles persists in the adult forms of lower vertebrates—fish, particularly—but not in the higher, terrestrial vertebrates. The heart, even as it beats, develops from a simple contracting tube into a four-chambered vessel.

During the second month, the embryo increases in mass about 500 times. By the end of this period, it weighs about 1 gram, slightly less than the weight of an aspirin tablet, and is about 3 centimeters long. Despite its small size, it is almost human-looking, and from this time on it is generally referred to as a fetus (Figure 31-19). Its head is still relatively large, because of the early and rapid development of the brain, but the size of the head in proportion to the body will continue to be reduced throughout gestation (and throughout childhood as well).

Arms, legs, elbows, knees, fingers, and toes are all forming during this time. As another reminder of our ancestry, there is a temporary tail. The tail reaches its greatest length in the second month and then gradually begins to disappear. The primitive organs have begun to form by this time. The liver now constitutes about 10 percent of the body of the fetus and is its main blood-forming organ. By the end of the second month, the major steps in organ development are more or less complete. In fact, the rest of development is mostly concerned with growth and the maturation of physiological processes.

The first two months are the most sensitive period as far as the possible influence of external factors is concerned. For example, when the arms and legs are mere rudiments (fourth and fifth weeks), a number of substances can upset the normal course of events and result in limb abnormalities. A large number of drugs have been linked to congenital deformities, and pregnant women and their doctors have become far more cautious about the use of any kind of medication during these critical first months. Similarly, exposure to x-rays at doses that would not affect an adult or even an older fetus may produce permanent abnormalities. Heavy alcohol consumption and smoking also affect normal development of organ systems.

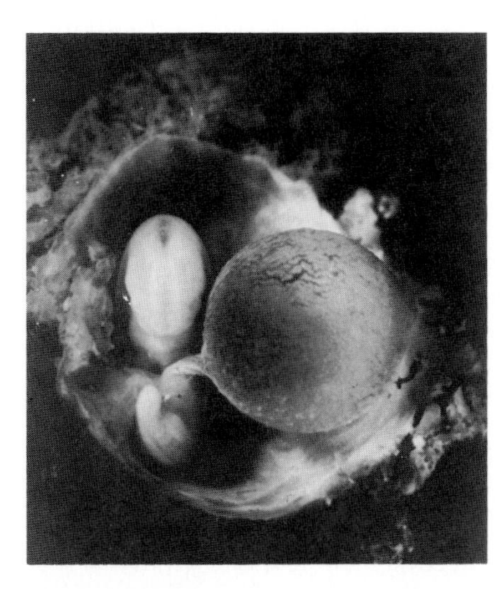

31-17 *A human embryo at 28 days. The balloon-like structure is the yolk sac. The embryo is curved toward you. At the top you can see the bulge of the rudimentary brain. The "seam" where the neural ridges closed is still visible.*

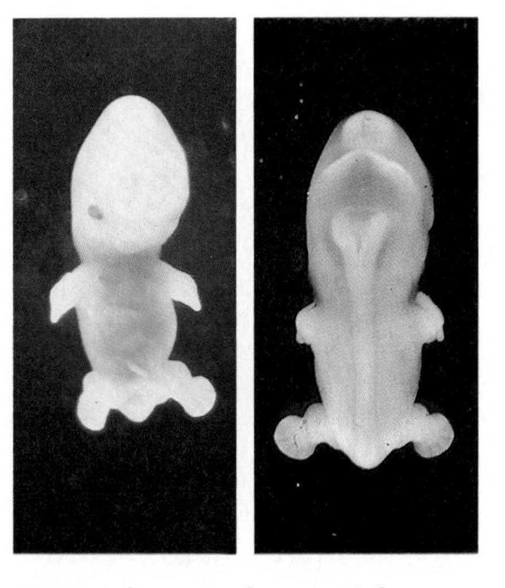

31-18 *A human embryo at 40 days, front and back views. Notice the spinal cord, brain, and paddlelike feet. The embryo is now about 16 millimeters long, little more than half an inch. (Rugh and Shettles, op. cit.)*

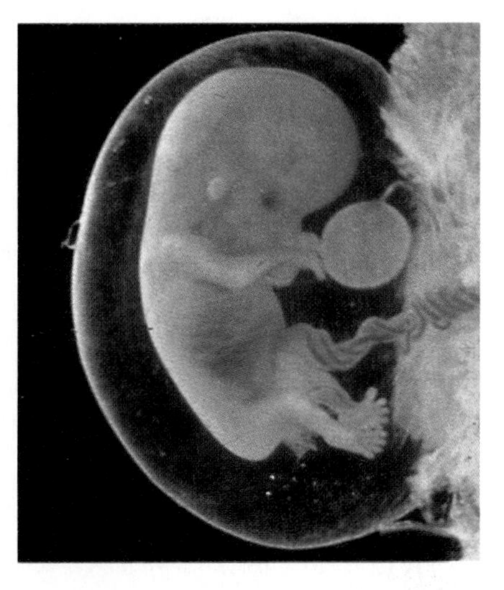

31-19 *Fetus at 2 months. The yolk sac is now smaller in comparison to the embryo but still persists. The umbilical cord, connecting the fetus to the placenta, contains both veins and arteries. (Rugh and Shettles, op. cit.)*

31-20 *Fetus at 2 months, 1 week, now almost 4 centimeters long. The eyes have lenses but are covered by lids that will fuse during the third month and remain closed for the next three months. (Rugh and Shettles, op. cit.)*

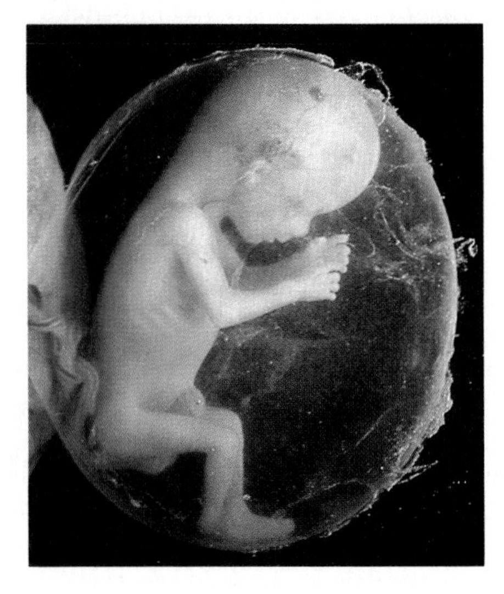

Infections may also affect the development of the embryo. Rubella (German measles) is a very mild disease in children and adults. Yet when contracted by the mother during the fourth through the twelfth weeks of pregnancy, it can have damaging effects on the formation of the heart, the lens of the eye, the inner ear, and the brain, depending on exactly when the infection occurs in relation to embryonic development.

During the third month, the fetus begins to move its arms and kick its legs, and the mother may become aware of its movements. Reflexes, such as the startle reflex and (by the end of the third month) sucking, first appear at this time. Its face becomes expressive; the fetus can squint, frown, or look surprised. Its respiratory organs are fairly well formed by this time but, of course, are not yet functional. The external sexual organs begin to develop.

By the end of the third month, the fetus is about 9 centimeters long from the top of its head to its buttocks and weighs about 15 grams ($\frac{1}{2}$ ounce). It can swallow and occasionally does swallow some of the fluid that surrounds it in the amniotic sac. The finger, palm, and toe prints are now so well developed that they can be clearly distinguished by ordinary fingerprinting methods. The kidneys and other structures of the urinary system develop rapidly, although waste products are still disposed of through the placenta. By the end of this period—the first trimester of development—all the major organ systems have been laid down.

The Second Trimester

During the fourth month, movements of the fetus become obvious to the mother. Its bony skeleton is forming and can be seen with x-rays. The body is becoming covered with a protective cheesy coating. The four-month-old fetus is about 14 centimeters long and weighs about 115 grams (4 ounces).

By the end of the fifth month, the placenta covers about 50 percent of the uterus. The fetus has grown to almost 20 centimeters and now weighs 250 grams. It has acquired hair on its head, and its body is covered with a fuzzy, soft hair called the lanugo, from the Latin word for "down." Its heart, which beats between 120 and 160 times per minute, can be heard with a

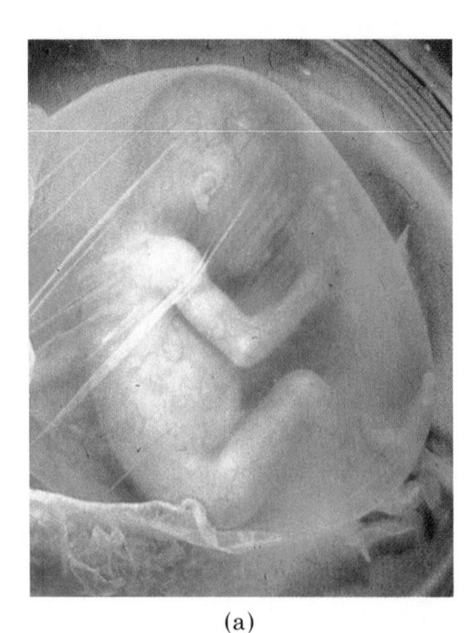

(a)

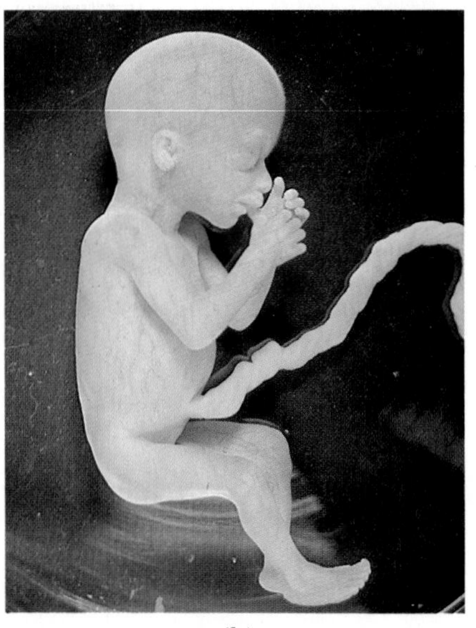

(b)

31-21 (a) *Fetus at 16 weeks. Blood vessels are visible through the translucent skin. The hands and feet are well formed; even the fingernails are visible. The fetus is now about 13 centimeters long and fills the uterus, which is expanding as the fetus grows.* (b) *A 17-week-old fetus, sucking its thumb. (Rugh and Shettles, op. cit.)*

stethoscope. The five-month-old fetus is already discarding some of its cells and replacing them with new ones, a process that will continue throughout its lifetime.

During the sixth month, the fetus has a sitting height of 30 to 36 centimeters and weighs about 680 grams. By the end of the sixth month, it could survive outside the mother's body, although probably only with respiratory assistance in an incubator. Its skin is red and wrinkled, and the cheesy body covering, which helps protect the fetus against abrasions, is now abundant. Reflexes are more vigorous. In the intestines is a pasty green mass of dead cells and bile, which will remain there until birth.

The Final Trimester

During the final trimester, the fetus increases greatly in size and weight. In fact, it normally doubles in size just during the last two months. During this period, many nerve tracts are forming and brain cells are being produced at a very rapid rate. By the seventh month, brain waves can be recorded, through the abdomen of the mother, from the cerebral cortex of the fetus. Some recent research indicates that the protein intake of the mother is important during this period if the child is to have full development of its brain. Infants weighing less than 2,000 grams (4 pounds, 6

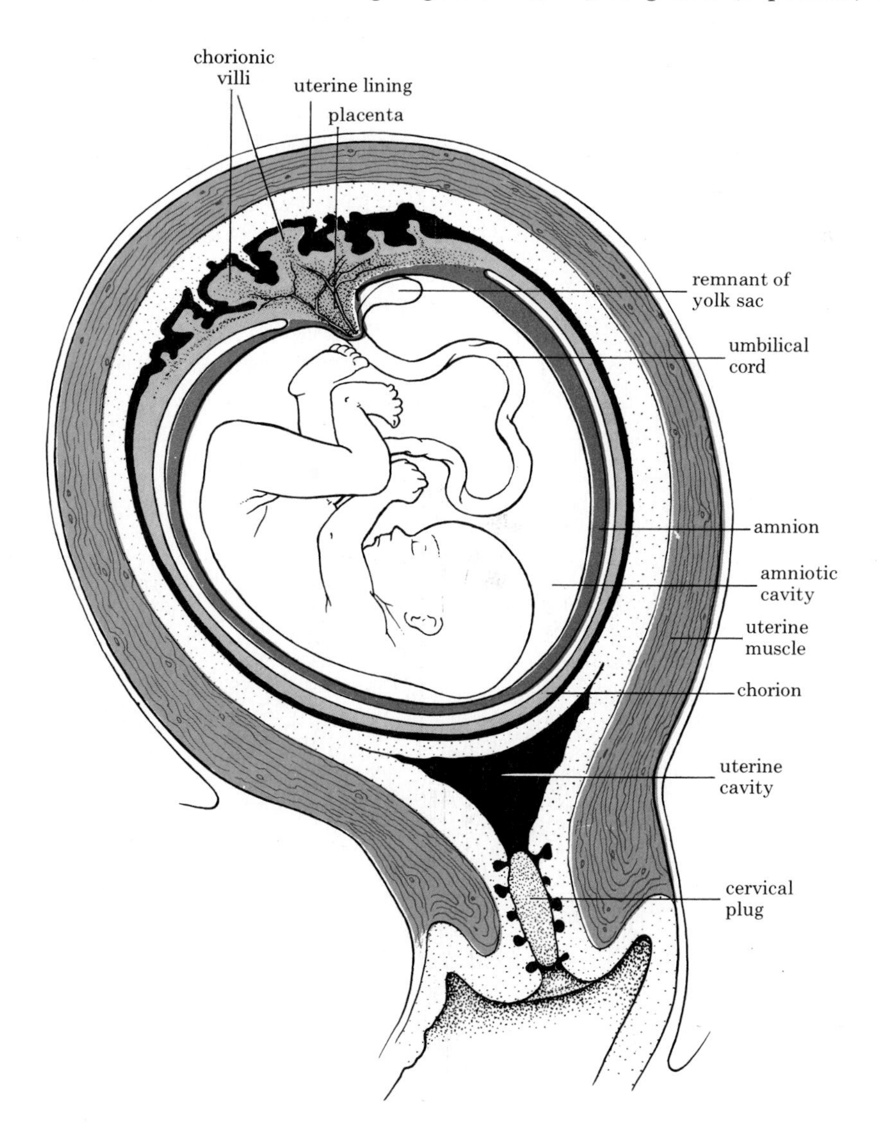

31-22 *A human fetus, shortly before birth, showing the protective membranes surrounding it and the uterine tissues. The cervical plug is composed largely of mucus. It develops under the influence of progesterone and serves to exclude bacteria and other infectious agents from the uterus. In 95 percent of all births, the fetus is in this head-down position.*

ounces) at birth are at high risk of death or severe brain damage. Alcohol consumed by the mother during this trimester is concentrated by and can lead to tissue damage in the fetus.

During the last month of pregnancy, the growth rate of the baby begins to slow down. (If it continued at the same rate, the child would weigh 90 kilograms—about 200 pounds—by its first birthday.) The placenta begins to regress and becomes tough and fibrous.

Birth

The date of birth is calculated as about 266 days after conception or 280 days after the beginning of the last regular menstrual period. Babies rarely are born on the scheduled day, but some 75 percent are born within two weeks of that day.

Labor is divided into three stages: dilation, expulsion, and placental stages. Dilation, which usually lasts from 2 to 16 hours (it is longer with the first baby than with subsequent births), begins with the onset of contractions of the uterus. It ends with the full dilation, or opening, of the cervix. At the beginning, uterine contractions occur at intervals of about 15 to 20 minutes and are relatively mild. By the end of the dilation stage, contractions are stronger and occur about every 1 to 2 minutes. At this point, the opening of the cervix is about 10 centimeters in diameter. Rupture of the amniotic sac, with the release of fluids, usually occurs during this stage.

The second, or expulsion, stage lasts 2 to 60 minutes. It begins with the full dilation of the cervix and the appearance of the head in the cervix, called crowning. Contractions at this stage last from 50 to 90 seconds and are 1 or 2 minutes apart.

31-23 *The birth of a baby.* (a) *The baby's head appears.* (b) *The parents see their newborn infant for the first time. When the umbilical cord—the baby's lifeline throughout its development—is severed, the infant will begin its separate existence.*

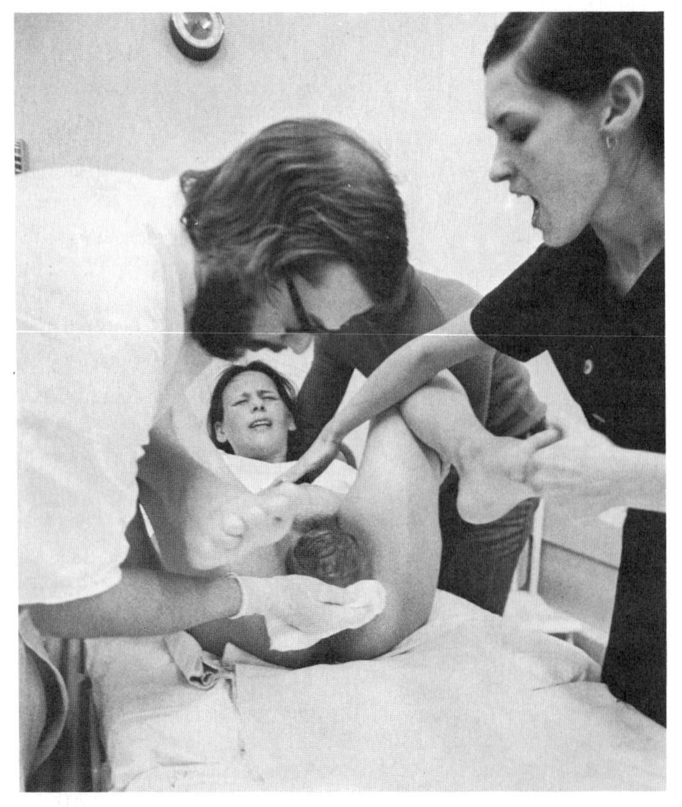

(a)

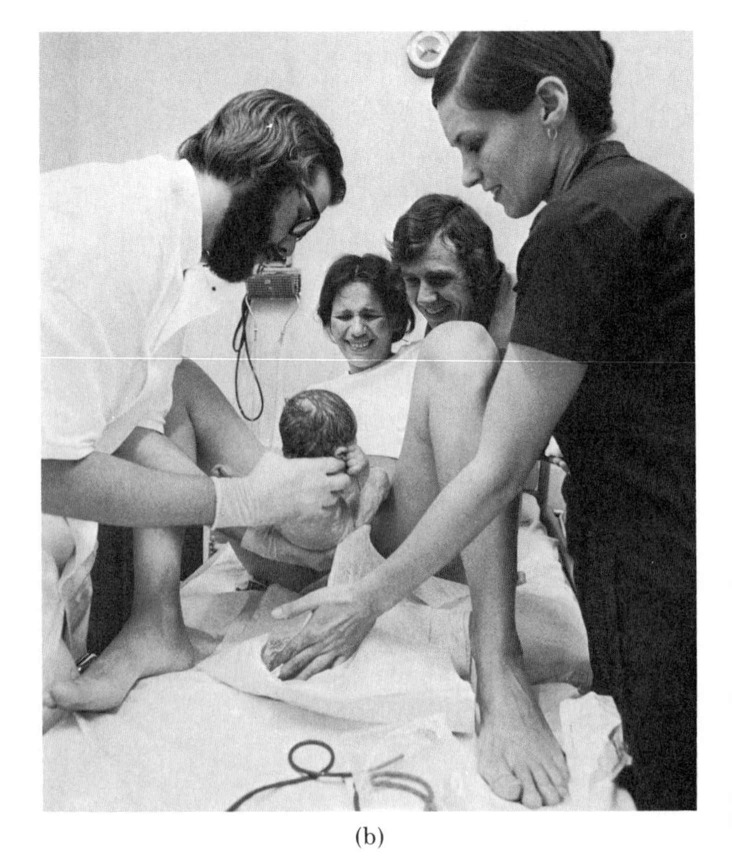

(b)

A Family of Potent Fatty Acids: Prostaglandins

Among the most potent of all biological materials is a group of related chemicals first detected in semen. These substances were thought at the time to be produced by the prostate gland, and so they were called prostaglandins. Later research revealed, however, that most of the prostaglandins in semen are synthesized in the seminal vesicles. Since their initial discovery, 16 natural prostaglandins have been found. One major source is menstrual fluid.

Although prostaglandins have hormonelike properties, they differ from hormones in four significant ways. First, unlike any other hormones, they are fatty acids. Second, their target tissues are generally either the same tissues in which they are produced or the tissues of another individual. (As you will recall from Chapter 26, most hormones are produced in one tissue and travel through the bloodstream to another tissue of the same individual where they exert their effects.) Third, they produce marked effects at extremely low concentrations. Finally, prostaglandins appear to be produced by cell membranes.

Among the many effects of prostaglandins, one of the most striking is their capacity to induce contractions in smooth muscle. This is believed to play several important roles in reproduction. The walls of the uterus, which are composed of smooth muscle, normally contract in continuous waves. After sexual intercourse, prostaglandins from semen are found in the female reproductive tract, where they increase the rhythmic contractions of the uterine wall and oviducts. It is believed that this action assists both the sperm on its journey to the oviduct and the oocyte as it travels from the oviduct to the uterus. The semen of some infertile males has been found to be poor in prostaglandins, and the uterus of infertile females is often unresponsive to prostaglandins.

The contractions of the uterus also increase when the endometrium is shed during a menstrual period and reach their greatest strength when a woman is in labor. Prostaglandins produced in the uterine lining are believed to play a key role in triggering the onset of both menstruation and labor. They may play a role in the effectiveness of IUDs.

Increased understanding of prostaglandins and their function has shed light on a long-perplexing medical problem. Between 30 and 50 percent of all women of child-bearing age experience painful abdominal cramps during the first day or two of each menstrual period, a condition known as dysmenorrhea. In most of these women, no abnormalities of the reproductive organs can be detected. Recent research has revealed, however, that the menstrual fluid of such women contains concentrations of prostaglandins two to three times higher than the levels found in the menstrual fluid of women without dysmenorrhea. Present evidence indicates that the increased prostaglandin levels not only cause stronger, more rapid contractions of the uterine walls but also reduce the blood supply to the tissue. As a result, less oxygen is available to the actively contracting muscles, creating an oxygen debt (page 108) and its accompanying pain. It is also hypothesized that prostaglandins may act directly on pain nerve endings, causing them to fire more rapidly.

Clinical studies in progress in this country and in Europe indicate that several compounds that inhibit prostaglandin synthesis are highly effective in reducing or, in some cases, completely eliminating the symptoms of dysmenorrhea. This alleviation is accompanied by a marked reduction in the levels of prostaglandins found in the menstrual fluid. Other studies are investigating the use of these compounds in preventing labor in women who are in danger of giving birth prematurely.

Prostaglandins have also been implicated in inflammatory and immune reactions and, by extension, in diseases such as rheumatoid arthritis and asthma. In the latter, they appear to cause the air passages in the bronchi to constrict. Aspirin inhibits prostaglandin activity, and this is thought to be the reason it reduces inflammation and fever. It is not surprising that the compounds that appear to be so effective in treating dysmenorrhea are also effective in the treatment of rheumatoid arthritis.

The third, or placental, stage begins immediately after the baby is born. It involves some contractions of the uterus and the expelling through the vagina of fluid, blood, and finally the placenta (also called the afterbirth), with the umbilical cord attached. The infant has begun his or her separate existence.

SUMMARY

The male and female gametes, sperm and ova, are formed by meiosis in the gonads (the testes and the ovaries). Sperm cells are produced in the seminiferous tubules of the testes. These sperm cells enter the epididymis, a tightly coiled tubule overlying the testis. The epididymis is continuous with the vas deferens, which leads through the abdominal cavity, around the bladder, and into the prostate. Within the prostate, the two vasa deferentia merge with the urethra, which leads out through the penis.

The penis is composed largely of spongy erectile tissue that can become engorged with blood, enlarging and hardening it. At the time of ejaculation, sperm are propelled along the vas deferens by contractions of smooth muscle in its walls. Secretions from the seminal vesicles and the prostate are added to the sperm as they pass to the urethra. Semen is expelled from the urethra by muscular contractions.

Production of sperm and the development of male secondary sex characteristics are under the control of hormones, including testosterone (an androgen) and two gonadotropins, luteinizing hormone (LH) and follicle-stimulating hormone (FSH). LH acts on the interstitial cells, located between the seminiferous tubules, to stimulate the production of testosterone. FSH and testosterone stimulate the production of sperm. Production of LH is inhibited by the presence of testosterone in the bloodstream, that is, through a negative feedback system.

The female gamete-producing organs are the ovaries. The primary oocytes develop within nests of cells called follicles. The first meiotic division of the oocyte begins in the female fetus and is not completed until ovulation. The second meiotic division is completed at fertilization. On the average, one oocyte is released every 28 days and travels down the oviduct to the uterus. If it is fertilized (which usually takes place in an oviduct), it becomes implanted in the lining of the uterus (the endometrium). If it is not fertilized, it degenerates and the endometrial lining is shed at menstruation.

The production of oocytes, the menstrual cycle, and the development at puberty of the uterus, vulva, breasts, and the secondary sex characteristics of the female are all controlled by hormones. Before ovulation, FSH and LH from the pituitary stimulate the ripening of the follicle and the secretion of estrogens. After ovulation, the corpus luteum, which forms from the emptied follicle, produces both estrogens and progesterone. Progesterone and estrogens both stimulate the growth of the endometrium.

The earliest stages in the development of a human embryo take place in the oviduct. There is a large increase in the number of cells but little or no increase in the total size of the embryo. Implantation of the embryo in the endometrium occurs about a week after fertilization. Subsequently, extraembryonic membranes combine with tissues of the endometrium to form the placenta, through which the fetus receives its food and oxygen and excretes carbon dioxide and other waste products. The fetus is attached to the placenta by the umbilical cord.

During the first three months, although the fetus remains very small (less than 8 centimeters in length), all the major organ systems are established and externally the fetus takes on a human appearance. This is the most critical period of development in terms of the risk of abnormalities.

During the second trimester, movements of the fetus can be felt by the mother, the heartbeat can be heard with a stethoscope placed on the mother's abdomen, and the skeleton, although still cartilaginous, can be seen in x-rays. By the end of six months, survival outside the uterus is possible, although an incubator would probably be required.

During the third trimester, there is a great increase in size and weight. New brain cells are formed rapidly during this period.

Birth occurs, on the average, 266 days after conception. Labor consists of three stages: dilation, during which the cervix opens and the muscular wall of the uterus contracts with increasing force; expulsion, during which the baby is expelled from the uterus; and the placental stage, during which fluids, blood, and the placenta are expelled.

QUESTIONS

1. Distinguish among the following: spermatocyte/spermatid/sperm cell; oocyte/ovum; ovarian follicle/corpus luteum; fertilization/implantation; embryo/fetus; dilation/expulsion/placental stage.

2. What would constitute the semen of a man who had undergone a vasectomy?

3. During which days in the menstrual cycle is a woman most likely to become pregnant? (Include data on the longevity of eggs and sperm in making this calculation.) Why is the use of the calendar method of birth control much less effective than other methods?

4. Describe the effects of luteinizing hormone (LH) and follicle-stimulating hormone (FSH) in the human male and female.

5. What would be an appropriate method of contraception for a couple who have infrequent intercourse (once a month)? For a couple who have frequent intercourse (three times a week), with plans to have children eventually? For a couple who have frequent intercourse, but do not wish to have any children? Under what circumstances, if any, would you elect to have a vasectomy or a tubal ligation?

32

The Brain and Its Functions

At birth a human infant is relatively helpless. It has been argued that we are born "too soon," and that the gestation period should be twice as long. However, the fetus remains in the uterus as long as possible, given the large size of its most unique organ—the human brain. This large and complex structure, which weighs about 1,400 grams (3 pounds) in an adult, has the consistency of semisoft cheese. As we saw in Chapter 25, the brain is made up of white matter (fiber tracts) and gray matter (nerve cell bodies and glial cells).

For at least 2,000 years, people have wondered about the relationship between the brain—this mass of semisoft substance—and the mind, the center of consciousness, thought, and emotion. René Descartes—"I think, therefore I am"—believed that the body is essentially a complex machine. He conceived of mind and body as two separate entities, whose meeting place was a small gland, the pineal, located within the skull. The persistence of this dualistic concept is reflected in such phrases, still in use today, as "mind over matter."

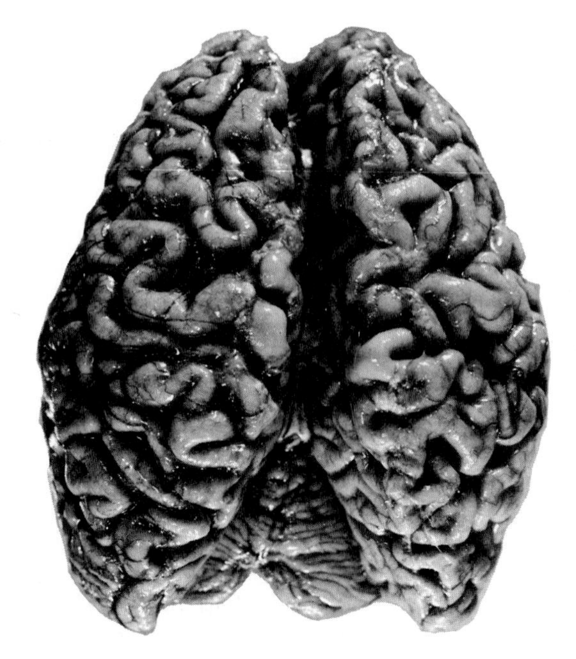

32-1 *A brain, viewed from above. A deep groove separates the two cerebral hemispheres, which are joined by a bridge of nerve tissue known as the corpus callosum. The cerebellum can be seen below the hemispheres at the base of the brain (see also page 362). Note the many convolutions of the cerebral cortex. By these, you can immediately distinguish this brain as human.*

Many years after the cell theory was accepted for all other parts of the body, the brain was considered to be an exception. One reason was that it was not possible to see neurons as readily as other types of cells because special staining techniques are required to reveal them. Another and undoubtedly more important reason is the feeling that mental processes are somehow special, different from other physiological functions.

Even today, many of us who accept without difficulty the fact that the tissues and cells of the digestive tract are responsible for digestion or that gases are exchanged across the surface of the lungs find it less easy to comprehend—to *really* believe—that the ideas, ideals, dreams, fantasies, thoughts, hypotheses, loves, hates, fears, and aspirations that make up the mind somehow can be explained by the interactions of the cells within our heads. Nor, indeed, can scientists begin to explain "mind" in terms of brain, despite the many intriguing discoveries that have been made in this field. Thus, the brain is the last remaining stronghold of the vitalists.

Probing Brain Functions

As we saw in Chapter 25 (page 363), the human brain consists of a number of different structures. Moreover, there is a vast, continuous series of interactions among the nerve cells of the many parts of the brain. The more that is learned about brain function, the more it becomes clear that the interactions—rather than the functions of the separate parts—determine how the brain works. Thus, the studies we are about to describe represent only the proverbial tip of the iceberg.

For example, when someone speaks to you, the sound waves produced by his or her voice activate receptors in your ear that stimulate the auditory nerve. This information is relayed through a network of nerve cells that lead to an area in the cerebral cortex; this is called the auditory area of the cortex or, more briefly, the auditory cortex. From the auditory cortex, the information, which is processed in some way that is not understood, is transmitted to some other part of the brain. It is not until this happens that you have "heard" what was said. The auditory cortex can be quite precisely defined: If it is destroyed—which can occur as a result of brain injury or disease—hearing is lost. Yet it is only a small part of a larger and much more complex picture.

The Cerebral Cortex

The cerebral cortex is a thin layer of gray matter that lies over the surface of the cerebrum. It is the most recent evolutionary development of the vertebrate nervous system. Fish and amphibians have no cerebral cortex, and reptiles and birds have only a rudimentary indication of a cortex. More primitive mammals, such as rats, have a relatively smooth cortex. Among the primates, the cortex becomes increasingly complex, reaching a culmination in the human cerebral cortex, in which the highly convoluted surface greatly increases the total area. Of the nearly 100 billion nerve cells in the human brain, about 10 billion are in the cerebral cortex.

Because of its relative accessibility, just below the surface of the skull, the cerebral cortex is the most thoroughly studied area of the human brain. In primates, each of the hemispheres is divided into lobes by two deep fissures, or grooves, in the surface. The principal fissures are the central sulcus and the lateral sulcus. There are four lobes: frontal, parietal, temporal, and occipital (Figure 32–2).

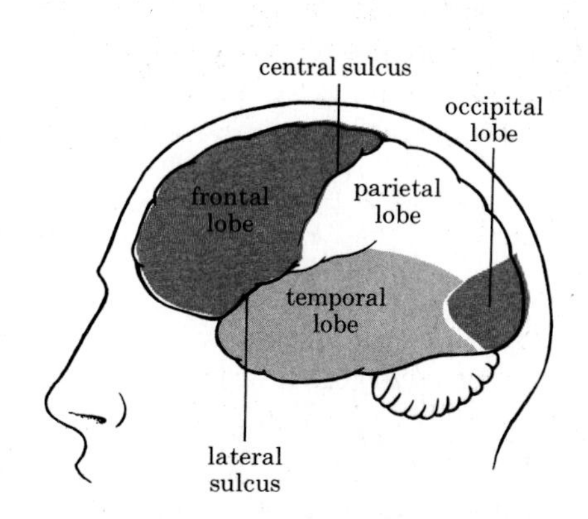

32–2 *The principal sulci (fissures) and lobes of the human cerebral cortex.*

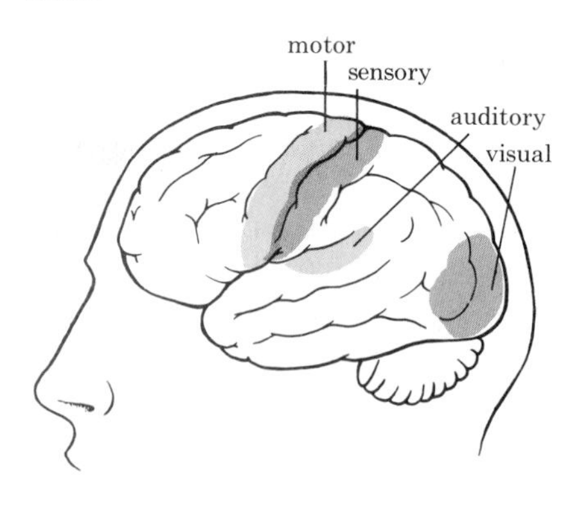

32-3 *The human cerebral cortex, showing the location of the motor and sensory areas, on either side of the central sulcus, and the auditory and visual zones.*

Motor and Sensory Functions

Certain areas of the cortex have been mapped in terms of the functions they perform (Figure 32-3). The area just anterior to the central sulcus, in the frontal lobe, is the motor cortex. It is involved in the integration of muscular activities. The area just posterior to the central sulcus, in the parietal lobe, is the sensory cortex. It is involved with the reception of tactile (touch) stimuli and taste. Many nerve fibers entering or leaving the brain cross in the brainstem; as a result, the motor and sensory areas for the left side of the body are in the right hemisphere, and vice versa.

In the temporal lobe, partially buried within the lateral sulcus, is the auditory cortex. By measuring electrical discharges in this area of the brain in dogs and cats exposed to sounds of varying frequencies, investigators have been able to show that different regions of the auditory cortex respond to different frequencies of sound.

The visual cortex occupies the occipital lobe. By using a tiny point of light to stimulate very small regions of the retina—the light-sensitive surface at the back of the eyeball—investigators have shown that each region of the retina is represented by a corresponding but larger region on the visual cortex.

In both experimental animals and humans, damage to specific areas of the parietal lobes seems to result in problems in discriminating specific stimuli. For example, if particular areas of the parietal lobes are damaged,

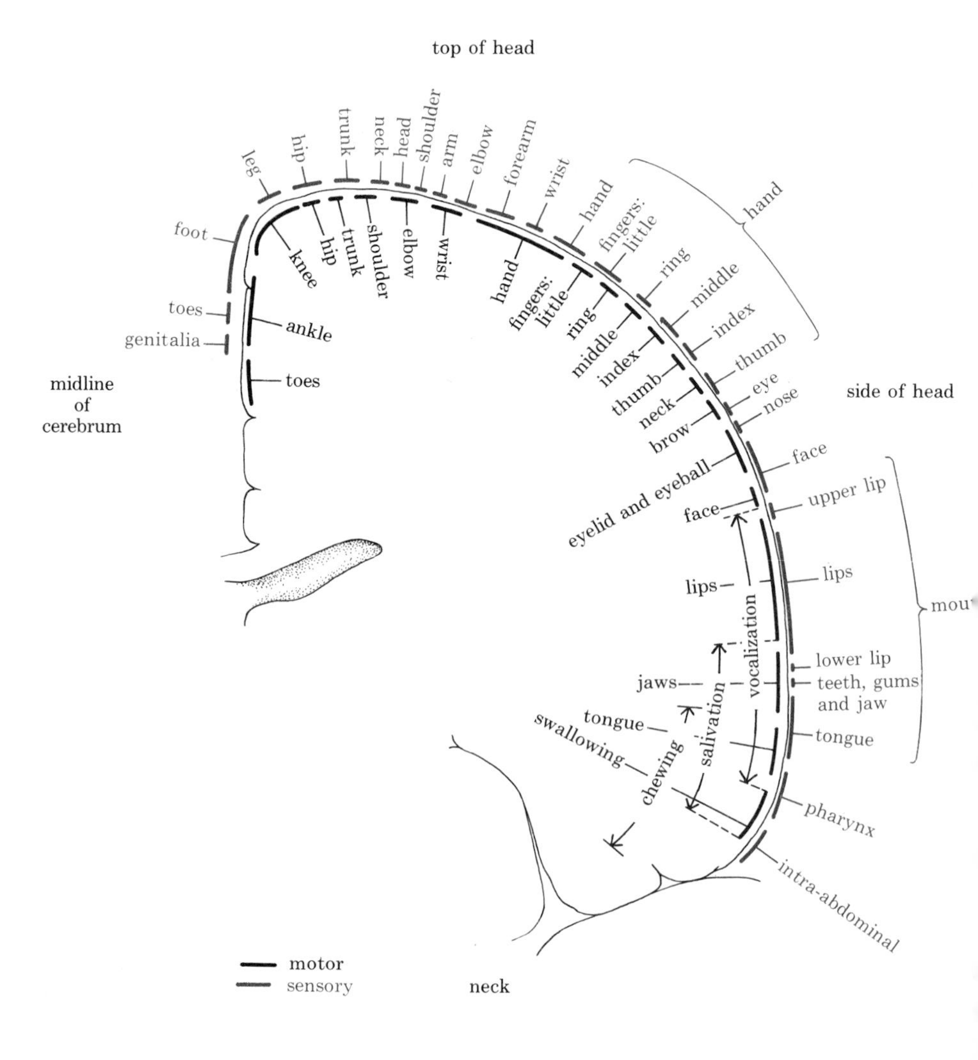

32-4 *A combined cross section of one hemisphere of the human cerebrum, indicating the functional areas of the motor and sensory cortices. The motor cortex is indicated in black; stimulation of these areas causes responses in corresponding parts of the body. The sensory cortex is in color; stimulation of various parts of the body produces electrical activity in corresponding parts of this cortex. Notice the relatively huge motor and sensory areas associated with the hand and the mouth. This map is based largely on studies done by neurosurgeon Wilder Penfield on patients undergoing surgical treatment for epilepsy.*

The motor and sensory cortices are located on either side of the central sulcus. Because nerve fibers cross in the brainstem, the motor and sensory cortices of the right hemisphere are associated with the left side of the body, and vice versa.

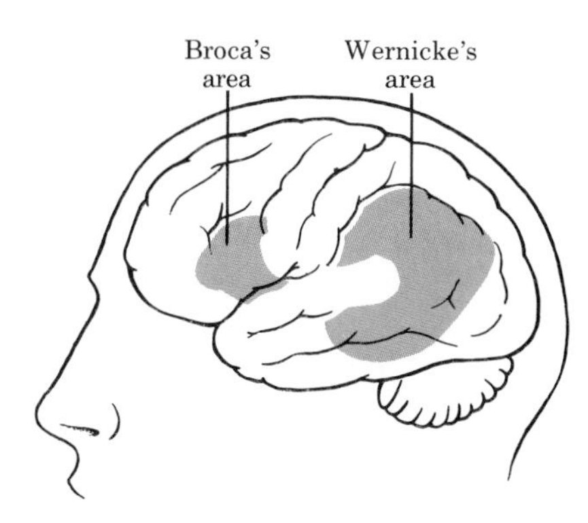

32–5 *The human cerebral cortex, showing the areas associated with speech. Damage to Broca's area—the more anterior area—affects motor control of speech. Damage to Wernicke's area affects conceptual aspects of language.*

monkeys can no longer tell the difference between patterns, objects, or sounds that they were previously able to distinguish.

In humans, damage to specific areas of the cortex on the left side of the brain (in most left-handed as well as all right-handed people) results in impairment of speech (Figure 32–5). The more anterior area (Broca's area) is adjacent to the region of the motor cortex that controls the movements of the muscles of the lips, tongue, jaw, and vocal cords. Damage to Broca's area results in slow and labored speech but does not affect comprehension. When the more posterior area (Wernicke's area) is damaged, speech is fluent but often meaningless, and comprehension of both spoken and written words is impaired. The two speech areas are joined by a nerve bundle. The corresponding areas in the right hemisphere are associated with music and spatial relations.

The Association Cortex

The unmapped areas of the cortex are sometimes known as the association, or "silent," cortex. The term "association" was introduced when this part of the cortex was thought to function as a sort of giant switchboard, interconnecting the motor and sensory zones. More recent studies suggest, however, that organization of the brain is more vertical than horizontal. For example, severing the motor cortex from the sensory areas by deep vertical cuts appears to have no effect at all on an animal's behavior. Anatomical studies support the interpretation that communication between the sensory and motor areas of the cortex takes place primarily via lower brain centers, particularly the thalamus.

Nor are the association areas actually "silent." Although sensory stimuli do not evoke discrete responses in these areas, and direct electrical stimulation of these areas does not elicit muscle movements, the association cortex shows steady background activity, apparent (but nonspecific) responses to stimuli, and increased activity following electrical stimulation of other areas of the brain. Clearly, something is taking place in these "silent" areas. The proportion of association area to motor and sensory areas is much higher in primates than in other mammals (even mammals such as cats), and is very large in humans. This suggests that these areas have something to do with what is special about the human mind.

About half of the association area of the cortex is in the frontal lobes, the part of the brain that has developed most rapidly during the recent evolution of *Homo sapiens*. The frontal lobes are responsible for our high forehead, as compared with the beetle brow of our most immediate ancestors. Public appraisal of their function is reflected in the terms "high brow" and "low brow."

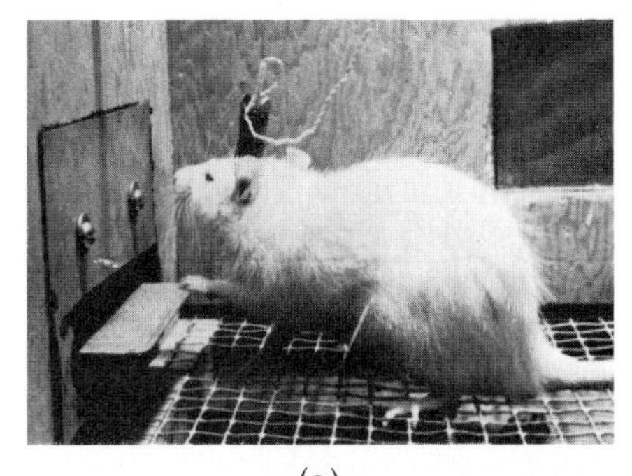

(a)

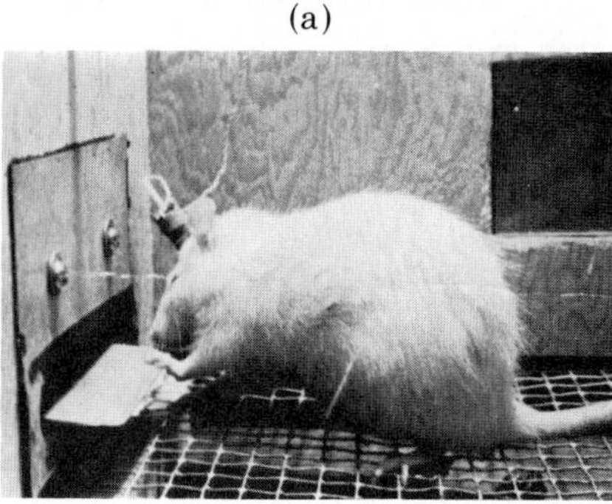

(b)

32–6 (a) *Rat with an electrode implanted in its hypothalamus.* (b) *By pressing a lever, the rat is able to stimulate particular cells in its own hypothalamus with a weak electric current. These nerve cells are components of the several "pleasure pathways" through the brain. Some animals press the lever as often as 8,000 times an hour and, even when hungry, prefer pressing the pleasure lever to pressing the one giving a food reward.*

The Hypothalamus

As we saw in Chapter 26, the hypothalamus regulates the activities of the pituitary gland and the several glands under pituitary feedback control. It is also the major central brain structure concerned with the functions of the autonomic nervous system, particularly with its sympathetic division. Thus it is, in effect, the chief coordinating center for the combined activities of the nervous and endocrine systems. The clusters of neurons that make up the hypothalamus—which are paradoxically very small—are critically involved in appetite, thirst, sexual behavior, sleep, temperature regulation, and emotional behavior in general (Figure 32–6).

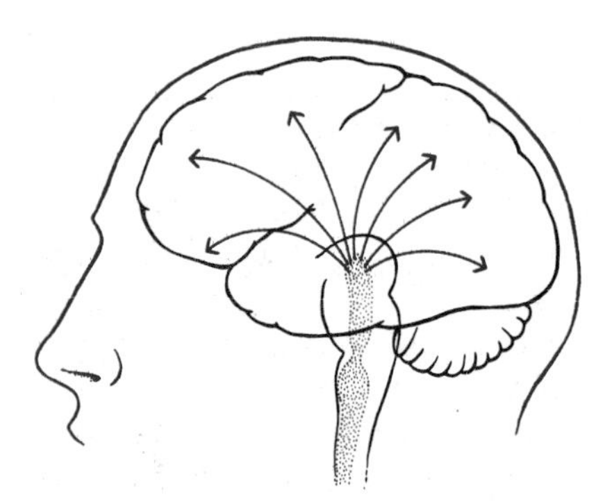

32-7 *The reticular system is a diffuse network of neurons in the brainstem. Here, incoming stimuli are monitored, analyzed, and relayed to other areas of the brain.*

Activating Systems of the Brain

Behavior, especially complex behavior, involves circuits connecting many different areas of the brain. These circuits are often difficult to identify anatomically. Two in particular, the reticular system and the limbic system, have commanded the attention of investigators.

The Reticular System

The reticular system, a diffuse network of neurons in the brainstem, is of particular interest to physiological psychologists because it is involved with arousal of that hard-to-define state we know as consciousness. All the sensory systems have fibers that feed into the reticular system, which apparently filters incoming stimuli and "decides" whether or not they may be important. Stimulation of this system, either artificially or by the incoming sensory impulses, results in increased activity in other areas of the brain.

The existence of such a filtering system is well verified by ordinary experience. A person may sleep through the familiar blare of a subway train or a loud radio or TV program but wake instantly at the cry of the baby or the stealthy turn of a doorknob. Similarly, we may be unaware of the contents of a dimly overhead conversation until something important—our own name, for instance—is mentioned, and then our degree of attention increases.

The Limbic System

The limbic system is a neuron network that forms a loop around the inside of the brain, connecting the hypothalamus to the cerebral cortex (Figure 32–8). It is thought to be the circuit by which drives and emotions, such as hunger, thirst, and desire for pleasure, are translated into complex actions, such as seeking food, drinking water, or pressing a lever.

Chemical Activity in the Brain

Transmission of nerve impulses across synapses in the human nervous system involves the release of chemicals into the synaptic cleft (page 361). In the peripheral nervous system, the chemical transmitters are acetylcholine and norepinephrine. Acetylcholine and norepinephrine are also

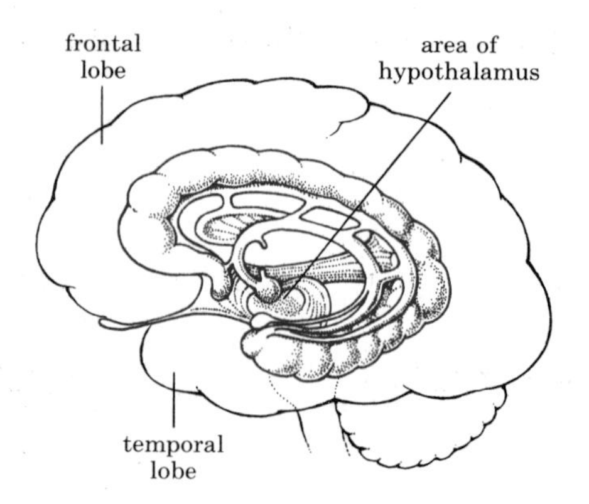

32-8 *The limbic system, a group of interconnected structures within the cerebrum, is associated with emotional and motivational behavior. The hypothalamus, which is connected to the cerebral cortex by the limbic system, is located more deeply in the interior of the brain than the structures shown here.*

Table 32-1 Drugs that affect mood and behavior

Drug	Effects	Possible Mechanisms of Action
Caffeine (coffee, tea, cola drinks)	Stimulant; promotes alertness; increases motor activities; insomnia	Facilitates synaptic transmission
Nicotine (tobacco)	Stimulant	Mimics acetylcholine; facilitates synaptic transmission
Amphetamines (Benzedrine, Methedrine, Dexedrine)	Stimulants; lessen fatigue and depression; suppress appetite; cause malnutrition, exhaustion, impairment of judgment, psychoses; strongly addictive	Increase activity of biogenic amines in brain; block uptake of norepinephrine and dopamine
Alcohol	Sedative (after preliminary excitatory effect); loss of motor coordination and alertness; euphoria; cirrhosis of liver	Depresses central nervous system (exact mechanism unknown)
Chlorpromazine (Thorazine, Compazine, Stelazine)	Tranquilizer; depresses reticular-system activity; suppresses hallucinations and delusions; antipsychotic	Blocks receptors of norepinephrine and dopamine
Barbiturates (Seconal, Amytal, Nembutal, Tuinal)	Sedatives; induce sleep, euphoria, irritability, hallucinations; highly addictive	Interfere with synthesis and release of norepinephrine and serotonin
Opium, morphine, heroin, methadone, Demerol, codeine	Relieve pain; induce muscle relaxation, drowsiness, lethargy, euphoria, sleep; produce constipation, loss of appetite; kill by depressing respiratory center; highly addictive	Depress central nervous system
Cocaine	Local anesthesia; euphoria; alertness; addiction	Inhibits uptake of norepinephrine
Psilocybin, mescaline	Increased sensory awareness; hallucinations; psychoses	Chemicals similar to biogenic amines; may enhance their effects
LSD	Rich visual imagery; sensory awareness; visual hallucinations; psychoses	Inhibits brain serotonin, perhaps by decreasing activity of serotonin-producing cells in reticular system
Marijuana, hashish	Mild euphoria; enhanced sensory perception	Unknown

transmitter substances in the brain. Some 30 additional substances found in the brain are believed to be transmitters. Two of the most important are dopamine and serotonin, which are closely related to norepinephrine. They belong to a group of chemicals known as the biogenic amines and are derived from amino acids.

Norepinephrine is a transmitter substance of the reticular system and thus may play a role in arousal and attention. It is also found in relatively high concentrations in the hypothalamus and the limbic system.

Dopamine is the transmitter substance for a relatively small group of neurons involved with muscular activity. Parkinson's disease, which is characterized by muscular tremor, is associated with a decrease in dopamine. L-Dopa, a closely related compound, has been used with great success in treating some persons with this condition. L-Dopa also appears to be of benefit in the treatment of some cases of depression.

Serotonin is found in many portions of the central nervous system, including, in particular, the hypothalamus and the limbic system. All of the cells producing serotonin, however, have their cell bodies in a single small area in the brainstem that is part of the reticular system. It thus appears that the serotonin synapses may characterize a very old portion of the brain dealing with basic functions such as sleep, consciousness, and emotions.

Table 32-1 lists drugs that act on the brain to alter mood and behavior. As you can see, many of these agents appear to exert their effects by enhancing or inhibiting the effects of chemical transmitters.

Internal Opiates

The word opium comes from the Greek *opion,* meaning "poppy juice." Since the time of the Greeks, poppy juice and its derivatives have been used for the control of pain. They are the most potent pain-killers known, and their psychological effects are greatly enhanced by the fact that they produce euphoria. They are also highly addictive.

Recent research has revealed that opiates act upon the brain by binding to specific receptors. These receptors are located primarily in the limbic system and in the spinal cord; their effect is to inhibit the production of nerve impulses by neurons in this circuit. Such receptors have been found not just in humans, but in all other vertebrates tested.

This discovery has triggered a search for naturally occurring substances with opiate activity. Four naturally occurring peptides that bind to the opiate receptors have now been isolated. All four have pain-relieving effects. Two, found predominantly in brain tissue and known as enkephalins, are believed to function as neurotransmitters with inhibitory effects. It is hypothesized that, under normal conditions, our opiate receptors are exposed to a relatively low level of enkephalins. Under conditions of stress or pain, the release of enkephalins increases, producing the pain-relieving effect and, probably, the mood-elevating state. Administering morphine or other opiates enhances these effects. However, by a negative feedback mechanism, administered opiates reduce the normal production of enkephalins, resulting in dependency on the artificial source (addiction).

Two other substances, isolated from the pituitary gland and known as endorphins, also have opiate activity. In fact, one has been found to be three times as powerful as morphine when injected intravenously, and 48 times as powerful when injected directly into the brain. Because endorphins are abundant in the pituitary and because they are not found in the brain, where the effects of pain are clearly mediated, it has been suggested that the endorphins may be hormones whose primary effects and target tissues are as yet unknown.

People who begin and maintain regular exercise programs (such as jogging, swimming, or bicycling) often notice a feeling of increased emotional well-being in addition to the physical benefits of the exercise program. Some recent biochemical research has suggested that regular exercise may increase the levels of internal opiates, with a consequent elevation of mood. Some researchers have speculated that individuals who seem to be addicted to exercise may actually be addicted to increased levels of these naturally occurring substances.

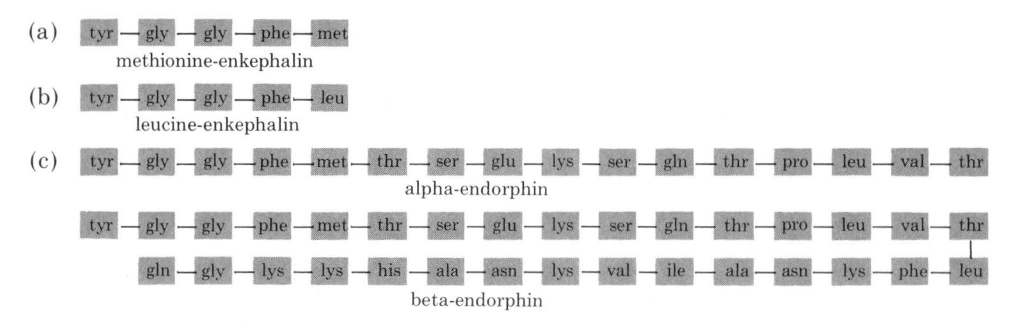

32-9 *The amino acid sequences of four naturally occurring substances with opiate activity:* (a) *methionine-enkephalin,* (b) *leucine-enkephalin,* (c) *alpha-endorphin, and* (d) *beta-endorphin. The enkephalins are found primarily in the brain tissue of vertebrates, and the endorphins are most abundant in the pituitary gland. The first four amino acids of each sequence are identical.*

The brain and the pituitary produce a number of substances in addition to the neurotransmitters, internal opiates, and hormones that we have described here and in Chapters 25 and 26. Intensive studies of these substances are being carried out at a number of laboratories around the world. It is hoped that these studies not only will increase our understanding of brain functions but also will lead to solutions to particularly difficult medical problems such as chronic pain.

Electrical Activity of the Brain

As you know, the transmission of a nerve impulse along a nerve fiber is an electrical phenomenon. The electrical activity of the brain can be detected by measuring the difference in electric potential between pairs of electrodes on the head or between an electrode placed on a specific area of the scalp and a "neutral" electrode placed elsewhere on the body. The voltages that arise are very weak—about 300 microvolts is the maximum in a normal adult—and extremely sensitive recording equipment is required. (A microvolt is a millionth of a volt.) A record of the electrical activity of the brain is known as an electroencephalogram (EEG).

EEG recordings obviously reflect only gross electrical activity. The electroencephalogram has been compared to measurements that an observer on Mars might make of the roars emanating from the Houston Astrodome during a baseball game. They would certainly indicate that something was going on in Houston, but it would be very difficult to infer the rules or determine the progress of the game from such recordings.

Nevertheless, there are characteristic EEG patterns that can be correlated with certain levels and types of brain activity.

Alpha and Beta Waves

One characteristic EEG pattern is the alpha wave, a slow variation in electric potential, with a frequency of about 8 to 12 cycles per second, recorded at the back of the head. Alpha waves are usually produced during periods of relaxation; in most persons they are conspicuous only when the eyes are shut. About two-thirds of the general population have alpha waves that are disrupted by attention. Of the remaining third, about one-half have almost no alpha rhythms at all and about one-half have persistent alpha rhythms that are not easily disrupted by attention.

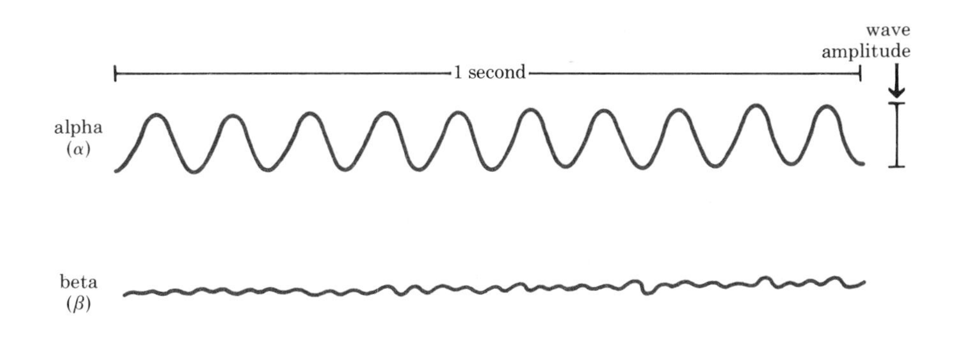

32-10 *The wave patterns of alpha and beta waves. Note that the frequency (number of waves per second) is much greater and the amplitude (voltage) is much lower for the beta waves.*

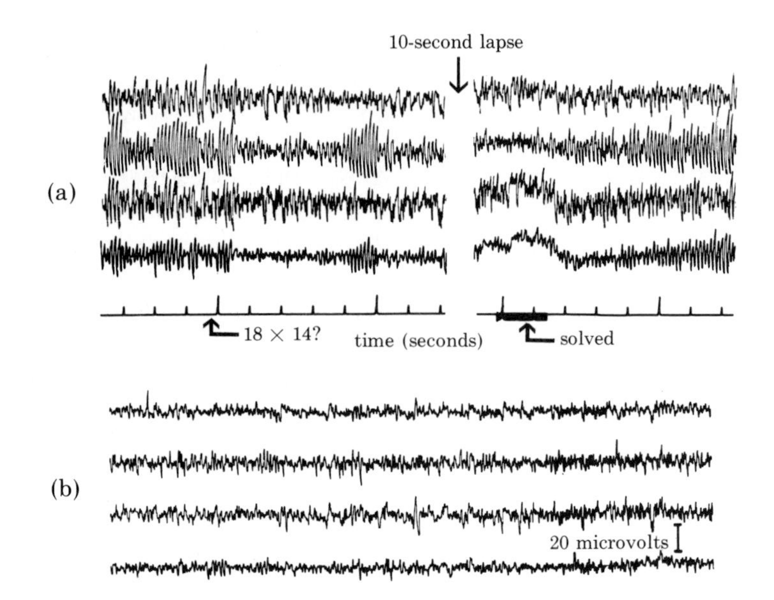

32-11 (a) *Brain wave recordings of a student with conspicuous alpha rhythm. Electrodes were placed at four positions on the head. When the student was asked to multiply 18 × 14, the alpha rhythm was suppressed and then resumed after the problem was solved (10 seconds of the recording are omitted in order to keep it on the page).*
(b) *Brain waves from a student with a complete absence of alpha rhythms. According to a recent study, people who have almost no alpha rhythms under any conditions think almost exclusively by visual imagery, whereas people with a persistent alpha rhythm tend to be abstract thinkers.*

Another EEG pattern is the beta wave. Beta waves are of smaller amplitude (lower voltage) than alpha waves, but their frequency is greater, being about 18 to 32 cycles per second. Beta waves occur in bursts in the anterior part of the brain and are associated with mental activity and excitement.

Sleep and Dreams

Other types of brain waves occur during sleep. On the basis of changes in the electroencephalogram, sleep can be divided into five stages: drowsiness, light sleep, intermediate sleep, deep sleep, and REM sleep. Sleepers pass through a regular sleep cycle, moving from one stage of sleep to another. The average young adult spends 50 percent of his or her sleeping time in light sleep and about 20 percent in REM sleep, divided into three to five episodes.

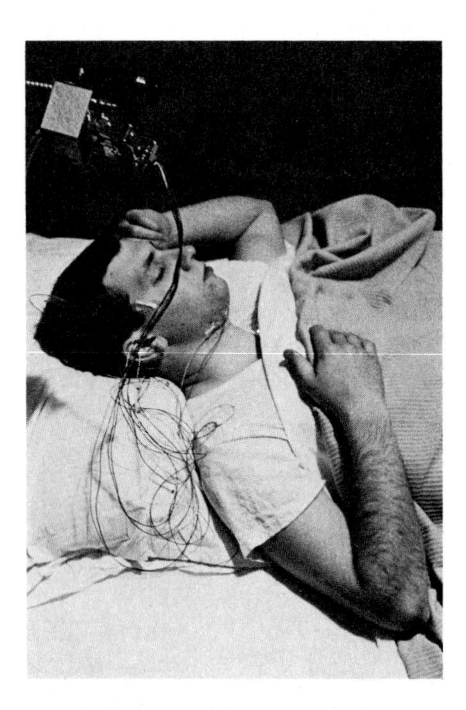

32-12 *Volunteer wired to record brain waves, eye movements, breathing rate, and other physical functions during sleep.*

REM sleep is characterized by rapid, low-voltage activity similar to that seen in alert persons and by rapid eye movements (REMs). The eyeballs move jerkily as if the sleeper were watching some scene of intense activity. Although the EEG shows evidence of alertness, the muscles are more relaxed than in ordinary deep sleep. Further, although the sleeper is harder to awaken from REM than from deep sleep, once awakened, he or she is more alert. Fluctuations in heart rate, blood pressure, and respiration are also seen during REM sleep. Many men experience erections of the penis during REM sleep, and women have similar erections of the clitoris with secretions of vaginal fluid. A person awakened during a period of REM sleep nearly always reports that he or she has been dreaming.

At one time, REM sleep was thought to be the period when all dreaming occurred. On the basis of more recent work, it has been suggested that dreaming can occur at any time during sleep but that conditions for recalling dreams are most favorable when subjects are awakened during REM sleep. Some alternative hypotheses have been suggested. One proposal is that REM sleep serves as an information-processing period during which data from the previous waking cycle are sorted, processed, and stored. If this is the case, then REM sleep is essential to the memory

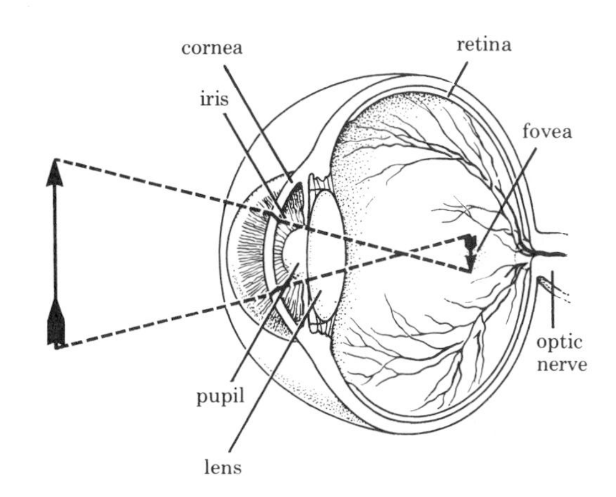

32-13 *The human eye. An image, represented by the arrow, is focused (and inverted) on the retina, which contains the photoreceptor cells. The fovea, near the center of the retina, is the area of greatest visual acuity. The pupil is a hole in the center of the iris, which is the colored part of the eye. The cornea is the transparent outer layer that covers the iris and pupil.*

Only the front of the eye is exposed; the rest of the eyeball is recessed in and protected by the bony socket of the skull.

process. Laboratory rats forget tasks they have learned if they are deprived of REM sleep. Similarly, evidence indicates that the student who stays up all night cramming for an exam will not have as good a recollection of the material as the student who studies and then sleeps.

According to other hypotheses, only the deep sleep stage is necessary for the restorative functions of sleep. REM sleep is thought to be a partial arousal, serving the same sort of sentinel function as the occasional arousals seen in hibernating animals. These studies and the controversies they engender are continuing.

The Eye and the Brain

Some of the most interesting studies on brain function have been concerned with interactions between brain and eye.

Structure of the Eye

The vertebrate type of eye is often called a camera eye. In fact, it has a number of features in common with an ordinary camera equipped with several expensive accessories, such as a built-in cleaning and lubricating system, an exposure meter, and an automatic focus. Light from the object being viewed passes through the transparent cornea and lens, which focus an inverted image of the object on the light-sensitive retina at the back of the eyeball (Figure 32-13).

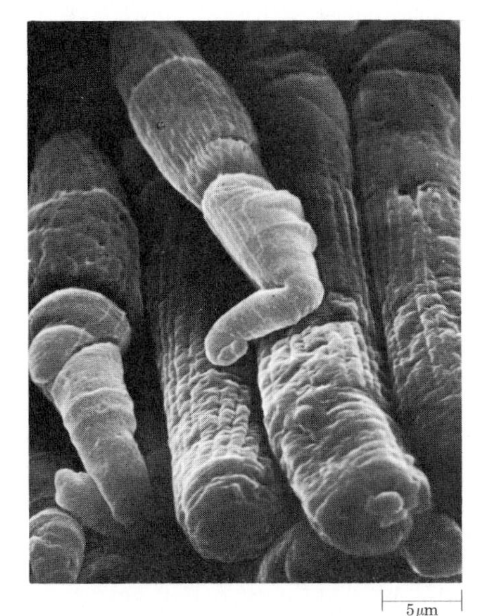

32-14 *Rods and cones as shown by the scanning electron microscope.*

The retina is made up of photoreceptor ("light-receiving") cells. In humans, these cells are of two types, named, because of their shapes, *rods* and *cones* (Figure 32-14). The cones, of which the human eye contains 5 to 10 million, provide greater resolution, giving a "crisper" picture. They are generally concentrated in the center of the retina in the *fovea*, which is the area of sharpest vision. We have about 160,000 cones per square millimeter of retina.

Rods do not provide as great a degree of resolution as cones do, but they are more light sensitive than cones. In dim light, our vision depends entirely on rods. In humans, rods are concentrated outside of the fovea. Some nocturnal animals, such as toads, mice, rats, and bats, have retinas made up almost entirely of rods, and some diurnal animals, such as some reptiles, have almost entirely cones.

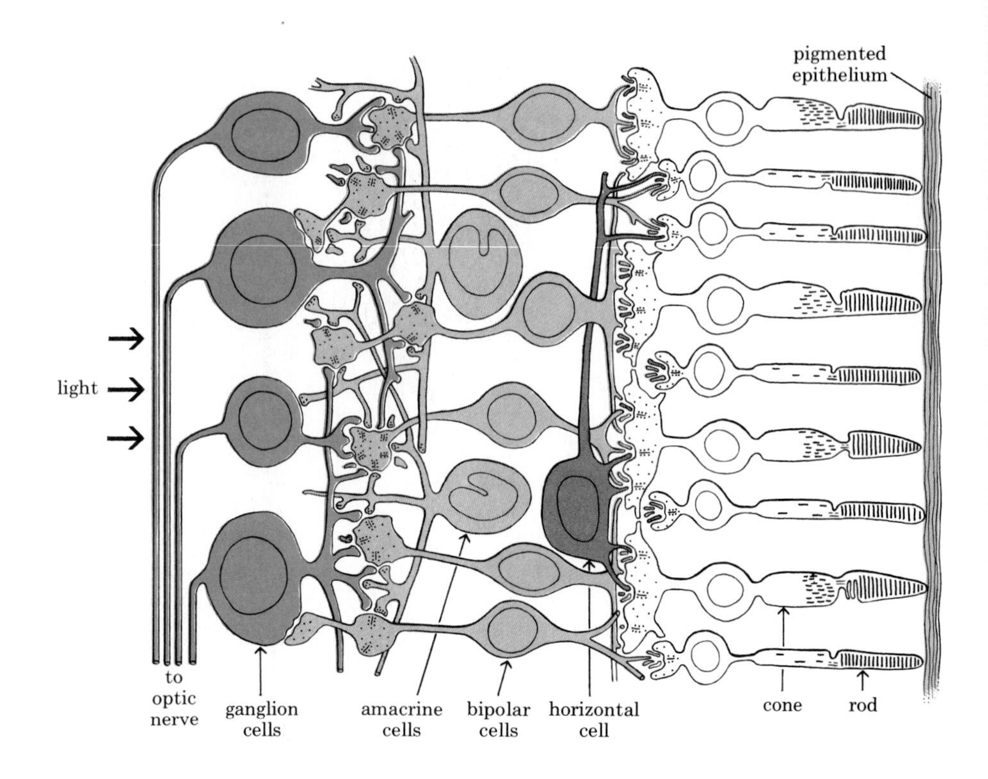

32-15 *The retina of the vertebrate eye. Light (shown here as entering from the left) must pass through a layer of cells to reach the photoreceptors (the rods and cones) at the back of the eye. Signals from the photoreceptor cells are then transmitted through the neurons known as bipolar cells to the ganglion cells, whose axons converge to become the optic nerve. The transmission paths involve elaborate interconnections.*

Cones are responsible for color vision, which is why the world becomes colorless to us at night—the light is too dim to stimulate the cone system and we must depend on the rods.

When the photoreceptor cells are stimulated by light, they transmit electrical signals to neurons known as bipolar cells, which pass the signals on to the ganglion cells, whose axons form the optic nerve (Figure 32-15). However, there are varied and elaborate interactions among these neurons and others even before nerve impulses leave the retina. This circuitry is involved in the preliminary analysis of visual information that takes place in the retina itself. The human retina contains about 125 million photoreceptors, and the optic nerve about 1 million axons.

The fibers of the optic nerve can be traced to a visual relay center in the thalamus, where the fibers from the two eyes come together and synapse with fibers leading to the visual cortex. Various experiments have shown that the spatial arrangement of neurons in the cortex corresponds topographically with the spatial arrangement of the image as it is received on the retina, except for "over-representation" of the fovea. The fovea, which occupies only about 1 percent of the area of the retina, projects to nearly 50 percent of the visual cortex.

Vision and Behavior

Vision is a direct stimulus to action in many of the vertebrates. Predatory fish are vision-stimulated, and many such fish literally "cannot help" striking at a moving object of appropriate size. The frog can learn not to strike at a small moving object, but only with difficulty. If it comes to associate an unpleasant experience with a particular shape, it can restrain itself, although apparently with effort, from striking at that shape. On the other hand, it can never learn to take an object that is not moving. Even a frog that has eaten thousands of live flies will starve to death surrounded by motionless ones.

Investigators have found that certain ganglion cell axons leaving the retina of the frog produce impulses only when certain photoreceptors in the retina are stimulated by a small, rounded, moving object. In the laboratory, the visual stimulus for these ganglion cells can be simulated by a variety of objects—such as paper clips and pencil erasers—provided they are in motion. In a frog's natural environment, however, the objects likely to elicit this response most frequently are flying insects. A large, moving object or a small, still one provokes no response. These ganglion cell axons terminate in a special layer of cells in the frog's midbrain, and, although no further connections have been traced, it is tempting to speculate that signals are relayed from this part of the midbrain to areas containing motor nerves that control the muscles involved in striking.

The Perception of Form

How do we perceive form? The logical answer is that we perceive form simply because of the way the visual image is arranged on the visual cortex. However, studies conducted over the past 20 years on cats and monkeys by David H. Hubel and Torsten N. Wiesel of Harvard University reveal that quite a different mechanism may be involved. They used microelectrodes to record from single cells of the visual cortex, recorded the responses of individual cells to visual stimuli, and came up with some very surprising results (Figure 32–16).

One class of cortical neurons, for example, was found to respond only to a horizontal bar; as the bar was tipped away from the horizontal, discharges from the cells slowed and ceased. Another class responded best to a vertical bar, ceasing to fire as the bar was oriented toward the horizontal. A third class fired only when the bar was moved from left to right; another

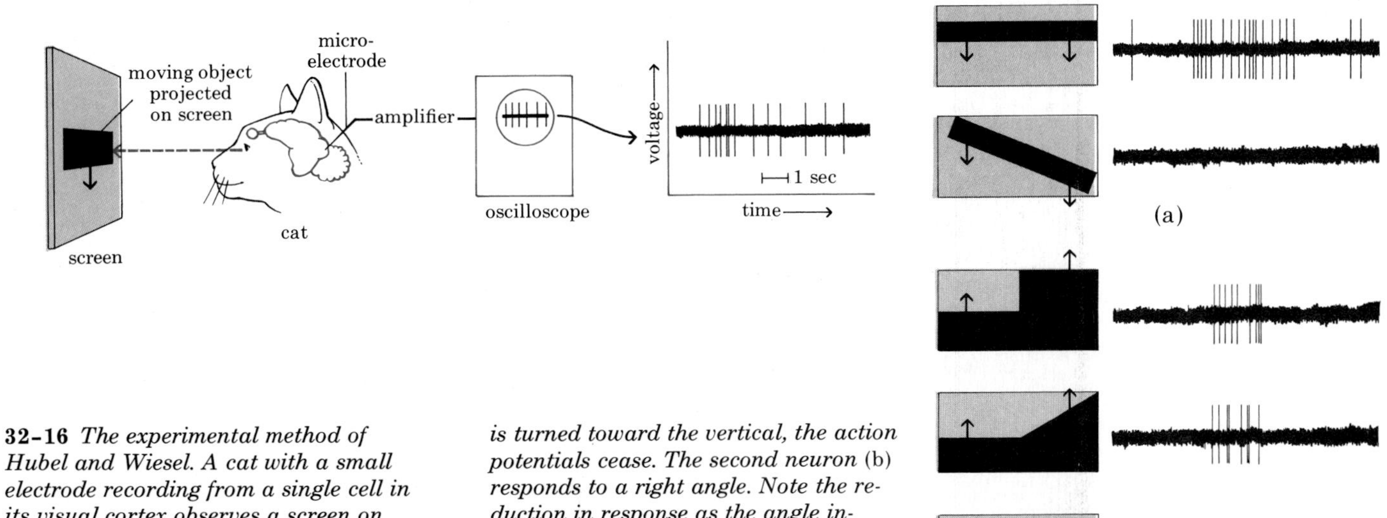

32–16 *The experimental method of Hubel and Wiesel. A cat with a small electrode recording from a single cell in its visual cortex observes a screen on which a moving shape is projected.* (a) *and* (b) *represent tracings from two different neurons. The first neuron* (a) *responds to a horizontal bar; as the bar is turned toward the vertical, the action potentials cease. The second neuron* (b) *responds to a right angle. Note the reduction in response as the angle increases. The arrows indicate the direction of movement of the shape on the screen.*

only to a right-to-left movement. Another class of cells turned out to be a right-angle detector; such cells responded only to an angle moving across the visual field and were most excited when the angle was a right angle. Thus an important component of the visual information-processing system appears to be a set of very specifically tuned neurons, each responsive to one aspect of the shape of an object in the visual field. If this hypothesis is correct, then there must be some further information-processing system synthesizing the final perception by integrating the information received from its individual elements.

The neurons in the frog's retina apparently carry out much this same kind of data processing. These very fundamental discoveries would seem to be leading to a new and far deeper understanding of the ancient dilemma about the relationship between the human mind and the outside world.

Split-Brain Studies

Additional interesting implications concerning processing of information in the cerebral cortex have come from split-brain studies carried out by Roger Sperry and co-workers at the California Institute of Technology. Sperry severed the connections between the two cerebral hemispheres in cats and monkeys by cutting through the corpus callosum and the optic chiasm (Figure 32–17). The animals' everyday behavior was normal, but by using specially contrived testing techniques, the investigators were able to show some remarkable and unexpected effects. The animals responded to test procedures as if they had two brains. When the right visual field of an

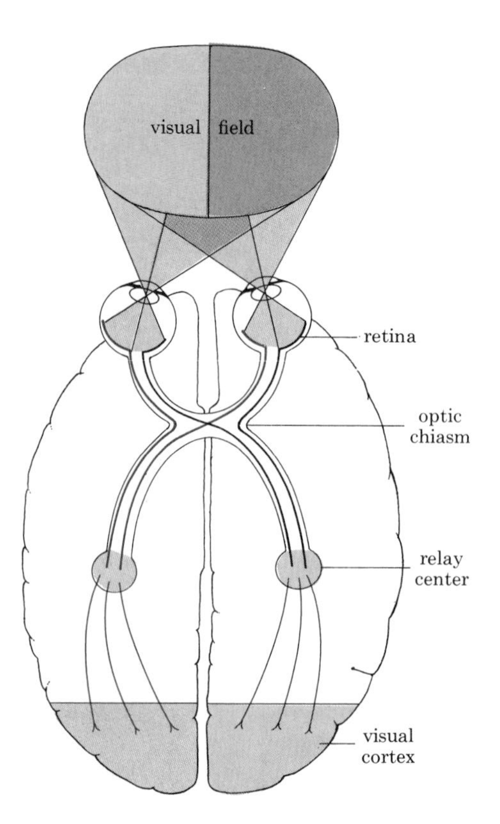

32–17 *The optic chiasm viewed from below. The chiasm is the structure formed by the crossing over of fibers traveling from the retina to the visual cortex. Because of this crossing-over, the right and left cortical areas each "see" only the opposite half of the visual field.*

animal was covered, for instance, it could learn tasks and discriminations using only the left visual field. But then if the left visual field were covered and the same tests were presented to the right visual field, the animal had to learn all over again. In fact, it was possible, without creating any confusion at all, to have the two sides of the brain learn opposite solutions to the same problem. Similar observations have been made in humans in whom the corpus callosum has been severed for the treatment of epilepsy. The patients are able to carry out their normal activities, but special tests indicate the presence of what Sperry calls "two realms of consciousness."

Under normal circumstances, the two cerebral hemispheres are integrated in their activities, and although one may be dominant in certain functions—as the left hemisphere is in language—the other retains at least the potential for this same function. The capacity of one hemisphere to compensate for the dysfunction of the other is well known to neurosurgeons who have observed patients functioning normally after removal of or damage to extensive areas of the right temporal lobe. Further, in a number of cases, children with left temporal lobe damage have, after an interval, learned to speak again, as a result of the right lobe's taking over the speech functions. The outlook for complete recovery varies with age.

More recent studies have revealed further differences in mental characteristics between the two hemispheres, with the left brain excelling in speech, logic, writing, and mathematics. The right brain excels in discriminations involving shape, form, tune, and texture, in nonverbal communication, and in general creativity.

The many convolutions of the cerebral cortex represent one way of increasing the working area of the brain. Specialization of function of the two cerebral hemispheres is another solution to the problem of increasing the capacity of the brain without enlarging its volume.

Learning and Memory

For scientists interested in brain research, the biggest present challenge is to understand the mechanisms of memory and learning. If we define learning as a change in behavior based on experience, and if the functions of the brain—the mind—are to be explained in terms of the atoms and molecules and the structures composed of them, then learning must involve changes in these atoms or molecules or structures. But what is this change, and where and how does it take place?

About 60 years ago, the late Karl Lashley set out to locate this physical change, the trace of memory, which he called the engram. He taught rats and other animals to solve particular problems and then performed operations to see if he could remove the portion of the cortex containing that particular engram. But he never found the engram. As long as he left enough brain tissue to enable the animal to respond to the test procedures at all, he left memory as well. And the amount of memory that remained was generally proportional to the amount of remaining brain tissue. Lashley concluded that memory is "nowhere and everywhere present."

Although much more sophisticated techniques are now available for brain research, modern neurobiologists have had little more success than Lashley in finding memory's hiding place. Moreover, the problem has become, if anything, more complex. It is now clear that there are two types of memory, short-term and long-term.

A simple example of short-term memory is looking up an unfamiliar number in a phone book; you usually remember it just long enough to dial it. A less common but also well-known phenomenon is a loss of memory for the events immediately preceding a blow to the head; only rarely (except in soap operas) is there any loss of the records of earlier experiences.

The transfer of information from short-term to long-term memory involves, according to the most widely held hypotheses, changes in synaptic pathways in the brain. The exact nature of these changes is unknown, but there is evidence that protein synthesis is involved as well as changes in the synthesis and release of neurotransmitters. The process can be compared to the establishment of a footpath. The more frequently the path is traveled, the better established it becomes. This concept is supported by the familiar experience of consciously retrieving a name, for instance, by seeking out related information that puts one "on the right track."

The human brain is the most complicated structure known. As we have noted, it contains about 100 billion neurons, and it has been calculated that the possible number of interconnections among them in a single brain is greater than the total number of atomic particles making up the universe. The human brain is capable of many remarkable activities. It remains to be seen if it can meet the most difficult challenge of all—understanding itself.

SUMMARY

The function of the brain is to process information received from the internal and external environments and to initiate appropriate responses. In both experimental animals and humans, it has been possible to correlate particular areas of the brain with particular functions.

The cerebral cortex, which is the outer layer of the cerebral hemispheres, is the area of the brain in which complex, conscious motor activities originate and to which sensory information is ultimately transmitted. Areas of the cortex that have been mapped include the motor cortex, sensory cortex, and parts of the cortex concerned with vision, hearing, and speech. Most of the human cortex is "silent"; that is, it has no direct sensory or motor function. It is known as the association cortex, about half of which is in the frontal lobes, and is the part of the brain that has developed most rapidly in human evolution.

The hypothalamus is a small group of nerve cell bodies with a great variety of regulatory functions. It is also apparently involved in basic drives and emotions, such as hunger, pleasure, and sex. Two important systems of the brain are the reticular system, concerned with arousal and attention, and the limbic system, which connects the hypothalamus with parts of the cerebral cortex.

Chemical transmitters in the brain include acetylcholine, norepinephrine, serotonin, and dopamine. A newly isolated group of chemical substances, the enkephalins and endorphins, resemble opiates in their function and may act as inhibitory transmitters. Many drugs appear to act on the brain by increasing or decreasing the effectiveness of particular neurotransmitters.

Wavelike variations in electric potential can be recorded from the brain surface in electroencephalograms. Two major types of brain waves can be recorded in normal waking subjects: beta waves, of relatively high frequency, associated with alertness and attention, and alpha waves, of lower frequency, associated with relaxation. Other types of brain waves characterize the five stages of sleep.

Studies of the eye and the brain have provided new insights into perception and the organization of the brain. Light passes into the eyeball to the retina where it is received by specialized cells, the rods and the cones. Cones provide greater resolution and are responsible for color vision. Rods are involved in night vision. Rods and cones transmit nerve impulses to overlying bipolar cells from which they are relayed to a network of ganglion cells whose axons form the optic nerve. Fibers from the left and right optic fields cross in the optic chiasm. Studies with cats and monkeys have revealed that visual form is coded by single neurons in the brain, each of which responds to one or two specific features of the visual image. Studies in which connections (the corpus callosum and the optic chiasm) between the two cerebral hemispheres have been severed indicate that the two hemispheres of the same individual can function independently and that they differ somewhat in their capacities.

Studies on the mechanisms of memory and learning involve the search for changes in the physical or chemical structure of brain cells or the connections among them.

QUESTIONS

1. Distinguish between the following: Broca's area/Wernicke's area; reticular system/limbic system; norepinephrine/dopamine; enkephalins/endorphins; alpha waves/beta waves; rods/cones; left brain/right brain.

2. Sketch the human cerebral cortex and indicate on it the areas that have been mapped.

3. What are the five stages of sleep? Which are thought to be most important if the sleeper is to waken alert and refreshed?

4. Draw a diagram of the human eye and label its parts.

5. On the basis of your own experience, define the differences between learning, memory, and habit.

SUGGESTIONS FOR FURTHER READING

BRECHER, EDWARD M., et al.: *Licit and Illicit Drugs: The Consumers Union Report on Narcotics, Stimulants, Depressants, Inhalants, Hallucinogens, and Marijuana—Including Caffeine, Nicotine, and Alcohol,* Little, Brown and Company, Boston, 1972.

A well-balanced, straightforward, and nonsensational review.

ECCLES, JOHN C.: *The Understanding of the Brain,* McGraw-Hill Book Company, New York, 1977.

A general survey, written for the layman, by one of the great leaders in the field of neurophysiology.

ECKERT, ROGER, and DAVID RANDALL: *Animal Physiology,* W. H. Freeman and Company, San Francisco, 1978.

The emphasis is on basic principles. Well-written and handsomely illustrated.

GREGORY, R. L.: *Eye and Brain: The Psychology of Seeing,* 3d ed., McGraw-Hill Book Company, New York, 1977.*

A vivid introduction to the science of vision.

* Available in paperback.

KESSEL, RICHARD G., and RANDY H. KARDON: *Tissues and Organs: A Text-Atlas of Scanning Electron Microscopy,* W. H. Freeman and Company, San Francisco, 1979.*

A collection of more than 700 outstanding scanning electron micrographs of vertebrate tissues and organs, beautifully reproduced. The accompanying text summarizes current knowledge of each organ system and its component parts.

KUFFLER, STEPHEN W., and JOHN G. NICHOLLS: *From Neuron to Brain,* Sinauer Associates, Inc., Sunderland, Mass., 1976.

A lively account of the cellular basis of the workings of the nervous system, based largely on work with which these two distinguished neurobiologists have had first-hand experience.

LUCIANO, DOROTHY S., A. J. VANDER, and J. H. SHERMAN: *Human Function and Structure,* McGraw-Hill Book Company, New York, 1978.

An introductory anatomy and physiology textbook. The text is clearly written, and the detailed illustrations—particularly of the skeletal, muscular, nervous, and circulatory systems—are outstanding.

MILLER, JONATHAN: *The Body in Question,* Random House, Inc., New York, 1978.

An informal history of the development of modern biology and medicine, originally prepared as a television series. The author, a physician, stresses that the medical achievements of the past 25 years are based on an "unprecedented understanding of what the healthy body is and how it survives and protects itself."

NOSSAL, G. J. V.: *Antibodies and Immunity,* 2d ed., Basic Books, Inc., New York, 1978.

An account for the general public of research in this important field by a scientist who has himself made many significant contributions.

ROMER, ALFRED: *The Vertebrate Story,* 4th ed., The University of Chicago Press, Chicago, 1971.*

The history of vertebrate evolution, written by an expert but as readable as a novel.

RUGH, ROBERTS, and LANDRUM B. SHETTLES: *From Conception to Birth: The Drama of Life's Beginnings,* Harper & Row, Publishers, Inc., New York, 1971.

This is an account of the history of life before birth. The book describes in detail the development of the unborn child from the moment of fertilization and also the changes in the mother during pregnancy. It discusses such related topics as birth control, congenital malformation, labor, and delivery of the baby. There are a large number of illustrations, including a group of magnificent color photographs of the developing fetus.

SCHMIDT-NIELSEN, KNUT: *Animal Physiology: Adaptations and Environment,* 2d ed., Cambridge University Press, New York, 1979.

Schmidt-Nielsen is concerned with underlying principles of animal physiology—the problems animals have to solve in order to survive. The emphasis is on comparative physiology, and the lucid exposition is illuminated by many interesting examples.

SCHMIDT-NIELSEN, KNUT: *Desert Animals,* Oxford University Press, New York, 1964.

Although considered the definitive work on the physiological problems relating to heat and water, this readable book also contains numerous anecdotes—such as that about Dr. Blagden—and many fascinating personal observations.

SCIENTIFIC AMERICAN: *The Brain,* W. H. Freeman and Company, San Francisco, 1979.*

A reprint of the September 1979 issue of Scientific American. *Its 11 chapters by different authors are devoted to current knowledge of neurons and their organization and functions within the brain. The many drawings and micrographs are excellent.*

SMITH, HOMER W.: *From Fish to Philosopher,* Doubleday & Company, Inc., Garden City, N.Y., 1959.*

Smith was an eminent specialist in the physiology of the kidney. Writing for the general public, he explains the role of this remarkable organ in the story of how, in the course of evolution, organisms have increasingly freed themselves from their environments.

* Available in paperback.

THOMPSON, R. F.: *Introduction to Physiological Psychology,* Harper & Row, Publishers, Inc., New York, 1975.

Intended for the undergraduate student, this text presents an up-to-date survey of the biological foundations of psychology.

VANDER, A. J., J. H. SHERMAN, and DOROTHY S. LUCIANO: *Human Physiology,* 3d ed., McGraw-Hill Book Company, New York, 1980.

Most highly recommended. The text is a model of clarity, and the diagrams, many of which we have borrowed, are splendid.

PART III

Biology of Populations

SECTION 7

Evolution

When Charles Darwin visited the Galapagos archipelago, he found that each major island had its own variety of tortoise, so distinct from the others that it was easily recognized by local sailors and fishermen. This was one of the clues that led him to the formulation of the theory of evolution.

The Galapagos consists of 13 volcanic islands that pushed up from the sea more than a million years ago. The major vegetation is thornbush and cactus, and the original black basaltic lava is often visible, as it is beneath the lumbering feet of this tortoise on Indefatigable Island—"what we might imagine the cultivated parts of the Infernal regions to be," young Darwin wrote in his diary.

33

Darwin and the Theory of Evolution

In 1831, the young Charles Darwin set sail from England on what was to prove the most consequential voyage in the history of biology. Not yet 23, Darwin had already abandoned a proposed career in medicine—he described himself as fleeing a surgical theater in which an operation was being performed on an unanesthetized child—and was a reluctant candidate for the clergy, a profession deemed suitable for the younger son of a wealthy English gentleman. An indifferent student, Darwin was an ardent hunter and horseman, a collector of beetles, mollusks, and shells, and an amateur botanist and geologist. When the captain of the surveying ship H.M.S. *Beagle,* himself only a little older than Darwin, offered passage to any young man who would volunteer to go without pay as a naturalist, Darwin eagerly seized the opportunity. This voyage, which lasted five years, shaped the course of Darwin's future work. He returned to an inherited fortune, an estate in the English countryside, and a lifetime of independent work and study that radically changed our view of life and of our place in the living world.

The Road to Evolutionary Theory

That Darwin was the founder of the modern theory of evolution is well known. In order to understand the meaning of his theory, however, it is useful to look briefly at the intellectual climate in which it was formulated. Aristotle (384–322 B.C.), the first great biologist, believed that all living things could be arranged in a hierarchy. This hierarchy became known as the *Scala Naturae,* or ladder of nature, in which the simplest creatures had a humble position on the bottommost rung, man occupied the top, and all other organisms had their proper places in between. Until the late nineteenth century, many biologists believed in such a natural hierarchy. But whereas to Aristotle living organisms had always existed, the later biologists believed, in harmony with the teachings of the Old Testament, that all living things were the products of a divine creation. They believed, moreover, that most were created for the service or pleasure of human beings. Indeed, it was pointed out, even the lengths of day and night were planned to coincide with the human need for sleep.

33–1 *"Afterwards, on becoming very intimate with Fitz Roy [the captain of the* Beagle], *I heard that I had run a very narrow risk of being rejected on account of the shape of my nose! He . . . was convinced that he could judge of a man's character by the outline of his features; and he doubted whether anyone with my nose could possess sufficient energy and determination for the voyage. But I think he was afterwards well satisfied that my nose had spoken falsely." (Charles Darwin,* The Voyage of the *Beagle.)*

33-2 *Linnaeus, the originator of the binomial system for naming species of organisms, is seen here wearing his collector's outfit. His ambition was to classify all the known kinds of plants and animals according to their genera. Linnaeus believed that each living thing corresponded more or less closely to some ideal model and that by classifying them, he was revealing the grand pattern of creation.*

That each type of living thing came into existence in its present form—specially and specifically created—was a compelling idea. How else could one explain the astonishing extent to which every living thing was fitted to its environment and to its role in nature? It was not only the authority of the church, but also, so it seemed, the evidence before one's own eyes that gave such strength to the concept of special creation.

Among those who believed in special creation was Carolus Linnaeus (1707–1778), the great Swedish naturalist, who devised our present system of biological nomenclature. All the time that Linnaeus was at work on his encyclopedic *Systema Naturae* and *Species Plantarum,* explorers of Africa and the New World were returning to Europe with new species of plants and animals and even, apparently, new kinds of human beings. Linnaeus revised edition after edition to accommodate these findings, but he did not change his opinion that all species then in existence were created on the sixth day of God's labor and had remained fixed ever since. During Linnaeus's time, however, it became clear that the pattern of creation was far more complex than had been previously imagined.

Evolution before Darwin

The French scientist Georges-Louis Leclerc de Buffon (1707–1788) was among the first to suggest that species undergo changes in the course of time. Buffon believed that these changes took place by a process of degeneration. He suggested that, in addition to the numerous creatures that were produced by divine creation at the beginning of the world, "there are lesser families conceived by Nature and produced by Time." Buffon's hypothesis, although vague as to the way in which changes might occur, did attempt to explain the bewildering variety of creatures in the modern world.

Another early doubter of fixed and unchanging species was Erasmus Darwin (1731–1802), Charles Darwin's grandfather. Erasmus Darwin was a physician, a gentleman naturalist, and a prolific writer, often in verse, on both botany and zoology. Erasmus Darwin suggested, largely in asides and footnotes, that species have historical connections with one another, that animals may change in response to their environment, and that their offspring may inherit these changes. He maintained, for instance, that a polar bear is an "ordinary" bear that, by living in the Arctic, became modified and passed the modifications along to its cubs. These ideas were never clearly formulated but are interesting because of their possible effects on Charles Darwin, although the latter, born after his grandfather died, did not profess to hold his grandfather's views in high esteem.

The Age of the Earth

It was geologists, more than biologists, who paved the way for the development of evolutionary theory. One of the most influential of these was James Hutton (1726–1797). Hutton proposed that the earth had been molded not by sudden, violent events but by slow and gradual processes—wind, weather, and the flow of water—the same processes that can be seen at work in the world today. This theory of Hutton's, which was known as *uniformitarianism,* was important for three reasons. First, it implied that the earth has a living history, and a long one. This was a new idea. Christian theologians, by counting the successive generations since Adam (as recorded in the Bible), had calculated the maximum age of the earth at about 6,000 years. No one, as far as we know, had ever thought in terms of a longer

(a)

(b)

(c)

33-3 *A fossil is a remnant or trace of a once-living organism.* (a) *Minerals may fill the hollows left by the decay of soft tissues, as, for example, in this fossilized* Sassafras *leaf.* (b) *A primitive wasplike ant caught in amber formed from the resin of a tree that lived some 100 million years ago. The ant is a worker (a sterile female), indicating that insect societies had evolved by this time.* (c) *Among the most common early fossils are the discarded exoskeletons of trilobites, a type of arthropod that flourished during the Cambrian period, which began 500 to 600 million years ago.*

period. And 6,000 years is not enough time for such major evolutionary changes as the formation of new species to have taken place. Second, the theory of uniformitarianism stated that change is the *normal* course of events, as opposed to the concept of a normally static system interrupted by an occasional unusual event, such as a flood. Third, although this was never explicit, uniformitarianism suggested that there might be interpretations of the Bible other than the literal one.

The Fossil Record

During the latter part of the eighteenth century, there was a revival of interest in fossils. In previous centuries, fossils had been collected as curiosities, but they had generally been regarded either as accidents of nature—stones that somehow looked like shells—or as evidence of great natural catastrophes, such as the Flood. The English surveyor William Smith (1769–1839) was the first to make a systematic study of fossils. Whenever his work took him down into a mine or along canals or cross-country, he carefully noted the order of the different layers of rock, which are called strata, and collected fossils from each layer. He eventually established that each stratum, no matter where he came across it in England, contained a characteristic group of specimens and that these fossils were actually the best way to identify a particular stratum. (The use of fossils to identify strata is still widely practiced, for instance, by geologists looking for oil.) Smith did not interpret his findings, but the implication that the present surface of the earth had been formed layer by layer over the course of time was an unavoidable one.

Like Hutton's world, the world seen and reported by William Smith was clearly a very ancient one. A revolution in geology was beginning; earth science was becoming a study of time and change rather than a mere cataloging of types of rocks. As a consequence, the history of the earth became inseparable from the history of living organisms, as revealed in the fossil record.

33-4 *The first ideas about evolution came from the work of geologists who discovered that different layers, or strata, of the earth's surface contain different types of fossils. This view of Olduvai Gorge, in Tanzania, clearly shows such strata, now seen as chapters in evolutionary history. This particular site has yielded a number of fossils important in our growing understanding of human evolution (to be discussed in Chapter 42).*

Catastrophism

Although the way was being prepared by the revolution in geology, the time was not yet ripe for a parallel revolution in biology. The dominating force in European science in the early nineteenth century was Georges Cuvier (1769–1832). Cuvier was the founder of vertebrate paleontology, the scientific study of the fossil record. An expert in anatomy and zoology, he applied his special knowledge of the way in which animals are constructed to the study of fossil animals, and he was able to make brilliant deductions about the form of an entire animal from a few fragments of bone. Today we think of paleontology and evolution as so closely connected that it is surprising to learn that Cuvier was a staunch and powerful opponent of evolutionary theories. He recognized the fact that many species that had once existed no longer did. (In fact, according to modern estimates, considerably less than one percent of all species that have ever lived are living on the earth today.) Cuvier explained the extinction of species by postulating a series of catastrophes. After each catastrophe, the most recent of which was the Flood, new species filled the vacancies.

Cuvier hedged somewhat on the source of the new animals and plants that appeared after the extinction of older forms; he was inclined to believe they moved in from parts unknown. A contemporary and supporter of his, the first professor of paleontology at the Paris museum, was more straightforward: He taught that all species were wiped out at each catastrophe and wholly new ones were specially created to take their places. (One symptom of the terminal illness of a scientific theory is the accumulation of new hypotheses required to bolster its shaking foundations.)

The Theories of Lamarck

The first scientist to work out a systematic theory of evolution was Jean Baptiste Lamarck (1744–1829). "This justly celebrated naturalist," as Darwin himself referred to him, boldly proposed in 1801 that all species, including *Homo sapiens,* are descended from other species. Lamarck, unlike

The Record in the Rocks

The earth's long history is recorded in the rocks that lie at or near its surface, layer piled upon layer, like the chapters in a book. These layers, or strata, are formed as rocks in upland areas are broken down to pebbles, sand, and clay and are carried to the lowlands and the seas. Once deposited, they slowly become compacted and cemented into a solid form as new material is deposited above them. As continents and ocean basins change shape, some strata sink below the surface of an ocean or a lake, others are forced upward into mountain ranges, and some are worn away, in turn, by water, wind, or ice or are deformed by heat or pressure.

Individual strata may be paper-thin or many meters thick. They can be distinguished from one another by the types of parent material from which they were laid down, by the way the material was transported, and by the environmental conditions under which the strata were formed, all of which leave their traces in the rock. They can be distinguished, moreover, by the types of fossils they contain. Small marine fossils, in particular, can be associated with specific periods in the earth's history. The fossil record is never complete in any one place, but because of the identifying characteristics of the strata, it is possible to piece together the evidence from many different sources. It is somewhat like having many copies of the same book, all with chapters missing—but different chapters, so it is possible to reconstruct the whole.

The geologic eras—Precambrian, Paleozoic, Mesozoic, and Cenozoic—which are the major subdivisions of the geologic "book," were established and named in the early nineteenth century. These eras were subdivided into periods, each named, quite simply, for the areas in which the particular strata were first studied or studied most completely: the Devonian for Devonshire in southern England, the Permian for the province of Perm in Russia, the Jurassic for the Jura Mountains between France and Switzerland, and so on.

Early attempts to date the various eras and periods were based simply on their relative ages compared to the age of the earth; obviously, a stratum occurring regularly above another was younger than the one below it. The first scientific estimate was made in the mid-1800s by the famous British physicist Lord Kelvin. On the basis of his calculations of the time necessary for the earth to have cooled from its original molten state, Kelvin maintained that the planet could not have been more than 25 million years old. This estimate greatly troubled Charles Darwin, who knew that it was too little time for evolution to have taken place. (Kelvin was not aware of the existence under the earth's surface of radioactive materials that heat the planet from within.) In the last 30 years, however, new methods for determining the ages of strata have been developed involving measurements of the decay of radioactive isotopes. As a result, the estimated age of the earth has increased, in little more than a century, from 25 million years to about 4.6 billion.

Darwin's concept of the time necessary for major evolutionary changes to take place has been more than borne out by subsequent studies. For instance, in the evolution of the horse (page 502), a major index of change is relation of tooth width to tooth height as horses gradually changed from browsers (leaf eaters) to grazers (grass eaters). The largest actual change—and this is a fast evolutionary line—was a difference of 1.61 millimeters in the height of a particular tooth in 1 million years. In the entire sequence of dawn horse to *Equus,* there were more than eight successive genera over 60 million years, or about 7.5 million years per genus.

Geologic strata (and the fossils within them) are now dated whenever possible by analysis of the radioactive isotopes they contain. Many naturally occurring isotopes are radioactive. All the heavier elements—atoms that have 84 or more protons in their nuclei—are unstable and, therefore, radioactive. All radioactive elements emit radiation (as particles or rays) at a fixed rate; this process is known as radioactive "decay." The rate of decay is measured in terms of half-life: The half-life of a radioactive element is defined as the time in which half the atoms in a sample of the element lose their radioactivity and become stable. Since the half-life of an element is constant, it is possible to calculate the fraction of decay that will take place for a given isotope in a given period of time.

Half-lives vary widely, depending on the isotope. The radioactive nitrogen isotope ^{13}N has a half-life of 10 minutes, and the most common isotope of uranium (^{238}U) has a half-life of $4\frac{1}{2}$ billion years. The uranium atom undergoes a series of decays, eventually being transformed to an isotope of lead (^{206}Pb).

The proportion of ^{238}U to ^{206}Pb in a given rock sample, for example, is a good indication of how long ago that rock was formed. However, because the daughter atoms (^{206}Pb, for instance) are fewer and so more likely than parent atoms (^{238}U, in this case) to go undetected, dates set by radioactive isotopes are more likely to be too recent. Hence, as methods become more precise and more samples are analyzed, geologic time stretches farther and farther backward.

most of the other zoologists of his time, was particularly interested in one-celled organisms and invertebrates. Undoubtedly it was his long study of these forms of life that led him to think of living things in terms of constantly increasing complexity, each species derived from an earlier, less complex one.

33-5 *According to Lamarck's hypothesis—now known to be in error—the necks of giraffes became longer when they stretched to reach high branches, and this acquired characteristic was transmitted to their offspring.*

Like Cuvier and others, Lamarck noted that older rocks generally contained fossils of simpler forms of life. Unlike Cuvier, however, Lamarck interpreted this as meaning that the more complex forms had arisen from the simpler forms by a kind of progression. According to his hypothesis, this progression, or "evolution," to use the modern term, is dependent on two main forces. The first is the inheritance of acquired characteristics. Organs in animals become stronger or weaker, more or less important, through use or disuse, and these changes, according to Lamarck's theory, are transmitted from the parents to the progeny. His most famous example was the evolution of the giraffe. According to Lamarck, the modern giraffe evolved from ancestors that stretched their necks to reach leaves on high branches. These ancestors transmitted the longer neck—acquired through stretching—to their offspring, which stretched their necks even longer, and so on.

The second important factor in Lamarck's theory of evolution was a universal creative principle, an unconscious striving upward on the *Scala Naturae* that moved every living creature toward greater complexity. Amoebas were on their way to *Homo sapiens*. Some might get waylaid—the orangutan, for instance, by being caught in an unfavorable environment, had been diverted off its course—but the will was always present. Life in its simplest forms was constantly emerging by spontaneous generation to fill the void left at the bottom of the ladder. In Lamarck's formulation, Aristotle's ladder of nature had been transformed into a steadily ascending escalator powered by a universal will.

Lamarck's contemporaries did not object to his ideas about the inheritance of acquired characteristics, which we, with our more advanced knowledge of genetics, now know to be false. Nor did they criticize his belief in a metaphysical force, which was actually a common element in many of the theories of the time. But these vague postulates provided a very shaky foundation for the radical proposal that more complex forms evolved from simpler forms. Moreover, Lamarck personally was no match for the brilliant and witty Cuvier, who relentlessly attacked his ideas. As a result, Lamarck's career was ruined, and both scientists and the public became even less prepared to accept any evolutionary doctrine.

Development of Darwin's Theory

The Earth Has a History

The person who most influenced Darwin, it is generally agreed, was Charles Lyell (1797–1875), a geologist who was Darwin's senior by 12 years. One of the books that Darwin took with him on his voyage was the first volume of Lyell's newly published *Principles of Geology,* and the second volume was sent to him while he was on the *Beagle*. On the basis of his own observations and those of his predecessors, Lyell opposed the theory of catastrophes. Instead, he produced new evidence in support of Hutton's earlier theory of uniformitarianism. According to Lyell, the slow, steady, and cumulative effect of natural forces had produced continuous change in the course of the earth's history. Since this process is demonstrably slow, its results being barely visible in a single lifetime, it must have been going on for a very long time. If the earth had a long continuous history and if no forces other than well-known, natural ones were needed to explain the events as they were recorded in the geologic record, might not living organisms have had a similar history?

The Voyage of the *Beagle*

This, then, was the intellectual equipment with which Charles Darwin set sail from England. As the *Beagle* moved down the Atlantic coast of South America, through the Straits of Magellan, and up the Pacific coast, Darwin traveled the interior, fished, hunted, and rode horseback. He explored the rich fossil beds of South America (with the theories of Lyell fresh in his mind) and collected specimens of the many new kinds of plant and animal life he encountered. He was impressed most strongly during his long, slow trip down the coast and up again by the constantly changing varieties of organisms he saw. The birds and other animals on the west coast, for example, were very different from those on the east coast, and even as he moved slowly up the western coast, one species would give way to another.

Most interesting to Darwin were the animals and plants that inhabited a small, barren group of islands, the Galapagos, which lie some 950 kilometers off the coast of Ecuador. The Galapagos were named after the islands' most striking inhabitants, the tortoises (*galápagos* in Spanish), some of which weigh 100 kilograms or more. Each island has its own type of tortoise; sailors who took these tortoises on board and kept them as convenient sources of fresh meat on their sea voyages could readily tell which island any particular tortoise had come from. Then there was a group of finchlike birds, 13 species in all, that differed from one another in the sizes and shapes of their bodies and beaks, and particularly in the type of food they ate. In fact, although clearly finches, they had many characteristics seen only in completely different types of birds on the mainland. One finch, for example, feeds by routing insects out of the bark of trees. It is not fully equipped for this, however, lacking the long tongue with which the woodpecker flicks out insects from under the bark. Instead, the woodpecker finch uses a small stick or cactus spine to pry the insects loose.

33-6 *The* Beagle's *voyage around South America. The ship left England in December of 1831 and arrived at Bahia, Brazil, in late February of 1832. About $3\frac{1}{2}$ years were spent along the coast of South America, surveying and making inland explorations. The stop at the Galapagos Islands was for slightly more than a month, and, during that brief time, Darwin made the wealth of observations that were to change the course of the science of biology. The remainder of the voyage, across the Pacific to New Zealand and Australia, across the Indian Ocean to the Cape of Good Hope, back to Bahia once more, and at last home to England, occupied another year.*

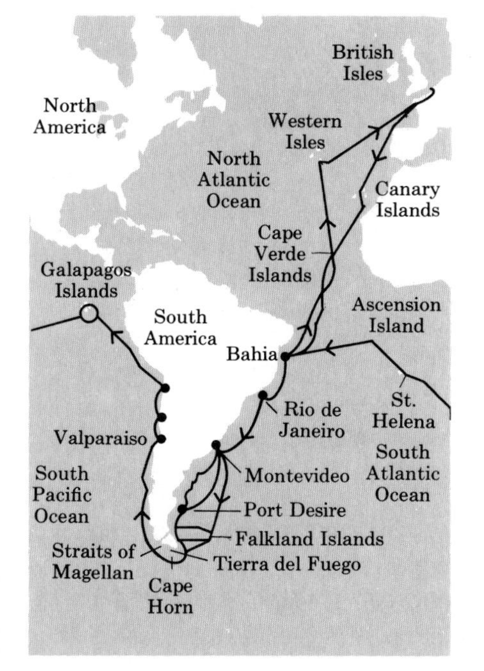

33-7 *The woodpecker finch is a rare phenomenon in the bird world because it is a tool user. Like the true woodpecker, it feeds on grubs, which it digs out of trees using its beak for a chisel. Lacking the woodpecker's long, barbed tongue, it resorts to an artificial probe, a twig or cactus spine, to dislodge the grub. The woodpecker finch shown has selected a cactus spine, which it inserts into the grub hole. The bird has succeeded in prying out the grub, which it eats. If the pick selected turns out to be an efficient tool, the bird will carry it from tree to tree in its search for grubs.*

Darwin's Long Delay

Darwin returned to England with the *Beagle* in 1836. Two years later, he read the essay by Malthus, and in 1842 he wrote a preliminary sketch of his theory, which he revised in 1844. On completing the revision, he wrote a formal letter to his wife requesting her, in the event of his death, to publish the manuscript (which was some 230 pages long). Then, with the manuscript and letter in safekeeping, he turned to other work, including a four-volume treatise on the classification and biology of barnacles. For more than 20 years following his return from the Galapagos, Darwin mentioned his ideas on evolution only in his private notebooks and in letters to his scientific colleagues.

In 1856, urged on by his friends Charles Lyell and botanist Joseph Hooker, Darwin set slowly to work preparing a manuscript for publication. In 1858, some 10 chapters later, Darwin received a letter from another English naturalist, Alfred Russel Wallace, presenting a theory of evolution that exactly paralleled Darwin's own. Like Darwin, Wallace had traveled extensively, both in South America (in the Amazon basin) and in the Malay archipelago, from which his letter had been mailed. He also had read Malthus's essay, and he had corresponded with Darwin on several previous occasions. Wallace, tossing in bed one night with a fever, had a sudden flash of insight. "Then I saw at once," Wallace recollected, "that the ever present variability of all living things would furnish the material from which, by the mere weeding out of those less adapted to the actual conditions, the fittest alone would continue the race." Within two days, Wallace's 20-page manuscript was completed and in the mail.

When Darwin received Wallace's letter, he turned to his friends for advice, and Lyell and Hooker, taking matters into their own hands, presented the theory of Darwin and Wallace at a scientific meeting just one month later. (Darwin described Wallace as "noble and generous," as indeed he was.) Lyell and Hooker read four papers from Darwin's notes of 1844, excerpts from two letters written by Darwin, and Wallace's manuscript. Their presentation received little attention, but for Darwin the floodgates were opened. He finished his long treatise in little more than a year, and the book was finally published. The first printing was a mere 1,250 copies, but they were sold out the same day.

Why Darwin's long delay? His own writings, voluminous though they are, shed little light on this question. However, it may be worth recalling that he had once been a divinity student. Perhaps more important, his wife, to whom he was deeply devoted, was extremely religious. It is difficult to avoid the speculation that Darwin, as has been the case with others, found the implications of his theory difficult to confront.

From his knowledge of geology, Darwin knew that these islands, clearly of volcanic origin, were much younger than the mainland. Yet the plants and animals of the islands were different from those of the mainland, and in fact the inhabitants of different islands in the archipelago differed from one another. Were the living things on each island the product of a separate special creation? "One might really fancy," Darwin mused at a later date, "that from an original paucity of birds in this archipelago one species had been taken and modified for different ends." After his return, this problem continued, in his own word, to "haunt" him.

The Darwinian Theory

Not long after Darwin's return, he read a book by the Reverend Thomas Malthus that had first appeared in 1798. In this book, Malthus warned, as economists have warned frequently ever since, that the human population was increasing so rapidly that it not only would soon outstrip the food supply, but also would leave "standing room only" on earth. Darwin saw that Malthus's conclusion—that food supply and other factors hold populations in check—is true for all species, not just the human one. For example, a single breeding pair of elephants, which are the slowest breeders of all animals, would, if all their progeny lived and reproduced the normal number of offspring over a normal life span, produce 19 million elephants in 750 years, yet the average number of elephants generally remains the same over the years. Where there might have been 19 million elephants in theory,

Alfred Russel Wallace (1823–1913).

there are, in fact, only two. But why those particular two? The process by which the two survivors are "chosen" Darwin called *natural selection*.

Darwin saw natural selection as a process analogous to the type of selection exercised by breeders of cattle, horses, or dogs. In the case of artificial selection, we humans choose individuals for breeding on the basis of characteristics that seem to us to be desirable. In the case of natural selection, environmental conditions are the principal forces that operate on the variations continually produced among individuals in all species of living organisms. As these forces select for some varieties and eliminate others, the characteristics of the population very slowly change. For example, if, due to anatomical variations, some individual horses were swifter than others, these individuals would be more likely to escape predators and survive; their progeny, in turn, might be swifter, which is exactly what happened over hundreds of thousands of years.

Where do the variations come from? According to Darwin's theory, variations occur by chance. They are not produced by the environment, by a "creative force," or by the unconscious striving of the organism. In themselves, they have no goal or direction. It is the operation of natural selection over a series of generations that gives direction to evolution. A variation that gives an animal even a slight advantage makes that animal more likely to leave surviving offspring. Thus, to return to Lamarck's giraffe, an animal with a slightly longer neck has an advantage in feeding and so is apt to leave more offspring than one with a shorter neck. If the longer neck is an inherited trait, some of these offspring will also have long necks, and, if the long-necked animals in this generation have an advantage, members of the next generation will include more long-necked individuals. Eventually, the population of short-necked giraffes will give way to a population of long-necked ones.

As you can see, the essential difference between Darwin's formulation and that of any of his predecessors is the central role he gave to variation. Others had thought of variations as mere disturbances in the overall design, whereas Darwin saw that variations among individuals are the real fabric of the evolutionary process. Species arise, he saw, when differences among individuals within a group are gradually converted into differences between groups as the groups become separated in both space and time.

The Origin of Species, which Darwin pondered for more than 20 years before its publication in 1859 (see essay), is, in his own words, "one long argument." Fact after fact, observation after observation, culled from the most remote Pacific island to a neighbor's pasture, is recorded, analyzed, and commented upon. Every objection is anticipated and countered. Because the process of evolution is so slow, Darwin did not believe that direct proof of his theory was possible. However, as we shall see in the next three chapters, the twentieth century has produced clear evidence of evolution in progress. Few scientists now doubt that species have originated in the past and are still originating, that species have become extinct in the past and are still becoming extinct, and that the different species living today share ancestral species in the past.

The real difficulty in accepting Darwin's theory has always been that it seems to diminish our significance. Earlier, astronomy had made it clear that the earth is not the center of the universe or even of our own solar system. Now the new biology asked us to accept the proposition that, like all other organisms, we too are the products of a random process and that, as far as science can show, we are not created for any special purpose or as part of any universal design.

SUMMARY

Until the eighteenth century, it was generally accepted that species were the products of special creation and remained unchanged. This concept, however, came into question as a result of several developments, including the discovery of many new species, as a consequence of New World exploration; studies in geology that indicated a gradual, constant change in the surface of the earth (uniformitarianism); the accompanying recognition that the earth has a long history (without which evolution would have been impossible); and the discovery of fossils and the recognition of their origin.

Darwin was not the first to propose a theory of evolution. His most notable predecessor was Lamarck, whose theory of evolution was based on the inheritance of characteristics acquired by an organism during its lifetime. Darwin's theory differed from others in that it envisioned evolution as a two-part process, depending upon (1) the existence in nature of inheritable variations among organisms and (2) the process of natural selection by which some organisms, by virtue of their inheritable variations, leave more offspring than others. Darwin's theory is rightfully regarded as the greatest unifying principle in biology.

QUESTIONS

1. What is the essential difference between Darwin's theory of evolution and that of Lamarck?

2. The chief predator of an English species of snail is the song thrush. Snails that inhabit woodland floors have dark shells, whereas those that live on grass have yellow shells, which are less clearly visible against the lighter background. Explain, in terms of Darwinian principles.

3. The phrase "chance and necessity" has been applied to the process by which species develop. Relate this to the fact that snails living on grass do not have green shells but there are, for example, green frogs and green insects.

4. What tool-using animals other than the woodpecker finch can you think of?

34

The Modern Theory of Evolution

Darwin's theory of evolution brought about a revolution, not only in biology but also in almost every aspect of Western culture. Now, the theory of evolution is so much a part of modern biological thought that today's discussions center around details of how the process takes place rather than the theory itself. As a result, one tends to lose sight of the grand design. Before we begin our own detailed discussion, let us take another brief look at the theory, beginning with the body of evidence from which the theory emerged.

First, the earth has a long history. It has been in existence, by present calculations, for 4.6 billion years. During this time it has experienced major geologic and climatic changes.

Second, in the course of the earth's history there has been a succession of living forms, as recorded in the fossil record, which now dates back more than 3 billion years. In this record, generally speaking, simpler forms precede more complex ones. In the case of many groups of organisms—vascular plants and vertebrates, for example—fossils can be found that exhibit a graded series of changes in anatomical characteristics, linking older forms with the modern forms and revealing pathways diverging from common ancestors.

Third, living members of the various groups of organisms share the same basic plans of organization, although the actual structures may vary considerably, both in form and in function. For example, Figure 34–1 (page 478) shows the forelimbs of various vertebrates. The bones in each are the same, but they have been modified in each type of organism. Such structures, which have a common origin but not necessarily a common function, are said to be homologous. (Analogous structures, by contrast, are superficially similar but have an entirely different evolutionary background—the wing of a bird and the wing of an insect, for example.)

Further evidence for evolution is seen in the great variety of existing organisms. Remember that it was not until comparatively recently in the history of human civilization that naturalists began to explore other countries, much less other continents. Such explorers, Darwin and Wallace among them, began to see that species were not distinct, but that as one traveled—up the western coast of South America, for instance—one could observe gradual changes in various characteristics of the plants and animals. These were interpreted as evidence against separate creation and for

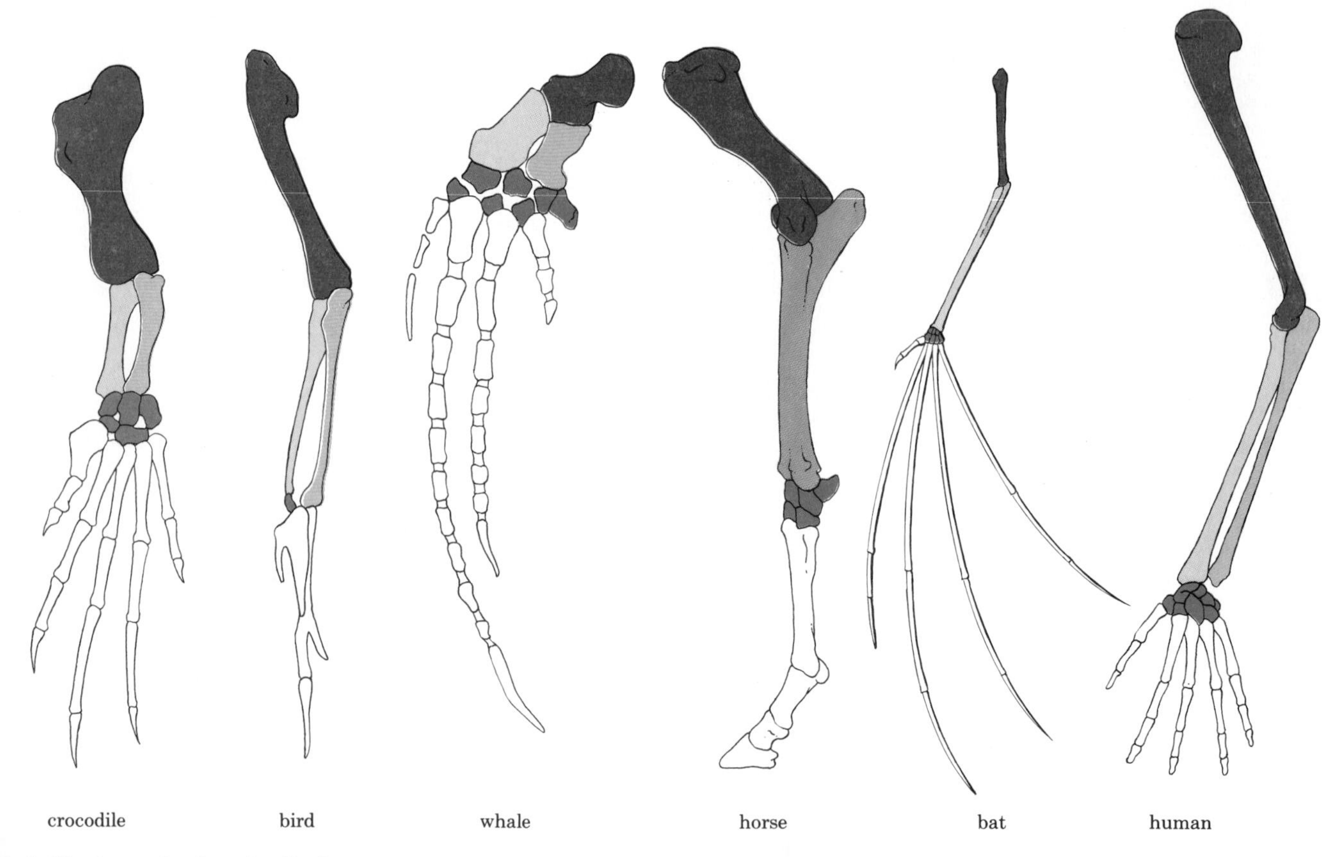

34-1 *The bones in these forelimbs are color-coded to indicate fundamental similarities of structure and organization. The crocodile is placed first because it is the closest to the ancestral type—the form from which all the others arose.*

the concept that organisms became modified with time, according to the different environments in which they lived.

The principle of evolution gains additional support from the remarkable similarity of living things. The cellular basis of life was demonstrated within a decade of the publication of the first edition of *The Origin of Species*. Since that time, almost every major discovery in biochemistry—glycolysis, the electron transport chain, the nature of ATP, the genetic code—has emphasized the fundamental similarity of living cells. Additional evidence for the unity of life has been uncovered by advances, first in light microscopy and more recently in electron microscopy, that have revealed, for example, the two-ply structure of cell membranes, the existence of ribosomes, and the internal organization of eukaryotic cilia and flagella.

Finally, there is the exquisite adaptedness of living things to their environment. To many, this seeming perfection of adaptation appeared to be strong support for the doctrine of special creation. Now, however, it is possible to show that such adaptations have taken place over a long period of time as a consequence of very small changes and, moreover, are still taking place at this very moment. The coevolution of insects and flowers, which we described in Chapter 22, is one example, and there are many others.

Evolution is inextricably intertwined with ecology, in what G. E. Hutchinson of Yale University aptly called "the ecological theater and the evolutionary play." In the remaining chapters of this section, we are going to examine the mechanisms by which the plot of the evolutionary play moves forward. In Section 8, with those mechanisms in mind, we shall place

34-2 *Within this long egg case, composed of a parchmentlike material, a whelk is depositing hundreds of eggs. When the baby mollusks hatch, they will break through thin areas in the individual capsules and emerge as miniature copies of their parents. Once the egg case is produced and the eggs laid, the mother whelk does not participate further in raising the young. Despite the large number of young produced, there is, in general, no increase in the number of whelks from generation to generation.*

ourselves in the ecological theater to observe the cast—all living things—in the variety of interrelationships that constitute the play in action.

The Theory Today

Although it is now more than 120 years since the first publication of *The Origin of Species,* Darwin's original concept of how evolution comes about still provides the basic framework for our understanding of the process. His concept rested on four premises:

1. Organisms beget like organisms—in other words, there is stability in the process of reproduction.
2. In any given population, there are variations among individual organisms, and some of the variations are inheritable.
3. In most species, the number of individuals that survive to reproductive age is small compared with the number produced.
4. Which individuals survive to reproduce and which do not is determined to a significant degree by inherited variations. Some variations provide an advantage that enables the individuals possessing them to survive and produce more offspring than other individuals. Darwin termed these "favorable" variations and argued that inherited favorable variations tend to become more common from one generation to the next. This is the process that Darwin called natural selection.

Modern evolutionary theory conceives of evolution as Darwin conceived of it, with the important addition of an understanding of the mechanisms of inheritance. Twentieth-century genetics answers three questions that Darwin was never able to resolve: (1) how genetic traits are transmitted from one generation to the next; (2) why genetic traits are not "blended out" but can disappear and then reappear in later generations (like whiteness in pea flowers); and (3) how the variations arise on which natural selection acts. The modern combination of evolutionary theory and genetics is known as the neo-Darwinian synthesis, or the synthetic theory of evolution. (Here "synthetic" does not mean artificial, which is the connotation it has for us in these days of synthetic fabrics and artificial colors and flavors, but has its original meaning of the putting together of two or more different elements.)

The branch of biology that emerged from the synthesis of Darwinian evolution and Mendelian principles is known as *population genetics.* A *population,* for the geneticist, is an interbreeding group of organisms. For instance, all the fish of one particular species in a pond are a population and so are all the fruit flies in one bottle. The population is defined and united by its *gene pool,* which is simply the total of all the alleles of all the genes of all the individuals in a population.

The relative proportion of alleles of certain genes in a gene pool may change from one generation to the next. In population genetics, *evolution* is defined as any change in the composition of the gene pool. In this view the individual is only a temporary vessel, holding a small portion of the gene pool for a short time. During that time, the interactions of the individual with its environment "test" the fitness of the particular combination of genes. *Fitness,* in the Darwinian sense, is measured by only one criterion: the reproductive success of one genotype measured relative to the reproductive success of another.

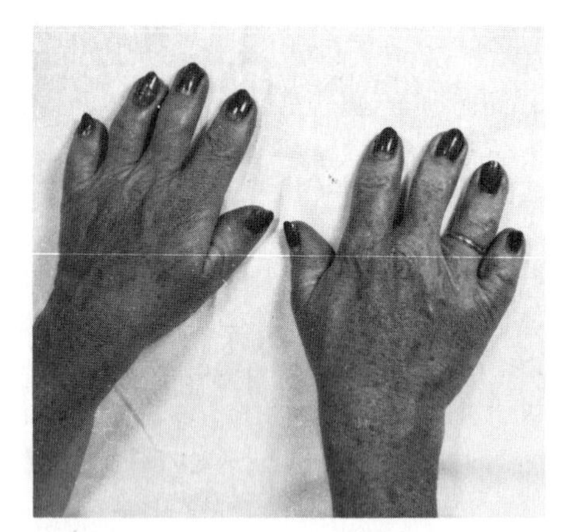

34-3 *A dominant gene is responsible for the trait known as brachydactylism (short fingers). In the brachydactylous hands shown here, the first bones of the fingers are of normal length, but the second and third bones are abnormally short. If brachydactylism is caused by a dominant gene, why is it such a rare trait?*

The Hardy-Weinberg Principle

Mendel's great contribution was to demonstrate that traits—even when they seem to disappear—are actually preserved, "hidden" in the heterozygote. But the laws of heredity had been demonstrated only for ideal conditions, in which simple, controlled matings were performed. For the purpose of evolutionary theory, it was necessary to establish how genes behave under the much more complex conditions that occur in nature. One needed to be able to detect and follow changes in the genetic makeup of a population. Population genetics is concerned with the sources and dynamics of variations in populations. Just as Mendel's laws form the cornerstone of classical genetics, the Hardy-Weinberg principle forms the cornerstone of population genetics.

In the early 1900s biologists raised an important question: How can both dominant and recessive genes remain in populations? Why don't dominants simply drive out recessives? For example, if brachydactylism (Figure 34–3) is caused by a dominant allele, why don't most or even all people have short, fat fingers? The question was answered in 1908 by G. H. Hardy, an English mathematician, and G. Weinberg, a German physician.

Working independently, Hardy and Weinberg examined the behavior of alleles in an idealized population. Four conditions govern the idealized population:

1. No mutations occur.
2. There is no net movement of individuals—with their genes—into the population (immigration) or out of it (emigration).
3. The population is large enough that the laws of probability apply—that is, it is highly unlikely that chance alone can alter gene frequencies.
4. Mating is random, and all alleles are equally viable. In other words, there is no difference in reproductive success. The offspring of all possible matings are equally likely to survive to reproduce in the next generation.

Now, let us look at just one gene, selecting one that has only two alleles, which we will call A and a. We are interested in the relative proportions of A and a from one generation to the next. By convention, the letter p is used to designate the frequency, or relative proportion, of one allele and q the frequency of the other. When there are only two alleles, p and q together must equal one:

$$p + q = 1$$

These proportions—p and q—could be expressed in terms of fractions, as done by Mendel, but since the proportions of the two alleles will probably not be equal, as they were in Mendel's carefully controlled experiments, it is more convenient to express the numbers as decimals. For example, suppose that in a particular population 80 percent of the alleles of the gene under study are allele A. The frequency of A is 0.8, or $p = 0.8$. Given that there are only two alleles, we then know that the frequency of allele a is 0.2 ($q = 1 - p$).

Now suppose that random mating takes place in this population. We can calculate the frequencies of the resulting genotypes by drawing a Punnett square (Figure 34–4). As you can see from the square, the population

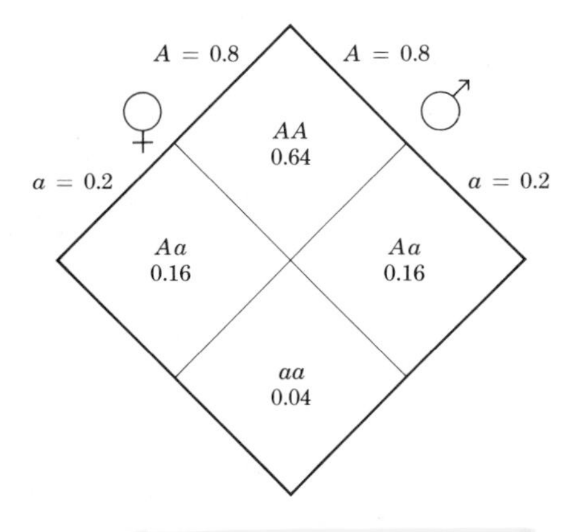

34-4 *Results of random breeding in a population in which the frequency* (p) *of allele* A *equals 0.8 and the frequency* (q) *of allele* a *equals 0.2. In setting up this Punnett square, we are assuming that the gene is not sex-linked and that the alleles appear in the same frequencies in males as in females.*

produced by random mating would consist of 64 percent *AA*, 32 percent *Aa*, and 4 percent *aa* genotypes.

Instead of drawing a Punnett square, we can do the same thing algebraically. Because $p + q = 1$, it follows that:

$$(p + q)(p + q) = 1 \times 1 = 1$$

Or, as you probably remember from high school algebra:

$$p^2 + 2pq + q^2 = 1$$

This algebraic expression of the allele frequencies is the Hardy-Weinberg formula.

Let us apply it to the random mating that just occurred in our population. Taking the initial values for the frequencies of the two alleles, we obtain the following results:

$p^2 = 0.8 \times 0.8 = 0.64$ (the frequency of *AA* genotypes)
$2pq = 2 \times 0.8 \times 0.2 = 0.32$ (the frequency of *Aa* genotypes)
$q^2 = 0.2 \times 0.2 = 0.04$ (the frequency of *aa* genotypes)

What has happened to the frequency of the alleles in the gene pool as a result of this round of mating? We know from our calculations that the frequency of *AA* is 0.64. In addition, half of the alleles in the heterozygotes *(Aa)* are *A*, so the total of allele *A* is 0.64 plus ½ of 0.32—that is, 0.64 plus 0.16, or 0.8. The frequency of allele *A (p)* has not changed. Similarly, the total of allele *a* is 0.04 (in the homozygotes) plus 0.16 (half the alleles in the heterozygotes), or 0.2. The frequency of allele *a (q)* has also remained the same.

If another round of mating occurs, the proportion of *AA*, of *Aa*, and of *aa* genotypes in our population will again be 64 percent, 32 percent, and 4 percent respectively. Again the frequency of allele *A* will be 0.8 and of allele *a* 0.2. And so on, and so on, generation after generation. *In an ideal population in which the four specified conditions are met, the allele frequencies do not change from generation to generation.*

Although normally any one diploid individual has no more than two alleles of the same gene, there may, of course, be more than two alleles for a given gene in the gene pool. For instance, in the case of the gene for coat color in rabbits, there are at least four alleles. Different diploid combinations of these alleles give rise to coat colors ranging from dark gray through medium and light gray to white, as well as a variety in which most of the coat is white but the feet, ears, and muzzle are black. The Hardy-Weinberg principle applies equally well to situations such as this, in which there are multiple alleles of the same gene, although the calculations become much more complex.

The Significance of the Hardy-Weinberg Principle

What the Hardy-Weinberg principle says, in effect, is that the genetic recombination that occurs as a result of meiosis and fertilization does not by itself change the overall composition of the gene pool. A dominant allele, for instance, does not tend to increase in a population, and a recessive one does not tend to decrease. A rare variant will continue to persist in the population, despite the shuffling and reshuffling of genes that take place every generation. The interbreeding of organisms does not change the frequencies of alleles.

As you will recall, however, the Hardy-Weinberg principle applies only to populations in which the four conditions mentioned earlier hold. What happens if one of these conditions is not met? Let us imagine another gene pair containing an allele a that has a damaging effect in the homozygous state, reducing the likelihood that a homozygous individual will survive to reproduce. We can estimate the number of aa genotypes in the population, perhaps by screening tests on newborn infants. Suppose, for instance, that the condition shows up in 1 in 10,000 infants. In other words, $q^2 = 1/10{,}000$, or 0.0001, so $q = \sqrt{0.0001}$, or 0.01. If $q = 0.01$, then $p = 0.99$ and $2pq = 0.0198$, or almost 0.02. Thus about 2 percent of the population—one person in every 50—can be estimated to be a carrier of this allele.

Now suppose that we do the same screening tests five years later and find that q is not 0.01 but 0.009, and then we repeat it a few years later and find it has once again decreased very slightly, perhaps to 0.008. In other words, evolution has occurred. One allele is decreasing while the other is increasing. Moreover, using Hardy-Weinberg as our yardstick, we know not only that the change has taken place but also that there must be some reason for it.

The Hardy-Weinberg principle gives us a way of detecting evolution, that is, changes in the gene pool of a population. When it does not hold from generation to generation, we know that evolution is occurring and we can then look for the factors causing the change.

The Agents of Evolution

Let us now look at the factors that can disturb the Hardy-Weinberg equilibrium and produce changes in the gene pool: mutations, gene flow, genetic drift, and natural selection. We shall consider each in turn.

Mutations

Mutations are defined as inheritable changes in the genotype. A mutation may involve a large part of a chromosome or only a single nucleotide in a DNA molecule. As we saw in Chapter 15, the change in a single nucleotide pair is responsible for the abnormal hemoglobin associated with sickle cell anemia. Mutations, as we noted previously, can be produced by exposure to x-rays, radioactive compounds, ultraviolet rays, and other agents. Most occur "spontaneously"—meaning simply that we do not know the reasons for them.

The rate of spontaneous mutation is low. For humans, the chance of a mutation occurring in any given gene has been estimated at less than 1 in 10,000. However, given the large number of genes in every gamete, there is an excellent possibility that any given gamete will carry a new mutation in at least one gene. Different genes have different rates of mutation. The reason for these differences is not known but probably has to do with the chemical nature of the gene itself or with its position in the chromosome.

At one time, many students of evolution viewed mutations as a most important element in determining the direction and extent of evolutionary change (Figure 34–5). Today, mutations are seen to be important as the original source of the variations within a population upon which selection then acts. Mutations *alone* do not effect major changes in the frequencies of alleles.

34-5 *An example of the sometimes dramatic effects of mutation. The ewe in the middle is an Ancon, an unusually short-legged strain of sheep. The first Ancon on record was born in the late nineteenth century into the flock of a New England farmer. By inbreeding (the trait was transmitted as a recessive), it was possible to produce a strain of animals with legs too short to jump the low stone walls that traditionally enclosed New England sheep pastures. A similar strain was produced in northern Europe as a result of an independent mutation. At one time, it was thought that evolution took place in sudden, large jumps such as this—a concept sometimes referred to as the "hopeful monster" theory. One reason this concept was abandoned is that nearly all mutations producing dramatic changes in the phenotype are harmful, as this one would be in a wild population.*

Gene Flow

Changes in the Hardy-Weinberg equilibrium of a population can, of course, be produced by the movement of breeding individuals (with their genes) into or out of the population or, as in the case of plants and many aquatic invertebrates, the movement of gametes (for example, in the form of pollen) from one population to another. This phenomenon is often referred to as *gene flow.*

In the human population, the effects of gene flow are particularly evident on the Hawaiian Islands, where the influx of missionaries and other Caucasians breached a once-formidable geographic barrier and ended the long genetic isolation of the Polynesians who originally inhabited the islands. The geographic isolation of a population can be, as we shall see in Chapter 36, very important in evolution by natural selection. Gene flow between populations can reunite previously separated gene pools and often counteracts the effects of natural selection.

Genetic Drift

As we stated previously, the Hardy-Weinberg principle holds true only if the population is large. This qualification is necessary because the Hardy-Weinberg equilibrium depends on the laws of probability. These laws—the laws of chance—apply equally well to flipping coins, rolling dice, or betting at roulette. In flipping coins, it is possible for heads to show up five times in a row, but, on the average, heads will show up half the time and tails half the time. The more times the coin is flipped, the more closely the expected frequencies of half (0.50) and half (0.50) are approached. Similarly, in a presidential poll, it is possible that the first five people interviewed will be Republicans, and it is not until a sufficiently large sample of the voting population is queried that accurate predictions become possible. In a small population, as in a small sample, chance plays a large role.

Consider, for example, an allele, say *a,* that has a frequency of 1 percent. In a population of 1 million, 20,000 *a* alleles would be present in the gene pool (remember that each diploid individual carries two alleles for any given gene). But in a population of 50, only one copy of allele *a* would be

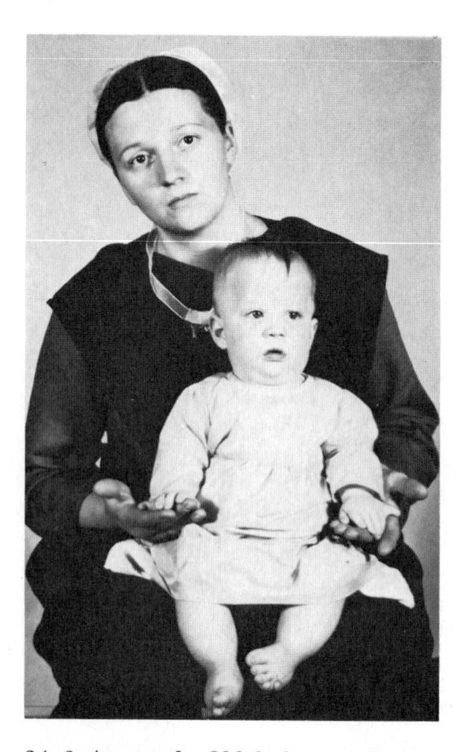

34-6 *Among the Old Order Amish, a group founded by only a few couples some 200 years ago, there is an unusually high frequency of a rare allele. In its homozygous state, the allele results in extra fingers and dwarfism. This Amish child is a six-fingered dwarf.*

present. If the individual carrying this allele failed to mate or were destroyed by chance before leaving offspring, allele *a* would be completely lost. Similarly, if 10 of the 49 individuals without allele *a* were lost, the frequency of *a* would jump from 1 in 100 to 1 in 80. This phenomenon, a change in the gene pool that takes place as a result of chance, is *genetic drift.*

The Founder Principle

A small population that branches off from a larger one may or may not be genetically representative of the larger population from which it was derived. Some rare alleles may be overrepresented or may be lost completely. As a consequence, even when and if the small population increases in size, it will have a different genetic composition—a different gene pool—from that of the parent group. This phenomenon, a type of genetic drift, is known as the *founder principle.* In a small population, there is likely to be less variation in the population. Thus, even though mating may be random, the mates will be more closely related to each other—more similar genetically—with a consequent increase in homozygosity.

An example of the founder principle is found in the Old Order Amish of Lancaster, Pennsylvania (Figure 34-6). Among these people, there is an unprecedented frequency of a recessive allele that, in the homozygous state, causes a combination of dwarfism and polydactylism (extra fingers). Since the group was founded in the early 1770s, some 61 cases of this rare congenital deformity have been reported, about as many as in all the rest of the world's population. Approximately 13 percent of the persons in the group, which numbers some 17,000, are estimated to carry this rare allele. The entire colony, which has kept itself virtually isolated from the rest of the world, is descended from only a few individuals. By chance, one of them must have been a carrier of the gene.

Genetic Drift and Non-Darwinian Evolution

Changes in the gene pool that originate as a result of random mutations and that increase or decrease as a result of genetic drift are sometimes referred to as non-Darwinian evolution. Now that scientists are beginning to analyze the genotype at the molecular level, more and more variations are being found that might be "neutral"—that is, that might have no effect on the fitness of the organism. Such variations will be described in more detail in the following chapter. Whether or not any variation can, in fact, be neutral and the extent to which changing frequencies of variation can be caused merely by limited sampling (genetic drift) are matters currently under debate by population biologists.

Natural Selection

Natural selection is the fourth and by far the most important agent of evolution. Darwin coined the term "natural selection" as an analogy to what he called "artificial selection," the process by which breeders of domestic animals or crops select certain individuals and discard others as they seek to develop desirable characteristics in their stock. Mating is not random but is instead determined by the breeder. In Darwin's time, pigeon fanciers had produced a number of different and often very exotic breeds by means of artificial selection. Among the examples more familiar to us are "purebred" dogs and cultivated plants (Figure 34-7).

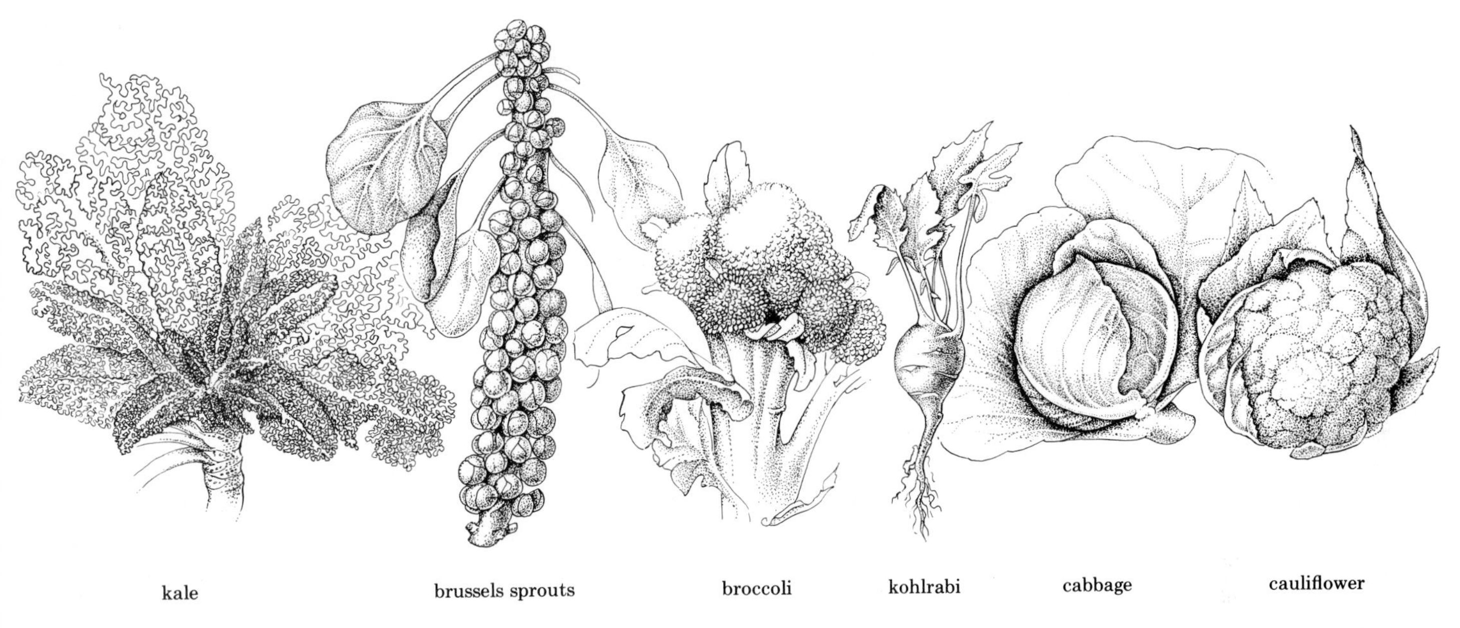

34-7 *Six vegetables produced from a single species of plant* (Brassica oleracea, *a member of the mustard family). They are the result of selection for leaves (kale), lateral buds (brussels sprouts), flowers and stem (broccoli), stem (kohlrabi), enlarged terminal buds (cabbage), and flower clusters (cauliflower). Kale most resembles the ancestral wild plant.*

In our discussion of the Hardy-Weinberg equilibrium and in most of the examples in the last few pages, we have considered changes in the frequencies of alleles at a single locus. That is, we limited ourselves to members of a single gene pair, in isolation from the rest of the genotype. It is therefore important to emphasize that *selection acts upon the phenotype.* The phenotype, you will recall, is the product of the interaction of the genes of an individual with one another and with the environment in the course of the individual's life (Figure 34-8).

34-8 *A pinyon pine growing out of a crack in a rock in Utah. Genotypically the tree is tall and straight; however, the constant, strong winds in which it has grown have produced this phenotype.*

Natural selection is the process of interaction between an organism and its environment that results in the differential rate of reproduction of different phenotypes in a population. A number of factors may contribute to the differential reproduction. These include nonrandom mate selection, more efficient mating by some phenotypes than by others, differing fertility of different phenotypes, and failure of some phenotypes to survive to reproductive maturity. Whatever the specific factor or factors involved, differential reproduction can result in changes in the relative frequencies of alleles in the gene pool—that is, in evolution.

Types of Selection

Three general types of selection operate within populations: stabilizing, disruptive, and directional (Figure 34-9).

Stabilizing selection, a process that goes on at all times in all populations, is the continual elimination of extreme individuals. Most mutant forms are probably immediately weeded out in this way, many of them in the zygote or the embryo. Clutch size in birds, for instance, is a result of stabilizing selection. Clutch size (the number of eggs a bird lays) is determined genetically. In a study of Swiss starlings (Table 34-1), the percentage of birds surviving increased for each clutch size up to five. With a clutch size larger than five, a smaller percentage of the birds survived—apparently because of inadequate nutrition. Female Swiss starlings whose genotypes dictate a clutch size of five will have more surviving young, on the average, than members of the same species that lay more or fewer eggs.

The second type of selection, *disruptive selection*, increases the proportions of two extreme types in a population at the expense of intermediate forms. As we shall see in the next chapter, disruptive selection can be important in maintaining variations within a population.

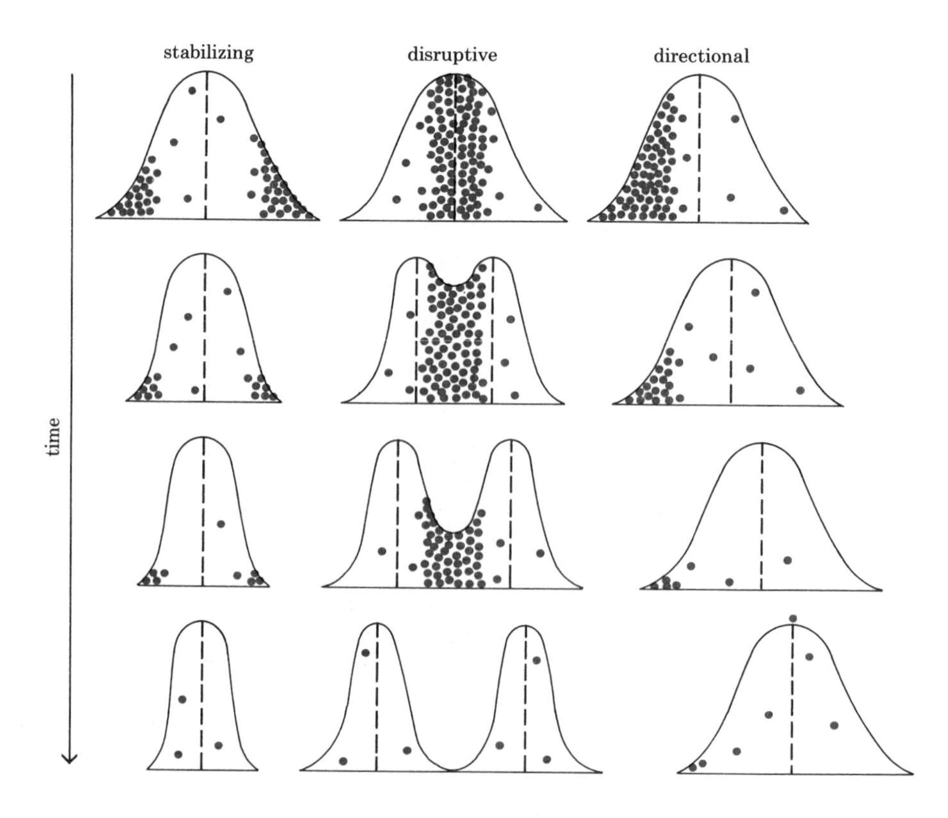

34-9 *The three different types of natural selection. The dots represent individuals that failed to reproduce or that left less than the average number of offspring. Stabilizing selection involves the elimination of extremes. In disruptive selection, intermediate forms are eliminated, producing two divergent gene pools. Directional selection, which is the gradual elimination of one phenotype in favor of another, produces adaptive change. In these graphs, the vertical axes denote the proportion of individuals in a population with a particular characteristic, and the horizontal axes, the varying dimensions of the characteristic being considered.*

Table 34-1 Survival in relation to number of young in Swiss starling

Number of young in brood	1	2	3	4	5	6	7	8
Number of young marked	65	328	1,278	3,956	6,175	3,156	651	120
Number of marked birds recaptured after 3 months	0	6	26	82	128	53	10	1
Percentage of marked birds recaptured after 3 months	0	1.8	2.0	2.1	2.1	1.7	1.5	0.8

The third type, which is the one we shall be most concerned with, is *directional selection*. Directional selection acts either for or against an extreme phenotypic characteristic and so is likely to result in the gradual replacement of one allele or group of alleles by another in the gene pool.

"Survival of the Fittest"

The phrase "survival of the fittest" is often used in describing the Darwinian theory. In the early twentieth century, the doctrine of survival of the fittest in natural populations was used by some individuals to defend gross social inequalities and ruthless competitive tactics in industry on the grounds that they were merely in accord with the "laws of nature." This philosophy was referred to by some as social Darwinism. However, in actual fact, very little in the process of evolutionary change fits the concept of "nature red in tooth and claw." A fuchsia plant with flowers a little brighter than those of its neighbors and so better able to catch the attention of a passing hummingbird is a more pertinent model of the struggle for survival. There is only one criterion of fitness as it is measured by population geneticists: the relative number of descendants of an individual in a future population.

SUMMARY

Population genetics is a synthesis of the Darwinian theory of evolution with the Mendelian principles of genetics. A population, for the population geneticist, is an interbreeding group of organisms, defined and united by its gene pool (the sum of all the alleles of all the genes of all the individuals in the population). Evolution is, by definition, a change in the composition of the gene pool.

The Hardy-Weinberg principle is the application of Mendelian genetics to population biology. It describes the gene frequencies in ideal populations in which four conditions are met: (1) no mutations occur; (2) there is no gene flow; (3) the population is large; and (4) there is no difference in the reproductive success of the offspring of random matings. The Hardy-Weinberg principle states that genetic recombinations alone do not change the frequencies of alleles in the gene pool. For a gene with two alleles, the mathematical expression of the Hardy-Weinberg equilibrium is $(p + q)^2 = 1$, where p equals the frequency of one allele and q equals the frequency of the other allele. Thus, in any given population, the homozygotes for each allele are represented by p^2 and q^2, and the heterozygotes by $2pq$. Changes in the Hardy-Weinberg equilibrium are caused by mutations, gene flow, genetic drift, and, particularly, natural selection.

The Hardy-Weinberg equilibrium applies only if the population is large. In a small population, chance may cause the frequency of certain alleles to increase or to decrease and perhaps even disappear. This phenomenon is known as genetic drift. A small population that has branched off from a larger population may not be a representative sample of the larger population. This phenomenon is known as the founder principle.

Natural selection is the cause of changes in the Hardy-Weinberg equilibrium resulting from differences in the number of offspring left by different organisms because of variations in their phenotypes. It may take the form of stabilizing selection (elimination of extreme individuals), disruptive selection (elimination of intermediate forms), or directional selection (replacement of one phenotype by another). Of these three, the first is the most frequent and the third is primarily responsible for the changes we usually associate with evolution.

QUESTIONS

1. Distinguish between the following terms: genotype/phenotype; natural selection/evolution; population/gene pool.

2. Why doesn't everyone in the human population have brown teeth and a cleft chin (both dominant traits)?

3. Suppose, in a breeding experiment, you combined 7,000 *AA* and 3,000 *aa*. After the first generation, what would be the value of p^2? Of q^2? Of $2pq$? What would be the values after the second generation?

4. Among black Americans, the frequency of sickle cell anemia (a homozygous recessive condition) is about 0.0025. What is the frequency of heterozygotes? When one black American marries another, what is the probability that both will be heterozygotes? If both are heterozygotes, what is the probability that one of their children will have sickle cell anemia?

5. What is the difference between gene flow and genetic drift? How does each affect the gene pool of a population?

6. How does stabilizing selection affect the gene pool? Directional selection?

7. What characteristics in the contemporary human population are probably being maintained by stabilizing selection? Which might be subject to directional selection?

35

Variability: Its Extent, Preservation, and Promotion

Like begets like, we know now, because of the remarkable precision with which the DNA is copied and transmitted from cell to cell. The DNA in the cells of any individual is, except for occasional mutations, a true replica of the DNA that individual received from its father and mother. This constancy is, of course, essential to the survival of the individual organisms of which the population is composed. However, if evolution is to occur, there must be variations among individuals. Such variations make it possible for species to change as conditions do; they provide the raw material on which natural selection acts.

As we emphasized previously, Darwin was the first to recognize the importance of widespread, inheritable variations in the process of evolution. Much of the research in modern population genetics is concerned with the extent of such variability (far greater than Darwin could have realized) and with the way variations are preserved and fostered in gene pools.

35-1 *Even in populations in which the individuals appear almost identical, such as the members of this group of king penguins, we know that variations exist because individuals have no difficulty in identifying their own mates or offspring. To penguins, all people probably look alike.*

Variations within populations may be continuous or polymorphic. Continuous variation, as we noted on page 160, is characteristic of a trait, such as height, that is governed by a number of different genes (polygenic inheritance). *Polymorphism* is the coexistence within a population of two or more phenotypically different forms with no intermediate forms connecting them. In the course of this chapter, we shall see examples of both types of variability.

The Extent of Variability

Darwin's awareness that variations exist among individuals in a population was based largely on his observations as a naturalist, both in England and during his voyage on the *Beagle*. He also recognized that the traits that emerged in the course of selective breeding of domestic animals and plants reflected variability somehow latent in the population. This important hypothesis of Darwin has now been confirmed by a variety of laboratory investigations.

Bristle Number in *Drosophila*

In one group of studies, for example, the extent of latent variability in a natural population was demonstrated in the laboratory by experiments with the fruit fly *Drosophila melanogaster*. An easily observable hereditary trait, the number of bristles on the ventral surface of the fourth and fifth abdominal segments, was chosen for study. In the starting stock, the average number of bristles was 36. Two groups were interbred, one selected for increase of bristles and one for decrease (an example of disruptive selection—in this case, artificial rather than natural). In every generation, individuals with the fewest bristles were selected and crossbred, and so were individuals with the highest number of bristles.

Selection for low bristle number resulted in a drop after 30 generations from 36 to an average of about 30 bristles. In the high-bristle-number line, progress was at first rapid and steady. In 21 generations, bristle number rose steadily from 36 to an average of about 56 (Figure 35-3). No

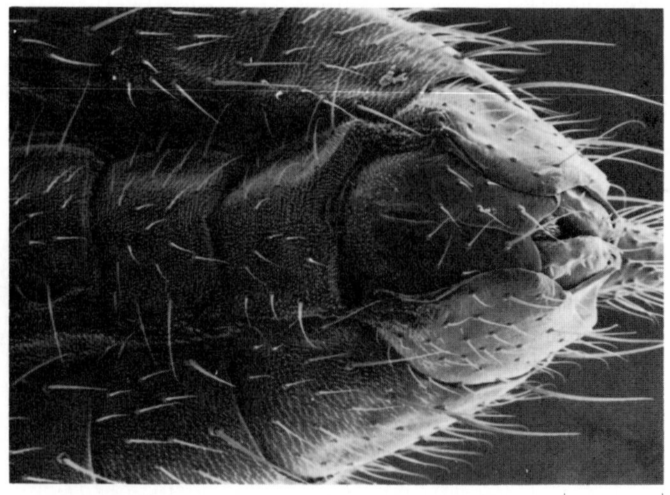

35-2 *Scanning electron micrographs of* (a) *a fruit fly* (Drosophila melonagaster) *and* (b) *the ventral surface of its posterior abdominal segments. The results of selection experiments to increase and decrease the number of bristles on the ventral surface of the fourth and fifth abdominal segments are shown in Figure 35-3.*

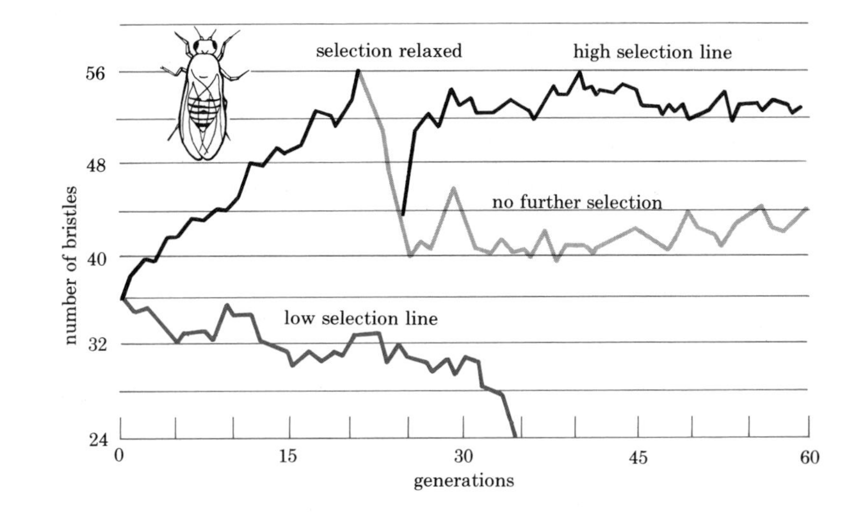

35-3 *The results of experiments with* Drosophila melanogaster, *demonstrating the extent of latent variability in a natural population. From a single parental stock, one group was selected for an increase in the number of bristles on the ventral surface (high selection line) and one for a decrease in the bristle number (low selection line).*

As you can see, the high selection line rapidly reached a peak of 56, but then the stock began to become sterile. Selection was abandoned at generation 21 and begun again at generation 24. This time, the previous high bristle number was regained, and there was no apparent loss in reproductive capacity. Note that after generation 24 the stock interbreeding without selection was also continued, as indicated by the line of gray. After 60 generations the freely breeding group from the high selection line had 45 bristles. The low selection line died out owing to sterility.

new genetic material had been introduced; within the single population the potential for a wide range of bristle numbers already existed. Subsequent experiments with *Drosophila* and other organisms have shown that no matter what single characteristic is selected for in breeding experiments, all reveal a comparable range of natural variability.

There is a second part to the bristle-number story. The low-bristle-number line soon died out, owing to sterility. Presumably, changes in factors affecting fertility had also taken place during selection. When sterility became severe in the high-bristle line, a mass culture was started; members of the high-bristle line were permitted to interbreed without selection. The average number of bristles fell sharply, and in five generations went from 56 to 40. Thereafter, as this line continued to breed without selection, the bristle number fluctuated up and down, usually between 40 and 45, which still was higher than the original 36. At generation 24, selection for high bristle number was begun again for a portion of this line. The previous high bristle number of 56 was regained, and this time there was no loss in reproductive capacity. Apparently, the genotype had rearranged and reintegrated itself so that the genes controlling bristle number were present in more favorable combinations.

Mapping studies have shown that bristle number is controlled by a large number of genes, at least one on every chromosome and sometimes several at different sites on the same chromosome. Therefore, although we do not know how important bristle number is to the survival of the animal, selection for this trait in some way disrupted the entire genotype. Livestock breeders are well aware of this consequence of artificial selection. Loss of fertility is a major problem in virtually all circumstances in which animals have been purposely inbred for particular traits. This result emphasizes the fact that in natural selection it is the entire phenotype that is selected rather than certain isolated traits, as is often the case in artificial selection.

Variability in Gene Products

Molecular biology provides another method for studying latent variability. Genes, as we saw in Section 3, code for and regulate the production of proteins. Proteins therefore provide clues about the genes coding for them. J. L. Hubby and R. C. Lewontin took a population of fruit flies, ground up the flies, and extracted proteins from them. From these proteins

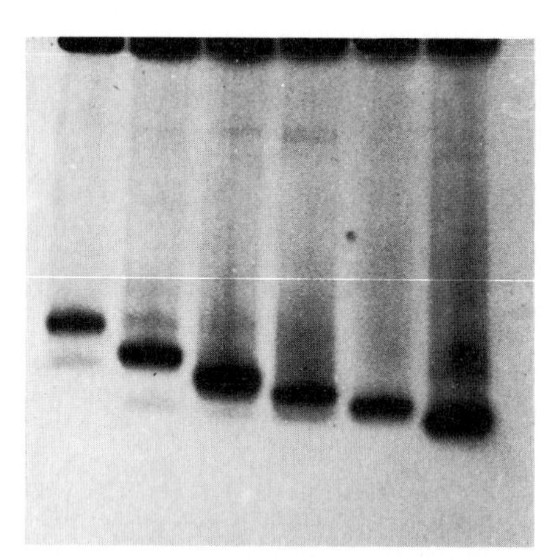

35-4 *The method used by Hubby and Lewontin to analyze* Drosophila *enzymes was electrophoresis. In this process, the proteins are dissolved, placed at the edge of a sheet of jellylike material, and exposed to a weak electric field. The rate at which the protein molecules move in the electric field is determined by their size and electric charge. As a result, protein molecules with even very slight structural differences can be separated.*

This photograph shows the separation of six different forms of one enzyme. The material in each column was obtained from flies homozygous for one of the six different alleles coding for the enzyme.

they were able to isolate 18 different enzymes, which were detected on the basis of their activity. Then they analyzed each of the 18 enzymes separately to see if its protein structure was identical in all the flies or if the structure differed from fly to fly (Figure 35-4).

Of the 18 enzymes studied in this way, nine were found to be composed of structurally indistinguishable proteins; in other words, the genes that had produced these proteins were the same throughout the entire population of fruit flies studied. However, nine of the enzyme groups contained two or more structurally different proteins. Therefore, without any direct analysis of the genes themselves, the investigators were able to conclude that among the fruit flies studied, there were two or more alleles of the gene responsible for each of the nine enzyme groups. In one group of enzymes, there were as many as six slightly different structural forms; that is, six alleles for the gene coding for that enzyme group were shown to exist in the population as a whole.

Each fruit-fly population examined was heterozygous for almost a third of the genes so tested. Each individual, it was estimated, was probably heterozygous for about 12 percent of its genes. Similar studies were performed on humans by H. Harris, using accessible tissues such as blood and placenta. These studies indicate that at least 25 percent of the genes in any given population are represented by two or more alleles, and individuals are heterozygous for at least 7 percent of their genes, on the average.

Explaining the Extent of Variation

The experiments of Hubby and Lewontin and those patterned after them disclosed a far greater degree of variability than had been previously imagined and, like most important scientific discoveries, raised major new questions. Most geneticists had previously thought that the individuals of a population should be close to genetic uniformity, as a result of a long history of selection for "optimal" genes. Yet, as these studies revealed, there is a great deal of genetic variation present in natural populations.

One school of geneticists, the "selectionists," claims that even such small variations as those in enzyme structure are maintained by a balance of forces of natural selection that favor some genotypes at some times and in some areas and others at other times or in other localities. An opposing school, the "neutralists," claims that the observed variations in the protein molecules are so slight that they do not make any difference in the function of the organism and so are not affected by natural selection. This latter group believes that such variations result from the accumulation over a long period of time of random mutations, many of which are lost from the population by genetic drift, but some of which persist. Continuing research may, in time, resolve the controversy between these two viewpoints.

Preservation of Variability

Diploidy

The most important factor in the preservation of variability in eukaryotes is diploidy. In a haploid organism, any genetic variations are immediately exposed to the selection process, whereas in the diploid organism, such variations may be stored as recessives, as with the allele for white flowers in Mendel's pea plants. The extent to which a rare allele is protected can be calculated with the Hardy-Weinberg formula (Table 35-1).

Table 35-1 Relation of allele frequency to genotype frequency

Frequency of Allele *a* in Gene Pool	Genotype Frequencies *AA*	*Aa*	*aa*	Percentage of *a* in Heterozygotes
0.9	0.01	0.18	0.81	10
0.1	0.81	0.18	0.01	90
0.01	0.9801	0.0198	0.0001	99

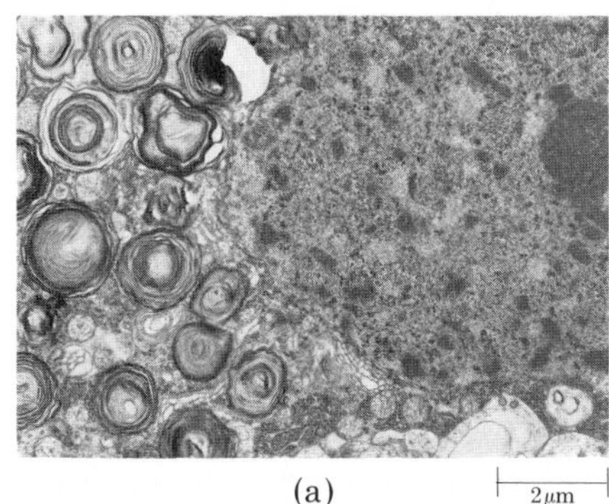

(a)

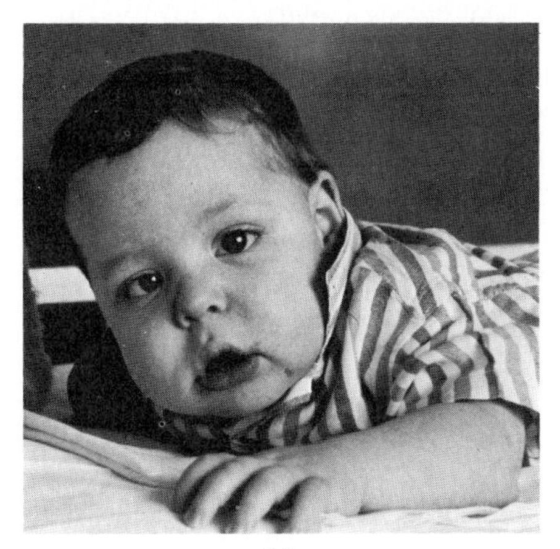

(b)

35-5 *Certain diseases, such as Tay-Sachs disease, are the result of the coming together of two recessive alleles. Even though this condition is fatal in the homozygote, the allele persists in the population because it is protected in the heterozygous state by diploidy. About 1 in 28 American Jews of Central European origin is heterozygous for the allele.*

(a) *The disease is caused by the absence of an enzyme involved with lipid metabolism. Without the enzyme, harmful fatty deposits accumulate in the lysosomes of brain cells, as shown in this micrograph.* (b) *In this one-year-old infant with Tay-Sachs disease, deterioration of the brain, already begun, will progress rapidly. The child will probably die before he is six years old.*

As the table reveals, the lower the frequency of allele *a*, the smaller the proportion of it exposed in the *aa* homozygote becomes. The removal of the allele by natural selection slows down accordingly. This result is of special interest to students of eugenics, the study of improvements in the human gene pool through controlled breeding. For instance, consider a genetic disorder, such as PKU (page 171), that is expressed only in the homozygous recessive. If the frequency of allele *a* is about 0.01, individuals with the *aa* genotype make up 0.0001 of the population (1 child for every 10,000 born). It would take 100 generations, roughly 2,500 years, of a program of sterilization of homozygote individuals with this condition to halve the allele frequency (to 0.005) and reduce the number born with this genetic disorder to 1 in 40,000.

Suppose, however, that a recessive allele in the homozygous state—harmful under ordinary circumstances—conferred an advantage under certain unusual environmental conditions or in conjunction with another particular set of genes. By appearing mostly in the heterozygous state, it would be protected against removal by selection. The allele would therefore persist in the gene pool and so would have a chance of being expressed under favorable selective circumstances.

The Evolution of Diploidy

Once sexual reproduction was established among the unicellular eukaryotes, the stage was set for the evolution of diploidy. By "accident"—an accident that apparently took place in a number of separate evolutionary lines—a zygote, originally the only diploid stage of the organism's life cycle, divided mitotically instead of meiotically, producing an organism with two complete sets of chromosomes. The fact that all the organisms we consider most highly evolved are diploid (or polyploid—see page 509) is an indication of the important role of diploidy in the evolutionary process.

Heterozygote Superiority

Recessive alleles, even ones that may be harmful in the homozygous state, may not only be sheltered in the heterozygous state, but sometimes may actually be selected for. This phenomenon, in which the heterozygotes have greater reproductive success than either type of homozygote, is known as *heterozygote superiority*. It is another way that variability is preserved.

A dramatic example of heterozygote superiority is found in association with sickle cell anemia. Until very recently, individuals homozygous for sickling almost never lived to maturity and thus did not become parents; in Darwinian terms, their fitness was zero. Therefore, almost every time two sickling alleles came together in a homozygous individual, two sickling alleles were removed from the gene pool. At one time, it was thought that the sickling allele was maintained in the population by a steady influx of new mutations. Yet in some African tribes, as much as 45 percent of the population is heterozygous for sickling, despite the loss of sickle alleles through homozygous individuals. To replace these alleles by mutations alone would require a rate of mutation about 5,000 times greater than any other known human mutation rate.

In the search for an alternative explanation, it was discovered that the sickling allele is maintained at high frequencies in certain areas where the heterozygote has a selective advantage. In many regions of Africa, malaria is one of the leading causes of illness and death, especially among young children. Studies of the incidence of malaria among young children showed that susceptibility to malaria is significantly lower in individuals heterozygous for sickling. Moreover, for reasons that are not known, women who carry the sickling allele are more fertile.

Among blacks in the United States, only about 9 percent are heterozygous for the sickling allele. Since no more than half of this loss of the sickling allele can be explained by the black-white admixture in America, the conclusion is that once selection pressure in favor of the heterozygote is relaxed, the mutant will tend to be eliminated slowly from the population.

Other alleles that are harmful in the homozygous state appear to be maintained in the population by heterozygote superiority. For example, again for reasons that are not known, women who are carriers of hemophilia (heterozygotes) have a fertility rate about 20 percent greater than that of other women.

The Role of Natural Selection

As we have just seen, natural selection can be an important factor in the preservation of variability in a population. There are many other examples, of which we shall consider two.

Color and Banding in Snails

One of the best-studied cases of polymorphism—the presence of two or more different phenotypic forms within a population—is found among land snails of the genus *Cepaea*. In one species *(Cepaea nemoralis),* for instance, the shell of the snail may be yellow, brown, or any shade from pale fawn through pink and orange to red. The lip of the shell may be black or dark brown (normally) or pink or white (rarely). Up to five black or dark-brown longitudinal bands may decorate it (Figure 35-6). Fossil evidence shows that these different types of shells have coexisted for more than 12,000 years.

Studies among English colonies of *Cepaea nemoralis* have revealed selective forces at work in some of the colonies. An important enemy of the snail is the song thrush. Song thrushes select snails from the colonies and take them to nearby rocks, where they break them open, eating the soft parts and leaving the shells. By comparing the proportions of types of shells around the thrush "anvils" with the proportions in the nearby colony, investigators have been able to show correlations between the shell patterns of snails seized by the thrushes and the habitats of the snails. For instance, of 560 individuals collected from a small bog near Oxford, 296 (52.8 percent) were unbanded; whereas of 863 broken shells collected from

(a)

(b)

35-6 (a) *A banded and an unbanded snail of the genus* Cepaea. (b) *An "anvil," where song thrushes break land snails open in order to obtain the soft, edible parts.*

From the evidence left by the empty shells, investigators have been able to show that in areas where the background is fairly uniform, unbanded snails have a survival advantage over the banded type. Conversely, in colonies of snails living on dark, mottled backgrounds, banded snails are preyed upon less frequently.

around the nearby rocks, only 377 (43.7 percent) were unbanded.* In other words, in bogs, where the background is fairly uniform, unbanded snails are less likely to be preyed upon than banded ones.

Studies of many different colonies have confirmed these correlations. In uniform environments, a higher proportion of snails is unbanded, whereas in rough, tangled habitats, such as woodland floors, far more tend to be banded. Similarly, the greenest habitats have the highest proportion of yellow shells, but among snails living on dark backgrounds, the yellow shells are much more visible and are clearly disadvantageous, judging from the evidence conveniently assembled by the thrushes.

Many of the snail colonies studied were at distances so great from one another that the possibility of migration between populations could be ruled out. Why, then, are both shell types still present in these colonies? One would expect populations living on uniform backgrounds to be composed almost entirely of unbanded snails, and colonies on dark, mottled backgrounds to lose most of their yellow-shelled individuals. The answer to this problem is not fully worked out, but it seems that there are physiological factors that are correlated with the particular shell patterns and that form a part of the same group of genes that control color and banding. Experiments have shown, for instance, that unbanded snails (especially yellow ones) are more heat-resistant and cold-resistant than banded snails. In other words, several different selection pressures are at work, and they appear to maintain the genetic variations of color and banding.

Geographic Variations: Clines

Sometimes variations within the same species follow a geographic distribution and are correlated with gradual changes in temperature, humidity, or some other environmental condition. Such a graded variation in a trait or a complex of traits within a particular species is known as a *cline*.

Many species exhibit north-south clines of various characteristics. House sparrows, for example, tend to have a smaller body size in the warmer parts of the range of the species and a larger body size in the cooler parts. In cooler regions, a larger body size is advantageous for heat conservation. Conversely, tails, ears, bills, and other extremities of animals are relatively longer in the warmer areas of a species' range. Such adaptations

* This difference (about 10 percent) may seem too small to be significant, but actually the number of shells counted was sufficiently large that the odds are only 1 in 1,000 that this difference could have occurred by chance alone.

Human Blood Groups: A Puzzle

The blood types A, B, AB, and O represent the most thoroughly studied polymorphism found in human populations. Apparently, the three alleles associated with these blood types are part of our ancestral legacy, since the same blood types are also found in other primates. A great deal is known about the chemistry of the different blood groups and about the allele frequencies in different populations. Yet we know very little about how this polymorphism has been maintained.

Some population biologists regard the blood types as probably neutral in selective value. Others maintain that polymorphism in human blood groups is a result of selection. For example, among Caucasian males, the life expectancy is greatest for those with group O and least with group B; exactly the opposite is true for Caucasian females with these blood types. People with type A blood run a relatively higher risk of cancer of the stomach and of pernicious anemia. People with type O blood have a higher risk of duodenal ulcers and are more likely to contract Asian flu. Most of these conditions, however, would not affect relative rates of reproduction since they generally occur in individuals who are past reproductive age.

It has been suggested that there are correlations between the different blood types and susceptibility to such diseases as plague, leprosy, tuberculosis, syphilis, and smallpox—all diseases that, in the past, could have been powerful selective forces. However, such correlations have not been substantiated.

The geographic distributions of the A, B, AB, and O groups are irregular. For example, there is a large predominance of O in the western hemisphere and an increase in the B allele as one moves from Europe toward Central Asia. The B allele is totally

allow for heat radiation from the animal. Plants growing in the south often have slightly different requirements for flowering or for ending dormancy than the same kind of plants growing in the north, although they all may belong to the same species. Most geographic variations, however, are too irregular to be classified into clines.

A species that occupies many different habitats may appear to be slightly different in each one. Are the differences in phenotypes determined entirely by the interaction of different environments with the same genotype, or are there genetic differences too? Figure 35-7 illustrates a series of experiments carried out with the perennial plant *Potentilla glandulosa,* a relative of the strawberry. Experimental gardens were established at various altitudes, and wild plants collected from near each of the experimental sites were grown in all the gardens. (*Potentilla glandulosa,* like the strawberry, reproduces asexually by runners, making it possible to study genetically uniform replicates.) Under these conditions it was possible to demonstrate that many of the phenotypic differences among the *P. glandulosa* plants from different altitudes were due to genetic variations. It is not surprising that in their very different environments, different traits were selected for and so, over time, genetic differences arose and were preserved.

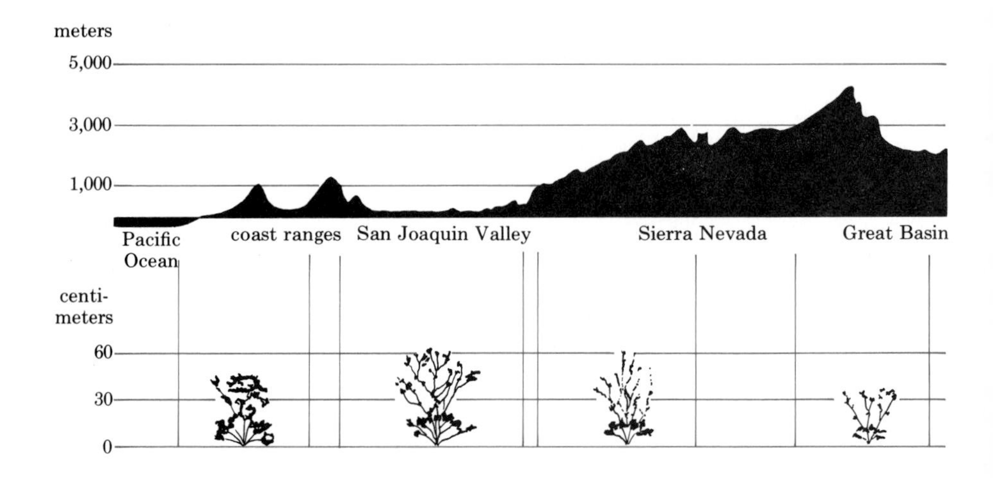

35-7 *Different phenotypes of* Potentilla glandulosa, *a relative of the strawberry. Notice the correlation between the height of the plant and the altitude at which it grows; there are other phenotypic differences among the plants as well. When plants from the four geographic areas are grown under identical conditions, many of these phenotypic differences persist and are passed on to the next generation, indicating that these plants are genotypically, as well as phenotypically, different.*

absent in American Indians and Australian aborigines who have not mixed with Europeans. Although most American Indians are type O, the Blackfoot tribe has the highest frequency of type A blood found anywhere in the world (55 percent). Even within an area as small as the British Isles, there are significant variations in allele frequency going both from north to south and from east to west. These differences may reflect some differences in the selective forces favoring particular blood groups under particular conditions, they may be the result of population migrations and genetic drift, or both. At the present time, we simply do not know.

Promotion of Variability

Sexual Reproduction

By far the most important method by which organisms promote variability in their offspring is by sexual reproduction. As we saw in Chapter 10, sexual reproduction combines parental genes in new ways so that wholly new genotypes appear in the next generation. The new genotypes, in turn, give rise to new phenotypes upon which natural selection can operate.

Assuming that the hypothesis is correct that, like other traits, the efficient utilization of energy and other resources is under selective pressure, consider the expenditure of energy and resources involved in sex. In the words of Edward O. Wilson:

> Sexual reproduction is in every way a consuming biological activity. Reproductive organs tend to be elaborate in structure, courtship activities lengthy and energetically expensive, and genetic mechanisms of sex determination finely tuned and easily wrecked. But more important, a female that elects to reproduce by sex cuts her genetic contribution to each gamete by one-half, without, in the vast majority of species, receiving any material aid from the male. If an egg develops . . . without meiosis—the simplest way to proceed—all of the genes in the resulting offspring will be identical to those of the mother. In sexual reproduction, which entails reduction division during meiosis, only half are identical. The female, in other words, has thrown away half of her investment.*

Note that the only selective advantage conferred by sexual reproduction is the production of totally new genetic combinations that can be tested in the next generation.

Some eukaryotic organisms reproduce only asexually. In some cases, it appears that the capacity for sexual reproduction was lost in the course of their evolutionary history. In effect, in balancing the costs, these organisms opted for the stability conferred by asexual reproduction. Many modern organisms, particularly members of the plant kingdom, hedge their bets, reproducing both sexually and asexually (Figure 35-8).

* E. O. Wilson, *Science,* vol. 188, page 139, 1975.

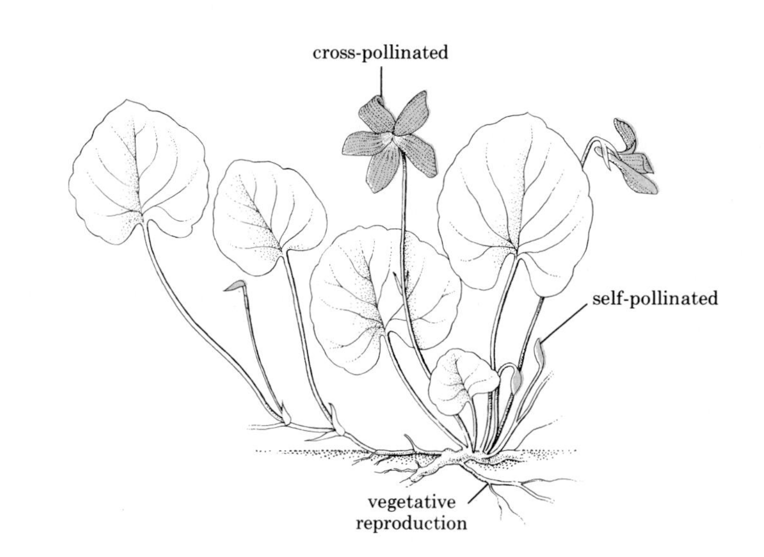

35-8 *Plants such as the violet shown here reproduce both sexually and asexually. The larger flowers are cross-pollinated (sexual reproduction) by insects, and the windborne seeds bearing new alleles in new combinations are carried some distance from the parent plant.*

The smaller flowers, closer to the ground, are self-pollinated and never open. Seeds from these flowers will carry only the alleles of the parent plant, although they may be present in new combinations. These seeds drop close to the parent plant, most likely in a very similar environment.

Underground stems can also produce a new series of genetically identical plants close to the parent by vegetative (asexual) reproduction.

35–9 *Two slugs mating, suspended from a branch by a cord of mucus. Like earthworms and snails, slugs are hermaphrodites. Hermaphroditism is advantageous for slow, solitary species because it doubles the chances of finding a mate. Every mature member of the same species that a hermaphrodite meets will always be of the opposite sex—and of the same sex too. As a result of the mating, each individual can produce new offspring.*

Mechanisms That Promote Outbreeding

Many ways have evolved by which new genetic combinations are promoted in sexually reproducing populations. Among animals, even in those that are hermaphrodites (each organism produces both egg and sperm cells), such as earthworms, snails, slugs, and some fish, an individual seldom fertilizes its own eggs (Figure 35–9).

Among plants, there are a variety of mechanisms that ensure that the sperm-bearing pollen is from a different individual than the stigma it lights upon. Many plants, such as the holly and the date palm, have male flowers on one tree and female on the other. In others, such as the avocado, the pollen of a particular plant matures at a time when its own stigma is not receptive. In some species, anatomical arrangements inhibit self-pollination (Figure 35–10).

Some plants have genes for self-sterility. Typically, such a gene has multiple alleles—s^1, s^2, s^3, and so on. A plant carrying the allele s^1 cannot pollinate another plant with an s^1 allele, one with an s^1/s^2 genotype cannot pollinate a plant with either of those alleles, and so forth. In one population of about 500 evening primrose plants, 37 different self-sterility alleles were found, and it has been estimated that there are more than 200 alleles for self-sterility in red clover. As a consequence, a plant with a rare self-sterility allele is more likely to be able to pollinate another plant than is a plant with a common self-sterility allele. Such a system strongly encourages variability in a population; selection for the rare allele makes it more common, whereas more common alleles become rarer.

The existence of such mechanisms that promote outbreeding, some of them very elaborate, reaffirms the selective value of genetic variability.

As we have seen, Darwin believed, as do modern students of evolution, that the differences between individuals within a population could, with time, be translated into the differences between species. We shall explore this concept further in the next chapter.

SUMMARY

Variations provide the raw material on which natural selection acts. They may be continuous, as in traits that are governed by a number of different genes (polygenic inheritance), or polymorphic. Polymorphism is the coexistence within a population of two or more distinct phenotypes.

Natural populations can be shown, by a variety of methods, to harbor a wide spectrum of latent variations. Variations can be demonstrated not only in morphological characteristics, such as bristle number in *Drosophila,* but also in the proteins directly coded by the genes. It has been estimated that at least 25 percent of the genes in any given human population are represented by two or more alleles.

Variability is preserved in populations primarily by diploidy, in which an organism has two sets of chromosomes; in the heterozygous state, rare alleles are protected from selection. Another factor, heterozygote superiority, is important in the case of certain recessive alleles that, while detrimental in the homozygous state, confer a selective advantage in the heterozygous state. Natural selection plays a role in maintaining polymorphisms such as color and banding in snails, gradual variations that follow a geographic distribution (clines), and distinct groups of phenotypes of the same species occupying different habitats.

35-10 *Flowers have various morphological adaptations that prevent self-pollination. Orchid pollen is bound together in a mass, called a pollinium, which contains a sticky extension that attaches the pollinium to the pollinating insect. In a bee-pollinated orchid, a bee with a pollinium on its back lands on the lip of the flower and crawls inside to gather the nectar. As it backs out, the flap separating the stigma from the pollen mass first scrapes the transported pollinium off its back onto the sticky surface of the stigma (from which the sperm enter the style and fertilize the egg cells in the ovules). Next, as it pushes the anther cap outward, the pollinium from the flower it is leaving becomes deposited on its back and is thus carried to another flower.*

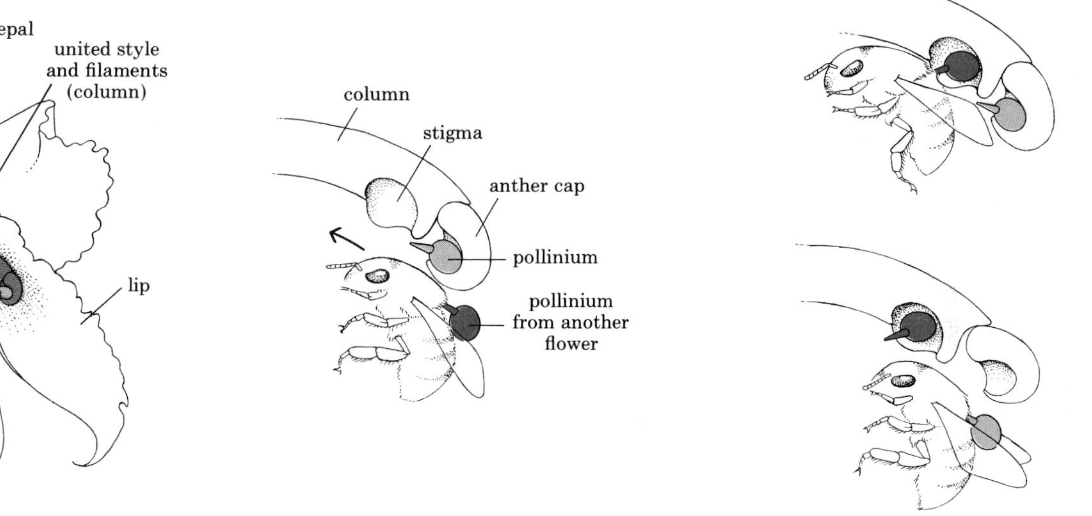

Sexual reproduction is the most important method by which variability is promoted. Mechanisms that promote outbreeding (and, in some cases, discourage or prevent self-fertilization) further promote variability. These include hermaphroditism in solitary, slow-moving invertebrates, the presence of male and female flowers on different plants of the same species, anatomic adaptations in plants, and self-sterility alleles.

QUESTIONS

1. Distinguish between the following terms: polygenic inheritance/polymorphism; homozygote/heterozygote; diploidy/heterozygote superiority.

2. At the end of the experiment on bristle number in *Drosophila,* the flies in the high selection line had an average number of 56 bristles. No fly at the beginning of the experiment had as many as 56, however. How do you explain this fact?

3. Describe in your own words the steps in the Hubby-Lewontin experiment and the significance of the results.

4. In what sense does a female reproducing sexually throw away half of her investment? If you are not sure of the answer, review the diagram of meiosis, page 144. Does the male also throw away part of his "investment"?

5. In bacteria, the exchange of genetic information and reproduction take place as separate events. In organisms other than prokaryotes, genetic recombination is generally linked with reproduction. Why do you think this might be so?

6. How do self-sterility alleles promote variability in a population?

36

Evolution in Action

Darwin believed evolution to be such a slow process that it could never be observed directly. However, the effects of human civilization have produced such extremely strong selection pressures on some organisms that it has been possible to observe not only the results but also the actual process of selection.

Evolution Observed

The Peppered Moth

One of the best-studied examples of natural selection is that of *Biston betularia,* the peppered moth. These moths were well known to British naturalists of the nineteenth century, who remarked that they were usually found on lichen-covered trees and rocks. Against this background, the light coloring of the moths made them practically invisible. Until 1845, all reported specimens of *Biston betularia* had been light-colored, but in that year one black moth of this species was captured at the growing industrial center of Manchester.

With the increasing industrialization of England, smoke particles began to pollute the foliage in the vicinity of industrial towns, killing the lichens and leaving the tree trunks bare. In heavily polluted districts, the trunks and even the rocks and ground became black. During this period, more and more black *Biston betularia* were found. Replacement of light-colored moths by dark ones proceeded briskly; by the 1950s, only a few of the light-colored population could be found, and these were far from industrial centers. Because of the prevailing westerly wind in England, the moths to the east of industrial towns tended to be of the black variety right up to the east coast of England. The few light-colored populations were concentrated in the west, where lichens still grew. A similar tendency for dark-colored forms to replace light-colored forms has been found among some 70 other moth species in England and some 100 species of moths in the Pittsburgh area of Pennsylvania.

Where did the black *Biston betularia* come from? Eventually, it was demonstrated that the black color was the result of a rare, recurring mutation. Why had they increased so rapidly? H. B. D. Kettlewell hypothesized that the color of the moths protected them from predators, notably birds (Figure 36–1). He marked a sample of moths of each color by carefully

(a)

(b)

36–1 *The two varieties of* Biston betularia, *the peppered moth, resting on* (a) *a lichen-covered tree trunk in an unpolluted English countryside and* (b) *a dark tree trunk, near Manchester. If you look carefully, you will notice that there are two moths in each photograph.*

Kettlewell demonstrated that on lichen-covered trees the black moths are more likely to be eaten by birds (which are visually oriented), whereas on polluted trees the light-colored moths are more likely to be eaten.

putting a spot on the underside of the wings, where it could not be seen by a predator. Then he released known numbers of marked individuals into a bird reserve near Birmingham, an industrial area where 90 percent of the local population consisted of black moths. Another sample was released into an unpolluted Dorset countryside, where no black moths ordinarily occurred. He returned at night with light traps to recapture his marked moths. From the area around Birmingham, he recovered 40 percent of the black moths, but only 19 percent of the light ones. From the area in Dorset, 6 percent of the black moths and 12.5 percent of the light moths were retaken.

To clinch the argument, Kettlewell placed moths on tree trunks in both locations, focused hidden movie cameras on them, and was able to record birds actually selecting and eating the moths. Near Birmingham, when equal numbers of dark and light moths were available, the birds seized 43 light-colored moths and only 15 black ones; in Dorset, they took 164 black moths but only 26 light-colored forms. Clearly, if you are a moth, it is advantageous to be black near Birmingham but light in Dorset.

Most recently, strong controls have been instituted in Great Britain on the particulate content of smoke, and the heavy soot pollution has begun to decrease. The light-colored moths are already increasing in proportion to the black forms, but it is not yet known whether a complete reversal either in pollution or in the selection process will come about.

Insecticide Resistance

Another example of evolution in process is the development of insecticide resistance. Chemicals poisonous to insects, such as DDT, were originally hailed as major saviors of human health and property. They have fallen into disfavor not only because of their tendency to accumulate in the environment, but also because of the extraordinary increase in resistant strains of insects. At least 225 species of insects are now resistant to one or more insecticides. One species is even able to remove a chlorine atom from a DDT molecule and use the remainder as food.

36-2 *A scale insect on an orchid. With their specialized mouthparts, these insects suck fluid from the phloem, debilitating the plant. If left unchecked, they eventually kill it.*

A particularly striking example of insecticide resistance has been found in the scale insects (Figure 36-2) that attack citrus trees in California. In the early 1900s, a concentration of hydrocyanic gas sufficient to kill nearly 100 percent of the insects was applied to orange groves at regular intervals with great success. By 1914, orange growers near Corona, California, began to notice that the standard dose of the fumigant was no longer sufficient to destroy one type of scale insect, the red scale. A concentration of the gas that had left fewer than 1 in 100 survivors in the nonresistant strain left 22 survivors out of 100 in the resistant strain. By crossing resistant and nonresistant strains, it was possible to show that the two differed in a single gene. The mechanism for this resistance is not known, but one group of experiments has shown that the resistant individual can keep its spiracles closed for 30 minutes under unfavorable conditions, whereas the nonresistant insect can do so for only 60 seconds.

Patterns of Evolution

Can relatively small changes, such as those observed in populations of peppered moths and scale insects, add up over the eons to the major changes associated with the emergence of entirely new forms? The leap is a large one, but the evidence that this is indeed the case is found in the fossil record.

Paleontologists—students of fossils—recognize three principal patterns of organic evolution: phyletic evolution, splitting evolution, and adaptive radiation. As we shall see, these three patterns are not really so distinct as they first appear to be.

Phyletic Evolution

Phyletic evolution describes the changes taking place in a single hereditary line of organisms over a long period of time. A well-known example is the evolution of the horse (Figure 36-3), whose history is abundantly represented in the fossil record.

At the beginning of the Tertiary period, some 65 million years ago, the horse lineage was represented by Eohippus ("dawn horse"). It was a small herbivore (25 to 50 centimeters high at the shoulders) with three toes on its hind feet, four toes on its front feet, and doglike footpads on which its weight was carried. Its eyes were halfway between the top of its head and the tip of its nose. Its teeth had small grinding surfaces and low crowns, which probably could not have stood much wear. The teeth indicate that Eohippus did not eat grass; in fact, there was probably not much grass to eat then. It most likely lived on succulent leaves. Eohippus evolved by stages into a group of larger, three-toed herbivores. Some of these lines survived and some did not. The molar teeth of some of the successful lines had large crowns that continued to grow as they were worn away (as do those of modern horses).

During the Oligocene, horses, of which there were several distinct species, became still larger. The middle or third digit of each foot expanded and became the weight-bearing part of the foot. Some of the species remained three-toed, while in others the non-weight-bearing digits became greatly reduced. The three-toed species eventually became extinct, but those in which the digits were greatly reduced—to one—survived, and were the direct ancestors of the modern horse, *Equus*.

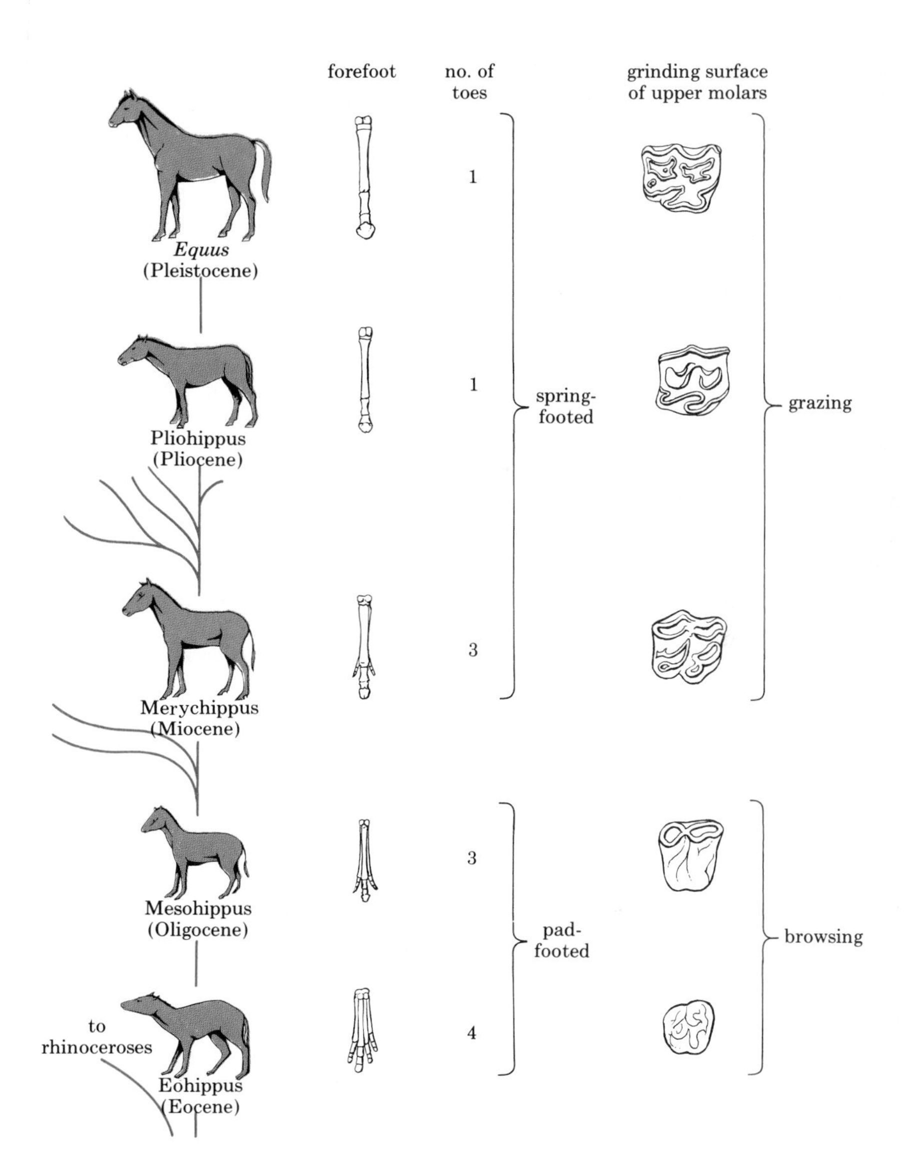

36-3 *Phyletic evolution: the modern horse and some of its ancestors. Over the past 60 million years, small several-toed browsers, such as Eohippus, were replaced in gradual stages by members of the genus* Equus. *The later horses were characterized by, among other features, a larger size, broad molars adapted to grinding coarse grass blades, a single toe surrounded by a tough, protective keratin hoof, and a leg in which the bones of the lower leg had fused, with joints becoming more pulley-like and motion restricted to a single plane. Only a few of the evolutionary lines are shown here.*

By hindsight, these changes can be correlated with changes in the environment and so interpreted as adaptive. In the time of little Eohippus, the land was marshy and the chief vegetation was leaves; the teeth of Eohippus were adapted for browsing. By the Miocene, the grasslands began to spread; groups of horses whose teeth became adapted to grinding grasses (which are very coarse and tough) survived, whereas those who remained browsers did not. The placement of the eye higher in the head may have facilitated watching for predators while grazing. The climate became drier and the ground became harder; reduction of the number of toes, with the development of the spring-footed gait characteristic of the modern horse, was an adaptation to harder ground and to larger size. An animal twice as high as another tends to weigh about eight times as much. Little Eohippus was probably as fast as the modern horse, but a larger, heavier horse with the foot and leg structure of Eohippus would have been too slow to escape from predators. (During this same period predators were developing adaptations that rendered them better able to catch large herbivores, including horses.)

Molecular Clocks

Evolutionary relationships may be established by comparisons of living species and by examination of fossil forms. Most recently a new method has become available: the study of macromolecules. In different species some of the amino acids that make up similar proteins may be different, and the degree of difference between the proteins appears to correspond well with the degree of evolutionary divergence between the species. For example, the protein chain of cytochrome *c* is exactly the same in human beings and chimpanzees, but it differs from the cytochrome *c* of *Neurospora* by 44 out of 100 amino acids.

The composition of cytochrome *c* from more than 60 species has now been determined, and it appears that about 20 million years are required to produce a change in 1 percent of the amino acids in this particular protein. The altered protein is able to function because the changes involve only certain amino acids; 35 of the amino acids are the same for all species. These are presumably the amino acids of the active site. Moreover, when a substitution is made, it almost always involves an amino acid of similar properties, so that the conformation of the chain is altered little if at all. Presumably mutations occur that involve other amino acids, but they are weeded out as disadvantageous.

Other proteins change at different rates. Hemoglobin, for example, appears to change at a rate of 1 percent in 6 million years, presumably because it is more tolerant of change than cytochrome *c*.

Although macromolecules may evolve at different rates, they are consistent in providing the same answer concerning the relative divergence of species. Moreover, these answers are generally, but not always, in accord with accepted interpretations of the fossil evidence. Perhaps most important, these ancient proteins may hold the key to unsolved evolutionary problems, such as the relationships of the invertebrate phyla to one another and to vertebrate origins.

Thus the evolution of *Equus,* viewed from the long retrospective of the geologic record, represents a fairly straightforward accumulation of adaptive changes related to pressures for grazing, increased size, and avoidance of predators.

Splitting Evolution

A second mode of evolution involves the splitting, or branching, of phylogenetic lineages. For example, *Ursus arctos*—the brown bear, a species that also includes the Kodiak and grizzly bears—is distributed throughout the northern hemisphere, ranging through the deciduous forests up through the coniferous forests and into the tundra, as it was some 1½ million years ago. As is characteristic of such widespread species, there are phenotypic differences between populations in different environments. During one of the massive glaciations of the Pleistocene, a population of *Ursus arctos* was split off from the main group, and, according to fossil evidence, this group, under extreme pressure from the harsh environment, evolved into the polar bear, *Ursus maritimus* (Figure 36-4).

Brown bears, although they are members of the order of carnivores and closely related to dogs, are mostly vegetarians, supplementing their diet only occasionally with fish and game. The polar bear, however, is almost entirely carnivorous, with seals being its staple diet. Unlike the brown bear, it does not hibernate. The polar bear differs physically from other bears in a number of ways, including, besides its white color, its carnivore-type teeth, its streamlined head and shoulders, and the stiff bristles that cover the soles of its feet, providing insulation and traction on the slippery ice.

The cave bear *(Ursus spelaeus),* which was twice as large as any existing bear, is pictured in ancient cave drawings but is now extinct. It also appears from fossil remains to have originated by splitting evolution from *Ursus arctos.*

Ernst Mayr of Harvard believes that the formation of new species by the splitting off of small populations from the parent stock (the founder principle) is responsible for almost all major evolutionary change. In small populations, favorable genetic combinations, if present, can increase rapidly in number and frequency without being diluted out by gene flow. Hence, evolution probably does not occur steadily and gradually but in spurts, accounting for the sudden appearance of new species in the fossil record.

Adaptive Radiation

Adaptive radiation is believed by many experts to be the major pattern of evolution. It involves the sudden (in geologic time) diversification of a group of organisms that share a common ancestor, often itself newly evolved. It also involves the opening up of a new biological frontier that may be as vast as the land or the air, or, as in the case of the Galapagos finches (page 512), as small as an archipelago.

The fossil record contains many examples of adaptive radiation. For example, some 300 million years ago, the reptiles, liberated from an amphibian existence by the "invention" of the amniote egg (page 277), diversified rapidly into terrestrial environments. A similar, even more rapid burst of evolution later gave rise to the birds. A number of biologists now believe that, contrary to long-held opinion, the birds are not descendants of the flying reptiles. More likely they are the result of adaptive radiation from a group of small, bipedal, carnivorous dinosaurs known as theropods. The

36-4 *An example of splitting evolution: the divergence of polar bears from the main line of brown bears. The white coloring of the polar bear is probably related to its need for camouflage while hunting, rather than a need to hide from predators, since it is one of the world's largest carnivores. Its greatest enemies are commercial fishermen who do not appreciate the polar bear's fishing activities and hunters who, prizing the skins for trophies, have taken to tracking the bears by helicopter.*

Females with young hibernate, but other polar bears roam the ice and icy waters all winter long in search of fish, seals, young walruses, and other game.

(a)

(b)

36–5 *While the placental mammals were undergoing adaptive radiation on the other continents, the marsupials, geographically isolated in Australia, went through a parallel pattern of adaptive radiation.*

(a) *The ecological role of large herbivore, usually filled by hooved mammals, is occupied in Australia by kangaroos. There are some 45 species of kangaroos; adults of the larger species weigh up to 90 kilograms and can jump 9 meters in a single leap.*

(b) *Though the koala is called a bear, it has no true placental equivalent. It is arboreal (that is, it lives in trees) and feeds exclusively on the oily aromatic leaves of certain* Eucalyptus *trees. As for the* Eucalyptus, *more than 600 different species are known in Australia; none are native to other continents.*

(c) *Wombats, of which there are four species, are the marsupial equivalent of large rodents. Nocturnal animals, they spend the day in tunnels and underground burrows dug with their shovel-like paws.*

theropods, which were ground-dwelling, may have been warm-blooded; in any case, they are thought to be the only dinosaurs whose descendants—the birds—have survived into modern times. As the dinosaurs became extinct, the mammals similarly burst forth on the evolutionary scene, with many different kinds appearing simultaneously in the fossil record (Figure 36–5).

Common Features in Patterns of Evolution

Actually, as we noted earlier, the three major patterns of evolution—phyletic evolution, splitting evolution, and adaptive radiation—have a great deal in common. Although we can trace an overall pattern of evolution in *Equus* and its progenitors, the evolution of the horse did not proceed in a straight line, and it involved many branches. Conversely, although a branching off was the crucial event in the creation of *Ursus maritimus,* the polar bear is itself a product of phyletic evolution, as is every modern organism. Adaptive radiation is a combination of both phyletic and splitting evolution.

Origin of Species

The crucial event in the origin of species is the branching of a single population into two or more populations that do not interbreed. For example, the polar bear could never have developed its distinctive characteristics had gene flow continued between it and the brown bear. The mechanisms of speciation have been a chief concern of twentieth-century evolutionists.

What is a species? According to Terrell H. Hamilton, "A species may be envisioned as an isolated pool of genes flowing through space and time,

(c)

constantly adapting to changes in its environment as well as to the new environments encountered by its extension into other geographic regions."*

As a working definition, a species is a group of populations whose members can interbreed with one another but cannot (or at least usually do not) interbreed with members of other species. The key concept in both these definitions is genetic isolation. How does one pool of genes split off from another and begin its solitary journey? How do two species, often very similar to one another, inhabit the same place at the same time and yet maintain the genetic isolation essential to their identity?

Population biologists agree that new species can arise only when populations are genetically isolated from one another. If, under such circumstances of genetic isolation, the populations are subject to different selection pressures, they will begin to diverge genetically. As a consequence of such genetic divergence, the populations will eventually be unable to interbreed. At this point, speciation is said to have occurred.

Speciation with Geographic Isolation

Every widespread species that has been carefully studied has been found to contain geographically representative populations that differ from each other to a greater or lesser extent, as with the different populations of *Potentilla glandulosa* (page 496). A species composed of geographic races of this sort is particularly susceptible to speciation if geographic barriers arise so that gene flow cannot occur.

36-6 *Depending on the species, islands, genetically speaking, may be* (a) *true islands,* (b) *mountaintops,* (c) *ponds, lakes, or even oceans, or* (d) *isolated clumps of vegetation.*

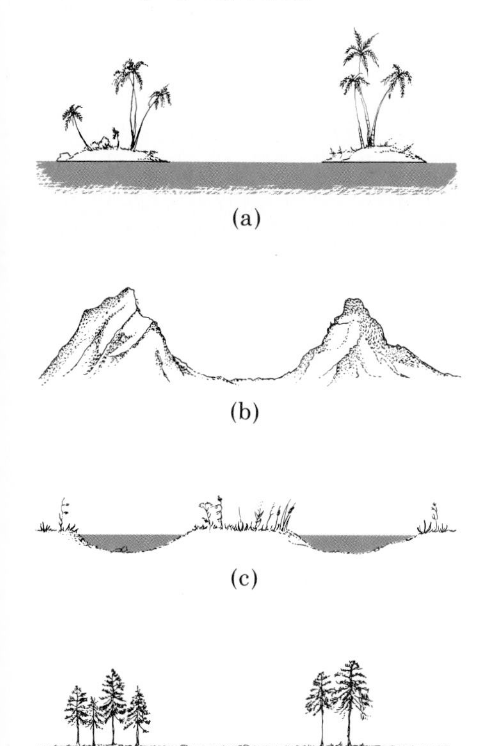

(a)

(b)

(c)

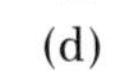

(d)

Geographic barriers are of many different types. Islands are frequent sites for the development of unusual species. The breakup of Pangaea (see essay on next page) profoundly altered the course of evolution, setting the island continent of Australia adrift as a veritable Noah's ark of marsupials (Figure 36-5). Populations of many organisms can become cut off from one another by barriers less obvious than oceans (Figure 36-6). For a plant, an island may be a mist-veiled mountaintop, and for a fish, a freshwater lake. A forest grove may be an island for a small mammal. A few meters of dry ground can isolate two populations of snails.

Because of the work of John A. Moore, the leopard frog *(Rana pipiens)* has become one of the best-known examples of geographic races and speciation. Leopard frogs are found in North America as far north as Quebec and as far south as Mexico. Moore has studied and compared a large number of characteristics in 29 separate populations of this species. He has found that there are marked variations, as might be expected, from population to population. Crossbreeding of individuals from different populations produces normal offspring when parents are drawn from populations that are geographically adjacent, such as those in central and southern Florida, or from populations that live at roughly the same latitude, such as those in Texas and central Florida. However, the greater the north-south gap separating the home populations of the parents, the greater also is the proportion of defective, often inviable, offspring. In Texas-Vermont hybrids, for example, mortality among the developing embryos may reach 100 percent. Thus what was clearly once a single species is becoming genetically fragmented. Some investigators believe that the genetic fragmentation has already reached the point at which *Rana pipiens* consists of several different species.

* Terrell H. Hamilton, *Process and Pattern in Evolution,* The Macmillan Company, New York, 1967.

The Breakup of Pangaea

In the past 15 years, the theory of continental drift has become firmly established. According to this theory, the outermost layer of the earth is divided into a number of segments, or plates. These plates, on which the continents rest, slide across the surface of the earth, moving in relation to one another. The geologic expression of this relative motion occurs mainly at the boundaries between the adjacent plates. Where plates collide, volcanic islands such as the Aleutians may be formed or mountain belts such as the Andes or Himalayas may be uplifted. At the boundaries where plates are separating, volcanic material wells up to fill the void. It is here that ocean basins are created. Plates may also move parallel to the boundary that joins them but in opposite directions, or in the same direction but at different speeds, as with the San Andreas fault.

About 200 million years ago, all the major continents were locked together in a supercontinent, Pangaea. Several reconstructions of Pangaea have been proposed, one of which is shown here. It is generally agreed that Pangaea began to break up about 190 million years ago, about the time the dinosaurs approached their zenith and the first mammals began to appear. First, the northern group of continents (Laurasia) split apart from the southern group (Gondwana). Subsequently, Gondwana broke into three parts: Africa–South America, Australia–Antarctica, and India. India drifted northward and collided with Asia about 50 million years ago. This collision initiated the uplift of the Himalayas, which continue to rise today as India still pushes northward into Asia.

By the end of the Cretaceous period—according to current reconstructions, about 65 million years ago—South America and Africa had separated sufficiently to have formed half the South Atlantic, and Europe, North America, and Greenland had begun to drift apart. However, final separation between Europe and North America–Greenland did not occur until the Eocene (43 million years ago). During the Cenozoic, Australia finally split from Antarctica and moved northward to its present position, and the two Americas were joined by the Isthmus of Panama, which was created by volcanic action.

From the outset, the theory of continental drift has been closely interwoven with that of evolution. One of the earliest and most impressive pieces of evidence in favor of continental drift was the discovery of fossil remains of a small, snaggle-toothed reptile, *Mesosaurus,* found in the coastal regions of Brazil and South Africa but nowhere else. Students of evolution have not yet felt the full impact of the new geology; however, it is already calling forth important new interpretations of fossil history.

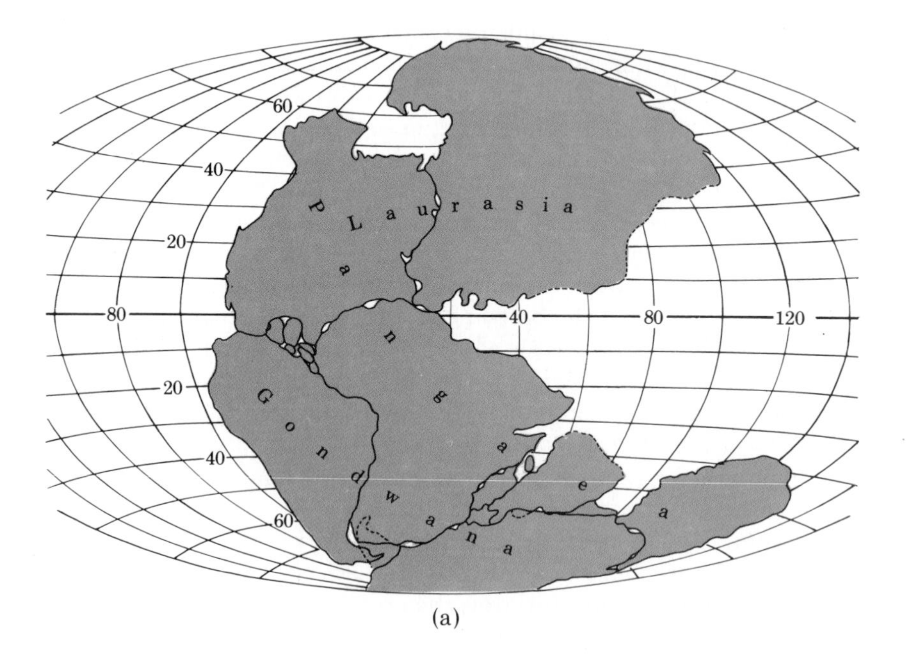

(a)

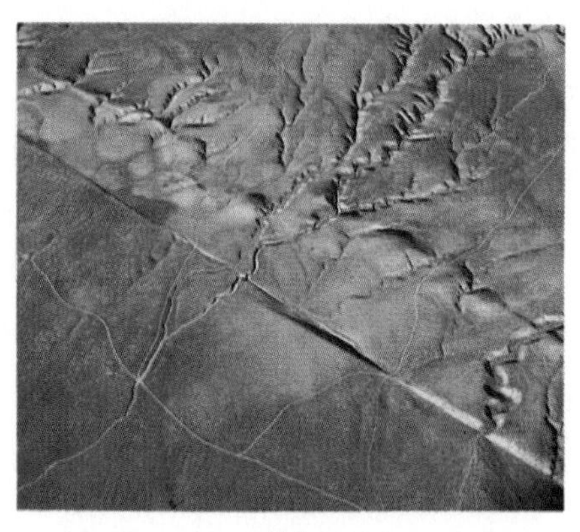

(b)

(a) *Pangaea.* (b) *The San Andreas fault is a boundary between two giant plates that are moving past one another. The fault, running through San Francisco and continuing southeast of Los Angeles, is responsible for California's notorious earthquakes.*

Speciation without Geographic Isolation

The only proven means of producing new species without geographic isolation occurs by hybridization or polyploidy. *Hybrids,* the offspring of parents of different species, occur in both animals and plants. In nature, they are far more frequent in plants. Kentucky bluegrass, for instance, occurs widely throughout the United States, is a promiscuous hybridizer, and has crossbred with many related species. Occasional offspring of these matings have been better adapted than either of the parents and have become successful in their own right. Some hybrid plants spread by runners and eventually cover large areas.

Hybrids in both plants and animals are nearly always sterile because the chromosomes cannot pair at meiosis (having no homologues), a necessary step for producing viable gametes (Figure 36–7a). In plants, however, fertile forms may arise from these infertile hybrids by the process of *polyploidy,* in which the cells acquire more than two sets of chromosomes. Polyploid cells arise at a low frequency as the result of a "mistake" in mitosis, so that the chromosomes divide but the cell does not. If such cells divide by further mitosis so that they eventually produce a new individual, either sexually or asexually, that individual will have twice the number of chromosomes of its parent. Polyploid individuals can be produced deliberately in the laboratory by the use of the drug colchicine, which prevents separation of chromosomes during mitosis.

If polyploidy occurs in a sterile hybrid, each chromosome from each parent will be present in duplicate. The duplicate chromosomes can then pair, meiosis will be normal, and fertility restored (Figure 36–7b). Approximately half of the 235,000 kinds of flowering plants have had a polyploid origin, and many important agricultural species, including wheat, are hybrid polyploids.

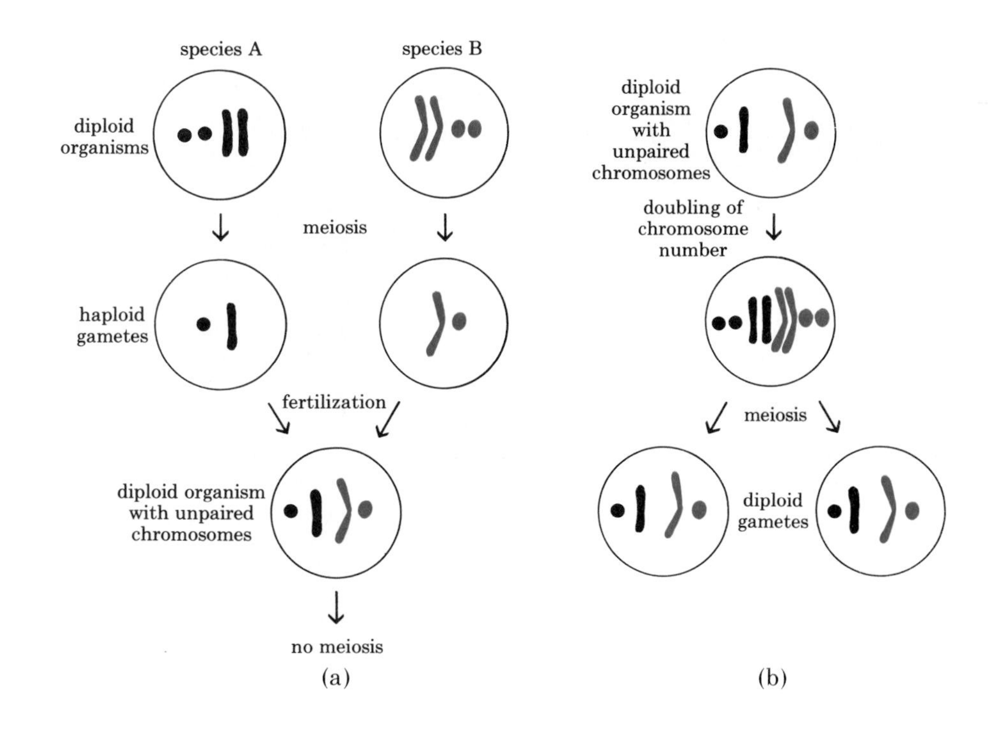

36–7 (a) *A hybrid organism (such as a mule) produced from two haploid* (n) *gametes can grow normally because mitosis is normal, but it cannot reproduce because the chromosomes cannot pair at meiosis.* (b) *If, however, polyploidy occurs and the chromosome number doubles, the hybrid can produce viable gametes. Since each chromosome will have a partner, the chromosomes can pair at meiosis. The resultant gametes will be diploid* (2n).

Isolating Mechanisms

Some species of organisms are so similar phenotypically that only an expert with a microscope can tell them apart—*Drosophila* offers several examples. Yet such species may live together in the same area without interbreeding. What factors operate to maintain genetic isolation of closely related species? Isolating mechanisms may be conveniently divided into two categories: premating mechanisms, which prevent mating between members of different species, and postmating mechanisms, which prevent the production of fertile offspring from such matings as do occur. One of the most significant things about the postmating isolating mechanisms is that, in nature, they are rarely tested. The premating mechanisms alone usually prevent interbreeding.

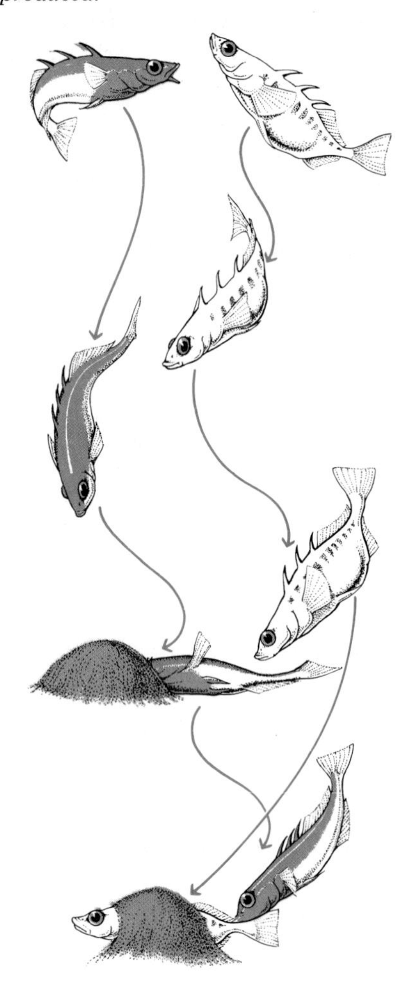

36-8 *Sticklebacks, small freshwater fish, have elaborate mating behavior. The male at breeding time, in response to increasing periods of sunlight, changes from dull brown to the radiant colors shown here. He builds a nest and begins to court females, zigging toward them and zagging away from them. A female ready to lay eggs responds by displaying her swollen belly. The male leads her down to the tunnel-like nest. He prods her tail and, in response, she lays her eggs and swims off. He follows her through, fertilizes the eggs, and stays to tend the brood. If either partner fails in any step of this quite elaborate ritual, no young are produced.*

Premating Isolating Mechanisms

In animals, premating mechanisms are often behavioral. They can take many forms, some of them quite elaborate (Figure 36–8).

Visual recognition is important, particularly among birds, which are visually oriented in all their behavior. According to observers, Galapagos finches recognize each other by their beaks. A male finch may mistakenly pursue a finch of another species—the finches often look very much alike from the rear—only to lose interest as soon as he sees the beak. He will not court a female of another species. The beak is a conspicuous feature in courtship, during which food is passed from the beak of the male to that of the female.

The flashings of fireflies (which are actually beetles) are sexual signals. Each species of firefly has its own particular flashing pattern, which is different from that of other species both in duration of flashes and in intervals between them (Figure 36–9). For example, one common flash pattern

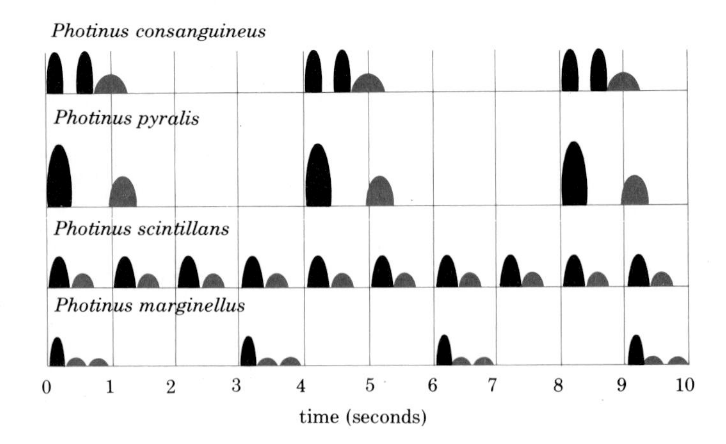

36-9 *Flashing patterns of various species of fireflies found in Delaware. Each division of the horizontal lines represents one second, and the height of each curve represents the brilliance of the flash. The black curves are male flashes, and the colored ones are the female responses. The male flashes first and the female answers, returning the species-specific signal. Firefly flashes differ from species to species, not only in duration, intensity, and timing, but also in color.*

consists of two short pulses of light separated by about 2 seconds, with the phrase repeated every 4 to 7 seconds. The lights also vary among the species in intensity and color. The male flashes first, and the female answers, returning the species-specific signal. By mimicking the signals with a flashlight, it is possible to attract the males of a given species.

Bird songs, frog calls, and the strident love notes of cicadas and crickets all serve to identify members of a species to one another. For example, studies of leopard frogs have recently revealed that some of the populations that can crossbreed under laboratory conditions do not do so in the wild. They are effectively isolated from one another by differences in their mating calls.

Among many invertebrates, pheromones (page 372) are the chief isolating mechanisms. They may serve as signals, as in the case of the cecropia moth, to attract the male or, with oysters, for instance, to trigger the release of gametes by the female.

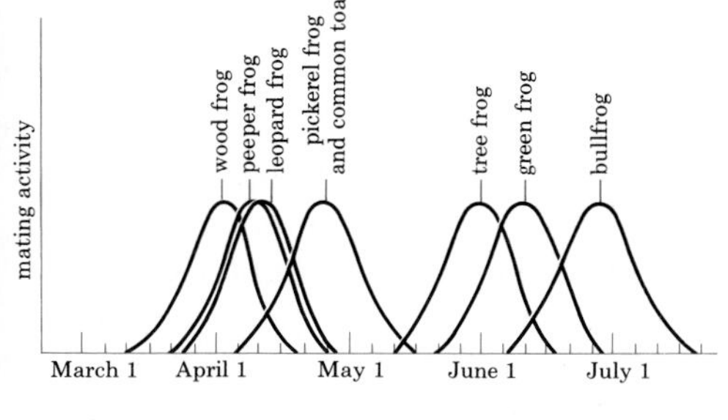

36-10 *Mating timetable for various frogs and toads that live near Ithaca, New York. In the two cases where two different species have mating seasons that coincide, the breeding sites differ. Peepers prefer woodland ponds and shallow water; leopard frogs breed in swamps; pickerel frogs mate in upland streams and ponds; and common toads use any ditch or puddle.*

Temporal differences also play an important role in sexual isolation. Species differences in flowering times are important isolating mechanisms in plants. Most mammals—we are a notable exception—have seasons for mating, often controlled by temperature or by day length. Figure 36-10 shows the mating calendar of species of frogs near Ithaca, New York. The spawning of freshwater fish is often regulated by water temperatures. Members of the species *Drosophila pseudoobscura* and *Drosophila persimilis,* which often are found in the same areas, congregate and mate almost every day but differ in their mating hour.

Postmating Isolating Mechanisms

On the rare occasions when members of two different species attempt to mate or succeed in doing so, anatomical or physiological incompatibilities maintain the genetic isolation of the species. These postmating isolating mechanisms are of several different types. For instance:

1. Differences in the shape of the genitalia may prevent insemination.
2. The sperm may not be able to survive in the reproductive tract of the female.
3. The pollen tube may not be able to grow on the stigma.
4. The sperm cell may not fuse with the ovum.
5. The ovum, once fertilized, may not develop.
6. The young, or some of the young, may survive but may not become reproductively mature.
7. The offspring may be hardy but sterile—the mule, for example. Sterility in such cases is apparently often caused by the fact that the dissimilar chromosomes cannot pair at meiosis.

The fact that offspring of matings between different species are often less viable serves to reinforce the premating isolating mechanisms. A female cricket that answers to the wrong song or a frog whose individual calendar is not synchronized with that of the rest of the species will contribute less to the gene pool. As a consequence, there will be a steady selection for premating isolating mechanisms.

An Example: Darwin's Finches

Admirers of Charles Darwin find it particularly appropriate that one of the best examples of speciation is provided by the finches observed by Darwin on his voyage to the Galapagos Islands. All the Galapagos finches are believed to have arisen from one common ancestral group—perhaps either a single pair or even a single pregnant female—transported from the South American mainland, some 950 kilometers away. How she or they got there is, of course, not known, but it may have been the result of some particularly severe storm. (Periodically, for example, some American birds and insects appear on the coasts of Ireland and England after having been blown across the North Atlantic.) It is very likely that finches were the first land birds to colonize the islands, which provide a highly diversified environment.

From the small ancestral group, 13 different species arose (plus one to the northeast on the Cocos Islands, 1,000 kilometers away). Apparently the various islands were near enough to one another so that, over the years, small founding groups could land. They were far enough apart, however, that once a group was established, there would be little or no gene flow between it and the parent group for a period of time long enough for genetic barriers to develop. (Finches are not very good fliers; if they had been, this

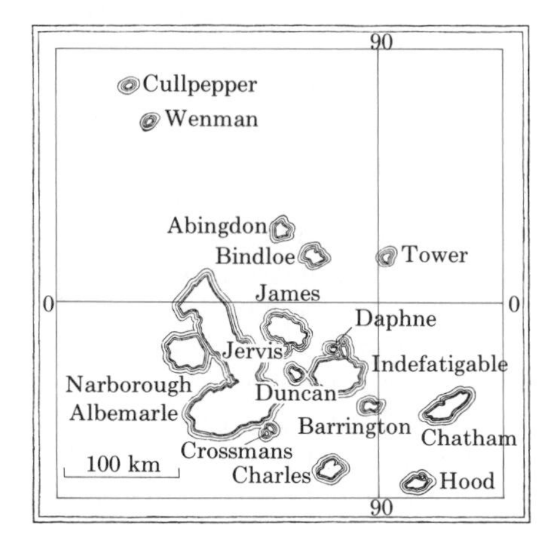

36-11 *The Galapagos Islands, some 950 kilometers west of the coast of Ecuador, have been called "a living laboratory of evolution." Species and subspecies of plants and animals that have been found nowhere else in the world inhabit these islands. "One is astonished," wrote Charles Darwin in 1837, "at the amount of creative force . . . displayed on these small, barren, and rocky islands. . . ."*

(a)

(b)

(c)

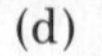

(d)

(e)

(f)

36-12 *Six of the 13 different species of Darwin's finches. There are six species of ground finch, six species of tree finch, and one warblerlike species—all derived from a single ancestral species. Except for the warbler finch, which resembles a warbler more than a finch, the species look very much alike; the birds are all small and dusky-brown or blackish, with stubby tails. The differences between them lie mainly in their bills, which vary from small, thin beaks to huge, thick ones.*

(a) *The small ground finch* (Geospiza fuliginosa) *and* (b) *the medium ground finch* (Geospiza fortis) *are both seed eaters.* Geospiza fortis, *with a somewhat larger beak than* G. fuliginosa, *is able to crack larger seeds.*

The cactus ground finches Geospiza scandens (c) *and* Geospiza conirostris (d) *live on cactus blooms and fruit. Notice that their beaks are much larger and more pointed than those of the other two ground finch species.*

The tree finches Camarhynchus parvulus (e) *and* Camarhynchus pauper (f), *both insectivorous, take prey of different sizes.*

natural experiment in evolution would have been a failure.) Development of a new species requires, it is estimated, at least 10,000 years of genetic isolation, but since there are many islands, several species could have been evolving at the same time.

The ancestral type was a finch, a smallish bird with a short, stout, conical bill especially adapted for seed-crushing; canaries and sparrows are finches and so are cardinals. The ancestor is believed to have been a ground-feeding finch, and six of the Galapagos finches are ground finches. Four species of ground finches live together on most of the islands. Three of them eat seeds and differ from one another mainly in the size of their beaks, which in turn, of course, influences the size of seeds they eat. The fourth lives largely on the prickly pear and has a much longer and more pointed beak. The two other species of ground finch are usually found only on outlying islands, where some supplement their diet with cactus.

In addition to the ground finches, there are six species of tree finch, also differing from one another mainly in beak size and shape. One has a parrotlike beak, suited to its diet of buds and fruit. Four of these tree finches have insect-eating beaks, each adapted to a different size range of insects. The sixth, and most remarkable of the insect eaters, is the woodpecker finch (Figure 33-7, page 473).

By all ordinary standards of external appearance and behavior, the thirteenth species of Galapagos finch would be classified as a warbler. Its beak is thin and pointed like a warbler's, it even has the warblerlike habit of flicking its wings partly open, and, warblerlike, it searches through leaves, twigs, and ground vegetation for small insects. However, its internal anatomy and other characteristics clearly place it among the finches. There is general agreement that it, too, is a descendant of the common ancestor or ancestors.

Notice also that the evolution of the Galapagos finches—the diversification of a single type to fill vacant ecological spaces—provides a good example of adaptive radiation.

36-13 *The woodpecker has a number of adaptations that enable it to obtain food. These include two toes pointing backward with which the woodpecker clings to the tree bark, strong tail feathers that prop it up, a strong beak that can chisel holes in the bark, strong neck muscles that make the beak work as a hammer, and a very long tongue that can reach insects under the bark.*

Adaptation

Adaptation is the tangible result of evolution. As we noted at the beginning of this book (page 3), all organisms are adapted. For example, note how a squirrel's tail serves as counterbalance as the squirrel leaps and turns; in addition, the same marvelous structure serves as a parasol, a blanket, an aerial rudder, and, in case of mishap, a parachute. Pluck a burr from your clothing and consider the ingenious devices by which it clings there. Consider the love and devotion characteristic of the domesticated dog. These are adaptations related to procurement of food and shelter as stringently selected for as the beak of a woodpecker (Figure 36-13). Thread a needle; your capacity to do so represents the cumulative effect of millions of years of selection pressures for digital dexterity and eye-hand coordination. (The needle itself made its appearance a mere 10,000 years ago.)

Two special types of adaptation deserve mention.

Preadaptation

Early critics of Darwin were fond of arguing that it was impossible by chance alone for so intricate and marvelous an organ as, for example, the vertebrate eye to be assembled. They were missing an important point:

Evolution is very conservative. Studies of comparative anatomy indicate that the eye began as a patch of light-sensitive epithelium and gradually accumulated, by mutation and recombination, modifications that increased its survival value to the organism. It did not spring into being, fully formed, without precursors.

Organic changes accumulate and are conserved because they are of immediate—not future—survival value. To take a simple example, the lungs and walking fins of certain Devonian fishes did not evolve in preparation for the transition to land; they improved the survival chances of the fishes when the temporary ponds of that time dried up. Similarly, the increased visual acuity and upright stance associated with life in the treetops fortuitously prepared a group of large primates for life on the open tropical grasslands. This phenomenon is often seen on the biochemical level as well, where a chemical such as AMP may be captured for another use in the course of evolution (see page 374).

Possession of specific traits that enable an organism to exploit a new or changing environment and on which wholly new adaptations can be constructed is called preadaptation. However, as with most aspects of evolution, preadaptations can be accurately identified only by hindsight.

Adaptations Related to Sexual Selection

One of the most powerful forces of natural selection is that exercised by individuals in choosing a mate.

Among most insects and vertebrates, males do the courting and females do the choosing. This division of labor makes sense. Sperm are less costly, energetically speaking, than eggs and are always more abundant. All other factors being equal, therefore, a male optimizes his contribution to the gene pool by mating as frequently as possible. On the other hand, a female seeks quality rather than quantity, since her energy investment in each act of reproduction is typically greater than the male's investment.

36-14 *A male frigate bird displays his crimson pouch. Throughout the courtship period, the pouch remains bright and inflated, even when the bird is flying or sleeping. The pouch's size and bright color are the result of sexual selection.*

As a consequence, the process of sexual selection is more apt to produce striking adaptations among males. Females are usually quick to reject suitors who deviate in any way from the "ideal." For instance, female fruit flies will not mate with a white-eyed male if a red-eyed male is present. Clearly, sexual selection is often stabilizing, but this is not necessarily the case. Sexual selection pressure led, for example, to the evolution of the elaborate plumage and huge tails of certain birds, the giant antlers of some deer, and the long, intricate mating rituals of many different types of organisms.

Convergent Evolution

Organisms that occupy similar environments often come to resemble one another even though they may be only very distantly related. When they are subjected to similar selection pressures, they show similar adaptations. This independent development of similar characteristics by different groups of organisms living in similar environments is known as *convergent evolution* (Figure 36–15). The whales, a group that includes the dolphin and the porpoises, are similar in many exterior features to sharks and other large fish, but the fins of whales conceal the remnants of a vertebrate hand (see Figure 34–1, page 478). Whales are warm-blooded, like their land-dwelling ancestors, and they have lungs rather than gills. Similarly, two families of plants invaded deserts in different parts of the world, giving rise to the cacti and the euphorbs. Both evolved large fleshy stems with water-storage tissues and protective spines, and they appear superficially similar. However, their quite different flowers reveal their widely separate evolutionary origins.

The biomes—the major groupings of organisms in the biosphere—contain many examples of convergent evolution. We shall discuss these in Chapter 40.

36–15 *An example of convergent evolution is provided by* (a) *the bull fur seal and* (b) *the king penguin shown here. Although one is a mammal and the other a bird, both have streamlined, fishlike bodies and a layer of insulating fat below the skin. The penguin swims by means of its flipperlike wings, using its webbed feet as rudders. The fur seal also swims primarily with its webbed forelimbs, using them like oars.*

(a)

(b)

36-16 *Darwin accurately predicted the discovery of an insect with a tongue 28 centimeters long in Madagascar, because that length would be required to reach the nectar of a species of orchid that blooms there. A similar, though slightly less startling, relationship resulted from the coevolution of the sphinx moth and the tobacco blossom, as shown here.*

Coevolution

When two or more populations interact so closely that each is a strong selective force on the other, simultaneous adjustments occur that result in coadaptation, or, as it is more commonly called, *coevolution*.

We have previously mentioned several examples of coevolution, such as the horse and its predators. One of the most important, in terms of sheer numbers of species and individuals involved, is the coevolution of flowers and their pollinators, described in Chapter 22. As we shall see in Section 8, when we turn our attention to the ecological theater, the natural world is filled with examples of coevolution, ranging from look-alike insects (page 542) to ants and acacia trees (page 545) to viruses and rabbits (page 546).

Extinction

Only a small fraction of all the species that have ever lived are presently in existence—certainly less than $^1/_{10}$ of 1 percent, perhaps less than $^1/_{1,000}$ of 1 percent. Extinction is very much a part of the evolutionary process, with the elimination of species making room for new ones. Certain times in the earth's history have seen the elimination of major life forms: the trilobites, which dominated the early Paleozoic seas; the dinosaurs, which disappeared at the end of the Mesozoic era; and the great land mammals that became extinct during the Pleistocene, victims of its harsh climatic changes or, perhaps, of the predations of a new life form, *Homo sapiens*.

At present, it has been estimated, extinction is proceeding at a far more rapid rate than ever before in evolutionary history, owing partly to human predation but mostly to the wholesale destruction of natural habitats. Conservationists try to indicate the value of endangered species, pointing out that seemingly worthless plants have been proven to contain chemicals with medicinal uses or that large mammals have "recreational" value and are the basis of revenue to national parks. Most endangered species, however, have no monetary value. Nor has it been proven that the "web of life" will be threatened by their disappearance. We know only that a species takes a very long time to evolve, and once it is gone it is gone forever. As David W. Ehrenfeld of Rutgers University points out, there has been, in all of Western culture, only one other major conservation effort, and not a single species was excluded as unworthy. "Of clean beasts, and of beasts that are not clean, and of fowls, and of everything that creepeth on the earth, there went in two and two. . . ." Not a bad precedent.

36-17 *The last known surviving member of the subspecies of tortoise found on the Galapagos island of Culpepper. With his death, this group of tortoises will be extinct.*

SUMMARY

Examples of the effects of natural selection on contemporary populations can be seen in the peppered moth *(Biston betularia)* and in the phenomenon of insecticide resistance.

The effects of natural selection on past populations can be seen in the fossil record. Three principal patterns of evolution are observed: phyletic, or "straight line," evolution; splitting, or "branching," evolution; and adaptive radiation. These are not distinct processes. Phyletic evolution always involves branching, splitting evolution always involves phyletic evolution, and adaptive radiation involves a combination of both patterns.

A species is defined in population biology as an isolated gene pool moving through time and space. For new species to originate, populations formerly within the same gene pool must be genetically isolated from one another and subsequently subjected to different selection pressures. When the genetic divergence between the populations becomes so great that they can no longer interbreed, speciation has occurred. Among animals, genetic isolation appears usually to require geographic isolation. Among plants, it may also be achieved by hybridization or polyploidy. Polyploidy is often a mechanism whereby hybrids become fertile. Genetic barriers among species are maintained by premating and postmating isolating mechanisms.

Adaptation, which is a property of all organisms, is the tangible result of evolution. Adaptations accumulate and are preserved because they are of immediate survival value. When, in retrospect, it is apparent that they enabled an organism to exploit changed circumstances and were the foundation for new adaptations, they are termed preadaptations. Many of the most striking adaptations are related to mate selection.

Convergent evolution is the development of similar characteristics by different organisms living in similar environments. Coevolution occurs when populations exert strong selection pressures on each other.

Extinction is a normal part of the evolutionary process, but it is occurring at an unprecedented rate in the modern world.

QUESTIONS

1. Distinguish among the following: phyletic evolution/splitting evolution/adaptive radiation; hybridization/polyploidy; premating isolating mechanisms/postmating isolating mechanisms; adaptation/evolution; convergent evolution/coevolution.

2. In your own words, define "species."

3. What are the three essential steps in speciation?

4. Describe the separate steps involved in the formation of the distinct species of Galapagos finches. How does this process of speciation illustrate both the founder principle and adaptive radiation?

5. In your opinion, why is there at present only one species in the genus *Homo*? Is a second *Homo* species likely to arise?

SUGGESTIONS FOR FURTHER READING

BATES, MARSTON, and PHILIP S. HUMPHREY, eds.: *The Darwin Reader,* Charles Scribner's Sons, New York, 1956.*

> *A collection of Darwin's writings, including* The Autobiography *and excerpts from* The Voyage of the Beagle, The Origin of Species, The Descent of Man, *and* The Expression of the Emotions. *Darwin was a fine writer, and you can discover here the wide range of his interests and concerns at different periods of his life.*

CALDER, NIGEL: *The Restless Earth,* Viking Press, Inc., New York, 1972.

> *A handsomely illustrated report on the new geology, written for the general reader.*

CARLQUIST, SHERWIN: *Island Life: A Natural History of the Islands of the World,* Natural History Press, Garden City, N.Y., 1965.

> *An exploration of the nature of island life and of the intricate and unexpected evolutionary patterns found in island plants and animals.*

DARWIN, CHARLES: *The Origin of Species by Means of Natural Selection, or The Preservation of Favored Races in the Struggle for Life,* Doubleday & Company, Inc., Garden City, N.Y., 1960.*

> *Darwin's "long argument." Every student of biology should, at the very least, browse through this book to catch its special flavor and to begin to understand its extraordinary force.*

DARWIN, CHARLES: *The Voyage of the* Beagle, Natural History Press, Garden City, N.Y., 1962.*

> *Darwin's own chronicle of the expedition on which he made the discoveries and observations that eventually led him to his theory of evolution. The sensitive, eager young Darwin that emerges from these pages is very unlike the image many of us have formed of him from his later portraits.*

EHRLICH, PAUL R., RICHARD W. HOLM, and DENNIS R. PARNELL: *The Process of Evolution,* 2d ed., McGraw-Hill Book Company, New York, 1974.

> *A good, short text for an undergraduate course in evolutionary theory, appropriate for students who have completed a year of general biology.*

EICHER, DON L.: *Geologic Time,* Prentice-Hall, Inc., Englewood Cliffs, N.J., 1968.*

> *A short, supplementary text that describes the fossil strata and their significance.*

GOULD, STEPHEN JAY: *Ever Since Darwin: Reflections in Natural History,* W. W. Norton & Company, New York, 1977.*

GOULD, STEPHEN JAY: *The Panda's Thumb: More Reflections in Natural History,* W. W. Norton & Company, New York, 1980.

> *Collections of thoughtful, witty, and well-written essays from* Natural History. *Evolution is the central theme.*

HAMILTON, TERRELL H.: *Process and Pattern in Evolution,* The Macmillan Company, New York, 1967.*

> *This short book, designed as a supplementary text, is outstanding for its clarity of definition and presentation of evolutionary concepts.*

LACK, DAVID: *Darwin's Finches,* Harper & Row, Publishers, Inc., New York, 1961.*

> *This short, readable book, first published in 1947, gives a marvelous account of the Galapagos, its finches and other inhabitants, and of the general process of evolution.*

LEWONTIN, R. C.: *The Genetic Basis of Evolutionary Change,* Columbia University Press, New York, 1974.*

> *This book is intended for the advanced student, but it is also highly recommended for any reader interested in the problems and perspectives of population genetics.*

MAYR, E.: *Populations, Species, and Evolution: An Abridgement of Animal Species and Evolution,* Harvard University Press, Cambridge, Mass., 1970.*

> *A masterly, authoritative, and illuminating statement of contemporary thinking about species—how they arise and their role as units of evolution.*

* Available in paperback.

MOOREHEAD, ALAN: *Darwin and the* Beagle, Harper & Row, Publishers, Inc., New York, 1969.*

A delightful narrative of Darwin's journey, beautifully illustrated with contemporary or near contemporary drawings, paintings, and lithographs.

SCIENTIFIC AMERICAN: *Evolution,* W. H. Freeman and Company, San Francisco, 1978.*

This reprint of the September 1978 issue of Scientific American *provides an overview of the history of life as understood by modern evolutionary theory. Such topics as the evolution of ecological systems and of behavior are discussed, in addition to the evolution of the major types of organisms that have inhabited our planet.*

SIMPSON, GEORGE GAYLORD: *Horses,* Oxford University Press, New York, 1951.

The story of the horse family in the modern world and through 60 million years of history, as recorded by a leading paleontologist.

SIMPSON, GEORGE GAYLORD: *Penguins: Past and Present, Here and There,* Yale University Press, New Haven, Conn., 1976.

As Simpson says, "Penguins are beautiful, interesting, inspiring, and funny." They are also excellent examples of evolution and adaptation, as this informal account demonstrates.

SMITH, JOHN MAYNARD: *The Theory of Evolution,* 3d ed., Penguin Books, Inc., New York, 1976.*

Written for the general public, this book is notable for its many concrete examples of evolution, past and present.

* Available in paperback.

SECTION 8

Ecology: Organisms in Nature

The African savanna, a tropical grassland with scattered clumps of trees, is one of the great biomes of the world. Among its more conspicuous residents are the wildebeest. Each year, some 1 million wildebeest migrate about 1,800 kilometers (1,100 miles) across the Serengeti Plain, following the path of the rainfall.

37

Population Dynamics: The Numbers of Organisms

Ecology (from the Greek word *oikos,* meaning "house" or "home") is the study of the interactions of organisms with their physical environment and with each other. As a science, it seeks to discover how organisms affect, and are affected by, their living and nonliving environment. Ecology further seeks to define how these interactions determine the kinds and numbers of organisms found in a particular place at a particular time.

Although few will quarrel with this definition of ecology as a science, the word ecology has, in fact, a broader social meaning. In 1962, Rachel Carson published her book *The Silent Spring* and heralded a new era in which "ecology" came to be associated with the environmental movement and with the preservation of wildlife and the conservation of natural resources. The environmental movement became both popular and productive. As a consequence, Lake Erie was brought back to life, strip mining was put under federal regulation, the smog was lifted from many of our major cities, and our springs were no longer silent. All the problems have not, of course, been solved, but these and other noteworthy achievements indicate that similar problems can be solved. More recently, the concerns of the environmental movement have come into conflict with other concerns in our society—economic development, preservation of jobs, and our need for abundant supplies of food and energy. The conflicts are very difficult ones to resolve.

During the period of environmental activism, a revolution was occurring in the science of ecology. At one time ecology was synonymous with natural history and with the observation (and enjoyment, as the early ecologists made clear) of organisms in their natural environment. Since the 1960s, however, ecologists have sought not only to quantify the many factors that may affect an organism in nature but even to manipulate these factors—that is, to perform experiments and to make predictions and construct testable hypotheses. An important tool in their work is the computer, which makes it possible to deal with vast quantities of data, involving many variable factors.

The basic unit of study in ecology is the population. In this chapter we shall consider some of the factors that determine the number of organisms in a particular population. In the chapters that follow, our attention will be focused on the interactions of populations with one another and their organization into communities and ecosystems.

37-1 *In the savanna, the major route of energy flow is from the grasses and trees, through grazing and browsing animals, to predators. The lion and the hyena are the great predators of the savanna.*

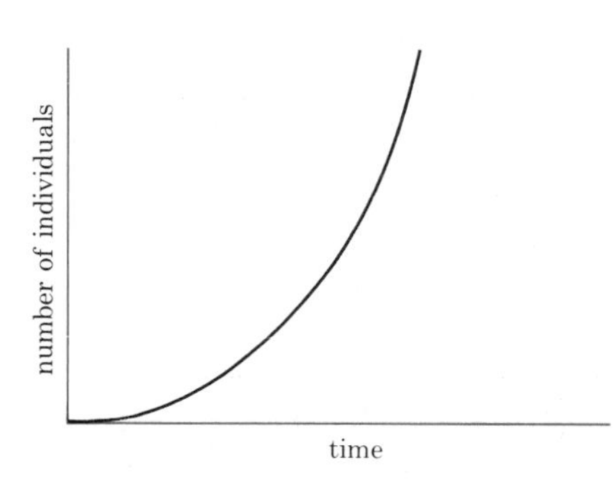

37-2 *An exponential growth curve. After an initial establishment phase, the population increases in the same fashion as a savings account with compound interest. Such growth is characteristic of a small population with access to abundant resources.*

The Rate of Increase

As you will recall from Chapter 34, geneticists define a population as an interbreeding group of organisms. Ecologists generally use the term in a more restricted sense to refer to organisms of the same species occupying a particular place at the same time. A population grows as the result of reproduction and immigration (the movement of other individuals of the species into the area from elsewhere) and decreases as a result of death and emigration (the departure of individuals).

Reproduction

As Darwin noted more than 120 years ago, the reproductive capacities of almost all species are very high. It can be calculated, for instance, that a single female housefly producing an average of 120 eggs per laying, with half developing into females and with seven generations per year per female, could produce 6,182,442,727,320 houseflies in the space of one year. At the other end of the reproductive scale is a pair of elephants, which have a gestation period of 20 to 22 months and usually give birth to only one offspring at a time. If all offspring and their offspring, in turn, survived and reproduced, this one pair of elephants would have 19 million descendants at the end of 750 years, as noted by Darwin.

This type of growth, in which essentially all individuals in each generation achieve their full reproductive potential, is known as *exponential growth* (Figure 37-2). It operates on the same principle as compound interest on a savings account; the more you have, the more you get. Exponential growth starts off slowly, and then it shoots up very rapidly. If we assume that immigration and emigration balance out, the rate at which the population increases is simply the reproduction rate minus the death rate; it is analogous to the basic interest rate on the savings account. The change in the number of individuals in the population over a given period of time, which is analogous to the change in your savings account balance over a given period of time (assuming you make no deposits or withdrawals), equals the rate of increase times the number of individuals already present. In other words, the greater the number of individuals present in the population, the greater the number of offspring that will be produced, leading to a greater number of individuals reproducing in the next generation.

Exponential growth occurs most often when a small population has access to abundant resources. Examples are bacteria or protists cultivated in the laboratory or seasonal blooms of algae in nutrient-rich waters. Exponential growth is also typical of insect outbreaks, such as the gypsy moth infestations (Figure 37-3) that recurrently plague the northeastern United States. Often, in nature, insect populations grow until they are overcome by an epidemic of disease organisms (another example of exponential growth) or until they reduce their food supply so that they are not able to complete their metamorphoses and reach the reproductive stages; the population then crashes to low levels.

37-3 *An infestation of gypsy moth caterpillars can rapidly strip the leaves from a tree, as shown in this photograph. Spraying the caterpillars with insecticides often reduces the size of the population just enough to delay the population "crash" that ultimately follows exponential growth.*

The Effect of Age

In organisms in which the life span exceeds the reproductive span, the average age of the population at any given time may profoundly affect the rate of increase. If a large proportion of the population is of reproductive age or younger, population growth will continue at a high rate even though

37-4 *Charting populations by age and sex permits predictions about future growth rates. In India, for example, nearly 45 percent of the population is under 15 years of age. Even if these young men and women limit their family size enough to produce only enough children to replace themselves (which means cutting the current birth rate in half), population growth will not level off until about the year 2040—and at a level of well over a billion. The population of Sweden, by contrast, will remain the same unless the birth rate is dramatically increased.*

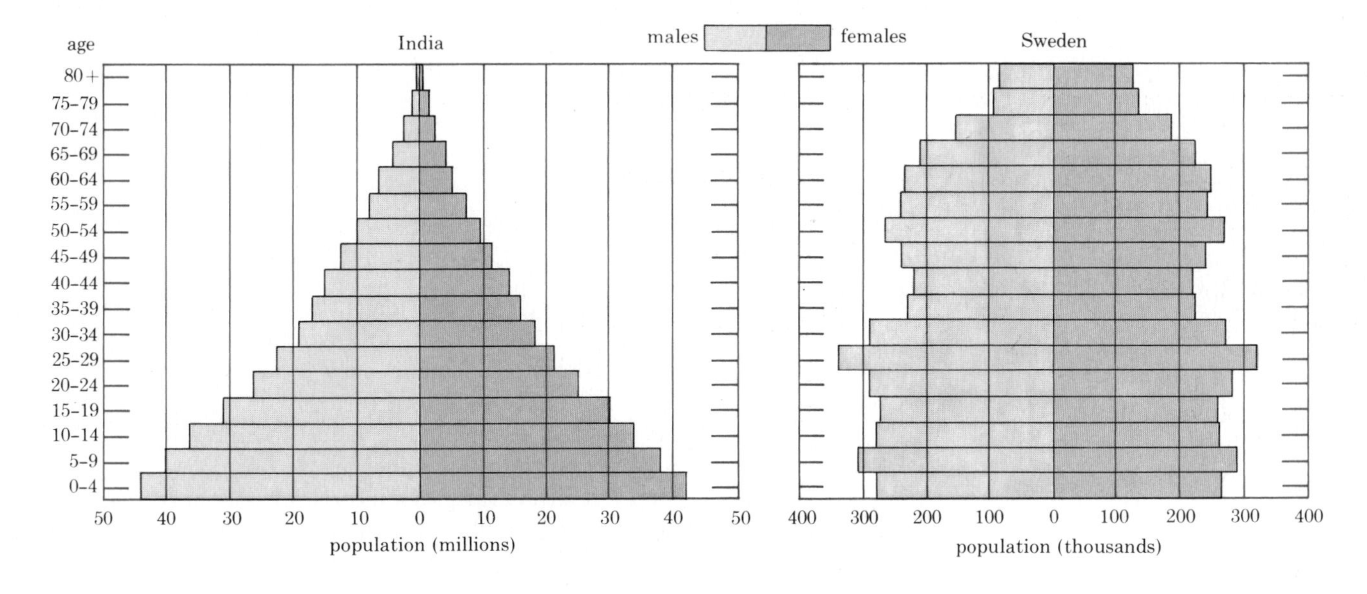

the birth rate may remain constant. This is of considerable importance in the growth of human populations (Figure 37-4). It is one of the reasons why the population of the United States has continued to increase despite the fact that, on the average, young couples are having slightly fewer than two children. (Another reason is continued immigration.)

Mortality

Another factor in population growth is, of course, mortality. In the so-called developing nations—the nations showing the most rapid population increases—gains in population are the result more of the decrease in mortality, especially among infants and young people, than of any increase in birth rates (Figure 37-5).

The rate of increase or decrease of a population is the difference between the birth rate and the death rate. Death rates are high in natural populations. For example, in a study of the saltmarsh song sparrows of San Francisco Bay, it was estimated that of every 100 eggs laid, 26 are lost before

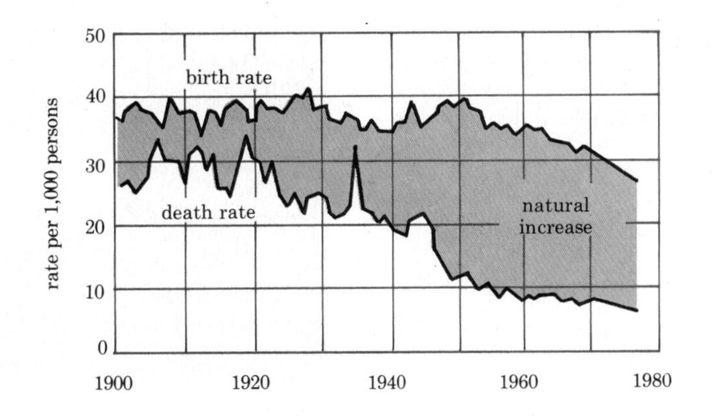

37-5 *In many tropical countries, the death rate has fallen rapidly since 1940, resulting in a rapid growth of the population. The drop in death rate is the result of increased medical services, control of malaria by DDT, and the availability of new antibacterial drugs, especially the antibiotics. Note that the birth rate has also begun to fall. The area on the graph marked in color indicates population growth. The data shown here are from Sri Lanka (formerly Ceylon).*

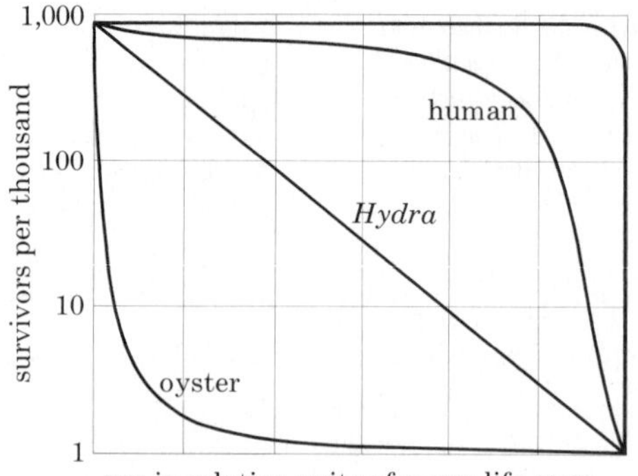

37-6 *Representative survivorship curves. In the oyster, mortality is extremely high during the free-swimming larval stage, but once the individual attaches itself to a favorable substrate, life expectancy levels off. Among* Hydra, *the mortality rate is the same at all ages.*

The curve at the top of the graph is for a hypothetical population in which all individuals live out the average life span of the species—a population, in other words, in which all individuals die at about the same age. The fact that the curve for humans approaches this hypothetical curve indicates that the human population as a whole is reaching a uniform age of mortality.

hatching. Of the 74 live nestlings, only 52 leave the nest, and of these, 42 die the first year. The remaining 10 breed the following season, but during the next year, 43 percent of these die, leaving only 6 out of the original 100. Each subsequent year, mortality among the survivors amounts to 43 percent. Once a bird survives its first, risk-laden year, the mortality rate for the subsequent years remains more or less constant.

In many species, the probability of survival for one more year (or one more period of time) seems to be highest during the middle of life and lowest for both very young and very old individuals. Patterns of mortality differ from species to species, however. Figure 37-6 shows the basic types of survivorship curves. Patterns of mortality affect the age structure of populations and may be important in population regulation.

Emigration

Emigration is the departure of individuals from a population. Arctic lemmings, among the most famous emigrants, increase in number and then periodically emigrate in large numbers, apparently in response to crowding. In northern Europe, they have been known to march by the millions across the countryside. Similar spectacular emigrations occur among locusts, resulting in the plagues of locusts recorded from Biblical times. Emigration may also occur continuously but less spectacularly, involving only small numbers of individuals at a time.

The stimuli that cause mass emigration undoubtedly differ from species to species, but the result is always the same: a reduction in the size of the local population.

37-7 *Greenland collared lemmings. Lemmings are inhabitants of the North American and Eurasian tundra (page 585). Lemming populations, like those of many other small rodents, undergo drastic cyclic fluctuations in numbers, resulting in mass emigrations.*

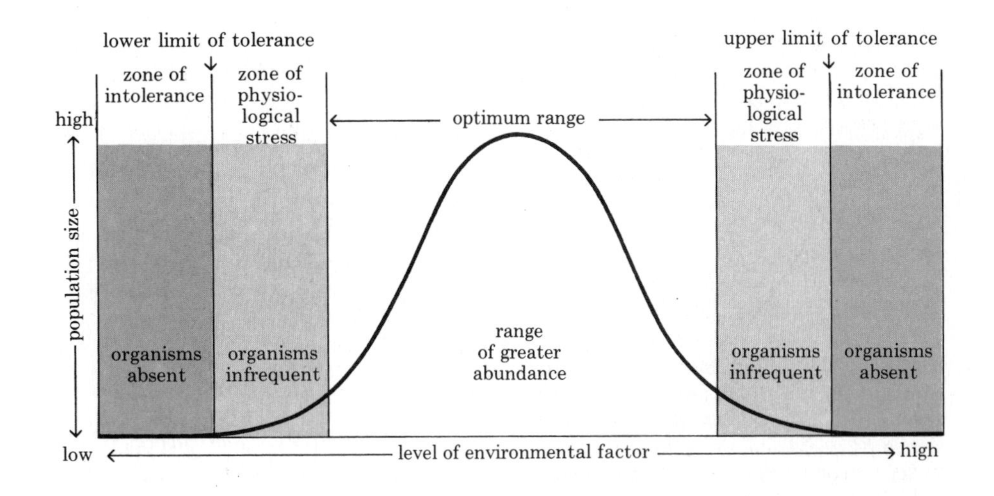

37-8 *The principle of limiting factors. Every species has a characteristic limiting-factor curve for each factor in its environment. In the zones of intolerance, individual organisms cannot survive. In the zones of physiological stress, some individuals are able to survive but the population cannot grow. In the optimum range, the population can flourish.*

The Limits to Growth

The size of a population depends on a number of factors that differ for different populations. Of critical importance is the organisms' range of tolerance for such factors as light, temperature, available water, salinity, nesting space, and shortages (or excesses) of required nutrients. If any essential requirement is in short supply or any environmental feature is too extreme, growth of the population is not possible, even though all other necessities are met. Whatever determines whether or not a population can grow in a given environment is known as a *limiting factor* (Figure 37-8).

The way limiting factors affect the growth of populations is illustrated by a common pollution problem. When phosphorus is the limiting factor in the growth of freshwater algae in a lake or slow-moving stream, phosphate-containing detergents added to the water from sewage systems will produce a spectacular bloom of algae. This period of rapid growth continues until the available supply of another essential element—perhaps calcium—is consumed. Then the algae begin to die, the decomposing bacteria take over, the respiration of the bacteria begins to use up the oxygen in the water, and oxygen becomes the limiting factor. Eventually the concentration of oxygen drops below the tolerance level of fish and other organisms, and these, too, begin to die. If the process is not interrupted, the lake becomes completely stagnant, and only anaerobic bacteria and other microorganisms can survive in it.

Types of Limiting Factors

Ecologists often divide the factors that affect the growth of a population into those that are density-dependent and those that are not. *Density-independent factors* are usually climatic and include such features of the environment as extremes of temperature, day length, and rainfall. *Density-dependent factors* exert effects that vary with the size of the population per unit area. In our previous example of water pollution, the limits to growth imposed by shortages of phosphorus, calcium, and then oxygen are all density-dependent.

Competition among members of the same population (that is, intraspecific competition) for requirements in short supply is one of the major forces of natural selection. For example, plants of a particular species compete with each other for sunlight, water, and nutrients from the soil. Birds

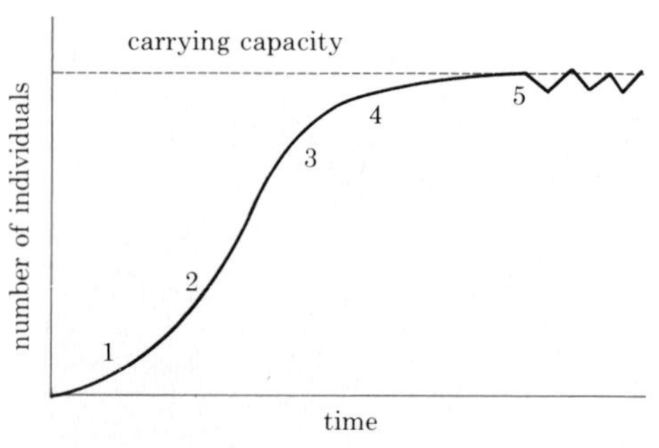

37–9 *The simplest growth pattern observed in most natural populations is known as logistic growth and is represented by a sigmoid, or S-shaped, curve. As with exponential growth, there is an initial establishment phase when population growth is relatively slow (1), and a phase of rapid acceleration (2). Then, as the population approaches the carrying capacity of the environment, the growth rate slows down (3 and 4), and finally stabilizes (5). Other growth patterns observed in natural populations are considerably more complex.*

of a given species compete with each other for nesting space and food for the young. All other factors being equal, those individuals that can obtain the needed resources most efficiently will make the greatest contribution to the gene pool of the next generation.

Interactions among populations of different species also function as density-dependent limiting factors. These interactions may involve competition for the same resources (in this case, interspecific competition), predation, or symbiosis. We shall consider their role in population regulation and as selective forces in the next chapter.

The Effect of the Carrying Capacity

The *carrying capacity* of the environment is the number of individuals in a particular population that the environment can support over an indefinite period of time. It is determined by the density-dependent limiting factors. Food resources are often the most important factor, but others, such as light, water, or nesting space, may also be limiting. The effect of the carrying capacity on the growth of a population is shown in Figure 37–9.

When the number of organisms is very small, the curve approximates the exponential growth curve of Figure 37–2. As the number of organisms increases, growth slows down. If the number of organisms exceeds the carrying capacity, the growth rate becomes negative and the population declines slightly. Finally, the population stabilizes and oscillates around the maximum size the environment can support. This type of growth, represented by an S-shaped curve, is called *logistic*.

This model is the simplest sort of growth pattern seen for most populations in nature. It holds true for microorganisms in a laboratory, for instance, or for organisms that have found their way to a new and favorable environment not being exploited by any other organism.

An understanding of the logistic model has practical utility. If, for instance, one wants to control a population of rats, killing half of them may merely reduce the population to the point at which it increases most rapidly. A more effective approach would be to reduce the carrying capacity, which, in the case of rats, usually means tighter control of garbage disposal. Similarly, if one wants to control the production of a particular type of economically valuable fish, the fish should not be harvested below the level of rapid growth unless one is willing to wait a long time for the population to recover.

Strategies for Growth

Given the goal of maximizing one's contribution to the gene pool, is it better to produce large numbers of offspring or to invest the same amount of energy in caring for a relative few? The answer seems to be that it depends (Figure 37–10). If one is a pioneer in a new environment, a rapid reproduction rate would appear to be the best strategy, capturing the available space and resources for one's own genes. An example would be grasses invading a recently abandoned field. Species that have evolved such a life-history tactic are called pioneer, or opportunistic, species. *Opportunistic species* are characterized, among other things, by high reproduction rates, rapid development, early reproduction, small body size, and uncertain adult survival. Under more constant conditions, when the population

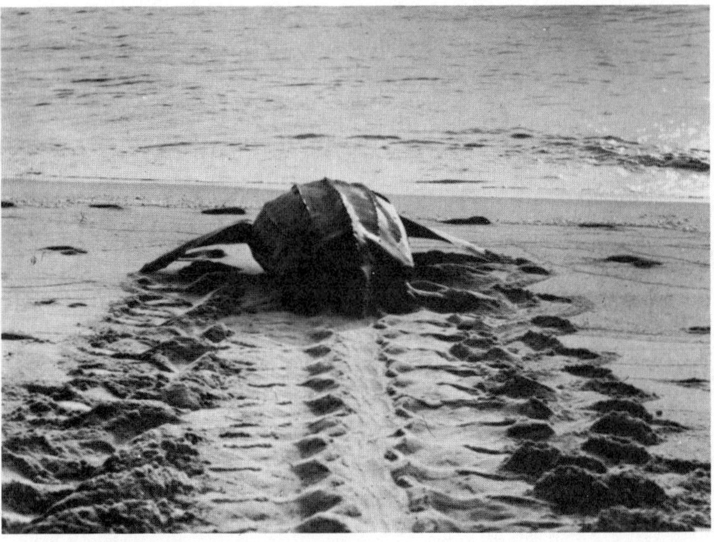

(a)

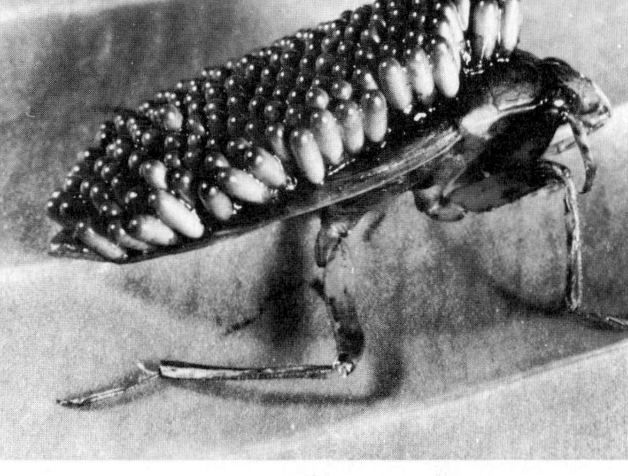

(b)

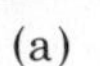

(c)

(d)

37-10 *Reproductive strategies.* (a) *A leatherback turtle returning to the sea after laying her eggs in a hole dug in the sand. The eggs, which she has covered over with sand, will be incubated by the warmth of the sun. Once the eggs are laid, the mother turtle does not participate further in raising the young.*

(b) *A female water bug lays her eggs on the back of the male. The male then carries the eggs, aerating them, until they hatch.*

(c) *Baby Canada geese remain with their parents through the summer after they are hatched and migrate with them to the winter feeding grounds. In proportion to life span, the time allotted to parental care by birds is longer than for any other animal group.*

(d) *Elephant calves are suckled by their mothers for at least two years, and the young are zealously guarded by the mother and the sisters and the aunts.*

is at or near its carrying capacity, the advantage appears to go to species with lower reproduction rates, longer development times, larger body size, and longer adult life with repeated reproductions. Such species are known as *equilibrium species*.

Whether a species is opportunistic or equilibrium—or somewhere in between—is a result, of course, of natural selection. In some cases, the result of selection has been different tactics for different members of the same population. Judith Myers and Charles Krebs studied emigration in a population of meadow mice. As with the lemmings, an increase in population size culminates in the movement of large numbers of individuals out of the population. What determines which individuals leave and which remain behind? The investigators found that the population consisted of both opportunistic- and equilibrium-type individuals. One group of mice had a higher birth rate, began to reproduce at an earlier age, and tended to disperse when population density increased. These individuals were the ones that emigrated, leaving the equilibrium-type individuals behind.

SUMMARY

Ecology is the study of the interactions of organisms with their physical environment and with each other. Ecologists are seeking to quantify the variables that affect organisms in nature, to make predictions and construct hypotheses, and to perform experiments that test the hypotheses.

The basic unit of study in ecology is the population—organisms of the same species living in the same place at the same time. A population increases as a result of reproduction and immigration and decreases as a result of death and emigration.

The reproductive potential of most populations is high. When the full reproductive potential of a population is achieved (a relatively infrequent occurrence in nature), exponential growth can occur. The change in the number of individuals in the population over a given period of time is the rate of increase times the number of individuals present at the beginning of the period.

The average age of a population affects its rate of increase. When a large proportion of a population is of or below reproductive age, population growth continues at a high rate even though the birth rate may be constant.

Mortality is generally high in natural populations. Different species, however, have different patterns of mortality. These patterns affect the age structure of the population and may contribute to population regulation. Emigration, the departure of individuals from a population, reduces the size of the local population.

A complex of environmental factors limits the growth of a population. Density-independent factors include extremes of temperature, day length, and rainfall. Density-dependent factors are generally resources such as moisture, mineral nutrients and sunlight (for plants), food (for animals), and space. Competition among members of a population for limited resources is a powerful selective force. Interactions among different populations also function as density-dependent limiting factors.

The density-dependent limiting factors determine the carrying capacity of the environment—the number of individuals in a particular population that the environment can support over an indefinite period of time.

The logistic growth curve, which takes the carrying capacity into account, represents the simplest growth pattern observed in most natural populations. Population growth is rapid when the population is small, gradually slows as it approaches the carrying capacity, and then oscillates as the population is maintained at or near its maximum size.

Different populations have different strategies for growth. Opportunistic species produce large numbers of young with small food supplies or little or no parental care, while equilibrium species generally produce fewer young and invest their energies in improved care for the existing young.

QUESTIONS

1. Distinguish between the following terms: immigration/emigration; exponential growth/logistic growth; limiting factors/carrying capacity; density-independent factors/density-dependent factors; opportunistic species/equilibrium species.

2. An old French riddle: "The pond lilies in a certain pond grow at a rate such that each day they cover twice as much of the pond as they did the day before. The pond is of a size that it will be completely covered at the end of 30 days. On what day is the pond half-covered? One-tenth covered? One-hundredth covered?"

What is the relevance of this riddle to human ecology?

3. Explain how each of the following factors would affect the growth rate of a population: age at first reproduction; time between generations; pre-reproductive mortality; post-reproductive mortality; length of period of parental care.

4. Suppose that you have a "farm" on which you grow, harvest, and sell edible freshwater fish. The growth of the fish population is logistic. You, of course, wish to obtain maximum yields from your "farm" over a number of years. To ensure this, how large should you allow the population to become before you begin harvesting? Identify the point on the logistic growth curve (Figure 37–9) at which you should begin harvesting.

How large a population should you leave unharvested? Identify the point on the logistic curve at which you should take no more fish from the population.

Factors in addition to the pattern of harvesting will affect the yields of fish obtained. What are some of these factors, and how might you adjust them to further increase the yields?

38

Interactions among Populations

In the previous chapter, we considered the factors that influence the size of a population in isolation from its interactions with other populations. In nature, of course, no population lives alone. A population is always part of a *community*—all of the organisms that inhabit a common environment and interact with one another. The interactions within a community influence the size of each population and, over long periods of time, its evolution.

The interactions among different populations are enormously varied and complex, but can generally be classified as competitive, predatory, or symbiotic.

Interspecific Competition: The Ecological Niche

Competition affects both the kinds and numbers of organisms found in a community. It generally occurs between organisms that have similar requirements and life styles. Plants compete with plants; herbivores, animals that eat plants and algae, compete with herbivores; carnivores, animals that eat other animals, compete with carnivores.

Closely linked to the phenomenon of competition is the concept of the ecological *niche*. The niche, which encompasses all of the roles and associations of a particular species in the community, includes but is more than the *habitat*, the portion of the environment in which the species lives. In the words of Richard Lewontin:

> The ecological niche is a multidimensional description of the total environment and way of life of an organism. Its description includes physical factors, such as temperature and moisture; biological factors, such as the nature and quantity of food sources and of predators, and factors of the behavior of the organism itself, such as its social organization, its pattern of movement, and its daily and seasonal activity cycles.*

* Richard Lewontin, "Adaptation," *Scientific American,* September 1978, page 215.

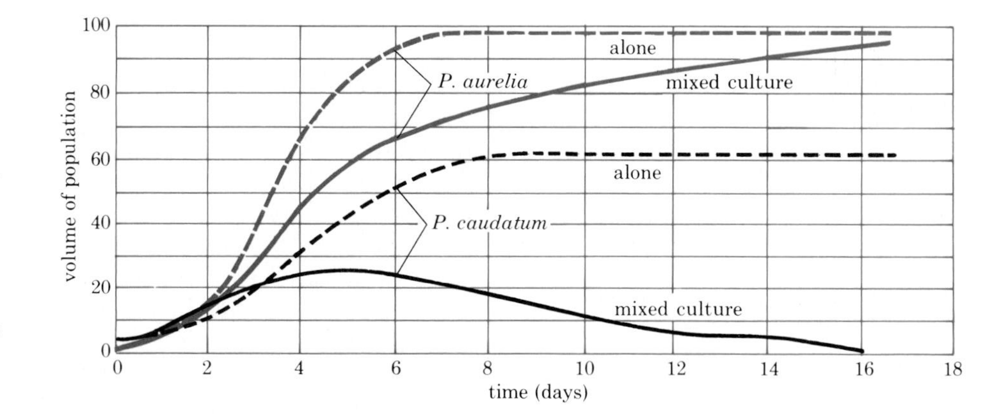

38–1 *Gause's experiment with* Paramecium *demonstrated that two species cannot occupy exactly the same niche indefinitely. In this experiment, two closely related species were in direct competition for the same resource—food.* Paramecium caudatum *and* Paramecium aurelia *were first grown separately under controlled conditions and with a constant food supply. As you can see,* P. aurelia *grew much more rapidly than* P. caudatum, *indicating that* P. aurelia *uses available food supplies more efficiently. When the two protozoans were grown together, the more rapidly growing species outmultiplied and eliminated the slower-growing species.*

According to the principle of competitive exclusion, formulated by the Russian biologist G. F. Gause, in any given community, no two species (that is, two populations) can occupy exactly the same niche for an extended period of time. Gause bolstered his hypothesis with a number of laboratory experiments. His simplest, now classic, experiment involved laboratory cultures of *Paramecium aurelia* and *Paramecium caudatum.* When the two species were grown under identical conditions in separate containers, *P. aurelia* grew much more rapidly than *P. caudatum,* indicating that the former used the available food supply more efficiently than the latter. When the two were grown together, the former rapidly outmultiplied the latter, which soon died out (Figure 38–1).

A similar situation occurs with two species of duckweed, *Lemna gibba* and *Lemna polyrrhiza. Lemna gibba* grows more slowly in pure culture than *L. polyrrhiza; L. gibba,* however, always replaces *L. polyrrhiza* when they are grown together. Again, evolution has provided one with an advantage. The plant bodies of *L. gibba* have air-filled sacs that serve as little pontoons, so that these plants form a mass over the other species, cutting off the light. As a consequence, the shaded *L. polyrrhiza* dies out (Figure 38–2).

It is possible to devise different culture conditions under which the outcomes of both the *Paramecium* and *Lemna* experiments can be reversed. However, as long as the conditions of the particular experiment are held constant, one species always wins.

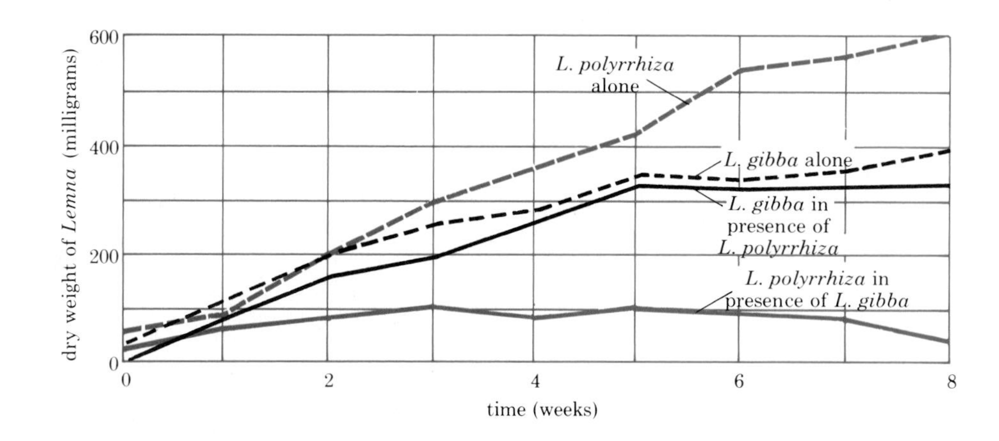

38–2 *An experiment with two species of floating duckweed, tiny angiosperms found in ponds and lakes. One species,* Lemna polyrrhiza, *grows more rapidly in pure culture than the other species,* Lemna gibba. *But* L. gibba *has tiny air-filled sacs that float it on the surface, and so it shades the other species and is the victor in the competition for light.*

The Niche in Nature

Gause's principle leads one to predict that in a situation in which two very similar species coexist in nature, it will be possible to demonstrate differences in their niches. The prediction has been fulfilled by a number of observations. For example, some New England forests are inhabited by five closely related species of warbler, all about the same size and all insect eaters. Why do these birds not eliminate one another by competitive exclusions? An analysis by Robert MacArthur showed that these warblers have different feeding zones in the trees (Figure 38–3). Because they exploit slightly different resources, they are able to coexist.

J. H. Connell studied competition between two species of barnacles in Scotland. Barnacles are crustaceans. When they change from their immature, larval forms, in which they are free-swimming, into their adult, feeding forms, they cement themselves to rocks and secrete shells. Once attached, barnacles remain fixed, so that by making careful records, one can determine the history of a particular population. One can identify exactly which barnacles have died and which new ones have arrived between visits to the study site.

One of the barnacle species studied, *Chthamalus stellatus,* occurs in the high part of the intertidal seashore, where conditions vary on a regular basis as the tide goes in and out. The other species, *Balanus balanoides,* occurs lower down, where conditions are more constant. Although *Chthamalus* larvae, after their period of drifting in the plankton, often attach to rocks in the lower, *Balanus*-occupied zone, adults are rarely found there. Connell was able to show that in the lower zone, *Balanus,* which grows faster, ousts *Chthamalus* by crowding it off the rocks and growing over it. Isolated from contact with *Balanus, Chthamalus* lived with no difficulty in the lower zone. However, when *Chthamalus* was removed, *Balanus* was unable to live in the higher intertidal zone. Although *Chthamalus* is driven from the lower zone by competition from the faster-growing *Balanus,* it survives in the intertidal community by special physiological adaptations that enable it to inhabit an area where *Balanus* cannot live.

A final example is afforded by Darwin's finches. As we noted in Chapter 36, the large, medium, and small ground finch species are very similar except for differences in overall body size and in the sizes of their beaks. These differences in beak size are correlated with the fact that they eat seeds of different sizes. Because of these different eating habits, competition for food is reduced. On islands such as Abingdon and Bindloe (see page 512), where all three species of ground finch exist together, there are

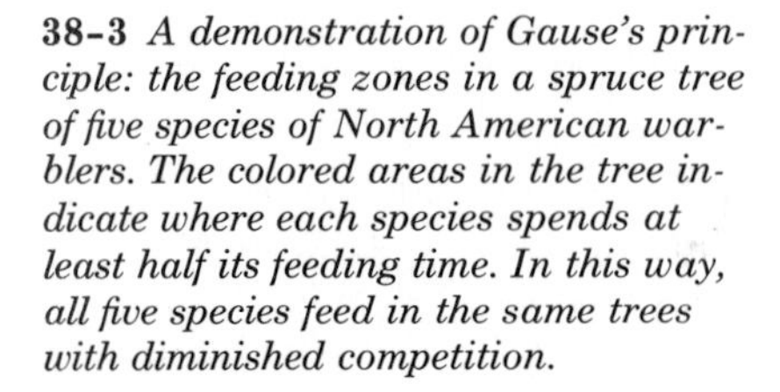

38–3 *A demonstration of Gause's principle: the feeding zones in a spruce tree of five species of North American warblers. The colored areas in the tree indicate where each species spends at least half its feeding time. In this way, all five species feed in the same trees with diminished competition.*

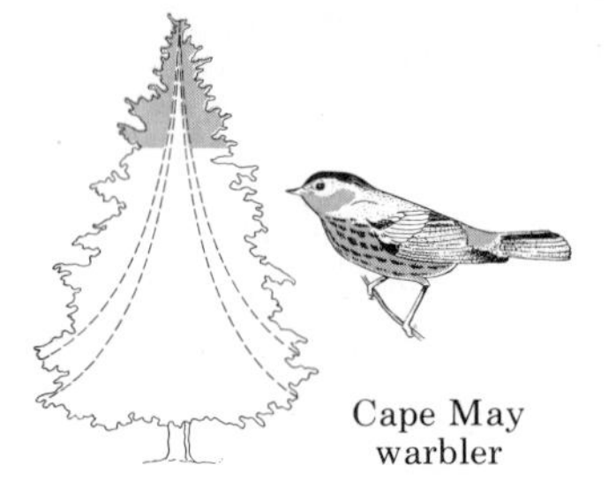

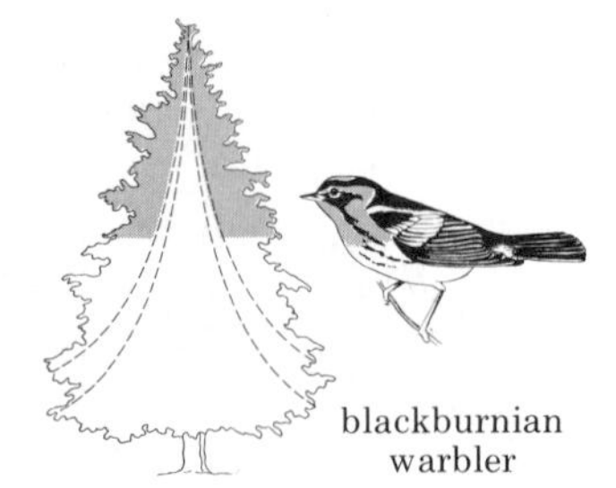

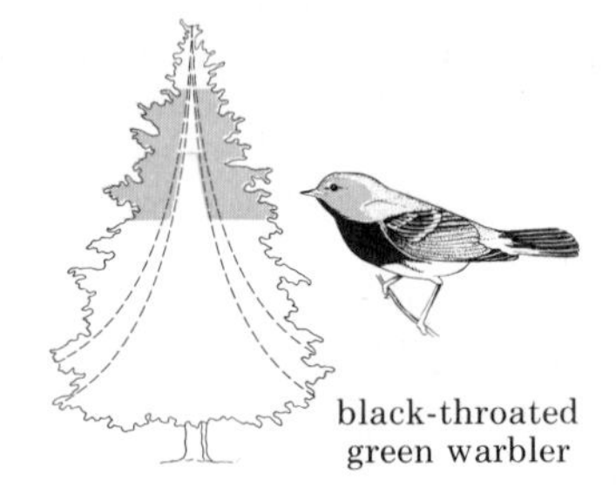

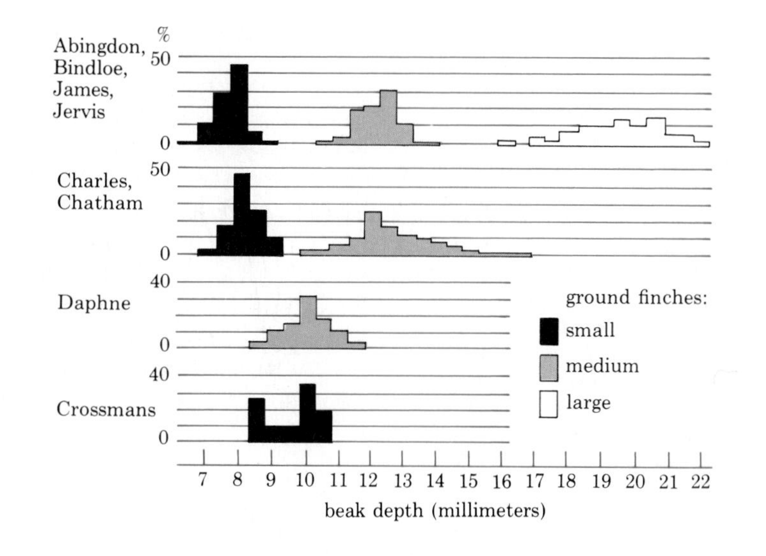

38-4 *Beak sizes in three species of ground finch found on the Galapagos Islands. Beak measurements are plotted horizontally, and the percentage of specimens of each species is shown vertically. Daphne and Crossmans, which are very small islands, each have only one species of ground finch. These species have beak sizes halfway between those of the medium-sized and small finches on the larger islands.*

clear-cut differences in beak size. On Charles and Chatham Islands, the large species is not found, and the beak sizes of the medium ground finches found on these islands overlap the beak sizes of the large finches found on Abingdon and Bindloe. Daphne and Crossmans, which are very small islands, each have only one species; Daphne has the medium-sized finch and Crossmans the small finch. These two populations have similar beak sizes, which are intermediate between those of the medium-sized and small finches on the larger islands (Figure 38-4). This phenomenon, in which species that live together in the same environment are more dissimilar than the same species living in separate environments, is known as *character displacement.*

As these examples indicate, it appears that although competing species may eliminate one another in the laboratory, in natural environments, which are far more complex, the pressure of competition may result in simply subdividing the niche; that is, the two species evolve in such a way as to minimize competition between them.

Predation

Predation is the eating of live organisms, including plants by animals, animals by animals, and even, as we saw on page 306, animals by plants. In many—but not all—populations, predation is an important density-dependent factor that strongly influences population size.

The classic example of the effect of predation on population size is that of the lynx and the snowshoe hare. The data (Figure 38-5, page 536) are based on pelts received yearly by the Hudson's Bay Company over almost 100 years. As you can see, there are oscillations in population density that occur about every 10 years. Generally speaking, a rise in the hare population is followed by a rise in the lynx population; the hare population then plummets, and the lynx population follows. Note that other factors also may be involved that were not recorded by the Hudson's Bay Company. For example, the fluctuations in the hare population might have resulted from overcropping of the vegetation by the hares rather than from predation by the lynxes.

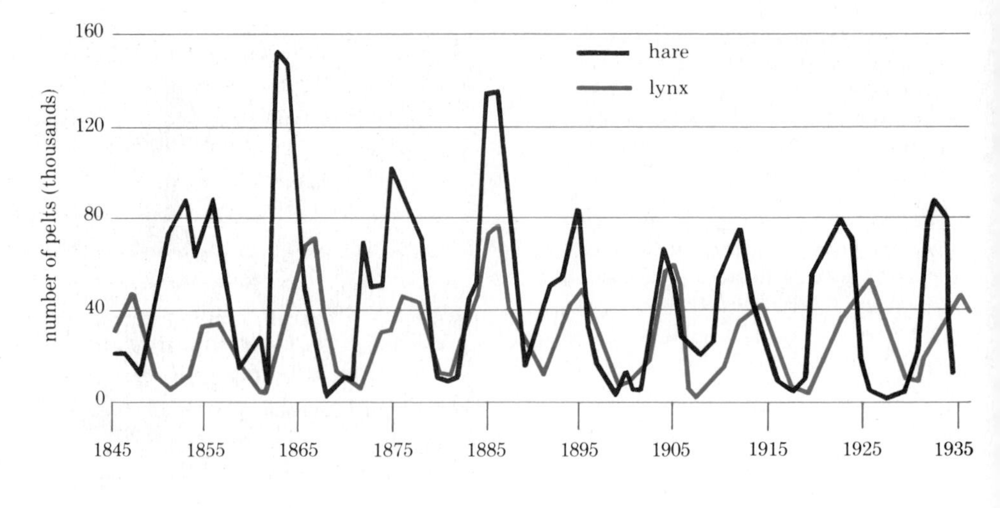

38-5 *The number of lynx and snowshoe hare pelts received yearly by the Hudson's Bay Company over a period of almost 100 years, indicating a pattern of 10-year oscillations in population density. The lynx reaches a population peak every 9 or 10 years, and these peaks are followed in each case by several years of sharp decline. The snowshoe hare follows the same cycle, with a peak abundance generally preceding that of the lynx by a year or more. Presumably a surge in the population of snowshoe hares, the prey, results in a corresponding increase in the predator population, with a resultant decrease in prey followed by a crash in the predator population.*

This type of pattern is characteristic of very simple communities containing a small number of species. In more complex communities, the switching of predators to other prey prevents such extreme oscillations in the population size of both predator and prey.

Many predators have more than one prey species, but they generally feed upon the most common prey species available. When it becomes less abundant, they switch to other, more abundant species so the relative sizes of the prey populations fluctuate. Thus, although oscillations occur in the population size of a prey species, they are usually not as striking as those seen with the lynx and the hare, which live in a community of relatively few species.

Under normal circumstances, predators tend to keep prey well within the carrying capacity of their environment. Moreover, predation, especially on large herbivores, tends to cull animals in poor physical condition. Wolves, for instance, have great difficulty overtaking healthy adult caribou or even healthy calves. A study of Isle Royale, an island in Lake Superior, showed that in some seasons more than 50 percent of the animals the wolves killed had lung disease, although the incidence of such individuals in the population was less than 2 percent. (Human hunters, however, with their superior weapons and their desire for a "prize" specimen, are more likely to injure or destroy strong, healthy animals.)

The Effect of Predation on Competition

Although predators often switch from one prey species to another as the relative sizes of prey populations fluctuate (a behavior that often enables very similar prey species to coexist in the same community), most predators have a preferred type of prey. Such preferences effectively partition resources among the predators, reducing competition.

The feeding habits of predators can also have important effects on the competitive balance among prey. For example, Jane Lubchenco, now of Oregon State University, showed that the herbivorous marine snail *Littorina littorea* controls the abundance and type of algae in the higher intertidal pools on the New England coast. In such pools, the snails' preferred food (the green alga *Enteromorpha*) is competitively superior, and its removal by the snail permits the growth of other algal species (Figure 38-6). However, in areas exposed at high tide, where *Enteromorpha* is competitively inferior, its removal by the snail facilitates the growth of the dominant algal species and reduces the number of species in the area.

38-6 *An example of the effects of predation on competition.* (a) *In this high intertidal pool on the New England coast, the density of the herbivorous marine snail* Littorina littorea *is very low (between one and five individuals per square meter). The competitively superior green alga* Enteromorpha *dominates the pool, excluding other algae.* (b) *In a neighboring pool, less than a meter away, the density of* Littorina *is much higher—more than 250 individuals per square meter. The snails have grazed heavily on* Enteromorpha, *permitting the growth of other, inedible algal species. The aluminum grid at the top of the photograph was used to estimate the density of the snails.*

The introduction of a new species in a community where it has no natural predators may sometimes completely upset existing competitive balances, with disastrous results. When prickly-pear cactus, for instance, was brought to Australia from South America, it escaped from the garden of the gentleman who imported it. It spread into fields and pastureland until more than 12 million hectares were so densely covered with prickly pears that they could support almost no other vegetation (Figure 38-7). The cactus then began to take over the rest of Australia at the rate of about 400,000 hectares a year. It was not brought under control until a natural predator was imported—a South American moth, whose caterpillars live only on the prickly-pear cactus. Now only an occasional cactus and a few moths can be found. (Note, however, that introduction of the moth was a risky experiment.)

38-7 *Prickly-pear cactus on a homestead in Australia. Such rapid and environmentally destructive spread is often seen among alien organisms introduced into a region where they have no natural enemies.*

38–8 *Like many other plants of arid regions, this species of acacia has thorns that protect it against vertebrate herbivores. The graceful gerenuk, however, which has coevolved in the savanna, is, in turn, specially adapted to circumvent these defenses. Acacias and other trees and shrubs provide the gerenuks with water as well as food; these giraffe-necked gazelles never need to drink.*

The Coevolution of Predators and Prey

Predation may exert strong selection pressures on both predator and prey populations. Natural selection favors the most efficient predator. Thus, over the course of generations, the big cats become swifter and more cunning, the necks of the giraffes become progressively longer, the grosbeaks' bills grow thicker and stronger, and eagles become more eagle-eyed. However, natural selection also strongly favors the prey that can escape predation. Thus, as predators have evolved, prey organisms have evolved their own defensive strategies. The coevolution of predators and prey has led to some of the most striking adaptations found in the natural world; an important result of these adaptations is that populations of predators and prey tend to remain in balance.

Natural Defenses in Plants

Plants have many adaptations to deter herbivores. Some natural defenses in plants are structural, such as the sharp-toothed edges of the holly leaf, the thorns of the rosebush, the spines of the cactus, and the stings of the nettle.

Plants also produce a number of chemical substances for which the plant itself has no apparent physiological use and which seem to function as defenses against leaf-eating insects and other predators. Eating foxgloves *(Digitalis purpurea)* can cause convulsive heart attacks in vertebrates. Other plants containing digitalis-like toxins include oleander, of which a single leaf may be fatal to humans, and the members of a large family of plants known as the milkweeds. Some plants contain other chemicals with insect-hormone activities that either arrest development or fatally accelerate it.

Plant defensive chemicals include many substances presently useful to us, among them digitalis, quinine, castor oil, and the tasty ingredients of peppercorns and other spices; some substances of more questionable value, such as nicotine, caffeine, and morphine; and others, such as the active principles in marijuana, mescaline, and peyote, whose desirability for the animal world is a matter of opinion. Plants appear to have been waging chemical warfare long before the coming of *Homo sapiens,* and it is possible that some small leafhopper was the first of all animals to have a psychedelic experience.

Natural Defenses in Animals

Natural defenses in animals include structural adaptations and the capacity to produce noxious chemicals, both of which are seen in plants, and behavioral strategies as well. Some animals are formidably armored—for example, the armadillo, the porcupine, and the sea urchin. Some have ingenious behavioral devices. The armadillo and the pillbug roll themselves up in tight armor-plated balls. Squids and octopuses vanish, jet-propelled, leaving behind only an ink cloud. Many lizards have brightly colored tails that break off when they are attacked and, conspicuous and wriggling, divert the predator from the prey—which is escaping, tailless but with all its vital organs intact.

(a)

(b)

(c)

38-9 (a) *When attacked, a blue-tailed skink leaves its tail to distract predators while it escapes.* (b) *When threatened, porcupines release barbed quills.* (c) *An arrow-poison frog from South America. Skin secretions from these frogs are used by local Indians to tip their arrows with a poison that produces a curare-like paralysis and death.*

Concealment and Camouflage

Hiding is one of the chief means of escape from predators. The young of many birds and of some other vertebrates respond to the warning cries of their parents by "freezing" or by running to cover. Small mammals often have nests, burrows, or makeshift residences in hollow logs or beneath tree roots in which they conceal themselves from their enemies.

Protective coloration is common. Mice, lizards, and arthropods that live on the sand are often light-colored, and such light-colored species, if removed from their home territories, will immediately return to them. Snails that live on mottled backgrounds are often banded (page 495). Grass snakes are grass-colored, as are many of the insects that live among the grasses.

Some animals are countershaded for camouflage. The next time you pass a fish market, look at the specimens laid out on view. Fish are nearly always darker on the top than on the bottom. Countershading reduces the contrast between the shaded and unshaded areas of the body when the sun is shining on the organism from overhead. A fish was once found—the Nile catfish—that was reverse-countershaded; that is, its dorsal surface was light and its ventral surface dark. The selective theory of camouflage was momentarily threatened, but scientific order was restored when it was discovered that the Nile catfish characteristically swims upside down.

Other organisms hide by looking like something else. One type of insect, the treehopper, looks like a thorn; another, the walkingstick, like a twig. Young larvae of some swallowtail butterflies look like bird droppings. In order for such disguises to work, animals must behave appropriately. *Biston betularia,* the peppered moth, lies very flat and motionless on its tree trunk, so that its colors blend with those of the tree (page 501). These and other moths that sit exposed on the bark of trees even habitually orient themselves so that the dark markings on their wings lie parallel to the dark cracks in the bark. Some desert succulent plants look like smooth stones, revealing their vegetable nature only once a year when they flower.

(a)

(b)

(c)

(d)

(e)

(f)

38-10 *Concealment and camouflage.* (a) *Bark katydid;* (b) *stone grasshopper in the desert;* (c) *horned lizard;* (d) *a francolin, an East African bird;* (e) *viper buried in sand;* (f) *a leaflike insect* (Anaea). *The physical adaptation of these animals is dependent, in large part, on a crucial behavioral adaptation. In times of danger, all of these animals remain absolutely still.*

Some insects manage through warning coloration to frighten off their would-be predators. Large spots that look like eyes are commonly found on the backs of butterflies or the bodies of caterpillars, where they will suddenly appear when the insect spreads its wings or arches its body. Small birds reported to flee at the sight of these insects are probably themselves the prey of larger, large-eyed birds, such as owls or hawks. Smaller eyespots, while probably not frightening, seem to have the effect of deflecting the point of attack away from the head. Examination of wounded butterflies has shown that if a part of the wings bears beak marks or is missing, it is most often the part that contains the eyespots. One investigator has tested and confirmed this conclusion by painting eyespots on the wings of living insects, releasing them, and later recapturing them for examination.

On Being Obnoxious

Some animals are protected from predators by a disagreeable taste, odor, or spray, often derived from distasteful chemicals in the plants they eat. Such animals are, generally, carefully avoided. Monarch and other related butterflies feed in their larval stages on milkweeds, and the digitalis-like compounds of the plants are concentrated in their tissues. The caterpillars are immune to these poisons, but birds that eat the caterpillars or adults become violently ill and learn to avoid these butterflies.

Obviously, tasting bad, while useful, is not an ideal defense from the point of view of the individual, since making this fact known may demand a certain amount of personal sacrifice. (Actually, birds drop the monarch butterfly after the first bite, but they often inflict fatal injury in the process.) Obnoxious sprays and odors have the advantage of warding off predators before they harm the prey. To us, the most familiar of such animals is the skunk, which advertises its malodorous threat by its distinctive color-

(a)

(b)

(c)

38-11 (a) *The beetle* Eleodes longicollis *has glands in its abdomen that secrete a foul-smelling liquid. When disturbed,* Eleodes *stands on its head and sprays the liquid at the potential predator.* (b) Eleodes longicollis *is on the right. On the left is* Megasida obliterata, *which, as you can see, resembles* Eleodes *and emphasizes this resemblance by also standing on its head. However,* Megasida *has no similar glands, no noxious secretion, and no spray.* (c) *A grasshopper mouse that has solved the problem of how to eat* Eleodes. *The mouse drives the posterior of the beetle into the ground and eats it head first. It would probably eat* Megasida *the same way.*

ation. Many insects and other arthropods have developed defensive secretions. In some millipedes, for example, the secretion oozes out of a gland onto the surface of the animal's body. Other arthropods are able to spray their secretion over a distance, in some instances even aiming it precisely at the attacker. Many arthropods possess a number of glands but discharge only from those closest to the point of attack, thus saving ammunition and gaining efficiency. The secretions usually act as topical irritants, especially to the mouth, nose, and eyes of the predator. Because of their relatively permeable skin, frogs and toads, common predators of arthropods, are sensitive to these irritants over their entire bodies.

Müllerian Mimicry: Advertising

For animals that have a highly effective protective device, such as a sting, a revolting smell, or a poisonous or bad-tasting secretion, it is advantageous to advertise. The more inconspicuous or rare such an animal is, the larger the number of individuals that must be sacrificed before the bird or other predator learns to avoid it. Müllerian mimics, named after F. Müller, who first described the phenomenon, are groups of organisms that, although not closely related, all have effective obnoxious defenses and all resemble one another. Bees, wasps, and hornets probably offer the most familiar example; even if we cannot tell which is which, we recognize them immediately as stinging insects and keep a respectful distance. Similarly, large numbers of bad-tasting butterflies are look-alikes. Müllerian mimicry is adaptive for all the individuals involved because each prospers from a predator's experience with another.

Batesian Mimicry: Deception

Batesian mimicry, first described by the British naturalist H. W. Bates in 1862, is deceptive mimicry. In Batesian mimicry, the innocuous mimic fools its predator by resembling a stinging or bad-tasting model that the predator has learned to avoid. Some species of stingless flies resemble bees or hornets, and many species of butterflies resemble monarchs or other unpalatable butterflies or moths.

Laboratory experiments have clearly demonstrated Batesian mimicry in operation. Jane Brower, working at Oxford, made artificial models by dipping mealworms in a solution of quinine, to give them a bitter taste, and then marking each one with a band of green cellulose paint. Other meal-

38-12 *Müllerian mimics:* (a) *a yellowjacket,* (b) *a sand wasp, and* (c) *a masarid wasp. All share the same warning coloration, and all sting.*

(a)

(b)

(c)

38-13 *A monarch butterfly (below) and one of its mimics, a viceroy butterfly. A Batesian mimic has been compared to an unscrupulous retailer who copies the advertisement of a successful firm. Müllerian mimics, by contrast, are reputable tradesmen who share a common advertisement and divide its costs. Viceroys are Batesian.*

worms, which had first been dipped in distilled water, were painted green like the models, so as to produce mimics, and still others were painted orange to indicate another species. These colors were chosen deliberately: Orange is a warning color, since it is clearly distinguishable, and green is usually found in species that are not repellent and for whom, therefore, advertising would be unwise.

The painted mealworms were fed to caged starlings, which ordinarily eat mealworms voraciously. Each of the nine birds tested received models and mimics in varying proportions. After initial tasting and violent rejection, the models were generally recognized by their appearance and avoided. In consequence, their mimics were protected also. Even when mimics made up as much as 60 percent of the green-banded worms, 80 percent of the mimics escaped the predators.

Batesian mimicry obviously works to the advantage only of the mimic. The model, on the other hand, suffers from attacks not only by inexperienced predators but also by predators who have had their first experience with mimic rather than with model. The mimetic pattern will be at its greatest advantage if the mimic is rare—that is, less likely to be encountered than the model—and also if the mimic makes its seasonal appearance after the model, thus reducing its chances of being encountered first. However, if the model is sufficiently distasteful—as in the case of the quinine-soaked mealworms—it may protect mimics even if the latter are very common.

Symbiosis

The competitive and predator-prey interactions within communities are, as we have seen, important in regulating population size, in determining the particular niche that an organism occupies, and in shaping the evolutionary history of species. Of equal importance are other, more subtle interactions.

Symbiosis ("living together") is a close and permanent association between organisms of different species. There are several types of symbiotic relationships. If the relationship is beneficial to both species, it is called *mutualism*. If one species benefits from the association while the other is neither harmed nor benefited, it is called *commensalism*. If one species benefits and the other is harmed, the relationship is known as *parasitism*. However, since all the details of symbiotic relationships are not understood, these distinctions are not always useful. We shall focus our attention on a few striking examples of mutualism and parasitism.

Mutualism

The examples of mutualism in nature are many and varied (Figure 38-14). A classic example is provided by the lichens, which are part alga, part fungus. The body of the lichen is composed largely of fungal mycelium; held within this mycelium are numerous photosynthetic algal cells. The two organisms together form a closely integrated unit that can grow under conditions where neither the fungus nor the alga alone could survive. Lichens occur from arid desert regions to the Arctic; they grow on bare soil, tree trunks, sunbaked rocks, and windswept alpine peaks all over the world. They are often the first colonists of bare rocky areas.

38-14 *Mutualism.* (a) *Sea anemones on the back of a snail shell occupied by a hermit crab. The anemone protects and camouflages the crab and, in turn, gains mobility—and so a wider feeding range—from its association with the crab. Hermit crabs, which periodically move into new, larger shells, will coax their anemones to move with them.*

(b) *Cleaner fish are permitted to approach larger fish with impunity because they feed off the algae, fungi, and other microorganisms on the fish's body. The fish recognize the cleaners by their distinctive markings. Other species of fish, by closely resembling the cleaners, are able to get close enough to the large fish to remove large bites of flesh. What would probably happen if cleaner mimics began to outnumber cleaners?*

(c) *Aphids suck phloem, removing certain amino acids, sugars, and other nutrients from it and excreting most of it as "honeydew," or "sugar-lerp," as it is called in Australia where it is harvested as food by the aborigines. Some species of aphids have been domesticated by some species of ants. These aphids do not excrete their honeydew at random, but only in response to caressing movements of the ant's antennae and forelimbs. The aphids involved in this symbiotic association have lost all their own natural defenses, including even their hard outer skeletons, relying upon their hosts for protection.*

(d) *Oxpeckers live on the ticks they remove from their hosts. An oxpecker forms an association with one particular animal, such as the wart hog shown here, conducting most of its activities, including courtship and mating, on the back of its host.*

(a)

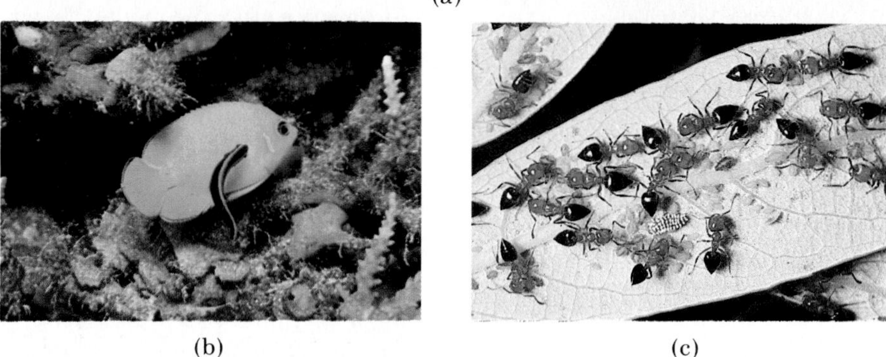

(b) (c)

(d)

(a)

(b)

(c)

(d)

38-15 *Ants and acacias.* (a) *The beginning. A queen ant cuts an entrance into a thorn on a seedling bull's-horn acacia. She will hollow out the thorn and raise her first brood inside it.* (b) *The tip of an acacia leaf. The orange structures at the tips of the leaflets are Beltian bodies. They are a source of food for the ants.* (c) *A worker ant drinking from the nectary.* (d) *Warriors in a battle for possession of an acacia. Such battles occur when the branches of acacias grow to touch each other. The largest colony usually wins.*

Ants and Acacias

Trees and shrubs of the genus *Acacia* grow throughout the tropical and subtropical regions of the world. In Africa and tropical America, acacia species are protected by thorns (Figure 38-8). The acacias of Australia, where there are few large grazing animals except recently introduced species, have no thorns.

On one of the African species of *Acacia,* ants of the genus *Crematogaster* gnaw entrance holes in the walls of the thorns and live permanently in them. Each colony of ants inhabits the thorns on one or more trees. The ants obtain food from nectar-secreting glands on the leaves of the acacias and eat caterpillars and other herbivores that they find on the trees. Both the ants and the acacias appear to benefit from this association.

In the lowlands of Mexico and Central America, the ant-acacia relationship has been extended to even greater lengths. The bull's-horn acacia, a common plant of that area, is found particularly frequently in cutover or disturbed areas, where the competition for light is intense. It grows extremely rapidly. This species of acacia has a pair of greatly swollen thorns several centimeters in length at the base of most leaves. The petioles bear nectaries, and at the very tip of each leaflet is a small structure rich in oils and proteins known as a Beltian body. Thomas Belt, the naturalist who first described these bodies, noted that their only apparent function was to nourish the ants. Ants live in the thorns, obtain sugars from the nectaries, eat the Beltian bodies, and feed them to their larvae.

Worker ants, which swarm over the surface of the plant, are very aggressive toward other insects and, indeed, toward animals of all sizes. For example, they become alert at the mere smell of a man or a cow. When their tree is brushed by an animal, they swarm out and attack at once, inflicting painful, burning stings. Moreover, and even more surprisingly, alien plants sprouting within as much as a meter of occupied acacias are chewed and mauled, and twigs and branches of other trees that touch an occupied acacia are similarly destroyed. Not so surprisingly, acacias inhabited by these ants grow very rapidly, soon overtopping other vegetation.

Daniel Janzen, who first analyzed the ant-acacia relationship in detail, removed ants from acacias artificially, by insecticides or by removing thorns or entire occupied branches. Acacias without their ants grew slowly and usually suffered severe damage from insect herbivores. Their stunted bodies were soon overshadowed by competing species of plants and vines. As for the ants, according to Janzen, these particular species can live only on acacias.

Parasitism

Parasitism is regarded by many ecologists as a special form of predation in which the predator is considerably smaller than the prey. The plants and animals in a natural community support hundreds of parasites of many species—in fact, perhaps millions, if one were to count viruses.

As with more obvious forms of predation, parasitic diseases are most likely to wipe out the very young, the very old, and the disabled—either directly or, more often, indirectly, by making them more susceptible to other predators or to the effects of climate or food shortages. It is logical that a parasite-caused disease should not be too virulent or too efficient. If a parasite were to kill all the hosts for which it is adapted, it too would perish. This principle is particularly well illustrated by a series of further misadventures on the continent of Australia.

There were no rabbits in Australia until 1859, when an English gentleman imported a dozen from Europe to grace his estate. Six years later he had killed a total of 20,000 on his own property and estimated he had 10,000 remaining. In 1887, in New South Wales alone, Australians killed 20 million rabbits. By 1950, Australia was being stripped of its vegetation by giant hordes of rabbits. In that year, rabbits infected with myxoma virus were released on the continent. (Myxoma virus causes only a mild disease in South American rabbits, its normal hosts, but is usually fatal to the European rabbit.) At first, the effects were spectacular, and the rabbit population steadily declined, yielding a share of pastureland once more to the sheep herds, on which much of the economy of the country depends. But then occasional rabbits began to survive, and their litters also showed resistance to the myxoma virus.

38-16 *Rabbits crowd a water hole in Australia. Imported from Europe as potentially valuable herbivores, they soon overran the countryside. The myxoma virus was introduced to control the rabbits, but now host and parasite are coexisting.*

A double process of selection had taken place. The original virus was so rapidly fatal that often a rabbit died before it could be bitten by a mosquito and thereby infect another rabbit; the virus strain then died with the rabbit. Strains less drastic in their effects, on the other hand, had a better chance of survival since they had a greater opportunity to spread to a new host. (After an initial infection, a rabbit is immune to the virus, just as we usually become immune to mumps or measles after one infection.) So, first, selection began to work in favor of a less virulent strain of myxoma virus. Almost simultaneously, rabbits that were resistant to the original virus began to appear. Now, as a result of coevolution, the virus and the rabbits are reaching an equilibrium, like most parasites and hosts.

Community Structure

In this chapter, we have examined the major types of interactions that occur among populations. Our examples, however, were limited to interactions involving only two or three species. Although some communities, such as those found in Arctic regions, contain few species, most com-

Equilibrium Disturbed: The Black Death

Epidemics of infectious disease occur when the equilibrium between a parasitic species and its host organism is nonexistent or is severely disturbed. The epidemic that swept Australian rabbits when myxoma virus was introduced occurred because there was no equilibrium between the two; as equilibrium was gradually established through coevolution, the incidence and severity of disease diminished greatly. The best-known epidemics in human populations—the outbreaks of bubonic plague in Europe in the fourteenth and seventeenth centuries, known as the Black Death—occurred when an existing equilibrium was upset.

The bacterium that causes bubonic plague is a parasite of rodents, such as rats and squirrels. In many natural populations the bacterium and its rodent hosts coexist in a stable equilibrium that also includes the fleas that live on the rodents. The fleas, particularly those of the rat, are capable of spreading the bacterium to humans. But, they, like most other parasites, have their preferred hosts—in this case, rats. Only when a flea population becomes so great that it exceeds the carrying capacity of its environment do the fleas emigrate to a new environment, humans, carrying the bacterium with them. Similarly, rats invade human habitations searching for food only when their numbers have become so great that the environment elsewhere cannot support them. As long as the rat population does not exceed the limits of its environment, rats do not generally come close enough to humans that fleas can get from one to the other.

Throughout most of history, rat populations have remained in balance with their environment, flea populations in balance with theirs, and bubonic plague bacteria in equilibrium with both fleas and rats. On a few occasions, however, the balances were upset, and great hordes of rats, bearing large populations of fleas, invaded cities and towns. This happened several times in Europe between 1348 and 1382, in London between 1603 and 1665, and, on a much smaller scale, in India between 1953 and 1959. In each case, the initial consequences to the human population were disastrous.

In the fourteenth century, the outbreak of the plague began in 1346 during the siege of Caffa, a small military post on the Crimean Straits. From there, it spread to Italy and the south of France, and by 1348 it had reached all of Europe. The disease did not disappear entirely until 1357, and then it reappeared three more times during the century. The incidence and mortality during the four outbreaks were as follows:

Year	Percent of Population Afflicted	Resultant Deaths
1348	67	Almost all
1361	50	Almost all
1371	10	Some
1382	5	Almost none

As these data reveal, the course of the disease was similar to that observed with myxoma virus and rabbits. Initially, a large proportion of the population was afflicted and the disease was so virulent that virtually everyone stricken died. As time passed, most susceptible individuals had already been killed, those still alive were either resistant or had developed immunity, and the bacterium became less virulent.

When the plague struck London in the seventeenth century, the same pattern was not observed. The outbreaks occurred in 1603, 1625, and 1665. In the first two epidemics approximately 13 percent of the population died, and in the third the mortality increased to 15 percent of the population. The explanation is found in the intervals between epidemics—22 years between the first two and 40 years between the second and third. Two things occurred in these long intervals: First, a significant proportion of the population lost the immunity it had acquired previously, and, second, a large proportion of the population was born after the preceding epidemic and had never acquired immunity to the bacterium. As a result, the disease could once more spread rapidly through the population with a high rate of mortality.

J. B. S. Haldane once remarked that disease probably has been the major agent of natural selection operating on human populations since they began to settle in dense concentrations about 10,000 years ago. When one considers that 67 percent of the population of Europe died from the Black Death in 1348 and that a mere 13 years later 50 percent of the population died from the same cause, the force of Haldane's remark becomes apparent.

A plague hospital in Vienna during a seventeenth-century epidemic.

munities are composed of a multitude of species, simultaneously interacting with each other. Even a brief reflection on one's own observations of nature indicates that these simultaneous interactions are anything but chaotic. A community, such as a meadow or a pond or a forest, has a recognizable structure that persists in time and yet can change in time. Moreover, the characteristics of a community can be correlated with its physical environment—for example, a community found in an area with plentiful rainfall is clearly different from one found in an arid region. In the next two chapters, we shall consider the structure of communities and their distribution on the earth's surface.

SUMMARY

Populations live as part of a community—a group of different organisms inhabiting a common environment and interacting with one another. These interactions involve competition, predation, and symbioses, all of which shape and affect the characteristics of the organisms that make up the community and the relative sizes of the different populations.

The ecological niche is a multidimensional description of an organism's way of life. It is determined by physical factors and also biological ones, including competition from other organisms. When niches overlap so that different populations are competing for the same resources, one population may be eliminated, or one or both populations may evolve in such a way as to reduce or eliminate the competition. The principal observable effect of competition is a subdivision of the resources of the community among the populations present.

Predation is a powerful influence on the size of many populations, often keeping the prey population well below the carrying capacity of the environment. Predators generally feed on the most common prey species available; when a prey population decreases in size, predation pressure relaxes and the population usually has a chance to recover. The feeding preferences of different predators not only reduce the competition among predators but also affect the competition among prey populations.

The interactions of predators and prey have profound evolutionary effects on the species involved. Plants and animals have developed a variety of defenses against predation. These include forms of structural protection, as seen in cacti, armadillos, and porcupines, behavioral strategies, and chemical weapons, such as plant poisons and the noxious secretions of insects. Many organisms are camouflaged. Some insects have come to resemble organisms of other species either to advertise an effective protective device that they possess in common (Müllerian mimicry) or to "pretend" they possess such a device when they actually do not (Batesian mimicry).

Symbiosis is a close and permanent association between organisms of different species. The association may be beneficial to both (mutualism), beneficial to one and harmless to the other (commensalism), or beneficial to one and harmful to the other (parasitism). As a result of coevolution, most parasites and hosts remain in equilibrium with each other.

QUESTIONS

1. Distinguish among the following terms: habitat/niche; competition/predation; Müllerian mimicry/Batesian mimicry; mutualism/commensalism/parasitism.

2. Compare the results of MacArthur's experiment with the warblers with Connell's experiment with barnacles. What step did Connell perform that MacArthur did not? Why is the step important?

3. Introducing a new species into a community can have a number of possible effects. Name some of these possible consequences both to the community and to the species. What types of studies should be made before the importing of an "alien" organism? Some states and many countries have laws restricting such importations. Has your own state adopted any such laws? Are they, in your opinion, ecologically sound?

4. In the American southwest, grasses and mesquite compete with each other for dominance. Mesquite, however, was rare before cattle were introduced into the western United States. How have cattle affected the competition between the two types of plant? Suppose all cattle were removed from a large area. What change would you predict in the competitive balance between grasses and mesquite?

5. In the opinion of some ecologists, animals that eat seeds, such as Galapagos ground finches, should be regarded as predators, while animals that eat leaves, such as deer, should be regarded as parasites. Justify this classification of herbivores as either predators or parasites.

39

The Organization of Communities and Ecosystems

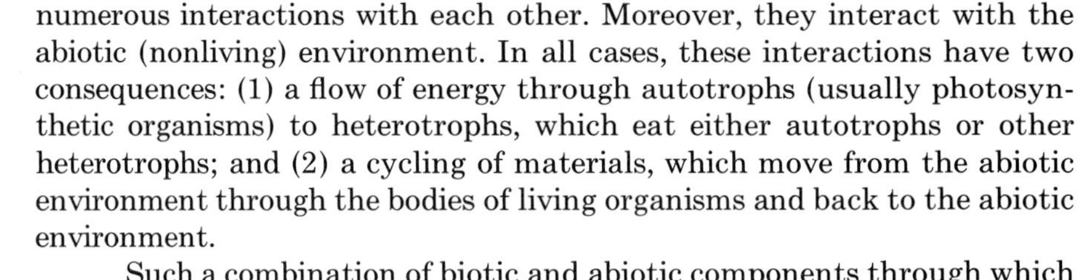

As we saw in the last chapter, the populations within a community have numerous interactions with each other. Moreover, they interact with the abiotic (nonliving) environment. In all cases, these interactions have two consequences: (1) a flow of energy through autotrophs (usually photosynthetic organisms) to heterotrophs, which eat either autotrophs or other heterotrophs; and (2) a cycling of materials, which move from the abiotic environment through the bodies of living organisms and back to the abiotic environment.

Such a combination of biotic and abiotic components through which energy flows and materials recycle is known as an *ecosystem*. Taking a large, astronautical view, the entire surface of the earth can be seen as a single ecosystem. This view is useful when studying materials that are circulated on a worldwide basis, such as carbon dioxide, oxygen, and water. A suitably stocked aquarium or terrarium is also an ecosystem, and such models may be useful in studying certain ecological problems, such as the details of transfer of a particular mineral element. Most studies of ecosystems and their component communities have been made, however, on more or less self-contained natural units—on a pond, for example, or a swamp, or a meadow.

39-1 *The energy on which life depends enters the living world in the form of light. It is converted to chemical energy by photosynthetic organisms, such as the wild barley shown here. Such photosynthetic organisms are, in turn, the energy source for all heterotrophs, including ourselves.*

The Flow of Energy

The flow of energy through ecosystems is the most important factor in the organization of biological communities. The ultimate source of energy is, of course, the sun. Every day, year in and year out, energy from the sun arrives at the upper surface of the earth's atmosphere at an average rate of 1.94 calories per square centimeter per minute, a total of about 10^{24} calories per year. This is known as the solar constant, and although it is only a tiny fraction of the total energy radiated by the sun, it is a tremendous quantity of energy. Of this energy, about 30 percent is reflected back into space from clouds and dust. Another 20 percent is absorbed in the various layers of the earth's atmosphere.

The remaining 50 percent of the incoming solar energy reaches the earth's surface. A small amount of this is reflected from bright areas, but most is absorbed. Energy absorbed by the oceans warms the surface of the water, evaporating water molecules and powering the water cycle (page 32). Energy absorbed by the ground is reradiated in the form of heat.

Of the solar energy that reaches the earth's surface, only a very small fraction is diverted into living systems—an estimated 1/10 of 1 percent on a worldwide basis. Even when the light falls where vegetation is abundant, as in a forest, a cornfield, or a marsh, only 1 or 2 percent of that light (calculated on an annual basis) is used in photosynthesis. Yet this fraction, as small as it is, may result in the production—from carbon dioxide, water, and a few minerals—of several thousand grams (dry weight) of organic matter per year in a single square meter of field or forest, a total of about 90 billion metric tons of such organic matter per year on a worldwide basis.

Trophic Levels

The passage of energy from one organism to another takes place along a particular food chain—that is, a sequence of organisms related to one another as prey and predator. The first is eaten by the second, the second by the third, and so on, in a series of *trophic levels*, or feeding levels. In most communities, food chains are linked together in complex *food webs*, with many branches and lateral connections and involving many different species (Figure 39–2). An organism's relative position in the food web is one of the principal distinguishing characteristics of its ecological niche.

Producers

The first trophic level of a food web is always occupied by a primary producer. On land, the primary producer is usually a plant; in aquatic ecosystems, it is usually a photosynthetic alga. These photosynthetic organisms use light energy to make carbohydrates and other compounds, which then become sources of chemical energy. Producers far outweigh consumers; 99 percent of all the organic matter in the living world is made up of plants and algae. All heterotrophs account for only 1 percent of the organic matter.

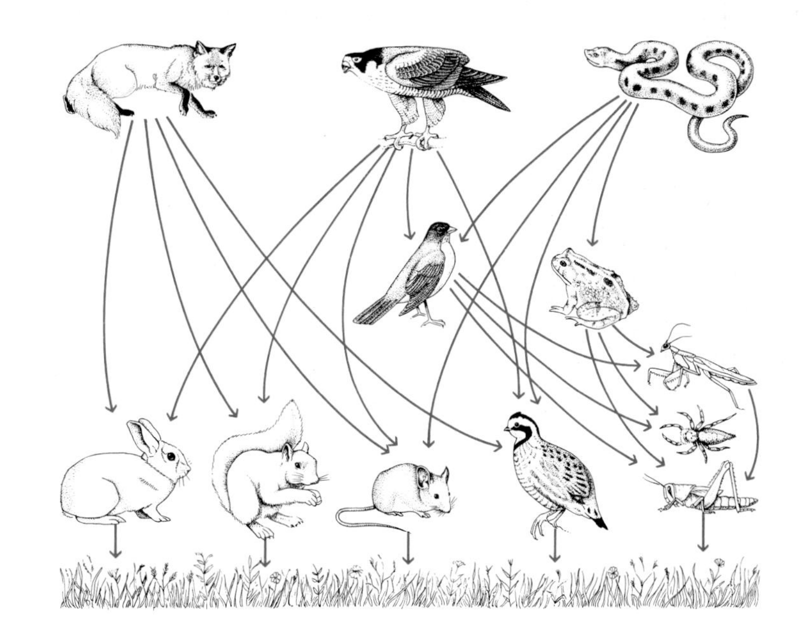

39–2 *Diagram of a food web. The arrows point from consumers to their energy source. This food web is much simplified; in reality, many more animals and plants would be involved.*

A Chemosynthetic Ecosystem

Throughout this text, we have been informing you that life on earth depends on radiant energy from the sun, which is converted by plants or algae to chemical energy by the process of photosynthesis. Not so! In 1977, a new type of ecosystem was discovered based on chemosynthesis.

Its discoverer was Alvin, the little research submarine from Woods Hole Oceanographic Institution, and its crew of scientific investigators, led by John B. Corliss of Oregon State University. The place was the Galapagos Rift, a boundary between adjacent plates (see page 508). Sea water seeps into the fissures in the volcanic rocks erupting along this boundary, becomes heated, and rises again. As a consequence, oases of warmth are created in the near-freezing waters 2.5 kilometers below the surface. More important, the water reacts with the rocks deep within the crust. Here, under high heat and pressure (300°C, 280 kilograms per square centimeter), chemical reactions take place. The crucial one, according to early evidence, is the conversion of sulfate in the sea water to hydrogen sulfide. Chemosynthetic bacteria use the hydrogen sulfide as an energy source, becoming the primary producers in a complex food web.

Five such oases have now been located, each with a somewhat different animal community. (According to the investigators, the types of animals that are present are probably the result of whatever floating larvae colonized the area during its early stages of development.) Among the most spectacular inhabitants are clams, measuring some 20 centimeters in diameter, which cover the lava floor. Other permanent inhabitants include crabs, mussels, and octopods. One oasis, known as the Garden of Eden, is dominated by huge tube worms. The unusual size of the organisms found at these warm springs of the rift is attributed to the high productivity of the chemosynthetic bacteria. Thus, although most life on earth is, indeed, based on radiant energy from the sun, there are clearly highly viable alternatives not only on this planet, but perhaps on others as well.

Deep in the Pacific Ocean, at the Garden of Eden oasis in the Galapagos Rift, giant tube worms flourish in the warmth of a sea-floor spring. Cancroid crabs hide in the crevices of the limpet-encrusted lava, while a pink brotulid fish swims by. Compared to the near-freezing temperatures normally found at these depths, it is 17°C (63°F).

Ecologists speak of the productivity of a trophic level, a community, or an ecosystem. *Gross productivity* is a measure of the rate at which energy is assimilated by the organisms in, for example, a particular trophic level. A more useful—and often more easily measured—quantity is the *net productivity*. It is defined as the amount of energy (measured in calories) stored in chemical compounds or as the increase in biomass (measured in grams or metric tons) in a particular period of time. (*Biomass* is a convenient shorthand term meaning the weight of all the living organisms in a given area.) Net productivity is the difference between the gross productivity and the energy used by the organisms in respiration. It is a measure of the rate at which organisms store energy, which then becomes available to the organisms of the next trophic level. In agricultural communities, the standing crop at the end of the growing season represents the *net primary production* for that season (Figure 39-3).

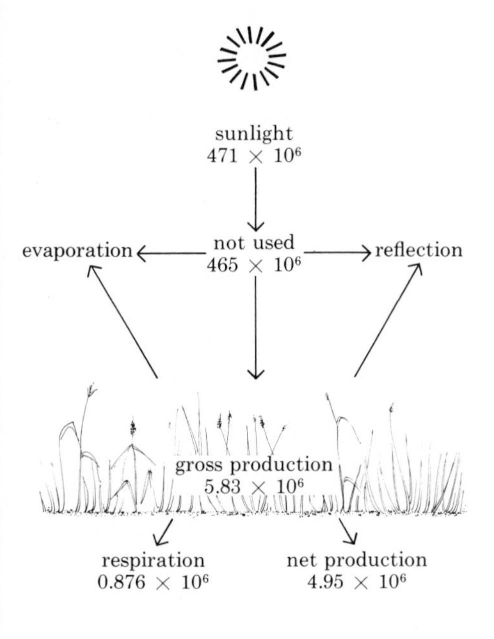

39-3 *Calculation of the productivity of a field in Michigan in which the vegetation was mostly perennial grasses and herbs. Measurements are in terms of calories per square meter per year. In this field, the net primary production—the amount of chemical energy stored in plant material—was 4,950,000 calories per square meter per year. Thus, slightly more than 1 percent of the 471,000,000 calories per square meter per year of sunlight reaching the field was converted to chemical energy and stored in the plant bodies.*

Consumers

Energy enters the animal world through the activity of herbivores, animals that eat plants or algae. An herbivore may be a caterpillar, an elephant, a sea urchin, a snail, or a field mouse; each type of community has its characteristic complement of herbivores. Of the organic material consumed by herbivores, much is eliminated undigested. Most of the chemical energy from digested food is used to maintain the metabolic processes of the animal and to power its daily activities: searching for food, eating and digesting it, mating and caring for the young, fleeing predators, and so on. Although this energy is generally described as "lost" through respiration, it is important to realize that for the individual organism this is the essential energy on which its life depends. A fraction of the chemical energy consumed by the herbivore is converted to new animal biomass. The increase in animal biomass is the sum of the increase in weight of individual animals plus the weight of new offspring. It represents energy available to the next trophic level.

(a)

(b)

39-4 *Representative consumers.* (a) *A whitetail buck, which is an herbivore, feeding on sumac.* (b) *This red-tailed hawk, a carnivore, has killed a prairie dog. Most hawks eat warmblooded vertebrates, such as small mammals and birds, but some prefer fish, amphibians, or reptiles.*

39-5 *The fate of food energy ingested by a carnivore. Only a fraction—between 2 and 15 percent—is converted to animal biomass, available for consumption by an organism at the next trophic level.*

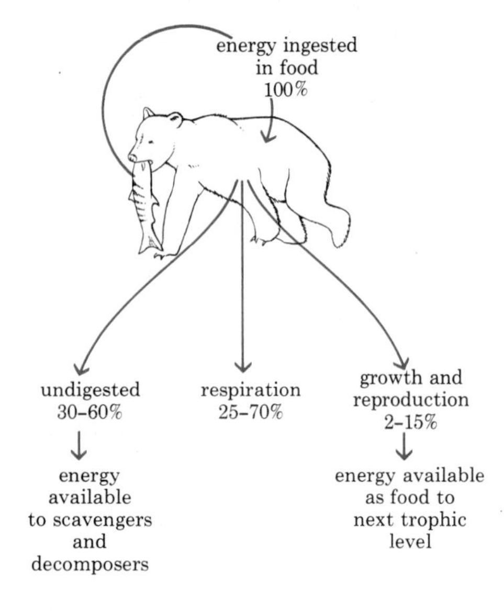

The next level in the food web, the secondary consumer level, involves a carnivore, an animal that eats other animals. The carnivore that devours the herbivore may be a lion, a minnow, a starfish, a robin, or a spider. In every case, only a small part of the organic substance present in the body of the herbivore becomes incorporated into the body of the carnivore (Figure 39-5).

Some food webs have third and fourth consumer levels, but five links are usually the limit. With each higher trophic level, there is a decrease in the total amount of energy stored in animal biomass and therefore available to other consumers.

Detritivores

Detritivores are organisms that live on the refuse, or detritus, of a community—dead leaves, branches, and tree trunks, the roots of annual plants, feces, carcasses, even the discarded exoskeletons of insects. They include large scavengers, such as vultures, jackals, and crabs, as well as smaller organisms, such as earthworms, fungi, and bacteria. Animal detritivores can be regarded as consumers that prefer dead prey to live prey. The detritivores that are commonly referred to as decomposers—fungi and bacteria—are also consumers, but with a difference: They have evolved specializations that enable them to exploit sources of chemical energy, such as cellulose and nitrogenous waste products, that cannot be used by animals.

In a forest community, more than 90 percent of the net primary production is consumed by detritivores rather than by herbivores. Some of this energy flows through the food web by way of consumers that feed on detritivores, while the rest of it is used in the metabolic processes of the detritivores themselves. As a result, essentially all of the energy stored by plants (and, in aquatic communities, by algae) is ultimately used to support life. Energy stored in organic matter goes unused only when it is trapped in an environment in which detritivores cannot live.

Ecological Pyramids

The flow of energy through a food web is often represented by a diagram of quantitative relationships among the various trophic levels. Because large amounts of energy and biomass are dissipated at every trophic level, with each level retaining a much smaller amount than the preceding level, these diagrams nearly always take the form of pyramids. An *ecological pyramid*, as such a diagram is called, may be (1) a pyramid of numbers showing the numbers of individual organisms at each level; (2) a pyramid of biomass, based either on the total weight of the organisms at each level or on the total number of calories at each level; or (3) a pyramid of energy flow showing the productivity of the different trophic levels.

39-6 *Pyramids of numbers for* (a) *a grassland ecosystem, in which the number of primary producers (grass plants) is large, and* (b) *a temperate forest, in which a single primary producer, a tree, can support a large number of herbivores.*

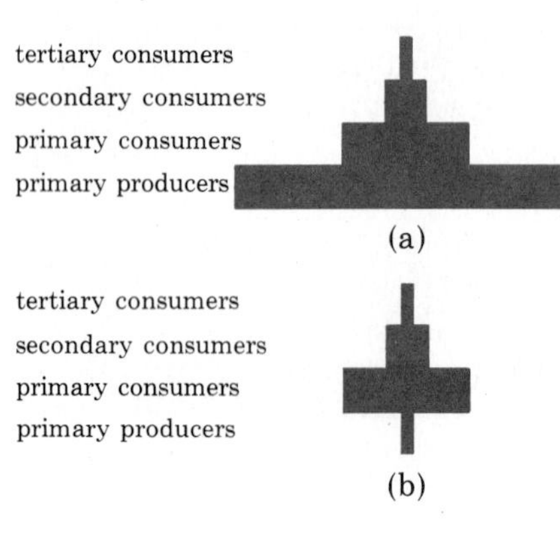

Pyramids of Numbers

The shape of any particular pyramid tells a great deal about the ecosystem it represents. Figure 39-6a, for example, shows a pyramid of numbers for a grassland ecosystem. In this type of ecosystem, the primary producers (the individual grass plants) are small, and so a large quantity of them is required to support the primary consumers (the herbivores). In an ecosystem in which the primary producers are large (for instance, trees), an individual primary producer may support many herbivores, as indicated in Figure 39-6b.

39-7 *Pyramids of biomass for* (a) *a field in Georgia and* (b) *the English Channel. Such pyramids reflect the mass present at any one time, hence the seemingly paradoxical relationship between phytoplankton and zooplankton.*

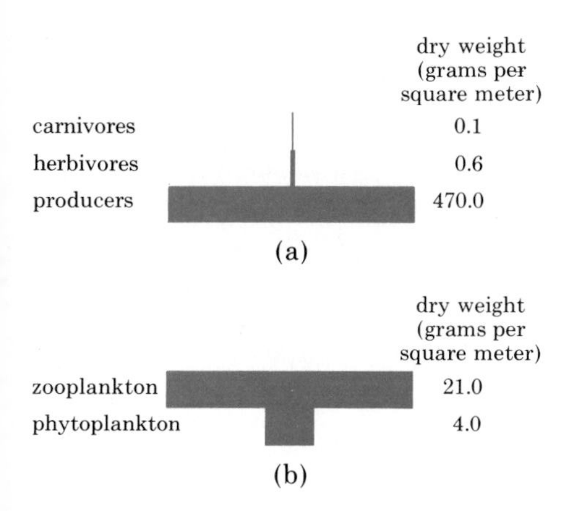

Pyramids of Biomass

A pyramid of biomass for a grassland ecosystem, like the pyramid of numbers for that system, takes the form of an upright pyramid, as shown in Figure 39-7a. Pyramids of biomass are inverted only when the producers have very high reproduction rates. For example, in the ocean, the standing crop of phytoplankton (primarily microscopic algae) may be smaller than the biomass of the zooplankton (animal plankton) that feeds upon it (Figure 39-7b). Because the growth rate of the phytoplankton population is much higher than that of the zooplankton population, a small biomass of phytoplankton can supply food for a larger biomass of zooplankton. Like pyramids of numbers, pyramids of biomass indicate only the quantity of organic material present at one time; they do not give the total amount of material produced or, as do pyramids of energy, the rate at which it is produced.

39-8 *Pyramid of energy flow for a river system in Florida. This type of pyramid, regardless of the ecosystem under analysis, is never inverted.*

net production
respiration
net prod. resp.
secondary carnivore 15 6
primary carnivore 316 67
herbivore 945 1,478 945
producer 5,988 8,833 5,988
10 8 6 4 2 0 2 4 6 8 10
energy flow ($\times 10^3$ kcal/cm^2/year)

Pyramids of Energy

Pyramids of energy, in keeping with the second law of thermodynamics (page 90), have the same basic shape for every ecosystem (Figure 39-8). In general, of the total amount of energy in the biomass that passes from one trophic level to another in a food chain, only about 10 percent is stored in body tissue. Approximately 90 percent of the total calories is either unassimilated or "burned" in respiration by the animals at that level. Hence, the tertiary consumers, the carnivores that eat the carnivores that eat the herbivores, are reduced to approximately $1/10 \times 1/10 \times 1/10 = 1/1{,}000$ the energy stored in the plants that are eaten. Supercarnivores, which eat these tertiary consumers, are reduced to one-tenth of this, or 1/10,000 the energy in the plant material. Individuals at the top of the ecological pyramid may have to be quite large, as individuals, to capture and eat other animals, but even if they are larger as individuals, they are usually smaller in total number and total biomass. And they are always lower in total captured energy than the organisms they eat.

This is the basis of the argument for living lower on the food chain. Beef, for instance, is energetically expensive. In the case of steer, only 10 percent of the energy present in the grasses or grain that fed the steer is converted to animal biomass, and even less to edible meat. Consequently, in terms of energy, those who eat meat are consuming 10 times as much of the world's energy resources as those who eat grains and other plant products.

Biogeochemical Cycles

Energy takes a one-way course through an ecosystem, but many substances cycle through the system. Such substances include water, nitrogen, carbon, phosphorus, potassium, sulfur, magnesium, calcium, sodium, chlorine, and also a number of other minerals, such as iron and cobalt, that are required by living systems in only very small amounts. The water cycle is shown on page 33, and the carbon cycle is on page 116.

Movements of inorganic substances are referred to as *biogeochemical cycles* because they involve geological as well as biological components of the ecosystem. The geological components are (1) the atmosphere, which is made up largely of gases, including water vapor; (2) the solid crust of the earth; and (3) the oceans, lakes, and rivers, which cover three-fourths of the earth's surface.

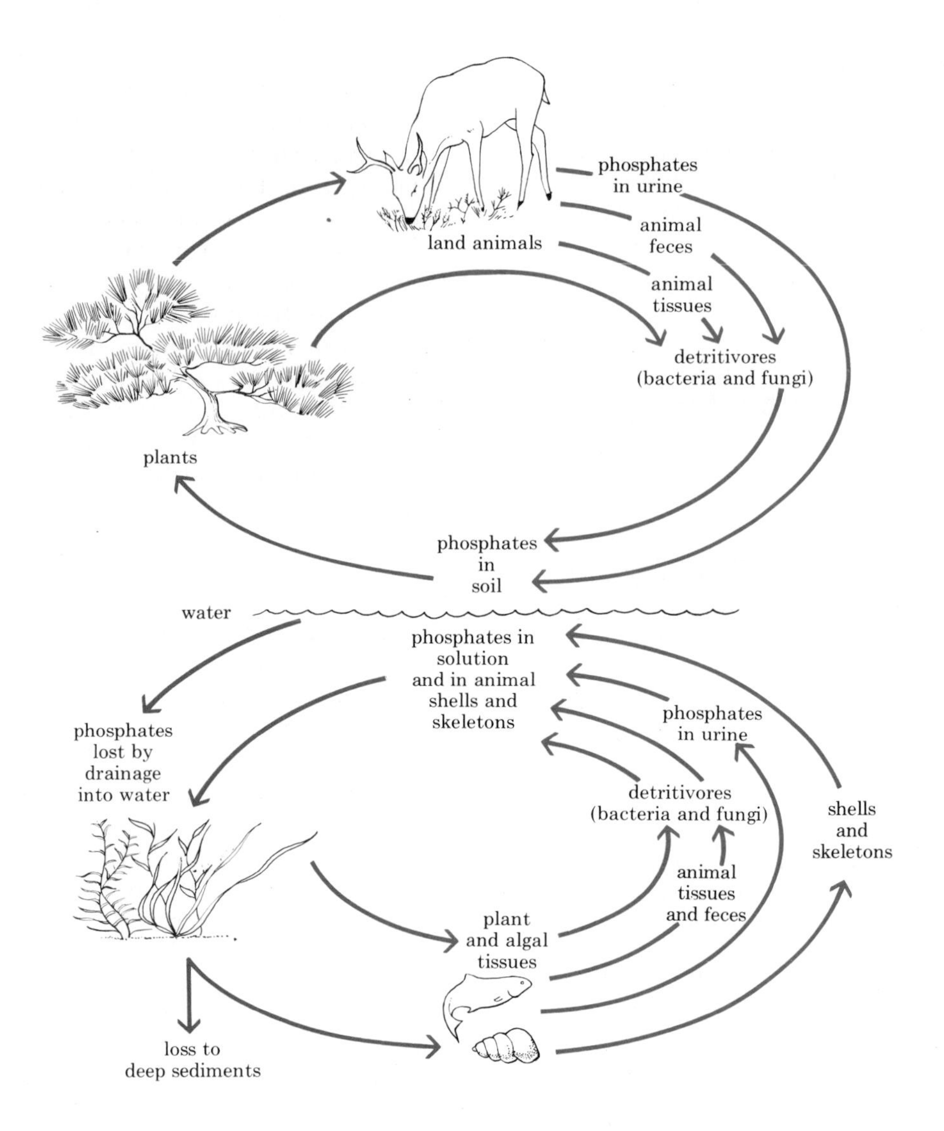

39-9 *The phosphorus cycle. Phosphorus is essential to all living systems as a component of the nucleotides of DNA and RNA and also of the energy-carrier molecules ATP and ADP. Like other minerals, it is released from dead tissues by the activities of detritivores, taken up from soil and water by plants and algae, and recycled through the ecosystem.*

The biological components of biogeochemical cycles include the producers, consumers, and detritivores. As a result of the metabolic work of the detritivores, inorganic substances are released from organic compounds and returned to the soil or water. From the soil or water, inorganic substances are again taken into the bodies of plants and enter their cycles again.

The cycling of one important mineral, phosphorus, through the ecosystem is shown in Figure 39-9. Other minerals in the ecosystem undergo similar cycles, differing only in details.

The Nitrogen Cycle

The chief reservoir of nitrogen is the atmosphere; in fact, nitrogen makes up 78 percent of the gases in the atmosphere. Since most living things, however, cannot use elemental atmospheric nitrogen to make amino acids and other nitrogen-containing compounds, they must depend on nitrogen present in soil minerals. The process by which this limited amount of nitrogen is circulated and recirculated throughout the world of living organisms is known as the *nitrogen cycle*. The three principal stages of this cycle are (1) ammonification, (2) nitrification, and (3) assimilation.

Much of the nitrogen found in the soil, a result of the decomposition of organic materials, is in the form of complex organic compounds, such as proteins, amino acids, nucleic acids, and nucleotides. However, these nitrogenous compounds are usually rapidly decomposed into simple compounds by detritivores, chiefly bacteria and fungi. These microorganisms use the proteins and amino acids as a source of their own needed proteins and release excess nitrogen in the form of ammonia (NH_3) or ammonium (NH_4^+). This process is known as *ammonification.*

Several species of bacteria common in soils are able to oxidize ammonia or ammonium. This oxidation, known as *nitrification,* is an energy-yielding process, and the energy released is used by these bacteria as their primary energy source. One group of bacteria oxidizes ammonia (or ammonium) to nitrite (NO_2^-):

$$\underset{\text{ammonia}}{2NH_3} + \underset{\text{oxygen}}{3O_2} \longrightarrow \underset{\text{nitrite ion}}{2NO_2^-} + \underset{\text{hydrogen ion}}{2H^+} + \underset{\text{water}}{2H_2O}$$

Nitrite is toxic to many plants, but it rarely accumulates. Members of another genus of bacteria oxidize the nitrite to nitrate (NO_3^-), again with a release of energy:

$$\underset{\text{nitrite ion}}{2NO_2^-} + \underset{\text{oxygen}}{O_2} \longrightarrow \underset{\text{nitrate ion}}{2NO_3^-}$$

Although plants can utilize ammonium directly, nitrate is the form in which most nitrogen moves from the soil into the roots.

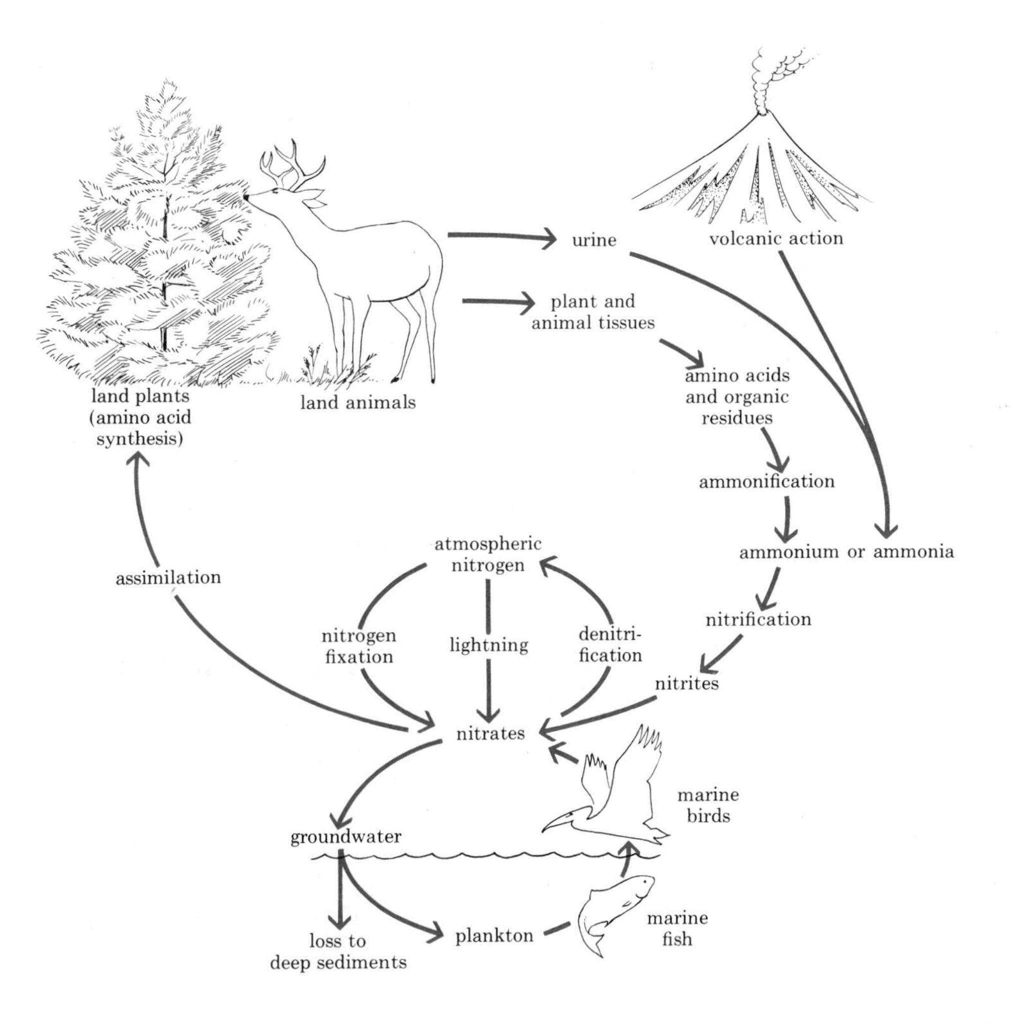

39-10 *The nitrogen cycle. Although the reservoir of nitrogen is in the atmosphere, where it makes up 78 percent of dry air, the movement of nitrogen in the ecosystem is more like that of a mineral than of a gas. Only a few microorganisms are capable of nitrogen fixation.*

Energy Costs of Food Gathering

How much does a calorie cost, in terms of calories? For the expenditure of 1 calorie, an organism in most natural populations obtains from 2 to 20 calories in food energy. This is true for organisms whose expenditures are very high—such as the hummingbird, which spends up to 330 calories a minute—as well as for organisms whose expenditures are very low—such as the damselfly, which uses less than a calorie a day.

In simple human societies, in which individuals obtain their food without fossil-fuel energy subsidies, the ratio of food calories gained to calories invested is similar to that which prevails for the rest of the animal kingdom. Hunter-gatherers average 5 to 10 calories for each calorie spent; shifting agriculture (which requires no fertilizer) yields about 20 calories per calorie spent.

As is true in most societies (the social insects and civilized man are among the exceptions), almost the entire adult population has a share in the business of getting food. In the United States, about 20 percent of the population is involved in the food supply system. (Only about 2 percent are actually farmers; the rest are involved in food processing, transportation, and marketing.) Thus 80 percent are, for better or worse, free for other pursuits.

On the surface, it would seem that we expend less energy than most animals on the mundane work of food supply. Not so. At the turn of the century, for each calorie expended, including human labor, fuel for farm machinery and food transport, and the energy cost of fertilizer, we received about a calorie in return. Today, in the United States, as in other "advanced" technological societies, for every calorie invested we get a return of 0.1 calorie. This cost figure does not include the energy used for heating or lighting or running private automobiles (even those that bring the food home from the market) or electric can openers.

Obviously, other animals cannot live so profligately; their energy income must exceed their expenditures. Like these other organisms, we also are dependent almost exclusively on solar energy. There is an important difference, however; because of our technology, we have been able to draw upon energy stored millions of years ago. It is only in the last decade that we have come to realize that not only are these resources finite but also they may soon be expended.

From Robert M. May, "Energy Costs of Food Gathering," *Nature*, vol. 225, page 669, 1975.

Once the nitrate is within the plant cell, it is reduced back to ammonium. In contrast to nitrification, this *assimilation* process requires energy. The ammonium ions thus formed are transferred to carbon-containing compounds to produce amino acids and other nitrogenous organic compounds needed by the plant.

Nitrogen Fixation

The nitrogen-containing compounds of green plants are returned to the soil with the death of the plants (or of the animals that have eaten the plants) and are reprocessed by detritivores, taken up by the plant roots in the form of nitrate dissolved in soil water, and reconverted to organic compounds. In the course of this cycle, a certain amount of nitrogen is always "lost," in the sense that it becomes unavailable to plants.

The main source of nitrogen loss is the removal of plants from the soil. Soils under cultivation often show a steady decline of nitrogen content. Nitrogen may also be lost when topsoil is carried off by soil erosion or when ground cover is destroyed by fire. Nitrogen is also leached away by water percolating down through the soil to the groundwater. In addition, numerous types of bacteria are present in the soil that, when oxygen is not present, can break down nitrates, releasing nitrogen into the air and using the oxygen for the oxidation of carbon compounds (respiration). This process, known as *denitrification*, takes place in poorly drained (hence, poorly aerated) soils.

If the nitrogen lost from the soil were not steadily replaced, virtually all life on this planet would finally flicker out. The "lost" nitrogen is re-

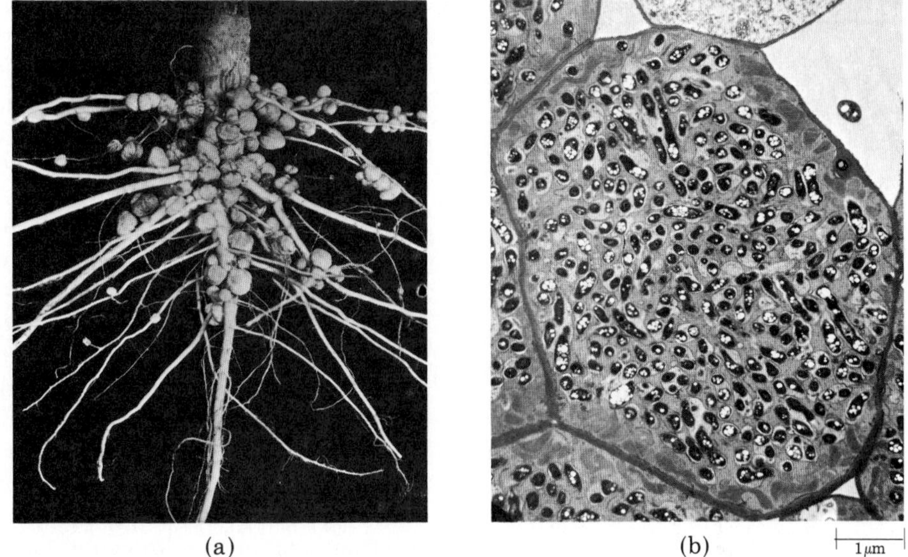

39-11 (a) *Nitrogen-fixing nodules on the roots of a soybean, a legume. These nodules are the result of a symbiotic relationship between a soil bacterium* (Rhizobium) *and root cells.* (b) *Cross section of an infected nodule, showing bacteria.*

The plant supplies the bacteria with an energy source; the bacteria supply the plant with fixed nitrogen.

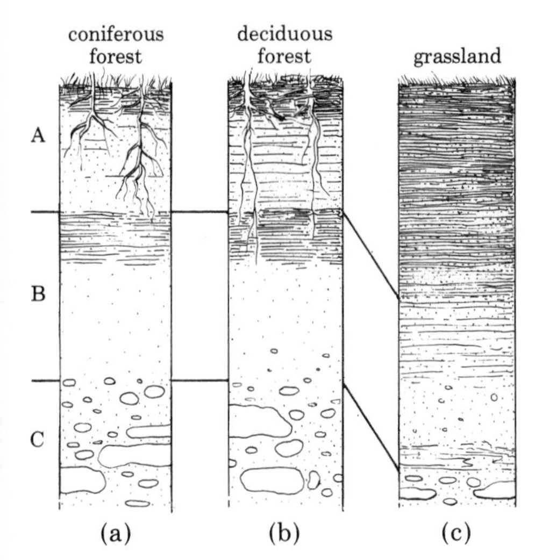

39-12 *Diagrams of soil layers of three major soil types.* (a) *The litter of the northern coniferous forest is acidic and slow to decay, and the soil has little accumulation of humus, is very acidic, and is leached of minerals.* (b) *In the cool, temperate deciduous forest, decay is somewhat more rapid, leaching less extensive, and the soil more fertile. Such soils have been used extensively for agriculture, but they need to be prepared by adding lime (to reduce acidity) and fertilizer.* (c) *In the grasslands, almost all of the plant material above the ground dies each year as do many of the roots, and so much organic matter is constantly returned to the soil. In addition, the finely divided roots penetrate the soil extensively. The result is highly fertile soil, often black in color, with a topsoil sometimes more than a meter in depth.*

turned to the soil by *nitrogen fixation*, the process by which gaseous nitrogen from the air is incorporated into organic nitrogen-containing compounds and thereby brought into the nitrogen cycle. It is carried out to a small extent by abiotic processes. Most nitrogen fixation, however, is carried out by a few types of microorganisms, including blue-green algae, some free-living bacteria, and some species of bacteria that live in symbiosis with plants. Of the various classes of nitrogen-fixing organisms, the symbiotic bacteria are among the most important in terms of total amounts of nitrogen fixed. The most common of the nitrogen-fixing symbiotic bacteria is *Rhizobium* (Figure 39-11), which invades the roots of leguminous plants, such as clover, peas, beans, and alfalfa.

Fifty million metric tons of nitrogen are added to the soil each year, of which 45 million tons are biological in origin. (The other 10 percent is largely in the form of chemical fertilizers.) Just as all organisms are ultimately dependent on photosynthesis for energy, they all depend on nitrogen fixation for their nitrogen.

Soils and Mineral Cycles

The composition of the soil influences the availability of minerals in an ecosystem. Soil, the uppermost layer of the earth's crust, is composed of weathered rock associated with organic material, both living and in various stages of decomposition. It typically has three layers: the A horizon, the B horizon, and the C horizon. The A horizon, or topsoil, is the zone of maximum organic accumulation (humus). The B horizon, or subsoil, consists of inorganic particles in combination with mineral nutrients that have leached down from the A horizon. The C horizon is made up of loose rock that extends down to the bedrock beneath it. Figure 39-12 shows profiles of three common soil types.

The mineral content of the soil depends in part on the parent rock from which the soil is formed. These differences in mineral content can be extremely localized, with sharp lines of demarcation. Geologists sometimes use types of vegetation or changes in color or growth patterns of plants as indicators of mineral deposits.

Table 39–1 Soil classification

	Diameter of Fragments (Micrometers)
Coarse sand	200–2,000 (0.2–2 millimeters)
Sand	20–200
Silt	2–20
Clay	Less than 2

In most soils, however, the mineral content is more dependent on the biotic component of the ecosystem. In an undisturbed environment, most of the mineral nutrients stay within the ecosystem. If, however, the vegetation is repeatedly removed, as when crops are harvested or grasslands are overgrazed, or the top, humus-rich layer of the A horizon is eroded, the soil rapidly becomes depleted and can be used for agricultural purposes only if it is heavily fertilized.

Another factor influencing the mineral content of soils is the soil composition. The smaller fragments of rock are classified as sand, silt, or clay, according to size (Table 39–1). Water and minerals drain rapidly through soil composed of large particles (sandy soil). Soil composed of small particles (clay) holds the water against gravity. Moreover, the small clay particles are negatively charged and so hold positively charged ions, such as calcium (Ca^{2+}) and potassium (K^+). However, a pure clay soil is not suitable for plant growth because it is usually too tightly packed to let in enough oxygen for the respiration of plant roots, soil animals, and most soil microorganisms. Clay soils that contain enough large particles to keep the soil from packing are known as loams, and these are generally the best soils for plant growth.

The pH of the soil also affects its capacity to retain minerals. In acidic soils, positively charged ions leach out of the soil. Soil pH also affects the solubility of certain nutrient elements. Calcium, for example, increases in solubility (and therefore in its availability to plant roots) as pH increases. Crops such as alfalfa and other legumes have high calcium requirements and so grow well only in alkaline soil. Rhododendrons and azaleas, on the other hand, have high requirements for iron, which is abundant only when the soil is acidic.

Soils and plant life interact. Plants constantly add to the humus, thereby changing not only the content of the soil but also its texture and its capacity to hold minerals and water. In turn, the plants are dependent upon the mineral content of the soil and its holding capacity. As these improve, plants increase in biomass and also often change in kind, thereby producing further changes in the soil. Thus, under natural conditions, the soil is constantly changing.

39–13 *Prairie soil. Note how the roots of the grasses bind the topsoil.*

Community Composition

An ecosystem consists of all of the organisms in a community plus the environment with which they interact. In all ecosystems, energy is captured by primary producers, flows through the various trophic levels of consumers and detritivores, and is ultimately dissipated as heat. Minerals cycle and recycle through living communities and the abiotic environment. These two phenomena—energy flow and mineral cycling—underlie the organization of every community. Despite the fundamental similarities of all communities, however, there is tremendous diversity among the communities found on the surface of the earth and in its waters.

Communities vary both in space and in time. Different organisms play comparable roles in communities in different locations or in communities in the same location at different times. Although competitive and symbiotic interactions occur in all communities, the species involved and the details and effects of their relationships differ. Moreover, in some communities, there are only a few species performing each role in the life of the community, but the populations of such species may be quite large. In other communities, a vast number of species are present, but each is represented by a relatively small population. In still other communities, a few species are represented by large populations, while many other species are present in smaller numbers.

The questions of most concern to many contemporary ecologists revolve around the matter of community composition. Why does a given community contain a particular "set" of species and not some other? What factors determine the number of species in a community and the relative sizes of their populations? What are the consequences for the composition of a community and its ongoing life when the environment is disturbed, either by natural forces (for example, a hurricane) or by human intervention? How is the rest of a community affected when the size of one or more of its populations changes significantly?

These and other related questions are of profound importance for our future and for that of our planet. Our knowledge of the answers, however, is very limited. As we have seen, the interactions within communities are multiple and complex, and only with the advent of the computer have ecologists had a way of dealing with the masses of data involved. Even collecting the necessary data remains a formidable task. Thus, although we shall describe some current views of community composition, you should bear in mind that this is an area of very active research in which new studies may significantly change our understanding.

Variation in Space

In all terrestrial communities, the basic structure is determined by the plant life. The kinds of plants that can flourish in a particular location are determined largely by abiotic factors: precipitation, temperature, sunlight, seasonal changes, and the mineral content of the soil. For example, the coast of northern California and southern Oregon is primarily redwood forest. The critical factor for the existence of this community is fog, which provides ample moisture during the otherwise dry summer. The redwoods dominate the community and alter the surrounding environment—creating gradients of shade and moisture, conditioning and changing the soil—so that only certain other plants can live in the same community.

39-14 *In the forest along the coast of northern California and southern Oregon, swordfern* (a) *and redwood sorrel* (b) *characteristically grow at the foot of the redwood trees. We call this particular association of plants the redwood community, rather than the swordfern community, because the redwoods so thoroughly dominate it.*

(a)

(b)

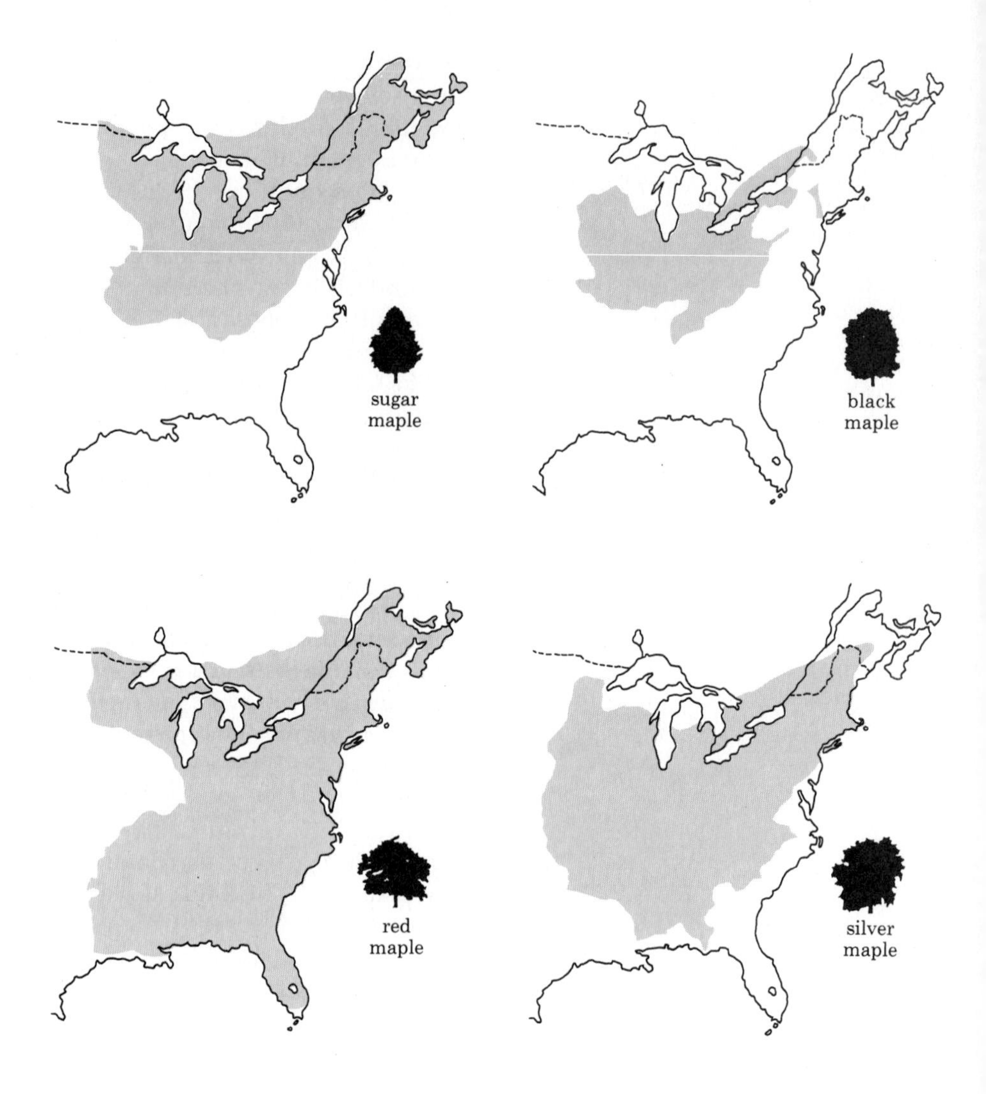

39-15 *Four species of maple trees are found in the forests of eastern North America. Although similar, the species differ somewhat in their requirements for optimum growth. In some areas, the needs of all four species are met and they are found in varying proportions in the same communities or in communities very close to one another. In other areas, only one or two of the species can grow.*

As we saw in Figure 37-8 (page 527), each species has a characteristic range of tolerance for each factor that limits its growth. Populations of a given species can grow only in areas where each essential environmental factor falls within its optimum range (Figure 39-15). This range differs from factor to factor for a particular species and from species to species for a particular factor. Moreover, the environment is not uniform. Even within a region of similar climate and topography, factors such as temperature, moisture, and available nutrients occur along gradients. As a result of environmental variation and variations in the requirements of different species, the species composition of a community changes through space. In locations where critical environmental factors change sharply, there may be clear boundaries separating different communities from one another. The more frequent pattern, however, is a gradual shift from one community to another (Figure 39-16).

The consumers found within a community are determined by their physiological tolerances for the particular environmental conditions and by the resources provided by the plants, such as food and nesting space. In general, the total number of individual consumers increases with the quantity of resources present. Similarly, the greater the variety of resources, such as types of nesting space and food, the greater the variety of consumer species present. But many factors—still poorly understood—are clearly in-

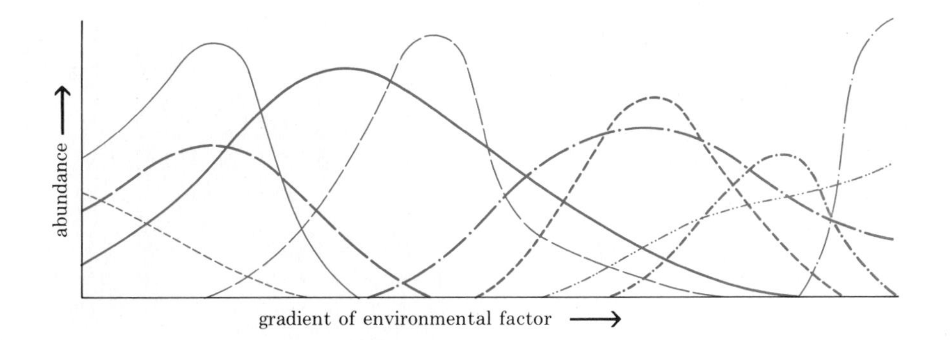

39-16 *Many ecologists believe that community composition is determined primarily by gradients of environmental factors, such as temperature and moisture, with interactions among species playing a lesser role. In this graph of the hypothetical distribution of species along such a gradient, each curve represents a different species. The species in any narrow region along the gradient would be members of a particular community that shifts gradually into the next community.*

volved in determining exactly how the resources are divided among the consumers and therefore what species are present. These factors include the adaptations of each species resulting from its particular evolutionary history; competitive, predatory, and symbiotic interactions; accidents of geography or history that enabled one species to establish itself securely in the community before competing species arrived; extinction of species and their replacement by newly evolved species or immigrants; and the recent history of the community itself.

Variation in Time

When land is laid bare, it will, if the environment is not too harsh, slowly become covered with vegetation and its accompanying animal life. The vegetation that initially colonizes the bare land is usually replaced in time by a second type, which gradually crowds out the first and which itself may eventually be replaced. In areas where these stages of replacement have been studied over a number of years, it has been found that communities of living organisms replace one another in a predictable and orderly sequence. This process is known as *ecological succession.*

The colonization by plants of an area not previously covered by vegetation is known as *primary succession*. It occurs on land made available by such natural forces as landslides, the eruption of a volcano, the retreat of a glacier, or the rising of a new island from the sea. When succession occurs on a site previously occupied by vegetation, it is known as *secondary succession.* Secondary succession commonly occurs on areas laid bare by human intervention—abandoned farmland and strip mines, roadsides, and landfills—but can also be the result of natural processes.

The process of ecological succession is carried out by living organisms themselves. According to the classic model, the organisms of each temporary community change the local conditions of temperature, light, humidity, soil content, and other abiotic factors, and so set up favorable conditions for the organisms of the next temporary community. Recently, however, ecologists have begun to question this model. They believe that factors other than modification of the physical environment by successive plant communities may also play a major role in succession. These factors include predation—especially that of herbivores on plants—and competition between opportunistic and equilibrium species (page 528). According to the emerging model, species employing opportunistic growth strategies are the first colonists of newly available areas. As time passes, species with equilibrium strategies gain a foothold and gradually the competitive balance shifts in their favor. Once a mature community is established, the opportunistic

39-17 *Secondary succession. A young stand of loblolly pine is taking over an abandoned field in Wake County, North Carolina. The plow furrows are still clearly visible. The field was abandoned perhaps 20 years ago.*

species are at a disadvantage, but enough survive to exploit areas that are subsequently disturbed.

When succession ceases in a particular area—or at least slows down considerably—the mature community is called the *climax community*. It, too, is in a state of constant change; individual organisms die, and their places are taken by new individuals. Usually these individuals are of the same sort and of the same species as the ones being replaced. In some areas, however, the environment is regularly disturbed by natural forces, such as fire, volcanic action, or severe weather disturbances, and the area is characterized by recurrent cycles of succession.

Some Examples of Succession

Primary Succession on Rock The first stage in primary succession on rocks or cliffs is the formation of soil. The solid rock is broken down by weathering processes, such as freezing and thawing or heating and cooling, which cause the rocks to expand and contract, thus splitting them. Water and wind exert a scouring action that breaks the fragmented rock into smaller particles, often carrying the fragments great distances. Water enters between the particles and dissolves soluble materials such as rock salt. Water in combination with carbon dioxide from the air forms a mild acid that dissolves additional substances. Chemical reactions that contribute to the disintegration of the rock begin to take place.

Soon, if other conditions such as light and temperature permit, bacteria, fungi, lichens, and then small plants begin to gain a foothold. Growing roots further split the rocks, and the disintegrating bodies of the plants and of the animals associated with them add to the accumulating material. Finally, larger plants move in, anchoring the soil in place with their root systems, and a new stage has begun. The patterns are similar worldwide, although the particular species at each stage differ from place to place.

(a)

(b)

39-18 (a) *Yapoah Crater, a volcanic cinder cone east of the Cascade Mountains in central Oregon. Succession leading to the establishment of climax forest on such a cone may take centuries and may often be interrupted by further volcanic activity long before it is complete.*
(b) *An early stage of succession on a rocky slope. Lichens have begun to accumulate soil, and a bladder fern has sprung up in a small crevice.*

Secondary Succession on an Abandoned Field An open area surrounded by other vegetation is bombarded by the seeds of numerous plants. An abandoned field, for example, is captured by those plants whose seeds can germinate most quickly. In an open field, these are the plants that can survive the sunlight and drying winds—weeds and grasses and trees such as cedars, white pines, poplars, and birch. For a while, these plants are dominant, but eventually they are replaced by other trees—oaks, red maples, white ash, and tulip trees. As the forest matures, the competitive edge goes to hemlock, beech, and sugar maple, and these ultimately take over the forest. Nothing else can compete with them in the conditions that have been established. This, in temperate regions, is the most common climax forest.

The Role of Fire In some areas, fire plays an important part in determining the final stage in forest succession. Young seedlings of deciduous trees are very susceptible to fire, while pines are resistant to it. Jack pines, in fact, open their cones and release their seeds only after they have been heated, so they tend to spread most readily after a fire. In the southern pine forests and the northern lake areas of the United States, recurrent ground fires maintain the pinewoods, keeping the forests perpetually young. Similarly, the forests of sugar pines and giant sequoias on the western coast of the United States are maintained by fires. The fires destroy competitive trees that are faster growing and less fire-resistant than the giant sequoias, and also bare the ground so the small sequoia seeds can germinate. These ground fires, characteristic of the fire-type ecosystem, are distinctly different from the uncontrolled crown fires of northern forests, which spread through the treetops, destroying entire communities of plants and animals. Also, there is now evidence that in prehistory grass fires played a significant role in maintaining grassland ecosystems by eliminating the early stages of forest succession.

39-19 *When fire sweeps through a forest, secondary succession—with regeneration from nearby unburned stands of vegetation—is initiated. Some plants produce sprouts from the stumps; others seed abundantly on the burn. In one group of pines, the closed-cone pines, the cones do not open to release their seeds until they have been heated by fire.*

Mature Ecosystems

As ecosystems pass through the various stages of succession, one type of community is replaced by another, distinctly different type. Although the changes that take place as an ecosystem matures differ in detail as to the exact species involved and the rates of change, many ecologists believe that they all have certain results in common:

1. There is an increase in total biomass. Compare, for example, a recently abandoned field, which is an immature ecosystem, with a deciduous forest, which is a more mature one.
2. There is a decrease in net productivity in relation to biomass (standing crop). In other words, the biomass of a mature system does not tend to increase as rapidly as that of a system in earlier stages of succession, since it is dominated by slower-growing species.
3. Mature systems have a greater capacity to entrap and hold nutrients.
4. During the early stages of succession, the diversity of species increases. There is also a general increase in the size of organisms, the length of their lives, and the complexity of their life cycles.
5. It is hypothesized that the ecosystem becomes more stable as it matures. Mature ecosystems, according to this hypothesis, show a greater capacity to recover from disturbance and are less likely to be drastically affected by biological changes—such as population explosions.

The physical characteristics of the environment determine the nature of the mature community. In regions where conditions are particularly unfavorable, such as the tundra, the process of succession involves relatively few stages and the climax community is correspondingly simple. Where physical conditions are less limiting, the mature community is rich and diverse.

Agricultural Ecosystems

An area under intensive cultivation is an immature ecosystem. It has high productivity, relatively little biomass, and few species. It is also a very unstable system, compared to a mature, complex ecosystem, which, with its complicated food webs, has many built-in checks and balances. Individual members of the plant and animal community may be sick or dying, but the natural, mature ecosystem itself is healthy, and species tend to endure in relatively stable numbers. In areas under cultivation, plants do not grow in complex communities, as they do in a forest, but in pure stands. A cornfield, for example, has little inherent stability. If not constantly guarded, it will be immediately overrun with insects and weeds. It is for this reason that insecticides and herbicides play such a large role in our modern life.

The susceptibility of modern crops to predators and parasites was tragically illustrated by the great potato famine of Ireland, which was caused by a fungus infection. The famine of 1845–1847 was responsible for more than a million deaths from starvation and initiated large-scale emigration from Ireland to the United States; within a decade, the population of Ireland dropped from 8 million to 4 million. Virtually the entire Irish potato crop was wiped out in a single week in the summer of 1846. A number of plant geneticists have warned that the new strains of wheat and rice, which have made major contributions toward feeding the growing human population, are particularly susceptible to such disasters because of their genetic uniformity and widespread distribution.

The primary producers of an agricultural ecosystem, such as this cornfield, must be constantly defended against disease, predation, and competition. Also, because organic matter is removed from the system, minerals must be constantly replaced, in the form of increasingly expensive fertilizer.

SUMMARY

An ecosystem is a unit of biological organization made up of all the organisms in a given area (that is, a community) and the environment in which they live. It is characterized by interactions between the living (biotic) and nonliving (abiotic) components that result in (1) a flow of energy from the sun through autotrophs to heterotrophs, and (2) a cycling of minerals and other inorganic materials.

Within an ecosystem, there are trophic (feeding) levels. All ecosystems have at least three such levels: primary producers, which are plants or photosynthetic algae; primary consumers, which are usually animals; and detritivores, which live on animal wastes and dead plant and animal tissues.

The primary producers (the autotrophs) convert a small proportion (above 1 percent) of the sun's energy into chemical energy. The primary consumers (herbivores) eat the primary producers. A carnivore that eats the herbivore is a secondary consumer, and so on. About 10 percent of the energy transferred at each link in a food chain is stored in body tissue; of the remaining 90 percent, part is used in the metabolism of the organism and part is unassimilated. This unassimilated energy is ultimately consumed by detritivores.

The movements of water, carbon, nitrogen, and minerals through ecosystems are known as biogeochemical cycles. In such cycles, inorganic materials from the air, water, or soil are taken up by primary producers, passed on to consumers, and eventually transferred to detritivores, chiefly bacteria and fungi. In the course of their metabolism, the detritivores release the inorganic materials to the soil or water in a form in which they can again be taken up by primary producers.

The nitrogen cycle is of critical importance to all organisms. It involves several stages: ammonification, the breakdown of nitrogenous organic compounds to ammonia or ammonium; nitrification, the oxidation of ammonia or ammonium to nitrates, which are taken up by plants; and assimilation, the conversion of nitrates to ammonia and its incorporation into organic compounds. Nitrogen-containing organic compounds are eventually returned to the soil, completing the cycle. Nitrogen is lost from the system by harvest, erosion, fire, leaching, and denitrification. It is added to the system by nitrogen fixation, the incorporation of elemental nitrogen into organic compounds. Biological nitrogen fixation is carried out entirely by microorganisms, especially bacteria *(Rhizobium)* in symbiotic associations with legumes.

Characteristics of soil affect the presence, retention, and recycling of minerals. These characteristics include the rock from which the soil was formed, the presence of humus on the soil surface, and soil composition and pH. Soils and plant life interact. Plants and detritivores increase the availability of minerals in the soil and affect its texture; these changes, in turn, improve the soil's capacity to maintain plant life, leading to a further increase in the humus content.

The fundamental organization of all communities is similar, but the species composition varies from location to location and over time. The basic structure of terrestrial communities is determined by the dominant plant life, which is, in turn, determined largely by abiotic factors. For a population to live in a particular location, the level of each factor limiting its growth must be within the optimum range for that species. Changes in the species composition of neighboring communities are associated with gradients of limiting factors. The animal populations of a community are determined by abiotic factors, the quantity and variety of resources provided by the producers, and the manner in which the resources are apportioned among the animal species.

Ecological succession is an orderly sequence of changes in the type of vegetation and other organisms on a particular site. Primary succession occurs in areas previously unoccupied by living organisms, whereas secondary succession occurs in disturbed areas where vegetation has been removed. Factors involved in succession are believed to include modifications of the physical environment caused by the organisms themselves, predation (especially by herbivores), and competition between opportunistic and equilibrium species. Succession culminates in the establishment of a climax community, which is the characteristic community for that area.

Maturation in ecosystems is accompanied by increases in biomass, the number of species present, and the retention of nutrients, as well as a decrease in net productivity.

QUESTIONS

1. Distinguish among the following: community/ecosystem; gross productivity/net productivity; producer/consumer/detritivore; A horizon/B horizon/C horizon; primary succession/secondary succession.

2. Describe what happens to the energy in light striking a temperate forest ecosystem. What happens when it strikes a cornfield? A pond? A field on which cattle are grazing?

3. Describe what happens to a nutrient mineral in each of the areas in Question 2.

4. Explain the different kinds of information provided by a pyramid of numbers, a pyramid of biomass, and a pyramid of energy. For what purposes might each type be more useful than the other two?

5. A leading ecologist has stated: "The plough is the most deadly agent of extinction ever devised; not even thermonuclear weapons pose such a threat to the beauty and diversity of life on earth." Explain.

40

The Biosphere

The biosphere, that part of the earth on which life exists, is only a thin film on the surface of our planet. It extends about 8 or 10 kilometers above sea level and a few meters down into the soil, as far as roots penetrate and microorganisms are found. It includes all of the surface waters and ocean depths. It is patchy, differing in both depth and density.

Factors Affecting the Distribution of Life

The sun, of course, powers the biosphere. It is responsible for the wind and the weather as well as for the energy flow that characterizes life. As we saw in the last chapter, the earth receives about 10^{24} calories of energy from the sun each year. Of this, approximately 50 percent is reflected from or absorbed by the atmosphere and 50 percent reaches the

40-1 *Photograph of earth taken from the Apollo 17 spacecraft during the final lunar landing mission in 1972. Virtually the entire continent of Africa is visible, as is the Arabian Peninsula. The Mediterranean Sea is at the top left of the photograph. At the bottom, Antarctica is blanketed under a heavy cloud cover.*

Different parts of the earth's surface receive different amounts of solar energy. These differences determine worldwide patterns of climate, wind, and weather, which are, in turn, major factors affecting the distribution of organisms in the biosphere.

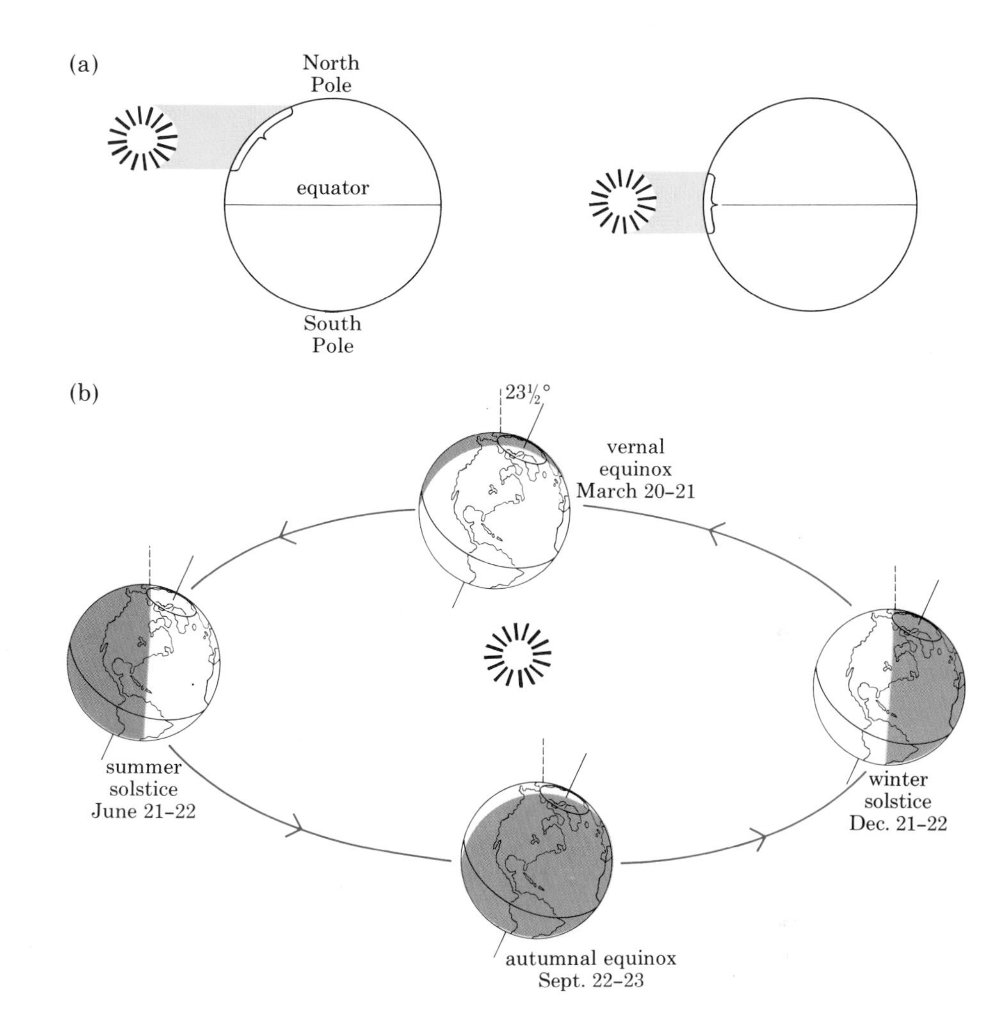

40–2 (a) *A beam of solar energy striking the earth near one of the poles is spread over a wider area of the earth's surface than is a similar beam striking the earth near the equator.*

(b) *In the northern and southern hemispheres, temperatures change in an annual cycle because the earth is slightly tilted on its axis in relation to its pathway around the sun. In winter, the northern hemisphere tilts away from the sun, which decreases the angle at which the sun's rays strike the surface and also decreases the duration of daylight, both of which result in lower temperatures. In the summer, the northern hemisphere tilts toward the sun. Note that the polar region of the northern hemisphere is continuously dark during the winter and continuously light during the summer.*

earth's surface. However, the amount of energy received by various parts of the earth's surface is not uniform. This is the major factor determining the distribution of life on earth.

Climate, Wind, and Weather

At the equator, the sun's rays are almost perpendicular to the earth's surface, and this sector receives more energy per unit area than the regions to the north and south, with the poles receiving the least (Figure 40–2a). Moreover, because the earth, which is tilted on its axis, rotates once every 24 hours and completes an orbit around the sun about once every 365 days, the amount of energy reaching different parts of the surface changes hour by hour and season by season (Figure 40–2b).

Temperature variations over the surface of the earth and the earth's rotation establish major patterns of air circulation and rainfall. These patterns depend, to a large extent, on the fact that cold air is denser than warm air. As a consequence, hot air rises and cold air falls. As air rises, it encounters lower pressure and consequently expands, and as a gas expands, it cools. Cooler air holds less moisture, so rising, cooling air tends to lose its moisture in the form of rain or snow.

The air is hottest along the equator, the region heated most by the sun. This air rises, creating a low-pressure area (the doldrums), moves away from the equator, cools, loses some of its moisture, and falls again at latitudes of about 30° north and south, the regions where most of the great

Acid Rain

Acid rain, an increasingly serious problem in North America and Europe, is the result of attempting to solve a pollution problem by dispersing the pollutant rather than by controlling its production. As we noted earlier (see page 293), sulfur oxides are produced by the combustion of sulfur-containing fossil fuels and by the smelting of sulfur-containing ores. These gases combine with water vapor to form sulfurous and sulfuric acids. Similarly, nitrogen oxides, a by-product of gasoline combustion, combine with water vapor to form nitrous and nitric acids.

As early as 1905, it was recognized that the gases produced by smelters and coal-burning factories had a devastating effect on the vegetation in surrounding areas. The solution devised for this problem—still used today—was to build very tall smokestacks, so the wind would carry the pollutants away from the immediate area. It was assumed that the pollutants would be so widely dispersed that they would be rendered harmless.

In the 1920s, the pH of rain and snow in Scandinavia began to drop, and by the 1950s, similar phenomena were observed elsewhere in Europe and in the northeastern United States. The average pH of normal rainfall is about 5.6 (mildly acidic), a result of the combination of carbon dioxide with water vapor to produce carbonic acid. As more data were collected, it was found that, in certain geographic areas, the average annual pH of precipitation was between 4.0 and 4.5. Occasional storms would release rain with a pH as low as 2.1, which is extremely acidic.

It was not until the 1960s that the cause of the increasingly acidic precipitation was understood. Sulfur oxides released from tall smokestacks are transported by the prevailing winds hundreds or thousands of miles to the east and then return to earth in rain and snow. The problem is compounded by nitrogen oxides released from automobiles and also carried off by the wind. What was once a local problem has become an international problem, in which the pollutants respect no boundaries.

The biological consequences of acid rain are numerous. In plants, it causes reduced germination of seeds, a decrease in the number of seedlings that mature, reduced photosynthesis, and lowered resistance to disease. Moreover, acid rain lowers the pH of soil, resulting in the leaching of essential plant nutrients from the soil and, perhaps, reducing nitrogen fixation by symbiotic bacteria.

The effects of acid rain on freshwater ecosystems, particularly lakes in mountainous regions, are also severe. A 1977 Cornell University study of the lakes at high elevations in the western Adirondack Mountains of New York found that 51 percent had a pH below 5.0 of which 90 percent were devoid of fish life. By contrast, a similar study performed between 1929 and 1937 found that only 4 percent of the lakes were acidic and without fish. Lakes at lower elevations, where the characteristics of the bedrock and soil are different, are generally not so vulnerable to the effects of acid rain.

Many people believe that acid rain is the most serious worldwide pollution problem confronting us in the

deserts of the world are found. This air warms, picks up moisture, and rises again at about 60° latitude (north and south); this is the polar front, another low-pressure area. The air rising at the polar front descends again at the poles, producing a region in which, as in other areas of descending air, there is virtually no rainfall. The earth's spinning movement twists the winds caused by the transfer of air from equator to poles, creating the major wind patterns (Figure 40–3).

40–3 *The earth's surface is covered by many belts of air currents, which determine the major patterns of rainfall and wind. Air rising at the equator loses moisture in the form of rain, and falling air at latitudes of 30° north and south is responsible for the great deserts found at these latitudes.*

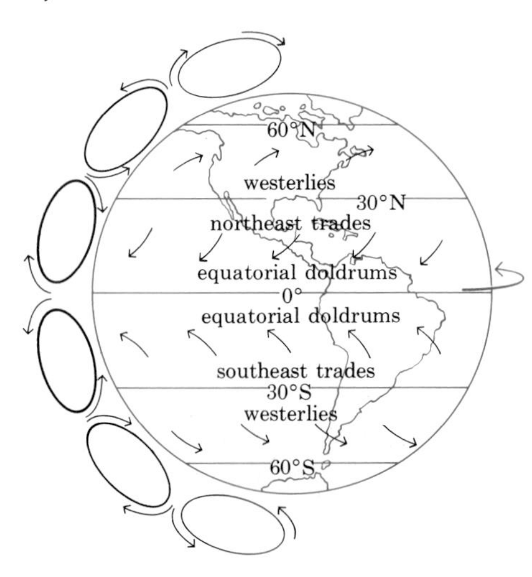

The worldwide patterns are modified locally by a variety of factors. For example, along our own West Coast, where the winds are prevailing westerlies, the western slopes of the Sierra Nevada have abundant rainfall, while the eastern slopes are dry and desertlike (Figure 40–4). As the air from the ocean hits the western slope, it rises, is cooled, and releases its water. Then, after passing the crest of the mountain range, the air descends again, becomes warmer, and its water-holding capacity increases, resulting in a so-called "rain shadow" on the eastern slope.

The Earth's Surface

Much of the iron and other heavier materials of which the earth is composed is collected in a dense core in the center. Surrounding the core is a lighter layer of solid and molten rock. The outermost layer is made up of overlapping, mobile plates on which the continents rest.

1980s. The potential consequences of its effects on biological systems are immense: lowered crop yields, decreased timber production, the loss of important freshwater fishing areas, and the need for greater amounts of increasingly expensive fertilizer to compensate for nutrient losses from leaching and decreased nitrogen fixation. The monetary and social costs of allowing the conditions that create acid rain to continue (or even to increase—a real possibility as we attempt to reduce our dependence on petroleum by increased use of coal) are potentially very great, as are the costs of available processes to remove the sulfur and nitrogen oxides at the source, before they enter the air.

In the last few years, scientists from many fields have begun new research projects to gain a greater understanding of the causes and effects of acid rain and the likely consequences of proposed solutions. Although scientists can provide information on which decisions can be based, the choices that lie ahead are essentially social and economic, to be made through political processes.

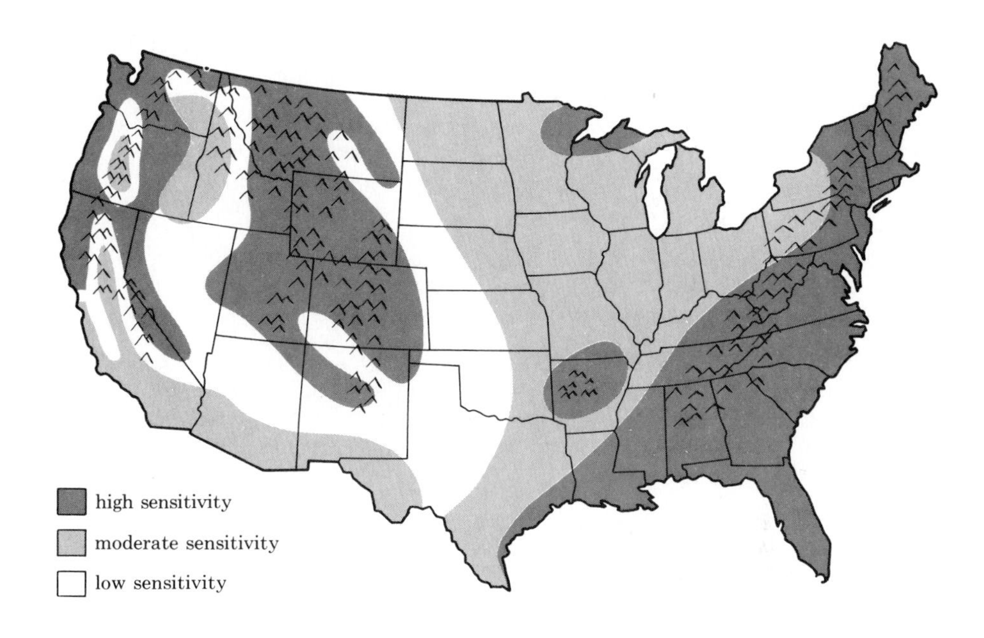

This map, based on estimates made by the Environmental Protection Agency, shows the sensitivity of different areas of the continental United States to acid rain. It takes into account such factors as major sources of sulfur and nitrogen oxides, weather patterns, altitude, and soil characteristics. Can you explain the factors that resulted in a prediction of low sensitivity for the vast grasslands in the interior of the continent?

The continents themselves are composed of granite, which is an igneous rock (from the Latin word *ignis,* for "fire"); igneous rock is formed directly from molten material. The surfaces of the continents change constantly. They are crumpled by contractions and collisions as the continents rise, sink, and collide because of the motion of the plates on which they are carried (page 508). As a consequence, the earth's surface is not at all uniform but varies widely from place to place in its composition and in its height above sea level. Both the mineral content of the earth's surface and the altitude affect the growth of plants and other living organisms, and, as we saw in the example of the Sierra Nevada, the mountain ranges of the continents do much to determine the patterns of rainfall.

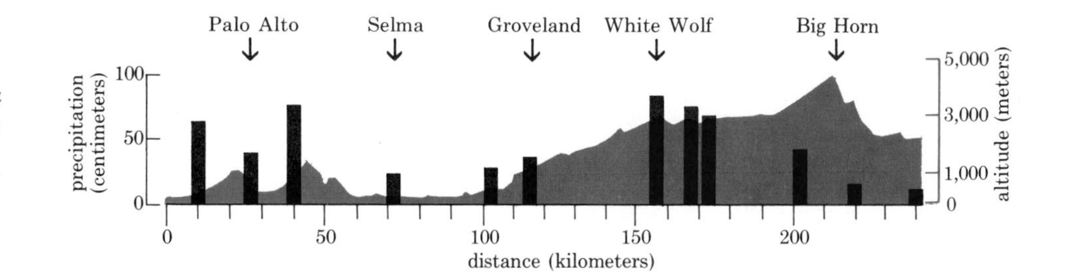

40-4 *The mean annual rainfall (vertical columns) in relation to altitude at a series of stations from Palo Alto on the Pacific Coast across the Coast Ranges and the Sierra Nevada. The prevailing winds are from the west, and there are rain shadows on the eastern slopes of the two mountain ranges.*

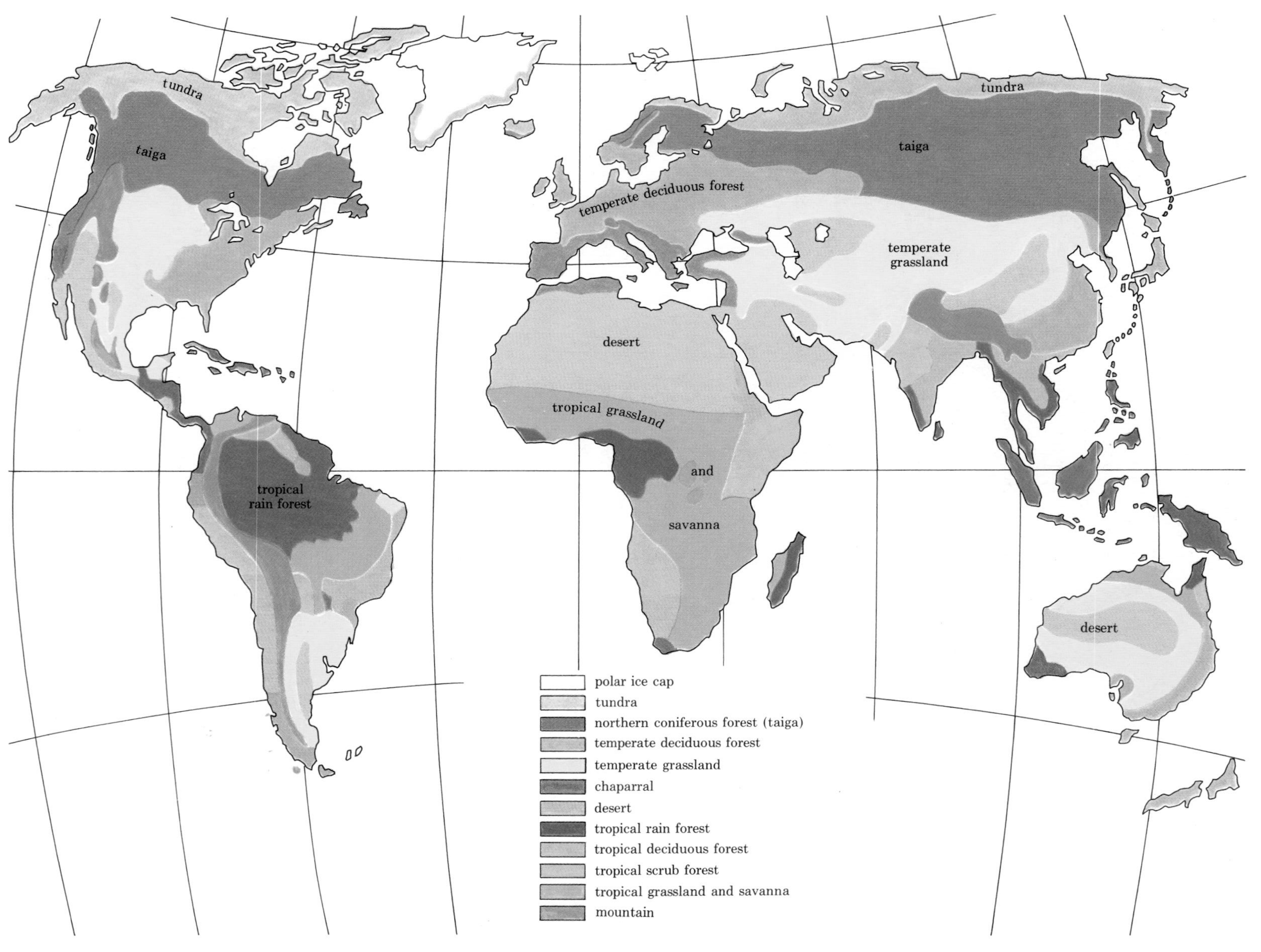

40–5 *Biomes of the world.*

Biomes

The varying patterns of temperature, seasonal change, and precipitation on the earth's surface are responsible for the local variations in the patterns of life. *Biomes*, the big patches into which the biosphere is divided, are large collections of communities (Figure 40-5). The patches are discontinuous, but a patch may closely resemble another patch on the other side of the planet. For instance, the deserts of the world look remarkably the same. However, when one looks more closely, one will see that although the physical features of the environment—temperature and rainfall—are the same, the organisms are not the same. But they look and act alike.

A biome is thus an abstraction. When we speak of the tropical forest biome, we are not speaking of a particular geographic region, but rather of all the tropical forests on the planet. As with most abstractions, important details are omitted. For example, the boundaries are not so sharp as shown on maps nor are all areas of the world easy to categorize. However, the biome concept emphasizes one important truth: Where the climate is the same, the organisms are also very similar, even though these organisms are not related genetically and are far apart in their evolutionary history. This phenomenon, as we have noted previously, is called convergent evolution.

Terrestrial Biomes

Tropical Forests

In the equatorial zone, where most of the tropical forest formations are found, the mean daily temperature is the same throughout the year, and the length of day varies by less than one hour. Rainfall is seasonal, however, with maxima at the time of the equinoxes (Figure 40-2). Variations in total rainfall from one area to another within these zones are caused principally by mountains and their rain shadows.

40-6 *A tropical rain forest, Rancho Grande National Park, in Venezuela. Notice the height of the trees and the huge, buttressed trunk of the tree in the center. The vine growing up it is a philodendron.*

Tropical Rain Forest

In the tropical rain forest, rainfall is abundant all year around; total rainfall is between 200 and 400 centimeters per year, and a month with less than 10 centimeters of rain is considered relatively dry. More species of plants and animals live in the tropical rain forest than in all the rest of the biomes of the world combined. As many as 100 species of trees can be counted on 1 hectare, but each species may be represented by only one tree. By contrast, a given area in a deciduous forest in the northeastern United States typically contains only a few tree species, but each species is represented by many individual trees.

Plants of the tropical forest compete for light. About 70 percent of all species of plants are trees. The upper tree story consists of solitary giants 50 to 60 meters tall. A lower story of trees characteristically forms a solid canopy. The trees are remarkably similar in appearance. Their trunks are usually slender and branch only near the crown. The crowns are high up and relatively small as a result of crowding. Because the soil is perpetually wet, their roots do not reach deep into it, and trunks often end in thick buttresses that provide firm, broad anchorage. Their leaves are large, leathery, and dark green; their bark is thin and smooth; and their flowers are generally inconspicuous and greenish or whitish in color.

(a) (b)

40-7 *Inhabitants of the tropical rain forest.* (a) *Epiphytes, such as this bromeliad, grow in the canopy of the tropical rain forest, obtaining water and minerals from the moist air. Bromeliads are members of the pineapple family that are especially adapted to the tropical forest biome.*
(b) *A great variety of birds, including this splendid parakeet* (Neophema splendida), *live in the tropical rain forest. In Colombia alone, there are at least 1,400 species of birds.*

Woody vines, or lianas, are abundant, especially where an opening has appeared in the forest, as a result, for example, of a tree's falling; vines as long as 240 meters have been measured. There are also many epiphytes, which are plants that grow on other plants, often high above the forest floor (Figure 40–7a). The epiphytes of the tropical rain forest germinate in the branches of trees and obtain water and minerals from the humid air of the canopy. Unlike the plants that have contact with the moist floor, epiphytes need to conserve water between rainfalls. Some epiphytes resemble desert succulents, having fleshy water-storing leaves and stems. Others have spongy roots or cup-shaped leaves that capture moisture and organic debris; many of these epiphytes can take up nutrients from decaying organisms in these storage tanks. A variety of plants, including ferns, orchids, mosses, and bromeliads, have exploited this life style.

An extraordinary variety of insects, birds, and other animals, including mammals, have moved into the treetops along with the vines and epiphytes to make it the most abundantly and diversely populated area of the tropical rain forest.

Little light reaches the forest floor (from 0.1 to 1 percent of the total), and the few plants that are found there are adapted to growing at low light intensities. Many of these, such as the African violet, are familiar to us as house plants.

There is almost no accumulation of leaf litter on the forest floor such as we find in our northern forests; decomposition is too rapid. Everything that touches the ground disappears almost immediately—carried off, consumed, or rapidly decomposed.

The soils of tropical rain forests are relatively infertile. Many are chiefly composed of a red clay; when they are cleared, in many cases they either erode rapidly or form thick, impenetrable crusts that cannot be cultivated after a season or two. Tropical soils are generally deficient in minerals; most of the nitrogen, phosphorus, calcium, and other nutrients are found in the plants rather than the soil. Also, those minerals that are present in the soil are leached out by the heavy rainfall. As a consequence, soils where tropical forests have stood are very poor agriculturally and will support crops for only a few years. The rapidly expanding human population in the tropics has made the traditional tropical agricultural practices of clearing and short-term cultivation immensely destructive because they are now carried out on such a wide scale. Although the tropical rain forest now forms about half of the forested area of the earth, some ecologists predict that, at the present rate of destruction, it may almost all have disappeared by the year 2000, and with it thousands of species of plants and animals found nowhere else in the world.

One of the questions ecologists would like to answer, before the tropical forests disappear, concerns their enormous diversity. One hypothesis holds that it is the result of past and present competition, which has produced many small ecological niches, each held by the best competitor. Other hypotheses have been proposed, involving such factors as the rate of speciation, the rate of extinction, and the relatively unchanging conditions of the environment as a whole, which have allowed the tropical forest to develop—until recently—without disruption. Still another hypothesis holds that localized disruptions and changes in the habitat have prevented the competition that would eliminate species and decrease diversity. Considerably more information is needed before ecologists will be able to determine which hypothesis—or combination of hypotheses—is correct.

Other Tropical Forests

Where the rainfall is more seasonal, tropical rain forests give way to tropical deciduous forests, which are dominated by trees that lose their leaves during the dry seasons. Many of the species flower before they put out new leaves (transpiration, or water loss, is extremely low in leafless, flowering trees because petals have few stomata). Tropical scrub forests, which have trees with small, water-conserving leaves, are found where rainfall is limited all year.

Savannas

Savannas are tropical grasslands with scattered clumps of trees (Figure 40-8). The transition from open forest with grassy undergrowth to savanna is gradual and is determined by the duration and severity of the dry season and, often, by fire and grazing and browsing animals.

In the savanna, the critical competition is for water, and grasses are the victors. Grasses are well suited to a fine, sandy soil with seasonal rain because their roots form a dense root system capable of extracting the maximum amount of water during the rainy period. During dry seasons, the aboveground portions of the plants die, but the roots are able to survive even many months of desiccation. The balance between woody plants and grasses is a delicate one. If rainfall decreases, the trees die entirely. If rainfall increases, the trees increase in number until they shade the grasses, which, in turn, die. If the grasses are overgrazed (which often happens when people begin to use the savanna for agricultural purposes and introduce livestock), enough water is left in the soil so that the woody plants can increase in number, and the grassland is eventually destroyed.

The best-known savannas are those of Africa, which are inhabited by the most abundant and diverse group of large herbivores in the world, including the gazelle, the impala, the eland, the buffalo, the giraffe, the zebra, and the wildebeest.

40-8 *A savanna in Kenya, with a giraffe, zebras, and an impala. The trees in the background are acacias. The African grasslands are able to support a great variety of herbivores because each eats a different type of vegetation.*

The Desert

The great deserts of the world are located at latitudes of about 30°, both north and south, and extend poleward in the interiors of the continents. These are areas of falling, warming air and, consequently, little rainfall. The Sahara, which stretches all the way from the Atlantic coast of Africa to Arabia, is the largest desert in the world, almost equal to the size of the United States, and is increasing in size, spreading along its southern boundaries. This spread is due in large part to the growth of the human population, resulting, in turn, in intensified grazing by domestic animals along the margins of the desert.

Desert regions are characterized by less than 25 centimeters of rain a year. Because there is little water vapor in the air to moderate the temperature, the nights are often extremely cold. The temperature may drop as much as 30°C at night, in contrast to the humid tropics, where day and night temperatures vary by only a few degrees.

Many of the plants in the desert are annuals that race from seed to flower to seed during periods when water is available, and during the brief growing seasons, the desert may be carpeted with flowers. Many of the perennials are succulents (adapted for water storage). Some, like cacti, have no leaves, some are drought-deciduous (dropping their leaves in dry seasons), and some have small, leathery water-conserving leaves.

The animals that live in the desert are also specially adapted to this extreme climate. Reptiles and insects, for example, have waterproof outer coverings and dry, and therefore water-conserving, excretions. The few mammals of the desert are small and nocturnal and get what little water they require from the plants they eat.

(a)

(b)

(c)

(d)

Great Basin desert

Mojave desert

Sonoran desert

Chihuahuan desert

40-9 *The North American deserts. The Sonoran stretches from southern California to western Arizona and down into Mexico. A dominant plant, the giant saguaro cactus* (a) *is often as much as 15 meters high, with a wide-spreading network of shallow roots. Water is stored in a thickened stem, which expands, accordionlike, after a rainfall.*

To the southeast is the Chihuahuan desert, one of whose principal plants is the agave (b), *or century plant, which is a monocot.*

North of the Sonoran is the Mojave, whose characteristic plant is the grotesque Joshua tree (c). *This plant was named by early Mormon colonists who thought that its strange, awkward form resembled a bearded patriarch gesticulating in prayer. The Mojave contains Death Valley, the lowest point on the continent (90 meters below sea level), only 130 kilometers from Mt. Whitney, whose elevation is more than 4,000 meters.*

The Mojave blends into the Great Basin, a cold desert bounded by the Sierra Nevada to the west and the Rockies to the east. It is the largest and bleakest of the American deserts. The dominant plant form is sagebrush (d), *shown here in the background. The large green plant in the foreground is shadscale, and the yellow-flowered plant to the left is rabbitbrush.*

(a)

(b)

(c)

(e)

(d)

(f)

(g)

40-10 *Inhabitants of the North American deserts:* (a) *Golden carpet, found only (and rarely) in Death Valley;* (b) *Gamble quail;* (c) *giant hairy scorpion;* (d) *roadrunner with a whiptail lizard;* (e) *horned lizard with a grasshopper;* (f) *bighorn sheep of the southwestern desert mountains;* (g) *gila monster.*

(a)

(b)

40-11 *The North American chaparral.* (a) *A cacomistle, or ring-tailed cat, a common inhabitant of the chaparral.* (b) *The bushy vegetation that characterizes the chaparral grows as dense as a mat on the foothills of southern California. It is the result of long, dry summers, during which much of the plant life is semidormant, followed by a brief, cool rainy season. The name comes from* chaparro, *the Indian word for the scrub oak that is one of the prominent components of the chaparral. Chaps, the leather leggings worn by cowboys making their way through this dense, dry growth, has the same derivation.*

Chaparral

Regions with mild, rainy winters and long, hot, dry summers, such as the southern coast of California, are dominated by small trees or often by spiny shrubs with broad, thick evergreen leaves. In the United States, such areas are known as *chaparral.* Similar communities are found in areas of the Mediterranean (where they are called the *maquis,* which became the name of the French underground in World War II), in Chile (where they are the *matorral*), in southern Africa, and along portions of the coast of Australia. Although the plants of these various areas are unrelated, they closely resemble one another in their growth patterns and appearance.

Mule deer live in the North American chaparral during the spring growing season, moving out to cooler regions during the summer. The resident vertebrates—brush rabbits, lizards, wren-tits, and brown towhees—are generally small and dull-colored, matching the dull-colored vegetation.

Grasslands

Grasslands, which are transitional areas between the deserts and the temperate forests, are found in the interior areas of continents. They are characterized by periodic droughts, rolling to flat terrain, and hot-cold seasons (rather than the wet-dry seasons of the savanna). The great grasslands of the world include the plains and prairies of North America, the steppes of the Soviet Union, the veld of South Africa, and the pampas of Argentina.

The vegetation is largely bunch or sod-forming grasses, often mixed with legumes (clover and alfalfa) and various annuals. In North America, there is a transition from the more desertlike, western short-grass prairie (the Great Plains), through the moister, richer tall-grass prairie (the Corn

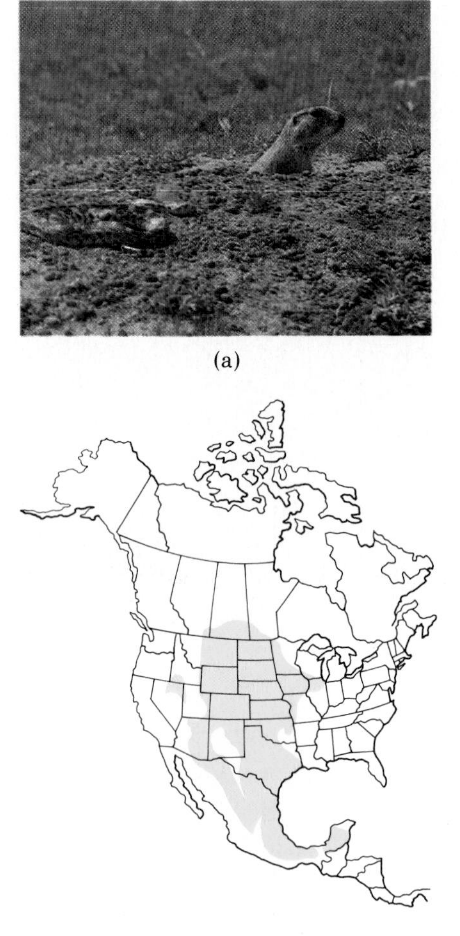

(a)

(b)

40-12 *The grasslands of North America.* (a) *Short-grass prairie with two familiar inhabitants, a black-tailed prairie dog peering out of its burrow and a bull snake.* (b) *A June day on a tall-grass prairie in North Dakota. The cottonwood grove by the prairie creek is characteristic of this biome. A thunderstorm is gathering on the horizon.*

Belt), to the eastern temperate deciduous woodland. Grasslands become drier and drier at increasing distances from the Atlantic Ocean and the Gulf of Mexico, which are the major sources of moisture-bearing winds in the eastern half of the continent.

The grasslands of the world support small, seed-eating rodents and also large herbivores, such as the bison of early America, the gazelles and zebras of the African veld, the wild horses, wild sheep, and ibex of the Asiatic steppes, and now the domestic herbivores. These large, grass-eating mammals, in turn, support the carnivores, such as lions, tigers, and wolves, as well as omnivorous humans. The herds of grazing animals serve, as do ground fires, to maintain the nature of the grassland landscape, destroying tree seedlings and preventing their encroachment.

Temperate Deciduous Forest

Deciduous forests occupy areas where there is a warm, mild growing season with moderate precipitation, followed by a colder period less suited to plant growth. Leaf-shedding in the temperate forest, as in the tropical forest, evolved as a protection against water loss.

The dominant trees of the deciduous forests vary from region to region, depending largely on the local rainfall. In the northern and upland regions, oak, birch, beech, and maple are the most prominent trees. Before the chestnut blight struck North America, an oak-chestnut forest extended from Cape Ann, Massachusetts, and the Mohawk River valley of New York to the southern end of the Appalachian highland. Maple and basswood predominate in Wisconsin and Minnesota, and maple and beech in southern Michigan, becoming mixed with hemlock and white pine as the forest moves northward. The southern and lowland regions have forests of oak and hickory. Along the southeastern coast of the United States, the wet, warm climate supports an evergreen forest of live oak and magnolia.

40-13 *The deciduous forest of North America.* (a) *Aspens growing on the Wasatch Mountains in Utah.* (b) *A raccoon fishing and* (c) *a red fox killing a cottontail rabbit.* (d) Trillium, *photographed in the Great Smoky Mountains National Park. All of these organisms are common inhabitants of the deciduous forests.*

The deciduous forest supports an abundant animal life. Smaller mammals, such as chipmunks, voles, squirrels, raccoons, opossums, and white-footed mice, live mainly on nuts and other fruits, mushrooms, and insects. The wolves, bobcats, gray foxes, and mountain lions, in the areas where they have not been driven out by the encroachment of civilization, feed on these smaller mammals. Deer live mainly on the forest borders, where they browse on shrubs and seedlings.

The soil of the forest is often a rich, gray-brown topsoil. Such soil is composed largely of organic material—decomposing leaves and other plant parts and decaying insects and other animals—and the bacteria, protozoans, fungi, worms, and arthropods that live on this organic matter. The roots of plants penetrate the soil to depths measurable in meters and add organic matter to the soil when they die. Carnivorous arthropods carry fragments of their prey to considerable depths in the soil. The myriad passageways left by dead roots and fungi and by the earthworms and other small animals that inhabit the forest make the soil into a sponge that holds the water and minerals. Land where deciduous forests have stood is good farmland.

Temperate deciduous forests once covered most of eastern North America, as well as most of Europe, part of Japan and Australia, and the tip of South America. In the United States, only scattered patches of the original forest still remain.

Coniferous Forests: The Taiga

Most conifers are evergreens, with small, compact leaves protected by a thick cuticle that guards against water loss. The dropping of leaves is a more efficient adaptation to a wet-dry season, and conifers generally cannot compete with deciduous trees in temperate zones with adequate summer rainfall. However, deciduous trees require a warm summer period of at least four months to permit their leaves to regrow. The northern boundary between the deciduous forest and the coniferous forest occurs where the summers are too short and the winters too long for deciduous trees to grow well.

The northern coniferous forest, called the taiga, is characterized by long, severe winters and a constant cover of winter snow. It is composed chiefly of evergreen needle-leaved trees such as pine, fir, spruce, and hemlock. A thick layer of needles and dead twigs, matted together by fungal mycelium, covers the ground. Along the stream banks grow deciduous trees, such as tamarack, willow, birch, alder, and poplar. There are virtually no annual plants.

The principal large animals of the North American coniferous forest are elk, moose, mule deer, black bears, and grizzlies. Among the smaller animals are porcupines, red-backed mice, snowshoe hares, shrews, wolverines, lynxes, warblers, and grouse. The small animals use the dense growths of the evergreens for breeding and for shelter. Wolves feed upon these mammals, particularly the larger ones. The black bear and grizzly bear eat everything—leaves, buds, fruits, nuts, berries, fish, the supplies of campers, and occasionally the flesh of other mammals. Porcupines are bark eaters, and many seriously damage trees by girdling them. Moose and mule deer are largely browsers. The ground layer of the coniferous forest is less richly populated by invertebrates than that of the deciduous forest because the accumulated litter is slower to decompose.

(a)

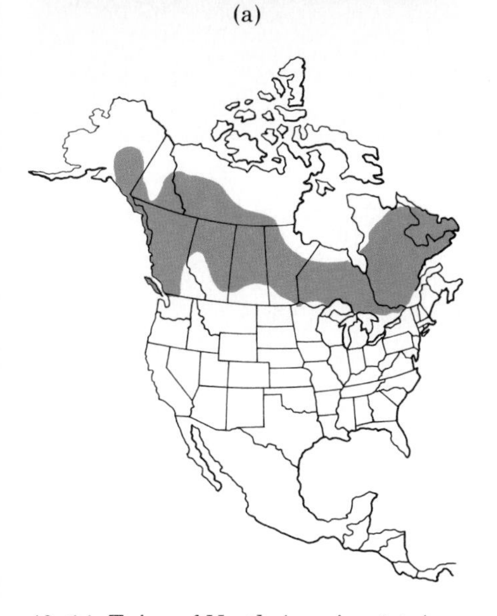

(b)

(c)

40-14 *Taiga of North America.* (a) *A bull moose, with strands of velvet hanging from his newly polished antlers, in Mount McKinley National Park, Alaska. The moose is browsing on willow, its staple food in this region.* (b) *A coniferous forest of balsam spruce and white pine, photographed near the Canadian border.* (c) *Floor of a virgin northern coniferous forest, carpeted with red pine needles. Decay is slower than on the warmer, wetter floors of the deciduous forest, and there is no undergrowth due to the shade cast by the mature trees.*

The Tundra

Where the climate is too cold and the winters too long even for the conifers, the taiga grades into the tundra. The tundra is a form of grassland that occupies one-tenth of the earth's land surface, forming a continuous belt across North America, Europe, and Asia. Its most characteristic feature is permafrost, a layer of permanently frozen subsoil. During the summer the ground thaws to a depth of a few centimeters and becomes wet and soggy; in winter it freezes again. This freeze-thaw process, which tears and crushes the roots, keeps the plants small and stunted. Drying winter winds and abrasive driven snow further reduce the growth of the Arctic tundra.

The virtually treeless vegetation of the tundra is dominated by herbaceous plants, such as grasses, sedges, rushes, and heather, beneath which is a well-developed ground layer of mosses and lichens, particularly the lichen known as reindeer moss. All the flowering plants are perennials.

The largest animals of the Arctic tundra are the musk oxen and caribou of North America and the reindeer of the Old World. Lemmings and ptarmigans (pigeon-sized grouse) feed off the tundra plants. The white fox and the snowy owl of the Arctic live largely on the lemmings.

During the brief Arctic summer, insects emerge in great numbers, and migratory birds visit, taking advantage of the insect hordes and the long periods of daylight. The growing season in many areas of the tundra is less than two months.

(a)

(b)

(c)

40-15 (a) *Tundra of North America on a long Arctic day. Cotton grass surrounds a kettle pond formed from a chunk of glacial ice.* (b) *The Arctic tern represents one of a number of bird species that breed in the tundra, taking advantage of the long summer days to gather food for their nestlings. The terns winter in the Antarctic, following migration routes of 13,000 to 18,000 kilometers. Within three months after hatching, the young must be ready to migrate.* (c) *A snowshoe, or varying, hare in its winter coat. Extra hair on its feet enables it to travel easily over the snow.*

Freshwater Biomes

In general, limnologists—scientists who study natural fresh waters in all their aspects—classify freshwater habitats into two groups: standing water (lakes and ponds) and running water (rivers and streams).

Lakes and Ponds

Lakes vary in size from very large bodies of water, covering thousands of square kilometers, to small ponds. Lakes contain three distinct zones: littoral, limnetic, and profundal.

The *littoral zone*, at the edge of the lake, is the most richly inhabited. Here the most conspicuous plants are angiosperms rooted to the bottom, such as cattails and rushes. Water lilies grow farther out from the shore. There is often a green blanket of duckweed, a small, free-floating angiosperm. Other pond weeds, which grow entirely beneath the water, lack a fatty outer cuticle and so can absorb minerals through their epidermis as well as through their roots. Their submerged plant surfaces harbor large numbers of small organisms. Snails, small arthropods, and mosquito larvae feed upon the plants. Other insects that live among the submerged plants, such as the larvae of the dragonfly and damselfly, and the water scorpion, are carnivorous. Clams, worms, snails, and still other insect larvae burrow in the mud. Frogs, salamanders, water turtles, and water snakes are found almost exclusively in the littoral zone, where they feed primarily upon the

40-16 *Some freshwater inhabitants.* (a) *A great blue heron at the edge of a lake. The water is covered with duckweed.* (b) *A pond with water lilies on the surface.*

insects. Fish, too, are found in greater numbers along the lake margins. Ducks, geese, and herons feed on the plants, insects, mollusks, fish, and amphibians abundant in this zone. The shallow margins of some lakes and ponds are marshy. Among the inhabitants of the marshes are such animals as snails, frogs, ducks, herons, bitterns, muskrats, otters, and beavers.

In the *limnetic zone*, the zone of open water, small, floating algae—phytoplankton—are usually the only photosynthetic organisms found. This zone, which extends down to the limits of light penetration, is the habitat, for example, of smallmouth bass, bluegills, and, in colder waters, of trout.

The deepwater *profundal zone*, extending down from the limnetic zone, has no plant life. Its principal occupants are detritivores—scavenging fish, fungi, and bacteria—that consume the organic debris filtering down from the overlying water.

Rivers and Streams

Rivers and streams are characterized by continuously moving water. They may begin as outlets of ponds or lakes, as runoffs from melting ice or snow, or they may arise from springs (flows of groundwater emerging from bedrock).

The character of life in a stream is determined to a large extent by the swiftness of the current, which characteristically changes as a stream moves downward from its source and, fed by tributaries, increases in volume and decreases in speed. In swift streams, most organisms live in the riffles, or shallows, where small photosynthetic algae and mosses cling to rock surfaces. Many insects, both adult and immature forms, live on the underside of rocks and gravel in the riffles. For those small organisms that can survive the swiftly moving current, there is an abundance of oxygen and of nutrients swept along by the flowing waters.

As the stream travels along its course, the riffles are often interrupted by quieter pools, where organic materials may collect and be decomposed. Few plants can gain footholds on the shifting bottoms of stream pools, but some invertebrates, such as dragonflies and water striders, are typically found in or about the pools. Some organisms, notably trout, move back and forth between the riffles and the pools.

As streams broaden and become slower, they begin to take on the characteristics of lakes and ponds.

The Oceans

The oceans cover almost three-fourths of the surface of the earth. Life extends to their deepest portions, but photosynthesizing organisms are restricted to the upper, lighted zones. The sea has an average depth of more than 3 kilometers and, except for a relatively small surface fraction, is dark and cold. Most of it, therefore, is inhabited by bacteria, fungi, and animals, rather than by plants.

Sea water absorbs light readily. Even in clear water, less than 40 percent of the sunlight reaches a depth of 1 meter, and less than 1 percent of the sunlight that reaches the surface penetrates below 50 meters. Red, orange, and yellow wavelengths of light are absorbed first, so that only the shorter wavelengths, specifically blue and green, penetrate deeply. Thus, below depths of a few meters, only those photosynthesizing organisms capable of utilizing the shorter wavelengths of light can grow.

There are two main divisions of life in the open ocean: *pelagic* (free-floating) and *benthic* (bottom-dwelling). A major component of the pelagic division is plankton. It is composed of photosynthetic algae (phytoplankton), intermingled with small shrimp and other crustaceans and the eggs and larval forms of many fish and invertebrates (zooplankton). These planktonic forms provide food for fish and other relatively large pelagic animals. The benthic division contains the sessile animals, such as sponges, sea anemones, and clams, and many motile animals, such as worms, starfish, snails, crustaceans, and fish. A variety of fungi and bacteria also inhabit the benthic zone, subsisting on the accumulation of debris steadily drifting down from the more populated levels of the ocean.

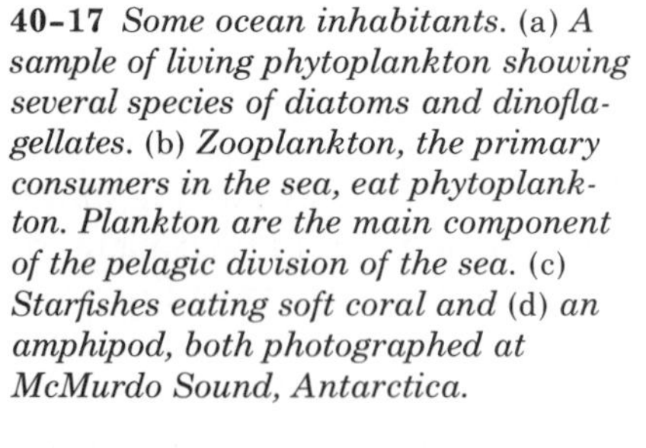

40-17 *Some ocean inhabitants.* (a) *A sample of living phytoplankton showing several species of diatoms and dinoflagellates.* (b) *Zooplankton, the primary consumers in the sea, eat phytoplankton. Plankton are the main component of the pelagic division of the sea.* (c) *Starfishes eating soft coral and* (d) *an amphipod, both photographed at McMurdo Sound, Antarctica.*

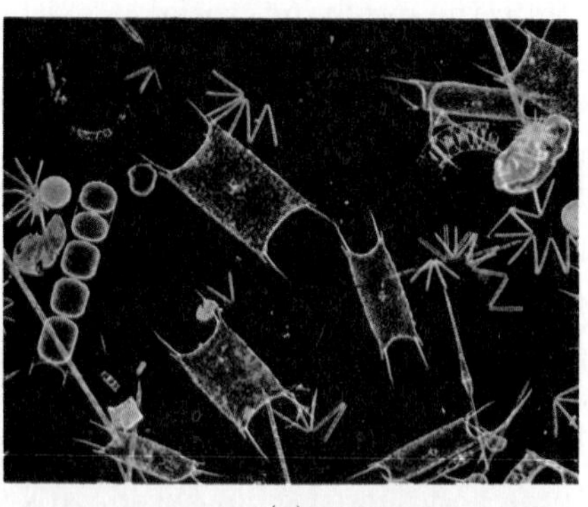

(a)

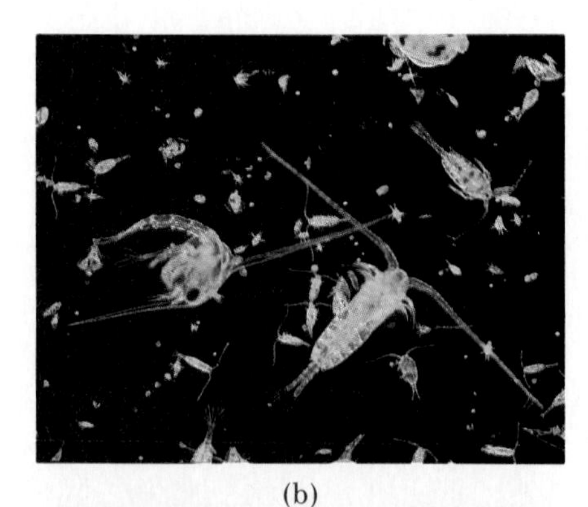

(b)

(c)

(d)

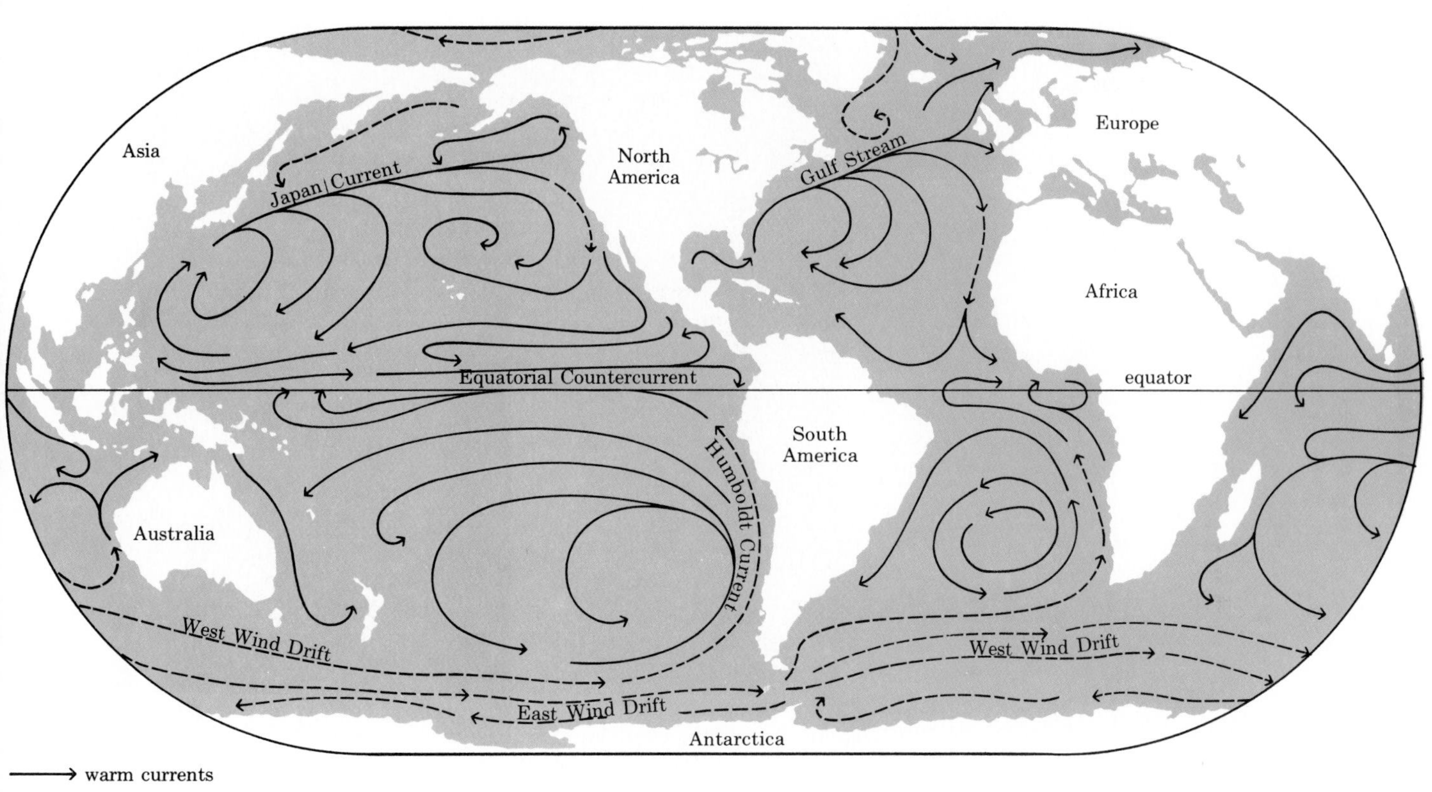

40-18 *The major currents of the ocean have profound effects on climate. Because of the warming effects of the Gulf Stream, Europe is milder in temperature than is North America at similar latitudes. The eastern coast of South America is warmed by water from the equator, and the Humboldt Current brings cooler weather to the western coast of South America.*

The major ocean currents, which are produced by a combination of winds and the earth's rotation, profoundly affect life in the oceans and alter the climate along the ocean coasts. These patterns of water circulation—clockwise in the northern hemisphere and counterclockwise in the southern hemisphere—move currents of warm water north and south from the equator (Figure 40-18). One such current, the Gulf Stream, warms a portion of the eastern coast of North America and the western shores of Europe, and another warms the eastern coast of South America. The same patterns of circulation bring cold waters to the western coasts of North and South America. Where the winds move the water continuously away from the shores, as off the coasts of Portugal and Peru, cold water rich in nutrients is brought to the surface from below, a process referred to as upwelling. Such areas contain high densities of pelagic life and traditionally have supported highly profitable fishing industries.

Despite the fact that oceans cover three times more surface area of the planet than does the land, the total productivity of the open ocean—as measured by the amount of carbon converted to organic compounds by photosynthesis—is only about one-third as great. In fact, the open ocean is only slightly more productive per square meter than the desert, presumably because of the low concentration of minerals in the areas of ocean where light penetrates and photosynthetic organisms can survive.

The Coral Reef

The coral reef is the most diverse of all marine communities. The reef structure is formed by colonial coelenterates and encrusting algae. Each polyp in the coelenterate colony secretes its own calcium-containing skeleton, which then becomes part of the reef. The photosynthetic activity of the

40-19 *A reef in the New Hebrides, islands in the Pacific, east of Australia. Staghorn coral is in the foreground.*

reef is carried out almost entirely by symbiotic algae living within the corals. Carbon, oxygen, and dissolved minerals flow over the reef as a result of the movement of waves and the ocean currents. The reef furnishes both food and shelter for other sea animals, including numerous species of reef fishes and a tremendous variety of invertebrates, such as sponges, sea urchins, polychaetes, and crustaceans. The coelenterates and algae that form the reef can grow only in warm, well-lighted surface water, where the temperature seldom falls below 21°C.

The longest reef in the world is the Great Barrier Reef of Australia, which extends some 2,000 kilometers. Other reefs are found throughout tropical waters and as far north as Bermuda, which is warmed by the Gulf Stream.

The Seashore

The edges of the continents extend 10 to 20 kilometers out into the sea. Along these edges, known as the continental shelves, nutrients are washed out from the land, and life is much denser than in the open seas. In temperate latitudes at the edge of the sea, where the large primary producers are rockweeds (such as *Fucus*) and kelps (*Laminaria*), net primary productivity is as high as anywhere else in the biosphere. Sessile animals, such as sponges and anemones, are found all over the ocean bottom, but they are abundant only in relatively shallow areas that are close to the shores. Predators, such as mollusks, echinoderms, crustaceans, and many kinds of fish, roam over the bottoms of the continental shelves. Eel grass, turtle grass, and seaweeds provide shelter for many animals and increase the supply of oxygen. Snails, sea slugs, and worms crawl over the surfaces of the plants and algae, eating off the encrusted growth.

Seashores—where the sea and the land join—are of three general types along most of the shores of the temperate zones: rocky, sandy, and muddy. Organisms that live on rocky coasts, like those that live in the riffles of fast-moving streams, often have special adaptations for clinging to rocks. The algae have strong holdfasts. The starfish of the rocky coasts lies spread-eagled on the rocks, clinging with its tube feet. The abalone holds tight with its well-developed muscular foot. Mussels secrete coarse, ropelike strands that anchor them to rocky surfaces.

The organisms of the rocky coast face the additional problem of the rising and falling tides. Life along the rocky coasts is highly stratified. The supratidal zone, which is wetted only by waves and spray, is a zone of dark algal and lichen growth. The intertidal zone, alternately submerged and exposed by the tides, is commonly characterized by rockweed (a brown alga), often intermixed with many other species of red and brown algae. Animal life includes barnacles, oysters, mussels, limpets, and periwinkles. The subtidal zone, always submerged, often contains forests of kelp (large brown algae such as *Laminaria* and *Macrocystis*), sea squirts, starfish, and various other invertebrates. The characteristic stratification is due, in part, to gradients of light, temperature, and wave action, partly to competitive interactions, and also to predation by both herbivores, such as sea urchins, and carnivores, such as starfish.

Sandy beaches have fewer bottom dwellers because of the constantly shifting sands. Clams, ghost crabs, sand fleas, lugworms, and other small invertebrates live below the surface of the sand, feeding on the debris washed in and out by the tide. Above the sandy beaches, beach grasses, which spread by means of underground stems, are important for stabilizing the shifting dunes.

The mud flat, while not so rich or diverse in species as the rocky coast, supports a large number of organisms with many animals living not only on but also beneath its surface. A mud flat can support tens of thousands of individuals per cubic meter.

Mud flats, salt marshes, and estuaries (areas where the fresh water of streams and rivers drains into the sea) are the receiving grounds for a constant flow of nutrients drained off from the land and so are extremely

40-20 *Some examples of life at the seashore.* (a) *Sea otter in a kelp bed off the coast of California.* (b) *Tidal pools at low tide on Alaska's rocky southeast coast.* (c) *Common inhabitants of sandy tropical beaches are ghost crabs. These crabs, also known as racing crabs, run sideways at high speeds across the sand.*

(a)

(b)

(c)

(a)

(b)

40-21 (a) *A salt marsh. The grasses are* Spartina. (b) *A mud flat, San Francisco Bay.*

rich in animal life. They serve an important function as the spawning places and nurseries for many forms of marine life.

In the tropics and subtropics (including parts of Florida, Puerto Rico, and Hawaii), mangrove forests are important tideland communities, serving as spawning grounds for marine organisms and exporting minerals and nutrients.

Because they are often located in prime recreational and commercial areas and because they cannot be directly exploited for agriculture or lumbering, thousands of kilometers of mud flats, salt marshes, and mangrove forests are destroyed each year as these wetlands are filled and paved and rendered sterile for human occupation. Their protection is of special importance because of their role in nurturing life in the oceans.

SUMMARY

The biosphere is that part of the earth that contains living organisms. It is a thin film on the surface of the planet. The biosphere is affected by the position and movements of the earth in relation to the sun and by the movements of air and water over the earth's surface. These factors can cause wide differences in temperature and rainfall from place to place and season to season on the earth. There are also differences in the surfaces of the continents, both in composition and in altitude.

As a result of these differences, the biosphere is divided into different types of communities, called biomes. The major terrestrial biomes include the tropical rain forests, the savannas, the deserts, the chaparrals, the grasslands, the temperate deciduous forests, the taiga, and the tundra. Life formations of fresh water include the standing waters (lakes and ponds) and running waters (rivers and streams). Marine environments comprise the oceans, the coral reefs, and the shores.

QUESTIONS

1. What are the eight major terrestrial biomes? Describe the principal abiotic features of each.

2. Name a plant and an animal associated with each biome, and describe the special adaptations of each.

3. Although each of the biomes we have considered is sufficiently different from all the others to warrant its identification as a distinct biome, there are important similarities among some of the biomes. Consider the following groups of biomes: tropical rain forest/tropical deciduous forest; tropical deciduous forest/temperate deciduous forest/taiga; savanna/temperate grasslands/tundra. Describe the essential similarities and the most significant differences in the environmental factors affecting the members of each group. How do these factors affect the types of plants characterizing each biome?

4. The rate of decomposition of plant litter, animal wastes, and dead plants and animals varies from biome to biome. Describe the differences in decomposition rates in the following biomes: tropical rain forest; temperate deciduous forest; taiga. What factors in each biome are important in causing these differences? What are the consequences of these differing decomposition rates for mineral recycling, soil quality, and the size and diversity of detritivore populations?

5. Mud flats are extremely rich in animal life, yet few plants are found there. What reasonable explanation can you give for the scarcity of plant life? How can the mud flats support such a profusion of animal life in the absence of plants?

41

Social Behavior

As we noted at the beginning of this section, ecology is the study of the interactions of organisms with their physical environment and with each other. In the preceding four chapters, we have explored population growth, the interactions of different populations with each other, the organization of populations into communities and their interaction with the abiotic environment in ecosystems, and the distribution of communities throughout the biosphere. Of all the interactions among organisms, perhaps the most fascinating are those that occur among individual members of populations in which there is a social structure. In this chapter, we shall consider a few of the best-studied examples of such interactions.

A *society* consists of individuals of the same species organized in a group in which there are divisions of labor and mutual dependence. It is thus something more than an aggregation of individuals. Often, similar

41–1 *Societies of animals are held together by various forms of communication. In insect societies, which are often very large, these communications are entirely impersonal, depending primarily on exchange of chemical signals (pheromones). Responses to these stimuli are genetically programmed. In nonhuman primate societies, by contrast, relationships are personal and are based on recognition of other members of the group as individuals. Awareness of the identity of other members of the group and one's behavior toward them depend on learning as well as instinct. The photograph shows two young baboons engaged in this learning process.*

organisms—bacteria, for instance, or *Paramecium,* or mealworms—are found gathered in the same place because of environmental conditions, such as humidity, shelter, temperature, food supply, and so on. These aggregations do not represent societies in the sense that the word is used by students of animal behavior. In a society, stimuli exchanged among members of the group hold the group together. These exchanges of stimuli, which are essentially forms of communication, result in what is defined as *social behavior.*

Insect Societies

We shall begin with a description of insect societies, since they are by far the most ancient of all societies, and also, with the single exception of modern human societies, by far the most complex. Of all animal organizations, the insect societies are probably the best understood, in terms both of their evolution and of the interplay of forces that keep them together. As with other animals, the social insects evolved from forms that were originally solitary. In fact, true sociality has evolved on at least eight separate occasions in bees and four times among wasps.

Stages of Socialization

Most species of bees and wasps are solitary. Among the solitary species, the female builds a small nest, lays her eggs in it, stocks it with a food supply, seals it off, and leaves it forever (Figure 41–2). She usually dies before the larvae mature.

Among subsocial or presocial species, the mother returns to feed the larvae for some period of time, and the emerging young may subsequently lay their eggs in the same nest or comb. However, the colony is not permanent (usually being destroyed over the winter), there is no division of labor, and all females are fertile.

Eusocial, or "truly social," insects are characterized by cooperation in caring for the young and a division of labor, with sterile individuals working on behalf of reproductive ones. All ants and termites and some species of wasps and bees—for example, honey bees—are eusocial.

(a)

(b)

41–2 *Wasps, most of which are not truly social insects, do not tend their young, but often provide for them.* (a) *Wasps of the* Apanteles, *a solitary and parasitic genus, inject their eggs under the skins of caterpillars. The resultant larvae eat the internal tissues of the caterpillar, chew their way to the surface, and spin the cocoons shown here. Adults emerge from these cocoons.* (b) *The potter wasp builds a graceful clay vase, lays the eggs inside, and stocks it with caterpillars for the larvae. A caterpillar, paralyzed but not dead, provides up to 40 days' food.*

(a)

(b)

41-3 *Honey-bee workers.* (a) *The first segment of each of the three pairs of legs has a patch of bristles on its inner surface. Those of the first and second pairs are pollen brushes, which gather the pollen that sticks to the bee's hairy body. On the third pair of legs, the bristles form a pollen comb that collects pollen from the brushes and the abdomen. From the comb, the pollen is forced up into the pollen basket, a concave surface fringed with hairs on the upper segment of the third pair of legs. Transfer of pollen to the pollen basket occurs in midflight. The sting is at the tip of the abdomen.*

(b) *The mouthparts are fused into a sucking tube containing a tongue with which the bee obtains nectar. The antennae, attached to the head by a ball-and-socket joint, contain receptors for touch and for odors. The large compound eyes cannot see red (which is black, or colorless, to them) but can see ultraviolet, which is invisible to human eyes.*

Honey Bees

A honey-bee society usually has a population of 30,000 to 40,000 workers and one adult queen. The life span of a worker is approximately six weeks. Each worker, always a female, begins life as a fertilized egg deposited by the queen in a separate wax cell. (Drones, or male bees, develop from unfertilized eggs.) The fertilized egg hatches to produce a white, grublike larva that is fed almost continuously by the nurse workers; each larval bee eats about 1,300 meals a day. After the larva has grown until it fills the cell, a matter of about six days, the nurses cover the cell with a wax lid, sealing it in. It pupates for about 12 days, after which the adult bee emerges.

The newly emerged adult worker rests for a day or two and then begins successive phases of employment. She is first a nurse, bringing honey and pollen from storage cells to the queen, drones, and larvae. This occupation usually lasts about a week, but it may be extended or shortened, depending on the conditions of the colony. Then she begins to produce wax, which is exuded from the abdomen, passed forward by the hind legs to the front legs, chewed thoroughly, and then used to enlarge the comb. During this stage of employment as a houseworking bee, she may also remove sick or dead comrades from the hive, clean emptied cells for reuse, or serve as a guard at the hive entrance. During this period, she begins to make brief trips outside, seemingly to become familiar with the immediate neighborhood of the hive. It is only in the third and final phase of her existence that the worker honey bee forages for nectar and pollen.

The Queen

The queen begins life as an egg genetically identical to that of the worker. The differences between the two depend on the substance fed the queen-to-be in the larval stage and on the pheromonal influence she, in turn, exerts upon her subjects.

Queens are raised in special cells larger than the ordinary cells and shaped somewhat like a peanut shell. Larvae in queen cells are fed special glandular secretions, known as royal jelly. Attempts have been made to identify the substance in royal jelly that confers queenhood, but so far they have not been successful.

If a hive loses its queen, workers will notice her absence very quickly and will become quite agitated. Very shortly, they begin enlarging worker cells to form emergency queen cells, and the larvae in the enlarged cells are then fed royal jelly. Any diploid larva so treated will become a queen.

The queen exerts influences on her subjects by means of pheromones (page 372), of which there appear to be several. As shown by the British entomologist C. G. Butler and his co-workers, the influence of one of the pheromones, known as queen substance, inhibits ovarian development in the worker bees and prevents them from becoming queens or producing rival queens. This pheromone passes among the workers of the hive orally. As the workers meet, they often exchange the contents of their stomachs. Studies in which queen substance has been tagged with a radioactive label have shown that, as a result of this activity, the pheromone travels through the hive with remarkable speed. Within only half an hour after removal of the queen, the shortage of the queen substance is already noticed, and the hive begins to grow restless. It is difficult to understand how a single queen can produce enough pheromone to influence the entire hive of as many as 30,000 to 40,000 workers, as well as tend to her stupendous egg-laying chore

(a)

(b)

(c)

41-4 *Honey bees.* (a) *Workers tending honey and pollen storage cells. The honey is made from nectar processed by special enzymes in the workers' bodies.* (b) *Workers feed the queen (top center) and lick queen substance from her body. Queen substance, a pheromone, prevents sexual maturation in the workers.* (c) *Four pupae in different stages of development. A fully developed worker at the far left has shed her pupal skin and is ready to emerge, which she will do by gnawing through the wax cap.*

(commonly more than 1,000 per day). It has been suggested that, after the pheromone is passed among the workers, it is fed back to the queen in a reduced form and she need simply oxidize it to reactivate it.

Note that the words "queen" and "royal" imply, by analogy, that the queen bee's life is more desirable than that of her sisters. However, she can also be viewed, from a slightly different perspective, as an egg-laying machine held captive and operated by the workers.

Winter Organization

The honey-bee colony differs from that of subsocial bees in that it survives the winter. This means that the bees must stay warm despite the cold. Honey bees cannot fly if the temperature falls below 10°C and cannot walk if the temperature is below 7°C. Within the wintering hive, bees maintain their temperature by clustering together in a dense ball; the lower the temperature, the denser the cluster. The clustered bees produce heat by constant muscular movements of their wings, legs, and abdomens. In very cold weather, the bees on the outside of the cluster keep moving toward the center, while those in the center move to the colder outside periphery. The entire cluster moves slowly about on the combs, eating the stored honey from the combs as it moves.

New Colonies

In the spring, when the nectar supplies are at their peak, so many young are raised that the group separates into colonies. The first new colony is always founded by the old queen, who leaves the hive, taking about half of the workers with her. The group stays together in a swarm for a few days, gathered around the queen, after which the swarm will settle in some suitable hollow tree or other shelter found by its scouts.

As the old queen is preparing to leave the hive, the new queens are getting ready to emerge. These two events are synchronized by sound signals transmitted through the comb. As these signals are exchanged, the workers remain motionless. During this period, ovarian development begins in some of the workers, a few of which lay eggs. The unfertilized eggs develop into drones. After the old queen leaves the hive, a new young queen emerges, and any other developing queens are destroyed. The young queen

41-5 *Until its scouts find a suitable location for a new hive, this honey-bee swarm is temporarily residing in a pine tree.*

then goes on her nuptial flight, exuding a pheromone (apparently also the queen substance) that entices the drones of neighboring colonies. She mates only on this one occasion (although she may mate with more than one male) and then returns to the hive to settle down to a life devoted to egg production.

During her nuptial flight, the queen receives enough sperm to last her entire life, which may be some five to seven years. These are stored in a special organ in her reproductive tract and are released, one at a time, to fertilize each egg as it is being laid. The queen usually lays unfertilized eggs only in the spring, at the time males are required to inseminate the new queens.

The drones' only contribution to the life of the hive is their participation in the nuptial flight. Since they are unable to feed themselves, they become an increasing liability to the social group. As nectar supplies decrease in the fall, they are stung to death by their sisters or are driven out to starve.

Organizational Principles of Insect Societies

Insect societies, although they resemble human societies in many ways, differ from them in at least two important respects. The first major point of difference is that in insect societies only a very few individuals reproduce. Because of the reproductive pattern, all of the insects in a particular hive are likely to be very similar genetically. The seeming altruism, or self-sacrifice, of the workers on behalf of the queen (their mother) and the larvae (their sisters) is apparently a way of ensuring that their own genes have maximum representation in the next generation. In some respects, an insect society is more like a superorganism with different specialized parts than a collection of individuals.

Second, recognition among members of the insect society is based entirely on chemical, tactile, visual, and auditory signals. There is no personal recognition of individual members as there is in primate or other mammalian societies. Stereotyped signals and responses are, of course, in keeping with (and demanded by) the relatively short life span of insects. A worker honey bee, for example, with her life span of about six weeks, has little time for complex learning. This impersonality is also in keeping with the large, sometimes enormous, size of the societies themselves.

Vertebrate Societies

Vertebrate societies range from small, often transient groups of closely related individuals, in which the social focus is on the raising of the young, to larger, permanent, quite stable groups, such as those found among some of the primates. Some of the societies—such as flocks of birds, schools of fish, and large migratory herds of herbivores—may number in the thousands, but they are not usually as large as the societies of most eusocial insects.

We are among the most social of the social animals, and so, from our point of view, social living would appear to be the norm. However, in fact, living in a society has decided disadvantages for the individual organism. The most important of these is the competition for food resources. Another is the much greater vulnerability to disease. Also, a large group may attract large predators, as when sharks attack a school of fish. Finally, an individual in a society may have to compete for a mate, thereby diminishing his or her chances of leaving viable offspring.

Generally, the advantage of social living seems to be related to avoidance of predation (Figure 41–6). For example, grazing animals feed in herds, and birds, when threatened, tend to tighten their flock. It is almost impossible for a hawk to catch a bird within the flock, since the hawk will be subjected to a constant bombardment of other bird bodies moving at high speeds. In some societies, such as those of wolves, social behavior also permits the hunting of large animals, such as moose or caribou, that could not be caught or killed by a solitary individual.

41–6 *Note that zebras standing together like this can not only scratch each other's back and whisk flies from each other but can also command a 360° view of the savanna.*

Communication among Vertebrates

As we noted earlier, societies are held together by reciprocal stimuli—communications. Among vertebrates, communication is often visual, involving signals of color and shape expressed in complex behavior patterns. As an indication of the complexities involved, we shall touch on two types of behavior that serve to keep animals in groups: imprinting and rituals.

Imprinting

Birds, such as swans, chickens, and turkeys, that are physiologically mature enough to leave the nest soon after they are hatched follow the first moving object they see. For ducklings, the effective objects can range from a matchbox on a string to a walking man. Once one of these newly hatched birds has become attached to a particular object, it will follow only that object. Under natural circumstances, of course, this object will be a parent, and it is this following response that keeps the young birds close behind and well within the protective range of the parent until the end of their juvenile period, when the response is lost.

The learning pattern involved in this act of recognition is called *imprinting*, and it differs from most other types of learning in that the period in which it can occur is very limited. Studies of the newly hatched mallard duckling, using a mechanically operated decoy, showed that imprinting was most effective between 12 and 16 hours after hatching. After 24 hours, 80 percent of the ducklings failed to follow moving objects; and after 30 hours, all the ducklings avoided them. Similarly, chicks will not follow a moving object when they are only a few hours old nor when they are several days old but only during the intervening period. It is believed that the end of the imprinting period is marked by the development of fear responses, which seem to arise in normally reared birds at about this time.

Imprinting seems to influence mate selection in the adult birds. Austrian zoologist Konrad Lorenz reports many examples of ducks and geese becoming sexually fixated on objects or on members of other species, including Lorenz himself (Figure 41–7). Controlled experiments with ducklings support these observations. Ducklings raised with females of another species courted females of that species when they reached maturity, while ducklings raised with females of their own species did not.

Nina Leen, Time-Life Picture Agency

41–7 *Many species of precocial birds (birds that are able to walk as soon as they are born) will follow the first moving object they see after hatching and will continue to show this following response as they mature. The phenomenon is known as imprinting.*

The birds shown here are goslings, and the object of their affection is Konrad Lorenz. The branch of biology known as ethology—the study of the behavior characteristic of species, with emphasis on the evolution and adaptive value of such behavior—had its origins in the work of Lorenz and his former student, Niko Tinbergen.

The term "imprinting" is usually applied only to phenomena associated with the following response in birds, but similar effects are seen in other animals. For instance, if a lamb is taken from its mother right after birth and bottle-raised, it will not follow other sheep when it is returned to pasture with the rest of the flock.

Rituals

Most vertebrates—fish and birds in particular—have a distinct preference for maintaining discrete distances between themselves and other animals, whether of their own or another species. Birds lined up on a telephone wire, for instance, will be found almost equidistant from each other, as though the spaces were measured off by an invisible meterstick. Even those animals that fly in flocks or swim in schools tend not to touch one another.

One function of the mating ceremonies of these animals is to allay the highly adaptive fear of contact so that mating can take place. As a consequence, many of these ceremonies incorporate both aggressive and appeasing behavior. In the mating behavior of the stickleback (page 510), for example, the male's "zig" toward the female is identical to a motion of attack, and when the male "zags," the motion entices her toward the nest. Most females flee the attack motion, but the one whose need to lay her eggs is sufficiently great stands still, turns sideways (an appeasing gesture on her part since an antagonist would zig back), and then follows the male's zag.

In mating ceremonies between birds, feelings of fear and aggression brought into play by the closeness of another individual may be handled by being redirected at either a real or an imaginary antagonist. Konrad Lorenz describes the so-called "triumph ceremony" in the graylag goose, one of his favorite research subjects and companions. As a form of greeting to his partner, the gander proceeds to attack an "enemy." This attack is performed, as is a real attack, with the head and neck pointing obliquely forward and upward and is accompanied by a raucous trumpeting. After the "enemy" is routed or defeated, the gander returns to his partner. On his return, the gander holds his head lowered and pointed forward, but instead of pointing directly at the goose, as he would at an enemy, he points obliquely past her. She comes forward to meet him, her head inverted submissively, and he cackles to her triumphantly.

This ceremony, which probably had a purely sexual origin, now serves to hold entire flocks together, and even small goslings participate in elements of the triumph ceremony. When a young male performs the ceremony with a strange young female, it usually marks the beginning of a mating bond that may last for the entire lifetime of the individuals. The ceremony is typically performed between young geese the year before mating and breeding begin, and it will continue to be performed by the partners throughout their lives whenever they encounter one another after even a short separation. By the intensity with which the ceremony is performed, an experienced observer can judge the length and strength of the bond between the partners.

Although few animals—except perhaps humans of ambassadorial rank—have such elaborate social rituals, animals of many species exhibit obligate social formalities upon meeting. Think, for example, of the greeting ceremonies between domestic dogs, and how they vary with sex, rank, and familiarity. Rituals serve the function both of allaying the anxieties of the individuals involved and of identifying them to one another as members of an "in" group.

41-8 *Rituals and ceremonies are common among birds.* (a) *Greeting ceremonies between storks at the nest. Pair bonding between storks characteristically endures for a lifetime, based, in part, on the repeated performance of this ritual.* (b) *Courtship display between two albatrosses of the Galapagos.* (c) *Female tern begging for a fish from a male tern. If he is ready to mate, he will give her the fish (although she may have to ask several times). If she is ready, she will eat it; otherwise she returns it. Such mating ceremonies serve to promote genetic isolation (page 510).*

(a)

(b)

(c)

41-9 *Wolves have dominance hierarchies of both males and females. Here a subordinate female is licking the muzzle of the dominant female. This same muzzle-nuzzle gesture is used by pups begging for food.*

Social Dominance

Vertebrate societies are often arranged in hierarchies. One type of social dominance that has been studied in some detail is the *pecking order* in chickens. A pecking order is established whenever a flock of hens is kept together over any period of time. In any one flock, one hen usually dominates all the others; she can peck any other hen without being pecked in return. A second hen can peck all hens but the first one; a third, all hens but the first two; and so on through the flock, down to the unfortunate pullet that is pecked by all and can peck none in return.

Hens that rank high in pecking order have privileges such as first chance at the food trough, the roost, and the nest boxes. As a consequence, they can usually be recognized at sight by their sleek appearance and confident demeanor. Low-ranking hens tend to look dowdy and unpreened and to hover timidly on the fringes of the group.

During the period when a pecking order is being established, frequent and sometimes bloody battles may ensue, but once rank is fixed in the group, a mere raising or lowering of the head is sufficient to acknowledge the dominance or submission of one hen in relation to another. Life then proceeds in harmony. If a number of new members are added to a flock, the entire pecking order must be reestablished, and the subsequent disorganization results in more fighting, less eating, and less tending to the essential business, from the poultry dealer's point of view, of growth and egg laying.

Pecking orders reduce the breeding population. Cocks and hens low in the pecking order copulate much less frequently than socially dominant chickens. Thus the final outcome is the same as if the social structure did not exist: The stronger and otherwise dominant animals eat better, sleep better, and leave the most offspring. However, because of the social hierarchy, this comes about with a minimum expenditure of lives and energy.

Territories and Territoriality

Most vertebrates stay close to their birthplaces, occupying a home range that is likely to be the same home range occupied by their parents. Even migratory birds that travel great distances are likely to return year after year to the same areas. Often these home ranges are defended, either by individuals (or more likely, mating pairs) or by groups against other individuals or groups of the same species or closely related species that use the same resources. Areas so defended are known as *territories*, and the behavior of defending an area against rivals is known as *territoriality*.

41-10 *A subordinate baboon turns his buttocks toward a superior. This gesture, known as presenting and used by females to indicate their readiness to mate, is also used between males and between females to signify submission or conciliation or to beg for special favors. The superior is reassuring the subordinate with a pat on the back.*

Territoriality in Birds

Territoriality was first recognized by an English amateur naturalist and bird watcher, Eliot Howard, who observed that the spring songs of male birds served not only to court the females but also to warn other males of the same species to stay away from the terrain that the prospective father had selected for his own. In general, a territory is established by a male. Courtship of the female, nest building, raising of the young, and often feeding are carried out within this territory. Frequently the female also participates in territory defense.

By virtue of territoriality, a mating pair is assured of a monopoly of food and nesting materials in the area and of a safe place to carry on all the activities associated with reproduction and care of the young. Some pairs carry out all their domestic activities within the territory. Others perform

(a)

(b)

(c)

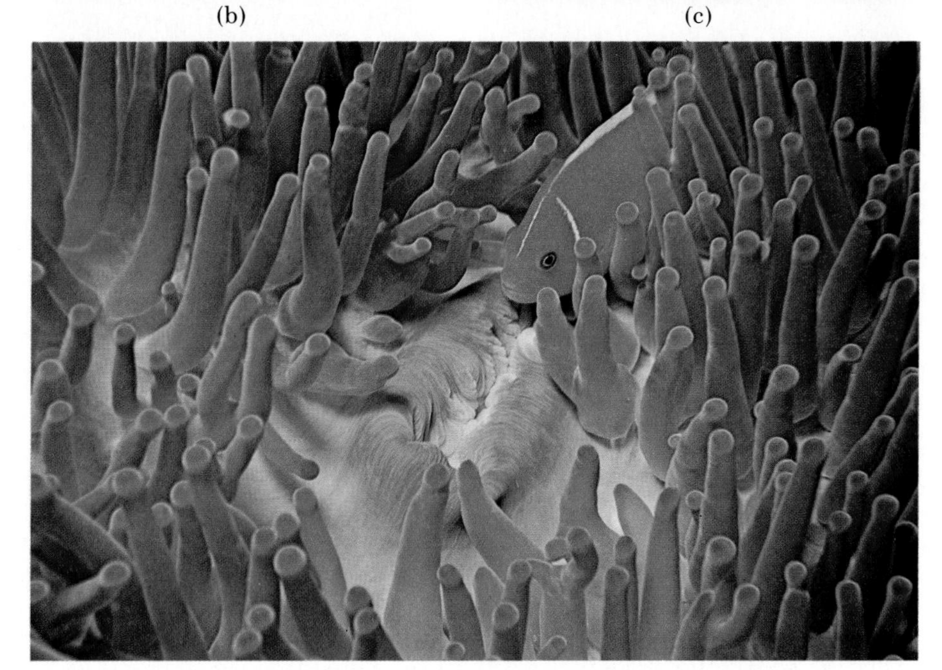

(d)

41-11 *Territories come in many shapes and sizes.* (a) *The male Uganda kob displays on his stamping ground, which is about 15 meters in diameter and is surrounded by similar stamping grounds on which other males display. A female signifies her choice by entering one of the stamping grounds and grazing there. Only a small proportion of males possess stamping grounds, and those that do are the only ones that breed.*

(b) *A fiddler crab's territory is a burrow, from which he signals with his large claw, beckoning females and warning off other males.*

(c) *Howler monkeys shift their territories as they move through the jungle canopy, but maintain spacing between groups by chorusing.*

(d) *Territoriality is common among reef fish. For many species, a territory is a crevice in the coral, but for others, such as the skunk clown fish shown here, the territory is a sea anemone. The fish is covered by thick slime that partially protects it from the poison of the tentacles, but its acceptance by the anemone is chiefly a consequence of behavioral adaptation of the fish, which even mates and raises its brood among the tentacles.*

mating and nesting activities in the territories, which are defended vigorously by the males, but do their food gathering on a nearby communal feeding ground, where the birds congregate amicably together. A third type of territory functions only for courtship and mating, as in the bower of the bowerbird or the arena of the prairie chicken. In these territories, the males prance, strut, and posture—but very rarely fight—while the females look on and eventually indicate their choice of a mate by entering his territory. Males that have not been able to secure a territory for themselves are not able to reproduce; in fact, there is evidence from studies of some territorial species, such as the Australian magpie, that adults that do not secure territories do not mature sexually.

Territorial Defense

Even though territorial boundaries may be invisible, they are clearly defined and recognized by the territory owner. With birds, for example, it is not the mere proximity of another bird of the same species that elicits aggression, but its presence within a part of a particular area. The territory owner patrols his territory by flying from tree to tree. He will ignore a nearby rival outside his territory, but he will fly off to attack a more distant one that has crossed the border. Animals of other species are generally ignored unless they are prey or predators.

Once an animal has taken possession of a territory, he is virtually undefeatable on it. Among territory owners, prancing, posturing, scent marking, and singing and other types of calls usually suffice to dispel intruders, which are at a great psychological disadvantage. For example, a male cichlid, a tropical freshwater fish, will dart toward a rival male within his territory, but, as he chases the rival back into its own territory, he begins to swim more slowly, the tail fin seemingly working harder and harder, just as if he were making his way against a current that increases in strength the farther he pushes into the other male's home ground. The fish know just where the boundaries are and, after chasing each other back and forth across them, will usually end up with each one trembling and victorious on his own side of the truce line.

Similarly, the expulsion from communal territories is typically accomplished by ritual rather than by force. For example, among the red grouse of Scotland, the males crow and threaten only very early in the morning, and then only when the weather is good. This ceremony may become so threatening that weaker members of the group leave the moor. Those that leave often starve or are killed by predators. Once the early-morning contest is over, the remaining birds flock together and feed side by side for the rest of the day.

Social Structure, Aggression, and Population Regulation

As we have seen, animals that are low in a social hierarchy, or that do not gain possession of territories, or that are excluded from the "home range" or breeding area do not produce young. As older animals die, younger ones will contend for their places, keeping the breeding population stabilized. If conditions are particularly favorable, or if the range is enlarged, more animals can gain access to the breeding community. However, many are never able to breed.

The submissive acceptance of an inferior social position would seem to offer little reward in terms of "fitness" to those low in a dominance hierarchy or deprived of a territory and therefore deprived of an opportunity to reproduce. Would it not be advisable for such individuals to fight with their last calorie of strength rather than to give up the struggle? What, after all, have they got to lose? Actual combat, however, is rare among members of the same species, and such fighting as does occur is often ritualized in such a way that both animals remain unharmed. For instance, male iguanas of the Galapagos Islands fight by pushing their heads against one another; the one that drops to its belly in submission is no longer attacked. Male cichlid fish of one species first display, presenting themselves head on and then side on, with their dorsal fins erected, and then beat water at each other with their tails. If this does not bring about a decision, each grasps the other by its thick, strong lips. They then pull and push with great force until one lets go and, unharmed but defeated, swims away. Rattlesnakes, which could kill each other with a single bite, never bite when they fight but instead glide along side by side, each pushing its head against the head of the other, trying to push it to the ground in a form of Indian wrestling. Many antlered animals, such as stags of the fallow deer, which have long and vicious horns, follow an equally careful ceremony and attack only when they are facing each other, so that their antlers are used only for dueling and not for goring.

Ritualized combat, which prevents killing and maiming, makes sense biologically. The winner is not injured, and the loser lives to fight another day. Waiting may be worthwhile. Some years ago, for example, a group of scientists undertook a study in Maine to determine whether or not the warblers nesting in the spruce trees exerted a controlling effect on the caterpillars of the spruce bud worm. They mapped the position of the singing males within an area of 16 hectares, found 148 pairs, and began to shoot them all. After 3 weeks, they had shot 302 male warblers (and a lesser number of females), and there were still male warblers singing throughout the 16 hectares. One moral of this story, among others, is that animals that do not have territories are always ready to claim them and sometimes succeed in doing so. Thus, in the long run, waiting may be a better evolutionary bet than engaging in a probably unsuccessful combat. Territoriality and other forms of social dominance are successful strategies, in short, because they ritualize competition and secure for the winner adequate resources for the breeding of young. The limitation of the size of the population is a side effect of these activities.

41-12 *Two Thomson's gazelles, sparring strictly according to the rules.*

Social Organization among Baboons

Baboons are large, quadrupedal African monkeys with protruding, doglike muzzles. One genus (*Papio*) and four species are commonly recognized: a savanna species (*Papio anubis*); two forest species (the drill and the mandrill), both short-tailed baboons; and a desert species, *Papio hamadryas,* the hamadryas baboon.

The baboons are of particular ecological interest for two reasons. First, unlike most monkeys, they live on the ground and have done so for several million years. For this reason, differences between baboons and other monkeys can be related to some degree to their environments. Second, there is a marked difference in social organization between the savanna and forest species, on the one hand, and the desert species, on the other. Since the groups are closely related genetically, obviously sharing a common ancestor in the not-too-distant past, it is very likely that the differences can be explained ecologically rather than genetically. Both of these statements must be tentative, however; field studies of baboons have been made only since the late 1950s, and most other monkey groups have been studied inadequately or not at all.

The Genus Papio

Baboons are large, with adult males weighing about 55 kilograms. Their size is clearly related to a terrestrial existence; among the primates, only large animals, such as baboons, some of the great apes, and *Homo sapiens,* are terrestrial.

Among all baboons, adult males are very different in appearance from adult females (Figure 41–13). This phenomenon, known as sexual dimorphism, is much less pronounced among most other primates, particularly those that are arboreal (tree-dwelling). It is believed to be the result of sexual selection (page 514).

41–13 *A male baboon and a female with an infant. The males are about twice as large as the females, with prominent canine teeth and long manes, or mantles. The young are born black, and their coats become lighter as they mature.*

41-14 *Baboons sleep in groups, either in tall trees, as shown here, or on ledges or steep cliffs. The size of the group and the choice of a sleeping site appear to depend more on the terrain than on the species.*

Like most primates, baboons are almost exclusively vegetarian, living on leaves, fruits, flowers, and young stems, all of which they pick and consume on the spot. They also dig for roots, bulbs, and tubers. This vegetable diet is supplemented by insects and occasionally by meat, such as snakes, lizards, fledgling birds, or young gazelles, when they are encountered accidentally. There is no organized hunting for prey animals and no sharing of prey. Baboons sleep in tall trees or on cliffs, returning to the nesting site every night (Figure 41-14). Hence they must often travel to find food, sometimes for great distances across open country.

The Structure of the Band

Baboons are organized into multimale bands; those observed have ranged in size from eight to more than 185 individuals. Adult females outnumber adult males by about 2 to 1. Much of this difference can be accounted for by the much slower maturation rate of the males. Although an adult male can probably produce viable sperm by the time he is about four years old, he does not reach full size nor are his canines fully erupted until he is eight or more, and therefore he is not counted as an adult male either in the social organization of the band or by the field observer peering through binoculars. Females, on the other hand, are adults by the age of two or three.

The bands are territorial, with fairly distinct home ranges, and bands rarely meet. When they do, the band farthest from the center of its customary range generally moves away, sometimes after an exchange of vocalizations, canine displays, and gestures, but more often with no visible reaction at all, except perhaps a slight display of nervousness. When the bands move in open country, females and infants characteristically are in the middle of the group, close to adult males. Mass counterattacks by adult males, who placed themselves between the group and the attackers, have been observed to rout a leopard and disperse a large dog pack.

Another function of the band is the pooling of information. Hans Kummer, one of the most experienced observers of baboons, describes a band at the beginning of a day's march as resembling a giant amoeba, stretching out and then withdrawing one pseudopod after another, as small segments begin to move off in one direction, only to rejoin the band if the others do not follow. Finally, a movement is initiated that meets with general approval, and the day's march begins.

The size of the group seems to be related to its habitat. Very small groups are correlated with small areas of vegetation, quite widely separated from one another, whereas larger groups may be found where a typical feeding site might be a clump of trees. Logically, group size must be a compromise determined by the availability of food resources and the need of the group to defend itself.

41-15 *Dominant males threaten subordinates by staring, yawning (with the ears laid back), raising the eyebrows, and, at close range, by grinding their teeth. The huge canines are found only in the males, indicating that they are related less to diet (male and female eating habits are the same) than to dominance and aggression.*

Social Dominance

In all baboon species except the hamadryas, the band is organized around a dominance hierarchy of adult males. As in other dominance hierarchies, the band is maintained mostly by behavioral conventions (Figure 41-15). The most reliable index of hierarchical position is the gesture of presentation, in which a subordinate animal demonstrates submissiveness by presenting its hindquarters. (This gesture is also used by females in

41-16 *As with all social primates (except* Homo sapiens), *grooming is one of the primary bonds among group members. The amount of grooming an individual receives is closely related to his or her position in the social hierarchy. Here an adult female grooms a dominant male. Being groomed is not only pleasurable but also removes insects and other parasites from the skin.*

estrus—the time of ovulation and increased sexual activity—as an invitation to copulate.) The top-ranking male presents to no other individual, the second-ranking male only to the top one, and so on down the social ladder. Adult females are subordinate to males but also have a separate dominance hierarchy, although not as clearly structured as that of the males. Adult females present to superior females.

Socially superior males mate more often than inferior males and are thus more likely to father offspring. Dominant males also have first choice of nesting sites. However, because there is no shared food supply, dominants are not allocated a larger share of food than subordinates, as they are in some other pecking orders.

Dominant males play a leading role in territorial disputes, thus maintaining the spacing among bands. They are also the most aggressive in defending the band against interspecific aggression.

Social Bonds

As among almost all higher primates, grooming is a prominent form of social behavior among baboons (Figure 41-16). Dominant males, females in estrus, and females with infants are groomed most frequently, but all members of the group receive some grooming attention. The function of grooming is clearly not only hygienic but also a continual reinforcement of social bonds.

Infants are another clear social bond. The infant, which is all black and therefore very conspicuous for its first few weeks of life (perhaps so it can be guarded more zealously), usually clings to its mother, who may support it with her arm. The infant or infant-mother pair are clearly attractive to other members of the group, who tend to cluster around, especially when the infant is newborn. Mothers with infants are groomed frequently, particularly by other females, and dominant males attend them closely as the band moves. An infant that loses its mother will be adopted, sometimes by a childless female, but often by a young male.

(a)

(b)

41-17 (a) *Mother and nursing infant.* (b) *A female baboon presenting, in social deference, to a mother and newborn infant because she wants to approach and touch the infant.*

The Hamadryas Baboon

The hamadryas baboon is unlike the other members of the genus *Papio* in that there are three distinct levels of social organization. The principal unit is a male with one or more females and their young, totaling up to seven or eight members. These one-male units group together into bands, similar in size and organization to the bands of the other baboon species. Finally, the bands come together in troops, some of which have been counted as containing 750 members.

Students of primate behavior hypothesize a direct relationship between these social organizations and the environmental resources of the hamadryas. The habitat, which is on the edges of deserts, is typically arid grassland, interspersed with thorny acacias and other small trees and bushes. There are no tall trees, and the baboons of this area sleep on ledges on the vertical slopes of steep cliffs. The baboons move across open country, often for long distances, between their sleeping cliffs and their feeding sites. They travel in bands, break up into one-male units for feeding—typically one unit to a tree—regroup into bands, return to the cliffs, where they often sleep in troops, and then recongregate each morning in the band. Thus each social unit serves a clear and important function. The one-male unit is an optimal foraging unit, provided with a protector; the band provides for mutual defense when traveling; and the troop makes possible maximum utilization of safe sleeping sites within the habitat.

The strongest bonds are those that hold the one-male unit together. The females mate exclusively with the unit leader, and almost all social interactions, such as grooming, are carried out within the unit. Juveniles sometimes leave to play with those of other units within the band.

The structure of the hamadryas band is generally similar to that of other baboon species, but the ties that hold the band together are less strong than those that hold the one-male unit together. The band, however, like the one-male unit, is a stable social group, always composed of the same members. The troop, on the other hand, is made up of varying groups of bands from the same territory that apparently recognize one another and are mutually tolerant.

41-18 *A hamadryas band. Note the grouping by one-male units.*

Because the male-dominated band is the primary social structure of the other *Papio* species (and also of the macaques, arboreal monkeys who are their closest relatives), the hamadryas band appears to be a legacy from their common ancestors. Similarly, the troop seems easy to understand as a direct outgrowth of environmental pressures; baboons of other species share water holes and large sleeping groves when they have to.

What, however, are the bonds that hold the one-male unit together? The cohesion of the unit turns out to depend primarily on two behavioral patterns, both those of the male. The first is a herding instinct toward the females. The females are taught to follow the male. He constantly watches them over his shoulder, and, if one drops back or attempts to slip away, he stares, threatens, and sometimes nips her on the back of the neck, after which she immediately follows. If a male hamadryas is presented with a female olive baboon (*Papio anubis*), he accepts her quite readily and immediately trains her to follow him.

The second behavioral pattern is a strong inhibition among baboon males against taking another male's females. This, too, was tested experimentally. Two adult males, who knew each other, and a strange adult female were trapped. One male was enclosed with the female while the other was permitted to watch from a cage 10 meters away. The male caged with the female immediately made grooming, mounting, and herding advances toward her, as he would toward a member of his harem. Fifteen minutes later, the second male was put in the same enclosure with the pair. He not only refrained from fighting over the female, but he even avoided looking at them. The animals were separated, and two days later the experiment was repeated again, this time switching the males. Once again, the male that had watched the activities between the male-female pair was strongly inhibited in his behavior.

Another adaptive feature of this inhibition is that it applies only to adult males and females. Young males are attracted to juvenile females, which they "kidnap" from their family groups before they are sexually mature and train to follow them. Thus the younger male has a detour around the inhibition barrier, and new family units can come into existence.

41–19 *A young male hamadryas with a juvenile female that he has stolen from another one-male unit. New family groups originate in this way.*

Lest you be tempted to extrapolate to human behavior from these accounts of baboon behavior, it is of some interest that, among the geladas (large, baboonlike monkeys that live in the mountainous grasslands of Ethiopia), groups are formed and held together by activities of both sexes. Dominant males pair with dominant females, wife number one is dominant over wife number two, and so on.

The baboons remind us that social behavior in animals is ecologically adaptive—that is, that it promotes the survival of members of the social organization (and particularly of their young) within a given habitat or range of environmental conditions.

Sociobiology

Sociobiology means, quite simply, the biology of social behavior. As such, it is a natural outgrowth of ethology—the study of the behavior characteristic of species. The evolution of social behavior began to command serious attention among modern biologists in 1962 when V. C. Wynne-Edwards proposed that natural selection might act on groups as well as individuals. This concept seemed to offer satisfying explanations for many previously puzzling phenomena, such as altruism in social insects and territorial exclusion, which seemed to serve the group but whose advantage to the individual was not immediately clear. In 1964, W. D. Hamilton offered a cogently argued response to Wynne-Edwards, based mainly on the genetic relationship of the social insects (page 598). He demonstrated, in these cases at any rate, that the concept of group selection was neither necessary nor useful—especially since no mechanism had been suggested as to how it might operate. Since that time, numerous publications have demonstrated the benefit in evolutionary terms to the individual of various types of group behavior, even, paradoxically, of those forms of behavior that lead to the failure to reproduce by the individual exhibiting the behavior. Most biologists are now in agreement that group selection, if it exists at all, is not a major adaptive force in social behavior.

The demonstrations by Hamilton and others of the extent to which evolutionary theory could explain social behavior, such as territoriality and altruistic acts, gave impetus to another group of studies and controversies that are still continuing. These center around attempts to explain human behavior on the same evolutionary bases used so successfully to explain the behavior of other social organisms. The most important serious publication in this area is the brilliant and beautiful *Sociobiology* by E. O. Wilson, a Harvard scientist whose work on insect societies, in particular, has won high esteem. The last chapter of *Sociobiology* (later expanded into another book, *On Human Nature*) is given over to speculations on the genetic basis of altruism, ethical values, "ease of doctrination," and other characteristics attributed to humankind.

Many attacks have been launched on the attempt to apply the theories of sociobiology to the human species. These attacks may seem contrary to the spirit of biology in general; over and over again, for instance, we have had occasion to point out that *Homo sapiens* is a member of the biological continuum, even sharing genes and enzymes with the prokaryotes. There are two major reasons for the objections, one biological and one political.

First, with the exception of some very simple reflexes—such as the suckling response in newborn infants—there is very little evidence for a genetic basis for any human behavior. In fact, little is known about the genetic basis of any complex behavior. There is, on the other hand, much evidence that human behavioral patterns are very flexible. This concept is supported by numerous observations that social behavior varies widely with culture and, within any culture, individual behavior varies widely with personal experience. Whatever may be programmed into our genes—and it is likely we shall never know what this is—would appear to be overridden by our cultural evolution and personal history.

Second, in the past, political and personal philosophies of biological determinism have proved particularly pernicious. They have offered a justification for economic exploitation (page 487), for slavery, for the racial policies of Nazi Germany, and for all the ugly remnants of racism that persist today. The attempts to apply sociobiology to *Homo sapiens* are highly speculative, and the deep concern of many people has been that such speculation might be misused—as other biological concepts have been misused in the past—to justify violence and aggression toward and exploitation and oppression of groups or individuals.

SUMMARY

A society is a group of individuals of the same species organized in a cooperative manner and communicating with one another in some way.

Among the most complex societies are those of insects. These societies are matriarchies, centering around the care of the queen and the raising of the brood. The behavior of members of the society is determined by chemical substances—pheromones—exchanged among members of the society. In eusocial insects, only one female is reproductive; the other females are workers, caring for the reproducing individual and the young.

Among vertebrates, the principal advantage of social living appears to involve either avoidance of predators or the communal hunting of prey. The factors that hold vertebrate societies together are varied and complex. Two such behavioral influences are imprinting and rituals. Imprinting involves a very specific sort of learning that takes place in a very narrow developmental time period and results, in nature, in the recognition of members of one's own species. Rituals are stylized behavior patterns, shared by members of a group, that allay anxieties and promote recognition of group members by one another.

Some animal societies are organized in social hierarchies. Socially inferior animals—those low in the pecking order—reproduce less frequently than their social superiors. They are also the first to starve if food is limited or to be driven out if shelter is limited.

Territories are areas defended by an individual or a society against others of the same species. Territories may be "real"—that is, they may be actual areas of land containing food and nesting material to support a mating pair and young—or they may be symbolic, such as an arena. In either case, the only animals that breed are those with territories.

Acts of aggression and conflicts establishing social dominance or territories are ritualized in animal societies, with the result that both victor and vanquished are unharmed. Animals without territories or low in the social hierarchy provide replacements for territory owners and a reserve for population expansion if the range or food supply available to a population is increased.

Baboons are large, quadrupedal terrestrial monkeys, with pronounced sexual dimorphism. They are organized into multimale bands. The males have a pronounced social hierarchy; superior males mate more often, have first choice of nesting sites, and play a more important role in defending the band than their social inferiors. The hamadryas baboon is unusual in that the band is subdivided into one-male family units. These units group into bands, and the bands come together in large troops, which utilize communal sleeping cliffs. Thus, social organization is hypothesized to be related to the habitat. The one-male family is an optimal foraging unit, the band is the traveling unit, and the troop provides for the most efficient use of the sleeping resources.

Sociobiology is the biology of social behavior.

QUESTIONS

1. Distinguish among the following: community/society; eusocial/subsocial; queen/worker/drone; social dominance/territoriality.

2. Territoriality and social dominance achieve the same results. What are they? How do the results differ in a population in which these forms of behavior are not present?

3. In what ways are human societies different from insect societies? How are they similar?

4. What behavioral conventions limit intraspecific aggression in *Homo sapiens?* Do any foster it?

42

Human Evolution and Ecology

According to the fossil record, the first mammals—the class to which *Homo sapiens* belongs—arose from a primitive reptilian stock about 200 million years ago, at about the time of the first dinosaurs. Our information about these mammals is very slight. From the scraps of fossil evidence—fragments of skulls and some teeth and jaws—we know that the first mammals were about the size of a house cat. They had sharp teeth, indicating that they were basically carnivorous, but since they were too small to attack most other vertebrates, they are assumed to have lived on insects and worms, supplementing their diet with tender buds, fruits, and perhaps eggs. These first mammals were probably nocturnal, judging by the large size of their eye sockets, and they were almost certainly warm-blooded. If such an animal were alive today, it would be classified as an insectivore, something like a ground shrew.

42-1 *Cave painting from Lascaux, France. This wild ox is the aurochs,* Bos primigenius, *which became extinct in Europe during the seventeenth century. These animals were much larger than modern cattle, with the bulls often as much as 2 meters high at the shoulders. The hunting of large game such as this bull was probably an important factor in human evolution.*

42-2 *A modern tree shrew, which the earliest primates probably resembled. If you look closely you will see five-digited paws. Although clawed, they can be spread out and used for grasping. In some classification systems, tree shrews are grouped with the primates, and in others, with the insectivores, which indicates the closeness of the two evolutionary lines.*

For about 130 million years, these small mammals led furtive existences in a land dominated by reptiles. Then suddenly, as geologic time is measured, the giant reptiles, the dinosaurs, disappeared. The cause of the dinosaur extinction is one of the great biological mysteries. The disappearance of the dinosaurs occurred at a time when, geologists believe, there was a drop in the average temperature and, perhaps more important, a marked increase in seasonal temperature variations. In any case, by the end of the Cretaceous period all of the dinosaurs had disappeared forever, and about 65 million years ago an explosive adaptive radiation of the mammals began.

The early mammals immediately diverged into the two dozen or so different lines that included (1) the monotremes, or egg-laying mammals, of which the duckbill platypus is one of the few remaining examples; (2) the marsupials, such as the kangaroos, opossums, koala bears, and others whose young are born in embryonic form and continue their development in pouches; and (3) the placentals, by far the largest group. Among the placentals are carnivores, ranging in size from the saber-toothed tiger down to small, weasel-like creatures; herbivores, which include not only the many wild grazing animals but also most of our domesticated farm animals; the omnipresent rodents; and such odd groups as the whales and dolphins, the bats, the modern insectivores, and the primates. We are placental mammals and members of the primate order, as are tarsiers, lemurs, monkeys, and apes, among others.

Trends in Primate Evolution

Primate evolution began when a group of small, shrewlike mammals took to the trees. Fossils of early primates indicate that they closely resembled modern-day tree shrews (Figure 42-2). Most trends in primate evolution seem to be related to various adaptations to arboreal life.

The Primate Hand and Arm

With a few exceptions, primates have five digits and a divergent thumb. The divergent thumb, which can be brought into opposition to the forefinger, greatly increases gripping powers and dexterity. There is an evolutionary trend among the primates toward finer manipulative ability that reaches its culmination in humans (Figure 42-3).

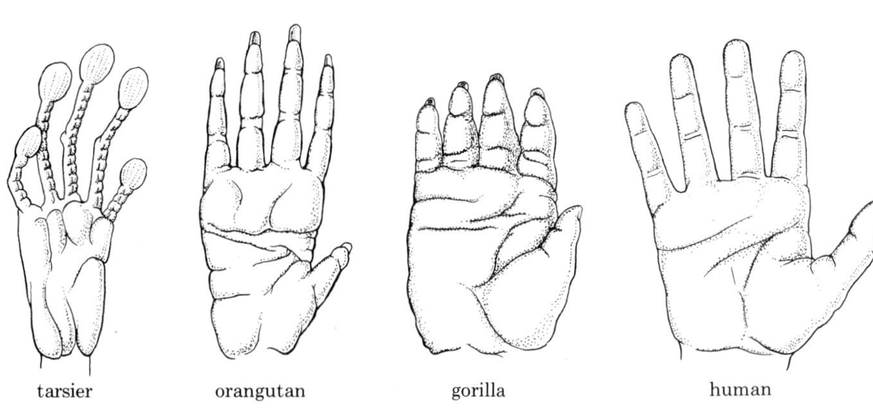

42-3 *Some primate hands. The hand of the tarsier has enlarged skin pads for grasping branches. In the orangutan, the fingers are lengthened and the thumb reduced, which provide for efficient brachiating (swinging arm over arm through the trees). The gorilla's hand, which is used in walking as well as handling, has shortened fingers. The human thumb is proportionately larger than that of any other primate, and opposition of thumb and fingers, on which the handling ability depends, is greatest in the human hand.*

(a)

(b)

42-4 *Life in the treetops made maternal care a major factor in infant survival. Also the necessity of carrying the young for long periods resulted in strong selection pressures for reduced numbers of offspring.* (a) *Anthropoids, such as this vervet monkey, usually have single births.* (b) *A mother chimpanzee with her infant. Field studies suggest that bonds between mother and offspring and perhaps also among siblings last well into adulthood, perhaps for a lifetime.*

Compared to other mammals, however, primates are relatively unspecialized. Their extremities resemble those of the primitive mammals—indeed, of the reptiles—more closely than do the extremities of mammals of most of the other major orders (see page 478). The first four-legged mammals all had five separate digits on each hand and foot, and each digit except the thumb and the first toe had three separate segments that made it flexible and capable of independent movement. In the course of evolution, most mammals developed hooves and paws more suited for running, seizing prey, and digging; other mammals developed flippers for swimming. The primates retained and elaborated on the primitive five-digited pattern.

In the basic quadrupedal structure of the early mammals and reptiles, the forelimb is supported by two bones (the radius and the ulna; see page 347), a pattern that provides for flexibility. Among mammals, it is the primates, in particular, that have retained the ability to twist the radius, the bone on the thumb side, over the ulna so that the hand can be rotated through a full semicircle without moving the elbow or the upper arm.

Similarly, only a few mammals have retained the ability to move the upper arm freely in the shoulder socket. A dog or horse, for instance, usually moves its legs in only one plane, forward and backward. Some South American monkeys, apes, and humans are among the few higher mammals that can rotate the arm widely in the socket.

Primates also have nails rather than claws. Nails leave the tactile surface of the digit free and so greatly increase the sensitivity of the digits for exploration and manipulation.

Visual Acuity

Another result of the move to the trees is the high premium placed on visual acuity, with a decreasing emphasis on the sense of smell, the most important of the senses among many of the other mammalian orders. (Flying produced similar evolutionary pressures among the birds, which were also evolving rapidly during this same period.) This shift from dependence on smell to dependence on sight has anatomical consequences. Tree shrews, like many other animals, have eyes that are directed laterally, but among the other primates there can be traced a steady evolutionary trend toward frontally directed eyes and stereoscopic vision.

Almost all primate retinas have cones as well as rods; cones, as we discussed on page 455, are concerned with color vision and with fine visual discrimination. All primate retinas have foveas, areas of closely packed cones that produce sharp visual images.

Care of the Young

Another principal trend in primate evolution is toward increased care of the young. Because mammals, by definition, nurse their young, they tend to have longer, stronger mother-child relationships than other vertebrates (with the exception, in some cases, of birds). In the larger primates, the young mature slowly and have long periods of dependency and learning.

Uprightness

Another adaptation to arboreal life is an upright posture. Even quadrupedal primates, such as monkeys, sit upright. One consequence of this posture is a change in the orientation of the head, allowing the animal to look straight ahead while in a vertical position; it is this characteristic, above all others, that makes primates look so "human" to us. Vertical posture was an important preadaptation for the upright stance of the genus *Homo.*

Major Lines of Primate Evolution

Primates are divided into two major groups: the *prosimians* (lorises, bush babies, tarsiers, and lemurs) and the higher primates, the *anthropoids* (monkeys, apes, and humans).

Prosimians

During the Paleocene and the Eocene (about 65 to 38 million years ago), a great abundance and variety of prosimians inhabited the tropical and subtropical forests that spread much farther north and south of the equator than they do today. Modern prosimians (Figure 42–5) are small, arboreal, furry, eat fruit and insects, and are almost all nocturnal. The prosimians are generally considered an unnatural group (as are the protists, for example) in the sense that they are not closely related.

(a)

(b)

(c)

42-5 *Some prosimians.* (a) *A lesser bush baby. Bush babies move over level ground by hopping like kangaroos, and sometimes leap 3 or 4 meters in the air from one tree to another. The fingers work together, not independently like human fingers.* (b) *A ring-tailed lemur, combing his tail. The second digit of each foot is a special grooming claw.* (c) *A native of Indonesia, the little tarsier (about the size of a kitten) has existed relatively unchanged for some 50 million years. Its upper lip, like ours, is free from the gum below it, giving it the ability to make faces. It has stereoscopic vision and a larger brain than the lemur's. Living entirely in trees, it has hands and feet with enlarged skin pads for grasping branches. As you may have guessed from the owl-like eyes, tarsiers are nocturnal.*

(a) (b)

Anthropoid Apes

The anthropoid apes include gibbons, orangutans, gorillas, and our closest living relatives, chimpanzees.

Gibbons are the smallest of the apes. They live in pairs or in small family groups, and, along with some human beings, they are the only monogamous anthropoids. Gibbons can stand and walk upright, but usually they move through the trees, swinging arm over arm as shown above (a).

The rarest and least studied of the large primates, the modern orangutan (b), is found only in the tropical rain forests of Borneo and Sumatra. Males, which can weigh as much as 100 kilograms (220 pounds), are usually found alone.

A male gorilla is about as tall as a tall man but weighs more than three times as much (some 180 to 275 kilograms). (c) Gorillas have a beetling brow, a strong, heavy jaw, and, on top of the skulls of adult males, a bony crest to which the strong jaw muscles are attached. Its legs are shorter in proportion to its body than are a man's, but its arms are much longer—with a span of about 3 meters—and extremely powerful.

Gorillas usually walk on all fours but they can stand erect, as they do when challenging an enemy, and they also walk erect for short distances. Gorillas have almost completely abandoned trees, with the exception of some of the smaller animals, which sleep in nests in the lower branches. The larger gorillas sleep on the ground, and all feed on ground plants. In (d), a mature male and family are feeding on a banana plant.

Gorillas live in groups ranging from eight to 24 individuals, with about twice as many females as males, and a number of juveniles and infants. Each gorilla troop has a large, mature (silver-backed) male as a leader. Al-

(c) (d) (e)

(f)

(g)

though reportedly placid by nature, when threatened, males will give fearsome exhibitions of power, thrashing with broken branches, standing erect, beating their chests, barking and roaring, even hitting themselves under the chin to make their teeth rattle. Given the massive size of an adult male, such as the silver-backed male lunging forward to charge (e), the display is impressive.

Chimpanzees are somewhat smaller than humans; an adult male weighs approximately 45 kilograms. They feed and sleep in the trees, but spend a large part of their waking hours on the ground. They walk on all fours, on the flats of their feet and the knuckles of their hands. Although their fingers are shorter than a man's, they have a fine precision grip. They use simple tools, such as a stick to pry out termites, a leaf as a blotter, or a stick as a weapon. Whereas the gorilla is shy by nature, chimps are gregarious, curious, boisterous, and extroverted. (f) A juvenile and infant wrestle in play; (g) a male pant-hoots; (h) a mother shares food with her young; (i) a young male relaxes in a tree; and (j) a chimpanzee group sits and grooms.

(h)

(i)

(j)

Anthropoids

The anthropoids arose from one prosimian line, apparently during the Eocene (53 to 38 million years ago). There are two major groups of anthropoids: the platyrrhines (meaning flat-nosed), which are the New World monkeys, and the catarrhines (downward-nosed), which include the Old World monkeys, apes, and humans (Figure 42-6). These two groups diverged early in their history, with the platyrrhines evolving in South America and the catarrhines in Africa, both apparently during the Oligocene, some 38 to 26 million years ago.

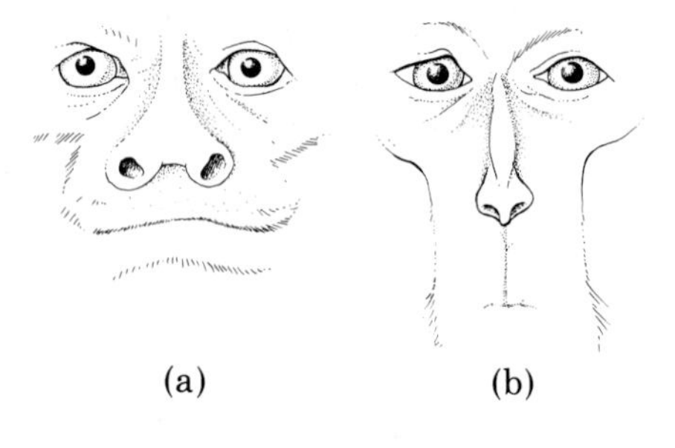

42-6 *Early in their evolution, the anthropoids split into two main lines, the platyrrhine, or flat-nosed* (a), *and the catarrhine, or downward-nosed* (b). *New World monkeys are platyrrhines; Old World monkeys, apes, and humans are catarrhines. There are other characteristic anatomical differences between the two groups.*

The Hominoids

A branching of the catarrhines about 35 million years ago gave rise to the *hominoids*, a group now comprising the anthropoid apes and ourselves. Hominoids are larger than the monkeys, have proportionately larger brain cases, and are tailless. They are adapted for brachiation—that is, swinging from one arm and then the other with their bodies upright. Although, among modern genera, only the gibbons move primarily in this way, brachiation may have played a role in the transition from the body structures associated with the crouching position characteristic of the lower primates to the body structure that makes possible our erect posture.

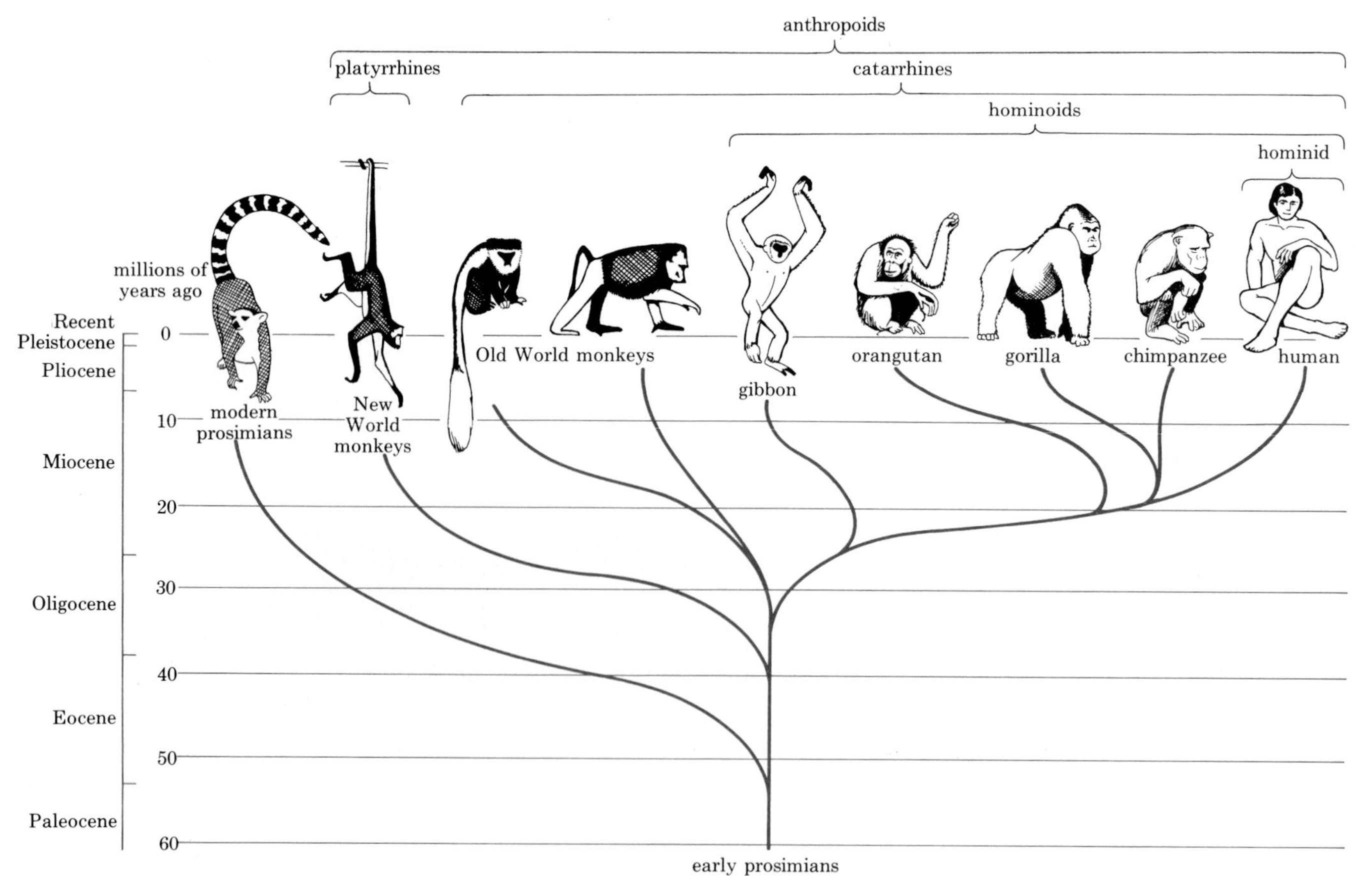

42-7 *A tentative phylogenetic tree of the primates, based on fossil evidence. Biochemical studies suggest that the divergence of the lines leading to the great apes and to humans may have occurred much more recently than is shown here.*

At some point, the hominoid line diverged, forming several major branches. One branch led to the modern gibbon, which is by far the smallest of the apes. Another line diverged into at least four branches, giving rise not only to the ancestors of the great apes (the orangutan, the gorilla, and the chimpanzee) but also to the ancestors of *Homo sapiens*. Until a few years ago, it was thought that these branchings occurred about 20 million years ago, as shown in Figure 42–7. However, studies of the similarities and differences in the proteins of the modern hominoids (see essay, page 504) indicate that gorillas, chimpanzees, and humans diverged from a common ancestor as recently as four to six million years ago. These results have triggered considerable controversy about the dating of key events in hominoid evolution.

The First Hominids

The earliest hominoid universally accepted as a *hominid*—a member of the human family, Hominidae—is the form generally known as *Australopithecus* ("southern ape"). The first *Australopithecus* specimen (Figure 42–8) was described in 1925 by anatomist Raymond Dart. Dart's evidence for its hominid status included the rounded appearance of the skull, the size of the brain case, and the roundness of the jaw. Also, the point of attachment of the vertebral column to the skull indicated that the young animal was a biped.

Subsequent fossil finds have built up a picture of a group of small, lightly built hominids, not more than $1\frac{1}{2}$ meters (5 feet) tall, with cranial capacities ranging from 450 to 700 cubic centimeters, as compared to 340 to 750 for male gorillas (a much larger animal) and 1,000 to 2,000 for modern humans. Their hands had a broad, flat thumb and the beginnings of a human-style grip. Their teeth were very much like our own, although the molars were larger. Australopithecines walked upright, possibly covering the ground with quick, short steps in a sort of jog trot, rather than by our far more efficient heel-and-toe striding movement.

42–8 *The skull of a child, found in a limestone quarry in Taung, South Africa, in 1924, was the first specimen discovered of the earliest known hominid, an australopithecine. The oldest australopithecine fossils, according to radioisotope dating methods, are specimens found near Lake Omo in eastern Africa, representing remains of creatures that lived about $3\frac{1}{2}$ million years ago. Shown here are* (a) *the skull and* (b) *a reconstruction drawing of the head of the child, which was about five or six years old when it died. The australopithecines were erect-walking creatures with brains somewhat larger proportionately than the gorilla's and with teeth that are more similar to ours than to the ape's.* Australopithecus *lived on the ground, and some of them probably ate meat and used tools of their own making.*

(a)

(b)

42-9 *Pebble tools such as these have been found at Olduvai and other sites in eastern Africa in fossil strata of 2 million or more years ago. Australopithecines (presumably) took pebbles of lava and quartz and either struck off flakes in two directions at one end, making a somewhat pointed implement, or in a row on one side, making a chopper. Pebble tools measure up to 10 centimeters in length and were probably used to prepare plant food and to butcher game. The flakes were also used, apparently as scrapers.*

They used simple tools (Figure 42-9) and ate meat; collections of bones of smaller animals—frogs, lizards, rats, and mice—and of the young of larger animals, such as antelopes, have been found at various australopithecine sites. The evidence indicates that they lived in small groups, like most anthropoids. In the past decade, owing largely to the efforts and extraordinary talents of the Leakey family, discoveries of australopithecine fossils have increased in number and in antiquity. According to radioisotope dating, the oldest remains were deposited about 3½ million years ago. The picture has also become much more complicated.

It is now beginning to appear that there was not one but several different lines of hominid evolution (which, of course, is in accord with evolutionary patterns of other organisms). The Leakeys now recognize at least two species of australopithecines, *Australopithecus africanus* and *Australopithecus boisei,* and a third species, which they call *Homo habilis. Homo habilis,* according to this interpretation, was the species that used tools, had the larger brain, and is our ancestor.

42-10 *Louis and Mary Leakey, the British anthropologists, with one of their sons at Olduvai Gorge in the Serengeti plain of Tanzania. Olduvai, from which many important australopithecine fossils have been recovered, represents a geologic sequence that is almost continuous from the present to nearly 2 million years ago. Like our Grand Canyon, it was cut by river action. Because it was by the side of a lake, it was visited by many animals and was apparently a popular hominid camping ground. Louis Leakey died in 1975, but Mary Leakey and their son Richard are still leading figures in the search for early hominids.*

How Did It Happen?

What prompted some primitive hominoids to take the first fateful step to bipedalism? We know from pollen specimens that grasslands were spreading and forests contracting during the Miocene, the epoch preceding the earliest hominid fossils. The grasslands may have presented a new kind of habitat. The horse and other grazing animals evolved rapidly during this time, filling the new niches available to herbivores. Baboons also seem to have become terrestrial in this period. We do not know why some hominoids and not others moved into the grasslands. One possible explanation is that they were driven out of the woodland by competition with other, perhaps larger, primates. Or, they may have possessed some important preadaptation that we do not know about.

Standing upright on the ground would have greatly increased the range of vision of a ground-dwelling species with highly developed eyesight; other mammals, including bears, squirrels, and prairie dogs, often raise themselves on their two hind legs to look around. Standing upright also, of course, frees the hands, which then can be employed for picking protein-rich grains from wheat and other grasses, for tool making and using, for carrying food, and for brandishing weapons. Use of weapons for defense correlates well with the reduction in the size of the canines among fossil hominids, as compared with baboons, for example, or gorillas, among which the males have large canines that are used in aggressive display. The advantages of bipedalism must have been very great for natural selection to have operated in its favor, because the first bipedal primates must have moved slowly and inefficiently compared with the swift movement of other large animals, either on the ground or in the trees. Among the consequences of bipedalism is the enormous human brain, the most characteristic feature of the genus *Homo*.

(a)

(b)

42-11 (a) *Eye-hand coordination is another characteristic of higher primates, such as the female olive baboon shown here with her young.* (b) *Bipedalism frees the upper extremities for a variety of activities. An adult male chimpanzee "fishes" for termites with a twig.*

The Emergence of *Homo sapiens*

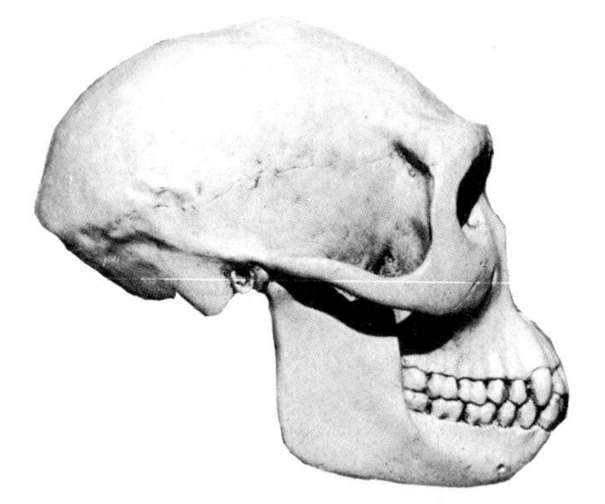

42-12 *A reconstruction of the skull of* H. erectus *from the Solo River, Java.* Homo erectus *was much larger than earlier hominids and had a much larger brain. The skull walls are thick and heavy, and the brow ridges are prominent. The protruding cheekbone was the point of attachment of strong, heavy neck muscles. The jaw is protruding and chinless.*

By about 1 million years ago, a group of hominids had evolved that are placed, without dispute, into the genus *Homo*. Within this genus are two species, the now-extinct *Homo erectus* and our own species, *Homo sapiens*.

Homo erectus

Homo erectus was distinctly different from even the most advanced australopithecine. Members of this species had body skeletons much like our own and were about our size. The bones of their legs indicate that they had a stride similar to our own. The chief differences between *H. erectus* and modern *H. sapiens* are in the skull. Skull specimens of *H. erectus* are thick and massive, with low foreheads. Their jaws are large and chinless, with large teeth. Brain capacity ranges from 700 to 1,100 cubic centimeters, overlapping ours.

Also, unlike the australopithecines, which have only been found in eastern and southern Africa, members of *H. erectus* seem to have occupied a much larger range. Fossil specimens have been found in Java (Java man), in the Choukoutien caves in China (Peking man), and in various sites in Africa, including Olduvai. There are also some skull fragments from Vérteszöllös, Hungary, and a jaw from Heidelberg that are believed to represent this species. The fossils cover a period of about 700,000 years, from about 1 million to 300,000 years ago. Along with the skeletal fragments have been uncovered a number of clues about the social life of *H. erectus*.

Hunting

Probably the most significant factor in the evolution of *H. erectus* was that they were hunters. As we mentioned earlier, recent field studies have shown that baboons and chimpanzees eat meat occasionally and obviously enjoy it (Figure 42-13), and the australopithecines were also meat eaters. Members of the species we call *H. erectus* became hunters on a different scale, however. During various stages of the Pleistocene, there seem to have been far more grasslands than now exist. Some areas that are now deserts were savannas, and to the north, much land that is now forest was prairie or steppe, which could support large numbers of grazing and browsing animals. *Homo erectus* was a rare species, greatly outnumbered by these large herbivores, and so would have had little difficulty finding game. And indeed, from sites as far apart as eastern Africa and China and dating back from 1 million to 350,000 years ago, the bones of very large animals—elephants, rhinoceroses, antelopes, bears, hippopotamuses, and giant baboons—are found in the debris of human campsites.

We do not know how these early hunters killed animals so much larger and more ferocious than themselves. Stone implements found at the sites were apparently made to be held in the hand and were used for chopping, cutting, or pounding. These tools therefore appear to have been used for butchering or preparing food rather than for killing game. There is some evidence that these early hunters killed animals by stampeding them into marshes or over cliffs, perhaps with the help of fire to frighten them. American Indians slaughtered game in this way until very recent times.

(a)

(b)

42–13 *The chimpanzee shows some behavioral traits also attributed to early humans.* (a) *Chimpanzee preparing to eat the forequarters of a young baboon.* (b) *Male chimpanzee throwing rocks and shouting at adult baboons.*

Certainly hundreds of thousands of years passed before these early humans developed weapons as sophisticated as spears with stone or metal tips or bows and arrows. Anthropologists who have studied primitive tribes have been impressed by the fact that, in general, their weapons are not efficient and that hunters take little interest in perfecting them. These modern hunters emphasize patience, endurance, skill in stalking, and highly specialized knowledge of animals and terrain, and perhaps the prehistoric hunters did also. Clearly the capacity to cooperate with other members of the hunting party would have been of extreme importance. It is not known whether or not *H. erectus* was capable of speech, but it would have been of great advantage ("Head off that hippopotamus!").

Tools

In the hands of *Homo erectus,* the pebble tool (Figure 42–9) became a new and highly distinctive implement, the hand ax (Figure 42–14). Hand axes were used extensively throughout Africa, India, and much of the Near East and Europe; tens of thousands have been found in the valleys of the Somme and Thames alone. All these hand axes closely resemble one another. Unlike the pebble tools, they were clearly fashioned according to a formal pattern. Thus we see the emergence of a cultural tradition, with skills and learning passed from one generation to another. Also, the wide distribution of the hand ax indicates communication among the groups of early humans roaming over these vast territories.

42–14 *The hand ax is a stone that has been worked on all its surfaces to provide what appears to be a gripping surface and various combinations of cutting edges, sometimes with a more or less sharp point. Making such an implement requires both skill and time, as reported by anthropologists who have tried. Hand axes came into use about 1 million years ago and are associated with* H. erectus.

Home and Hearth

All higher primates except modern *Homo sapiens* are continuously on the move, and though they range back and forth over the same territory, they sleep in a different place every night. Any member of the group who

The Ice Ages

During most of our planet's history, its climate appears to have been warmer than it is at the present time. However, these long periods of milder temperatures have been interrupted periodically by Ice Ages, so called because they are characterized by glaciations, or persistent accumulations of ice and snow. Such glaciations occur whenever the summers are not hot enough and long enough to melt ice that has accumulated during the winter. In many parts of the world, an alteration of only a few degrees in temperature is enough to begin or end a glaciation.

An early Ice Age appears to have occurred at the beginning of the Paleozoic era, some 600 million years ago. Another, marked by extensive glaciations in the Southern Hemisphere, closed the Paleozoic, some 250 million years ago. The conifers evolved during this period and possibly the angiosperms, as older forest types disappeared. A more recent, less severe cold, dry period occurred at the end of the Mesozoic, about 65 million years ago, and was perhaps a principal cause of the extinction of the dinosaurs.

The most recent Ice Age began during the Pleistocene epoch, about $1\frac{1}{2}$ million years ago. The Pleistocene has been marked by four extensive glaciations that have covered large areas of North America, England, and northern Europe. Between the glaciations there have been intervals, called interglacials, during which the climate has become warmer. In each of these four Pleistocene glaciations, sheets of ice, thicker than 3 kilometers in some regions, spread out locally from the poles, scraped their way over much of the continents—reaching as far south as southern Illinois in North America and covering Scandinavia, most of Great Britain, northern Germany, and northern Russia—and then receded again. We are living at the end of the fourth glaciation, which began its retreat only some 10,000 years ago.

During these periods of violent climatic changes, the fossil record shows that the populations of these regions were under extraordinary evolutionary pressures. Plant and animal populations moved, changed, or became extinct. In the interglacial periods, during which the average temperatures were at times warmer than those of today, the tropical forests and their inhabitants spread up through today's temperate zones. During the periods of glaciations, only animals of the northern tundra could survive in these same locations. Rhinoceroses, great herds of horses, large bears, and lions roamed Europe in the interglacial periods, and in North America, as the fossil record shows, there were camels and horses, saber-toothed cats, and great ground sloths, one species as large as an elephant. In the colder periods, reindeer ranged as far south as southern France, while during the warmer periods, the hippopotamus reached England. It was during this most recent period of glaciations and interglacials that hominids of the genus *Homo* evolved.

The reason for these large changes in temperature is one of the most controversial issues in modern science. They have been variously ascribed to changes in the earth's orbit, variations in the earth's angle of inclination toward the sun, migration of the magnetic poles, fluctuations in the solar energy, migration of the continents, higher elevations of the continental masses, continental drift, and combinations of these and other causes. Also under debate is the question of whether this period of glaciations is over—perhaps for another 200 million years—or whether we are merely enjoying a brief interglacial before the ice sheets begin to creep toward the equator once more.

Glaciers are accumulations of snow and ice that flow across a land surface as a result of their own weight. This glacier is flowing into the sea. As the glacier moves forward, its edges melt; the rocks carried within it have become concentrated in the dark band at the right. This is part of the Greenland ice sheet, which has an area of about 1,726,000 square kilometers and at some places is more than 3 kilometers thick. Glaciers covered much of North America and Europe during the period of the evolution of the genus Homo.

42–15 *This large cave near the little village of Shanidar in northern Iraq has been almost continuously inhabited for more than 100,000 years. Nine Neanderthal skeletons have been found in the cave, including one who was, according to analysis of fossil pollen, buried on a bed of woody branches and June flowers gathered from the hillside.*

cannot keep up is left behind. Early humans were also nomads, as are most modern hunter-gatherers, but there is evidence that they maintained base camps. Some of these camps appear to have been fairly permanent, in the sense that they were returned to year after year.

The australopithecines and the earliest humans probably lived mostly in the open, although one australopithecine site is in the natural shelter of overhanging rocks. The campsites found in southeastern Africa are near streams or lakes, which would not only have provided water for the campers but would also have attracted game.

The Choukoutien caves, near Peking, China, inhabited 350,000 years ago, are among the earliest caves known to have been used for human habitation. It is probably no coincidence that within these same caves are also the first traces of the human use of fire. Before fire, caves were inhabited by bears and other large carnivores and were too dangerous for humans to occupy. Fire, however, made them available for human habitation, and they have been so used up to modern times (Figure 42–15).

Increase in Intelligence

In mammals, the proportion of brain size to body size can be correlated with the intelligence of the species under consideration. It is therefore reasonable to assume that the enormous increase in cranial capacity that took place in the course of human evolution signifies strong selection pressures for intelligence. In fact, insofar as one can compare the evolution of the human brain with the evolution, for instance, of the foreleg of the horse (which is considered very rapid), the human brain evolved 100 times more quickly. The increase in cranial capacity (Figure 42–16) took place well after bipedalism and the early use of tools and well before the development of sophisticated tools or weapons, which did not take place until a mere 30,000 to 40,000 years ago.

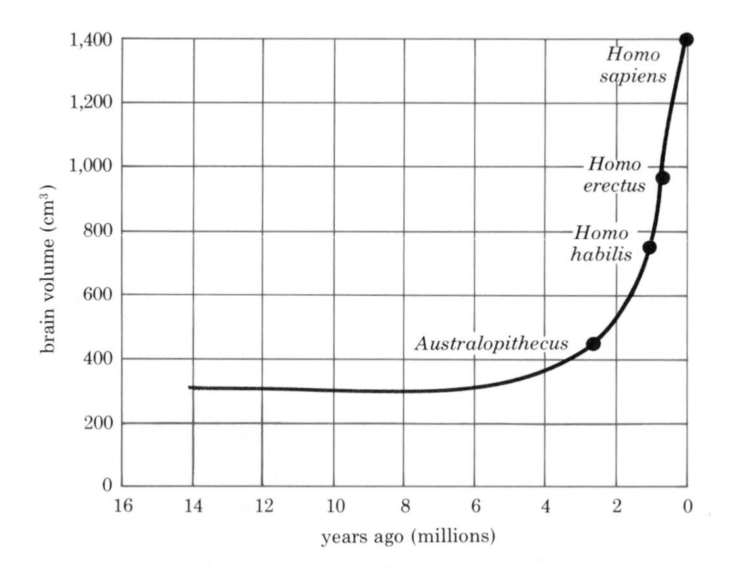

42-16 *Increase in brain volume in the course of hominid evolution. Some of the increase can be correlated with the increase in body size that was also taking place during this period, but most of it is believed to represent the results of strong selection pressures for greater intelligence.*

Homo sapiens

The earliest fossils that are classified as *Homo sapiens* come from Swanscombe in England and Steinheim in Germany. They consist only of some skull fragments, but these clearly indicate that the brains of these humans were larger than those of *Homo erectus* and that their skulls were less massive. These skull fragments are dated at about 250,000 to 150,000 years ago, during an interglacial period when Europe was warm, probably warmer than it is today.

Neanderthal Man

Then we have another of those large and frustrating gaps in the fossil record. The next specimens that are known to us date from around 80,000 to about 40,000 years ago, the period of the last glaciation. This period abounds with specimens of what we have come to call Neanderthal man. They have been found largely in Europe but also in the Near East and Central Asia. Neanderthals stood erect, had a brain capacity somewhat larger than ours, a thick skull, a protruding muzzle, a low forehead, and heavy brow ridges. They are now classified as a variety of *Homo sapiens*.

Neanderthals also used simple, hand-held stone tools (Figure 42-17). Some of the stone tools appear to have been used for scraping hides, suggesting that Neanderthals wore clothing made of animal skins, which would certainly have been in keeping with the climate in which they lived.

flake struck from core

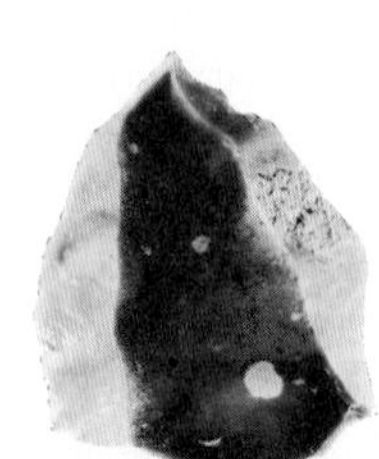

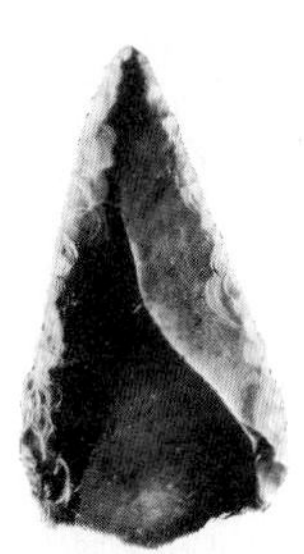

42-17 *Fossils of Neanderthal man are associated with a decrease in the use of core tools, such as the hand ax, and a predominance of flake tools. Multiple flakes were struck off from a central core and then retouched. They are known as points and scrapers, which also presumably describes their functions. They are usually 5 to 8 centimeters long. Some may have been used as points for spears or javelins, but, in general, it appears that they were hand-held. Spears, which were in use at that time, were of fire-hardened wood.*

Neanderthals buried their dead, often with food and weapons and, in at least one instance, with spring flowers. Formal burials such as these suggest a belief in life after death. Whether or not Neanderthals were able to speak is a matter of current debate, but it is difficult to understand how a society could hold such abstract concepts without some means of exchanging ideas among its members.

Cro-Magnon Man

About 30,000 to 40,000 years ago, specimens of Neanderthal man disappear abruptly from the fossil record and are replaced by what is known as Cro-Magnon man, or sometimes as Upper Pleistocene man, who is physically indistinguishable from modern *Homo sapiens*. We do not know what became of the Neanderthals. Perhaps they were exterminated in warfare, although there is no evidence of this. Perhaps they were simply unable to compete for food and living space with the better-equipped Cro-Magnon type. Perhaps they interbred, although there are no clear traces of intermediate forms. Some have suggested that Cro-Magnon men brought with them some disease to which they themselves were resistant and the Neanderthals were not. In any case, soon after the appearance of Cro-Magnon man, there were no other hominids in Europe, and within the course of 10,000 to 20,000 years, this new variety of primate had spread over the face of the planet.

Cro-Magnon man, when he first appeared in Europe, came bearing a new, quite different, and far better tool kit (Figure 42–18). The stone tools were essentially flakes—which, of course, had been in use for more than 2½ million years—but they were struck from a carefully prepared core with the aid of a punch (a tool made to make another tool). These flakes, usually referred to as blades, were smaller, flatter, and narrower, and, most important, they could be and were shaped in a large variety of ways. From the beginning they included various scraping and piercing tools, flat-backed knives, awls, chisels, and a number of different engraving tools. Using these tools to work other materials, especially bone and ivory, Cro-Magnon man made a variety of projectile points, barbed points for spears and harpoons, fishing hooks, and needles. Thus, although Cro-Magnon man lived much the same sort of existence as that of his forebears, he apparently lived it with more possessions, more comfort, and more style.

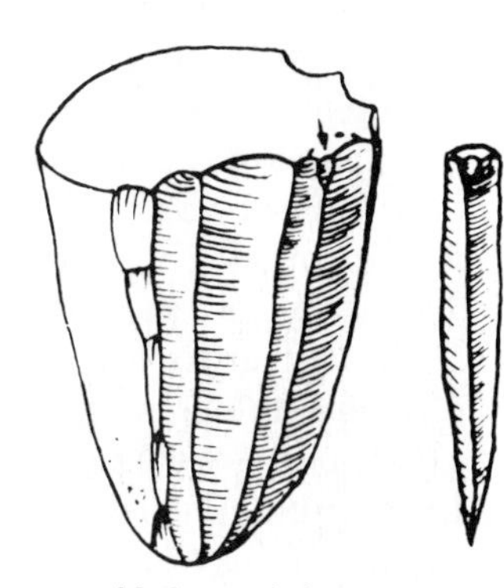

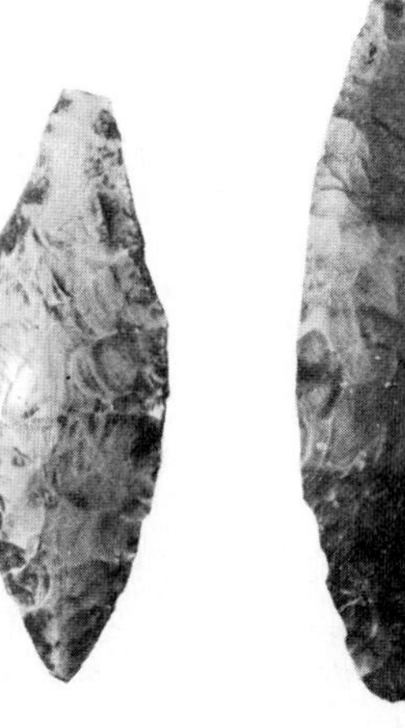

42–18 *Cro-Magnon culture was characterized by a great increase in the types of specialized tools made from long and relatively thin flakes with parallel sides, called blades. Beautifully worked "laurel leaf" blades, such as the two shown at the right, are often more than 30 centimeters long and only ½ centimeter thick.*

42-19 *In almost every cave, a small number of the animals show wounds, as seen in this steppe wisent from the Cave of Niaux in France. Although the animals are shown in realistic detail, the weapons and the hunters, if shown at all, are abstractions, perhaps to keep their identity a secret. The illusion of movement is greatly enhanced by the patterns of light and shadow in the dark, narrow recesses of the cave.*

Perhaps our closest emotional links to this most immediate ancestor are the cave paintings of western Spain and southern France (Figure 42-19). The examples that remain to us, many surprisingly untouched by time, clearly form a part of a rich artistic tradition that endured for at least 10,000 years. The cave drawings are almost entirely of animals, nearly all game animals, and they are deep within the caves, so they must have been viewed (as they must have been painted) by the light of crude lamps or torches.

The meaning of these drawings and paintings has long been a matter of debate. Some of the animals are marked with darts or wounds (although very few appear to be seriously injured or dying). Such markings have led to the suggestion that the figures are examples of sympathetic magic, in which there is the notion that one can do harm to one's enemy by sticking needles in his image. The fact that many of the animals appear to be pregnant suggests that they may symbolize fertility. Many appear also to be in motion. Perhaps these animals, so vital to the hunters' welfare, were migratory in these areas, and they may have seemed to vanish at certain times in the year, mysteriously returning, heavy with young, in the springtime. This return of the animals might have been an event to be solicited or celebrated in much the same spirit as the rites of spring of more recent peoples or Easter are celebrated.

Cave art came to an end perhaps 8,000 or 10,000 years ago. Not only were the tools and pigments laid aside, but the sacred places—for such they seem to have been—were no longer visited. New forces were at work to mold the course of human existence, and the end of this first great era in human art is our clearest line of demarcation between the old life of the hunter and the new life that was to come.

The Agricultural Revolution

The most important event in human cultural evolution was the change to an agricultural way of life. The reasons for such a transition are not clear, but one factor seems to have been a change in the climate. The most recent of the glaciations began to retreat about 18,000 years ago, withdrawing slowly for about 10,000 years. As the glaciers retreated, the plains of northern Europe and of North America, once cold grasslands, or steppes, gave way to forest. Many of the great herbivores that roamed these steppes retreated northward and eventually vanished; the woolly mammoth was last seen in Siberia more than 10,000 years ago. While some animals became extinct, humans adapted, as they had during the entire period of violent climatic changes that have marked their evolutionary history. With the migratory animals gone, attention shifted to smaller game. The hunting and gathering of small animals—instead of large migratory herbivores—undoubtedly resulted in a less nomadic existence for the hunters and formed a prelude to the agricultural revolution. (However, it should be noted that similar fluctuations had occurred previously in human history without leading to a revolution in the way of life.)

The earliest traces of agriculture, dating back about 11,000 years, are found in areas of the Near East, which are now parts of Iran, Iraq, and Turkey. Here were the raw materials required by an agricultural economy: cereals, which are grasses with seeds capable of being stored for long periods without serious deterioration, and herbivorous herd animals, which can be readily domesticated. The grasses in these areas were wild wheats and barley, which still grow wild in the foothills. The animals were wild sheep and goats.

By 8,000 years ago, agricultural communities were established in eastern Europe. By 7,000 years ago (about 5000 B.C.), agriculture had spread to the western Mediterranean and up the Danube River into central Europe and, by 4000 B.C., to Britain. During this same period, agriculture originated separately in Central and South America, and perhaps slightly later in the Far East.

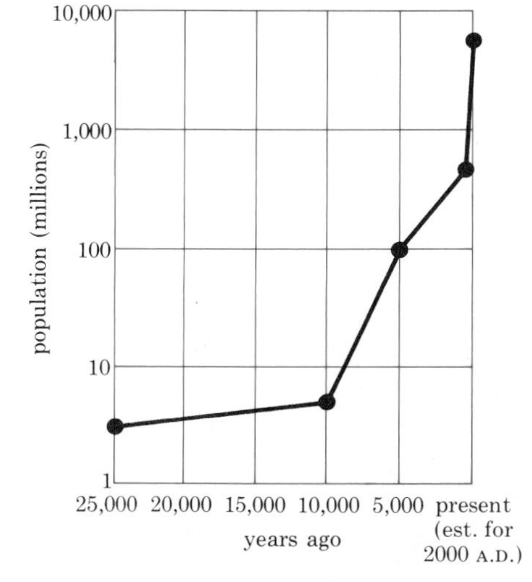

42-20 *When population is plotted on a logarithmic scale (vertical axis) against time on a normal arithmetic scale (horizontal axis), the slope of the line between any two points represents the rate at which the population increased during that time interval.*

In the period from 25,000 to 10,000 years ago, when Homo sapiens *lived as a hunter-gatherer, the rate of increase was low. During the next 5,000 years, when the agricultural revolution was spreading around the world, population growth increased dramatically. Population growth continued after an agricultural way of life had become established, but at a significantly slower rate. However, with the advent of the scientific and industrial revolution, about 1650, population growth again increased dramatically and is continuing at a very rapid rate. If present trends continue, there will be about 6 billion people on earth in the year 2000.*

The Population Explosion

One immediate and direct consequence of the agricultural revolution was an increase in population. It is estimated that about 25,000 years ago there may have been as many as 3 million people. Some 15,000 years later, at the close of the Pleistocene epoch, the population probably numbered a little more than 5 million, spread over the entire world. At this point, some 10,000 years ago, the establishment of agricultural communities began. In the next 5,000 years, agriculture spread throughout the world and was accompanied by a twentyfold increase in the human population—to about 100 million in 3000 B.C. (Figure 42-20).

From 3000 B.C. until about 1650, human population growth slowed considerably. During this period—slightly less than 5,000 years—the population increased only fivefold, to about 500 million. At about this time, the rapid development of science, technology, and industrialization began, bringing profound changes in human life and its relationship to nature. In the 200 years between 1650 and 1850, the population doubled, to 1,000 million (1 billion), and then it doubled again by 1930, to 2 billion.

By 1980, there were more than 4½ billion people on our planet, and the number is increasing at a rapid rate. Although the birth rate in the United States plunged dramatically from 18.4 babies born per 1,000 people in the population in 1970 to 15 in 1973, and is now about 14.8, the world's population is growing at 1.7 percent per year. This means that about 139 people are added to the world population every minute, about 200,000 each day, and 74 million every year. If this rate of increase is sustained, there will be close to 6 billion people on earth by the year 2000.

Factors Affecting Population Growth

As we saw earlier (page 524), the rate at which a population increases is the result of both its birth rate and its death rate. The relative influence of these two factors has varied at different times in human history. In the period between 25,000 and 10,000 years ago, when humans lived as hunter-gatherers, the low rate of increase is believed to have resulted from a low birth rate, in turn the result of physiological factors. For example, in some primitive hunter-gatherer societies, a woman is unable to conceive a child until she is about 19½; moreover, throughout the period when she is breast-feeding a child (which may be as long as three or four years), ovulation does not occur. Compared to other contemporary human societies, the age at first reproduction is later (and thus the total reproductive span of the woman is shorter) and the interval between children is longer; these factors combine to produce a much lower birth rate in the population. It has also been suggested that the difficulties of carrying small children in a nomadic

42–21 *A nomadic !Kung woman with her two children, photographed while visiting relatives who live a settled life in a cattle camp. Among the nomadic !Kung people, the age of first menstruation is about 16 to 16½, the first child is born when the mother is about 20, and there is an interval of 4 years between children. The completed family is about five children, three of whom survive to reproductive age. Among the settled !Kung people, the women are heavier, children are weaned earlier, and the birth interval is less than 3 years. In recent years, the growth rate of the population has increased dramatically. It has been suggested that similar changes in human fertility occurred when hunter-gatherers first settled into agricultural communities some 10,000 years ago.*

society that lacked domesticated pack animals would have created social pressures for a long interval between children.

It is thought that the changes in culture and nutrition that accompanied the shift to an agricultural way of life may have led to a breakdown in the mechanisms that controlled the birth rate, resulting in the dramatic population increases that occurred between 10,000 and 5,000 years ago. The reasons for the subsequent slowing of population growth are not known with certainty. One contributing factor was probably an increased vulnerability to infectious diseases as people lived in closer proximity to each other within agricultural communities. Although the population continued to grow, increased mortality—particularly among the young—had a damping effect on the growth rate. Social customs, such as postpartum taboos, may have reduced the birth rate also.

The enormous recent increase in the world's population is primarily a product of the decline in death rates, especially among the young (Figure 37-5, page 525). For example, in Pakistan, the population was 20 million in 1911, 35 million in 1950, and over 80 million in 1979. During this period, birth rates seem to have remained the same—about 45 per 1,000—but the death rate is one-third of what it was only 20 years ago.

As we saw earlier (Figure 37-4, page 525), population growth also depends on the age of the population. In Pakistan, again, almost half of the population is under 15, so if the ideal of a two-child family were to be obtained in the near future, Pakistan's population could still double in the next half century.

Birth Rates, Death Rates, and Social Security

Despite the fact that the most recent phase of the population increase is related to a reduction in death rates, particularly from infectious diseases, some experts believe that further reductions in death rates and a general increase in the standard of living will reduce the rate of population growth. Other experts, however, disagree. They point out that for many countries a significant rise in the standard of living seems an almost unattainable goal and, in some cases, a rapidly receding one. Moreover, the age structure in many countries is such that even if birth rates were to drop dramatically, it would be many years before population size would begin to level off.

There is, however, a correlation between economic deprivation and high birth rates. It is in the developing countries that the greatest rates of population growth are found. In India, for example, where large numbers of people are now starving, the annual rate of population increase is 2.0 percent. Yet, most Indian women do not seek help in birth control until they have three or four children. The desire for large families is deeply rooted in the Indian culture. Moreover, in Indian tradition, children provide security for their parents in old age. Two recent studies of life expectancy in India show that, with the high death rate among children, it is necessary for a mother to bear five children if the couple is to be 95 percent certain that one son will survive the father's sixty-fifth birthday.

It is clear that a reduction in birth rate is not directly correlated with the availability of birth-control measures. United States' birth rates are very low now, but they were nearly as low in the 1930s during the Depression. The Caribbean island of Jamaica has a Family Planning Center in almost every small town, dispensing contraceptives at very low cost, yet the current birth rate is 29 per 1,000 population.

Within the last few years, the rate of population growth worldwide has begun to decline, but it is too early to know if this decline represents a long-term trend or is only temporary. What has become clear, however, is that bringing the growth of the human population under control is an urgent and complex task involving biological, economic, political, and social factors.

Our Hungry Planet

The consequences of the increase in the human population include the problems of pollution, depletion of fossil fuels, destruction of natural resources, and extinction of other species. By far the most difficult problem and also the most urgent is hunger and starvation. Of the world's $4\frac{1}{2}$ billion people, at least 1 billion are inadequately nourished. It is estimated that about one-third of the deaths that now occur worldwide are due directly or indirectly to malnutrition. Perhaps even more important, in terms of the world's future, are the effects, both physical and psychological, of the prolonged chronic hunger of so large a proportion of our population.

The effort to increase agriculture by the development of new crop plants—especially grains—is called the Green Revolution. Enormous progress has been made. From 1950 to 1970, the production of wheat in Mexico increased from 270,000 metric tons per year to 2.35 million; the corn harvest increased a more modest 250 percent, with yields per hectare almost doubling. Between 1950 and the present, India has increased its production of food grains about 2.8 percent a year. (Its population during this period has increased about 2.1 percent a year, which is significantly less.) Since 1971, China, the most populous nation in the world, has become agriculturally self-sustaining. Most of this success has come about as a result of improved varieties of crop plants, combined with better techniques of irrigation and fertilization. Moreover, and perhaps most important, the full potential of the Green Revolution has not yet been realized.

42-22 *Field workers weeding a rice plot at the International Rice Research Institute in the Philippines. The new strains of rice are dwarf varieties with short, stiff stalks and respond well to heavy fertilization. Older, nondwarf strains become too tall when heavily fertilized and tend to fall over, spilling the grain.*

Despite its acknowledged success, this massive effort has come under criticism in recent years. One reason is the increasing cost of fertilizer; these new grains require intensive cultivation. Because the large landowners are able to afford the investment in fertilizer and farming equipment that the small-scale farmers cannot, these new agricultural developments are seen as accelerating the consolidation of farm lands into a few large holdings by the very wealthy. Most serious of all, although food production is still outstripping population growth, the Green Revolution, at its most productive, will not be able to keep pace indefinitely with the rapid growth of the world's population. (If the population had remained at its 1950 levels, there would be enough food to feed everyone on the planet adequately today.)

Finally, there is a more fundamental though more elusive reason for the dissatisfaction with the Green Revolution. When it was first introduced, it appeared to many to be an almost magical solution to problems so enormous and distressing that they had seemed insoluble. It is now clear, however, that poverty and famine and the unrest and violence they may bring will not be solved by a "technological fix." The Green Revolution must of course go forward. At the same time, we must recognize that the broader solutions are social, political, and ethical, involving not only the growth of crops but their distribution, not only the limiting of populations but the raising of living standards of these populations to tolerable levels.

The cultural evolution of *Homo sapiens* in the past 10,000 years—and especially in the past 350 years—has given us the capacity to thwart the effects of natural selection on ourselves and to disrupt and, in many cases, to control its effects on other species. The freeing of human population growth from natural biological controls has created, directly and indirectly, enormous pressures on other species and on the intricate interactions that characterize ecosystems. Our survival as part of the living world depends, however, on those interactions. In the age of the dinosaurs, the earliest primates survived, it would appear, largely by their wits. Now, if we are to survive the consequences of our own evolution, we will have to do it again, by the contents of our own skulls. For within the human brain—that complex collection of neurons and synapses—resides the uniquely human capacity to accumulate knowledge, to plan with foresight, and so to act with enlightened self-interest and even, on occasion, with compassion and reverence.

SUMMARY

The first mammals arose from primitive reptilian stock about 200 million years ago and coexisted with the dinosaurs for 130 million years. The extinction of the dinosaurs was followed by a rapid adaptive radiation of the mammals. The primates are an order of mammals that became adapted to arboreal life. Primates are characterized by five-digited extremities adapted for grasping with nails rather than claws and by freely movable limbs. They are dependent more upon vision than upon smell, and the higher primates all have stereoscopic vision with foveas for fine focus and cones for color vision.

The two principal groups of living primates are the prosimians and the anthropoids. Prosimians were widespread and abundant during the Paleocene and Eocene epochs. Modern prosimians include lorises, bush babies, lemurs, and tarsiers.

The anthropoids include the New World monkeys, the Old World monkeys, and the hominoids (apes and ourselves). At some point in hominoid evolution—perhaps in the Miocene and perhaps more recently—the hominoids diverged into the lines that gave rise to the modern apes and to hominids, members of the human family.

From about 3½ million to 1 million years ago, groups of hominids lived that were bipedal, had larger brains than apes, and, some of whom at least, used simple tools and ate meat. Evidence now indicates the coexistence of at least three species, which have been called *Australopithecus africanus, Australopithecus boisei,* and *Homo habilis.*

Homo erectus lived about 1 million to 300,000 years ago. Individuals of the species were successful hunters of very large animals, indicating that they lived in fairly large groups, shared food, and worked together cooperatively. The hand ax is associated with *Homo erectus,* and its widespread use indicates cultural exchanges between groups. Some groups lived in caves and had fire, two developments that are probably related.

Neanderthal man is considered a variety of *Homo sapiens.* Neanderthal fossils date from about 80,000 to 30,000 years ago, the period of the last glacial advance. The majority of specimens have been found in Europe. Neanderthals had fire, inhabited caves, hunted large animals, and probably wore clothing. They used stone tools of a characteristic type. They buried their dead, sometimes with food and weapons. Neanderthals disappeared about 30,000 to 40,000 years ago.

Cro-Magnon men, or Upper Pleistocene men, who replaced the Neanderthal, were nomads and hunters, like their predecessors. They had tools distinctly different, smaller, and far more varied than had been seen previously. Their art, which covers a period of some 10,000 years, is evidence of a culture rich in beauty, mystery, magic, and tradition.

The most important event in human cultural evolution was the advent of agriculture, about 10,000 years ago. About 8000 B.C., when agriculture had its beginning, there were probably about 5 million people in the world. By 3000 B.C., the population had increased to about 100 million, and by 1650, there were about 500 million. By 1980, there were more than 4½ billion, and the population continues to increase, especially in the developing countries. A most serious consequence of this rapid increase in world population is widespread malnutrition and starvation.

QUESTIONS

1. Distinguish among the following: primate/prosimian; monkey/ape; hominid/hominoid/anthropoid; *Homo habilis*/*Australopithecus*.

2. If you were to meet one of the following, how would you distinguish him (or her): *Australopithecus? Homo habilis? Homo erectus?* Cro-Magnon man?

3. Name five evolutionary trends among primates, and discuss the probable selective value for each.

4. Describe a possible relation between these two facts: (1) the only modern lemurs are a widely diversified group found on the island of Madagascar; (2) the only other primate on Madagascar is *Homo sapiens,* a late arrival.

5. Buckminster Fuller, the architect, has said, "Pollution is resources we are not harvesting." Give some examples to support this statement. Do you agree with his definition?

6. In an editorial entitled "Food, Overpopulation, and Irresponsibility" that appeared recently in a scientific journal, the authors concluded: "Because it creates a vicious cycle that compounds human suffering at a high rate, the provision of food to the malnourished nations of the world that cannot, or will not, take very substantial measures to control their own reproductive rates is inhuman, immoral and irresponsible." What is your opinion?

SUGGESTIONS FOR FURTHER READING

CARSON, RACHEL: *Silent Spring,* Houghton Mifflin Company, Boston, 1962.*

This is the book that awakened the nation to the dangers of pesticides. Once seen as highly controversial, it is now regarded as a classic of modern ecology.

COLINVAUX, PAUL: *Introduction to Ecology,* John Wiley & Sons, Inc., New York, 1973.

A good, general text by a teacher who clearly enjoys his subject, designed primarily for second-year students and science majors. It is notable for its stress on and criticism of the concepts of ecology.

COLINVAUX, PAUL: *Why Big Fierce Animals Are Rare,* Princeton University Press, Princeton, N.J., 1978.*

A series of essays that form an excellent introduction to modern ecology.

* Available in paperback.

COMMONER, BARRY: *The Closing Circle,* Alfred A. Knopf, Inc., New York, 1971.*

The author traces the relationship between the present ecological crisis and technological and social factors in our society. His title, The Closing Circle, *refers to his conviction that if we are to survive, "we must learn how to restore to nature the wealth we borrow from it." Commoner's book is criticized by Ehrlich and some other ecologists and demographers as presenting an oversimplified and too optimistic solution to environmental problems.*

EHRLICH, PAUL R., ANNE H. EHRLICH, and JOHN P. HOLDREN: *Ecoscience: Population, Resources, Environment,* W. H. Freeman and Company, San Francisco, 1977.*

Required, although not cheerful, reading for all concerned with the biological, economic, and social consequences of human population growth. Topics covered include basic geology and ecology, human population dynamics, food supplies, energy and material resources, environmental disruption, and approaches to solutions for the many problems confronting us. The authors present a wealth of useful data, even though you may not agree with all their conclusions.

EMMEL, THOMAS C.: *An Introduction to Ecology and Populations,* W. W. Norton & Company, New York, 1973.*

A short, sound text for the beginner, logically presented and sensibly organized.

KRUTCH, J. W.: *The Desert Year,* Viking Press, Inc., New York, 1960.*

A description by one of the best American nature writers of the animal and plant life of the American desert.

KUMMER, HANS: *Primate Societies: Group Techniques of Ecological Adaptation,* Aldine, Atherton, Inc., Chicago, 1971.

Most of this book is concerned with behavior in baboons. This book is greatly enriched both by the author's personal experience in observing baboons and other primates and by his interpretation of primate behavior patterns in terms of evolutionary and ecological principles. Some wonderful and unusual photographs, also by the author.

LAPPÉ, FRANCES MOORE: *Diet for a Small Planet,* Ballantine Books, Inc., New York, 1971.*

Lappé clearly establishes the feasibility of eating "low on the food chain," thereby greatly increasing the availability of proteins to other human populations.

LEAKEY, RICHARD, and ROGER LEWIN: *Origins,* E. P. Dutton & Co., Inc., New York, 1977.*

A beautifully illustrated, thoughtful, and imaginative book about 3 million years of human history, and how we know what we do about it, with some speculations on human nature.

LE GROS CLARK, W. E.: *Antecedents of Man: An Introduction to the Evolution of the Primates,* Harper & Row, Publishers, Inc., New York, 1963.*

Le Gros Clark's emphasis is on the anatomy and physiology of the entire primate order.

MORAN, JOSEPH M., MICHAEL D. MORGAN, and JAMES H. WIERSMA: *An Introduction to Environmental Sciences,* 2d ed., W. H. Freeman and Company, San Francisco, 1980.

A clear, matter-of-fact presentation of the basic features of our planet and, in particular, of the biosphere. It provides a good background for the study of ecology in general and problems of environmental pollution and technological change in particular.

PILBEAM, DAVID: *The Ascent of Man: An Introduction to Human Evolution,* The Macmillan Company, New York, 1972.

A thorough survey of primate evolution, including evolution of the hominids.

RICKLEFS, ROBERT E.: *The Economy of Nature,* Chiron Press, New York, 1976.

This outstanding textbook for the beginning student provides a comprehensive introduction to the basic concepts of modern ecology. Beautifully written and rich with examples, it is highly recommended.

* Available in paperback.

SANDERS, N. K.: *Prehistoric Art in Europe,* Penguin Books, Ltd., Harmondsworth, England, 1968.

An illustrated history of art's first 30,000 years.

SCIENTIFIC AMERICAN: *The Biosphere,* W. H. Freeman and Company, San Francisco, 1970.*

A reprint of the September 1970 issue of Scientific American. *Its 11 chapters by different authors are devoted entirely to energy flow and biogeochemical cycles and their relationship to current human problems of food consumption and pollution.*

SMITH, ROBERT L.: *Ecology and Field Biology,* 3d ed., Harper & Row, Publishers, Inc., New York, 1980.

The outstanding sections of this text are those that deal with descriptions of biomes and communities and the plants and animals within them. Although meant to accompany a course in field biology, this book would be an asset and a pleasure to the amateur naturalist.

STORER, JOHN H.: *The Web of Life,* New American Library, Inc., New York, 1966.*

One of the first books ever written on ecology for the general public. In its simple presentation of the interdependence of living things, it remains a classic.

TINBERGEN, NIKO: *Curious Naturalists,* Natural History Library, Doubleday & Company, Inc., Garden City, N.Y., 1968.*

Some charming descriptions of the activities and discoveries of scientists studying the behavior of animals in their natural environment.

UCKO, PETER J., and ANDRÉE ROSENFELD: *Paleolithic Cave Art,* McGraw-Hill Book Company, New York, 1967.*

This book not only presents the cave art, in photos and drawings, but also examines its contents and context and discusses the various interpretations of it.

VAN LAWICK-GOODALL, JANE: *In the Shadow of Man,* Houghton Mifflin Company, Boston, 1971.

An absorbing personal account of 11 years spent observing the complex social organization of a single chimpanzee community in Tanzania.

WICKLER, WOLFGANG: *Mimicry in Plants and Animals,* World University Library, London, 1968.*

Many examples and illustrations of a delightful subject.

WILSON, E. O.: *The Insect Societies,* Harvard University Press, Cambridge, Mass., 1971.

A comprehensive and fascinating account of the social insects. An unusual nominee for the National Book Award.

WILSON, E. O.: *Sociobiology: The New Synthesis,* Harvard University Press, Cambridge, Mass., 1975.*

In this extremely interesting, beautifully written, and beautifully illustrated book, Wilson undertakes to set forth the biological principles that govern social behavior in all kinds of animals. The last chapter, which concerns human sociobiology, has become a focus of the current controversy about the inheritance of behavioral traits in Homo sapiens. *Also available in an abridged, less technical version.*

* Available in paperback.

Appendixes

APPENDIX A

Metric Table

	Quantity	Numerical Value	English Equivalent	Converting English to Metric
Length	kilometer (km)	1,000 (10^3) meters	1 km = 0.62 mile	1 mile = 1.609 km
	meter (m)	100 centimeters	1 m = 1.09 yards	1 yard = 0.914 m
			= 3.28 feet	1 foot = 0.305 m
	centimeter (cm)	0.01 (10^{-2}) meter	1 cm = 0.394 inch	= 30.5 cm
	millimeter (mm)	0.001 (10^{-3}) meter	1 mm = 0.039 inch	1 inch =2.54 cm
	micrometer (μm)	0.000001 (10^{-6}) meter		
	nanometer (nm)	0.000000001 (10^{-9}) meter		
	angstrom (Å)	0.0000000001 (10^{-10}) meter		
Area	square kilometer (km^2)	100 hectares	1 km^2 = 0.3861 square mile	1 square mile = 2.590 km^2
	hectare (ha)	10,000 square meters	1 ha = 2.471 acres	1 acre = 0.4047 ha
	square meter (m^2)	10,000 square centimeters	1 m^2 = 1.1960 square yards	1 square yard = 0.8361 m^2
			= 10.764 square feet	1 square foot = 0.0929 m^2
	square centimeter (cm^2)	100 square millimeters	1 cm^2 = 0.155 square inch	1 square inch = 6.4516 cm^2
Mass	metric ton (t)	1,000 kilograms =	1 t = 1.103 tons	1 ton = 0.907 t
		1,000,000 grams		
	kilogram (kg)	1,000 grams	1 kg = 2.205 pounds	1 pound = 0.4536 kg
	gram (g)	1,000 milligrams	1 g = 0.0353 ounce	1 ounce = 28.35 g
	milligram (mg)	0.001 gram		
	microgram (μg)	0.000001 gram		
Time	second (sec)	1,000 milliseconds		
	millisecond	0.001 second		
	microsecond	0.000001 second		
Volume (Solids)	1 cubic meter (m^3)	1,000,000 cubic centimeters	1 m^3 = 1.3080 cubic yards	1 cubic yard = 0.7646 m^3
			= 35.315 cubic feet	1 cubic foot = 0.0283 m^3
	1 cubic centimeter (cm^3)	1,000 cubic millimeters	1 cm^3 = 0.0610 cubic inch	1 cubic inch = 16.387 cm^3
Volume (Liquids)	kiloliter (kl)	1,000 liters	1 kl = 264.17 gallons	1 gal = 3.785 l
	liter (l)	1,000 milliliters	1 l = 1.06 quarts	1 qt = 0.94 l
	milliliter (ml)	0.001 liter	1 ml = 0.034 fluid ounce	1 pt = 0.47 l
	microliter (μl)	0.000001 liter		1 fluid ounce = 29.57 ml

APPENDIX B

Temperature Conversion Scale

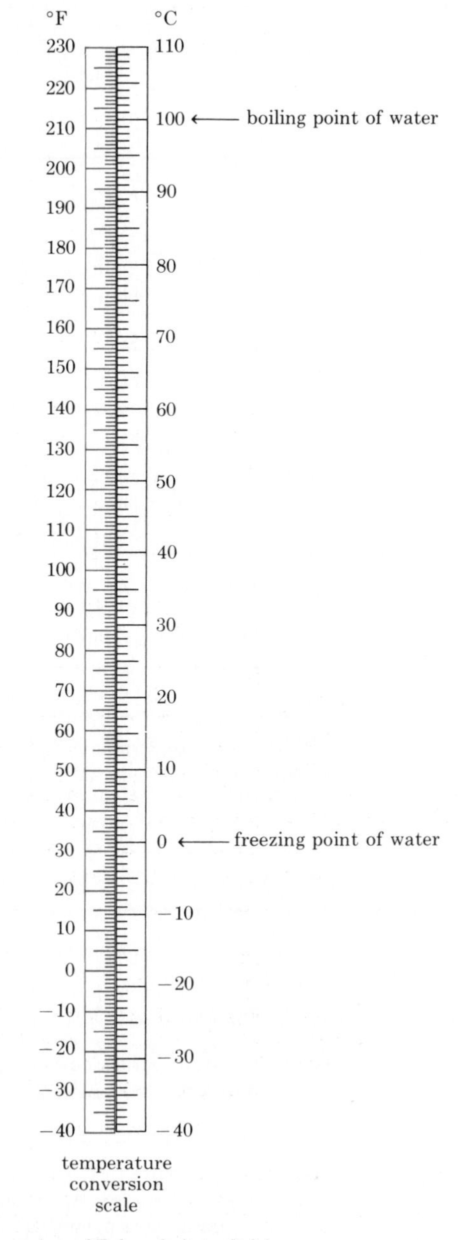

temperature
conversion
scale

for conversion of Fahrenheit to Celsius,
the following formula can be used:

$$°C = \frac{5}{9}(°F - 32)$$

for conversion of Celsius to Fahrenheit,
the following formula can be used:

$$°F = \frac{9}{5}°C + 32$$

APPENDIX C

Classification of Organisms

There are several ways to classify organisms. The one presented here is based closely upon that of Whittaker (in *Science,* vol. 163, pages 150–160, 1969). Organisms are divided into five major groups, or kingdoms: Prokaryotae, Protista, Fungi, Plantae, and Animalia.

The chief taxonomic divisions are kingdom, phylum, class, order, family, genus, species. The following classification includes all of the major phyla. Certain classes and orders, particularly those mentioned in this book, are also included, but the listing is far from complete. The number of species given for each group is the estimated number of living species described and named.

KINGDOM PROKARYOTAE

Prokaryotes are cells that lack a nuclear envelope, chloroplasts and other plastids, mitochondria, and 9 + 2 flagella. Prokaryotes are unicellular but sometimes aggregate into filaments or other superficially multicellular bodies. Their predominant mode of nutrition is absorption, but some groups are photosynthetic or chemosynthetic. Reproduction is primarily asexual, by mitotic division or budding, but conjugation occurs in some species.

PHYLUM

PHYLUM SCHIZOPHYTA: the bacteria. Unicellular prokaryotes; reproduction usually asexual by cell division; nutrition usually heterotrophic. About 1,600 species.

PHYLUM CYANOPHYTA: blue-green algae. Unicellular or colonial; prokaryotic; chlorophyll, but no chloroplasts; nutrition usually autotrophic; reproduction by fission. Common on damp soil and rocks and in fresh and salt water. About 200 distinct, nonsymbiotic species.

KINGDOM PROTISTA

Eukaryotic organisms, including unicellular heterotrophs (protozoans) and unicellular or multicellular algae. Their modes of nutrition include ingestion, photosynthesis, and sometimes absorption. Reproduction both sexual (in some forms) and asexual. They move by 9 + 2 flagella or pseudopods or are nonmotile.

PHYLUM

PHYLUM PROTOZOA: microscopic, unicellular or simple colonial heterotrophic organisms; reproduction usually asexual by mitotic division; classified by type of locomotion. About 30,000 species.

CLASS

Class Mastigophora: protozoans with flagella, including a number of symbiotic forms such as *Trichonympha* and *Trypanosoma,* the cause of sleeping sickness.

Class Sarcodina: protozoans with pseudopods, such as amoebas. No stiffening pellicle; some secrete shells.

Class Ciliophora: protozoans with cilia, including *Paramecium* and *Stentor.*

Class Sporozoa: parasitic protozoans; usually without locomotive organs during a major part of their life cycle. Includes *Plasmodium,* several species of which cause malaria.

PHYLUM

PHYLUM EUGLENOPHYTA: euglenoids. Unicellular photosynthetic (or sometimes secondarily heterotrophic) organisms with chlorophylls *a* and *b.* They store food as paramylon, an unusual carbohydrate. Euglenoids usually have a single apical flagellum and a contractile vacuole. Sexual reproduction is unknown. Euglenoids occur mostly in fresh water. There are some 800 species.

PHYLUM CHRYSOPHYTA: golden algae and diatoms. Unicellular photosynthetic organisms with chlorophylls *a* and *c* and the accessory pigment fucoxanthin. Food stored as the carbohydrate leucosin or as large oil droplets. Cell walls consisting mainly of pectic compounds, sometimes heavily impregnated with siliceous materials. Some 6,000 to 10,000 living species.

CLASS

Class Bacillariophyceae: diatoms. Chrysophyta with double siliceous shells, the two halves of which fit together like a pillbox. They are sometimes motile by the secretion of mucilage fibrils along a specialized groove, the raphe. There are many extinct species and 5,000 to 9,000 living species.

Class Chrysophyceae: golden algae. A diverse group of organisms, including flagellated, amoeboid, and nonmotile forms, some naked and others with a cell wall that may be ornamented with siliceous scales. At least 1,000 species.

PHYLUM

PHYLUM PYRROPHYTA: "fire" algae, sometimes called golden-brown algae. Unicellular photosynthetic organisms with chlorophylls *a* and *c*. Food is stored as starch. Cell walls contain cellulose. The phylum contains some 1,100 species, mostly biflagellated organisms, of which the great majority belong to the following class:

CLASS

Class Dinophyceae: dinoflagellates. Pyrrophyta with lateral flagella, one of which beats in a groove that encircles the organism. They probably have no form of sexual reproduction, and their mitosis is unlike that in any other organism. More than 1,000 species.

PHYLUM

PHYLUM CHLOROPHYTA: green algae. Unicellular or multicellular, characterized by chlorophylls *a* and *b* and various carotenoids. The carbohydrate food reserve is starch. Motile cells have two whiplash flagella at the apical end. True multicellular genera do not exhibit complex patterns of differentiation. Multicellularity has arisen at least three times, and quite possibly more often. There are about 7,000 known species and possibly many more.

PHYLUM PHAEOPHYTA: brown algae. Multicellular marine organisms characterized by the presence of chlorophylls *a* and *c* and the pigment fucoxanthin. Their food reserve is a carbohydrate called laminarin. Motile cells are biflagellate, with one forward flagellum of the tinsel type and one trailing one of the whiplash type. A considerable amount of differentiation is found in some of the kelps, with specialized conducting cells for transporting photosynthate to dimly lighted regions of the alga present in some genera. There is, however, no differentiation into leaves, roots, and stem, as in the plants. About 1,100 species.

PHYLUM RHODOPHYTA: red algae. Primarily marine organisms characterized by the presence of chlorophyll *a* and pigments known as phycobilins. Their carbohydrate reserve is a special type of starch (floridian). No motile cells are present at any stage in the complex life cycle. The algal body is built up of closely packed filaments in a gelatinous matrix and is not differentiated into leaves, roots, and stem. It lacks specialized conducting cells. There are some 4,000 species.

PHYLUM GYMNOMYCOTA: the slime molds. Heterotrophic amoeboid organisms that mostly lack a cell wall but form sporangia at some stage in their life cycle. Predominant mode of nutrition is by ingestion. There are three classes:

CLASS

Class Myxomycetes: plasmodial slime molds. Slime molds with multinucleate plasmodium that creeps along as a mass and eventually differentiates into sporangia, each of which is multinucleate and eventually gives rise to many spores. About 450 species.

Class Acrasiomycetes: cellular slime molds. Slime molds in which there are separate amoebas that eventually swarm together to form a mass but retain their identity within this mass, which eventually differentiates into a compound sporangium. Seven genera and about 26 species.

Class Protostelidomycetes: In this recently discovered group, the amoebas may remain separate or mass together, but each one eventually differentiates into a simple stalked sporangium with one or two spores at its apex. Five genera and more than a dozen species.

KINGDOM FUNGI

Eukaryotic unicellular or multinucleate organisms in which the nuclei occur in a basically continuous mycelium; this mycelium becomes septate (partitioned off) in certain groups and at certain stages of the life cycle. They are heterotrophic, with nutrition by absorption. Reproductive cycles often include both sexual and asexual phases. Some 100,000 species of fungi have been named.

CLASS

Class Oomycetes: Mostly aquatic fungi with motile cells characteristic of certain stages of the life cycle; their cell walls are composed of glucose polymers, including cellulose. There are several hundred species.

Class Zygomycetes: terrestrial fungi, such as black bread mold, with the hyphae septate only during the formation of reproductive bodies; chitin predominant in the cell walls. The class includes several hundred species.

Class Ascomycetes: terrestrial and aquatic fungi, including *Neurospora,* powdery mildews, morels, and truffles. The hyphae are septate but the septa perforated; complete septa cut off the reproductive bodies, such as spores or gametangia. Chitin is predominant in the cell walls. Sexual reproduction involves the formation of a characteristic cell, the ascus, in which meiosis takes place and within which spores are formed. The hyphae in many Ascomycetes are packed together into complex "fruiting bodies." Yeasts are unicellular Ascomycetes that reproduce asexually by budding. About 30,000 species.

Class Basidiomycetes: terrestrial fungi, including the mushrooms and toadstools, with the hyphae septate but the septa perforated; complete septa cut off reproductive bodies, such as spores or gametangia. Chitin is predominant in the cell walls. Sexual reproduction involves formation of basidia, in which meiosis takes place and on which the spores are borne. There are some 25,000 species.

Fungi Imperfecti: Mainly fungi with the characteristics of Ascomycetes but in which the sexual cycle has not been observed; a few probably belong to other classes. The Fungi Imperfecti are classified by their asexual spore-bearing organs. There are some 25,000 species, including a *Penicillium,* the original source of penicillin, fungi that cause athlete's foot and other skin diseases, and many of the molds that give cheese, such as Roquefort and Camembert, their special flavor.

KINGDOM PLANTAE

Multicellular photosynthetic terrestrial eukaryotes and closely related forms. The photosynthetic pigment is chlorophyll *a,* with chlorophyll *b* and a number of carotenoids serving as accessory pigments. The cell walls contain cellulose. There is considerable differentiation of organs and tissues. Their reproduction is primarily sexual with alternating gametophytic and sporophytic phases; the gametophytic phase has been progressively reduced in the course of evolution.

PHYLUM

PHYLUM BRYOPHYTA: mosses, hornworts, and liverworts. Multicellular plants with photosynthetic pigments and food reserves similar to those of the green algae. They have gametangia with a multicellular sterile jacket one cell layer thick. The sperm are biflagellate and motile. Gametophytes and sporophytes both exhibit complex multicellular patterns of development, but the conducting tissues are usually completely absent and not well differentiated when present. Most of the photosynthesis in these primarily terrestrial plants is carried out by the gametophyte, upon which the sporophyte is initially dependent. More than 23,500 species.

CLASS

Class Hepaticae: liverworts. The gametophytes are either thallose (not differentiated into roots, leaves, and stem) or leafy, and the sporophytes relatively simple in construction. About 9,000 species.

Class Antherocerotae: hornworts. The gametophytes are thallose. The sporophyte grows from a basal meristem for as long as conditions are favorable. Stomata are present on the sporophyte. About 100 species.

Class Musci: mosses. The gametophytes are leafy. The sporophytes have complex patterns of spore discharge. Stomata are present on the sporophyte. About 14,500 species.

PHYLUM

PHYLUM TRACHEOPHYTA: vascular plants. Plants with complex differentiation of organs into leaves, roots, and stems. The only motile cells are the male gametes of some species, which are propelled by many cilia. The vascular plants have well-developed strands of conducting tissue for the transport of water and organic materials. The main trends of evolution in the vascular plants involve a progressive reduction in the gametophyte, which is green and free-living in ferns but heterotrophic and more or less enclosed by sporophytic tissue in the others; the loss of multicellular gametangia and motile sperm; and the evolution of the seed. The phylum includes the following subphyla with living representatives:

SUBPHYLUM

SUBPHYLUM LYCOPHYTINA: lycophytes. Homosporous and heterosporous vascular plants with microphylls; extremely diverse in appearance. All lycophytes have motile sperm. There are five genera and about 1,000 species.

SUBPHYLUM SPHENOPHYTINA: horsetails. Homosporous vascular plants with jointed stems marked by conspicuous nodes and elevated siliceous ribs and sporangia borne in a strobilus at the apex of the stem. Leaves are scalelike. Sperm are motile. Although now thought to have evolved from a megaphyll, the leaves of the horsetails are structurally indistinguishable from microphylls. There is one genus, *Equisetum,* with about two dozen living species.

SUBPHYLUM PTEROPHYTINA: ferns, gymnosperms, and flowering plants. Although diverse, these groups possess in common the megaphyll, which in certain genera has become much reduced. About 260,000 species.

CLASS

Class Filicineae: the ferns. They are mostly homosporous, although some are heterosporous. The gametophyte is more or less free-living and usually photosynthetic. Multicellular gametangia and free-swimming sperm are present. About 11,000 species.

Class Coniferinae: the conifers. Seed plants with active cambial growth and simple leaves, in which the ovules are not enclosed and the sperm are not flagellated. There are some 50 genera and about 550 species; the most familiar group of gymnosperms.

Class Cycadinae: cycads. Seed plants with sluggish cambial growth and pinnately compound, palmlike or fernlike leaves. The ovules are not enclosed. The sperm are flagellated and motile, but are carried to the vicinity of the ovule in a pollen tube. Cycads are gymnosperms. There are nine genera and about 100 species.

Class Ginkgoinae: ginkgo. Seed plants with active cambial growth and fan-shaped leaves with open dichotomous venation. The ovules are not enclosed and are fleshy at maturity. Sperm are carried to the vicinity of the ovule in a pollen tube, but are flagellated and motile. They are gymnosperms. There is one species only.

Class Angiospermae: flowering plants. Seed plants in which the ovules are enclosed in a carpel (in all but a very few genera), and the seeds at maturity are borne within fruits. They are extremely diverse vegetatively but characterized by the flower, which is basically insect-pollinated. Other modes of pollination, such as wind pollination, have been derived in a number of different lines. The gametophytes are much reduced, with the female gametophyte often consisting of only eight cells or nuclei at maturity. Double fertilization involving two of the three nuclei from the mature male gametophyte gives rise to the zygote and to the primary endosperm nucleus; the former becomes the embryo and the latter a special nutritive tissue, the endosperm. About 250,000 species.

SUBCLASS

Subclass Dicotyledoneae: dicots. Flower parts are usually in fours or fives; leaf venation is usually netlike, pinnate, or palmate; there is true secondary growth with vascular cambium commonly present; there are two cotyledons; and the vascular bundles in the stem are in a ring. About 190,000 species.

Subclass Monocotyledoneae: monocots. Flower parts are usually in threes, leaf venation is usually parallel, true secondary growth is not present, there is one cotyledon, and vascular bundles in the stem are scattered. About 60,000 species.

KINGDOM ANIMALIA

Eukaryotic multicellular organisms. Their principal mode of nutrition is by ingestion. Many animals are motile, and they generally lack the rigid cell walls characteristic of plants. Considerable cellular migration and reorganization of tissues often occur during the course of embryonic development. Their reproduction is primarily sexual, with male and female diploid organisms producing haploid gametes that fuse to form the zygote. More than a million species have been described and the actual number may be close to 10 million.

PHYLUM

PHYLUM PORIFERA: sponges. Simple multicellular animals, largely marine, with stiff skeletons, and bodies perforated by many pores that admit water containing food particles. All have choanocytes, "collar cells." About 4,200 species.

PHYLUM COELENTERATA: coelenterates. Animals with radially symmetrical, "two-layered" bodies of a jellylike consistency. Reproduction is asexual or sexual. They are the only organisms with cnidoblasts, special stinging cells. All are aquatic and most are marine. About 11,000 species.

CLASS

Class Hydrozoa: Hydra, Obelia, and other *Hydra*-like animals. They are often colonial and frequently have a regular alternation of asexual and sexual generations. The polyp form is dominant.

Class Scyphozoa: marine jellyfishes or "cup animals," including *Aurelia.* The medusa form is dominant. They have true muscle cells.

Class Anthozoa: sea anemones ("flower animals") and colonial corals. They have no medusa stage.

PHYLUM

PHYLUM CTENOPHORA: comb jellies and sea walnuts. They are free-swimming, often almost spherical animals. They are translucent, gelatinous, delicately colored, and often bioluminescent. They possess eight bands of cilia, for locomotion. About 80 species.

PHYLUM PLATYHELMINTHES: flatworms. Bilaterally symmetrical with three tissue layers. The gut has only one opening. They have no coelom or circulatory system. They have complex hermaphroditic reproductive systems and excrete by means of special (flame) cells. About 15,000 species.

CLASS

Class Turbellaria: planaria and other nonparasitic flatworms. They are ciliated, carnivorous, and have ocelli ("eyespots").

Class Trematoda: flukes. They are parasitic flatworms with digestive tracts.

Class Cestoidea: tapeworms. They are parasitic flatworms with no digestive tracts; they absorb nourishment through body surfaces.

PHYLUM

PHYLUM RHYNCHOCOELA: proboscis, nemertine, or ribbon worms. They are nonparasitic, usually marine, and have a tubelike gut with mouth and anus, a protrusible proboscis armed with a hook for capturing prey, and simple circulatory and reproductive systems. About 600 species.

PHYLUM NEMATODA: roundworms. The phylum includes minute free-living forms, such as vinegar eels, and plant and animal parasites, such as hookworms. They are characterized by elongated, cylindrical, bilaterally symmetrical bodies. About 80,000 species have been described.

PHYLUM ACANTHOCEPHALA: spiny-headed worms. They are parasitic worms with no digestive tract and a head armed with many recurved spines. About 300 species.

PHYLUM CHAETOGNATHA: arrow worms. Free-swimming, planktonic marine worms, they have a coelom, a complete digestive tract, and a mouth with strong sickle-shaped hooks on each side. About 50 species.

PHYLUM NEMATOMORPHA: horsehair worms. They are extremely slender, brown or black worms up to 1 meter long. Adults are free-living, but the larvae are parasitic in insects. About 250 species.

PHYLUM ROTIFERA: microscopic, wormlike or spherical animals, "wheel animalcules." They have a complete digestive tract, flame cells, and a circle of cilia on the head, the beating of which suggests a wheel; males are minute and either degenerate or unknown in many species. About 1,500 species.

PHYLUM GASTROTRICHA: These are microscopic, wormlike animals that move by longitudinal bands of cilia. About 140 species.

PHYLUM BRYOZOA: "moss" animals. These microscopic aquatic organisms are characterized by a U-shaped row of ciliated tentacles, with which they feed. They usually form fixed and branching colonies. Bryozoans superficially resemble hydroid coelenterates but are much more complex, having anus and coelom; they retain larvae in special brood pouch. About 4,000 species.

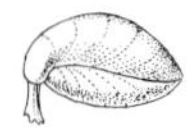

PHYLUM BRACHIOPODA: lamp shells. Marine animals with two hard shells (one dorsal and one ventral), they superficially resemble clams. Fixed by a stalk or one shell in adult life, they obtain food by means of ciliated tentacles. About 260 living species; 3,000 extinct.

PHYLUM PHORONIDEA: sedentary, elongated, wormlike animals that secrete and live in a leathery tube. They have a U-shaped digestive tract and a ring of ciliated tentacles with which they feed. Marine. About 15 species.

PHYLUM ANNELIDA: ringed or segmented worms. They usually have a well-developed coelom, a one-way digestive tract, head, and circulatory system, nephridia, and well-defined nervous system. About 8,800 species.

CLASS

Class Archiannelida: small, simple, probably primitive, marine worms. About 35 species.

Class Polychaeta: mainly marine worms, such as *Nereis.* They have a distinct head with tentacles, antennae, and specialized mouthparts. Parapodia are often brightly colored. About 4,000 species.

Class Oligochaeta: soil, freshwater, and marine annelids, including the earthworm *(Lumbricus).* They have scanty bristles and usually a poorly differentiated head. About 2,500 species.

Class Hirudinea: leeches. They have a posterior sucker and usually an anterior sucker surrounding the mouth. They are freshwater, marine, and terrestrial; either free-living or parasitic. About 300 species.

PHYLUM

PHYLUM MOLLUSCA: unsegmented animals, with a head, a mantle, and a muscular foot, variously modified. They are mostly aquatic; soft-bodied, often with one or more hard shells, and a three-chambered heart. All mollusks, except bivalves, have a radula (rasplike organ used for scraping or marine drilling). About 110,000 species.

CLASS

Class Amphineura: chitons. The simplest type of mollusks, they have an elongated body covered with a mantle in which are embedded eight dorsal shell plates. About 700 species.

Class Pelecypoda: two-shelled mollusks, including clams, oysters, mussels, scallops. They usually have a hatchet-shaped foot and no distinct head. Generally sessile. About 15,000 species.

Class Scaphopoda: tooth or tusk shells. They are marine mollusks with a conical tubular shell. About 350 species.

Class Gastropoda: asymmetrical mollusks, including snails, whelks, slugs. They usually have a spiral shell and a head with one or two pairs of tentacles. About 80,000 species.

Class Cephalopoda: octopus, squid, *Nautilus.* They are characterized by a "head-foot" with eight or ten arms or many tentacles, mouth with two horny jaws, and well-developed eyes and nervous system. The shell is external *(Nautilus),* internal (squid), or absent (octopus). All except *Nautilus* have ink glands. About 400 species.

PHYLUM

PHYLUM ARTHROPODA: The largest phylum in the animal kingdom, arthropods are segmented animals with paired jointed appendages, a hard jointed exoskeleton, a complete digestive tract, reduced coelom, no nephridia, a dorsal brain, and a ventral nerve cord with paired ganglia in each segment. About 1 million species are known.

CLASS

Class Merostomata: horseshoe crabs. They are aquatic, with book gills. About 5 species.

Class Crustacea: lobsters, crabs, crayfish, shrimps. Crustaceans are mostly aquatic, with two pairs of antennae, one pair of mandibles, and typically two pairs of maxillae. The thoracic segments have appendages, and the abdominal segments are with or without appendages. About 30,000 species.

Class Arachnida: spiders, mites, ticks, scorpions. Most members are terrestrial, air-breathing; usually have four or five pairs of legs; first pair of appendages used for grasping; have chelicerae (pincers or fangs) in place of jaws or antennae. About 35,000 species.

Class Onychophora: simple, terrestrial arthropods. All belong to one genus, *Peripatus*. They have many short, unjointed pairs of legs. About 73 species.

Class Insecta: insects, including bees, ants, beetles, butterflies, fleas, lice, flies, etc. Most insects are terrestrial, and most breathe by means of tracheae. They have one pair of antennae, three pairs of legs, and three distinct parts of the body (head, thorax, and abdomen). Most have two pairs of wings. More than 700,000 species.

Class Chilopoda: centipedes. They have 15 to 173 trunk segments, each with one pair of jointed appendages. About 2,000 species.

Class Diplopoda: millipedes. They have an abdomen with 20 to 100 segments, each with two pairs of appendages. About 7,000 species.

PHYLUM

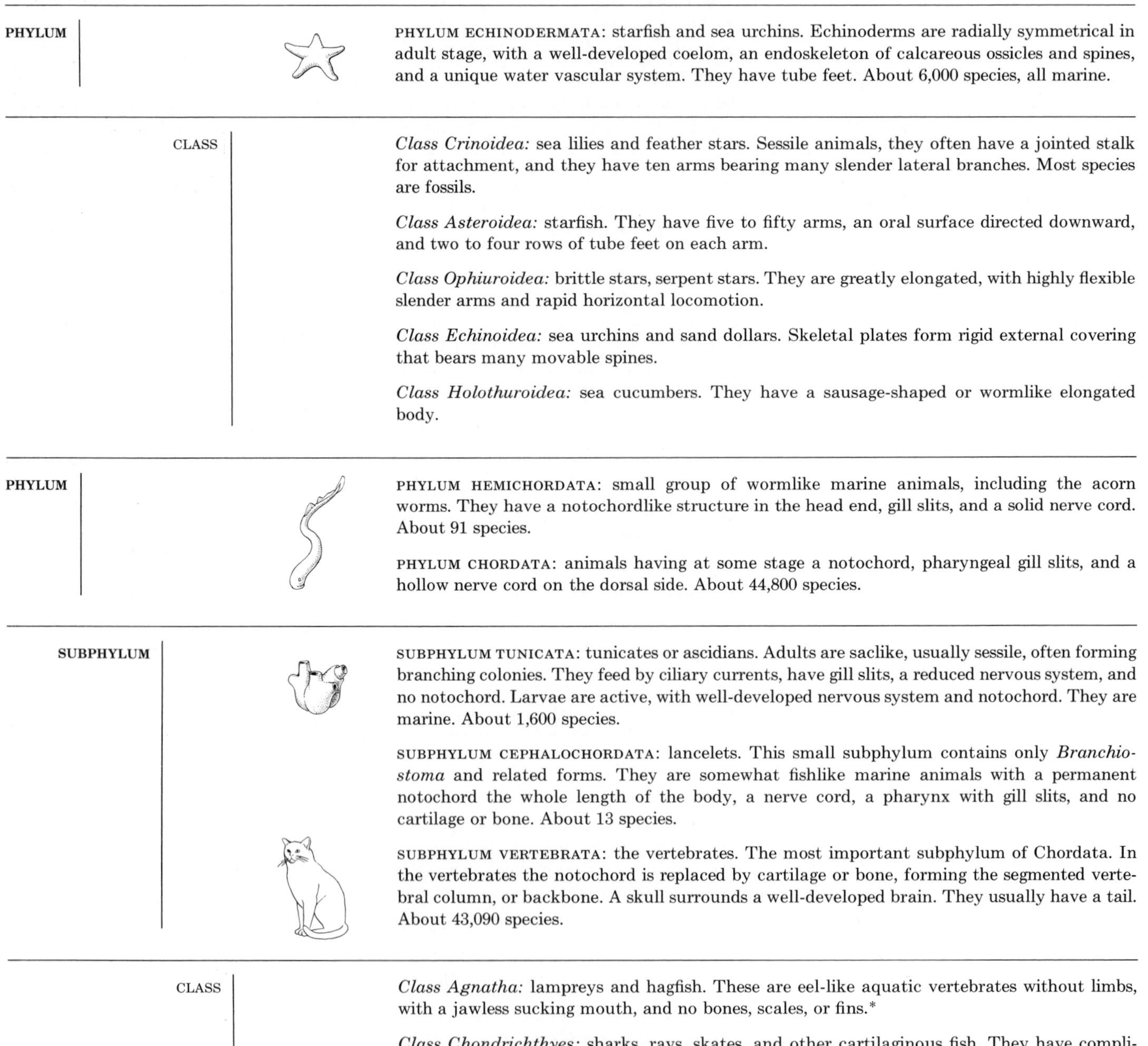

PHYLUM ECHINODERMATA: starfish and sea urchins. Echinoderms are radially symmetrical in adult stage, with a well-developed coelom, an endoskeleton of calcareous ossicles and spines, and a unique water vascular system. They have tube feet. About 6,000 species, all marine.

CLASS

Class Crinoidea: sea lilies and feather stars. Sessile animals, they often have a jointed stalk for attachment, and they have ten arms bearing many slender lateral branches. Most species are fossils.

Class Asteroidea: starfish. They have five to fifty arms, an oral surface directed downward, and two to four rows of tube feet on each arm.

Class Ophiuroidea: brittle stars, serpent stars. They are greatly elongated, with highly flexible slender arms and rapid horizontal locomotion.

Class Echinoidea: sea urchins and sand dollars. Skeletal plates form rigid external covering that bears many movable spines.

Class Holothuroidea: sea cucumbers. They have a sausage-shaped or wormlike elongated body.

PHYLUM

PHYLUM HEMICHORDATA: small group of wormlike marine animals, including the acorn worms. They have a notochordlike structure in the head end, gill slits, and a solid nerve cord. About 91 species.

PHYLUM CHORDATA: animals having at some stage a notochord, pharyngeal gill slits, and a hollow nerve cord on the dorsal side. About 44,800 species.

SUBPHYLUM

SUBPHYLUM TUNICATA: tunicates or ascidians. Adults are saclike, usually sessile, often forming branching colonies. They feed by ciliary currents, have gill slits, a reduced nervous system, and no notochord. Larvae are active, with well-developed nervous system and notochord. They are marine. About 1,600 species.

SUBPHYLUM CEPHALOCHORDATA: lancelets. This small subphylum contains only *Branchiostoma* and related forms. They are somewhat fishlike marine animals with a permanent notochord the whole length of the body, a nerve cord, a pharynx with gill slits, and no cartilage or bone. About 13 species.

SUBPHYLUM VERTEBRATA: the vertebrates. The most important subphylum of Chordata. In the vertebrates the notochord is replaced by cartilage or bone, forming the segmented vertebral column, or backbone. A skull surrounds a well-developed brain. They usually have a tail. About 43,090 species.

CLASS

Class Agnatha: lampreys and hagfish. These are eel-like aquatic vertebrates without limbs, with a jawless sucking mouth, and no bones, scales, or fins.*

Class Chondrichthyes: sharks, rays, skates, and other cartilaginous fish. They have complicated copulatory organs, scales, and no air bladders. They are almost exclusively marine.*

Class Osteichthyes: the bony fish, including nearly all modern freshwater fish, such as sturgeon, trout, perch, anglerfish, lungfish, and some almost extinct groups. They usually have an air bladder or (rarely) a lung.*

Class Amphibia: salamanders, frogs, and toads. They usually breathe by gills in the larval stage and by lungs in the adult stage. They have incomplete double circulation and a usually naked skin. The limbs are legs. They were the first vertebrates to inhabit the land and are ancestors of the reptiles. Their eggs are unprotected by a shell and embryonic membranes. About 2,000 species.

Class Reptilia: turtles, lizards, snakes, crocodiles; includes extinct species such as the dinosaurs. Reptiles breathe by lungs and have incomplete double circulation. Their skin is usually covered with scales. The four limbs are legs (absent in snakes). They are ectotherms. Most live and reproduce on land, although some are aquatic. Their embryo is enclosed in an egg shell and has protective membranes. About 5,000 species.

Class Aves: birds. Birds are homeothermic animals with complete double circulation and a skin covered with feathers. The forelimbs are wings. Their embryo is enclosed in an egg shell with protective membranes. Includes the extinct *Archaeopteryx*. About 8,590 species.

Class Mammalia: mammals. Mammals are homeothermic animals with complete double circulation. Their skin is usually covered with hair. The young are nourished with milk secreted by the mother. They have four limbs, usually legs (forelimbs sometimes arms, wings, or fins), a diaphragm used in respiration, a lower jaw made up of a single pair of bones, three bones in each middle ear connecting eardrum and inner ear, and seven vertebrae in the neck. About 4,500 species.

* The total number of species of fish is estimated to be about 23,000.

SUBCLASS

Subclass Prototheria: monotremes. These are the oviparous (egg-laying) mammals with imperfect temperature regulation. There are only two living species: the duckbill platypus and spiny anteater of Australia and New Guinea.

Subclass Metatheria: marsupials, including kangaroos, opossums, and others. Marsupials are viviparous mammals without a placenta (or with a poorly developed one); the young are born in an undeveloped state and are carried in an external pouch of the mother for some time after birth. They are found chiefly in Australia.

Subclass Eutheria: mammals with a well-developed placenta. This subclass comprises the great majority of living mammals. There are 12 principal orders of Eutheria:

Order

Insectivora: shrews, moles, hedgehogs, etc.

Edentata: toothless mammals—anteaters, sloths, armadillos, etc.

Rodentia: the rodents—rats, mice, squirrels, etc.

Artiodactyla: even-toed ungulates (hoofed mammals)—cattle, deer, camels, hippopotamuses, etc.

Perissodactyla: odd-toed ungulates—horses, zebras, rhinoceroses, etc.

Proboscidea: elephants.

Lagomorpha: rabbits and hares.

Sirenia: the manatee, dugong, and sea cows. Large aquatic mammals with the forelimbs finlike, the hind limbs absent.

Carnivora: carnivorous animals—cats, dogs, bears, weasels, seals, etc.

Cetacea: the whales, dolphins, and porpoises. Aquatic mammals with the forelimbs fins, the hind limbs absent.

Chiroptera: the bats. Aerial mammals with the forelimbs wings.

Primates: the lemurs, monkeys, apes, and humans.

GLOSSARY

This list does not include units of measure or names of taxonomic groups, which can be found in Appendixes A and C, or terms that are used only once in the text and defined there.

abdomen: In vertebrates, the portion of the trunk containing visceral organs other than heart and lungs; in arthropods, the posterior portion of the body, made up of similar segments and containing the reproductive organs and part of the digestive tract.

abscission [L. *ab,* away, off + *scissio,* dividing]: In plants, the dropping of leaves, flowers, fruits, or stems at the end of a growing season, as the result of formation of a layer of specialized cells (the abscission layer) and the action of a hormone (abscisic acid).

absorption [L. *absorbere,* to swallow down]: The movement of water and dissolved substances into a cell, tissue, or organism.

absorption spectrum: A graph or other display of the wavelengths (colors) of light absorbed by a particular pigment.

acetylcholine (a-**sea**-tell-**co**-leen): One of the chemicals (neurotransmitters) responsible for the passing of nerve impulses across synaptic junctions.

acid [L. *acidus,* sour]: A substance that causes an increase in the relative number of hydrogen ions (H^+) in a solution and a decrease in the relative number of hydroxide ions (OH^-); having a pH of less than 7; the opposite of a base.

actin [Gk. *aktis,* a ray]: One of the two major proteins of muscle (the other is myosin); the principal constituent of thin filaments.

action potential: A transient change in electric potential across a membrane; in nerve cells, results in transmission of a nerve impulse; in muscle cells, results in contraction.

activation energy: The initial input of energy that must be supplied from an outside source before a chemical reaction can begin; used to break existing chemical bonds.

active site: That part of an enzyme molecule into which the substrate fits during the reaction catalyzed by the enzyme.

active transport: The pumping of a substance across a cell membrane from a region of lower concentration to a region of higher concentration (that is, against a concentration gradient); an energy-requiring process.

adaptation [L. *adaptare,* to fit]: (1) The evolution of features that make a group of organisms better suited to live and reproduce in their environment. (2) A peculiarity of structure, physiology, or behavior that aids the organism in its environment.

adaptive radiation: The evolution from a primitive and unspecialized ancestor to several divergent forms, each specialized to fit a distinct niche.

adenosine diphosphate (ADP): A nucleotide consisting of adenine, ribose, and two phosphate groups; formed by the removal of one phosphate from an ATP molecule.

adenosine monophosphate (AMP): A nucleotide consisting of adenine, ribose, and one phosphate group; formed by the removal of two phosphates from an ATP molecule; in its cyclic form, functions as a "second messenger" for many vertebrate hormones.

adenosine triphosphate (ATP): The nucleotide that provides the energy currency for cell metabolism; composed of adenine, ribose, and three phosphate groups. On hydrolysis, ATP loses one phosphate group to become adenosine diphosphate (ADP), releasing energy in the process. ATP is formed from ADP + P_i in an enzymatic reaction that traps the energy released by the breakdown of glucose and other molecules or the energy captured in photosynthesis.

ADH: Abbreviation of antidiuretic hormone.

adhesion [L. *adhaerrere,* to stick to]: The holding together of unlike substances.

ADP: Abbreviation of adenosine diphosphate.

adrenal gland [L. *ad,* near + *renes,* kidney]: A vertebrate endocrine gland. The cortex (outer surface) is the source of cortisol, aldosterone, and other steroid hormones; the medulla (inner core) secretes epinephrine and norepinephrine.

adrenaline: *See* Epinephrine.

aerobic [Gk. *aēr,* air + *bios,* life]: Any biological process that can occur in the presence of molecular oxygen (O_2).

afferent [L. *ad,* near + *ferre,* to carry]: Bringing inward to a central part, applied to nerves and blood vessels.

aldosterone [Gk. *aldainō,* to nourish + *stereō,* solid]: A hormone produced by the adrenal cortex that affects the concentration of ions in the blood; it stimulates the resorption of sodium and the excretion of potassium by the kidney.

alga, *pl.* **algae** (**al**-gah, **al**-jee): A unicellular or simple multicellular photosynthetic organism lacking multicellular sex organs.

alkaline: Pertaining to substances that increase the relative number of hydroxide ions (OH^-) in water; having a pH greater than 7; basic; opposite of acidic.

alleles (al-**eels**) [Gk. *allelon,* of one another]: Two or more genes that occupy the same position (locus) on homologous chromosomes (and so are separated from each other at meiosis); alleles can mutate one to the other.

alternation of generations: A sexual life cycle in which a haploid (n) phase alternates with a diploid ($2n$) phase. The gametophyte (n) produces gametes (n) by mitosis. Fertilization of gametes yields zygotes ($2n$). Each zygote develops into a sporophyte ($2n$) that forms haploid spores (n) by meiosis. Each haploid spore forms a new gametophyte, completing the cycle.

alveolus, *pl.* **alveoli** [L. dim. of *alveus,* cavity, hollow]: One of the many small air sacs within the lungs in which the bronchioles terminate. The thin walls of the alveoli contain numerous capillaries and are the site of gas exchange between the air in the alveoli and the blood in the capillaries.

amino acids (am-**ee**-no) [Gk. *Ammon,* referring to the Egyptian sun god, near whose temple ammonium salts were first prepared from camel dung]: Organic molecules containing nitrogen in the form of NH_2 and an acidic carbon atom, COOH, bonded to the same carbon atom; the "building blocks" of protein molecules.

amnion (**am**-neon) [Gk. dim. of *amnos,* lamb]: Fluid-filled sac that surrounds the embryo in reptiles, birds, and mammals.

amniote egg: An egg that is isolated and protected from the environment by a more or less impervious shell during the period of its development and that is completely self-sufficient, requiring only oxygen from the outside.

AMP: Abbreviation of adenosine monophosphate.

amphibian [Gk. *amphibios,* living a double life]: A class of vertebrates intermediate in many characteristics between fish and reptiles, which live part or most of the time on land but must return to water to reproduce because fertilization is external.

anaerobic [Gk. *an,* without + *aēr,* air + *bios,* life]: Applied to a process that can occur without oxygen, as anaerobic fermentation.

analogous [Gk. *analogos,* proportionate]: Applied to structures similar in function but different in evolutionary origin, such as the wing of a bird and the wing of an insect.

anaphase (**anna**-phase) [Gk. *ana,* up + *phasis,* form]: A stage in mitosis or meiosis in which the chromatids of each chromosome separate and move to opposite poles.

androgens [Gk. *andros,* man + *genos,* origin, descent]: Male sex hormones.

angiosperms (**an**-jee-o-sperms) [Gk. *angeion,* vessel + *sperma,* seed]: The flowering plants. Literally, a seed borne in a vessel; thus, any plant whose seeds are borne within a matured ovary (fruit).

annual plant [L. *annus,* year]: A plant that completes its life cycle (from seed germination to seed production) and dies within a single growing season.

antennae: Long, paired sensory appendages on the head of many arthropods.

anterior [L. *ante,* before, toward, in front of]: The front end of an organism; in human anatomy, the ventral surface.

anther [Gk. *anthos,* flower]: In flowering plants, the pollen-bearing portion of a stamen.

antheridium, *pl.* **antheridia:** In bryophytes and some vascular plants, the multicellular sperm-producing organ.

anthropoid [Gk. *anthropos,* man, human]: A higher primate; includes monkeys, apes, and humans.

antibiotic [Gk. *anti,* against + *bios,* life]: An organic compound formed and secreted by an organism that is inhibitory or toxic to other species.

antibody [Gk. *anti,* against]: A globular protein produced in response to a foreign substance (antigen), with which it reacts specifically.

anticodon: In a tRNA molecule, the three-nucleotide sequence that "plugs in" to the mRNA codon for the amino acid carried by that particular tRNA; the anticodon is complementary to the mRNA codon.

antidiuretic hormone (ADH) [Gk. *anti,* against + *diurgos,* thoroughly wet; *hormon,* excite, stimulate]: A hormone secreted by the hypothalamus that inhibits urine excretion by inducing the resorption of water from the nephrons of the kidneys; also called vasopressin.

antigen [Gk. *anti,* against + *genos,* origin, descent]: A foreign substance, usually a protein or polysaccharide, that stimulates the formation of specific antibodies.

aorta (a-**ore**-ta) [Gk. *aeirein,* to lift, heave]: The major artery in blood-circulating systems; the aorta sends blood to the other body tissues.

apical meristem [L. *apex,* top + Gk. *meristos,* divided]: In vascular plants, the growing point at the tip of the root or stem.

arboreal [L. *arbor,* tree]: Tree-dwelling.

archegonium, *pl.* **archegonia** [Gk. *archegonos,* first of a race]: In bryophytes and some vascular plants, the multicellular egg-producing organ.

artery: A vessel carrying blood from the heart to the tissues; arteries are usually thick-walled, elastic, and muscular. A small artery is known as an arteriole.

arthropod [Gk. *arthron,* joint + *pous, podos,* foot]: Any invertebrate animal with jointed appendages; a member of the phylum Arthropoda.

artificial selection: The breeding of special strains of organisms for the purpose of enhancing desired traits.

asexual reproduction: Any reproductive process, such as budding or the division of a cell or body into two or more approximately equal parts, that does not involve the union of gametes.

atmospheric pressure [Gk. *atmos,* vapor + *sphaira,* globe]: The weight of the earth's atmosphere over a unit area of the earth's surface.

atom [Gk. *atomos,* indivisible]: The smallest particle into which a chemical element can be divided and still retain the properties characteristic of the element; consists of a central core, the nucleus, containing protons and neutrons, and electrons that move around the nucleus.

atomic number: The number of protons in the nucleus of an atom; equal to the number of electrons in the neutral atom.

atomic weight: The number of protons plus neutrons in the nucleus of an atom.

ATP: Abbreviation of adenosine triphosphate, the principal energy-carrying compound of the cell.

atrium, *pl.* **atria** (**a**-tree-um) [L., yard, court, hall]: A thin-walled chamber of the heart that receives blood and passes it on to a thick, muscular ventricle.

autonomic [Gk. *autos,* self + *nomos,* usage, law]: Self-controlling, independent of outside influences.

autonomic nervous system [Gk. *autos,* self + *nomos,* usage, law]: A system of motor nerves and ganglia in vertebrates that is not under voluntary control; innervates the heart, glands, visceral organs, and smooth muscle; subdivided into the sympathetic and the parasympathetic divisions.

autosome [Gk. *autos,* self + *soma,* body]: Any chromosome other than the sex chromosomes. Humans have 22 pairs of autosomes and 1 pair of sex chromosomes.

autotroph [Gk. *autos,* self + *trophos,* feeder]: An organism that is able to synthesize all needed organic molecules from simple inorganic substances (e.g., H_2O, CO_2, NH_3) and some energy source (e.g., sunlight); in contrast to heterotroph. Plants, algae, and some bacteria are autotrophs.

auxin [Gk. *auxein,* to increase + *in,* of, or belonging to]: One of a group of plant hormones with a variety of growth-regulating effects, including promotion of cell elongation.

axis: An imaginary line passing through a body or organ around which parts are symmetrically aligned.

axon [Gk. *axon,* axle]: The part of a neuron that carries impulses away from the cell body.

bacteriophage [L. *bacterium* + Gk. *phagein,* to eat]: A virus that parasitizes a bacterial cell.

bacterium [Gk. dim. of *baktron,* staff]: A unicellular prokaryote.

bark: In plants, all tissues outside the vascular cambium in a woody stem.

basal body [Gk. *basis,* foundation]: A cytoplasmic organelle of protozoa and animals that organizes cilia or flagella; identical in structure to the centriole, which is involved in mitosis and meiosis in most protozoa and animals and some plants.

base: A substance that causes an increase in the relative number of hydroxide ions (OH^-) in a solution and a decrease in the relative number of hydrogen ions (H^+); having a pH of more than 7; the opposite of an acid. *See* Alkaline.

base-pairing rule: In the formation of nucleic acids, the requirement that adenine must always pair with thymine (or uracil) and guanine with cytosine.

biennial [L. *biennium,* a space of two years; *bi,* twice + *annus,* year]: Occurring once in two years; a plant that requires two years to complete its reproductive cycle; vegetative growth occurs in the first year, sexual reproduction and death in the second.

bilateral symmetry [L. *bi,* twice, two + *lateris,* side; Gk. *summetria,* symmetry]: A body form in which the right and left halves of an organism are approximate mirror images of each other.

bile: A yellow secretion of the vertebrate liver, temporarily stored in the gallbladder and composed of organic salts that emulsify fats in the small intestine.

binomial system [L. *bi,* twice, two + Gk. *nomos,* usage, law]: A system of naming organisms in which the "scientific" name consists of two parts (the first designates genus and the second, species).

biogeochemical cycle [Gk. *bios,* life + *geō,* earth + *chēmeia,* alchemy; *kyklos,* circle, wheel]: The cyclic path of an inorganic substance, such as carbon or nitrogen, through an ecosystem. Its geological components are the atmosphere, the crust of the earth, and the oceans, lakes, and rivers; its biological components are producers, consumers, and detritivores.

biological clock [Gk. *bios,* life + *logos,* discourse]: Proposed internal factor(s) in plants and animals that governs what seem to be the innate biological rhythms (growth and activity patterns) of some organisms.

biomass [Gk. *bios,* life]: Total weight of all organisms (or some group of organisms) living in a particular habitat or place.

biomes: Communities of organisms recognized by common patterns of distinctive vegetation and climate; for example, the grassland biome, the tropical rain forest biome, etc.

biosphere [Gk. *bios,* life + *sphaira,* globe]: The zones of air, land, and water at the surface of the earth occupied by living things.

biosynthesis [Gk. *bios,* life + *synthesis,* a putting together]: Formation by living organisms of organic compounds from elements or simple compounds.

blade: The broad, expanded part of a leaf or leaflike organ.

blastula [Gk. *blastos,* sprout]: An early stage in the development of an animal embryo; usually consists of a hollow sphere, the walls of which are composed of a single layer of cells.

botany [Gk. *botanikos,* of herbs]: The study of plants.

Bowman's capsule: In the vertebrate kidney, the bulbous unit of the nephron, which surrounds the glomerulus. In filtration, the initial process in urine formation, blood plasma is forced from the glomerular capillaries into Bowman's capsule.

brainstem: The most posterior portion of the vertebrate brain; includes medulla, pons, and midbrain.

bronchus, *pl.* **bronchi** (**bronk**-us, **bronk**-eye) [Gk. *bronchos,* windpipe]: One of a pair of respiratory tubes branching into either lung at the lower end of the trachea; it subdivides into progressively finer passageways, the bronchioles, culminating in the alveoli.

bud: (1) In plants, an embryonic shoot, including rudimentary leaves, often protected by specialized bud scales. (2) In animals, an asexually produced outgrowth that develops into a new individual.

bulb: A modified bud with thickened leaves adapted for underground food storage.

bulk flow: The overall movement of a liquid induced by gravity, pressure, or an interplay of both.

calorie [L. *calor,* heat]: The amount of energy in the form of heat required to raise the temperature of 1 gram of water 1°C; in making metabolic measurements the kilocalorie (Calorie) is generally used. A Calorie is the amount of heat required to raise the temperature of 1 kilogram of water 1°C.

Calvin cycle: The set of reactions through which carbon dioxide is reduced to carbohydrates during the dark phase of photosynthesis.

capillaries [L. *capillaris,* relating to hair]: Smallest thin-walled blood vessels through which exchanges between blood and the tissues occur; connect arteries with veins.

capillary action: The movement of water or any liquid along a surface; results from the combined effect of cohesion and adhesion.

capsule (**kap**-sul) [L. *capsula,* a little chest]: (1) A slimy layer that surrounds the cells of certain bacteria. (2) The sporangium of Bryophyta.

carbohydrate [L. *carbo,* charcoal + *hydro,* water]: An organic compound consisting of a chain or ring of carbon atoms to which hydrogen and oxygen are attached in a ratio of approximately 2:1; carbohydrates include sugars, starch, glycogen, cellulose, etc.

carbon cycle: Worldwide circulation and reutilization of carbon atoms, chiefly due to metabolic processes of living organisms. Inorganic carbon, in the form of carbon dioxide, is incorporated into organic compounds by photosynthetic organisms; when the organic compounds are broken down in respiration, carbon dioxide is released. Large quantities of carbon are "stored" in the seas and the atmosphere, as well as in fossil fuel deposits.

carnivore [L. *caro, carnis,* flesh + *voro,* to devour]: Predator that obtains its nutrients and energy by eating animals.

carotenoids [L. *carota,* carrot]: A class of pigments that includes the carotenes (yellows, oranges, and reds) and the xanthophylls (yellow); accessory pigments in photosynthesis.

carpel: A leaflike floral structure enclosing the ovule or ovules of angiosperms, typically divided into ovary, style, and stigma; a flower may have one or more carpels, either single or fused.

carrying capacity: In ecology, the largest number of organisms of a particular species that can be maintained indefinitely in a given part of the environment.

cartilage [L. *cartilago,* gristle]: A connective tissue in skeletons of vertebrates; forms much of the skeleton of adult lower vertebrates and immature higher vertebrates.

Casparian strip (after Robert Caspary, German botanist): In plants, a thickened, waxy strip that extends around and seals the walls of endodermal root cells, thus restricting the diffusion of solutes across the endodermis into the vascular tissues of the root.

catalyst [Gk. *katalysis,* dissolution]: A substance that accelerates the rate of a chemical reaction but is not used up in the reaction; enzymes are catalysts.

cell [L. *cella,* a chamber]: The structural unit of organisms, surrounded by a membrane and composed of cytoplasm and, in eukaryotes, one or more nuclei. In most plants, fungi, and bacteria, there is a cell wall outside the membrane.

cell cycle: A regular, timed sequence of the events of cell growth and division through which dividing cells pass.

cell membrane: The outermost membrane of the cell; also called the plasma membrane.

cell plate: In the dividing cells of most plants (and in some fungi and algae), a flattened structure that forms at the equator of the mitotic spindle in early telophase; gives rise to the middle lamella.

cell theory: All living things are composed of cells; cells arise only from other cells. No exception has been found to these two rules since they were first proposed well over a century ago.

cellulose [L. *cellula,* a little cell]: The chief constituent of the cell wall in all green plants; an insoluble complex carbohydrate formed of microfibrils of glucose molecules.

cell wall: A plastic or rigid structure, produced by the cell and located outside the cell membrane in most plants, algae, fungi, and bacteria; in plant cells, it consists mostly of cellulose.

central nervous system: In vertebrates, the brain and spinal cord; in invertebrates it usually consists of one or more cords of nervous tissue plus their associated ganglia.

centriole (**sen**-tree-ole) [Gk. *kentron,* center]: A cytoplasmic organelle identical in structure to a basal body; flagellated cells and all animal cells, including those without flagella, have centrioles at the spindle poles during division.

centromere (**sen**-tro-mere) [Gk. *kentron,* center + *meros,* a part]: Region of constriction of chromosome that holds sister chromatids together.

cerebellum [L. dim. of *cerebrum,* brain]: An enlarged part of the dorsal side of the vertebrate brain; functions in coordinating muscular activities and maintaining equilibrium.

cerebral cortex [L. *cerebrum,* brain]: A layer of neurons (gray matter) forming the upper surface of the cerebrum, well developed only in mammals; the seat of conscious sensations and voluntary muscular activity.

cerebrum [L., brain]: The portion of the vertebrate brain occupying the upper part of the skull, consisting of two cerebral hemispheres united by the corpus callosum.

character displacement: A phenomenon in which species that live together in the same environment are more dissimilar than the same species living in separate environments; exemplified by Darwin's finches.

chemical reaction: An interaction among atoms, ions, or molecules that results in the formation of new combinations of atoms, ions, or molecules; the making or breaking of chemical bonds.

chitin (**kye**-tin) [Gk. *chitōn,* a tunic, undergarment]: A tough, resistant, nitrogen-containing polysaccharide present in the exoskeleton of arthropods, the epidermal cuticle or other surface structures of many other invertebrates, and in the cell walls of certain fungi.

chlorophyll [Gk. *chloros,* green + *phyllon,* leaf]: A class of green pigments found in chloroplasts; necessary for photosynthesis.

chloroplast [Gk. *chloros,* green + *plastos,* formed]: A membrane-bound, chlorophyll-containing organelle in eukaryotes (algae and green plants) that is the site of photosynthesis.

chordate [L. *chorda,* cord, string]: Member of the animal phylum Chordata of which all members possess a notochord, dorsal nerve cord, and pharyngeal gill slits, at least at some stage of the life cycle.

chorion (**core**-ee-on) [Gk. *chorion,* skin, leather]: The outermost membrane of the embryos of reptiles, birds, and mammals; in placental mammals it contributes to the structure of the placenta.

chromatid (**crow**-ma-tid) [Gk. *chrōma,* color]: Either of the two strands of a replicated chromosome, which are joined at the centromere.

chromatin [Gk. *chrōma,* color]: The deeply staining complex of DNA and proteins of which the chromosomes are composed.

chromosome [Gk. *chrōma,* color + *soma,* body]: One of the bodies in the cell nucleus along which the genes are located; visualized as threads or rods of chromatin, which appear in a contracted form during mitosis and meiosis.

chromosome map: A diagram of the linear order of the genes on a chromosome; determined from the frequency of crossing over between pairs of genes.

cilium, *pl.* **cilia** (**silly**-um) [L., eyelash]: A short, thin structure embedded in the surface of some eukaryotic cells, usually in large numbers and arranged in rows; has a highly characteristic internal structure of two inner microtubules surrounded by nine pairs of outer microtubules; involved in locomotion and the movement of substances across the cell surface.

circadian rhythms [L. *circa,* about + *dies,* day]: Regular rhythms of growth or activity that occur on an approximately 24-hour cycle.

class: A taxonomic grouping of related, similar orders; category above order and below phylum.

climax community: Final, relatively stable community in a successional series.

cline [Gk. *klinein,* to lean]: A graded series of changes in some characteristic within a species, correlated with some gradual change in climate or another geographical factor.

clitoris (**klit**-o-ris) [Gk. *kleitoris,* a small hill]: A small erectile body at the anterior part of the vulva; homologous to the penis.

clone [Gk. *klon,* twig]: A line of cells, all of which have arisen from the same single cell by mitotic division; a population of individuals derived by asexual reproduction from a single ancestor.

codon (**code**-on): Basic unit ("letter") of the genetic code; three adjacent nucleotides in a molecule of mRNA that form the code for a specific amino acid.

coelenteron (see-**len**-t-ron) [Gk. *koilos,* hollow + *enteron,* gut]: A digestive cavity with only one opening, characteristic of the phylum Coelenterata (jellyfish, hydra, corals, etc.).

coelom (**see**-loam) [Gk. *koilos,* a hollow]: A body cavity within which the viscera of higher animals are suspended; formed between layers of mesoderm.

coenzyme [L. *co,* together + Gk. *en,* in + *zyme,* leaven]: An organic molecule that plays an accessory role in enzyme-catalyzed processes, often by acting as a donor or acceptor of a substance involved in the reaction. NAD^+, FAD, and coenzyme A are common coenzymes.

coevolution [L. *co,* together + *e-,* out + *volvere,* to roll]: The simultaneous development of adaptations in two or more populations that interact so closely that each is a strong selective force on the other.

cohesion [L. *cohaerere,* to stick together]: The attraction or holding together of like molecules or like substances.

cohesion-tension theory: A theory accounting for the upward movement of water in plants. According to this theory, transpiration of a water molecule results in a negative (below 1 atmosphere) pressure in the leaf cells, inducing the entrance from the vascular tissue of another water molecule, which, because of the cohesive property of water, pulls with it a chain of water molecules extending up from the cells of the root tip.

coleoptile (coal-ee-**op**-tile) [Gk. *koleon,* sheath + *ptilon,* feather]: The sheath enclosing the apical meristem and leaf primordia of the germinating monocot.

collagen [Gk. *kolla,* glue]: A fibrous protein in bones, tendons, and other connective tissues.

colony: A group of unicellular or multicellular organisms living together in close association.

commensalism [L. *com,* together + *mensa,* table]: *See* Symbiosis.

community: All of the organisms inhabiting a common environment and interacting with one another.

competition: Interaction between members of the same population or of two or more populations in order to obtain a mutually required resource.

competitive exclusion: The hypothesis that two species with identical ecological requirements cannot coexist stably in the same locality and the species that is more efficient in utilizing the available resources will exclude the other; also known as Gause's principle, after the Russian biologist G. F. Gause.

compound [L. *componere,* to put together]: A molecule composed of two or more kinds of atoms in definite ratios, held together by chemical bonds.

compound eye: In arthropods, a complex eye composed of many separate elements, each with light-sensitive cells and a lens that can form an image.

condensation [L. *co,* together + *densare,* to make dense]: A type of chemical reaction in which two molecules join to form one larger molecule, simultaneously splitting out a molecule of water. The biosynthetic reactions in which monomers (e.g., monosaccharides, amino acids) are joined to form polymers (e.g., polysaccharides, polypeptides) are condensation reactions.

cone: (1) In plants, the reproductive structure of a conifer. (2) In vertebrates, a type of light-sensitive neuron in the retina, concerned with the perception of color and with the most acute discrimination of detail.

conifer [Gk. *konos,* cone + *phero,* carry]: One of the cone-bearing plants, such as pines and firs.

conjugation [L. *conjugatio,* a joining, connection]: The sexual process in some unicellular organisms by which genetic material is passed from one cell to another across a cytoplasmic bridge.

connective tissues: Supporting or packing tissues that lie between groups of nerves, glands, and muscle cells, and beneath epithelial cells, in which the cells are irregularly distributed through a relatively large amount of intercellular material; include bone, cartilage, blood, and lymph.

consumer, in ecological pyramids: A heterotroph that derives its energy from living or freshly killed organisms or parts thereof. Primary consumers are herbivores; higher-level consumers are carnivores.

continuous variation: A gradation of small differences in a particular trait, such as height, within a population; occurs in traits that are controlled by a number of genes.

convergent evolution [L. *convergere,* to turn together; *evolutio,* to unfold]: The independent development of similarities between unrelated groups, such as porpoises and sharks, resulting from adaptation to similar environments.

cork [L. *cortex,* bark]: A secondary tissue that is a major constituent of bark in woody and some herbaceous plants; made up of flattened cells, dead at maturity; restricts gas and water exchange and protects the vascular tissues from injury.

cork cambium [L. *cortex,* bark + *cambium,* exchange]: The lateral meristem that produces cork.

corolla (ko-**role**-a) [L. dim. of *corona,* wreath, crown]: Petals, collectively; usually the conspicuously colored flower parts.

corpus luteum [L., yellowish body]: An ovarian structure that secretes estrogens and progesterone, which maintain the uterus during pregnancy. It develops from the remaining cells of the ruptured follicle following ovulation.

cortex [L., bark]: (1) The outer layer. (2) In a stem or root, the primary tissue bounded externally by the epidermis and internally by the central cylinder of vascular tissue.

cotyledon (cottle-**ee**-don) [Gk. *kotyledon,* a cup-shaped hollow]: A leaflike structure of the embryo of a seed plant; contains stored food used during germination.

covalent bond [L. *con,* together + *valere,* to be strong]: A chemical bond formed between atoms as a result of the sharing of one or more pairs of electrons.

crossing over: During meiosis, the exchange of genetic material between paired chromatids of homologous chromosomes.

cuticle (**ku**-tik-l) [L. *cuticula,* dim. of *cutis,* the skin]: (1) In plants, a layer of fatty substance (cutin) on outer surface of epidermal cell walls. (2) In animals, the noncellular, outermost layer of many invertebrates.

cytochromes [Gk. *kytos,* vessel + *chrōma,* color]: Iron-containing proteins that participate in electron transport chains; involved in cellular respiration and photosynthesis.

cytokinesis [Gk. *kytos,* vessel + *kinesis,* motion]: Division of the cytoplasm of a cell following nuclear division.

cytokinin [Gk. *kytos,* vessel + *kinesis,* motion]: One of a group of chemically related plant hormones that promote cell division, among other effects.

cytoplasm (**sight**-o-plazm) [Gk. *kytos,* vessel + *plasma,* anything molded]: The living matter within a cell, excluding the nucleus.

dark reactions: The second stage of photosynthesis, which does not require light; energy stored in ATP and NADPH by the light reactions is used to reduce carbon from carbon dioxide to simple sugars.

deciduous [L. *decidere,* to fall off]: Refers to plants that shed their leaves at a certain season.

decomposers: Specialized detritivores, usually bacteria or fungi, that consume such substances as cellulose and nitrogenous waste products. Their metabolic processes release inorganic nutrients, which are then available for reuse by plants and other organisms.

dendrite [Gk. *dendron,* tree]: Nerve fiber, typically branched, that conducts impulses toward the cell body of a neuron.

density-dependent factors: Factors affecting the size of a population, the effects of which vary with the density (number of individuals per unit area) of the population; include resources for which members of the same or different populations compete, predation, and symbiotic interactions.

density-independent factors: Factors affecting the size of a population, the effects of which are independent of the density of the population; are usually climatic, such as extremes of temperature, day length, and rainfall.

deoxyribonucleic acid (DNA) (dee-ox-y-rye-bo-new-**klee**-ick): The carrier of genetic information in cells, composed of two complementary chains of nucleotides wound in a double helix; capable of self-replication as well as coding for RNA synthesis.

dermis [Gk. *derma,* skin]: The inner layer of the skin, beneath the epidermis.

detritivores [L. *detritus,* worn down, worn away + *voro,* to devour]: Organisms that live on dead and discarded organic matter; include large scavengers, smaller animals such as earthworms and some insects, as well as decomposers (fungi and bacteria).

development: The progressive production of the phenotypic characteristics of a multicellular organism, beginning with the fertilization of an egg.

diaphragm [Gk. *diaphrassein,* to barricade]: In mammals, a sheetlike tissue (tendon and muscle) forming the partition between the abdominal and thoracic cavities; functions in breathing.

dicotyledon (**dye**-cottle-**ee**-don) [Gk. *di,* double, two + *kotyledon,* a cup-shaped hollow]: A member of a subclass of angiosperms having two seed leaves, or cotyledons, among other distinguishing features; often abbreviated as dicot.

differentiation: The developmental process by which a relatively unspecialized cell or tissue undergoes a progressive (usually irreversible) change to a more specialized cell or tissue.

diffusion [L. *diffundere,* to pour out]: The net movement of suspended or dissolved particles down a concentration gradient as a result of the random spontaneous movements of individual particles; the process tends to distribute the particles uniformly throughout a medium.

digestion [L. *digestio,* separating out, dividing]: The breakdown of complex, usually insoluble foods into simple, usually soluble forms by means of enzymes.

dioecious (dye-**ee**-shus) [Gk. *di,* two + *oikos,* house]: In angiosperms, having the male and female flowers on different individuals of the same species.

diploid [Gk. *di,* double, two + *ploion,* vessel]: The condition in which each autosome is represented twice ($2n$); in contrast to haploid (n). Characteristic of some microorganisms, the sporophyte generation in plants and algae, and the somatic cells of animals.

DNA: Abbreviation of deoxyribonucleic acid.

dominance: (1) In genetics, the ability of one allelic form of a gene to determine the phenotype of a heterozygous individual. The homologous chromosome carries a different allele, which is said to be recessive. (2) In ecology, the capacity of a group of organisms, by virtue of size, number, or behavior, to exert a controlling influence on their environment and, as a result, to determine what other kinds of organisms exist in that ecosystem.

dominant allele: An allele whose phenotypic effect is the same in both the heterozygous and homozygous conditions.

dormancy [L. *dormire,* to sleep]: A period during which growth ceases and is resumed only if certain requirements, as of temperature or day length, have been fulfilled.

dorsal [L. *dorsum,* the back]: Pertaining to or situated near the back; opposite of ventral.

double fertilization: A phenomenon unique to the angiosperms, in which the egg and one sperm nucleus fuse (resulting in a $2n$ fertilized egg, the zygote) and simultaneously the second sperm nucleus fuses with the two polar nuclei (resulting in a $3n$ endosperm nucleus).

duodenum (duo-**dee**-num) [L. *duodeni,* twelve each—from its length, about 12 fingers' breadth]: The upper portion of the small intestine in vertebrates, where food is digested into molecules that can be absorbed by intestinal cells.

ecological niche: A description of the roles and associations of a particular species in the community of which it is a part; the way in which an organism interacts with the biotic and abiotic parts of its environment.

ecological pyramid: A graphic representation of the quantitative relationships of number of organisms, biomass, or energy flow between the trophic levels of a food chain. Because large amounts of energy and biomass are dissipated at every trophic level, these diagrams nearly always take the form of pyramids.

ecological succession: The gradual process by which the species composition of a community changes, culminating in a stable composition, known as a climax community.

ecology [Gk. *oikos,* home + *logos,* a discourse]: The study of the interactions of organisms with their physical environment and with each other and of the results of such interactions.

ecosystem [Gk. *oikos,* home + *systema,* that which is put together]: All organisms in a community plus the associated abiotic environmental factors with which they interact.

ectoderm [Gk. *ecto,* outside + *derma,* skin]: In animals, the outermost layer of body tissue.

ectotherm [Gk. *ecto,* outside + *therme,* heat]: An organism that regulates its body temperature by taking in heat from the environment or giving it off to the environment. Reptiles are ectotherms.

effector [L. *ex,* out of + *facere,* to make]: Cell, tissue, or organ (such as muscle or gland) capable of producing a response to a stimulus.

efferent [L. *ex,* out of + *ferre,* to bear]: Carrying away from a center, applied to nerves and blood vessels.

egg: A female gamete, which usually contains abundant cytoplasm and yolk; usually immotile, often larger than a male gamete.

electric potential: The difference in the amount of electric charge between an area of positive charge and an area of negative charge. The electric potential across the membrane of nerve cells makes possible the transmission of nerve impulses.

electron: A subatomic particle with a negative electric charge equal in magnitude to the positive charge of the proton but with a much smaller mass; normally found within orbitals surrounding the atom's positively charged nucleus.

electron acceptor: Substance that accepts or receives electrons in an oxidation-reduction reaction, becoming reduced in the process.

electron carrier: A specialized molecule, such as a cytochrome, that can lose and gain electrons reversibly, alternately becoming oxidized and reduced.

electron donor: Substance that donates or gives up electrons in an oxidation-reduction reaction, becoming oxidized in the process.

electron transport chain: A series of electron-carrier molecules that hold electrons at slightly different energy levels; as electrons move down the chain, the energy released is used to form ATP from ADP and phosphate. Electron transport chains play essential roles in the final stage of cellular respiration and in the light reactions of photosynthesis.

element: A substance composed only of atoms of the same atomic number and which cannot be decomposed by ordinary chemical means.

embryo [Gk. *en,* in + *bryein,* to swell]: The early developmental stage of an organism produced from a fertilized egg; a young organism before it emerges from the seed, egg, or body of its mother. In humans, the embryo is called a fetus from the third month of pregnancy until birth.

endocrine gland [Gk. *endon,* within + *krinein,* to separate]: Ductless gland whose secretions (hormones) are released into the circulatory system; in vertebrates, includes pituitary, sex glands, adrenal, thyroid, and others.

endoderm [Gk. *endon,* within + *derma,* skin]: In animals, the innermost layer of body tissue.

endodermis [Gk. *endon,* within + *derma,* skin]: In plants, a one-celled layer of specialized cells that lies between the cortex and the vascular tissues in young roots. The Casparian strip of the endodermis prevents diffusion of solutes across the root.

endometrium [Gk. *endon,* within + *metrios,* of the womb]: The glandular lining of the uterus in mammals; thickens in response to secretion of estrogens and progesterone and is sloughed off in menstruation.

endoplasmic reticulum [Gk. *endon,* within + *plasma,* from cytoplasm; L. *reticulum,* network]: An extensive system of membranes present in most eukaryotic cells, dividing the cytoplasm into compartments and channels, often coated with ribosomes.

endosperm [Gk. *endon,* within + *sperma,* seed]: In plants, a $3n$ tissue containing stored food that develops from the union of a sperm nucleus and the polar nuclei of the egg; found only in angiosperms.

endotherm [Gk. *endon,* within + *therme,* heat]: An organism that regulates its body temperature internally through metabolic processes. Birds and mammals are endotherms.

enzyme [Gk. *en,* in + *zyme,* leaven]: A protein molecule that catalyzes a chemical reaction.

epidermis [Gk. *epi,* on or over + *derma,* skin]: In plants and animals, the outermost layers of cells.

epinephrine: A hormone produced by the medulla of the adrenal gland, which increases the concentration of sugar in the blood, raises blood pressure and heartbeat rate, and increases muscular power and resistance to fatigue; also a chemical transmitter across synaptic junctions. Also called adrenaline.

epithelial tissue [Gk. *epi,* on or over + *thele,* nipple]: In animals, a type of tissue that covers a body or structure or lines a cavity; epithelial cells form one or more regular layers with little intercellular material.

equilibrium [L. *aequus,* equal + *libra,* balance]: The state of a system in which no further net change is occurring; result of counterbalancing forward and backward processes.

equilibrium species [L. *aequus,* equal + *libra,* balance + *species,* kind, sort]: Species characterized by low reproduction rates, long development times, large body size, and long adult life with repeated reproductions. Such species appear to have a competitive advantage when environmental conditions are relatively constant.

erythrocyte (eh-**rith**-ro-site) [Gk. *erythros,* red + *kytos,* vessel]: Red blood cell, the carrier of hemoglobin.

estrogens: Female sex hormones, which are the predominant secretions of the ovarian follicle during the preovulatory phase of the menstrual cycle.

ethology [Gk. *ethos,* habit, custom + *logos,* discourse]: The study of whole patterns of animal behavior in natural environments, stressing adaptation and evolution.

eukaryote (you-**car**-ry-oat) [Gk. *eu,* good + *karyon,* nut, kernel]: A cell having a membrane-bound nucleus, membrane-bound organelles, and chromosomes in which DNA is combined with special proteins; an organism composed of such cells.

eusocial [Gk. *eu,* good + L. *socius,* companion]: Applied to insect societies in which sterile individuals work on behalf of individuals involved in reproduction.

evolution [L. *e-,* out + *volvere,* to roll]: Any change in the gene pool from one generation to the next; Darwinian evolution is the result of natural selection operating on random genetic variations.

exocrine glands [Gk. *ex,* out of + *krinein,* to separate]: Glands, such as digestive glands and sweat glands, that secrete their products into ducts.

exocytosis [Gk. *ex,* out of + *kytos,* vessel]: A cellular process in which particles are wrapped in a vacuole and transported to the cell surface; there, the membrane of the vacuole fuses with the cell membrane, expelling the vacuole's contents to the outside.

exoskeleton: The outer supporting covering of the body; common in arthropods.

exponential growth: In populations, the increasingly accelerated rate of growth due to the increasing number of individuals being added to the reproductive base. In the absence of control factors, populations experience exponential growth.

extraembryonic membranes: Membranes formed of embryonic tissues that lie outside the embryo, protecting it and aiding metabolism; include amnion, chorion, allantois, and yolk sac.

F_1 (first filial generation): The offspring resulting from the crossing of plants or animals of a parental generation.

F_2 (second filial generation): The offspring resulting from crossing members of the F_1 generation among themselves.

Fallopian tubes: *See* Oviduct.

family: A taxonomic grouping of related, similar genera; the category below order and above genus.

feedback systems: Control mechanisms whereby an increase or decrease in the level of a particular factor inhibits or stimulates the production, utilization, or release from the body of that factor; important in the regulation of hormone levels, ion concentrations, temperature, and many other factors.

fertilization: The fusion of two haploid gamete nuclei to form a diploid zygote nucleus.

fetus [L., pregnant]: An unborn or unhatched vertebrate that has passed through the earliest developmental stages; a developing human from about the third month after conception until birth.

fibril [L. *fibra,* fiber]: Any minute, threadlike organelle within a cell.

fibrous protein: Insoluble structural protein in which the polypeptide chain is coiled along one dimension; constitutes the main structural elements of many animal tissues.

filament [L. *filare,* to spin]: (1) A chain of cells. (2) In plants, the stalk of a stamen.

filtration: The first stage of kidney function; blood plasma is forced, under pressure, out of the glomerular capillaries into Bowman's capsule, through which it enters the renal tubule.

fitness: The relative ability to leave offspring.

flagellum, *pl.* **flagella** (fla-**jell**-um) [L. *flagellum,* whip]: A long, threadlike organelle found in eukaryotes and used in locomotion and feeding; has an internal structure of nine pairs of microtubules encircling two central microtubules.

flower: The reproductive structure of angiosperms; a complete (perfect) flower includes sepals, petals, stamens (male sex organs), and carpels (female sex organs).

food chain, food web: A set of interactions among organisms, including producers, herbivores, and carnivores, through which energy and materials move within a community or ecosystem.

fossil [L. *fossilis,* dug up]: The remains of an organism, or direct evidence of its presence (such as tracks). May be an unaltered hard part (tooth or bone), a mold in a rock, petrification (wood or bone), unaltered or partially altered soft parts (a frozen mammoth).

fossil fuels: The remains of once-living organisms that are burned to release energy. Examples are coal, oil, and natural gas.

founder principle: Type of genetic drift that occurs when a small, genetically nonrepresentative population branches off from a larger population; results in an altered gene pool, of lower heterozygosity.

fovea [L., pit]: A small area in the center of the retina in which cones are concentrated; the area of sharpest vision.

fruit [L. *fructus,* fruit]: In angiosperms, a matured, ripened ovary or group of ovaries and associated structures; contains the seeds.

function [L. *fungor,* to busy oneself]: Characteristic role or action of a structure or process in the normal metabolism or behavior of an organism.

gametangium, *pl.* **gametangia** [Gk. *gamein,* to marry + L. *tangere,* to touch]: A unicellular or multicellular structure in which gametes are produced.

gamete (**gam**-meet) [Gk., wife]: The haploid reproductive cell whose nucleus fuses with that of another gamete of an opposite sex (fertilization); the resulting cell (zygote) may develop into a new diploid individual or, in some species, may undergo meiosis to form haploid somatic cells.

gametophyte: In plants, which have alternation of haploid and diploid generations, the haploid (n) gamete-producing generation.

ganglion, *pl.* **ganglia** (**gang**-lee-on) [Gk. *ganglion,* a swelling]: Aggregate of nerve cell bodies.

Gause's principle: *See* Competitive exclusion.

gene [Gk. *genos,* birth, race; L. *genus,* birth, race, origin]: A unit of heredity in the chromosome; a sequence of nucleotides in a DNA molecule that codes for a protein.

gene flow: The exchange of genes between different populations.

gene frequency: In a population, the proportion of a particular allele at a given genetic locus (at which two or more allelic forms may occur).

gene pool: All the alleles of all the genes of all the individuals in a population.

genetic code: The system of nucleotide triplets in DNA and RNA that carries genetic information; referred to as a code because it determines the amino acid sequence in the enzymes and other protein components synthesized by the organism.

genetic drift: Random fluctuations in the relative allele frequencies of a small breeding population.

genetic isolation: The absence of genetic exchange between populations or species as a result of geographic separation or as a result of premating or postmating mechanisms (behavioral, morphological, or physiological) that prevent reproduction.

genome: The genetic identity of an individual; the complete set of chromosomes, with their associated genes.

genotype (**jean**-o-type): The genetic constitution, latent or expressed, of a cell or an organism, as contrasted with the phenotype; the sum total of all the genes present in an individual.

genus, *pl.* **genera** (**jean**-us) [L. *genus,* race, origin]: A taxonomic grouping of closely related species.

geologic eras: See Table 19–1, page 249.

germination [L. *germinare,* to bud]: In plants, the resumption of growth or the development from seed or spore.

gibberellins (jibb-e-**rell**-ins) [Fr. *gibberella,* genus of fungi]: A group of chemically related plant growth hormones, whose most characteristic effect is stem elongation in dwarf plants and bolting.

gill: The respiratory organ of aquatic animals, usually a thin-walled projection from some part of the external body surface or, in vertebrates, from some part of the digestive tract.

gland [L. *glans, glandis,* acorn]: An organ specialized to produce one or more secretions that are discharged to the outside of the gland.

globular protein [L. dim. of *globus,* a ball]: A polypeptide chain folded into a roughly spherical shape.

glomerulus (glom-**mare**-u-lus) [L. *glomus,* ball]: In the vertebrate kidney, a cluster of capillaries enclosed by Bowman's capsule; blood plasma minus large molecules filters through the walls of the glomerular capillaries into the renal tubules.

glucagon [Gk. *glukus,* sweet + *agō,* to lead toward]: Hormone produced in the pancreas that acts to raise the concentration of blood sugar.

glucose [Gk. *glukus,* sweet]: A six-carbon sugar ($C_6H_{12}O_6$); the most common monosaccharide in animals.

glycogen [Gk. *glukus,* sweet + *genos,* race or descent]: A complex carbohydrate (polysaccharide); one of the main stored food substances of most animals and fungi; it is converted into glucose by hydrolysis.

glycolysis (gly-**coll**-y-sis) [Gk. *glukus,* sweet + *lysis,* loosening]: The process by which a glucose molecule is changed anaerobically to two molecules of pyruvic acid with the liberation of a small amount of useful energy; catalyzed by soluble cytoplasmic enzymes.

Golgi body (**goal**-jee): An organelle present in many eukaryotic cells; consists of flat, disk-shaped sacs, tubules, and vesicles. It functions as a collecting and packaging center for substances that the cell manufactures for export.

gonad [Gk. *gone,* seed]: Gamete-producing organ of multicellular animals; ovary or testis.

granum, *pl.* **grana** [L., grain or seed]: In chloroplasts, stacked membrane-bound disks (thylakoids) that contain chlorophylls and carotenoids and are the sites of the light reactions of photosynthesis.

gross productivity: A measure of the rate at which energy is assimilated by the organisms in a trophic level, a community, or an ecosystem.

groundwater: Water in the zone of saturation where all openings in rocks and soil are filled, the upper surface of which forms the water table.

guard cells: Specialized epidermal cells surrounding a pore, or stoma, in a leaf or green stem; changes in turgor of a pair of guard cells cause opening and closing of the pore.

habitat [L. *habitare,* to live in]: The surroundings in which individuals of a particular species can usually be found.

haploid [Gk. *haploos,* single + *ploion,* vessel]: Having only one set of chromosomes (n), in contrast to diploid ($2n$); characteristic of all eukaryotic gametes, of gametophytes in plants, and of some microorganisms.

Hardy-Weinberg principle: The mathematical expression of the relationship between relative frequencies of two or more alleles in an idealized population; it states that both the allele frequencies and the genotype frequencies will remain constant in the absence of evolutionary forces.

heat of vaporization: The amount of heat required to change a given amount of a liquid into a gas; 540 calories are required to change 1 gram of liquid water into vapor.

heme [Gk. *haima,* blood]: The iron-containing group of heme proteins such as hemoglobin and the cytochromes.

hemoglobin [Gk. *haima,* blood + L. *globus,* a ball]: The iron-containing protein in vertebrate blood that carries oxygen.

hemophilia [Gk. *haima,* blood + *philios,* friendly]: A group of hereditary diseases characterized by failure of the blood to clot and consequent excessive bleeding from even minor wounds.

herbaceous (her-**bay**-shus) [L. *herba,* grass]: In plants, nonwoody.

herbivore [L. *herba,* grass + *vorare,* to devour]: A consumer that eats plants or other photosynthetic organisms to obtain its food and energy.

heredity [L. *herres, heredis,* heir]: The transmission of characteristics from parent to offspring.

hermaphrodite [Gk. *Hermes* and *Aphrodite*]: An organism possessing both male and female reproductive organs; hermaphrodites may or may not be self-fertilizing.

heterotroph [Gk. *heteros,* other, different + *trophos,* feeder]: An organism that must feed on organic materials formed by other organisms in order to obtain energy and small building-block molecules; in contrast to autotroph. Animals, fungi, and many unicellular organisms are heterotrophs.

heterozygote [Gk. *heteros,* other + *zugōtos,* a pair]: A diploid organism that carries two different alleles at one or more genetic loci.

heterozygote superiority: The greater fitness of a heterozygote as compared with the two homozygotes at a given genetic locus.

hibernation [L. *hiberna,* winter]: A period of dormancy and inactivity, varying in length depending on the species and occurring in dry or cold seasons. During hibernation, metabolic processes are greatly slowed and, even in mammals, body temperature may drop to just above freezing.

homeostasis (home-e-o-**stay**-sis) [Gk. *homos,* same or similar + *stasis,* standing]: Maintenance of a relatively stable internal physiological environment or internal equilibrium in an organism.

homeotherm [Gk. *homos,* same or similar + *therme,* heat]: Organism capable of maintaining a stable body temperature independent of the environment; warm-blooded. Birds and mammals are homeotherms.

hominid [L. *homo,* man]: Humans and closely related primates; includes modern and fossil hominoids but not the apes.

hominoid [L. *homo,* man]: Hominids and the great apes.

homologues [Gk. *homologia,* agreement]: Chromosomes that carry corresponding genes and associate in pairs in the first stage of meiosis; each member of the pair is derived from a different parent.

homology [Gk. *homologia,* agreement]: Similarity in structure and/or position, assumed to result from a common ancestry, regardless of function, such as the wing of a bird and the foreleg of a mammal.

homozygote [Gk. *homos,* same or similar + *zugōtos,* a pair]: A diploid organism that carries identical alleles at one or more genetic loci.

hormone [Gk. *hormaein,* to excite]: An organic molecule secreted, usually in minute amounts, in one part of an organism and transported to another part of that organism where it has a specific effect on some target organ or tissue.

host: (1) An organism on or in which a parasite lives. (2) A recipient of grafted tissue.

hybrid [L. *hybrida,* the offspring of a tame sow and a wild boar]: (1) Offspring of two parents that differ in one or more heritable characteristics. (2) Offspring of two different varieties or of two different species.

hydrocarbon [L. *hydro,* water + *carbo,* charcoal]: An organic compound consisting of only carbon and hydrogen.

hydrogen bond: A weak molecular bond linking a hydrogen atom that is covalently bonded to another atom (usually oxygen, nitrogen, or fluorine) to another oxygen, nitrogen, or fluorine atom of the same or another molecule.

hydrolysis [L. *hydro,* water + Gk. *lysis,* loosening]: Splitting of one molecule into two by addition of H^+ and OH^- ions of water.

hydrophilic [L. *hydro,* water + Gk. *philios,* friendly]: Having an affinity for water; applied to polar molecules or polar regions of large molecules.

hydrophobic [L. *hydro,* water + Gk. *phobos,* fearing]: Having no affinity for water; applied to nonpolar molecules or nonpolar regions of molecules.

hypertonic [Gk. *hyper,* above + *tonos,* tension]: Of two solutions of different concentration, the solution that contains the higher concentration of solute particles; water moves across a semipermeable membrane into a hypertonic solution.

hypha [Gk. *hyphe,* web]: A single tubular filament of a fungus; the hyphae together make up the mycelium, the matlike "body" of a fungus.

hypothalamus [Gk. *hypo,* under + *thalamos,* inner room]: The floor and sides of the vertebrate brain just below the cerebral hemispheres; controls the autonomic nervous system and the pituitary gland and contains centers that regulate body temperature and appetite.

hypothesis [Gk. *hypo,* under + *tithenai,* to put]: A temporary working explanation or supposition based on accumulated facts and suggesting some general principle or relation of cause and effect; a postulated solution to a scientific problem that must be tested by experimentation and, if not validated, discarded.

hypotonic [Gk. *hypo,* under + *tonos,* tension]: Of two solutions of different concentration, the solution that contains the lower concentration of solute particles; water moves across a semipermeable membrane from a hypotonic solution.

immune response: A highly specific defensive reaction of the body to invasion by a foreign substance or organism; consists of a primary response in which the invader is recognized as foreign, or "not-self," and eliminated and a secondary response to subsequent attacks by the same invader; is mediated by two types of lymphocytes—B-cells, which play a key role in antibody production, and T-cells.

imprinting: A rapid and extremely narrow form of learning to respond to a specific stimulus, which occurs during a very short, critical period in the early life of an organism, such as the following response in certain birds.

inbreeding: The mating of individuals closely related genetically.

incomplete dominance: In genetics, the phenomenon in which the effects of both alleles at a particular locus are apparent in the phenotype of the heterozygote.

independent assortment: *See* Mendel's second law.

inflammatory response: A nonspecific defensive reaction of the body to invasion by a foreign substance or organism; involves phagocytosis by white blood cells and is often accompanied by accumulation of pus and an increase in the local temperature.

instinct: An inflexible, unlearned behavior or response triggered by specific stimuli; thought to be genetically determined.

insulin: A peptide hormone produced by the vertebrate pancreas, the action of which results in the lowering of the concentration of sugar in the blood.

intelligence: The capacity to profit by experience, through reasoning and the analysis and association of ideas.

interneuron: Neuron that transmits nerve impulses from one neuron to another; may receive impulses from and transmit impulses to many different neurons.

interphase: The stage between two mitotic or meiotic cycles.

intervening sequence: In a segment of DNA coding for a particular protein (that is, in a gene), a long sequence of nucleotides that is removed enzymatically from the transcribed mRNA molecule before it enters the cytoplasm and is translated.

ion (**eye**-on): An atom or molecule that has lost or gained one or more electrons. By the process known as ionization, the atom or molecule becomes electrically charged.

ionic bond: A chemical bond formed as a result of the mutual attraction of ions of opposite charge.

isolating mechanisms: Mechanisms that prevent genetic exchange between different populations or species; they prevent mating or successful reproduction even when mating occurs; may be behavioral, morphological, or physiological.

isotonic [Gk. *isos,* equal + *tonos,* tension]: Having the same concentration of solutes as another solution. If two isotonic solutions are separated by a semipermeable membrane, there will be no net flow of water across the membrane.

isotope [Gk. *isos,* equal + *topos,* place]: Atom of an element that differs from other atoms of the same element in the number of neutrons in the atomic nucleus; isotopes thus differ in atomic weight. Some isotopes are unstable and emit radiation.

karyotype [Gk. *kara,* the head + *typos,* stamp or print]: The general appearance of the chromosomes in a genome with regard to number, size, and shape.

keratin [Gk. *karas,* horn]: One of a group of tough, fibrous proteins formed by certain epidermal tissues and especially abundant in skin, claws, hair, feathers, and hooves.

kidney: In vertebrates, the organ that regulates the balance of water and solutes in the blood and the excretion of nitrogenous wastes in the form of urine.

kinetic energy: Energy of motion.

kinetochore [Gk. *kinetikos,* putting in motion + *choros,* chorus]: Region within the centromere to which the spindle fibers attach.

Krebs cycle: Stage of cellular respiration in which pyruvate fragments are completely broken down into carbon dioxide; molecules reduced in the process can be used in ATP formation.

lamella (lah-**mell**-ah) [L. dim. of *lamina,* plate or leaf]: Layer, thin sheet.

larva [L., ghost]: An immature animal that is morphologically very different from the adult; examples are caterpillars and tadpoles.

lateral meristem [L. *latus, lateris,* side + Gk. *meristos,* divided]: In vascular plants, one of the two rings of tissue (vascular cambium and cork cambium) that produce new cells for secondary growth.

leaching: The dissolving of minerals and other elements in soil or rocks by the downward movement of water.

learning: The process that leads to adaptive change in individual behavior as the result of experience.

leukocyte [Gk. *leukos,* white + *kytos,* vessel]: White blood cell. Two principal types are phagocytic cells, involved in the inflammatory response, and lymphocytes, involved in immune reactions.

lichen: Organism composed of an alga and a fungus that are symbiotically associated.

life cycle: The entire span of existence of any organism from time of zygote formation (or asexual reproduction) until it itself reproduces.

light reactions: The first stage of photosynthesis, in which light energy trapped by the pigments of the chloroplast is converted to chemical energy stored in molecules of ATP and NADPH.

limbic system [L. *limbus,* border]: Neuron network forming a loop around the inside of the brain and connecting the hypothalamus to the cerebral cortex; thought to be circuit by which drives and emotions are translated into complex actions.

limiting factor: Any environmental factor the level or concentration of which limits the growth or some other activity of an organism or population.

linkage: The tendency for certain genes to be inherited together because they are located on the same chromosome.

linkage group: A pair of homologous chromosomes.

lipid [Gk. *lipos,* fat]: One of a large variety of organic fats or fatlike compounds; includes waxes, steroids, phospholipids, and carotenes.

locus, *pl.* **loci** [L., place]: In genetics, the position of a gene in a chromosome. For any given locus, there may be several possible alleles.

logistic growth: A pattern of population growth in which growth is rapid when the population is small, gradually slows as the population approaches the carrying capacity of its environment, and then oscillates as the population stabilizes at or near its maximum size; it is the simplest growth pattern observed for most populations in nature.

loop of Henle (after F. G. J. Henle, German pathologist): A hairpin-shaped portion of the renal tubule found in birds and mammals in which a hypertonic urine is formed by processes of osmosis and active transport.

lymph [L. *lympha,* water]: Colorless fluid derived from blood by filtration through capillary walls in the tissues; carried in special lymph ducts.

lymphatic system: In higher vertebrates, the system that returns lymph to the bloodstream; consists of lymph capillaries, which begin blindly in the tissues, and a network of progressively larger vessels that empty into the vena cava; also includes the lymph nodes, spleen, thymus, tonsils, and adenoids.

lymphocyte [L. *lympha,* water + Gk. *kytos,* vessel]: A type of white blood cell involved in the immune response; B-cells differentiate into antibody-producing plasma cells, whereas T-cells interact directly with the foreign invader.

lysis [Gk. *lysis,* a loosening]: Disintegration of a cell by breaking its cell membrane.

lysogenic bacteria (lye-so-**jenn**-ick) [Gk. *lysis,* a loosening + *genos,* race or descent]: Bacteria carrying a bacteriophage genome integrated into the bacterial chromosome. The viral genome may subsequently begin to replicate independently and set up an active cycle of infection, causing lysis of the bacterial cells.

lysosome [Gk. *lysis,* loosening + *soma,* body]: A membrane-bound organelle in which destructive enzymes (e.g., digestive enzymes) are segregated.

macromolecule [Gk. *makros,* large + L. dim. of *moles,* mass]: An extremely large molecule; refers specifically to proteins, nucleic acids, polysaccharides, and complexes of these.

mantle: In mollusks, the outermost layer of the body wall or a soft extension of it; usually secretes a shell.

marine [L. *marini(us),* from *mare,* the sea]: Living in salt water.

marsupial [Gk. *marsypos,* pouch, little bag]: A nonplacental mammal in which the female has a ventral pouch or folds surrounding the nipples; the premature young leave the uterus and crawl into the pouch, where each one attaches itself by the mouth to a nipple until development is completed.

medulla (med-**dull**-a) [L., the innermost part]: (1) The inner as opposed to the outer part of an organ, as in the adrenal gland. (2) The most posterior region of the vertebrate brain; connects with the spinal cord.

medusa: The free-swimming, bell- or umbrella-shaped stage in the life cycle of many coelenterates; a jellyfish.

megaspore [Gk. *megas,* great, large + *spora,* a sowing]: In plants, a haploid (n) spore that develops into a female gametophyte.

meiosis (my-**o**-sis) [Gk. *meioun,* to make smaller]: The two successive nuclear divisions in which a single diploid ($2n$) cell forms four haploid (n) cells, and segregation, crossing over, and reassortment of the genes occur; gametes or spores may be produced as a result of meiosis.

Mendel's first law: The factors for a pair of alternative characters are separate and only one may be carried in a particular gamete (genetic segregation). In modern form: Alleles segregate in meiosis.

Mendel's second law: The inheritance of a pair of factors for one trait is independent of the simultaneous inheritance of factors for other traits, such factors "assorting independently" as though there were no other factors present (later modified by the discovery of linkage). Modern form: Unlinked genes assort independently.

menstrual cycle [L. *mensis,* month]: In humans and certain other primates, the cyclic, hormone-regulated changes in the condition of the uterine lining; marked by the periodic discharge of blood and disintegrated uterine lining through the vagina.

meristem [Gk. *meris,* part of, portion + *stamon,* the warp of a loom]: The undifferentiated plant tissue, including a mass of rapidly dividing cells, from which new tissues arise.

mesoderm [Gk. *mesos,* middle + *derma,* skin]: In animals, the middle layer of body tissue.

mesophyll [Gk. *mesos,* middle + *phyllon,* leaf]: The internal tissue of a leaf, sandwiched between two layers of epidermal cells; consists of palisade parenchyma and spongy parenchyma cells.

messenger RNA (mRNA): The single-stranded RNA that carries genetic information from the gene to the ribosome, where it determines the order of the amino acids in the formation of a polypeptide; mRNA is formed by transcription and functions in translation.

metabolism [Gk. *metabole,* change]: The sum of all chemical reactions occurring within a cell or organism.

metamorphosis [Gk. *metamorphoun,* to transform]: Abrupt transition from larval to adult form, such as the transition from tadpole to adult frog.

metaphase [Gk. *meta,* middle + *phasis,* form]: The stage of mitosis or meiosis during which the chromosomes lie in the equatorial plane of the spindle.

microbe [Gk. *mikros,* small + *bios,* life]: A microscopic organism.

micronutrient [Gk. *mikros,* small + L. *nutrire,* to nourish]: A mineral required in only minute amounts for plant growth, such as iron, chlorine, copper, manganese, zinc, molybdenum, and boron.

microspore [Gk. *mikros,* small + *spora,* a sowing of seeds]: In plants, a spore that develops into a male gametophyte; in seed plants, it becomes a pollen grain.

microtubule [Gk. *mikros,* small + L. dim. of *tubus,* tube]: An extremely small hollow tube composed of two types of globular protein subunits. Among their many functions, microtubules make up the internal structure of cilia and flagella.

middle lamella: In plants, distinct layer between adjacent cell walls, rich in pectins and other polysaccharides; derived from the cell plate.

mimicry [Gk. *mimos,* mime]: The superficial resemblance in form, color, or behavior of certain organisms (mimics) to other more powerful or more protected ones (models), resulting in protection, concealment, or some other advantage for the mimic.

mineral: A naturally occurring element or inorganic compound.

mitochondrion, *pl.* **mitochondria** [Gk. *mitos,* thread + *chondros,* cartilage or grain]: An organelle bound by a double membrane in which the reactions of the Krebs cycle and electron transport chain take place, resulting in the formation of ATP, CO_2, and H_2O from acetyl CoA and ADP. Mitochondria are the organelles in which most of the ATP of the eukaryotic cell is produced.

mitosis [Gk. *mitos,* thread]: Nuclear division characterized by chromosome replication and formation of two identical daughter nuclei.

molecule [L. dim. of *moles,* mass]: A particle consisting of two or more atoms held together by chemical bonds; the smallest unit of a compound that displays the properties of the compound.

molting: Shedding of all or part of an organism's outer covering; in arthropods, periodic shedding of the exoskeleton to permit an increase in size.

monocotyledon [Gk. *monos,* single + *kotyledon,* a cup-shaped hollow]: A member of a subclass of angiosperms, characterized by a variety of features, among which is the presence of a single seed leaf (cotyledon); abbreviated as monocot.

monoecious (mo-**nee**-shuss) [Gk. *monos,* single + *oikos,* house]: In angiosperms, having the male and female organs (the stamens and the carpels, respectively) on the same individual but on different flowers.

monomer [Gk. *monos,* single + *meros,* part]: A simple, relatively small molecule that can be linked to others to form a polymer.

monosaccharide [Gk. *monos,* single + *sakcharon,* sugar]: A simple sugar, such as glucose, fructose, ribose.

monotreme [Gk. *monos,* single + *trēma,* hole]: A nonplacental mammal, such as the duckbill platypus, in which the female lays shelled eggs and nurses the young.

morphogenesis [Gk. *morphe,* form + *genesis,* origin]: The development of size, form, and other structural features of organisms.

morphological [Gk. *morphe,* form + *logos,* discourse]: Pertaining to form and structure, at any level of organization.

motor neuron: Neuron that transmits nerve impulses from the central nervous system to an effector, which is typically a muscle or a gland.

muscle fiber: Muscle cell; a long, cylindrical, multinucleated cell containing numerous myofibrils, which is capable of contraction when stimulated.

mutagen [L. *mutare,* to change + *genus,* source or origin]: A chemical or physical agent that increases the mutation rate.

mutant [L. *mutare,* to change]: A mutated gene or an organism carrying a gene that has undergone a mutation.

mutation [L. *mutare,* to change]: The change of a gene from one allelic form to another; an inheritable change in the DNA sequence of a chromosome not resulting from recombination.

mutualism [L. *mutuus,* lent, borrowed]: *See* Symbiosis.

mycelium [Gk. *mykes,* fungus]: The mass of hyphae forming the body of a fungus.

myelin sheath [Gk. *myelinos,* full of marrow]: A fatty layer surrounding the axons of some nerve cells in vertebrates; made up of the membranes of Schwann cells.

myofibril [Gk. *mys,* muscle + L. *fibra,* fiber]: Contractile element of a muscle fiber, made up of thick and thin filaments arranged in sarcomeres.

myosin [Gk. *mys,* muscle]: One of the principal proteins in muscle; makes up the thick filaments.

NAD: Abbreviation of nicotinamide adenine dinucleotide, a coenzyme that functions as an electron acceptor.

natural selection: A process of interaction between organisms and their environment that results in a differential rate of reproduction of different phenotypes in a population; can result in changes in the relative frequencies of alleles in the gene pool—that is, in evolution.

nectar [Gk. *nektar,* the drink of the gods]: A sugary fluid that attracts insects to plants.

negative feedback: A control mechanism whereby an increase in some substance inhibits the process leading to the increase; also known as feedback inhibition.

nephridium [Gk. *nephros,* kidney]: A type of excretory organ found in many invertebrates.

nephron [Gk. *nephros,* kidney]: The functional unit of the kidney in reptiles, birds, and mammals; a human kidney contains about 1 million nephrons.

nerve: A group or bundle of nerve fibers with accompanying connective tissue.

nerve fiber: A filamentous process of a neuron; either dendrite or axon.

nerve impulse: A rapid, transient change in electric potential across the membrane of a nerve cell; it is propagated along a nerve fiber from one part of an animal to another.

nervous system: All the nerve cells of an animal; the receptor-conductor-effector system; in humans, the nervous system consists of brain, spinal cord, and all nerves.

net productivity: In a trophic level, a community, or an ecosystem, the amount of energy (in calories) stored in chemical compounds or the increase in biomass (in grams or metric tons) in a particular period of time; it is the difference between gross productivity and the energy used by the organisms in respiration.

neuron [Gk. *neuron,* nerve]: Nerve cell, including cell body, dendrites, and axon.

neutron (**new**-tron): An uncharged particle with a mass slightly greater than that of a proton. Found in the atomic nucleus of all elements except hydrogen, in which the nucleus consists of a single proton.

niche: *See* Ecological niche.

nitrification: The oxidation of ammonia or ammonium to nitrites and nitrates, as by nitrifying bacteria.

nitrogen cycle: Worldwide circulation and reutilization of nitrogen atoms, chiefly due to metabolic processes of living organisms; plants take up inorganic nitrogen and convert it into organic compounds (chiefly proteins), which are assimilated into the bodies of one or more animals; excretion and bacterial and fungal action on dead organisms return nitrogen atoms to the inorganic state.

nitrogen fixation: Incorporation of atmospheric nitrogen into inorganic nitrogen compounds available to plants, a process that can be carried out only by certain soil bacteria and many blue-green algae, or by certain symbiotic microorganisms in association with legumes.

nitrogenous base: A nitrogen-containing molecule having basic properties (tendency to acquire an H^+ ion); a purine or pyrimidine.

node [L. *nodus,* knot]: In plants, a joint of a stem; the place where branches and leaves are joined to the stem.

nondisjunction [L. *non,* not + *disjungere,* to separate]: The failure of chromatids to separate during meiosis, resulting in one or more extra chromosomes in some gametes and correspondingly fewer in others.

norepinephrine: A hormone produced by the medulla of the adrenal gland, which increases the concentration of sugar in the blood, raises blood pressure and heartbeat rate, and increases muscular power and resistance to fatigue; also one of the principal chemical transmitters across synaptic junctions.

notochord [Gk. *noto,* back + L. *chorda,* cord]: A dorsal rodlike structure that runs the length of the body and serves as the internal skeleton in the embryos of all chordates; in most adult chordates the notochord is replaced by a vertebral column that forms around (but not from) the notochord.

nuclear envelope [L. *nucleus,* a kernel]: The double membrane surrounding the nucleus within a eukaryotic cell.

nucleic acid: A macromolecule consisting of nucleotides; the principal types are deoxyribonucleic acid (DNA) and ribonucleic acid (RNA).

nucleolus (new-**klee**-o-lus) [L. *nucleolus,* a small kernel]: A small, dense body, containing DNA, RNA, and protein, present in the nucleus of eukaryotic cells; site of production of ribosomal RNA.

nucleotide [L. *nucleus,* a kernel]: A building block of nucleic acid composed of phosphate, a five-carbon sugar (either ribose or deoxyribose), and a purine or pyrimidine base.

nucleus [L., a kernel]: (1) The central part of an atom. (2) The cellular organelle that contains the genetic information in the form of DNA. In the eukaryotic cell, a specialized body bound by a double membrane, containing the chromosomes. (3) A group of nerve cell bodies in the central nervous system.

omnivorous [L. *omnis,* all + *vorare,* devour]: Eating "everything," for example, using both plants and animals as food.

oocyte (**o**-uh-sight) [Gk. *oion,* egg + *kytos,* vessel]: A cell that gives rise by meiosis to an ovum.

operator gene: A segment of DNA to which a repressor protein may bind, thus regulating the transcription of an operon.

operon [L. *opus, operis,* work]: In bacterial cells, a unit of genetic transcription; a group of adjacent structural genes whose functions are related to a particular biochemical pathway and which are regulated by a single repressor protein.

opportunistic species: Species characterized by high reproduction rates, rapid development, early reproduction, small body size, and uncertain adult survival. Such species appear to have a competitive advantage in colonizing new environments or when environmental conditions are unstable.

order: A taxonomic grouping of related, similar families; the category below class and above family.

organ [Gk. *organon,* tool]: A body part composed of several tissues grouped together in a structural and functional unit.

organelle [Gk. *organon,* instrument, tool]: A formed body in the cytoplasm of a cell.

organic [Gk. *organon,* instrument, tool]: Pertaining to (1) organisms or living things generally, or (2) compounds formed by living organisms, or (3) the chemistry of compounds containing carbon.

organism [Gk. *organon,* instrument, tool]: Any living creature, either unicellular or multicellular.

osmosis [Gk. *osmos,* impulse, thrust]: The net movement of water across a selectively permeable membrane (a membrane that permits the free passage of water but prevents or retards the passage of a solute). The net movement of water is from the side containing a lesser concentration of solute toward the side containing a greater concentration.

osmotic pressure [Gk. *osmos,* impulse, thrust]: A measure of the difference in solute concentrations on either side of a semipermeable membrane; pressure generated by the osmotic flow of water.

ovary [L. *ovum,* egg]: (1) In animals, the egg-producing organ. (2) In flowering plants, the enlarged basal portion of a carpel or a fused carpel, containing the ovule or ovules; the ovary matures to become the fruit.

oviduct [L. *ovum,* egg + *ductus,* duct]: The tube serving to transport the eggs to the uterus or to the outside; Fallopian tubes (in humans).

ovulation: In animals, release of an egg or eggs from the ovary.

ovule [L. dim. of *ovum,* egg]: In seed plants, a structure composed of a protective outer coat, a tissue specialized for food storage, and a female gametophyte with egg cell; becomes a seed after fertilization.

ovum, *pl.* **ova** [L., egg]: The egg cell; female gamete.

oxidation: Gain of oxygen, loss of hydrogen, or loss of an electron by an atom, ion, or molecule. Oxidation and reduction take place simultaneously, with the electron lost by one reactant being transferred to another reactant. Oxidation-reduction reactions are an important means of energy transfer within living systems.

pacemaker: Area of the vertebrate heart that initiates the heartbeat; located where the superior vena cava enters the right atrium; sinoatrial node.

paleontology [Gk. *palaios,* old + *onta,* things that exist + *logos,* discourse]: The study of the life of past geologic times, principally by means of fossils.

palisade cells [L. *palus,* stake + *cella,* a chamber]: In plant leaves, the columnar, chloroplast-containing parenchyma cells of the mesophyll.

pancreas (pang-kree-us) [Gk. *pan,* all + *kreas,* meat, flesh]: In vertebrates, a small complex gland located between the stomach and the duodenum, which produces digestive enzymes and the hormones insulin and glucagon; sweetbread.

parasite [Gk. *para,* beside, akin to + *sitos,* food]: An organism that lives on or in an organism of a different species and derives nutrients from it. *See* Symbiosis.

parasympathetic division [Gk. *para,* beside, akin to]: A subdivision of the autonomic nervous system of vertebrates with centers located in the brain and in the most anterior part and in the most posterior parts of the spinal cord; stimulates digestion; generally inhibits other functions and restores the body to normal following emergencies.

parenchyma (pah-**renk**-ee-ma) [Gk. *para,* beside, akin to + *en,* in + *chein,* to pour]: A plant tissue composed of living, thin-walled, randomly arranged cells with large vacuoles; usually photosynthetic or storage tissue.

pecking order: A social ranking in poultry flocks in which each bird dominates (can peck) others lower in rank, but is dominated by birds higher in rank.

peptide [Gk. *pepto,* to soften, digest]: Two or more amino acids linked together. Molecules made up of a relatively small number of amino acids are called peptides, while those formed of a larger number of amino acids are called polypeptides.

peptide bond [Gk. *pepto,* to soften, digest]: The type of bond formed when two amino acids are joined end to end; the acidic group ($-COOH$) of one amino acid is linked covalently to the basic group ($-NH_2$) of the next, and a molecule of water (H_2O) is removed.

perennial [L. *per,* through + *annus,* year]: A plant that persists in whole or in part from year to year and usually produces reproductive structures in more than one year.

pericycle [Gk. *peri,* around + *kyklos,* circle]: One or more layers of cells completely surrounding the vascular tissues of the root; branch roots arise from the pericycle.

peripheral nervous system [Gk. *peripherein,* to carry around]: All of the neurons and nerve fibers outside the central nervous system, including both motor neurons and sensory neurons; consists of the somatic nervous system and the autonomic nervous system.

peristalsis [Gk. *peristellein,* to wrap around]: Successive waves of muscular contraction in the walls of a tubular structure, such as the digestive tract or an oviduct; moves the contents, such as food or an egg cell, through the tube.

permeable [L. *permeare,* to pass through]: Penetrable by molecules, ions, or atoms; usually applied to membranes that let given solutes pass through.

petiole (**pet**-ee-ole) [Fr., from L. *petiolus,* dim. of *pes, pedis,* a foot]: The stalk of a leaf, connecting the blade of the leaf with the branch or stem.

pH: A symbol denoting the relative concentration of hydrogen ions in a solution; pH values range from 0 to 14; the lower the value, the more acidic a solution, that is, the more hydrogen ions it contains; pH 7 is neutral, less than 7 is acidic, more than 7 is alkaline.

phagocytosis [Gk. *phagein,* to eat + *kytos,* vessel]: Cell "eating"; the intake of solid particles by a cell, by flowing over and engulfing them; characteristic of amoebas, the digestive cells of some invertebrates, and vertebrate white blood cells.

phenotype [Gk. *phainein,* to show + *typos,* stamp, print]: Observable properties of an organism (as distinguished from genotype).

pheromone (**fair**-o-moan) [Gk. *phero,* to bear, carry]: Substance secreted by an animal that influences the behavior or morphological development of other animals of the same species, such as the sex attractants of moths, the odor trail of ants.

phloem (**flow**-em) [Gk. *phloos,* bark]: Vascular tissue of higher plants; conducts sugars and other organic molecules from the leaves to other parts of the plant; in angiosperms, composed of sieve tubes and companion cells, parenchyma, and fibers.

phospholipids: Organic molecules similar in structure to fats, but in which a phosphate group rather than a fatty acid is attached to the third carbon of the glycerol molecule; as a result, the molecule has a hydrophilic "head" and a hydrophobic "tail." Phospholipids form the basic structure of cellular membranes.

photoperiodism [Gk. *photos,* light]: The response to relative day and night length, a mechanism by which organisms measure seasonal change.

photoreceptor [Gk. *photos,* light]: A cell or organ capable of detecting light.

photosynthesis [Gk. *photos,* light + *syn,* together + *tithenai,* to place]: The conversion of light energy to chemical energy; the synthesis of organic compounds from carbon dioxide and water in the presence of chlorophyll, using light energy.

phototropism [Gk. *photos,* light + *trope,* turning]: Movement in which the direction of the light is the determining factor, such as the growth of a plant toward a light source; turning or bending response to light.

phyletic evolution [Gk. *phylon,* race, tribe + L. *e-,* out + *volvere,* to roll]: The changes taking place in a single hereditary line of organisms over a long period of time; one of the three principal patterns of organic evolution.

phylogeny [Gk. *phylon,* race, tribe]: Evolutionary history of a taxonomic group. Phylogenies are often depicted as "evolutionary trees."

phylum [Gk. *phylon,* tribe, stock]: A taxonomic grouping of related, similar classes; a high-level category beneath kingdom and above class.

physiology [Gk. *physis,* nature + *logos,* a discourse]: The study of function in cells, organs, or entire organisms; the processes of life.

phytochrome [Gk. *phyton,* plant + *chrōma,* color]: A plant pigment that is a photoreceptor for red or far-red light and is involved with a number of developmental processes, such as flowering, dormancy, leaf formation, and seed germination.

phytoplankton [Gk. *phyton,* plant + *planktos,* wandering]: Aquatic, free-floating, microscopic, photosynthetic organisms.

pigment [L. *pigmentum,* paint]: A colored substance that absorbs light over a narrow band of wavelengths.

pinocytosis [Gk. *pinein,* to drink + *kytos,* vessel]: Cell "drinking"; the intake of fluid droplets by a cell, probably by a mechanism similar to that of phagocytosis but triggered by different stimuli.

pituitary [L. *pituita,* phlegm]: Endocrine gland in vertebrates; the anterior lobe is the source of tropic hormones, growth hormone, and prolactin and is stimulated by secretions of the hypothalamus; the posterior lobe stores and releases oxytocin and ADH produced by the hypothalamus.

placenta [Gk. *plax,* a flat object]: A tissue formed in part from the inner lining of the mammalian uterus and in part from the extraembryonic membranes; serves as the connection through which exchanges of nutrients and wastes occur between the blood of the mother and that of the embryo.

plankton [Gk. *planktos,* wandering]: Small (mostly microscopic) aquatic and marine organisms found in the upper levels of the water, where light is abundant; includes both photosynthetic (phytoplankton) and heterotrophic (zooplankton) forms.

planula [L. dim. of *planus,* a wanderer]: The ciliated, free-swimming type of larva formed by many coelenterates.

plasma [Gk. *plasma,* form or mold]: The clear, colorless fluid component of vertebrate blood, containing dissolved salts and proteins; blood minus the blood cells.

plasma cell: An antibody-producing cell resulting from the multiplication and differentiation of a B-lymphocyte that has interacted with an antigen; a single plasma cell can produce about 2,000 molecules of antibody per second.

plasmid: In bacteria, a DNA molecule much smaller than the chromosome; may exist free in the cell or may be incorporated into the chromosome.

plasmodesma, *pl.* **plasmodesmata** [Gk. *plassein,* to mold + *desmos,* band, bond]: In plants, a minute, cytoplasmic thread that extends through pores in cell walls and connects the cytoplasm of adjacent cells.

platelet (**plate**-let) [Gk. *platus,* flat]: In mammals, a minute, granular body suspended in the blood and involved in the formation of blood clots.

pleiotropy (**plee**-o-trope-ee) [Gk. *pleios,* more + *trope,* a turning]: The capacity of a gene to affect a number of different phenotypic characteristics.

polar [L. *polus,* end of axis]: Having parts or areas with opposed or contrasting properties, such as positive and negative charges, head and tail.

polar body: Minute, nonfunctioning cell produced during those meiotic divisions that lead to egg cells; contains a nucleus but very little cytoplasm.

polar covalent bond: A covalent bond in which the electrons are shared unequally between the two atoms; the resulting polar molecule has regions of slightly negative and slightly positive charge.

pollen [L., fine dust]: In seed plants, spores consisting of an immature male gametophyte and a protective outer covering.

pollination [L. *pollen,* fine dust]: The transfer of pollen from the anther to a receptive surface of a flower.

polygenic inheritance [Gk. *polus,* many + *genos,* race, descent]: The determination of a given characteristic, such as weight or height, by the interaction of many genes.

polymer [Gk. *polus,* many + *meris,* part or portion]: A large molecule composed of many molecular subunits.

polymorphism [Gk. *polus,* many + *morphe,* form]: Occurrence together in a single population of two or more morphologically distinct forms.

polyp [Gk. *polus,* many + *pous,* foot]: The sessile stage in the life cycle of coelenterates.

polypeptide [Gk. *polus,* many + *pepto,* to soften, digest]: A molecule consisting of many amino acids linked together by peptide bonds.

polyploid [Gk. *polus,* many + *ploion,* vessel]: Cell with more than two complete sets of chromosomes per nucleus.

polysaccharide [Gk. *polus,* many + *sakcharon,* sugar]: A carbohydrate polymer composed of monosaccharide monomers in long chains; includes starch, cellulose.

population: Any group of individuals of one species that occupy a given area at the same time; in genetic terms, an interbreeding group of organisms.

posterior: Of or pertaining to the rear end. In humans, the back of the body is said to be posterior.

potential energy: Energy in a potentially usable form that is not, for the moment, being used; often called "energy of position."

preadaptation [L. *pre,* before + *adaptare,* to fit]: (1) The evolution in one environment of a structure or function that turns out to be advantageous when later generations shift to a new environment. (2) A feature of an organism that suits the organism for survival in new circumstances.

predator [L. *praedari,* to prey upon; from *prehendere,* to grasp, seize]: An organism that eats other living organisms.

pressure-flow hypothesis: A hypothesis accounting for sap flow through the phloem system. According to this hypothesis, the solution containing nutrient sugars is moved by active transport and osmosis from the sieve tubes to the cells of the rest of the plant.

prey [L. *prehendere,* to grasp, seize]: An organism eaten by another organism.

primary growth: In plants, growth originating in the apical meristem of the shoots and roots, as contrasted with secondary growth; results in an increase in length.

primary structure of proteins: The amino acid sequence of a protein.

primary succession: Ecological succession in an area not previously covered by vegetation.

primate: A member of the order of mammals that includes anthropoids and prosimians.

primitive [L. *primus,* first]: Not specialized; at an early stage of evolution or development.

producer, in ecological pyramids: An autotrophic organism, usually a photosynthesizer, that contributes to the net primary productivity of a community.

progesterone [L. *progerere,* to carry forth or out + *steiras,* barren]: In mammals, a steroid hormone produced by the corpus luteum that prepares the uterus for implantation of the ovum.

prokaryote [L. *pro,* before + Gk. *karyon,* nut, kernel]: A cell lacking a membrane-bound nucleus and membrane-bound organelles; a bacterium or a blue-green alga.

promoter: Specific segment of DNA to which RNA polymerase attaches to initiate transcription of mRNA from an operon.

prophase [Gk. *pro,* before + *phasis,* form]: An early stage in nuclear division, characterized by the condensing of the chromosomes and their movement toward the equator of the spindle. Homologous chromosomes pair up during meiotic prophase.

prosimian [L. *pro,* before + *simia,* ape]: A lower primate; includes lemurs, lorises, tarsiers, and tree shrews, as well as many fossil forms.

prostaglandins [Gk. *prostas,* a porch or vestibule + L. *glaus,* acorn]: A group of fatty acid hormones discovered in semen, now found to be present in many other tissues; believed to play a role in fertilization.

prostate gland [Gk. *prostas,* a porch or vestibule + L. *glaus,* acorn]: A mass of muscle and glandular tissue surrounding the base of the urethra in male mammals; the vasa deferentia merge with ducts from the seminal vesicles, enter the prostate gland, and there merge with the urethra. The prostate gland secretes an alkaline fluid that has a stimulating effect on the sperm as they are released.

protein [Gk. *proteios,* primary]: A complex organic compound composed of one or more polypeptide chains, each made up of many (about 100 or more) amino acids linked together by peptide bonds.

proton: A subatomic particle with a single positive charge equal in magnitude to the charge of an electron and with a mass slightly less than that of a neutron; a component of every atomic nucleus.

protoplasm: *See* Cytoplasm.

pseudopod [Gk. *pseudes,* false + *pous, pod-,* foot]: A temporary cytoplasmic protrusion from an amoeboid cell, which functions in locomotion or in feeding by phagocytosis.

pulmonary artery [L. *pulmonis,* lung]: In birds and mammals, an artery that carries blood from the heart to the lungs, where it is oxygenated.

pulmonary vein [L. *pulmonis,* lung]: In birds and mammals, a vein carrying oxygenated blood from the lungs to the left atrium of the heart, from which blood is pumped into the left ventricle and from there to the body tissues.

Punnett square: The checkerboard diagram used for analysis of gene segregation.

pupa [L., girl, doll]: A developmental stage of some insects, in which the organism is nonfeeding, immotile, and sometimes encapsulated or in a cocoon; the pupal stage occurs between the larval and adult phases.

purine [Gk. *purinos,* fiery, sparkling]: A nitrogenous base such as adenine or guanine; one of the components of nucleic acids.

pyramid, ecological: *See* Ecological pyramid.

pyramid of energy: A diagram of the energy flow between the trophic levels of a food chain; plants (at the base of the pyramid) represent the greatest amount of energy, herbivores next, then primary carnivores, secondary carnivores, etc.

pyrimidine: A nitrogenous base such as cytosine, thymine, or uracil; one of the components of nucleic acids.

quaternary structure of proteins: The overall structure of a globular protein molecule that consists of two or more polypeptide chains.

queen: In social insects (ants, termites, and some species of bees and wasps), the fertile, or fully developed, female whose function is to lay eggs.

radial symmetry [L. *radius,* a spoke of a wheel + Gk. *summetros,* symmetry]: The regular arrangement of body parts, like the spokes of a wheel, around a longitudinal axis; seen in coelenterates and adult echinoderms.

radiation [L. *radius,* a spoke of a wheel, hence, a ray]: Energy emitted in the form of waves or particles.

radioactive isotope: An isotope with an unstable nucleus that stabilizes itself by emitting radiation.

recessive allele [L. *recedere,* to recede]: An allele that is not expressed phenotypically in the heterozygous condition.

recombination: In genetics, the formation of gene combinations that differ from the combinations present in the parents; in eukaryotes, may be accomplished through assortment of unlinked genes during sexual reproduction, crossing over between linked genes, or both; in prokaryotes, may be accomplished through transformation, conjugation, or transduction.

reduction [L. *reducere,* to lead back]: Loss of oxygen, gain of hydrogen, or gain of an electron by an atom, ion, or molecule; oxidation and reduction take place simultaneously with the electron lost by one reactant being transferred to another. Oxidation-reduction reactions are an important means of energy transfer within living systems.

reflex [L. *reflectere,* to bend back]: Unit of action of the nervous system involving a sensory neuron, often an interneuron or -neurons, and one or more motor neurons.

regulator gene: A gene that codes for a specific protein, a repressor, that controls the rate of transcription of the structural genes of an operon; often located some distance from the operon on the chromosome.

renal [L. *renes,* kidneys]: Pertaining to the kidney.

repressor [L. *reprimere,* to press back, keep back]: A protein produced by a regulator gene; it binds to the operator gene of an operon, preventing RNA polymerase from attaching to the promoter and transcribing the structural genes.

resolving power [L. *resolvere,* to loosen, unbind]: The ability of a lens to distinguish two lines as separate.

respiration [L. *respirare,* to breathe]: (1) In aerobic organisms, the intake of oxygen and the liberation of carbon dioxide. (2) In cells, the oxygen-requiring stage in the breakdown and release of energy from fuel molecules.

resting potential: The difference in electric potential (about 70 millivolts) across the membrane of an axon at rest.

reticular system [L. *reticulum,* a network]: A core of tissue that runs centrally through the entire brainstem; a weblike network of fibers and neurons; involved with consciousness.

reticulum [L., network]: A fine network (e.g., endoplasmic reticulum).

retina [L. dim. of *rete,* net]: The photosensitive layer of the vertebrate eye; contains several layers of neurons and light-receptors (rods and cones); receives the image formed by the lens and transmits it to the brain via the optic nerve.

rhizoid [Gk. *rhiza,* root]: Rootlike anchoring tissue in nonvascular plants.

rhizome [Gk. *rhizoma,* mass of roots]: In vascular plants, an underground stem; may be enlarged for storage, or may function in vegetative reproduction.

ribonucleic acid (RNA) (rye-bo-new-**klee**-ick): A class of nucleic acids characterized by the presence of the sugar ribose and the pyrimidine uracil; includes mRNA, tRNA, and rRNA. RNA is the genetic material of many viruses.

ribosomal RNA (rRNA): A class of RNA molecules found, along with characteristic proteins, in ribosomes; transcribed from the DNA of the nucleolus.

ribosome: A small organelle composed of protein and ribonucleic acid; the site of translation in protein synthesis; in eukaryotic cells, often bound to the endoplasmic reticulum. Many ribosomes attached to a single strand of mRNA comprise a polyribosome.

RNA: Abbreviation of ribonucleic acid.

rod: Light-sensitive nerve cell found in the vertebrate retina; sensitive to very dim light, responsible for "night vision."

root: The descending axis of a plant, normally below ground and serving both to anchor the plant and to take up and conduct water and minerals.

sarcomere [Gk. *sarx,* the flesh + *meris,* part of, portion]: Functional and structural unit of contraction in striated muscle.

secondary sex characteristics: Characteristics of animals that distinguish between the two sexes but that do not produce or convey gametes; includes facial hair of the human male and enlarged hips and breasts of the human female.

secondary structure of proteins: The simple structure (often a helix, a sheet, or a cable) resulting from the spontaneous folding of a polypeptide chain as it is formed; maintained by hydrogen bonds and other weak forces.

secondary succession: Ecological succession in an area that was previously covered by vegetation but has been laid bare by human activities or natural processes.

secretion [L. *secermere,* to sever, separate]: (1) Product of any cell, gland, or tissue that is released through the cell membrane and that performs its function outside the cell that produced it. (2) The second stage of kidney function; through active transport processes, molecules remaining in the blood plasma are selectively removed from the peritubular capillaries and pumped into the filtrate in the renal tubule.

seed: A complex organ formed by the maturation of the ovule of seed plants following fertilization; upon germination, a seed develops into a new sporophyte; generally consists of seed coat, embryo, and a food reserve.

segregation: The separation of alleles into different gametes during meiosis. *See* Mendel's first law.

selectively permeable [L. *seligere,* to gather apart + *permeare,* to go through]: Applied to membranes that permit passage of water and some solutes but block passage of most solutes; semipermeable.

self-pollination: The transfer of pollen from anther to stigma in the same flower or to another flower of the same plant, leading to self-fertilization.

semen [L., seed]: Product of the male reproductive system; includes sperm and the sperm-carrying fluids.

seminal vesicles [L. *semen,* seed + *vesicula,* a little bladder]: In male mammals, small vesicles, the ducts of which merge with the vasa deferentia as they enter the prostate gland; they produce an alkaline, fructose-containing fluid that suspends and nourishes the sperm cells.

sensory neuron: A neuron that carries impulses from a receptor to the central nervous system or central ganglion.

sensory receptor: A cell, tissue, or organ that detects internal or external stimuli.

sessile [L. *sedere,* to sit]: Attached; not free to move about.

sex chromosomes: Pairs of chromosomes that are similar in one sex but not in the other—for example, human females are *XX,* males are *XY;* chromosomes that determine sexual characteristics.

sex-linked characteristic: A genetic characteristic, such as color blindness, determined by a gene located on the *X* or *Y* chromosome.

sexual reproduction: Reproduction involving meiosis and fertilization.

shoot: The aboveground portions, such as the stem and leaves, of a vascular plant.

sieve tube: A series of sugar-conducting cells (sieve-tube elements) found in the phloem of angiosperms.

sinoatrial node: *See* Pacemaker.

smooth muscle: Nonstriated muscle; lines the walls of internal organs and arteries and is under involuntary control.

social dominance: A hierarchical pattern of social organization involving domination of some members of a group by other members in a relatively orderly and long-lasting pattern.

society [L. *socius,* companion]: An organization of individuals of the same species in which there are divisions of labor and mutual dependence; a society is held together by stimuli exchanged among members of the group.

sociobiology: The study of the biological basis of social behavior.

solution: A homogeneous mixture (usually liquid) in which one or more substances (the solutes) are uniformly dispersed throughout a solvent.

somatic cells [Gk. *soma,* body]: The differentiated cells composing body tissues of multicellular plants and animals; all body cells except those giving rise to gametes.

somatic nervous system [Gk. *soma,* body]: In vertebrates, the motor, sensory, and interneurons; the "voluntary" system, as contrasted with the "involuntary," or autonomic, nervous system.

somite: One of the segments into which the body of many animals is divided, especially arthropods, annelids, and vertebrate embryos.

specialized: (1) Of organisms, having special adaptations to a particular habitat or mode of life. (2) Of cells, having particular functions in a multicellular organism.

species, *pl.* **species** [L., kind, sort]: A group of organisms that actually (or potentially) interbreed in nature and are reproductively isolated from all other such groups; a taxonomic grouping of morphologically similar individuals (the category beneath genus).

specific: Unique; for example, the proteins in a given organism, the enzyme catalyzing a given reaction, or the antibody to a given antigen.

specific heat: The amount of heat (in calories) required to raise the temperature of 1 gram of a substance 1°C. The specific heat of water is 1 calorie per gram.

sperm [Gk. *sperma,* seed]: A mature male sex cell, or gamete, usually motile and smaller than the female gamete.

spermatid [Gk. *sperma,* seed]: Each of four haploid (n) cells resulting from the meiotic divisions of a spermatocyte; each spermatid becomes differentiated into a sperm cell.

spermatocytes [Gk. *sperma,* seed + *kytos,* vessel]: The diploid ($2n$) cells formed by the enlargement of the spermatogonia; they give rise by meiotic division to the spermatids.

spermatogonia [Gk. *sperma,* seed + *gonos,* a child, the young]: The unspecialized diploid ($2n$) cells on the walls of the testes that, by meiotic division, become spermatocytes, then spermatids, then sperm cells.

spinal cord: Part of the vertebrate central nervous system; consists of a thick, dorsal, longitudinal bundle of nerve fibers extending posteriorly from the brain.

spiracle [L. *spirare,* to breathe]: One of the external openings of the respiratory system in terrestrial arthropods.

splitting evolution: A pattern of evolutionary change characterized by the splitting, or branching, of phylogenetic lineages; one of the three principal patterns of organic evolution.

sporangiophore (spo-**ran**-ji-o-for) [Gk. *spora,* seed + *phore,* from *phorein,* to bear]: A branch bearing one or more sporangia.

sporangium, *pl.* **sporangia** [Gk. *spora,* seed]: A unicellular or multicellular structure in which spores are produced.

spore [Gk. *spora,* seed]: An asexual reproductive cell capable of developing into an adult without fusion with another cell; in contrast to a gamete.

sporophyll [Gk. *spora,* seed + *phyllon,* the leaves]: Spore-bearing leaf. The carpels and stamens of flowers are modified sporophylls.

sporophyte [Gk. *spora,* seed + *phytos,* growing]: The spore-producing diploid ($2n$) phase in the life cycle of a plant having alternation of generations.

stamen [L., a thread]: The male organ of a flower, which produces pollen; usually consists of a stalk, the filament, bearing a pollen-producing anther at its tip.

starch [M.E. *sterchen,* to stiffen]: A class of complex, insoluble carbohydrates, the chief food-storage substances of plants; composed of up to several hundred sugar units and readily broken down enzymatically into these units.

stem: The aboveground part of the axis of vascular plants, as well as anatomically similar portions below ground (such as rhizomes).

stigma [Gk. *stigme,* a prick mark, puncture]: In plants, the region of a carpel serving as a receptive surface for pollen grains, which germinate on it.

stimulus [L., goad, incentive]: Any internal or external change or signal that influences the activity of an organism or of part of an organism.

stoma, *pl.* **stomata** (**sto**-ma) [Gk. *stoma,* mouth]: A minute opening bordered by guard cells in the epidermis of leaves and stems through which gases pass. Also used to refer to the entire stomatal apparatus, the guard cells plus their included pore.

striated muscle [L., from *striare,* to groove]: Skeletal voluntary muscle and cardiac muscle. The name derives from the striped appearance, which reflects the arrangement of contractile elements.

stroma [Gk. *stroma,* a bed, from *stronnymi,* to spread out]: A dense solution in the interior of the chloroplast; the dark reactions of photosynthesis take place in the stroma.

structural gene: Any gene that produces a protein in distinction to other genes of an operon (operator and regulator genes).

style [L. *stilus,* stake, stalk]: In angiosperms, the stalk of a carpel, down which the pollen tube grows.

substrate [L. *substratus,* strewn under]: (1) The foundation to which an organism is attached. (2) A substance on which an enzyme acts.

succession: *See* Ecological succession.

sucrose: Cane sugar; a common disaccharide found in many plants; a molecule of glucose linked to a molecule of fructose.

sugar: Any monosaccharide or disaccharide.

surface tension: A tautness of the surface of a liquid, caused by the cohesion of the molecules of liquid. Water has an extremely high surface tension.

symbiosis [Gk. *syn,* together with + *bioonai,* to live]: An intimate and protracted association between two or more organisms of different species. Includes mutualism, in which the association is beneficial to both; commensalism, in which one benefits and the other is neither harmed nor benefited; and parasitism, in which one benefits and the other is harmed.

sympathetic division: A subdivision of the autonomic nervous system, with centers in the midportion of the spinal cord; slows digestion; generally excites other functions.

synapse [Gk. *synapsis,* a union]: The region of nerve-impulse transfer between two neurons.

synthesis [Gk. *syntheke,* a putting together]: The formation of a more complex substance from simpler ones.

taxonomy [Gk. *taxis,* arrange, put in order + *nomos,* law]: The study of the classification of organisms, the ordering of organisms into a hierarchy that reflects on their essential similarities and differences.

telophase [Gk. *telos,* end + *phasis,* form]: The last stage in mitosis and meiosis, during which the chromosomes become reorganized into two new nuclei.

template: A pattern or mold guiding the formation of a negative or complement.

tentacles [L. *tentare,* to touch]: Long, flexible protrusions located about the mouth in many invertebrates; usually prehensile or tactile.

territory: An area or space occupied and defended by an individual or a group; trespassers are attacked (and usually defeated); may be the site of breeding, nesting, food gathering, or any combination thereof.

tertiary structure of proteins: A complex structure, usually globular, resulting from further folding of the helical secondary structure of a protein; forms spontaneously due to attractions and repulsions among amino acids with different charges on their R groups.

testcross: A mating between a phenotypically dominant individual and a homozygous recessive "tester" to determine the genetic constitution of the dominant phenotype, that is, whether it is homozygous or heterozygous for the relevant gene.

testis, *pl.* **testes** [L., witness]: The sperm-producing organ; also the source of male sex hormone.

testosterone [Gk. *testis,* testicle + *steiras,* barren]: A hormone secreted by the testes in higher vertebrates and stimulating the development and maintenance of male sex characteristics and the production of sperm; an androgen.

tetrad [Gk. *tetras,* four]: A pair of homologous chromosomes that have replicated and come together in prophase I of meiosis; consists of four chromatids.

thalamus [Gk. *thalamos,* chamber]: A part of the vertebrate forebrain just posterior to the cerebrum; an important intermediary between all other parts of the nervous system and the cerebrum.

theory [Gk. *theorein,* to look at]: A generalization based on many observations and experiments.

thermodynamics [Gk. *therme,* heat + *dynamis,* power]: The study of transformations of energy, especially heat. The first law of thermodynamics states that in all processes, the total energy of an isolated system remains constant. The second law states that in all processes, randomness, or disorder, tends to increase.

thorax [Gk., breastplate]: (1) In vertebrates, that portion of the trunk containing the heart and lungs. (2) In insects, the three leg-bearing segments between head and abdomen.

thylakoid [Gk. *thylakos,* a small bag]: A flattened sac, or vesicle, that forms part of the internal membrane structure of the chloroplast and is the site of the light reactions of photosynthesis; stacks of thylakoids collectively form the grana.

thyroid [Gk. *thyra,* a door]: An endocrine gland of vertebrates, located in the neck; source of an iodine-containing hormone (thyroxine) that increases the metabolic rate and affects growth.

tissue [L. *texere,* to weave]: A group of similar cells organized into a structural and functional unit.

trachea, *pl.* **tracheae** (**trake**-ee-a) [Gk. *tracheia,* rough]: An air-conducting tube, such as the windpipe of mammals and the breathing systems of insects.

tracheid (**tray**-key-idd) [Gk. *tracheia,* rough]: In vascular plants, an elongated, thick-walled conducting and supporting cell of xylem, characterized by tapering ends and pitted walls without true perforations.

transcription [L. *trans,* across + *scribere,* to write]: The biosynthesis of a coded strand of RNA; nucleotides available in the cell are linked together in a sequence dictated by the DNA strand from which the RNA is transcribed; the RNA strand is complementary to the DNA strand.

transduction [L. *trans,* across + *ducere,* to lead]: The transfer of genetic material (DNA) from one cell to another by a virus.

transfer RNA (tRNA) [L. *trans,* across + *ferre,* to bear or carry]: A class of small RNAs (about 80 nucleotides) with two functional sites; one recognizes a specific activated amino acid; the other carries the nucleotide triplet (anticodon) for that amino acid. Each type of tRNA accepts a specific activated amino acid and transfers it to a growing polypeptide chain as specified by the nucleotide sequence of the messenger RNA being translated.

transformation [L. *trans,* across + *formare,* to shape]: A genetic change produced by the incorporation into a cell of DNA from another cell.

translation [L. *trans,* across + *latus,* that which is carried]: The synthesis of a polypeptide from amino acids carried by tRNAs to the ribosome-bound mRNA specifying the polypeptide.

translocation [L. *trans,* across + *locare,* to put or place]: (1) In plants, the transport of the products of photosynthesis from a leaf to another part of the plant. (2) In genetics, the breaking off of a piece of chromosome and reattachment to a nonhomologous chromosome.

transpiration [L. *trans,* across + *spirare,* to breathe]: In plants, the loss of water vapor from the stomata.

trophic level [Gk. *trophos,* feeder]: The position of a species in the food web or chain; a step in the movement of biomass or energy through a system.

tropic [Gk. *trope,* a turning]: Pertaining to behavior or action brought about by specific stimuli, for example, phototropic ("light-oriented") motion, gonadotropic ("stimulating the gonads") hormone.

tuber [L. *tuber,* bump, swelling]: A much-enlarged, short, fleshy underground stem, such as that of the potato.

turgor [L. *turgere,* to swell]: The pressure exerted on the inside of a plant cell wall by the fluid contents of the cell; the interior of the cell is hypertonic in relation to the fluids surrounding it and so gains water by osmosis.

urea [Gk. *ouron,* urine]: An organic compound formed in the vertebrate liver; the principal form of disposal of nitrogenous wastes by mammals.

ureter [Gk. from *ourein,* to urinate]: The tube carrying urine from the kidney; in mammals, it empties into the bladder.

urethra [Gk. from *ourein,* to urinate]: The tube carrying urine from the bladder to the exterior of mammals.

uric acid [Gk. *ouron,* urine]: An insoluble nitrogenous waste product that is the principal excretory product in birds, reptiles, and insects.

urine [Gk. *ouron,* urine]: The liquid waste filtered from the blood by the kidney and stored in the bladder pending elimination through the urethra.

uterus [L., womb]: The muscular, expanded portion of the female reproductive tract modified for the storage of eggs or for housing and nourishing the developing embryo.

vacuole [L. *vacuus,* empty]: A membrane-bound, fluid-filled sac within the cytoplasm of a cell.

vagina [L., a sheath]: The part of the female reproductive tract in mammals that receives the male penis during copulation; leads to the uterus.

vagus nerve [L. *vagus,* wandering]: A nerve arising from the medulla of the vertebrate brain that innervates the heart and visceral organs; carries parasympathetic fibers.

vaporization [L. *vapor,* steam]: The change from a liquid to a gas; evaporation.

vascular [L. *vasculum,* a small vessel]: Containing or concerning vessels that conduct fluid.

vascular bundle: In plants, a group of longitudinal supporting and conducting tissues (xylem and phloem).

vascular cambium [L. *vasculum,* a small vessel + *cambium,* exchange]: A cylindrical sheath of meristematic cells that divide mitotically, producing secondary phloem to one side and secondary xylem to the other, but always with a cambial cell remaining.

vas deferens (vass **deff**-er-ens) [L. *vas,* a vessel + *deferre,* to carry down]: In mammals, the tube carrying sperm from the testes to the urethra.

vein [L. *vena,* a blood vessel]: (1) In plants, a vascular bundle forming a part of the framework of the conducting and supporting tissue of a leaf or other expanded organ. (2) In animals, a blood vessel carrying blood from the tissues to the heart. A small vein is known as a venule.

vena cava (**vee**-na **cah**-va) [L., blood vessel + hollow]: A large vein that brings blood from the tissues to the right atrium of the four-chambered vertebrate heart. The superior vena cava collects blood from the forelimbs, head, and anterior or upper trunk; the inferior vena cava collects blood from the posterior body region.

ventilation lungs: Lungs in which the exchange of air with the atmosphere takes place as a result of changes in lung volume; vertebrate lungs.

ventral [L. *venter,* belly]: Pertaining to the undersurface of an animal that moves on all fours; to the front surface of an animal that holds its body erect.

ventricle [L. *ventriculus,* the stomach]: A muscular chamber of the heart that receives blood from an atrium and pumps blood out of the heart, either to the lungs or to the body tissues.

vertebral column [L. *vertebra,* joint]: The backbone; in nearly all vertebrates, it forms the supporting axis of the body and it protects the spinal cord.

vesicle [L. *vesicula,* a little bladder]: A small, intracellular membrane-bound sac.

vessel [L. *vas,* a vessel]: A tubelike element of the xylem of angiosperms; composed of dead cells (vessel elements) arranged end to end. Its function is to conduct water and minerals from the soil.

viability [L. *vita,* life]: Ability to live.

villus, *pl.* **villi** [L., a tuft of hair]: In vertebrates, one of the minute, fingerlike projections lining the small intestine that serve to increase the absorptive surface area of the intestine.

virus [L., slimy, liquid, poison]: A submicroscopic, noncellular particle composed of a nucleic acid core and a protein coat; parasitic; reproduces only within a host cell.

viscera [L., internal organs]: The collective term for the internal organs of an animal.

vitamin [L. *vita,* life]: Any of a number of unrelated organic substances that cannot be synthesized by a particular organism and are essential in minute quantities for normal growth and function.

vulva [L., the womb]: External genital organs of the female mammal; in humans, includes the clitoris and the labia.

water cycle: Worldwide circulation of water molecules, powered by the sun. Water evaporates from oceans, lakes, rivers, and, in smaller amounts, soil surfaces and bodies of organisms; water returns to the earth in the form of rain and snow. Of the water falling on land, some flows into rivers that pour water back into the oceans and some percolates down through the soil until it reaches a zone where all pores and cracks in the rock are filled with water (groundwater); the deep groundwater eventually reaches the oceans, completing the cycle.

water potential: The potential energy of water molecules; regardless of the reason (e.g., gravity, pressure, concentration of solute particles) for the water potential, water moves from a region where water potential is greater to a region where water potential is lower.

water table: The upper limit of permanently water-saturated soil; the top layer of the groundwater.

wild type: In genetics, the phenotype that is characteristic of the vast majority of individuals of a species in a natural environment.

worker: A member of the nonreproductive laboring caste of social insects.

xylem [Gk. *xylon,* wood]: A complex vascular tissue through which most of the water and minerals are conducted from the roots to other parts of the plant; consists of tracheids or vessel elements, parenchyma cells, and fibers; constitutes the wood of trees and shrubs.

yolk: The stored food in egg cells that nourishes the embryo.

zoology [Gk. *zoe,* life + *logos,* a discourse]: The study of animals.

zooplankton [Gk. *zoe,* life + *plankton,* wanderer]: A collective term for the nonphotosynthetic organisms present in plankton.

zygote (**zi**-got) [Gk. *zygon,* yolk, pair]: The diploid ($2n$) cell resulting from the fusion of male and female gametes (fertilization); a zygote may either develop into a diploid individual by mitotic divisions or may undergo meiosis to form haploid (n) individuals that divide mitotically to form a population of cells.

ILLUSTRATION ACKNOWLEDGMENTS

Page 1 Francisco Erize, Bruce Coleman
I-2 Runk/Schoenberger, Grant Heilman Photography
I-3 T. E. Adams, Peter Arnold
I-4 Tom Bledsoe, Photo Researchers
I-5 D. R. Specker, Animals Animals Enterprises
I-6 Margot Conte, Animals Animals Enterprises
I-7 (b) Zig Leszczynski, Animals Animals Enterprises
I-8 (a) Field Museum, Photo Researchers; (b) Y. J. Rey-Millet, Photo Researchers
I-9 (a) Rare Book Division, The New York Public Library; (b) John Mais
I-10 John Dominis, LIFE © 1971
I-11 Heather Angel

Pages 12, 13 T. E. Adams, Peter Arnold
1-1 California Institute of Technology and The Carnegie Institution of Washington
Page 18 John Reader, © National Geographic Society
1-4 Elihu Blotnick
1-6 (b) Runk/Schoenberger, Grant Heilman Photography
1-7 Barbara F. Reese and Thomas S. Reese, National Institute of Neurological and Communicative Disorders and Stroke

2-1 Runk/Schoenberger, Grant Heilman Photography
2-4 Steven J. Krasemann, Photo Researchers
2-5 Larry Pringle, Photo Researchers
2-6 Jack Dermid
2-8 Shirley Baty
2-9 (b) Grant Heilman
Page 34 Grant Heilman

3-4 (a), (b) After Albert L. Lehninger, *Biochemistry,* 2d ed., Worth Publishers, Inc., New York, 1975; (c) L. M. Beidler
3-5 Don Fawcett
3-6 (a) R. D. Preston; (b) Eric V. Gravé
3-7 Larry West
Page 41 N. E. Beck, Jr., National Audubon Society Collection/PR
3-11 B. E. Juniper
3-16 After E. O. Wilson et al., *Life: Cells, Organisms, Populations,* Sinauer Associates, Inc., Sunderland, Mass., 1977
3-17 Dan Friend
3-18 (a) Emil Bernstein and Eila Kairinen, Gillette Company Research Institute, *Science,* vol. 173, 1971, © 1971 by AAAS; (b) John Mais

4-1 National Center for Atmospheric Research
4-2 NASA
4-3 (a) S. Jonasson, from R. Anderson et al., *Science,* vol. 148, pages 1179–1190, 1965, © 1965 by AAAS
4-4 E. S. Barghoorn, *Science,* vol. 152, pages 758–763, 1966, © 1966 by AAAS
4-5 Micrograph by A. Ryter
4-6 Micrograph by Norma J. Lang, *Journal of Phycology,* vol. 1, pages 127–134, 1965
4-7 Micrograph by George Palade
4-9 Micrograph by Michael A. Walsh
Pages 60, 61 David M. Phillips
4-11 J. David Robertson
4-12 (a) M. C. Ledbetter; (b) Peter Albersham, *Scientific American,* April 1975
4-13 Daniel Branton
4-15 Don Fawcett, from William Bloom and D. W. Fawcett, *A Textbook of Histology,* 10th ed., Saunders College/Holt, Rinehart and Winston, Philadelphia, 1975
4-16 C. J. Flickinger, *Journal of Cell Biology,* vol. 49, pages 221–226, 1971
4-17 After Stephen Wolfe, *Biology of the Cell,* Wadsworth Publishing Co., Inc., Belmont, Calif., 1972
4-18 (a) D. Henderson; (b) Mary Osborn
4-19 Gregory Antipa
4-20 After Gerald Karp, *Cell Biology,* McGraw-Hill Book Company, New York, 1979
4-21 (a) After E. J. DuPraw, *Cell and Molecular Biology,* Academic Press, Inc., New York, 1968; (b) A. V. Grimstone

5-1 Keith R. Porter
5-3 K. Weidmann, National Audubon Society Collection/PR
5-5 After Lehninger, *op. cit.*
5-6 (a) Eric V. Gravé; (b), (c) Thomas Eisner
5-10 Gregory Antipa
5-11 Birgit H. Satir

Pages 86, 87 M. C. Ledbetter
6-1 R. D. Estes
6-3 (a) Jack Dermid; (b) L. L. Rue III, Animals Animals Enterprises (c) Larry West; (d) Y. Haneda
6-6 After Lehninger, *op. cit.*

7-1 Jen and Des Bartlett, Bruce Coleman
7-2 Keith R. Porter
7-6 (b) L. J. Reed, from *The Enzymes,* 3rd ed., vol. 1, page 213, © 1970 by Academic Press, New York
7-8 After Lehninger, *op. cit.*
7-9 (b) Grant Heilman
Page 107 Charles S. Lieber, "The Metabolism of Alcohol," *Scientific American,* March 1976

8-1 J. Robert Waaland
8-6 (a) A. D. Greenwood; (b) L. K. Shumway
8-12 J. Heslop-Harrison

Pages 124, 125 Gernsheim Collection, Humanities Research Center, The University of Texas
9-1 Kurt Hirschhorn, M.D., from *Birth Defects: Original Article Series,* vol. 4, no. 4, *Guide to Human Chromosome Defects,* by Audrey Redding and Kurt Hirschhorn, M.D., The National Foundation
9-2 Ursula Goodenough
9-3 Mia Tegner and David Epel
9-5 (b) William Tai
9-6 (a) Etienne de Harven, *The Nucleus,* Academic Press, Inc., New York, 1968; (b) Michael Friedlander
9-7 (a) G. Östergren; (b) J. T. Pickett-Heaps
9-8 Eric V. Gravé
9-9 H. W. Beams and R. G. Kessel, *American Scientist,* vol. 64, page 279, 1976
9-10 James Cronshaw
9-12 Hugh Spencer, National Audubon Society Collection/PR

10-7 B. John
10-8 Arnold Sparrow
10-9 After DuPraw, *op. cit.*
Page 147 William Tai

11-2 After Karl von Frisch, *Biology,* translated by Jane Oppenheimer, Harper & Row, Publishers, Inc., New York, 1964
11-7 Dr. V. Orel, The Moravian Museum, Brno

12-1 C. G. G. J. van Steenis
12-3 F. B. Hutt, *Journal of Genetics,* vol. 22, page 126, 1930
12-4 After E. D. Merrell, 1964
12-5 B. John
Page 164 Shirley Baty
12-10 Micrograph by B. John
12-13 B. P. Kaufmann

13-1 Marjorie Guthrie
13-2 Margaret Clark
13-5 After I. Michael Lerner, *Heredity, Evolution, and Society,* W. H. Freeman and Company, San Francisco, 1968

13-6 The Bettmann Archive
13-7 Paediatric Research Unit, Guy's Hospital Medical School
13-10 George Ancona

14-1 A. K. Kleinschmidt, D. Land, D. Jacherts, and R. K. Zahn, *Biochemica Biophysica Acta,* vol. 61, pages 857–864, 1962
14-3 Robert Austrian, *Journal of Experimental Medicine,* vol. 92, page 21, 1953
14-5 After Derry D. Koob and William E. Boggs, *The Nature of Life,* Addison-Wesley Publishing Co., Inc., Reading, Mass., 1972
14-8 Lee D. Simon
14-10 (a) After Lehninger, *op. cit.*
Page 193 From James D. Watson, *The Double Helix,* Atheneum Publishers, New York, 1968

15-5 O. L. Miller, Jr., and Barbara R. Beatty, *Journal of Cell Physiology,* vol. 74, Supplement 1, pages 252–232, 1969
15-6 After sketch by Keith Roberts
15-8 Hans Ris

16-1 John Cairns
16-5 Jack Griffith
16-7 Ada L. Olins, *The American Scientist,* vol. 66, page 6, November–December, 1978

17-1 Charles C. Brinton, Jr., and Judith Carnahan
Page 217 Stanley N. Cohen, *Nature,* December 27, 1969
Page 220 Jack Griffith
17-6 David A. Jackson
17-7 After *City of Hope Quarterly,* vol. 7, page 2, Winter 1978

Pages 226, 227 Edward S. Ross
18-1 Victor Lorian
18-2 (a) John S. Flannery, Bruce Coleman; (b) Ken Brate, Photo Researchers; (c) M. and B. Reed, Animals Animals Enterprises
18-3 (a), (b) Eric V. Gravé; (c) Edward S. Ross; (d) Les Blacklock; (e) Alvin E. Staffan
18-4 (a) J. P. Grilione and J. Pangborn, *Journal of Bacteriology,* vol. 124, page 1558, 1975; (b) Centers for Disease Control; (c) Virus Laboratory, Parke, Davis & Co.
18-5 (a) H. Forest; (b) Eric V. Gravé
18-6 (a), (b) J. D. Almeida and A. F. Howatson, *Journal of Cell Biology,* vol. 16, page 616, 1963; (c) Frederick A. Murphy; (d) Lee D. Simon
18-8 Richard W. Greene
Page 238 Ny Carlsberg Glyptotek
18-9 Eric V. Gravé
18-10 John Mais
18-11 (a), (c) Douglas P. Wilson; (b) Eric V. Gravé
18-12 Douglas P. Wilson
18-13 (a) Grant Heilman; (b), (d) Oxford Scientific Films, Animals Animals Enterprises; (c) Douglas P. Wilson
18-15 Larry West
18-16 (a) Alvin E. Staffan; (b) Edward S. Ross; (c) Charlie Ott, Photo Researchers
18-18 (a) Jack Dermid; (b) James R. Leard, National Audubon Society Collection/PR

19-1 Edward S. Ross
19-2 Alvin E. Staffan
19-3 (a) Edward S. Ross; (b) Robert Carr; (c) Gene Ahrens, Bruce Coleman
19-4 (a) William M. Harlow; (b) Larry West
Page 253 Jack Dermid
19-8 (a) H. N. Darrow, Bruce Coleman; (b) Heather Angel
19-9 Carleton Ray, Photo Researchers
19-11 Douglas P. Wilson
19-14 Steve Earley, Animals Animals Enterprises
19-17 (a) Maria Wimmer; (b) Raymond K. Mendez, Animals Animals Enterprises; (c) J. E. Sulston
Page 263 Centers for Disease Control
19-19 Maria Wimmer
19-22 (a) Runk/Schoenberger, Grant Heilman Photography; (b) Peter Ward, Bruce Coleman; (c) Tom McHugh, Steinhart Aquarium, Photo Researchers; (d) Jane Burton, Bruce Coleman
19-24 Douglas P. Wilson
19-25 (a) Zig Leszczynski, Animals Animals Enterprises; (b) Breck P. Kent, Animals Animals Enterprises; (c), (d), (e) George K. Bryce, Animals Animals Enterprises; (f) Neville Fox-Davies, Bruce Coleman
19-28 Grant Heilman
19-29 John Mais
19-30 (a), (b) Larry West; (c), (d) E. R. Degginger, Bruce Coleman
19-31 Paul E. Taylor, National Audubon Society Collection/PR
19-35 Donald H. Fritts
19-36 Russ Kinne, Photo Researchers
19-37 Runk/Schoenberger, Grant Heilman Photography
19-38 Lynwood M. Chace, National Audubon Society Collection/PR
19-39 (a) The American Museum of Natural History; (b) Allen Rokach
19-40 (a) The New York Zoological Society; (b) Jack Dermid
19-42 (a) Karl H. Maslowski, National Audubon Society Collection/PR, (b) L. L. Rue III, National Audubon Society Collection/PR, (c) L. L. Rue III, Bruce Coleman, (d) Norman Myers, Bruce Coleman, (e) Karen Tweedy-Holmes, Animals Animals Enterprises, (f) Mark Boulton, National Audubon Society Collection/PR

Pages 284, 285 Grant Heilman
20-1 Jack Dermid
20-5 Grant Heilman
20-6 (a) L. M. Beidler; (b) John H. Troughton
20-8 (a), (b) Jack Dermid; (c) Grant Heilman
Page 293 Ray Evert
20-9 Ray Evert

21-2 (a) Jack Dermid; (b) Grant Heilman
21-3 Ray Evert
21-4 Jack Dermid
21-5 After Peter Ray, *The Living Planet,* 2d ed., Holt, Rinehart and Winston, Inc., New York, 1971
21-6 Ray Evert
21-7 Ray Evert
21-8 M. C. Ledbetter, from Myron C. Ledbetter and Keith R. Porter, *Introduction to the Fine Structure of Plant Cells,* Springer-Verlag, New York, 1970
21-10 M. C. Ledbetter, from Ledbetter and Porter, *op. cit.*
21-11 (a), (b) Ray Evert; (c) Carolina Biological Supply
21-12 (a) John Colwell, Grant Heilman Photography; (b) Jack Dermid
21-13 After A. C. Leopold, *Plant Growth and Development,* McGraw-Hill Book Company, New York, 1964
21-14 George Whiteley, Photo Researchers
21-15 After M. Richardson, *Translocation in Plants,* St. Martin's Press, Inc., New York, 1968 (after Stoat and Hoagland, 1939)
Page 306 (a) Jack Dermid; (b) Ross Hutchins; (c), (d) Runk/Schoenberger, Grant Heilman Photography; (e) Carolina Biological Supply
Page 307 S. A. Wilde
21-17 (a) Martin H. Zimmermann; (b) George A. Schaefers
21-18 After W. A. Jensen and F. B. Salisbury, *Botany: An Ecological Approach,* Wadsworth Publishing Co., Inc., Belmont, Calif., 1972
Page 312 Grant Heilman

22-2 Larry West
22-4 Edward S. Ross
22-5 (a), (b), (e) Edward S. Ross; (c) W. G. Calder; (d) D. J. Howell
22-6 (b) Grant Heilman
22-7 (b) Anita Sabarese
22-8 (a) John Mais; (b) L. Anderson
Page 320 Ray Evert
22-12 After Peter Ray, *op. cit.*
22-13 Peter Ray
22-15 From Richard Ketchum, *The Secret Life of the Forest,* American Heritage Press, 1970
Page 328 Grant Heilman

23-1 Jack Dermid
23-5 J. P. Nitsch, *American Journal of Botany,* vol. 37, page 3, 1950
23-6 S. H. Wittwer and The Michigan Agricultural Experiment Station
23-7 J. E. Varner
23-8 H. R. Chen
23-10 After Aubrey W. Naylor, "The Control of Flowering," *Scientific American,* May 1952
23-11 Hugh Spencer, National Audubon Society Collection/PR
23-14 Frank Salisbury
23-15 Richard F. Trump, Photo Researchers

23-16 Jack Dermid
23-17 Jack Dermid

Pages 342, 343 Ken Heyman
24-1 D. M. Phillips
24-4 From Richard G. Kessel and Randy H. Kardon, *Tissues and Organs: A Text-Atlas of Scanning Electron Microscopy,* W. H. Freeman and Company, San Francisco, copyright © 1979
24-5 (a) Keith R. Porter
24-6 M. H. Ross and Edward J. Reith
24-8 Micrograph by Hugh E. Huxley
24-10 Don Fawcett
Page 354 Hugh E. Huxley

25-2 (b) C. S. Raine
25-6 After Wilson et al., *op. cit.*
25-7 (b) David Hubel
25-8 After S. W. Kuffler and J. G. Nicholls, *From Neuron to Brain,* Sinauer Associates, Inc., Sunderland, Mass., 1976
25-9 After Kuffler and Nicholls, *op. cit.*
25-10 After Kuffler and Nicholls, *op. cit.*
25-11 John Heuser
25-12 Arthur Jacques
25-14 After S. S. Mader, *Inquiry into Life,* William C. Brown Publishers, Dubuque, Iowa, 1976

26-1 Don Fawcett
Page 372 Grant Heilman
26-7 After Wilson et al., *op. cit.*
26-8 (b) J. T. Bonner

27-1 After M. B. V. Roberts, *Biology: A Functional Approach,* The Ronald Press Company, New York, 1971
27-2 Don Fawcett
27-3 Kessel and Kardon, *op. cit.*
27-9 After Roberts, *op. cit.*
27-10 Thomas L. Hayes and L. McDonald, *Experimental and Molecular Pathology,* vol. 10, Academic Press, Inc., New York, 1969
27-11 P. Armstrong and J. Lackie, *Journal of Cell Biology,* vol. 65, page 439, 1975
27-13 Joseph Feldman
27-14 (a) After Lehninger, *op. cit.*

28-3 M. P. L. Fogden, Bruce Coleman
Page 395 Gary W. Griffen, Animals Animals Enterprises
28-5 (b) H. R. Duncker, Department of Anatomy, Justus-Liebig University, Giessen, Germany
28-7 Micrograph by Keith R. Porter
Page 399 Myron Melamed
Page 400 Anker Odum, from *Scenic Wonders of Canada,* © 1976 by Reader's Digest Association (Canada) Ltd.

29-2 (a) I. Kaufman Arenburg
29-5 Micrograph by Jeanne M. Riddle
29-7 Don Fawcett
29-8 Keith R. Porter
29-9 (b) L. L. Rue III, Photo Researchers

30-1 Jack Dermid
30-2 (a) Ron Garrison, San Diego Zoo; (b) L. L. Rue III, Bruce Coleman; (c) L. L. Rue III, National Audubon Society Collection/PR
30-4 After Arthur J. Vander, James H. Sherman, and Dorothy S. Luciano, *Human Physiology: The Mechanisms of Body Function,* 3d ed., McGraw-Hill Book Company, New York, 1980
30-5 (a) Nicholas Mrosovsky; (b) L. L. Rue III, National Audubon Society Collection/PR; (c) S. C. Bisserôt, Bruce Coleman
30-9 Kessel and Kardon, *op. cit.*
30-11 From Lewis Carroll, *Alice's Adventures in Wonderland,* illustrated by John Tenniel
30-14 Bob Leatherman, National Audubon Society Collection/PR
30-15 Pat Kirkpatrick, National Audubon Society Collection/PR

31-2 After Roberts, *op. cit.*
31-6 Top: left, Charles G. Summers, Jr., Bruce Coleman; middle, Jane Burton, Bruce Coleman; right, Russ Kinne, Photo Researchers. Bottom: left, David Hughes, Bruce Coleman; right, L. L. Rue III, Bruce Coleman
31-13 Roberts Rugh and Landrum B. Shettles, M.D., *From Conception to Birth: The Drama of Life's Beginnings,* Harper & Row, Publishers, Inc., New York, 1971
31-14 Carnegie Institution of Washington
31-16 Rugh and Shettles, *op. cit.*
31-17 Rugh and Shettles, *op. cit.*
31-18 Rugh and Shettles, *op. cit.*
31-19 Rugh and Shettles, *op. cit.*
31-20 Rugh and Shettles, *op. cit.*
31-21 Rugh and Shettles, *op. cit.*
31-23 Suzanne Arms, Jeroboam

32-1 Arthur Jacques
32-6 James Olds
32-11 W. R. Klemm, *Science, the Brain, and Our Future,* Pegasus, New York, 1972
32-12 Jerry Hecht, National Institute of Mental Health
32-14 E. R. Lewis

Pages 464, 465 Jen and Des Bartlett, Photo Researchers
33-1 Radio Times Hulton Picture Library
33-2 Burndy Library
33-3 (a) Grant Heilman; (b) Frank Carpenter; (c) Simon C. Morris
33-4 Shirley Baty
33-5 Jen and Des Bartlett, Photo Researchers
33-7 I. Eibl-Eibesfeldt
Page 475 The American Museum of Natural History

34-2 Marineland of Florida
34-3 Victor McKusick
34-5 Edmund B. Gerard
34-6 Victor McKusick
34-7 After Jack R. Harlow, "The Plants and Animals That Nourish Man," *Scientific American,* September 1976. Vegetables courtesy of The Green Thumb, Water Mill, N.Y.
34-8 Tony Gauba, Photo Researchers
34-9 After Charles W. Brown, Santa Rosa Junior College

35-1 Niall Rankin, Eric and David Hosking Photographs
35-3 After Mather and Harrison, *Heredity,* vol. 3, 1949
35-4 R. C. Lewontin, *The Genetic Basis of Evolutionary Change,* Columbia University Press, New York, 1974
35-5 (a) John S. O'Brien; (b) Herbert A. Fischler, Isaac Albert Research Institute of the Kingsbrook Jewish Medical Center
35-6 P. M. Sheppard
35-7 After J. Clausen and W. M. Hiesey, Carnegie Institution of Washington, publication 615, 1958
35-9 Lynwood M. Chace, National Audubon Society Collection/PR
35-10 After Grover C. Stephens and Barbara Best North, *Biology,* John Wiley & Sons, Inc., New York, 1974

36-1 H. B. D. Kettlewell
36-2 Jerome Wexler, National Audubon Soeiety Collection/PR
36-3 After Stephens and North, *op. cit.*
36-4 Department of Indian Affairs and Northern Development
36-5 (a) J. R. Brownlie, Bruce Coleman; (b) Fritz Prenzel, Bruce Coleman; (c) J. Wallis, Bruce Coleman
36-6 After E. O. Wilson and William H. Bossert, *A Primer of Population Biology,* Sinauer Associates, Inc., Sunderland, Mass., 1971
Page 508 John S. Shelton
36-9 After Bruce Wallace and Adrian Srb, *Adaptation,* Prentice-Hall, Inc., Englewood Cliffs, N.J., 1964
36-10 After Wallace and Srb, *op. cit.*
36-12 M. P. Harris
36-13 Allan D. Cruickshank, Photo Researchers
36-14 Jen and Des Bartlett, Bruce Coleman
36-15 (a) C. Haagner, Bruce Coleman; (b) Robert Clark, Photo Researchers
36-16 Treat Davidson, National Audubon Society Collection/PR
36-17 George Holton, Photo Researchers

Pages 520, 521 Tom Nebbia
37-1 David C. Fritts, Animals Animals Enterprises
37-3 L. L. Rue III, Photo Researchers
37-7 Tom McHugh, Photo Researchers
37-8 After S. Charles Kendeigh, *Animal Ecology,* Prentice-Hall, Inc., Englewood Cliffs, N.J., 1961 (after Shelford, 1911)
37-10 (a) Jane Burton, Bruce Coleman; (b) Lynwood M. Chace, National Audubon Society Collection/PR; (c) Margot Conte, Animals Animals Enterprises; (d) Masud Quraishy, Bruce Coleman

38-2 After John L. Harper, *Symposia Society for Experimental Biology,* vol. 15, page 25, 1961

38-6 Jane Lubchenco
38-7 Biological Section Labs, Australian Department of Lands
38-8 N. Myers, Bruce Coleman
38-9 (a) Edward S. Ross; (b) Doug Fulton, Photo Researchers; (c) S. C. Bisserôt, Bruce Coleman
38-10 Edward S. Ross
38-11 Thomas Eisner
38-12 Edward S. Ross
38-13 Edward S. Ross
38-14 (a) R. Marischal, Bruce Coleman; (b) Douglas Faulkner; (c) Edward S. Ross; (d) L. L. Rue III, Bruce Coleman
38-15 Daniel Janzen
38-16 Australian News and Information Service
Page 547 The Bettmann Archive

39-1 Jack Dermid
39-4 (a) L. L. Rue III, Bruce Coleman; (b) Charles G. Summers Jr., Photo Researchers
Page 553 John M. Edmond, from John B. Corliss and Robert D. Ballard, *National Geographic,* vol. 152, no. 4, October 1977
39-11 (a) Raymond C. Valentine; (b) R. R. Hebert, J. G. Griswell, and R. W. F. Hardy, Central Research and Development Department, E. I. duPont de Nemours & Co., Wilmington, Del.
39-13 R. H. Wright, National Audubon Society Collection/PR
39-14 Dennis Brokaw
39-15 From Robert E. Ricklefs, *Ecology,* 2d ed., Chiron Press, New York, 1979, after H. A. Fowells, *Silvics of Forest Trees of the United States,* Agricultural Handbook No. 271, Washington, D.C., U.S. Department of Agriculture
39-16 From Ricklefs, *op. cit.*
39-17 Jack Dermid
39-18 (a) U.S. Forest Service; (b) Larry West
39-19 Jack Dermid
Page 567 Grant Heilman

40-1 NASA
Page 573 After *Science News,* October 13, 1979, page 244; based on data from the U.S. Environmental Protection Agency
40-6 Karl Weidmann, National Audubon Society Collection/PR
40-7 (a) Edward S. Ross; (b) John Markham, Bruce Coleman
40-8 Bruce Coleman
40-9 (a) L. L. Rue III, Bruce Coleman; (b) Max Thompson, National Audubon Society Collection/PR; (c) N. Myers, Bruce Coleman; (d) Verna R. Johnston, Photo Researchers
40-10 (a) Bill Ratcliffe; (b) Jen and Des Bartlett, Bruce Coleman; (c) Robert H. Wright, National Audubon Society Collection/PR; (d) Grace Thompson, Photo Researchers; (e) Anthony Mercieca, National Audubon Society Collection/PR; (f) Willis Peterson; (g) Alan Blank, Bruce Coleman
40-11 (a) Jen and Des Bartlett, Bruce Coleman; (b) Dennis Brokaw
40-12 (a) Tom McHugh, Photo Researchers; (b) Patricia Caulfield
40-13 (a) Ed Cooper; (b) Les Blacklock; (c) L. L. Rue III, Bruce Coleman; (d) James Amos, Photo Researchers
40-14 (a) Craig Blacklock; (b), (c) Les Blacklock
40-15 (a) Les Blacklock; (b) Alvin E. Staffan; (c) Bill Ruth, Bruce Coleman
40-16 Les Blacklock
40-17 (a), (b) Douglas P. Wilson; (c) William R. Curtsinger, Rapho/PR; (d) Kjell B. Sandved, Photo Researchers
40-19 Allan Power, Bruce Coleman
40-20 (a) Jeff Foote, Bruce Coleman; (b) Charlie Ott, National Audubon Society Collection/PR; (c) Stephen J. Kraseman, Photo Researchers
40-21 (a) Jack Dermid; (b) Al Lowry, Photo Researchers

41-1 Masud Quraishy, Bruce Coleman
41-2 (a) Edward S. Ross; (b) Alan Blank, Bruce Coleman
41-3 (a) Edward S. Ross; (b) Treat Davidson, National Audubon Society Collection/PR
41-4 (a) Edward S. Ross; (b) Colin G. Butler, Bruce Coleman; (c) Stephen Dalton, Bruce Coleman
41-5 Grant Heilman
41-6 Shirley Baty
41-8 (a) M. P. Kahl, Photo Researchers; (b) George Holton, Photo Researchers; (c) Jeff Foote, Bruce Coleman
41-9 Patricia Caulfield
41-10 T. W. Ransom
41-11 (a) L. L. Rue III, Bruce Coleman; (b) R. R. Pawlaski, Bruce Coleman; (c) S. Gaulin, Anthro-Photo; (d) Douglas Faulkner
41-12 Tanaka Kojo, Animals Animals Enterprises
41-13 Ron Garrison, San Diego Zoo
41-14 Irven DeVore, Anthro-Photo
41-15 Hans Kummer
41-16 Irven DeVore, Anthro-Photo
41-17 (a) Tom McHugh, Photo Researchers; (b) Irven DeVore, Anthro-Photo
41-18 J. Popp, Anthro-Photo
41-19 J. Popp, Anthro-Photo

42-1 Editorial Photocolor Archives
42-2 Ron Garrison, San Diego Zoo
42-4 (a) Edward S. Ross; (b) H. Albrecht, Bruce Coleman
42-5 (a) George Holton, Photo Researchers; (b) Tom McHugh, Photo Researchers; (c) M. P. L. Fogden, Bruce Coleman
42-6 After W. W. Howells, *Mankind in the Making,* rev. ed., Doubleday & Company, Inc., Garden City, N. Y., 1967
Pages 620, 621 (a) Ralph Morse, Time-Life; (b) G. D. Plage, Bruce Coleman; (c) George Holton, Photo Researchers; (d), (e) Lee Lyon, Bruce Coleman; (f), (h) Teleki/Baldwin; (g), (i), (j) Richard Wrangham, Anthro-Photo
42-8 The American Museum of Natural History
42-9 The American Museum of Natural History
42-10 Robert F. Sisson, © National Geographic Society
42-11 (a) L. L. Rue III, National Audubon Society Collection/PR; (b) Geza Teleki
42-12 Tom McHugh, Photo Researchers
42-13 Geza Teleki
Page 626 George Holton, Photo Researchers
42-14 Lee Boltin
42-15 Ralph S. Solecki
42-17 Lee Boltin. Illustrations from *Tools of the Old and New Stone Age,* by Jacques Bordaz. Copyright © 1970 by Jacques Bordaz; copyright © 1958, 1959 by The American Museum of Natural History. Used by permission of Doubleday and Company, Inc.
42-18 Peabody Museum. Illustrations from Bordaz, *op. cit.*
42-19 Tom McHugh, Photo Researchers
42-21 Mel Korner, Anthro-Photo
42-22 International Rice Research Institute

Contents: page xiv, Jen and Des Bartlett, Photo Researchers; page xv, Kurt Hirschhorn, M.D., from Redding and Hirschhorn, *op. cit.;* page xvi, Karen Tweedy-Holmes, Animals Animals Enterprises; page xvii, Jack Dermid; page xviii, Russ Kinne, Photo Researchers; page xix, D. M. Phillips; page xx, Lynwood M. Chace, National Audubon Society Collection/PR; page xxi, Masud Quraishy, Bruce Coleman

INDEX

Page numbers in italic type indicate references to illustrations, tables, and boxed essays; boldface type is used to indicate major text discussions.